4

The SAGE Handbook of
Qualitative
Research

INTERNATIONAL ADVISORY BOARD

4 | The SAGE Handbook of Qualitative Research

Edited by

Norman K. Denzin
University of Illinois, Urbana-Champaign

Yvonna S. Lincoln
Texas A&M University

Los Angeles | London | New Delhi
Singapore | Washington DC

Los Angeles | London | New Delhi
Singapore | Washington DC

FOR INFORMATION:

SAGE Publications, Inc.
2455 Teller Road
Thousand Oaks, California 91320
E-mail: order@sagepub.com

SAGE Publications Ltd.
1 Oliver's Yard
55 City Road
London EC1Y 1SP
United Kingdom

SAGE Publications India Pvt. Ltd.
B 1/I 1 Mohan Cooperative Industrial Area
Mathura Road, New Delhi 110 044
India

SAGE Publications Asia-Pacific Pte. Ltd.
33 Pekin Street #02-01
Far East Square
Singapore 048763

Acquisitions Editor: Vicki Knight
Associate Editor: Lauren Habib
Editorial Assistant: Kalie Koscielak
Production Editor: Astrid Virding
Copy Editor: Jackie Tasch, Taryn Bigelow,
    Robin Gold, and Teresa Herlinger
Typesetter: C&M Digitals (P) Ltd.
Proofreader: Dennis Webb
Indexer: Kathy Paparchontis
Cover Designer: Candice Harman
Marketing Manager: Helen Salmon
Permissions Editor: Adele Hutchinson

Printed in the United States of America

Library of Congress Cataloging-in-Publication Data

The Sage handbook of qualitative research / editors, Norman K. Denzin, Yvonna S. Lincoln. — 4th ed.

p. cm.
Includes bibliographical references and index.

ISBN 978-1-4129-7417-2 (cloth)

1. Social sciences--Research. 2. Qualitative research. I. Denzin, Norman K. II. Lincoln, Yvonna S. III. Title: Handbook of qualitative research.

H62.H2455 2011 001.4´2—dc22 2010052892

This book is printed on acid-free paper.

11 12 13 14 15 10 9 8 7 6 5 4 3 2

CONTENTS

PREFACE

The fourth edition of the *Handbook of Qualitative Research,* like the third edition, is virtually a new volume. Nearly two thirds of the authors from the third edition have been replaced by new contributors. Indeed, there are 53 new chapters, authors, and/or coauthors. There are 18 totally new chapter topics, including contributions on critical social science, endarkened transnational feminist praxis, critical pedagogy, Asian epistemologies, disability communities and transformative research for social justice, human rights, oral history, indigenous inquiry, evidence, politics, science and government, criteria for assessing interpretive validity, models of representation, varieties of validity, qualitative research and technology, queer theory, performance ethnography, narrative inquiry, arts-based inquiry, the politics and ethics of online ethnography, analytic methodologies, writing strategies, policy and qualitative evaluation, the future of qualitative inquiry, teaching qualitative research, talk and text, focus groups, critical pedagogy, and models, issues, and controversies in mixed methods research,. All returning authors have substantially revised their original contributions, in many cases producing a totally new and different chapter.

There were and continue to be multiple social science and humanities audiences for the *Handbook*: graduate students who want to learn how to do qualitative research; interested faculty hoping to become better informed about the field; persons in policy settings, who understand the value of qualitative research methodologies and want to learn about the latest developments in the field; and faculty who are experts in one of more areas of the *Handbook,* but who also want to be informed about the most recent developments in the field. We never imagined this audience would be so large. Nor did we imagine that the *Handbook* would become a text used in undergraduate and graduate research methods courses, but it did. In 2008, we created three new paperback volumes for classroom use: *The Landscape of Qualitative Research, Strategies of Qualitative Inquiry,* and *Collecting and Interpreting Qualitative Materials.*

The fourth edition of the *Handbook of Qualitative Research* continues where the third edition ended. Sometime during the last two decades, critical qualitative inquiry came of age, or more accurately moved through another historical phase.[1] Out of the qualitative-quantitative paradigm wars of the 1980s, there appeared, seemingly overnight, journals,[2] handbooks,[3] textbooks,[4] dissertation awards,[5] annual distinguished lectures,[6] and scholarly associations.[7] All of these formations were dedicated to some version of qualitative inquiry (see the Erickson, Chapter 3, this volume). Scholars were in the midst of a social movement of sorts, a new field of inquiry; a new discourse had arrived, or so it seemed, and it flourished.

Qualitative researchers proudly took their place at the table. Students flocked to graduate programs for study and mentoring. Instruction in qualitative and mixed methods models became commonplace. Now there were QUAN and QUAL programs (see the chapter by Eisenhart & Jurow, Chapter 43, this volume). Paradigm proliferation prevailed, a rainbow coalition of racialized and queered post-isms, from feminism, to structuralism, postmodernism, postcolonialism, poststructuralism, postpostivism, post-scientism, Marxism, and postconstructivism (Erickson, Chapter 3, this volume).

All of this took place within and against a complex historical field, a global war on terror, a third methodological movement (Teddlie & Tashakkori, Chapter 16, this volume), the beginning or end of the eighth moment (Denzin & Lincoln, 2005, p. 3).[8] In the *methodologically contested present,* qualitative researchers confronted and then went beyond the scientific backlash associated with the evidence-based social movement connected in North American education with the No Child Left Behind legislation (see Hatch, 2006). At the same time, they embraced multiple and mixed methods approaches to inquiry (see Teddlie & Tashakkori, Chapter 16, and Creswell, Chapter 15, this volume).

So at the beginning of the second decade of the 21st century, it is time to move forward. It time to open up new spaces, time to explore new discourses. We need to find new ways of connecting persons and their personal troubles with social justice methodologies. We need to become better accomplished in linking these interventions to those institutional sites where troubles are turned into public issues and public issues transformed into social policy.

A critical framework is central to this project. It privileges practice, politics, action, consequences, performances, discourses, methodologies of the heart, and pedagogies of hope, love, care, forgiveness, and healing (Pelias, Chapter 40, this volume; Dillard & Okpalaoka, Chapter 8, this volume). It speaks for and with those who are on the margins. As a liberationist philosophy, it is committed to examining the consequences of racism, poverty, and sexism on the lives of interacting individuals.

Moving forward, it is necessary to confront and work through the criticisms that continue to be directed to qualitative inquiry. Each generation must draw its line in the sand and take a stance toward the past. Each generation must articulate its epistemological, methodological, and ethical stance toward critical inquiry. Each generation must offer its responses to current and past criticisms. In the spirit of inclusion, let us listen to our critics. But in doing so, we must renew our efforts to de-colonize the academy, to honor the voices of those who have been silenced by dominant paradigms. Let us do this in a spirit of cooperation and collaboration and mutual self-respect.

There is a pressing need to show how the practices of qualitative research can help change the world in positive ways. It is necessary to continue to engage the pedagogical, theoretical, and practical promise of qualitative research as a form of radical democratic practice.

In our invitation letter to authors and editorial boards members, we stated that

> As with the third edition, which was published by Sage in 2005, we regard the *Handbook* as a major benchmark for future work in this field. One measure of a benchmark work is its status in graduate education. We want the fourth edition to be a work that all doctoral students in your field will continue to want to study as they prepare for their exams and their dissertations. We have also been gratified to discover that many faculty use the *Handbook* as a class textbook; we hope that the fourth edition fulfills the same teaching needs. The new edition should advance a democratic project committed to social justice in an age of uncertainty. We are working with authors who can write chapters that will address practical, concrete issues of implementation while critiquing the field and mapping key current and emergent themes, debates, and developments.

This is the three-sided agenda of the fourth edition, to show how the discourses of qualitative research, inside and outside the classroom, can be used to help create and imagine a free democratic society. Each of the chapters that follow is defined by these commitments, in one way or another.

◫ ◫ ◫

We ask of a handbook that it do many things. A handbook should ideally represent the distillation of knowledge of a field; it should be a benchmark volume that synthesizes an existing literature, helping to define and shape the present and future of that discipline. A handbook charts the past, the present, and the future of the discourses at hand. It represents the very best thinking of the very best scholars in the world. It is reflexive, comprehensive, dialogical, accessible. It is authoritative and definitive. Its subject matter is clearly defined. Its authors work within a shared framework. Its authors and editors seek to impose an order on a field and a discipline. Yet they respect and attempt to honor diversity across disciplinary and paradigmatic perspectives.

A handbook is more than a review of the literature. It speaks to graduate students, to established scholars, and to scholars who wish to learn about the field. It has hands-on information. It shows persons how to move from ideas to inquiry, from inquiry to interpretation, from interpretation to praxis to action in the world. It locates its project within larger disciplinary and historical formations. It takes a stand on social justice issues; it is not just about pure scholarship. It is humble. It is indispensable.

These understandings organized the first three editions of this *Handbook*. In metaphorical terms, if you were to take one book on qualitative research with you to a desert island (or for a comprehensive graduate examination), a handbook would be the book.

A critical social science seeks its external grounding not in science, in any of its revisionist postpositivist forms, but rather in a commitment to critical pedagogy and communitarian feminism with hope but no guarantees. It seeks to understand how power and ideology operate through and across systems of discourse, cultural commodities, and cultural texts. It asks how words and texts and their meanings play a pivotal part in the culture's "decisive performances of race, class [and] gender" (Downing 1987, p. 80).

We no longer just write culture. We perform culture. We have many different forms of qualitative inquiry today. We have multiple criteria for evaluating our work (see Appendix B). It is a new day for our generation. We have drawn our line in the sand, and we may redraw it. But we stand firmly behind the belief that critical qualitative inquiry inspired by the sociological imagination can make the world a better place.

◫ ORGANIZATION OF THIS VOLUME

The organization of the *Handbook* moves from the general to the specific, the past to the present. Part I locates the field, starting with applied qualitative research traditions in the academy, then takes up the history of qualitative inquiry in social and educational research, ethics, politics, and critical social science traditions. Part II isolates what we regard as the major historical and contemporary paradigms now structuring and influencing qualitative research in the human disciplines. The chapters move from competing paradigms (positivist, postpositivist, constructivist, critical theory) to specific interpretive perspectives (critical ethnography, feminist and endarkened transnational discourse, critical race theory, cultural studies, critical humanism and queer theory, Asian epistemologies, and disability studies).

Part III isolates the major strategies of inquiry—historically, the research methods—a researcher can use in a concrete study. The contributors in this section embed their discussions of specific strategies of inquiry (mixed methods, case study, performance ethnography, narrative ethnography, interpretive practice, grounded theory, *testimonio,* participatory action research, clinical research) in social justice topics. The history and uses of these strategies are extensively explored in the 11 chapters in Part III.

Still, the question of methods begins with the design of the qualitative research project. This always begins with a socially situated researcher who moves from a research question, to a paradigm or perspective, to the empirical world. So located, the researcher then addresses the range of methods that can be employed in any study. In Chapter 14 of this volume, Julianne Cheek wisely observes that questions surrounding the practice and politics of funding qualitative research are often paramount at this point in any study. Globally, funding for qualitative research becomes more difficult as methodological conservatism gains momentum in neoliberal political regimes.

Part IV examines methods of collecting and analyzing empirical materials. It moves from narrative inquiry to chapters on arts-based inquiry, oral history, observation, visual methodology, performative autoethnography, the politics, ethics, and forms of online ethnography, and analyses of talk and text, then on to focus groups, pedagogy, and politics.

Part V takes up the art and practices of interpretation, evaluation, and presentation, including criteria for judging the adequacy of qualitative materials in an age of relativism, the interpretive process, writing as a method of inquiry, the poetics of place, cultural *poesis,* investigative poetry and the politics of witnessing, and qualitative evaluation and changing social policy. The three chapters in Part VI speculate on the future and promise of the social sciences and qualitative research in an age of global uncertainty.

◨ PREPARATION OF THE REVISED *HANDBOOK*

In preparation of a revised *Handbook,* it again became clear in our lengthy discussions that we needed input from perspectives other than our own. To accomplish this, we assembled a highly prestigious, international, and interdisciplinary editorial board (listed at the front of this volume), who assisted us in the selection of equally prestigious authors, the preparation of the Table of Contents, and the reading of (often multiple drafts) of each chapter. We used editorial board members as windows into their respective disciplines. We sought information on key topics, perspectives, and controversies that needed to be addressed. In our selection of editorial board members and chapter authors, we attempted to crosscut disciplinary, gender, race, paradigm, and national boundaries. Our hope was to use the authors' views to minimize our own disciplinary blinders.

Extensive feedback was received from the editorial board, including suggestions for new chapters, different slants to take on each of the chapters, and suggestions of authors for different chapters. In addition to considering social justice issues, each *Handbook* author—internationally recognized in his or her subject matter—was asked to treat such topics as history, epistemology, ontology, exemplary texts, key controversies, competing paradigms, and predictions about the future.

◨ RESPONDING TO CRITICS

We were gratified by the tremendous response from the field; especially gratifying were the hundreds of professors from around the world who choose the *Handbook* (in one form or another) as an assigned reading for their students. We were also gratified by the critical responses to the work. The *Handbook* has helped open a space for dialogue. This dialogue was long overdue. Many found problems with our approach to the field, and these problems indicate places where more conversations need to take place.

Among the criticisms of the first three editions were the following: our framework was unwieldy; we did not give enough attention to the Chicago School; there was too much emphasis on the postmodern period; we had an arbitrary historical model (Alasuutari, 2004; Atkinson, Coffey, & Delamont, 2003); we were too eclectic; we overemphasized the contemporary period and the crisis of representation; we gave too much attention to political correctness and not enough to knowledge for its own sake; there was not enough on how to do it. Some felt that a revolution had not occurred and wondered, too, how we proposed to evaluate qualitative research, now that the narrative turn has been taken. Others contended that our framework exposed the social sciences to unnecessary criticism and indeed threatened the entire project of social inquiry.

We cannot speak for the more than 160 chapter authors from the first, second, and third editions. Each person has taken a stance on these issues. As editors, we have attempted to represent a number of competing or at least contesting ideologies and frames of reference. This *Handbook* is not nor is it intended to be the view from the bridge of Denzin or Lincoln. We are not saying that there is only one way to do research, or that our way is best, or that the so-called old ways are bad. We are just saying this is one way to conceptualize this field, and it is a way that we find useful.

Of course, the *Handbook* is not a single thing. It even transcends the sum of its parts, and there is enormous diversity within and between every chapter. It is our hope that readers find spaces within these spaces that work for them. It is our desire that new dialogue take place within these spaces. This will be a gentle, probing, neighborly, and critical conversation, a conversation that bridges the many diverse interpretive communities that today make up this field called qualitative research. We value passion, we invite criticism, and we seek to initiate a discourse of resistance. Internationally, qualitative researchers must struggle against neoliberal regimes of truth, science, and justice.

◧ DEFINING THE FIELD

The qualitative research community consist of groups of globally dispersed persons who are attempting to implement a critical interpretive approach that will help them (and others) make sense of the terrifying conditions that define daily life at the first decade of this new century. These individuals employ constructivist, critical theory, feminist, queer, and critical race theory, as well as cultural studies models of interpretation. They locate themselves on the borders between postpositivism and poststructuralism. They use any and all of the research strategies (case study, ethnography, phenomenology, grounded theory, biographical, historical, participatory, and clinical) discussed in Part III of the *Handbook*. As interpretive *bricoleurs* (see Harper, 1987, pp. 9, 74; Kincheloe, 2008), the members of this group are adept at using all of the methods of collecting and analyzing empirical materials discussed by the authors of the chapters in Part IV of the *Handbook*. And, as writers and interpreters, these individuals wrestle with positivist, postpositivist, poststructural, and postmodern criteria for evaluating their written work.[9]

These scholars constitute a loosely defined international interpretive community. They are slowly coming to agreement on what constitutes a "good" and "bad," or banal, or an emancipatory, troubling analysis and interpretation. They are constantly challenging the distinction between the "real" and that which is constructed, understanding that all events and understandings are mediated and made real through interactional and material practices, through discourse, conversation, writing, narrative, scientific articles, realist, postrealist, and performance tales from the field.

This group works at both the centers and the margins of those emerging interdisciplinary, transnational formations that crisscross the borders between communications; race, ethnic, religious, and women's studies; sociology; history; anthropology; literary criticism; political science; economics; social work; health care; and education. This work is characterized by a quiet change in outlook, a transdisciplinary conversation, and a pragmatic change in practices, politics, and habits.

At this juncture—the uneasy, troubled crossroads between neoliberalism, pragmatism, and postmodernism—a quiet revolution is occurring. This revolution is defined by the politics of representation, which asks what is represented in a text and how should it be judged. We have left the world of naïve realism, knowing now that a text does not mirror the world, it creates the world. Furthermore, there is no external world or final arbiter—lived experience, for example—against which a text is judged.

Pragmatism is central to this conversation, for it is itself a theoretical and philosophical concern, firmly rooted in the postrealist tradition. As such, it is a theoretical position that privileges practice and method over reflection and deliberative action. Indeed, postmodernism itself has no predisposition to privilege discourse or text over observation. Instead, postmodernism (and poststructuralism) would simply have us attend to discourse and performance as seriously as we attend to observation (or any other fieldwork methods) and to recognize that our discourses are the vehicles for sharing our observations with those who were not in the field with us.

The angst attending our recognition of the hidden powers of discourses is precisely what leaves us now at the threshold of postmodernism and signals the advent of questions that will leave none of us untouched. It is true that contemporary qualitative, interpretive research exists within competing fields of discourse. Our present history of the field locates seven moments—and an eighth—the future. These moments all circulate in the present, competing with and defining one another. This discourse is moving in several directions at the same time. This has the effect of simultaneously creating new spaces, new possibilities, and new formations for qualitative research methods while closing down others.

There are those who would marginalize and politicize the postmodern, poststructural versions of qualitative research, equating them with political correctness, with radical relativism, narratives of the self, and armchair commentary. Some would chastise this *Handbook* for not paying adequate homage to the hands-on, nuts-and-bolts approach to fieldwork, to texts that tell us how to study the "real" world. Still others would seek a preferred, canonical, but flexible version of this project, returning to the Chicago School or to more recent formal, analytic, realist versions. Some would criticize the formation from within, contending that the privileging of discourse over observation does not yield adequate criteria for evaluating interpretive work, wondering what to do when left with only voice and interpretation. Many ask for a normative framework for evaluating their own work. None of these desires are likely to be satisfied anytime soon, however. Contestation, contradiction, and philosophical tensions make the achievement of consensus on any of these issues less than imminent.

We are not collating history here, although every chapter describes the history in a subfield. Our intention, which our contributors share, is to point to the future, where the field of qualitative research methods will be 10 years from now. Of course, much of the field still works within frameworks defined by earlier historical moments. This is how it should be. There is no one way to do interpretive, qualitative inquiry. We are all interpretive *bricoleurs* stuck in the present, working against the past, as we move into a politically charged and challenging future.

▣ COMPETING DEFINITIONS OF QUALITATIVE RESEARCH METHODS

The open-ended nature of the qualitative research project leads to a perpetual resistance against attempts to impose a single, umbrella-like paradigm over the entire project. There are multiple interpretive projects, including the decolonizing methodological project of indigenous scholars, theories of critical pedagogy, performance [auto] ethnographies; standpoint epistemologies, critical race theory; critical, public, poetic, queer, materialist, feminist, reflexive, ethnographies; projects connected to the British cultural studies and Frankfurt schools; grounded theorists of several varieties; multiple strands of ethnomethodology; African American, prophetic, postmodern, and neo-pragmatic Marxism; an American-based critical cultural studies model; and transnational cultural studies projects.

The generic focus of each of these versions of qualitative research moves in five directions at the same time: (1) the "detour through interpretive theory" and a politics of the local, linked to (2) the analysis of the politics of representation and the textual analyses of literary and cultural forms, including their production, distribution, and consumption; (3) the ethnographic qualitative study and representation of these forms in everyday life; (4) the investigation of new pedagogical and interpretive practices that interactively engage critical cultural analysis in the classroom and the local community; and (5) a utopian politics of possibility (Madison, 1998) that redresses social injustices and imagines a radical democracy that is not yet (Weems, 2002, p. 3)

▣ WHOSE REVOLUTION?

To summarize, a single, several-part thesis organizes our reading of where the field of qualitative research methodology is today. First, this project has changed because the world that qualitative research confronts, within and outside the academy, has changed. It has also changed because of the increasing sophistication—both theoretical and methodological—of interpretivist researchers everywhere. Disjuncture and difference, violence and terror, define the global political economy. This is a post- or neo-colonial world. It is necessary to think beyond the nation or the local group as the focus of inquiry.

Second, this is a world where ethnographic texts circulate like other commodities in an electronic world economy. It may be that ethnography is one of the major discourses of the neomodern world. But if this is so, it is no longer possible to take for granted what is meant by ethnography, even by traditional, realist qualitative research (see Snow, 1999, p. 97).[10] Global and local legal processes have erased the personal and institutional distance between the ethnographer and those he or she writes about. We do not "own" the fieldnotes we make about those we study. We do not have an undisputed warrant to study anyone or anything. Subjects now challenge how they have been written about, and more than one ethnographer has been taken to court.

Third, this is a gendered project. Feminist, postcolonial, and queer theorists question the traditional logic of the heterosexual, narrative ethnographic text, which reflexively positions the ethnographer's gender-neutral (or masculine) self within a realist story. Today there is no solidified ethnographic identity. The ethnographer works within a hybrid reality. Experience, discourse, and self-understandings collide against larger cultural assumptions concerning race, ethnicity, nationality, gender, class, and age. A certain identity is never possible; the ethnographer must always ask, "not *who* am I?" but "*when, where, how* am I?" (Trinh, 1992, p. 157).

Fourth, qualitative research is an inquiry project, but it is also a moral, allegorical, and therapeutic project. Ethnography is more than the record of human experience. The ethnographer writes tiny moral tales, tales that do more than celebrate cultural difference or bring another culture alive. The researcher's story is written as a prop, a pillar that, to paraphrase William Faulkner (1967, p. 724), will help men and women endure and prevail in the opening years of the 21st century.

Fifth, while the field of qualitative research is defined by constant breaks and ruptures, there is a shifting center to the project: the avowed humanistic and social justice commitment to study the social world from the perspective of the interacting individual. From this principle flow the liberal and radical politics of action that are held by feminist, clinical, ethnic, critical, queer, critical race theory, and cultural studies researchers. While multiple interpretive communities now circulate within the field of qualitative research, they are all united on this single point.

Sixth, qualitative research's seventh and eighth moments will be defined by the work that interpretive scholars do as they implement the above assumptions. These situations set the stage for qualitative research's transformations in the 21st century. Finally, we anticipate a continued performance turn in qualitative inquiry, with writers performing their texts for others.

▣ TALES OF THE *HANDBOOK*

Many of the difficulties in developing a volume such as this are common to any project of this magnitude. Others were set by the essential tensions and contradictions that operate in this field at this historical moment. As with the first, second, and third editions, the "right" chapter author was unavailable, too busy, or overcommitted. Consequently, we sought out others, who turned out to be more "right" than we imagined possible. Few overlapping networks cut across the many disciplines we were attempting to cover. We were fortunate, in more than one instance, when an editorial board member pointed us in a direction of which we were not even aware. We are grateful to Michelle Fine for connecting us with the international community of indigenous scholars. We have attempted to represent some of the best work available in the North and South American, European, Asian, South African, Australian, and New Zealand traditions of qualitative research.

Although we knew the territory somewhat better this time around, there were still spaces we blundered into with little knowledge about who should be asked to do what. We confronted disciplinary and generational blinders—including our own—and discovered there were separate traditions surrounding each of our topics within distinct interpretive communities. It was often difficult to know how to bridge these differences, and our bridges were often makeshift constructions. We also had to cope with vastly different styles of thinking about a variety of different topics based on disciplinary, epistemological, gender, racial, ethnic, cultural, and national beliefs, boundaries, and ideologies.

In many instances, we unwittingly entered into political battles over who should write a chapter or over how a chapter should be written or evaluated. These disputes clearly pointed to the political nature of this project and to the fact that each chapter was a potential if not real site for multiple interpretations. Many times, the politics of meaning came into play, as we attempted to negotiate and navigate our way through areas fraught with high emotion. On more than one occasion, we disagreed with both an author and an editorial board member. We often found ourselves adjudicating between competing editorial reviews, working the hyphens between meaning-making and diplomacy. Regrettably, in some cases, we hurt feelings and perhaps even damaged long-standing friendships. In such moments, we sought forgiveness. With the clarity of hindsight, there are many things we would do differently today, and we apologize for the damage we have done.

We, as well as our authors and advisers, struggled with the meanings we wanted to bring to such terms as theory, paradigm, epistemology, interpretive framework, empirical materials versus data, research strategies, and so on. We discovered that the very term *qualitative research* means different things to many different people.

We abandoned the goal of being comprehensive, even with 1,500-manuscript pages. We fought with authors over deadlines, and the number of pages we would give them. We also fought with authors over how to conceptualize their chapters and found that what was clear to us was not necessarily clear to anyone else. We fought, too, over when a chapter was done and constantly sought the forbearance of our authors as we requested yet another revision.

◨ READING THE *HANDBOOK*

Were we to write our own critique of this book, we would point to the shortcomings we see in it, and in many senses, these are the same as those in previous editions. They include an over-reliance on the perspectives of our respective disciplines (sociology, communications, and education), as well as a failure to involve more scholars from the international indigenous community. We do not have a detailed treatment of the intersection of critical and indigenous inquiry, nor do we have a comprehensive chapter on human subject research and institutional review boards (IRBs). We worked hard to avoid all of these problems. On the other hand, we have addressed some of the problems present in the third edition. We have made a greater effort to cover more areas of applied qualitative work. We have helped initiate dialogue between different chapter authors. We have created spaces for more voices from other disciplines, especially anthropology and communications, but we still have a shortfall of voices representing people of color and of the Third World. We would have liked to include more non-English speakers from outside Europe and North America. You, the reader, will certainly have your own response to this book, which may highlight other issues that we do not see.

This is all in the nature of the *Handbook* and in the nature of doing qualitative research. This handbook is a social construction, a socially enacted, co-created entity, and though it exists in a material form, it will no doubt be re-created in subsequent iterations as generations of scholars and graduate students use it, adapt it, and launch from it additional methodological paradigmatic, theoretical, and practical work. It is not a final statement. It is a starting point, a springboard for new thought and new work, work that is fresh and sensitive and that blurs the boundaries of our disciplines, but always sharpens our understandings of the larger human project.

With all its strengths and all its flaws, it is our hope that this project, in its fourth edition, will contribute to the growing maturity and global influence of qualitative research in the human disciplines. And, following our original intent, we hope this convinces you, the reader, that qualitative research now constitutes a field of study in its own right, allowing you to better anchor and locate your own work in the qualitative research tradition and its central place in a radical democratic project. If this happens, we will have succeeded in building a bridge that serves all of us well.

◨ ACKNOWLEDGMENTS

This *Handbook* would not exist without its authors, as well as the editorial board members, who gave freely, often on very short notice, of their time, advice, and ever-courteous suggestions. We acknowledge en masse the support of the authors and the editorial board members, whose names are listed facing the title page. These individuals were able to offer both long-term, sustained commitments to the project and short-term emergency assistance.

There are other debts, intensely personal and closer to home. The *Handbook* would never have been possible without the ever-present help, support, wisdom, and encouragement of our editors and publishers at Sage: Michele Sordi, Vicki Knight, Sean Connelly, and Lauren Habib. Their grasp of this field, its history, and diversity is extraordinary. Their conceptions of what this project should look like were extremely valuable. Their energy kept us moving forward. Furthermore, whenever we confronted a problem, Michele, Vicki, and Lauren were there with their assistance and good-natured humor.

We would also like to thank the following individuals and institutions for their assistance, support, insights, and patience: our respective universities, administrations, and departments. In Urbana, James Salvo, Melba Velez, Koeli Goel, and Katia Curbelo were the sine qua non. Their good humor and grace kept our ever-growing files in order and everyone on the same timetable. Without them, this project would never have been completed.

The following individuals at Sage Publications helped move this project through production: Astrid Virding, Jackie Tasch, Taryn Bigelow, Robin Gold, and Teresa Herlinger. We are extremely grateful to them, as well as to Dennis Webb and Kathy Paparchontis for their excellent work during the proofreading and indexing phases of production. Our spouses, Katherine Ryan and Egon Guba, helped keep us on track, listened to our complaints, and generally displayed extraordinary patience, forbearance, and support.

Finally, there is another group of individuals who gave unstintingly of their time and energy to provide us with their expertise and thoughtful reviews when we needed additional guidance. Without the help of these individuals, we would often have found ourselves with less than complete understandings of the various traditions, perspectives, and methods represented in this volume. We would also like to acknowledge the important contributions of the following special readers to this project: Bryant Alexander, Susan Chase, Michele Fine, Susan Finley, Andrea Fontana, Jaber Gubrium, James Holstein, Alison Jones, Stacy Holman Jones, Tony Kuzel, Luis Miron, Ron Pelias, John Prosser, Johnny Saldana, and Harry Torrance.

Norman K. Denzin
University of Illinois at Urbana-Champaign
Yvonna S. Lincoln
Texas A & M University

▣ NOTES

1. Qualitative inquiry in North America has passed through several historical moments or phases: the traditional (1900–1950), the modernist or golden age (1950–1970), blurred genres (1970–1986), the crisis of representation (1986–1990), the postmodern (1990–1995), postexperimental inquiry (1995–2000), the methodologically contested present (2000–2004), and the fractured future (2005-). These moments overlap and coexist in the present (see Denzin & Lincoln, 2005, pp. 2–3). This model has been termed a progress narrative by Alasuutari (2004, pp. 599–600); Seale, Gobo, Gubrium, and Silverman (2004, p. 2); and Atkinson, Coffey, and Delamont (2003). The critics assert that we believe that the most recent moment is the most up-to-date, the avant-garde, the cutting edge (Alasuutari, 2004, p. 601). Naturally, we dispute this reading. Teddlie and Tashakkori (Chapter 16, this volume) have modified our historical periods to fit their historical analysis of the major moments in the emergence of mixed methods in the last century.

2. Today the list for the United States (and England) is very, very long, many of the journals are published by Sage, including *Qualitative Inquiry, Qualitative Health Research, Qualitative Research, Qualitative Social Work, Cultural Studies <=> Critical Methodologies, Journal of Contemporary Ethnography, Discourse Studies, Discourse and Society, Ethnography,* and *Field Methods.* Other important journals include *International Journal of Qualitative Studies in Education, Anthropology and Education, Communication and Critical/Cultural Studies, Text and Performance Quarterly,* and *The International Review of Qualitative Research.*

3. Again, from Sage—the *Handbooks of: Qualitative Research, Grounded Theory, Ethnography, Interviewing, Narrative Inquiry, Performance Studies,* and *Critical and Indigenous Methodologies.*

4. Sage seemingly has dozens of these texts, including those focused on case study, interviewing, Internet inquiry, ethnography, focus groups, visual data, conversation analysis, observation, participatory action research, ethics, qualitative design and analysis, life history, and interpretive biography (see Staller, Block, & Horner, 2008, for a review of Sage's place in this discourse.

5. Including the distinguished qualitative dissertation awards of the International Association of Qualitative Inquiry and the American Educational Research Association (AERA).

6. Including the Annual Egon Guba Distinguished Lecture for the QUALSIG of AERA.

7. On May 7, 2005, the last day of the First International Congress of Qualitative Inquiry, the International Association of Qualitative Inquiry (IAQI) was founded in Urbana, Illinois. IAQI is the first international association solely dedicated to the scholarly promotion, representation, and global development of qualitative research. At present, IAQI has 3,500 delegates representing 60 nations worldwide. It has established professional affiliations with more than 150 collaborating sites in Oceana, Africa, North and South America, the Caribbean, Europe, the Middle East, Japan, Korea, and China (see icqi. org). The *IAQI Newsletter* appears quarterly, as does a new journal, *The International Review of Qualitative Research.*

8. Mixed methods research is Teddlie and Tashakkori's third movement or moment. The first movement is quantitative research, and the second is qualitative inquiry. The third moment offers a middle ground that mediates quantitative and qualitative disputes (Teddlie and Tashakkori, Chapter 16, this volume).

9. These criteria range from those endorsed by postpositivists (variations on validity and reliability, including credibility and trustworthiness), to poststructural feminist standpoint concerns emphasizing collaborative, evocative performance texts that create ethically responsible relations between researchers and those they study.

10. The realist text, Jameson (1990) argues, constructed its version of the world by "programming. . . . readers; by training them in new habits and practices. . . . such narratives must ultimately produce that very category of Reality. . . . of the real, of the 'objective' or 'external' world, which itself historical, may undergo decisive modification in other modes of production, if not in later stages of this one" (p. 166). The new ethnographic text is producing its versions of reality and teaching readers how to engage this view of the social world.

◨ REFERENCES

Alasuutari, P. (2004). The globalization of qualitative research. In C. Seale, G. Gobo, J. F. Gubrium, & D. Silverman (Eds.), *Qualitative research practice* (pp. 595–608). London: Sage.

Atkinson, P., Coffey, A., & Delamont, S. (2003). *Key themes in qualitative research: Continuities and change.* Walnut Creek, CA: AltaMira.

Denzin, N. K., & Lincoln, Y. S. (2005). Introduction: The discipline and practice of qualitative research. In N. K. Denzin & Y. S. Lincoln (Eds.), *The SAGE handbook of qualitative research* (3rd ed., pp. 1–32). Thousand Oaks, CA: Sage.

Downing, D. B. (1987). Deconstruction's scruples: The politics of enlightened critique. *Diacritics, 17,* 66–81.

Faulkner, W. (1967). Address upon receiving the Nobel Prize for Literature. In M. Cowley (Ed.), *The portable Faulkner* (Rev. and expanded ed., pp. 723–724). New York: Viking.

Harper, D. (1987). *Working knowledge: Skill and community in a small shop.* Chicago: University of Chicago Press.

Hatch, A. (2006). Qualitative studies in the era of scientifically-based research: Musings of a former QSE editor. *International Journal of Qualitative Studies in Education, 19*(July-August), 403–409.

Jameson, F. (1990). *Signatures of the visible.* New York: Routledge.

Kincheloe, J. (2008). *Critical pedagogy primer.* New York: Peter Lang.

Madison, D. S. (1998). Performances, personal narratives, and the politics of possibility." In S. J. Dailey (Ed.), *The future of performance studies: Visions and revisions* (pp. 276–286). Washington, DC: National Communication Association.

Seale, C., Gobo, G., Gubrium, J. F., & Silverman, D. (Eds.). (2004). *Qualitative research practice.* London: Sage.

Snow, D. (1999). Assessing the ways in which qualitative/ethnographic research contributes to social psychology: Introduction to special issues. *Social Psychology Quarterly, 62,* 97–100.

Staller, K. M., Block, E., & Horner, P. S. (2008). History of methods in social science research. In S. N. Hesse-Biber & P. Leavy (Eds.), *Handbook of emergent methods* (pp. 25–54). New York: Guilford Press.

Trinh, T. M. (1992). *Framer framed.* New York: Routledge.

Weems, M. (2002). *I speak from the wound that is my mouth.* New York: Peter Lang.

1

INTRODUCTION

The Discipline and Practice of Qualitative Research

Norman K. Denzin and Yvonna S. Lincoln

The global community of qualitative researchers is midway between two extremes, searching for a new middle, moving in several different directions at the same time.[1] Mixed methodologies and calls for scientifically based research, on the one side, renewed calls for social justice inquiry from the critical social science tradition on the other. In the methodological struggles of the 1970s and 1980s, the very existence of qualitative research was at issue. In the new paradigm war, "every overtly social justice-oriented approach to research . . . is threatened with de-legitimization by the government-sanctioned, exclusivist assertion of positivism . . . as the 'gold standard' of educational research" (Wright, 2006, pp. 799–800).

The evidence-based research movement, with its fixed standards and guidelines for conducting and evaluating qualitative inquiry, sought total domination: one shoe fits all (Cannella & Lincoln, Chapter 5, this volume; Lincoln, 2010). The heart of the matter turns on issues surrounding the politics and ethics of evidence and the value of qualitative work in addressing matters of equity and social justice (Torrance, Chapter 34, this volume).

In this introductory chapter, we define the field of qualitative research, then navigate, chart, and review the history of qualitative research in the human disciplines. This will allow us to locate this handbook and its contents within their historical moments. (These historical moments are somewhat artificial; they are socially constructed, quasi-historical, and overlapping conventions. Nevertheless, they permit a "performance" of developing ideas. They also facilitate an increasing sensitivity to and sophistication about the pitfalls and promises of ethnography and qualitative research.) A conceptual framework for reading the qualitative research act as a multicultural, gendered process is presented.

We then provide a brief introduction to the chapters, concluding with a brief discussion of qualitative research. We will also discuss the threats to qualitative human-subject research from the methodological conservatism movement, which was noted in our Preface. As indicated there, we use the metaphor of the bridge to structure what follows. This volume provides a bridge between historical moments, politics, the decolonization project, research methods, paradigms, and communities of interpretive scholars.

HISTORY, POLITICS, AND PARADIGMS

To better understand where we are today and to better grasp current criticisms, it is useful to return to the so-called paradigm wars of the 1980s, which resulted in the serious crippling of quantitative research in education. Critical pedagogy, critical theorists, and feminist analyses fostered struggles to acquire power and cultural capital for the poor, non-whites, women, and gays (Gage, 1989).

Charles Teddlie and Abbas Tashakkori's history is helpful here. They expand the time frame of the 1980s war to embrace at least three paradigm wars, or periods of conflict: the postpositivist-constructivist war against positivism (1970–1990); the conflict between competing postpositivist, constructivist, and critical theory paradigms (1990–2005); and the current conflict between evidence-based methodologists and the mixed methods, interpretive, and critical theory schools (2005–present).[2]

Egon Guba's (1990a) *The Paradigm Dialog* signaled an end to the 1980s wars. Postpositivists, constructivists, and critical theorists talked to one another, working through issues connected to ethics, field studies, praxis, criteria, knowledge accumulation, truth, significance, graduate training, values, and politics. By the early 1990s, there was an explosion of published work on qualitative research; handbooks and new journals appeared. Special interest groups committed to particular paradigms appeared, some with their own journals.[3]

The second paradigm conflict occurred within the mixed methods community and involved disputes "between individuals convinced of the 'paradigm purity' of their own position" (Teddlie & Tashakkori, 2003b, p. 7). Purists extended and repeated the argument that quantitative and qualitative methods and postpositivism and the other "isms" cannot be combined because of the differences between their underlying paradigm assumptions. On the methodological front, the incompatibility thesis was challenged by those who invoked triangulation as a way of combining multiple methods to study the same phenomenon (Teddlie & Tashakkori, 2003a, p. 7). This ushered in a new round of arguments and debates over paradigm superiority.

A soft, apolitical pragmatic paradigm emerged in the post-1990 period. Suddenly, quantitative and qualitative methods became compatible, and researchers could use both in their empirical inquiries (Teddlie & Tashakkori, 2003a, p. 7). Proponents made appeals to a "what works" pragmatic argument, contending that "no incompatibility between quantitative and qualitative methods exists at either the level of practice or that of epistemology . . . there are thus no good reasons for educational researchers to fear forging ahead with 'what works'" (Howe, 1988, p. 16). Of course, what works is more than an empirical question. It involves the politics of evidence.

This is the space that evidence-based research entered. It became the battleground of the third war, "the current upheaval and argument about 'scientific' research in the scholarly world of education" (Clark & Scheurich, 2008; Scheurich & Clark, 2006, p. 401). Enter Teddlie and Tashakkori's third moment: Mixed methods and evidence-based inquiry meet one another in a soft center. C. Wright Mills (1959) would say this is a space for abstracted empiricism. Inquiry is cut off from politics. Biography and history recede into the background. Technological rationality prevails.

Resistances to Qualitative Studies

The academic and disciplinary resistances to qualitative research illustrate the politics embedded in this field of discourse. The challenges to qualitative research are many. To better understand these criticisms, it is necessary to "distinguish analytically the political (or external) role of [qualitative] methodology from the procedural (or internal) one" (Seale, Gobo, Gubrium, & Silverman, 2004, p. 7). Politics situate methodology within and outside the academy. Procedural issues define how qualitative methodology is used to produce knowledge about the world (Seale et al., 2004, p. 7).

Often, the political and the procedural intersect. Politicians and hard scientists call qualitative researchers *journalists* or "soft" scientists. Their work is termed unscientific, only exploratory, or subjective. It is called criticism and not theory, or it is interpreted politically, as a disguised version of Marxism or secular humanism (see Huber, 1995; also Denzin, 1997, pp. 258–261).

These political and procedural resistances reflect an uneasy awareness that the interpretive traditions of qualitative research commit one to a critique of the positivist or post-positivist project. But the positivist resistance to qualitative research goes beyond the "ever-present desire to maintain a distinction between hard science and soft scholarship" (Carey, 1989, p. 99). The experimental (positivist) sciences (physics, chemistry, economics, and psychology, for example) are often seen as the crowning achievements of Western civilization, and in their practices, it is assumed that "truth" can transcend opinion and personal bias (Carey, 1989, p. 99; Schwandt, 1997b, p. 309). Qualitative research is seen as an assault on this tradition, whose adherents often retreat into a "value-free objectivist science" (Carey, 1989, p. 104) model to defend their position. The positivists seldom attempt to make explicit, and critique the "moral and political commitments in their own contingent work" (Carey, 1989, p. 104; Lincoln, Lynham, & Guba, Chapter 6, this volume).

Positivists further allege that the so-called new experimental qualitative researchers write fiction, not science, and have no way of verifying their truth statements. Ethnographic poetry and fiction signal the death of empirical science, and there is little to be gained by attempting to engage in moral criticism. These critics presume a stable, unchanging reality that can be studied with the empirical methods of objective social science (see Huber, 1995). The province of qualitative research, accordingly, is the world of lived experience, for this is where individual belief and action intersect with culture. Under this model, there is no preoccupation with discourse and method as material interpretive practices that constitute representation and description. This is the textual, narrative turn rejected by the positivists.

The opposition to positive science by the poststructuralists is seen, then, as an attack on reason and truth. At the same time, the positivist science attack on qualitative research is regarded as an attempt to legislate one version of truth over another.

The Legacies of Scientific Research

Writing about scientific research, including qualitative research, from the vantage point of the colonized, a position that she chooses to privilege, Linda Tuhiwai Smith states that "the term 'research' is inextricably linked to European imperialism and colonialism." She continues, "the word itself is probably one of the dirtiest words in the indigenous world's vocabulary . . .

It is "implicated in the worst excesses of colonialism" (p. 1), with the ways in which "knowledge about indigenous peoples was collected, classified, and then represented back to the West" (Smith, 1999, p. 1). This dirty word stirs up anger, silence, distrust. "It is so powerful that indigenous people even write poetry about research " (Smith, 1999, p. 1). It is one of colonialism's most sordid legacies, she says.

Frederick Erickson's Chapter 3 of this volume charts many key features of this painful history. He notes with some irony that qualitative research in sociology and anthropology was born out of concern to understand the exotic, often dark-skinned "other." Of course, there were colonialists long before there were anthropologists and ethnographers. Nonetheless, there would be no colonial—and now no neo-colonial—history, were it not for this investigative mentality that turned the dark-skinned other into the object of the ethnographer's gaze. From the very beginning, qualitative research was implicated in a racist project.[4]

▣   DEFINITIONAL ISSUES

Qualitative research is a field of inquiry in its own right. It crosscuts disciplines, fields, and subject matter.[5] A complex, interconnected family of terms, concepts, and assumptions surrounds the term. These include the traditions associated with foundationalism, positivism, postfoundationalism, postpositivism, poststructuralism, postmodernism, post-humanism, and the many qualitative research perspectives and methods connected to cultural and interpretive studies (the chapters in Part II of this volume take up these paradigms).[6] There are separate and detailed literatures on the many methods and approaches that fall under the category of qualitative research, such as case study, politics and ethics, participatory inquiry, interviewing, participant observation, visual methods, and interpretive analysis.

In North America, qualitative research operates in a complex historical field that crosscuts at least eight historical moments. These moments overlap and simultaneously operate in the present.[7] We define them as the traditional (1900–1950), the modernist or golden age (1950–1970), blurred genres (1970–1986), the crisis of representation (1986–1990), the postmodern, a period of experimental and new ethnographies (1990–1995), postexperimental inquiry (1995–2000), the methodologically contested present (2000–2010), and the future (2010–), which is now. The future, the eighth moment, confronts the methodological backlash associated with the evidence-based social movement. It is concerned with moral discourse, with the development of sacred textualities. The eighth moment asks that the social sciences and the humanities become sites for critical conversations about democracy, race, gender, class, nation-states, globalization, freedom, and community.[8]

The postmodern and postexperimental moments were defined in part by a concern for literary and rhetorical tropes and the narrative turn, a concern for storytelling, for composing ethnographies in new ways (Ellis, 2009; and in this volume, Hamera, Chapter 18; Tedlock, Chapter 19; Spry, Chapter 30; Ellingson, Chapter 36; St.Pierre, Chapter 37; and Pelias, Chapter 40).

Successive waves of epistemological theorizing move across these eight moments. The traditional period is associated with the positivist, foundational paradigm. The modernist or golden age and blurred genres moments are connected to the appearance of postpositivist arguments. At the same time, a variety of new interpretive, qualitative perspectives were taken up, including hermeneutics, structuralism, semiotics, phenomenology, cultural studies, and feminism.[9] In the blurred genre phase, the humanities became central resources for critical, interpretive theory and the qualitative research project broadly conceived. The researcher became a *bricoleur* (as discussed later), learning how to borrow from many different disciplines.

The blurred genres phase produced the next stage, the crisis of representation. Here researchers struggled with how to locate themselves and their subjects in reflexive texts. A kind of methodological diaspora took place, a two-way exodus. Humanists migrated to the social sciences, searching for new social theory and new ways to study popular culture and its local ethnographic contexts. Social scientists turned to the humanities, hoping to learn how to do complex structural and poststructural readings of social texts. From the humanities, social scientists also learned how to produce texts that refused to be read in simplistic, linear, incontrovertible terms. The line between a text and a context blurred. In the postmodern experimental moment, researchers continued to move away from foundational and quasifoundational criteria (in this volume, see Altheide & Johnson, Chapter 35; St.Pierre, Chapter 37). Alternative evaluative criteria were sought, ones that might prove evocative, moral, critical, and rooted in local understandings.

Any definition of qualitative research must work within this complex historical field. Qualitative research means different things in each of these moments. Nonetheless, an initial, generic definition can be offered. *Qualitative research* is a situated activity that locates the observer in the world. Qualitative research consists of a set of interpretive, material practices that make the world visible. These practices transform the world. They turn the world into a series of representations, including fieldnotes, interviews, conversations, photographs, recordings, and memos to the self. At this level, qualitative research involves an interpretive, naturalistic approach to the world. This means that qualitative researchers study things in their natural settings, attempting to make sense of or interpret phenomena in terms of the meanings people bring to them.[10]

Qualitative research involves the studied use and collection of a variety of empirical materials—case study, personal experience, introspection, life story, interview, artifacts, and cultural texts and productions, along with observational, historical, interactional, and visual texts—that describe routine and problematic moments and meanings in individuals' lives. Accordingly, qualitative researchers deploy a wide-range of interconnected interpretive practices, hoping always to get a better understanding of the subject matter at hand. It is understood, however, that each practice makes the world visible in a different way. Hence, there is frequently a commitment to using more than one interpretive practice in any study.

◼ THE QUALITATIVE RESEARCHER-AS-BRICOLEUR AND QUILT MAKER

Multiple gendered images may be brought to the qualitative researcher: scientist, naturalist, fieldworker, journalist, social critic, artist, performer, jazz musician, filmmaker, quilt maker, essayist. The many methodological practices of qualitative research may be viewed as soft science, journalism, ethnography, *bricolage*, quilt making, or montage. The researcher, in turn, may be seen as a *bricoleur*, as a maker of quilts, or in filmmaking, a person who assembles images into montages (on montage, see Cook, 1981, pp. 171–177; Monaco, 1981, pp. 322–328; and discussion below; on quilting, see hooks, 1990, pp. 115–122; Wolcott, 1995, pp. 31–33).

Douglas Harper (1987, pp. 9, 74–75, 92); Michel de Certeau (1984, p. xv); Cary Nelson, Paula A. Treichler, and Lawrence Grossberg (1992, p. 2); Claude Lévi-Strauss (1962/1966, p. 17); Deena and Michael Weinstein (1991, p. 161); and Joe L. Kincheloe (2001) clarify the meaning of bricolage and bricoleur.[11] A bricoleur makes do by "adapting the bricoles of the world. Bricolage is 'the poetic making do'" (de Certeau, 1984, p. xv), with "such bricoles—the odds and ends, the bits left over" (Harper, 1987, p. 74). The bricoleur is a "Jack of all trades, a kind of professional do-it-yourself[er]" (Lévi-Strauss, 1962/1966, p. 17). In Harper's (1987) work, the bricoleur defines herself and extends herself (p. 75). Indeed, her life story, her biography, "may be thought of as bricolage" (Harper, 1987, p. 92).

There are many kinds of bricoleurs—interpretive, narrative, theoretical, political. The interpretive bricoleur produces a bricolage; that is, a pieced-together set of representations that are fitted to the specifics of a complex situation. "The solution (bricolage) which is the result of the bricoleur's method is an [emergent] construction" (Weinstein & Weinstein, 1991, p. 161), which changes and takes new forms as different tools, methods, and techniques of representation and interpretation are added to the puzzle. Nelson et al. (1992) describe the methodology of cultural studies "as a bricolage. Its choice of practice, that is, is pragmatic, strategic, and self-reflexive" (p. 2). This understanding can be applied, with qualifications, to qualitative research.

The qualitative-researcher-as-bricoleur or a maker of quilts uses the aesthetic and material tools of his or her craft, deploying whatever strategies, methods, or empirical materials are at hand (Becker, 1998, p. 2). If new tools or techniques have to be invented or pieced together, then the researcher will do this. The choice of which interpretive practices to employ is not necessarily set in advance. The "choice of research practices depends upon the questions that are asked, and the questions depend on their context" (Nelson et al., 1992, p. 2), what is available in the context, and what the researcher can do in that setting.

These interpretive practices involve aesthetic issues, an aesthetics of representation that goes beyond the pragmatic or the practical. Here the concept of *montage* is useful (see Cook, 1981, p. 323; Monaco, 1981, pp. 171–172). Montage is a method of editing cinematic images. In the history of cinematography, montage is associated with the work of Sergei Eisenstein, especially his film, *The Battleship Potemkin* (1925). In montage, a picture is made by superimposing several different images on one another. In a sense, montage is like *pentimento*, where something painted out of a picture (an image the painter "repented," or denied) now becomes visible again, creating something new. What is new is what had been obscured by a previous image.

Montage and pentimento, like jazz, which is improvisation, create the sense that images, sounds, and understandings are blending together, overlapping, and forming a composite, a new creation. The images seem to shape and define one another; an emotional gestalt effect is produced. Often, these images are combined in a swiftly run sequence. When done, this produces a dizzily revolving collection of several images around a central or focused picture or sequence; such effects signify the passage of time.

Perhaps the most famous instance of montage is given in the Odessa Steps sequence in *The Battleship Potemkin*.[12] In the climax of the film, the citizens of Odessa are being massacred by tsarist troops on the stone steps leading down to the city's harbor. Eisenstein cuts to a young mother as she pushes her baby's carriage across the landing in front of the firing troops. Citizens rush past her, jolting the carriage, which she is afraid to push down to the next flight of stairs. The troops are above her firing at the citizens. She is trapped between the troops and the steps. She screams. A line of rifles pointing to the sky erupts in smoke. The mother's head sways back. The wheels of the carriage teeter on the edge of the steps. The mother's hand clutches the silver buckle of her belt. Below her, people are being beaten by soldiers. Blood drips over the mother's white gloves. The baby's hand reaches out of the carriage. The mother sways back and forth. The troops advance. The mother falls back against the carriage. A woman watches in horror as the rear wheels of the carriage roll off the edge of the landing. With accelerating speed, the carriage bounces down the steps, past the dead citizens. The baby is jostled from side to side inside the carriage. The soldiers fire their rifles into a group of wounded citizens. A student screams, as the carriage leaps across the steps, tilts, and overturns (Cook, 1981, p. 167).[13]

Montage uses sparse images to create a clearly defined sense of urgency and complexity. Montage invites viewers to construct interpretations that build on one another as a scene unfolds. These interpretations are built on associations based on the contrasting images that blend into one another. The underlying assumption of montage is that viewers perceive and interpret the shots in a "montage sequence not *sequentially,* or one at a time, but rather *simultaneously*" (Cook, 1981, p. 172, italics in original). The viewer puts the sequences together into a meaningful emotional whole, as if at a glance, all at once.

The qualitative researcher who uses montage is like a quilt maker or a jazz improviser. The quilter stitches, edits, and puts slices of reality together. This process creates and brings psychological and emotional unity to an interpretive experience. There are many examples of montage in current qualitative research. Using multiple voices and different textual formations, voices, and narrative styles, Marcelo Diversi and Claudio Moreira (2009) weave a complex text about race, identity, nation, class, sexuality, intimacy, and family. As in quilt making and jazz improvisation, many different things are going on at the same time: different voices, different perspectives, points of views, angles of vision. Autoethnographic performance texts use montage simultaneously to create and enact moral meaning. They move from the personal to the political, the local to the historical and the cultural. These are dialogical texts. They presume an active audience. They create spaces for give and take between reader and writer. They do more than turn the other into the object of the social science gaze (in this volume, see Spry, Chapter 30; Pelias, Chapter 40).

Of course, qualitative research is inherently multimethod in focus (Flick, 2002, pp. 226–227; 2007). However, the use of multiple methods, or triangulation, reflects an attempt to secure an in-depth understanding of the phenomenon in question. Objective reality can never be captured. We know a thing only through its representations. Triangulation is not a tool or a strategy of validation but an alternative to validation (Flick, 2002, p. 227; 2007). The combination of multiple methodological practices, empirical materials, perspectives, and observers in a single study is best understood, then, as a strategy that adds rigor, breadth complexity, richness, and depth to any inquiry (see Flick, 2002, p. 229; 2007, pp. 102–104).

Laura L. Ellingson (Chapter 36, this volume; also 2009) disputes a narrow conception of triangulation, endorsing instead a postmodern form (2009, p. 190). It asserts that the central image for qualitative inquiry is the crystal—multiple lenses—not the triangle. She sees crystallization as embodying an energizing, unruly discourse, drawing raw energy from artful science and scientific artwork (p. 190). Mixed-genre texts in the postexperimental moment have more than three sides.

Like crystals, Eisenstein's montage, the jazz solo, or the pieces in a quilt, the mixed-genre text combines "symmetry and substance with an infinite variety of shapes, substances, transmutations . . . crystals grow, change, alter . . . crystals are prisms that reflect externalities and refract within themselves, creating different colors, patterns, arrays, casting off in different directions" (Richardson, 2000, p. 934).

In the crystallization process, the writer tells the same tale from different points of view. Crystallized projects mix genres and writing formats, offering partial, situated, open-ended conclusions. In *Fires in the Mirror* (1993) Anna Deavere Smith presents a series of performance pieces based on interviews with people involved in a racial conflict in Crown Heights, Brooklyn, on August 19, 1991. Her play has multiple speaking parts, including conversations with gang members, the police, and anonymous young girls and boys. There is no correct telling of this event. Each telling, like light hitting a crystal, gives a different reflection of the racial incident.

Viewed as a crystalline form, as a montage, or as a creative performance around a central theme, triangulation as a form of, or alternative to, validity thus can be extended. Triangulation is the display of multiple, refracted realities simultaneously. Each of the metaphors "works" to create simultaneity rather than the sequential or linear. Readers and audiences are then invited to explore competing visions of the context, to become immersed in and merge with new realities to comprehend.

The methodological bricoleur is adept at performing a large number of diverse tasks, ranging from interviewing to intensive self-reflection and introspection. The theoretical bricoleur reads widely and is knowledgeable about the many interpretive paradigms (feminism, Marxism, cultural studies, constructivism, queer theory) that can be brought to any particular problem. He or she may not, however, feel that paradigms can be mingled or synthesized. If paradigms are overarching philosophical systems denoting particular ontologies, epistemologies, and methodologies, one cannot move easily from one to the other. Paradigms represent belief systems that attach the user to a particular worldview. Perspectives, in contrast, are less well developed systems, and it can be easier to move between them. The researcher-as-bricoleur-theorist works between and within competing and overlapping perspectives and paradigms.

The interpretive bricoleur understands that research is an interactive process shaped by one's personal history, biography, gender, social class, race, and ethnicity and those of the people in the setting. Critical bricoleurs stress the dialectical and hermeneutic nature of interdisciplinary inquiry, knowing that the boundaries between traditional disciplines no longer hold (Kincheloe, 2001, p. 683). The political bricoleur knows that science is power, for all research findings have political implications. There is no value-free science. A civic social science based on a politics of hope is sought (Lincoln, 1999). The gendered, narrative bricoleur also knows that researchers all tell stories about the worlds they have studied. Thus, the narratives or stories scientists tell are accounts couched and framed within specific storytelling traditions, often defined as paradigms (e.g., positivism, postpositivism, constructivism).

The product of the interpretive bricoleur's labor is a complex, quilt-like bricolage, a reflexive collage or montage; a set of fluid, interconnected images and representations. This interpretive structure is like a quilt, a performance text, or a sequence of representations connecting the parts to the whole.

◨ QUALITATIVE RESEARCH AS A SITE OF MULTIPLE INTERPRETIVE PRACTICES

Qualitative research, as a set of interpretive activities, privileges no single methodological practice over another. As a site of discussion or discourse, qualitative research is difficult to define clearly. It has no theory or paradigm that is distinctly its own. As Part II of this volume reveals, multiple theoretical paradigms claim use of qualitative research methods and strategies, from constructivism to cultural studies, feminism, Marxism, and ethnic models of study. Qualitative research is used in many separate disciplines, as we will discuss below. It does not belong to a single discipline.

Nor does qualitative research have a distinct set of methods or practices that are entirely its own. Qualitative researchers use semiotics, narrative, content, discourse, archival, and phonemic analysis—even statistics, tables, graphs, and numbers. They also draw on and use the approaches, methods, and techniques of ethnomethodology, phenomenology, hermeneutics, feminism, rhizomatics, deconstructionism, ethnographies, interviews, psychoanalysis, cultural studies, survey research, and participant observation, among others.[14] All of these research practices "can provide important insights and knowledge" (Nelson et al., 1992, p. 2). No specific method or practice can be privileged over another.

Many of these methods or research practices are used in other contexts in the human disciplines. Each bears the traces of its own disciplinary history. Thus, there is an extensive history of the uses and meanings of ethnography and ethnology in education (Erickson, Chapter 3, this volume); of participant observation and ethnography in anthropology (Tedlock, Chapter 19, this volume); sociology (Holstein & Gubrium, Chapter 20, this volume); communications (in this volume, Hamera, Chapter 18; Spry, Chapter 30); cultural studies (Giardina & Newman, Chapter 10, this volume); textual, hermeneutic, feminist, psychoanalytic, arts-based, semiotic, and narrative analysis in cinema and literary studies (in this volume, Olesen, Chapter 7; Chase, Chapter 25; Finley, Chapter 26); and narrative, discourse, and conversational analysis in sociology, medicine, communications, and education (in this volume, Chase, Chapter 25; Peräkylä & Ruusuvuori, Chapter 32).

The many histories that surround each method or research strategy reveal how multiple uses and meanings are brought to each practice. Textual analyses in literary studies, for example, often treat texts as self-contained systems. On the other hand, a cultural studies or feminist perspective reads a text in terms of its location within a historical moment marked by a particular gender, race, or class ideology. A cultural studies use of ethnography would bring a set of understandings from feminism, postmodernism, and poststructuralism to the project. These understandings would not be shared by mainstream postpositivist sociologists. Similarly, postpositivist and poststructural historians bring different understandings and uses to the methods and findings of historical research. These tensions and contradictions are evident in many of the chapters in this handbook.

These separate and multiple uses and meanings of the methods of qualitative research make it difficult to agree on any essential definition of the field, for it is never just one thing.[15] Still, a definition must be made. We borrow from and paraphrase Nelson et al.'s (1992, p. 4) attempt to define cultural studies:

> Qualitative research is an interdisciplinary, transdiciplinary, and sometimes counterdisciplinary field. It crosscuts the humanities, as well as the social and the physical sciences. Qualitative research is many things at the same time. It is multiparadigmatic in focus. Its practitioners are sensitive to the value of the multimethod approach. They are committed to the naturalistic perspective and to the interpretive understanding of human experience. At the same time, the field is inherently political and shaped by multiple ethical and political positions.
>
> Qualitative research embraces two tensions at the same time. On the one hand, it is drawn to a broad, interpretive, postexperimental, postmodern, feminist, and critical sensibility. On the other hand, it is drawn to more narrowly defined positivist, postpositivist, humanistic, and naturalistic conceptions of human experience and its analysis. Furthermore, these tensions can be combined in the same project, bringing both postmodern and naturalistic, or both critical and humanistic, perspectives to bear.

This rather awkward statement means that qualitative research is a set of complex interpretive practices. As a constantly shifting historical formation, it embraces tensions and contradictions, including disputes over its methods and the forms its findings and interpretations take. The field sprawls between and crosscuts all of the human disciplines, even including, in some cases, the physical sciences. Its practitioners are variously committed to modern, postmodern, and postexperimental sensibilities and the approaches to social research that these sensibilities imply.

Politics and Reemergent Scientism

In the first decade of this new century, the scientifically based research movement (SBR) initiated by the National Research

Council (NRC) created a new and hostile political environment for qualitative research (Howe, 2009). Connected to the No Child Left Behind Act of 2001 (NCLB), SBR embodied a reemergent scientism (Maxwell, 2004), a positivist evidence-based epistemology. Researchers are encouraged to employ "rigorous, systematic, and objective methodology to obtain reliable and valid knowledge" (Ryan & Hood, 2004, p. 80). The preferred methodology has well-defined causal models using independent and dependent variables. Causal models are examined in the context of randomized controlled experiments, which allow replication and generalization (Ryan & Hood, 2004, p. 81).

Under this framework, qualitative research becomes suspect. There are no well-defined variables or causal models. Observations and measurements are not based on random assignment to experimental groups. Hard evidence is not generated by these methods. At best, case study, interview, and ethnographic methods offer descriptive materials that can be tested with experimental methods. The epistemologies of critical race, queer, postcolonial, feminist, and postmodern theories are rendered useless, relegated at best to the category of scholarship, not science (Ryan & Hood, 2004, p. 81; St.Pierre & Roulston, 2006, p. 132).

Critics of the evidence movement are united on the following points. The movement endorses a narrow view of science (Lather, 2004; Maxwell, 2004), celebrating a "neoclassical experimentalism that is a throwback to the Campbell-Stanley era and its dogmatic adherence to an exclusive reliance on quantitative methods" (Howe, 2004, p. 42). There is "nostalgia for a simple and ordered universe of science that never was" (Popkewitz, 2004, p. 62). With its emphasis on only one form of scientific rigor, the NRC ignores the need for and value of complex historical, contextual, and political criteria for evaluating inquiry (Bloch, 2004).

Neoclassical experimentalists extol evidence-based "medical research as the model for educational research, particularly the random clinical trial" (Howe, 2004, p. 48). But the random clinical trial—dispensing a pill—is quite unlike "dispensing a curriculum" (Howe, 2004, p. 48), nor can the "effects" of the educational experiment be easily measured, unlike a "10-point reduction in diastolic blood pressure" (Howe, 2004, p. 48).

Qualitative researchers must learn to think outside the box as they critique the NRC and its methodological guidelines (Atkinson, 2004). We must apply our critical imaginations to the meaning of such terms as *randomized design, causal model, policy studies,* and *public science* (Cannella & Lincoln, 2004; Weinstein, 2004). At a deeper level, we must resist conservative attempts to discredit qualitative inquiry by placing it back inside the box of positivism.

Contesting Mixed Methods Experimentalism

Kenneth R. Howe (2004) observes that the NRC finds a place for qualitative methods in mixed methods experimental designs. In such designs, qualitative methods may be "employed either singly or in combination with quantitative methods, including the use of randomized experimental designs" (Howe, 2004, p. 49; also Clark & Creswell, 2008; Hesse-Biber & Leavy, 2008). Clark, Creswell, Green, and Shope (2008) define mixed methods research "as a design for collecting, analyzing, and mixing both quantitative and qualitative data in a study in order to understand a research problem" (p. 364).[16] Mixed methods are direct descendants of classical experimentalism and the triangulation movement of the 1970s (Denzin, 1989b). They presume a methodological hierarchy, with quantitative methods at the top, relegating qualitative methods to "a largely auxiliary role in pursuit of the *technocratic* aim of accumulating knowledge of 'what works'" (Howe, 2004, pp. 53–54).

The *incompatibility thesis* disputes the key claim of the mixed methods movement, namely that methods and perspectives can be combined. Recalling the paradigm wars of the 1980s, this thesis argues that "compatibility between quantitative and qualitative methods is impossible due to incompatibility of the paradigms that underlie the methods" (Teddlie & Tashakkori 2003a, pp. 14–15; 2003b). Others disagree with this conclusion, and some contend that the incompatibility thesis has been largely discredited because researchers have demonstrated that it is possible to successfully use a mixed methods approach.

There are several schools of thought on this thesis, including the four identified by Teddlie and Tashakkori (2003a); that is, the complementary, single paradigm, dialectical, and multiple paradigm models. There is by no means consensus on these issues. Morse and Niehaus (2009) warn that ad hoc mixing of methods can be a serious threat to validity. Pragmatists and transformative emancipatory action researchers posit a dialectical model, working back and forth between a variety of tension points, such as etic–emic, value neutrality–value committed. Others (Guba & Lincoln, 2005; Lather, 1993) deconstruct validity as an operative term. Sharlene Nagy Hesse-Biber and Patricia Leavy's (2008) emphasis on emergent methods pushes and blurs the methodological boundaries between quantitative and qualitative methods.[17] Their model seeks to recover subjugated knowledges hidden from everyday view.

The traditional mixed methods movement takes qualitative methods out of their natural home, which is within the critical interpretive framework (Howe, 2004, p. 54; but see Teddlie and Tashakkori, 2003a, p. 15; also Chapter 16 in this volume). It divides inquiry into dichotomous categories, exploration versus confirmation. Qualitative work is assigned to the first category, quantitative research to the second (Teddlie & Tashakkori, 2003a, p. 15). Like the classic experimental model, this movement excludes stakeholders from dialogue and active participation in the research process. Doing so weakens its democratic and dialogical dimensions and decreases the likelihood that previously silenced voices will be heard (Howe, 2004, pp. 56–57).

Howe (2004) cautions that it is not just

[the] "methodological fundamentalists" who have bought into [this] approach. A sizeable number of rather influential . . . educational researchers . . . have also signed on. This might be a compromise to the current political climate; it might be a backlash against the perceived excesses of postmodernism; it might be both. It is an ominous development, whatever the explanation. (p. 57; also 2009, p. 438; Lincoln, 2010, p. 7)

The hybrid dialogical model, in contrast, directly confronts these criticisms.

The Pragmatic Criticisms of Anti-Foundationalism

Clive Seale et al. (2004) contest what they regard as the excesses of an antimethodological, "anything goes," romantic postmodernism that is associated with our project. They assert that too often the approach we value produces "low quality qualitative research and research results that are quite stereotypical and close to common sense" (p. 2). In contrast they propose a practice-based, pragmatic approach that places research practice at the center. Research involves an engagement "with a variety of things and people: research materials . . . social theories, philosophical debates, values, methods, tests . . . research participants" (p. 2). (Actually this approach is quite close to our own, especially our view of the bricoleur and bricolage).

Their situated methodology rejects the antifoundational claim that there are only partial truths, that the dividing line between fact and fiction has broken down (Seale et al., 2004, p. 3). They believe that this dividing line has not collapsed and that we should not accept stories if they do not accord with the best available facts (p. 6). Oddly, these pragmatic procedural arguments reproduce a variant of the evidence-based model and its criticisms of poststructural performative sensibilities. They can be used to provide political support for the methodological marginalization of many of the positions advanced in this handbook.

This complex political terrain defines the many traditions and strands of qualitative research: the British and its presence in other national contexts; the American pragmatic, naturalistic, and interpretive traditions in sociology, anthropology, communications, and education; the German and French phenomenological, hermeneutic, semiotic, Marxist, structural, and poststructural perspectives; feminist, African American, Latino, and queer studies; and studies of indigenous and aboriginal cultures. The politics of qualitative research create a tension that informs each of the above traditions. This tension itself is constantly being reexamined and interrogated, as qualitative research confronts a changing historical world, new intellectual positions, and its own institutional and academic conditions.

To summarize, qualitative research is many things to many people. Its essence is two-fold: (1) a commitment to some version of the naturalistic, interpretive approach to its subject matter and (2) an ongoing critique of the politics and methods of postpositivism. We turn now to a brief discussion of the major differences between qualitative and quantitative approaches to research. We will then discuss ongoing differences and tensions within qualitative inquiry.

Qualitative Versus Quantitative Research

The word *qualitative* implies an emphasis on the qualities of entities and on processes and meanings that are not experimentally examined or measured (if measured at all) in terms of quantity, amount, intensity, or frequency. Qualitative researchers stress the socially constructed nature of reality, the intimate relationship between the researcher and what is studied, and the situational constraints that shape inquiry. Such researchers emphasize the value-laden nature of inquiry. They seek answers to questions that stress *how* social experience is created and given meaning. In contrast, quantitative studies emphasize the measurement and analysis of causal relationships between variables, not processes. Proponents claim that their work is done from within a value-free framework.

Research Styles: Doing the Same Things Differently?

Of course, both qualitative and quantitative researchers "think they know something about society worth telling to others, and they use a variety of forms, media, and means to communicate their ideas and findings" (Becker, 1986, p. 122). Qualitative research differs from quantitative research in five significant ways (Becker, 1996). These points of difference turn on different ways of addressing the same set of issues. They return always to the politics of research and who has the power to legislate correct solutions to these problems.

Using Positivism and Postpositivism: First, both perspectives are shaped by the positivist and postpositivist traditions in the physical and social sciences (see discussion below). These two positivist science traditions hold to naïve and critical realist positions concerning reality and its perception. Proponents of the positivist version contend that there is a reality out there to be studied, captured, and understood, whereas the postpositivists argue that reality can never be fully apprehended, only approximated (Guba, 1990a, p. 22). Postpositivism relies on multiple methods as a way of capturing as much of reality as possible. At the same time, emphasis is placed on the discovery and verification of theories. Traditional evaluation criteria like internal and external validity are stressed, as are the use of qualitative procedures that lend themselves to structured (sometimes statistical) analysis. Computer-assisted methods of analysis, which permit frequency counts, tabulations, and low-level statistical analyses, may also be employed.

The positivist and postpositivist traditions linger like long shadows over the qualitative research project. Historically, qualitative research was defined within the positivist paradigm, where qualitative researchers attempted to do good positivist research with less rigorous methods and procedures. Some mid-century qualitative researchers (Becker, Geer, Hughes, & Strauss, 1961) reported findings from participant observations in terms of quasi-statistics. As recently as 1999 (Strauss & Corbin, 1999), two leaders of the grounded theory approach to qualitative research attempted to modify the usual canons of good (positivistic) science to fit their own postpositivist conception of rigorous research (but see Charmaz, Chapter 21, this volume; also see Glaser, 1992). Some applied researchers, while claiming to be atheoretical, often fit within the positivist or postpositivist framework by default.

Uwe Flick (2002, pp. 2–3) usefully summarizes the differences between these two approaches to inquiry. He observes that the quantitative approach has been used for purposes of isolating "causes and effects . . . operationalizing theoretical relations . . . [and] measuring and . . . quantifying phenomena . . . allowing the generalization of findings" (p. 3). But today, doubt is cast on such projects.

> Rapid social change and the resulting diversification of life worlds are increasingly confronting social researchers with new social contexts and perspectives . . . traditional deductive methodologies . . . are failing . . . thus research is increasingly forced to make use of inductive strategies instead of starting from theories and testing them . . . knowledge and practice are studied as local knowledge and practice. (Flick, 2002, p. 2)

George and Louise Spindler (1992) summarize their qualitative approach to quantitative materials.

> Instrumentation and quantification are simply procedures employed to extend and reinforce certain kinds of data, interpretations and test hypotheses across samples. Both must be kept in their place. One must avoid their premature or overly extensive use as a security mechanism. (p. 69)

While many qualitative researchers in the postpositivist tradition will use statistical measures, methods, and documents as a way of locating a group of subjects within a larger population, they will seldom report their findings in terms of the kinds of complex statistical measures or methods that quantitative researchers are drawn to (i.e., path, regression, log-linear analyses).

Accepting Postmodern Sensibilities: The use of quantitative, positivist methods and assumptions has been rejected by a new generation of qualitative researchers who are attached to poststructural or postmodern sensibilities. These researchers argue that positivist methods are but one way of telling a story about society or the social world. They may be no better or no worse than any other method; they just tell a different kind of story.

This tolerant view is not shared by everyone. Many members of the critical theory, constructivist, poststructural, and postmodern schools of thought reject positivist and postpositivist criteria when evaluating their own work. They see these criteria as being irrelevant to their work and contend that positivist and postpositivist research reproduces only a certain kind of science, a science that silences too many voices. These researchers seek alternative methods for evaluating their work, including verisimilitude, emotionality, personal responsibility, an ethic of caring, political praxis, multivoiced texts, dialogues with subjects, and so on. In response, positivist and postpositivists argue that what they do is good science, free of individual bias and subjectivity. As noted above, they see postmodernism and poststructuralism as attacks on reason and truth.

Capturing the Individual's Point of View: Both qualitative and quantitative researchers are concerned with the individual's point of view. However, qualitative investigators think they can get closer to the actor's perspective by detailed interviewing and observation. They argue that quantitative researchers are seldom able to capture the subject's perspective because they have to rely on more remote, inferential empirical methods and materials. Many quantitative researchers regard empirical materials produced by interpretive methods as unreliable, impressionistic, and not objective.

Examining the Constraints of Everyday Life: Qualitative researchers are more likely to confront and come up against the constraints of the everyday social world. They see this world in action and embed their findings in it. Quantitative researchers abstract from this world and seldom study it directly. They seek a nomothetic or etic science based on probabilities derived from the study of large numbers of randomly selected cases. These kinds of statements stand above and outside the constraints of everyday life. Qualitative researchers, on the other hand, are committed to an emic, ideographic, case-based position, which directs their attention to the specifics of particular cases.

Securing Rich Descriptions: Qualitative researchers believe that rich descriptions of the social world are valuable, whereas quantitative researchers, with their etic, nomothetic commitments, are less concerned with such detail. They are deliberately unconcerned with such descriptions because such detail interrupts the process of developing generalizations.

These five points of difference described above (using positivism and postpositivism, accepting postmodern sensibilities, capturing the individual's point of view, examining the constraints of everyday life, securing thick descriptions) reflect commitments to different styles of research, different epistemologies, and different forms of representation. Each work tradition is governed by a different set of genres, and each has its own classics and its own preferred forms of representation, interpretation, trustworthiness, and textual evaluation (see Becker, 1986, pp. 134–135). Qualitative researchers use ethnographic prose, historical narratives, first-person accounts, still photographs, life history, fictionalized "facts," and biographical and autobiographical materials, among others. Quantitative researchers use mathematical models, statistical tables, and graphs and usually write in an impersonal, third-person prose.

◨ TENSIONS WITHIN QUALITATIVE RESEARCH

It is erroneous to presume that qualitative researchers share the same assumptions about these five points of difference. As the discussion below will reveal, positivist, postpositivist, and poststructural differences define and shape the discourses of qualitative research. Realists and postpositivists within the interpretive, qualitative research tradition criticize poststructuralists for taking the textual, narrative turn. These critics contend that such work is navel-gazing. It produces the conditions "for a dialogue of the deaf between itself and the community" (Silverman, 1997, p. 240). Those who attempt to capture the point of view of the interacting subject in the world are accused of naïve humanism, of reproducing a Romantic impulse that elevates the experiential to the level of the authentic (Silverman, 1997, p. 248).

Still others argue that lived experience is ignored by those who take the textual, performance turn. David Snow and Calvin Morrill (1995) argue that

> This performance turn, like the preoccupation with discourse and storytelling, will take us further from the field of social action and the real dramas of everyday life and thus signal the death knell of ethnography as an empirically grounded enterprise. (p. 361)

Of course, we disagree.

According to Martyn Hammersley (2008, p. 1), qualitative research is currently facing a crisis symbolized by an ill-conceived postmodernist image of qualitative research, which is dismissive of traditional forms of inquiry. He feels that "unless this dynamic can be interrupted the future of qualitative research is endangered" (p. 11).

Paul Atkinson and Sara Delamont (2006), two qualitative scholars in the traditional, classic Chicago School tradition,[18] offer a corrective. They remain committed to qualitative (and quantitative) research *provided that they are conducted rigorously and contribute to robustly useful knowledge* (p. 749, italics in original). Of course, these scholars are committed to social policy initiatives at some level. But, for them, the postmodern image of qualitative inquiry threatens and undermines the

value of traditional qualitative inquiry. Atkinson and Delamont exhort qualitative researchers to "think hard about whether their investigations are the best social science they could be" (p. 749). Patricia and Peter Adler (2008) implore the radical postmodernists to "give up the project for the good of the discipline and for the good of society" (p. 23).

Hammersley (2008, pp. 134–136, 144), extends the traditional critique, finding little value in the work of ethnographic postmodernists and literary ethnographers.[19] This new tradition, he asserts, legitimates speculative theorizing, celebrates obscurity, and abandons the primary task of inquiry, which is to produce truthful knowledge about the world (p. 144). Poststructural inquirers get it from all sides. The criticisms, Carolyn Ellis (2009, p. 231) observes, fall into three overlapping categories. Our work (1) is too aesthetic and not sufficiently realistic; it does not provide hard data; (2) is too realistic and not mindful of poststructural criticisms concerning the "real" self and its place in the text; and (3) is not sufficiently aesthetic, or literary; that is, we are second-rate writers and poets (p. 232).

The Politics of Evidence

The critics' model of science is anchored in the belief that there is an empirical world that is obdurate and talks back to investigators. This is an empirical science based on evidence that corroborates interpretations. This is a science that returns to and is lodged in the real, a science that stands outside nearly all of the turns listed above; this is Chicago School neo-postpositivism.

Contrast this certain science to the position of those who are preoccupied with the politics of evidence. Jan Morse (2006), for example, says: "Evidence is not just something that is out there. Evidence has to be produced, constructed, represented. Furthermore, the politics of evidence cannot be separated from the ethics of evidence" (pp. 415–416). Under the Jan Morse model, representations of empirical reality become problematic. Objective representation of reality is impossible. Each representation calls into place a different set of ethical questions regarding evidence, including how it is obtained and what it means. But surely a middle ground can be found. If there is a return to the spirit of the paradigm dialogues of the 1980s, then multiple representations of a situation should be encouraged, perhaps placed alongside one another.

Indeed, the interpretive camp is not antiscience, per se. We do something different. We believe in multiple forms of science: soft, hard, strong, feminist, interpretive, critical, realist, postrealist, and post-humanist. In a sense, the traditional and postmodern projects are incommensurate. We interpret, we perform, we interrupt, we challenge, and we believe nothing is ever certain. We want performance texts that quote history back to itself, texts that focus on epiphanies; on the intersection of biography, history, culture, and politics; on turning point moments in people's lives. The critics are correct on this point. We have a political orientation that is radical, democratic, and interventionist. Many postpositivists share these politics.

Critical Realism

For some, there is a third stream between naïve positivism and poststructuralism. Critical realism is an antipositivist movement in the social sciences closely associated with the works of Roy Bhaskar and Rom Harré (Danermark, Ekstrom, Jakobsen, & Karlsson, 2002). Critical realists use the word *critical* in a particular way. This is not Frankfurt School critical theory, although there are traces of social criticism here and there (Danermark et al., 2002, p. 201). *Critical,* instead, refers to a transcendental realism that rejects methodological individualism and universal claims to truth. Critical realists oppose logical positivist, relativist, and antifoundational epistemologies. Critical realists agree with the positivists that there is a world of events out there that is observable and independent of human consciousness. Knowledge about this world is socially constructed. Society is made up of feeling, thinking human beings, and their interpretations of the world must be studied (Danermark et al., 2002, p. 200). A correspondence theory of truth is rejected. Critical realists believe that reality is arranged in levels. Scientific work must go beyond statements of regularity to the analysis of the mechanisms, processes, and structures that account for the patterns that are observed.

Still, as postempiricist, antifoundational, critical theorists, we reject much of what is advocated here. Throughout the last century, social science and philosophy were continually tangled up with one another. Various "isms" and philosophical movements criss-crossed sociological and educational discourse, from positivism to postpositivism to analytic and linguistic philosophy, to hermeneutics, structuralism, and poststructuralism; to Marxism, feminism, and current post-post-versions of all of the above. Some have said that the logical positivists steered the social sciences on a rigorous course of self-destruction.

We do not think critical realism will keep the social science ship afloat. The social sciences are normative disciplines, always already embedded in issues of value, ideology, power, desire, sexism, racism, domination, repression, and control. We want a social science committed up front to issues of social justice, equity, nonviolence, peace, and universal human rights. We do not want a social science that says it can address these issues if it wants to do so. For us, this is no longer an option.

▣   QUALITATIVE RESEARCH AS PROCESS

Three interconnected, generic activities define the qualitative research process. They go by a variety of different labels, including theory, method, and analysis; or ontology, epistemology, and methodology. Behind these terms stands the personal biography of the researcher, who speaks from a particular class, gendered, racial, cultural, and ethnic community perspective. The gendered, multiculturally situated researcher approaches the world with a set of ideas, a framework (theory, ontology) that specifies a set of questions (epistemology), which are then examined (methodology, analysis) in specific ways. That is, empirical materials bearing on the question are collected and then analyzed and written about. Every researcher speaks from within a distinct interpretive community, which configures, in its special way, the multicultural, gendered components of the research act.

In this volume, we treat these generic activities under five headings or phases: the researcher and the researched as multicultural subjects, major paradigms and interpretive perspectives, research strategies, methods of collecting and analyzing empirical materials, and the art of interpretation. Behind and within each of these phases stands the biographically situated researcher. This individual enters the research process from inside an interpretive community. This community has its own historical research traditions, which constitute a distinct point of view. This perspective leads the researcher to adopt particular views of the "other" who is studied. At the same time, the politics and the ethics of research must also be considered, for these concerns permeate every phase of the research process.

▣   THE OTHER AS RESEARCH SUBJECT

From its turn-of-the-century birth in modern, interpretive form, qualitative research has been haunted by a double-faced ghost. On the one hand, qualitative researchers have assumed that qualified, competent observers could, with objectivity, clarity, and precision, report on their own observations of the social world, including the experiences of others. Second, researchers have held to the belief in a real subject or real individual who is present in the world and able, in some form, to report on his or her experiences. So armed, researchers could blend their own observations with the self-reports provided by subjects through interviews, life story, personal experience, and case study documents.

These two beliefs have led qualitative researchers across disciplines to seek a method that would allow them to record accurately their own observations while also uncovering the meanings their subjects brought to their life experiences. This method would rely on the subjective verbal and written expressions of meaning given by the individuals, which are studied as windows into the inner life of the person. Since Wilhelm Dilthey (1900/1976), this search for a method has led to a perennial focus in the human disciplines on qualitative, interpretive methods.

Recently, as noted above, this position and its beliefs have come under assault. Poststructuralists and postmodernists have contributed to the understanding that there is no clear window into the inner life of an individual. Any gaze is always filtered through the lenses of language, gender, social class, race, and ethnicity. There are no objective observations, only observations socially situated in the worlds of—and between—the observer and the observed. Subjects, or individuals, are seldom able to give full explanations of their actions or intentions; all they can offer are accounts or stories about what they did and why. No single method can grasp the subtle variations in ongoing human experience. Consequently, qualitative researchers deploy a wide-range of interconnected interpretive methods, always seeking better ways to make more understandable the worlds of experience that have been studied.

Table 1.1 depicts the relationships we see among the five phases that define the research process (the researcher; major paradigms; research strategies; methods of collecting and analyzing empirical materials; and the art, practices, and politics of interpretation). Behind all but one of these phases stands the biographically situated researcher. These five levels of activity, or practice, work their way through the biography of the researcher. We take them up in brief order here, for each phase is more fully discussed in the transition sections between the various parts of this volume.

Phase 1: The Researcher

Our remarks above indicate the depth and complexity of the traditional and applied qualitative research perspectives into which a socially situated researcher enters. These traditions locate the researcher in history, simultaneously guiding and constraining work that will be done in any specific study. This field has been constantly characterized by diversity and conflict, and these are its most enduring traditions (see Levin & Greenwood, Chapter 2, this volume). As a carrier of this complex and contradictory history, the researcher must also confront the ethics and politics of research (Christians, Chapter 4, this volume). It is no longer possible for the human disciplines to research the native, the indigenous other, in a spirit of value-free inquiry. Today researchers struggle to develop situational and transsituational ethics that apply to all forms of the research act and its human-to-human relationships. We no longer have the option of deferring the decolonization project.

Phase 2: Interpretive Paradigms

All qualitative researchers are philosophers in that "universal sense in which all human beings . . . are guided by highly abstract principles" (Bateson, 1972, p. 320). These principles combine beliefs about *ontology* (What kind of being is the human being? What is the nature of reality?), *epistemology* (What is the relationship between the inquirer and the known?), and *methodology* (How do we know the world or gain knowledge of it?)

Table 1.1   The Research Process

Phase 1: The Researcher as a Multicultural Subject

History and research traditions
Conceptions of self and the other
The ethics and politics of research

Phase 2: Theoretical Paradigms and Perspectives

Positivism, postpositivism
Interpretivism, constructivism, hermeneutics
Feminism(s)
Racialized discourses
Critical theory and Marxist models
Cultural studies models
Queer theory
Post-colonialism

Phase 3: Research Strategies

Design
Case study
Ethnography, participant observation, performance ethnography
Phenomenology, ethnomethodology
Grounded theory
Life history, *testimonio*
Historical method
Action and applied research
Clinical research

Phase 4: Methods of Collection and Analysis

Interviewing
Observing
Artifacts, documents, and records
Visual methods
Autoethnography
Data management methods
Computer-assisted analysis
Textual analysis
Focus groups
Applied ethnography

*Phase 5: The Art, Practices, and
Politics of Interpretation and Evaluation*

Criteria for judging adequacy
Practices and politics of interpretation
Writing as interpretation
Policy analysis
Evaluation traditions
Applied research

(see Guba, 1990a, p. 18; Lincoln & Guba, 1985, pp. 14–15; and Lincoln, Lynham, & Guba in Chapter 6 of this volume). These beliefs shape how the qualitative researcher sees the world and acts in it. The researcher is "bound within a net of epistemological and ontological premises which—regardless of ultimate truth or falsity—become partially self-validating" (Bateson, 1972, p. 314).

The net that contains the researcher's epistemological, ontological, and methodological premises may be termed a *paradigm* (Guba, 1990a, p. 17) or interpretive framework, a "basic set of beliefs that guides action" (Guba, 1990a, p. 17). All research is interpretive: guided by a set of beliefs and feelings about the world and how it should be understood and studied. Some beliefs may be taken for granted, invisible, or only assumed, whereas others are highly problematic and controversial. Each interpretive paradigm makes particular demands on the researcher, including the questions that are asked and the interpretations that are brought to them.

At the most general level, four major interpretive paradigms structure qualitative research: positivist and postpositivist, constructivist-interpretive, critical (Marxist, emancipatory), and feminist-poststructural. These four abstract paradigms become more complicated at the level of concrete specific interpretive communities. At this level, it is possible to identify not only the constructivist but also multiple versions of feminism (Afrocentric and poststructural),[20] as well as specific ethnic, feminist, endarkened, social justice, Marxist, cultural studies, disability, and non-Western-Asian paradigms. These perspectives or paradigms are examined in Part II of this volume.

The paradigms examined in Part II work against or alongside (and some within) the positivist and postpositivist models. They all work within relativist ontologies (multiple constructed realities), interpretive epistemologies (the knower and known interact and shape one another), and interpretive, naturalistic methods.

Table 1.2 presents these paradigms and their assumptions, including their criteria for evaluating research, and the typical form that an interpretive or theoretical statement assumes in the paradigm.[21]

Each paradigm is explored in considerable detail in chapters 6 through 10. The positivist and postpositivist paradigms were discussed above. They work from within a realist and critical realist ontology and objective epistemologies, and they rely on experimental, quasi-experimental, survey, and rigorously defined qualitative methodologies.

The *constructivist paradigm* assumes a relativist ontology (there are multiple realities), a subjectivist epistemology (knower and respondent co-create understandings), and a naturalistic (in the natural world) set of methodological procedures. Findings are usually presented in terms of the criteria of grounded theory or pattern theories (in this volume, see Lincoln, Lynham, & Guba, Chapter 6; Creswell, Chapter 15; Teddlie & Tashakkori, Chapter 16; Charmaz, Chapter 21; Morse, Chapter 24; Altheide & Johnson, Chapter 35; and St.Pierre, Chapter 37). Terms like credibility, transferability, dependability, and confirmability replace the usual positivist criteria of internal and external validity, reliability, and objectivity.

Table 1.2   Interpretive Paradigms

Paradigm/Theory	Criteria	Form of Theory	Type of Narration
Positivist/ postpositivist	Internal, external validity	Logical-deductive, grounded	Scientific report
Constructivist	Trustworthiness, credibility, transferability, confirmability	Substantive-formal, standpoint	Interpretive case studies, ethnographic fiction
Feminist	Afrocentric, lived experience, dialogue, caring, accountability, race, class, gender, reflexivity, praxis, emotion, concrete grounding, embodied	Critical, standpoint	Essays, stories, experimental writing
Ethnic	Afrocentric, lived experience, dialogue, caring, accountability, race, class, gender	Standpoint, critical, historical	Essays, fables, dramas
Marxist	Emancipatory theory, falsifiability, dialogical, race, class, gender	Critical, historical, economic	Historical, economic, sociocultural analyses
Cultural studies	Cultural practices, praxis, social texts, subjectivities	Social criticism	Cultural theory-as-criticism
Queer theory	Reflexivity, deconstruction	Social criticism, historical analysis	Theory-as-criticism, autobiography

Feminist, ethnic, Marxist, cultural studies, queer theory, Asian, and disability models privilege a materialist-realist ontology; that is, the real world makes a material difference in terms of race, class, and gender. Subjectivist epistemologies and naturalistic methodologies (usually ethnographies) are also employed. Empirical materials and theoretical arguments are evaluated in terms of their emancipatory implications. Criteria from gender and racial communities (e.g., African American) may be applied (emotionality and feeling, caring, personal accountability, dialogue).

Poststructural feminist theories emphasize problems with the social text, its logic, and its inability to ever represent the world of lived experience fully. Positivist and postpositivist criteria of evaluation are replaced by other terms, including the reflexive, multivoiced text, which is grounded in the experiences of oppressed people.

The cultural studies and queer theory paradigms are multi-focused, with many different strands drawing from Marxism, feminism, and the postmodern sensibility (in this volume, Giardina & Newman, Chapter 10; Plummer, Chapter 11; St.Pierre, Chapter 37). There is a tension between a humanistic cultural studies, which stresses lived experiences (meaning), and a more structural cultural studies project, which stresses the structural and material determinants and effects (race, class, gender) of experience. Of course, there are two sides to every coin; both sides are needed and are indeed critical. The cultural studies and queer theory paradigms use methods strategically, that is, as resources for understanding and for producing resistances to local structures of domination. Such scholars may do close textual readings and discourse analysis of cultural texts (in this volume, Olesen, Chapter 7; Chase, Chapter 25), as well as local, online, reflexive, and critical ethnographies; open-ended interviewing; and participant observation. The focus is on how race, class, and gender are produced and enacted in historically specific situations.

Paradigm and personal history in hand, focused on a concrete empirical problem to examine, the researcher now moves to the next stage of the research process, namely working with a specific strategy of inquiry.

Phase 3: Strategies of Inquiry and Interpretive Paradigms

Table 1.1 presents some of the major strategies of inquiry a researcher may use. Phase 3 begins with research design, which broadly conceived involves a clear focus on the research question, the purposes of the study, "what information most appropriately will answer specific research questions, and which strategies are most effective for obtaining it" (LeCompte & Preissle with Tesch, 1993, p. 30; see also Cheek, Chapter 14, this volume). A research design describes a flexible set of guidelines that connect theoretical paradigms, first, to strategies of inquiry

and, second, to methods for collecting empirical material. A research design situates researchers in the empirical world and connects them to specific sites, people, groups, institutions, and bodies of relevant interpretive material, including documents and archives. A research design also specifies how the investigator will address the two critical issues of representation and legitimation.

A strategy of inquiry refers to a bundle of skills, assumptions, and practices that researchers employ as they move from their paradigm to the empirical world. Strategies of inquiry put paradigms of interpretation into motion. At the same time, strategies of inquiry also connect the researcher to specific methods of collecting and analyzing empirical materials. For example, the case study relies on interviewing, observing, and document analysis. Research strategies implement and anchor paradigms in specific empirical sites or in specific methodological practices, for example, making a case an object of study. These strategies include the case study, phenomenological and ethnomethodological techniques, the use of grounded theory, and biographical, autoethnographic, historical, action, and clinical methods. Each of these strategies is connected to a complex literature; each has a separate history, exemplary works, and preferred ways for putting the strategy into motion.

Phase 4: Methods of Collecting and Analyzing Empirical Materials

The researcher has several methods for collecting empirical materials.[22] These methods are taken up in Part IV. They range from the interview to direct observation, the use of visual materials or personal experience. The researcher may also use a variety of different methods of reading and analyzing interviews or cultural texts, including content, narrative, and semiotic strategies. Faced with large amounts of qualitative materials, the investigator seeks ways of managing and interpreting these documents, and here data management methods and computer-assisted models of analysis may be of use. In this volume, David L. Altheide and John M. Johnson (Chapter 35), Laura L. Ellingson (Chapter 36), and Judith Davidson and Silvana diGregorio (Chapter 38) take up these techniques.

Phase 5: The Art and Politics of Interpretation and Evaluation

Qualitative research is endlessly creative and interpretive. The researcher does not just leave the field with mountains of empirical materials and easily write up his or her findings. Qualitative interpretations are constructed. The researcher first creates a field text consisting of fieldnotes and documents from the field, what Roger Sanjek (1992, p. 386) calls "indexing" and David Plath (1990, p. 374) "filework." The writer-as-interpreter moves from this text to a research text; notes and interpretations based on the field text. This text is then re-created as a working interpretive document that contains the writer's initial attempts to make sense out of what has been learned. Finally, the writer produces the public text that comes to the reader. This final tale from the field may assume several forms: confessional, realist, impressionistic, critical, formal, literary, analytic, grounded theory, and so on (see Van Maanen, 1988).

The interpretive practice of making sense of one's findings is both artistic and political. Multiple criteria for evaluating qualitative research now exist, and those we emphasize stress the situated, relational, and textual structures of the ethnographic experience. There is no single interpretive truth. As argued earlier, there are multiple interpretive communities, each having its own criteria for evaluating an interpretation.

Program evaluation is a major site of qualitative research, and qualitative researchers can influence social policy in important ways. Applied, qualitative research in the social sciences has a rich history (discussed in this volume by Levin & Greenwood, Chapter 2; Cheek, Chapter 14; Brydon-Miller, Kral, Maguire, Noffke, & Sablok, Chapter 23; Morse, Chapter 24; Torrance, Chapter 34; Abma & Widdershoven, Chapter 41). This is the critical site where theory, method, praxis, action, and policy all come together. Qualitative researchers can isolate target populations, show the immediate effects of certain programs on such groups, and isolate the constraints that operate against policy changes in such settings. Action and clinically oriented qualitative researchers can also create spaces for those who are studied (the other) to speak. The evaluator becomes the conduit for making such voices heard.

Bridging the Historical Moments: What Comes Next?

St. Pierre (2004) argues that we are already in the post "post" period—post-poststructuralism, post-postmodernism, post-experimental. What this means for interpretive, ethnographic practices is still not clear. But it is certain that things will never again be the same. We are in a new age where messy, uncertain multivoiced texts, cultural criticism, and new experimental works will become more common, as will more reflexive forms of fieldwork, analysis, and intertextual representation. In a complex space like this, pedagogy becomes critical—that is, How do we teach qualitative methods? Judith Preissle (Chapter 42) and Margaret Eisenhart and S. Jurow (Chapter 43) offer insights on the future. It is true, as the poet said, the center no longer holds. We can reflect on what should be in this new center.

Thus, we come full circle. And returning to our bridge metaphor, the chapters that follow take the researcher back and forth through every phase of the research act. Like a good bridge, the chapters provide for two-way traffic, coming and going between moments, formations, and interpretive communities. Each chapter examines the relevant histories, controversies, and current practices that are associated with each paradigm, strategy,

and method. Each chapter also offers projections for the future, where a specific paradigm, strategy, or method will be 10 years from now, deep into the formative years of the next century.

In reading this volume, it is important to remember that the field of qualitative research is defined by a series of tensions, contradictions, and hesitations. This tension works back and forth between and among (1) the broad, doubting, postmodern sensibility; (2) the more certain, more traditional positivist, postpositivist, and naturalistic conceptions of this project; and (3) an increasingly conservative, neoliberal global environment. All of the chapters that follow are caught in and articulate these tensions.

▣ Notes

1. The following paragraphs draw from Denzin (2010, pp. 19–25).

2. They contend that our second moment, the Golden Age (1950–1970), was marked by the debunking of positivism, the emergence of postpositivism, and the development of designs that used mixed quantitative and qualitative methods. Full-scale conflict developed throughout the 1970–1990 period, the time of the first "paradigm war."

3. Conflict broke out between the many different empowerment pedagogies: feminist, anti-racist, radical, Freirean, liberation theology, postmodernists, poststructuralists, cultural studies, and so on (see Guba & Lincoln, 2005; also, Erickson, Chapter 3, this volume).

4. Recall bell hooks's reading of the famous cover photo on *Writing Culture* (Clifford & Marcus, 1986), which consists of a picture of Stephen Tyler doing fieldwork in India. Tyler is seated some distance from three dark-skinned people. A child is poking its head out of a basket. A woman is hidden in the shadows of the hut. A male, a checkered white and black shawl across his shoulder, elbow propped on his knee, hand resting along the side of his face, is staring at Tyler. Tyler is writing in a field journal. A piece of white cloth is attached to his glasses, perhaps shielding him from the sun. This patch of whiteness marks Tyler as the white male writer studying these passive brown and black people. Indeed, the brown male's gaze signals some desire or some attachment to Tyler. In contrast, the female's gaze is completely hidden by the shadows and by the words in the book's title, which cross her face (hooks, 1990, p. 127).

5. Qualitative research has separate and distinguished histories in education, social work, communications, psychology, history, organizational studies, medical science, anthropology, and sociology.

6. Definitions: *positivism:* Objective accounts of the real world can be given; *postpositivism:* Only partially objective accounts of the world can be produced, for all methods are flawed; *foundationalism:* We can have an ultimate grounding for our knowledge claims about the world, and this involves the use of empiricist and positivist epistemologies (Schwandt, 1997a, p. 103); *nonfoundationalism:* We can make statements about the world without "recourse to ultimate proof or foundations for that knowing" (Schwandt, 1997a, p. 102); *quasifoundationalism:* Certain knowledge claims about the world based on neorealist criteria can be made, including the correspondence concept of truth. There is an independent reality that can be mapped.

7. Jameson (1991, pp. 3–4) reminds us that any periodization hypothesis is always suspect, even one that rejects linear, stage-like models. It is never clear to what reality a stage refers. What divides one stage from another is always debatable. Our seven moments are meant to mark discernible shifts in style, genre, epistemology, ethics, politics, and aesthetics.

8. See Denzin and Lincoln (2005, pp. 13–21) for an extended discussion of each of these phases. This model has been termed a progress narrative by Alasuutari (2004, pp. 599–600) and Seale, Gobo, Gubrium, and Silverman (2004, p. 2). The critics assert that we believe that the most recent moment is the most up-to-date, the avant-garde, the cutting edge (Alasuutari, 2004, p. 601). Naturally, we dispute this reading. Teddlie and Tashakkori (2003a, pp. 5–8) have modified our historical periods to fit their historical analysis of the major moments in the emergence of mixed methods in the last century.

9. Definitions: *structuralism*: Any system is made up of a set of oppositional categories embedded in language; *semiotics*: the science of signs or sign systems—a structuralist project; *poststructuralism*: Language is an unstable system of referents, making it impossible to ever completely capture the meaning or an action, text, or intention; *postmodernism*: a contemporary sensibility, developing since World War II, which privileges no single authority, method, or paradigm; *hermeneutics*: An approach to the analysis of texts that stresses how prior understandings and prejudices shape the interpretive process; *phenomenology*: A complex system of ideas associated with the works of Edmund Husserl, Martin Heidegger, Jean-Paul Sartre, Maurice Merleau-Ponty, and Alfred Schutz; *cultural studies*: a complex, interdisciplinary field that merges with critical theory, feminism, and poststructuralism.

10. Of course, all settings are natural, that is, places where everyday experience takes place. Qualitative researchers study people doing things together in the places where these things are done (Becker, 1986). There is no field site or natural place where one goes to do this kind of work (see also Gupta & Ferguson, 1997, p. 8). The site is constituted through our interpretive practices. Historically, analysts have distinguished between experimental (laboratory) and field (natural) research settings; hence the argument that qualitative research is naturalistic. Activity theory erases this distinction (Keller & Keller, 1996, p. 20; Vygotsky, 1978).

11. "The meaning of bricoleur in French popular speech is 'someone who works with his (or her) hands and uses devious means compared to those of the craftsman . . . the bricoleur is practical and gets the job done" (Weinstein & Weinstein, 1991, p. 161). These authors provide a history of this term, connecting it to the works of the German sociologist and social theorist Georg Simmel, and by implication to Charles Baudelaire. Martyn Hammersley (2000) disputes our use of this term. Following Claude Lévi-Strauss, he reads the bricoleur as a myth maker. He suggests it be replaced with the notion of the boat builder. Hammersley also quarrels with our "moments" model of qualitative research, contending it implies some sense of progress.

12. Brian De Palma reproduces this baby carriage scene in his 1987 film, *The Untouchables*.

13. In the harbor, the muzzles of the Potemkin's two huge guns swing slowly into the camera. Words on screen inform us: "The brutal military power answered by guns of the battleship." A final famous three-shot montage sequence shows, first, a sculptured sleeping lion, then the lion rising from his sleep, and finally the lion roaring, symbolizing the rage of the Russian people (Cook, 1981, p. 167). In this sequence, Eisenstein uses montage to expand time, creating a psychological duration for this horrible event. By drawing out this sequence, by showing the baby in the carriage, the soldiers firing on the citizens, the blood on the mother's glove, the descending carriage on the steps, he suggests a level of destruction of great magnitude.

14. Here it is relevant to make a distinction between techniques that are used across disciplines and methods that are used within disciplines. Ethnomethodologists, for example, employ their approach as a method, whereas others selectively borrow that method-as-technique for their own applications. Harry Wolcott (in conversation) suggests this distinction. It is also relevant to make a distinction between topic, method, and resource. Methods can be studied as topics of inquiry; that is how a case study gets done. In this ironic, ethnomethodological sense, method is both a resource and a topic of inquiry.

15. Indeed any attempt to give an essential definition of qualitative research requires a qualitative analysis of the circumstances that produce such a definition.

16. They identify four major mixed methods designs: triangulation, embedded, explanatory, and exploratory (Clark et al., 2008, p. 371).

17. Their emergent model focuses on methods that break out of traditional frameworks and exploit new technologies and innovations; this is a process model that works between politics, epistemology, theory, and methodology.

18. There are several generations of the Chicago School, from Robert Park and Ernest Burgess, Herbert Blumer, and Everett Hughes (1920–1950) period, to second (Becker, Strauss, Goffman), to third (Hammersley, Atkinson, Delamont, Snow, Anderson, Fine, Adler and Adler, Prus, Maines, Flaherty, Sanders et al).

19. His blanket term for auto, performance, poststructural ethnography.

20. Olesen (Chapter 7, this volume) identifies three strands of feminist research: mainstream empirical; standpoint and cultural studies; and poststructural, postmodern; placing Afrocentric and other models of color under the cultural studies and postmodern categories.

21. These, of course, are our interpretations of these paradigms and interpretive styles.

22. *Empirical materials* is the preferred term for what are traditionally described as data.

▣ REFERENCES

Adler, P. A., & Adler, P. (2008). Of rhetoric and representation: The four faces of ethnography. *Sociological Quarterly, 49*(4), 1–30.

Alasuutari, P. (2004). The globalization of qualitative research. In C. Seale, G. Gobo, J. F. Gubrium, & D. Silverman (Eds.), *Qualitative research practice* (pp. 595–608). London: Sage.

Atkinson, E. (2004). Thinking outside the box: An exercise in heresy. *Qualitative Inquiry, 10*(1), 111–129.

Atkinson, P., & Delamont, S. (2006). In the roiling smoke: Qualitative inquiry and contested fields. *International Journal of Qualitative Studies in Education, 19*(6), 747–755.

Bateson, G. (1972). *Steps to an ecology of mind.* New York: Ballantine.

Becker, H. S. (1986). *Doing things together.* Evanston, IL: Northwestern University Press.

Becker, H. S. (1996). The epistemology of qualitative research. In R. Jessor, A. Colby, & R. A. Schweder (Eds.), *Ethnography and human development* (pp. 53–71). Chicago: University of Chicago Press.

Becker, H. S. (1998). *Tricks of the trade.* Chicago: University of Chicago Press.

Becker, H S., Geer, B., Hughes, E. C., & Strauss, A. L. (1961). *Boys in white.* Chicago: University of Chicago Press.

Bloch, M. (2004). A discourse that disciplines, governs, and regulates: On scientific research in education. *Qualitative Inquiry, 10*(1), 96–110.

Cannella, G. S. (2004). Regulatory power: Can a feminist poststructuralist engage in research oversight? *Qualitative Inquiry, 10*(2), 235–245.

Cannella, G. S., & Lincoln, Y. S. (2004a). Dangerous discourses II: Comprehending and countering the redeployment of discourses (and resources) in the generation of liberatory inquiry. *Qualitative Inquiry, 10*(2), 165–174.

Cannella, G. S., & Lincoln, Y. S. (2004b). Epilogue: Claiming a critical public social science—reconceptualizing and redeploying research. *Qualitative Inquiry, 10*(2), 298–309.

Carey, J. W. (1989). *Culture as communication.* Boston: Unwin Hyman.

Cicourel, A. V. 1964. *Method and measurement in sociology.* New York: Free Press.

Clark, C., & Scheurich, J. (2008). Editorial: The state of qualitative research in the early twenty-first century. *International Journal of Qualitative Research in Education, 21*(4), 313.

Clark, V. L. P., & Creswell, J. W. (2008). Introduction. In V. L. Plano Clark & J. W. Creswell (Eds.), *The mixed methods reader* (pp. xv–xviii). Thousand Oaks: Sage.

Clark, V. L. P., Creswell, J. W., Green, D. O., & Shope, R. J. (2008). Mixing quantitative and qualitative approaches: An introduction to emergent mixed methods research. In S. N. Hesse-Biber & P. Leavy (Eds.), *Handbook of emergent methods* (pp. 363–388). New York: Guilford.

Clifford, J. (1988). *Predicament of culture.* Cambridge: Harvard University Press.

Clifford, J. (1997). *Routes: Travel and translation in the late twentieth century.* Cambridge: Harvard University Press.

Clifford, J., & Marcus, G. E. (Eds.). (1986). *Writing culture.* Berkeley: University of California Press.

Clough, P. T. (1992). *The end(s) of ethnography.* Newbury Park, CA: Sage.

Clough, P. T. (1998). *The end(s) of ethnography* (2nd ed.). New York: Peter Lang.

Clough, P. T. (2000). Comments on setting criteria for experimental writing. *Qualitative Inquiry, 6,* 278–291.

Cook, D. A. (1981). *A history of narrative film.* New York: W. W. Norton.

Creswell, J. W. (1998). *Qualitative inquiry and research design: Choosing among five traditions.* Thousand Oaks, CA: Sage.

Danermark, B., Ekstrom, M., Jakobsen, L., & Karlsson, J. C. (2002). *Explaining society: Critical realism in the social sciences.* London: Routledge.

de Certeau, M. (1984). *The practice of everyday life.* Berkeley: University of California Press.

Denzin, N. K. (1970). *The research act.* Chicago: Aldine.

Denzin, N. K. (1978). *The research act* (2nd ed.). New York: McGraw-Hill.

Denzin, N. K. (1989a). *Interpretive interactionism.* Newbury Park, CA: Sage.

Denzin, N. K. (1989b). *The research act* (3rd ed.). Englewood Cliffs, NJ: Prentice Hall.

Denzin, N. K. (1997). *Interpretive ethnography.* Thousand Oaks, CA: Sage.

Denzin, N. K. (2003). *Performance ethnography: Critical pedagogy and the politics of culture.* Thousand Oaks, CA: Sage.

Denzin, N. K. (2009). *Qualitative inquiry under fire: Toward a new paradigm dialogue.* Walnut Creek, CA: Left Coast Press.

Denzin, N. K. (2010). *The qualitative manifesto: A call to arms.* Walnut Creek, CA: Left Coast Press.

Denzin, N. K., & Lincoln, Y. S. (2005). Introduction: The discipline and practice of qualitative research. In N. K. Denzin & Y. S. Lincoln (Eds.), *The SAGE handbook of qualitative research* (3rd ed., pp. 1–32). Thousand Oaks, CA: Sage.

Dilthey, W. L. (1976). *Selected writings.* Cambridge, UK: Cambridge University Press. (Original work published 1900)

Diversi, M. (1998). Glimpses of street life: Representing lived experience through short stories. *Qualitative Inquiry, 4,* 131–137.

Diversi, M., & Moreira, C. (2009). *Betweener talk: Decolonizing knowledge production, pedagogy, and praxis.* Walnut Creek, CA: Left Coast Press.

Ellingson, L. L. (2009). *Engaging crystallization in qualitative research.* Thousand Oaks, CA: Sage.

Ellis, C. (2009). *Revision: Autoethnographic reflections on life and work.* Walnut Creek, CA: Left Coast Press.

Ellis, C., & Bochner, A. P. (Eds.). (2000). *Ethnographically speaking: Autoethnography, literature, and aesthetics.* Walnut Creek, CA: AltaMira Press.

Filstead, W. J. (Ed.). (1970). *Qualitative methodology.* Chicago: Markham.

Flick, U. (1998). *An introduction to qualitative research.* London: Sage.

Flick, U. (2002). *An introduction to qualitative research* (2nd ed.). London: Sage.

Flick, U. (2007). *Designing qualitative research.* London: Sage

Gage, N. L. (1989). The paradigm wars and their aftermath: A "historical" sketch of research and teaching since 1989. *Educational Researcher, 18*(7), 4–10.

Geertz, C. (1973). *Interpreting cultures.* New York: Basic Books.

Geertz, C. (1983). *Local knowledge.* New York: Basic Books.

Geertz, C. (1988). *Works and lives.* Stanford, CA: Stanford University Press.

Geertz, C. (1995). *After the fact: Two countries, four decades, one anthropologist.* Cambridge: Harvard University Press.

Glaser, B. G. (1992). *Emergence vs. forcing: Basics of grounded theory.* Mill Valley, CA: Sociology Press.

Glaser, B., & Strauss, A. (1967). *The discovery of grounded theory.* Chicago: Aldine.

Goodall, H. L., Jr. (2000). *Writing the new ethnography.* Walnut Creek, CA: AltaMira.

Gordon, D. A. (1988). Writing culture, writing feminism: The poetics and politics of experimental ethnography. *Inscriptions, 3/4* (8), 21–31.

Gordon, D. A. (1995). Conclusion: Culture writing women: Inscribing feminist anthropology. In R. Behar & D. A. Gordon (Eds.), *Women writing culture* (pp. 429–441). Berkeley: University of California Press.

Greenblatt, S. (1997). The touch of the real. In S. B. Ortner (Ed.), The fate of "culture": Geertz and beyond [Special issue]. *Representations, 59*, 14–29.

Grossberg, L., Nelson, C., & Treichler, P. (Eds.) (1992). *Cultural studies.* New York: Routledge.

Guba, E. G. (1990a). The alternative paradigm dialog. In E. G. Guba (Ed.), *The paradigm dialog* (pp. 17–30). Newbury Park, CA: Sage.

Guba, E. G. (1990b). Carrying on the dialog. In Egon G. Guba (Ed.), *The paradigm dialog* (pp. 368–378). Newbury Park, CA: Sage.

Guba, E., & Lincoln, Y. S. (1989). *Fourth generation evaluation.* Newbury Park, CA: Sage.

Guba, E., & Lincoln, Y. S. (2005). Paradigmatic controversies and emerging confluences. In N. K. Denzin & Y. S. Lincoln (Eds.), *The SAGE handbook of qualitative research* (3rd ed., pp. 191–216). Thousand Oaks, CA: Sage.

Gupta, A., & Ferguson, J. (Eds.). (1997). Discipline and practice: "The field" as site, method, and location in anthropology. In A. Gupta & J. Ferguson (Eds.), *Anthropological locations: Boundaries and grounds of a field science* (pp. 1–46). Berkeley: University of California Press.

Hammersley, M. (1992). *What's wrong with ethnography?* London: Routledge.

Hammersley, M. (2000). Not bricolage but boatbuilding. *Journal of Contemporary Ethnography, 28*, 5.

Hammersley, M. (2008). *Questioning qualitative inquiry: Critical essays.* London: Sage.

Harper, D. (1987). *Working knowledge: Skill and community in a small shop.* Chicago: University of Chicago Press.

Hesse-Biber, S. N., & Leavy, P. (2008). Introduction: Pushing on the methodological boundaries: The growing need for emergent methods within and across the disciplines. In S. N. Hesse-Biber & P. Leavy (Eds.), *Handbook of emergent methods* (pp. 1–15). New York: Guilford Press.

Holman-Jones, S. H. (1999). Torch. *Qualitative Inquiry, 5*, 235–250.

hooks, b. (1990). *Yearning: Race, gender, and cultural politics.* Boston: South End Press.

Howe, K. (1988). Against the quantitative-qualitative incompatibility thesis (Or dogmas die hard). *Educational Researcher, 17*(8), 10–16.

Howe, K. R. (2004). A critique of experimentalism. *Qualitative Inquiry, 10*(1), 42–61.

Howe, K. R. (2009). Positivist dogmas, rhetoric, and the education science question. *Education Researcher, 38* (August/September), 428–440.

Huber, J. (1995). Centennial essay: Institutional perspectives on sociology. *American Journal of Sociology, 101,* 194–216.

Jackson, M. (1998). *Minima ethnographica.* Chicago: University of Chicago Press.

Jameson, F. (1991). *Postmodernism, or the cultural logic of late capitalism.* Durham, NC: Duke University Press.

Keller, C. M., & Keller, J. D. (1996). *Cognition and tool use: The blacksmith at work.* New York: Cambridge University Press.

Kincheloe, J. L. (2001). Describing the bricolage: Conceptualizing a new rigor in qualitative research. *Qualitative Inquiry, 7*(6), 679–692.

Lather, P. (1993). Fertile obsession: Validity after poststructuralism. *Sociological Quarterly, 35,* 673–694.

Lather, P. (2004). This *is* your father's paradigm: Government intrusion and the case of qualitative research in education. *Qualitative Inquiry, 10*(1), 15–34.

Lather, P., & Smithies, C. (1997). *Troubling the angels: Women living with HIV/AIDS.* Boulder, CO: Westview Press.

LeCompte, M. D., & Preissle, J. with R. Tesch. (1993). *Ethnography and qualitative design in educational research* (2nd ed.). New York: Academic Press.

Lévi-Strauss, C. (1966). *The savage mind.* Chicago: University of Chicago Press. (Original work published 1962)

Lincoln, Y. S. (1997). Self, subject, audience, text: Living at the edge, writing in the margins. In W. G. Tierney & Y. S. Lincoln (Eds.), *Representation and the text: Re-framing the narrative voice* (pp. 37–56). Albany: SUNY Press.

Lincoln, Y. S. (1999, June 3–6). *Courage, vulnerability, and truth.* Paper presented to the Reclaiming Voice II Conference, University of California-Irvine, Irvine, CA.

Lincoln, Y. S. (2010). What a long, strange trip it's been . . . : Twenty-five years of qualitative and new paradigm research. *Qualitative Inquiry, 16*(1), 3–9.

Lincoln, Y. S., & Cannella, G. S. (2004a). Dangerous discourses: Methodological conservatism and governmental regimes of truth. *Qualitative Inquiry, 10*(1), 5–14.

Lincoln, Y. S., & Cannella, G. S. (2004b). Qualitative research, power, and the radical right. *Qualitative Inquiry, 10*(2), 175–201.

Lincoln, Y. S., & Guba, E. G. (1985). *Naturalistic inquiry.* Beverly Hills, CA: Sage.

Lincoln, Y. S., & Tierney, W. G. (2004). Qualitative research and institutional review boards. *Qualitative Inquiry, 10*(2), 219–234.

Lofland, J. (1971). *Analyzing social settings.* Belmont, CA: Wadsworth.

Lofland, J. (1995). Analytic ethnography: Features, failings, and futures. *Journal of Contemporary Ethnography, 24*, 30–67.

Lofland, J., & Lofland, L. H. (1984). *Analyzing social settings.* Belmont, CA: Wadsworth.

Lofland, J., & Lofland, L. H. (1995). *Analyzing social settings* (3rd ed.). Belmont, CA: Wadsworth.

Lofland, L. (1980). The 1969 Blumer-Hughes talk. *Urban Life and Culture, 8*, 248–260.

Malinowski, B. (1948). *Magic, science and religion, and other essays.* New York: Natural History Press. (Original work published 1916)

Malinowski, B. (1967). *A diary in the strict sense of the term.* New York: Harcourt.

Marcus, G., & Fischer, M. (1986). *Anthropology as cultural critique.* Chicago: University of Chicago Press.

Maxwell, J. A. (2004). Reemergent scientism, postmodernism, and dialogue across differences. *Qualitative Inquiry, 10*(1), 35–41.

Mills, C. W. (1959). *The sociological imagination.* New York: Oxford University Press.

Monaco, J. (1981). *How to read a film: The art, technology, language, history and theory of film* (Rev. ed.). New York: Oxford University Press.

Morse, J. M. (2006). The politics of evidence. In N. Denzin & M. Giardina (Eds.), *Qualitative inquiry and the conservative challenge* (pp. 79–92). Walnut Creek, CA: Left Coast Press.

Morse, J. M., & Niehaus, L. (2009). *Mixed method design: Principles and procedures.* Walnut Creek, CA: Left Coast Press.

Nelson. C., Treichler, P. A., & Grossberg, L. (1992). Cultural studies. In L. Grossberg, C. Nelson, & P. A. Treichler (Eds.), *Cultural studies* (pp. 1–16). New York: Routledge.

Ortner, S. B. (1997). Introduction. In S. B. Ortner (Ed.), The fate of "culture": Clifford Geertz and beyond [Special issue]. *representations, 59,* 1–13.

Pelias, R. J. (2004). *A methodology of the heart: Evoking academic & daily life.* Walnut Creek, CA: AltaMira.

Plath, David. (1990). Fieldnotes, filed notes, and the conferring of note. In R. Sanjek (Ed.), *Fieldnotes* (pp. 371–384). Albany: SUNY Press.

Popkewitz, T. S. (2004). Is the National Research Council committee's report on scientific research in education scientific? On trusting the manifesto. *Qualitative Inquiry, 10*(1), 62–78.

Richardson, L. (1991). Postmodern social theory. *Sociological Theory, 9,* 173–179.

Richardson, L. (1992). The consequences of poetic representation: Writing the other, rewriting the self. In C. Ellis & M. G. Flaherty (Eds.), *Investigating subjectivity: Research on lived experience.* Newbury Park, CA: Sage.

Richardson, L. (1997). *Fields of play.* New Brunswick, NJ: Rutgers University Press.

Richardson, L. (2000). Writing: A method of inquiry. In N. K. Denzin & Y. S. Lincoln (Eds.), *Handbook of qualitative research* (2nd ed., pp. 923–948). Thousand Oaks, CA: Sage.

Richardson, L., & Lockridge, E. (2004). *Travels with Ernest: Crossing the literary/sociological divide.* Walnut Creek, CA: AltaMira.

Roffman, P., & Purdy, J. (1981). *The Hollywood social problem film.* Bloomington: Indiana University Press.

Ronai, C. R. (1998). Sketching with Derrida: An ethnography of a researcher/erotic dancer. *Qualitative Inquiry, 4,* 405–420.

Rosaldo, R. (1989). *Culture & truth.* Boston: Beacon.

Ryan, K. E., & Hood, L. K. (2004). Guarding the castle and opening the gates. *Qualitative Inquiry, 10*(1): 79–95.

Sanjek, R. (1992). *Fieldnotes.* Albany: SUNY Press.

Scheurich, J. & Clark, M. C. (2006). Qualitative studies in education at the beginning of the twenty-first century. *International Journal of Qualitative Studies in Education, 19*(4), 401.

Schwandt, T. A. (1997a). *Qualitative inquiry.* Thousand Oaks, CA: Sage.

Schwandt, T. A. (1997b). Textual gymnastics, ethics, angst. In W. G. Tierney & Y. S. Lincoln (Eds*.), Representation and the text: Re-framing the narrative voice* (pp. 305–313). Albany: SUNY Press.

Seale, C., Gobo, G., Gubrium, J. F., & Silverman, D. (2004). Introduction: Inside qualitative research. In C. Seale, G. Gobo, J. F. Gubrium, & D. Silverman (Eds.), *Qualitative research practice* (pp. 1–11). London: Sage.

Semaili, L. M., & Kincheloe, J. L. (1999). Introduction: What is indigenous knowledge and why should we study it? In L. M. Semaili & J. L. Kincheloe (Eds.), *What is indigenous knowledge? Voices from the academy* (pp. 3–57). New York: Falmer Press.

Silverman, D. (1997). Towards an aesthetics of research. In D. Silverman (Ed.), *Qualitative research: Theory, method, and practice* (pp. 239–253). London: Sage.

Smith, A. D. (1993). *Fires in the mirror.* New York: Anchor Books.

Smith, L. T. (1999). *Decolonizing methodologies: Research and indigenous peoples.* Dunedin, NZ: University of Otago Press.

Snow, D., & Morrill, C. (1995). Ironies, puzzles, and contradictions in Denzin and Lincoln's vision of qualitative research. *Journal of Contemporary Ethnography, 22,* 358–362.

Spindler, G., & Spindler, L. (1992). Cultural process and ethnography: An anthropological perspective. In M. D. LeCompte, W. L. Millroy, &

J. Preissle (Eds.), *The handbook of qualitative research in education* (pp. 53–92). New York: Academic Press.

Stocking, G. W., Jr. (1986). Anthropology and the science of the irrational: Malinowski's encounter with Freudian psychoanalysis. In *History of anthropology: Vol. 4. Malinowski, Rivers, Benedict, and others: Essays on culture and personality* (pp. 13–49). Madison: University of Wisconsin Press.

Stocking, G. W., Jr. (1989). The ethnographic sensibility of the 1920s and the dualism of the anthropological tradition. In *History of anthropology: Vol. 6. Romantic Motives: Essays on anthropological sensibility* (pp. 208–276). Madison: University of Wisconsin Press.

Stoller, P., & Olkes, C. (1987). *In sorcery's shadow.* Chicago: University of Chicago Press.

St.Pierre, E. A. (2004). Refusing alternatives: A science of contestation. *Qualitative Inquiry, 10*(1), 130–139.

St.Pierre, E. A., & Roulston, K. (2006). The state of qualitative inquiry: A contested science. *International Jouranl of Qualitative Studies in Education, 19*(6), 673–684.

Strauss, A. (1987). *Qualitative analysis for social scientists.* New York: Cambridge.

Strauss, A., & Corbin, J. (1999). *Basics of qualitative research* (2nd ed.). Thousand Oaks, CA: Sage.

Taylor, S. J., & Bogdan, R. (1998). *Introduction to qualitative research methods: A phenomenological approach to the social sciences* (3rd ed.). New York: Wiley.

Teddlie, C., & Tashakkori, A. (2003a). Major issues and controversies in the use of mixed methods in the social and behavioral sciences. In A. Tashakkori & C. Teddlie (Eds.), *Handbook of mixed-methods in social and behavioral research* (pp. 3–50). Thousand Oaks, CA: Sage.

Teddlie, C., & Tashakkori, A. (2003b). Preface. In A. Tashakkori & C. Teddlie (Eds.), *Handbook of mixed-methods in social and behavioral research* (pp. ix-xv). Thousand Oaks, CA: Sage.

Turner, V., & Bruner, E. (Eds.). (1986). *The anthropology of experience.* Urbana: University of Illinois Press.

Van Maanen, J. (1988). *Tales of the field.* Chicago: University of Chicago Press.

Vygotsky, L. S. (1978). *Mind in society.* Cambridge, MA: Harvard University Press.

Weinstein, D., & Weinstein, M. A. (1991). Georg Simmel: Sociological flaneur bricoleur. *Theory, Culture & Society, 8,* 151–168.

Weinstein, M. (2004). Randomized design and the myth of certain knowledge: Guinea pig narratives and cultural critique. *Qualitative Inquiry, 10*(2), 246–260.

West, C. (1989). *The American evasion of philosophy.* Madison: University of Wisconsin Press.

Wolcott, H. F. (1990). *Writing up qualitative research.* Newbury Park, CA: Sage.

Wolcott, H. F. (1992). Posturing in qualitative research. In M. D. LeCompte, W. L. Millroy, & J. Preissle (Eds.), *The handbook of qualitative research in education* (pp. 3–52). New York: Academic Press, Inc.

Wolcott, H. F. (1995). *The art of fieldwork.* Walnut Creek, CA: AltaMira Press.

Wolfe, M. (1992). *A thrice-told tale.* Stanford, CA: Stanford University Press.

Wright, H. K. (2006). Are we there yet? Qualitative research in education's profuse and contested present. *International Journal of Qualitative Studies in Education, 19*(6), 793–802.

Part I

LOCATING THE FIELD

Part I of the *Handbook* begins by locating qualitative research within the academy. It then turns to the history of qualitative inquiry in social and educational research. The last two chapters take up the ethics, politics, and moral responsibilities of the qualitative researcher.

▣ THE ACADEMY AND THE PARTICIPATORY ACTION TRADITION

The opening chapter, by Morten Levin and Davydd Greenwood, calls for a reinvention of the social sciences. Their chapter reveals the depth and complexity of the traditional and applied qualitative research perspectives that are consciously and unconsciously inherited by the researcher-as-interpretive-bricoleur.[1] These traditions locate the investigator in academic systems of historical (and organizational) discourse. This system guides and constrains the interpretive work that is done in any specific study. The academy is in a state of crisis. Traditional funding connections to stakeholders no longer hold. Radical change is required, and action research can help lead the way.

Levin and Greenwood argue that action researchers have a responsibility to do work that is socially meaningful and socially responsible. The relationship between researchers, universities, and society must change. Politically informed action research, inquiry committed to praxis and social change, is the vehicle for accomplishing this transformation.

Action researchers are committed to a set of disciplined, material practices that produce radical, democratizing transformations in the civic sphere. These practices involve collaborative dialogue, participatory decision-making, inclusive democratic deliberation, and the maximal participation and representation of all relevant parties (Ryan & Destefano, 2000, p. 1). Action researchers literally help transform inquiry into praxis or action. Research subjects become co-participants and stakeholders in the process of inquiry. Research becomes praxis—practical, reflective, pragmatic action—directed to solving problems in the world.

These problems originate in the lives of the research co-participants; they do not come down from on high by way of grand theory. Together, stakeholders and action researchers co-create knowledge that is pragmatically useful and grounded in local knowledge. In the process, they jointly define research objectives and political goals, co-construct research questions, pool knowledge, hone shared research skills, fashion interpretations and performance texts that implement specific strategies for social change, and measure validity and credibility by the willingness of local stakeholders to act on the basis of the results of the action research.

Academic science has a history of not being able to accomplish goals such as these consistently. Levin and Greenwood offer several reasons for this failure, including the inability of a so-called positivistic, value-free social science to produce useful social research; the increasing tendency of outside corporations to define the needs and values of the university; the loss of research funds to entrepreneurial and private-sector research organizations; and bloated, inefficient internal administrative infrastructures.

Levin and Greenwood are not renouncing the practices of science; rather, they are calling for a reformulation of what science and the academy are all about. Their model of pragmatically grounded action research is not a retreat from disciplined scientific inquiry.[2] This form of inquiry reconceptualizes science as a multiperspective, methodologically diverse, collaborative, communicative, communitarian, context-centered, moral project. Levin and Greenwood want to locate action research at the center of the contemporary university. Their chapter is a call for a civic social science, a pragmatic science that will lead to the radical reconstruction of the university's relationships with society, state, and community in this new century.

▣ HISTORY

In their monumental chapter ("Qualitative Methods: Their History in Sociology and Anthropology"), reprinted in the second edition of the *Handbook,* Arthur Vidich and Stanford

Lyman (2000) show how the ethnographic tradition extends from the Greeks through the 15th- and 16th-century interests of Westerners in the origins of primitive cultures; to colonial ethnology connected to the empires of Spain, England, France, and Holland; to several 20th-century transformations in the United States and Europe. Throughout this history, the users of qualitative research have displayed commitments to a small set of beliefs, including objectivism, the desire to contextualize experience, and a willingness to interpret theoretically what has been observed.

In Chapter 3 of this volume, Frederick Erickson shows that these beliefs supplement the positivist tradition of complicity with colonialism, the commitments to monumentalism, and the production of timeless texts. The colonial model located qualitative inquiry in racial and sexual discourses that privileged white patriarchy. Of course, as indicated in our Introduction, these beliefs have recently come under considerable attack.

Erickson, building on Vidich and Lyman, documents the extent to which early as well as contemporary qualitative researchers were (and remain) implicated in these systems of oppression. His history extends Vidich-Lyman's, focusing on five foundational footings: disciplinary perspectives on qualitative research—especially sociology and anthropology; the participant observer as observer/author; the people observed during fieldwork; the rhetorical and substantive content of the qualitative research report; and the audiences for such texts.

He offers a trenchant review of recent disciplinary efforts (by the American Educational Research Association) to impose fixed criteria of evaluation on qualitative inquiry. He carefully reviews recent criticisms of the classic ethnographic text. He argues that the realist ethnographic text—the text with its omniscient narrator—is no longer a genre of reporting that can be responsibly practiced.

◧ THE ETHICS OF INQUIRY

Clifford Christians locates the ethics and politics of qualitative inquiry within a broader historical and intellectual framework. He first examines the Enlightenment model of positivism, value-free inquiry, utilitarianism, and utilitarian ethics. In a value-free social science, codes of ethics for professional societies become the conventional format for moral principles. By the 1980s, each of the major social science associations (contemporaneous with passage of federal laws and promulgation of national guidelines) had developed its own ethical code with an emphasis on several guidelines: informed consent, nondeception, the absence of psychological or physical harm, privacy and confidentiality, and a commitment to collecting and presenting reliable and valid empirical materials. Institutional review boards (IRBs) implemented these guidelines, including ensuring that informed consent is always obtained in human subject

research. However, Christians notes that in reality IRBs protect institutions and not individuals.

Several events challenged the Enlightenment model, including the Nazi medical experiments, the Tuskegee syphilis study, Project Camelot in the 1960s, Stanley Milgram's deception of subjects in his psychology experiments, Laud Humphrey's deceptive study of homosexuals, and the complicity of social scientists with military initiatives in Vietnam. In addition, charges of fraud, plagiarism, data tampering, and misrepresentation continue to the present day.

Christians details the poverty of the Enlightenment model. It creates the conditions for deception, for the invasion of private spaces, for duping subjects, and for challenges to the subject's moral worth and dignity (see also Angrosino & Rosenberg, Chapter 28, this volume; also Guba & Lincoln, 1989, pp. 120–141). Christians calls for its replacement with an ethics based on the values of a feminist communitarianism.

This is an evolving, emerging ethical framework that serves as a powerful antidote to the deception-based, utilitarian IRB system. The new framework presumes a community that is ontologically and axiologically prior to the person. This community has common moral values, and research is rooted in a concept of care, of shared governance, of neighborliness, or of love, kindness, and the moral good. Accounts of social life should display these values and be based on interpretive sufficiency. They should have sufficient depth to allow the reader to form a critical understanding about the world studied. These texts should exhibit an absence of racial, class, and gender stereotyping. These texts should generate social criticism and lead to resistance, empowerment, social action, and positive change in the social world.

In the feminist communitarian model, as with the model of participatory action research advocated by Levin and Greenwood, participants have a co-equal say in how research should be conducted, what should be studied, which methods should be used, which findings are valid and acceptable, how the findings are to be implemented, and how the consequences of such action are to be assessed. Spaces for disagreement are recognized, while discourse aims for mutual understanding and the honoring of moral commitments.

A sacred, existential epistemology places us in a noncompetitive, nonhierarchical relationship to the earth, to nature, and to the larger world (Bateson, 1972, p. 335). This sacred epistemology stresses the values of empowerment, shared governance, care, solidarity, love, community, covenant, morally involved observers, and civic transformation. As Christians observes, this ethical epistemology recovers the moral values that were excluded by the rational Enlightenment science project. This sacred epistemology is based on a philosophical anthropology that declares that "all humans are worthy of dignity and sacred status without exception for class or ethnicity" (Christians, 1995, p. 129). A universal human ethic, stressing the sacredness of life, human dignity, truth telling, and nonviolence, derives from this position (Christians, 1997, pp. 12–15). This ethic is based on locally experienced, culturally prescribed pro-tonorms (Christians, 1995, p. 129). These primal norms provide a defensible "conception of good rooted in universal human solidarity" (Christians, 1995, p. 129; also 1997, 1998). This sacred epistemology recognizes and interrogates the ways in which race, class, and gender operate as important systems of oppression in the world today.

In this way, Christians outlines a radical ethical path for the future. He transcends the usual middle-of-the-road ethical mod-els, which focus on the problems associated with betrayal, decep-tion, and harm in qualitative research. Christians's call for a collaborative social science research model makes the researcher responsible, not to a removed discipline (or institution), but rather to those studied. This implements critical, action, and feminist traditions, which forcefully align the ethics of research with a politics of the oppressed. Christians's framework reorga-nizes existing discourses on ethics and the social sciences.[3]

Clearly the existing, Belmont and Common Rule defini-tions have little, if anything, to do with a human rights and social justice ethical agenda. Regrettably, these principles have been informed by notions of value-free experimenta-tion and utilitarian concepts of justice. They do not concep-tualize research in participatory terms. In reality, these rules protect institutions and not people, although they were originally created to protect human subjects from unethical biomedical research. The application of these regulations is an instance of mission or ethics creep, or the overzealous extension of IRB regulations to interpretive forms of social science research. This has been criticized by many, including Kevin Haggerty (2004), C. K. Gunsalus et al. (2007), Leon Dash (2007), and the American Association of University Professors (AAUP, 2001, 2002, 2006a, 2006b).[4]

Oral historians (see Shopes, Chapter 27, this volume) have contested the narrow view of science and research contained in current reports (American Historical Association, 2008; Shopes & Ritchie, 2004). Anthropologists and archaeologists have chal-lenged the concept of informed consent as it impacts ethno-graphic inquiry (see Fluehr-Lobban, 2003a, 2003b; also Miller & Bell, 2002). Journalists argue that IRB insistence on anonymity reduces the credibility of journalistic reporting, which rests on naming the sources used in a news account. Dash (2007, p. 871) contends that IRB oversight interferes with the First Amend-ment rights of journalists and the public's right to know. Indig-enous scholars Marie Battiste (2008) and Linda Tuhiwai Smith (2005) assert that Western conceptions of ethical inquiry have "severely eroded and damaged indigenous knowledge" and indigenous communities (Battiste, 2008, p. 497).[5]

As currently deployed, these practices close down critical ethical dialogue. They create the impression that if proper IRB procedures are followed, then one's ethical house is in order. But this is ethics in a cul de sac.

◙ DISCIPLINING AND CONSTRAINING ETHICAL CONDUCT

The consequence of these restrictions is a disciplining of quali-tative inquiry that extends from granting agencies to qualitative research seminars and even the conduct of qualitative disserta-tions (Lincoln & Cannella, 2004a, 2004b). In some cases, lines of critical inquiry have not been funded and have not gone for-ward because of criticisms from local IRBs. Pressures from the right discredit critical interpretive inquiry. From the federal to the local levels, a trend seems to be emerging. In too many instances, there seems to be a move away from protecting human subjects to an increased monitoring, censuring, and policing of projects that are critical of the right and its politics.

Yvonna S. Lincoln and William G. Tierney (2004) observe that these policing activities have at least five important impli-cations for critical social justice inquiry. First, the widespread rejection of alternative forms of research means that qualitative inquiry will be heard less and less in federal and state policy forums. Second, it appears that qualitative researchers are being deliberately excluded from this national dialogue. Consequently, third, young researchers trained in the critical tradition are not being heard. Fourth, the definition of research has not changed to fit newer models of inquiry. Fifth, in rejecting qualitative inquiry, traditional researchers are endorsing a more distanced form of research, one that is compatible with existing stereo-types concerning people of color.

These developments threaten academic freedom in four ways: (1) they lead to increased scrutiny of human subjects research and (2) new scrutiny of classroom research and training in qualitative research involving human subjects; (3) they connect to evidence-based discourses, which define qualitative research as unscientific; and (4) by endorsing methodological conservatism, they reinforce the status quo on many campuses. This conservatism produces new constraints on graduate training, leads to the improper review of faculty research, and creates conditions for politicizing the IRB review process, while protecting institutions and not individuals from risk and harm.

◙ A PATH FORWARD

Since 2004, many scholarly and professional societies have fol-lowed the Oral History and American Historical Associations in challenging the underlying assumptions in the standard campus IRB model. A transdisciplinary, global, counter-IRB discourse has emerged (Battiste, 2008; Christians, 2007; Ginsberg & Mertens, 2009; Lincoln, 2009). This discourse has called for the blanket exclusion of non-federally funded research from IRB review. The AAUP (2006a, 2006b) has gone so far as to recommend that

> exemptions based on methodology, namely research on autonomous adults whose methodology consists entirely of collecting data by surveys, conducting interviews, or observing behavior in public places should be exempt from the requirement of IRB review, with no provisos, and no requirement of IRB approval of the exemption. (p. 4)

The executive council of the Oral History Association endorsed the AAUP recommendations at its October 2006 annual meeting. They were quite clear: "Institutions consider as straightforwardly exempt from IRB review any 'research whose methodology consists entirely of collecting data by surveys, conducting interviews, or observing behavior in public places'" (Howard, 2006, p. 9). This recommendation can be extended: Neither the Office for Human Resource Protection, nor a campus IRB has the authority to define what constitutes legitimate research in any field, only what research is covered by federal regulations.

We agree.

◩ ETHICS AND CRITICAL SOCIAL SCIENCE

In Chapter 5, Gaile Cannella and Yvonna S. Lincoln, building on the work of Michel Foucault, argue that a critical social science requires a radical ethics, an "ethics that is always/already concerned about power and oppression even as it avoids constructing 'power' as a new truth" (p. 97). A critical ethical stance works outward from the core of the person. A critical social science incorporates feminist, postcolonial, and even postmodern challenges to oppressive power. It is aligned with a critical pedagogy and a politics of resistance, hope, and freedom.

A critical social science focuses on structures of power and systems of domination. It creates spaces for a decolonizing project. It opens the doors of the academy so that the voices of oppressed people can be heard and honored and so that others can learn from them.

◩ CONCLUSION

Thus do the chapters in Part I of the *Handbook* come together over the topics of ethics, power, politics, social justice, and the academy. We endorse a radical, participatory ethic, one that is communitarian and feminist, an ethic that calls for trusting, collaborative nonoppressive relationships between researchers and those studied, an ethic that makes the world a more just place (Collins, 1990, p. 216).

◩ NOTES

1. Any distinction between applied and nonapplied qualitative research traditions is somewhat arbitrary. Both traditions are scholarly. Each has a long tradition and a long history, and each carries basic implications for theory and social change. Good theoretical research should also have applied relevance and implications. On occasion, it is argued that applied and action research are nontheoretical, but even this conclusion can be disputed.

2. We will develop a notion of a sacred science below and in our concluding chapter.

3. Given Christians's framework, there are primarily two ethical models: utilitarian and nonutilitarian. However, historically, and most recently, one of five ethical stances (absolutist, consequentialist, feminist, relativist, deceptive) has been followed, although often these stances merge with one another. The *absolutist* position argues that any method that contributes to a society's self-understanding is acceptable, but only conduct in the public sphere should be studied. The *deception* model says any method, including the use of lies and misrepresentation, is justified in the name of truth. The *relativist* stance says researchers have absolute freedom to study what they want; ethical standards are a matter of individual conscience. Christians's feminist-communitarian framework elaborates a *contextual-consequential framework*, which stresses mutual respect, noncoercion, nonmanipulation, and the support of democratic values (see Guba & Lincoln, 1989, pp. 120–141; Smith, 1990; also Collins, 1990, p. 216; Mitchell, 1993).

4. Mission creep includes these issues and threats: rewarding wrong behaviors, focusing on procedures and not difficult ethical issues, enforcing unwieldy federal regulations, and involving threats to academic freedom and the First Amendment (Becker, 2004; Gunsalus et al., 2007; also Haggerty, 2004). Perhaps the most extreme form of IRB mission is the 2002 State of Maryland Code, Title 13—Miscellaneous Health Care Program, Subtitle 20—Human Subject Research § 13–2001, 13–2002:Compliance with Federal Regulations: A person may not conduct research using a human subject unless the person conducts the research in accordance with the federal regulations on the protection of human subjects (see Shamoo & Schwartz, 2007).

5. There is a large Canadian project on indigenous intellectual property rights—Intellectual Property Issues in Cultural Heritage. This project represents an international, interdisciplinary collaboration among more than 50 scholars and 25 partnering organizations embarking on an unprecedented and timely investigation of intellectual property (IP) issues in cultural heritage that represent emergent local and global interpretations of culture, rights, and knowledge. Their objectives are:

- to document the diversity of principles, interpretations, and actions arising in response to IP issues in cultural heritage worldwide;
- to analyze the many implications of these situations;

- to generate more robust theoretical understandings as well as exemplars of good practice; and
- to make these findings available to stakeholders—from Aboriginal communities to professional organizations to government agencies—to develop and refine their own theories, principles, policies, and practices.

Left Coast is their publisher. See their website: http://www.sfu.ca/ipinch/

▣   REFERENCES

American Association of University Professors. (2001). Protecting human beings: Institutional review boards and social science research. *Academe, 87*(3), 55–67.

American Association of University Professors. (2002). Should all disciplines be subject to the common rule? Human subjects of social science research. *Academe, 88*(1), 1–15.

American Association of University Professors, Committee A. (2006a). *Report on human subjects: Academic freedom and the institutional review boards.* Available at http://www.aaup.org/AAUP/About/committees/committee+repts/CommA/

American Association of University Professors (AAUP). (2006b). *Research on human subjects: Academic freedom and the institutional review board.* Available at www.aaup.org/AAUP/comm./rep/A/humansub.htm

American Historical Association. (2008, February). AHA statement on IRBs and oral history research. *Perspectives on History.*

Bateson, G. (1972). *Steps to an ecology of mind.* New York: Ballantine.

Battiste, M. (2008). Research ethics for protecting indigenous knowledge and heritage: Institutional and researcher responsibilities. In N. K. Denzin, Y. S. Lincoln, & L. T. Smith (Eds.), *Handbook of critical and indigenous methodologies* (pp. 497–510). Thousand Oaks, CA: Sage.

Christians, G. C. (1995). The naturalistic fallacy in contemporary interactionist-interpretive research. *Studies in Symbolic Interaction, 19,* 125–130.

Christians, G. C. (1997). The ethics of being in a communications context. In C. Christians & M. Traber (Eds.), *Communication ethics and universal values* (pp. 3–23). Thousand Oaks, CA: Sage.

Christians, G. C. (1998). The sacredness of life. *Media Development, 2,* 3–7.

Christians, C. G. (2007). Neutral science and the ethics of resistance. In N. K. Denzin & M. D. Giardina (Eds.), *Ethical futures in qualitative research* (pp. 47–66). Walnut Creek, CA: Left Coast Press.

Collins, P. H. (1990). *Black feminist thought.* New York: Routledge.

Dash, L. (2007). Journalism and institutional review boards. *Qualitative Inquiry, 13*(6), 871–874.

Fluehr-Lobban, C. (Ed.). (2003a). *Ethics and the profession of anthropology* (2nd ed.). Walnut Creek, CA: AltaMira.

Fluehr-Lobban C. (2003b). Informed consent in anthropological research. In C. Fluehr-Lobban (Ed.), *Ethics and the profession of anthropology* (2nd ed., pp. 159–177). Walnut Creek, CA: AltaMira.

Ginsberg, P. E., & Mertens, D. M. (2009). Frontiers in social research ethics: Fertile ground for evolution. In D. M. Mertens & P. E. Ginsberg (Eds.), *The handbook of social research ethics* (pp. 580–613). Thousand Oaks, CA: Sage.

Guba, E. S., & Lincoln, Y. S. (1989). *Fourth generation evaluation.* Newbury Park, CA: Sage.

Gunsalus, C. K., Bruner, E. M., Burbules, N. C., Dash, L., Finkin, M., Goldberg, J. P., Greenough, W. T., Miller, G. A., Pratt, M. G., Iriye, M., & Aronson, D. (2007). The Illinois white paper: Improving the system for protecting human subjects: Counteracting IRB "mission creep." *Qualitative Inquiry, 13*(5), 617–649.

Haggerty, K. D. (2004). Ethics creep: Governing social science research in the name of ethics. *Qualitative Sociology, 27*(4), 391–414.

Howard, J. (2006, November 10). Oral history under review. *Chronicle of Higher Education.* Available at http:///chronicle.com/free/v53/112/12a01401.htm

Lincoln, Y. S. (2009). Ethical practices in qualitative research. In D. M. Mertens & P. E. Ginsberg (Eds.), *The handbook of social research ethics* (pp. 150–170). Thousand Oaks, CA: Sage.

Lincoln, Y. S., & Cannella, G. S. (2004a). Dangerous discourses: Methodological conservatism and governmental regimes of truth. *Qualitative Inquiry, 10*(1), 5–14.

Lincoln, Y. S., & Cannella, G. S. (2004b). Qualitative research, power, and the radical right. *Qualitative Inquiry, 10*(2), 175–201.

Lincoln, Y. S., & Tierney, W. G. (2004). Qualitative research and institutional review boards. *Qualitative Inquiry, 10*(2), 219–234.

Miller, T., & Bell, L. (2002). Consenting to what? Issues of access, gatekeeping, and "informed consent." In M. Mauthner, M. Birtch, J. Jessop, & T. Miller (Eds.), *Ethics in qualitative research* (pp. 70–89). London: Sage.

Mitchell, Richard J. Jr. (1993). *Secrecy and fieldwork.* Newbury Park: Sage.

Ryan, K., & Destefano, L. (2000). Introduction. In K. Ryan & L. Destefano (Eds.), *Evaluation in a democratic society: Deliberation, dialogue, and inclusion* (pp. 1–20). New Directions in Evaluation Series. San Francisco: Jossey-Bass.

Shopes, L., & Ritchie, D. (2004, March). Exclusion of oral history from IRB review: An update. *Perspectives online.* Available at htttp://www.historians.org/Perspecxtives/Issues'2004/0403new1.cfn

Smith, L. M. (1990). Ethics, field studies, and the paradigm crisis. In E. G. Guba (Ed.), *The paradigm dialog* (pp. 139–157). Newbury Park, CA: Sage.

Smith, L. T. (2005). On tricky ground: Researching the native in the age of uncertainty. In N. K. Denzin & Y. S. Lincoln (Eds.), *The SAGE handbook of qualitative research* (3rd ed., pp. 85–107). Thousand Oaks, CA: Sage.

Vidich, A., & Lyman, S. (2000). Qualitative methods: Their history in sociology and anthropology. In N. K. Denzin & Y. S. Lincoln (Eds.), *Handbook of qualitative research* (2nd ed., pp. 37–84). Thousand Oaks, CA: Sage.

2

REVITALIZING UNIVERSITIES BY REINVENTING THE SOCIAL SCIENCES

Bildung *and Action Research*

Morten Levin and Davydd Greenwood

Doing social science is, among other things, a form of contextualized institutional social practice. This banality, taken to its obvious conclusion and set in the context of contemporary academic social science, yields a number of consequences that most academic social scientists will not like. One implication is that theoretical and methodological approaches must be interpreted within the institutional contexts and social practices where they are embedded and practiced. If the desire for theoretical and methodological development is genuine, then this means the social sciences cannot proceed without developing and advocating an understanding of how universities, research institutions, and disciplinary structures shape the contexts and practices of their activities. Academic social scientists' engagement in autopoetic theoretical and methodological efforts disconnects them from society at large. Research and teaching agendas are motivated more by what is fashionable in the professionalized arenas of institutionalized social science than by the aim of addressing pertinent societal problems. Since the larger organizational structures and processes of universities, campus administrative structures, national and international professional societies, and national and international ranking systems currently are inimical to the development of socially meaningful theories/practices in social sciences, then those structures have to be analyzed and changed as well.

We make a situated, pragmatist analysis that examines university organizational structures, power relations, discourses, and external relations as they affect social research methodologies and practices. Doing this creates an epistemological,

political, methodological, theoretical, and ethical necessity to go beyond conventional organizational analyses of the academic professions and analyze actual social science behavior in concrete contexts. Academic social scientists have to confront existing choices about university organizational structures and the larger extra-university context in which social science research operates. Social scientists have the tools to reveal the contours of these problems and the obligation to use them in playing a role in the pro-social reform of those structures. Leaving the changes to professional administrators, their consultants, and outside policymakers has already undermined universities in significant ways.

We pretend no neutrality on these matters. We believe that universities as something more than vocational schools and research shops are in real jeopardy. Current methods, professional practices, and organizational structures make the academic social sciences almost impossible to justify to increasingly hostile publics, funders, and policymakers. Since the Tayloristic structures of university organization are inimical to more than cosmetic institutional reform (e.g., strategic planning without any significant organizational change), we challenge them directly. We believe that universities matter and are therefore worth reforming, but only as loci for the formation of citizens; the analysis of complex technical, social, and ethical issues; and the support of meaningful efforts toward the solution of society's most pressing problems. Such universities could thrive only by means of fluid, multidimensional relationships within their own structures and with the nonuniversity worlds that are the source of their legitimacy and funding. We believe that the social sciences should have a privileged position and a core responsibility in bringing about the necessary changes.

Four important elements in practicing social science emerge as fundamental issues to be addressed if social science were to regain a solid foothold at universities and in society as a whole.

Multiperspective research. Social science research at universities has to include relevant social science, humanistic, and scientific professional expertise in multiperspective research on key societal problems. This multiperspective cogeneration of knowledge is vital in mobilizing the array of expertise found within the existing disciplines to generate meaningful and useful social knowledge and reform and to develop valid theories and methods in the fields of the participating academic partners. Fundamental reforms in teaching are also required to engage students, early and often, in multidisciplinary team research on complex problems. Doing so requires a significant reorganization of university operations and a revised set of ways of connecting intra-university worlds.

Methodological diversity. We necessarily support disciplinary and methodological diversity. For example, we believe that qualitative and quantitative methods are mutually necessary in the study of any important social problem (See, e.g., Creswell, 2003; Creswell & Clark, 2007). Significant problems do not come neatly divided into quantitative and qualitative dimensions. It is up to the researchers to combine these dimensions whenever necessary in comprehensive and actionable frameworks.

Academic social scientists want to believe that theory and practice can be neatly separated and that they should be (Eikeland, 2008). We disagree, and we assert that theory can best be generated in practice and can be properly tested only in practice. This means that the comfortable campus office/library/laboratory life is an insufficient context for the practice and further development of the social sciences. This is problematic since many academic social scientists have become academics precisely to withdraw from direct encounters with the nonacademic world.

Inclusiveness of stakeholders. While it increases both theoretical and methodological demands, nonuniversity stakeholders should be included in social science research. Contrary to the widespread view within academia, creating mutual learning opportunities between universities and nonacademic stakeholders does not lower the expectations for theoretical and methodological rigor in the social sciences. Rather, it increases those demands because the researchers are forced to deal with more complex, multidimensional problems than most academics want to address (and are rewarded professionally for studying), and they must do so in ways that are persuasive to nonacademic stakeholders whose personal well-being is at stake.

Changes in social science teaching. Much social science teaching has become antisocial. Lecturing on general theory and method to passive students, equating social science development with theoretical and methodological elaboration in the absence of practice, and privileging the critique of the latest journal articles rather than evaluating the substantive contributions to understanding and managing social problems are standard practices. They sever the connection between the social sciences and everyday social problems. Our own experiences have shown us that sustained linkages between social science theory and method and work on concrete social problems with local stakeholders help students and their teachers become more competent theorists and practitioners.

To achieve this, teaching must depart from the abstract presentations of lectures on theory and methods. Formal presentation has its place but must be accompanied by supervised social research practice in multidisciplinary team situations with multiple stakeholders who are internal and external to the university. Teaching must create learning opportunities built on real-life problems where theory and method are challenged and also used to broaden understandings.

These changes are directly opposed to the hegemonic Tayloristic logic of academic organizational structures. To meet the tests of complexity, applicability, and trustworthiness in social research requires multidisciplinary research and teaching that redefines departmental boundaries and professional identities and that recontextualizes the relationships between universities and extra-university stakeholders. This involves radical changes in universities, in the social sciences, and in the ways these interact with society at large.

▣ ACTION RESEARCH

The changes proposed here form the core elements in the kind of social research we have practiced for decades, action research. We know that the approaches we recommend work. If they are possible and they work effectively, it is scientifically unacceptable to ignore them, particularly when the social sciences are at risk with all but their internal professional constituencies.

This chapter is our third contribution to the *Handbook of Qualitative Inquiry*. The main thrust in the previous two contributions was to argue that action research is a viable research strategy enabling a balance between rigor and relevance and that it has great transformative potential. In our first contribution, "Reconstructing the Relationships Between Universities and Society Through Action Research" (Greenwood & Levin, 2000, pp. 85–106), we began with a limited presentation of the problems created by the disconnectedness between social sciences and society at large and advocated addressing it through action research as the core approach to university social research. In doing this, we briefly laid out the basic elements in action research built on a pragmatic philosophical position.

In our second contribution, "Reform of the Social Sciences and of Universities Through Action Research" (Greenwood & Levin, 2005, pp. 43–64), we tightened the focus on what counts as scientific knowledge in universities and developed arguments for action research as a genuinely scientific practice. We claimed that action research could be institutionalized as the principal model for research and teaching in universities. Action research would, we believed, support a closer linkage between academic knowledge creation and enhancement of concrete problem solving for all engaged stakeholders. We argued that action research is a research and teaching strategy that both could reform social science knowledge production and create a closer link between social research and society. The core idea was the creation of a research and teaching praxis that integrated researchers (teachers) and relevant stakeholders in the same knowledge acquisition process.

During the time since our first contribution and especially since 1998, when we published a synthetic introduction to action research (Greenwood & Levin, 1998a, 2007), we noticed that these arguments about action research have had no visible effect on university social science behavior. Parochial academic professionalism, ranking by peer review within disciplinary specialties, the separation of qualitative and quantitative research as methodological specialties, the separation of theory and practice—all of these continue. An understanding of action research as a major alternative strategy for social research is nowhere visible. The current financial crises of higher education have resulted in even more bunkerism among the disciplines, subdisciplines, and specialties. The standard administrative approach to financial problems has been to distribute the cuts according to the strengths and weaknesses of different departments, which creates even fewer incentives for cooperation across disciplines. As a result, we are once again "introducing" action research before we can proceed to our core arguments about universities. However, we do so more briefly than in the previous contributions. The reader can turn to those or the second edition of our *Introduction to Action Research* (Greenwood & Levin, 2007) for a more extended treatment.

In our book, *Introduction to Action Research*, we defined action research as follows:

> AR is a set of self-consciously collaborative and democratic strategies for generating knowledge and designing action in which trained experts in social and other forms of research and local stakeholders work together. The research focus is chosen collaboratively between the local stakeholders and the action researchers and the relationships among the participants are organized as joint learning processes. AR centers on doing "with" rather than doing "for" stakeholders and credits local stakeholders with the richness of experience and reflective possibilities that long experience living in complex situations brings with it. (Greenwood & Levin, 2007, p. 1)

Action researchers link praxis and theory in social research. Social research that is not applied cannot meaningfully be called research, we believe, because theories not tested in context are merely speculations. We reject the notion that there can be applied research practices that are not explicitly connected to theories and methods. So action research rejects the theory/practice dichotomy on which most conventional social research relies (Greenwood & Levin, 1998a, 1998b, 2000, 2001a, 2001b; Levin & Greenwood, 1998).

To many social scientists, action research is "mere" activism and is viewed as a retreat from rigorous theories and methods. The justification given for this position is that greater relevance requires less rigor (an extensive counterposition is found in Argyris & Schön, 1978, 1996). Our experience shows us that this view is wrong, although it conveniently allows conventional social researchers to reside in their universities without the "rigors" of connecting with engaged social actors in the world beyond.

The philosophical foundations for our action research position come from the pragmatism of John Dewey, William James, Charles Sanders Peirce, and others (Diggins, 1994). We have laid out this position in other publications and will not repeat the arguments here. Pragmatism builds a direct link between theory and praxis. Reflection proceeds from acting in a real context, reflecting on the results, and then acting again. This is necessarily a group process involving diverse stakeholders with different experiences and knowledge of the problems at hand. Pragmatic inquiry results in "warranted" assertions that guide both action and theory/method developments.

Pragmatism is intimately connected to democracy; it is the social science approach to democratic deliberation and action. We take the betterment of democratic societies to be a core mission of the "social" sciences. We believe that action research is "scientific" (Greenwood & Levin, 2007, Chapter 5) because it leads to results tested in action and evaluated by professional social researchers and the relevant local stakeholders.

Central to action research is a collaborative relationship we call *cogenerative inquiry.* This brings the experience and training of professional social researchers together with the depth of experience and commitment of the local stakeholders for the benefit of all. Both the professional researchers and the local stakeholders have needed knowledge to contribute to the process.

Action research produces significant generalizations, methodological developments, and empirical findings, as a reading of any issue of the journals *Action Research, Systemic Practice and Action Research,* and the *International Journal of Action Research* will show.

Critique of Action Research

In making these arguments, we realize that our perspectives are idealized. Our definitional arguments regarding the potential of action research do not pay attention to problems and pitfalls that assail the everyday practice of action research, both inside and outside the university. We have published a number of articles and papers that deal critically with action research as it is actually practiced. (Greenwood, 2002, 2004; Levin, 2003). Our critique centers on seeing how little action researchers have contributed to theoretical and methodological debates in the social sciences. Much action research writing involves endless case reporting without a sharp intellectual focus, often unlinked to any particular scientific discourse. These writings are often hard to distinguish from work done in any of the applied social science fields. We believe that, like the conventional social sciences, action research has not lived up to its potential for the same reason: a lack of integration between solving relevant practical problems and a well-developed theoretical and methodological agenda.

However, there are enough good examples of action research that bridge practical problem solving and have significant theoretical and methodological ambitions to make our positive case. (For examples, see Eikeland, 2008; Emery & Thorsrud, 1976; Emery & Trist, 1973; and these exemplary doctoral dissertations: Aslaksen, 1999; Crane, 2000; Hittleman, 2007; Kassam, 2005; Klemsdal, 2008; Raymer, 2007; Ruiz-Casares, 2006; Skule, 1994; Vasily, 2006.)

Action research is not reducible to "public scholarship." The notion that there are legitimate "public" and "private" spheres of scholarship runs directly against our understanding of social research as a process that engages simultaneous understanding and social action as the way to produce reliable theories, methods, and knowledge.

Pedagogy

Action research pedagogy runs directly against the passive "banking method" (Freire, 1970; Giroux & Giroux, 2004; McLaren & Farahmandpur, 2005; McLaren & Jaramillo, 2007). Training action researchers cannot be done from a lectern or in university seminars alone. Students need to work collaboratively with the faculty, with their fellow students, and on real projects in order to learn. They need to develop theoretical and methodological competencies but also organizational, coordinating, leadership, and ethnographic skills that arise from experience sharing responsibility for both their own learning and the learning and welfare of others.

This kind of pedagogy is possible. Morten Levin and Davydd Greenwood have both been able to practice it in their universities, and so have others (Levin & Martin, 2007). It is not common because of the hierarchical, compartmentalized, and authoritarian structures that dominate higher education. Paradoxically, it is more likely for advanced science, engineering, medical, and law students to learn these kinds of skills than for social scientists and humanists. In those fields, some teaching involves structured

teamwork and the struggle to apply knowledge to concrete situations, which is rarely the case in the non-performance oriented humanities and the conventional social sciences.

Thus, we argue that the social sciences, including action research, all must be moved to address both the intellectual and practical challenges of social knowledge creation and competence development with and for the key stakeholders.

▣ UNIVERSITY REFORM—THE BALANCING ACT

The position on university reform we articulated in our prior contributions to this Handbook has shifted significantly. When we first wrote, we believed the possibilities of reform were sufficient to warrant our effort. Since then, we have watched the juggernaut of neoliberal policies; the vocationalization of higher education; the Bologna Process (the reform of European universities); and the deepening crisis of confidence in the value of university education as a source of social mobility, citizen formation, and meaningful social reform undermine the university systems of the world. In the current climate of economic panic, energy for much of anything other than cost-cutting Tayloristic exercises has dried up. We believe we understand how very significant reforms in universities could be undertaken, which would dramatically improve them as teaching, research, and social reform institutions, but we also believe that such reforms are unlikely. Rather, the crisis has emphasized the worst features of universities as organizations. So we have moved from writing in the voice of hopeful reformers to writing in the mode of "what if" arguments, the "what if" being "what if society and academics really wanted to recreate a meaningful university system"?

In our vision of the university, the core organizational processes are a multifaceted integration of the generation of research strategies, methods, and findings in social research; the reform of university organizational structures and processes; and the engagement of that research with the multidimensional, urgent, and dynamic problems in the world. This scholarship would depend on the collaborative engagement of students and teachers as learners and actors in these processes. Linking these tasks is a daunting challenge. After all, the conventional social sciences reject the linkage. For many academics, research has to be rigorous or relevant, theoretical or applied, and so on. No point of encounter seems to exist and it is a major challenge to support the creation of a different conceptualization.

Our proposed conceptualization is what we call the "balancing act." It might appear that such a concept is lame because it could suggest an uninteresting compromise in which every involved actor and conceptualization is juxtaposed within its previous frames of reference to create a compromise position, a common ground of "consensus" where issues and people involved give a little and get a little in return. But this is not at all what we mean. For us, the balancing act is a radical and trans-formative vision of the future of the social sciences and of universities because it involves creating new points of encounter arising as everyone involved moves away from their former positions and institutional bunkers, taking on new theoretical, methodological, and institutional positions.

Thus, our model is based on Jürgen Habermas's (1984) discourse ethics. The balancing act is a reasoned way to let arguments and positions confront each other, not in a win-lose competition, but in a collaborative learning process where good arguments support transformative learning for all (see, e.g., Freire, 1970; Mezirow, 1991). We also build on Ronald Barnett´s (2003) arguments that the essence of academic life is to demand the exercise of reason to support or to reject any position. We assert not only that there is a middle ground but also that meaningful social research must take place precisely on that middle ground. By forcing us to strive both to be relevant for practical problem-solving and rigorous enough to make an intellectual contribution to the ongoing development of social research approaches, the balancing act requires us to stand on this middle ground and justify our work in both practical and epistemological terms and then to struggle to reorganize the work environments and the external links of universities to make this possible and sustainable. This is the first dimension of the balancing act.

We argue for multiperspective research and teaching as prerequisites for connected knowledge generation. In the research arenas where different disciplines must contribute, it is evident that a middle ground has to be shaped to facilitate transdisciplinary research and teaching (Gibbons et al., 1994; Nowotny, Scott, & Gibbons, 2001). This is the second dimension of the balancing act.

Action research embodies this middle ground because it accepts the challenge to serve two "masters"—the demand for practical solutions and the scientific demands for intellectual focus and linkages resulting in publications that expand the understandings of professional peers. To do so, social scientists must have integrity, as they can neither operate fully in the world of abstract academic communication nor in the world of practical solutions to social problems. The integrity of the action researcher, moving continuously between these potentially contradictory demands, is key. The action researcher's self-imposed demand to maintain integrity in searching for the best possible theoretical, methodological, and practical outcomes is the only guarantee.

Practicing this integrity, action researchers also model scientific and social integrity for their students. The integrity of the university as an institution depends on facilitating these processes and protecting all parties from internal or external coercion caused by sensitive issues involving multiple stakeholders. This is the third element of the balancing act.

Because action research is built on a commitment to democratic dialogue and social processes, the further obligation of action researchers is to weigh the fairness and democratic implications of their research and teaching processes and of the practical solutions they propose (Flood & Romm, 1996). The power and interests of the relevant stakeholders affect these processes, and the researchers seek to balance these interests through open processes characterized by integrity throughout.

Another balancing act within universities is an institutional challenge to mediate between the development and promotion of deep expertise and high skill levels in many fields and the deployment of that capability around important transdisciplinary projects within and beyond the university. This is the fourth element of the balancing act. Disciplinary silos and autonomy oppose such a change project, but doing away with the ongoing development and teaching of deep expert knowledge would also be destructive to the future of the university and society at large. As important as this is, we see little evidence of a meaningful role played by university management structures in achieving and protecting this balance. Current evidence points in the opposite direction, toward academic commodity production in a fee-for-service environment.

Action research teaching is the fifth element in the balancing act. This teaching balances conveying social theories and methods drawn from the social sciences and connecting these theories and methods practically with everyday social life. Telling students how to think and act is not successful in giving students the ability to evaluate theories and methods, gather and analyze social research data, and work with diverse actors to bring about social change. Nothing short of balancing theory and practice in the classroom and taking the professor and students out of the classroom in the company of other colleagues from other fields and nonuniversity stakeholders constitutes "teaching social science." If the teaching activity does not bridge theory and praxis, then the students are not learning social science. Instead, they are becoming experts in academic commodity production for the benefit of their own careers.

This kind of engaged reflective research is impossible in the conventional academic social sciences or in the existing organizational structures of universities, despite the depth of the crisis in the funding of higher education and the loss of public confidence in the academic social sciences. To explain this, we provide a perspective on Tayloristic organization and management in universities, organizational dynamics that create the disconnected social sciences, which cannot deliver meaningful social formation (*Bildung*) and which have created a marketized teaching system where "shopping" for courses substitutes for a well-reasoned course plan that creates personal formation (*Bildung*).

Social Scientists' Antisocial Self-Understandings

Deep expertise in particular topics and approaches is essential to research about and understanding of broader systemic relationships. Disciplinary specialization accompanied by the organizational isolation of disciplines does not promote good social research. It makes important social problems impossible to understand and resolve and promotes poor quality higher education. Students are forced to walk around from academic department to department to "get an education," while the faculty who are not in intellectual communication generally have only stereotypical ideas of fields outside their own.

The lack of understanding of the contexts of the social scientists' own practices is paralleled by social scientists' lack of understanding of universities' organizational dynamics and their uneasy position in society. After working for generations to separate universities as producers of social science from nonacademic stakeholders engaged in the problems under study, academic social scientists have made meaningful and valid social research difficult and often professionally suicidal.

We have long been struck that most of our academic social science colleagues, whose specialties involve understanding the pervasive, complexly structured ways humans live in institutional and cultural worlds, conduct themselves personally and academically as if they were suprasocial and supracultural individualists whose behavior is not subject to their own theories and analytical methods. This self-estrangement from local organizational life and from society results in an absence of individual and collective self-understanding among social scientists and humanists, who claim to understand society and culture better than nonprofessionals.

This shows that many academics do not really believe or have not reflected on the ways that their theories and methods apply to themselves. They intuitively place themselves in a suprasocial position, adopting a modernist view, even though generations of social theory and philosophy have demonstrated the impossibility and inadvisability of pretending to take such a superhuman position. Social scientists have positioned themselves as "spectator" analysts (Eikeland, 2008; Skjervheim, 1974) and not as participants in their institutions and society.

Many academic social scientists combine professional hyperindividualism as academic entrepreneurs with a lack of understanding of organizational contexts in which academic social science knowledge is generated and communicated. As these contexts are rapidly changing, even the most un-self-conscious academics now are aware of some changes. But this lack of organizational self-understanding and reflection leaves them unprepared to develop and defend their own narrow academic interests and those of their students in the emerging academic regimes of "marketized" global higher education.

Instead, there is collective self-denial regarding the impact of the changing institutional and societal environments. Disturbing signals from external stakeholders are often met by retrenching: continuing to teach as always, admitting graduate students and training them for nonexistent academic jobs, and doing research on subjects of interest only to immediate professional peers This is an example of single-loop learning and Model O-I behavior (Argyris & Schön, 1978, 1996). This historical naiveté and resistance to confronting the challenge of a critical examination of their own research and teaching practices, institutional working contexts, and the roles they could play in society means that they do not exhibit behavior that could result in greater interest and respect for the social sciences.

These regressive behaviors play directly into the hands of "marketizing" (Slaughter & Leslie, 1997) managerial ideologies, using professional ranking systems to measure excellence in academic fields, and attempting to convert higher education into a fee-for-service training enterprise rather than a research and education effort to improve the quality of democratic societies. Rather, universities are organized according to an antiquated, dysfunctional Tayloristic model of a hierarchical, bureaucratic division of labor resulting in managerialism, authoritarianism, internal competition, and alienation from key external constituencies.

For generations, the more qualitatively oriented and interpretivist social scientists have been freer of constraints other than a lack of financial resources. They were not free in the political economy of university life or free from political attacks on their theories. We think they have confused being marginal with being free. Now both the public who pays the bills and the policymakers expect academic research and teaching to deliver value for the money. Nonacademics logically imagine that the importance of the social sciences is the light they shed on how society works in order to improve our lives. The distance between this public view of the social sciences and the lifeworld of a great many academic social scientists is significant.

University Management as Taylorism

Tayloristic organization, with its hermetic organizational units and command and control structures, makes a bad situation worse by concentrating the communication with the outside world and strategic planning in the central administration. Senior administrators, while privy to the overall "bottom line" of the institutions, generally do not have a good understanding of what real activities and processes produce this bottom line or how the bottom line is likely to change as new discoveries are made, new demands emerge, and so on. Unaware of the details of the research, teaching, and other work contexts that many individual faculty members know far better, the administrators and accountants see the results of the work but do not understand the contexts that make these results positive or negative. They thus elaborate policies and plans that often obstruct or undermine important developments while favoring the interests of the incumbent players and organizations they already know.

These systems unleash dynamics by which senior administrators maintain their power by controlling access to certain kinds of information and causing those who report to them to compete. Those lieutenants, in turn, do the same down the chain. It is a recipe for the already visible disaster. Like the bosses in Tayloristic factories, they are remote from the point of value creation and increasingly surrounded by accountants, finance managers, human resource professionals, advertising and public relations experts, lawyers, and risk managers. Thus, senior administrators have authority without having the relevant information. Not surprisingly, they routinely make decisions that are either counterproductive or impossible to implement.

When administrators argue that the crucial role of universities now is to prepare students for the knowledge society of the 21st century, neither they nor the faculty have a clear idea what they are talking about because they are divorced from the real worlds of work outside the university. Only a small portion of the faculty on most university campuses have relevant or current experience of the extra-university world to which their educational and research supposedly links. Lacking this experience, they cannot teach or provide research of value to those whose lives will be lived mainly outside the university.

This problem is less acute in the sciences and engineering. In these fields, there is a more constant and fluid link between the private sector, the public sector, and the university and more opportunity for sensible compromises between basic and applied research, all supported by the process of gaining external funding. While the process is not perfect, there are more external linkages, and work organization takes account of them.

In the social sciences, opportunities abound for academics to engage in disciplinary debates and set priorities for research and teaching having little linkage to what happens outside their disciplinary structures. The modesty of funding for all but some forms of quantitative positivistic research "liberates" many academic social scientists to pursue whatever topics and methods interest them or are currently in vogue. Doing irrelevant research, they do not receive much funding, and this allows them to continue the irrelevant work.

Should anyone care? We believe so because, as counterproductive as this situation is, the current direction of university reform results in a society of even more narrowly professionalized people, who neither understand their societies nor how to play a social solidary and self-conscious role in them. What academic Taylorism and adversarial politics with the outside world undermines are theoretically ambitious, methodologically sophisticated, and socially relevant social sciences dealing with the interests of students, policymakers, scientists, engineers, humanists, and society at large.

There is nothing to recommend in current university organizational systems that are ineffectual, costly, isolated, and prey to neoliberal accountability pressures. Only a radical transformation will prevent universities from continuing down the path that is converting them into technical (vocational) training

institutes and fee-for-service research shops under direct external control.

The Disconnected Social Sciences

University relationships with key external constituencies often are handled in pecuniary and selfish ways or in aggressive and self-destructive ways. Universities often claim a service mission or an expertise-producing mission. However, the lack of fluid communication between universities and the taxpayers and the irrelevance of much university social research and teaching to nonuniversity people suggests that this mission is not real. The decline of the public land grant universities in the United States is a harbinger of things to come for all universities.

Calling attention to the disconnection between the social sciences, their own organizational and cultural environments, and the larger society sounds like the tired cliché about the "ivory tower." What interests us is not the isolation of social scientists but the radical contradiction between the stated missions of the social sciences and the organizational behavior of most academic social scientists.

It is tempting to treat this tension between irrelevant ivory towers and the "real world" in a stereotypical and moralizing way, but this is not our purpose. Enough jeremiads do this already (Giroux & Giroux, 2004; Kirp, 2003; Washburn, 2005). Our purpose is to highlight how unsustainable the social science practices and methods of academic social scientists are.

The Role of Neoliberal Higher Education Reforms

The issues we describe are not endogenous to universities. They express broader processes of political economic change that go under the headings of neoliberalism and globalization. These processes are real and menacing to organizations producing public goods, which do not fully follow the supposed logic of commodity production market processes. Davydd Greenwood (2009) has published on neoliberal reforms in higher education elsewhere, and so we will be brief here.

Neoliberalism is not conservatism, which believes that some values are not market negotiable, and it is not liberalism, which believes that human beings have basic rights beyond those allocated by market processes. Neoliberalism is based on a utopian belief that the market will allocate all goods and services to those who deserve them and away from those who do not, if left to do its work. Since it has not been left to do its work, neoliberals intervene constantly claiming to free up the market by destroying public goods and reallocating them (energy production, environmental protection, education, etc.) to private actors. These private actors are generally the sponsors of the neoliberal politicians, and the charade results in a rapid increase in socioeconomic inequality accompanied by increased corporatist governmental bureaucracy.

Higher education is among the public goods neoliberal policies have focused on heavily. Beginning with the Thatcher reforms in Britain, followed by the Bologna Process, and the work of the Department of Education in the United States (starting with George W. Bush but continuing now under Barack Obama), neoliberal reforms are decimating the independence and finances of public higher education. The metrics applied are customer satisfaction of students, transparency and accountability for resources expended, and the "flexibilization" of the academic workforce.

The conventional social sciences and universities as organizations are not faring well under these conditions, except for neoliberal economics and other forms of quantitative research. Support for social science that does not seem to be about anything of importance to nonsocial scientists is evaporating. By not studying and illuminating problems of immediate interest to external constituencies, social scientists have separated themselves from sources of support. By not studying social world problems in context, they do not challenge themselves theoretically or methodologically with complex problems. They substitute complex language and baroque methodologies for engagement with real social and cultural complexity. And, without application, they rarely discover if the theories and methods they produce have any value.

The study of social problems in context is more challenging theoretically and methodologically than disciplinary work because social world problems are multidimensional, dynamic, and puzzling "messes" (Ackoff, 1974). These messes are part of large-scale systems that include dimensions relevant to all the social sciences and humanities. Studying them out of context and in bits to fit disciplinary boundaries yields academic commodity production but rarely actionable understandings. Studying pieces of messes allows social scientists to acquiesce to the Tayloristic organization of universities and avoid facing the urgent challenges of university reorganization.

Frustrated with the waste of resources and lack of attention to salient social issues, policymakers and administrators make these problems worse. Rather than demanding fundamental structural reorganizations of universities, these constituencies have been persuaded to demand accountability and transparency within existing structures. They impose discipline-based accountability and ranking schemes, deepening the antiquated Tayloristic organization of the academy and creating less rather than more change. The ranking systems by discipline and the beauty contest rankings of universities conserve existing structures because they take the disciplines and organizational structures of universities for granted.

Academic Taylorism

Looking at the mission statements of some of the professional associations in United States makes clear the autopoetic professional orientations of the academic social sciences (see, e.g., those of the American Anthropological Association, http://www.aaanet.org/about/WhatisAnthropology.cfm; the American Political Science Association, http://www.apsanet.org/content_4403.cfm?navID=733; or the American Economic Association, http://www.vanderbilt.edu/AEA/gen_info.htm).

They take for granted the existence and rationality of the boundaries of the disciplines and then proceed to occupy and protect those turfs, with bows in the direction of being socially valuable. A greater emphasis on social value is absent not because most academic social scientists don't believe what they do is valuable but because they view their value as being beyond debate. Until now, they have rarely been asked to defend what they do outside of their disciplinary confines.

Compatible with the broad Taylorism of academia, this results in the compartmentalization of knowledge and a unit-based command-and-control structure in which powerful central administrative figures distribute resources among disciplinary departments according to the politics of the institution, the ranking of the departments in national and international league tables, and the research monies they gather. No one asks if this overall organization makes any sense. In what way do a set of departments side-by-side add up to a university? How does travel to and between these units amount to an education? Occasionally, when at an event a president or provost is asked to say something generic about the overall university, the emptiness of their pronouncements is evident.

The kind of discussion we are promoting rarely finds its way into the arena of open debate. Anthropologists, for example, who, in their own internal discussions, routinely describe the methodological poverty and ethnocentrism of fields like economics and political science, rarely stand up in a university forum and state that political science or economics needs anthropology's help. The Tayloristic rules are live and let live: Compete for resources by fighting for enrollments and majors, office space, and budget allocations, and improve your ranking in the national and international ranking system. The rules are clear and the professional consequences of "coloring outside the lines" for all but a few great practitioners (e.g., Claude Lévi-Strauss, Clifford Geertz, Jürgen Habermas, Michel Foucault, etc.) are harsh—failed promotions, low salaries, few students, isolation, and opprobrium. The everydayness of these ways of living and thinking prevents them from becoming a subject of conversation and analysis.

Another dimension of this organizational dynamic is that academic faculty members generally are competitive, individualist entrepreneurs (see Wright, Brenneis, & Shore, 2005). The process of getting into a university, graduating, doing a postgraduate degree, getting an academic job, securing grants, publishing, teaching effectively, and providing sufficient institutional service is a calculated career process. The individual is engaged in building curricula vitae that will lead to permanent appointment, advancement through the ranks, salary increases, increased influence, and eventually greater personal autonomy.

The whole process is based on individualistic competition. Disciplinary solidarity may be asserted when competing with other disciplines, but within the disciplinary department, the ethos is competitive and individualistic. What one academic gets often is gained by doing better than other colleagues in the same unit.

People who spend their professional lives operating according to these rules are unlikely to think of themselves as deeply connected to the structures within which they operate except when they look up the chain of command. Those who succeed within their disciplines nationally and internationally do sometimes become senior statesmen locally, taking on tasks for the collectivity, but they rarely arrive at the position of senior statesmen without first having won a competition with colleagues in their earlier years within their departments.

This behavior is amply supported by the intellectual property regimes current in academia. The ownership of ideas and the authorship of manuscripts are taken for granted as the property of individuals and disciplinary research teams. Ideas are supposed to be original, and the fiction is that an academic's original ideas belong to her or him alone. He or she communicates them and tries to get others to use some of her or his language and to refer to her or his work in the process. If the ideas result in useful inventions, an all-out struggle between the faculty member and the university administration often ensues over the distribution of the rights to the profits between the individual and the university (Marginson & Considine, 2000; Kirp, 2003; Slaughter & Leslie, 1997; Washburn, 2005).

We could multiply examples and arguments, but we have said enough to show how the organizational environment of the social sciences encourages anti- or at least nonsocial thought and action. The relevant social life is within the discipline and department, and even there, it is generally competitive. It is rare for an academic social scientist to think of her- or himself as a part of a university collectivity with shared cultural norms, a worldview, and preferred methods and as a person whose behavior is largely explained by the social and cultural context in which he or she operates. Instead, it is the "others," the informants, the people the social scientists study outside the university, who have culture, roles, and values and who live in a sociocultural context, not the social scientists. Taylorism is firmly backed up by modernism.

We provide a concrete example from anthropology. For generations, it was assumed that the ability of anthropologists to see and understand the cultures of others was based on their unquestioned rationality and training as Western intellectuals. This was a perverse legacy. Culture and society are claimed to have a pervasive causal influence on the behavior of humans, but the anthropologists making this claim operated professionally as

if this general human condition did not apply to them. In anthropology, this tension was long hidden by giving up the study of North America and Europe as part of anthropology. It is telling that the Society for the Anthropology of Europe was not founded until 1987 and that the Society for North American Anthropology was founded at nearly the same time. By not treating these areas as suitable for anthropology, anthropologists removed themselves from the study of their own societies (also reducing competition with economics, political science, and sociology) and steered clearer of political repression like that suffered in the era of the House Un-American Activities Committee and Senator Joseph McCarthy (Price, 2004). They could also engage the modernist fiction of the unquestioned superiority of Western knowledge systems.

This untenable position became more paradoxical when the combination of feminism and cultural studies made positionality, the impossibility of neutral stances, the politics of research, and other previously obscured issues open to discussion. Taking on these perspectives at a discursive level and representing them in the bibliographies of manuscripts and course syllabi, anthropologists and other social scientists still generally have resisted studying themselves, their own institutions, and their own practices. Talking about positionality and reflexivity is not the same as understanding one's positions and being reflexive.

Social science teaching shows the same kind of dynamic. Typically, the general introductory courses are taught as lectures, sometimes with discussion sections, but mainly as passive learning activities. The lecturers state their understanding of what the discipline is about, how professionals operate, and what the key lessons from generations of research are. Students do not learn how to act as social scientists, why the disciplines exist, how they are similar and different from each other, or how research is done. These practices change some in upper-level courses, where enrollments are smaller and more interaction is possible, but many social science majors after 3 years cannot conduct research nor explain how or why the discipline they majored in differs from other disciplines.

At the graduate level, at least in the United States, the situation is more extreme. Graduate students are mentored more individually and must learn to "talk the talk" and "walk the walk" of their professors as a condition for getting a PhD. Taking a particularly egregious example from anthropology, fewer than 10% of the graduate departments of anthropology in the United States require a methodology course as part of graduate training. Students who want to learn how to do anthropological field research often find themselves doing their doctoral research without training on how to proceed.

Other disciplines offer more methodological training. Graduates in sociology, political science, and economics know the main techniques associated with their disciplines. Are they trained, therefore, as researchers? Do they know what their discipline "is," why it exists, and how it relates to others? Our experience is that they do not. Nor do many of these students have practice collaborating with academics from other disciplines on joint research projects and/or as members of teams doing research outside the university. A few get this experience in fields like archaeology and landscape architecture, but it is a rare practice in the conventional core of the social sciences.

The market-competitive model exacerbates the worst features of these teaching situations. Universities are presented to beginning students as giant educational cafeterias. They are told to go down the line among the offered dishes, picking and choosing within the limits set by curricular requirements, administrative rules, and course enrollment limitations. Departments that attract many students get more university resources. Departments with popular majors get more university resources. Departments with lots of students get graduate teaching assistantships and so are able to admit more graduate students. Departments with large and highly ranked graduate programs get more university resources. The "business model" is clear.

This system already is based on the student market model, although the market language long remained hidden. Students are supposed to make choices according to how attractive they find the courses offered, the campus reputation of the lecturers, and what they take to be the market value of one discipline over another. It is no accident that fields like classics have low enrollments and applied economics and management have high ones.

The departments or disciplines compete rather than cooperate because the Tayloristic system demands it. How these fields relate to each other is a not question addressed in such systems, nor is it possible to make any serious analysis of the mix and relative sizes of the constituent units. What the students learn and why is much less relevant than how many students enroll. How the student as a whole person emerges changed from the exposure to multiple fields and agendas is a question not asked. Students graduate, are given a degree, and are defined as having an "education."

Research suffers a similar fate. Research topics commanding the most external resources contribute most to the internal prestige and power of a unit and a faculty member and to the ranking of the university at large. Universities play little or no role in setting these funding agendas. National governments, private foundations, national research councils, and private sector funders call the tune, and thus the research market is controlled by powerful nonacademic forces that channel research into their areas of military, industrial, and economic development. Most universities do not have their own means of linking to the external world, engaging important external groups, or working toward university activities that are both intellectually challenging and socially desired. Most such relationships pass through individual faculty members and the research groups and centers they create.

Some successful university researchers may work in teams, and these teams are driven by grants. Often the team leader spends most of her or his time, not in the laboratory or classroom, but writing and administering grants, applying for patents, and working out negotiations over the sharing of resources created by the research projects. Many well-funded university researchers barely teach at all, although they might have some students working in their labs and getting some valuable mentoring by observation and participation.

Poorly funded university researchers are either working on unpopular topics or are not very good at getting research funding. They cobble together modest resources or simply do research on weekends and evenings or during the summers. They are less likely to work in teams, and without funding, few work with students as apprentices. They also end up doing the bulk of the teaching at the university.

What is the research agenda of a given university? Every institution now has a research statement and mission statements, but most employees and students have no idea what they are or how they affect them. What the "mission" of the institution might be is a question not asked in a way to produce organizational innovation and change. Rather, it is mainly a public relations instrument.

The research agenda of most universities, like their teaching, is the sum total of the activity going on that year. Whether it adds up to anything, whether the whole is more than the sum of its parts, is mostly irrelevant. Senior administrators answer questions about the mission of their institutions either with vague statements about the "knowledge society," "environmental crisis," and so on or with a table showing the research rankings of their university vis-à-vis peers.

The Missing *Bildung* in the "Marketized" Model of Academic Teaching and Research

There are a host of attempts worldwide to use what are wrongly imagined to be conventional business management strategies to direct higher education institutions (Barnett, 2003; Birnbaum, 2000; Kirp, 2003; Newfield, 2008). Because this management approach has not been countered successfully at most universities, it threatens the public and private university systems around the world.

There are ideological antimarket thinkers in academia (McLaren & Farahmandpur, 2005; McLaren & Jaramillo, 2007), but antimarket ideology does not motivate our critique. We see that pseudo-market ideologies and management practices are imposed on universities with the justification of the need to "rationalize" operations for efficiency and improved quality. However, our observations and much of the analytical literature, both on the left (Ehrenberg, 2006; Kirp, 2003; Newfield, 2008; Slaughter & Leslie, 1997) and among conservatives (MacMahon, 2009), show that what is claimed as rational economic management is not economically rational. Sheila Slaughter and Larry L. Leslie call this pseudo-economic management "marketizing" higher education, by which they mean claiming economic rationality while actually imposing authoritarian command-and-control management systems on faculty, students, and staff members. The rational allocation of resources, efficiency, quality, and transparency are the supposed motives, but the results are authoritarianism, suppression of information, maladaptive behavior, lowering of quality and transparency, and the creation of scores of new administrative positions to run an unwieldy, inefficient system.

A powerful critique of this pseudo-economics of university management comes from a well-known conservative economist, William MacMahon, whose *Higher Learning, Greater Good* (2009) uses social capital theory to attack the neoliberal "new public management" of higher education. MacMahon shows that the neoliberal management models in higher education and higher education policy underestimate by at least half the value of the goods produced by higher education because half of these goods are "public goods" that accrue to individuals and society over longer periods. They are not simple academic commodities for sale at a given instant.

This argument is key because it shows that, by failing to count the public goods and failing to understand that universities are capable of producing vast stores of public goods, the current models of management by the numbers actually undermine the ability of higher education to produce public goods and drastically reduce their productivity and contribution to the economy and to positive social change. In effect, MacMahon (2009) argues that many higher education managers and policymakers actually have no idea of the breadth and complexity of the goods their institutions produce and thus make economically irrational decisions, undermining their institutions' ability to operate efficiently and effectively. A related argument has been made for conventional manufacturing businesses by H. Thomas Johnson and Robert S. Kaplan in *Relevance Lost* (1987).

Another way of framing this is to state that these managers *do not know what higher education is*. In the language of many, students are customers, faculty are employees, tuition is the payment of a fee-for-service, and research is a profit/loss effort to be analyzed according to cost-benefit criteria. While there are senses in which all this is true, such views miss a significant part of what goes on in higher education. The focus is on the vocational training, content transmission activities, and research profit/loss ratios that obtain, but these by themselves are not the defining characteristics of university teaching, research, and service. They do not constitute an acceptable definition of higher education.

University teaching and research efforts, while satisfying some preferences from the "customers," involve processes aimed at shaping and reshaping the preferences of students, colleagues, administrators, and external constituencies, that is, at *Bildung*. It is the way we link scientific approaches to the social functions of higher education (Prange, 2004).

No unified understanding of *Bildung* exists, but the concept is widely used. The central meanings focus on an ongoing process of formation, of enhancing the intellectual and ethical strengths of individuals and thus preparing them to play meaningful roles in democratic societies (Bruford, 1975). *Bildung* creates critical, well informed, and reflective intellectuals able to address societal problems with integrity. The effects of *Bildung* are precisely what MacMahon (2009) captures in accounting for the value of higher education. These are reasoned affirmations about values, definable processes with measurable outcomes.

Generations after the Manhattan Project and sending a man to the moon, we are still living off the public goods created in efforts that remade (for good and ill) our world (and the industrial economies) fundamentally. University research and education can do this, but proving it, as MacMahon (2009) points out, involves a longer temporal perspective and deeper analysis than the single-year balance sheet favored by university administrators and marketizing policymakers.

Promoting *Bildung* in university teaching, research, and service is demanding. It is much harder than having students read a number of classic texts (the approach favored in the Norwegian curricular debates and in the U.S. "great books" culture wars). It requires vision, an ability to take a long-term time perspective to achieve a greater gain or a greater good. It involves conceptualizing university teaching, research, and service as knowledge creation and transmission activities with outcomes that are both immediate and long-term and conceptualizing them in relationship to a concept of what education and knowledge are. To strengthen these processes requires the autonomy and support to permit students and faculty take up unpopular, divisive topics and to study complex, multisystem problems. It also requires flexible, open-minded administrators to allocate resources wisely in substantive terms rather than to paint by the numbers. In a word, it requires creative and reflective leadership.

This leadership is essential to the integrity of knowledge creation and transmission processes, but it is inhibited by the marketization of university teaching and research. In the case of social science research, inhibition means that social research is unlikely to occur if it involves examining controversial problems in their complex scope, in collaboration with important outsiders or over sustained periods of time without immediate results. The result has been the destruction of the classical *Bildung*, the conversion of education into vocational training, the disconnection of faculty teachers and researchers from the complexities of the real-world contexts, and the disconnection of most of the university from the ongoing core processes in democratic societies. These developments limit the academic freedom of both students and faculty to follow subjects wherever the teaching and research processes take them. Stated more briefly, the lack of collective reflective practice to address these challenges is a direct consequence of academic Taylorism. We argue for a new

Bildung that includes individual formation and also collective efforts and shared responsibility.

We are not backward-looking romantics about *Bildung*. This notion was used to found the Humboldtian university and to justify the creation of the U.S. research universities. It is historically connected to the current dreary scene we portray. The conventional understanding of *Bildung* pointed to the formation of the individual in the perspective of accessing classical virtues of philosophy, history, and literature. Through reading Greek and Roman texts in philosophy, studying history, and seeking the beauty in poetry, prose, and theater, *Bildung* would automatically emerge in the mind and body of the student. But as Prange (2004) argues, "The concept of *Bildung* is a latecomer in a long line of spiritual independence versus material circumstances [tearing sic.] us down to considerations of earthly well-being. Education is for now, *Bildung* is forever" (p. 506). The eternal perspective is matched with the earlier argument that *Bildung* links the scientific ideal to the social impact of higher education. Our own understanding of the social function of education relates to the role of the academically trained person. We see genuine humanism as the expected consequence of a broader understanding of forces and processes that have created our societies and cultures.

But we believe that more than this is needed. As a consequence of our being members of society, *Bildung* has to deal with integrity, equality, and democracy. These virtues can be taught about, but basically they should emerge as a by-product of participating in a learning community, the university. Through active engagement in discourses among students, between students and teachers, and with citizens at large, integrity, equality, and democracy can be nourished. So, *Bildung* cannot simply be taught in class; it emerges as an effect of having joined the university's larger learning community, a community that is open to society itself.

▣ THE ROLE OF ACTION RESEARCH IN ADDRESSING REFORM OF UNIVERSITIES

We argue that there must now be a *New Bildung*, one that readdresses the meaning of university education and knowledge in the 21st century, not one that looks backward. We have indicated a few meanings of *Bildung* in an age of globalization, marketization, and increasing inequality. How, then, do universities contribute to addressing these problems? Part of our answer is that action research itself can be a significant source of the *New Bildung*.

We have addressed some major challenges for social science in universities: antisocial behavior by academics, the Tayloristic leadership and organizational models, the disconnection from society at large, and the evaporation of *Bildung* as a unique mission of universities. What, then, would the application of action research to academic organization and behavior look like, and how could this lead to a positive change that could regenerate vigorous universities for the 21st century?

Almost all universities are subject to pressure related to economic resources, whether they are private or publicly funded. Clearly no university on either side of the Atlantic will see the glory years of the last century again soon. Universities also now face the dilution of the public trust in university education and research as a major driver of economic prosperity and trustworthy knowledge generation. Universities can either adapt passively by further tweaking of the Tayloristic organizational structure to streamline themselves as marketplaces for knowledge commodities, or they can fundamentally reorganize through collective, participative engagement (professors, students, administrators, and support staff).

Fundamental reorganization is the only feasible way out of this economic and social crisis. It involves both a bottom-up process and a top-down process because we must create a common space for collective reflection where different points of view and sources of expertise and experience are confronted and where the reflections that emerge can lead directly to changes. This involves a balancing act between what changes are possible and what forces are counteracting change processes.

The change process we describe is relevant both for teaching and research. In fact, we would argue that the same type of knowledge generation process permeates both arenas. Reflection and experimentation would form the kind of continuous learning spiral central to action research (Greenwood & Levin, 2007; Heron 1996; Kolb 1984; Reason, 1994). This reform process has to be multidisciplinary, multiperspective, and transorganizational. It is clear that no single branch of social science is capable of encapsulating the reform process.

What we advocate is controversial, and we know from long experience that this perspective is not welcome in the conventional practices of the social sciences. Among the central problems that would arise are several ontological, epistemological, and methodological clashes right on the horizon. The ontological and epistemological fractures between modernism, realism, positivism, hermeneutics, structuralism, and so on have divided the social sciences for many years. Discussions of ontology and epistemology are ongoing, but they have had little impact on daily life among social scientists, where methodological approaches are a constant source of debate and academic commodity production. This dynamic is already well known and has been well discussed in the theory of science literature (Berger & Luckmann, 1966; Skjervheim, 1974; Toulmin, 1990).

We prefer to concentrate on methodological clashes that would have to be dealt with in any process of linking the social sciences. Only when we engage in praxis in the social sciences do different perspectives productively confront each other, and we also know that any change activity in universities ultimately involves changes in work life as it is lived daily. To engage the change process, we believe it is necessary to confront the conventional social sciences with the direct challenge of creating a new and different social science praxis.

Action research, well practiced, offers a way to accomplish this because it links all disciplines, the university, and its external stakeholders in a cogenerative social research process that tests theories and methods for validity in the form of concrete solutions to problems in real-world contexts. Action research also involves collaborative research teams in which new learners from within and outside the university are welcome and contribute their energy and experiences to the process. In this way, action research necessarily develops the democratization of knowledge generation, transmission, and application.

The action research process is based on making concrete organizational and behavioral changes, and these change processes are used as a systematic tool for learning. As such, action research forms a spiral of experimentation and reflection where all involved take part in the learning activities. This is a democratic and engaged activity giving a voice to everyone involved; it is what we have labeled cogenerative learning.

Obviously, this runs counter to the disciplinary, proprietary, commodity view of research and teaching. In recommending action research, we are insisting that the way forward is to reconfigure universities, particularly public universities, as central institutions in the further development of democracy through participative processes.

What would such a change activity look like at universities? Action research activity would have to address the antisocial behavior of academics that we have alluded to earlier, the Tayloristic organizational structure and leadership systems of universities, and universities' disconnectedness from society; finally, it would be oriented around a core *Bildung* process for all involved parties.

Where is there both energy and possibility for such a process? It is fairly clear where it is not. Attacking the bunkers of the professionalized disciplines and departments directly is a recipe for failure. Making demands on senior administrators and policymakers to give up their Taylorist, marketized addictions is routinely advocated and ignored. Insisting that universities serve society democratically at a time when the only service that counts is service to powerful economic and political players is not promising.

In this challenging environment, we are left with the re-creation of the university as a center of *Bildung*. The one place where we think it might be possible to imagine reform through *Bildung* managed by action research is in teaching and research activities. For centuries, university teaching has meant learning that is a top-down, passive process, where the teacher knows what the students need to come to know. By contrast, in line with a long history in adult education and with the principles advocated by Dewey, we see learning as an active process in which the

students are presented problems, raise questions, and are assisted in gaining the skills to seek answers for themselves. In this perspective, the teacher, who is also a learner, is a mentor and participant in the same learning process. We see the relationship between students and teachers as a genuine cogenerative process where each participant contributes her or his knowledge and insight as a collaborator in this joint learning activity.

But this kind of learning works only when the students and the teachers see the problems being dealt with as important. Thus, this kind of education can and should make solving practical problems its point of entry—for example, learning what it means to be "green" by working with multidisciplinary teams of inside and external stakeholders to clean up the local water supply, learning administrative skills by helping a local group set up a volunteer health clinic, and so on. Such projects, which work equally well at the beginning university level and the postgraduate level, connect universities to the outside society and necessarily include those who own the local problem in the same learning activity. Because the focus of learning is real problems that are too complex for single discipline approaches, such projects are necessarily multidisciplinary and multiperspective ones.

We are not advocating the impossible. The best way to prove that something is possible is to show that it has already been done somewhere. What we present here are two modest efforts to push the boundaries of what can be possible, even within the current *modus operandi* of universities.

Levin provides one example from a class in organizational development at the Norwegian University of Science and Technology (NTNU). When the class began, there was no clear problem focus. Instead, the students began by visiting a company and meeting with managers, trade union representatives, and workers. Students had the option of interviewing local people, or they could have access to videotaped interviews done by Levin that were later subjected to analysis. The next stage was for the students to interpret the situation and develop perspectives on a meaningful problem focus. In this phase, they met for the second time with the local company people. The students worked in groups of three to five members.

They created a plan for a developmental process in the company, which was presented in writing to the class, and they got feedback from Levin. This feedback shaped a dynamic that effectively simulates a real-life dynamic on organizational development processes. Finally, representatives from the company were invited to the presentation of the students' work, and the company people also participated in the grading process. The companies found this process useful in helping them think through organizational dilemmas, and it has been relatively easy to get companies to volunteer for it.

The *Bildung* elements are clear. Students are receiving formation by interacting with each other, with the professor, and with external stakeholders over real-world problems with real data and real consequences. They work with real people and experience the social responsibility involved in interacting with other human beings. What the students do in their activity can have real impact on people in the company. The stakes are high, and the problems are complex enough to require collaboration among the members of the class and the acquisition of relevant knowledge and coaching from other parts of the university as well. The students, the faculty member, and the company partners all improve their skills, knowledge, and understanding and learn to share their thinking in a cogenerative environment in which all are stakeholders.

Greenwood offers another example, this one drawn from an English composition class he teaches to a group of 14 students who are in their first year at Cornell University and are about 18 years old. This is part of a Cornell system for teaching freshman writing through small intensive writing seminars on topics of interest to faculty members in many disciplines. Greenwood's course focuses on the anthropological study of universities as its topic and introduces students to action research in the process.

This particular edition of the course began conventionally with a short essay from each student on the process of application to universities and the experiences he or she had. In addition to the writing corrections and revision, the course dealt with developing an ability to study and conceptualize organizational processes such as applications and admissions, residential living, dining, physical organization, structure of requirements, and many other topics through combined ethnographic work by the students (some in teams) and readings.

Early in the course, the students (with widely varied intellectual interests, ethnic backgrounds, and social interests) read Paolo Freire's (1970) *The Pedagogy of the Oppressed* and then began making connections to their experiences of the passive, banking model of education. One visionary student said that if they really believed what Freire wrote, they should convert the seminar into a group project and take their education into their own hands. After brief negotiations about the process and requirements, they developed a project to eliminate contaminants in the water coming from and passing through the university campus.

The class, including Greenwood, worked out an overall plan, and they divided up into teams according to interest and skills; they spread out all over the campus, dealing with the central administration, the water plant, the city's water treatment system, the conservation biologists who had ways of cleaning the water by means of the use of plants, among others. The dynamic was intense and resulted in a submission to a national competition for green campus projects sponsored by General Electric and MTV. Thousands of e-mails were exchanged, and the collaborative website grew to hundreds of pages.

Their motivation, work, solidarity, and sophistication grew beyond Greenwood's expectations, and he ended up spending time trying to prevent them from ignoring their other courses

while they worked for a month on this project. Greenwood is convinced that these students developed the kinds of capacities summarized by the *Bildung* ideal, did so happily and willingly, and received support and approval from people around the university.

Thus, it is possible to work in existing universities within the action research mode. We also know from experience that such learning arenas often become pivotal in the development of individual students, whose academic and life choices thereafter are strongly affected by these experiences.

In previous versions of this Handbook, we have shown how research also can be re-organized on university campuses. Rather than recapitulate these here, we refer you to those chapters (Greenwood & Levin, 2000, 2005). The lesson from the examples is that both teaching and research can be based on the principles of action research. As the professors who engaged in those efforts, we could be in a state of great optimism because we see a way to make a modest contribution to students' *Bildung*. However, we are not because our experiments have shown that it is possible to reconstruct teaching and research, but we have seen little or no diffusion to other classes or research arenas. Business as usual prevails.

▣ CONCLUSION

Our arguments about teaching in this chapter, as well as our arguments in earlier editions of this Handbook about the epistemology and methodology of action research, indicate that this is a superior way to link teaching, research, and real-world engagement. If this is true, why does it not dominate the research universities of the world? Most action research activities take place outside the boundaries of higher education.

What we recommend requires academic social scientists to change their behavior radically—away from hyperprofessional internal debates, away from individualism and entrepreneurialism, and toward multidisciplinary research and action that takes them beyond the university. It requires the Tayloristic organization of the university to be, if not abandoned, transformed to permit easy collaborative work across internal institutional boundaries and across the boundaries between the university and society without a commodity production view of external linkages. And it would require a recommitment of universities to *Bildung* and democracy as core values. All of these changes seem quite unlikely.

Action research disconnected from universities, as some advocate, breaks the link to educating and forming new generations of social scientists. This permits universities to continue training people who lack the knowledge and skills to make the contributions we advocate to democratic society. And, if future researchers are not trained in action research at universities, they are not likely to develop this capability after graduation. Equally important, the potential contribution of action research to the redevelopment of the social sciences is greatly hampered. Thus, we believe that universities as centers of *Bildung* cannot survive without reorganizing teaching and research along action research lines and that action research will not survive unless it develops a key position within *Bildung*-oriented higher education.

Is action research likely to achieve this key position any time soon? No. Still, we know that the fit between the problems we have identified and action research is real, and we know that it is possible because we have done it on an admittedly limited basis. We have seen the power of the results.

Perhaps the current crisis in higher education will create a propitious environment on a few campuses for the changes we advocate. If so, we have provided a map of one road toward a better future for both the university and democratic society.

▣ REFERENCES

Ackoff, R. L. (1974). *Redesigning the future: A systems approach to societal problems.* New York: John Wiley.

Argyris, C., & Schön, D. A. (1978). *Organizational learning: A theory of action perspective.* New York: Addison-Wesley.

Argyris, C., & Schön, D. A. (1996). *Organizational learning II: Theory, method, and practice.* New York: Addison-Wesley.

Aslaksen, K. (1999). *Strategies for change in corporate settings: A study of diffusion of organizational innovations in a Norwegian corporation.* Unpublished doctoral dissertation, Norwegian University of Science and Technology, Department of Industrial Economics and Technology Management, Trondheim.

Barnett, R. (2003). *Beyond all reason: Living with ideology in the university.* Buckingham, UK: Society for Research in Higher Education and Open University Press.

Berger, P., & Luckmann, T. (1966). *The social construction of reality.* Garden City, NY: Doubleday.

Birnbaum, R. (2000). *Management fads in higher education: Where they come from, what they do, why they fail.* San Francisco: Jossey-Bass.

Bruford, W. H. (1975). *The German tradition of self-cultivation: Bildung from Humboldt to Thomas Mann.* Cambridge, UK: Cambridge University Press.

Crane, B. (2000). *Building a theory of change and a logic model for an empowerment-based family support training and credentialing program.* Unpublished doctoral dissertation, Cornell University, Ithaca, NY.

Cresswell, J. W. (2003). *Research design: Qualitative, quantitative, and mixed methods approaches* (2nd ed.). Thousand Oaks, CA: Sage.

Cresswell, J. W., & Clark, V. L. P. (2007). *Mixed methods research.* Thousand Oaks, CA: Sage.

Diggins, J. (1994). *The promise of pragmatism.* Chicago: University of Chicago Press.

Ehrenberg, R. G. (Ed.). (2006). *What is happening to public higher education?* Westport, CT: American Council on Education and Praeger.

Eikeland, O. (2008). *The ways of Aristotle: Aristotelian phronesis, Aristotelian philosophy of dialogue, and action research.* Bern, Switzerland: Peter Lang.

Emery, F., & Thorsrud, E. (1976). *Democracy at work.* Leiden, Netherlands: Martinus Nijhoff.

Emery, F., & Trist, E. (1973). *Towards a social ecology.* London: Plenum Press.

Flood, R., & Romm, N. R. A. (1996). *Diversity management: Triple-loop learning.* Chichester, UK: Wiley.

Freire, P. (1970). *The pedagogy of the oppressed.* New York: Herder & Herder.

Gibbons, M., Limoges, C., Nowotny, H., Schwartzman, S., Scott, P., & Trow, M. (1994). *The new production of knowledge: The dynamics of science and research in contemporary society.* London: Sage.

Giroux, H. A., & Giroux, S. S. (2004). *Take back higher education.* New York: Palgrave.

Greenwood, D. J. (2002). Action research: Unfulfilled promises and unmet challenges. *Concepts and Transformation, 7*(2), 117–139.

Greenwood, D. J. (2004). Action research: Collegial responses fulfilled. *Concepts and Transformation, 9*(1), 80–93.

Greenwood, D. J. (2009). Bologna in America: The Spellings Commission and neoliberal higher education policy. *Learning and Teaching, 2*(1), 1–38.

Greenwood, D. J., & Levin, M. (1998a). *Introduction to action research: Social research for social change.* Thousand Oaks, CA: Sage.

Greenwood, D. J., & Levin, M. (1998b). The reconstruction of universities: Seeking a different integration into knowledge development processes. *Concepts and Transformation, 21*(2), 145–163.

Greenwood, D. J., & Levin, M. (2000). Reconstructing the relationships between universities and society through action research. In N. K. Denzin & Y. S. Lincoln (Eds.), *Handbook of qualitative research* (2nd ed., pp. 85–106). Thousand Oaks, CA: Sage.

Greenwood, D. J., & Levin, M. (2001a). Pragmatic action research and the struggle to transform universities into learning communities. In P. Reason & H. Bradbury (Eds.), *Handbook of action research* (pp. 103–114). Thousand Oaks, CA: Sage.

Greenwood, D. J., & Levin, M. (2001b). Reorganizing universities and "knowing how": University restructuring and knowledge creation for the twenty-first century. *Organization, 8*(2), 433–440.

Greenwood, D. J., & Levin, M. (2005). Reform of the social sciences and of universities through action research. In N. K. Denzin & Y. S. Lincoln (Eds.), *The SAGE handbook of qualitative research* (3rd ed., pp. 43–64). Thousand Oaks, CA: Sage.

Greenwood, D. J., & Levin, M. (2007). *Introduction to action research: Social research for social change* (2nd ed.). Thousand Oaks, CA: Sage.

Habermas, J. (1984). *The theory of communicative action: Reason and the rationality of society.* Boston: Beacon.

Heron, J. (1996). *Co-operative inquiry: Research into the human condition.* London: Sage.

Hittleman, M. (2007). *Counting caring: Accountability, performance, and learning at the Greater Ithaca Activities Center.* Unpublished doctoral dissertation, Cornell University, Ithaca, NY.

Johnson, H. T., & Kaplan, R. (1987). *Relevance lost: The rise and fall of management accounting.* Cambridge, MA: Harvard Business Press.

Kassam, K.-A. (2005). *Diversity, ways of knowing, and validity—a demonstration of relations between the biological and the cultural among indigenous peoples of the circumpolar north.* Unpublished doctoral dissertation, Cornell University, Ithaca, NY.

Kirp, D. L. (2003). *Shakespeare, Einstein, and the bottom line: The marketing of higher education.* Cambridge, MA: Harvard University Press.

Klemsdal, L. (2008). *Making sense of the "new way of organizing": Managing the micro processes of planned change in a municipality.* Unpublished doctoral dissertation, Norwegian University of Science and Technology, Department of Industrial Economics and Technology Management, Trondheim.

Kolb, D. (1984). *Experiential learning.* Englewood Cliffs, NJ: Prentice Hall.

Levin, M. (2003). Action research and the research community. *Concepts and Transformation, 8*(3), 275–280.

Levin, M., & Greenwood, D. J. (1998). Action research, science, and co-optation of social research. *Studies in Cultures, Organizations, and Societies, 4*(2), 237–261.

Levin, M., & Martin, A. W. (Eds.). (2007). The praxis of education action researchers [Special issue]. *Action Research, 5,* 249–264.

MacMahon, W. W. (2009). *Higher learning, greater good: The private and social benefits of higher education.* Baltimore. Johns Hopkins University Press.

Marginson, S., & Considine, M. (2000). *The enterprise university: Power, governance, and reinvention in Australia.* Cambridge, UK: Cambridge University Press.

McLaren, P., & Farahmandpur, R. (2005). *Teaching against global capitalism and the new imperialism.* Lanham, MD: Rowman & Littlefield.

McLaren, P., & Jaramillo, N. (2007). *Pedagogy and praxis.* Boston: Sense Publishers.

Mezirow, J. (1991). *Transformative dimensions of adult learning.* San Francisco: Jossey-Bass.

Newfield, C. (2008). *Unmaking the public university: The forty-year assault on the middle class.* Cambridge, MA: Harvard University Press.

Nowotny, H., Scott, P., & Gibbons, M. (2001). *Re-thinking science: Knowledge and the public in the age of uncertainty.* London: Sage.

Prange, K. (2004, November). Bildung: A paradigm regained? *European Educational Research Journal, 3,* 501–509.

Price, D. (2004). *Threatening anthropology.* Durham, NC: Duke University Press.

Raymer, A. L. (2007). *Democratic places through democratic means with participatory evaluative action research (PEAR), a model of inquiry for habits and habitats where public life matters.* Unpublished doctoral dissertation, Cornell University, Ithaca, NY.

Reason, P. (Ed.). (1994). *Participation in human inquiry.* London: Sage.

Ruiz-Casares, M. (2006). *Strengthening the capacity of child-headed households in Namibia to meet their own needs: A social networks approach.* Unpublished doctoral dissertation, Cornell University, Ithaca, NY.

Skjervheim, H. (1974). *Objektivismen og studiet av mennesket* [Objectivity and the study of man]. Oslo, Norway: Gyldendal.

Skule, S. (1994). *From skills to organizational practice: A study of the relation between vocational education and organizational learning in the food-processing industry.* Unpublished doctoral dissertation, Norwegian University of Science and Technology, Department of Industrial Management and Work Science, Trondheim.

Slaughter, S., & Leslie, L. L. (1997). *Academic capitalism: Politics, policies, and the entrepreneurial university.* Baltimore: Johns Hopkins University Press.

Toulmin, S. (1990). *Cosmopolis: The hidden agenda of modernity.* Chicago: University of Chicago Press.

Vasily, L. (2006). *Reading one's life: A case study of an adult educational participatory action research curriculum development project for Nepali Dalit social justice.* Unpublished doctoral dissertation, Cornell University, Ithaca, NY.

Washburn, J. (2005). *University, Inc.: The corporate corruption of American higher education.* New York: Basic Books.

Wright, S., Brenneis, D., & Shore, C. (Eds.). (2005). Universities and the politics of accountability [Special issue]. *Anthropology in Action, 12*(1).

3

A HISTORY OF QUALITATIVE INQUIRY IN SOCIAL AND EDUCATIONAL RESEARCH[1]

Frederick Erickson

Qualitative inquiry seeks to discover and to describe in narrative reporting what particular people do in their everyday lives and what their actions mean to them. It identifies meaning-relevant *kinds* of things in the world—kinds of people, kinds of actions, kinds of beliefs and interests—focusing on differences in forms of things that make a difference for meaning. (From Latin, *qualitas* refers to a primary focus on the qualities, the features, of entities—to distinctions in kind—while the contrasting term *quantitas* refers to a primary focus on differences in amount.) The qualitative researcher first asks, "What are the kinds of things (material and symbolic) to which people in this setting orient as they conduct everyday life?" The quantitative researcher first asks, "How many instances of a certain kind are there here?" In these terms, quantitative inquiry can be seen as always being preceded by foundational qualitative inquiry, and in social research, quantitative analysis goes haywire when it tries to shortcut the qualitative foundations of such research—it then ends up counting the wrong kinds of things in its attempts to answer the questions it is asking.

This chapter will consider major phases in the development of qualitative inquiry. Because of the scale of published studies using qualitative methods, the citations of literature present illustrative examples of work in each successive phase of qualitative inquiry's development rather than an exhaustive review of literature in any particular phase. I have referred the reader at various points to additional literature reviews and historical accounts of qualitative methods, and at the outset, I want to acknowledge the comprehensive historical chapter by Arthur Vidich and Stanford Lyman (1994, pp. 23–59), which was published in the first edition of this *Handbook*. Our discussion here takes a somewhat different perspective concerning the crisis in authority that has developed in qualitative inquiry over the last 30 years.

This chapter is organized both chronologically and thematically. It considers relationships evolving over time between five foundational "footings" for qualitative research: (1) disciplinary perspectives in social science, particularly in sociology and anthropology; (2) the participant-observational fieldworker as an observer/author; (3) the people who are observed during the fieldwork; (4) the rhetorical and substantive content of the qualitative research report as a text; and (5) the audiences to which such texts have been addressed. The character and legitimacy of each of these "footings," have been debated over the entire course of qualitative social inquiry's development, and these debates have increased in intensity in the recent past.

I. ORIGINS OF QUALITATIVE RESEARCH

In the ancient world, there were precursors to qualitative social inquiry. Herodotus, a Greek scholar writing in the 5th century B.C.E., had interests that were cross-cultural as well as historical. Writing in the 2nd century C.E., the Greek skeptical philosopher Sextus Empiricus conducted a cross-cultural survey of morality, showing that what was considered right in one society was considered wrong in others. Both he and Herodotus worked from the accounts of travelers, which provided the primary basis for comparative knowledge about human lifeways until the late 19th century. Knowledge of nature also was reported descriptively, as in the physics of Aristotle and the medicine of Galen.

Descriptive reporting of everyday social practices flourished again in the Renaissance and Baroque eras in the publication of "how to do it books" such as Baldassar Castiglione's *The Book of the Courtier* and the writing of Thoinot Arbeau (*Orchésographie*) on courtly dancing, of Johann Comenius (*Didactica Magna*) on pedagogy, of Isaak Walton (*The Compleat Angler*) on fishing, and of John Playford (*The Division Viol)* on how to improvise in playing the viola da gamba. The treatises on dancing and music especially were descriptive accounts of very particular practices—step-by-step description at molecular grain size. Narrative descriptive reports were also written in broader terms, such as the accounts of the situation of Native Americans under early Spanish colonial rule in Latin America, written by Bartolomeo de las Casas in the 16th century, and the 17th-century reports French Jesuits submitted to superiors regarding their missionary work in North America (*Relations)*. A tension between scope and specificity of description remains in contemporary qualitative inquiry and reporting.

Simultaneously with the 17th-century writing on everyday practices, the quantitative physics of Galileo Galilei and Isaac Newton was being established. As the Enlightenment developed, quantitatively based inquiry became the standard for physical science. The search was for general laws that would apply uniformly throughout the physical world and for causal relations that would obtain universally. Could there be an equivalent to this in the study of social life—a "social physics"—in which social processes were monitored by means of frequency tabulation and generalizations about social processes could be derived from the analysis of frequency data? In England, William Petty's *Political Arithmetic* was one such attempt, published in 1690. In France and Germany, the term *statistics* began to be used to refer to quantitative information collected for purposes of the state—information about finance, population, disease, and mortality. Some of the French Enlightenment philosophers of the 18th century saw the possibility that social processes could be mathematically modeled and that theories of the state and of political economy could be formulated and empirically verified in ways that would parallel physics, chemistry, and astronomy.

As time went on, a change of focus occurred in published narrative descriptive accounts of daily practices. In the 16th and 17th centuries, the activities of the leisured classes were described, while the lower classes were portrayed patronizingly at the edges of the action, as greedy, lascivious, and deceitful, albeit clever. (A late example can be found in the portrayal of the lusty, pragmatic countrymen and women in Picander's libretto for J. S. Bach's *Peasant Cantata,* written and performed in 1742.) By the end of the 18th century, the everyday lives of servants and rustics were being portrayed in a more sympathetic way. Pierre Beaumarchais's play, *The Marriage of Figaro,* is an example. Written in 1778, it was initially banned in both Paris and Vienna on the grounds that by valorizing its servant characters and satirizing its aristocratic characters, it was dangerously subversive and incited insubordination. By the early 19th century, the Brothers Grimm were collecting the tales of German peasants, and documentation of folklore and folklife of commoners became a general practice.

By the mid-19th century, attempts were being made to define foundations for the systematic conduct of social inquiry. A fundamental disagreement developed over what kind of a "science" the study of society should be. Should such inquiry be modeled after the physical sciences, as Enlightenment philosophers had hoped? That is what Auguste Comte (1822/2001) claimed as he developed a science of society he would come to call *sociology*; his contemporary, Adolphe Quetelet (1835/2010) advocated the use of statistics to accomplish a "social physics." Early anthropologists with foundational interests in social and cultural evolution also aimed their inquiry toward generalization (e.g., Morgan, 1877; Tylor, 1871); they saw the comparative study of humans as aiming for general knowledge, in their case, an understanding of processes of change across time in physical and cultural ways of being human—of universal stages of development from barbarism to contemporary (European) civilization—comparative study that came to be called *ethnology*. Like Comte, they saw the purposes of social inquiry as the discovery of causal laws that applied to all cases, laws akin to those of physics and chemistry.

In contrast, the German social philosopher Wilhelm Dilthey (1883/1989) advocated an approach that differed from that of natural sciences (which he called *Naturwissenschaften*). He advocated conducting social inquiry as *Geisteswissenschaften*— literally "sciences of the spirit" and more freely translated as "human sciences" or, better, "human studies." Such inquiry was common to both the humanities and what we would now call the social sciences. It focused on the particulars of meaning and action taken in everyday life. The purpose of inquiry in the human sciences was understanding (*verstehen*) rather than proof or prediction. Dilthey's ideas influenced younger scholars—in particular Max Weber and Georg Simmel in sociology and early phenomenologists in philosophy such as Edmund Husserl and Martin Heidegger. His ideas became even more influential in the mid-20th century "hermeneutical turn" taken by philosophers such as Hans-Georg Gadamer and Jürgen Habermas and by anthropologists such as Ernest Gellner and Clifford Geertz.

The emergence of ethnography. In the last quarter of the 19th century, anthropologists began to use the term *ethnography* for descriptive accounts of the lifeways of particular local sets of people who lived in colonial situations around the world. These accounts, it was claimed, were more accurate and comprehensive than the reports of travelers and colonial administrators. In an attempt to improve the information quality and comprehensiveness of description in traveler's accounts, as well as to support the fieldwork of scholars in the emerging field of anthropology, the British Society for the Advancement of Science published in 1874 a manual to guide data collection in observation and interviewing, titled *Notes and Queries on Anthropology for the Use of Travelers and Residents in Uncivilized Lands* (available at http://www.archive.org/details/notesandqueries00readgoog). The editorial committee for the 1874 edition of *Notes and Queries* included George Lane-Fox Pitt-Rivers, Edward Tylor, and Francis Galton, the latter being one of the founders of modern statistics. The *Notes and Queries* manual continued to be reissued in further editions by the Royal Anthropological Society, with the sixth and last edition appearing in 1951.

At 6 ½ by 4 inches, the book could be carried to field settings in a large pocket, such as that of a bush jacket or suit coat. Rulers in both inches and centimeters are stamped on the edge of the cover to allow the observer to readily measure objects encountered in the field. The volume contains a broad range of questions and observation topics for what later became the distinct branches of physical anthropology and social/cultural anthropology: topics include anatomical and medical observations, clothing, navigation, food, religion, laws, and "contact with civilized races," among others. The goal was an accurate collection of facts and a comprehensive description of the whole way of life of those who were being studied.

This encyclopedic approach to fieldwork and information collection characterized late 19th-century qualitative research, for example, the early fieldwork of Franz Boas on the northwest coast of North America and the two expeditions to the Torres Straits in Oceania led by Alfred Haddon. The second Haddon expedition involved fieldworkers who would teach the next generation of British anthropologists—for example, W. H. R. Rivers and C. G. Seligman, with whom A. R. Radcliffe Brown and B. Malinowski later studied. (For further discussion of the early history of field methods in anthropology, see Urrey, 1984, pp. 33–61.)

This kind of data collection and reporting in overseas settings was called *ethnography,* combining two Greek words: *graphein,* the verb for "to write," and *ethnoi,* a plural noun for "the nations—the others." For the ancient Greeks, the *ethnoi* were people who were not Greek—Thracians, Persians, Egyptians, and so on—contrasting with *Ellenoi* or Hellenes, as us versus them. The Greeks were more than a little xenophobic, so that ethnoi carries pejorative implications. In the Greek translation of the Hebrew scriptures, *ethnoi* was the translation for the Hebrew term for "them"—*goyim*—which is not a compliment. Given its etymology and its initial use in the 19th century for descriptive accounts of non-Western people, the best definition for ethnography is "writing about other people."

Perhaps the first monograph of the kind that would become modern realist ethnography was *The Philadelphia Negro,* by W. E. B. DuBois (1899). His study of a particular African American census tract combined demographic data, area maps, recent community history, surveys of local institutions and community groups, and some descriptive accounts of the conduct of daily life in the neighborhood. His purpose was to make visible the lives—and the orderliness in those lives—of people who had been heretofore invisible and voiceless in the discourses of

middle class white society and academia. A similar purpose and descriptive approach, combining demography and health statistics with narrative accounts, was taken in the reports of working class life in East London by Charles Booth (1891), whose collaborators included Sidney and Beatrice Webb. Even more emphasis on narrative description was found in *How the Other Half Lives,* an account of the everyday life of immigrants on the lower East Side of New York City, written by the journalist Jacob Riis (1890) and illustrated with photographs. All of these authors—and especially Booth and DuBois—aimed for factual accuracy and holistic scope. Moreover, these authors were social reformers—Booth and the Webbs within the Fabian Socialist movement in England, Riis as a founder of "muckraking" journalism and popular sociology, and DuBois as an academic sociologist who turned increasingly to activism, becoming a leader of the early 20th-century African American civil rights movement. Beyond description for its own sake, their purpose was to advocate for and to inform social change.

None of these early practitioners claimed to be describing everyday life from the points of view of those who lived it. They were outsider observers. DuBois, although an African American, grew up in a small New England town, not Philadelphia, and he had a Harvard education. Booth and the Webbs were upper middle class, and so was Riis. They intended to provide accurate descriptions of "facts" about behavior, presented as self-evidently accurate and "objective," but not about their functional significance in use, or as Clifford Geertz (1973) said, what distinguishes an eye blink from a wink (p. 6). To use terms that developed later in linguistics and metaphorically applied to ethnography, their descriptions were *etic* rather than *emic* in content and epistemological status.

Adding point of view. Portraying social action (as wink) rather than behavior (as eye blink)—that is, describing the conduct of everyday life in ways that make contact with the subjective orientations and meaning perspectives of those whose conduct is being reported—is the fundamental shift in interpretive (hermeneutical) stance within ethnography that Bronislaw Malinowski claimed to have accomplished a generation later. In his groundbreaking monograph, *Argonauts of the Western Pacific* (Malinowski, 1922), he said that ethnographic description should not only be holistic and factually accurate, but should aim "to grasp the native's point of view, his relation to life, his vision of his world" (p. 25).

During World War I, Malinowski, a Pole who had studied anthropology in England, was interned by British colonial authorities during his fieldwork in the Trobriand Islands of Melanesia because they were concerned that, as a subject of the Austro-Hungarian Empire, he might be a spy. He was not allowed to return home until the war had ended. Malinowski later made a virtue of necessity and claimed that his 4 years of enforced fieldwork and knowledge of the local language enabled him to write a report that encompassed the system of everyday life in its entirety and accurately represented nuances of local meaning in its daily conduct. After Malinowski, this became a hallmark of ethnography in anthropology—reporting that included the meaning perspectives of those whose daily actions were being described.

Interpretively oriented (i.e., hermeneutic) realist ethnography presumed that *local meaning* is causal in social life and that local meaning varies fundamentally (albeit sometimes subtly) from one local setting to another. One way this manifested in anthropology was through cultural relativism—a position that Franz Boas had taken before Malinowski. By the late 1920s, anthropologists were presuming that because human societies were very different culturally, careful ethnographic case study documentation was necessary before valid ethnological comparison could take place—the previous armchair speculations of scholars like Edward Tylor and Lewis Henry Morgan were seen as having been premature.

What is implied in the overall emphasis on the distinctive differences in local meaning from one setting to another is a presumption that stands in sharp contrast to a basic presumption in natural science. There one assumes a fundamental *uniformity of nature* in the physical universe. For example, one can assume that a unit measurement of heat, or of force, or a particular chemical element is the same entity in Mexico City and Tokyo as it is in London—and also on the face of the sun and in a far distant galaxy. The presumption of uniformity of natural elements and processes permitted the statement of general laws of nature in physics, chemistry, and astronomy, and to a lesser extent in biology. In contrast, a human science focus on locally constructed meaning and its variability in construction presumes, in effect, a fundamental *nonuniformity of nature in social life.* That assumption was anathema to those who were searching for a social physics. But qualitative social inquiry is not aiming to be a social physics. Or is it? Within anthropology, sociology, and educational research, researchers disagreed about this, even as they did ethnographic case studies in traditional and modern societies.

A basic, mainstream approach was developing in qualitative social inquiry. We can see that approach as resting on five foundational grounds or footings: the disciplinary enterprise of social science, the social scientific observer, those who are observed, the research report as a text, and the research audience to which that text is addressed. Each of these five was considered as an entity whose nature was simple and whose legitimacy was self-evident. In current qualitative inquiry, the nature and the legitimacy of each of those footings have been called into question.

First, the *enterprise of social science.* By the late 19th century, sociology and anthropology were developing as new disciplines, beginning to achieve acceptance within universities. Physical sciences had made great progress since the 17th century, and social scientists were hoping for similar success.

Next, the *social scientist as observer.* His (and these were men) professional warrant for paying research attention to other humans was the social scientific enterprise in which he was engaged—that engagement gave him the right to watch other people and question them. It was assumed that he would and should be systematic and disinterestedly open-minded in the exercise of research attention. The process of looking closely and carefully at another human was seen as being no more ethically or epistemologically problematic than looking closely and carefully at a rock or a bird. Collecting specimens of human activity was justifiable because it would lead to new knowledge about social life. (Unlike the field biologist the social scientist was not justified to kill those he studied or to capture them for later observation in a zoological museum—although some non-Western people were exhibited at world expositions and the anthropologist Alfred Kroeber had housed a Native American, Ishi, at the anthropological museum of the University of California, Berkeley, making him available for observation and interview there—but artifact collecting and the writing of field-notes were the functional equivalent of the specimen collection and analysis methods of biologists and geologists.) Moreover, research attention in social inquiry was a one-way matter—just as the field biologist dissected an animal specimen and not the other way around, it was the researcher's watching and asking that counted in social inquiry, not the attending to and questioning of the researcher by the people whose daily lives were being studied.

Those who were observed as research objects (not as subjects but as objects) were thus considered as essentially passive participants in the research enterprise—patients rather than agents—there to be acted upon by observing and questioning, not there to affect the direction taken in the inquiry. Thus, in the division of labor within the process of qualitative social inquiry, a fundamental line of distinction and asymmetry was drawn between the observer and the observed, with control over the inquiry maximized for the observer and minimized for the observed.

That asymmetry extended to the process of *producing the text of a research report,* which was entirely the responsibility of the social scientist as author. Such reports were not written in collaboration with those whose lives were studied, nor were they accompanied by parallel reports produced by those who were studied (just as the finches of the Galapagos islands had not published a report of Darwin's visit to them). In reports of the results of social inquiry by means of firsthand participant observation, the portrayal of everyday life of the people studied was done by the researcher.

The asymmetry in text production extended further to text consumption. The written report of social inquiry was addressed to *an audience consisting of people other than those who had been studied*—the community of the researcher's fellow social scientists (and perhaps, of policymakers who might commission the research work). This audience had as its primary interests the substantive significance of the research topic and the technical quality of the conduct of the study. The success of the report (and of the author's status as a reporter) was a matter of judgment residing in the scholarly community. The research objects' existential experience of being scrutinized during the researcher's fieldwork and then described in the researcher's report was not a primary consideration for the readers of the report, nor for its author. Indeed those who had been studied were not expected to read the research report, since many were not literate.

For a time, each of these five footings had the stability of canonical authority in the "normal science" practice of qualitative inquiry. That was a period that could be called a "golden age," but with a twinge of irony in such a designation, given what we now know about the intense contestation that has developed recently concerning each of the footings.

▣   II. A "Golden Age" of Realist Ethnography

From the mid 1920s to the early 1950s, the basic approach in qualitative inquiry was realist general ethnography—at the time it was just called *ethnography*. More recently, such work has been called *realist* because of its literary quality of "you are there" reporting, in which the narrator presents description as if it were plain fact, and *general* because it attempted a comprehensive description of a whole way of life in the particular setting that was being described—a setting (such as a village or an island or, later, an urban neighborhood or workplace within a formal organization) that was seen as being distinctly bounded. Typically, the narrator wrote in third person and did not portray him- or herself as being present in the scenes of daily life that were described. A slightly distanced authorial voice was intended to convey an impression of even-handedness—conveying "the native's point of view" without either overt advocacy of customary practices or explicit critique of them. (For a discussion of the stance of detachment, see Vidich & Lyman, 1994, p. 23.) Usually, the social theory perspective underlying such work was some form of functionalism, and this led authors to focus less on conflict as a driving force in society and more on the complementarity of various social institutions and processes within the local setting.

Ethnographic monographs in anthropology during this time followed the overall approach found in Bronislaw Malinowski's (1922) *Argonauts,* where he said that an adequate ethnography should report three primary bodies of evidence:

1. *The organisation of the tribe, and the anatomy of its culture* must be recorded in firm, clear outline. The method of *concrete, statistical documentation* is the means through which such an outline has to be given.

2. Within this frame *the imponderabilia of actual life,* and the *type of behaviour* must be filled in. They have to be collected through minute, detailed observations, in the form of some sort of ethnographic diary, made possible by close contact with native life.

3. A collection of ethnographic statements, characteristic narratives, typical utterances, items of folk-lore, and magical formulae has to be given as a *corpus inscriptionem,* as documents of native mentality. (p. 24)

What was studied was a certain village or region in which a named ethnic/linguistic group resided. The monograph usually began with an overall description of the physical setting (and often of subsistence activities). This was followed by a chapter on an annual cycle of life, one on a typical day, one on kinship and other aspects of "social organization," one on child rearing, and then chapters on certain features of the setting that were distinctive to it. (Thus, for example, Evans-Pritchard's 1940 monograph on a herding people, *The Nuer,* contains detailed description of the aesthetics of appreciation of color patterns in cowhide.) Narrative vignettes describing the actions of particular people in an actual event were sometimes provided, or typical actions were described more synoptically. These vignettes and quotes from informants were linked in the text by narrating commentary. Often maps, frequency tables, and analytic charts (including kinship diagrams) were included.

Notable examples in British and American anthropology during this period include volumes by students of Franz Boas, such as Margaret Mead's (1928) semipopular account, *Coming of Age in Samoa.* Raymond Firth, a student of Malinowski, produced *We the Tikopia* (1936/2004), E. E. Evans-Pritchard, a student of Malinowski's contemporary, Alfred Radcliffe-Brown (who himself had published a monograph *The Andaman Islanders* in the same year as Malinowski's *Argonauts,* 1922) published *The Nuer* in 1940. David Holmberg (1950) published a study of the Siriono, titled *Nomads of the Longbow.* In addition to American work on indigenous peoples of the Western Hemisphere, there were monograph series published on British colonial areas—from Australia, studies of New Guinea, Micronesia, and Melanesia, and from England, studies of East Africa, West Africa, and South Africa.

In the United States, community studies in an anthropologically ethnographic vein were encouraged by Robert Park and Ernest Burgess at the department of sociology of the University of Chicago. On the basis of hunches about geographic determinism in the founding and maintenance of distinct social areas within cities, various Chicago neighborhoods were treated as if they were bounded communities, for example, Louis Wirth's (1928) study of the West Side Jewish ghetto and Harvey Warren Zorbaugh's (1929) study of contiguous working-class Italian and upper class "mainstream American" neighborhoods on the near North Side. A tradition of community study followed in American sociology. Robert and Helen Lynd (1929, 1937) conducted a two-volume study of a small Midwestern city, Muncie, Indiana, which they called Middletown. The anthropologist W. Lloyd Warner studied Newburyport, Massachusetts (1941), the Italian neighborhood of Boston's North End was described by William F. Whyte (1943/1955), and the anthropologists Conrad Arensberg and Solon Kimball studied a rural Irish village (1940).

The urban community studies efforts continued after World War II, with St Clair Drake and Horace Cayton's (1945) description of the African American neighborhoods of Chicago's South Side and Herbert J. Gans's report (1962) on an Italian American neighborhood in New York, among others. Gerald Suttles (1968) revisited the "social areas" orientation of Chicago School sociology in a study of interethnic relations in a multi-ethnic neighborhood on Chicago's Near West Side, and Elijah Anderson (1992) described a multiracial West Philadelphia neighborhood in a somewhat similar vein. Some studies narrowed the scope of community studies from a whole neighborhood to a particular setting within it, as in the case of bars as sites for friendship networks among African American men in the reports (e.g., Liebow, 1967). Rural sociology in America during the 1930s had also produced ethnographic accounts. (For an extensive review and listing of American community studies, see the discussion in Vidich & Lyman, 1994.)

Institutional and workplace studies began to be done ethnographically, especially in the postwar era. Labor-management relations were studied by means of participant observation (e.g., Roy, 1959). Chris Argyris published descriptive accounts of daily work in a bank department (1954a, 1954b) and of the worklife of a business executive (1953). Ethnographic accounts of socialization into professions began to appear (e.g., Becker & Geer, 1961; Glaser & Strauss, 1965). Workplace accounts, as in community studies, began to focus more closely on immediate scenes of everyday social interaction, a trend that continued into the future (see, e.g., Vaught & Smith, 1980; Fine, 1990).

Journal-length reports of workplace studies (as well as accounts of overseas development interventions by applied anthropologists) appeared in the interdisciplinary journal *Human Organization,* which began publication under that title in 1948, sponsored by the Society for Applied Anthropology.

Ethnographic documentary film developed in the 1950s and 1960s as field recording of sound became easier, with more portable equipment—audiotape and the 16-mm camera. Boas had used silent film in the 1920s to document Kwakiutl life on the Northwest Coast of Canada, and Gregory Bateson and Mead used silent film in the late 1930s in their study of dance instruction in Bali. Robert Flaherty produced semifictional, partially staged films of Canadian Inuit in the 1920s, notably *Nanook of the North.*

The new ethnographic documentaries were shot in naturalistic field situations, using for the most part hand-held cameras and microphones in order to move with the action. John Marshall's film, *The Hunters,* featured Kalahari Bushmen of southern Africa; Napoleon Chagnon's "The Ax Fight" and Tim Asch's "The Feast" were filmed in the Amazon River Delta in Brazil, among the Yanomamo. John Adair and Sol Worth gave 16 mm hand-held cameras to Navaho informants in a project that tried to identify differences in ways of seeing between the Navaho and Western European cinematographers. They produced film footage and a monograph on the project titled, "Through Navaho Eyes" (Worth & Adair, 1972). John Collier Jr. shot extensive silent film footage showing Native American school classrooms in Alaska. He also published a book on the use of still photographs for ethnographic documentation (Collier, 1967)—a practice that Mead had pioneered a generation earlier (see Byers, 1966, 1968). The Society for Visual Anthropology, a network of ethnographic filmmakers and scholars of documentary film semiotics, was founded in 1984.

U.S. sociologists made institutionally focused documentary films during the same time period, notably the films produced in the 1960s and 1970s by Frederick Wiseman. These interpretive film essays, through the editing of footage of naturally occurring events, bridge fiction and more literal documentary depiction. They include "Titicut Follies" (1967), a portrayal of a mental hospital; "High School" (1968); "Hospital" (1970); and "Essene" (1972), a portrayal of conflict and community in a monastery (for further discussion, see Barnouw, 1993; Benson & Anderson 2002; deBrigard, 1995; Heider, 1982; Ruby, 2000).

▣ III. CRISES IN ETHNOGRAPHIC AUTHORITY

A gathering storm. Even in the postwar heyday of realist ethnography, some cracks in its footings were beginning to appear. In American anthropology, a bitter controversy developed over accuracy and validity of competing ethnographic descriptions of a village on the outskirts of Mexico City, Tepoztlán. Robert Redfield (1930) at the University of Chicago had published an account of everyday life in Tepoztlán; in keeping with a functionalist perspective in social theory, he characterized the community as harmonious and internally consistent, a place where people led predictable, happy lives. Beginning fieldwork in the same village 17 years after Redfield and viewing everyday life in the community through a lens of Marxist conflict theory, Oscar Lewis (1951) saw life in Tepoztlán as fraught with tension and individual villagers as tending toward continual anger, jealousy, and anxiety; in his monograph, he harshly criticized Redfield's portrayal. Two fieldworkers had gone to the "same" place and collected very different evidence. Which one was right?

Concern was developing over texts that reported the general ethnography of a whole community—those reports seemed increasingly to be hazy in terms of evidence: Description flowed a mile wide but an inch deep. One way to address this limitation was to narrow the scope of research description and to focus on a particular setting within a larger community or institution. Another way was to become more careful in handling evidence. Within American anthropology, specialized "hyphenated" sub-fields of sociocultural study developed, such as cognitive anthropology, economic anthropology, anthropology of law, ethnography of communication, and interactional sociolinguistics. Studies in those subfields were often published as tightly focused journal-length articles in which evidence was presented deliberately and specifically. Careful elicitation techniques and increasing use of audio and audiovisual recording were used in attempts to get "better data." An interdisciplinary field called sociolinguistics developed across the disciplines of linguistics, anthropology, sociology, and social psychology.

In sociology first and then increasingly in anthropology, methods texts were published—becoming more explicit about methods of participant observation as another route to "better data." Notable examples are McCall and Simmons (1969), Glaser and Strauss (1967), Denzin (1970), Pelto and Pelto (1970), Hammersley and Atkinson (1983), Ellen (1984), and Sanjek (1990).

Autobiographical accounts of fieldwork also began to be published. The second edition of Whyte's *Street Corner Society* (1943/1955) and subsequent editions contained an extensive appendix in which Whyte described, in first person, his field experience. Hortense Powdermaker (1966) described her field experience in White and Black Southern U.S. rural communities in the 1930s. Even earlier, Laura Bohannon had published a fictionalized memoir of fieldwork, writing a quasi-novel under the pseudonym Elenore Smith Bowen (1954) because frank revelations of ambivalence, ethical dilemmas, the intense emotionality of fieldwork, and tendencies toward self-deception were not considered proper topics of "academic" discourse at the time. Rosalie Wax (1971) candidly recalled the difficulties of her fieldwork as a White woman in Japanese internment camps during World War II. These accounts showed that actual fieldwork was not so consistently guided by detached, means-ends rationality as ethnographic monographs had sometimes suggested. In 1967, Malinowski's Trobriand Island field diary was published posthumously. Over the next 15 years, the diary came to occupy a central place in what became a firestorm of criticism of realist general ethnography.

After World War II, the accuracy of ethnography began to draw challenges from the "natives" whose lives were portrayed in them. Thirty years after Malinowski left the Trobriands, Father Baldwin, a Roman Catholic missionary who succeeded him there, reported in a master's thesis how the "natives" had reacted to the text of *Argonauts*. Baldwin had lived on the island of Boyowa longer than Malinowski had done and learned the local language more thoroughly. To check the validity of Malinowski's portrayal of the "native's point of view," Father Baldwin translated large portions of *Argonauts* and read those texts with the Boyowans he knew, some of whom remembered Malinowski's presence among them:

> He seems to have left nothing unexplained and his explanations are enlightening, even to the people who live there. It is curious, then, that this exhaustive research, and patient, wise, and honest explanation, should leave a sense of incompleteness. But it does. I feel that his material is still not properly digested, that Malinowski would be regarded in some ways naive by the people he was studying . . .
>
> I was surprised at the number of times informants helping me with checking Malinowski would bridle. Usually when a passage has been gone over more than once, they would say it was not like that. They did not quarrel with facts or explanations, but with the coloring as it were. The sense expressed was not the sense they had of themselves or of things Boyowan. (Baldwin, n.d., pp. 17–18, as cited in Young, 1979, pp. 15–16)

Vine deLoria, a Native American, was more harsh in his criticism of American anthropologists, in a book evocatively titled *Custer Died for Your Sins* (1969). He characterized Amerindian studies done by American anthropologists as ethnocentric and implicitly colonialist. Sociological community studies also drew negative reactions from the "natives." Some small-town residents in rural New York were deeply offended by the monograph titled, *Small Town in Mass Society* (Vidich & Bensman, 1958; see Vidich's discussion of this reaction in Vidich & Bensman, 2000, and in Young, 1996). They castigated the authors for inaccuracy, for taking sides in local disputes, and for violating the confidentiality of individuals (e.g., there being only one mayor, his anonymity was compromised even though his name was not used; this later became a classic example of ethical difficulties in the conduct of qualitative research and its reporting.) The rise of Black nationalism in African American communities in the late 1960s (and the reaction of African American scholars to the "blame the victim" tone of studies about inner-city families such as that of Moynihan, 1965) gave further impetus to the contention that only "insiders" could study fellow insiders in ways that would be unbiased and accurate.

This directly contradicted the traditional view that an outsider researcher, with enough time to develop close acquaintance, could accurately observe and interpret meaning, without being limited by the insider's tendency to overlook phenomena so familiar they were taken for granted and had become invisible. As the anthropologist Clyde Kluckhohn (1949) put it in a vivid metaphor: "It would hardly be fish who discovered the existence of water" (p. 11).

This was not only a matter of inaccurate conclusions—it also had to do with the power relations that obtained in the conduct of "participant observation" itself. Various feminist authors, in a distinct yet related critique of standard anthropology and sociology, pointed out that fieldworkers should attend to their own mentality/subjectivity as a perceiving *subject* trying to make sense of others' lives, especially when power relations between the observer and the observed were asymmetric. An early instantiation of these perspectives was Jean Briggs's (1970) study of her conflicting relationships with an Inuit (Canadian Eskimo) nuclear family with whom she lived during fieldwork. Titled *Never in Anger,* her monograph reported in first person and placed her self and her reactions to her "informants" centrally in the narrative picture her monograph presented.

The notion that the researcher always sees from within (and is also blinded by) the power relationships between her and those she studies was pointedly explicated in Dorothy Smith's 1974 essay "Women's Perspective as a Radical Critique of Sociology." That idea continues to evolve in feminist criticism (see, e.g., Harding, 1991; Lather, 1991) that advocates reflexivity regarding the personal standpoints, the positionality, through which the fieldworker perceives—gendered, classed, age-graded, and raced/ethnicized ways of seeing and feeling in the world, especially as these are in part mutually constructed in the interaction that takes place between the observer and observed.

George Marcus and James Clifford (1986; Clifford, 1988) extended this line of criticism in the mid 1980s, a period when Malinowski became a prime target for those who considered conventional "participant observation" to be deeply flawed. With the publication of his *Diary,* Malinowski had become an easy target. The diary had unmasked power relationships that his ethnographic reporting had disguised. Thus, Malinowski's portrayal of the "native's point of view" in *Argonauts* may have had to do with the power relationships of his fieldwork. He does not mention this in his discussion of his fieldwork method; rather, he portrays himself simply as a detective, a Sherlock Holmes searching avidly for clues concerning native customs and character (Malinowski, 1922, p. 51):

> It is difficult to convey the feelings of intense interest and suspense with which an ethnographer enters for the first time the district that is to be the future scene of his field work. Certain salient features characteristic of the place had once riveted attention and filled him with hopes or apprehensions. The appearance of the natives, their manner, their types of behavior, may augur well or ill for the possibilities of rapid and easy research. One is on the lookout for the symptoms of deeper sociological facts. One suspects many hidden and mysterious ethnographic phenomena behind the commonplace aspect of things. Perhaps that queer looking, intelligent native is a renowned sorcerer. Perhaps between those two groups of men there exists some important rivalry or vendetta, which may throw much light on the customs and character of the people if one can only lay a hand upon it.

From the diary (Malinowski, 1967), a very different voice sounds—boredom, frustration, hostility, lust.

December 14, 1917: "When I look at women I think of their breasts and figure in terms of ERM [an Australian woman who he later married]." (pp. 151–152)

December 17, 1917: "I was fed up with the niggers and with my work." (p. 154)

December 18, 1917: "I thought about my present attitude toward ethnographic work and the natives, my dislike of them, my longing for civilization." (p. 154)

What went without mention was the asymmetry in power relationships between Malinowski and those he studied. He was the primary initiator of actions toward those around him. Years later, working with the same informants, Father Baldwin (n.d.) reported:

> It was a surprise to me to find that Malinowski was mostly remembered by the natives as a champion ass at asking damn fool questions, like "You bury the seed tuber root end or sprout end down?... They said of him that he made of his profession a sacred cow. You had to defer though you did not see why. (p. 41, as cited in Young, 1979, p. 15)

In contrast, Malinowski's tone in the original monograph suggests a certain smugness and lack of self-awareness: "In fact, as they knew that I would thrust my nose into everything, even where a well-mannered native would not dream of intruding, they finished by regarding me as part and parcel of their life, unnecessary evil or nuisance, mitigated by donations of tobacco" (Malinowski, 1922, p. 8).

Admittedly, the alienation Malinowski revealed in the diary was not unique to him. As Young (1979) puts it,

> It is only fair to point out that the chronic sense of alienation which permeates the diary is a common psychic experience of anthropologists in the field, and it is intensified by homesickness, nostalgia, loneliness, and sexual frustration, all of which Malinowski suffered in full measure. (p. 13)

That is humanly true, but it does not square with the popular image of the scientist—rather, it puts the professional social scientist on the same plane as the practical social actor, the "man in the street." Furthermore, it makes one distrust the dispassionate tenor of what Rosaldo (1989, p. 60) called "distanced normalizing description" in ethnographic research reporting.

Malinowski—and the overall credibility of ethnographic research reporting—was further undermined by similar criticism of Margaret Mead. Her first published study, titled *Coming of Age in Samoa* (Mead, 1928), had considered the experience of adolescence from the culturally relativist perspective of her teacher, Boas. Interviewing young Samoan girls and women, Mead concluded that their adolescent years were not emotionally turbulent and that, unlike American teenagers, they were able to engage in sexual experimentation without guilt. Her book attracted a wide popular audience and, together with subsequent popular writing, established Mead's reputation in the United States as a public intellectual. Derek Freeman (1983), an Australian anthropologist, waited until after Mead's death to publish a scathing critique of Mead's research in Samoa. He claimed that Mead had been naive in believing what her informants told her; that they had exaggerated their stories in the direction she had signaled that she wanted to hear. Subsequent consideration suggests that Mead's interpretation was correct overall (see, e.g., Shankman, 1996), but the highly authoritative style of Mead's text (and the lack of systematic presentation of evidence to support the claims she was making) left her vulnerable to the accusation that she had got her findings wrong.

Were all ethnographers self-deceived—or worse, were many of them "just making things up?" The Redfield-Lewis controversy—two vastly different descriptions of the same group—raised an even deeper question: Do the perspective, politics, and ideology of the observer so powerfully influence what he or she notices and reflects on that it overdetermines the conclusions drawn? Realist general ethnography was experiencing heavy weather indeed.

One line of response to these doubts was the "better evidence" movement already discussed. Somewhat earlier, another stream of work had developed that led to participatory action research or collaborative action research. In this approach, outside researchers worked with members of a setting to effect change that was presumed to be of benefit there—for example, improvements in public health, agricultural production, the formation of cooperatives for marketing, and the organization of work in factories. Research efforts accompanied attempts at instituting change, as in the study of local community health practices and beliefs within a project aimed to prevent cholera and dysentery by providing clean water. The social psychologist Kurt Lewin (1946) was one of the pioneers of these attempts, focusing especially on labor-management relations in England. The attempts in England spread through trade union channels into Scandinavia (see Emery & Thorsrud, 1969). Another pioneer was Whyte, working in industrial settings in the United States (see Whyte, Greenwood, & Lazes, 1989).

Also in the period immediately before and after World War II, anthropologists were undertaking change-oriented research overseas, and the Society for Applied Anthropology was founded in 1948. During the 1960s and 1970s, applied anthropologists and linguists worked in action projects in the United States and England in ethnic and racial minority communities (e.g., Gumperz, Roberts, & Jupp, 1979; Schensul & Schensul, 1992).

One line of justification for applied research harked back to the "better evidence" movement: Through a researcher's "involvement in the action" (Schensul, 1974), the accuracy and validity of evidence collection and analysis are tested in conditions of natural experimentation.

Another justification for applied research had to do with the explicit adoption of value positions by action researchers and their community partners. This is similar to the "critical" position in social research that especially took hold in the 1970s and 1980s, and as action research progressed, it combined increasingly with the various critical approaches discussed in the previous section (for elaboration, see Kemmis & McTaggart, 2005).

This aspect of action research led away from the stance of cultural relativism itself—from even the appearance of value neutrality—toward value affirmation. In research efforts to effect social change, explicit value commitments had to be adopted if the work was to make change in specific directions. This was called critical ethnography, related to the "critical theory" perspective articulated by the Frankfurt School. Theodor Adorno and Max Horkheimer had developed a critique, based in neo-Marxist social analysis, of both capitalism and fascism. The point was to criticize whatever material or cultural influences might lead people to take actions or support actions that resulted in limiting their own life chances—that is, their collusion in their own oppression. In Marxist terms one could say that critical theory made visible social processes that worked against the class interests of those being dominated—for example U.S. white workers supporting an oligarchy that oppressed both them and Black workers. Culturally relativist ethnography had not called domination by that name, nor had it named suffering as an object of attention and of description. Critical ethnography claimed to do just that, and in so doing, the ethnographer stepped out of a defended position of value neutrality to one of vulnerability, shifting from distanced relations with informants to relations of solidarity. This was to engage in social inquiry as ethnography "that breaks your heart" (Behar, 1996).

The adoption of an explicit value position created a fixed fulcrum from which analytic leverage could be exerted in distinguishing between which everyday practices led to increase or decrease in life chances (see Bredo & Feinberg 1982). As the critical ethnography movement developed, the focus shifted somewhat from careful explication of the value yardsticks used to judge habitual practices to claims about domination and oppression as if the inequity involved was self-evident. There was a push back from the earlier generation of scholars, who accused critical ethnographers of letting their values so drive their fieldwork that they were able to see only what they expected to see, ignoring disconfirming evidence.

As critical ethnographers identified more and more kinds of inequity, it became apparent that social criticism itself was relative depending on which dimension of superordination/subordination was the locus for analysis. If it was economic relations, then processes of class-based oppression appeared most salient; if gender relations, then patriarchal processes of domination; if postcolonial relations, the survivals of "colonized" status; if sexual identification, then heterosexual domination. And if race became the primary fulcrum for critical social analysis—race, as distinct from, yet as linked to class, gender, colonization, or sexuality—then racial privilege and disprivilege occupied the foreground of attention, with other dimensions of inequity less prominent. Arguments over whose oppression was more heinous or more fundamental—"oppresseder than thou"—took on a sectarian character.

There was also a new relativity in the considerations of the seats of power itself, its manifestations in various aspects or domains, and the ways in which existing patterns of life (including patterns of domination) are reproduced within and across successive generations. Marxism had explained social order as a forcefield of countervailing tensions that were the result of macro-social economic forces. Structural functionalism in anthropology and sociology had explained social order as the result of socialization of individuals, who followed systems of cultural rules. Structuralism in anthropology and linguistics had identified cultural rule systems, which appeared to operate according to inner logics that could be identified and specified by the social scientist. All these approaches treated macro-social structures as determining factors that constrained local social actors. Poststructuralist critiques of this top-down determinism developed. One line of critique stressed the opportunistic character of the everyday practices of local social actors, who as agents made choices of conduct within sets constrained by social processes (i.e. "structures") operating at the macro-social level (for example, Bourdieu's 1977 critique of Lévi-Strauss's structuralism). Another line of critique (Foucault, 1977) showed how power could be exercised over local social actors without physical coercion through the knowledge systems that were maintained discursively and through surveillance by secular "helping" professions—the modern successors of pre-modern religion—whose ideologically ratified purpose was to benefit the clients they "served" by controlling them—medicine, psychiatry, education, and modern prisons. Michel Foucault's notion of discourse as embodied in the conventional common sense of institutions is akin to Gramsci's (1988) notion of "cultural hegemony"—again, an ideological means by which control can be exercised nonviolently through commonsense rationalization justifying the exercise of such power. Power and social structure are thus seen to be strongly influential processes, even though the influence is partial, indirect, and contested—local actors are considered to be agents, not simply passive rule followers, yet they are agents who must swim in rivers that have strong currents.

At the same time, historians began to look away from the accounts of the past that were produced by the powerful (rich, literate, Caucasian, male, or any combination of those traits) and began to focus more centrally on the daily life practices of people whose subaltern "unwritten" lives could fly, as it were, below the radar of history. (This was a challenge to the accounts of orthodox historians who stuck to the conventional primary source materials.) An additional line of criticism of the authoritativeness of texts, which was once taken for granted, came from postmodern scholars (e.g., Derrida, Lyotard, Deleuze) who questioned the entire Enlightenment project of authoritative academic discourse concerning human activity, whether this discourse manifested in the arts, in history, or in social science. With roots in the early modernism of the Enlightenment, all these discourses attempted to construct "master narratives" whose credibility would be robust because they were based on reason and evidence. For the postmodernists, the rhetorical strategies that scholarly authors used to persuade readers of their text's accuracy and truthfulness could be unmasked through a textual analysis called *deconstruction*. Critical ethnography had challenged the authority of realist narrative accounts that left out explicit mention of processes of conflict and struggles over power; the postmodern line of criticism challenged the fundamental authoritativeness of texts per se. Moreover, lines of demarcation between qualitative social inquiry and scholarship in the humanities were dissolving. Approaches from literary criticism—outside the boundaries of mainstream social science—were used both in the interpretist (hermeneutic) orientation in ethnography and in the critical scrutiny of scholarly texts by means of deconstruction.

One of the ways to demystify the text of a qualitative research report is to include the author (and the author's "standpoint" perspectives) as an explicit presence in the fieldwork. The author becomes a character in the story being told—perhaps a primary one—and much or all of the text is written in first-person narration using past tense rather than the earlier ethnographic convention of present-tense narration, which to critics of realist ethnography seemed to connote timelessness—weightless social action in a gravitationless world outside history and apart from struggle. This autobiographical reporting approach came to be called *autoethnography*. Early examples of the approach have already been mentioned: the fiction of Bohannon (Bowen, 1954) and the first confessional ethnographic monograph by Jean Briggs (1970). Later examples of autoethnographic reporting include Rabinow (1977) and Kondo (1990)—see also the recent comprehensive discussion in Bochner and Ellis (2002).

Another approach toward alteration in the text of reports came from attempts to heighten the dramatic force of those texts, making full use of the rhetorics of performance to produce vivid kinds of narration, for example, breaking through from prose into poetry or adopting the means of "street theater," in which scripted or improvised dramatic performances were presented. Ethnographers have sometimes been invidiously called failed novelists and poets because their monographs typically did not make for compelling reading. By analogy with performance art, the new performance ethnography sought to employ more audience-engaging means of representation (see Conquergood, 1989, 2000; Denzin, 2003; Madison & Hamera, 2006.) Examples of arts-based representation approaches are also found in the recent work of Richardson (2004, 2007; see also the discussion in Richardson, 1999).

Classic and more innovative approaches to qualitative inquiry were extensively reviewed in the three successive handbooks on qualitative research methods edited by Denzin and Lincoln (1994, 2000, 2005).

Bent Flyvbjerg (2001), a member of the urban planning faculty at the University of Aalborg, Denmark, made an important statement in the edgily titled *Making social science matter: How social inquiry fails and how it can succeed again*. The book argued for the use of case study to address matters of value, power, and local detail, as these are pertinent to policy decision making. What policymakers need in making decisions is not general knowledge, says Flyvbjerg, but rather the specific circumstances of the local situation. He uses as an example the planning of auto parking and pedestrian mall arrangements in the city of Aalborg. To achieve the best traffic solution for Aalborg, one cannot make a composite of what was done in Limerick, Bruges, Genoa, Tokyo, and Minneapolis. To know what is good for Aalborg involves detailed understanding of Aalborg itself. Such understanding comes from a kind of knowledge that Flyvbjerg calls *phronesis*, action-oriented knowledge of a local social ecosystem.

Qualitative inquiry in educational research. The authority of realist ethnography was beginning to be challenged at the very time when qualitative research approaches developed in certain fields of human services delivery, especially in education. By the 1950s, a subfield of anthropology of education was forming (Spindler 1955, 1963). Henry (1963) published chapter-length accounts of elementary school classrooms that were highly critical of the practices used to encourage competition among students. The first book-length reports, modeled after the writing of ethnologists and anthropologists, were Smith and Geoffrey's (1968) *The Complexities of an Urban Classroom,* and Jackson's (1968) *Life in Classrooms.* Also in 1968, the Council on Anthropology and Education was founded within the American Anthropological Association. Its newsletter developed into a journal in 1973, the *Anthropology and Education Quarterly,* and for a time, this was the primary journal outlet for qualitative studies in education in the United States. Spindler became the editor of a series of overseas ethnographic studies of educational settings, published from the 1960s to the late 1980s by Holt, Rinehart, and Winston.

In England, qualitative inquiry was pioneered by educational evaluation researchers with an orientation from sociology and action research. At CARE, Laurence Stenhouse formed a generation of evaluators who studied schools and classrooms by means of participant observation and who wrote narrative research reports (see, for example, in chronological order, Walker & Adelman, 1975; Adelman, 1981; Kushner, Brisk, & MacDonald, 1982; Kushner, 1991; Torrance, 1995). Various sociologists also engaged in qualitative educational research. In 1977, Willis published *Learning to Labour.* See also Delamont

(1984, 1989, 1992) and Walkerdine (1998). Following in the tradition of Henry and Spindler in the United States and the "new sociology of education" in England, many of these studies focused on aspects of the "hidden curriculum" of social relations and values socialization in classrooms.

Because of the "objectivist" postpositivist tenor of mainstream educational research, this early work in education anticipated to some extent the criticisms of ethnographic authority that developed in anthropology in the late 1970s and early 1980s. In defense, the early qualitative researchers in education took pains to present explicit evidence; indeed, some of them had come out of the 'better data" and "hyphenated subfields" movements in anthropology or the ethnomethodological critique of mainstream work in sociology.

In the United States, qualitative approaches began to be adopted within research on subject matter instruction—initially in literacy studies (Heath, 1983) and social studies. Some of this work derived from the ethnography of communication/sociolinguistics work begun in the 1960s. As portable video equipment became available, classroom participant observation research was augmented by audiovisual recording (Erickson & Shultz, 1977/1997; McDermott, Gospodinoff, & Aron, 1978; Mehan, 1978). A literature on classroom discourse analysis developed, involving transcriptions of recordings of speech (see Cazden, 2001). Initially focused on literacy instruction, after the mid 1980s, this approach was increasingly used in studies of "teaching for understanding" in mathematics and science that were funded by the National Science Foundation in the United States, and that tendency has increased up to the present time.

Methods texts began to appear, explaining to postpositivist audiences of educational researchers how qualitative research could be rigorous and systematic: Guba (1978), Bogdan and Biklen (1982), and Guba and Lincoln (1985); see also Schensul, LeCompte, and Schensul (1999). Erickson's (1986) essay on interpretive qualitative research on teaching appeared in a handbook sponsored by the American Educational Research Association, and that discussion came to be widely cited in educational research. Preceded by a meeting in 1978 at which Mead was the keynote speaker, shortly before her death, and established as an annual meeting 2 years later, the Ethnography in Education conference at the University of Pennsylvania soon became the largest gathering of qualitative educational researchers in the world, surpassed in scale only recently by the International Congress of Qualitative Inquiry at the University of Illinois, Urbana. Also in the 1980s, a movement of practitioner research in education developed in the United States, principally as teachers began to write narrative accounts of their classroom practice (see Cochran-Smith & Lytle, 1993). This was related to participatory action research (see the discussion in Erickson, 2006).

By the early 1990s, qualitative research on subject areas in both the humanities and in science/mathematics had become commonplace, where 20 years earlier it had been very rare.

Video documentation was especially useful in the study of "hands on" instruction in science and in the use of manipulables in teaching mathematics instruction (see Goldman, Barron, Pea, & Derry, 2007). Increasingly, the subject matter studies—especially those supported by NSF funds—focused on the "manifest curriculum" rather uncritically. This tendency was counterbalanced by the adoption of "critical ethnography" by some educational researchers (e.g., Fine, 1991; Kincheloe, 1993; Lather, 1991; McLaren, 1986).

In a number of ways qualitative inquiry in education anticipated and later ran in parallel with the shifts taking place within recent qualitative work in anthropology and sociology. From the outset of qualitative inquiry in education, its research subjects—school teachers, administrators, parents—were literate, fully able to read the research reports that were written about them, and capable of talking back to researchers using the researchers' own terms. The "gaze" of educational researchers—its potential for distorted perception and its status as an exercise of power over those observed—had been identified as problematic in qualitative educational inquiry before critics such as Clifford and Marcus (1986) had published on those matters. Also action research and practitioner research—involving "insiders" in studying and reflecting on their own customary practices—had been done by educational researchers before such approaches were attempted by scholars from social science disciplines.

Today there is a bifurcation in qualitative educational studies—with subject matter-oriented studies on the one hand and critical or postmodern studies on the other. In effect, this results in a split between attention to issues of manifest curriculum and hidden curriculum. Ironically, as the authority of realist ethnography was increasingly challenged within sociology and anthropology, "realist" work in applied research in education, medicine, nursing, and business came to be the most valued.

IV. THE CURRENT SCENE

At this writing, there appear to be seven major streams of qualitative inquiry: a continuation of realist ethnographic case study, a continuation of "critical" ethnography, a continuation of collaborative action research, "indigenous" studies done by "insiders" (including practitioner research in education), autoethnography, performance ethnography, and further efforts along postmodern lines, including literary and other arts-based approaches.

The differences go beyond technique to basic assumptions. A question arises: Is it more useful to consider these as differing "paradigms," as Guba and Lincoln have done among many others (e.g., Guba 1990), or as a more mundane phenomenon? As Hammersley (2008) has put it, "differences among qualitative researchers are embedded in diverse forms of situated practices that incorporate characteristic ways of thinking and feeling about the research process" (p. 167). Whatever terms one uses to characterize this divergence, it is apparent that major differences in purposes, value positions, and ontological and epistemological assumptions obtain.

At the outset of this chapter, I mentioned five foundational "footings" for qualitative inquiry, each of which has been contested across the course of the development of such inquiry: (1) disciplinary perspectives in social science, (2) the participant-observational fieldworker as an observer/author, (3) the people who are observed during the fieldwork, (4) the rhetorical and substantive content of the research report, and (5) the audiences to which such reports have been addressed.

As the social sciences began to develop along lines of natural science models, its social theory orientations (social evolution, then functionalism combined with cultural relativism) were seen to justify data collection and analysis as a "value-neutral" enterprise. That stance was challenged by conflict-oriented social theory, with the research enterprise redefined as social criticism. Today the possibility of valid social critique is itself questioned by postmodern skepticism about the authoritativeness of scholarly inquiry in general, and core organizing notions taken from arts and humanities disciplines inform much new qualitative research. Sociology and anthropology are no longer the foundational "homes" for social and cultural studies.

Formerly, an "expert knowledge" model of the social scientist was seen as justifying long-term firsthand observation and interviewing—"fieldwork"—that was conducted autonomously by a researcher, who operated in ways akin to those of a field biologist. Today the adequacy and legitimacy of that researcher stance has been seriously challenged, with many researchers allying themselves as advocates (collaborators/joint authors/editors) with the people who are studied, or working with researchers coming from the ranks of the "studied." Thus, the roles of "researcher" and "researched" have been blended in recent work.

The research report was formerly considered to be an accurate, realistic, and comprehensive portrayal of the lifeways of those who were studied, with an underlying rhetoric of persuasion as to the realism of the account. Today qualitative research reports are often considered to be partial—renderings done from within the standpoints of the life experience of the researcher. The "validity" of these accounts can be compared to that of novels and poetry—a pointing toward "truths" that are not literal; fiction may be employed as a means of illuminating interpretive points in a report.

Initially, the audiences of such reports were the author's scholarly peers—fellow social scientists, and rarely those who were studied. Today those who are studied are expected to read the report—and they may also participate in writing it.

Moreover, in action research and other kinds of advocacy research, research may also address popular audiences.

This is a story of decentering and jockeying for position as qualitative inquiry has evolved over the last 120 years. Today there is an uneven pattern of adoption and rejection of the newer approaches in qualitative inquiry. In applied fields, such as education, medicine, and business, "realist" ethnography has gained wide acceptance, while more recently developed approaches have sometimes been adopted (especially in education) and sometimes met with skepticism or with outright rejection. In anthropology, heroic "lone ethnographer" fieldwork and reporting, after the self-valorizing model of Malinowski, has generally gone out of fashion. In sociology, the detached stance of professional researcher has also been seriously questioned, together with the realist mode of research reporting.

Yet there has also been push back. In education, for example, while realist ethnography was officially accepted as legitimately scientific in an influential report issued by the National Research Council (Shavelson & Towne, 2002), postmodern approaches were singled out for harsh criticism. The report also took the position that science is a seamless enterprise, with social scientific inquiry being continuous in its fundamental aims and procedures with that of natural science. This position was reinforced by a statement by the primary professional society of researchers in education, the American Educational Research Association. Quoting from the AERA website:

> The following definition of scientifically based research (SBR) was developed by an expert working group convened by the American Educational Research Association (AERA) ... AERA provided this definition in response to congressional staff requests for an SBR definition that was grounded in scientific standards and principles. The request derived from an interest in averting the inconsistencies and at times narrowness of other SBR definitions used in legislation in recent years.

Alternate Definition of Scientifically Based Research (SBR) Supported by AERA Council, July 11, 2008

The term "principles of scientific research" means the use of rigorous, systematic, and objective methodologies to obtain reliable and valid knowledge. Specifically, such research requires

development of a logical, evidence-based chain of reasoning;

methods appropriate to the questions posed;

observational or experimental designs and instruments that provide reliable and generalizable findings;

data and analysis adequate to support findings;

explication of procedures and results clearly and in detail, including specification of the population to which the findings can be generalized;

adherence to professional norms of peer review;

dissemination of findings to contribute to scientific knowledge; and

access to data for reanalysis, replication, and the opportunity to build on findings.

The statements by the NRC panel and the AERA Council claimed to provide a more broadly ecumenical definition of scientific research than that which some members of the U.S. Congress and their staffs were trying to insist on in developing criteria of eligibility for federal funding. Some legislators proposed that funding should be restricted to experimental designs with random assignment of subjects to treatment or control conditions. However, AERA's adoption of the "seamless" view of science means that many of the recent approaches to qualitative inquiry are declared beyond the boundaries of legitimate research. Moreover, the statements by the NRC and by AERA show no awareness of an intellectual history of social and cultural research in which, across many generations of scholars, serious doubts have been raised as to the possibility that inquiry in the human sciences should be, or could be, conducted in ways that were continuous with the natural sciences.

Geertz warned against the "broad umbrella" conception of science in his favorable review of Flyvbjerg's (2001) book, *Making Social Science Matter*:

> Using the term "science" to cover everything from string theory to psychoanalysis is not a happy idea because doing so elides the difficult fact that the ways in which we try to understand and deal with the physical world and those in which we try to understand and deal with the social one are not altogether the same. The methods of research, the aims of inquiry, and the standards of judgment all differ, and nothing but confusion, scorn, and accusation—relativism! Platonism! reductionism! verbalism!—results from failing to see this. (Geertz, 2001, p. 53)

In addition to external critique from the advocates of social inquiry as "hard science," there is also a conservative reaction from within the community of qualitative researchers. One such statement appears in a recent collection of essays by Martin Hammersley (2008):

> I have argued that this postmodern approach is founded on some false assumptions that undermine the distinctive nature of social research . . . one consequence of this has been a legitimization of speculative theorizing; another has been a celebration of obscurity, and associated denunciations of clarity . . . [this] leads toward an abdication of the responsibility for clear and careful argument aimed at discovering what truths qualitative inquiry is capable of providing. (p. 144)

> We must work to overcome, or at least to reduce, methodological pluralism. It is not that all research can or should be done in the same standardised way. Rather, my point is that any approach to methodological thinking needs to engage with the same general issues. (p. 181)

This postmodernist image of qualitative inquiry is not only ill-conceived but . . . its prominence at the present time, not least in arguments against what it dismisses as methodological conservatism, is potentially very damaging—not just to qualitative research but to social science more generally. (p. 11)

The postmodern range within the current spectrum of qualitative inquiry approaches was also criticized in a recent presidential address at the Midwest Sociological Society's annual meeting in 2006 (Adler & Adler, 2008). Reviewing past and current practices, Adler and Adler contrasted mainstream, realist interpretive ethnography and its literary tropes of verisimilitude with the postmodern approaches: "With its focus on the exploration of new forms, it [i.e., postmodern ethnography] offers great possibilities for continuing innovation. There is increasing unlikeliness, however, that it will ever be legitimized beyond its own rather narrow orbit" (p. 29).

In a response to these and other critics, Denzin (2009) wrote a rejoinder in the form of an imaginary play in which various characters—some of whom are scholars, living or dead, some of whom are fictional—engage in dispute on either side of this argument. Many of the lines the characters "speak" come from published work by the various scholars, and the form of the rejoinder as a performance text—a blending of scholarly quotation with novelistic creation of new utterances—mocks the high seriousness of the critics.

Mark Twain is said to have said, "History doesn't repeat itself—at best it sometimes rhymes." If he was correct, then the proponents of postpositivist social science are in serious trouble. Such inquiry, grounded in what is assumed to be a seamless whole of science, aims to discover general laws of social process that are akin to the laws of physics, that is, an enterprise firmly grounded in prose and in literal meanings of things. It will continue to be controverted by the stubborn poetics of everyday social life—its rhyming, the nonliteral, labile meanings inherent in social action, the unexpected twists and turns that belie prediction and control. It may well be that social science will at last give up on its perennially failing attempts to assume that history actually repeats itself and therefore can be studied as if it did. One might think that contemporary qualitative social inquiry would be better equipped than such a prosaic social physics to take account of the poetics of social and cultural processes, and yet qualitative social inquiry expends considerable energy on internecine dispute, with "classical" and "anticlassical" approaches vying for dominance. It seems too soon to know whether this situation is more an opportunity than a liability.

Let me conclude in first person. It seems to me that the full-blown realist ethnographic monograph, with its omniscient narrator speaking to the reader with an apparent neutrality as if from nowhere and nowhen, is no longer a genre of reporting that can responsibly be practiced, given the duration and force of the critique that has been leveled against it. Some adaptation, some deviation from the classic form seems warranted. It also

seems to me that there should be a viable place within qualitative inquiry between harsh critique and self-satisfied nostalgia—and that this need not involve what Hammersley (2008) has called overcoming methodological pluralism. But it does require adopting a certain degree of humility as we consider what any of our work is capable of accomplishing.

It is only 86 years since Malinowski set foot in the Trobriand Islands. I want to say that Malinowski's overall aim for ethnography was a noble one, especially as amended in the words that follow: "to grasp the points of view of those who are studied and of those who are studying, their relations to life, their visions of their worlds." I think it is fair to say that we have learned over the past 60 years how hard it is to achieve such an aim partially, even to move in the direction of that aim. We know now that this is far more difficult than Malinowski and his contemporaries had anticipated. Yet it could still orient our continuing reach.

◪ NOTE

1. Some discussion here is adapted from my own previous writing on these topics, drawing especially on Erickson 1986 and 2006. Because the literature on qualitative research methods is huge, the reader is also referred to Vidich and Lyman (1994) for an extensive review of classic realist ethnography in American sociology and anthropology; to Urrey (1994) for an extensive review of field research methods, primarily in British social anthropology; and to Heider (1982) for an extensive review of ethnographic film.

◪ REFERENCES

Adelman, C. (1981). *Uttering, muttering: Collecting, using, and reporting talk for social and educational research.* London: McIntyre.

Adler, P., & Adler, P. (2008). Of rhetoric and representation: The four faces of ethnography. *Sociological Quarterly, 49*(1), 1–30.

Anderson, E. (1992). *Streetwise: Race, class, and change in an urban community.* Chicago: University of Chicago Press.

Arensberg, C., & Kimball, S. (1940). *Family and community in Ireland.* Cambridge MA: Harvard University Press.

Argyris, C. (1953). *Executive leadership: An appraisal of a manager in action.* New York: Harper.

Argyris, C. (1954a). Human relations in a bank. *Harvard Business Review, 32*(5), 63–72.

Argyris, C. (1954b). *Organization of a bank: A study of the nature of organization and the fusion process.* New Haven, CT: Yale University Labor and Management Center.

Baldwin, B. (n.d.). Traditional and cultural aspects of Trobriand Island chiefs. Unpublished MS thesis. Canberra: Australia National University, Anthropology Department, Royal Society of Pacific Studies.

Barnouw, E. (1993). *Documentary: A history of the non-fiction film* (2nd Rev. ed.). New York: Oxford University Press.

Becker, H., & Geer, B. (1961). *Boys in white: Student culture in medical school*. Chicago: University of Chicago Press.

Behar, R. (1996). *The vulnerable observer: Anthropology that breaks your heart*. Boston: Beacon.

Benson, T., & Anderson C. (2002). *Reality fictions: The films of Frederick Wiseman* (2nd ed.). Carbondale: Southern Illinois University Press.

Bochner, A., & Ellis, C. (Eds.). (2002). *Ethnographically speaking: Autoethnography, literature, and aesthetics*. Walnut Creek, CA: Alta Mira.

Bogdan, R., & Biklen, S. (1982). *Qualitative research for education: An introduction to theory and methods*. Boston: Allyn & Bacon.

Booth, C. (1891). *Labour and life of the people of London*. London and Edinburgh: Williams and Nargate.

Bourdieu, P. (1977). *Outline of a theory of practice* (R. Nice, Trans.). Cambridge, UK: Cambridge University Press.

Bowen, E. (1954). *Return to laughter*. Garden City, NY: Doubleday.

Bredo, E., & Feinberg, W. (1982). *Knowledge and values in social and educational research*. Philadelphia: Temple University Press.

Briggs, J. L. (1970). *Never in anger: Portrait of an Eskimo family*. Cambridge, MA: Harvard University Press.

Byers, P. (1966). Cameras don't take pictures. *The Columbia University Forum, 9*(1), Winter. Reprinted in *Afterimage*, Vol. 4, No. 10, April 1977.

Byers, P. (with Mead, M.). (1968). *The small conference: An innovation in communication*. The Hague: Mouton.

Cazden, C. (2001). *Classroom discourse: The language of teaching and learning*. Portsmouth, NH: Heineman.

Clifford, J. (1988). *The predicament of culture: Twentieth century ethnography, literature, and art*. Cambridge, MA: Harvard University Press.

Clifford, J., & Marcus, G. (1986). *Writing culture: The poetics and politics of ethnography*. Berkeley: University of California Press.

Cochran-Smith, M., & Lytle, S. (1993). *Inside/outside: Teacher research and knowledge*. New York: Teachers College Press.

Collier, J., Jr. (1967). *Visual anthropology: Photography as a research method*. New York: Holt, Rinehart, & Winston.

Comte, A. (2001). *Plan des travaux scientifiques nécessaires pour reorganizer la societe*. Paris: L'Harmattan. (Original work published 1822)

Conquergood, D. (1989). *I am a shaman: A Hmong life story with ethnographic commentary*. Minneapolis: University of Minnesota, Center for Urban and Regional Affairs.

Conquergood, D. (2000). Rethinking elocution: The trope of the talking book and other figures of speech. *Text and Performance Quarterly, 20*(4), 325–341.

deBrigard, E. (1995). The history of ethnographic film. In P. Hockings (Ed.), *Principles of visual anthropology* (2nd ed., pp. 13–44). New York: Mouton de Gruyter.

deLoria, V. (1969). *Custer died for your sins: An Indian manifesto*. New York: Macmillan.

Delamont, S. (1984). The old girl network. In R. Burgess (Ed.), *The research process in educational settings*. London: Falmer.

Delamont, S. (1989). *Knowledgeable women: Structuralism and the reproduction of elites*. London: Routledge.

Delamont, S. (1992). *Fieldwork in educational settings: Methods, pitfalls, and perspectives*. London: Falmer.

Denzin, N. (1970). *The research act in sociology: A theoretical introduction to sociological methods*. London: Butterworths.

Denzin, N. (2003). *Performance ethnography: Critical pedagogy and the politics of culture*. Thousand Oaks, CA: Sage.

Denzin, N. K. (2009). Apocalypse now: Overcoming resistances to qualitative inquiry. *International Review of Qualitative Inquiry, 2*(3), 331–344.

Denzin, N. K., & Lincoln, Y. S. (Eds.). (1994). *The handbook of qualitative research*. Thousand Oaks, CA: Sage.

Denzin, N. K., & Lincoln, Y. S. (Eds.). (2000). *The handbook of qualitative research* (2nd ed.). Thousand Oaks, CA: Sage.

Denzin, N. K., & Lincoln, Y. S. (Eds.). (2005). *The SAGE handbook of qualitative research* (3rd ed.). Thousand Oaks, CA: Sage.

Dilthey, W. (1989). *Einleitung in die Geisteswissenschaften—Introduction to the human sciences* (R. Makkreel & F. Rodi, Ed. & Trans.). Princeton, NJ: Princeton University Press. (Original work published 1883)

Drake, S. C., & Cayton, H. (1945). *Black metropolis: A study of Negro life in a northern city*. Chicago: University of Chicago Press.

DuBois, W. E. B. (1899). *The Philadelphia negro: A social study*. New York: Schocken.

Ellen, R. (1984). *Ethnographic research: A guide to general conduct*. London and San Diego: Academic Press.

Emery, F., & Thorsrud, E. (1969). *Form and content of industrial democracy: Some experiments from Norway and other European countries*. Assen, The Netherlands: Van Gorcum.

Erickson, F. (1986). Qualitative methods in research on teaching. In M. C. Wittrock (Ed.), *Handbook of research on teaching* (3rd ed., pp. 119–161). New York: Macmillan.

Erickson, F. (2006). Studying side by side: Collaborative action ethnography in educational research. In G. Spindler & L. Hammond (Eds.), *New horizons for ethnography in education* (pp. 235–257). Mahwah, NJ: Lawrence Erlbaum.

Erickson, F., & Shultz, J. (1997). When is a context?: Some issues and methods in the analysis of social competence. Reprinted in M. Cole, M. Engeström, & O. Vasquez (Eds.), *Mind, culture, and activity: Seminal papers from the Laboratory of Comparative Human Cognition* (pp. 22–31). Cambridge, UK: Cambridge University Press. (Original work published 1977)

Evans-Pritchard, E. (1940). *The Nuer: A description of the modes of livelihood and political institutions of a Nilotic people*. Oxford, UK: Oxford University Press.

Fine, G. (1990). Organizational time: Temporal demands and the experience of work in restaurant kitchens. *Social Forces, 69*(1), 95–114.

Fine, M. (1991). *Framing dropouts*. Albany: SUNY Press.

Firth, R. (2004). *We the Tikopia*. New York: Routledge. (Original work published 1936)

Flyvbjerg, B. (2001). *Making social science matter: How social inquiry fails and how it can succeed again*. Cambridge and New York: Cambridge University Press.

Flyvbjerg, B. (2006). Five misunderstandings about case study research. *Qualitative Inquiry, 12*(2), 219–245.

Foucault, M. (1977). *Discipline and punish: The birth of the prison*. London: Penguin Books.

Freeman, D. (1983). *Margaret Mead and Samoa*. Cambridge, UK: Harvard University Press.

Gans, H. (1962). *The urban villagers*. New York: The Free Press.

Geertz, C. (1973). *The interpretation of cultures: Selected essays*. New York: Basic Books.

Geertz, C. (2001). Empowering Aristotle. *Science, 293*, 53.

Glaser, B., & Strauss, A. (1965). *Awareness of dying*. Chicago: Aldine.

Glaser, B., & Strauss, A. (1967). *The discovery of grounded theory: Strategies for qualitative research*. Chicago: Aldine.

Goldman, R., Barron, B., Pea, R., & Derry, S. (Eds.). (2007). *Video research in the learning sciences*. Mahwah, NJ: Lawrence Erlbaum.

Gramsci, A. (1988). *A Gramsci reader*. London: Lawrence & Wishart.

Guba, E. (1978). *Toward a methodology of naturalistic inquiry in educational evaluation*. Los Angeles: UCLA, Center for the Study of Evaluation.

Guba, E. (1990). *The paradigm dialogue*. Thousand Oaks, CA: Sage.

Guba, E., & Lincoln, Y. (1985). *Naturalistic inquiry*. Beverly Hills, CA: Sage.

Gumperz, J., Roberts, C., & Jupp, T. (1979). *Culture and communication: Background and notes to accompany the BBC film "Crosstalk."* London: British Broadcasting Company.

Hammersley, M. (2008). *Questioning qualitative inquiry: Critical essays*. London: Sage.

Hammersley, M., & Atkinson, P. (1983). *Ethnography: Principles in practice*. London: Tavistock.

Harding, S. (1991). *Whose science? Whose knowledge? Thinking from women's lives*. Ithaca, NY: Cornell University Press.

Heath, S. (1983). *Ways with words: Language, life, and work in communities and classrooms*. Cambridge: Cambridge University Press.

Heider, K. (1982). *Ethnographic film* (3rd ed.). Austin: University of Texas Press.

Henry, J. (1963). *Culture against man*. New York: Random House.

Hollingshead, A. (1949). *Elmtown's youth: The impact of social classes on adolescents*. New York: John Wiley.

Holmberg, A. (1950). *Nomads of the long bow: The Siriono of Eastern Bolivia*. Garden City, NY: Natural History Press.

Jackson, P. (1968). *Life in classrooms*. New York: Holt, Rinehart, & Winston.

Kemmis, S., & McTaggart, R. (2005). Participatory action research: Communicative action and the public sphere. In N. K. Denzin & Y. S. Lincoln (Eds.), *The SAGE handbook of qualitative research* (3rd ed., pp. 559–603). Thousand Oaks, CA: Sage.

Kincheloe, J. (1993). *Toward a critical politics of teacher thinking*. S. Hadley, MA: Bergin & Garvey.

Kluckhohn, C. (1949). *Mirror for man*. New York: McGraw-Hill.

Kondo, D. (1990). *Crafting selves: Power, gender, and discourses of identity in a Japanese workplace*. Chicago: Chicago University Press.

Kushner, S. (1991). *The children's music book: Performing musicians in schools*. London: Calouste Gulbenkian Foundation.

Kushner, S., Brisk, M., & MacDonald, B. (1982). *Bread and dreams: A case study of bilingual schooling in the U.S.* Norwich, UK: University of East Anglia, Centre for Applied Research in Education.

Lather, P. (1991). *Getting smart: Feminist research and pedagogy with/in the postmodern*. New York: Routledge.

Latour, B., & Woolgar, S. (1979). *Laboratory life: The social construction of scientific facts*. Beverly Hills: Sage.

Lewin, K. (1946). Action research and minority problems. *Journal of Social Issues, 24*(1), 34–46.

Lewis, O. (1951). *Life in a Mexican village: Tepoztlán restudied*. Urbana: University of Illinois Press.

Liebow, E. (1967). *Tally's corner: A study of Negro streetcorner men*. Boston: Little, Brown.

Lynch, M. (1993). *Scientific practice and ordinary action: Ethnomethodology and social studies of science*. Cambridge, UK: Cambridge University Press.

Lynd, R., & Lynd, H. (1929). *Middletown: A study in contemporary American culture*. New York: Harcourt, Brace.

Lynd, R., & Lynd, H. (1937). *Middletown in transition: A study in cultural conflicts*. New York: Harcourt, Brace.

Madison, D. S., & Hamera, J. (Eds.). (2006). *The SAGE handbook of performance studies*. Thousand Oaks, CA: Sage.

Malinowski, B. (1922). *Argonauts of the Western Pacific: An account of native enterprise and adventure in the archipelagoes of Melanesian New Guinea*. London and New York: G. Routledge and E. P. Dutton.

Malinowski, B. (1967). *A diary in the strict sense of the term*. New York: Harcourt, Brace.

McCall, G., & Simmons, J. (1969). *Issues in participant observation: A text and reader*. Reading, MA: Addison-Wesley.

McLaren, P. (1986). *Schooling as a ritual performance*. London: Routledge and Kegan Paul.

McDermott, R., Gospodinoff, K., & Aron, J. (1978). Criteria for an ethnographically adequate description of concerted activities and their contexts. *Semiotica, 24*(3–4), 245–276.

Mead, M. (1928). *Coming of age in Samoa: A psychological study of primitive youth for Western civilization*. New York: William Morrow.

Mehan, H. (1978). *Learning lessons: Social organization in the classroom*. Cambridge, MA: Harvard University Press.

Morgan, L. H. (1877). *Ancient society: Researches in the lines of human progress from savagery through barbarism to civilization*. New York: MacMillan.

Moynihan, D. (1965). *The Negro family: The case for national action*. Washington, DC: U.S. Department of Labor, Office of Policy Planning and Research.

Munhall, P. (Ed.). (2001). *Nursing research: A qualitative perspective*. Sudbury MA: Jones and Bartlett.

Nash, J. (1979). *We eat the mines and the mines eat us*. New York: Columbia University Press National Research Council.

Pelto, P. J., & Pelto, G. H. (1970). *Anthropological research: The structure of inquiry*. New York: Harper & Row.

Powdermaker, H. (1966). *Stranger and friend: The way of an anthropologist*. New York: W. W. Norton.

Quetelet, L. A. (2010). *A treatise on man and the development of his faculties* (T. Smibert, Ed). Charlestown, SC: Nabu Press. (Original work published 1835)

Rabinow, P. (1977). *Reflections on fieldwork in Morocco*. Berkeley: University of California Press.

Radcliffe-Brown, A. (1922). *The Andaman islanders: A study in social anthropology*. Cambridge, UK: Cambridge University Press.

Redfield, R. (1930). *Tepoztlán, a Mexican village: A study in folk life*. Chicago: University of Chicago Press.

Richardson, L. (1999). Feathers in our CAP. *Journal of Contemporary Ethnography, 28*, 660–668.

Richardson, L. (2004). *Travels with Ernest: Crossing the literary/sociological divide*. Walnut Creek, CA: AltaMira.

Richardson, L. (2007). *Last writes: A daybook for a dying friend.* Thousand Oaks, CA: Left Coast Press.

Riis, J. (1890). *How the other half lives: Studies among the tenements of New York.* New York: Charles Scribner's Sons.

Rosaldo, R. (1989). *Culture and truth: The remaking of social analysis.* Boston: Beacon.

Roy, D. (1959). "Banana Time": Job satisfaction and informal interaction. *Human Organization, 18*(04), 158–168.

Ruby, J. (2000). *Picturing culture: Explorations of film and anthropology.* Chicago: University of Chicago Press.

Sanjek, R. (1990). *Fieldnotes: The makings of anthropology.* Ithaca, NY: Cornell University Press.

Schensul, J., LeCompte, M., & Schensul, S. (1999). *The ethnographer's toolkit* (Vols. 1–5). Walnut Creek, CA: AltaMira Press.

Schensul, J., & Schensul, S. (1992). Collaborative research: Methods of inquiry for social change. In M. LeCompte, W. Milroy, & J. Preissle (Eds.), *The handbook of qualitative research in education.* San Diego and New York: Academic Press.

Schensul, S. (1974). Skills needed in action anthropology: Lessons learned from El Centro de la Causa. *Human Organization, 33,* 203–209.

Shankman, P. (1996). The history of Samoan sexual conduct and the Mead-Freeman controversy. *American Anthropologist, 98*(3), 555–567.

Smith, D. (1974). Women's perspective as a radical critique of sociology. *Sociological Inquiry, 44,* 7–13.

Smith, L., & Geoffrey, W. (1968). *The complexities of an urban classroom.* New York: Holt, Rinehart, & Winston.

Spindler, G. (1955). *Education and anthropology.* Stanford, CA: Stanford University Press.

Spindler, G. (1963). *Education and culture: Anthropological approaches.* New York: Holt, Rinehart, & Winston.

Stenhouse, L. (1975). *An introduction to curriculum research and development.* London: Heineman.

Torrance, H. (1995). *Evaluating authentic assessment: Problems and possibilities in new approaches to assessment.* Buckingham, UK: Open University Press.

Tylor, E. B. (1871). *Primitive culture.* London: John Murray.

Urrey, J. (1984). A history of field methods. In R. Ellen (Ed.), *Ethnographic research: A guide to general conduct* (pp. 33–61). London and San Diego: Academic Press.

Van Maanen, J. (1988). *Tales of the field: On writing ethnography.* Chicago: University of Chicago Press.

Van Maanen, J. (2006). Ethnography then and now. *Qualitative Research in Organizations and Management: An International Journal, 1*(1), 13–21.

Vaught, C., & Smith, D. L. (1980). Incorporation & mechanical solidarity in an underground coal mine. *Sociology of Work and Occupations, 7*(2), 159–187.

Vidich, A., & Bensman, J. (1958). *Small town in mass society: Class, power, and religion in a rural community.* Garden City, NY: Doubleday.

Vidich, A., & Bensman, J. (2000). *Small town in mass society: Class, power, and religion in a rural community* (Rev. ed.). Urbana: University of Illinois Press.

Vidich, A., & Lyman, S. (1994). Qualitative methods: Their history in sociology and anthropology. In N. K. Denzin & Y. S. Lincoln (Eds.), *Handbook of qualitative research* (pp. 23–59). Thousand Oaks, CA: Sage.

Walker, R., & Adelman, C. (1975). *A guide to classroom observation.* London: Routledge.

Walkerdine, V. (1998). *Counting girls out: Girls and mathematics.* London: Falmer.

Warner, W. L. (1941). *Yankee city.* New Haven, CT: Yale University Press.

Wax, R. (1971). *Doing fieldwork: Warnings and advice.* Chicago: University of Chicago Press.

Whyte, W. F. (1955). *Street corner society: The social structure of an Italian slum.* Chicago: University of Chicago Press. (Original work published 1943)

Whyte, W. F., Greenwood, D. J., & Lazes, P. (1989). Participatory action research: Through practice to science in social research. *American Behavioral Scientist, 32*(5), 513–551.

Willis, P. (1977). *Learning to labour: How working class kids get working class jobs.* Westemead, UK: Saxon House.

Wirth, L. (1928). *The ghetto.* Chicago: University of Chicago Press.

Worth, S., & Adair, J. (1972). *Through Navaho eyes: An exploration of film communication and anthropology.* Bloomington: Indiana University Press.

Young, F. (1996). Small town in mass society revisited. *Rural Sociology, 61*(4), 630–648.

Young, M. (1979). *The ethnography of Malinowski: The Trobriand Islands 1915–18.* London: Routledge and K. Paul.

Zorbaugh, H. (1929). *The gold coast and the slum: A sociological study of Chicago's Near North Side.* Chicago: University of Chicago Press.

ETHICS AND POLITICS IN QUALITATIVE RESEARCH

Clifford G. Christians

Getting straight on ethics in qualitative research is not an internal matter only. Putting ethics and politics together is the right move intellectually, but it engages a major agenda beyond adjustments in qualitative theory and methods. The overall issue is the Enlightenment mind and its progeny. Only when the Enlightenment's epistemology is contradicted will there be conceptual space for a moral-political order in distinctively qualitative terms. The Enlightenment's dichotomy between freedom and morality fostered a tradition of value-free social science and, out of this tradition, a means-ends utilitarianism. Qualitative research insists on starting over philosophically, without the Enlightenment dualism as its foundation. The result is an ethical-political framework that is multicultural, gender inclusive, pluralistic, and international in scope.

▣ ENLIGHTENMENT DUALISMS

The Enlightenment mind clustered around an extraordinary dichotomy. Intellectual historians usually summarize this split in terms of subject/object, fact/value, or material/spiritual dualisms. All three of these are legitimate interpretations of the cosmology inherited from Galileo Galilei, René Descartes, and Isaac Newton. None of them puts the Enlightenment into its sharpest focus, however. Its deepest root was a pervasive autonomy. The cult of human personality prevailed in all its freedom. Human beings were declared a law unto themselves, set loose from every faith that claimed their allegiance. Proudly self-conscious of human autonomy, the 18th-century mind saw nature as an arena of limitless possibilities in which human sovereignty is master over the natural order. Release from nature spawned autonomous individuals, who considered themselves independent of any authority. The freedom motif was the deepest driving force, first released by the Renaissance and achieving maturity during the Enlightenment.

Obviously, one can reach autonomy by starting with the subject/object dualism. In constructing the Enlightenment worldview, the prestige of natural science played a key role in setting people free. Achievements in mathematics, physics, and astronomy allowed humans to dominate nature, which formerly had dominated them. Science provided unmistakable evidence that by applying reason to nature and human beings in fairly obvious ways, people could live progressively happier lives. Crime and insanity, for example, no longer needed repressive theological explanations but were deemed capable of mundane empirical solutions.

Likewise, one can get to the autonomous self by casting the question in terms of a radical discontinuity between hard facts and subjective values. The Enlightenment pushed values to the fringe through its disjunction between knowledge of what is and what ought to be. And Enlightenment materialism in all its forms isolated reason from faith, knowledge from belief. As Robert Hooke insisted three centuries ago, when he helped found London's Royal Society: "This Society will eschew any discussion of religion, rhetoric, morals, and politics." With factuality gaining a stranglehold on the Enlightenment mind, those regions of human interest that implied oughts, constraints, and imperatives simply ceased to appear. Certainly those who see the Enlightenment as separating facts and values have identified a cardinal difficulty. Likewise, the realm of the spirit can easily dissolve into mystery and intuition. If the spiritual world contains no binding force, it is surrendered to speculation by the divines, many of whom accepted the Enlightenment belief that their pursuit was ephemeral.

But the Enlightenment's autonomy doctrine created the greatest mischief. Individual self-determination stands as the centerpiece, bequeathing to us the universal problem of integrating human freedom with moral order. In struggling with the complexities and conundrums of this relationship, the Enlightenment, in effect, refused to sacrifice personal freedom. Even though the problem had a particular urgency in the 18th century, its response was not resolution but a categorical insistence on autonomy. Given the despotic political regimes and oppressive ecclesiastical systems of the period, such an uncompromising stance for freedom at this juncture is understandable. The Enlightenment began and ended with the assumption that human liberty ought to be cut away from the moral order, never integrated meaningfully with it (cf. Taylor, 2007, Chapter 10).

Jean-Jacques Rousseau was the most outspoken advocate of this radical freedom. He gave intellectual substance to free self-determination of the human personality as the highest good. Rousseau is a complicated figure. He refused to be co-opted by Descartes' rationalism, Newton's mechanistic cosmology, or John Locke's egoistic selves. He was not content merely to isolate and sacralize freedom either, at least not in his *Discourse on Inequality* or in the *Social Contract,* where he answers Thomas Hobbes.

Rousseau represented the romantic wing of the Enlightenment, revolting against its rationalism. He won a wide following well into the 19th century for advocating immanent and emergent values rather than transcendent and given ones. While admitting that humans were finite and limited, he nonetheless promoted a freedom of breathtaking scope—not just disengagement from God or the church, but freedom from culture and from any authority. Autonomy became the core of the human being and the center of the universe. Rousseau's understanding of equality, social systems, axiology, and language were anchored in it. He recognized the consequences more astutely than those comfortable with a shrunken negative freedom. The only solution that he found tolerable was a noble human nature that enjoyed freedom beneficently and therefore, one could presume, lived compatibly in some vague sense with a moral order.

◙ VALUE-FREE EXPERIMENTALISM

Typically, debates over the character of the social sciences revolve around the theory and methodology of the natural sciences. However, the argument here is not how they resemble natural science, but their inscription into the dominant Enlightenment worldview. In political theory, the liberal state as it developed in 17th- and 18th-century Europe left citizens free to lead their own lives without obeisance to the church or the feudal order. Psychology, sociology, and economics—known as the human or moral sciences in the 18th and 19th centuries—were conceived as "liberal arts" that opened minds and freed the imagination. As the social sciences and liberal state emerged and overlapped historically, Enlightenment thinkers in Europe advocated the "facts, skills, and techniques" of experimental reasoning to support the state and citizenry (Root, 1993, pp. 14–15).

Consistent with the presumed priority of individual liberty over the moral order, the basic institutions of society were designed to ensure "neutrality between different conceptions of the good" (Root, 1993, p. 12). The state was prohibited "from requiring or even encouraging citizens to subscribe to one religious tradition, form of family life, or manner of personal or artistic expression over another" (Root, 1993, p. 12). Given the historical circumstances in which shared conceptions of the good were no longer broad and deeply entrenched, taking sides on moral issues and insisting on social ideals were considered counterproductive. Value neutrality appeared to be the logical alternative "for a society whose members practiced many religions, pursued many different occupations, and identified with many different customs and traditions" (Root, 1993, p. 11). The theory and practice of mainstream social science reflect liberal Enlightenment philosophy, as do education, science, and politics. Only a reintegration of autonomy and the moral order provides an alternative paradigm for the social sciences today.[1]

Mill's Philosophy of Social Science

For John Stuart Mill,

> neutrality is necessary in order to promote autonomy. . . . A person cannot be forced to be good, and the state should not dictate the kind of life a citizen should lead; it would be better for citizens to choose badly than for them to be forced by the state to choose well. (Root, 1993, pp. 12–13)

Planning our lives according to our own ideas and purposes is sine qua non for autonomous beings in Mill's *On Liberty* (1859/1978): "The free development of individuality is one of the principal ingredients of human happiness, and quite the chief ingredient of individual and social progress" (p. 50; see also Copleston, 1966, p. 303, note 32). This neutrality, based on the supremacy of individual autonomy, is the foundational principle in Mill's *Utilitarianism* (1861/1957) and in *A System of Logic* (1843/1893) as well. For Mill, "the principle of utility demands that the individual should enjoy full liberty, except the liberty to harm others" (Copleston, 1966, p. 54). In addition to bringing classical utilitarianism to its maximum development and establishing with Locke the liberal state, Mill delineated the foundations of inductive inquiry as social scientific method. In terms of the principles of empiricism, he perfected the inductive techniques of Francis Bacon as a problem-solving methodology to replace Aristotelian deductive logic.

According to Mill, syllogisms contribute nothing new to human knowledge. If we conclude that because "all men are mortal," the Duke of Wellington is mortal by virtue of his manhood, then the conclusion does not advance the premise (see Mill, 1843/1893, II.3.2, p. 140). The crucial issue is not reordering the conceptual world but discriminating genuine knowledge from superstition. In the pursuit of truth, generalizing and synthesizing are necessary to advance inductively from the known to the unknown. Mill seeks to establish this function of logic as inference from the known, rather than certifying the rules for formal consistency in reasoning (Mill, 1843/1893, III). Scientific certitude can be approximated when induction is followed rigorously, with propositions empirically derived and the material of all our knowledge provided by experience.[2] For the physical sciences, Mill establishes four modes of experimental inquiry: agreement, disagreement, residues, and the principle of concomitant variations (1843/1893, III.8, pp. 278–288). He considers them the only possible methods of proof for experimentation, as long as one presumes the realist position that nature is structured by uniformities.[3]

In Book 6 of *A System of Logic,* "On the Logic of the Moral Sciences," Mill (1843/1893) develops an inductive experimentalism as the scientific method for studying "the various phenomena which constitute social life" (VI.6.1, p. 606). Although he conceived of social science as explaining human behavior in terms of causal laws, he warned against the fatalism of full predictability. "Social laws are hypothetical, and statistically-based generalizations that by their very nature admit of exceptions" (Copleston, 1966, p. 101; see also Mill, 1843/1893, VI.5.1, p. 596). Empirically confirmed instrumental knowledge about human behavior has greater predictive power when it deals with collective masses than when it concerns individual agents.

Mill's positivism is obvious throughout his work on experimental inquiry.[4] Based on Auguste Comte's *Cours de Philosophie Positive* (1830), he defined matter as the "permanent possibility of sensation" (Mill, 1865b, p. 198) and believed that nothing else can be said about the metaphysical.[5] Social research is amoral, speaking to questions of means only. Ends are outside its purview. In developing precise methods of indication and verification, Mill established a theory of knowledge in empirical terms. Truth is not something in itself but "depends on the past history and habits of our own minds" (Mill, 1843/1893, II, Vol. 6, p. 181). Methods for investigating society must be rigorously limited to the risks and benefits of possible courses of action. With David Hume and Comte, Mill insisted that metaphysical substances are not real; only the facts of sense phenomena exist. There are no essences or ultimate reality behind sensations; therefore, Mill (1865/1907, 1865a, 1865b) and Comte (1848/1910) argued that social scientists should limit themselves to particular data as a factual source out of which experimentally valid laws can be derived. For both, this is the only kind of knowledge that yields practical benefits (Mill, 1865b, p. 242); in fact, society's salvation is contingent on such scientific knowledge (p. 241).[6]

Like his consequentialist ethics, Mill's philosophy of social science is built on a dualism of means and ends. Citizens and politicians are responsible for articulating ends in a free society and science for providing the know-how to achieve them. Science is amoral, speaking to questions of means but with no wherewithal or authority to dictate ends. Methods in the social sciences must be disinterested regarding substance and content. Protocols for practicing liberal science "should be prescriptive, but not morally

or politically prescriptive and should direct against bad science but not bad conduct" (Root, 1993, p. 129). Research cannot be judged right or wrong, only true or false. "Science is political only in its applications" (Root, 1993, p. 213). Given his democratic liberalism, Mill advocates neutrality "out of concern for the autonomy of the individuals or groups" social science seeks to serve. It should "treat them as thinking, willing, active beings who bear responsibility for their choices and are free to choose" their own conception of the good life by majority rule (Root, 1993, p. 19).

Value Neutrality in Max Weber

When 21st-century mainstream social scientists contend that ethics is not their business, they typically invoke Max Weber's essays written between 1904 and 1917. Given Weber's importance methodologically and theoretically for sociology and economics, his distinction between political judgments and scientific neutrality is given canonical status.

Weber distinguishes between value freedom and value relevance. He recognizes that in the discovery phase, "personal, cultural, moral, or political values cannot be eliminated; . . . what social scientists choose to investigate . . . they choose on the basis of the values" they expect their research to advance (Root, 1993, p. 33). But he insists that social science be value-free in the presentation phase. Findings ought not to express any judgments of a moral or political character. Professors should hang up their values along with their coats as they enter their lecture halls.

"An attitude of moral indifference," Weber (1904/1949b) writes, "has no connection with scientific objectivity" (p. 60). His meaning is clear from the value-freedom/value-relevance distinction. For the social sciences to be purposeful and rational, they must serve the "values of relevance."

> The problems of the social sciences are selected by the value relevance of the phenomena treated. . . . The expression "relevance to values" refers simply to the philosophical interpretation of that specifically scientific "interest" which determines the selection of a given subject matter and problems of empirical analysis. (Weber, 1917/1949a, pp. 21–22)

> In the social sciences the stimulus to the posing of scientific problems is in actuality always given by practical "questions." Hence, the very recognition of the existence of a scientific problem coincides personally with the possession of specifically oriented motives and values. . . .
>
> Without the investigator's evaluative ideas, there would be no principle of selection of subject matter and no meaningful knowledge of the concrete reality. Without the investigator's conviction regarding the significance of particular cultural facts, every attempt to analyze concrete reality is absolutely meaningless. (Weber, 1904/1949b, pp. 61, 82)

Whereas the natural sciences, in Weber's (1904/1949b, p. 72) view, seek general laws that govern all empirical phenomena, the social sciences study those realities that our values consider significant. Whereas the natural world itself indicates what reality to investigate, the infinite possibilities of the social world are ordered in terms of "the cultural values with which we approach reality" (1904/1949b, p.78).[7] However, even though value relevance directs the social sciences, as with the natural sciences, Weber considers the former value-free. The subject matter in natural science makes value judgments unnecessary, and social scientists by a conscious decision can exclude judgments of "desirability or undesirability" from their publications and lectures (1904/1949b, p. 52). "What is really at issue is the intrinsically simple demand that the investigator and teacher should keep unconditionally separate the establishment of empirical facts . . . and his own political evaluations" (Weber, 1917/1949a, p. 11).

Weber's opposition to value judgments in the social sciences was driven by practical circumstances (Brunn, 2007). Academic freedom for the universities of Prussia was more likely if professors limited their professional work to scientific know-how. With university hiring controlled by political officials, only if the faculty refrained from policy commitments and criticism would officials relinquish their control.

> Few of the offices in government or industry in Germany were held by people who were well trained to solve questions of means. Weber thought that the best way to increase the power and economic prosperity of Germany was to train a new managerial class learned about means and silent about ends. The mission of the university, on Weber's view, should be to offer such training.[8] (Root, 1993, p. 41; see also Weber, 1973, pp. 4–8)

Weber's practical argument for value freedom and his apparent limitation of it to the reporting phase have made his version of value neutrality attractive to 21st-century social science. He is not a positivist like Comte or a thoroughgoing empiricist in the tradition of Mill. He disavowed the positivist's overwrought disjunction between discovery and justification and developed no systematic epistemology comparable to Mill's. His nationalism was partisan compared to Mill's liberal political philosophy. Nevertheless, Weber's value neutrality reflects Enlightenment autonomy in a fundamentally similar fashion. In the process of maintaining his distinction between value relevance and value freedom, he separates facts from values and means from ends. He appeals to empirical evidence and logical reasoning rooted in human rationality. "The validity of a practical imperative as a norm," he writes, "and the truth-value of an empirical proposition are absolutely heterogeneous in character" (Weber, 1904/1949b, p. 52). "A systematically correct scientific proof in the social sciences" may not be completely attainable, but that is most likely "due to faulty data" not because it is conceptually impossible (1904/1949b, p. 58).[9] For Weber, like Mill, empirical science deals with questions of means, and his warning against inculcating political and moral values presumes a means-ends dichotomy (see Weber, 1917/1949a, pp. 18–19; 1904/1949b, p. 52; cf. Lassman, 2004).

As Michael Root (1993) concludes, "John Stuart Mill's call for neutrality in the social sciences is based on his belief" that the language of science "takes cognizance of a phenomenon and endeavors to discover its laws." Max Weber likewise "takes it for granted that there can be a language of science—a collection of truths—that excludes all value-judgments, rules, or directions for conduct" (p. 205). In both cases, scientific knowledge exists for its own sake as morally neutral. For both, neutrality is desirable "because questions of value are not rationally resolvable" and neutrality in the social sciences is presumed to contribute "to political and personal autonomy" (p. 229). In Weber's argument for value relevance in social science, he did not contradict the larger Enlightenment ideal of scientific neutrality between competing conceptions of the good.

Utilitarian Ethics

In addition to its this-worldly humanism, utilitarian ethics has been attractive for its compatibility with scientific thought. It fit the canons of rational calculation as they were nourished by the Enlightenment's intellectual culture.

> In the utilitarian perspective, one validated an ethical position by hard evidence. You count the consequences for human happiness of one or another course, and you go with the one with the highest favorable total. What counts as human happiness was thought to be something conceptually unproblematic, a scientifically establishable domain of facts. One could abandon all the metaphysical or theological factors which made ethical questions scientifically undecidable. (Taylor, 1982, p. 129)

Utilitarian ethics replaces metaphysical distinctions with the calculation of empirical quantities, reflecting the inductive processes Mill delineated in his *System of Logic*. Utilitarianism favors specific actions or policies based on evidence. It follows the procedural demand that if "the happiness of each agent counts for one . . . the right course of action should be what satisfies all, or the largest number possible" (Taylor, 1982, p. 131). Autonomous reason is the arbiter of moral disputes.

With moral reasoning equivalent to calculating consequences for human happiness, utilitarianism presumes there is "a single consistent domain of the moral, that there is one set of considerations which determines what we ought morally to do" (Taylor, 1982, p. 132). This "epistemologically-motivated reduction and homogenization of the moral" marginalizes the qualitative languages of admiration and contempt—integrity, healing, liberation, conviction, dishonesty, and self-indulgence, for example (Taylor, 1982, p. 133). In utilitarian terms, these languages designate subjective factors that "correspond to nothing in reality. . . . They express the way we feel, not the way things are" (Taylor, 1982, p. 141).[10] This single-consideration theory not only demands that we maximize general happiness, but considers irrelevant other moral imperatives that conflict with it, such as equal distribution. One-factor models appeal to the "epistemological squeamishness" of value-neutral social science, which "dislikes contrastive languages." Moreover, utilitarianism appealingly offers "the prospect of exact calculation of policy through . . . rational choice theory" (Taylor, 1982, p. 143). "It portrays all moral issues as discrete problems amenable to largely technical solutions" (Euben, 1981, p. 117). However, to its critics, this kind of exactness represents "a semblance of validity" by leaving out whatever cannot be calculated (Taylor, 1982, p. 143).[11]

Another influential critique of utilitarianism was developed earlier by W. David Ross.[12] Ross (1930) argued against the utilitarian claim that others are morally significant to us only when our actions impact them pro or con (pp. 17–21). We usually find ourselves confronting more than one moral claim at the same time involving different ethical principles. Asking only what produces the most good is too limiting. It does not cover the ordinary range of human relationships and circumstances. People recognize promise-keeping, equal distribution, nonviolence, and prevention of injury as moral principles. In various situations, any of them might be the most stringent.

> Ordinary moral sensitivities suggest that when someone fulfills a promise because he thinks he ought to do so, it seems clear that he does so with no thought of its total consequences. . . . What makes him think it's right to act in a certain way is the fact that he has promised to do so—that and, usually, nothing more. (Ross, 1930, p. 17)

For both Taylor and Ross, the domain of the good in utilitarian theory is extrinsic. Given its dualism of means and ends, all that is worth valuing is a function of the consequences. Prima facie duties are literally inconceivable. "The degree to which my actions and statements" truly express what is important to someone does not count. Ethical and political thinking in consequentialist terms legislate[s] intrinsic valuing out of existence" (Taylor, 1982, p. 144). The exteriority of ethics is seen to guarantee the value neutrality of experimental procedures.[13]

Codes of Ethics

In value-free social science, codes of ethics for professional and academic associations are the conventional format for moral principles. By the 1980s, each of the major scholarly associations had adopted its own code, with an overlapping emphasis on four guidelines for directing an inductive science of means toward majoritarian ends.

1. *Informed consent.* Consistent with its commitment to individual autonomy, social science in the Mill and Weber tradition insists that research subjects have the right to be informed about the nature and consequences of experiments in which they are involved. Proper respect for human freedom generally includes two necessary conditions. Subjects must agree voluntarily to participate—that is, without physical or psychological coercion. In addition, their agreement must be based on full and open information. "The Articles of the Nuremberg Tribunal and the Declaration of Helsinki both state that subjects must be told the duration, methods, possible risks, and the purpose or aim of the experiment" (Soble, 1978, p. 40).

The self-evident character of this principle is not disputed in rationalist ethics. Meaningful application, however, generates ongoing disputes. As Punch (1998) observes, "In much fieldwork there seems to be no way around the predicament that informed consent—divulging one's identity and research purpose to all and sundry—will kill many a project stone dead" (p. 171). True to the privileging of means in a means-ends model, Punch reflects the general conclusion that codes of ethics should serve as a guideline prior to fieldwork but not intrude directly on the research process itself. "A strict application of codes" may "restrain and restrict" a great deal of "innocuous" and "unproblematic" research (p. 171).

2. *Deception.* In emphasizing informed consent, social science codes of ethics uniformly oppose deception. Even paternalistic arguments for possible deception of criminals, children in elementary schools, or the mentally incapacitated are no longer credible. The ongoing exposé of deceptive practices since Stanley Milgram's experiments have given this moral principle special status; that is, deliberate misrepresentation is not ethically justified. In Kai Erikson's (1967) classic formulation:

> The practice of using masks in social research compromises both the people who wear them and the people for whom they are worn, and in doing so violates the terms of a contract which the sociologist should be ready to honor in his dealings with others. (pp. 367–368)

The straightforward application of this principle suggests that researchers design experiments free of active deception. But with ethical constructions exterior to the scientific enterprise, no unambiguous application is possible. Within both psychological experimentation and medical research, some information cannot be obtained without at least deception by omission. Given that the search for knowledge is obligatory and deception is codified as morally unacceptable, in some situations, both criteria cannot be satisfied. The standard resolution for this dilemma is to permit a modicum of deception when there are explicit utilitarian reasons for doing so. Opposition to deception in the codes is de facto redefined in these terms: "The crux of the matter is that some deception, passive or active, enables you to get at data not obtainable by other means" (Punch, 1998, p. 172). As Bulmer (2008) contends,

> As a general principle, the use of deception in research has been condemned. But there are many situations in which it is not possible to be completely open to all participants and sometimes a full explanation of one's purposes would overwhelm the listener. (p. 154)

The general exhortations of codes are considered far removed from the interactional complexities of the field.

3. *Privacy and confidentiality.* Codes of ethics insist on safeguards to protect people's identities and those of the research locations. Confidentiality must be assured as the primary safeguard against unwanted exposure. All personal data ought to be secured or concealed and made public only behind a shield of anonymity. Professional etiquette uniformly concurs that no one deserves harm or embarrassment as a result of insensitive research practices. "The single most likely source of harm in social science inquiry" is the disclosure of private knowledge considered damaging by experimental subjects (Reiss, 1979, p. 73).

As Enlightenment autonomy was developed in philosophical anthropology, a sacred innermost self became essential to the construction of unique personhood. Already in Locke, this private domain received nonnegotiable status. Democratic life was articulated outside these atomistic units, a secondary domain of negotiated contracts and problematic communication. In the logic of social science inquiry revolving around the same understanding of autonomy, invading people's fragile but distinctive privacy is intolerable.

Despite the signature status of privacy protection, watertight confidentiality has proved to be impossible. Pseudonyms and disguised locations are often recognized by insiders. What researchers consider innocent is perceived by participants as misleading or even betrayal. What appears neutral on paper is often conflictual in practice. When government agencies or educational institutions or health organizations are studied, what private parts ought not be exposed? And who is blameworthy if aggressive media carry the research further? Encoding privacy protection is meaningless when there is no distinction between public and private that has consensus any longer (Punch, 1998, p. 175).

4. *Accuracy.* Ensuring that data are accurate is a cardinal principle in social science codes as well. Fabrications, fraudulent materials, omissions, and contrivances are both nonscientific and unethical. Data that are internally and externally valid are the coin of the realm, experimentally and morally. In an instrumentalist, value-neutral social science, the definitions entailed by the procedures themselves establish the ends by which they are evaluated as moral.

Accuracy defined in scientist terms and included in codes of ethics represents a version of Alfred North Whitehead's fallacy of misplaced concreteness. The moral domain becomes equivalent to the epistemological. The unspecifiable abstract is said to have existence in the rigorous concrete. A set of methodological operations becomes normative, and this confusion of categories is both illogical and stale.

Institutional Review Boards

As a condition of funding, government agencies in various countries have insisted that review and monitoring bodies be established by institutions engaged in research involving human subjects. Institutional review boards (IRBs) embody the utilitarian agenda in terms of scope, assumptions, and procedural guidelines.

In 1978, the U.S. National Commission for the Protection of Human Subjects in Biomedical and Behavioral Research was established. As a result, three principles, published in what became known as the Belmont Report, were developed as the moral standards for research involving human subjects: respect for persons, beneficence, and justice.

1. The commitment to respect for persons reiterates the codes' demands that subjects enter the research voluntarily and with adequate information about the experiment's procedures and possible consequences. On a deeper level, respect for persons incorporates two basic ethical tenets: "First, that individuals should be treated as autonomous agents, and second, that persons with diminished autonomy [the immature and incapacitated] are entitled to protection" (University of Illinois, *Investigator Handbook,* 2009).

2. Under the principle of beneficence, researchers are enjoined to secure the well-being of their subjects. Beneficent actions are understood in a double sense as avoiding harm altogether and, if risks are involved for achieving substantial benefits, minimizing as much harm as possible:

> In the case of particular projects, investigators and members of their institutions are obliged to give forethought to the maximization of benefits and the reduction of risks that might occur from the research investigation. In the case of scientific research in general, members of the larger society are obliged to recognize the longer term benefits and risks that may result from the improvement of knowledge and from the development of novel medical, psychotherapeutic, and social procedures. (University of Illinois, *Investigator Handbook,* 2009)

3. The principle of justice insists on fair distribution of both the benefits and burdens of research. An injustice occurs when some groups (e.g., welfare recipients, the institutionalized, or particular ethnic minorities) are overused as research subjects because of easy manipulation or their availability. When research supported by public funds leads to "therapeutic devices and procedures, justice demands that these not provide advantages only to those who can afford them" (University of Illinois, *Investigator Handbook,* 2009).

These principles reiterate the basic themes of value-neutral experimentalism—individual autonomy, maximum benefits with minimal risks, and ethical ends exterior to scientific means. The policy procedures based on them reflect the same guidelines that dominate the codes of ethics: informed consent, protection of privacy, and nondeception. The authority of IRBs was enhanced in 1989 when Congress passed the NIH Revitalization Act and formed the Commission on Research Integrity. The emphasis at that point was on the invention, fudging, and distortion of data. Falsification, fabrication, and plagiarism continue as federal categories of misconduct, with a revised report in 1996 adding warnings against unauthorized use of confidential information, omission of important data, and interference (that is, physical damage to the materials of others).

With IRBs, the legacy of Mill, Comte, and Weber comes into its own. Value-neutral science is accountable to ethical standards through rational procedures controlled by value-neutral academic institutions in the service of an impartial government. Consistent with the way anonymous bureaucratic regimes become refined and streamlined toward greater efficiency, the regulations rooted in scientific and medical experiments now extend to humanistic inquiry. Protecting subjects from physical harm in laboratories has grown to encompass human behavior, history, and ethnography in natural settings. In Jonathon Church's (2002) metaphor, "a biomedical paradigm is used like some threshing machine with ethnographic research the resulting chaff" (p. 2). Whereas Title 45/Part 46 of the Code of Federal Regulations (45 CFR 46) designed protocols for research funded by 17 federal agencies, at present, most universities have multiple project agreements that consign all research to a campus IRB under the terms of 45 CFR 46 (cf. Shopes & Ritchie, 2004).

While this bureaucratic expansion has gone on unremittingly, most IRBs have not changed the composition of their membership. Medical and behavioral scientists under the aegis of value-free neutrality continue to dominate. And the changes in procedures have generally stayed within the biomedical model also. Expedited review under the common rule, for social research with no risk of physical or psychological harm, depends on enlightened IRB chairs and organizational flexibility. Informed consent, mandatory before medical experiments, is simply incongruent with interpretive research that does not reduce humans to subjects but sees itself as collaboration among human beings (Denzin & Giardina, 2007, pp. 20–28).[14] Despite technical improvements,

> Intellectual curiosity remains actively discouraged by the IRB. Research projects must ask only surface questions and must not deviate from a path approved by a remote group of people.... Often the review process seems to be more about gamesmanship than anything else. A better formula for stultifying research could not be imagined. (Blanchard, 2002, p. 11)

In its conceptual structure, IRB utilitarian policy is designed to produce the best ratio of benefits to costs (McIntosh & Morse, 2009, pp. 99–100). IRBs ostensibly protect the subjects who fall under the protocols they approve. However, given the interlocking utilitarian functions of social science, the academy, and the state that Mill identified and promoted, IRBs in reality protect their own institutions rather than subject populations in society at large (see Vanderpool, 1996, Chapters 2 to 6). Only when professional associations like the American Anthropological Association create their own best practices for ethnographic research is the IRB structure pushed in the right direction. Such renovations, however, are contrary to the centralizing homogeneity of closed systems such as the IRBs.

Current Crisis

Mill and Comte, each in his own way, presumed that experimental social science benefited society by uncovering facts about the human condition. Durkheim and Weber believed that a scientific study of society could help people come to grips with the development of big-business monopolies and industrialism. The American Social Science Association was created in 1865 to link "real elements of the truth" with "the great social problems of the day" (Lazarsfeld & Reitz, 1975, p. 1). This myth of beneficence was destroyed with "the revelations at the Nuremberg trials (recounting the Nazis' 'medical experiments' on concentration camp inmates) and with the role of leading scientists in the Manhattan Project" (Punch, 1998, pp. 166–167).

The crisis of confidence multiplied with the exposure of actual physical harm in the Tuskegee Syphilis Study and the Willowbrook Hepatitis Experiment. In the 1960s, Project Camelot, a U.S. Army attempt to use social science to measure and forecast revolutions and insurgency, was bitterly opposed around the world and had to be canceled. Milgram's (1974) deception of unwitting subjects and Laud Humphreys's (1970, 1972) deceptive research on homosexuals in a public toilet and later in their homes, were considered scandalous for psychologically abusing research subjects. Noam Chomsky (1969/2002) exposed the complicity of social scientists with military initiatives in Vietnam.

Vigorous concern for research ethics since the 1980s, support from foundations, and the development of ethics codes and the IRB apparatus are credited by their advocates with curbing outrageous abuses. However, the charges of fraud, plagiarism, and misrepresentation continue on a lesser scale, with dilemmas, conundrums, and controversies unabated over the meaning and application of ethical guidelines. Entrepreneurial faculty competing for scarce research dollars are generally compliant with institutional control, but the vastness of social science activity in universities and research entities makes full supervision impossible.[15]

Underneath the pros and cons of administering a responsible social science, the structural deficiencies in its epistemology have become transparent (Mantzavinos, 2009). A positivistic philosophy of social inquiry insists on neutrality regarding definitions of the good, and this worldview has been discredited. The understanding of society it entails and promotes is inadequate (Winch, 2007). The dominant Enlightenment model, setting human freedom at odds with the moral order, is bankrupt. Even Weber's weaker version of contrastive languages rather than oppositional entities is not up to the task. Reworking the ethics codes so that they are more explicit and less hortatory will make no fundamental difference. Requiring ethics workshops for graduate students and strengthening government policy are desirable but of marginal significance. Refining the IRB process and exhorting IRBs to account for the pluralistic nature of academic research are insufficient.

In utilitarianism, moral thinking and experimental procedures are homogenized into a unidimensional model of rational validation. Autonomous human beings are clairvoyant about aligning means and goals, presuming that they can objectify the mechanisms for understanding themselves and the social world surrounding them (see Winch, 2007, Chapters 3 and 4).[16] This restrictive definition of ethics accounts for some of the goods we seek, such as minimal harm, but those outside a utility calculus are excluded. "Emotionality and intuition" are relegated "to a secondary position" in the decision-making process, for example, and no attention is paid to an "ethics of caring" grounded in "concrete particularities" (Denzin, 1997, p. 273; see also Ryan, 1995, p. 147). The way power and ideology influence social and political institutions is largely ignored. Under a rhetorical patina of deliberate choice and the illusion of autonomous creativity, a means-ends system operates in fundamentally its own terms.

This constricted environment no longer addresses adequately the complicated issues we face in studying the social world. But failure in the War on Poverty, contradictions over welfare, ill-fated studies of urban housing, and the thinness of medical science in health care reform have dramatized the limitations of a utility calculus that occupies the entire moral domain. Certainly, levels of success and failure are open to dispute even within the social science disciplines themselves. More unsettling and threatening to the empirical mainstream than disappointing performance is the recognition that neutrality is not pluralistic but imperialistic. Reflecting on past experience, disinterested research under presumed conditions of value freedom is increasingly seen as de facto reinscribing the agenda in its own terms. Empiricism is procedurally committed to equal reckoning, regardless of how research subjects may constitute the substantive ends of life. But experimentalism is not a neutral meeting ground for all ideas; rather, it is a "fighting creed" that imposes its own ideas on others while uncritically assuming the very "superiority that powers this imposition" (Taylor et al., 1994, pp. 62–63).[17] In Foucault's (1979, pp. 170–195) more decisive terms, social science is a regime of power that helps maintain social order by normalizing subjects into categories designed by political authorities. A liberalism of equality is not

neutral but represents only one range of ideals and is itself incompatible with other goods.

This noncontextual, nonsituational model that assumes "a morally neutral, objective observer will get the facts right" ignores "the situatedness of power relations associated with gender, sexual orientation, class, ethnicity, race, and nationality" (Denzin, 1997, p. 272). It is hierarchical (scientist-subject) and biased toward patriarchy. "It glosses the ways in which the observer-ethnographer is implicated and embedded in the 'ruling apparatus' of the society and the culture" (p. 272). Scientists "carry the mantle" of university-based authority as they venture out into "local community to do research" (Denzin, 1997, p. 272; see also Ryan, 1995, pp. 144–145).[18] There is no sustained questioning of expertise itself in democratic societies that belong in principle to citizens who do not share this specialized knowledge (Pacey, 1996, Chapter 3).

◫ FEMINIST COMMUNITARIANISM

Social Ethics

Over the past decade, social and feminist ethics have made a radical break with the individual autonomy and rationalist presumption of canonical ethics (see Koehn, 1998). The social ethics of Agnes Heller (1988, 1990, 1996, 1999, 2009), Charles Taylor (1989, 1991, 1995, 2007; Taylor et al., 1994), Carole Pateman (1985, 1988, 1989; Pateman & Mills, 2007), Edith Wyschogrod (1985, 1990, 1998, 2002), Kwasi Wiredu (1996), and Cornel West (1989, 1991, 1993/2001) and the feminist ethics of Carol Gilligan (1982, 1983; Gilligan, Ward, & Taylor, 1988), Nel Noddings (1984, 1989, 1990, 2002), Virginia Held (1993), and Seyla Benhabib (1992, 1994, 2002, 2008) are fundamentally reconstructing ethical theory (see Code, 1991; Steiner, 2009). Rather than searching for neutral principles to which all parties can appeal, social ethics rests on a complex view of moral judgments as integrating into an organic whole everyday experience, beliefs about the good, and feelings of approval and shame, in terms of human relations and social structures. This is a philosophical approach that situates the moral domain within the general purposes of human life that people share contextually and across cultural, racial, and historical boundaries (Christians, Glasser, McQuail, Nordenstreng, & White, 2009, Chapters 2 and 3). Ideally, it engenders a new occupational role and normative core for social science research (Gunzenhauser, 2006; White, 1995).

Carol Gilligan (1982, 1983; Gilligan et al., 1988) characterizes the female moral voice as an ethic of care. This dimension of moral development is rooted in the primacy of human relationships. Compassion and nurturance resolve conflicting responsibilities among people, standards totally opposite of merely avoiding harm.[19] In Caring, Nel Noddings (1984) rejects outright the "ethics of principle as ambiguous and unstable" (p. 5), insisting that human care should play the central role in moral decision making. Feminism in Linda Steiner's work critiques the conventions of impartiality and formality in ethics while giving precision to affection, intimacy, nurturing, egalitarian and collaborative processes, and empathy. Feminists' ethical self-consciousness also identifies subtle forms of oppression and imbalance and teaches us to "address questions about whose interests are regarded as worthy of debate" (Steiner, 1989, p. 158; see also Steiner, 1997).

> Feminist approaches to ethics challenge women's subordination, pre-scribe morally justifiable ways of resisting oppressive practices, and envision morally desirable alternatives that promote emancipation.... Fully feminist ethics, far more than their feminine and maternal counterparts, are distinctively political.... A feminist approach to ethics asks questions about power even before it asks questions about good and evil, care and justice, or maternal and paternal thinking. With feminism's persuasive critique of the disembodied ethical subject generating a healthy respect for difference, a multiculturalist feminism may yet construct a non-sexist theory that respects difference of all sorts. (Steiner, 2009, p. 377)

While sharing in the turn away from an abstract ethics of calculation, Charlene Seigfried (1996) argues against the Gilligan-Noddings tradition (cf. Held, 2006). Linking feminism to pragmatism, in which gender is socially constructed, she contradicts "the simplistic equation of women with care and nurturance and men with justice and autonomy" (Seigfried, 1996, p. 206). Gender-based moralities de facto make one gender subservient to another. In her social ethics, gender is replaced with engendering: "To be female or male is not to instantiate an unchangeable nature but to participate in an ongoing process of negotiating cultural expectations of femininity and masculinity" (p. 206). Seigfried challenges us to a social morality in which caring values are central but contextualized in webs of relationships and constructed toward communities with "more autonomy for women and more connectedness for men" (p. 219). Heller and Wyschogrod are two promising examples of proponents of social ethics that meet Seigfried's challenge while confronting forthrightly today's contingency, mass murder, conceptual upheavals in ethics, and hyperreality (cf. Noddings, 2002).

Heller is a former student of Georg Lukács and a dissident in Hungary, who became the Hannah Arendt Professor of Philosophy (Emerita) at the New School for Social Research. Her trilogy developing a contemporary theory of social ethics (Heller, 1988, 1990, 1996) revolves around what she calls the one decisive question: "Good persons exist—how are they possible?" (1988, p. 7). She disavows an ethics of norms, rules, and ideals external to human beings. Only exceptional acts of responsibility under duress and predicaments, each in its own way, are "worthy of theoretical interest" (1996, p. 3). Accumulated wisdom, moral

meaning from our own choices of decency, and the ongoing summons of the Other together reintroduce love, happiness, sympathy, and beauty into a modern, nonabsolutist, but principled theory of morals.

In *Saints and Postmodernism,* Edith Wyschogrod (1990) asserts that anti-authority struggles are possible without assuming that our choices are voluntary. She represents a social ethics of self and Other in the tradition of Emmanuel Levinas (see Wyschogrod, 1974).[20] "The other person opens the venue of ethics, the place where ethical existence occurs" (Wyschogrod, 1990, p. xxi). The Other, "the touchstone of moral existence, is not a conceptual anchorage but a living force." Others function "as a critical solvent;" their existence carries "compelling moral weight" (p. xxi). As a professor of philosophy and religious thought at Rice University, with a commitment to moral narrative, Wyschogrod believes that one venue for Otherness is the saintly life, defined as one in "which compassion for the Other, irrespective of cost to the saint, is the primary trait." Saints put their own "bodies and material goods at the disposal of the Other. . . . Not only do saints contest the practices and beliefs of institutions, but in a more subtle way they contest the order of narrativity itself" (1990, pp. xxii–xxiii).

In addition to the Other, directed across a broad spectrum of belief systems who have "lived, suffered, and worked in actuality," Wyschogrod (1990, p. 7) examines historical narratives for illustrations of how the Other's self-manifestation is depicted. Her primary concern is the way communities shape shared experience in the face of cataclysms and calamities, arguing for historians who situate themselves "in dynamic relationship to them" (1998, p. 218). The overriding challenge for ethics, in Wyschogrod's view, is how researchers enter into communities that create and sustain hope in terms of immediacy—"a presence here and now" but "a presence that must be deferred" to the future (1998, p. 248). Unless it is tangible and actionable, hope serves those in control. Hope that merely projects a future redemption obscures abuses of power and human need in the present.

Martin Buber (1958) calls the human relation a primal notion in his famous lines, "in the beginning is the relation" and "the relation is the cradle of life" (pp. 69, 60). Social relationships are preeminent. "The one primary word is the combination I-Thou" (p. 3). This irreducible phenomenon—the relational reality, the in-between, the reciprocal bond, the interpersonal—cannot be decomposed into simpler elements without destroying it.[21] Given the primacy of relationships, unless we use our freedom to help others flourish, we deny our own well-being (cf. Verlinden, 2008, pp. 201–210).

Rather than privileging an abstract rationalism, the moral order is positioned close to the bone, in the creaturely and corporeal rather than the conceptual. "In this way, ethics . . . is as old as creation. Being ethical is a primordial movement in the beckoning force of life itself" (Olthuis, 1997, p. 141). The ethics of Levinas is one example:

> The human face is the epiphany of the nakedness of the Other, a visitation, a meeting, a saying which comes in the passivity of the face, not threatening, but obligating. My world is ruptured, my contentment interrupted. I am already obligated. Here is an appeal from which there is no escape, a responsibility, a state of being hostage. It is looking into the face of the Other that reveals the call to a responsibility that is before any beginning, decision or initiative on my part. (Olthuis, 1997, p. 139)

Humans are defined as communicative beings within the fabric of everyday life. Through dialogic encounter, subjects create life together and nurture one another's moral obligation to it. Levinas's ethics presumes and articulates a radical ontology of social beings in relation (see, e.g., Levinas, 1985, 1991).

Moreover, in Levinasian terms, when I turn to the face of the Other, I not only see flesh and blood, but a third party arrives— the whole of humanity. In responding to the Other's need, a baseline is established across the human race. For Benhabib (1992, cf. 1994), this is interactive universalism.[22] Our universal solidarity is rooted in the principle that "we have inescapable claims on one another which cannot be renounced except at the cost of our humanity" (Peukert, 1981, p.11).

A Feminist Communitarian Model

Feminist communitarianism is Norman Denzin's (1997, pp. 274–287; 2003, pp. 242–258; 2009, pp. 155–162) label for the ethical theory to lead us forward at this juncture (Christians, 2002b).[23] This is a normative model that serves as an antidote to individualist utilitarianism. It presumes that the community is ontologically and axiologically prior to people. Human identity is constituted through the social realm, and human bonding is the epicenter of social formation. We are born into a sociocultural universe where values, moral commitments, and existential meanings are negotiated dialogically. Feminist communitarianism "embodies a sacred, existential epistemology that locates persons in a noncompetitive, nonhierarchical relationship to the larger moral universe" (Denzin, 2009, p. 158). Moral reasoning does not depend on formal consensus but goes forward because reciprocal care and understanding make moral discourse possible. Every communal act is measured against the ideals of a universal respect for the dignity of all human beings, regardless of gender, age, race, or religion (see Benhabib, 1992, Chapter 1).

For communitarians, the liberalism of Locke and Mill confuses an aggregate of individual pursuits with the common good (Christians, Ferre, & Fackler, 1993, Chapter 1). Moral agents need a context of social commitments and community ties for assessing what is valuable. What is worth preserving as a good cannot be self-determined in isolation; it can be ascertained only within specific social situations where human identity is nurtured. The public sphere is conceived as a mosaic of particular communities, a pluralism of ethnic identities and worldviews intersecting to form a social bond but each seriously held and competitive as well. Rather than pay lip service to the social nature of the self while presuming a dualism of two orders, communitarianism interlocks personal autonomy with communal well-being. Morally appropriate action intends community. Common moral values are intrinsic to a community's ongoing existence and identity.

Therefore, the mission of social science research is enabling community life to prosper—equipping people to come to mutually held conclusions. The aim is not fulsome data per se, but community transformation. The received view assumes that research advances society's interests by feeding our individual capacity to reason and make calculated decisions. Instead of moving forward with IRB approval of human subjects, research is intended to be collaborative in its design and participatory in its execution. Rather than having their concerns defined by ethics codes in the files of academic offices and distributed in research reports prepared for clients, the participants themselves are given a forum to activate the polis mutually. In contrast to utilitarian experimentalism, the substantive conceptions of the good that drive the problems reflect the conceptions of the community rather than the expertise of researchers or funding agencies.

In the feminist communitarian model, participants have a say in how the research should be conducted and are involved in actually conducting it. Participants offer "a voice or hand in deciding which problems should be studied, what methods should be used to study them, whether the findings are valid or acceptable, and how the findings are to be used or implemented" (Root, 1993, p. 245). This research is rooted in "community, shared governance . . . and neighborliness." Given its cooperative mutuality, it serves "the community in which it is carried out, rather than the community of knowledge producers and policymakers" (Lincoln, 1995, pp. 280, 287). It finds its genius in the maxim that "persons are arbitrators of their own presence in the world" (Denzin, 1989, p. 81).

For feminist communitarians, research becomes "a civic, participatory, collaborative project. It uses democratically arrived at, participant-driven criteria of evaluation" (Denzin, 2009, p. 158). Researchers and subjects become "coparticipants in a common moral project." Ethnographic inquiry is "characterized by shared ownership of the research project, community-based analyses, an emancipatory, dialectical, and transformative commitment" to social action (Denzin, 2009, p. 158; see also Denzin, 1984, p. 145; Reinharz, 1993). This collaborative research model "makes the researcher responsible not to a removed discipline (or institution), but to those he or she studies." It aligns the ethics of research "with a politics of resistance, hope and freedom" (Denzin, 2003, p. 258).

▣ INTERPRETIVE SUFFICIENCY

Within a feminist communitarian model, the mission of social science research is interpretive sufficiency. In contrast to an experimentalism of instrumental efficiency, this paradigm seeks to open up the social world in all its dynamic dimensions. The thick notion of sufficiency supplants the thinness of the technical, exterior, and statistically precise received view. Rather than reducing social issues to financial and administrative problems for politicians, social science research enables people to come to terms with their everyday experience themselves.

Interpretive sufficiency means taking seriously lives that are loaded with multiple interpretations and grounded in cultural complexity. Ethnographic accounts should, therefore, "possess that amount of depth, detail, emotionality, nuance, and coherence that will permit a critical consciousness to be formed by the reader. Such texts should also exhibit representational adequacy, including the absence of racial, class, and gender stereotyping" (Denzin, 1997, p. 283; see 1989, pp. 77–81).

From the perspective of a feminist communitarian ethics, interpretive discourse is authentically sufficient when it fulfills three conditions: represents multiple voices, enhances moral discernment, and promotes social transformation. Consistent with the community-based norms advocated here, the focus is not on professional ethics per se but on the general morality. When feminist communitarianism is integrated with non-Enlightenment communal concepts such as *ubuntu* (from the Zulu maxim *umuntu ngumuntu ngabantu,* "a person is a person through other persons" or "I am because of others"), a dialogic ethics is formed that expands the general morality to the human race as a whole (Christians, 2004).

Multivocal and Cross-Cultural Representation

Within social and political entities are multiple spaces that exist as ongoing constructions of everyday life. The dialogical self is situated and articulated within these decisive contexts of gender, race, class, and religion. In contrast to contractarianism, where tacit consent or obligation is given to the state, promises are made and sustained to one another. Research narratives reflect a community's multiple voices through which promise-keeping takes place.

In Carole Pateman's communitarian philosophy, sociopolitical entities are not to be understood first of all in terms of contracts. Making promises is one of the basic ways in which consenting human beings "freely create their own social relationships" (Pateman, 1989, p. 61; see also Pateman, 1985, pp. 26–29). We assume an obligation by making a promise. When individuals promise, they are obliged to act accordingly. But promises are primarily made not to authorities through political contracts, but to fellow citizens. If obligations are rooted in promises, obligations are owed to other colleagues in institutions and to participants in community practices. Therefore, only under conditions of participatory democracy can there be self-assumed moral obligation.

Pateman understands the nature of moral agency. We know ourselves primarily in relation and derivatively as thinkers withdrawn from action. Only by overcoming the traditional dualisms between thinker and agent, mind and body, reason and will, can we conceive of being as "the mutuality of personal relationships" (MacMurray, 1961a, p. 38). Moral commitments arise out of action and return to action for their incarnation and verification. From a dialogical perspective, promise-keeping through action and everyday language is not a supercilious pursuit because our way of being is not inwardly generated but socially derived.

> We become full human agents, capable of understanding ourselves, and hence of defining our identity, through . . . rich modes of expression we learn through exchange with others. . . .
>
> My discovering my own identity doesn't mean that I work it out in isolation, but that I negotiate it through dialogue, partly overt, partly internal, with others. My own identity crucially depends on my dialogical relations with others. . . .
>
> In the culture of authenticity, relationships are seen as the key loci of self-discovery and self-affirmation. (Taylor et al., 1994, pp. 32, 34, 36)

If moral bondedness flows horizontally and obligation is reciprocal in character, the affirming and sustaining of promises occurs cross-culturally. But the contemporary challenge of cultural diversity has raised the stakes and made easy solutions impossible. One of the most urgent and vexing issues on the democratic agenda at present is not just how to meet the moral obligation to treat ethnic differences with fairness but how to recognize explicit cultural groups politically (Benhabib, 2002, 2008).

Communitarianism as the basis for ethnic plurality rejects melting pot homogeneity and replaces it with the politics of recognition. The basic issue is whether democracies are discriminating against their citizens in an unethical manner when major institutions fail to account for the identities of their members (Taylor et al., 1994, p. 3). In what sense should the specific cultural and social features of African Americans, Asian Americans, Native Americans, Buddhists, Jews, the physically disabled, or children matter publicly? Should not public institutions ensure only that democratic citizens share an equal right to political liberties and due process without regard to race, gender, or religion? Beneath the rhetoric is a fundamental philosophical dispute that Taylor calls the "politics of recognition." As he puts it, "Nonrecognition or misrecognition can inflict harm, can be a form of oppression, imprisoning someone in a false, distorted, and reduced mode of being. Due recognition is not just a courtesy we owe people. It is a vital human need" (Taylor et al., 1994, p. 26). This foundational issue regarding the character of cultural identity needs to be resolved for cultural pluralism to come into its own. Feminist communitarianism is a nonassimilationist framework in which such resolution can occur.

However, liberal proceduralism cannot meet this vital human need. Emphasizing equal rights with no particular substantive view of the good life "gives only a very restricted acknowledgement of distinct cultural identities" (Taylor et al., 1994, p. 52). Insisting on neutrality, and without collective goals, produces at best personal freedom, safety, and economic security understood homogeneously. As Bunge (1996) puts it, "Contractualism is a code of behavior for the powerful and the hard—those who write contracts, not those who sign on the dotted line" (p. 230). However, in promise-based communal formation the flourishing of particular cultures, religions, and ethnic groups is the substantive goal to which we are morally committed as human beings.

Denzin (2002) demonstrates how multicultural representation ought to operate in the media's construction of the American racial order. An ethnic cinema that honors racial difference is not assimilationist, nor does it "celebrate exceptional blackness" supporting white values; and it refuses to pit "the ethnic other against a mainstream white America" as well as "dark skin against dark skin" (p. 6). Rather than "a didactic film aesthetic based on social problems realism"—one that is "trapped by the modernist agenda"—Denzin follows Hal Foster and bell hooks in arguing for an anti-aesthetic or postmodern aesthetic that is cross-disciplinary, is oriented to the vernacular, and denies "the idea of a privileged aesthetic realm" (pp. 11, 180). A "feminist, Chicana/o and black performance-based aesthetic" creates "a critical counter-hegemonic race consciousness" and implements critical race theory (p. 180).

In feminist communitarian terms, this aesthetic is simultaneously political and ethical. Racial difference is imbricated in social theories and in conceptions of the human being, of justice, and the common good. It requires an aesthetic that "in generating social criticism ... also engenders resistance" (p. 181). It is not a "protest or integrationist initiative" aimed at "informing a white audience of racial injustice," but instead "offers new forms of representation that create the space for new forms of critical race consciousness" (p. 182). The overarching standard made possible by this aesthetic is enhancing moral agency, that is, serving as a catalyst for moral discernment (Christians, 2002a, p. 409).

With the starting hypothesis that all human cultures have something important to say, social science research recognizes particular cultural values consistent with universal human dignity (Christians, 1997, pp. 11–14; 2008, pp. 16–17). Interpretive sufficiency in its multicultural dimension helps people in their home territory see how life could be different. This framework "imagines new forms of human transformation and emancipation" (Denzin, 2009, p. 158). These transformations are enacted "through dialogue. If necessary, it sanctions nonviolent forms of civil disobedience." In its "asking that interpretive work provide the foundations for social criticism and social action, this ethic represents a call to action" (Denzin, 2009, p. 158).

Moral Discernment

Societies are embodiments of institutions, practices, and structures recognized internally as legitimate. Without allegiance to a web of ordering relations, society becomes, as a matter of fact, inconceivable. Communities not only constitute linguistic entities but also require at least a minimal moral commitment to the common good. Because social entities are moral orders and not merely functional arrangements, moral commitment constitutes the self-in-relation. Our identity is defined by what we consider good or worth opposing. Only through the moral dimension can we make sense of human agency. As Stephen Mulhall and Adam Swift (1996) write:

> Developing, maintaining and articulating [our moral intuitions and reactions] is not something humans could easily or even conceivably dispense with. . . . We can no more imagine a human life that fails to address the matter of its bearings in moral space than we can imagine one in which developing a sense of up and down, right and left is regarded as an optional human task. . . . A moral orientation is inescapable because the questions to which the framework provides answers are themselves inescapable. (pp. 106–108; see also Taylor, 1989, pp. 27–29)

A self exists only within "webs of interlocution," and all self-interpretation implicitly or explicitly "acknowledges the necessarily social origin of any and all their conceptions of the good and so of themselves" (Mulhall & Swift, 1996). Moral frameworks are as fundamental for orienting us in social space as the need to establish ourselves in physical space. The moral dimension must, therefore, be considered intrinsic to human beings, not a system of rules, norms, and ideals external to society. Moral duty is nurtured by the demands of social linkage and not produced by abstract theory.

The core of a society's common morality is pretheoretical agreement. However, "what counts as common morality is not only imprecise but variable . . . and a difficult practical problem" (Bok, 1995, p. 99). Moral obligation must be articulated within the fallible and irresolute voices of everyday life. Among disagreements and uncertainty, we look for criteria and wisdom in settling disputes and clarifying confusions; and normative theories of an interactive sort can invigorate our common moral discourse. But generally accepted theories are not necessary for the common good to prosper. The common good is not "the complete morality of every participant . . . but a set of agreements among people who typically hold other, less widely shared ethical beliefs" (Bok, 1995, p. 99). Instead of expecting more theoretical coherence than history warrants, Reinhold Niebuhr inspires us to work through inevitable social conflicts while maintaining "an untheoretical jumble of agreements" called here the common good (Barry, 1967, pp. 190–191). Through a common morality, we can approximate consensus on issues and settle disputes interactively. In Jürgen Habermas's (1993) terms, discourse in the public sphere must be oriented "toward mutual understanding" while allowing participants "the communicative freedom to take positions" on claims to final validity (p. 66; see also Habermas, 1990).

Communitarians challenge researchers to participate in a community's ongoing process of moral articulation. Conceptions of the good are shared by researchers and subjects, both of them self-reflexive and collaborating to bring moral issues to clarity. In fact, culture's continued existence depends on this type of identification and defense of its normative base (Fackler, 2009, pp. 312–315). Therefore, ethnographic texts must enable us "to discover moral truths about ourselves"; narratives ought to "bring a moral compass into readers' lives" by accounting for things that matter to them (Denzin, 1997, p. 284). Feminist communitarianism seeks to engender moral reasoning internally. Communities are woven together by narratives that invigorate their common understanding of good and evil, happiness and reward, the meaning of life and death. Recovering and refashioning moral vocabulary helps to amplify our deepest humanness. Researchers are not constituted as ethical selves antecedently, but moral discernment unfolds dialectically between researchers and the researched who collaborate with them.

Our widely shared moral convictions are developed through discourse within a community. These communities, where moral discourse is nurtured and shared, are a radical alternative to the utilitarian individualism of modernity. But in feminist communitarianism, communities are entered from the universal. The total opposite of an ethics of individual autonomy is universal human solidarity. Our obligation to sustain one another defines our existence. The primal sacredness of all without exception is the heart of the moral order and the new starting point for our theorizing (Christians, 1998, 2008).

The rationale for human action is reverence for life on earth. Living nature reproduces itself as its very character. Embedded in the animate world is the purposiveness of bringing forth life. Therefore, within the natural order is a moral claim on us for its own sake and in its own right. Nurturing life has a taken-for-granted character outside subjective preferences. Reverence for life on earth is a pretheoretical given that makes the moral order possible. The sacredness of life is not an abstract imperative but the ground of human action.[24] It is a primordial generality that underlies reification into ethical principles, an organic bond that everyone shares inescapably. In our systematic reflection on this protonorm, we recognize that it entails such basic ethical principles as human dignity and nonviolence (Christians, Rao, Ward, & Wasserman, 2009, pp. 143–145).

Reverence for life on earth establishes a level playing floor for cross-cultural collaboration in ethics. It represents a universalism from the ground up. Various societies articulate this protonorm in different terms and illustrate it locally, but every culture can bring to the table this fundamental norm for ordering political relationships and social institutions. We live out our values in a community setting where the moral life is experienced and a moral vocabulary articulated. Such protonorms as reverence for life can be recovered only locally. Language situates them in history. The sacredness of life reflects our common condition as a species, but we act on it through the immediate reality of geography, ethnicity, and ideology (Fackler, 2003). But according to feminist communitarianism, if we enter this communal arena not from individual decision making but from a universal commonness, we have the basis for believing that researchers and the researched can collaborate on the moral domain. Researchers do not bring a set of prescriptions into which they school their subjects. Instead they find ways interactively to bring the sacredness of life into its own—each culture and all circumstances providing an abundance of meaning and application.

How the moral order works itself out in community formation is the issue, not, first of all, what researchers consider virtuous. The challenge for those writing culture is not to limit their moral perspectives to their own generic and neutral principles but to engage the same moral space as the people they study. In this perspective, research strategies are not assessed, first of all, in terms of statistical sophistication, but for their vigor in illuminating how communities can flourish.

Politics of Resistance

Ethics in the feminist communitarian mode generates social criticism, leads to resistance, and empowers to action those who are interacting (see Habermas, 1971, pp. 301–317). Thus a basic norm for interpretive research is enabling the humane transformation of the multiple spheres of community life, such as religion, politics, ethnicity, and gender.

From his own dialogic perspective, Paulo Freire speaks of the need to reinvent the meaning of power:

> For me the principal, real transformation, the radical transformation of society in this part of the century demands not getting power from those who have it today, or merely to make some reforms, some changes in it. . . . The question, from my point of view, is not just to take power but to reinvent it. That is, to create a different kind of power, to deny the need power has as if it were metaphysics, bureaucratized, and anti-democratic. (quoted in Evans, Evans, & Kennedy, 1987, p. 229)

Certainly, oppressive power blocs and monopolies—economic, technological, and political—need the scrutiny of researchers and their collaborators. Given Freire's political-institutional bearing, power for him is a central notion in social analysis. But, in concert with him, feminist communitarian research refuses to deal with power in cognitive terms only. The issue is how people can empower themselves instead.

The dominant understanding of power is grounded in nonmutuality; it is interventionist power, exercised competitively and seeking control. In the communitarian alternative, power is relational, characterized by mutuality rather than sovereignty. Power from this perspective is reciprocity between two subjects, a relationship not of domination but of intimacy and vulnerability—power akin to that of Alcoholics Anonymous, in which surrender to the community enables the individual to gain mastery. In these terms, Cannella and Lincoln (2009) challenge us to "construct critical research that does not simultaneously create new forms of oppressive power for itself or for its practitioners" (p. 54). The indigenous Kaupapa Maori approach to research meets this standard: "The researcher is led by the members of the community and does not presume to be a leader, or to have any power that he or she can relinquish" (Denzin, 2003, p. 243).

Dialogue is the key element in an emancipatory strategy that liberates rather than imprisons us in manipulation or antagonistic relationships. Although the control version of power considers mutuality weakness, the empowerment mode maximizes our humanity and thereby banishes powerlessness. In the research process, power is unmasked and engaged through solidarity as a researched-researcher team. There is certainly no monologic "assumption that the researcher is giving the group power" (Denzin, 2003, p. 243). Rather than play semantic games with power, researchers themselves are willing to march against the barricades. As Freire insists, only with everyone filling his or her own political space, to the point of civil disobedience as necessary, will empowerment mean anything revolutionary (in McLaren & Leonard, 1993, Chapters 8, 10).

What is nonnegotiable in Freire's theory of power is participation of the oppressed in directing cultural formation (Stefanos, 1997). If an important social issue needs resolution, the most vulnerable will have to lead the way: "Revolutionary praxis cannot tolerate an absurd dichotomy in which the praxis of the people is merely that of following the [dominant elite's] decisions" (Freire, 1970a, p.120; see also Freire, 1978, pp. 17ff.).[25] Arrogant politicians—supported by a bevy of accountants, lawyers, economists, and social science researchers—trivialize the nonexpert's voice as irrelevant to the problem or its solution. On the contrary, transformative action from the inside out is impossible unless the oppressed are active participants rather than a leader's object. "Only power that springs from the weakness of the oppressed will be sufficiently strong to free both" (Freire, 1970b, p. 28).[26]

In Freire's (1973) terms, the goal is conscientization, that is, a critical consciousness that directs the ongoing flow of praxis and reflection in everyday life. In a culture of silence, the oppressor's language and way of being are fatalistically accepted without contradiction. But a critical consciousness enables us to exercise the uniquely human capacity of "speaking a true word" (Freire, 1970b, p. 75). Under conditions of sociopolitical control, "the vanquished are dispossessed of their word, their expressiveness, their culture" (1970b, p. 134). Through conscientization, the oppressed gain their own voice and collaborate in transforming their culture (1970a, pp. 212–213). Therefore, research is not the transmission of specialized data but, in style and content, a catalyst for critical consciousness. Without what Freire (1970b, p. 47) calls "a critical comprehension of reality" (that is, the oppressed "grasping with their minds the truth of their reality"), there is only acquiescence in the status quo.

The resistance of the empowered is more productive at the interstices—at the fissures in social institutions where authentic action is possible. Effective resistance is nurtured in the backyards, the open spaces, and voluntary associations, among neighborhoods, schools, and interactive settings of mutual struggle without elites. Since only nonviolence is morally acceptable for sociopolitical change, there is no other option except an educational one—having people movements gain their own voice and nurturing a critical conscience through dialogic means. People-based development from below is not merely an end in itself, but a fundamental condition of social transformation. "We are no longer called to just interpret the world"; rather than be limited to this mandate of traditional ethnography, "we are called to change the world and to change it in ways that resist injustice" (Denzin & Giardina, 2009, p. 23). In seeking research strategies of this kind, Guba (1990) insists correctly on a dialogic framework, a conversation of peace and hope "that will move us to new, more informed, and more sophisticated empowerment paradigms" (p. 27).

◙ CONCLUSION

As Guba and Lincoln (1994) argue, the issues in social science ultimately must be engaged at the worldview level. "Questions of method are secondary to questions of paradigm, which we define as the basic belief system or worldview that guides the investigator, not only in choices of method but in ontologically and epistemologically fundamental ways" (p. 105). The conventional view, with its extrinsic ethics, gives us a truncated and unsophisticated paradigm that needs to be ontologically transformed. This historical overview of theory and practice points to the need for an entirely new model of research ethics in which human action and conceptions of the good are interactive.

"Since the relation of persons constitutes their existence as persons, . . . morally right action is [one] which intends community" (MacMurray, 1961b, p. 119). In feminist communitarianism, personal being is cut into the very heart of the social universe. The common good is accessible to us only in personal form; it has its ground and inspiration in a social ontology of the human.[27] "Ontology must be rescued from submersion in things by being thought out entirely from the viewpoint of person and thus of Being" (Lotz, 1963, p. 294). "Ontology is truly itself only when it is personal, and persons are truly themselves only as ontological" (Lotz, 1963, p. 297).

When rooted in a positivist or postpositivist worldview, explanations of social life are considered incompatible with the renderings offered by the participants themselves. In problematics, lingual form, and content, research production presumes greater mastery and clearer illumination than the nonexperts who are the targeted beneficiaries. Protecting and promoting individual autonomy has been the philosophical rationale for value neutrality since its origins in Mill. But the incoherence in that view of social science is now transparent. By limiting the active involvement of rational beings or judging their self-understanding to be false, empiricist models contradict the ideal of rational beings who "choose between competing conceptions of the good" and make choices "deserving of respect" (Root, 1993, p. 198). The verification standards of an instrumentalist system "take away what neutrality aims to protect: a community of free and equal rational beings legislating their own principles of conduct" (Root, 1993, p. 198). The social ontology of feminist communitarianism escapes this contradiction by reintegrating human life with the moral order.

Freed from neutrality and a superficial instrumentalism, the ethics of feminist communitarianism participates in the revolutionary social science advocated by Cannella and Lincoln (2009):

Research conceptualizations, purposes, and practices would be grounded in critical ethical challenges to social (therefore science) systems, supports for egalitarian struggle, and revolutionary ethical awareness and activism from within the context of community. Research would be relational (often as related to community) and grounded within critique of systems, egalitarian struggle, and revolutionary ethics. (p. 68)

In this form, the positivist paradigm is turned upside down intellectually, and qualitative research advances social justice and is grounded in hope (Denzin & Giardina, 2009, pp. 41–42). Denzin, Yvonna Lincoln, and Linda Tuhiwai Smith (2008) correctly locate the politics and ethics of this chapter in global terms. For them, Occidental social scientists advocating alternative interpretive research "and indigenous communities alike have been moving toward the same goals." They both "seek a set of ethical principles that are feminist, caring, communitarian, holistic, respectful, mutual (rather than power imbalanced), sacred, and ecologically sound" (p. 569).

▣ Notes

1. Michael Root (1993) is unique among philosophers of the social sciences in linking social science to the ideals and practices of the liberal state on the grounds that both institutions "attempt to be neutral between competing conceptions of the good" (p. xv). As he elaborates: "Though liberalism is primarily a theory of the state, its principles can be applied to any of the basic institutions of a society; for one can argue that the role of the clinic, the corporation, the scholarly associations, or professions is not to dictate or even recommend the kind of life a person should aim at. Neutrality can serve as an ideal for the operations of these institutions as much as it can for the state. Their role, one can argue, should be to facilitate whatever kind of life a student, patient, client, customer, or member is aiming at and not promote one kind of life over another" (p. 13). Root's interpretations of Mill and Weber are crucial to my own formulation.

2. Although committed to what he called "the logic of the moral sciences" in delineating the canons or methods for induction, Mill shared with natural science a belief in the uniformity of nature and the presumption that all phenomena are subject to cause-and-effect relationships. His five principles of induction reflect a Newtonian cosmology.

3. Utilitarianism in John Stuart Mill was essentially an amalgamation of Jeremy Bentham's greatest happiness principle, David Hume's empirical philosophy and concept of utility as a moral good, and Comte's positivist tenets that things-in-themselves cannot be known and knowledge is restricted to sensations. In his influential *A System of Logic,* Mill (1843/1893) is typically characterized as combining the principles of French positivism (as developed by Comte) and British empiricism into a single system.

4. For an elaboration of the complexities in positivism—including reference to its Millian connections—see Lincoln and Guba (1985, pp. 19–28).

5. Mill's realism is most explicitly developed in his *Examination of Sir William Hamilton's Philosophy* (1865b). Our belief in a common external world, in his view, is rooted in the fact that our sensations of physical reality "belong as much to other human or sentient beings as to ourselves" (p. 196; see also Copleston, 1966, p. 306, note 97).

6. Mill (1873/1969) specifically credits Comte for his use of the inverse deductive or historical method: "This was an idea entirely new to me when I found it in Comte; and but for him I might not soon (if ever) have arrived at it" (p. 126). Mill explicitly follows Comte in distinguishing social statics and social dynamics. He published two essays on Comte's influence in the *Westminster Review,* which were reprinted as *Auguste Comte and Positivism* (Mill, 1865a; see also Mill, 1873/1969, p. 165).

7. Emile Durkheim is more explicit and direct about causality in both the natural and the social worlds. While he argues for sociological over psychological causes of behavior and did not believe intention could cause action, he unequivocally sees the task of social science as discovering the causal links between social facts and personal behavior (see, e.g., Durkheim, 1966, pp. 44, 297–306).

8. As one example of the abuse Weber resisted, Root (1993, pp. 41–42) refers to the appointment of Ludwig Bernhard to a professorship of economics at the University of Berlin. Although he had no academic credentials, the Ministry of Education gave Bernhard this position without a faculty vote (see Weber, 1973, pp. 4–30). In Shils's (1949) terms, "A mass of particular, concrete concerns underlies [his 1917] essay—his recurrent effort to penetrate to the postulates of economic theory, his ethical passion for academic freedom, his fervent nationalist political convictions, and his own perpetual demand for intellectual integrity" (p. v).

9. The rationale for the Social Science Research Council in 1923 is multilayered, but in its attempt to link academic expertise with policy research, and in its preference for rigorous social scientific methodology, the SSRC reflects and implements Weber.

10. In *Utilitarianism,* Chapter 4, Mill (1861/1957) drew an analogy between visibility and desirability to prove the utilitarian moral standard. He argued that the proof an object is visible is the fact that people in real life actually see it. By analogy, the proof that something is desirable is people actually desiring it. Therefore, since people do in fact desire happiness, happiness must be desirable or good. As Harris (2006, p. 142) and others have argued, although visibility/desirability illustrates Mill's empiricism, his intended proof is not convincing. Insisting that something is actually desired by people does not mean it should be desired. People often desire what they should not. My desiring happiness does not itself make the promotion of happiness a moral obligation for me or in general.

11. Often in professional ethics at present, we isolate consequentialism from a full-scale utilitarianism. We give up on the idea of maximizing happiness, but "still try to evaluate different courses of action purely in terms of their consequences, hoping to state everything worth considering in our consequence-descriptions" (Taylor, 1982, p. 144). However, even this broad version of utilitarianism, in Taylor's terms, "still legislates certain goods out of existence" (p. 144). It is likewise a restrictive definition of the good that favors the mode of reasoned calculation and prevents us from taking seriously all facets of moral and normative political thinking (p. 144). As Lincoln observes, utilitarianism's inescapable problem is that "in advocating the greatest good for the greatest number, small groups of people (all minority groups, for example) experience the political regime of the 'tyranny of the majority.'" She refers correctly to "liberalism's tendency to reinscribe oppression by virtue of the utilitarian principle" (personal communication, February 16, 1999).

12. John Rawls's (1971) justice-based moral theory is also a compelling critique of utilitarianism. Utilitarianism is a teleological theory and Rawls's justice-as-fairness is deontological. Rawls needs to be elaborated in debates over moral theory itself. Taylor and Ross are included here since they are more explicitly epistemological, interlacing Mill's empiricism and utilitarianism.

13. Given the nature of positivist inquiry, Jennings and Callahan (1983) conclude that only a short list of ethical questions are considered, and they "tend to merge with the canons of professional scientific methodology.... Intellectual honesty, the suppression of personal bias, careful collection and accurate reporting of data, and candid admission of the limits of the scientific reliability of empirical studies— these were essentially the only questions that could arise. And, since these ethical responsibilities are not particularly controversial (at least in principle), it is not surprising that during this period [the 1960s] neither those concerned with ethics nor social scientists devoted much time to analyzing or discussing them" (p. 6).

14. Most biomedical research occurs in a laboratory. Researchers are obliged to inform participants of potential risk and obtain consent before the research takes place. Ethnographic research occurs in settings where subjects live, and informed consent is a process of "ongoing interaction between the researcher and the members of the community being studied. . . . One must establish bonds of trust and negotiate consent . . . taking place over weeks or months—not prior to a structured interview" (Church, 2002, p. 3).

15. As Taylor (1982) puts it, "The modern dispute about utilitarianism is not about whether it occupies some of the space of moral reason, but whether it fills the whole space." "Comfort the dying" is a moral imperative in contemporary Calcutta, even though "the dying are in an extremity that makes [utilitarian] calculation irrelevant" (p. 134).

16. While rejecting this utilitarian articulation of means to ends, a philosophical critique of the means-ends trajectory is necessary for this rejection to have long-term credibility. Drescher (2006, pp. 183–188) represents a recent review of the means-ends relation, establishing criteria in rationalist terms.

17. This restates the well-known objection to a democratic liberalism of individual rights: "Liberalism is not a possible meeting ground for all cultures, but is the political expression of one range of cultures, and quite incompatible with other ranges. Liberalism can't and shouldn't claim complete cultural neutrality. Liberalism is also a fighting creed. Multiculturalism as it is often debated today has a lot to do with the imposition of some cultures on others, and with the assumed superiority that powers this imposition. Western liberal societies are thought to be supremely guilty in this regard, partly because of their colonial past, and partly because of their marginalization of segments of their populations that stem from other cultures" (Taylor et al., 1994, pp. 62–63).

18. Denzin in this passage credits Smith (1987, p. 107) with the concept of a "ruling apparatus."

19. Gilligan's research methods and conclusions have been debated by a diverse range of scholars. For this debate and related issues, see Brabeck (1990), Card (1991), Tong (1989, pp. 161–168; 1993, pp. 80–157), Seigfried (1996), and Wood(1994).

20. Levinas (b. 1905) was a professor of philosophy at the University of Paris (Nanterre) and head of the Israelite Normal School in Paris. In Wyschogrod's (1974) terms, "He continues the tradition of Martin Buber and Franz Rosenweig" and was "the first to introduce Husserl's work into . . . the French phenomenological school" (pp. vii–viii). Although Wyschogrod is a student of Martin Heidegger, Georg Wilhelm Friedrich Hegel, and Edmund Husserl (see, e.g., Wyschogrod, 1985)— and engages Jacques Derrida, Jean-François Lyotard, Michel Foucault, and Gilles Deleuze—her work on ethics appeals not to traditional philosophical discourse but to concrete expressions of self-Other transactions in the visual arts, literary narrative, historiography, and the normalization of death in the news.

21. Levinas sees the irreducibility of the I-Thou relation as a critical contribution to the history of ideas: "The dialogical relation and its phenomenological irreducibility . . . will remain the unforgettable contribution of Martin Buber's philosophical labours. . . . Any reflection of the alterity of the others in his or her irreducibility to the objectivity of objects and the being of beings must recognize the new perspective Buber opened" (Levinas, as cited in Friedman, 2002, p. 338).

22. Martha Nussbaum (1993) argues for a version of virtue ethics in these terms, contending for a model rooted in Aristotle that has cross-cultural application without being detached from particular forms of social life. In her model, various spheres of human experience that are found in all cultures represent questions to answer and choices to make—attitudes toward the ill or good fortune of others, how to treat strangers, management of property, control over bodily appetites, and so forth. Our experiences in these areas "fix a subject for further inquiry" (p. 247), and our reflection on each sphere will give us a "thin or nominal definition" of a virtue relevant to this

sphere. On this basis, we can talk across cultures about behavior appropriate in each sphere (see Nussbaum, 1999).

23. Root (1993, Chapter 10) also chooses a communitarian alternative to the dominant paradigm. In his version, critical theory, participatory research, and feminist social science are three examples of the communitarian approach. This chapter offers a more complex view of communitarianism developed in political philosophy and intellectual history, rather than limiting it to social theory and practical politics. Among the philosophical communitarians (Sandel, 1982/1998; Taylor, 1989; Walzer, 1983, 1987), Pateman (1985, 1989) is explicitly feminist, and her promise motif forms the axis for the principle of multivocal representation outlined below. In this chapter's feminist communitarian model, critical theory is integrated into the third ethical imperative—empowerment and resistance. In spite of that difference in emphasis, I agree with Root's (1993) conclusion: "Critical theories are always critical for a particular community, and the values they seek to advance are the values of that community. In that respect, critical theories are communitarian. . . . For critical theorists, the standard for choosing or accepting a social theory is the reflective acceptability of the theory by members of the community for whom the theory is critical" (pp. 233–234). For a review of communitarian motifs in terms of Foucault, see Olssen (2002).

24. The sacredness of life as a protonorm differs fundamentally from the Enlightenment's monocultural ethical rationalism in which universal imperatives were considered obligatory for all nations and epochs. Cartesian foundationalism and Immanuel Kant's formalism presumed noncontingent starting points. Universal human solidarity does not. Nor does it flow from Platonism, that is, the finite participating in the infinite and receiving its essence from it (see Christians, 2008, pp. 10–12). In addition to the sacredness of life as a protonorm, there are other appeals to universals that are not Western or do not presume a Newtonian cosmology; for a summary, see Cooper & Christians (2008, pp. 296–300).

25. Mutuality is a cardinal feature of the feminist communitarian model generally and therefore crucial to the principle of empowerment. For this reason, critical theory is inscribed into the third principle here, rather than following Root (see note 18, above), allowing it to stand by itself as an illustration of communitarianism. Root (1993, p. 238) himself observes that critical theorists often fail to transfer the "ideals of expertise" to their research subjects or give them little say in the research design and interpretation. Without a fundamental shift to communitarian interactivity, research in all modes is prone to the distributive fallacy.

26. Because of his fundamental commitment to dialogue, empowering for Freire avoids the weaknesses of monologic concepts of empowerment in which researchers are seen to free up the weak and unfortunate (summarized by Denzin, 2003, pp. 242–245, citing Bishop, 1998). While Freire represents a radical perspective, he does not claim, "as more radical theorists" do, that "only they and their theories can lead" the researched into freedom (Denzin, 2003, p. 246; citing Bishop, 1998).

27. Michael Theunissen (1984) argues that Buber's relational self (and therefore its legacy in Levinas, Freire, Heller, Wyschogrod, and Taylor) is distinct from the subjectivity of continental existentialism. The subjective sphere of Husserl and Jean-Paul Sartre, for example, "stands in no relation to a Thou and is not a member of a We" (p. 20; see also p. 276). "According to Heidegger the self can only come to itself in a voluntary separation from other selves; according to Buber, it has its being solely in the relation" (p. 284).

◨ REFERENCES

Barry, B. (1967). Justice and the common good. In A. Quinton (Ed.), *Political philosophy* (pp. 190–191). Oxford, UK: Oxford University Press.

Benhabib, S. (1992). *Situating the self: Gender, community, and post-modernism in contemporary ethics.* Cambridge, UK: Polity.

Benhabib, S. (1994). *Feminist contentions: A philosophical exchange.* New York: Routledge.

Benhabib, S. (2002). *The claims of culture: Equality and diversity in the global era.* Princeton, NJ: Princeton University Press.

Benhabib, S. (2008). *Democracy and difference.* New York: Oxford University Press.

Bishop, R. (1998). Freeing ourselves from neo-colonial domination in research: A Maori approach to creating knowledge. *International Journal of Qualitative Studies in Education, 11,* 199–219.

Blanchard, M. A. (2002, January). *Should all disciplines be subject to the common rule?* Washington, DC: U.S. Department of Health and Human Services.

Bok, S. (1995). *Common values.* Columbia: University of Missouri Press.

Brabeck, M. M. (Ed.). (1990). *Who cares? Theory, research, and educational implications of the ethic of care.* New York: Praeger.

Brunn, H. H. (2007). *Science, values, and politics in Max Weber's methodology.* Surrey, UK: Ashgate.

Buber, M. (1958). *I and thou* (2nd ed.; R. G. Smith, Trans.). New York: Scribner's.

Bulmer, M. (2008). The ethics of social research. In N. Gilbert (Ed.), *Researching social life* (3rd ed., pp. 145–161). London: Sage.

Bunge, M. (1996). *Finding philosophy in social science.* New Haven, CT: Yale University Press.

Cannella, G. S., & Lincoln, Y. S. (2009). Deploying qualitative methods for critical social purposes. In N. K. Denzin & M. D. Giardina (Eds.), *Qualitative inquiry and social justice* (pp. 53–72). Walnut Creek, CA: Left Coast Press.

Card, C. (Ed.). (1991). *Feminist ethics.* Lawrence: University of Kansas Press.

Chomsky, N. (2002). *American power and the new mandarins.* New York: The Free Press. (Original work published 1969)

Christians, C. G. (1997). The ethics of being. In C. G. Christians & M. Traber (Eds.), *Communication ethics and universal values* (pp. 3–23). Thousand Oaks, CA: Sage.

Christians, C. G. (1998). The sacredness of life. *Media Development, 45*(2), 3–7.

Christians, C. G. (2002a). Introduction. In C. G. Christians (Ed.), Ethical theorists and qualitative research [Special issue]. *Qualitative Inquiry, 8*(1), 407–410.

Christians, C. G. (2002b). Norman Denzin's feminist communitarianism. *Studies in Symbolic Interactionism, 25,* 167–177.

Christians, C. G. (2004). *Ubuntu* and communitarianism in media ethics. *Ecquid Novi, 25*(2), 235–256.

Christians, C. G. (2008). The ethics of universal being. In S. J. A. Ward & H. Wasserman (Eds.), *Media ethics beyond borders: A global perspective* (pp. 6–23). Johannesburg, South Africa: Heinemann.

Christians, C., Ferre, J., & Fackler, M. (1993). *Good news: Social ethics and the press.* New York: Oxford University Press.

Christians, C. G., Glasser, T. L., McQuail, D., Nordenstreng, K., & White, R. (2009). *Normative theories of the media: Journalism in democratic societies.* Urbana: University of Illinois Press.

Christians, C., Rao, S., Ward, S. J. A., & Wasserman, H. (2009). Toward a global media ethics: Theoretical perspectives. *Ecquid Novi: African Journalism Studies, 29*(2), 135–172.

Church, J. T. (2002, January). *Should all disciplines be subject to the common rule?* Washington, DC: U. S. Department of Health and Human Services.

Code, L. (1991). *What can she know? Feminist theory and the construction of knowledge.* Ithaca, NY: Cornell University Press.

Comte, A. (1830). *Cours de Philosophie Positive.* Paris: Bachelier Librarie pour les Mathematiques.

Comte, A. (1910). *A general view of positivism* (J. H. Bridges, Trans.). London: Routledge. (Original work published 1848)

Cooper, T. W., & Christians, C. G. (2008). On the need and requirements for a global ethic of communication. In J. V. Ciprut (Ed.), *Ethics, politics, and democracy: From primordial principles to prospective practices* (pp. 293–318). Cambridge, MA: MIT Press.

Copleston, F. (1966). *A history of philosophy: Vol. 8. Modern philosophy: Bentham to Russell.* Garden City, NY: Doubleday.

Denzin, N. K. (1984). *On understanding emotion.* San Francisco: Jossey-Bass.

Denzin, N. K. (1989). *Interpretive biography.* Newbury Park, CA: Sage.

Denzin, N. K. (1997). *Interpretive ethnography: Ethnographic practices for the 21st century.* Thousand Oaks, CA: Sage.

Denzin, N. K. (2002). *Reading race: Hollywood and the cinema of racial violence.* Thousand Oaks, CA: Sage.

Denzin, N. K. (2003). *Performance ethnography: Critical pedagogy and the politics of culture.* Thousand, Oaks, CA: Sage.

Denzin, N. K. (2009). *Qualitative inquiry under fire: Toward a new paradigm dialogue.* Walnut Creek, CA: Left Coast Press.

Denzin, N. K., & Giardina, M. D. (Eds.). (2007). *Ethical futures in qualitative research,* Walnut Creek, CA: Left Coast Press.

Denzin, N. K., & Giardina, M. D. (Eds.). (2009). *Qualitative inquiry and social justice.* Walnut Creek, CA: Left Coast Press.

Denzin, N. K., Lincoln, Y. S., & Smith, L. T. (Eds.). (2008). *Handbook of critical and indigenous methodologies.* Thousand Oaks, CA: Sage.

Drescher, G. L. (2006). *Good and real: Demystifying paradoxes from physics to ethics.* Cambridge, MA: MIT Press.

Durkheim, E. (1966). *Suicide: A study of sociology.* New York: Free Press.

Erikson, K. (1967). Disguised observation in sociology. *Social Problems, 14,* 366–373.

Euben, J. P. (1981). Philosophy and the professions. *Democracy, 1*(2), 112–127.

Evans, A. F., Evans, R. A., & Kennedy, W. B. (1987). *Pedagogies for the non-poor.* Maryknoll, NY: Orbis.

Fackler, M. (2003). Communitarian theory with an African flexion. In J. Mitchell & S. Marriage (Eds.), *Mediating religion: Conversations in media, religion, and culture* (pp. 317–327). London: T & T Clark.

Fackler, M. (2009). Communitarianism. In L. Wilkins & C. Christians (Eds.), *The handbook of mass media ethics* (pp. 305–316). New York: Routledge.

Foucault, M. (1979). *Discipline and punish: The birth of the prison* (A. Sheridan, Trans.). New York: Random House.

Freire, P. (1970a). *Education as the practice of freedom: Cultural action for freedom.* Cambridge, MA: Harvard Educational Review/Center for the Study of Development.

Freire, P. (1970b). *Pedagogy of the oppressed.* New York: Seabury.

Freire, P. (1973). *Education for critical consciousness.* New York: Seabury.

Freire, P. (1978). *Pedagogy in process: The letters of Guinea-Bissau.* New York: Seabury.

Friedman, M. S. (2002). *Martin Buber: The life of dialogue.* New York: Routledge.

Gilligan, C. (1982). *In a different voice: Psychological theory and women's development.* Cambridge, MA: Harvard University Press.

Gilligan, C. (1983). Do the social sciences have an adequate theory of moral development? In N. Haan, R. N. Bellah, P. Rabinow, & W. N. M. Sullivan (Eds.), *Social science as moral inquiry* (pp. 33–51). New York: Columbia University Press.

Gilligan, C., Ward, J. V., & Taylor, J. M. (1988). *Mapping the moral domain.* Cambridge, MA: Harvard University, Graduate School of Education.

Guba, E. G. (1990). The alternative paradigm dialog. In E. Guba (Ed.), *The paradigm dialog* (pp. 17–30). Thousand Oaks, CA: Sage.

Guba, E. G., & Lincoln, Y. S. (1994). Competing paradigms in qualitative research. In N. K. Denzin & Y. S. Lincoln (Eds.), *Handbook of qualitative research* (pp. 105–117). Thousand Oaks, CA: Sage.

Gunzenhauser, M. G. (2006). A moral epistemology of knowing subjects: Theorizing a relational turn for qualitative research. *Qualitative Inquiry, 12*(3), 621–647.

Habermas, J. (1971). *Knowledge and human interests* (J. J. Shapiro, Trans.). Boston: Beacon.

Habermas, J. (1990). *Moral consciousness and communicative action* (C. Lenhardt & S. W. Nicholson, Trans.). Cambridge, MA: MIT Press.

Habermas, J. (1993). *Justification and application: Remarks on discourse ethics* (C. Cronin, Trans.). Cambridge, MA: MIT Press.

Harris, C. E. (2006). *Applying moral theories* (5th ed.). Stamford, CT: Wadsworth.

Held, V. (1993). *Feminist morality: Transforming culture, society, and politics.* Chicago: University of Chicago Press.

Held, V. (2006). *The ethics of care: Personal, political, and global.* New York: Oxford University Press.

Heller, A. (1988). *General ethics.* Oxford, UK: Blackwell.

Heller, A. (1990). *A philosophy of morals.* Oxford, UK: Blackwell.

Heller, A. (1996). *An ethics of personality.* Oxford, UK: Blackwell.

Heller, A. (1999). *A theory of modernity.* Oxford, UK: Blackwell.

Heller, A. (2009). *A theory of feelings* (2nd ed.). Lanham, MD: Lexington Books.

Humphreys, L. (1970). *Tearoom trade: Impersonal sex in public places.* Chicago: Aldine.

Humphreys, L. (1972). *Out of the closet.* Englewood Cliffs, NJ: Prentice Hall.

Jennings, B., & Callahan, D. (1983, February). Social science and the policy-making process. *Hastings Center Report,* pp. 3–8.

Koehn, D. (1998). *Rethinking feminist ethics: Care, trust, and empathy.* New York: Routledge.

Lassman, P. (2004). Political theory in an age of disenchantment: The problem of value pluralism—Weber, Berlin, Rawls. *Max Weber Studies, 4*(2), pp. 251–269.

Lazarsfeld, P., & Reitz, J. G. (1975). *An introduction to applied sociology.* New York: Elsevier.

Levinas, E. (1985). *Ethics and infinity* (R. A. Cohen, Trans.). Pittsburgh, PA: Duquesne University Press.

Levinas, E. (1991). *Otherwise than being or beyond essence* (A. Lingis, Trans.). Dordrecht, Netherlands: Kluwer Academe.

Lincoln, Y. S. (1995). Emerging criteria for quality in qualitative and interpretive inquiry. *Qualitative Inquiry, 1,* 275–289.

Lincoln, Y. S., & Guba, E. G. (1985). *Naturalistic inquiry.* Beverly Hills, CA: Sage.

Lotz, J. B. (1963). Person and ontology. *Philosophy Today, 7,* 294–297.

MacMurray, J. (1961a). *The form of the personal: Vol. 1. The self as agent.* London: Faber & Faber.

MacMurray, J. (1961b). *The form of the personal: Vol. 2. Persons in relation.* London: Faber & Faber.

Mantzavinos, C. (Ed.). (2009). *Philosophy of the social sciences: Philosophical theory and scientific practice.* Cambridge, UK: Cambridge University Press.

McIntosh, M. J., & Morse, J. M. (2009). Institutional review boards and the ethics of emotion. In N. K. Denzin & M. D. Giardina (Eds.), *Qualitative inquiry and social justice* (pp. 81–107). Walnut Creek, CA: Left Coast Press.

McLaren, P., & Leonard, P. (Eds.). (1993). *Paulo Freire: A critical encounter.* London: Routledge.

Milgram, S. (1974). *Obedience to authority.* New York: Harper & Row.

Mill, J. S. (1865a). *Auguste Comte and positivism.* London.

Mill, J. S. (1865b). *Examination of Sir William Hamilton's philosophy and of the principal philosophical questions discussed in his writings.* London: Longman, Green, Roberts & Green.

Mill, J. S. (1893). *A system of logic, ratiocinative and inductive: Being a connected view of the principles of evidence and the methods of scientific investigation* (8th ed.). New York: Harper & Brothers. (Original work published 1843)

Mill, J. S. (1957). *Utilitarianism.* Indianapolis, IN: Bobbs-Merrill. (Original work published 1861)

Mill, J. S. (1969). *Autobiography.* Boston: Houghton Mifflin. (Original work published posthumously 1873)

Mill, J. S. (1978). *On liberty.* Indianapolis: Hackett. (Original work published 1859)

Mulhall, S., & Swift, A. (1996). *Liberals and communitarians* (2nd ed.). Oxford, UK: Blackwell.

Noddings, N. (1984). *Caring: A feminine approach to ethics and moral education.* Berkeley: University of California Press.

Noddings, N. (1989). *Women and evil.* Berkeley: University of California Press.

Noddings, N. (1990). Ethics from the standpoint of women. In D. L. Rhode (Ed.), *Theoretical perspectives on sexual difference* (pp. 160–173). New Haven, CT: Yale University Press.

Noddings, N. (2002). *Starting at home: Caring and social policy.* Berkeley: University of California Press.

Nussbaum, M. (1993). Non-relative virtues: An Aristotelian approach. In M. Nussbaum & A. Sen (Eds.), *The quality of life* (pp. 242–269). Oxford, UK: Clarendon.

Nussbaum, M. (1999). *Sex and social justice.* New York: Oxford University Press.

Olssen, M. (2002). Michel Foucault as "thin" communitarian: Difference, community, democracy." *Cultural Studies <=> Critical Methodologies, 2*(4), 483–513.

Olthuis, J. (1997). Face-to-face: Ethical asymmetry or the symmetry of mutuality? In J. Olthuis (Ed.), *Knowing other-wise* (pp. 134–164). New York: Fordham University Press.

Pacey, A. (1996). *The culture of technology.* Cambridge, MA: MIT Press.

Pateman, C. (1985). *The problem of political obligation: A critique of liberal theory.* Cambridge, UK: Polity.

Pateman, C. (1988). *The sexual contract.* Stanford, CA: Stanford University Press.

Pateman, C. (1989). *The disorder of women: Democracy, feminism and political theory.* Stanford, CA: Stanford University Press.

Pateman, C., & Mills, C. W. (2007). *Contract and domination.* Cambridge, UK: Polity Press.

Peukert, H. (1981). Universal solidarity as the goal of ethics. *Media Development, 28*(4), 10–12.

Punch, M. (1998). Politics and ethics in qualitative research. In N. K. Denzin & Y. S. Lincoln (Eds.), *The landscape of qualitative research* (pp. 156–184). Thousand Oaks, CA: Sage.

Rawls, J. (1971). *A theory of justice.* Cambridge, MA: Harvard University Press.

Reinharz, S. (1993). *Social research methods: Feminist perspectives.* New York: Elsevier.

Reiss, A. J., Jr. (1979). Governmental regulation of scientific inquiry: Some paradoxical consequences. In C. B. Klockars & F. W. O'Connor (Eds.), *Deviance and decency: The ethics of research with human subjects* (pp. 61–95). Beverly Hills, CA: Sage.

Root, M. (1993). *Philosophy of social science: The methods, ideals, and politics of social inquiry.* Oxford, UK: Blackwell.

Ross, W. D. (1930). *The right and the good.* Oxford, UK: Clarendon.

Ryan, K. E. (1995). Evaluation ethics and issues of social justice: Contributions from female moral thinking. In N. K. Denzin (Ed.), *Studies in symbolic interaction: A research annual* (Vol. 19, pp. 143–151). Greenwich, CT: JAI.

Sandel, M. J. (1998). *Liberalism and the limits of justice* (2nd ed.). Cambridge, UK: Cambridge University Press. (Original work published 1982)

Seigfried, C. H. (1996). *Pragmatism and feminism: Reweaving the social fabric.* Chicago: University of Chicago Press.

Shils, E. A. (1949). Foreword. In M. Weber, *The methodology of the social sciences* (pp. iii-x). New York: Free Press.

Shopes, L., & Ritchie, D. (2004). Exclusion of oral history from IRB review: An update. *Perspectives on History.* Available at http://www.historians.org/Perspectives/Issues/2004/0403/0403new1.cfm

Smith, D. E. (1987). *The everyday world as problematic: A feminist sociology.* Boston: Northeastern University Press.

Soble, A. (1978, October). Deception in social science research: Is informed consent possible? *Hastings Center Report,* pp. 40–46.

Stefanos, A. (1997). African women and revolutionary change: A Freirian and feminist perspective. In P. Freire (Ed.), *Mentoring the mentor: A critical dialogue with Paulo Freire* (pp. 243–271). New York: Peter Lang.

Steiner, L. (1989). Feminist theorizing and communication ethics. *Communication, 12*(3), 157–174.

Steiner, L. (1997). A feminist schema for analysis of ethical dilemmas. In F. L. Casmir (Ed.), *Ethics in intercultural and international communication* (pp. 59–88). Mahwah, NJ: Lawrence Erlbaum.

Steiner, L. (2009). Feminist media ethics. In L. Wilkins & C. Christians (Eds.), *The handbook of mass media ethics* (pp. 366–381). New York: Routledge.

Taylor, C. (1982). The diversity of goods. In A. Sen & B. Williams (Eds.), *Utilitarianism and beyond* (pp. 129–144). Cambridge, UK: Cambridge University Press.

Taylor, C. (1989). *Sources of the self: The making of the modern identity.* Cambridge, MA: Harvard University Press.

Taylor, C. (1991). *The ethics of authenticity.* Cambridge, MA: Harvard University Press.

Taylor, C. (1995). *Philosophical arguments.* Cambridge, MA: Harvard University Press.

Taylor, C. (2007). *A secular age.* Cambridge, MA: Harvard University Press.

Taylor, C., Appiah, K. A., Habermas, J., Rockefeller, S. C., Walzer, M., & Wolf, S. (1994). *Multiculturalism: Examining the politics of recognition* (A. Gutmann, Ed.). Princeton, NJ: Princeton University Press.

Theunissen, M. (1984). *The other: Studies in the social ontology of Husserl, Heidegger, Sartre, and Buber* (C. Macann, Trans.). Cambridge: MIT Press.

Tong, R. (1989). *Feminist thought.* Boulder, CO: Westview.

Tong, R. (1993). *Feminine and feminist ethics.* Belmont, CA: Wadsworth.

University of Illinois at Urbana-Champaign, Institutional Review Board. (2009). Part I: Fundamental principles for the use of human subjects in research. In *Investigator handbook.* Available at http://irb.illinois.edu/?q=investigator-handbook/index.html

Vanderpool, H. Y. (Ed.). (1996). *The ethics of research involving human subjects: Facing the 21st century.* Frederick, MD: University Publishing Group.

Verlinden, A. (2008). Global ethics as dialogism. In M. S. Comers, W. Vanderkerchove, & A. Verlinden (Eds.), *Ethics in an era of globalization* (pp. 187–215). Aldershot, UK: Ashgate.

Walzer, M. (1983). *Spheres of justice: A defense of pluralism and equality.* New York: Basic Books.

Walzer, M. (1987). *Interpretation and social criticism.* Cambridge, MA: Harvard University Press.

Weber, M. (1949a). The meaning of ethical neutrality in sociology and economics. In M. Weber, *The methodology of the social sciences* (E. A. Shils & H. A. Finch, Eds. & Trans.). New York: Free Press. (Original work published 1917)

Weber, M. (1949b). Objectivity in social science and social policy. In M. Weber, *The methodology of the social sciences* (E. A. Shils & H. A. Finch, Eds. & Trans.). New York: Free Press. (Original work published 1904)

Weber, M. (1973). *Max Weber on universities* (E. A. Shils, Ed. & Trans.). Chicago: University of Chicago Press.

West, C. (1989). *The American evasion of philosophy: A genealogy of pragmatism.* Madison: University of Wisconsin Press.

West, C. (1991). *The ethical dimensions of Marxist thought.* New York: Monthly Review Books.

West, C. (2001). *Race matters.* Boston: Beacon. (Original work published 1993)

White, R. (1995). From codes of ethics to public cultural truth. *European Journal of Communication, 10,* 441–460.

Winch, P. (2007). *The idea of a social science and its relation to philosophy* (2nd ed.). New York: Routledge. (Original work published 1958)

Wiredu, K. (1996). *Cultural universals: An African perspective.* Bloomington: Indiana University Press.

Wood, J. T. (1994). *Who cares? Women, care, and culture.* Carbondale: Southern Illinois University Press.

Wyschogrod, E. (1974). *Emmanuel Levinas: The problem of ethical metaphysics.* The Hague: Martinus Nijhoff.

Wyschogrod, E. (1985). *Spirit in ashes: Hegel, Heidegger, and man-made death.* Chicago: University of Chicago Press.

Wyschogrod, E. (1990). *Saints and post-modernism: Revisioning moral philosophy.* Chicago: University of Chicago Press.

Wyschogrod, E. (1998). *An ethics of remembering: History, heterology, and the nameless others.* Chicago: University of Chicago Press.

Wyschogrod, E. (2002). *Emmanuel Levinas: The problem of ethical metaphysics* (2nd ed.). New York: Fordham University Press.

5

ETHICS, RESEARCH REGULATIONS, AND CRITICAL SOCIAL SCIENCE

Gaile S. Cannella and Yvonna S. Lincoln

The social, intellectual, and even political positions from which the notion of research ethics can be defined have certainly emanated from diverse knowledges and ways of experiencing the world, as well as from a range of historical locations. The regulation of research ethics (especially legislated regulation) has, however, most often been influenced by traditional, postpositivist orientations. Clifford G. Christians (2005) discusses the histories of research ethics, from a value-free scientific neutrality that constructs science as "political only in its application" (Mill, 1859/1978; Root, 1993, p. 129; Weber, 1904/1949) to communitarian perspectives that challenge researchers to join with communities in new forms of moral articulation (Benhabib, 1992; Denzin, 1997, 2003).

In 2007, in a special issue of *Qualitative Inquiry* dedicated to research ethics and regulation, we discussed these multiple locations as well as contemporary power orientations from which diverse perspectives emanate. We focused on legislation imposed on researchers regarding the ethical conduct of research; ethical perspectives practiced, taught, or denied by those who teach and perform research methods; contemporary concerns that research is legitimated through market philosophies; and voices of the marginalized, created as the Other by or through research practices. Intertwined throughout our discussion was the recognition that regulation in its multiple forms results in an illusion of ethical practice and that any universalist ethic would be "catastrophic" (Foucault, 1985, p. 12). Furthermore, diversity of theoretical positions and perspectives within the field of qualitative inquiry has already generated rich and profound possibilities for reflexive ethics. From within these diverse perspectives, authors in the special issue reconceptualized research ethics as particularized, infused throughout inquiry, and requiring a continued moral dialogue—as calling for the development of a critical consciousness that would challenge the contemporary predatory ethical policies facilitated through neoliberalism (Christians, 2007; Clark & Sharf, 2007).

We who identify ourselves as *critical* in some form (whether hybrid–other–subject–feminist–scholar) have attempted to engage with the multiplicities embedded within notions of ethical scholarship. Being critical requires a radical ethics, an *ethics that is always/already concerned about power and oppression even as it avoids constructing 'power' as a new truth*. The intersection of power, oppression, and privilege with issues of human suffering, equity, social justice, and radical democracy results in a critical ethical foundation. Furthermore, ethical orientations are believed to be played out within the personal core of the researcher as she or he examines and makes decisions about the conceptualization and conduct of research as either oppressive or emancipatory practice.

A conceptualization of what some have called a *critical social science* incorporates the range of feminist, postcolonial, and even postmodern challenges to oppressive power, as well as the various interpretations of critical theory and critical pedagogies that are radically democratic, multilogical, and publicly, centrally concerned with human suffering and oppression. Traditional social science tends to address research ethics as following particular methodological rules in practices that are designed in advance and would reveal universalist results identified as ethical from within an imperative that would generalize to "save" humankind. For criticalists, however, this "will to save" is an imperialist imperative. Rather, critical radical ethics is relational and collaborative; it aligns with resistance and marginality. In *Ethical Futures in Qualitative Research*, Norman K. Denzin and Michael D. Giardina (2007) describe the range of scholars who have called for a collaborative critical social science model that "aligns the ethics of research with a politics of the oppressed, with a politics of resistance, hope, and freedom" (p. 35).

A critical social science literally requires that the researcher reconstruct the purposes of inquiry to engage with the struggle for equity and justice, while at the same time examining (and countering) individual power created for the researcher within the context of inquiry. The ethics of critical social science require that scholars "take up moral projects that decolonize, honor, and reclaim indigenous cultural practices" (Denzin & Giardina, 2007, p. 35), as well as engage with research that mobilizes collective actions that result in "a radical politics of possibility, of hope, of love, care, and equality for all humanity" (p. 35).[1] Researcher actions must avoid the perpetuation or maintenance of inquirer-oriented power (as savior, decolonizer, or one that would empower).

A critical social science reconceptualizes everything, from the embeddedness of ethics (and what that means) to the role of ethics in constructing research questions, methodologies, and possibilities for transformation. The major focus of this chapter is to examine the complexities of creating an ethical critical social science within our contemporary sociopolitical condition, a condition that has reinvigorated the privilege of empire through neoliberal Western discourses and regulatory technologies that would intervene into the lives of and literally create the Other and that continues redistribution of resources for neoliberal purposes (even from within a new administration in the United States that we believe is concerned with equity, anti-oppression, and social justice). We have previously discussed the positions from which research ethics tend to have been drawn, ranging from government regulation to voices of peoples who have not benefited and have often been damaged by research (Cannella & Lincoln, 2007; Cannella & Manuelito, 2008; Viruru & Cannella, 2006). In this chapter, we use these various standpoints to further explore a radical ethics as necessary for critical social science. We focus on constructing dialogic critical foundations (that we hope are anticolonial and even countercolonial) as well as reconceptualizing inquiry and forms of research (and researcher) regulation. Critical perspectives are located in the continuous alliance (and attempts at solidarity) with countercolonial positions and bodies and with the always/already historical acknowledgment of intersecting forms of privilege/oppression within contemporary contexts.

Furthermore, an evolving critical pedagogy (Kincheloe, 2007, 2008) is employed as a lens from which to generate forms of critical ethics that would transform academic (and public) spaces. This evolving criticality reconfigures the purposes of inquiry to focus on the dynamics and intersections of power relations between competing interests. Inquiry becomes the examination of contemporary forms of domination, as well as studies of what "could be"—of equitable and socially just futures. In addition, governmentality is addressed as produced by and producing forms of regulation interwoven with individual technologies of desire and accepted institutional practices. Finally, research regulation as ethical construct is rethought as reconfigured through the voices of those who have been traditionally marginalized as well as through the deployment of a critical social science whose purposes are to "join with," rather than "know and save."

▣ CONSTRUCTING CRITICAL WAYS OF BEING

Although not without conflicting beliefs, the range of critical perspectives (whether feminisms, poststructuralist work, queer theories, postcolonial critique, or other forms of knowledge that would address power) all tend to recognize the ways that particular groups of people have historically and continually been denied access to sites of power and have been systematically disenfranchised. These critical viewpoints have increasingly identified with marginalized peoples and have recognized the need to avoid forms of representation that maintain power in traditional locations. Furthermore, critical perspectives have called for the formation of alliances and attempts to join the struggle for solidarity with those who have been oppressed and inequitably treated. Patriarchal, racist, and colonializing forms of power are understood as historically grounded and recognized as never independent of cultural, political, and social context. For these reasons, we begin with a discussion of the need for critical ethical alliances that are always cognizant of the historical grounding and dominant power structures within the present.

Ethics and countercolonial alliance. An ethical perspective that would always address human suffering and life conditions, align with politics of the oppressed, and move to reclaim multiple knowledges and ways of being certainly involves complexity, openness to uncertainty, fluidity, and continued reflexive insight. Diverse conceptualizations of critical social science have reintroduced multiple knowledges, logics, ways of being in the world, and ethical orientations that have been historically marginalized and brutally discredited, facing violent attempts at erasure. As examples, Linda Tuhiwai Smith (1999) proposes four research processes that represent Maori collective ethics—decolonization, healing, transformation, and mobilization. Lester-Irabinna Rigney recommends that research methods privilege indigenous voices, resistance, and political integrity (1999). Sandy Grande (2007) puts forward Red pedagogy, an indigenous methodology that requires critique of democracy and indigenous sovereignty, functions as a pedagogy of hope that is contingent with the past, cultivates collective agency, is concerned with the dehumanizing effects of colonization on both the colonized and the colonizer, and is boldly and unabashedly political. Using Emmanuel Levinas's focus on the primacy of the well-being of the Other (1988), Jenny Ritchie and

Cheryl Rau (2010) construct a countercolonial ethics, labeled an *ethics of alterity,* which would shift the focus from "us" or "them" to "a collective reconfiguring of who 'we' are" (p. 364). Corrine Glesne (2007) even suggests that the purpose of research should be solidarity: "If you want to research us, you can go home. If you have come to accompany us, if you think our struggle is also your struggle, we have plenty of things to talk about" (p. 171). Critical pedagogues focus on the underpinnings of power in whatever context they find themselves and the ways that power performs or is performed to create injustice.

These are just a few of the ethical locations from which a critical social science has been proposed, introducing multiplicities, complexities, and ambiguities that would be part of any moral conceptualization and practice of research focusing on human suffering and oppression, radical democracy, and the struggle for equity and social justice. Furthermore, those of us who have been privileged through our connection with the dominant (e.g., education, economic level, race, gender) and may at least appear as the face of the oppressor must always avoid actions or interpretations that appropriate. We must struggle to "join with," and "learn from" rather than "speak for" or "intervene into." Voices from the margins demonstrate the range of knowledges, perspectives, languages, and ways of being that should become foundational to our actions, that should become a new center.

At various points, we have attempted to stand for a critical, transformative social science, for example: with Viruru (Viruru & Cannella, 2006) the critique of the construction of the ethnographic subject and the examination of privilege created by language in research practices; with Manuelito (Cannella & Manuelito, 2008) in proposing that social science be constructed in ways that are egalitarian, anticolonial, and ethically embedded within the nonviolent revolutionary consciousness proposed by hooks (1990). Recognizing that ethics as a construct is always and already essentializing, we have suggested that a revolutionary ethical conscience would be anticolonial and ask questions like: How are groups being used politically to perpetuate power within systems? How can we enlarge the research imaginary (e.g., regarding gender, race, childhood) to reveal the possibilities that our preoccupations have obscured? Can we cultivate ourselves as those who can desire and inhabit unthought spaces regarding research (about childhood, diverse views of the world)? (Lincoln & Cannella, 2007). Can we critique our own privilege? Can we join the struggle for social justice in ways that support multiple knowledges and multiple logics? These diverse perspectives and the underlying moral foundations from which they are generated are basic to the construction of an ethical, critical, even anticolonial social science. The ethics and the science must be understood as complex, must always be fluid, and must continually employ self-examination.

Furthermore, using the scholarship of Michel Foucault, Frantz Fanon, Judith Butler, and Gayatri Chakravorty Spivak,

Anthony C. Alessandrini (2009) calls for an ethics without subjects that is a new concept of ethical relationships, a responsible ethics that can be considered "after" humanism (p. 78). This postcolonial ethics would not be between people; rather in its future-oriented construction, an ethical relationship would occur with "would-be subjects that have not yet come into existence" (p. 78). The ethical relations would address contemporary political and power orientations by recognizing that the investigator and investigated (whether people, institutions, or systems) are subjects of the presence or aftermath of colonialism (Spivak, 1987). The tautology of humanist piety that would "save" others through science, religion, or politics would be avoided (Fanon, 1967; Foucault, 1984a). Yet, the Enlightenment blackmail that insists on a declaration of acceptance or rejection would be circumvented, while at the same time a critical flexibility is maintained (Butler, 2002; Foucault, 1984b). Ethics would involve being responsive and responsible to, while both trusting and avoiding construction of the Other. Ethical responsibility would be to a future, which can be accepted as unknowable (Attridge, 1994).

Drawing from Ritchie and Rau (2010), we would also support a *critical research ethics* that would counter colonialism. This critical ethics would value and recognize the need to

- Expose the diversity of realities
- Engage with the webs of interaction that construct problems in ways that lead to power/privilege for particular groups
- Reposition problems and decisions toward social justice
- Join in solidarity with the traditionally oppressed to create new ways of functioning

The magnitude and history of contemporary power. The ethics of a critical social science cannot avoid involvement with contemporary, everyday life and dominant societal discourses influencing that life. Research that would challenge oppression and foster social justice must acknowledge the gravity of context and the history of power within that context.

In the 21st century, this life has been constructed by the "Imperial Court of Corporate Greed and Knowledge Control" (Kincheloe, 2008, p. 15). Interpretations of knowledge and literally all human activity have been judged as valid and reliable if they fit the entrepreneurial imperative, if they foster privatization, competition, corporatization, and profiteering. In recent years, many of us have expressed outrage regarding this hypercapitalist influence, the free market illusion, over everything from definitions of public and higher education as benchmarked and measureable to privatization of services for the public good, to war mongering as a vehicle for corporatization to technologies that produce human desires that value self and others only as economic, measured, and entrepreneurial performers (Cannella & Miller, 2008; Cannella & Viruru, 2004; Chomsky, 1999; Horwitz, 1992).

Many of us would hope that a different administration in Washington, D.C., combined with the current financial crisis around the world, would result in confrontation with and transformation of capitalist imperialism. However, contemporary corporate fundamentalism is so foundational to dominant discourses that questioning failing corporations is not at all synonymous with contesting corporate forms of intellectual colonization. Examples abound in the early 21st century, like the discourse that labeled AIG as "too big" to fail, attempts to convince European governments to create stimulus packages, or presidential admonitions regarding "raising standards" in public schools (rather than the recognition of structural inequities in the system and taking actions to broaden definitions of public education as related to critical democracy and social justice).

Actually, the economic crisis may have created a new urgency within which critical scholars and others must take action. Living within a context in which "corporate-produced images" (Kincheloe, 2007, p. 30) have created new ideological templates for both affect and intellect, the need to accept corporate constructs and align with business interests is assumed. Corporate discourses have been so infused into the fabric of everyday life that most are not even recognized as such (for example, the construction of elitist public schools, which had been previously denied as not equitable or benefiting the common good—for example, by Lusher (Klein, 2007) and others—immediately following Hurricane Katrina in New Orleans). This illustrates what Klein (2007) has identified as "disaster capitalism." In the current economic crisis, even as big business is criticized, an unquestioned language of hypercapitalism (e.g., competition, free market, choice) results in further depoliticization of corporate colonization of the mind (both the mind of society and the mind of the individual) and of societal institutions (e.g., acceptance of privatized public services, education, even the armed forces). The Obama administration's unquestioned implementation of the Bush administration's charter school agenda for public education in the United States is an excellent example. The charter school concept has been used to reawaken the "free market" notion of public school choice (which was originally rejected when put forward as vouchers) and reinvigorate the power of the business roundtable, corporate turnaround models, and profiteering in public education.

"Western knowledge producers" (Kincheloe, 2008, p. 10) have held that their various forms of information were universal and enlightened (and as the progress that all should embrace, whether tied to the Christian religion or Cartesian science), in all conditions a risky circumstance for those who do not produce that knowledge. However, the politics of knowledge is even more dangerous when embedded within hypercapitalism and the power generated by capital and those that control resources. The acceptance of corporate perspectives that would invoke capitalist accountability constructs like evidenced-based research or scores on particular achievement tests (created by multinational companies) decontextualize and further subjectify and objectify students and children, their teachers, and their families. Human beings are treated as if their bodies (defined as achievement test scores) were the measure of "what works" within a particular discourse, just as financial success is used as the measure of a supposedly free-market, competitive, successful enterprise. Definitions are not questioned because the measured and measurement language and discourses of neoliberalism are accepted as correct, efficient, indisputable, universal, and even just. This contemporary condition constructs particular views of morality and equity, and thus expectations for what can be defined as ethical. From within this context, conceptualizing ethics and ethical practices as independent from (and necessary challenges to) hypercapitalism is very difficult but absolutely necessary.

The ethics of a critical social science requires the cultivation of a consciousness that is aware of both the sociopolitical condition of the times and one's own self-productive reactions to dominant disciplinary and regulatory technologies. This awareness involves engaging with the complexities of power and how it operates in the social order. Critical ethics would recognize the dominant (in our contemporary condition economics) but would never accept the truth of a superstructure (like economics) as always dictating human existence. Finally, a critical immanence would be necessary to move beyond ethnocentrism or egocentrism and construct new, previously unthought-of relationships and societal possibilities (Kincheloe, 2007).

▣ **ETHICS, CRITICAL SOCIAL SCIENCE, AND INSTITUTIONALIZED FORMS OF GOVERNMENTALITY**

In recent years, research ethics have been most often tied to one of the following:

- An ethics of entitlement (Glesne, 2007) that legitimizes engagement in research and the right to "know" the other
- Qualitative research methods, which require and employ ethical considerations like reflexive ethics (Guilleman & Gillam, 2004)
- Communitarian ethics through which values and moral commitments are negotiated socially (Christians, 2007; Denzin, 1997, 2003)
- Forms of legislated research regulation (e.g., institutional reviews of projects) that create an illusion of ethical concern (Lincoln & Tierney, 2004)

All are embedded within the notion of governmentality, either the construction of technologies that govern by producing control of populations (regulatory power) or the internalized discipline of bodies of individuals (researchers) based on the desire (from a range of value perspectives) to construct a particular self within the context (Foucault, 1978). The reader can consider *govern* as the action and *mentality* as the way people think about accepting control, the internalization of beliefs that allow regulation (Dean, 1999).

Research regulation that is legislated is most often recognized (and critiqued) as an institutionalized form of governmentality, a technology of power that constructs, produces, and limits and is thus tied to the generation of intersecting oppressions. However, Foucault (1986) also discusses the construction of self-governance, "political technologies of individuals" (p. 87), that are entirely internalized. There is a range of examples of this individual governmentality, from technologies of the "free citizen" (Rose, 1999), to the "well-educated person," to the "good teacher," even to the "transformative activist" or the "dialogically engaged researcher." We believe that our discussion of ethics within critical social science can be interpreted as a form of governmentality; most likely, any construction of ethics (however flexible) represents a form of governance. To construct a critical ethics regarding research is to address mentality. Any belief structure, however emergent or flexible, certainly serves as discipline and regulation of the self.

Since research has traditionally been a predominantly individual project and research regulation is legislated practice, both forms of governmentality (self and researcher population) must be considered in constructing an ethical critical social science. While a critical social science would always examine and challenge the notion of governmentality as "truth structure," the construction of a critical desire for countercolonial solidarity, the embeddedness within institutional expectations regarding research, and the contemporary regulatory context within which research is practiced cannot be denied as themselves forms of governmentality.

Individual desire and forms of governmentality. Critical and qualitative researchers have for some time critiqued the power orientations of research methods, have discussed practices that facilitate a reflexive ethical orientation throughout the research process, and have certainly rethought the purposes of research as construct. As examples, Walkerdine (1997) warns against the "voyeuristic thrill" of observation that constructs researcher as expert in what people are "really like" (p. 67). Feminists, poststructuralists, constructivists, and other scholars associated with postmodern concerns with oppression and power have engaged in principled struggles concerning the conceptualization of research itself, from the purposes of research, to forms of representation (Fine, Weis, Weseen, & Wong, 2000; Tedlock, 2005), to the role of the researcher. Questions like the following have been asked: "How are forms of exclusion being produced? Is transformative and liberatory research possible that also examines its own will to emancipate? . . . How does the practice of research reinscribe our own privilege?" (Cannella & Lincoln, 2007, p. 321). These ethical positions and concerns are certainly being incorporated into constructions of research projects and publications, as well as in new forms of education and coursework for graduate students. These positions are critical forms of governmentality.

However, the interconnected structures that characterize the dominant (noncritical) research community and the institutions that support research are not critical and tend to support modernist forms of governmentality. Ethics are likely to be legislated or constructed by individual researchers from within value structures that either maintain that science can solve all problems, therefore legitimating intervention into the lives of others in the name of science, or that free-market capitalism will improve life conditions for all, also used as the ethical justification for research choices and actions. These conceptualizations of ethics (for individuals and institutions) remain modernist, male-oriented, and imperialist (especially as related to labeling individuals, supporting particular forms of knowledge, and underpinning the dominance of neoliberal economics generally). These structures are interconnected (Collins, 2000) and invasive, have a long history, and will likely dominate into the foreseeable future.

Even though we support a critical social science that would be relational, collaborative, and less individualistically oriented, the contemporary context continues to be oriented toward power for the individual researcher. Therefore, while we would continually critique the privileging of the individual as construct, we also believe that perspectives that avoid universalist ethical codes yet address individual ethical frameworks are necessary. We hope that from the perspective of an ethical critical social science, individual governmentality as construct can always be challenged. However, we would also avoid the Enlightenment blackmail (Butler, 2002; Foucault, 1984b) that either accepts or rejects individualism and would submit that the individual is conceptually a useful master's tool (Lorde, 1984) as well as a critical agent. We would, therefore, propose the development of the desire to be critical, of a form of doubled individual governmentality through which the researcher is both instrument in the critique of power and collaborative agent in joining with traditionally marginalized communities.

The work of Foucault (1985), which challenges the individual to counter his or her own fascist orientations that would yield to the love of power and domination, is an illustration of this doubled conceptualization, even a doubled identity. An ethical framework is proposed that avoids the inscription of universalist moral codes but rather constructs "an intensification of the relation to oneself by which one constitutes oneself as the subject of one's acts" (Foucault, 1986, p. 41). The purpose of this use of the individually oriented master's tools is to suggest a critical framework through which self-absorption could be avoided, as the researcher conducts a continuous genealogy of the self along the axes of truth, power, and ethics (Foucault, 1985; Rabinow, 1994). Our focus in this discussion is on the ethical axis through which the self acts on itself, although the self's construction of both truth and power are not unrelated. Four components are included within the ethical axis of self: (1) ethical substance, (2) mode of subjectification, (3) ethical work, and (4) *telos* or disassembly of oneself. These components can be pondered from an individualistic rationalist perspective that also attempts to incorporate critical pedagogies and postcolonial critique.

Ethical substance is the way in which the researcher legitimates self morally. This substance is not a given but is constituted as relational to the self as a creative agent. To some extent, we can describe ethical substance as that which is important to the researcher, as that which facilitates or disallows self-deception and is the grounding for ethics. The ethical substance is "that which enables one to get free from oneself" (Foucault, 1985, p. 9), and it varies for everyone. As examples, the unification of pleasure and desire served as the ethical substance for many in ancient Greece; for some, collective existence and communal decision making is ethical substance (Ritchie & Rau, 2010); for some, addressing equity and social justice in solidarity with those who have most likely been oppressed may be the ethical substance. Foucault (1985) suggests genealogical questions to determine the substance of the self that we believe can be applied to the researcher, focusing on circumstances in which research is constituted as a moral activity—whether circumstances related to research as construct, interpretation of the meaning of research, or circumstances under which the researcher defines his or her scholarship as a moral or ethical act.

We propose (and we are not the first) that the belief in critical social science that would address oppression and construct alliances and solidarity with those who have traditionally been excluded constitutes ethical substance. Recognizing that governmentality and technologies of the self are more often subconscious (but acknowledging conscious possibilities), we would further suggest that those who choose such critical mentalities join in the broader reconceptualizations that are literally creating a new ethical substance for research. An example of this is the work of critical pedagogues. In describing the "ever-evolving conceptual matrix" of criticality, Joe Kincheloe (2007, p. 21) provides us with content for both ethical substance and the further creation of domains of critical social science that can be the content of ethical substance. These critical domains can even construct the foundations for research. They include:

1. Analysis of the dynamics of competing power interests

2. Exposure of forces that inhibit the ability of individuals and groups to determine the direction of their own lives

3. Research into the intersection of various forms of domination

4. Analysis of contemporary forms of technical rationality and the impact on diverse forms of knowledge and ways of being

5. Examination of forms of self-governmentality, always recognizing the sociopolitical and sociocultural context

6. Inquiry into what "could be," into ways of constructing a critical immanence that moves toward new, more equitable relationships between diverse peoples (yet always avoids utopian, humanist rationalities)

7. Exploration of the continually emerging, complex exercise of power, as hegemonic, ideological, or discursive

8. Examination of the role of culture in the contested production and transmission of knowledge(s)

9. Studies of interpretation, perception, and diverse vantage points from which meaning is constructed

10. Analysis of the role of cultural pedagogy as education, as producing hegemonic forms of interpretation

As ethical substance, this critical content can lead to specific inquiry like historical problematizations (of the present) that refuse to either blame or endorse; examinations of policy discourses, networks, or resources; or research that exposes power while refusing to co-opt the knowledge(s), skills, and resources of the other.

The *mode of subjectification* is probably the ethical component most illustrative of governmentality. The notion that the individual submits the self to particular rules and obligations is included; the rules are constructed and accepted dependent on the ethical substance. For example, Immanuel Kant (whose ethical substance focused on intention as embedded within reason) valued the obligation to know and the use of reason as the method of self-governance (Foucault, 1985). Critical social scientists may construct an ethical obligation (and resultant related rules) to a critical, historical disposition that is flexible and responds to issues of oppression. As Glesne (2007) implies, this critical mode of subjectification would most likely reject the sense of entitlement that would "know" others and would further recognize the alienation created when one is placed under the observational gaze of the researcher. A criticalist's ethical rules might be more likely to accept communal decision making rather than rationalist forms of negotiation.

From within the ethical axis, researchers can ask questions of themselves related to the rules that are constructed within particular constructions of ethical substance and used to determine the existence of moral activity. "How are these rules acted on in research activities to conceptualize/legitimate and implement moral obligations" (e.g., for an individual researcher in choice of study, in choice of population, in collaborations with others, as I educate other researchers) (Cannella & Lincoln, 2007, p. 325)?

Ethical work is the method used to transform self into the form that one defines as ethical. Foucault (1994) proposes that this work requires a self-criticism that historically examines the constitution of the self. The work is expected to reveal the conditions under which one questions the self, invents new ways of forming relationships, and constructs new ways of being. This form of self-governance involves examination of the ways one can change oneself (as person and/or as researcher). An evolving critical pedagogy can be used to illustrate the ethics of an ontological transformation that goes beyond Western constructions of the self. Kincheloe (2007) illustrates the central critical features that can be related to ethical identity development. These features include constructs like socioindividual imagination, challenges to the boundaries of abstract individualism, socioindividual analysis of power, alternatives to the alienation of the individual, mobilizing desire, and critical consciousness that acknowledges self-production. To illustrate, socioindividual imagination is the ability to conceptualize new forms of collaboration, rethinking subjectivities and acknowledging that the professional and personal are critical social projects; institutions like education are thus constructed as emphasizing social justice and democratic community as the facilitator of human development. Another example, mobilizing desire, is constructed as a radical democratization, joining continued efforts of the excluded to gain access and input into civic life.

Finally, *telos* is the willingness to disassemble self, to deconstruct one's world (and one's research practices if a researcher) in ways that demonstrate commitment to an ethical practice that would avoid the construction of power over any individual or group of others (even unpredictable, yet to be determined others located in the future). Telos is a form of self-bricolage, slowly elaborating and establishing a self that is committed to think differently, that welcomes the unknown and can function flexibly (Foucault, 1994). As critical pedagogy again suggests, alternatives to alienation of the individual are created, forms of domination that construct isolation are rejected, and unthought-of ways to be with and for others are constructed (Kincheloe, 2007). Furthermore, telos can construct new pathways through which individual researchers, as well as groups of scholars, can consider notions like an ethics without subjects that combines critical and postcolonial perspectives that are committed to the future and to avoiding the continued colonialist construction of the Other (Alessandrini, 2009).

Although certainly consistent with modernist approaches to individual rationality, the examination of an individual ethical axis demonstrates the ways that even the master's tools can be used for critique and transformation.

Currently, researchers must both engage in their own individual ethical decisions regarding research and function within institutional forms of regulation. From a range of critical locations, we are continuously reminded that different disciplinary strategies are enacted by institutions dependent on the historical moment and context (Foucault, 1977). Certainly, individual critically ethical selves (in our modernist academic community, which privileges the scientific individual) will be more prepared to engage with the conflicting ethical messages within institutions, whether academic expectations or legislated regulation; to take hold of our own existence as researchers, to transform academic spaces, and to redefine discourses (Denzin & Giardina, 2007).

▣ TRANSFORMING REGULATIONS: REDEFINING THE TECHNOLOGIES THAT GOVERN US

Qualitative and critical qualitative researchers have continued to "take hold" of their academic spaces as they have clashed with legislated research regulation (especially, for example, as practiced by particular institutional review boards in the United States). This conflict has been much discussed and will not end any time soon. This work has demonstrated not only that legislated attempts to regulate research ethics are an illusion, but that regulation is culturally grounded and can even lead to ways of functioning that are damaging to research participants and collaborators. As examples, Marzano (2007) demonstrates the ways that following Anglo-Saxon ethical research regulation in an Italian setting with medical patients involved in qualitative research can be detrimental to the participant patients. Susan Tilley and Louise Gormley (2007) illustrate the ways that the construction of confidentiality represents challenges to understandings of individual integrity in a Mexican setting. Furthermore, a range of scholarship demonstrates that research ethics is particularized, must be infused throughout the process, and requires a continued dialogue with self (Christians, 2007; Clark & Sharf, 2007). Legislated forms of governmentality can certainly not address these particulars.

If researchers accompany communities, rather than "test/know/judge" them, perhaps community members will want to address review boards and legislators themselves concerning collaborative practices. In describing the Mi'kmaw Ethics Watch, Marie Battiste and James (Sa'ke'j) Youngblood Henderson (2000; Battiste, 2008) demonstrate just such a practice, as Mi'kmaw people have constructed research guidelines in which research is always to be an equal partnership in which the Mi'kmaw people are the guardians and interpreters of their intellectual and cultural property and review research conclusions for accuracy and sensitivity.

Aligned with the ethics of the traditionally marginalized, which could ultimately reconceptualize the questions and practices of research, a critical social science would no longer accept the notion that one group of people can "know" and define (or even represent) "others." This perspective would certainly change the research purposes and designs that are submitted for human subjects review, perhaps even eliminating the need for "human subjects" in many cases. This change could result in research questions and forms of data collection that do not require researchers to interpret the meaning making or constructions of participants. Rather, research questions could address the intersections of power across systems, institutions, and societal practices. As examples, assumptions underlying the conceptualizations of public policy, dominant knowledges, and dominant ideologies (in particular areas); actions that would protect and celebrate diverse knowledges; and analyses of forms of representation privileged by those in power, can all become research purposes without constructing human subjects as objects of data collection. If societal structures, institutions, and oppressions become the subjects of our research (rather than human beings), perhaps we can avoid further creation and subjectification of an or the Other. Denzin (2009) even suggests that we "abandon the dirty word called research" and take up a "critical, interpretive approach to the world" (p. 298), a practice that could benefit us all and would require major forms of activism within our academic settings.

This section on the legislated regulation of research is noticeably and purposely brief. We would suggest that, first, critical qualitative researchers make all efforts to move to the center the reconceptualized, broad-based critical social science that addresses institutionalized, policy-based, intersecting forms of power. This critical social science can even include studies of regulation from an ethics-without-subjects perspective. And, it would undoubtedly include alliances with countercolonial positions, as well as critical historical recognitions of context and ethical examinations of the researcher self. Until this critical social science is accepted as an important form of practice (perhaps even vital enough to be threatening to the mainstream), modernist research regulation will most likely change very little. We will simply (although it is not at all simple, or any less important) continue our attempts to educate those who have not learned about qualitative research as a field or the methods associated with it. However, if a critical social science aligns with the oppressed, demonstrating solidarity with the traditionally marginalized and constructing research that addresses power, our constructions of and concerns about legislated research regulations will be of a different nature. Perhaps our critical research ethics can anticipate and facilitate that change.

◨ NOTE

1. Recognizing that we could be accused of assuming that postpositivist science has no ethical base, we must absolutely acknowledge that we understand that researchers from a range of philosophical perspectives believe that their research questions and practices are grounded in the ethical attempt to improve life for everyone, and following an Enlightenment, rational science orientation, we would agree. However, very often, these postpositivist forms of legitimation and scientific intentions do not acknowledge embeddedness within the Euro American "error" (Jaimes, 1992). This error is the unquestioned belief in modernist, progressive (both U.S. liberal and conservative) views of the world that would "unveil" universalist interpretations of all human

experience; it assumes the omnipotent ability (and right) to "know" and interpret "others." Unfortunately, these ethical good intentions have most often denied the multiple knowledges, logics, and ways of being in the world that have characterized a large number of human beings. Furthermore, focusing on the individual and the discovery of theories and universals has masked societal, institutional, and structural practices that perpetuate injustices. Finally, an ethics that would help others "be like us" has created power for "us." This ethics of good intentions has tended to support power for those who construct the research and the furthering of oppressive conditions for the subjects of that research.

◨ REFERENCES

Alessandrini, A. C. (2009). The humanism effect: Fanon, Foucault, and ethics without subjects. *Foucault Studies, 7*, 64–80.

Attridge, D. (1994). Trusting the other: Ethics and politics in J. M. Coetzee's Age of Iron. *South Atlantic Quarterly, 93*, 70–71.

Battiste, M. (2008). Research ethics for protecting indigenous knowledge and heritage: Institutional and researcher responsibilities. In N. K. Denzin, Y. S. Lincoln, & L. T. Smith (Eds.), *Handbook of critical indigenous methodologies* (pp. 600–625). Thousand Oaks, CA: Sage.

Battiste, M., & Youngblood Henderson, J. (Sa'ke'j). (2000). *Protecting indigenous knowledge and heritage.* Saskatoon, Saskatchewan, Canada: Purich.

Benhabib, S. (1992). *Situating the self: Gender, community, and postmodernism in contemporary ethics.* Cambridge, UK: Polity.

Butler, J. (2002). What is critique? An essay on Foucault's virtue. In D. Ingram (Ed.), *The political* (pp. 212–227). Cambridge, MA: Blackwell.

Cannella, G. S., & Lincoln, Y. S. (2007). Predatory vs. dialogic ethics: Constructing an illusion or ethical practice as the core of research methods. *Qualitative Inquiry, 13*(3), 315–335.

Cannella, G. S., & Manuelito, K. (2008). Feminisms from unthought locations: Indigenous worldviews, marginalized feminisms, and revisioning an anticolonial social science. In N. K. Denzin, Y. S. Lincoln, & L. T. Smith (Eds.), *Handbook of critical and indigenous methodologies* (pp. 45–59). Thousand Oaks, CA: Sage.

Cannella, G. S., & Miller, L. L. (2008). Constructing corporatist science: Reconstituting the soul of American higher education. *Cultural Studies <=> Critical Methodologies, 8*(1), 24–38.

Cannella, G. S., & Viruru, R. (2004). *Childhood and postcolonization: Power, education, and contemporary practice.* New York: RoutledgeFalmer.

Chomsky, N. (1999). *Profit over people: Neoliberalism and global order.* New York: Seven Stories Press.

Christians, C. G. (2007). Cultural continuity as an ethical imperative. *Qualitative Inquiry, 13*(3), 437–444.

Clark, M. C., & Sharf, B. F. (2007). The dark side of truth(s): Ethical dilemmas in researching the personal. *Qualitative Inquiry, 13*(3), 399–416.

Collins, P. H. (2000). *Black feminist thought: Knowledge, consciousness, and the politics of empowerment.* New York: Routledge.

Dean, M. (1999). *Governmentality: Power and rule in modern society.* London: Sage.

Denzin, N. K. (1997). *Interpretive ethnography: Ethnographic practices for the 21st century.* Thousand Oaks, CA: Sage.

Denzin, N. K. (2003). *Performance ethnography: Critical pedagogy and the politics of culture.* Thousand Oaks, CA: Sage.

Denzin, N. K. (2009). *Qualitative inquiry under fire: Toward a new paradigm dialogue.* Walnut Creek, CA: Left Coast Press.

Denzin, N. K., & Giardina, M. D. (2007). Introduction: Ethical futures in qualitative research. In N. K. Denzin & M. D. Giardina (Eds.), *Ethical futures in qualitative research: Decolonizing the politics of knowledge* (pp. 9–44). Walnut Creek, CA: Left Coast Press.

Fanon, F. (1967). *Black skin, white masks* (C. L Markmann, Trans.). New York: Grove.

Fine, M., Weis, L., Weseen, S., & Wong, L (2000). For whom? Qualitative research, representation, and social responsibilities. In N. K. Denzin & Y. S. Lincoln (Eds.), *Handbook of qualitative research* (2nd ed., pp. 107–131). Thousand Oaks, CA: Sage.

Foucault, M. (1977). *Discipline and punish: The birth of the prison.* London: Allen Lane.

Foucault, M. (1978). Governmentality. In B. Burchell, C. Gordon, & P. Miller (Eds.), *The Foucault effect: Studies in governmentality* (pp. 87–104). Chicago: University of Chicago Press.

Foucault, M. (1984a). Nietzsche, genealogy, history (D. F. Bouchard & S. Simon, Trans.). In P. Rabinow (Ed.), *The Foucault reader* (pp. 76–100). New York: Vintage.

Foucault, M. (1984b). What is enlightenment? (C. Porter, Trans.). In P. Rabinow (Ed.), *The Foucault reader* (pp 32–50). New York: Vintage.

Foucault, M. (1985). *History of sexuality: Vol. 2. The use of pleasure* (R. Hurley, Trans.). New York: Pantheon.

Foucault, M. (1986). *History of sexuality: Vol. 3. The care of the self* (R. Hurley, Trans.). New York: Pantheon.

Foucault, M. (1994). On the genealogy of ethics: An overview of work in progress. In P. Rabinow (Ed.). *Michel Foucault: Ethics, subjectivity, and truth, 1954–1984* (Vol. 1, pp. 253–280). New York: The New York Press.

Glesne, C. (2007). Research as solidarity. In N. K. Denzin & M. D. Giardina (Eds.), *Ethical futures in qualitative research: Decolonizing the politics of knowledge* (pp. 169–178). Walnut Creek, CA: Left Coast Press.

Grande, S. (2007). Red pedagogy: Indigenizing inquiry or the un-methodology. In N. K. Denzin & M. D. Giardina (Eds.), *Ethical futures in qualitative research: Decolonizing the politics of knowledge* (pp. 133–144). Walnut Creek, CA: Left Coast Press.

Guilleman, M., & Gillam, L. (2004). Ethics, reflexivity, and "ethically important moments" in research. *Qualitative Inquiry, 10*(2), 261–280.

hooks, b. (1990). *Yearning: Race, gender, and cultural politics.* Boston: South End Press.

Horwitz, M. (1992). *The transformation of American law, 1870–1960.* Cambridge, MA: Harvard University Press.

Jaimes, M. A. (1992). La raza and indigenism: Alternatives to autogenocide in North America. *Global Justice, 3*(2–3), 4–19.

Kincheloe, J. L (2007). Critical pedagogy in the twenty-first century. In P. McLaren & J. L. Kincheloe (Eds.), *Critical pedagogy: Where are we now?* (pp. 9–42). New York: Peter Lang.

Kincheloe, J. L. (2008). Critical pedagogy and the knowledge wars of the twenty-first century. *International Journal of Critical Pedagogy, 1*(1), 1–22.

Klein, N. (2007). *The shock doctrine: The rise of disaster capitalism.* New York: Metropolitan Books.

Levinas, E. (1988). Useless suffering. In R. Bernasconi & D. Wood (Eds.), *The provocation of Levinas: Rethinking the Other* (pp. 156–167). London & New York: Routledge.

Lincoln, Y. S., & Cannella, G. S. (2007). Ethics and the broader rethinking/reconceptualization of research as construct. In N. K. Denzin & M. D. Giardina (Eds.), *Ethical futures in qualitative research: Decolonizing the politics of knowledge* (pp. 67–84). Walnut Creek, CA: Left Coast Press.

Lincoln, Y. S., & Tierney, W. G. (2004). Qualitative research and institutional review boards. *Qualitative Inquiry, 10,* 219–234.

Lorde, A. (1984). *Sister outsider.* Langhorne, PA: Crossing Press.

Marzano, M. (2007). Informed consent, deception, and research freedom in qualitative research: A cross-cultural comparison. *Qualitative Inquiry, 12*(3), 417–436.

Mill, J. S. (1978). *On liberty.* Indianapolis, IN: Hackett. (Original work published 1859)

Rabinow, P. (1994). *Michel Foucault: Ethics, subjectivity, and truth, 1954–1984* (Vol. 1). New York: The New York Press.

Rigney, L.-I. (1999). Internationalization of an indigenous anticolonial cultural critique of research methodologies. *Wicazo Sa Review, 14*(2), 109–121.

Ritchie, J., & Rau, C. (2010). Kia mau ki te wairuatanga: Counter narratives of early childhood education in Aotearoa. In G. S. Cannella & L. D. Soto (Eds.), *Childhoods: A handbook* (pp. 355–373). New York: Peter Lang.

Root, M. (1993). *Philosophy of social science: The methods, ideals, and politics of social inquiry.* Oxford, UK: Blackwell.

Rose, N. (1999). *Powers of freedom: Reframing political thought.* Cambridge, UK: Cambridge University Press.

Smith, L. T. (1999). *Decolonizing methodologies: Research and indigenous peoples.* London: Zed Books.

Spivak, G. C. (1987). *In other worlds: Essays in cultural politics.* New York: Routledge.

Tedlock, B. (2005). The observation of participation and the emergence of public ethnography. In N. K. Denzin & Y. S. Lincoln (Eds.), *The SAGE handbook of qualitative research* (3rd ed., pp. 467–482), Thousand Oaks, CA: Sage.

Tilley, S., & Gormley, L. (2007). Canadian university ethics review: Cultural complications translating principles into practice. *Qualitative Inquiry, 13*(3), 368–387.

Viruru, R., & Cannella, G. S. (2006). A postcolonial critique of the ethnographic interview: Research analyzes research. In N. K. Denzin & M. D. Giardina (Eds.), *Qualitative inquiry and the conservative challenge* (pp. 175–192). Walnut Creek, CA: Left Coast Press.

Walkerdine, V. (1997). *Daddy's girl: Young girls and popular culture.* Cambridge, MA: Harvard University Press.

Weber, M. (1949). Objectivity in social science and social policy. In E. A. Shils & H. A. Finch (Eds. & Trans.), *The methodology of the social sciences* (pp. 50–112). New York: Free Press. (Original work published 1904)

Part II

PARADIGMS AND PERSPECTIVES IN CONTENTION

In our introductory chapter, following Egon G. Guba (1990, p. 17), we defined a paradigm as a basic set of beliefs that guide action. Paradigms deal with first principles or ultimates. They are human constructions. They define the worldview of the researcher-as-interpretive-*bricoleur*. These beliefs can never be established in terms of their ultimate truthfulness. Perspectives, in contrast, are not as solidified nor as well unified as paradigms, although a perspective may share many elements with a paradigm, for example, a common set of methodological assumptions or a particular epistemology.

A paradigm encompasses four terms: ethics (axiology), epistemology, ontology, and methodology. Ethics ask, "How will I be as a moral person in the world?" Epistemology asks, "How do I know the world?" "What is the relationship between the inquirer and the known?" Every epistemology, as Christians indicates (Chapter 4, this volume) implies an ethical-moral stance toward the world and the self of the researcher. Ontology raises basic questions about the nature of reality and the nature of the human being in the world. Methodology focuses on the best means for gaining knowledge about the world.

Part II of the *Handbook* examines the major paradigms and perspectives that now structure and organize qualitative research. These paradigms and perspectives are positivism, postpositivism, critical theory, constructivism, and participatory action frameworks. Alongside these paradigms are the perspectives of feminism (in its multiple forms), critical race theory, critical pedagogy, cultural studies, queer theory, Asian epistemologies, and disability theories, coupled with transformative, social justice paradigms. Each of these perspectives has developed its own criteria, assumptions, and methodological practices. These practices are then applied to disciplined inquiry within that framework. The tables in Chapter 6 by Guba & Yvonna Lincoln, with Susan A. Lynham outline the major differences between the positivist, postpositivist, critical theory (feminism + race), constructivism, and participatory (+ postmodern) paradigms.

We provided a brief discussion of each paradigm and perspective in Chapter 1; here we elaborate them in somewhat more

detail. However, before turning to this discussion, it is important to note three interconnected events. Within the last decade, the borders and boundary lines between these paradigms and perspectives have begun to blur. As Lincoln and Guba observe, the "pedigrees" of various paradigms are themselves beginning to "interbreed." However, although the borders have blurred, perceptions of differences between perspectives have hardened. Even as this occurs, the discourses of methodological conservatism, discussed in our Preface and in Chapter 1, threaten to narrow the range and effectiveness of qualitative research practices. Hence, the title of this part, Paradigms and Perspectives in Contention.

▣ MAJOR ISSUES CONFRONTING ALL PARADIGMS

In Chapter 6, Lincoln, Lynham, and Guba suggest that, in the present moment, all paradigms must confront seven basic, critical issues. These issues involve (1) axiology (ethics and values), (2) accommodation and commensurability (can paradigms be fitted into one another), (3) action (what the researcher does in the world), (4) control (who initiates inquiry, who asks questions), (5) foundations of truth (foundationalism vs. anti- and nonfoundationalism), (6) validity (traditional positivist models vs. poststructural-constructionist criteria), and (7) voice, reflexivity, and postmodern representation (single vs. multivoiced).

Each paradigm takes a different stance on these topics. Of course, the positivist and postpositivist paradigms provide the backdrop against which these other paradigms and perspectives operate. Lincoln and Guba analyze these two traditions in considerable detail, including their reliance on naive realism; their dualistic epistemologies; their verificational approach to inquiry; and their emphasis on reliability, validity, prediction, control, and a building block approach to knowledge. Lincoln and Guba discuss the inability of these paradigms to address adequately issues surrounding voice, empowerment, and praxis. They also allude to the failure to satisfactorily address the theory- and value-laden nature of facts, the interactive nature of inquiry, and the fact that the same set of "facts" can support more than one theory.

▣ CONSTRUCTIVISM, INTERPRETIVISM AND HERMENEUTICS

According to Lincoln and Guba, constructivism adopts a relativist ontology (relativism), a transactional epistemology, and a hermeneutic, dialectical methodology. Users of this paradigm are oriented to the production of reconstructed understandings of the social world. The traditional positivist criteria of internal and external validity are replaced by such terms as trustworthiness and authenticity. Constructivists value transactional knowledge. Their work overlaps with the several different participatory action approaches discussed by Morten Levin and Davydd Greenwood (Chapter 2), and Mary Brydon-Miller, Michael Kral, Patricia Maguire, Susan Noffke, and Anu Sabhlok (Chapter 23) in this volume. Constructivism connects action to praxis and builds on antifoundational arguments, while encouraging experimental and multivoiced texts.

In the third edition of the *Handbook*, Douglas Foley and Angela Valenzuela (2005) offered a history and analysis of critical ethnography, giving special attention to critical ethnographers who study applied policy and also involve themselves in political movements. Foley and Valenzuela observe that post-1960s critical ethnographers began advocating cultural critiques of modern society. These scholars revolted against positivism and sought to pursue a politically progressive agenda using multiple standpoint epistemologies. Various approaches were taken up in this time period, including action anthropology; global, neo-Marxist, Marxist feminist, and critical ethnography; and participatory action research.

▣ THE FEMINISMS

In Chapter 7, Virginia Olesen observes that feminist qualitative research, at the dawn of the second decade of this new century, is a highly diversified and contested site. Already we see multiple articulations of gender and its enactment in post-9/11 spaces. Competing models blur together on a global scale. But beneath the fray and the debate, there is agreement that feminist inquiry in the new millennium is committed to action in the world. Feminists insist that a social justice agenda address the needs of men and women of color because gender, class, and race are intimately interconnected. Olesen's is an impassioned feminism. "Rage is not enough," she exclaims. We need "incisive scholarship to frame, direct, and harness passion in the interests of redressing grievous problems in the many areas of women's health" (Olesen, 2000, p. 215).

In 1994, Olesen identified three major strands of feminist inquiry (standpoint epistemology, empiricist, postmodernism-cultural studies). A decade later, these strands continued to multiply. Today separate feminisms are associated with specific disciplines and with the writings of women of color; women problematizing whiteness; postcolonial, transnational discourse; decolonizing arguments of indigenous women; lesbian research and queer theory; disabled women; standpoint theory; and postmodern and deconstructive theory. Two critical trends emerge from these developments: (1) endarkening, decolonizing, indigenizing inquiry and (2) an expanding and maturing intersectionality as a critical approach. This complexity has made the researcher-participant relationship more complicated. It has destablized the insider-outsider model of inquiry. Within indigenous spaces, it has produced a call for the decolonization of the academy. This is linked to a deconstruction of such traditional terms as experience, difference, and gender.

A gendered decolonizing discourse focuses on the concepts of experience, difference, bias and objectivity, validity and trustworthiness, voice, performance, and feminist ethics. On this last point, Olesen's masterful chapter elaborates the frameworks presented by Cannella and Lincoln (Chapter 5) and Christians (Chapter 4) in Part I.

▣ THE ENDARKENED FEMINIST PRAXIS

In their chapter in the third edition of the *Handbook*, Gloria Ladson-Billings and Jamel Donnor presented an activist version of critical race theory (CRT) committed to social justice and a revolutionary habitus. They focused their analysis on the meaning of the "call," those epiphanic moments when people of color are reminded that they are locked into a hierarchical racial structure. Critical race theorists experiment with multiple interpretive strategies, ranging from storytelling to autoethnography, case studies, textual and narrative analyses, traditional fieldwork, and most important, collaborative, action-based inquiries and studies of race, gender, law, education, and racial oppression in daily life. Inquiry for social justice is the goal. For justice to happen, the academy must change; it must embrace the principles of decolonization. A reconstructed university will become a home for racialized others, a place where indigenous, liberating empowering pedagogies have become commonplace.

In Chapter 8, Cynthia B. Dillard and Chinwe Okpalaoka radically extend the spaces of CRT by opening up a paradigm that embodies cultural and spiritual understandings. Their endarkened framework foregrounds spirituality, with links to Africa and the African diaspora. An endarkened feminist epistemology intersects with the historical and contemporary contexts of oppression for African ascendant woman. Under this model, research is a moral responsibility. It honors the wisdom, spirituality, and critical interventions of transnational Black women. These are powerful recipes for action.

▣ CRITICAL PEDAGOGY AND CRITICAL THEORY

Multiple critical theories and Marxist or neo-Marxist models now circulate within the discourses of qualitative research In Lincoln and Guba's (2000) framework, this paradigm, in its many formulations, articulates an ontology based on historical realism, an epistemology that is transactional and a methodology that is both dialogic and dialectical. In Chapter 9, Joe L. Kincheloe, Peter McLaren, and Shirley Steinberg trace the history of critical research (and Marxist theory) from the Frankfurt School through more recent transformations in poststructural, postmodern, feminist, critical pedagogy, and cultural studies theory.

They outline a critical theory, a bricolage, which they call *critical humility,* an evolving criticality for the new millennium, beginning with the assumption that the societies of the West are not unproblematically democratic and free. Their version of critical theory rejects economic determinism and focuses on the media, culture, language, power, desire, critical enlightenment, and critical emancipation. Their framework embraces a critical hermeneutics. They read instrumental rationality as one of the most repressive features of contemporary society. Building on Paulo Freire, Karl Marx, Max Weber, Mikhail Bakhtin, and Jürgen Habermas, they present a critical, pragmatic approach to texts and their relationships to lived experience. This leads to a "resistance" version of critical theory, a version connected to critical ethnography, and partisan, critical inquiry committed to social criticism and the empowerment of individuals. As bricoleurs, critical theorists seek to produce practical, pragmatic knowledge, a bricolage that is cultural and structural, judged by its degree of historical situatedness and its ability to produce praxis or action.

Like Olesen's Chapter 7, this chapter is a call to arms. Getting mad is no longer enough. We must learn how to act in the world in ways that allow us to expose the workings of an invisible empire that leaves even more children behind.

▣ CULTURAL STUDIES

Cultural studies cannot be contained within a single framework. There are multiple cultural studies projects, including those connected to the Birmingham School and the work of Stuart Hall and his associates (see Hall, 1996). Cultural studies research is historically self-reflective, critical, interdisciplinary, conversant with high theory, and focused on the global and the local; it takes into account historical, political, economic, cultural, and everyday discourses. It focuses on "questions of community, identity, agency, and change" (Grossberg & Pollock, 1998).

In its generic form, cultural studies involves an examination of how the history people live is produced by structures that have been handed down from the past. Each version of cultural studies is joined by a three-fold concern with cultural texts, lived experience, and the articulated relationship between texts and everyday life. Within the cultural text tradition, some scholars examine the mass media and popular culture as sites where history, ideology, and subjective experiences come together. These scholars produce critical ethnographies of the audience in relation to particular historical moments. Other scholars read texts as sites where hegemonic meanings are produced, distributed, and consumed. Within the ethnographic tradition, there is a postmodern concern for the social text and its production.

The disciplinary boundaries that define cultural studies keep shifting, and there is no agreed upon standard genealogy of its emergence as a serious academic discipline. Nonetheless, there are certain prevailing tendencies, including feminist understandings of the politics of the everyday and the personal; disputes between proponents of textualism, ethnography, and autoethnography; and continued debates surrounding the dreams of modern citizenship.

The open-ended nature of the cultural studies project leads to a perpetual resistance against attempts to impose a single definition over the entire project. There are critical-Marxist, constructionist, and postpositivist paradigmatic strands within the formation, as well as emergent feminist and ethnic models. Scholars within the cultural studies project are drawn to historical realism and relativism as their ontology, to transactional epistemologies and dialogic methodologies, while remaining committed to a historical and structural framework that is action-oriented.

In Chapter 10, Michael D. Giardina and Josh L. Newman outline a performative, embodied, poststructural, contextualist, and globalized cultural studies project. They locate the bodies of cultural studies within a post-9/11 militarization of culture, a destabilized Middle East, and endless wars in Iraq and Afghanistan. Cultural studies' bodies are under duress, assailed by heteronormative logics of consumption, racism, and gender oppression. Drawing on their own research, Giardina and Newman outline a methodological program for a radically embodied cultural studies that is defined by its interest in lived, discursive, and contextual dimensions of reality, weaving back and forth between culturalist and realist agendas.

Theirs is a historically embodied, physical cultural studies. It works outward from the politically located body, locating that body in those historical structures that overdetermine meaning, identity, and opportunity. They seek a performative cultural studies that makes the world visible in ways that implement the goals of social justice and radical, progressive democracy. Thus do they move back and forth between the local and the global, the cultural and the real, the personal and the political, the embodied and the performative.

▣ CRITICAL HUMANISM AND QUEER THEORY

Critical race theory brought race and the concept of a complex racial subject squarely into qualitative inquiry. It remained for queer theory to do the same; namely, to question and deconstruct the concept of a unified sexual (and racialized) subject. In Chapter 11, Ken Plummer takes queer theory in a new direction. He writes from his own biography, a post-gay humanist, a sort of feminist, a little queer, a critical humanist who wants to move on. He thinks that in the postmodern moment certain terms, like *family* and much of our research methodology language, are obsolete. He calls them zombie categories. They are no longer needed. They are dead.

With the arrival of queer theory, the social sciences are in a new space. This is the age of postmodern fragmentation, globalization, and post-humanism. This is a time for new research styles, styles that take up the reflexive queer, polyphonic, narrative, ethical turn. Plummer's critical humanism, with its emphasis on symbolic interactionism, pragmatism, democratic thinking, storytelling, moral progress, and social justice, enters this space. It is committed to reducing human suffering, to an ethics of care and compassion, a politics of respect, and the importance of trust.

His queer theory is radical. It encourages the postmodernization of sexual and gender studies. It deconstructs all conventional categories of sexuality and gender. It is transgressive, gothic, and romantic. It challenges the heterosexual/homosexual binary; the deviance paradigm is abandoned. His queer methodology takes the textual turn seriously, endorses subversive ethnographies, scavenger methodologies, ethnographic performances, and queered case studies.

By troubling the place of the homo-heterosexual binary in everyday life, queer theory has created spaces for multiple discourses on gay, bisexual, transgendered, and lesbian subjects. This means researchers must examine how any social arena is structured, in part, by this homo-hetero dichotomy. They must ask how the epistemology of the closet is central to the sexual and material practices of everyday life. Queer theory challenges this epistemology, just as it deconstructs the notion of unified subjects. Queerness becomes a topic and a resource for investigating the way group boundaries are created, negotiated, and changed. Institutional and historical analyses are central to this project, for they shed light on how the self and its identities are embedded in institutional and cultural practices.

In a short postscript to his 2005 chapter, Plummer asks, in this current moment, "Is a global critical humanism possible—Is it possible to generate a transnational queer studies?" And, if so, what would it look like? He calls for a cosmopolitan methodology, a methodological open mindedness, a respect, and a willingness to listen, learn, and dialogue across the spaces of intimate citizenship.

▣ ASIAN EPISTEMOLOGIES

In Chapter 12, James H. Liu analyzes Asian epistemologies and their influence on contemporary social psychological research in Asia and elsewhere. He thus extends the qualitative research project into non-Western cultures, noting that the social sciences in Asia continue to be shaped by an imported Western logical positivism.

Recently, there has been a rise of indigenous Asian epistemologies and indigenous psychologies—from Japan to China, India, Taiwan, the Philippines, and Korea. Chinese indigenous psychology, for example, has its own journal, regular conferences, and preferred research methods. A highly pragmatic approach, rooted in research practices rather than epistemology, characterizes the modal Asian indigenous psychology. Hermeneutic and empiricist schools of thought compete for attention.

Much of Asian ontology is holistic; culture is embedded within the processes and objects of inquiry. Culture is understood to be historically constructed. Culture is constitutive of mind. People develop a theory of mind and construe themselves in and through language.

Tensions persist, and the positivist paradigm is still in ascendency. Asian scholars are in a transition period, and an epistemological breakthrough may be imminent. Even if the breakthrough does not occur, as Professor Liu observes, "even if the sky does not fall down, it may still be useful to work on the margins to connect the centers of an increasingly interconnected world of parallel and distributed cultural values" (p. 224).

▣ DISABILITY COMMUNITIES: TRANSFORMATIVE RESEARCH FOR SOCIAL JUSTICE

In Chapter 13, Donna Mertens, Martin Sullivan, and Hilary Stace outline the major interpretive contours of a social disability paradigm. Following the examples set by antiracist, feminist, and gay rights movements of the 1960s, disabled people throughout the Western world began to organize to challenge the oppressive stereotypes of disabled people and to propose ways to conduct nonoppressive, empowering disability research. Positivist, emancipatory, and transformative inquiry paradigms are compared and contrasted. The transformative model focuses, not on disability per se (as in the emancipatory model), but on the historical, cultural contexts surrounding disability, gender, race, sexual orientation, and social class. Interpretive and mixed methods of inquiry are employed under both paradigms. The transformative paradigm is activist, critical, and constructivist, embedded in social justice and human rights agendas. It combines the transformative lens with the arguments of disability rights theorists. Examples from work with indigenous peoples are offered, understanding that there is no single, homogenous, indigenous or disability community. Mertens et al. end on a powerful note: "The pathway to full realization of human rights and social justice for people with disability is not smooth . . . the transformative paradigm provides a way forward" (p. 237).

▣ CONCLUSION

The researcher-as-interpretive-bricoleur cannot afford to be a stranger to any of the paradigms and perspectives discussed in Part II of the *Handbook*. The researcher must understand the basic ethical, ontological, epistemological, and methodological assumptions of each and be able to engage them in dialogue. The differences between paradigms and perspectives have significant and important implications at the practical, material, everyday level. The blurring of paradigm differences is likely to continue, as long as proponents continue to come together to discuss their differences, while seeking to build on those areas where they are in agreement.

It is also clear that there is no single "truth." All truths are partial and incomplete. There will be no single conventional paradigm, as Lincoln and Guba (2000) argue, to which all social scientists might ascribe. We occupy a historical moment marked by multivocality, contested meanings, paradigmatic controversies, and new textual forms. This is an age of emancipation, freedom from the confines of a single regime of truth, emancipation from seeing the world in one color.

▣ REFERENCES

Foley, D., & Valenzuela, A. (2005). Critical ethnography: The politics of collaboration. In N. K. Denzin & Y. S. Lincoln (Eds.), *The SAGE handbook of qualitative research* (3rd ed., pp. 217–234). Thousand Oaks, CA: Sage.

Grossberg, L., & Pollock, D. (1998). Editorial statement. *Cultural Studies, 12*(2), 114.

Guba, E. (1990). The alternative paradigm dialog. In E. Guba (Ed.), *The paradigm dialog* (pp. 17–30). Newbury Park, CA: Sage.

Hall, S. (1996). Gramsci's relevance for the study of race and ethnicity. In D. Morley & K.-H. Chen (Eds.), *Stuart Hall: Critical dialogues in cultural studies* (pp. 411–444). London: Routledge.

Ladson-Bilings, G., & Donner, J. (2005). The moral activist role of critical race theory scholarship. In N. K. Denzin & Y. S. Lincoln (Eds.), *The SAGE handbook of qualitative research* (3rd ed., pp. 279–302). Thousand Oaks, CA: Sage.

Lincoln, Y. S., & Guba, E. (2000). Paradigmatic controversies, contradictions, and emerging confluences. In N. K. Denzin & Y. S. Lincoln (Eds.), *Handbook of qualitative research* (2nd ed., pp. 163–188). Thousand Oaks, CA: Sage.

6

PARADIGMATIC CONTROVERSIES, CONTRADICTIONS, AND EMERGING CONFLUENCES, REVISITED

Yvonna S. Lincoln, Susan A. Lynham, and Egon G. Guba

In our chapter for the first edition of the *Handbook of Qualitative Research* (Guba & Lincoln, 1994), we focused on the contention among various research paradigms for legitimacy and intellectual and paradigmatic hegemony. The postmodern paradigms that we discussed (postmodernist, critical theory, and constructivism)[1] were in contention with the received positivist and postpositivist paradigms for legitimacy and with one another for intellectual legitimacy. In the 15 years that have elapsed since that chapter was published, substantial changes have occurred in the landscape of social scientific inquiry. On the matter of legitimacy, we observe that readers familiar with the literature on methods and paradigms reflect a high interest in ontologies and epistemologies that differ sharply from those undergirding conventional social science, including, but not limited to, feminist theories, critical race and ethnic studies, queer theory, border theories, postcolonial ontologies and epistemologies, and poststructural and postmodern work. Second, even those established professionals trained in quantitative social science (including the two of *us)* want to learn more about qualitative approaches because new professionals being mentored in graduate schools are asking serious questions about and looking for guidance in qualitatively oriented studies and dissertations. Third, the number of qualitative texts, research papers, workshops, and training materials has exploded. Indeed, it would be difficult to miss the distinct turn of the social sciences toward more interpretive, postmodern, and critical practices and theorizing (Bloland, 1989, 1995). This nonpositivist orientation has created a context (surround) in which virtually no study can go unchallenged by proponents of contending paradigms. Furthermore, it is obvious that the number of practitioners of new paradigm inquiry is growing daily. The legitimacy of postpositivist and postmodern paradigms is well established and at least equal to the legitimacy of received and conventional paradigms (Denzin & Lincoln, 1994).

On the matter of hegemony, or supremacy, among postmodern paradigms, it is clear that Clifford Geertz's (1988, 1993) prophecy about the "blurring of genres" is rapidly being fulfilled. Inquiry methodology can no longer be treated as a set of universally applicable rules or abstractions.

Methodology is inevitably interwoven with and emerges from the nature of particular disciplines (such as sociology and psychology) and particular perspectives (such as Marxism, feminist theory, and queer theory). So, for instance, we can read feminist critical theorists such as Virginia Olesen (2000; Chapter 7, this volume) and Patricia Lather (2007) or queer theorists such as Joshua Gamson (2000), or we can follow arguments about teachers as researchers (Kincheloe, 1991) while we understand the secondary text to be teacher empowerment and democratization of schooling practices. Indeed, the various paradigms are beginning to "interbreed" such that two theorists previously thought to be in irreconcilable conflict may now appear, under a different theoretical rubric, to be informing one another's arguments. A personal example is our own work, which has been heavily influenced by action research practitioners and postmodern and poststructural critical theorists. Consequently, to argue that it is paradigms that are in contention is probably less useful than to probe where and how paradigms exhibit confluence and where and how they exhibit differences, controversies, and contradictions. As the field or fields of qualitative research mature and continue to add both methodological and epistemological as well as political sophistication, new linkages will, we believe, be found, and emerging similarities in interpretive power and focus will be discovered.

In our chapter in the first edition of this *Handbook,* we presented two tables that summarized our positions, first, on the axiomatic nature of paradigms (the paradigms we considered at that time were positivism, postpositivism, critical theory, and constructivism; Guba & Lincoln, 1994, p. 109, Table 6.1); and second, on the issues we believed were most fundamental to differentiating the four paradigms (p. 112, Table 6.2). These tables are reproduced here in slightly different form as a way of reminding our readers of our previous statements. The axioms defined the ontological, epistemological, and methodological bases for both established and emergent paradigms; these are shown here in Table 6.1. The issues most often in contention were inquiry aim, nature of knowledge, the way knowledge is accumulated, goodness (rigor and validity) or quality criteria, values, ethics, voice, training (the nature of preparatory work that goes into preparing a researcher to engage in responsible and reflective fieldwork), accommodation, and hegemony; these are shown in Table 6.2. An examination of these two tables will reacquaint the reader with our original *Handbook* treatment; more detailed information is, of course, available in our original chapter. Readers will notice that in the interim, Susan Lynham has joined us in creating a new and more substantial version of one of the tables, one that takes into account both our own increasing understandings and her work with us and students in enlarging the frames of reference for new paradigm work.

Since publication of that chapter, at least one set of authors, John Heron and Peter Reason, has elaborated on our tables to include the *participatory/cooperative* paradigm (Heron, 1996; Heron & Reason, 1997, pp. 289–290). Thus, in addition to the paradigms of positivism, postpositivism, critical theory, and constructivism, we add the participatory paradigm in the present chapter (this is an excellent example, we might add, of the hermeneutic elaboration so embedded in our own view, constructivism; see, e.g., Guba 1990, 1996). Our aim here is to extend the analysis further by building on Heron and Reason's additions and by rearranging the issues to reflect current thought. The issues we have chosen include our original formulations and the additions, revisions, and amplifications made by Heron and Reason (1997) as well as by Lynham, and we have also chosen what we believe to be the issues most important today. We should note that *important* means several things to us. An important topic may be one that is widely debated (or even hotly contested)—validity is one such issue. An important issue may be one that bespeaks a new awareness (an issue such as recognition of the role of values). An important issue may be one that illustrates the influence of one paradigm on another (such as the influence of feminist, action research, critical theory, and participatory models on researcher conceptions of action within and with the community in which research is carried out). Or issues may be important because new or extended theoretical or field-oriented treatments for them are newly available—voice and reflexivity are two such issues. Important may also indicate that new or emerging treatments contradict earlier formulations in such a way that debates about method, paradigms, or ethics take the forefront once again, resulting in rich and fruitful conversations about what it means to do qualitative work. *Important* sometimes foregrounds larger social movements that undermine qualitative research in the name of science or that declare there is only one form of science that deserves the name (National Research Council, 2002).

Table 6.1 Basic Beliefs (Metaphysics) of Alternative Inquiry Paradigms

Item	*Positivism*	*Postpositivism*	*Critical Theory et al.*	*Constructivism*
Ontology	Naïve realism—"real" reality but apprehendible	Critical realism—"real" reality but only imperfectly and probabilistically apprehendible	Historical realism—virtual reality shaped by social, political, cultural, economic, ethnic, and gender values; crystallized over time	Relativism—local and specific constructed and co-constructed realities
Epistemology	Dualist/objectivist; findings true	Modified dualist/objectivist; critical tradition/community; findings probably true	Transactional/subjectivist; value-mediated findings	Transactional/subjectivist; created findings
Methodology	Experimental/manipulative; verification of hypotheses; chiefly quantitative methods	Modified experimental/manipulative; critical multiplism; falsification of hypotheses; may include qualitative methods	Dialogic/dialectical	Hermeneutical/dialectical

Table 6.2 Paradigm Positions on Selected Practical Issues

Item	Positivism	Postpositivism	Critical Theory et al.	Constructivism
Inquiry aim	Explanation: prediction and control		Critique and transformation; restitution and emancipation	Understanding; reconstruction
Nature of knowledge	Verified hypotheses established as facts or laws	Nonfalsified hypotheses that are probable facts or laws	Structural/historical insights	Individual or collective reconstructions coalescing around consensus
Knowledge accumulation	Accretion—"building blocks" adding to "edifice of knowledge"; generalizations and cause-effect linkages		Historical revisionism; generalization by similarity	More informed and sophisticated reconstructions; vicarious experience
Goodness or quality criteria	Conventional benchmarks of "rigor": internal and external validity, reliability, and objectivity		Historical situatedness; erosion of ignorance and misapprehension; action stimulus	Trustworthiness and authenticity, including catalyst for action
Values	Excluded—influence denied		Included—formative	Included—formative
Ethics	Extrinsic: tilt toward deception		Intrinsic: moral tilt toward revelation	Intrinsic: process tilt toward revelation; special problems
Voice	"Disinterested scientist" as informer of decision makers, policy makers, and change agents		"Transformative intellectual" as advocate and activist	"Passionate participant" as facilitator of multivoice reconstruction
Training	Technical and quantitative; substantive theories	Technical; quantitative and qualitative; substantive theories	Resocialization; qualitative and quantitative; history; values of altruism, empowerment, and liberation	
Accommodation	Commensurable		Incommensurable with previous two	
Hegemony	In control of publication, funding, promotion, and tenure		Seeking recognition and input; offering challenges to predecessor paradigms, aligned with postcolonial aspirations	

Table 6.3 Basic Beliefs of Alternative Inquiry Paradigms—Updated

Issue	Positivism	Postpositivism	Critical Theory et al.	Constructivism	Participatory[a]
Ontology	Naïve realism—"real" reality but apprehendible	Critical realism—"real" reality but only imperfectly and probabilistically apprehendible	Historical realism—virtual reality shaped by social, political, cultural, economic, ethnic, and gender values; crystallized over time	Relativism— local and specific co-constructed realities	Participative reality— subjective-objective reality, co-created by mind and given cosmos
Epistemology	Dualist/objectivist; findings true	Modified dualist/objectivist; critical tradition/community; findings probably true	Transactional/subjectivist; value-mediated findings	Transactional/subjectivist; co-created findings	Critical subjectivity in participatory transaction with cosmos; extended epistemology of experiential, propositional, and practical knowing; co-created findings
Methodology	Experimental/manipulative; verification of hypotheses; chiefly quantitative methods	Modified experimental/manipulative; critical multiplism; falsification of hypotheses; may include qualitative methods	Dialogic/dialectical	Hermeneutical/dialectical	Political participation in collaborative action inquiry; primacy of the practical; use of language grounded in shared experiential context

a. Entries in this column are based on Heron and Reason (1997).

Table 6.3 reprises the original Table 8.3 but adds the axioms of the participatory paradigm proposed by Heron and Reason (1997). Table 6.4 deals with seven issues and represents an update of selected issues first presented in the old Table 8.4. *Voice* in the 1994 version of Table 6.2 has been renamed *inquirer posture,* and we have inserted a redefined *voice* in the current table.

In all cases except inquirer posture, the entries for the participatory paradigm are those proposed by Heron and Reason; in the one case not covered by them, we have added a notation that we believe captures their intention. We make no attempt here to reprise the material well discussed in our earlier handbook chapter. Instead, we focus primarily on the issues in Table 6.4: axiology; accommodation and commensurability; action; control; foundations of truth and knowledge; validity; and voice, reflexivity, and postmodern textual representation. In addition, we take up the issues of cumulation and mixed methods since both prompt some controversy and friendly debate within the qualitative camp. We believe these issues to be the most important at this time. While we believe these issues to be the most contentious, we also believe they create the intellectual, theoretical, and practical space for

dialogue, consensus, and confluence to occur. There is great potential for interweaving of viewpoints, for the incorporation of multiple perspectives, and for borrowing, or *bricolage,* where borrowing seems useful, richness-enhancing, or theoretically heuristic. For instance, even though we are ourselves social constructivists or constructionists, our call to action embedded in the authenticity criteria we elaborated in *Fourth Generation Evaluation* (Guba & Lincoln, 1989) reflects strongly the bent to action embodied in critical theorists' and participatory action research perspectives well outlined in the earlier editions (Kemmis & McTaggart, 2000; Kincheloe & McLaren, 2000). And although Heron and Reason have elaborated a model they call the *cooperative paradigm,* careful reading of their proposal reveals a form of inquiry that is postpostpositive, postmodern, and criticalist in orientation.

As a result, the reader familiar with several theoretical and paradigmatic strands of research will find that echoes of many streams of thought come together in the extended table. What this means is that the categories, as Laurel Richardson (personal communication, September 12, 1998) has pointed out, "are fluid, indeed what should be a category keeps altering, enlarging." She notes that "even as [we] write, the boundaries between the paradigms are

(Text Continued on page 116)

Table 6.4 Paradigm Positions on Selected Issues—Updated

Issue	Positivism	Postpositivism	Critical Theories	Constructivism	Participatory[a]
Nature of knowledge	Verified hypotheses established as facts or laws	Nonfalsified hypotheses that are probable facts or laws	Structural/historical insights	Individual and collective reconstructions sometimes coalescing around consensus	Extended epistemology: primacy of practical knowing; critical subjectivity; living knowledge
Knowledge accumulation	Accretion—"building blocks" adding to "edifice of knowledge"; generalizations and cause-effect linkages		Historical revisionism; generalization by similarity	More informed and sophisticated reconstructions; vicarious experience	In communities of inquiry embedded in communities of practice
Goodness or quality criteria	Conventional benchmarks of "rigor": internal and external validity, reliability, and objectivity		Historical situatedness; erosion of ignorance and misapprehensions; action stimulus	Trustworthiness and authenticity including catalyst for action	Congruence of experiential, presentational, propositional, and practical knowing; leads to action to transform the world in the service of human flourishing
Values	Excluded—influence denied		Included—formative		
Ethics	Extrinsic—tilt toward deception		Intrinsic—moral tilt toward revelation	Intrinsic—process tilt toward revelation	
Inquirer posture	"Disinterested scientist" as informer of decision makers, policy makers, and change agents		"Transformative intellectual" as advocate and activist	"Passionate participant" as facilitator of multivoice reconstruction	Primary voice manifest through aware self-reflective action; secondary voices in illuminating theory, narrative, movement, song, dance, and other presentational forms
Training	Technical and quantitative; substantive theories	Technical; quantitative and qualitative; substantive theories	Resocialization; qualitative and quantitative; history; values of altruism, empowerment and liberation		Coresearchers are initiated into the inquiry process by facilitator/researcher and learn through active engagement in the process; facilitator/researcher requires emotional competence, democratic personality and skills

a. Entries in this column are based on Heron and Reason (1997), except for "ethics" and "values."

Table 6.5 Themes of Knowledge: An Heuristic Schema of Inquiry, Thought, and Practice*

	Positivism	Postpositivism	Critical (+ Feminism + Race)	Constructivism (or Interpretivist)	Participatory (+ Postmodern)
THEMES OF KNOWLEDGE: Inquiry Aims, Ideals, Design, Procedures, and Methods					
	Realists, "hard science" researchers	*A modified form of positivism*	*Create change, to the benefit of those oppressed by power*	*Gain understanding by interpreting subject perceptions*	*Transformation based on democratic participation between researcher and subject*
A: BASIC BELIEFS (METAPHYSICS) OF ALTERNATIVE INQUIRY PARADIGMS					
Ontology *The worldviews and assumptions in which researchers operate in their search for new knowledge* (Schwandt, 2007, p. 190). *The study of things that exist and the study of what exists* (Latsis, Lawson, & Martins, 2007). What is the nature of reality? (Creswell, 2007).	Belief in a single identifiable reality. There is a single truth that can be measured and studied. The purpose of research is to predict and control nature (Guba & Lincoln, 2005; Merriam, 1991; Merriam, Caffarella, & Baumgartner, 2007).	Recognize that nature can never fully be understood. There is a single reality, but we may not be able to fully understand what it is or how to get to it because of the hidden variables and a lack of absolutes in nature (Guba & Lincoln, 2005; Merriam, 1991; Merriam et al., 2007).	Human nature operates in a world that is based on a struggle for power. This leads to interactions of privilege and oppression that can be based on race or ethnicity, socioeconomic class, gender, mental or physical abilities, or sexual preference (Bernal, 2002; Giroux, 1982; Kilgore, 2001).	Relativist: Realities exist in the form of multiple mental constructions, socially and experientially based, local and specific, dependent for their form and content on the persons who hold them (Guba, 1990, p. 27). Relativism: local and specific constructed and co-constructed realities (Guba & Lincoln, 2005, p. 193). "Our individual personal reality—the way we think life is and the part we are to play in it is—*self-created. We put together our own personal reality*" (Guba & Lincoln, 1985, p. 73). Multiple realities exist and are dependent on the individual (Guba, 1996). "Metaphysics that embraces relativity" (Josselson, 1995, p. 29).	Participative reality: subjective-objective reality, co-created by mind and the surrounding cosmos (Guba & Lincoln, 2005, p. 195). Freedom from objectivity with a new understanding of relation between self and other (Heshusius, 1994, p. 15). Socially constructed: similar to constructive, but do not assume that rationality is a means to better knowledge (Kilgore, 2001, p. 54). Subjective–objective reality: Knowers can only be knowers when known by other knowers. Worldview based on participation and participative realities (Heron & Reason, 1997).

Epistemology						
Epistemology *The process of thinking. The relationship between what we know and what we see.* *The truths we seek and believe as researchers* (Bernal, 2002; Guba & Lincoln, 2005; Lynham & Webb-Johnson, 2008; Pallas, 2001). What is the relationship between the researcher and that being researched? (Creswell, 2007).	Belief in total objectivity. There is no reason to interact with who or what researchers study. Researchers should value only the scientific rigor and not its impact on society or research subjects (Guba & Lincoln, 2005; Merriam, 1991; Merriam et al., 2007).	Assume we can only approximate nature. Research and the statistics it produces provide a way to make a decision using incomplete data. Interaction with research subjects should be kept to a minimum. The validity of research comes from peers (the research community), not from the subjects being studied (Guba & Lincoln, 2005; Merriam, 1991; Merriam et al., 2007).	"We practice inquiries that make sense to the public and to those we study" (Preissle, 2006, p. 636). Assumes that reality as we know it is constructed intersubjectively through the meanings and understandings developed socially and experientially (Guba & Lincoln, 1994). *To me this means that we construct knowledge through our lived experiences and through our interactions with other members of society. As such, as researchers, we must participate in the research process with our subjects to ensure we are producing knowledge that is reflective of their reality.*	Research is driven by the study of social structures, freedom and oppression, and power and control. Researchers believe that the knowledge that is produced can change existing oppressive structures and remove oppression through empowerment (Merriam, 1991).	Subjectivist: Inquirer and inquired into are fused into a single entity. Findings are literally the creation of the process of interaction between the two (Guba, 1990, p. 27). Transactional/subjectivist: co-created findings (Guba & Lincoln, 2005, p. 195). The philosophical belief that people construct their own understanding of reality; we construct meaning based on our interactions with our surroundings (Guba & Lincoln, 1985). "Social reality is a construction based upon the actor's frame of reference within the setting" (Guba & Lincoln, 1985, p. 80).	Holistic: "Replaces traditional relation between 'truth' and 'interpretation' in which the idea of truth antedates the idea of interpretation" (Heshusius, 1994, p. 15). Critical subjectivity in participatory transaction with cosmos; extended epistemology of experiential, propositional, and practical knowing; co-created findings (Guba & Lincoln, 2005, p. 195). Critical subjectivity: Understanding how we know what we know and the knowledge's consuming relations. Four ways of knowing: (1) experiential, (2) presentational, (3) propositional, and (4) practical (Heron & Reason, 1997).

(Continued)

Table 6.5 (Continued)

	Findings are due to the interaction between the researcher and the subject (Guba, 1996). "We cannot know the real without recognizing our own role as knowers" (Flax, 1990). "Simultaneously empirical, intersubjective, and process-oriented" (Flax, 1990). "We are studying ourselves studying ourselves and others" (Preissle, 2006, p. 691). Assumes that we cannot separate ourselves from what we know. The investigator and the object of investigation are linked such that who we are and how we understand the world is a central part of how we understand ourselves, others, and the world (Guba & Lincoln, 1994). *This means we are shaped by our lived experiences, and these will always come out in the knowledge we generate as researchers and in the data generated by our subjects.*				Political participation in collaborative action inquiry, primacy of the practical; use of language grounded in shared experiential context (Guba & Lincoln, 2005, p. 195). Use deconstruction as a tool for questioning prevailing representations of learners and learning in the adult education literature; this discredits the false binaries that structure a
				Hermeneutic, dialectic: Individual constructions are elicited and refined hermeneutically, and compared and contrasted dialectically, with aim of generating one or a few constructions on which there is substantial consensus (Guba, 1990, p. 27). Hermeneutical; dialectical (Guba & Lincoln, 1985, p. 195). Hermeneutical discussion (Geertz, 1973).	
			Dialogic/Dialectical (Guba & Lincoln, 2005) Search for participatory research, which empowers the oppressed and supports social transformation and revolution (Merriam, 1991, p. 56).		
		Researchers should attempt to approximate reality. Use of statistics is important to visually interpret our findings. Belief in the scientific method. Research is the effort to create new knowledge, seek scientific discovery. There is an attempt to ask more questions than positivists			
Methodology *The process of how we seek out new knowledge. The principles of our inquiry and how inquiry should proceed* (Schwandt, 2007, p. 190). What is the process of research? (Creswell, 2007).	Belief in the scientific method. Value a "gold standard" for making decisions. Grounded in the conventional hard sciences. Belief in the falsification principle (results and findings are true until disproved). Value data produced by studies that can be replicated (Merriam, 1991).				

(Continued)

because of the unknown variables involved in research. There is a unifying method. Distance the researcher to gain objectivity. Use the hypothetical deductive method—hypothesize, deduce, and generalize (Guba & Lincoln, 2005; Merriam, 1991; Merriam et al., 2007).	Hermeneutics (interpretation, i.e., recognition and explanation of metaphors) and comparing and contrasting dialectics (resolving disagreements through rational discussion) (Guba, 1996). "Everyday consciousness of reality and its chameleonlike quality pervade politics, the media, and literature" (Guba & Lincoln, 1985, p. 70). "The construction of realities must depend on some form of consensual language" (Guba & Lincoln, 1985, p. 71). "Stock taking and speculations regarding the future nevertheless help us comprehend the past and the present and aid our choices for the futures we desire" (Preissle, 2006, p. 686). Interpretive approaches rely heavily on naturalistic methods (interviewing and observation and analysis of existing texts (Angen, 2000). These methods ensure an adequate dialog between the researchers and those with whom they interact in order to collaboratively construct a meaningful reality (Angen, 2000). Generally, meanings are emergent from the research process (Angen, 2000). Typically, qualitative methods are used (Angen, 2000). *Hermeneutic Cycle: Actions lead to collection of data, which leads to interpretation of data which spurs action based on data. (Class notes, 2008)*	communication and challenges the assertions of what is to be included or excluded as normal, right, or good (Kilgore, 2001, p. 56). Experiential knowing is through face-to-face learning, learning new knowledge through the application of the knowledge. Democratization and co-creation of both content and method. Engage together in democratic dialogue as co-researchers and as co-subjects (Heron & Reason, 1997).

Table 6.5 (Continued)

THEMES OF KNOWLEDGE: Inquiry Aims, Ideals, Design, Procedures, and Methods

	Positivism	Postpositivism	Critical (+ Feminist + Race)	Constructivism (or Interpretivist)	Participatory (+ Postmodern)
B: PARADIGM POSITIONS ON SELECTED PRACTICAL ISSUES					
Inquiry aim *The goals of research and the reason why inquiry is conducted. What are the goals and the knowledge we seek?* (Guba & Lincoln, 2005).	Research should be geared toward the prediction and control of natural phenomena. Demonstrate laws that can be applied to natural order.	Researchers attempt to get as close to the answer as possible. Cannot fully attain reality but can approximate it.	Aim of inquiry is to find the social power structure in an attempt to discover the truth as it relates to social power struggles (Giroux, 1982; Merriam, 1991). Transformation (Guba & Lincoln, 2005). Stimulate oppressed people to rationally scrutinize all aspects of their lives to reorder their collective existence on the basis of the understanding it provides, which will ultimately change social policy and practice (Fay, 1987).	To understand and interpret through meaning of phenomena (obtained from the joint construction/reconstruction of meaning of lived experience); such understanding is sought to inform praxis (improved practice). Understanding/reconstruction (Guba & Lincoln, 2005, p. 194). Consensus toward understanding of culture (Geertz, 1973). Scientific generalizations may not fit in solving all problems (Guba, 1996). An approach needed to fill in the gaps between theory and practice (Guba, 1996). The essential message of hermeneutics is that to be human is to mean, and only by investigating the multifaceted nature of human meaning can we approach the understanding of people (Josselson, 1995).	What is the form and nature of reality and, therefore, what is there that can be known about it? What is the relationship between the knower or would-be knower and what can be known? How can the inquirer . . . go about finding out whatever he or she believes can be known about? What is intrinsically valuable in human life, in particular what sort of knowledge, if any, is intrinsically valuable? (Heron & Reason, 1997).
Nature of knowledge *How researchers view the knowledge that is generated through inquiry research* (Guba & Lincoln, 2005).	Hypothesis is verified as fact.	There is a correct single truth, which may have multiple hidden values and variables that prevent ever fully knowing the answer.	Knowledge is viewed as "subjective, emancipatory, and productive of fundamental social change" (Merriam, 1991, p. 53). Rationality is a means to better knowledge. Knowledge is a logical outcome of human interests (Kilgore, 2001).	The constructed meanings of actors are the foundation of knowledge. Individual and collective reconstructions sometimes coalescing around consensus (Guba & Lincoln, 2005, p. 196). Collective reconstruction coalescing around consensus on meaning of culture (Geertz, 1973).	Believe knowledge is socially constructed and takes the form in the eyes of the knower rather than being formulated from an existing reality (Kilgore, 2001, p. 51). Extended epistemology: primacy of practical knowing; living knowledge subjectivity; living knowledge (Guba & Lincoln, 2005, p. 196).

Structural/historical insights (Guba & Lincoln, 2005). Believe knowledge is socially constructed and takes the form in the eyes of the knower rather than being formulated from an existing reality (Kilgore, 2001, p. 51).	People construct their own understanding of reality (Guba, 1990). "Realities are taken to exist in the form of multiple mental constructions that are socially and experientially based, local and specific, and dependent for their form and content on the persons who hold them" (Guba, 1990, p. 27). Knowledge is cognitively constructed from experience and interaction of the individual with others and the environment (Class Notes, 2008). Subjective and co-created through the process of interaction between the *inquirer* and the *inquired into* (Class Notes, 2008). Knowledge is socially constructed, not discovered (Class Notes, 2008). "Observing dialogue allows us to construct a meta-narrative of whole people, not reducing people to parts, but recognizing in the interplay of parts the essence of wholeness. Only then can we begin to imagine the real" (Josselson, 1995, p. 42).	Experiential participation. Propositional knowing. Subjective-objective reality. Practical knowing is knowing how to do something, demonstrated in a skill or competence (Heron & Reason, 1997). The constructed meanings of actors are the foundation of knowledge. Individual and collective reconstructions sometimes coalescing around consensus (Guba & Lincoln, 2005, p. 196). Collective reconstruction coalescing around consensus on meaning of culture (Geertz, 1973). People construct their own understanding of reality (Guba, 1990). "Realities are taken to exist in the form of multiple mental constructions that are socially and experientially based, local and specific, and dependent for their form and content on the persons who hold them" (Guba, 1990, p. 27). Knowledge is cognitively constructed from experience and interaction of the individual with others and the environment (Epistemology Class Notes). Subjective and co-created through the process of interaction between the inquirer and the inquired into (Epistemology Class Notes) Knowledge is socially constructed, not discovered (Epistemology class notes)

(Continued)

Table 6.5 (Continued)

Knowledge accumulation					
How does knowledge build off of prior knowledge to develop a better understanding of the subject or field? (Guba & Lincoln, 2005).	Seek to find cause-and-effect linkages that can build into a better understanding of the field. This can become law over time through use of the scientific method (Merriam, 1991).	Use statistics and other techniques to get as close as possible to reality. Although it can never be attained, approximations of reality can be made to develop further understanding.	Knowledge accumulation is based on historical perspective and revision of how history is viewed so that it no longer serves as an oppressive tool by those with structural power (Guba & Lincoln, 2005).	More informed and sophisticated reconstructions; vicarious experience (Guba & Lincoln, 2005, p. 196). "Since the 1980s, for example, qualitative inquiry has been much influenced by the poststructural and postmodern developments from the arts and the humanities. These bring a sensitivity to language, especially to linguistic assumptions embedded in disciplinary terminology (e.g. Scheurich, 1996) that has challenged scholars working in post-positivist, interpretive, and critical traditions" (Preissle, 2006, p. 688).	In communities of inquiry embedded in communities of practice (Guba & Lincoln, 2005, p. 196). "Mind's conceptual articulation of the world is grounded in its experiential participation in what is present, in what there is" … Experiential knowing consists of conceptual, symbolic frameworks of conceptual, propositional knowing" (Heron & Reason, 1997, pp. 277–278).
Goodness or quality criteria					
How researchers judge the quality of inquiry (Guba & Lincoln, 2005).	Rigorous data produced through scientific research.	Statistical confidence level and objectivity in data produced through inquiry.	The value is found in the erosion of unearned privileges and its ability to impart action for the creation of a more fair society (Giroux, 1982; Guba & Lincoln, 2005).	Intersubjective agreement and reasoning among actors, reached through dialogue; shared conversation and construction. Trustworthiness and authenticity, including catalyst for action (Guba & Lincoln, 2005, p. 196). Credibility, transferability, dependability, and confirmability (Guba & Lincoln, 2005). "To interrogate objectivity and subjectivity and their relationship to one another" (Preissle, 2006, p. 691).	Congruence of experiential, presentational, and practical knowing; leads to action to transform the world in the service of human flourishing (Guba & Lincoln, 2005, p. 196). Intersubjective agreement and reasoning among actors, reached through dialogue; shared conversation and construction. Trustworthiness and authenticity, including catalyst for action (Guba & Lincoln, 2005, p. 196). Credibility, transferability, dependability, and confirmability (Guba & Lincoln, 2005). "To interrogate objectivity and subjectivity and their relationship to one another" (Preissle, 2006, p. 691).

B: PARADIGM POSITIONS ON SELECTED PRACTICAL ISSUES continued

	Positivism	Postpositivism	Critical (+ Feminist + Race)	Constructivism (or Interpretivist)	Participatory (+ Postmodern)
Values *What do researchers seek as important products within inquiry research? (Guba & Lincoln, 2005).*	Standards-based research. Value is found in the scientific method. Gold standard is scientific rigor.	Can find useful information even if data are incomplete and contain hidden values.	Included, formative (Guba & Lincoln, 2005). Researchers seek data that can be transformative and useful in imparting social justice (Giroux, 1982). Value is found in the reasoned reflection and the change in practice (Creswell, 2007). Values of research produced should include: rational self-clarity, collective autonomy, happiness, justice, bodily pleasure, play, love, aesthetic self-expression, and other values within these primary values (Fay, 1987).	Are personally relative and need to be understood. Inseparable from the inquiry and outcomes. (Class Notes, 2008). Included; formative (Guba & Lincoln, 2005, p. 194).	Included, formative (Guba & Lincoln, 2005, p. 196). Values are personally relative and need to be understood (Epistemology Class Notes).
Ethics *The interaction and relationship between the researcher and the subject as well as the effect inquiry research has on populations (Schwandt, 2007).*	Belief that the data drive the side effects of any research. The effort is to study nature, not to influence how nature affects populations (Guba & Lincoln, 2005).	Attempt to be as statistically accurate in their interpretation of reality as possible. Effect on others is not taken into account because research is driven to gain accuracy, not influence populations.	Frankfurt School of thought: Research is tied to a specific interest in the development of a society without injustice (Giroux, 1982).	Intrinsic: process tilt toward revelation; special problems (Guba & Lincoln, 2005, p. 196). Included in all aspects of inquiry and examination of culture (Geertz, 1973).	Intrinsic: process tilt toward revelation (Guba & Lincoln, 2005, p. 196). Included in all aspects of inquiry and examination of culture (Geertz, 1973).

(Continued)

Table 6.5 (Continued)

Voice *Who narrates the research that is produced? Qualitative approach: The ability to present the researcher's material along with the story of the research subject (Guba & Lincoln, 2005).* *What is the language of research? (Creswell, 2007).*	The data speak for themselves. Consistent findings from inquiry leads to the researcher being disinterested in effect (Guba & Lincoln, 2005).	Researchers are to inform populations using the data produced through their inquiry (Guba & Lincoln, 2005).	The data are created with the intent of producing social change and imparting a social justice that leads to equal rights for all (Giroux, 1982). (Advocate/Activist).	"Passionate participant" as facilitator of multivoice reconstruction (Guba & Lincoln, 2005). Facilitator of multivoice reconstruction of culture (Geertz, 1973). *This means that while critical theorists attempt to get involved in their research to change the power structure, researchers in this paradigm attempt to gain increased knowledge regarding their study and subjects by interpreting how the subjects perceive and interact within a social context.*	"Passionate participant" as facilitator of multivoice reconstruction (Guba & Lincoln, 2005). Facilitator of multivoice reconstruction of culture (Geertz, 1973).
Training *How are researchers prepared to conduct inquiry research?*	Researchers are trained in a technical and very quantitative way (Guba & Lincoln, 2005). Prescribe scientific method.	Researchers are trained in a technical and very quantitative way but also have the ability to conduct mixed-methods research (Guba & Lincoln, 2005).	Researchers are trained using both qualitative and quantitative approaches. They study history and social science to understand empowerment and liberation (Guba & Lincoln, 2005).	Resocialization; qualitative and quantitative; history, values of altruism, empowerment, and liberation (Guba & Lincoln, 2005, p. 196).	Co-researchers are initiated into the inquiry process by facilitator/researcher and learn through active engagement in the process; facilitator/researcher requires emotional competence, democratic personality, and skills (Guba & Lincoln, 2005, p. 196).
Inquirer posture *The point of view in which the researcher operates. How does the researcher approach the inquiry process? (Guba & Lincoln, 2005).*	Disinterested scientist. Researchers should remain distant from the change process and should not attempt to influence decisions (Guba & Lincoln, 2005).	Researchers are removed from the process, but concerned about its results (Guba & Lincoln, 2005).	The researcher serves as an activist and a transformative intellectual. The researcher understands a way of producing a fair society through social justice (Bernal, 2002; Giroux, 1982; Guba & Lincoln, 2005; Merriam, 1991).	A co-constructor of knowledge, of understanding and interpretation of the meaning of lived experiences (Guba & Lincoln, 2005, p. 196).	Primary voice manifested through aware self-reflective action; secondary voices in illuminating theory, narrative, movement, song, dance, and other presentational forms (Guba & Lincoln, 2005, p. 196). Can include alternative forms of data representation including film and ethnography (Eisner, 1997).

THEMES OF KNOWLEDGE: Inquiry Aims, Ideals, Design, Procedures, and Methods

	Positivism	Postpositivism	Critical (+ Feminist + Race)	Constructivism (or Interpretivist)	Participatory (+ Postmodern)
Accommodation *What needs are provided by the inquiry research? (Guba & Lincoln, 2005).*	Commensurable: Research has a common unit for study and analysis (Guba & Lincoln, 2005, p. 194).	Commensurable: Research has a common unit for study and analysis (Guba & Lincoln, 2005, p. 194).	Incommensurable: Data produced do not have to be from a common unit of measurement. Approaches research with different styles and methods that can produce multiple forms of data (Guba & Lincoln, 2005).	Incommensurable with positivism and postpositivism; commensurable with critical and participatory inquiry (Guba & Lincoln, 2005, p. 194). Some accommodation with criticalist and participatory methods of examining culture (Geertz, 1973). Incommensurable: Data produced do not have to be from a common unit of measurement. Approaches research with different styles and methods that can produce multiple forms of data (Guba & Lincoln, 2005).	Incommensurable: Data produced does not have to be from a common unit of measurement. Approaches research with different styles and methods that can produce multiple forms of data (Guba, & Lincoln, 2005). Some accommodation with criticalist and participatory methods of examining culture (Geertz, 1973).
Hegemony *The influence researchers have on others. Who has the power in inquiry and what is inquired. Presenting definition of reality (Kilgore, 2001).*	Belief that research should have the influence – not the person conducting the inquiry. Aim is to produce truth, not provide ways for that reality to affect others.	Statistical analysis of reality will produce data from which decisions can be made. Ultimately, the researcher is in charge of the inquiry process (Guba & Lincoln, 2005, p.194).	Research demonstrates the interactions of privilege and oppression as they relate to race/ethnicity, gender, class, sexual orientation, physical or mental ability, and age (Kilgore, 2001).	Seeks recognition and input; offers challenges to predecessor paradigms, aligned with postcolonial aspirations (Guba & Lincoln, 2005, p. 196). *Postcolonial is in reference to theories that deal with the cultural legacy of colonial rule (Gandhi, 1998).*	Power is a factor in what and how we know (Kilgore, 2001, p. 51).

C: CRITICAL ISSUES OF THE TIME

	Positivism	Postpositivism	Critical (+ Feminist + Race)	Constructivism (or Interpretivist)	Participatory (+ Postmodern)
Axiology *How researchers act based on the research they produce—also the criteria of values and value judgments especially in ethics (Merriam-Webster, 1997).*	Researchers should remain distant from the subject so their actions are to not have influence on populations—only the laws their inquiry produces (Guba & Lincoln, 2005).	Researchers should attempt to gain a better understanding of reality and as close as possible to truth through the use of statistics that explains and describes what is known as	Researchers seek to change existing education as well as other social institutions' policies and practice (Bernal, 2002).	Propositional, transactional knowing is instrumentally valuable as a means to social emancipation, which is an end in itself, is intrinsically valuable (Guba & Lincoln, 2005, p. 198).	Practical knowing how to flourish with a balance of autonomy, co-operation and hierarchy in a culture is an end in itself, is intrinsically valuable (Heron & Reason, 1997).

(Continued)

Table 6.5 (Continued)

What is the role of values? (Creswell, 2007).	reality (Guba & Lincoln, 2005).	Attempt to conduct research to improve social justice and remove barriers and other negative influences associated with social oppression (Giroux, 1982).	Emancipatory, but longer term, more reflective versus critical theory's desire for immediate results. "Intellectual digestion"	What is the purpose for which we create reality? To change the world or participation implies engagement, which implies responsibility. In terms of human flourishing, social practices and institutions need to enhance human associations by integration of these three principles; deciding for others with others and for ones self (Heron & Reason, 1997).
Accommodation and commensurability *Can the paradigm accommodate other types of inquiry?* (Guba & Lincoln, 2005). Can the results of inquiry accommodate each other? (Guba & Lincoln, 1989). Can the paradigms be merged together to make an overarching paradigm? (Guba & Lincoln, 1989).	According to Guba and Lincoln, all positivist forms are commensurable. The data produced are equal in measure to all other data created (Guba & Lincoln, 2005).	There is a priority or rank order to data created by different forms of research. Because critical researchers want to transform society, critical theory data must come before all other forms. (Incommensurable with empirical-analytical epistemologies and accommodates different forms of research paradigms) (Guba & Lincoln, 2005; Skrtic, 1990).	Incommensurable with positivistic forms; some commensurability with constructivist, criticalist, and participatory approaches, especially as they merge in liberationist approaches outside the West (Guba & Lincoln, 2005, p. 198). Commensurable with other modern paradigms; exception: attempt to understand a problem, but not transform (effect a change). Accommodates critical and participatory approaches to understanding of culture (Geertz, 1973). "Qualitative inquiry is composed of multiple and overlapping communities of practice. Many qualitative inquirers are members of several of these communities" (Preissle, 2006, p. 692).	Incommensurable with positivistic forms; some commensurability with constructivist, criticalist, and participatory approaches, especially as they merge in liberationist approaches outside the West (Guba & Lincoln, 2005, p. 198).

Action					
What is produced as a result of the inquiry process beyond the data? How does society use the knowledge generated? (Guba & Lincoln, 2005).	Researchers are to remain strictly objective, therefore do not concern themselves with the action that is produced as a result of inquiry research (Guba & Lincoln, 2005, p. 198).	Researchers are to remain strictly objective, therefore do not concern themselves with the action that is produced as a result of inquiry research (Guba & Lincoln, 2005, p. 198).	The research produced is to impart social change, change how people think, or serve as an examination of human existence (Creswell, 2007).	Intertwined with validity; inquiry often incomplete without action on the part of participants; constructivist formulation mandates training in political action if participants do not understand political systems (Guba & Lincoln, 2005, p. 198). Must act to be valid or trustworthy. If do not educate participants to act appropriately politically, could actually cause harm to them (accountability in research). Encourages readers to consider the findings presented and understanding of culture that is offered (Geertz, 1973). *According to my understanding of the readings, researchers must understand the social context and the culture in which the data are produced to accurately reflect what the data actually mean to the study.*	Intertwined with validity; inquiry often incomplete without action on the part of participants; constructivist formulation mandates training in political action if participants do not understand political systems (Guba & Lincoln, 2005, p. 198).
Control					
Who dictates how the research is produced and used? (Guba & Lincoln, 2005).	According to Guba and Lincoln (2005), the control is conducted by the researchers without the input and/or concern of the participants and/or society as a whole.	According to Guba and Lincoln (2005), the control is conducted by the researchers without the input and/or concern of the participants and/or society as a whole.	Critical race theory and critical raced-gendered epistemologies demonstrate that within the critical paradigm, control can be shared by the researcher and the subject, and ultimately the subject can have a say in how the research is conducted (Bernal, 2002).	Shared between inquirer and participants (Guba & Lincoln, 2005, p. 198). Without equal or co-equal control, research cannot be carried out.	Shared between inquirer and participants (Guba & Lincoln, 2005, p. 198). Without equal or co-equal control, research cannot be carried out. Knowledge is an expression of power (Kilgore, 2001, p. 59).

(Continued)

Table 6.5 (Continued)

THEMES OF KNOWLEDGE: Inquiry Aims, Ideals, Design, Procedures, and Methods					
	Positivism	Postpositivism	Critical (+ Feminist + Race)	Constructivism (or Interpretivist)	Participatory (+ Postmodern)
C: CRITICAL ISSUES OF THE TIME continued					
Relationships to foundations of truth and knowledge Helps make meaning and significance of components explicit (Guba & Lincoln, 2005).	Positivists believe there is only one truth or reality. Knowledge is the understanding and control over nature.	Postpositivists believe in a single reality; however, they also believe it will never fully be understood. Knowledge is the attempt to approximate reality and get as close to truth as possible.	The foundation of the critical paradigm is found in the struggle for equality and social justice, and social science demonstrates the oppression of people. Knowledge is an attempt to emancipate the oppressed and improve human condition (Fay, 1987).	Antifoundational (Guba & Lincoln, 2005, p. 198). Refusal to adopt any permanent standards by which truth can be universally known. *According to the readings, to approach inquiry from a constructivist viewpoint is to yield to multiple perspectives of the same data.*	Knowledge is founded in transformation and experience as demonstrated through shared research inquiry between the researcher and subject(s) (Epistemology Class Notes). Knowledge is tentative, multifaceted, not necessarily rational (Kilgore, 2001, p. 59).
Extended considerations of validity (goodness criteria) Bringing ethics and epistemology together (the moral trajectory) (Guba & Lincoln, 2005).	Validity is found in "gold standard" data, data that can be proven and replicated.	Validity is found in data that can be analyzed and studied using statistical tests. Data can be an approximation of reality.	Validity is found when research creates action (or action research) or participatory research, which creates the capacity for positive social change and emancipatory community action (Guba, & Lincoln, 2005; Merriam, 1991)	Extended constructions of validity (Guba & Lincoln, 2005, p. 198). Validity is a construct of the development of consensus. Based on participants and inquirer. "Assessment of any particular piece of research, then, may depend on very general expectations, on criteria tailored to the subcategory of approach and on emergent expectations that vary in all areas as the methodology itself changes" (Preissle, 2006, p. 691). *Based on this assessment of validity, can it be argued that all data are valid because what may not have meaning to one person could be the foundation of all truth to another? Taking this*	Extended constructions of validity (Guba & Lincoln, 2005, p. 198). Validity is found in the ability of the knowledge to become transformative according to the findings of the experiences of the subjects (Epistemology Class Notes).

Voice, reflexivity, postmodern textural representations				*approach, could we say that there is no such thing as invalidity of data or method if someone can find it to be an accurate reflection of their interpretation of reality?*
Voice, reflexivity, postmodern textural representations Voice: Can include the voice of the author, the voice of the respondents (subjects), and the voice of the researcher through their inquiry (Guba & Lincoln, 2005). Reflexivity: The process of reflecting critically on the self as researcher, "the human instrument" (Guba & Lincoln, 2005). Postmodern texual representations: The approach researchers take in understanding how social science is written and presented to avoid "dangerous illusions" which may exist in text (Guba & Lincoln, 2005). *Whose voices are heard in the research produced through the inquiry process? Whose views are presenting and/or producing the data?* (Guba & Lincoln, 2005).	Only the researcher has a voice; any effort to include the voice of the participants would impact objectivity (Guba & Lincoln, 2005).	The researcher has a voice, but also imparts the voice of the subjects. The researcher is careful to present knowledge through his or her own paradigm while being sensitive to the views of others (Bernal, 2002; Guba & Lincoln, 2005).	Voices mixed with participants' voices sometimes dominant; reflexivity serious and problematic; textual representation and extended issue (Guba & Lincoln, 2005, p. 198). Voices mixed, with participants' voices sometimes dominant. Reflexivity is serious and problematic. Researchers do not wish to give direction to study. Must use reflection as a researcher: "A few issues seem to be perennial: combining research approaches, assessing research quality, and the researcher's relationship to theory and philosophy, on the one hand, and participants and the public, on the other hand" (Preissle, 2006, p. 689).	Voices mixed; textual representation rarely discussed but problematic; reflexivity relies on critical subjectivity and self-awareness (Guba & Lincoln, 2005, p. 199). Textural: Must be within the context of who or what (for institutions or organizations) is being studied. The subject(s) voice must be present in the research (Epistemology Class Notes).

* Table originally developed by Guba and Lincoln, later expanded and extended by Susan A. Lynham as a teaching tool. The columns were filled in by David Byrd, a Ph.D. student in Dr. Lynham's epistemology class, 2008 Texas A&M University.

shifting." This is the paradigmatic equivalent of the Geertzian "blurring of genres" to which we referred earlier, and we regard this blurring and shifting as emblematic of a dynamism that is critical if we are to see qualitative research begin to have an impact on policy formulation or on the redress of social ills.

Our own position is that of the constructionist camp, loosely defined. We do not believe that criteria for judging either "reality" or validity are absolutist (Bradley & Schaefer, 1998); rather, they are derived from community consensus regarding what is "real": what is useful and what has meaning (especially meaning for action and further steps) within that community, as well as for that particular piece of research (Lather, 2007; Lather & Smithies, 1997). We believe that a goodly portion of social phenomena consists of the meaning-making activities of groups and individuals around those phenomena. The meaning-making activities themselves are of central interest to social constructionists and constructivists simply because it is the meaning-making, sense-making, attributional activities that shape action (or inaction). The meaning-making activities themselves can be changed when they are found to be incomplete, faulty (e.g., discriminatory, oppressive, or nonliberatory), or malformed (created from data that can be shown to be false). We have tried, however, to incorporate perspectives from other major nonpositivist paradigms. This is not a complete summation; space constraints prevent that. What we hope to do in this chapter is to acquaint readers with the larger currents, arguments, dialogues, and provocative writings and theorizing, the better to see perhaps what we ourselves do not even yet see: where and when confluence is possible, where constructive rapprochement might be negotiated, where voices are beginning to achieve some harmony.

▣ AXIOLOGY

Earlier, we placed values on the table as an "issue" on which positivists or phenomenologists might have a "posture" (Guba & Lincoln, 1989, 1994; Lincoln & Guba, 1985). Fortunately, we reserved for ourselves the right to either get smarter or just change our minds. We did both. Now, we suspect that *axiology* should be grouped with basic beliefs. In *Naturalistic Inquiry* (Lincoln & Guba, 1985), we covered some of the ways in which values feed into the inquiry process: choice of the problem, choice of paradigm to guide the problem, choice of theoretical framework, choice of major data-gathering and data-analytic methods, choice of context, treatment of values already resident issue within the context, and choice of format(s) for presenting findings. We believed those were strong enough reasons to argue for the inclusion of values as a major point of departure between positivist, conventional modes of inquiry and interpretive forms of inquiry. A second reading of the burgeoning literature and subsequent rethinking of our own rationale have led us to conclude that the issue is much larger than we first conceived. If we had it to do all over again, we would make values or, more

correctly, axiology (the branch of philosophy dealing with ethics, aesthetics, and religion) a part of the basic foundational philosophical dimensions of paradigm proposal. Doing so would, in our opinion, begin to help us see the embeddedness of ethics within, not external to, paradigms (see, e.g., Christians, 2000) and would contribute to the consideration of and dialogue about the role of spirituality in human inquiry. Arguably, axiology has been "defined out" of scientific inquiry for no larger a reason than that it also concerns religion. But defining religion broadly to encompass spirituality would move constructivists closer to participative inquirers and would move critical theorists closer to both (owing to their concern with liberation from oppression and freeing of the human spirit, both profoundly spiritual concerns). The expansion of basic issues to include axiology, then, is one way of achieving greater confluence among the various interpretivist inquiry models. This is the place, for example, where Peter Reason's (1993) profound concerns with "sacred science" and human functioning find legitimacy; it is a place where Richardson's (1994) "sacred spaces" become authoritative sites for human inquiry; it is a place—or *the* place—where the spiritual meets social inquiry, as Reason (1993), and later Lincoln and Denzin (1994), proposed some years earlier.

▣ ACCOMMODATION, COMMENSURABILITY, AND CUMULATION

Positivists and postpositivists alike still occasionally argue that paradigms are, in some ways, commensurable; that is, they can be retrofitted to each other in ways that make the simultaneous practice of both possible. We have argued that at the paradigmatic or philosophical level, commensurability between positivist and constructivist worldviews is not possible, but that within each paradigm, mixed methodologies (strategies) may make perfectly good sense (Guba & Lincoln, 1981, 1982, 1989, 1994; Lincoln & Guba, 1985). So, for instance, in *Effective Evaluation* (Guba & Lincoln, 1981), we argued:

> The guiding inquiry paradigm most appropriate to responsive evaluation is . . . the naturalistic, phenomenological, or ethnographic paradigm. It will be seen that qualitative techniques are typically most appropriate to support this approach. There are times, however, when the issues and concerns voiced by audiences require information that is best generated by more conventional methods, especially quantitative methods. . . . In such cases, the responsive conventional evaluator will not shrink from the appropriate application. (p. 36)

As we tried to make clear, the "argument" arising in the social sciences was *not about method*, although many critics of the new naturalistic, ethnographic, phenomenological, or case study approaches assumed it was.[2] As late as 1998, Weiss could be found to claim that "some evaluation theorists, notably Guba and Lincoln (1989), hold that it is impossible to combine qualitative and quantitative approaches responsibly within an evaluation" (p. 268), even though we stated early on in *Fourth Generation Evaluation* (1989) that those claims, concerns, and issues that have *not* been resolved become the advance organizers for information collection by the evaluator: "The information may be quantitative or qualitative. Responsive evaluation does not rule out quantitative modes, as is mistakenly believed by many, but deals with whatever information is responsive to the unresolved claim, concern, or issue" (p. 43).

We had also strongly asserted earlier, in *Naturalistic Inquiry* (1985), that

> qualitative methods are stressed within the naturalistic paradigm not because the paradigm is antiquantitative but because qualitative methods come more easily to the human-as-instrument. *The reader should particularly note the absence of an antiquantitative stance*, precisely because the naturalistic and conventional paradigms are so often—mistakenly—equated with the qualitative and quantitative paradigms, respectively. Indeed, *there are many opportunities for the naturalistic investigator to utilize quantitative data— probably more than are appreciated.* (pp. 198–199, emphases added)

Having demonstrated that we were not then (and are not now) talking about an antiquantitative posture or the exclusivity of *methods*, but rather about the philosophies of which paradigms are constructed, we can ask the question again regarding commensurability: Are paradigms commensurable? Is it possible to blend elements of one paradigm into another, so that one is engaging in research that represents the best of both worldviews? The answer, from our perspective, has to be a cautious *yes*. This is so if the models (paradigms, integrated philosophical systems) share axiomatic elements that are similar or that resonate strongly. So, for instance, *positivism* and *postpositivism* (as proposed by Phillips, 2006) are clearly commensurable. In the same vein, elements of *interpretivist/ postmodern*, critical theory, constructivist, and participative inquiry fit comfortably together. Commensurability is an issue only when researchers want to "pick and choose" among the axioms of positivist and interpretivist models because the axioms are contradictory and mutually exclusive. Ironically enough, the National Research Council's 2002 report, when defining their take on science, made this very point clearly and forcefully for us. Positivism (their stance) and interpretivism (our stance) are not commensurable.

Cumulation. The argument is frequently made that one of the problems with qualitative research is that it is not cumulative, that is, it cannot be aggregated in such a way as to make larger understandings or policy formulations possible. We would argue this is not the case. Beginning with the Lucas (1974, 1976) case study aggregation analyses, developed at Rand Corporation in the 1970s, researchers have begun to think about ways in which similar studies, carried out via qualitative methods with similar populations or in similar contexts, might be cumulated into meta-analyses, especially for policy purposes. This is now a far more readily available methodology with the advent of large databases manageable on computers. Although the techniques have not, we would argue, been tested extensively, it would seem that cumulation of a growing body of qualitative research is now within our grasp. That makes the criticisms of the non-cumulativeness of qualitative research less viable now, or even meaningless.

◩ THE CALL TO ACTION

One of the clearest ways in which the paradigmatic controversies can be demonstrated is to compare the positivist and postpositivist adherents, who view action as a form of contamination of research results and processes, and the interpretivists, who see action on research results as a meaningful and important outcome of inquiry processes. Positivist adherents believe action to be either a form of advocacy or a form of subjectivity, either or both of which undermine the aim of objectivity. Critical theorists, on the other hand, have always advocated varying degrees of social action, from the overturning of specific unjust practices to radical transformation of entire societies (Giroux, 1982). The call for action—whether in terms of internal transformation, such as ridding oneself of false consciousness, or of external social transformation (in the form, for instance, of extended social justice)—differentiates between positivist and postmodern criticalist theorists (including feminist and queer theorists). The sharpest shift, however, has been in the constructivist and participatory phenomenological models, where a step beyond interpretation and *verstehen*, or understanding, toward social action is probably one of the most conceptually interesting of the shifts (Lincoln, 1997, 1998a, 1998b).

For some theorists, the shift toward action came in response to widespread nonutilization of evaluation findings and the desire to create forms of evaluation that would attract champions who might follow through on recommendations with meaningful action plans (Guba & Lincoln, 1981, 1989). For others, embracing action came as both a political and an ethical commitment (see, e.g., Carr & Kemmis, 1986; Christians, 2000; Greenwood & Levin, 2000; Schratz & Walker, 1995; Tierney, 2000). Whatever the source of the problem to which inquirers were responding, the shift toward connecting action with research, policy analysis, evaluation, and social deconstruction (e.g., deconstruction of the patriarchal forms of oppression in social structures, which is the project informing much feminist theorizing, or deconstruction of the homophobia embedded in public policies) has come to characterize much new-paradigm inquiry work, both at the theoretical and at the practice and *praxis-oriented* levels. Action has become a major controversy that limns the ongoing debates among practitioners of the various paradigms. The mandate for social action, especially action designed and created by and for research participants with the aid and cooperation of researchers, can be most sharply delineated between positivist/postpositivist and new-paradigm inquirers. Many positivist and postpositivist inquirers still consider action the domain of communities other than researchers and research participants: those of policy personnel, legislators, and civic and political officials. Hard-line foundationalists presume that the taint of action will interfere with or even negate the objectivity that is a (presumed) characteristic of rigorous scientific method inquiry.

◫ CONTROL

Another controversy that has tended to become problematic centers on *control* of the study: Who initiates? Who determines salient questions? Who determines what constitutes findings? Who determines how data will be collected? Who determines in what forms the findings will be made public, if at all? Who determines what representations will be made of participants in the research? Let us be very clear: The issue of control is deeply embedded in the questions of voice, reflexivity, and issues of postmodern textual representation, which we shall take up later, *but only for new-paradigm inquirers.* For more conventional inquirers, the issue of control is effectively walled off from voice, reflexivity, and issues of textual representation because each of those issues in some way threatens claims to rigor (particularly objectivity and validity). For new-paradigm inquirers who have seen the preeminent paradigm issues of ontology and epistemology effectively folded into one another, and who have watched as methodology and axiology logically folded into one another (Lincoln, 1995, 1997), control of an inquiry seems far less problematic, except insofar as inquirers seek to obtain participants' genuine participation (see, e.g., Guba & Lincoln, 1981, on contracting and attempts to get some stakeholding groups to do more than stand by while an evaluation is in progress). Critical theorists, especially those who work in community organizing programs, are painfully aware of the necessity for members of the community or research participants to take control of their futures (see, e.g., Lather, 2007). Constructivists desire participants to take an increasingly active role in nominating questions of interest for any inquiry and in designing outlets for findings to be shared more widely within and outside the community. Participatory inquirers understand action controlled by the local context members to be the aim of inquiry within a community. For none of these paradigmatic adherents is control an issue of advocacy, a somewhat deceptive term usually used as a code within a larger metanarrative to attack an inquiry's rigor, objectivity, or fairness.

Rather, for new-paradigm researchers, control is a means of fostering emancipation, democracy, and community empowerment and of redressing power imbalances such that those who

were previously marginalized now achieve voice (Mertens, 1998) or "human flourishing" (Heron & Reason, 1997). Control as a controversy is an excellent place to observe the phenomenon that we have always termed "Catholic questions directed to a Methodist audience:" We use this description—given to us by a workshop participant in the early 1980s—to refer to the ongoing problem of illegitimate questions: questions that have no meaning because the frames of reference are those for which they were never intended. (We could as well call these "Hindu questions to a Muslim" to give another sense of how paradigms, or overarching philosophies—or theologies—are incommensurable, and how questions in one framework make little, if any, sense in another.) Paradigmatic formulations interact such that control becomes inextricably intertwined with mandates for objectivity. Objectivity derives from the Enlightenment prescription for knowledge of the physical world, which is postulated to be separate and distinct from those who would know (Polkinghorne, 1989). But if knowledge of the social (as opposed to the physical) world resides in meaning-making mechanisms of the social, mental, and linguistic worlds that individuals inhabit, then knowledge cannot be separate from the knower but rather is rooted in his or her mental or linguistic designations of that world (Polkinghorne, 1989; Salner, 1989).

◫ FOUNDATIONS OF TRUTH AND KNOWLEDGE IN PARADIGMS

Whether or not the world has a "real" existence outside of human experience of that world is an open question. For modernist (i.e., Enlightenment, scientific method, conventional, positivist) researchers, most assuredly there is a "real" reality "out there," apart from the flawed human apprehension of it. Furthermore, that reality can be approached (approximated) only through the utilization of methods that prevent human contamination of its apprehension or comprehension. For foundationalists in the empiricist tradition, the foundations of scientific truth and knowledge about reality reside in rigorous application of testing phenomena against a template as devoid as instrumentally possible of human bias, misperception, and other "idols" (Francis Bacon, cited in Polkinghorne, 1989). As Donald Polkinghorne (1989) makes clear:

> The idea that the objective realm is independent of the knower's subjective experiences of it can be found in Descartes's dual substance theory, with its distinction between the objective and subjective realms. . . . In the splitting of reality into subject and object realms, what can be known "objectively" is only the objective realm. True knowledge is limited to the objects and the relationships between them that exist in the realm of time and space. Human consciousness, which is subjective, is not accessible to science, and thus not truly knowable. (p. 23)

Now, templates of truth and knowledge can be defined in a variety of ways—as the end product of rational processes, as the result of experiential sensing, as the result of empirical observation, and others. In all cases, however, the referent is the physical or empirical world: rational engagement with it, experience of it, and empirical observation of it. Realists, who work on the assumption that there is a "real" world "out there" may in individual cases also be foundationalists, taking the view that all of these ways of defining are rooted in phenomena existing outside the human mind.

Although we can think about them, experience them, or observe them, the elements of the physical world are nevertheless transcendent, referred to but beyond direct apprehension. Realism is an ontological question, whereas foundationalism is a criterial question. Some foundationalists argue that having real phenomena necessarily implies certain final, ultimate criteria for testing them as truthful (although we may have great difficulty in determining what those criteria are); nonfoundationalists tend to argue that there are no such ultimate criteria, only those that we can agree on at a certain time, within a certain community (Kuhn, 1967) and under certain conditions. Foundational criteria are discovered; nonfoundational criteria are negotiated. It is the case, however, that most realists are also foundationalists, and many nonfoundationalists or antifoundationalists are relativists.

An ontological formulation that connects realism and foundationalism within the same "collapse" of categories that characterizes the ontological-epistemological collapse is one that exhibits good fit with the other assumptions of constructivism. That state of affairs suits new-paradigm inquirers well. Critical theorists, constructivists, and participatory/cooperative inquirers take their primary field of interest to be precisely that subjective and intersubjective, critical social knowledge and the active construction and co-creation of such knowledge by human agents, which is produced by human consciousness. Furthermore, new-paradigm inquirers take to the social knowledge field with zest, informed by a variety of social, intellectual, and theoretical explorations. These theoretical excursions include

■ Saussurian linguistic theory, which views all relationships between words and what those words signify as the function of an internal relationship within some linguistic system;

■ Literary theory's deconstructive contributions, which seek to disconnect texts from any *essentialist* or transcendental meaning and resituate them within both author's and reader's historical and social contexts (Hutcheon, 1989; Leitch, 1996);

■ Feminist (Addelson, 1993; Alpern, Antler, Perry, & Scobie, 1992; Babbitt, 1993; Harding, 1993), race and ethnic (Kondo, 1990, 1997; Trinh, 1991), and queer theorizing (Gamson, 2000), which seeks to uncover and explore varieties of oppression and

historical colonizing between dominant and subaltern genders, identities, races, and social worlds;

■ The postmodern historical moment (Michael, 1996), which problematizes truth as partial, identity as fluid, language as an unclear referent system, and method and criteria as potentially coercive (Ellis & Bochner, 1996); and

■ Criticalist theories of social change (Carspecken, 1996; Schratz & Walker, 1995).

The realization of the richness of the mental, social, psychological, and linguistic worlds that individuals and social groups create and constantly re-create and co-create gives rise, in the minds of new-paradigm postmodern and poststructural inquirers, to endlessly fertile fields of inquiry rigidly walled off from conventional inquirers. Unfettered from the pursuit of transcendental scientific truth, inquirers are now free to resituate themselves within texts, to reconstruct their relationships with research participants in less constricted fashions, and to create representations (Tierney & Lincoln, 1997) that grapple openly with problems of inscription, reinscription, metanarratives, and other rhetorical devices that obscure the extent to which human action is locally and temporally shaped. The processes of uncovering forms of inscription and the rhetoric of metanarratives are *genealogical*—"*expos[ing]* the origins of the view that have become *sedimented and accepted as truths*" (Polkinghorne, 1989, p. 42; emphasis added)—or *archaeological* (Foucault, 1971; Scheurich, 1997).

New-paradigm inquirers engage the foundational controversy in quite different ways. Critical theorists, particularly critical theorists who are more positivist in orientation, who lean toward Marxian interpretations, tend toward foundational perspectives, with an important difference. Rather than locating foundational truth and knowledge in some external reality "out there," such critical theorists tend to locate the foundations of truth in specific historical, economic, racial, gendered, and social infrastructures of oppression, injustice, and marginalization. Knowers are not portrayed as *separate from* some objective reality, but they may be cast as unaware actors in such historical realities ("false consciousness") or as aware of historical forms of oppression but unable or unwilling, because of conflicts, to act on those historical forms to alter specific conditions in this historical moment ("divided consciousness"). Thus, the "foundation" for critical theorists is a duality: social critique tied in turn to raised consciousness of the possibility of positive and liberating social change. Social critique may exist apart from social change, but both are necessary for most critical perspectives.

Constructivists, on the other hand, tend toward the antifoundational (Lincoln, 1995, 1998b; Schwandt, 1996). *Antifoundational* is the term used to denote a refusal to adopt any permanent, unvarying (or "foundational") standards by which truth can be universally known. As one of us has argued, truth—and any agreement regarding what is valid knowledge—arises from the relationship between members of some stakeholding community (Lincoln, 1995). Agreements about truth may be the subject of community *negotiations* regarding what will be accepted as truth (although there are difficulties with that formulation as well; Guba & Lincoln, 1989). Or agreements may eventuate as the result of a *dialogue* that moves arguments about truth claims or validity past the warring camps of objectivity and relativity toward "a communal test of validity through the argumentation of the participants in a discourse" (Bernstein, 1983; Polkinghorne, 1989; Schwandt, 1996). This "communicative and pragmatic concept" of validity (Rorty, 1979) is never fixed or unvarying. Rather, it is created by means of a community narrative, itself subject to the temporal and historical conditions that gave rise to the community. Thomas A. Schwandt (1989) has also argued that these discourses, or community narratives, can and should be bounded by moral considerations, a premise grounded in the emancipatory narratives of the critical theorists, the philosophical pragmatism of Richard Rorty, the democratic focus of constructivist inquiry, and the "human flourishing" goals of participatory and cooperative inquiry.

The controversies around foundationalism (and, to a lesser extent, essentialism) are not likely to be resolved through dialogue between paradigm adherents. The likelier event is that the "postmodern turn" (Best & Kellner, 1997), with its emphasis on the social construction of social reality, fluid as opposed to fixed identities of the self, and the partiality of all truths, will simply overtake modernist assumptions of an objective reality, as indeed, to some extent, it has already done in the physical sciences. We might predict that, if not in our lifetimes, at some later time, the dualist idea of an objective reality suborned by limited human subjective realities will seem as quaint as flat-earth theories do to us today.

▣ VALIDITY: AN EXTENDED AGENDA

Nowhere can the conversation about paradigm differences be more fertile than in the extended controversy about validity (Howe & Eisenhart, 1990; Kvale, 1989, 1994; Ryan, Greene, Lincoln, Mathison, & Mertens, 1998; Scheurich, 1994, 1996). Validity is not like objectivity. There are fairly strong theoretical, philosophical, and pragmatic rationales for examining the concept of objectivity and finding it wanting. Even within positivist frameworks, it is viewed as conceptually flawed. But validity is a more irritating construct, one neither easily dismissed nor readily configured by new-paradigm practitioners (Angen, 2000; Enerstvedt, 1989; Tschudi, 1989). Validity cannot be dismissed simply because it points to a question that has to be answered in one way or another: Are these findings

sufficiently authentic (isomorphic to some reality, trustworthy, related to the way others construct their social worlds) that I may trust myself in acting on their implications? More to the point, would I feel sufficiently secure about these findings to construct social policy or legislation based on them? At the same time, radical reconfigurations of validity leave researchers with multiple, sometimes conflicting, mandates for what constitutes rigorous research. One of the issues around validity is the conflation between method and interpretation. The postmodern turn suggests that no method can deliver on ultimate truth and, in fact, "suspects all methods," the more so the larger their claims to delivering on truth (Richardson, 1994). Thus, although one might argue that some methods are more suited than others for conducting research on human construction of social realities (Lincoln & Guba, 1985), no one would argue that a single method—or collection of methods—is the royal road to ultimate knowledge. In new-paradigm inquiry, however, it is not merely method that promises to deliver on some set of local or context-grounded truths; it is also the processes of interpretation.

Thus, we have two arguments proceeding simultaneously. The first, borrowed from positivism, argues for a kind of rigor in the application of method, whereas the second argues for both a community consent and a form of rigor-defensible reasoning, plausible alongside some other reality that is known to author and reader in ascribing salience to one interpretation over another and in framing and bounding the interpretive study itself. Prior to our understanding that there were, indeed, two forms of rigor, we assembled a set of methodological criteria, largely borrowed from an earlier generation of thoughtful anthropological and sociological methodological theorists. Those methodological criteria are still useful for a variety of reasons, not the least of which is that they ensure that such issues as prolonged engagement and persistent observation are attended to with some seriousness.

It is the second kind of rigor, however, that has received the most attention in recent writings: Are we *interpretively* rigorous? Can our co-created constructions be trusted to provide some purchase on some important human phenomenon? Do our findings point to action that can be taken on the part of research participants to benefit themselves or their particular social contexts?

Human phenomena are themselves the subject of controversy. Classical social scientists would like to see *human phenomena* limited to those social experiences from which (scientific) generalizations may be drawn. New-paradigm inquirers, however, are increasingly concerned with the single experience, the individual crisis, the epiphany or moment of discovery, with that most powerful of all threats to conventional objectivity, feeling, and emotion and to action. Social scientists concerned with the expansion of what count as social data rely increasingly on the experiential, the embodied, the emotive qualities of human experience, which contribute the narrative quality to a life. Sociologists such as Carolyn Ellis and Arthur P. Bochner (2000) and Richardson (2000), qualitative researchers such as Ronald Pelias (1999, 2004), and psychologists such as Michelle Fine (see Fine, Weis, Weseen, & Wong, 2000) and Ellis (2009) concern themselves with various forms of autoethnography and personal experience and performance methods, both to overcome the abstractions of a social science far gone with quantitative descriptions of human life and to capture those elements that make life conflictual, moving, and problematic. For purposes of this discussion, we believe the adoption of the most radical definitions of social science is appropriate because the paradigmatic controversies are often taking place at the edges of those conversations. Those edges are where the border work is occurring, and accordingly, they are the places that show the most promise for projecting where qualitative methods will be in the near and far future.

Whither and Whether Criteria

At those edges, several conversations are occurring around validity. The first and most radical is a conversation opened by Schwandt (1996), who suggests that we say "farewell to criteriology" or the "regulative norms for removing doubt and settling disputes about what is correct or incorrect, true or false" (p. 59); this has created a virtual cult around criteria. Schwandt does not, however, himself say farewell to criteria forever; rather, he resituates and resuscitates social inquiry, with other contemporary philosophical pragmatists, within a framework that transforms professional social inquiry into a form of practical philosophy, characterized by "aesthetic, prudential, and moral considerations as well as more conventionally scientific ones" (p. 68). When social inquiry becomes the practice of a form of practical philosophy—a deep questioning about how we shall get on in the world and what we conceive to be the potentials and limits of human knowledge and functioning—then we have some preliminary understanding of what entirely different criteria might be for judging social inquiry.

Schwandt (1996) proposes three such criteria. First, he argues, we should search for a social inquiry that "generate[s] knowledge that complements or supplements rather than displac[ing] lay probing of social problems," a form of knowledge for which we do not yet have the *content,* but from which we might seek to understand the aims of practice from a variety of perspectives, or with different lenses. Second, he proposes a "social inquiry as practical philosophy" that has as its aim "enhancing or cultivating *critical* intelligence in parties to the research encounter," critical intelligence being defined as "the capacity to engage in moral critique." And finally, he proposes a third way in which we might judge social inquiry as practical philosophy: We might make judgments about the social inquirer-as-practical-philosopher. He or she might be "evaluated on the

success to which his or her reports of the inquiry enable the training or calibration of human judgment" (p. 69) or "the capacity for practical wisdom" (p. 70). Schwandt is not alone, however, in wishing to say "farewell to criteriology," at least as it has been previously conceived. Scheurich (1997) makes a similar plea, and in the same vein, Smith (1993) also argues that validity, if it is to survive at all, must be radically reformulated if it is ever to serve phenomenological research well (see also Smith & Deemer, 2000).

At issue here is not whether we shall have criteria, or whose criteria we as a scientific community might adopt, but rather what the nature of social inquiry ought to be, whether it ought to undergo a transformation, and what might be the basis for criteria within a projected transformation. Schwandt (1989; also personal communication, August 21, 1998) is quite clear that both the transformation and the criteria are rooted in dialogic efforts. These dialogic efforts are quite clearly themselves forms of "moral discourse:" Through the specific connections of the dialogic, the idea of practical wisdom, and moral discourses, much of Schwandt's work can be seen to be related to, and reflective of, critical theorist and participatory paradigms, as well as constructivism, although Schwandt specifically denies the relativity of truth. (For a more sophisticated explication and critique of forms of constructivism, hermeneutics, and interpretivism, see Schwandt, 2000. In that chapter, Schwandt spells out distinctions between realists and nonrealists and between foundationalists and nonfoundationalists far more clearly than it is possible for us to do in this chapter.) To return to the central question embedded in validity: How do we know when we have specific social inquiries that are faithful enough to some human construction that we may feel safe in acting on them, or, more important, that members of the community in which the research is conducted may act on them? To that question, there is no final answer. There are, however, several discussions of what we might use to make both professional and lay judgments regarding any piece of work. It is to those versions of validity that we now turn.

Validity as Authenticity

Perhaps the first nonfoundational criteria were those we developed in response to a challenge by John K. Smith (see Smith & Deemer, 2000). In those criteria, we attempted to locate criteria for judging the processes and *outcomes* of naturalistic or constructivist inquiries (rather than the application of methods; see Guba & Lincoln, 1989). We described five potential outcomes of a social constructionist inquiry (evaluation is one form of disciplined inquiry, alongside research and policy analyses; see Guba & Lincoln, 1981), each grounded in concerns specific to the paradigm we had tried to describe and construct and apart from any concerns carried over from the positivist legacy. The criteria were instead rooted in the axioms and assumptions of the constructivist paradigm, insofar as we could extrapolate and infer them. Those authenticity criteria—so called because we believed them to be hallmarks of authentic, trustworthy, rigorous, or "valid" constructivist or phenomenological inquiry—were fairness, ontological authenticity, educative authenticity, catalytic authenticity, and tactical authenticity (Guba & Lincoln, 1989, pp. 245–251). *Fairness* was thought to be a quality of balance; that is, all stakeholder views, perspectives, values, claims, concerns, and voices should be apparent in the text. Omission of stakeholder or participant voices reflects, we believe, a form of bias.

This bias, however, was and is not related directly to the concerns of objectivity that flow from positivist inquiry and that are reflective of inquirer blindness or subjectivity. Rather, this fairness was defined by deliberate attempts to prevent marginalization, to act affirmatively with respect to inclusion, and to act with energy to ensure that all voices in the inquiry effort had a chance to be represented in any texts and to have their stories treated fairly and with balance. *Ontological and educative authenticity* were designated as criteria for determining a raised level of awareness, in the first instance, by individual research participants and, in the second, by individuals about those who surround them or with whom they come into contact for some social or organizational purpose. Although we failed to see it at that particular historical moment (1989), there is no reason these criteria cannot be—at this point in time, with many miles under our theoretic and practice feet—reflective also of Schwandt's (1996) "critical intelligence," or capacity to engage in moral critique. In fact, the authenticity criteria we originally proposed had strong moral and ethical overtones, a point to which we later returned (see, e.g., Lincoln, 1995, 1998a, 1998b). It was a point to which our critics strongly objected before we were sufficiently self-aware to realize the implications of what we had proposed (see, e.g., Sechrest, 1993).

Catalytic and tactical authenticities refer to the ability of a given inquiry to prompt, first, action on the part of research participants and, second, the involvement of the researcher/ evaluator in training participants in specific forms of social and political action if participants desire such training. It is here that constructivist inquiry practice begins to resemble forms of critical theorist action, action research, or participative or cooperative inquiry, each of which is predicated on creating the capacity in research participants for positive social change and forms of emancipatory community action. It is also at this specific point that practitioners of positivist and postpositivist social inquiry are the most critical because any action on the part of the inquirer is thought to destabilize objectivity and introduce subjectivity, resulting in bias. The problem of subjectivity and bias has a long theoretical history, and this chapter is simply too brief for us to enter into the various formulations that either take account of subjectivity or posit it as a positive learning experience, practical, embodied, gendered, and emotive. For purposes of this discussion, it is enough to say that we are persuaded that objectivity is a chimera: a mythological creature that never existed, save in the imaginations of those who believe that knowing can be separated from the knower.

Validity as Resistance and as Poststructural Transgression

Richardson (1994, 1997) has proposed another form of validity, a deliberately "transgressive" form, the *crystalline*. In writing experimental (i.e., nonauthoritative, nonpositivist) texts, particularly poems and plays, Richardson (1997) has sought to "problematize reliability, validity, and truth" (p. 165) in an effort to create new relationships: to her research participants, to her work, to other women, to herself (see also Lather, who seeks the same ends, 2007). Richardson says that transgressive forms permit a social scientist to "conjure a different kind of social science ... [which] means changing one's relationship to one's work, *how* one knows and tells about the sociological" (p. 166). To see "how transgression looks and how it feels," it is necessary to "find and deploy methods that allow us to uncover the hidden assumptions and life-denying repressions of sociology; resee/refeel sociology. Reseeing and retelling are inseparable" (p. 167). The way to achieve such validity is by examining the properties of a crystal in a metaphoric sense. Here we present an extended quotation to give some flavor of how such validity might be described and deployed:

> I propose that the central imaginary for "validity" for postmodernist texts is not the triangle—a rigid, fixed, two-dimensional object. Rather the central imaginary is the crystal, which combines symmetry and substance with an infinite variety of shapes, substances, transmutations, multidimensionalities, and angles of approach. Crystals grow, change, alter, but are not amorphous. Crystals are prisms that reflect externalities *and* refract within themselves, creating different colors, patterns, arrays, casting off in different directions. What we *see* depends upon our angle of repose. Not triangulation, crystallization. In postmodernist mixed-genre texts, we have moved from plane geometry to light theory, where light can be *both* waves *and* particles. Crystallization, without losing structure, deconstructs the traditional idea of "validity" (we feel how there is no single truth, we see how texts validate themselves); and crystallization provides us with a deepened, complex, thoroughly partial understanding of the topic. Paradoxically, we know more and doubt what we know. (Richardson, 1997, p. 92)

The metaphoric "solid object" (crystal/text), which can be turned many ways, which reflects and refracts light (light/ multiple layers of meaning), through which we can see both "wave" (light wave/human currents) and "particle" (light as "chunks" of energy/elements of truth, feeling, connection, processes of the research that "flow" together) is an attractive metaphor for validity. The properties of the crystal-as-metaphor help writers and readers alike see the interweaving of processes in the research: discovery, seeing, telling, storying, representation.

Other "Transgressive" Validities

Richardson is not alone in calling for forms of validity that are "transgressive" and disruptive of the status quo. Patti Lather (1993) seeks "an incitement to discourse," the purpose of which is "to rupture validity as a regime of truth, to displace its historical inscription . . . via a dispersion, circulation and proliferation of counterpractices of authority that take the crisis of representation into account" (p. 674). In addition to catalytic validity (Lather, 1986), Lather (1993) poses *validity as simulacra/ironic validity; Lyotardian paralogy/neopragmatic validity,* a form of validity that "foster[s] heterogeneity, refusing disclosure" (p. 679); *Derridean rigor/rhizomatic validity,* a form of behaving "via relay, circuit, multiple openings" (p. 680); and *voluptuous/ situated validity,* which "embodies a situated, partial tentativeness" and "brings ethics and epistemology together . . . via practices of engagement and self reflexivity" (p. 686). Together, these form a way of interrupting, disrupting, and transforming "pure" presence into a disturbing, fluid, partial, and problematic presence—a poststructural and decidedly postmodern form of discourse theory, hence textual revelation (see also Lather, 2007, for further reflections and disquisitions on validity).

Validity as an Ethical Relationship

As Lather (1993) points out, poststructural forms for validities "bring ethics and epistemology together" (p. 686); indeed, as Parker Palmer (1987) also notes, "every way of knowing contains its own moral trajectory" (p. 24). Alan Peshkin reflects on Nel Noddings's (1984) observation that "the search for justification often carries us farther and farther from the heart of morality" (p. 105; quoted in Peshkin, 1993, p. 24). The *way* in which we know is most assuredly tied up with both *what* we know and our *relationships with our research participants.* Accordingly, one of us worked on trying to understand the ways in which the ethical intersects both the interpersonal and the epistemological (as a form of authentic or valid knowing; Lincoln, 1995). The result was the first set of understandings about emerging criteria for quality that were also rooted in the epistemology/ethics nexus. Seven new standards were derived from that search: positionality, or standpoint, judgments; specific discourse communities and research sites as arbiters of quality; voice, or the extent to which a text has the quality of polyvocality; critical subjectivity (or what might be termed intense self-reflexivity; see, for instance, Heron & Reason, 1997); reciprocity, or the extent to which the research relationship becomes reciprocal rather than hierarchical; sacredness, or the profound regard for how science can (and does) contribute to human flourishing; and sharing of the perquisites of privilege that accrue to our positions as academics with university positions. Each of these standards was extracted from a body of research, often from disciplines as disparate as management, philosophy, and women's studies (Lincoln, 1995).

▣ VOICE, REFLEXIVITY, AND POSTMODERN TEXTUAL REPRESENTATION

Texts have to do a lot more work these days than in the past. Even as they are charged by poststructuralists and postmodernists to reflect on their representational practices, those practices become more problematic. Three of the most engaging, but painful issues are voice, the status of reflexivity, and postmodern/poststructural textual representation, especially as those problematics are displayed in the shift toward narrative and literary forms that directly and openly deal with human emotion.

Voice

Voice is a multilayered problem, simply because it has come to mean many things to different researchers. In former eras, the only appropriate voice was the "voice from nowhere"—the "pure presence" of representation, as Lather (2007) terms it. As researchers became more conscious of the abstracted realities their texts created (Lather 2007), they became simultaneously more conscious of having readers "hear" their informants— permitting readers to hear the exact words (and, occasionally, the paralinguistic cues, the lapses, pauses, stops, starts, and reformulations) of the informants. Today, especially in more participatory forms of research, voice can mean not only having a real researcher—and a researcher's voice—in the text, but also letting research participants speak for themselves, either in text form or through plays, forums, "town meetings," or other oral and performance-oriented media or communication forms designed by research participants themselves (Bernal, 1998, 2002). Performance texts, in particular, give an emotional immediacy to the voices of researchers and research participants far beyond their own sites and locales (see McCall, 2000). Rosanna Hertz (1997) describes voice as

> a struggle to figure out how to present the author's self while simultaneously writing the respondents' accounts and representing their selves. Voice has multiple dimensions: First, there is the voice of the author. Second, there is the presentation of the voices of one's respondents within the text. A third dimension appears when the self is the subject of the inquiry. . . . Voice is how authors express themselves within an ethnography. (pp. xi–xii)

But knowing how to express ourselves goes far beyond the commonsense understanding of "expressing ourselves." Generations of ethnographers trained in the "cooled-out, stripped-down rhetoric" of positivist inquiry (Firestone, 1987) find it difficult, if not nearly impossible, to "locate" themselves deliberately and squarely within their texts (even though, as Geertz, 1988, has demonstrated finally and without doubt, the authorial voice is rarely genuinely absent, or even hidden).

Specific textual experimentation can help; that is, composing ethnographic work in various literary forms—Richardson's poetry and plays are good examples, or Lather and Chris Smithies's (1997) *Troubling the Angels*—can help a researcher to overcome the tendency to write in the distanced and abstracted voice of the disembodied "I." But such writing exercises are hard work. This is also work that is embedded in the practices of reflexivity and narrativity, without which achieving a voice of (partial) truth is impossible.

Reflexivity

Reflexivity is the process of reflecting critically on the self as researcher, the "human as instrument" (Guba & Lincoln, 1981). It is, we would assert, the critical subjectivity discussed early on in Peter Reason and John Rowan's edited volume, *Human Inquiry* (1981). It is a conscious experiencing of the self as both inquirer and respondent, as teacher and learner, as the one coming to know the self within the processes of research itself. Reflexivity forces us to come to terms not only with our choice of research problem and with those with whom we engage in the research process, but with ourselves and with the multiple identities that represent the fluid self in the research setting (Alcoff & Potter, 1993). Shulamit Reinharz (1997), for example, argues that we not only "*bring* the self to the field . . . [we also] *create* the self in the field" (p. 3). She suggests that although we all have many selves we bring with us, those selves fall into three categories: research-based selves, brought selves (the selves that historically, socially, and personally create our standpoints), and situationally created selves (p. 5). Each of those selves comes into play in the research setting and consequently has a distinctive voice.

Reflexivity—as well as the poststructural and postmodern sensibilities concerning quality in qualitative research—demands that we interrogate each of our selves regarding the ways in which research efforts are shaped and staged around the binaries, contradictions, and paradoxes that form our own lives. We must question ourselves, too, regarding how those binaries and paradoxes shape not only the identities called forth in the field and later in the discovery processes of writing, but also our interactions with respondents, in who we become to them in the process of *becoming* to ourselves (Mayan, 2009). Someone once characterized qualitative research as the twin processes of "writing up" (fieldnotes) and "writing down" (the narrative). But D. Jean Clandinin and F. Michael Connelly (1994) have made clear that this bitextual reading of the processes of qualitative research is far too simplistic. In fact, many texts are created in the process of engaging in fieldwork.

As Richardson (1994, 1997, 2000) makes clear, writing is not merely the transcribing of some reality. Rather, writing—of all the texts, notes, presentations, and possibilities—is also a process of discovery: discovery of the subject (and sometimes of the problem itself) and discovery of the self.[3]

There is good news and bad news with the most contemporary of formulations. The good news is that the multiple selves—ourselves and our respondents—of postmodern inquiries may give rise to more dynamic, problematic, open-ended, and complex forms of writing and representation. The bad news is that the multiple selves we create and encounter give rise to more dynamic, problematic, open-ended, and complex forms of writing and representation. Among the various proposals for textual presentations, it is occasionally difficult to know to which proposals we should be attending; while it is often a matter of specific model (e.g., critical feminist studies, queer theories, hybrid theorists, postcolonial theorists, and the like) to which we are theoretically, philosophically, and morally inclined, it is nevertheless a buffet of wildly rich fare, and some choices must be made. Often such choices are made on the basis of both the needs of our research participants and coresearchers and the needs of our intended audiences.

Postmodern Textual Representations

There are two dangers inherent in the conventional texts of scientific method: They may lead us to believe the world is rather simpler than it is, and they may reinscribe enduring forms of historical oppression. Put another way, we are confronted with a crisis of authority (which tells us the world is "this way" when perhaps it is some other way, or many other ways) and a crisis of representation (which serves to silence those whose lives we appropriate for our social sciences, and which may also serve subtly to re-create *this* world, rather than some other, perhaps more complex, but just one; Eisner, 1997). Catherine Stimpson (1988) has observed:

> Like every great word, "representation/s" is a stew. A scrambled menu, it serves up several meanings at once. For a representation can be an image visual, verbal, or aural. . . . A representation can also be a narrative, a sequence of images and ideas. . . . Or, a representation can be the product of ideology, that vast scheme for showing forth the world and justifying its dealings. (p. 223)

One way to confront the dangerous illusions (and their underlying ideologies) that texts may foster is through the creation of new texts that break boundaries; that move from the center to the margins to comment on and decenter the center; that forgo closed, bounded worlds for those more open-ended and less conveniently encompassed; that transgress the boundaries of conventional social science; and that seek to create a social science about human life rather than *on* subjects.

Experiments with how to do this have produced "messy texts" (Marcus & Fischer, 1986). Messy texts are not typographic nightmares (although they may be typographically nonlinear); rather, they are texts that seek to break the binary between science and literature; to portray the contradiction and truth of human experience; to break the rules in the service of showing, even partially (Flax, 1990), how real human beings cope with both the eternal verities of human existence and the daily irritations and tragedies of living that existence. Postmodern representations search out and experiment with narratives that expand the range of understanding, voice, and storied variations in human experience. As much as they are social scientists, inquirers also become storytellers, poets, and playwrights, experimenting with personal narratives, first-person accounts, reflexive interrogations, and deconstruction of the forms of tyranny embedded in representational practices (see Richardson, 2000; Tierney & Lincoln, 1997).

Representation may be arguably the most open-ended of the controversies surrounding phenomenological research today because the ideas of what constitutes legitimate inquiry are expanding and, at the same time, the forms of narrative, dramatic, and rhetorical structure are far from being either explored or exploited fully and because we know that there is extensive slippage between life as lived and experienced and our ability to cast that life into words that exhibit perfect one-to-one correspondence with that experience. Words, and therefore any and all representations, fail us. Because, too, each inquiry, each inquirer, brings a unique perspective to our understanding, the possibilities for variation and exploration are limited only by the number of those engaged in inquiry and the realms of social and intrapersonal life that become interesting to researchers. The only thing that can be said for certain about postmodern representational practices is that they will proliferate as forms and they will seek and demand much of audiences, many of whom may be outside the scholarly and academic world. In fact, some forms of inquiry may never show up in the academic world because their purpose will be use in the immediate context, for the consumption, reflection, and use of local or indigenous audiences. Those that are produced for scholarly audiences will, however, continue to be untidy, experimental, and driven by the need to communicate social worlds that have remained private and "nonscientific" until now.

▣ A GLIMPSE OF THE FUTURE

The issues raised in this chapter are by no means the only ones under discussion for the near and far future. But they are some of the critical ones, and discussion, dialogue, and even controversies are bound to continue as practitioners of the various new and emergent paradigms continue either to look for common ground or to find ways in which to distinguish their forms of inquiry from others.

Some time ago, we expressed our hope that practitioners of both positivist and new-paradigm forms of inquiry might find some way of resolving their differences, such that all social scientists could work within a common discourse—and perhaps even several traditions—once again. In retrospect, such a resolution

appears highly unlikely and would probably even be less than useful. This is not, however, because neither positivists nor phenomenologists will budge an inch (although that, too, is unlikely), or because the reinscription of stern positivist "science" abounds, with even more rancorous pronouncements about qualitative research than we have heard in previous decades. Rather, it is because, in the postmodern (and post-postmodern) moment, and in the wake of poststructuralism, the assumption that there is no single "truth"—that all truths are but partial truths; that the slippage between signifier and signified in linguistic and textual terms creates representations that are only and always shadows of the actual people, events, and places; that identities are fluid rather than fixed—leads us ineluctably toward the insight that there will be no single "conventional" paradigm to which all social scientists might ascribe in some common terms and with mutual understanding. Rather, we stand at the threshold of a history marked by multivocality, contested meanings, paradigmatic controversies, and new textual forms. At some distance down this conjectural path, when its history is written, we will find that this has been the era of emancipation: emancipation from what Hannah Arendt calls "the coerciveness of Truth," emancipation from hearing only the voices of Western Europe, emancipation from generations of silence, and emancipation from seeing the world in one color.

We may also be entering an age of greater spirituality within research efforts. The emphasis on inquiry that reflects ecological values, on inquiry that respects communal forms of living that are not Western, on inquiry involving intense reflexivity regarding how our inquiries are shaped by our own historical and gendered locations, and on inquiry into "human flourishing," as Heron and Reason (1997) call it, may yet reintegrate the sacred with the secular in ways that promote freedom and self-determination. Egon Brunswik, the organizational theorist, wrote of "tied" and "untied" variables—variables that are linked, or clearly not linked, with other variables—when studying human forms of organization. We may be in a period of exploring the ways in which our inquiries are both tied and untied, as a means of finding where our interests cross and where we can both be and promote others' being, as whole human beings.

▣ NOTES

1. There are several versions of critical theory, just as there are several varieties of postmodernism, including classical critical theory, which is most closely related to neo-Marxist theory; postpositivist formulations, which divorce themselves from Marxist theory but are positivist in their insistence on conventional rigor criteria; and postmodernist, poststructuralist, or constructivist-oriented varieties. See, for instance, Fay (1987), Carr and Kemmis (1986), and Lather (1991). See also Kemmis and McTaggart (2000) and Kincheloe and McLaren (2000).

2. For a clearer understanding of how methods came to stand in for paradigms, or how our initial (and, we thought, quite clear) positions came to be misconstrued, see Lancy (1993) or, even more currently, Weiss (1998, esp. p. 268).

3. For example, compare this chapter with, say, the work of Richardson (2000) and Ellis and Bochner (2000), where the authorial voices are clear, personal, vocal, and interior, interacting subjectivities. Although some colleagues have surprised us by correctly identifying which chapters each of us has written in given books, nevertheless, the style of this chapter more closely approximates the more distanced forms of "realist" writing rather than the intimate, personal "feeling tone" (to borrow a phrase from Studs Terkel) of other chapters. Voices also arise as a function of the material being covered. The material we chose as most important for this chapter seemed to demand a less personal tone, probably because there appears to be much more "contention" than calm dialogue concerning these issues. The "cool" tone likely stems from our psychological response to trying to create a quieter space for discussion around controversial issues. What can we say?

▣ REFERENCES

Addelson, K. P. (1993). Knowers/doers and their moral problems. In L. Alcoff & E. Potter (Eds.), *Feminist epistemologies* (pp. 265–294). New York: Routledge.

Alcoff, L., &Potter, E. (Eds.). (1993). *Feminist epistemologies.* New York: Routledge.

Alpern, S., Antler, J., Perry, E. I., & Scobie, I. W. (Eds.). (1992). *The challenge of feminist biography: Writing the lives of modern American women.* Urbana: University of Illinois Press.

Angen, M. J. (2000). Evaluating interpretive inquiry: Reviewing the validity debate and opening the dialogue. *Qualitative Health Research, 10*(3), 378–395.

Babbitt, S. (1993). Feminism and objective interests: The role of transformation experiences in rational deliberation. In L. Alcoff & E. Potter (Eds.), *Feminist epistemologies* (pp. 245–264). New York: Routledge.

Bernal, D. D. (1998). Using a Chicana feminist epistemology in educational research. *Harvard Educational Review, 68*(4), 1–19.

Bernal, D. D. (2002). Critical race theory, Latino critical theory, and critical race-gendered epistemologies; Recognizing students of color as holders and creators of knowledge. *Qualitative Inquiry, 9*(1), 105–126.

Bernstein, R. J. (1983). *Beyond objectivism and relativism: Science, hermeneutics, and praxis.* Oxford, UK: Blackwell.

Best, S., & Kellner, D. (1997). *The postmodern turn.* New York: Guilford.

Bloland, H. (1989). Higher education and high anxiety: Objectivism, relativism, and irony. *Journal of Higher Education, 60,* 519–543.

Bloland, H. (1995). Postmodernism and higher education. *Journal of Higher Education, 66,* 521–559.

Bradley, J., & Schaefer, K. (1998). *The uses and misuses of data and models.* Thousand Oaks, CA: Sage.

Carr, W. L., & Kemmis, S. (1986). *Becoming critical: Education, knowledge, and action research.* London: Falmer.

Carspecken, P. F. (1996). *Critical ethnography in educational research: A theoretical and practical guide.* New York: Routledge.

Christians, C. G. (2000). Ethics and politics in qualitative research. In N. K. Denzin & Y. S. Lincoln (Eds.), *Handbook of qualitative research* (2nd ed., pp.133–155). Thousand Oaks, CA: Sage.

Clandinin, D. J., & Connelly, F. M. (1994). Personal experience methods. In N. K. Denzin & Y. S. Lincoln (Eds.), *Handbook of qualitative research* (pp. 413–427). Thousand Oaks, CA: Sage.

Creswell, J. W. (2007). *Qualitative inquiry and research design: Choosing among five approaches.* Thousand Oaks, CA: Sage.

Denzin, N. K., & Lincoln, Y. S. (Eds.). (1994). *Handbook of qualitative research.* Thousand Oaks, CA: Sage.

Eisner, E. W. (1997). The promise and perils of alternative forms of data representation. *Educational Researcher, 26*(6), 4–10.

Ellis, C. (2009). *Autoethnographic reflections on life and work.* Walnut Creek, CA: Left Coast Press.

Ellis, C., & Bochner, A. P. (Eds.). (1996). *Composing ethnography: Alternative forms of qualitative writing.* Walnut Creek, CA: AltaMira.

Ellis, C., & Bochner, A. P. (2000). Autoethnography, personal narrative, reflexivity: Researcher as subject. In N. K. Denzin & Y. S. Lincoln (Eds.), *Handbook of qualitative research* (2nd ed., pp. 733–768). Thousand Oaks, CA: Sage.

Enerstvedt, R. (1989). The problem of validity in social science. In S. Kvale (Ed.), *Issues of validity in qualitative research* (pp. 135–173). Lund, Sweden: Studentlitteratur.

Fay, B. (1987). *Critical social science.* Ithaca, NY: Cornell University Press.

Fine, M., Weis, 1., Weseen, S., & Wong, 1. (2000). For whom? Qualitative research, representations, and social responsibilities. In N. K. Denzin & Y. S. Lincoln (Eds.), *Handbook of qualitative research* (2nd ed., pp. 107–131). Thousand Oaks, CA: Sage.

Firestone, W. (1987). Meaning in method: The rhetoric of quantitative and qualitative research. *Educational Researcher, 16*(7), 16–21.

Flax, J. (1990). *Thinking fragments.* Berkeley: University of California Press.

Foucault, M. (1971). *The order of things: An archaeology of the human sciences.* New York: Pantheon.

Gamson, J. (2000). Sexualities, queer theory, and qualitative research. In N. K. Denzin & Y. S. Lincoln (Eds.), *Handbook of qualitative research* (2nd ed., pp. 347–365). Thousand Oaks, CA: Sage.

Gandhi, L. (1998). *Postcolonial theory: A critical introduction.* St. Leonards, N.S.W.: Allen & Unwin.

Geertz, C. (1973). Thick description: Toward an interpretive theory of culture. In C. Geertz, *The interpretation of cultures* (pp. 2–30). New York: Basic Books.

Geertz, C. (1988). *Works and lives: The anthropologist as author.* Cambridge, UK: Polity.

Geertz, C. (1993). *Local knowledge: Further essays in interpretive anthropology.* London: Fontana.

Giroux, H. A. (1982). *Theory and resistance in education: A pedagogy for the opposition.* Boston: Bergin & Garvey.

Greenwood, D. J., &Levin, M. (2000). Reconstructing the relationships between universities and society through action research. In N. K. Denzin & Y. S. Lincoln (Eds.), *Handbook of qualitative research* (2nd ed., pp. 85–106). Thousand Oaks, CA: Sage.

Guba, E. G., (1990). *The paradigm dialog.* Newbury Park, CA: Sage.

Guba, E. G., (1996). What happened to me on the road to Damascus. In L. Heshusius & K. Ballard (Eds.), *From positivism to interpretivism and beyond: Tales of transformation in educational and social research* (pp. 43–49). New York: Teachers College Press.

Guba, E. G., & Lincoln, Y. S. (1981). *Effective evaluation: Improving the usefulness of evaluation results through responsive and naturalistic approaches.* San Francisco: Jossey-Bass.

Guba, E. G., & Lincoln, Y. S. (1982). Epistemological and methodological bases for naturalistic inquiry. *Educational Communications and Technology Journal, 31,* 233–252.

Guba, E. G., & Lincoln, Y. S. (1985). *Naturalistic inquiry.* Newbury Park, CA: Sage.

Guba, E. G., & Lincoln, Y. S. (1989). *Fourth generation evaluation.* Newbury Park, CA: Sage.

Guba, E. G., & Lincoln, Y. S. (1994). Competing paradigms in qualitative research. In N. K. Denzin & Y. S. Lincoln (Eds.), *Handbook of qualitative research* (pp. 105–117). Thousand Oaks, CA: Sage.

Guba, E. G., & Lincoln, Y. S. (2005). Paradigmatic controversies, contradictions, and emerging confluences. In N. K. Denzin & Y. S. Lincoln (Eds.), *The SAGE handbook of qualitative research* (3rd ed., pp. 191–215). Thousand Oaks, CA: Sage.

Harding, S. (1993). Rethinking standpoint epistemology: What is "strong objectivity"? In L. Alcoff & E. Potter (Eds.), *Feminist epistemologies* (pp. 49–82). New York: Routledge.

Heron, J. (1996). *Cooperative inquiry: Research into the human condition.* London: Sage.

Heron, J., & Reason, P. (1997). A participatory inquiry paradigm. *Qualitative Inquiry, 3,* 274–294.

Hertz, R. (1997). Introduction: Reflexivity and voice. In R. Hertz (Ed.), *Reflexivity and voice.* Thousand Oaks, CA: Sage.

Heshusius, L. (1994). Freeing ourselves from objectivity: Managing subjectivity or turning toward a participatory mode of consciousness? *Educational Researcher, 23*(3), 15–22.

Howe, K., & Eisenhart, M. (1990). Standards for qualitative (and quantitative) research: A prolegomenon. *Educational Researcher, 19*(4), 2–9.

Hutcheon, L. (1989). *The politics of postmodernism.* New York: Routledge.

Josselson, R. (1995). Imagining the real. *Interpreting experience. The narrative study of lives* (Vol. 3.). Thousand Oaks, CA: Sage.

Kemmis, S., & McTaggart, R. (2000). Participatory action research. In N. K. Denzin & Y. S. Lincoln (Eds.), *Handbook of qualitative research* (2nd ed., pp. 567–605). Thousand Oaks, CA: Sage.

Kilgore, D. W. (2001). Critical and postmodern perspectives in learning. In S. Merriam (Ed.), *The new update of education theory: New directions in adult and continuing education.* San Francisco: Jossey-Bass.

Kincheloe, J. L. (1991). *Teachers as researchers: Qualitative inquiry as a path to empowerment.* London: Falmer.

Kincheloe, J. L., & McLaren, P. (2000). Rethinking critical theory and qualitative research. In N. K. Denzin & Y. S. Lincoln (Eds.), *Handbook of qualitative research* (2nd ed., pp. 279–313). Thousand Oaks, CA: Sage.

Kondo, D. K. (1990). *Crafting selves: Power, gender, and discourses of identity in a Japanese workplace.* Chicago: University of Chicago Press.

Kondo, D. K. (1997). *About face: Performing race in fashion and theater.* New York: Routledge.

Kuhn, T. (1967). *The structure of scientific revolutions* (2nd ed.). Chicago: University of Chicago Press.

Kvale, S. (Ed.). (1989). *Issues of validity in qualitative research.* Lund, Sweden: Studentlitteratur.

Kvale, S. (1994, April). *Validation as communication and action.* Paper presented at the annual meeting of the American Educational Research Association, New Orleans.

Lancy, D. F. (1993). *Qualitative research in education: An introduction to the major traditions.* New York: Longman.

Lather, P. (1986). Issues of validity in openly ideological research: Between a rock and a soft place. *Interchange, 17*(4), 63–84.

Lather, P. (1991). *Getting smart: Feminist research and pedagogy within the postmodern.* New York: Routledge.

Lather, P. (1993). Fertile obsession: Validity after poststructuralism. *Sociological Quarterly, 34,* 673–693.

Lather, P. (2007). *Getting lost: Feminist efforts toward a double(d) science.* Albany: State University of New York Press.

Lather, P., & Smithies, C. (1997). *Troubling the angels: Women living with HIV/AIDS.* Boulder, CO: Westview/HarperCollins.

Latsis, J., Lawson, C., & Martins, N. (2007). Introduction: Ontology, philosophy, and the social sciences. In C. Lawson, J. Latsis, & N. Martins (Eds.), *Contributions to social ontology.* New York: Routledge.

Leitch, Y. B. (1996). *Postmodern: Local effects, global flows.* Albany: State University of New York Press.

Lincoln, Y. S. (1995). Emerging criteria for quality in qualitative and interpretive research. *Qualitative Inquiry, 1,* 275–289.

Lincoln, Y. S. (1997). What constitutes quality in interpretive research? In C. K. Kinzer, K. A. Hinchman, & D. J. Leu (Eds.), *Inquiries in literacy: Theory and practice* (pp. 54–68). Chicago: National Reading Conference.

Lincoln, Y. S. (1998a). The ethics of teaching qualitative research. *Qualitative Inquiry, 4,* 305–317.

Lincoln, Y. S. (1998b). From understanding to action: New imperatives, new criteria, new methods for interpretive researchers. *Theory and Research in Social Education, 26*(1), 12–29.

Lincoln, Y. S., & Denzin, N. K. (1994). The fifth moment. In N. K. Denzin & Y. S. Lincoln (Eds.), *Handbook of qualitative research* (pp. 575–586). Thousand Oaks, CA: Sage.

Lincoln, Y. S., & Guba, E. G. (1985). *Naturalistic inquiry.* Beverly Hills, CA: Sage.

Lucas, J. (1974, May). *The case survey and alternative methods for research aggregation.* Paper presented at the Conference on Design and Measurement Standards in Political Science, Delavan, WI.

Lucas, J. (1976). *The case survey method: Aggregating case experience* (R-1515-RC). Santa Monica, CA: The Rand Corporation.

Lynham, S. A., & Webb-Johnson, G. W. (2008). Models of Epistemology and Inquiry Class Notes. Texas A&M University.

Marcus, G. E., & Fischer, M. M. J. (1986). *Anthropology as cultural critique: An experimental moment in the human sciences.* Chicago: University of Chicago Press.

Mayan, M. J. (2009). *Essentials of qualitative inquiry.* Walnut Creek, CA: Left Coast Press.

McCall, M. M. (2000). Performance ethnography: A brief history and some advice. In N. K. Denzin & Y. S. Lincoln (Eds.), *Handbook of qualitative research* (2nd ed., pp. 421–433). Thousand Oaks, CA: Sage.

Merriam, S. B. (1991). How research produces knowledge. In J. M. Peters & P. Jarvis (Eds.), *Adult education*. San Francisco: Jossey-Bass.

Merriam, S. B., Caffarella, R. S., & Baumgartner, L. M. (2007). *Learning in adulthood: A comprehensive guide*. San Francisco: Jossey-Bass.

Mertens, D. (1998). *Research methods in education and psychology: Integrating diversity with quantitative and qualitative methods*. Thousand Oaks, CA: Sage.

Michael, M. C. (1996). *Feminism and the postmodern impulse: Post-World War II fiction*. Albany: State University of New York Press.

Noddings, N. (1984). *Caring: A feminine approach to ethics and moral education*. Berkeley: University of California Press.

Olesen, Y. L. (2000). Feminisms and qualitative research at and into the millennium. In N. K. Denzin & Y. S. Lincoln (Eds.), *Handbook of qualitative research* (2nd ed., pp. 215–255). Thousand Oaks, CA: Sage.

Pallas, A. M. (2001). Preparing education doctoral students for epistemological diversity. *Educational Researcher, 30*(5), 6–11.

Palmer, P. J. (1987, September-October). Community, conflict, and ways of knowing. *Change, 19,* 20–25.

Pelias, R. J. (1999). *Writing performance: Poeticizing the researcher's body*. Carbondale: Southern Illinois University Press.

Pelias, R. J. (2004). *A methodology of the heart*. Walnut Creek, CA: AltaMira Press.

Peshkin, A. (1993). The goodness of qualitative research. *Educational Researcher, 22*(2), 24–30.

Phillips, D. C. (2006). A guide for the perplexed: Scientific educational research, methodolatry, and the gold versus the platinum standards. *Educational Research Review, 1*(1), 15–26.

Polkinghorne, D. E. (1989). Changing conversations about human science. In S. Kvale (Ed.), *Issues of validity in qualitative research* (pp. 13–46). Lund, Sweden: Studentlitteratur.

Preissle, J. (2006). Envisioning qualitative inquiry: A view across four decades. *International Journal of Qualitative Studies in Education 19*(6), 685–695.

Reason, P. (1993). Sacred experience and sacred science. *Journal of Management Inquiry, 2,* 10–27.

Reason, P., & Rowan, J. (Eds.). (1981). *Human inquiry*. London: John Wiley.

Reinharz, S. (1997). Who am I? The need for a variety of selves in the field. In R. Hertz (Ed.), *Reflexivity and voice* (pp. 3–20). Thousand Oaks, CA: Sage.

Richardson, L. (1994). Writing: A method of inquiry. In N. K. Denzin & Y. S. Lincoln (Eds.), *Handbook of qualitative research* (pp. 516–529). Thousand Oaks, CA: Sage.

Richardson, L. (1997). *Fields of play: Constructing an academic life*. New Brunswick, NJ: Rutgers University Press.

Richardson, L. (2000). Writing: A method of inquiry. In N. K. Denzin & Y. S. Lincoln (Eds.), *Handbook of qualitative research* (2nd ed., pp. 923–948). Thousand Oaks, CA: Sage.

Rorty, R. (1979). *Philosophy and the mirror of nature*. Princeton, NJ: Princeton University Press.

Ryan, K. E., Greene, J. C., Lincoln, Y. S., Mathison, S., & Mertens, D. (1998). Advantages and challenges of using inclusive evaluation approaches in evaluation practice. *American Journal of Evaluation, 19,* 101–122.

Salner, M. (1989). Validity in human science research. In S. Kvale (Ed.), *Issues of validity in qualitative research* (pp. 47–72). Lund, Sweden: Studentlitteratur.

Scheurich, J. J. (1994). Policy archaeology. *Journal of Educational Policy, 9,* 297–316.

Scheurich, J. J. (1996). Validity. *International Journal of Qualitative Studies in Education, 9,* 49–60.

Scheurich, J. J. (1997). *Research method in the postmodern*. London: Falmer.

Schratz, M., & Walker, R. (1995). *Research as social change: New opportunities for qualitative research*. New York: Routledge.

Schwandt, T. A. (1989). Recapturing moral discourse in evaluation. *Educational Researcher, 18*(8), 11–16, 34.

Schwandt, T. A. (1996). Farewell to criteriology. *Qualitative Inquiry, 2,* 58–72.

Schwandt, T. A. (2000). Three epistemological stances for qualitative inquiry: Interpretivism, hermeneutics, and social constructionism. In N. K. Denzin & Y. S. Lincoln (Eds.), *Handbook of qualitative research* (2nd ed., pp. 189–213). Thousand Oaks, CA: Sage.

Schwandt, T. A. (2007). *The SAGE dictionary of qualitative inquiry* (3rd ed.). Thousand Oaks, CA: Sage.

Sechrest, 1. (1993). *Program evaluation: A pluralistic enterprise*. San Francisco: Jossey-Bass.

Skrtic, T. M. (1990). Social accommodation: Toward a dialogical discourse in educational inquiry. In E. Guba (Ed.), *The paradigm dialog*. Newbury Park, CA: Sage.

Smith, J. K. (1993). *After the demise of empiricism: The problem of judging social and educational inquiry*. Norwood, NJ: Ablex.

Smith, J. K., & Deemer, D. K. (2000). The problem of criteria in the age of relativism. In N. K. Denzin & Y. S. Lincoln (Eds.), *Handbook of qualitative research* (2nd ed., pp. 877–896). Thousand Oaks, CA: Sage.

Stimpson, C. R. (1988). Nancy Reagan wears a hat: Feminism and its cultural consensus. *Critical Inquiry, 14,* 223–243.

Tierney, W. G. (2000). Undaunted courage: Life history and the postmodern challenge. In N. K. Denzin & Y. S. Lincoln (Eds.), *Handbook of qualitative research* (2nd ed., pp. 537–553). Thousand Oaks, CA: Sage.

Tierney, W. G., & Lincoln, Y. S. (Eds.). (1997). *Representation and the text: Re-framing the narrative voice*. Albany: State University of New York Press.

Trinh, T. M. (1991). *When the moon waxes red: Representation, gender, and cultural politics*. New York: Routledge.

Tschudi, F. (1989). Do qualitative and quantitative methods require different approaches to validity? In S. Kvale (Ed.), *Issues of validity in qualitative research* (pp. 109–134). Lund, Sweden: Studentlitteratur.

Weiss, C. H. (1998). *Evaluation* (2nd ed.). Upper Saddle River, NJ: Prentice Hall.

7

FEMINIST QUALITATIVE RESEARCH IN THE MILLENNIUM'S FIRST DECADE

Developments, Challenges, Prospects[1]

Virginia Olesen

There are many discourses of feminism in circulation, and we need, at times, to deploy them all.

—Susanne Gannon and Bronwyn Davies (2007, p. 100)

Feminisms and qualitative research practices continue to be highly diversified, contentious, dynamic, and challenging. Disparate orientations to both theoretical issues and research practices exist as new ideas and practices emerge, old ones ossify or fade (Fonow & Cook, 2005). Amid the multiple complexities, maturing and deepened developments in theory and research on intersectionality, participatory action research and transnational feminist work, insights, and practices expand, even as they destabilize some foundations. Energizing these developments is the growing importance of "endarkened"/ decolonized feminist research. These position feminist qualitative researchers to address enduring and emergent questions of gendered social justice. This does not assume a global, homogeneous feminism. Feminists draw from different theoretical and pragmatic orientations that reflect national contexts where feminist agendas differ widely (Evans, 2002; Franks, 2002; Howard & Allen, 2000). Ideas of once dominant groups in the northern hemisphere are no longer the standard (Alexander, 2005; Arat-Koc, 2007; Harding & Norberg, 2005; Mohanty, 2003). Replicating whiteness is a major concern (Evans, Hole, Berg, Hutchinson, & Sookraj, 2009).

This chapter derives from the sharpening and focusing of my own research sensibilities since my 1975 chapter, "Rage Is Not Enough." That chapter called for incisive feminist scholarship relevant for policy to frame and harness passion to challenge injustice around women's health, one of my enduring concerns. Feminist postcolonial and deconstructive thought later substantially expanded my groundedness in constructionist symbolic interaction. Postmodern research that addresses social justice issues has also influenced me, as has the work of feminists of color and lesbian feminists.

A brief review of diverse feminist qualitative research will introduce a discussion of transformative themes and developments. A short exploration follows of some enduring concerns. A review of unresolved and emergent issues introduces discussion of new opportunities and an examination of realizing social justice in difficult times.

Breadth

Feminist qualitative researchers continue to explore topics that range from interpersonal issues, that is, domestic violence (Jiwani, 2005; Renzetti, 2005), body and health (Dworkin & Wachs, 2009), health and illness (Schulz & Mullings, 2006), medical knowledge (Shim, 2000), and social movements (Bell, 2009; Klawiter, 2008; Kuumba, 2002).

Policy research, once erroneously thought impossible with qualitative approaches, increasingly draws feminist attention (Fonow & Cook, 2005), although the area is a challenge (Campbell, 2000; Harding & Norberg, 2005; Mazur, 2002; Priyadharshini, 2003).

If there is a dominant theme in feminist qualitative research, it is the issue of knowledges. Whose knowledges? Where and how obtained, by whom, from whom, and for what purposes?

It moved feminist research from the lack of or flawed attention to marginalized women, usually nonwhite, homosexual, or disabled, to recognition of differences among women and within the same groups of women and the recognition that multiple identities and subjectivities are constructed in particular historical and social contexts. It opened discussion of critical epistemological issues, the researcher's characteristics and relationships to the research participants.

Transformative Developments

Transformative developments continue to emerge from approaches (postcolonial, globalization, transnational feminism), conceptual and theoretical shifts (standpoint theory, poststructural thought) and research by and about specific groups of women (gay, lesbian, and queer; disabled; women of color).

Postcolonial feminist thought. If the criticisms of an unremitting whiteness in feminist research in Western, industrialized societies unsettled feminist research frames, powerful and sophisticated research and feminist thought from postcolonial theorists continued to shift grounds of feminist research with regard to "woman" and "women," the very definitions of feminism itself, and constructions of color. Feminism takes many different forms depending on the context of contemporary nationalism. Concerned about the invidious effects of "othering" (invidious, oppressive definitions of the people with whom research is done), postcolonial feminists claimed that Western feminist models were inappropriate for thinking of research with women in postcolonial sites.

Postcolonial feminists raised incisive questions whether subordinates can speak or are forever silenced by virtue of representation within elite thought (Mohanty, 1988, 2003; Spivak, 1988). They also asked whether all women could be conceptualized as unified subjectivities located in the category of woman. They argued that subjectivity and identity are constructed in many different ways in any historical moment (Kim, 2007) and undercut the concept of woman, the assumptions of subjectivity and objectivity, and the utility of the interview (Trinh, 1989, 1992). Postcolonial feminist thought demands decolonizing self and other (Kim, 2007).

Globalization and transnational feminism. Globalization, the relentless, neoliberal flow of capitalism across national borders, destabilizes labor markets, induces movements of workers (Kim-Puri, 2005, pp. 139–142), and creates new sites of inquiry beyond the nation-state and new interpretations of power as multisited and shifting (Mendez & Wolf, 2007, pp. 652). Feminists have complicated the nature and characteristics of globalization (Desai, 2007). Globalization is rife with contradictions and the potential to produce multiple subjectivities (Kim-Puri, 2005;

Naples, 2002a, 2002b). Research examines the tension between the dominance of the state and economic forces and women's potential resistance (Thayer, 2001) and dialectic between "new" opportunities and oppressions (Chang, 2001; Lan, 2006).

Others have examined women's lives and working conditions in diverse international contexts: sex workers (Gulcur & Ilkkaracan, 2002; Katsulis, 2009); the international sex trade (Dewey, 2008; Hanochi, 2001); care work (Zimmerman, Litt, & Bose, 2006); domestic servants (Parrenas, 2008); and laborers (Keough, 2009) as well as how governments create "heroic" migrant labor (Guevarra, 2009).

This work invokes the efficacy of postmodern thinking (Lacsamana, 1999); the risk of reproducing Eurocentric concepts of feminism (Grewal & Kaplan, 1994; Kempadoo, 2001); questions of female agency (Doezema, 2000); and the inadequacies of cultural analyses to understand oppressions rooted in material conditions under globalization (Fraser, 2005; Kim-Puri, 2005; Mendoza, 2002).

Closely related, transnational feminism analyzes national and cross-national feminist organizing and action (Davis, 2007; Mendez & Wolf, 2007; Mendoza, 2002). This work examines bases of feminist mobilization, for example, class, race, ethnicity, religion, and regional struggles; it sidesteps imposing a Westernized version of feminism. It poses substantial critical challenges (Mendez & Wolf, 2007).

Transnational feminists also examine sex trafficking (DeRiviere, 2006; Firdous, 2005; Stout, 2008), violence against women (Jiwani, 2005), and reproductive technologies (Gupta, 2006).

Standpoint research. Standpoint research flourishes in the early years of the millennium (Harding, 2008). Sociologist Dorothy Smith,[2] sociologist Patricia Hill Collins,[3] philosopher Sandra Harding,[4] and political scientist Nancy Hartsock[5] replaced the concept of essentialized, universalized woman with the idea of a situated woman with experiences and knowledge specific to her place in the material division of labor and the racial stratification systems. Standpoint theorists are not identical; they offer divergent approaches for qualitative researchers.[6] Moreover, feminist qualitative researchers must read these theorists in their latest version—for example, Harding's plea to start with women's lives in households (2008)—if they are to avoid misinterpretation. Standpoint theories came in for extensive criticisms,[7] which evoked vigorous responses (Collins, 1997; Harding, 1997; Hartsock, 1997; D. E. Smith, 1997).

Regarding the relationship of standpoint theory to postmodern and poststructural thinking, "poststructural approaches have been especially helpful in enabling standpoint theories systematically to examine critically pluralities of power relations" (Harding, 1996, p. 451). Collins (1998b) warns about the corrosive effects of postmodern and deconstructive thought for Black women's group authority and social action, but she also argues that postmodernism's powerful analytic tools can challenge

Table 7.1

I. Transformative Developments		
Approaches	Postcolonial feminist thought	Kim, 2007; Mohanty, 1988, 2003; Spivak, 1988; Trinh, 1989, 1992
	Globalization	Chang, 2001; Dewey, 2008; Fraser, 2005; Guevara, 2009; Kim-Puri, 2005; Lan, 2006; Naples, 2002a,b; Parrenas, 2008; Zimmerman, Litt, & Bose, 2006
	Transnational feminism	Davis, 2007; DeRiviere, 2006; Firdous, 2005; Mendez & Wolf, 2007; Stout, 2008
	Standpoint theory	Collins, 1992, 1998 a,b; Haraway, 1991; Harding, 1987, 1993, 2008; Hartsock, 1983, 1997; Naples, 2007; Smith, 1987, 1997; Weeks, 2004
	Postmodern and poststructural deconstructive theory	Clough, 2000; Collins, 1998b; Flax, 1987, 1990; Gannon & Davies, 2007; Haraway, 1991; Hekman, 1990; Lacsamana, 1999; Lather, 2007; Mazzei, 2003, 2004; Pillow, 2003; St.Pierre, 1997b, 2009
Work By and About Specific Groups of Women	Lesbian research	Anzaldúa, 1987, 1990; Connolly, 2006; Kennedy & Davis, 1993; Lewin, 1993, 2009; Mamo, 2007; Merlis & Linville, 2006; Mezey, 2008; Weston, 1991
	Queer theory	Butler, 1990, 1993, 2004; Rupp & Taylor, 2003
	Disabled women	Fine, 1992; Garland-Thompson, 2005; Lubelska & Mathews, 1997; Meekosha, 2005; Mertens, 2009; Petersen, 2006 ; Tregaskis & Goodley, 2005
	Women of color	Acosta, 2008; Anzaldúa, 1990; Chow, 1987; Collins, 1986; Cummins & Lehman, 2007; Davis, 1981; Dill, 1979; Espiritu, 2007; Few, 2007; Glenn, 2002; Green, 1990; hooks, 1990; Majumdar, 2007; Mihesuah, 2003 ; Moore, 2008; Tellez, 2008
	Problematizing unremitting whiteness	Frankenberg, 1994; Hurtado & Stewart, 1997
II. Critical Trends		
Endarkening, Decolonizing, Indigenizing Feminist Research		Anzaldúa, 1987; Battiste, 2008; Collins, 2000; Dillard, 2008; Gardiner & Meyer, 2008a; Saavedra & Nymark, 2008; Segura & Zavella, 2008; Smith, 1999, 2005
Intersectionality		Andersen 2005, 2008; Bhavnani, 2007; Bowleg, 2008; Brah & Phoenix, 2004; Collins, 2000, 2008, 2009; Crenshaw, 1989, 1991; Davis, 2008; Denis, 2008; Dill, McLaughlin, & Nieves, 2007; Dill & Zambrana, 2009; Glenn, 2002; Hancock, 2007a,b; McCall, 2005; Risman, 2004; Shields, 2008; Stewart & McDermott, 2004; Warner, 2008; Yuval-Davis, 2006
III. Continuing Issues		
Problematizing Researcher and Participant		Kahn, 2005; Lather & Smithies, 1997; Lather, 2007; Lincoln, 1993, 1997
Destabilizing Insider-Outsider		Kondo, 1990; Lewin, 1993; Naples, 1996; Narayan, 1997; Ong, 1995; Weston, 1991; Zavella, 1996
Troubling Traditional Concepts	Experience	Scott, 1991

(Continued)

Table 7.1 (Continued)

	Difference	Felski, 1997; hooks, 1990
	Gender	Baravosa-Carter, 2001; Butler, 1990, 1993, 2004; Jurik & Siemsen, 2009; Lorber, 1994; West & Zimmerman, 1987
IV. Enduring Concerns		
"Bias" and Objectivity		Diaz, 2002; Fine, 1992; Harding, 1993, 1996, 1998; Phoenix, 1994; Scheper-Hughes, 1983
Reflexivity		Few, 2007; Guilleman & Gillam, 2004; Hesse-Biber & Piatelli, 2007; Pillow, 2003
"Validity" and Trustworthiness		Lather, 1993, 2007; Manning, 1997; Richardson, 1993; St.Pierre (Ch. 37, this handbook)
Participants' Voices		Behar, 1993; Ellis & Bochner, 1996, 2000; Fine, 1992; Gray & Sinding, 2002; Kincheloe, 1997; Kondo, 1995; Lather & Smithies, 1997; Lincoln, 1993, 1997; Phoenix, 1994; Richardson, 1997; Stacey, 1998
Deconstructing Voice		Jackson, 2003; MacLure, 2009; Mazzei, 2009; Mazzei & Jackson, 2009; Lather & Smithies, 1997
Performance Ethnography		Alexander, 2005; Battacharya, 2009; Case & Abbitt, 2004; Cho & Trent, 2009; Denzin, 2005; Gray & Sinding, 2002; Kondo, 1995; Madison, 2005, 2006; Valentine, 2006
Ethics in Feminist Research		Battacharya, 2007; Battiste, 2008; Corrigan, 2003; Edwards & Mauthner, 2002; Ellis, 2009a; Fine, Weis, Weseem, & Wong, 2000; Guilleman & Gillam, 2004; Halsey & Honey, 2005; Lincoln, 2005; Llewelyn, 2007; Mauthner, Birch, Jessop, & Miller, 2002; Miller & Bell, 2002; Morse, 2005, 2007; L. T. Smith, 1999, 2005; Stacey, 1988; Thapar-Bjorkert & Henry, 2004; Wolf, 1996
Participatory Action Research		Cancian, 1996; Etowa, Bernard, Oynisan, & Clow, 2007; Evans, Hole, Berg, Hutchinson, & Sookraj, 2009; Fine & Torre, 2006; Reid, Tom, & Frisby, 2008
V. Influences on Feminist Work		
Contexts	The Academy	Dever, 2004; Laslett & Brenner, 2001; Messer-Davidow, 2002; Shields, 2008
	Publishing and Eurocentric Parochialism	Messer-Davidow, 2002
VI. Into the Future		
Challenges: Making Feminist Work Count		Cook & Fonow, 2007; Davis & Craven, 2011; Hesse-Biber, 2007; Laslett & Brenner, 2001; Stacey, 2003

dominant discourses and the very rules of the game. Nancy A. Naples (2007) argues for a multidimensional approach to standpoint research, which recognizes both the embodied aspects and the multiplicity of researcher and participant perspectives.

Poststructural Postmodern Thought

Postmodern and poststructural/deconstructive thinking continues to be controversial, yet energizes other feminist researchers (Gannon & Davies, 2007; Lather, 2007).

Concerned that it is impossible to produce more than a partial story of women's lives in oppressive contexts, postmodern feminists regard "truth" as a destructive illusion. They see the world as a series of stories or texts that sustain the integration of power and oppression and actually "constitute us as subjects in a determinant order" (Hawkesworth, 1989, p. 549). Influenced by French feminists (Luce Irigaray, Hélène Cixous) and theorists (Michel Foucault, Gilles Deleuze, Jean-François Lytoard, Jacques Derrida, and Jean Baudrillard) and American theorist Judith Butler, postmodern/deconstructive feminist research studies focus on representation and text. Some scholars also use Marxist theory from Louis Althusser, and psychoanalytic views (Flax, 1987, 1990; Gannon & Davies, 2007).

Taking the position that text is central to incisive analysis as a fundamental mode of social criticism, these inquiries typically analyze cultural objects (film, etc.) and their meanings (Balsamo, 1993; Clough, 2000; deLauretis, 1987; Denzin, 1992; Morris, 1998). Included are textual analyses of these objects and the discourses surrounding them (Denzin, 1992) and the "study of lived cultures and experiences which are shaped by the cultural meanings that circulate in everyday life" (Denzin, 1992, p. 81).

Here, too, will be found sophisticated feminist work in gender and science, wherein science is deconstructed to reveal its practices, discourses, and implications for control of women's lives (Haraway, 1991; Martin, 1999), including their health (Clarke & Olesen, 1999), and to suggest avenues for resistance or intervention. Research about women's reproductive issues also moved into this area (Clarke, 1998; Mamo, 2007; Rapp, 1999). These productions discomfort not only male-dominated institutions, such as science, but feminism itself by complicating where and how "women" are controlled, how multiple, shifting identities and selves are produced.

In particular, poststructural deconstructive feminists question the very nature and limits of qualitative research (Lather, 1991, 2007; St.Pierre, 2009). They argue that traditional empirical research, imbedded in regimes of power, merely replicates oppressive structures while fruitlessly seeking the impossible, namely a full, complete account of whatever is investigated with inadequate strategies. They do not seek "a method" but attempt to exploit these shortcomings with centripetal strategies that reach outward, "strategies, approaches, and tactics that defy definition or closure" (Gannon & Davies, 2007, p. 81), rather than centrifugal (leaning inward toward one, stable interpretation).

Poststructural deconstructive feminists question taken-for-granted terms such as data, arguing for "transgressive data" (emotional, dreams, sensual response) (St.Pierre, 1997b) and for analysis of silences (Mazzei, 2003, 2004). They have also deconstructed validity (Lather, 1993), reflexivity (Pillow, 2003), and voice, to be discussed shortly. They but point to "a less comfortable science" (Lather, 2007, p. 4) wherein researchers trouble their own categories, while recognizing the uncertainties and the absence of absolute frames of reference (Lather, 2007).

Critics of the postmodern/poststructural position alleged that it left no grounds for reform-oriented research, reinforced the status quo, erased structural power, and failed to address problems or to represent a cultural system.[8] However, as already noted, standpoint theorists Collins and Harding see the possibility of deconstructing power and opening new spaces for social action.

Poststructural feminist work offers the potential for thinking differently about obdurate problems (Gannon & Davies, 2007), which appears useful for feminist policy research. These feminists have done work oriented to social justice (Lather & Smithies, 1997; Mazzei, 2004; Scheurich & Foley, 1997; St.Pierre, 1997a). Transformative developments continue in work by and about groups of women.

Lesbian research. Research dissolved homogeneous views of lesbians (Lewin, 1993; Weston, 1991).[9] Other work revealed multiple bases of lesbian identity to further differentiate these views and destabilize notions of heteronormativity (Anzaldúa, 1987, 1990; Kennedy & Davis, 1993). Early millennial lesbian research continued this trend (Connolly, 2006; Lewin, 2009; Mamo, 2007; Merlis & Linville, 2006; Mezey, 2008). *Queer theory*, loosely used as a cover term for gay and lesbian studies, also refers to a more precise political stance and the push against "disciplinary legitimation and rigid categorization" (Adams & Jones, 2008, p. 381). Disruption of normalizing ideologies is the key to queer theory, which is oriented to a politics of change (Alexander, 2008).

Research shows how gay and lesbian marriage ceremonies simultaneously reflect accommodation and subversion (Lewin, 1998), and it questions the very stability of "man" and "woman" (Rupp & Taylor, 2003).

Disabled women. Disabled women were depersonalized and degendered, sometimes even, regrettably, within feminist circles (Lubelska & Mathews, 1997), when researchers overlooked their multiple statuses and viewed them solely in terms of their disability (Asch & Fine, 1992). Feminist scholars, both disabled and abled, began to problematize disability (Garland-Thompson, 2005).

In the new millennium, their work ranges widely (Meekosha, 2005; Mertens, 2009; Mertens, Sullivan, & Stace, Chapter 13, this volume; Petersen, 2006; Tregaskis & Goodley, 2005).

Women of color. That there *are* multiple knowledges, that women of color were frequently overlooked or interpreted in terms of white women has been forcefully argued (Anzaldúa, 1987, 1990; Chow, 1987; Collins, 2000; Davis, 1981; Dill, 1979; Green, 1990; hooks, 1990). This continues with exploration of Black families (Few, 2007; Moore, 2008), AIDS and Black women (Foster, 2007; Latino critical theory (Delgado Bernal, 2002), diversities among American Indian women (Mihesuah, 2003), Asian American men and women (Espiritu, 2007), eating disorders among Asian women (Cummins & Lehman, 2007), marriage among Southeast Asian women (Majumdar, 2007), and Chicana experiences on the U.S.-Mexican border (Acosta, 2008; Tellez, 2008). Important theoretical contributions examined interlocking influences of gender and race on citizenship (Glenn, 2002) and the argument that Blacks are a monolithic group (Collins, 2008).

Parallel investigations problematized the construction of women of color in relationship to whiteness (Puar, 1996) and whiteness itself (Frankenberg, 1994; Hurtado & Stewart, 1997). As Yen Le Espiritu has noted, "Racism affects not only people of color but organizes and shapes experiences of all women" (personal communication, September 15, 2003). To untangle whiteness and the existence of a global color line, Chandra Mohanty (2003) noted the necessity to think relationally about questions of power, equality, and justice, to make thinking and organizing contextual, and to root questions of history and experience.

Critical Trends

Two critical trends emerged from these developments: (1) "endarkening," decolonizing, indigenizing feminist research and (2) expansion and maturing intersectionality as a critical approach.

▣ ENDARKENING, DECOLONIZING, INDIGENIZING FEMINIST RESEARCH

Feminist scholars of color deepened thought and research to move away from colonial legacies, wherever found, and stressed the critical nature of subordinated women's (and men's) knowledge as legitimate foundations for attempts to realize social justice. Influential work on decolonizing methodologies (L. T. Smith, 1999, 2005) and on protecting indigenous knowledge (Battiste, 2008) spurred these developments, as did writing by African American and Mexican American feminists (Cannella & Manuelito, 2008).

Anzaldúa's (1987) experimental writing and work decenters Western thinking and theorizing to emphasize decolonizing research (Saavedra & Nymark, 2008). More specifically, her conceptualization of borderlands posed "dynamic processes deployed for specific purposes—fluctuating, permeable, and rife with possibilities and consequences" (Gardiner & Meyer, 2008b, p. 10). (See Gardiner & Meyer, 2008a; Segura & Zavella, 2008).

Anzaldúa's innovative thinking also emphasized spirituality as requisite to the political (Gardiner & Meyer, 2008a). A similar proposal, but more specifically directed to feminist research and action, is Dillard's (2008) call to locate spirituality and qualitative research in endarkening feminist research (see also Dillard and Okpalaoka, Chapter 8, this volume).

Intersectionality. Intersectionality (Crenshaw, 1989, 1991) denotes how social divisions are constructed and intermeshed with one another in specific historical conditions to contribute to the oppression of women not in mainstream white, heterosexual, middle-class, able-bodied America. By the early years of the new millennium, intersectional analysis had spread to numerous disciplines and professions (Brah & Phoenix, 2004; Davis, 2008; Denis, 2008; Yuval-Davis, 2006) and prompted special journal issues (Phoenix & Pettynama, 2006).

Not surprisingly, different views emerged. Some preferred interconnections, which configure one another, to intersectionality, which was seen as too static and at risk of overlooking agency (Bhavnani, 2007). Other worries include that intersectionality applies to all groups, not just the marginalized (Warner, 2008); is empirically weak (Nash, 2008); and does not attend to narrative accounts (Prins, 2006). Working only within an intersectional framework fails to acknowledge how structural mechanisms produce different inequalities (Risman, 2004).

Feminists should not overlook "the broader, political, economic, and social processes that constitute and buttress inequality" (Acker, 2006; Andersen, 2008, p. 121). However, others claim that intersectionality addresses the very meaning of power (Collins, 2009; Dill & Zambrana, 2009; Hancock, 2007a) and is useful in political struggles (Davis, 2008).

Although there is agreement that categories are not additive but interactive and mutually constructed (Acker, 2008; Andersen, 2005; Collins, 2009; Hancock, 2007b; Shields, 2008; Yuval-Davis, 2006), debates about which combinations to use continue. Related is the criticism that intersectionality courts problems of "infinite regress" of categories (Hancock, 2007b). Three observations responded to this: (1) Judgments *can* be made about which categories to use (Stewart & McDermott, 2004); (2) researchers must be explicit as to which are chosen (Warner, 2008); and (3) in specific situations and for specific people, some social divisions are more important than others (Dill & Zambrana, 2009).

Running through these arguments are questions of dynamic interactions between individual identities and institutional factors (Hancock, 2007a, 2007b) that locate any group in socially stratified systems. This pushes feminist researchers to articulate ways to analyze simultaneously identities at structural and political levels (Dill, McLaughlin, & Nieves, 2007) and necessitates placing "social structural and narrative/interpretive approaches to social reality in dialogue with one another" (Collins, 2009, p. xi). This daunting challenge implicates research design, methods (Hancock, 2007b), and interpretation (Bowleg, 2008).

It also questions how complexity in categories is viewed: Anticategorical complexity holds that the completeness of any category can be challenged; for example, sexuality is no longer merely gay or straight but more complicated (McCall, 2005). Intracategorical complexity posits range of diversity and experiences within the same social category, for example, working-class men and working-class women (McCall, 2005).

Intercategorical complexity centers comparison of groups across analytical categories (McCall, 2005). Intersectionality analysis is a "field of cognitive land mines" (Collins, 2008, p. 73), one that is "typically partial" (one cannot handle race, class, gender, sexuality, able-bodiness, and age simultaneously) and is inherently comparative (Collins, 2008).

How, then, to do manageable intersectionality analyses? Collins (2008) finds dynamic centering and relational thinking useful. Dynamic centering places two or more entities at the center of analysis to get a closer look at their mutual construction (Collins, 2008). Relational thinking asks how categories mutually construct one another as systems of power.

Intersectional research promises to address complex feminist issues (Bredstrom, 2006; Dworkin, 2005; Morgen, 2006) to yield new insights, but much remains to be done to handle earlier criticisms (Luft & Ward, 2009). Thanks to new developments in qualitative analysis (Clarke, 2004) and the maturing of institutional ethnography (Smith, 2006), feminist qualitative research *in its own right* is well positioned to undertake these challenges. Blended with quantitative research approaches, it is a powerful way to analyze mechanisms of intersectionality in play (Weber, 2007).

▣ CONTINUING ISSUES

Problematizing researcher and participants. Recognition grew that the researcher's attributes also enter the research interaction. History and context position both researcher and participant (Andrews, 2002). The subjectivity of the researcher, as much as that of the researched, became foregrounded, blurring phenomenological and epistemological boundaries between the researcher and the researched. This questioned whether being an "insider" gave feminist researchers access to inside knowledge (Collins, 1986; Kondo, 1990; Lewin, 1993; Naples, 1996; Narayan, 1997; Ong, 1995; Williams, 1996; Zavella, 1996). Also questioned were the views that insider knowledge and insider/outsider positions are fixed and unchanging (Kahn, 2005).

Troubling traditional concepts. Also under critical scrutiny were concepts key to feminist thought and research, experience, difference, and the workhorse concept, gender.

Experience. Recognition continues to grow that merely focusing on experience does not account for how that experience emerged (Scott, 1991) and the characteristics of the material, historical, and social circumstances. (For early millennial feminist research that does attend to those circumstances, see Garcia-Lopez, 2008; Higginbotham, 2009). Taking experience in an unproblematic way replicates rather than criticizes oppressive systems and carries a note of essentialism. Moreover, personal experience is not a self-authenticating claim to knowledge (O'Leary, 1997).

Difference. The recognition of difference pulled feminist thinkers and researchers away from the view of a shared gynocentric identity but gave way to concerns about the nature of the concept and whether its use led to an androcentric or imperialistic "othering" (Felski, 1997; hooks, 1990). Some wanted it replaced by such concepts as *hybridity, creolization,* and *metissage,* which "not only recognize differences within the subject but also address connections between subjects" (Felski, 1997, p. 12). Others argued that identity cannot be dropped entirely (hooks, 1990). They see differences as autonomous, not fragmented, producing knowledge that accepts "the existence of and possible solidarity with knowledges from other standpoints" (O'Leary, 1997, p. 63).

Gender. Influential reformulations of gender as performative rather than static (Butler, 1990, 1993; West & Zimmerman, 1987)

or wholly constructed (Lorber, 1994) have shifted views away from gender as an individual attribute or biological characteristic. Gender is conceptualized as "done" and "undone" in everyday social interaction (Butler, 2004).[10]

Vigorous criticisms highlight conceptual problems. Some argued that Butler's performative conceptualizations draw attention away from practical interventions (Barvosa-Carter, 2001, p. 129), a point echoed in some criticisms of Candace West and Don Zimmerman (Jurik & Siemsen, 2009). Another critique examines whether the "doing gender" perspective obscures inequality in social relations (Smith, 2009).

▣ ENDURING CONCERNS

Concerns about bias, validity, voice, the text, and ethical conduct, well explored in an earlier era, continue to produce thoughtful uneasiness. Feminist empiricists and standpoint researchers share these worries, while deconstructionists focus on voice and text. All feminist researchers worry about replicating oppression and privilege.

Bias. Foregoing rigid ideas about objectivity, feminist theorists and researchers earlier opened new spaces around the enduring question of bias. Sandra Harding suggested "strong objectivity," which takes researchers as well as those researched as the focus of critical, causal, scientific explanations (1993, 1996, 1998). Donna Haraway (1997) urged going beyond strong objectivity to diffracting, which turns the researchers' lenses to show fresh combinations and possibilities of phenomena.

Reflexivity. This recognizes that both participants and researcher produce interpretations that are "the data" (Diaz, 2002) and goes beyond mere reflection on the conduct of the research. Reflexivity demands steady, uncomfortable assessment about the interpersonal and interstitial knowledge-producing dynamics of qualitative research, in particular, acute awareness as to what unrecognized elements in the researchers' background contribute (Gorelick, 1991; Scheper-Hughes, 1983).

Some have reservations; for example, reflexivity may only generate a rehearsal of the familiar, which reproduces hegemonic structures (Pillow, 2003). However, others argue that it facilitates preventing perpetuation of racial and ethnic stereotypes (Few, 2007). Finally, there remain difficult questions of how much and what kinds of reflexivity are possible and how they are realized (Hesse-Biber & Piatelli, 2007).

Validity. Feminist qualitative researchers address validity, also called "trustworthiness," in different ways depending on how they frame their approaches. Those who work in a traditional vein, reflecting the positivist origins of social science (there is a reality to be discovered), will use established techniques. Others disdain positivistic origins and use techniques that reflect their postpositivist views but do not hold out hard and fast criteria for according "authenticity" (Lincoln & Guba, 1985; Manning, 1997). Other feminist qualitative researchers "challenge different kinds of validity and call for different kinds of science practices" (Richardson, 1993, p. 65).

Lather's (1993) transgressive validity remains the most completely worked out feminist model; it calls for a subversive move, "retaining the term to circulate and break with the signs that code it" (p. 674) in a feminist deconstuctionist mode. This formulation and the articulation of a transgressive validity checklist (Lather, 2007, pp. 128–129) firmly retain a feminist emancipatory stance while working out problems in validity.

Voice(s) and text. How to avoid exploiting or distorting women's voices has long worried feminists (Hertz, 1997). In the new millennium, poststructural feminists raise critical questions about the very nature of voice.

Researchers earlier explored ideology, hegemonic pressures, or interpretation (Fine, personal communication). In the end, whoever writes up the account also has responsibility for the text, selects the audience that shapes voice (Kincheloe, 1997; Lincoln, 1993), and remains in a powerful position (Lincoln, 1997; Phoenix, 1994; Stacey, 1998).

To address this, researchers have outlined various strategies: using voice-centered relational methods (Mauthner & Doucet, 1998) or reconstructed research narratives (Birch, 1998), writing the less powerful voices (Standing, 1998), and presenting versions of voices (Wolf, 1992). Feminist researchers should articulate how, how not, and within what limits voices are framed and used (Fine, 1992).

Other feminist researchers blend respondent voices with their own in various formats: a doubled-voice ethnographic text (Behar, 1993), split-page textual format (Lather & Smithies, 1997), or sociological poetry and tales (Richardson, 1997). Autoethnography foregrounds deeply personal researcher experiences and participants' voices interwoven with political and social issues (Ellingson, 1998, 2009a, 2009b; Ellis, 1995; Ellis & Bochner, 1996, 2000; Gatson, Chapter 31, this volume; Holman Jones 2005). Autoethnographic work links the personal and the political to refute criticisms that such personal reflections are merely solipsistic.

Autoethnography is a way to understand and change the world (Ellis, 2009a). Reflecting that it unsettles ideas of research, social scientists and poststructural feminists and scholars with literary perspectives have criticized the approach (Ellis, 2009a). There are ways to evaluate it (Richardson, 2000).

Deconstructing voice. Poststructural feminists question what constitutes voice (Jackson, 2003, 2009; Mazzei & Jackson, 2009). Their research problematizes voice to yield examples for others: laughter, silence, irony (MacLure, 2009); silent narratives (Mazzei, 2009); and HIV-positive women (Lather & Smithies, 1997).

Performance ethnography. Performance ethnography shifts from conventional prose and reporting findings to dramatic representations (Kondo, 1995). These pieces dramatize feminist subversions (Case & Abbitt, 2004): the experience of metastatic breast cancer (Gray & Sinding, 2002), lives of imprisoned women (Valentine, 2006), and issues of human rights (Madison, 2006). (See also Alexander, 2005; Denzin, 2005; Madison, 2005). Performance ethnography could be useful in taking feminist research public (Stacey, 2003). Work continues on how to evaluate these inquiries (Alexander, 2005, pp. 428–430; Battacharya, 2009; Cho & Trent, 2009; Madison, 2005, 2006).

Ethics. Feminist research ethics moved beyond universalist positions in moral philosophy (duty ethics of principles, utilitarian ethics of consequences) to recognize relationships with research participants as an ethical issue, called *relational ethics* (Edwards & Mauther, 2002; Ellis, 2009a; Mauthner, Birch, Jessop, & Miller, 2002; Preissle, 2007). This necessitates critical reflection to recognize, analyze, and act on ethically important research moments (Guilleman & Gillam, 2004; Halsey & Honey, 2005; Llewelyn, 2007).

Indigenous scholars continue to raise critical elements in feminist ethics. They see the dreary history of research as "a corporate, deeply colonial institution" (L. T. Smith, 2005, p. 101) that exploits indigenous peoples and commodifies indigenous knowledge (Battiste, 2008; Smith, 1999). They conceptualize indigenous research as a seedbed for ethical standards that reference not just the individual but the collective (Battiste, 2008; L. T. Smith, 2005) and, above all, stress respectful relationships and reflect mutual understanding (L. T. Smith, 2005).

Scrutiny of informed consent, destabilized as unproblematic, continues (Battacharya, 2007; Corrigan, 2003; Fine, Weis, Weseem, & Wong, 2000; Miller & Bell, 2002). Carolyn Ellis proposes *process consent,* the practice of continually checking with participants to accommodate changing research relationships and respondents' willingness to continue participating since what is outlined on an institutional review board (IRB) protocol will not necessarily reflect later events (2009a).

Other ethical dilemmas abound (Bell & Nutt 2002; Kirsch, 2005; Morse, 2005, 2007; Stacey, 1988). The view that the researcher occupies a more powerful position has been tempered by realization that the researcher's power is often partial (Ong, 1995), tenuous (Stacey, 1998; Wolf, 1996), and confused with researcher responsibility (Bloom, 1998); also, respondents manipulate or exploit shifts of power (Thapar-Bjorkert & Henry, 2004.)

Feminist qualitative researchers face, along with all qualitative researchers, a pinched, conservative era in which many IRB review practices are not sympathetic to even the most traditional qualitative research, never mind the complex approaches discussed in this chapter (Lincoln, 2005). The restrictive effects of "these politics of evidence" (Morse, 2005, 2006) add another level of struggle to the feminist qualitative search for social justice (Lincoln, 2005) and reflect an enduring climate of positivism. The challenge is to influence local IRBs and to seek changes in legislation and policy (Lincoln, 2005).

Feminists have also examined ethics qua ethics as a research topic. The view shifted from ethical or moral behavior as inherent in gender to the position that an ethics of caring emerges from an interaction between the individual and the milieu (Seigfried, 1996). These positions reach to concerns with the just community (Seigfried, 1996) and the potential to transform society in the public sphere (DesAutels & Wright, 2001; Fiore & Nelson, 2003). Long-standing concerns about ethical (or nonethical) treatment of women in health care systems carried into inquiries on new technologies, such as assisted reproduction, genetic screening and the regrettably enduring problems of equitable care for elderly, poor, deprived women in all ethnic groups.

Participatory action research. In participatory action research (PAR), "researchers" and "participants" fully share aspects of the research process to undertake emancipatory projects. Earlier PAR explored research-related matters: power (Cancian, 1996; Lykes, 1989); data (Acker, Barry, & Esseveld, 1991); and corrections of researchers' and participants' distortions (Skeggs, 1994). These continue with inquiries into participant vulnerability (Fine & Torre, 2006), risks for marginalized individuals (Reid, Tom, & Frisby, 2008), and ethical questions (Rice, 2009). In the new millennium, PAR examined health issues (Etowa, Bernard, Oyinsan, & Clow, 2007; Evans, Hole, Berg, Hutchinson, & Sookraj, 2009) and imprisoned women (Fine & Torre, 2006).

▣ CONTEXTS' INFLUENCE ON
   QUALITATIVE FEMINIST WORK, AGENDAS

Academic sites. Structures of traditional academic life—at least in the United States—have influenced feminist qualitative research and not always to transform the university or realize reform more generally (Dever, 2004; Messer-Davidow, 2002). Continued emphasis on positivism in the social and behavioral sciences has also blunted reform efforts, but feminist scholars continue to argue for transformative scholarship (Shields, 2008).

To realize transformative scholarship, feminist researchers need to recognize the way higher education institutions work while generating "new strategies that correspond to new opportunities as well as the difficulties of these times" (Laslett & Brenner 2001, pp. 1233–1234). (For analyses of difficulties with Black women's scholarship and transformation of the academy, see "Black Women's Studies," 2010).

Publishing and Anglo/Eurocentric parochialism. Publishers bring out increasing numbers of feminist works—theoretical, empirical, experimental, and methodological (Messer-Davidow, 2002).

More international scholars are being published, but in English because of translation difficulties and marketing pressures (Meaghan Morris, personal communication). Fortunately, these publications foreground different perspectives and postcolonial, endarkened feminist research, to undercut Westernizing and homogenizing assumptions about "women" anywhere and everywhere. Feminist talk lists and websites offer information about international feminist work, conferences, and publications, for example, those run by the Sociologists for Women in Society and the Anthropology Feminist Association. Some that are outside the United States or Britain, for instance, http://www.qualitative-research.net/, agi-feministafrica@act.cu.za, regularly cite international researchers.

Feminist research has yet to extensively explore Internet communication resources such as Twitter and Facebook, but their growing popularity has implications for dissemination of reform-oriented inquiries.

▣ INTO THE FUTURE

Challenges. Challenges to feminist qualitative research in all its complexity, diversity, and contentiousness will continue. Notable among these is deeper exploration and extension of intersectionality, using mature methodological approaches (Choo & Ferree, 2010). These explorations position feminist efforts to examine more incisively the interplay of multiple factors in all women's lives. They sharpen understandings of and the potential to generate action and policy in the pursuit of social justice. They link to emergent methods and new knowledge from critical work in "borderlands" and endarkening feminist research.

Also necessary is continued close attention to representation, voice, and text to avoid replication of the researcher and hidden or not so hidden oppressions and instead display participants' representations.

Feminist qualitative research grows stronger because theorists and researchers critically examine foundations; try new research approaches, experimental and traditional; and search for unexamined equity issues. They are more self-conscious and aware of and sensitive to issues in the formulation and conduct of research, as well as the nature of a feminist science. More sophisticated approaches position feminists to examine material social and cultural dynamics, for example, globalization and neoliberalism, which shape women's lives and their contexts (see Davis & Craven, 2011). The hope is for, if not emancipation, at least modest intervention and transformation without replicating oppression.

Making feminist work count. Feminist researchers have articulated thoughtful and realistic suggestions for change or transformation. "We must take our work public with extraordinary levels of reflexivity, caution, and semiotic and rhetorical sophistication" (Stacey, 2003, p. 28). Sociologists for Women in Society (www .socwomen.org) reports mainstream critical feminist research on urgent topics. Feminists have yet to explore the potential of cyberspace to intervene for social justice or disseminate research findings.

I believe that "it is important to recognize that knowledge production is continually dynamic—new frames open which give way to others which in turn open again and again. Moreover, knowledges are only partial" (Olesen & Clarke, 1999, p. 356). (See Cook & Fonow, 2007; Hesse-Biber, 2007.) Early millennial feminist qualitative research, outlined far too sketchily here, lays foundations to realize social justice in different feminist versions: "Our mission . . . must be nothing short of rethinking and reworking our future" (Randall, 2004, p. 23).

Recalling my 1975 paper, "Rage Is Not Enough" (1975), I contend that rage is *still* not enough, but developments in feminist qualitative scholarship, in whatever style or framework, are harnessing passion to realize social justice in more incisive ways. Much more, however, remains to be done to grapple with enduring and emerging issues of equity and social justice.

◙ NOTES

1. I am grateful for incisive criticisms from Norman Denzin, Yvonna Lincoln, Patricia Clough, Michelle Fine, Meaghan Morris, and Yen Le Espiritu and to Adele Clarke for continuing, stimulating feminist dialogue.

2. Dorothy Smith conceptualizes the everyday world as problematic, continually created, shaped, and known by women within it; its organization is shaped by external material factors or textually mediated relations (1987). She has fully explicated this approach, *institutional ethnography* (Smith, 2005, 2006), which she and others are developing (Campbell, 2002; Campbell & Gregor, 2002).

3. Collins (2000) grounds her Black women's standpoint in Black women's material circumstances and political situation. She refuses to abandon situated standpoints and links the standpoint of Black women with intersectionality, while she amplifies standpoint theory (1998a) *always* with keen consideration for power and structural relations (1998a).

4. Harding, a philosopher, early recognized three types of feminist inquiry (1987): (1) *Feminist empiricism,* which is of two types: (a) "spontaneous feminist empiricism" (rigorous adherence to existing research norms and standards) and (b) "contextual empiricism" (recognition of the influence of social values and interests in science) (1993); (2) *standpoint theory,* which recognizes that all knowledge attempts are socially situated and that some of these objective social locations are better than others for knowledge projects" (1993, 1998); and (3) *postmodern theories,* which void the possibility of a feminist science in favor of the many and multiple stories women tell about the knowledge they have (1987).

5. Key to Hartsock's (1983) Marxist standpoint theory is her view that women's circumstances in the material order provide them with experiences that generate particular and privileged knowledge, which

both reflects oppression and women's resistance. Such knowledge is not innately essential nor do all women have the same experiences or the same knowledge. Rather, there is the possibility of a "concrete multiplicity" of perspectives (1990). "The subjects who matter are not individual subjects, but collective subjects, or groups" (Hartsock, 1997, p. 371).

6. See Harding, 1997; Weeks, 2004; Naples, 2003, 2007; Ramazanoglu and Holland, 2002.

7. See Clough, 1993a, 1993b, 1994; Collins, 1992; Harding, 1987; Hawkesworth, 1987; Hekman, 1990, 1997a 1997b; Kim, 2007; Maynard, 1994; Scott, 1997; Smith, 1992, 1993; Welton, 1997.

8. See Benhabib, 1995; Collins, 1998b; Ebert 1996.

9. It is useful to differentiate studies that focus on sexuality as an object of study from those that make sexuality a central concept (Yen Le Espiritu, personal communication, September 15, 2003). The former includes research that dissolved a homogeneous view of lesbians just noted in text. Alexander's work in the second category conceptualizes sexuality as fundamental to gender inequality and as a salient marker of otherness that has been central to racist and colonial ideologies (Alexander & Mohanty, 1997).

10. Differences among women as well as similarities between men and women were acknowledged (Brabeck, 1996; Lykes, 1994). For gender as causal explanation and analytic category and research implications for research, see Connell, 1997; Hawkesworth, 1997a, 1997b; McKenna & Kessler, 1997; S. G. Smith, 1997).

◙ REFERENCES

Acker, J. (2006). Inequality regimes: Gender, class and race in organizations. *Gender & Society, 4,* 441–464.

Acker, J. (2008). Feminist theory's unfinished business. *Gender & Society, 22,* 104–108.

Acker, J., Barry, K., & Esseveld, J. (1991). Objectivity and truth: Problems in doing feminist research. In M. M. Fonow & J. A. Cook (Eds.), *Beyond methodology: Feminist scholarship as lived research* (pp. 133–153). Bloomington: University of Indiana Press.

Acosta, K. L. (2008). Lesbians in the borderlands: Shifting identities and imagined communities. *Gender & Society, 22,* 639–659.

Adams, T. E., & Jones, S. H. (2008). Autoethnography is queer. In N. K. Denzin, Y. S. Lincoln, & L. T. Smith (Eds.), *Handbook of critical and indigenous methodologies* (pp. 373-390). Thousand Oaks, CA: Sage.

Alexander, B. K. (2005). Performance ethnography: The reenacting and citing of culture. In N. K. Denzin & Y. S. Lincoln (Eds.), *The SAGE handbook of qualitative research* (3rd ed., pp. 411–442). Thousand Oaks, CA: Sage.

Alexander, B. K. (2008). Queer(y)ing the post-colonial through the West(ern). In N. K. Denzin, Y. L. Lincoln, & L. T. Smith (Eds.), *Handbook of critical and indigenous methodologies* (pp. 101–134). Thousand Oaks, CA: Sage.

Alexander, M. J., & Mohanty, C. T. (1997). *Feminist geneaologies, colonial legacies, democratic futures.* New York: Routledge.

Andersen, M. L. (2005). Thinking about women: A quarter century's view. *Gender & Society, 19,* 437–455.

Andersen, M. L. (2008). Thinking about women some more: A new century's view. *Gender & Society, 22,* 120–125.

Andrews, M. (2002). Feminist research with non-feminist and anti-feminist women: Meeting the challenge. *Feminism and Psychology, 12,* 55–77.

Anzaldúa, G. (1987). *Borderlands/La frontera.* San Francisco: Auntie Lute.

Anzaldúa, G. (1990). *Making soul, Haciendo caras.* San Francisco: Auntie Lute.

Arat-Koc, S. (2007). (Some) Turkish transnationalisms in an age of capitalist globalization and empire: "White Turk" discourse, the new geopolitics and implications for feminist transnationalism. *Journal of Middle East Women's Studies, 3,* 35–57.

Asch, A., & Fine, M. (1992). Beyond the pedestals: Revisiting the lives of women with disabilities. In M. Fine (Ed.), *Disruptive voices: The possibilities of feminist research* (pp.139–174). Ann Arbor: University of Michigan Press.

Balsamo A. (1993). On the cutting edge: Cosmetic surgery and the technological production of the gendered body. *Camera Obscura, 28,* 207–237.

Barvosa-Carter, E. (2001). Strange tempest: Agency, structuralism and the shape of feminist politics to come. *International Journal of Sexuality and Gender Studies, 6,* 123–137.

Battacharya, K. (2007). Consenting to the consent form: What are the fixed and fluid understandings between the researcher and the researched. *Qualitative Inquiry, 13,* 1095–1115.

Battacharya, K. (2009). Negotiating shuttling between transnational experiences: A de/colonizing approach to performance ethnography. *Qualitative Inquiry, 15,* 1061–1083.

Battiste, M. (2008). Research ethics for protecting indigenous knowledge and heritage, In N. K. Denzin, Y. S. Lincoln, & L. T. Smith (Eds.), *Handbook of critical and indigenous methodologies* (pp. 497–510). Thousand Oaks, CA: Sage.

Behar, R. (1993). *Translated woman: Crossing the border with Esparanza's story.* Boston: Beacon.

Bell, L., & Nutt, L. (2002). Divided loyalties, divided expectations: Research ethics, professional and occupational responsibilities. In M. Mauthner, M. Birch, J. Jessop, & T. Miller (Eds.), *Ethics in qualitative research* (pp. 70–90). Thousand Oaks, CA: Sage.

Bell, S. E. (2009). *DES daughters, embodied knowledge, and the transformation of women's health politics.* Philadelphia: Temple University Press.

Benhabib, S. (1995). Feminism and post-modernism: An uneasy alliance. In S. Benhabib, J. Butler, D. Cornell, & N. Fraser (Eds.), *Feminist contentions: A philosophical exchange* (pp. 17–34). New York: Routledge.

Bhavnani, K.-K. (2007). Interconnections and configurations: Toward a global feminist ethnography, In S. N. Hesse-Biber (Ed.), *Handbook of feminist research: Theory and praxis* (pp. 639–650). Thousand Oaks, CA: Sage.

Birch, M. (1998). Reconstructing research narratives: Self and sociological identity in alternative settings. In J. Ribbens & R. Edwards (Eds.), *Feminist dilemmas in qualitative research: Public knowledge and private lives* (pp. 171–185). Thousand Oaks, CA: Sage.

Black women's studies and the transformation of the academy [Symposium]. (2010). *Signs, 35*(4).

Bloom, L. R. (1998). *Under the sign of hope: Feminist methodology and narrative interpretation.* Albany: State University of New York Press.

Bowleg, L. (2008). When Black + lesbian + woman = Black lesbian woman: The methodological challenges of qualitative and quantitative intersectionality research, *Sex Roles, 59,* 312–325.

Brabeck, M. M. (1996). The moral self, values, and circles of belonging. In K. F. Wyche & F. J. Crosby (Eds.), *Women's ethnicities: Journeys through psychology* (pp. 145–165). Boulder, CO: Westview Press.

Brah, A., & Phoenix, A. (2004). Ain't I a woman? Revisiting intersectionality. *International Journal of Women's Studies, 5,* 75–86.

Bredstrom, A. (2006). Intersectionality: A challenge for feminist HIV/AIDS Research? *European Journal of Women's Studies, 13,* 229–243.

Butler, J. (1990). *Gender trouble: Feminism and the subversion of identity.* London: Routledge.

Butler, J. (1993). *Bodies that matter: On the discursive limits of "sex."* London: Routledge.

Butler, J. (2004). *Undoing gender.* New York: Routledge.

Campbell, N. D. (2000). *Using women: Gender, policy, and social justice.* New York: Routledge.

Campbell, N. D. (2002). Textual accounts, ruling action: The intersection of knowledge and power in the routine conduct of nursing work. *Studies in Cultures, Organizations and Societies, 7,* 231–250.

Campbell, N. D., & Gregor, F. (2002). *Mapping social relations: A primer in doing institutional ethnography.* Toronto: Garamond.

Cancian, F. M. (1996). Participatory research and alternative strategies for activist sociology. In H. Gottfried (Ed.), *Feminism and social change* (pp. 187–205). Urbana: University of Illinois Press.

Cannella, G. S., & Manuelito, K. D. (2008). Indigenous world views, marginalized feminisms and revisioning an anticolonial social science. In N. K. Denzin, Y. S. Lincoln, & L. T. Smith (Eds.), *Handbook of critical and indigenous methodologies* (pp. 45–59). Thousand Oaks, CA: Sage.

Case, S-E., & Abbitt, E. S. (2004). Disidentifications, diaspora, and desire: Questions on the future of the feminist critique of performance. *Signs, 29,* 925–938.

Casper, M. J., & Talley, H. L. (2007). Feminist disability studies. In G. Ritzer (Ed.), *Blackwell encyclopedia of sociology* (pp. 15–30). London: Blackwell.

Chang, G. (2001). *Disposable domestics: Immigrant women workers in the global economy.* Cambridge, MA: South End Press.

Cho, J., & Trent, A. (2009). Validity criteria for performance-related qualitative work: Toward a reflexive, evaluative, and co-constructive framework for performance in/as. *Qualitative Inquiry, 15,* 1013–1041.

Choo, H. Y., & Ferree, M. M. (2010). Practicing intersectionality in sociological research: A critical analysis of inclusions, interactions and institutions in the study of inequalities. *Sociological Theory, 28,* 129–149.

Chow, E. N. (1987). The development of feminist consciousness among Asian American women. *Gender & Society, 1,* 284–299.

Clarke, A. (1998). *Disciplining reproduction: Modernity, American life sciences, and the problems of "sex."* Berkeley: University of California Press.

Clarke, A. (2004). *Grounded theory after the postmodern turn: Situational maps and analyses.* Thousand Oaks, CA: Sage.

Clarke, A., & Olesen, V. L. (Eds.). (1999). *Revisioning women, health, and healing. Feminist, cultural, and technoscience perspectives.* New York: Routledge.

Clough, P. T. (1993a). On the brink of deconstructing sociology: Critical reading of Dorothy Smith's standpoint epistemology. *The Sociological Quarterly, 34,* 169–182.

Clough, P. T. (1993b). Response to Smith's response. *The Sociological Quarterly, 34,* 193–194.

Clough, P. T. (1994). *Feminist thought: Desire, power, and academic discourse.* London: Basil Blackwell.

Clough, P. T. (2000). *Autoaffection: The unconscious in the age of teletechnology.* Minneapolis: University of Minnesota Press.

Collins, P. H. (1986). Learning from the outsider within: The sociological significance of Black feminist thought. *Social Problems, 33,* 14–32.

Collins, P. H. (1992). Transforming the inner circle: Dorothy Smith's challenge to sociological theory. *Sociological Theory, 10,* 73–80.

Collins, P. H. (1997). Comment on Hekman's "Truth and method: Feminist standpoint theory revisited." *Signs, 22,* 375–381.

Collins, P. H. (1998a). *Fighting words: Black women and the search for justice.* Minneapolis: University of Minnesota Press.

Collins, P. H. (1998b). What's going on? Black feminist thought and the politics of postmodernism. In P. H. Collins, *Fighting words, Black women and the search for justice* (Ch. 4, pp. 124–154). Minneapolis: University of Minnesota Press.

Collins, P. H. (2000). *Black feminist thought. Knowledge, consciousness and the politics of empowerment* (2nd ed.). Boston: Unwin Hyman.

Collins, P. H. (2008). Reply to commentaries: *Black sexual politics* revisited. *Studies in Gender and Sexuality, 9,* 68–85.

Collins, P. H. (2009). Foreword: Emerging intersections—Building knowledge and transforming institutions. In B. T. Dill & R. E. Zambrana (Eds.), *Emerging intersections. Race, class, and gender in theory, policy, and practice,* (pp. vii–xiii). New Brunswick, NJ: Rutgers University Press.

Connell, R. W. (1997). Comment on Hawkesworth's "Confounding Gender." *Signs, 22,* 702–706.

Connolly, C. M. (2006). A feminist perspective of resilience in Lesbian couples, *Journal of Family Therapy, 18,* 137–162.

Cook, J. A., & Fonow, M. M. (2007). A passion for knowledge: The teaching of feminist methodology. In S. N. Hesse-Biber (Ed.), *Handbook of feminist research, theory, and praxis* (pp. 705–712). Thousand Oaks, CA: Sage.

Corrigan, O. (2003). Empty ethics: The problem with informed consent. *Sociology of Health and Illness, 25,* 768–792.

Crenshaw, K. (1989). Demarginalizing the intersection of race and sex: A Black feminist critique of antidiscrimination doctrine, feminist theory, and antiracist politics. *University of Chicago Legal Forum,* pp. 139–167.

Crenshaw, K. (1991). Mapping the margins: Intersectionality, identity politics, and violence against women of color. *Stanford Law Review, 43,* 1241–1299.

Cummins, L. H., & Lehman, J. (2007). Eating disorders and body image concerns in Asian American women: Assessment and treatment from a multi-cultural and feminist perspective. *Eating Disorders: The Journal of Treatment And Prevention, 15,* 217–230.

Davis, A. Y. (1981). *Women, race and class.* London: The Women's Press.

Davis, D.-A., & Craven, C. (2011). Revisiting feminist ethnography: Methods and activism at the intersection of neoliberal policy in the U.S. *Feminist Formations, 23.*

Davis, K. (2007). *The making of* Our Bodies, Ourselves: *How feminism travels across borders.* Durham, NC: Duke University Press.

Davis, K. (2008). Intersectionality as buzzword: A sociology of science perspective on what makes a feminist theory successful. *Feminist Theory, 9,* 67–85.

deLauretis, T. (1987). *Technologies of gender: Essays on theory, film, and fiction.* Bloomington: Indiana University Press.

Delgado Bernal, D. (2002). Critical race theory, Latino critical theory, and critical raced-gendered epistemologies: Recognizing students as creators and holders of knowledge. *Qualitative Inquiry, 8,* 105–126.

Denis, A. (2008). Intersectional analysis: A contribution of feminism to sociology, *International Sociology, 23,* 677–694.

Denzin, N. K. (1992). *Symbolic interaction and cultural studies.* Oxford, UK: Basil Blackwell.

Denzin, N. K. (2005). *Performance ethnography: Critical pedagogy and the politics of culture.* Thousand Oaks, CA: Sage.

DeRiviere, L. (2006). A human capital methodology for estimating the lifelong personal costs of young women leaving the sex trade. *Feminist Economics, 12,* 367–402.

Desai, M. (2007). The messy relationship between feminisms and globalization. *Gender & Society, 21,* 797–803.

DesAutels, P., & Wright, J. (2001). *Feminists doing ethics.* Boulder, CO: Rowan & Littlefield.

Dever, C. (2004). *Skeptical feminism, activist theory, activist practice.* Minneapolis: University of Minnesota Press.

Dewey, S. (2008). *Hollow bodies: Institutional responses to the traffic in women in Armenia, Bosnia, and India.* Sterling, VA: Kumarian Press.

Diaz, C. (2002). Conversational heuristic as a reflexive method for feminist research. *International Review of Sociology, 2,* 249–255.

Dill, B. T. (1979). The dialectics of Black womanhood. *Signs, 4,* 543–555.

Dill, B. T., McLaughlin, A. E., & Nieves, A. D. (2007). Future directions of feminist research: Intersectionality. In S. N. Hesse-Biber (Ed.), *Handbook of feminist research, theory and praxis* (pp. 629–638). Thousand Oaks, CA: Sage.

Dill, B. T., & Zambrana, R. E. (2009). Critical thinking about inequality: An emerging lens. In B. T. Dill & R. E. Zambrana (Eds.), *Emerging intersections: Race, class, and gender in theory, policy, and practice* (pp. 1–22). New Brunswick, NJ: Rutgers University Press.

Dillard, C. B. (2008). When the ground is black, the ground is fertile. In N. K. Denzin, Y. S. Lincoln, & L. T. Smith (Eds.), *Handbook of critical and indigenous methodologies* (pp. 277–292). Thousand Oaks, CA: Sage.

Doezema, J. (2000). Loose women or lost women? The re-emergence of the myth of white slavery in contemporary discourses in trafficking in women. *Gender Issues, 18,* 23–50.

Dworkin, S. L. (2005). Who is epidemiologically fathomable in the HIV-AIDS epidemic? Gender, sexuality, and intersectionality in public health. *Culture, Health, & Sexuality, 7,* 615–623.

Dworkin, S. L., & Wachs, F. L. (2009). *Body panic: Gender, health and the selling of fitness.* New York: New York University Press.

Ebert, T. (1996). *Ludic feminism and after: Postmodernism, desire and labor in late capitalism.* Ann Arbor: University of Michigan Press.

Edwards, R., & Mauthner, M. (2002). Ethics and feminist research: Theory and practice. In M. Mauthner, M. Birch, J. Jessop, &

T. Miller (Eds.), *Ethics in qualitative research* (pp. 14–31). Thousand Oaks, CA: Sage.

Ellingson, L. L. (1998). Then you know how I feel: Empathy, identity, and reflexivity in fieldwork. *Qualitative Inquiry, 4,* 492–514.

Ellis, C. (1995). *Final negotiations: A story of love, loss and chronic illness.* Philadelphia: Temple University Press.

Ellis, C. (2009a). Fighting back or moving on: An autoethnographic response to critics. *International Review of Qualitative Research, 3*(2).

Ellis, C. (2009b). *Revision: Autoethnographic reflections on life and work.* Walnut Creek, CA: Left Coast Press.

Ellis, C., & Bochner, A. P. (1996). *Composing ethnography, Alternative forms of qualitative writing.* Walnut Creek, CA: AltaMira Press.

Ellis, C., & Bochner, A. P. (2000). Autoethnography, personal narrative, reflexivity: Researcher as subject. In N. K. Denzin & Y. S. Lincoln (Eds.), *Handbook of qualitative research* (2nd ed., pp. 733–768). Thousand Oaks, CA: Sage.

Espiritu, Y. L. (2007). *Asian American women and men: Labor, laws, and love.* Thousand Oaks, CA: Sage.

Etowa, J. B., Bernard, W. T., Oyinsan, B., & Clow, B. (2007). Participatory action research (PAR): An approach for improving Black women's health in rural and remote communities. *Journal of Transcultural Nursing, 18,* 349–357.

Evans, M., Hole, R., Berg, L. C., Hutchinson, P., & Sookraj, D. (2009). Common insights, differing methodologies: Toward a fusion of indigenous methodologies, participatory action research, and white studies in an urban aboriginal research agenda. *Qualitative Inquiry, 15,* 893–910.

Evans, S. M. (2002). Re-viewing the second wave. *Feminist Studies, 28,* 259–267.

Felski, R. (1997). The doxa of difference. *Signs, 23,* 1–22.

Few, A. L. (2007). Integrating Black consciousness and critical race feminism into family studies research. *Journal of Family Issues, 28,* 452–473.

Fine, M. (1992). Passions, politics, and power: Feminist research possibilities. In M. Fine (Ed.), *Disruptive voices* (pp. 205–232). Ann Arbor: University of Michigan Press.

Fine, M., Weis, L., Weseem, S., & Wong, L. (2000). For whom? Qualitative research, representations and social responsibilities. In N. K. Denzin & Yvonna S. Lincoln (Eds.), *Handbook of qualitative research* (2nd ed., pp. 107–132). Thousand Oaks, CA: Sage.

Fine, M., & Torre, M. E. (2006). Intimate details. Participatory research in prison. *Action Research, 4,* 253–269.

Fiore, R. N., & Nelson, H. L. (2003). *Recognition, responsibility, and rights: Feminist ethics and social theory.* Boulder, CO: Rowan & Littlefield.

Firdous, A. (2005). Feminist struggles in Bangladesh. *Feminist Review, 80,* 194–197.

Flax, J. (1987). Postmodernism and gender relations in feminist theory. *Signs, 14,* 621–643.

Flax, J. (1990). *Thinking fragments: Psychoanalysis, feminism, and postmodernism in the contemporary West.* Berkeley: University of California Press.

Fonow, M. M., & Cook, J. A. (2005). Feminist methodology: New applications in the academy and public policy. *Signs, 30,* 2211–2236.

Foster, N. (2007). Reinscribing Black women's position within HIV and AIDS discourses. *Feminism and Psychology, 17,* 323–329.

Frankenberg, R. (1994). *White women, race matters: The social construction of whiteness.* Minneapolis: University of Minnesota Press.

Franks, M. (2002). Feminisms and cross ideological feminist social research: Standpoint, situatedness, and positionality: Developing cross-ideological feminist research. *Journal of International Women's Studies, 3.* Available at http://www.bridgew.edu/SoAS/jiws/

Fraser, N. (2005). Mapping the feminist imagination: From redistribution to recognition to representation. *Constellations, 12,* 295–307.

Gannon, S., & Davies, B. (2007). Postmodern, poststructural, and critical theories. In S. N. Hesse-Biber (Ed.), *Handbook of feminist research: Theory and praxis* (pp. 71–106). Thousand Oaks, CA: Sage.

Garcia-Lopez, G. (2008). "*Nunca te toman en cuenta* [They never take you into account]": The challenges of inclusion and strategies for success of Chicana attorneys. In B. T. Dill & R. E. Zambrana (Eds.), *Emerging intersections: Race, class, and gender in theory, policy, and practice* (pp. 22–49). New Brunswick, NJ: Rutgers University Press.

Gardiner, J. K., & Meyer, L. D. (Eds.). (2008a). *Chicana studies* [Special issue] *34*(1),

Gardiner, J. K., & Meyer, L. D., for the editorial collective. (2008b). Preface, Chicana Studies. *Feminist Studies, 34,* 10–22.

Garland-Thomson, R. (2005). Feminist disability studies. *Signs, 30,* 1557–1587.

Glenn, E. N. (2002). *Unequal freedom. How race and gender shaped American citizenship and labor.* Cambridge, MA: Harvard University Press.

Gorelick, S. (1991). Contradictions of feminist methodology. *Gender & Society, 5,* 459–477.

Gray, R., & Sinding, C. (2002). *Standing ovation, Performing social science research about cancer.* Boulder, CO: Rowan & Littlefield.

Green, R. (1990). The Pocahontas perplex: The image of Indian women in American culture. In E. C. DuBois & V. L. Ruiz (Eds.), *Unequal sisters: A multi-cultural reader in U.S. women's history* (pp. 15–21). London: Routledge.

Grewal, I., & Caplan, K. (1994). *Scattered hegemonies: Postmodernity and trans-national practices.* Minneapolis: University of Minnesota Press.

Guevarra, A. (2009). *Marketing dreams, manufacturing heroes: The transnational labor brokering of Filipino workers.* New Brunswick, NJ: Rutgers University Press.

Guilleman, M., & Gillam, L. (2004). Ethics, reflexivity, and "ethically important" moments in research. *Qualitative Inquiry, 10,* 261–280.

Gulcur, L., & Ilkkaracan, P. (2002). The 'Natasha' experience: Migrant sex workers from the former Soviet Union and Eastern Europe in Turkey. *Women's Studies International Forum, 25,* 411–421.

Gupta, J. S. (2006). Toward transnational feminisms: Some reflections and concerns in relation to the globalization of reproductive technologies. *European Journal of Women's Studies, 13,* 23–38.

Halsey, C., & Honey, A. (2005). Unravelling ethics: Illuminating the moral dilemmas of research. *Signs, 30,* 2141–2162.

Hancock, A-M. (2007a). Intersectionality as a normative and empirical paradigm. *Politics and Gender, 3,* 248–254.

Hancock, A-M. (2007b). When multiplication doesn't equal quick addition: Examining intersectionality as a research paradigm. *Perspectives on Politics, 5,* 63–78.

Hanochi, S. (2001). Japan and the global sex industry. In R. M. Kelly, J. H. Hayes, M. H. Hawkesworth, & B. Young (Eds.), *Gender, globalization, and democratization.* Lanham, MD: Rowan & Littlefield.

Haraway, D. J. (1991). *Simians, cyborgs, and women. The reinvention of nature.* London: Routledge.

Haraway, D. J. (1997). Modest_witness@second millenium.FemaleMan©_ Meets_On coMouse™. New York: Routledge.

Harding, S. (1987). Conclusion: Epistemological questions. In S. Harding (Ed.), *Feminism and methodology* (pp. 181–90). Bloomington: University of Indiana Press.

Harding, S. (1993). Rethinking standpoint epistemology: What is "strong objectivity?" In L. Alcoff & E. Potter (Eds.), *Feminist epistemologies* (pp. 49–82). New York: Routledge.

Harding, S. (1996). Gendered ways of knowing and the "epistemological crisis" of the West. In N. R. Goldberger, J. M. Tarule, B. M. Clinchy, & M. F. Belenky (Eds.), *Knowledge, difference, and power: Essays inspired by women's ways of knowing* (pp. 431–454). New York: Basic Books.

Harding, S. (1997). Comment on Hekman's "Truth and method: Feminist standpoint theory revisited." *Signs, 22,* 382–391.

Harding, S. (1998). *Is science multicultural? Postcolonialisms, feminisms, and epistemologies.* Bloomington: Indiana University Press.

Harding, S. (2008). *Sciences from below: Feminisms, postcolonialities, and modernities.* Durham, NC: Duke University Press.

Harding, S., & Norberg, K. (2005). New feminist approaches to social science methodologies: An introduction. *Signs, 30,* 2009–2019.

Hartsock, N. (1983). The feminist standpoint: Developing the ground for a specifically feminist historical materialism. In S. Harding & M. B. Hintikka (Eds.), *Discovering reality* (pp. 283–310). Amsterdam: D. Reidel.

Hartsock, N. (1997). Comment on Hekman's "Truth and method: Feminist standpoint theory revisited": Truth or justice? *Signs, 22,* 367–374.

Hawkesworth, M. E. (1987). Feminist epistemology: A survey of the field. *Women and Politics, 7,* 115–127.

Hawkesworth, M. E. (1989). Knowing, knowers, known: Feminist theory and claims of Truth. *Signs, 14,* 553–557.

Hawkesworth, M. E. (1997a). Confounding gender. *Signs, 22,* 649–686.

Hawkesworth, M. E. (1997b). Reply to McKenna and Kessler, Smith, Scott and Connell: Interrogating gender. *Signs, 22,* 707–713.

Hekman, S. (1990). *Gender and knowledge: Elements of a post-modern feminism.* Boston: Northeastern University Press.

Hekman, S. (1997a). Truth and method: Feminist standpoint theory revisited. *Signs, 22,* 341–365.

Hekman, S. (1997b). Reply to Hartsock, Collins, Harding and Smith. *Signs, 22,* 399–402.

Hertz, R. (Ed.). (1997). *Reflexivity and voice.* Thousand Oaks, CA: Sage.

Hesse-Biber, S. N. (2007). Dialoguing about future directions in feminist theory, research, and pedagogy. In S. N. Hesse-Biber (Ed.), *Handbook of feminist research: Theory and praxis* (pp. 535–545). Thousand Oaks, CA: Sage.

Hesse-Biber, S. N., & Piatelli, D. (2007). Holistic reflexivity: The feminist practice of reflexivity. In S. N. Hesse-Biber (Ed.), *Handbook of feminist research: Theory and praxis* (pp. 493–544). Thousand Oaks, CA: Sage.

Higginbotham, E. (2009). Entering a profession: Race, gender, and class in the lives of black women attorneys. In B. T. Dill & R. E. Zambrana (Eds.), *Emerging intersections: Race, class, and gender in theory, policy, and practice* (pp. 22–49). New Brunswick, NJ: Rutgers University Press.

Holman Jones, S. (2005). Autoethnography: Making the personal political. In N. K. Denzin & Y. S. Lincoln (Eds.), *The SAGE handbook of qualitative research* (3rd ed., pp. 763–791). Thousand Oaks, CA: Sage.

hooks, b. (1990). Culture to culture: Ethnography and cultural studies as critical intervention. In b. hooks (Ed.), *Yearning: Race, gender, and cultural politics* (pp. 123–133). Boston: South End Press.

Howard, J. A., & Allen, C. (Eds.). (2000). *Women at a millennium* [Special issue]. *Signs, 25*(4).

Hurtado, A., & Stewart, A. J. (1997). Through the looking glass: Implications of studying whiteness for feminist methods. In M. Fine, L. Weis, L. C. Powell, & L. M. Wong (Eds.), *Off white: Readings on race, power, and society* (pp. 297–311). New York: Routledge.

Jackson, A. Y. (2003). Rhizovocality. *Qualitative Studies in Education, 16,* 693–710.

Jackson, A. Y. (2009). "What am I doing when I speak of this present?" Voice, power, and desire in truth-telling, In A. Y. Jackson & L. A. Mazzei (Eds.), *Voice in qualitative inquiry: Challenging conventional, interpretive, and critical consequences in qualitative research* (pp. 165–174). New York: Routledge.

Jiwani, Y. (2005). Walking a tightrope: The many faces of violence in the lives of immigrant girls and young women. *Violence Against Women, 11,* 846–875.

Jurik, N. C., & Siemsen, C. (2009). "Doing gender" as canon or agenda: A symposium on West and Zimmerman. *Gender & Society, 23,* 72–75.

Kahn, S, (2005). Reconfiguring the native informant: Positionality in the golden age. *Signs, 30,* 2017–2055.

Katsulis, Y. (2009). *Sex work and the city: The social geography of health and safety in Tijuana, Mexico.* Austin: University of Texas.

Kempadoo, K. (2001). Women of color and the global sex trade: Transnational feminist perspectives. *Meridians: Feminism, Race, Transnationalism, 1,* 28–51.

Kennedy, E. L., & Davis, M. (1993). *Boots of leather, slippers of gold: The history of a lesbian community.* New York: Routledge.

Keough, L. J. (2009). "Driven women": Gendered moral economies of women's migrant labor in postsocialist Europe's peripheries. *Dissertation Abstracts International, Section A: Humanities and Social Sciences, 69*(9-A), p. 3602.

Kim, H. S. (2007). The politics of border crossings: Black, postcolonial, and transnational feminist perspectives. In S. N. Hesse-Biber (Ed.), *Handbook of feminist research: Theory and praxis* (pp. 107–122). Thousand Oaks, CA: Sage.

Kim-Puri, H. J. (2005). Conceptualizing gender-sexuality-state-nation, An introduction. *Gender & Society, 19,* 137–159.

Kincheloe, J. (1997). Fiction formulas: Critical constructivism and the representation of reality. In W. G. Tierney & Y. S. Lincoln (Eds.), *Representation and the text: Reframng the narrative voice* (pp. 57–80). Albany: State University of New York Press.

Kirsch, G. E. (2005). Friendship, friendliness, and feminist fieldwork. *Signs, 30,* 2162–2172.

Klawiter, M. (2008). *The biopolitics of breast cancer: Changing cultures of disease and activism.* Minneapolis: University of Minnesota Press.

Kondo, D. K. (1990). *Crafting selves, power, gender, and discourses of identity in a Japanese workplace.* Chicago: University of Chicago Press.

Kondo, D. K. (1995). Bad girls: Theater, women of color, and the politics of representation. In R. Behar & D. Gordon (Eds.), *Women writing culture* (pp. 49–64). Berkeley: University of California Press.

Kuumba, M. B. (2002). "You've struck a rock": Comparing gender, social movements, and transformation in the United States and South Africa. *Gender & Society, 4,* 504–523.

Lacsamana, A. E. (1999). Colonizing the South: Postmodernism, desire, and agency, *Socialist Review, 27,* 95–106.

Lan, P-C. (2006). *Global Cinderellas: Migrant domestics and newly rich employers in Taiwan.* Durham, NC: Duke University Press.

Laslett, B., & Brenner, B. (2001). Twenty-first academic feminism in the United States: Utopian visions and practical actions. *Signs, 25,* 1231–1236.

Lather, P. (1991). *Getting smart: Feminist research and pedagogy within the postmodern.* New York: Routledge.

Lather, P. (1993). Fertile obsession: Validity after post-structuralism. *The Sociological Quarterly, 34,* 673–694.

Lather, P. (2007). *Getting lost: Feminist efforts toward a double(d) science.* Albany: State University of New York Press.

Lather, P., & Smithies, C. (1997). *Troubling the angels: Women living with AIDS.* Boulder, CO: Westview.

Lewin, L. (1993). *Lesbian mothers.* Ithaca, NY: Cornell University Press.

Lewin, L. (1998). *Recognizing ourselves: Ceremonies of lesbian and gay committment.* New York: Columbia University Press.

Lewin, L. (2009). *Gay fathers: Narratives of family and citizenship.* Chicago: University of Chicago Press.

Lincoln, Y. S. (1993). I and thou: Method, voice, and roles in research with the silenced. In D. McLaughlin & W. G. Tierney (Eds.), *Naming silenced lives: Personal narratives and processes of educational change* (pp. 20–27). New York: Routledge.

Lincoln, Y. S. (1997). Self, subject, audience, text: Living at the edge, writing at the margins. In W. G. Tierney & Y. S. Lincoln (Eds.), *Representation and the text* (pp. 37–55). Albany: State University of New York Press.

Lincoln, Y. S. (2005). Institutional review boards and methodological conservatism: The challenge to and from phenomenological paradigms. In N. K. Denzin & Y. S. Lincoln (Eds.), *The SAGE handbook of qualitative research* (3rd ed., pp. 165–182). Thousand Oaks, CA: Sage.

Lincoln, Y. S., & Guba, E G. (1985). *Naturalistic inquiry.* Beverly Hills, CA: Sage.

Llewelyn, S. (2007). A neutral feminist observer? Observation-based research and the politics of feminist knowledge making. *Gender and Development, 15,* 299–310.

Lorber, J. (1994). *Paradoxes of gender.* New Haven, CT: Yale University Press.

Lubelska, C., & Mathews, J. (1997). Disability issues in the politics and processes of feminist studies. In M. Ang-Lygate, C. Corrin, & M. S. Henry (Eds.), *Desperately seeking sisterhood: Still challenging and building* (pp. 117–137). London: Taylor & Francis.

Luft, R. E., & Ward, J. (2009). Toward an intersectionality just out of reach: Confronting challenges to intersectional practice. *Advances in Gender Research. Special Volume: Intersectionality, 13,* 9–37.

Lykes, M. B. (1989). Dialogue with Guatemalan Indian women: Critical perspectives on constructing collaborative research. In R. Unger (Ed.), *Representations: Social constructions of gender* (pp. 167–184). Amityville, NY: Baywood.

Lykes, M. B. (1994). Whose meeting at which crossroads? A response to Brown and Gilligan. *Feminism and Psychology, 4,* 345–349.

MacLure, M. (2009). Broken voices, dirty words: On the productive insufficiency of voice. In A. Y. Jackson & L. A. Mazzie(Eds.), *Voice in qualitative inquiry: Challenging conventional, interpretive, and critical conceptions in qualitative research* (pp. 98–113). New York: Routledge.

Madison, D. S. (2005). *Critical ethnography: Methods and performance.* Thousand Oaks, CA: Sage.

Madison, D. S. (2006). Staging fieldwork/performing human rights. In D. S. Madison & J. Hameva (Eds.), *The SAGE handbook of performance studies* (pp. 397–418). Thousand Oaks, CA: Sage.

Majumdar, A. (2007). Researching South Asian women's experiences of marriage: Resisting stereotypes through an exploration of "space" and "embodiment." *Feminism and Psychology, 17,* 316–322.

Mamo, L. (2007). *Queering reproduction: Achieving pregnancy in the age of technoscience.* Durham, NC: Duke University Press.

Manning, K. (1997). Authenticity in constructivist inquiry: Methodological considerations without prescription. *Qualitative Inquiry, 3,* 93–116.

Martin, E. (1999). The woman in the flexible body. In A. E. Clarke & V. L. Olesen (Eds.), *Revisioning women, health, and healing: Feminist, cultural, and technoscience perspectives* (pp. 97–118). New York: Routledge.

Mauthner, M., Birch, M., Jessop, J., & Miller, T. (2002). *Ethics in qualitative research.* Thousand Oaks, CA: Sage.

Mauthner, N., & Doucet, A. (1998). Reflections on a voice-centered relational method: Analyzing maternal and domestic voices. In J. Ribbens & R. Edwards (Eds.), *Feminist dilemmas in qualitative research: Public knowledge and private lives* (pp. 119–146). Thousand Oaks, CA: Sage.

Maynard, M. (1994). Race, gender, and the concept of "difference" in feminist thought. In H. Afshar & M. Maynard (Eds.), *The dynamics of "race" and gender: Some feminist interventions* (pp. 9–25). London: Taylor & Francis.

Mazur, A. G. (2002). *Theorizing feminist policy.* Oxford, UK: Oxford University Press.

Mazzei, L. A. (2003). Inhibited silences: In pursuit of a muffled subtext. *Qualitative Inquiry, 9,* 355–366.

Mazzei, L. A. (2004). Silent listenings: Deconstructive practices in discourse-based research. *Educational Researcher, 33,* 26–34.

Mazzei, L. A. (2009). An impossibly full voice. In A. Y. Jackson & L. A. Mazzei (Eds.), *Voice in qualitative inquiry: Challenging conventional, interpretive, and critical concepts in qualitative research* (pp. 45–62). New York: Routledge.

Mazzei, L. A., & Jackson, A. Y. (2009). Introduction: The limit of voice. In A. Y. Jackson & L. A. Mazzei (Eds.), *Voice in qualitative inquiry: Challenging conventional, interpretive, and critical concepts in qualitative research* (pp. 1–13). New York: Routledge.

McCall, L. (2005). The complexity of intersectionality. *Signs, 30,* 1771–1800.

McKenna, W., & Kessler, S. (1997). Comment on Hawkesworth's "Confounding gender": Who needs gender theory? *Signs, 22,* 687–691.

Meekosha, H. (2005). *Body battles: Disability, representation, and participation.* Thousand Oaks, CA: Sage.

Mendez, J. B., & Wolf, C. L. (2007). Feminizing global research/globalizing feminist research. In S. N. Hesse-Biber (Ed.), *Handbook of feminist research: Theory and practice* (pp. 651–662). Thousand Oaks, CA: Sage.

Mendoza, B. (2002). Transnational feminisms in question. *Feminist Theory, 3,* 295–314.

Merlis, S. R., & Linville, D. (2006). Exploring a community's response to Lesbian domestic violence through the voices of providers: A qualitative study. *Journal of Feminist Family Therapy, 18,* 97–136.

Mertens, D. B. (2009). *Transforming research and evaluation.* New York: Guilford.

Messer-Davidow, E. (2002). *Disciplining feminism: From social activism to academic discourse.* Durham, NC: Duke University Press.

Mezey, N. J. (2008). *New choices, new families: How lesbians decide about motherhood.* Baltimore, MD: Johns Hopkins University Press.

Mihesuah, D. A. (2003). *American indigenous women: Decolonization, empowerment, activism.* Lincoln: University of Nebraska Press.

Miller, T., & Bell, L. (2002). Consenting to what? Issues of access, gate-keeping and 'informed' consent." In M. Mauthner, M. Birch, J. Jessop, & T. Miller (Eds.), *Ethics in qualitative research* (pp. 37–54). New York: Routledge.

Mohanty, C. (1988). Under Western eyes: Feminist scholarship and colonial discourses, *Feminist Review, 30,* 60–88.

Mohanty, C. (2003). *Feminism without borders: Decolonizing theory, practicing solidarity.* Durham, NC: Duke University Press.

Moore, M. R. (2008). Gendered power relations among women: A study of household decision making in Black, lesbian stepfamilies. *American Sociological Review, 73,* 335–336.

Morgen, S. (2006). Movement-grounded theory: Intersectional analysis of health inequities in the United States. In A. Schulz & L. Mullings (Eds.), *Race, class, gender, and health* (pp. 394–423). San Francisco: Jossey-Bass.

Morris, M. (1998). *Too soon, too late: History in popular culture.* Bloomington: University of Indiana Press.

Morse, J. (2005). Ethical issues in institutional research, *Qualitative Health Research, 15,* 435–437.

Morse, J. (2006). The politics of evidence, *Qualitative Health Research, 16,* 395–404.

Morse, J. (2007). Ethics in action: Ethical principles for doing qualitative health research, *Qualitative Health Research, 17,* 1003–1005.

Naples, N. A. (1996). A feminist revisiting of the insider/outsider debate: The 'outsider phenomenon' in rural Iowa. *Qualitative Sociology, 19,* 83–106.

Naples, N. A. (2002a). The challenges and possibilities of transnational feminist praxis. In N. A. Naples & M. Desai, M. (Eds.), *Women's activism and globalization: Linking local struggles and transnational politics* (pp. 267–282). New York: Routledge

Naples, N. A. (2002b). Changing the terms: Community activism, globalization, and the dilemmas of traditional feminist praxis. In N. A. Naples& M. Desai (Eds.), *Women's activism and globalization: Linking local struggles and transnational politics* (pp. 3–14). New York: Routledge.

Naples, N. A. (2007). Standpoint epistemology and beyond. In S. N. Hesse-Biber (Ed.), *Handbook of feminist research: Theory and praxis* (pp. 579–589). Thousand Oaks, CA: Sage.

Narayan, U. (1997). *Dislocating cultures: Identities and third world feminism.* New York: Routledge.

Nash, J. C. (2008). Re-thinking intersectionality. *Feminist Review, 89,* 1–15.

O'Leary, C. M. (1997). Counteridentification or counterhegemony? Transforming feminist theory. In S. J. Kenney & H. Kinsella (Eds.), *Politics and standpoint theories* (pp. 45–72). New York: Haworth Press.

Olesen, V. L. (1975). Rage is not enough: Scholarly feminism and research in women's health. In V. L. Olesen (Ed.), *Women and their health: Research implications for a new era* (DHEW Publication No. HRA-3138, pp. 1–2). Washington, DC: U.S. Department of Health, Education and Welfare, Public Health Service.

Olesen, V. L. (2005). Early millennial feminist qualitative research: Challenges and contours. In N. K. Denzin & Y. S. Lincoln (Eds.), *The SAGE handbook of qualitative Research* (3rd ed., pp. 235–278). Thousand Oaks, CA: Sage.

Olesen, V. L., & Clarke, A. E. (1999). Resisting closure, embracing uncertainties, creating agendas. In A. E. Clarke and V. L. Olesen (Eds.), *Revisioning women, health and healing: Feminist, cultural studies and technoscience perspectives (355–357).* New York:Routledge.

Ong, A. (1995). Women out of China: Traveling tales and traveling theories in postcolonial feminism. In R. Behar & D. Gordon (Eds.), *Women writing culture* (pp. 350–372). Berkeley: University of California Press.

Parrenas, R. S. (2008). *The force of domesticity: Filipina migrants and globalization.* New York: New York University Press.

Petersen, A. (2006). An African-American woman with disabilities: The intersection of gender, race and disability. *Disability and Society, 21,* 721–734.

Phoenix, A. (1994). Practicing feminist research: The intersection of gender and "race" in the research process. In M. Maynard & J. Purvis (Eds.), *Researching women's lives from a feminist perspective* (pp. 35–45). London: Taylor & Francis.

Phoenix, A., & Pettynama, P. (2006). Intersectionality [Special issue]. *European Journal of Women's Studies, 13*(3).

Pillow, W. S. (2003). Confession, catharsis, or cure. The use of reflexivity as methodological power in qualitative research. *International Journal of Qualitative Studies in Education, 16,* 175–196.

Preissle, J. (2007). Feminist research ethics. In S. N. Hesse-Biber (Ed.), *Handbook of feminist research: Theory and praxis* (pp. 515–534). Thousand Oaks, CA: Sage.

Priyadharshini, E. (2003). Coming unstuck: Thinking otherwise about "studying up." *Anthropology and Education Quarterly, 34,* 420–437.

Prins, B. (2006). Narrative accounts of origins: A blind spot in the intersectional approach? *European Journal of Women's Studies, 13,* 277–290.

Puar, J. K. (1996). Resituating discourses of "Whiteness" and "Asianness" in northern England: Second-generation Sikh women and constructions of identity. In M. Maynard & J. Purvis (Eds.), *New frontiers in women's studies* (pp. 125–150). London: Taylor & Francis.

Randall, M. (2004). Know your place: The activist scholar in today's political culture. *SWS Network News, 21,* 20–23.

Rapp, R. (1999). *Testing women, testing the foetus: The social impact of amnio-centesis in America.* New York: Routledge.

Reid, C., Tom, A., & Frisby, W. (2008). Finding the "action" in feminist participatory research. *Action Research, 4,* 315–322.

Renzetti, C (2005). Editor's introduction. *Violence Against Women, 11,* 839–841.

Rice, C. (2009). Imagining the other? Ethical challenges of researching and writing women's embodied lives. *Feminism & Psychology, 19*(2), 245–266.

Richardson, L. (1993). Poetics, dramatics, and transgressive validity: The case of the skipped line. *The Sociological Quarterly, 34,* 695–710.

Richardson, L. (1997). *Fields of play: Constructing an academic life.* New Brunswick, NJ: Rutgers University Press.

Richardson, L. (2000). Introduction: Assessing alternative modes of qualitative and ethnographic research: How do we judge? Who judges? *Qualitative Inquiry, 6,* 251–252.

Risman, B. (2004). Gender as a social structure: Theory wrestling with activism, *Gender & Society, 18,* 429–450.

Rohrer, J. (2005). Toward a full-inclusion feminism: A feminist deployment of disability analysis. *Feminist Studies, 31,* 34–63.

Rupp, L. J., & Taylor, V. (2003). *Drag queens at the 801 cabaret.* Chicago: University of Chicago Press.

Saavedra, C. M., & Nymark, E. D. (2008). Borderland-Mestija feminism, The new tribalism. In N. K. Denzin, Y. S. Lincoln, & L. T. Smith (Eds.), *Handbook of critical and indigenous methodologies* (pp. 255–276). Thousand Oaks, CA: Sage.

Scheper-Hughes, N. (1983). Introduction: The problem of bias in androcentric and feminist anthropology. *Women's Studies,19,* 109–116.

Scheurich, J. J., & Foley, D. (Eds.). (1997). Feminist poststructuralism [Special issue]. *International Journal of Qualitative Studies in Education, 10*(3).

Scott, J. (1991). The evidence of experience. *Critical Inquiry, 17,* 773–779.

Scott, J. (1997). Comment on Hawkesworth's "Confounding Gender." *Signs, 22,* 697–702.

Schulz, A., & Mullings, L. (Eds.). (2006). *Race, class, gender, and health.* San Francisco: Jossey-Bass.

Segura, D. A., & Zavella, P. (2008). Introduction: Gendered borderlands. *Gender & Society, 22,* 537–544.

Seigfried, C. H. (1996). *Pragmatism and feminism: Reweaving the social fabric.* Chicago: University of Chicago Press.

Shields, S. A. (2008). Gender: An intersectionality perspective. *Sex Roles, 59,* 301–311.

Shim, J. K. (2000). Bio-power and racial, class, and gender formation in biomedical Knowledge production. In J. J. Kronenefield (Ed.), *Research in the sociology of health care* (pp. 173–95). Stamford, CT: JAI Press.

Skeggs, B. (1994). Situating the production of feminist ethnography. In M. Maynard & J. Purvis (Eds.), *Researching women's lives from a feminist perspective* (pp. 72–92). London: Taylor & Francis.

Smith, D. E. (1987). *The everyday world as problematic.* Boston: Northeastern University Press.

Smith, D. E. (1992). Sociology from women's experience: A reaffirmation. *Sociological Theory, 10,* 88–98.

Smith, D. E. (1993). High noon in textland: A critique of Clough. *The Sociological Quarterly, 34,* 183–192.

Smith, D. E. (1997). Telling the truth after postmodernism. *Symbolic Interaction, 19,* 171–202.

Smith, D. E. (2005). *Institutional ethnography: A sociology for people.* Walnut Creek, CA: AltaMira Press.

Smith, D. E. (2006). *Institutional ethnography as practice.* Lanham, MD: Rowan & Littlefield.

Smith, D. E. (2009). Categories are not enough. *Gender & Society, 23,* 76–80.

Smith, L. T. (1999). *Decolonizing methodologies: Research and indigenous peoples.* London: Zed Books.

Smith L. T. (2005). On tricky ground: Researching the native in an age of uncertainty. In N. K. Denzin & Y. S. Lincoln (Eds.), *The SAGE handbook of qualitative research* (3rd ed., pp. 85–107). Thousand Oaks, CA: Sage.

Smith, S. G. (1997). Comment on Hawkesworth's "Confounding Gender." *Signs, 22,* 691–697.

Spivak, G. C. (1988). Subaltern studies: Deconstructing historiography. In G. C. Spivak, *In other worlds: Essays in cultural politics* (pp. 197–221). London: Routledge.

Stacey, J. (1988). Can there be a feminist ethnography? *Women's Studies International Forum, 11,* 21–27.

Stacey, J. (1998*). Brave new families: Stories of domestic upheaval in late twentieth century America.* Berkeley: University of California Press.

Stacey, J. (2003). Taking feminist sociology public can prove less progressive than you wish. *SWS Network News, 20,* 27–28.

Standing, K. (1998). Writing the voices of the less powerful. In J. Ribbens & R. Edwards (Eds.), *Feminist dilemmas in qualitative research: Public knowledge and private lives* (pp. 186–202). Thousand Oaks, CA: Sage.

Stewart, A. J., & McDermott, C. (2004). Gender in psychology. *Annual Review of Psychology, 55,* 519–544.

Stout, N. M. (2008). Feminists, queers, and critics: Debating the Cuban sex trade. *Journal of Latin American Studies, 40,* 721–742.

St.Pierre, E. A. (1997a). Guest editorial: An introduction to figurations—a post-structural practice of inquiry. *International Journal of Qualitative Studies in Education, 19,* 279–284.

St.Pierre, E. A. (1997b). Methodology in the fold and the irruption of transgressive data. *International Journal of Qualitative Studies in Education, 19,* 175–179.

St.Pierre, E. A. (2009). Afterword: Decentering voice in qualitative inquiry. In Y. Jackson & L. A. Mazzei (Eds.), *Voice in qualitative inquiry: Challenging conventional, interpretive, and critical conceptions in qualitative research* (pp. 221–236). New York: Routledge.

Tellez, M. (2008). Community of struggle: Gender, violence, and resistance on the U.S./Mexican border. *Gender & Society, 22,* 545–567.

Thapar-Bjorkert, S., & Henry, M. (2004). Reassessing the research relationship: Location, position, and power in fieldwork accounts. *International Journal of Research Methodology, 7,* 363–381.

Thayer, M. (2001). Transnational feminism: Reading Joan Scott in the Brazilian Sertao. *Ethnography, 2,* 243–271.

Tregaskis, C., & Goodley, D. (2005). Disability research by disabled and non-disabled people: Towards a relational methdology of research production. *International Journal of Social Research Methodology, 8,* 363–374.

Trinh, T. M-ha. (1989). *Woman, native, other: Writing post-coloniality and feminism.* Bloomington: University of Indiana Press.

Trinh, T. M-ha. (1992). *Framer framed.* New York: Routledge.

Valentine, K. B. (2006). Unlocking the doors for incarcerated women through performance and creating writing. In D. S. Madison & J. Hameva (Eds.), *The SAGE handbook of performance studies* (pp. 309–324). Thousand Oaks, CA: Sage.

Warner, L. R. (2008). A best practices guide to intersectional approaches in psychological research. *Sex Roles, 59,* 454–463.

Weber, L. (2007). Future directions of feminist research: New directions in social policy—the case of women's health. In S. N. Hesse-Biber (Ed.), *Handbook of feminist research: Theory and praxis* (pp. 669–679). Thousand Oaks, CA: Sage.

Weeks, K. (2004). Labor, standpoints, and feminist subjects. In S. G. Harding (Ed.), *The feminist standpoint theory reader* (pp. 181–195). New York: Routledge.

Welton, K. (1997). Nancy Hartsock's standpoint theory: From content to "concrete multiplicity." In S. J. Kenney & H. Kinsella (Eds.), *Politics and feminist standpoint theories* (pp. 7–24). New York: Haworth Press.

West, C., & Zimmerman, D. (1987). Doing gender. *Gender & Society, 1,* 125–151.

Weston, K. (1991). *Families we chose: Lesbians, gays, kinship.* New York: Columbia.

Williams, B. (1996). Skinfolk, not kinfolk: Comparative reflections on identity and participant observation in two field situations. In D. Wolf (Ed.), *Feminist dilemmas in field work* (pp. 72–95). Boulder, CO: Westview Press.

Wolf, D. (1996). *Feminist dilemmas in fieldwork.* Boulder, CO: Westview Press.

Wolf, M. (1992). *A thrice-told tale. Feminism, postmodernism, and ethnographic responsibility.* Stanford, CA: Stanford University Press.

Yuval-Davis, N. (2006). Intersectionality and feminist politics. *European Journal of Women's Studies, 13,* 193–209.

Zavella, P. (1996). Feminist insider dilemmas: Constructing ethnic identity with Chicana informants. In D. Wolf (Ed.), *Feminist dilemmas in field work* (pp. 138–159). Boulder, CO: Westview Press.

Zimmerman, M. K., Litt, J. S., & Bose, C. E. (2006). *Global dimensions of gender and carework.* Stanford, CA: Stanford University Press.

8

THE SACRED AND SPIRITUAL NATURE OF ENDARKENED TRANSNATIONAL FEMINIST PRAXIS IN QUALITATIVE RESEARCH[1]

Cynthia B. Dillard and Chinwe Okpalaoka

I. SANKOFA (GO BACK TO FETCH IT)[2]

History is sacred because it is the only chance that you have of knowing who you are outside of what's been rained down upon you from a hostile environment. And when you go to the documents created inside the culture, you get another story. You get another history. The history is sacred and the highest, most hallowed songs in tones are pulled into service to deliver that story (Latta, 1992).

Revisiting "Paradigms"

Several years ago, responding to James J. Scheurich and Michelle D. Young's (1997) *Educational Researcher* article, a number of researchers presented sessions at national meetings, wrote papers, and responded to the challenge inherent in Scheurich and Young's rather provocative title, "Coloring Epistemologies: Are Our Research Epistemologies Racially Biased?"[3] Among other writings, Cynthia Dillard's (2006a) modest contribution to this paradigm talk became a chapter in her book, *On Spiritual Strivings: Transforming an African American Woman's Academic Life*. In this chapter, as in the aforementioned discussions, she explored the cultural, political, and spiritual nature of the entire conversation about paradigms and the way that the swirling assumptions and conclusions about their proliferation were mostly carried out at a level of abstraction (and distraction), absent any examination of the ways that racism, power, and politics profoundly shape our research and representations, especially as scholars of color. She spoke to how such exclusion

brings a particularized paradox for scholars of color as we seek to imagine, create and embrace new and useful paradigms from and through which we engage educational research ... [as] there are deep and serious implications in choosing to embrace paradigms that resonate with our spirit as well as our intellect, regardless of issues of "proliferation" (Dillard, 2006a, pp. 29–30).

She raised up the all too common absence of Black voices and voices of scholars of color in the discussions of the meanings and outcomes of the "coloring" of epistemologies, a discussion that had been carried out as if we did not exist as subjects within the conversation but solely as objects of it, invisible, silent. However well intentioned this discussion may have been, Black people and our thoughts about paradigms were the focus of the steady and often distorted gaze and descriptions of White researchers.

The part of the discussion that still resonates with Dillard most deeply today—and with many students of qualitative research—is the call for scholars of color to turn our attention and desires away from "belonging" to a particular paradigm (or even to the discussion of paradigm proliferation that still often swirls around us but does not include us), but instead to construct and nurture paradigms that encompass and embody our cultural and spiritual understandings and histories and that shape our epistemologies and ways of being.

We see evidence of the same call echoed throughout the literature on qualitative research. Gloria Ladson-Billings's (2000) handbook chapter, "Racialized Discourses and Ethnic Epistemologies," contrasts the concept of individualism and the elevation of the individual mind prevalent in Western thought with the African notion of *Ubuntu* (I am because we are). This notion of individual well-being that is predicated on the wholeness of the community speaks to the same spiritual and epistemological stance of endarkened feminist epistemology. Ladson-Billings's reference to the necessity of using different discourses and epistemologies, ones that can disrupt Western epistemological discourse and the dominant worldview, can be interpreted as an echo of an endarkened transnational feminism, whose notions of the sacred and the spiritual in research disrupts the Western tendency to bifurcate the mind and the spirit. Furthermore, she suggests that we must research and better understand "well developed systems of knowledge . . . that stand in contrast to the dominant Euro-American epistemology" (p. 258) in order to address the critical questions of the world today. While critical race theory is the framework used by Ladson-Billings and Donner (2005), like the endarkened transnational epistemologies put forward in this chapter, it is part of a larger effort by scholars of color and of conscience to address the ravages of racism and other discriminations and to create a space of freedom for all humanity. It is important also to note that a major revision in Virginia Olesen's Chapter 7 on feminism in this volume included an entirely new overview of endarkened, decolonizing, and indigenous feminist qualitative research and the weighty recognition of the breadth of discourses of feminisms that must be strategically used at this moment. Hence, the need for our exploration of endarkened and transnational feminisms here is an extension of the same need that encouraged Norman Denzin, Yvonna Lincoln, and Linda Tuhiwai Smith to edit the extant *Handbook of Critical and Indigenous Methodologies* (HCIM) (2008). Central to HCIM is the embrace of indigenous and critical research epistemologies that foreground spirituality, including feminist, native, indigenous, endarkened and Black feminist, spiritual, hybrid, Chicana, and border/ *mestizaje* among others (see chapters by Cannella & Manuelito, 2008; Dillard, 2008; Meyer, 2008; Saavedra & Nymark, 2008, for examples).

We see in these works gestures toward the sacred nature of science, an idea that Peter Reason (1993) forwarded more than a decade ago. In this work, his call is for researchers to consider how spirituality and the sacred can be brought to bear on pressing human and environmental problems. This chapter is a response to Reason's call. Our attempt here is to examine the complexities of Black and endarkened feminisms and epistemologies, which link the continent of Africa and the African diaspora, bringing discourses of the spiritual and the sacred to bear in this discussion in a way that is fundamental. We also recognize a discussion that is still missing in our examinations of multiple epistemologies and theories of research.

So, this Handbook chapter emerges from a paradigm and worldview that is spirit filled, endarkened, and centered in Blackness and international womanhood. However, with the publication of *On Spiritual Strivings* (2006a), and the global reaction to it, Dillard found herself (echoed by her students) wondering more deeply about the way that knowledge travels and moves in the world, enlarging and engaging the discussions and constructions of what it means to be a Black woman (and thus, what it will mean for *us,* as Black women scholars). As the work continued to travel, new opportunities for dialogue and cooperation also arose, including one with the coauthor of this chapter, Chinwe Okpalaoka, who ascends from Nigeria. As sister scholars, we have begun to recognize not just the spiritual but the *sacred* nature of research that African ascendant women have always done and are continuing to do all over the globe. We believe it is the same sacred and divine energy that has brought us together in this writing.

Some definitions of key terms may be important here. An *endarkened feminist epistemology* (Dillard, 2000, 2006a, 2006b) articulates how reality is known when based in the historical roots of global Black feminist thought. More specifically, such an epistemology embodies a distinguishable difference in cultural standpoint from mainstream (White) feminism in that it is located in the intersection or overlap of the culturally constructed notions of race, gender, class, and national and other identities. Maybe most important, it arises from and informs the historical and contemporary contexts of oppressions and resistance for African ascendant women. From an endarkened feminist epistemological space, we are encouraged to move away from the traditional metaphor of research as recipe to fix a "problem" to a metaphor that centers on reciprocity and relationship between the researcher and the researched, between knowing and the production of knowledge. Thus, Dillard (2000, 2006b) suggests that a more useful metaphor of research from an endarkened feminist epistemological stance is *research as a responsibility,* answerable and obligated to the very people and communities being engaged in the inquiry.

Our use of the term *transnational* is a literal one. We are simply meaning a way of looking at endarkened feminism that is beyond or through (*trans*) the boundaries of nations. But we also believe that such a look brings to bear the possibility of a *change* in our viewpoints as well.

An endarkened feminist epistemology is also an approach to research that honors the wisdom, spirituality, and critical interventions of transnational Black woman's ways of knowing and being in research, with the sacred serving as a way to describe the doing of it, the way that we approach the work. Noting the distinction between spirituality and the sacred is important here. What we mean by *spirituality* is to have a consciousness of the realm of the spirit in one's work and to recognize that consciousness as a transformative force in research and teaching (Alexander, 2005; Dillard, 2006a; Dillard, Tyson, & Abdur-Rashid, 2000; Fernandes, 2003; hooks, 1994; Hull, 2001; Moraga & Anzaldúa, 1981; Ryan, 2005; Wade-Gayles, 1995). However, when we speak of the *sacred* in endarkened feminist its research, we are referring to *the way the work is honored and embraced as it carried out.* Said another way, work that is sacred is worthy of being held with *reverence* as it is done. The idea here is that, as we consider paradigms and epistemology from endarkened or Black feminist positions, the work embodies and engages spirituality and is carried out in sacred ways. Thus, we are using the notions of both spirituality and sacredness to explore more globally the meanings, articulations, and possibilities of an endarkened feminist epistemology and research as sacred, spiritual, and relevant practices of inquiry for Black women on the continent of Africa and throughout her diaspora. Mostly, we are suggesting that both spirituality and the sacred are embedded fundamentally in the very ground of inquiry, knowledge, and cultural production of Black women's lives and experiences and that it is this understanding that helps us to understand the radical activism of Black feminism transnationally. However, as we look back to the earlier articulations of the cultural and spiritual nature and work of paradigms for scholars of color, what is missing is an explicit attention to the epistemologies of Black or endarkened feminism in an interconnected, intersubjective, and transnational way that renders *visible* the work of research as *sacred* work, centered in the spiritual notions constructed by Black women on the continent and in the diaspora.

Whether in the United States, Africa, or elsewhere in the African diaspora, women of African ascent share experiences with some form of oppression characterized and related by our class, race, or gender, by our existence as women. And often, it is some version of or belief in spirit that has allowed us to stand in the face of hostility and degradation, however severe (Akyeampong & Obeng, 2005; Alexander, 2005; Dillard, 2006a, 2006b; hooks, 1994; Hull, 2001; Keating, 2008; Moraga & Anzaldúa, 1981; Walker, 2006). Most arguments that have arisen around similarities and differences in transnational Black women's experience with interlocking oppressions have focused on whether there exists a hierarchy of oppressions for women on the African continent versus in the African diaspora and the issue of appropriate naming of the struggle (Hudson-Weems, 1998b; Nnaemeka, 1998; Steady, 1981; Walker, 1983). This is a discussion we take up later in this chapter.

However, such ground leads us to several questions that guide our examination here. What are the contours of Black or endarkened feminist epistemology and paradigms that emerge from African women's voices and spirits transnationally? What are the tensions, the cultural and historical experiences, diversities, nuances, and relationships that have created visions and versions of Black women's thinking, theorizing, and praxis that include or exclude Black feminism, Africana womanism, and other theoretical frameworks for Black women's thinking and being with and against one another? Where is spirituality in these global discussions, and how does it matter to the work of research? To its methods, methodologies,

and representations? Most important, we seek to further explore the profound question—and the equally profound response—put forth by M. Jacqui Alexander (2005) in her groundbreaking work, *Pedagogies of Crossings: Meditations on Feminism, Sexual Politics, Memory, and the sacred*:

> What would taking the Sacred seriously mean for transnational feminism and related radical projects beyond an institutionalized use value of theorizing marginalization? *It would mean wrestling with the praxis of the Sacred.* (p. 326, emphasis added)

Taking the sacred seriously means exploring and creating sacred versions and approaches to research and a critical revisioning of the very meanings of Black feminist inquiry and paradigms. It means taking up spirituality and the sacred as a place "from which to launch a critique of the status quo" (Wright, 2003, p. 209) from a Black-eyed female gaze.

▣ II. Nyame Nti (Since God Exists)

Spirituality as Necessity: Exploring the Cultural and Historical Landscape of Endarkened/Black Feminist Perspectives of Research in Africa and the Diaspora

> *Everything we engage in our lives is primarily a practice ordered by spirit, or authorized by spirit and executed by someone who recognizes that [she] cannot, by herself, make happen what she has been invited towards (Some, 1994).*

The concepts of intersecting oppressions and domination, although universal in practice, take on varying forms of expression from one society to another (Collins, 2000). Patricia Hill Collins further suggests that the shape of domination changes as it takes on specific forms across temporal, historical, and geographical contexts. The key difference lies in the organization of particular oppressions. Said another way, although contexts of domination might be similar across the globe (in that there is some combination of interlocking systems of oppression), the differences arise in the ways these particular oppressions manifest and the historical roots of said oppressions. The type of clothing that oppression is dressed in (that is, apartheid, colonialism, imperialism, enslavement) may vary. However, they are all systems of oppression, intersecting in various combinations and contexts. Ultimately, Collins (2000) suggests that it is fundamentally about whose knowledge and agenda are front, center, and definitive. However, to think about and work through the differences across the continent and diaspora, we must find our "common agenda," a *transnational* Black feminism.

Collins (2000) was not alone in her clarion call for a transnational Black feminism to confront intersecting oppressions of race, class, gender, and sexuality across the globe. She was joined by many, including Obioma Nnaemeka (1998), a groundbreaking

Black feminist scholar from Nigeria. Yet, a closer look at the historically and sociological manifestations of oppression and domination of endarkened or Black feminist thought in the African diaspora reveals a sort of dynamism, a constantly changing nature of oppressions for African ascendant women, very particularly within and across national contexts. For example, although interlocking oppressions of race, class, gender, and sexuality characterize Black women's experiences in parts of the African diaspora, the particular oppression that dominates might differ from one geographical and national context to another. Nnaemeka (1998) speaks of multiple feminisms within the countries of Africa and even between Africa and other continents as an indication of multiplicity of perspectives. She further explains that the multiplicity of perspectives must include cultural and historical forces that have fueled women's movements in Africa. Nnaemeka describes the African feminist spirit as both "complex and diffused" (p. 5):

> The much bandied-about intersection of class, race, sexual orientation, etc., in Western feminist discourse does not ring with the same urgency for most African women, for whom other basic issues of everyday life are intersecting in most oppressive ways. This is not to say that issues of race and class are not important to African women in the continent . . . African women see and address such issues first as they configure in and relate to their own lives and immediate surroundings. (p. 7)

Collins (2000) urges us to think globally when we consider the shared legacy of struggle and oppressions and remember that the experiences of women of African ascent have been shaped by varying forms of domination including slavery and colonialism. The oppression of continental African women cannot be isolated from the persistent consequences of colonialism. In other words, as Nnaemeka (1998) frames it, "to meaningfully explain . . . African feminism, it is . . . to the African environment that one must refer" (p. 9). Likewise, the oppressions of African ascendant women in the diaspora cannot be isolated from the persistent consequences of centuries of enslavement. It is noteworthy that the 1960s were turbulent times for African ascendant people on the continent in the fight for independence and for those in the diaspora engaged in various civil rights movements. While the former fought to gain independence from colonial rulers, the latter marched for the rights of Black Americans, women, the disabled, and other marginalized groups.

We believe that a brief historical look at the roots of Black feminism in the United States and on the continent of Africa may help situate Black women's experiences with oppression on the globe, the ground from which interventions and transformations of research must arise. While we begin this discussion with a brief history of Black feminism, we do not believe that U.S. Black feminism marked the beginning of feminism for women of color all over the world. In fact, there are ongoing critiques of U.S (and European) Black feminisms and the dangers of the

cultural hegemony throughout the world, particularly from continental African feminist scholars.[4] Instead, we begin here because of the important role U.S. Black feminism continues to play in global discussions on Black women's experiences with oppression. While the holistic nature of interlocking systems of oppression has not been particular to U.S. Black feminists, it has provided the stage for reconceptualizing new relationships within and between African women's spirits and experiences in the diaspora and the African continent and for shaping research paradigms and methodology as well. And while Black feminists in the U.S. context may have strong understandings of our experiences and struggles as African ascendant women in the United States, we often know precious little of the experiences and struggles of African ascendant women throughout the diaspora and on the continent of Africa.

Given our positionalities, the authors attempt to situate our experiential knowings into this historical survey, speaking directly to the material ways that these histories were experienced and are enacted in our lives: Dillard speaking primarily to experiences with Black feminism in the United States and Okpalaoka speaking primarily to her experiences with Black feminism in Nigeria. This is our attempt to bring our herstories to bear across the globe in ways that allow us to create paradigms across our differences "that resonate with our spirit as well as our intellect, regardless of issues of proliferation" (Dillard, 2006a, p. 29–30) and that highlight the sacred and spiritual embodiment of the praisesong that is our story as Black women everywhere.

▣ U.S. BLACK FEMINISM IN BRIEF

By Cynthia Dillard

I remember being very powerfully influenced by the image of Angela Davis in the 1970s. And it wasn't simply the perfect afro that framed her face like a crown that moved me: It was the powerful sound of her voice as she talked about freedom and truth and Black women's struggles in the United States and beyond. I remember my own desires to be a part of the Black Panther Party, but I wasn't quite of age. However, when I saw those brothers (and an occasional sister) walking into my former elementary school to serve lunches to children who needed them most, that act was transformational for me. I realized in that moment that whatever "Black power" meant, it included the commitment to knowing our history, enacting our culture with spirit, and engaging in social and sacred action on behalf of Black people, especially the young and those most needy.

While my parents were involved with more mainstream Black organizations (the Links and Omega Psi Phi), I became more interested in what they deemed the more "radical" Black organizations. And I was especially interested in the places where Black *women* were organizing, marching, making their voices heard. The National Black Feminist Organization (NBFO) was one of the first Black feminist organizations with an explicit commitment to confronting the interlocking systems of racism, sexism, and heterosexism that plagued Black women in the United States. Emerging in 1973, the organization was also a forceful response to the lack of attention and regard for Black women's experiences within both the women's movement and within Black power movements witnessed above (Hull, Bell-Scott, & Smith, 1982; Wallace, 1982). By 1974, a spin-off group of U.S. Black feminists formed the Combahee River Collective, focusing on a more radical commitment to the oppressions that Black women still faced in the United States. The mission of this group of women, in comparison to the NBFO's, was to confront these complex systems of oppression through a Black feminist *political movement* (Combahee River Collective, 1982). Rather than project themselves as "firsts" or as pioneers of Black feminism, the collective's members historically acknowledged their work as an extension of the earlier work of Black women activists like Sojourner Truth, Harriet Tubman, Frances E.W. Harper, Ida B. Wells Barnett, and Mary Church Terrell, whose intellectual and activist work flourished during the antislavery era (Combahee River Collective, 1982). There was also a very strong commitment to spiritually center the work of the Combahee River Collective, both in the sacred approach to seeing and acknowledging the above Black women ancestors and in setting a purpose and vision that sought to transform the social and political milieu away from oppression and toward equality and justice, particularly for U.S. Black women.

By the early 1970s, we witnessed a critical intervention of theorizing and knowledge production, as Black feminist literature (including anthologies and fiction) began to be published and find their way to bookstores and bookshelves, both in the U.S. and abroad. This was not simply publishing as an economic intervention in the lives and knowledges of Black women: This was a radical intervention, as these literatures fundamentally shifted and shaped the foundations of Black feminist thought and actions. Toni Cade Bambara's *The Black Woman: An Anthology* (1970), Toni Morrison's *The Bluest Eye* (1970), Audre Lorde's *Cables to Rage* (1970), Alice Walker's *In Search of Our Mothers' Gardens: Womanist Prose* (1983), and a reissue of Zora Neale Hurston's *Their Eyes Were Watching God* (1978) are examples of landmark literary texts that defined and theorized the early Black feminist movement in the United States. As an adolescent African American girl, I felt these early works profoundly, as I sought desperately to define what it meant to be both Black and female in the predominately White schooling contexts of my youth. All of the texts we were required to read centered images of White womanhood as virtuous and worthy of emulation. Louisa May Alcott's *Little Women* was the standard by which we were asked to aspire, and watching "The Brady Bunch" was the free time text of the day. But my mother's version of Black womanhood (albeit similarly tethered to homemaking and child rearing as Mrs. Brady) was

tied to a simple and explicit truth, manifest in her strict attention to our school lives, homework, and consistent trips to the public library: Education and learning to read the word and the world were the *only* ways to create options for Black women's lives. In her precious free time, my mother read these texts along with me, opening me to a world that in some cases highlighted the harsh realities of her own life as a Black woman, growing up in poverty and during segregation in the United States. In other cases, these words on the page opened something that could exist only in her imagination and our own but that always also existed as *possibilities.* These texts also stirred significant debates and controversies *within* the Black community, especially for Black men, who often resented what they interpreted as direct accusations that they were perpetrators of gender and sexual oppression. Regardless of the consequences, my Mom and I continued to read every story of Black womanhood we could. And I learned how powerful words could be: Black women's literature helped define ourselves for ourselves, and as an oral tradition, it goes back generations. Now, through the voices of Walker, Hurston, and others, as well as the words on the page, we could *see* our definitions and return to them over and over again.

The 1980s brought more radical overtly political texts, responding in part to the birth of woman's studies and specifically Black women's thought and knowledge production "in public." We came to know, through their writings, major Black feminist scholars *and* activists like Gloria (Akasha) Hull, Barbara Smith, and Patricia Bell-Scott, whose co-edited text (Hull, Bell-Scott, & Smith, 1982), *All the Women Are White, All the Blacks Are Men, But Some of Us Are Brave,* became a pioneering text for Black feminist studies across the United States. This relative proliferation of Black feminist writing in the 1980s also included works like Barbara Smith's *Home Girls: A Black Feminist Anthology* (1983) and bell hooks's (1981) *Ain't I a Woman: Black Women and Feminism,* which focused on the impact of sexism on Black women. But these women also began to bring questions and concerns of sexual identities and spirituality within Black feminism to the forefront. Lorde's *Sister Outsider* (1984) spoke directly to the need for integration and wholeness in Black women's multifaceted identities, including our sexualities. Paule Marshall's *Praisesong for the Widow* (1984) brought to the fore the ways that remembering culture and history as a Black woman is truly a transformative act, particularly from a spiritual perspective.

Cherie Moraga and Gloria Anzaldúa's *This Bridge Called My Back: Radical Writings by Women of Color* (1981) was one of the earliest attempts to link the underlying oppressions of women across differences of race, class, sexuality, and culture. Equally important, Anzaldúa brought the scholarship and voices of women of color together in an edited volume that began to speak explicitly about the importance of spirituality, healing, and self-recovery as necessities for women of color across our ethnicities and identities.

The proliferation of scholarship of the 1990s and beyond picked up Anzaldúa's call to recognize the sacred and spiritual ethos of Black and endarkened feminisms. From hooks's (1993) *Sisters of the Yam: Black Women and Self-Recovery* and Bambara's *The Salt Eaters* (1992) to Collins's (1990) landmark, *Black Feminist Thought: Knowledge, Consciousness, and the Politics of Empowerment,* which literally transformed our understandings of gender, race, and class, centering it firmly in the epistemologies and theories of Black womanhood. It is interesting to note that one of the most radical revisions in Collins's second edition of *Black Feminist Thought* (2000) was her explorations of the limits of a Black feminism bounded by nations: The revisions provided direction for how to place U.S. Black feminist thought into coalition with the voices and efforts of African ascendant women worldwide. And whether on the continent of Africa, in the United States, or all of the spaces between, around, and among them both, I see that the creation of Black feminism in the United States is only a part of a living, breathing legacy. As Pearl Cleage (2005) suggests, Black/endarkened feminist thought is *itself* a praisesong:

the flesh and blood of our collective dreaming,

[through which] we realize with a knowing deeper than the flow of human blood in human veins

that we are part of something *better, truer, deeper.* (p. 15)

At least part of the "better, truer, deeper" is found in the connections between the Black feminism in the United States and efforts of African feminists on the continent, toward which we now turn.

◧ BLACK FEMINISM IN AFRICA IN BRIEF

By Chinwe Okpalaoka

While the struggle of U.S. Black feminists in the 1960s included the fight for the rights of women and people of color, Africans on the continent of Africa witnessed a decade that ushered in the end of colonial rule in many African nations. As former colonizers retreated to their countries of origin, newly formed African nations began what has and continues to be an arduous and complicated journey toward independence. Oyeronke Oyewumi (1997) asserts that colonization should not be understood solely in the context of the period of the actual colonization. For many, she suggests, the period of the Atlantic slave trade and colonization "were logically *one* process unfolding over many centuries" (p. xi, emphasis added). This argument is critical to understanding the spiritual connection between African women on the continent and throughout the diaspora, in our knowledge production and in praxis. And as Dillard (2006a) asserts, these connections are *always* present, whether

we are conscious of them or not. It is an understanding and embrace of this connection that drew me to Dillard shortly after our first meeting. Soon after I began my doctoral studies, I was introduced to the histories of African Americans and theories about their lived experiences. As Oyewumi (1997) argues above, I quickly identified a connection between me (as "representative" of the sister who never left the continent) and my African American sister ("representative" of those who were enslaved and taken from the continent of Africa). My dilemma, then, was how to gain legitimacy to speak on behalf of women from both sides of the Atlantic Ocean. Was I now estranged from my sister because I did not make that trip with her? Was I no longer a part of her story because we were now divided by history, distance, and experience? Did she not understand that I, too, knew the pain of oppression, albeit in a different form and intensity? That at the same time she was fighting for human rights in the United States, I was fighting to end centuries of oppression that began with slavery and continues today as neocolonialism? It is this connection of struggle and spirit that we speak in this chapter.

Amina Mama (2007) connects the advent of African independence struggles with the emergence of feminist activism in Africa, most notably in Nigeria and Egypt. However, many argue that gender, as a political category, was not necessarily a salient category for women in many African societies, especially in comparison to Black American feminists (Aina, 1998; Oyewumi, 1997; Taiwo, 2003). However, although African nations were beginning the slow process of achieving independence from colonial regimes, women's issues were not foregrounded in this independence struggle. Continental African women quickly learned that the fight for independence did not necessarily place women's rights front and center in the fight for independence. Speaking specifically about Nigerian women, Molara Ogundipe-Leslie (1994) explains that the seeming lack of focus on women's issues postindependence could be explained by the preexistence and availability of economic opportunities for women in precolonial Nigeria. This is echoed by Ifi Amadiume (1987) in her precolonial, colonial, and postindependence analysis of the ways that women enjoyed relative power and influence, which diminished in eastern Nigeria only with the advent of British colonialism and its own versions of gender roles. This study counters the master narrative among Western feminists that portrays African women as having had limited political and economic power in comparison to their male counterparts. Amadiume demonstrates the way that colonialism actually disempowered women by limiting the economic freedom that they enjoyed in precolonial times. We see this echoed in prominent African feminist literature that theorizes women's lives, such as Kenyan Margaret Ogola's (1994) novel, *The River and the Source;* Nigerian Flora Nwapa's (1966) *Efuru;* and Ghanaian writer Ama Ata Aidoo's (1970) collection of short stories, entitled *No Sweetness Here.* Having spent the first 25 years of my life in eastern

Nigeria, I witnessed, firsthand, the spirit of enterprise and economic independence that characterized women's efforts at running their households. Traditionally, men were the heads of households, but it was apparent that women were the glue that held the homefront together. I understood, even as a young girl, why many women set up their own businesses, even when said business was a small table strategically placed in front of her home from which she hawked basic household and food items, while remaining within eyesight of her home and children. Although I wondered how much profit the women made from selling such small items, I understood that it was their gesture toward economic independence and empowerment. A woman who depended entirely on her husband for financial help was usually perceived as lazy. So, the struggle for independence for the African woman was not necessarily a struggle for her economic independence, but a struggle for the independence of her local and national community at large. Several scholars claim that the struggle for economic opportunities and the right to work, which characterized the struggles of women in the diaspora and elsewhere, could not easily be applied to women on the continent of Africa (Bray, 2008; Mohanty, Russo, & Torres, 1991; Nnaemeka, 1998; Ogundipe-Leslie, 1994). Instead, it was and continues to be neocolonialism, oppressive regimes, and marriage and cultural norms that we must unpack to understand the African woman's experience with oppression in Africa and her feminisms. Aidoo (1998) and Zulu Sofola (1998) concur that the African woman's burdens of oppression can be traced to both internal influences from sociocultural and patriarchal structures and external influences stemming from colonialism and postcolonial crises of leadership. Aidoo (1998) echoes these sentiments when she states:

> Three major historical factors have influenced the position of the African woman today: Indigenous African social patterns; the conquest of the continent by Europe; and the apparent lack of vision, or courage, in the leadership of the post-colonial period. (p. 42)

Like her African ascendant sisters in the diaspora, the African woman on the continent of Africa has had to fight the voicelessness caused by centuries of domination through slavery, colonialism, and imperialism. She, too, has had to confront intersecting oppressions of racism, sexism, and classism. But the multiplicity of manifestations of the particular set of oppressions that plague women within and outside of the African continent has caused African feminists like Aidoo, Abena Busia, Sofola, and Ogundipe-Leslie to advocate for the consideration of *culture* as a form of oppression for the African woman. According to Ogundipe-Leslie (1994) culture, much more than race, more aptly determines African women's identity. We understand that the three major axes of oppression (race, class, and gender), which may plague Black women in the diaspora (and within and against which we have theorized our versions of feminisms), must be expanded to include oppressive cultural norms; we will avoid thinking of African womanhood in universal terms, a tension that is apparent in many discussions of transnational Black feminisms (Collins, 2000; Guy-Sheftall, 1995; Ogundipe-Leslie, 1994; Omolade, 1994; Oyewumi, 1997; Steady, 1981). Nowhere is this wrestling more apparent than in the naming of what Nnaemeka (1998) calls the "feminist spirit" across the globe, to which we now turn.

Call Me by My True Names:[5] Naming Black Feminism in the United States, the Diaspora and the Continent of Africa

Within African culture, naming is a sacred practice, one that is not only important to the continuation of the group's heritage and work, but also to the purpose and future work of the individual being named. Through this issue of naming, we can begin to see the interconnected nature of Black feminist struggles in the United States with those of Black women throughout the diaspora and the continent of Africa.

Given the too often exclusionary spaces for U.S. Black feminists within the broader conversations of feminism, early Black feminists in the United States began to create names that more carefully honored and described a collective *Black* feminism. Walker (1983) first introduced the term *womanism* into the ongoing debates by White feminists, who seemed to quickly forget that their Black counterparts had been their allies nearly a decade before in the fight for civil rights. According to Walker, a womanist is

> a black feminist or feminist of color . . . a woman who loves other women, sexually and/or nonsexually . . . committed to survival and wholeness of entire people, male *and* female . . . Womanist is to feminist as purple is to lavender." (emphasis in the original, pp. xi–xii)

However, in direct criticism of Walker's definition, Clenora Hudson-Weems (1998a, 1998b) argues that, regardless of where women of African heritage exist, we should not adopt the label of feminist because, in comparison to our Western counterparts, gender is not primary in the struggle for equity and recognition. Hudson-Weems (1995) prefers the term *Africana womanism* to womanism, Black feminism, and African feminism. She believes that Africana womanism more succinctly captures the family centeredness in an African framework, rather than the female centeredness of Western feminism. This also resonates with the stance that many continental African feminists have taken, suggesting the crucial need for Africana men and women to come together to confront all oppression, given what is seen as the interdependency of men and women in equally worthwhile albeit different roles in an African cosmology and worldview (Nzegwu, 2006; Oyewumi, 1997; Richards, 1980). In other words, African ascendant people must take control of our struggles for the sake of collective justice for African people (Hudson-Weems, 1998b). While Hudson-Weems suggests there are strong and fundamental differences between her notion of African womanism and womanism as defined by Walker, one can also argue that her version of naming does at least partly align with Walker's original definition of womanism, in that it includes a commitment "to survival and wholeness of entire people, male and female. Not a separatist [movement]" (Walker, 1983, pp. xi–xii). Hudson-Weems's (1998a) main concern is with the issue of self-naming or what she calls "a reclamation of Africana women through properly identifying our own collective struggle and acting on it" (p. 160). She further believes that the agenda of the Africana woman must be "shaped by the dictates of their past and present cultural reality. No one can be accurately defined outside of one's historical and cultural context" (Hudson-Weems, 1998b, p. 450). She goes on to claim that Africana scholars are sometimes forced to identify as feminists, to either gain legitimacy in the academy or because of a lack of a more appropriate framework that is suitable for their particular experiences. However, Busia (1993), in speaking of the need to negotiate multiple and transnational identities, calls for a more fluid, layered, and particular naming that more aptly describes the crossing of national and international boundaries in the act of naming self and other. Busia typifies the sort of complexity and dynamism within which Black women in Africa and in the diaspora wrestle, in her self-identification as

a Ghanaian-born poet, educated in the United Kingdom, teaching in the universities of the United States of America (p. 204) . . . [or] as scholar, as poet, as Black, as female, as African, as an exile, as an Afro-Saxon living in Afro-America. (p. 209)

These arguments over naming Black feminism are not simply about the act of naming: They are also about defining and constructing the boundaries and possibilities for relationships *across* Black feminisms, across racial, tribal, ethnic, and national differences, as well as advocating for fundamental *human* rights. In her well-known text, *The Black Woman Cross-Culturally,* Filomina Steady (1981) called for a redefinition of concepts, perspectives, and methodologies that position the transnational Black feminist researcher as an advocate for basic *human* rights throughout the world and not solely an advocate for women's rights or the rights of those in her local community. We hear echoes of that same call throughout this chapter, of African ascendant feminist acknowledgment of the oneness of male and female energy in the struggle against oppression (Wekker, 1997). These arguments are also spiritual in nature, seeing African feminism as a standpoint on human life from "a total, rather than a dichotomous and exclusive perspective" (Steady, 1981, p. 7). Steady goes on to echo a common cosmological concept in African thought that "for African women, the male is not 'the other' but part of the human same. Each gender constitutes the critical half that makes the human whole. Neither sex is totally complete in itself to constitute a unit by itself" (p. 7).

hooks (1994) welcomed these contestations in naming, perspectives, positions, and language, seeing these confrontations as less about naming and more about how these "differences [mean] that we must change ideas about how we learn" (p. 113). She continues, suggesting that "rather than fearing conflict [in naming], we have to find ways to use it as a catalyst for new thinking, for growth" (p. 113). Walker (2006) also cautions against arguments that suggest a lack of unity of purpose and proffers an alternative "combined energy" through which we can "scrutinize an oncoming foe" (p. 4). According to Walker (2006), this coming together has the potential to "rebalance the world" (p. 4) and, in the case of these contestations, help us refocus on the task before us. This task— whether historical or contemporary—requires us to remember that the struggle for injustice, regardless of geographical location, must include *an awareness of the specific historical and cultural contexts within which oppressions are taking place* in order to identify effective frameworks with which to do the necessary work to dismantle them.

The rumblings of dissatisfaction with naming and creating an organized and collective transnational Black feminist response to oppression continue today. Led by activist scholars and writers such as Steady, Collins, and Beverly Guy Sheftall, Black feminists across the globe are troubling the boundaries of the definition of feminism to describe differences in African ascendant women's experiences with racial, sexual, class, and cultural oppression. This stands to produce what Steady (1996) has called

a more inclusive brand of feminism through which women are viewed primarily as human beings and not simply women . . . [that] emphasizes the totality of human experience and [is] optimistic for the total liberation of humanity . . . African feminism is *humanistic* feminism. (p. 4, emphasis added)

This holistic view of the African woman, in relation to her community, echoes precolonial African practices and values regarding the physical as well as the spiritual well-being of the community. Therefore, in contrast to the Western tendency to dichotomize the material and the spiritual, male and female, the emotional and logical, a transnational African feminism merely reflects an age-old concept of a human oneness or human wholeness, where the male is not the enemy but a coparticipator in the struggle for human survival. It is important to note that this concept of oneness existed prior to the European invasion of the continent of Africa and defines the nature of relationships and life both historically and contemporarily. Consequently, the African spiritual concept of communal well-being is more highly valued than the individualism that marks Western feminist thought. This was at least part of the tension that existed between African American feminist scholars and White American feminist scholars as well. We argue here that, whether conscious of this African moral value as a carryover from cultural ways of being and ways of thinking prior to the trans-Atlantic slave trade, the pursuit of the well-being of the whole on both the continent and in the African diaspora as a means of meeting the needs of the individual and the community is a spiritual concept. Although the spiritual concept of communal wholeness and wellness prevalent in African feminism sharply contradicts the historical split between the spiritual and material so pervasive in the academy, it has been a critical part of Africana and Black feminist thought historically and continues to be pushed to the fore by scholars like Alexander (2005), Dillard (2006a, 2006b), Guy-Sheftall, (1995), hooks, (2000), Hull, (2001), and Oyewumi, (1997), to name a few.

Thus, like sisters in the diaspora, continental African women scholars have not escaped the tensions present within and among African feminist scholars in defining the connections, however contentious, between versions and visions of Black feminism on the continent and in the diaspora. The well-known Ghanaian feminist writer and activist Aidoo (1998) has been criticized primarily by continental African women scholars for identifying as a feminist. Like the struggles of Black women in the United States, the accusations have to do both with the issue of naming and the critique at the epistemological level that the term *feminism* is a Western construct and that African women should seek empowerment through their own self-naming. The premise of the argument is that the historical realities of Western feminism do not mirror the reality of the continental African woman's historical struggles, particularly as they relate to the sufferings and current realities during and after colonialization. This argument mirrors the one made by Hudson-Weems (1995) about the need to marshal more appropriate terminology to capture Black women's experiences in the diaspora. Aidoo's defense, similar to that of her Sierra Leonean counterpart Steady, is that all men and women are feminists if they believe that the struggle for liberation for all Africans cannot be isolated from the struggle for the well-being of the African woman. Here we see again echoes of the communal versus the individual or the collective versus the self, fundamental to an African-centered cosmology.

While this discussion is by no means complete, the ethos and spirit of transnational Black feminist thought is clear: Black feminist scholars have *always* talked back to the exclusion of Black women's experiences in feminist research, paving the way for more global and diasporic conversations about Black women and the specialized angle of visions that we bring to the question of knowledges, knowledge production, and the practices of research. The call for a naming and marshalling of Black feminism arises from a place—epistemologically, spiritually, paradigmatically—that both acknowledges and addresses the complex intersections of culture, race, class, sexualities, nation, and gender in Black women's experiences and in a way that is historically and sufficiently grounded in African

ascendant ways of knowing and being. This call is reminiscent of one made nearly two centuries earlier by Anna Julia Cooper, one of only two Black women to address the Pan African Congress organized by W. E. B. DuBois in 1919 in Paris, France. An acclaimed forerunner of Black feminism in the United States and abroad, Cooper (1892) cautioned against the expectation within Black communities generally and White feminist communities as well that the Black woman be required to fracture her identity by uplifting her gender identity over race or class. Instead, the work was and is about establishing linkages with African ascendant women globally in the struggle for elimination of all oppressions, wherever Black women are. It is about finding the sacred ground between us.

◼ III. Nkyimkyim (Devotion to Service and Willingness to Withstand Hardship): The Ethos of an Endarkened Transnational Feminism

Re-member what is dark and ancient and divine within yourself that aids your speaking. As outsiders, we need each other for support and connection and all the other necessities of living on the borders . . . The oppression of women knows no ethnic nor racial boundaries, true, but that does not mean it is identical within those boundaries. (Lorde, 1984, p. 69–71)

Our politics initially sprang from the shared belief that black women are inherently valuable, that our liberation is a necessity not as an adjunct to somebody else's but because of our need as human persons for autonomy . . . We realize that only people who care enough about us to work consistently for our liberation are us . . . we have a very definite revolutionary task to perform, and we are ready for the lifetime of work and struggle before us. (Combahee River Collective, 1982)

The quotations above speak volumes to the possibilities of both living and creating spaces for epistemologies and methodologies that arise from the Black or endarkened feminist voices gathered here, including those voices representing multiple forms of migration to the far reaches of globe, those who have transitioned to the spirit world, and those yet to come. What we are suggesting here is that a more globally attentive Black or endarkened feminism and its methodologies would be less about traditional academic notions of research practices and more about a sort of radical spiritual activism that encompasses the collective diversity of Black women's knowings and doings, that defines and describes our collective *ethos*, particularly given that previous definitions of *Black feminism, womanism, Africana womanism,* or *Third World feminism* may no longer hold as bridges (if they ever did) across our differences in paradigms, practices, and purposes. We have also come to know that whatever descriptions and definitions of transnational or global Black feminisms *are,* they must necessarily be "simultaneously historically specific and dynamic, not frozen in time in the form of a spectacle" (Mohanty, 1991, p. 6). That what fundamentally defines and shapes an "in common" ethos and experience of Black women in the world across culture, ethnicity, national affiliation, sexual affinity, economic class and condition, and other forms of identity can be articulated by two core experiences of African ascendant women wherever you find us on the globe. These may seem both obvious and common sense, given the body of literature around Black/African/Africana feminism. However, this is our attempt to make explicit two salient knowings which all Black women experience our lives.

1. Black women work and live within a context of struggle against systems of oppression and exploitation, both large and small.

2. Black women work and live within a context of spirituality and the sacred, holding beliefs in something larger than ourselves.

Such spiritual consciousness is what enables us as Black women to both work against that which oppresses and to find strength and sometimes even joy in the process of the work as well. *This is our collective ethos or spirit as African ascendant women.* What is needed is an approach to research and inquiry that honors the wisdom, spirituality, and critical interventions of Black women's ways of knowing and being, with spirituality and sacredness being central to the work. But what is the nature or character of an endarkened feminist approach to research that can work within and against these struggles, that can transcend our present differences, and that embraces spirituality and the sacred nature of inquiry? What might we need to consider and question as we think (and feel) our way into and toward epistemologies and methodologies that might be useful wherever we find ourselves on this globe? That is the focus of this final section.

▣ SECTION IV. FUNTUMMIREKU-DENKYEMMIREKU (WE HAVE A COMMON DESTINY, A UNITY THROUGH DIVERSITY)

Sacred Practice, Sacred Dialogues: Some Considerations for an Endarkened Transnational Feminist Methodology

An important component of African indigenous pedagogy is the vision of the teacher [and researcher] as a selfless healer intent on inspiring, transforming, and propelling students to a higher spiritual level. (Hilliard, 1995, pp. 69–70)

Sacredness and spirituality are central to endarkened feminism. From Cooper's (1892) advocacy for the well-being of the African American community, to Steady's (1996) call

for a feminism that attends to the total liberation of humanity. From Walker's (1983) definition of womanism, which includes a commitment to the survival and wholeness of entire people, male and female, to hooks's (1994) concept of the basic interdependency of life. From these voices, we can clearly see both the expectation and the relative requirement that endarkened feminist scholars bring a spiritual vision and sacred practice to bear within whatever version of Black feminism we might ascribe to. Similar to hooks (1994) and Walker (1983), Dillard (2006a) asserts that "a spiritual life is first and foremost about commitment to a way of thinking and behaving that honors principles of inter-being and interconnectedness" (p. 77). This suggests that bringing sacredness and spirituality to bear in any exploration of an endarkened transnational feminist methodology is not a frivolous exercise: It is a radical response to the need for an approach to research that honors the wisdom, spirituality, and critical interventions of transnational Black women's ways of knowing and being in research. We note again that the distinction between spirituality and the sacred is important here, particularly as it relates to research. Again, what is meant by *spirituality* is to have a consciousness of the spiritual in one's work, recognizing the ways that consciousness can transform research and teaching. However, the *sacred* in endarkened feminist research refers to *the way the work is honored and embraced as it carried out.* Said another way, work that is sacred is worthy of being held with *reverence* as it is done.

Recently, our department offered the first doctoral seminar in education on Black feminist thought. In addition to other goals, the course was designed to be a space where an endarkened and transnational feminist epistemology and pedagogy (Dillard, 2006a, 2006b) would be created, engaged, and experienced. It was about enacting a radical humanism as intervention in higher education, about becoming more fully human in all of our variations as African ascendant women (who made up the entire class). However, we were a very diverse group of Black women, representing identities, histories, and cultural affiliations that enriched our ability to engage the discussion of Black feminisms and endarkened epistemologies not simply with our minds but embodied in our methodology as well. Nationally, we had deep roots in the United States, Japan, Ghana, Nigeria, and Kenya. Economically, we had experienced the economic range from poverty to middle- and upper middle-class wealth. We had migratory and immigration experiences that included a third of the class growing up in countries outside the United States and "becoming" citizens of the United States in adolescence, with the other two thirds having grown up in the United States (in the Pacific Northwest, the South, the North, and in both rural and urban environments). We spanned the continuum of sexualities and partnering. Most important, we were deeply committed to ourselves and to doing our work *as Black women* in the academy.

One of the first assignments in the class was a creative autobiography, where we shared stories of who we are and why we are.[6] The course reading list represented Black and endarkened feminisms that were transnational and historical, as well as multiple in genre including films, poetry, visual art, letters, narrative, research studies, and other course syllabi. This corpus is what Bell-Scott (1994) calls *life notes*. Many of these readings are represented in the bibliography of this chapter; others were suggested by the students and faculty as the course progressed. Our weekly class sessions were mostly dialogues about the readings and short presentations about content. As we were preparing this chapter (a text that we desired at the time of the class but could not find in the literature), the class became a space to engage what bell hooks and Cornel West (1991) call critical affirmation, the humanizing process of critique that "cuts to heal not to bleed."[7] More than that, we found ourselves raising questions and critique that explicitly showed us the difficulty of talking across our different versions of Black womanhood, even given our common "texts" and deep commitments to dialogue as sacred praxis. As the course continued, we began to see more clearly the nature and character of an endarkened transnational feminist dialogue as we engaged with each other. We also experienced the tensions and intelligibilities that still existed between us, despite our good intentions and deep commitments to dialogue. We also had the opportunity to engage in a cross-class dialogue with other doctoral students enrolled in a qualitative course on feminist methodologies that was being offered at the same time. These interactions further exposed the tensions and challenges of such a dialogue, particularly the racialized spaces of feminisms, which often exclude feminisms of color (and certainly African ascendant feminisms) from the consciousness of White researchers.

What would it take to productively enter into these tensions and differences? And could explicating the character of such a dialogue help us to speak to what an endarkened transnational feminist methodology in research might be (especially given our new understandings of the history, culture, and contestations within and among African ascendant feminisms)? A methodology that would be "historically specific and dynamic" (Mohanty, 1991)? One that would be a useful way forward in the spirit of sacred praxis that Alexander (2005) called for?

The following section [Table 8.1] is our attempt to share what we learned *and* some questions we learned to raise, as authors, as members of the course, and as researchers and sisters of the yam.[8] We hope it is not seen as a checklist to legitimate one's identity or research position ("I do these things and now I'm an endarkened transnational feminist researcher!"). Instead, we see it as an offering to the research community of the ways that an endarkened transnational feminist methodology may have the potential to

Table 8.1   Engaging Transnational Endarkened Feminist Research: Some Considerations and Questions

Considerations for Endarkened Transnational Feminist Research	Some Relevant Questions for the Researcher
On the meaning of African womanhood . . .	
Seeks to examine the multiple intersections of oppressions of Black women relationally *and* historically	What is or was going on here/there? What is or was my/our relation to the lives of Black women here/there?
Sees the way that temporality shapes relative relationships between and among versions of Black womanhood, personhood	Whose story will I tell and from what time period of "African womanhood?" How do I struggle with the tension of the African "continent" and the "diaspora" and their relative and multiple meanings? Have I dealt with questions of the timing (and manner) of im/migrations and the relationship to "authenticity" and naming oneself "African"?
Seeks to know Black women's experiences, contributions, cultures, and "feminisms" in all of their varieties/versions	What do I know about African ascendant woman? How did I gain that knowledge and what would enrich my understanding of specific versions of Black womanhood or endarkened feminisms? How can this knowledge get in the way of my seeing and understanding the vision and version of African ascendant feminism under study?

(Continued)

Table 8.1 (Continued)

Considerations for Endarkened Transnational Feminist Research	Some Relevant Questions for the Researcher
Embodies responsibility and respect, different than the cult of womanhood	How have I prepared to study the lives of Black women differently than I would for other women? What would show that I respect the particularities of her understandings and embodiment of cultural norms, geographies, and traditions?
On the sacred nature of experience . . .[9]	
Seeks to recognize multiple experiences outside of one's own	In what ways does the story I'm hearing (or the text I'm reading) map on to my experience and knowings? In what ways is it different? How do I hold those differences as sacred (with reverence), without judgment or denial in their difference?
Recognizes that you can never be the "expert" on another's experience and, thus, must move yourself out of the way to make room for the liberation of others	What does their experience *mean* to them? Can I *hear* and *imagine* the depth of the meaning of their experiences and empathize without trying to "save" another? What does their story mean to me and what emotions/memories does it evoke? How do my emotions mediate (or distort) *their* intended meaning?
On recognizing African community and landscapes . . .	
Shares the need for alliance and reliance: *I am because we are*	Where are the recognitions and engagements in this work of an endarkened womanhood that moves between and even beyond nation, culture, sexualities, economic class, language, and so on?
Recognizes the dynamic and shifting landscapes and configurations of identity and social location of groups	How does what I *thought* I knew about this individual/group match what I am hearing from engagements with him/her/them? Where are the places and people who could provide disconfirming data? Have I sought this out?
Is committed to knowing one another's stories through sustained relationship for the purposes of bettering conditions that may not mirror our own	Can I rest in that place where it is not all/always about me? Are humility, sacrifice, and selflessness at the center of my desire to "know"?
On engaging body, mind, and spirit . . .	
Makes *space* for mind, body, and spirit to be a part of the work	How have I sought knowledge at a level of intimacy and wholeness (beyond the mind), at the level of the senses, the sensual, and the spiritual? What questions have I asked of myself and another that move toward connections of our spirit? What would happen if I "went there?"
Is reciprocal, as every person is both teacher and taught, changing as we know the other and the other knows us	In what ways are my views of research shifting as a result of my research? What "lessons" have I learned from others in this inquiry? What are the lessons they've learned from me? When someone reads this work, how will they know that I approached this project with reverence?
Requires radical openness and vulnerability	In what ways have I "shown up" for this inquiry? How am I hiding in fear of what I am, what I don't know or misunderstand, or who the other is or what they know?

shape a more reverent and sacred approach to inquiry that transcends our differences, our feminisms, and our lives.

These considerations and questions suggest that, as we work to live and theorize from and through endarkened transnational epistemologies, we must also shift our gaze and engagement to embrace a more sacred (reverent) understanding of our relationships with and in endarkened spaces of womanhood and feminisms. We must go beyond employing or engaging our methodologies: We must *be* differently, asking relevant (and reverent) questions of our practice and of ourselves. Some considerations are discussed in brief here.

On the meaning of African womanhood. Endarkened transnational research acknowledges that the lives of African ascendant women are intertwined and interconnected, given our shared legacy of oppressions on the African continent and in the African diaspora. This awareness does not discount the ways that temporality shapes Black women's experiences (Okpalaoka, 2009). Neither does this awareness discount the notion that there are variations of feminisms that reflect the varied nuances of oppression manifested in women's specific historical, cultural, and geographical locations. The disruption of African ascendant peoples' lives through enslavement, colonization, and apartheid across temporal and geographical boundaries only serves to connect us across these boundaries. A respect for the particularities of Black women's understandings and embodiment of cultural norms, geographies, and traditions must be reflected in the research and work of inquiry.

On the sacred nature of experience. At the core of Black feminism (Collins, 2000; Steady, 1996) and endarkened feminism (Dillard, 2006a) is the recognition of the expertise that Black women acquire through our lived experiences and specific to our lived conditions. An approach to endarkened transnational feminist research is one in which the researcher and the researched are engaged in a mutually *humbling* experience, where each understands our limitations in speaking *for the* other. An endarkened transnational feminist epistemology and methodology recognizes that there are multiple experiences outside of one's own. Therefore, the role of researcher as expert will serve only to hinder the liberation of those with whom we engage in research and the cultural and spiritual knowledge that is inherently valuable to both of us as human and spiritual beings.

On recognizing African community and landscapes. The South African concept of *Ubuntu* ("I am because we are") and the Ghanaian (Akan) concept of *Funtummireku-denkyemmireku* ("We have a common destiny") embody the need to recognize the powerful and omnipresent role of community from an endarkened transnational perspective. Contrary to Western thought that seeks to elevate the individual above the community, researchers committed to an endarkened

transnational feminist praxis are also committed to knowing another's stories through both telling one's own and through the sustained relationship that such dialogue requires. From this standpoint, our work as researchers has as part of its purpose to make better conditions that may not mirror our own. In other words, while we recognize the specifics of the oppressions within and amongst African ascendant women, as long as some form of oppression is present within our collective reality, we all must engage in the struggle for freedom from oppression and full humanhood. We are in a collective struggle for liberation regardless of the specifics of our conditions. An endarkened transnational feminist praxis works beyond self to recognize the dynamic and shifting landscapes and configurations of identity and social location of groups.

On engaging body, mind, and spirit. Endarkened transnational feminist research is research that makes space for mind, body, and spirit to be a part of the work. It invites the whole person of the researcher and the whole person of the researched into the work, knowing that the mind, body, and spirit are intertwined in their functions of maintaining the well-being of the individual and community. The place of the sacred in endarkened and transnational feminisms requires radical openness, especially on the part of the researcher, who understands deeply that her or his humanity is linked with that of the people with whom he or she studies. The act of sharing with those who have been silenced and marginalized is a spiritual task that embodies a sense of humility and intimacy. Furthermore, a sense of reciprocity is fundamental from this epistemological space, a sense that the researcher and the researched are changed in the process of mutual teaching and learning the world together.

As we end this chapter, two things are clear. First, this exploration of endarkened transnational feminisms affirms the sacred praxis of Black women on the continent and throughout the diaspora. This chapter is our contribution to the collective legacy of struggle and spirit, to "write all the things I should have been able to read" (Walker, 1983, p. 13). However, this exploration of endarkened transnational feminisms also points to the ways that the paradigms and epistemologies that have been marshaled in qualitative research have still not answered the deeper, spiritual questions that undergird many cultural phenomenon, the persistent social problems of equity and justice, the difficulties of community and solidarity, and the complex nature of identity and personhood.

Given an African cosmology and epistemology, the authors strongly resist attempts to predict the "future" of the field of Black/endarkened feminist thought, as past, present, and future are implicated and embraced in our current existence, not as separate but as part of the same. As we have examined the history of Black feminisms on both sides of the water to arrive at a place where we believe a more transnational endarkened feminism is a necessity, one thing is clear: That attention to epistemologies and praxis that also center the spiritual and sacred nature of qualitative research are the necessary way forward.

◻ Notes

1. The coauthors would like to acknowledge the "sisters of the yam" who were participants in the first *Black Feminist Thought* course in the School of Teaching and Learning at The Ohio State University: Charlotte Bell, Tanikka Price, Detra Price-Dennis, Jacquie Scott, Samatha Wahome, and Ann Waliaula, for their very thoughtful feedback on drafts of this chapter, as well as their willingness to help us answer Alice Walker's call to write the texts that we wish we could read. We are grateful and honored by your wisdom and love. The coauthors would also like to acknowledge Norm Denzin and Yvonna Lincoln for their thoughtful reviews of this chapter.

2. In honor to the long traditions of proverbs in the African and African diasporic communities, we introduce each section of this chapter with an adinkra symbol from Ghana and its corresponding proverb, which represents the focus of the section. For further reading on the language of adinkra symbols, see Willis (1998).

3. "Coloring epistemologies" was a descriptive termed used by Scheurich and Young (1997) for the ways that traditional research epistemologies were being created; centered with/in identity markers such as race, culture, class, gender, national origin, and religion; and marshaled in research projects, primarily by people of color. Tyson's (1998) response in *Educational Researcher* critiques this notion on the basis of its unexamined assumptions and implicit racist implications. The idea of "paradigm proliferation" was an extension of the same argument, that is, that paradigms represented what is known and upheld as legitimizing knowledge in research. Thus, the discussion of proliferation of paradigms (again particularly given that they are being advanced by people of color) continued to also advance the same racist assumptions. See Dillard (2006b) for an examination and critique of the paradigm proliferation.

4. See Oyewumi (2004) *Conceptualizing Gender: Eurocentric Foundations of Feminist Concepts and the Challenge of African Epistemologies* for an in-depth discussion of the expansion of Euro/American feminism and its imperialistic outcomes for a one-sided radicalization of knowledge production, the dismissal of African realities, and consequent distortion of the human condition. See also Steady (2004), *An Investigative Framework for Gender Research in Africa in the New Millennium,* for her critical study of the ways that corporate globalization (led by the United States and Europe) and the persistence of Eurocentric concepts and paradigms continue to shape one-dimensional examinations of Africa and African women.

5. This is a borrowed title from Thich Nhat Hanh's (1999) book of the same name.

6. See Dillard (1994, 1996) for in-depth discussions of the pedagogical power of creative autobiography within teacher education, particularly for people of color.

7. See Wright (2003) for an excellent example of critical affirmation and Dillard (2003) for an explanation and response to that critical affirmation.

8. bell hooks used this as the title for her 1993 book, in which she names her Black women's support group by this name saying: "I felt the yam was a life-sustaining symbol of black kinship and community. Everywhere black women live in the world, we eat yam. It is a symbol of our diasporic connections. Yams provide nourishment for the body as food yet they are also used medicinally—to heal the body" (p. 13). The sisters in the Black feminist thought course also referred to ourselves by this name, in the spirit of solidarity and life-affirming connections to the African continent and diasporic communities we were connected to and responsible for.

9. Experience has and continues to be a space of contestation, particularly for poststructural scholars. As example, Jackson and Mazzei (2008) argue for the need for experience in autoethnography "that acknowledges the constraints of 'one' telling, that theorizes the ethics of such tellings, and that works the limits of the narrative 'I'" (p. 299), seeking instead a deconstructive autoethnography that puts experience under deconstruction, that confronts experience as questionable, incomplete, and problematic "rather than as a foundation for truth" (p. 304)." The authors further speak of the need, in a deconstructive autoethnography, to engage "a critique of the *relations of power* in the production of meaning from experience" (p. 304, emphasis in original). In this need to critique the relations of power, we see Jackson and Mazzei's call for a different sort of autoethnography (and by extension, qualitative research generally) as useful and important. However, even in their call for a critique of power, the framing of experience remains something that is strikingly singular and somewhat personal (even in their notion of a performative versus narrative "I"). This is counter to African-centered understandings of experience as *collective*, sacredly imbued with a connected spirituality to all that exists past, present, and future. As Dillard (2006b) suggests, we remain in agreement with Lubiano's (1991) notion that an African feminist "post" position must "be politically nuanced in a radical way, focusing on such differences and implications especially in moments of oppositional transgressions" (p. 160). Dillard goes on to suggest that one way that the African American presence in postmodernism can offer a critique is to engage in an alternative cultural discourse in keeping with the spirit of an African ethos. Hence, our notion of experience here is centered in the African notions of the collective, spiritual, and sacred.

◻ References

Aidoo, A. A. (1970). *No sweetness here and other stories.* Harlow, UK: Longman.

Aidoo, A.A. (1998). The African woman today. In O. Nnaemeka (Ed.). *Sisterhood: Feminisms and power from Africa to the diaspora.* Trenton, NJ: Africa World Press.

Aina, O. (1998). African women at the grassroots. In O. Nnaemeka (Ed.), *Sisterhood: Feminisms and power from Africa to the diaspora.* Trenton, NJ: Africa World Press.

Akyeampong, E., & Obeng, P. (2005). Spirituality, gender, and power in Asante history. In O. Oyewumi (Ed.), *African gender studies: A reader.* New York: Palgrave.

Alexander, M. J. (2005). *Pedagogies of crossing: Meditations on feminism, sexual politics, memory, and the sacred.* Durham, NC: Duke University Press.

Amadiume, I. (1987). *Male daughters, female husbands: Gender and sex in an African society.* London: Zed Books.

Bambara, T.C. (1970). *The Black woman: An anthology.* New York: New American Library.

Bambara, T. C. (1992). *The salt eaters.* New York: Random House. (Original work published 1980)

Bell-Scott, P. (1994). *Life notes: Personal writings by contemporary Black women.* New York: W. W. Norton.

Bray, Y. A. (2008). All the "Africans" are male, all the "sistas" are "American," but some of us resist: Realizing African feminism(s) as an Afrological research methodology. *The Journal of Pan African Studies, 2*(2), 58–73.

Busia, A. P. A. (1993). Languages of the self. In S. M. James & A. P. A. Busia (Eds.), *Theorizing Black feminisms: The visionary pragmatism of Black women* (pp. 204–209). London: Routledge.

Cannella, G. S., & Manuelito, K.D. (2008). Feminisms from unthought locations: Indigenous worldviews, marginalized feminisms, and revisioning an anticolonial social science. In N. K Denzin, Y. S. Lincoln, & L. T. Smith (Eds), *Handbook of critical and indigenous methodologies* (pp. 45–59). Thousand Oaks, CA: Sage.

Cleage, P. (2005). *We speak your names: A celebration.* New York: One World Books.

Collins, P. H. (1990). *Black feminist thought: Knowledge, consciousness, and the politics of empowerment.* New York: Routledge.

Collins, P. H. (2000). *Black feminist thought: Knowledge, consciousness, and the politics of empowerment* (2nd ed.). New York: Routledge.

Combahee River Collective. (1982). A black feminist statement. In G. T. Hull, P. B. Scott, & B. Smith (Eds.). *All the women are white, all the blacks are men, but some of us are brave: Black women's studies.* New York: The Feminist Press.

Cooper, A. J. (1892). *A voice from the South: By a woman from the South.* Xenia, OH: Aldine.

Denzin, N. K., Lincoln, Y. S., & Smith, L. T. (Eds.). (2008). *Handbook of critical and indigenous methodologies.* Thousand Oaks, CA: Sage.

Dillard, C. B. (1994). Beyond supply and demand: Critical pedagogy, ethnicity, and empowerment in recruiting teachers of color. *Journal of Teacher Education, 45,* 1–9.

Dillard, C. B. (1996). From lessons of self to lessons of others: Exploring creative autobiography in multicultural learning and teaching. *Multicultural Education, 4*(2), 33–37.

Dillard, C. B. (2000). The substance of things hoped for, the evidence of things not seen: Examining an endarkened feminist epistemology in educational research and leadership. *The International Journal of Qualitative Studies in Education, 13,* 661–681.

Dillard, C. B. (2003). Cut to heal, not to bleed: A response to Handel Wright's "An endarkened feminist epistemology? Identity, difference, and the politics of representation in educational research". *International Journal of Qualitative Studies in Education, 16*(2), 227–232.

Dillard, C. B. (2006a). *On spiritual strivings: Transforming an African American woman's academic life.* New York: SUNY.

Dillard, C. B. (2006b). When the music changes, so should the dance: Cultural and spiritual considerations in paradigm "proliferation." *International Journal of Qualitative Studies in Education, 19*(1), 59–76.

Dillard, C. B. (2008). When the ground is black, the ground is fertile: Exploring endarkened feminist epistemology and healing methodologies in the spirit. In N. K. Denzin, Y. S. Lincoln, & L. T. Smith (Eds.). *Handbook of critical sand indigenous methodologies* (pp. 277–292). Thousand Oaks, CA: Sage.

Dillard, C. B., Tyson, C. A., & Abdur-Rashid, D. (2000b). My soul is a witness: Affirming pedagogies of the spirit. *International Journal of Qualitative Studies in Education, 13,* 447–462.

Fernandes, L. (2003). *Transforming feminist practice: Non-violence, social justice, and the possibilities of a spiritualized feminism.* San Francisco: Aunt Lute Books.

Guy-Sheftall, B. (1995). *Words of fire: An anthology of African-American feminist thought.* New York: The New Press.

Hanh, T. N. (1999). *Call me by my true names: The collected poems of Thich Nhat Hanh.* Berkeley, CA: Parallax Press.

Hilliard, A. G. (1995). *The maroon within us: Selected essays on African American community socialization.* Baltimore: Black Classic Press.

hooks, b. (1981). *Ain't I a woman? Black women and feminism.* Cambridge, MA: South End Press.

hooks, b. (1993). *Sisters of the yam: Black women and self-recovery.* Cambridge, MA: South End Press.

hooks, b. (1994). *Teaching to transgress.* New York: Routledge.

hooks, b. (2000). *All about love: New visions.* New York: William Morrow.

hook, b. (2008). *Belonging: A culture of place.* New York: Routledge.

hooks, b., & West, C. (1991). *Breaking bread: Insurgent Black intellectual life.* Cambridge, MA: South End Press.

Hudson-Weems, C. (1995). *Africana womanism: Reclaiming ourselves* (3rd ed.). Troy: MI: Bedford.

Hudson-Weems, C. (1998a). Africana womanism. In O. Nnaemeka (Ed.), *Sisterhood: Feminisms and power from Africa to the diaspora* (pp. 149–162). Trenton, NJ: Africa World Press.

Hudson-Weems, C. (1998b). Self naming and self definition: An agenda for survival (pp. 450–452). In O. Nnaemeka (Ed.), *Sisterhood: Feminisms and power from Africa to the diaspora* (pp.149–162). Trenton, NJ: Africa World Press.

Hull, A. G. (2001). *Soul talk: The new spirituality of African American women.* Rochester, VT: Inner Traditions.

Hull, G., Bell-Scott, P., & Smith, B. (1982). *All the women are white, all the blacks are men, but some of us are brave: Black women's studies.* New York: The Feminist Press.

Hurston, Z. N. (1978). *Their eyes were watching God.* Urbana: University of Illinois Press.

Jackson, A. Y., & Mazzei, L. A. (2008). Experience and "I" in autoethnography: A deconstruction. *International Review of Qualitative Research, 1*(3), 299–318.

Keating, A. (2008). "I'm a citizen of the universe": Gloria Anzaldúa's spiritual activism as catalyst for social change. *Feminist Studies, 34*(1/2), 53–69.

Ladson-Billings, G. (2000). Racialized discourses and ethnic epistemologies. In N. K. Denzin & Y. S. Lincoln (Eds.), *Handbook of qualitative research* (2nd ed., pp. 257–277). Thousand Oaks, CA: Sage.

Ladson-Billings, G., & Donner, J. (2005). The moral activist role of critical race theory. In N. K. Denzin & Y. S. Lincoln (Eds.), *The SAGE handbook of qualitative research* (pp. 279–302). Thousand Oaks, CA: Sage.

Latta, J. M. (1992). *Sacred songs as history* (Interview with Bernice Johnson Reagon). Recorded August 4, 1992. Washington, DC: National Public Radio Archives, Wade in the Water Program.

Lorde, A. (1970). *Cables to rage.* London: Paul Breman Limited.

Lorde, A. (1984). *Sister outsider.* Freedom, CA: The Crossing Press.

Lubiano, W. (1991). Shuckin' off the African-American native other: What's "po-mo" got to do with it? *Cultural Critique, 18,* 149–186.

Mama, A. (2007). Critical connections: Feminist studies in African contexts. In A. Cornwall, E. Harrison, & A. Whitehead (Eds.), *Feminisms in development: Contradictions, contestations and challenge* (p. 152). London: Zeal Books.

Marshall, P. (1984). *Praisesong for the widow.* New York: E. P. Dutton.

Meyer, M. A. (2008). Indigenous and authentic: Hawaiian epistemology and the triangulation of meaning. In N. K. Denzin, Y. S. Lincoln, & L. T. Smith (Eds.). *Handbook of critical and indigenous methodologies* (pp. 217–232). Thousand Oaks, CA: Sage.

Mohanty, C. T. (1991). Cartographies of struggle: Third world women and the politics of feminism. In C. T. Mohanty, A Russo, & L. Torres (Eds.), *Third world women and the politics of feminism* (pp. 1–50). Bloomington: Indiana University Press.

Mohanty, C. T., Russo, A., & Torres, L. (Eds.). (1991). *Third world women and the politics of feminism.* Bloomington: Indiana University Press.

Moraga, C., & Anzaldúa, G. (1981). *This bridge called my back: Writings by radical women of color.* Watertown, MA: Persephone Press.

Morrison, T. (1970). *The bluest eye.* New York: Vintage Books.

Nnaemeka, O. (1998). *Sisterhood: Feminisms and power—from Africa to the diaspora.* Trenton, NJ: Africa World Press.

Nwapa, F. (1966). *Efuru.* London: Cox & Wyman.

Nzegwu, N. (2006). *Family matters: Feminist concepts in African philosophy of culture.* Albany: SUNY Press.

Ogala, M. (1994). *The river and the source.* Nairobi, Kenya: Focus Publications.

Ogundipe-Leslie, O. (1994). *Re-creating ourselves: African women & critical transformations.* Trenton: NJ: Africa World Press.

Okpalaoka, C. L. (2009). *Endarkened feminism and qualitative research: Colonization and connectedness in Black women's experiences.* Unpublished manuscript.

Omolade, B. (1994). *The rising song of African American women.* New York: Routledge.

Oyewumi, O. (1997). *The invention of women: Making an African sense of Western gender discourses.* Minneapolis: University of Minnesota Press.

Oyewumi, O. (2004). Conceptualizing gender: Eurocentric foundations of feminist concepts and the challenge of African epistemologies. In *CODESRIA, African gender scholarship: Concept, methodologies, and paradigms.* Dakar, Senegal: CODESRIA.

Reason, P. (1993). Reflections on sacred experience and sacred sciences. *Journal of Management Inquiry, 2,* 10–27.

Richards, D. (1980). *Let the circle be unbroken: The implications of African spirituality in the diaspora.* Lawrenceville, NJ: The Red Sea Press.

Ryan, J. S. (2005). *Spirituality as ideology in Black women's film and literature.* Charlottesville: University of Virginia Press.

Saavedra, C. M., & Nymark, E. D. (2008). Borderland-Mestizaje feminism: The new tradition. In N. K. Denzin, Y. S. Lincoln, & L. T. Smith (Eds.), *Handbook of critical and indigenous methodologies* (pp. 255–276). Thousand Oaks, CA: Sage.

Scheurich, J., & Young, M. (1997). Coloring epistemologies: Are our research epistemologies racially biased? *Educational Researcher, 26*(4), 4–16.

Smith, B. (1983). *Home girls: A Black feminist anthology.* New York: Kitchen Table: Women of Color Press.

Sofola, Z. (1998). Feminism and African womanhood. In O. Nnaemeka (Ed.), *Sisterhood: Feminisms and power—from Africa to the diaspora.* Trenton, NJ: Africa World Press.

Some, M. P. (1994). *Of water and the spirit: Ritual, magic, and initiation in the life of an African shaman.* New York: G. P. Putnam.

Steady, F. C. (1981). The Black woman cross-culturally: An overview. In F. C. Steady (Ed.), *The Black woman cross-culturally.* Cambridge, MA: Schenkman.

Steady, F. C. (1996). African feminism: A worldwide perspective. In R. Terbog-Penn & R. Benton (Eds.), *Women in Africa: A reader* (2nd ed.). Washington, DC: Howard University Press.

Steady, F. C. (2004). An investigative framework for gender research in Africa in the new millennium. In *CODESRIA, African gender scholarship: Concepts, methodologies, and paradigms* (pp. 42–60). Dakar, Senegal: CODESRIA.

Taiwo, O. (2003). Reflections on the poverty of theory. In O. Oyewumi (Ed.), *African women and feminism: Reflecting on the politics of sisterhood* (pp. 45–66). Trenton, NJ: Africa World Press.

Tyson, C. A. (1998). A response to "Coloring epistemologies: Are our qualitative research epistemologies racially biased?" *Educational Researcher, 27,* 21–22.

Wade-Gayles, G. (Ed.). (1995). *My soul is a witness: African-American women's spirituality.* Boston: Beacon.

Walker, A. (1983). *In search of our mother's gardens: Womanist prose.* San Diego: Harvest Books.

Walker, A. (2006). *We are the ones we have been waiting for: Inner light in a time of darkness.* New York: The New Press.

Wallace, M. (1982). A black feminist's search for sisterhood. In G. T. Hull, P. B. Scott, & B. Smith (Eds.). *All the women are white, all the blacks are men, but some of us are brave: Black women's studies.* New York: The Feminist Press.

Wekker, G. (1997). One finger does not drink okra soup: Afro-Surinamese women and critical agency. In M. J. Alexander & C. T. Mohanty (Eds.), *Feminist geneologies, colonial legacies, democratic futures* (pp. 330–352). New York: Routledge.

Willis, B. (1998). *The Adinkra dictionary: A visual primer on the language of Adinkra.* Washington, DC: The Pyramid Complex.

Wright, H. K. (2003). An endarkened feminist epistemology? Identity, difference, and the politics of respresentation in educational research. *International Journal of Qualitative Research, 16*(2), 197–214.

9

CRITICAL PEDAGOGY AND QUALITATIVE RESEARCH

Moving to the Bricolage

Joe L. Kincheloe, Peter McLaren, and Shirley R. Steinberg[1]

CRITICALITY AND RESEARCH

Over the past 35 years of our involvement in critical theory, critical pedagogy, and critical research, we have been asked to explain how critical theory relates to pedagogy. We find that question difficult to answer because (1) there are many critical theories; (2) the critical tradition is always changing and evolving; and (3) critical theory attempts to avoid too much specificity, as there is room for disagreement among critical theorists. To lay out a set of fixed characteristics of the position is contrary to the desire of such theorists to avoid the production of blueprints of sociopolitical and epistemological beliefs. Given these disclaimers, we will now attempt to provide one idiosyncratic "take" on the nature of critical theory and critical research in the second decade of the 21st century. Please note that this is our subjective analysis and that there are many brilliant critical theorists who disagree with our pronouncements. We tender a description of an ever-evolving criticality, a reconceptualized critical theory that was critiqued and overhauled by the "post-discourses" of the last quarter of the 20th century and has been further extended in the 21st century (Collins, 1995; Giroux, 1997; Kellner, 1995; Kincheloe, 2008b; McLaren & Kincheloe, 2007; Roman & Eyre, 1997; Ryoo & McLaren, 2010; Steinberg & Kincheloe, 1998; Tobin, 2009; Weil & Kincheloe, 2004).

A reconceptualized critical theory questions the assumption that societies such as Australia, Canada, Great Britain, New Zealand, and the United States, along with some nations in the European Union and Asia, are unproblematically democratic and free (Steinberg, 2010). Over the 20th century, especially after the early 1960s, individuals in these societies were acculturated to feel comfortable in relations of domination and subordination rather than equality and independence. Given the social and technological changes of the last half of the century, which led to new forms of information production and access, critical theorists argued that questions of self-direction and democratic egalitarianism should be reassessed. Researchers informed by the postdiscourses (e.g., postmodernism, critical feminism, poststructuralism) came to understand that individuals' view of themselves and the world were even more influenced by social and historical forces than previously believed. Given the changing social and informational conditions of late-20th century and early-21st century, media-saturated Western culture (Steinberg, 2004a, 2004b), critical theorists have needed new ways of researching and analyzing the construction of individuals (Agger, 1992; Flossner & Otto, 1998; Giroux, 2010; Hammer & Kellner, 2009; Hinchey, 2009; Kincheloe, 2007; Leistyna, Woodrum, & Sherblom, 1996; Quail, Razzano, & Skalli, 2004; Skalli, 2004; Steinberg, 2007, 2009; Wesson & Weaver, 2001).

Partisan Research in a "Neutral" Academic Culture

In the space available here, it is impossible to do justice to all of the critical traditions that have drawn inspiration from Karl Marx; Immanuel Kant; Georg Wilhelm Friedrich Hegel; Max Weber; the Frankfurt School theorists; Continental social theorists such as Jean Baudrillard, Michel Foucault, Jürgen Habermas, and Jacques Derrida; Latin American thinkers such as Paulo Freire; French feminists such as Luce Irigaray, Julia Kristeva, and Hélène Cixous; or Russian socio-sociolinguists such as Mikhail Bakhtin and Lev Vygotsky—most of whom regularly find their way into the reference lists of contemporary critical researchers. Today, there are criticalist schools in many fields, and even a superficial discussion of the most prominent of these schools would demand much more space than we have available (Chapman, 2010; Flecha, Gomez, & Puigvert, 2003).

The fact that numerous books have been written about the often-virulent disagreements among members of the Frankfurt School only heightens our concern with the "packaging" of the different criticalist schools. Critical theory should not be treated as a universal grammar of revolutionary thought objectified and reduced to discrete formulaic pronouncements or strategies. Obviously, in presenting our version of a reconceptualized critical theory or an evolving criticality, we have defined the critical tradition broadly for the purpose of generating understanding; as we asserted earlier, this will trouble many critical researchers. In this move, we decided to focus on the underlying commonality among critical schools of thought at the cost of focusing on differences. This is always risky business in terms of suggesting a false unity or consensus where none exists, but such concerns are unavoidable in a survey chapter such as this.

We are defining a criticalist as a researcher, teacher, or theorist who attempts to use her or his work as a form of social or cultural criticism and who accepts certain basic assumptions:

- All thought is fundamentally mediated by power relations that are social and historically constituted;
- Facts can never be isolated from the domain of values or removed from some form of ideological inscription;
- The relationship between concept and object and between signifier and signified is never stable or fixed and is often mediated by the social relations of capitalist production and consumption;
- Language is central to the formation of subjectivity (conscious and unconscious awareness);
- Certain groups in any society and particular societies are privileged over others and, although the reasons for this privileging may vary widely, the oppression that characterizes contemporary societies is most forcefully reproduced when subordinates accept their social status as natural, necessary, or inevitable;
- Oppression has many faces, and focusing on only one at the expense of others (e.g., class oppression versus racism) often elides the interconnections among them; and finally
- Mainstream research practices are generally, although most often unwittingly, implicated in the reproduction of systems of class, race, and gender oppression (De Lissovoy & McLaren, 2003; Gresson, 2006; Kincheloe & Steinberg, 1997; Rodriguez and Villaverde, 2000; Steinberg, 2009; Villaverde, 2007; Watts, 2008, 2009a).

In today's climate of blurred disciplinary genres, it is not uncommon to find literary theorists doing anthropology and anthropologists writing about literary theory, political scientists trying their hand at ethnomethodological analysis, or philosophers doing Lacanian film criticism. All of these inter- and cross-disciplinary moves are examples of what has been referred to as *bricolage*—a key innovation, we argue, in an evolving criticality. We will explore this dynamic in relation to critical research later in this chapter. We offer this observation about blurred genres, not as an excuse to be wantonly eclectic in our treatment of the critical tradition but to make the point that any attempts to delineate critical theory as discrete schools of analysis will fail to capture the evolving hybridity endemic to contemporary critical analysis (Denzin, 1994; Denzin & Lincoln, 2000; Kincheloe, 2001a, 2008b; Kincheloe & Berry, 2004; Steinberg, 2008, 2010, 2011).

Critical research can be understood best in the context of the empowerment of individuals. Inquiry that aspires to the name "critical" must be connected to an attempt to confront the injustice of a particular society or public sphere within the society. Research becomes a transformative endeavor unembarrassed by the label "political" and unafraid to consummate a relationship with emancipatory consciousness. Whereas traditional researchers cling to the guardrail of neutrality, critical researchers frequently announce their partisanship in the struggle for a better world (Chapman, 2010; Grinberg, 2003; Horn, 2004; Kincheloe, 2001b, 2008b).

Critical Pedagogy Informing Social Research

The work of Brazilian educator Paulo Freire is instructive in relation to constructing research that contributes to the struggle for a better world. The research of the authors of this chapter has been influenced profoundly by the work of Freire (1970, 1972, 1978, 1985). Concerned with human suffering and the pedagogical and knowledge work that helped expose the genesis of it, Freire modeled critical theoretical research throughout his career. In his writings about research, Freire maintained that there were no traditionally defined objects of his research—he insisted on involving the people he studied as *partners* in the research process. He immersed himself in their ways of thinking and modes of perception, encouraging them to begin thinking about their own thinking. Everyone involved in Freire's critical research, not just the researcher, joined in the process of investigation, examination, criticism, and reinvestigation—all participants and researchers learned to see more critically, think at a more critical level, and to recognize the forces that subtly shape their lives. Critiquing traditional methods of research in schools, Freire took a critical pedagogical approach to research that serves to highlight its difference from traditional research (Kirylo, 2011; Mayo, 2009; Tobin & Llena, 2010).

After exploring the community around the school and engaging in conversations with community members, Freire constructed generative themes designed to tap into issues that were important to various students in his class. As data on these issues were brought into the class, Freire became a problem poser. In this capacity, Freire used the knowledge he and his students had produced around the generative themes to construct questions. The questions he constructed were designed to teach the lesson that no curriculum or knowledge in general was beyond examination. We need to ask questions of all knowledge, Freire argued, because all data are shaped by the context and by the individuals that produced them. Knowledge, contrary to the pronouncements of many educational leaders, does not transcend culture or history.

In the context of reading the word and the world and problem-posing existing knowledge, critical educators reconceptualize the notion of literacy. Myles Horton spoke of the way he read books with students in order "to give testimony to the students about what it means to read a text" (Horton & Freire, 1990). Reading is not an easy endeavor, Horton continued, for to be a good reader is to view reading as a form of research. Reading becomes a mode of finding something, and finding something, he concluded, brings a joy that is directly connected to the acts of creation and re-creation. One finds in this reading that the word and world process typically goes beyond the given, the common sense of everyday life. Critical pedagogical research must have a mandate to represent a form of reading that understood not only the words on the page but the unstated dominant ideologies hidden between the sentences as well.

Going beyond is central to Freirean problem posing. Such a position contends that the school curriculum should in part be shaped by problems that face teachers and students in their effort to live just and ethical lives (Kincheloe, 2004). Such a curriculum promotes students as researchers (Steinberg & Kincheloe, 1998) who engage in critical analysis of the forces that shape the world. Such critical analysis engenders a healthy and creative skepticism on the part of students. It moves them to problem pose and to be suspicious of neutrality claims in textbooks; it induces them to look askance at, for example, oil companies' claims in their TV commercials that they are and have always been environmentally friendly organizations. Students and teachers who are problem posers reject the traditional student request to the teacher: "just give us the facts, the truth, and we'll give it back to you." On the contrary, critical students and teachers ask in the spirit of Freire and Horton: "Please support us in our explorations of the world."

By promoting problem posing and student research, teachers do not relinquish their authority in the classroom. Over the last couple of decades, several teachers and students have misunderstood the subtlety of the nature of teacher authority in a critical pedagogy. In the last years of his life, Freire was very concerned with this issue and its misinterpretation by those operating in his name. Teachers, he told us, cannot deny their position of authority in such a classroom. It is the teacher, not the students, who evaluates student work, who is responsible for the health, safety, and learning of students. To deny the role of authority the teacher occupies is insincere at best, dishonest at worst. Critical teachers, therefore, must admit that they are

in a position of authority and then demonstrate that authority in their actions in support of students. One action involves the ability to conduct research and produce knowledge. The authority of the critical teacher is dialectical; as teachers relinquish the authority of truth providers, they assume the mature authority of facilitators of student inquiry and problem posing. In relation to such teacher authority, students gain their freedom—they gain the ability to become self-directed human beings capable of producing their own knowledge (Kirylo, 2011; Siry & Lang, 2010).

Freire's own work was rooted in both liberation theology and a dialectical materialist epistemology (Au, 2007), both of which were indebted to Marx's own writings and various Marxist theorists. Standard judgments against Marxism as economistic, productivist, and deterministic betray an egregious and scattershot understanding of Marxist epistemology, his critique of political economy, and Marx's dialectical method of analyzing the development of capitalism and capitalist society. We assert that the insights of Marx and those working within the broad parameters of the Marxist tradition are foundational for any critical research (Lund & Carr, 2008); Marxism is a powerful theoretical approach to explaining, for instance, the origins of racism and the reasons for its resiliency (McLaren, 2002). Many on the left today talk about class as if it is one of many oppressions, often describing it as "classism." But class is not an "ism." It is true that class intersects with race, and gender, and other antagonisms. And while clearly those relations of oppression can reinforce and compound each other, they are grounded in the material relations shaped by capitalism and the economic exploitation that is the motor force of any capitalist society (Dale & Hyslop-Margison, 2010; Macrine, McLaren, & Hill, 2009).

To seriously put an end to racism, and shatter the hegemony of race, racial formations, the racial state, and so on, we need to understand class as an objective process that interacts upon multiple groups and sectors in various historically specific ways. When conjoined with an insightful class analysis, the concept of race and the workings of racism can be more fully understood and racism more forcefully contested and as a result more powerful transformative practices can be mobilized. Class and race are viewed here as co-constitutive and must be understood as dialectically interrelated (McLaren & Jarramillo, 2010).

Teachers as Researchers

In the conservative educational order of mainstream schooling, knowledge is something that is produced far away from the school by experts in an exalted domain. This must change if a critical reform of schooling is to exist. Teachers must have more voice and more respect in the culture of education. Teachers must join the culture of researchers if a new level of educational rigor and quality is ever to be achieved. In such a democratized culture, critical teachers are scholars who understand the power implications of various educational reforms. In this context, they appreciate the benefits of research, especially as they relate to understanding the forces shaping education that fall outside their immediate experience and perception. As these insights are constructed, teachers begin to understand what they know from experience. With this in mind they gain heightened awareness of how they can contribute to the research on education. Indeed, they realize that they have access to understandings that go far beyond what the expert researchers have produced. In the critical school culture, teachers are viewed as learners—not as functionaries who follow top-down orders without question. Teachers are seen as researchers and knowledge workers who reflect on their professional needs and current understandings. They are aware of the complexity of the educational process and how schooling cannot be understood outside of the social, historical, philosophical, cultural, economic, political, and psychological contexts that shape it. Scholar teachers understand that curriculum development responsive to student needs is not possible when it fails to account for these contexts.

Critical teacher/researchers explore and attempt to interpret the learning processes that take place in their classrooms. "What are its psychological, sociological, and ideological effects?" they ask. Thus, critical scholar teachers research their own professional practice. With empowered scholar teachers working in schools, things begin to change. The oppressive culture created in our schools by top-down content standards, for example, is challenged. In-service staff development no longer takes the form of "this is what the expert researchers found—now go implement it." Such staff development in the critical culture of schooling gives way to teachers who analyze and contemplate the power of each other's ideas. Thus, the new critical culture of school takes on the form of a "think tank that teaches students," a learning community. School administrators are amazed by what can happen when they support learning activities for both students and teachers. Principals and curriculum developers watch as teachers develop projects that encourage collaboration and shared research. There is an alternative, advocates of critical pedagogy argue, to top-down standards with their deskilling of teachers and the dumbing-down of students (Jardine, 1998; Kincheloe, 2003a, 2003b, 2003c; Macedo, 2006).

Promoting teachers as researchers is a fundamental way of cleaning up the damage of deskilled models of teaching that infantilize teachers by giving them scripts to read to their students. Deskilling of teachers and the stupidification (Macedo, 2006) of the curriculum take place when teachers are seen as receivers, rather than producers, of knowledge. A vibrant professional culture depends on a group of practitioners who have the freedom to continuously reinvent themselves via their research and knowledge production. Teachers engaged in critical practice find it difficult to allow top-down content standards and their poisonous effects to go unchallenged. Such teachers cannot abide the deskilling and reduction in professional status that accompany these top-down reforms. Advocates of critical pedagogy understand that teacher empowerment does not occur just because we wish it to do so. Instead, it takes place when teachers develop the knowledge-work skills, the power literacy, and the pedagogical abilities befitting the calling of teaching. Teacher research is a central dimension of a critical pedagogy (Porfilio & Carr, 2010).

Teachers as Researchers of Their Students

A central aspect of critical teacher research involves studying students so they can be better understood and taught. Freire argued that all teachers need to engage in a constant dialogue with students, a dialogue that questions existing knowledge and problematizes the traditional power relations that have served to marginalize specific groups and individuals. In these research dialogues with students, critical teachers listen carefully to what students have to say about their communities and the problems that confront them. Teachers help students frame these problems in a larger social, cultural, and political context in order to solve them.

In this context, Freire argued that teachers uncover materials and generative themes based on their emerging knowledge of students and their sociocultural backgrounds (Mayo, 2009; Souto-Manning, 2009). Teachers come to understand the ways students perceive themselves and their interrelationships with other people and their social reality. This information is essential to the critical pedagogical act, as it helps teachers understand how they make sense of schooling and their lived worlds. With these understandings in mind, critical teachers come to know what and how students make meaning. This enables teachers to construct pedagogies that engage the impassioned spirit of students in ways that move them to learn what they do not know and to identify what they want to know (A. Freire, 2000; Freire & Faundez, 1989; Janesick, 2010; Kincheloe, 2008b; Steinberg & Kincheloe, 1998; Tobin, in press).

It is not an exaggeration to say that before critical pedagogical research can work, teachers must understand what is happening in the minds of their students. Advocates of various forms of critical teaching recognize the importance of understanding the social construction of student consciousness, focusing on motives, values, and emotions. Operating within this critical context, the teacher-researcher studies students as living texts to be deciphered. The teacher-researcher approaches them with an active imagination and a willingness to view students as socially constructed beings. When critical teachers have approached research on students from this perspective, they have uncovered some interesting information. In a British action research project, for example, teachers used student diaries, interviews, dialogues, and shadowing (following students as they pursue their daily routines at school) to uncover a student preoccupation with what was labeled a second-order curriculum. This curriculum involved matters of student dress, conformance to school rules, strategies of coping with boredom and failure, and methods of assuming their respective roles in the school pecking order. Teacher-researchers found that much of this second-order curriculum worked to contradict the stated aims of the school to respect the individuality of students, to encourage sophisticated thinking, and to engender positive self-images. Students often perceived that the daily lessons of teachers (the intentional curriculum) were based on a set of assumptions quite different from those guiding out-of-class teacher interactions with students. Teachers consistently misread the anger and hostility resulting from such inconsistency. Only in an action research context that values the perceptions of students could such student emotions be understood and addressed (Hooley, 2009; Kincheloe, 2001a; Sikes, 2008; Steinberg, 2000, 2009; Vicars, 2008).

By using IQ tests and developmental theories derived from research on students from dominant cultural backgrounds, schools not only reflect social stratification but also extend it. This is an example of school as an institution designed for social benefit actually exerting hurtful influences. Teachers involved in the harmful processes most often do not intentionally hurt students; they are merely following the dictates of their superiors and the rules of the system. Countless good teachers work every day to subvert the negative effects of the system but need help from like-minded colleagues and organizations. Critical pedagogical research works to provide such assistance to teachers who want to mitigate the effects of power on their students. Here schools as political institutions merge with critical pedagogy's concern with creating a social and educational vision to help teachers direct their own professional practice. Anytime teachers develop a pedagogy, they are concurrently constructing a political vision. The two acts are inseparable (Kincheloe, 2008b; Wright & Lather, 2006).

Unfortunately, those who develop noncritical pedagogical research can be unconscious of the political inscriptions embedded within them. A district supervisor who writes a curriculum in social studies, for example, that demands the simple transference of a body of established facts about the great men and great events of American history is also teaching a political lesson that upholds the status quo (Keesing-Styles, 2003; McLaren & Farahmandpur, 2003, 2006). There is no room for teacher-researchers in such a curriculum to explore alternate sources, to compare diverse historical interpretations, or to do research of their own and produce knowledge that may conflict with prevailing interpretations. Such acts of democratic citizenship may be viewed as subversive and anti-American by the supervisor and the district education office. Indeed, such personnel may be under pressure from the state department of education to construct a history curriculum that is inflexible, based on the status quo, unquestioning in its approach, "fact-based," and teacher-centered. Dominant power operates in numerous and often hidden ways (Nocella, Best, & McLaren, 2010; Watts, 2006, 2009a, 2009b).

Traditional researchers see their task as the description, interpretation, or reanimation of a slice of reality; critical pedagogical researchers often regard their work as a first step toward forms of political action that can redress the injustices found in the field site or constructed in the very act of research itself. Horkheimer (1972) puts it succinctly when he argues that critical theory and research are never satisfied with merely increasing knowledge (see also Agger, 1998; Britzman, 1991; Giroux, 1983, 1988, 1997; Kincheloe, 2003c, 2008a, 2008b; Kincheloe & Steinberg, 1993; Quantz, 1992; Shor, 1996; Villaverde & Kincheloe, 1998; Wexler, 2008). Research in the critical tradition takes the form of self-conscious criticism—self-conscious in the sense that researchers try to become aware of the ideological imperatives and epistemological presuppositions that inform their research as well as their own subjective, intersubjective, and normative reference claims. Critical pedagogical researchers enter into an investigation with their assumptions on the table, so no one is confused concerning the epistemological and political baggage they bring with them to the research site.

On detailed analysis, critical researchers may change these assumptions. Stimulus for change may come from the critical researchers' recognition that such assumptions are not leading to emancipatory actions. The source of this emancipatory action involves the researchers' ability to expose the contradictions of the world of appearances accepted by the dominant culture as natural and inviolable (Giroux, 1983, 1988, 1997; Kincheloe, 2008b; McLaren, 1992, 1997; San Juan, 1992; Zizek, 1990). Such appearances may, critical researchers contend, conceal social relationships of inequality, injustice, and exploitation. If we view the violence we find in classrooms not as random or isolated incidents created by aberrant individuals willfully stepping out of line in accordance with a particular form of social pathology, but as possible narratives of transgression and resistance, then this could indicate that the "political unconscious" lurking beneath the surface of everyday classroom life is not unrelated to practices of race, class, and gender oppression but rather intimately connected to them. By applying a critical pedagogical lens within research, we create an empowering qualitative research, which expands, contracts, grows, and questions itself within the theory and practice examined.

The Bricolage

It is with our understanding of critical theory and our commitment to critical social research and critical pedagogy that we identify the bricolage as an emancipatory research construct. Ideologically grounded, the bricolage reflects an evolving criticality in research. Norman K. Denzin and Yvonna S. Lincoln (2000) use the term in the spirit of Claude Lévi-Strauss (1968 and his lengthy discussion of it in *The Savage Mind)*. The French word *bricoleur* describes a handyman or handywoman who makes use of the tools available to complete a task (Harper, 1987; Steinberg, 2011). Bricolage implies the fictive and imaginative elements of the presentation of all formal research. The bricolage can be described as the process of getting down to the nuts and bolts of multidisciplinary research. Research knowledges such as ethnography, textual analysis, semiotics, hermeneutics, psychoanalysis, phenomenology, historiography, discourse analysis combined with philosophical analysis, literary analysis, aesthetic criticism, and theatrical and dramatic ways of observing and making meaning constitute the methodological bricolage. In this way, bricoleurs move beyond the blinders of particular disciplines and peer through a conceptual window to a new world of research and knowledge production (Denzin, 2003; Kincheloe & Berry, 2004; Steinberg, 2011).

Bricolage, in a contemporary sense, is understood to involve the process of employing these methodological processes as they are needed in the unfolding context of the research situation. While this interdisciplinary feature is central to any notion of the bricolage, critical qualitative researchers must go beyond this dynamic. Pushing to a new conceptual terrain, such an eclectic process raises numerous issues that researchers must deal with to maintain theoretical coherence and epistemological innovation. Such multidisciplinarity demands a new level of research self-consciousness and awareness of the numerous contexts in which any researcher is operating. As one labors to expose the various structures that covertly shape our own and other scholars' research narratives, the bricolage highlights the relationship between a researcher's ways of seeing and the social location of his or her personal history. Appreciating research as a power-driven act, the critical researcher-as-bricoleur abandons the quest for some naïve concept of realism, focusing instead on the clarification of his or her position in the web of reality and the social locations of other researchers and the ways they shape the production and interpretation of knowledge.

In this context, bricoleurs move into the domain of complexity. The bricolage exists out of respect for the complexity of the lived world and the complications of power. Indeed, it is grounded on an epistemology of complexity. One dimension of this complexity can be illustrated by the relationship between research and the domain of social theory. All observations of the world are shaped either consciously or unconsciously by social theory—such theory provides the framework that highlights or erases what might be observed. Theory in a modernist empiricist mode is a way of understanding that operates without variation in every context. Because theory is a cultural and linguistic artifact, its interpretation of the object of its observation is inseparable from the historical dynamics that have shaped it (Austin & Hickey, 2008). The task of the bricoleur is to attack this complexity, uncovering the invisible artifacts of power and culture and documenting the nature of their influence not only on their own works, but on scholarship in general.

In this process, bricoleurs act on the concept that theory is not an explanation of nature—it is more an explanation of our relation to nature.

In its hard labors in the domain of complexity, the bricoleur views research methods actively rather than passively, meaning that we actively construct our research methods from the tools at hand rather than passively receiving the "correct," universally applicable methodologies. Avoiding modes of reasoning that come from certified processes of logical analysis, bricoleurs also steer clear of preexisting guidelines and checklists developed outside the specific demands of the inquiry at hand. In its embrace of complexity, the bricolage constructs a far more active role for humans both in shaping reality and in creating the research processes and narratives that represent it. Such an active agency rejects deterministic views of social reality that assume the effects of particular social, political, economic, and educational processes. At the same time and in the same conceptual context, this belief in active human agency refuses standardized modes of knowledge production (Bresler & Ardichvili, 2002; Kincheloe & Berry, 2004; McLeod, 2000; Selfe & Selfe, 1994; Steinberg, 2010, 2011; Wright, 2003a).

Some of the best work in the study of social complexity is now taking place in the qualitative inquiry of numerous fields including sociology, cultural studies, anthropology, literary studies, marketing, geography, media studies, nursing, informatics, library studies, women's studies, various ethnic studies, education, and nursing. Denzin and Lincoln (2000) are acutely aware of these dynamics and refer to them in the context of their delineation of the bricolage. Yvonna Lincoln (2001), in her response to Joe L. Kincheloe's development of the bricolage, maintains that the most important border work between disciplines is taking place in feminism and race-ethnic studies.

In many ways, there is a form of instrumental reason, of rational irrationality, in the use of passive, external, monological research methods. In the active bricolage, we bring our understanding of the research context together with our previous experience with research methods. Using these knowledges, we *tinker* in the Lévi-Straussian sense with our research methods in field-based and interpretive contexts (Steinberg, in press). This tinkering is a high-level cognitive process involving construction and reconstruction, contextual diagnosis, negotiation, and readjustment. Researchers' interaction with the objects of their inquiries, bricoleurs understand, are always complicated, mercurial, unpredictable, and, of course, complex. Such conditions negate the practice of planning research strategies in advance. In lieu of such rationalization of the process, bricoleurs enter into the research act as methodological negotiators. Always respecting the demands of the task at hand, the bricolage, as conceptualized here, resists its placement in concrete as it promotes its elasticity. In light of Lincoln's (2001) discussion of two types of bricoleurs, (1) those who are committed to research eclecticism, allowing circumstance to shape methods employed, and (2) those who want to engage in the genealogy/ archeology of the disciplines with some grander purpose in mind, critical researchers are better informed as to the power of the bricolage. Our purpose entails both of Lincoln's articulations of the role of the bricoleur (Steinberg & Kincheloe, 2011).

Research method in the bricolage is a concept that receives more respect than in more rationalistic articulations of the term. The rationalistic articulation of method subverts the deconstruction of wide varieties of unanalyzed assumptions embedded in passive methods. Bricoleurs, in their appreciation of the complexity of the research process, view research method as involving far more than procedure. In this mode of analysis, bricoleurs come to understand research method as also a technology of justification, meaning a way of defending what we assert we know and the process by which we know it. Thus, the education of critical researchers demands that everyone take a step back from the process of learning research methods. Such a step back allows us a conceptual distance that produces a critical consciousness. Such a consciousness refuses the passive acceptance of externally imposed research methods that tacitly certify modes justifying knowledges that are decontextualized, reductionistic, and inscribed by dominant modes of power (Denzin & Lincoln, 2000; Foster, 1997; Kincheloe & Berry, 2004; McLeod, 2000).

In its critical concern for just social change, the bricolage seeks insight from the margins of Western societies and the knowledge and ways of knowing of non-Western peoples. Such insight helps bricoleurs reshape and sophisticate social theory, research methods, and interpretive strategies, as they discern new topics to be researched. This confrontation with difference so basic to the concept of the bricolage enables researchers to produce new forms of knowledge that inform policy decisions and political action in general. In gaining this insight from the margins, bricoleurs display once again the blurred boundary between the hermeneutical search for understanding and the critical concern with social change for social justice (Jardine, 2006a). Kincheloe has taken seriously Peter McLaren's (2001) important concern—offered in his response to Kincheloe's (2001a) first delineation of his conception of the bricolage— that merely focusing on the production of meanings may not lead to "resisting and transforming the existing conditions of exploitation" (McLaren, 2001, p. 702). In response, Kincheloe maintained that in the critical hermeneutical dimension of the bricolage, the act of understanding power and its effects is merely one part—albeit an inseparable part—of counterhegemonic *action*. Not only are the two orientations not in conflict, they are synergistic (DeVault, 1996; Lutz, Jones, & Kendall, 1997; Soto, 2000; Steinberg, 2001, 2007; Tobin, 2010).

To contribute to social transformation, bricoleurs seek to better understand both the forces of domination that affect the lives of individuals from race, class, gender, sexual, ethnic, and religious backgrounds outside of dominant culture(s) and the worldviews of such diverse peoples. In this context, bricoleurs attempt to remove knowledge production and its benefits from the control of elite groups. Such control consistently operates to reinforce elite privilege while pushing marginalized groups farther away from the center of dominant power. Rejecting this normalized state of affairs, bricoleurs commit their knowledge work to helping address the ideological and informational needs of marginalized groups and individuals. As detectives of subjugated insight, bricoleurs eagerly learn from labor struggles, women's marginalization, the "double consciousness" of the racially oppressed, and insurrections against colonialism (Kincheloe & Steinberg, 1993; Kincheloe, Steinberg, & Hinchey, 1999; Kincheloe & Berry, 2004). In this way, the bricolage hopes to contribute to an evolving criticality.

The bricolage is dedicated to a form of rigor that is conversant with numerous modes of meaning making and knowledge production—modes that originate in diverse social locations. These alternative modes of reasoning and researching always consider the relationships, the resonances, and the disjunctions between formal and rationalistic modes of Western epistemology and ontology and different cultural, philosophical, paradigmatic, and subjugated expressions. In these latter expressions, bricoleurs often uncover ways of accessing a concept without resorting to a conventional validated set of prespecified procedures that provide the distance of objectivity (Thayer-Bacon, 2003). This notion of distance fails to take into account the rigor of the hermeneutical understanding of the way meaning is preinscribed in the act of being in the world, the research process, and objects of research. This absence of hermeneutical awareness undermines the researcher's quest for a thick description and contributes to the production of reduced understandings of the complexity of social life (Jardine, 2006b; Selfe & Selfe, 1994).

The multiple perspectives delivered by the concept of difference provide bricoleurs with many benefits. Confrontation with difference helps us to see anew, to move toward the light of epiphany. A basic dimension of an evolving criticality involves a comfort with the existence of alternative ways of analyzing and producing knowledge. This is why it's so important for a historian, for example, to develop an understanding of phenomenology and hermeneutics. It is why it is so important for a social researcher from a metropolitan center to understand forms of indigenous knowledge, urban knowledge, and youth knowledge production (Darder, 2010; Dei, 2011; Grande, 2006; Hooley, 2009; Porfilio & Carr, 2010). The incongruities between such cultural modes of inquiry are quite valuable, for within the tensions of difference rest insights into multiple dimensions of the research act. Such insights move us to new levels of understanding of the subjects, purposes, and nature of inquiry (Gadamer, 1989; Kincheloe & Berry, 2004; Kincheloe & Steinberg, 2008; Mayers, 2001; Semali & Kincheloe, 1999; Watts, 2009a, 2009b; Willinsky, 2001).

Difference in the bricolage pushes us into the hermeneutic circle as we are induced to deal with parts in their diversity in relation to the whole. Difference may involve culture, class, language, discipline, epistemology, cosmology, ad infinitum. Bricoleurs use one dimension of these multiple diversities to explore others, to generate questions previously unimagined. As we examine these multiple perspectives, we attend to which ones are validated and which ones have been dismissed. Studying such differences, we begin to understand how dominant power operates to exclude and certify particular forms of knowledge production and why. In the criticality of the bricolage, this focus on power and difference always leads us to an awareness of the multiple dimensions of the social. Freire (1970) referred to this as the need for perceiving social structures and social systems that undermine equal access to resources and power. As bricoleurs answer such questions, we gain new appreciations of the way power tacitly shapes what we know and how we come to know it.

Ontologically Speaking

A central dimension of the bricolage that holds profound implications for critical research is the notion of a critical ontology (Kincheloe, 2003a). As bricoleurs prepare to explore that which is not readily apparent to the ethnographic eye, that realm of complexity in knowledge production that insists on initiating a conversation about what it is that qualitative researchers are observing and interpreting in the world, this clarification of a complex ontology is needed. This conversation is especially important because it has not generally taken place. Bricoleurs maintain that this object of inquiry is ontologically complex in that it cannot be described as an encapsulated entity. In this more open view, the object of inquiry is always a part of many contexts and processes; it is culturally inscribed and historically situated. The complex view of the object of inquiry accounts for the historical efforts to interpret its meaning in the world and how such efforts continue to define its social, cultural, political, psychological, and educational effects.

In the domain of the qualitative research process, for example, this ontological complexity undermines traditional notions of triangulation. Because of its in-process (processual) nature, interresearcher reliability becomes far more difficult to achieve. Process-sensitive scholars watch the world flow by like a river in which the exact contents of the water are never the same. Because all observers view an object of inquiry from their own vantage points in the web of reality, no portrait of a social phenomenon is ever exactly the same as another. Because all physical, social, cultural, psychological, and educational dynamics are connected in a larger fabric, researchers will produce different descriptions of an object of inquiry depending on what part of the fabric they have focused on—what part of the river they have seen. The more unaware observers are of this type of complexity, the more reductionistic the knowledge they produce

about it. Bricoleurs attempt to understand this fabric and the processes that shape it in as thick a way as possible (Kincheloe & Berry, 2004).

The design and methods used to analyze this social fabric cannot be separated from the way reality is construed. Thus, ontology and epistemology are linked inextricably in ways that shape the task of the researcher. The bricoleur must understand these features in the pursuit of rigor. A deep interdisciplinarity is justified by an understanding of the complexity of the object of inquiry and the demands such complications place on the research act. As parts of complex systems and intricate processes, objects of inquiry are far too mercurial to be viewed by a single way of seeing or as a snapshot of a particular phenomenon at a specific moment in time.

This deep interdisciplinarity seeks to modify the disciplines and the view of research brought to the negotiating table constructed by the bricolage (Jardine, 1992). Everyone leaves the table informed by the dialogue in a way that idiosyncratically influences the research methods they subsequently employ. The point of the interaction is not standardized agreement as to some reductionistic notion of "the proper interdisciplinary research method" but awareness of the diverse tools in the researcher's toolbox. The form such deep interdisciplinarity may take is shaped by the object of inquiry in question. Thus, in the bricolage, the context in which research takes place always affects the nature of the deep interdisciplinarity employed. In the spirit of the dialectic of disciplinarity, the ways these context-driven articulations of interdisciplinarity are constructed must be examined in light of the power literacy previously mentioned (Friedman, 1998; Kincheloe & Berry, 2004; Lemke, 1998; Pryse, 1998; Quintero & Rummel, 2003).

In social research, the relationship between individuals and their contexts is a central dynamic to be investigated. This relationship is a key ontological and epistemological concern of the bricolage; it is a connection that shapes the identities of human beings and the nature of the complex social fabric. Bricoleurs use multiple methods to analyze the multidimensionality of this type of connection. The ways bricoleurs engage in this process of putting together the pieces of the relationship may provide a different interpretation of its meaning and effects. Recognizing the complex ontological importance of relationships alters the basic foundations of the research act and knowledge production process. Thin reductionistic descriptions of isolated things-in-themselves are no longer sufficient in critical research (Foster, 1997; Wright, 2003b).

The bricolage is dealing with a double ontology of complexity: first, the complexity of objects of inquiry and their being-in-the-world; second, the nature of the social construction of human subjectivity, the production of human "being." Such understandings open a new era of social research where the process of becoming human agents is appreciated with a new level of sophistication. The complex feedback loop between an unstable social structure and the individual can be charted in a way that grants human beings insight into the means by which power operates and the democratic process is subverted. In this complex ontological view, bricoleurs understand that social structures do not *determine* individual subjectivity but *constrain* it in remarkably intricate ways. The bricolage is acutely interested in developing and employing a variety of strategies to help specify these ways subjectivity is shaped.

The recognitions that emerge from such a multiperspectival process get analysts beyond the determinism of reductionistic notions of macrosocial structures. The intent of a usable social or educational research is subverted in this reductionistic context, as human agency is erased by the "laws" of society. Structures do not simply "exist" as objective entities whose influence can be predicted or "not exist" with no influence over the cosmos of human affairs. Here fractals enter the stage with their loosely structured characteristics of irregular shape—fractal structures. While not *determining* human behavior, for example, fractal structures possess sufficient order to affect other systems and entities within their environment. Such structures are never stable or universally present in some uniform manifestation (Slee, 2011; Varenne, 1996). The more we study such dynamics, the more diversity of expression we find. Taking this ontological and epistemological diversity into account, bricoleurs understand there are numerous dimensions to the bricolage (Denzin & Lincoln, 2000). As with all aspects of the bricolage, no description is fixed and final, and all features of the bricolage come with an elastic clause.

▣   EMPLOYING A "METHOD" WITHIN BRICOLAGE: ETHNOGRAPHY AS AN EXAMPLE

As critical researchers attempt to get behind the curtain, to move beyond assimilated experience, to expose the way ideology constrains the desire for self-direction, and to confront the way power reproduces itself in the construction of human consciousness, they employ a plethora of research methodologies (Hyslop-Margison, 2009). We are looking at the degree to which research moves those it studies to understand the world and the way it is shaped in order for them to transform it. Noncritical researchers who operate within an empiricist framework will perhaps find catalytic validity to be a strange concept. Research that possesses catalytic validity displays the reality-altering impact of the inquiry process and directs this impact so that those under study will gain self-understanding and self-direction.

Theory that falls under the rubric of postcolonialism (see McLaren, 1999; Semali & Kincheloe, 1999; Wright 2003a, 2003b) involves important debates over the knowing subject and object of analysis. Such works have initiated important new modes of analysis, especially in relation to questions of imperialism,

colonialism, and neocolonialism. Recent attempts by critical researchers to move beyond the objectifying and imperialist gaze associated with the Western anthropological tradition (which fixes the image of the so-called informant from the colonizing perspective of the knowing subject), although laudatory and well-intentioned, are not without their shortcomings (Bourdieu & Wacquant, 1992). As Fuchs (1993) has so presciently observed, serious limitations plague recent efforts to develop a more reflective approach to ethnographic writing. The challenge here can be summarized in the following questions: How does the knowing subject come to know the Other? How can researchers respect the perspective of the Other and invite the Other to speak (Ashcroft, Griffiths, & Tiffin, 1995; Brock-Utne, 1996; Goldie, 1995; Gresson, 2006; Macedo, 2006; Myrsiades & Myrsiades, 1998; Pieterse & Parekh, 1995; Prakash & Esteva, 2008; Scheurich & Young, 1997; Semali & Kincheloe, 1999; Steinberg, 2009; Viergever, 1999)?

Although recent confessional modes of ethnographic writing, for example, attempt to treat so-called informants as "participants" in an attempt to avoid the objectification of the Other (usually referring to the relationship between Western anthropologists and non-Western culture), there is a risk that uncovering colonial and postcolonial structures of domination may, in fact, unintentionally validate and consolidate such structures as well as reassert liberal values through a type of covert ethnocentrism. Fuchs (1993) warns that the attempt to subject researchers to the same approach to which other societies are subjected could lead to an "'othering' of one's own world" (p. 108). Such an attempt often fails to question existing ethnographic methodologies and therefore unwittingly extends their validity and applicability while further objectifying the world of the researcher. Foucault's approach to this dilemma is to "detach" social theory from the epistemology of his own culture by criticizing the traditional philosophy of reflection. However, Foucault falls into the trap of ontologizing his own methodological argumentation and erasing the notion of prior understanding that is linked to the idea of an "inside" view (Fuchs, 1993). Louis Dumont fares somewhat better by arguing that cultural texts need to be viewed simultaneously from the inside and from the outside.

However, in trying to affirm a "reciprocal interpretation of various societies among themselves" (Fuchs, 1993, p. 113) through identifying both transindividual structures of consciousness and transsubjective social structures, Dumont aspires to a universal framework for the comparative analysis of societies. Whereas Foucault and Dumont attempt to "transcend the categorical foundations of their own world" (Fuchs, 1993, p. 118) by refusing to include themselves in the process of objectification, Pierre Bourdieu integrates himself as a social actor into the social field under analysis. Bourdieu achieves such integration by "epistemologizing the ethnological content of his own presuppositions" (Fuchs, 1993, p. 121). But the self-objectification of the observer (anthropologist) is not unproblematic. Fuchs (1993) notes, after Bourdieu, that the chief difficulty is "forgetting the difference between the theoretical and the practical relationship with the world and . . . imposing on the object the theoretical relationship one maintains with it" (p. 120). Bourdieu's approach to research does not fully escape becoming, to a certain extent, a "confirmation of objectivism," but at least there is an earnest attempt by the researcher to reflect on the preconditions of his or her own self-understanding—an attempt to engage in an "ethnography of ethnographers" (p. 122). As an example, critical ethnography, in a bricolage context, often intersects—to varying degrees—with the concerns of postcolonialist researchers, but the degree to which it fully addresses issues of exploitation and the social relations of capitalist exploitation remains questionable. Critical ethnography shares the conviction articulated by Marc Manganaro (1990):

> No anthropology is apolitical, removed from ideology and hence from the capacity to be affected by or, as crucially, to effect social formations. The question ought not to be if an anthropological text is political, but rather, what kind of sociopolitical affiliations are tied to particular anthropological texts. (p. 35)

This critical ethnographic writing faces the challenge of moving beyond simply the reanimation of local experience, an uncritical celebration of cultural difference (including figural differentiations within the ethnographer's own culture), and the employment of a framework that espouses universal values and a global role for interpretivist anthropology (Silverman, 1990). Criticalism can help qualitative researchers challenge dominant Western research practices that are underwritten by a foundational epistemology and a claim to universally valid knowledge at the expense of local, subjugated knowledges (Peters, 1993). The issue is to challenge the presuppositions that inform the normalizing judgments one makes as a researcher.

Although critical ethnography (Hickey & Austin, 2009) allows, in a way conventional ethnography does not, for the relationship of liberation and history, and although its hermeneutical task is to call into question the social and cultural conditioning of human activity and the prevailing sociopolitical structures, we do not claim that this is enough to restructure the social system. But it is certainly, in our view, a necessary beginning (Trueba & McLaren, 2000). Clough (1998) argues that "realist narrativity has allowed empirical social science to be the platform and horizon of social criticism" (p. 135). Ethnography needs to be analyzed critically not only in terms of its field methods but also as reading and writing practices. Data collection must give way to "rereadings of representations in every form" (p. 137). In the narrative construction of its authority as empirical science, ethnography needs to face the unconscious processes on which it justifies its canonical formulations, processes that often involve the disavowal of oedipal or authorial

desire and the reduction of differences to binary oppositions. Within these processes of binary reduction, the male ethnographer is most often privileged as the guardian of "the factual representation of empirical positivities" (Clough, 1998).

Critical research traditions have arrived at the point where they recognize that claims to truth are always discursively situated and implicated in relations of power. We do not suggest that because we cannot know truth absolutely, truth can simply be equated with an effect of power. We say this because truth involves regulative rules that must be met for some statements to be more meaningful than others. Otherwise, truth becomes meaningless and, if that is the case, liberatory praxis has no purpose other than to win for the sake of winning. As Phil Carspecken (1993, 1999) remarks, every time we act, in every instance of our behavior, we presuppose some normative or universal relation to truth. Truth is internally related to meaning in a pragmatic way through normative referenced claims, intersubjective referenced claims, subjective referenced claims, and the way we deictically ground or anchor meaning in our daily lives. Carspecken explains that researchers are able to articulate the normative evaluative claims of others when they begin to see them in the same way as their participants by living inside the cultural and discursive positionalities that inform such claims.

While a researcher can use, as in this example, critical ethnography (Willis, 1977, 2000) as a focus within a project, she or he, as a bricoleur (Steinberg, 2011) employs the additional use of narrative (Janesick, 2010; Park, 2005), hermeneutic interpretation (Jardine, 2006a), phenomenological reading (Kincheloe, 2008b), content analysis (Steinberg, 2008), historiography (Kincheloe, 2008b), autoethnography (Kress, 2010), social media analysis (Cucinelli, 2010; Kress, 2008; Kress & Silva, 2009), anthropology (Marcus & Fischer, 1986), quantitative analysis (Hyslop-Margison & Naseem, 2007), and so on; and the bricoleur creates a polysemic read and multiple ways of both approaching and using research. The bricolage, with its multiple lenses allows necessary fluidity and goes beyond a traditional triangulated approach for verification. The lenses expand the research and prevent a normalized methodology from creating a scientistic approach to the research. Bricolage becomes a failsafe way in which to ensure that the multiple reads create new dialogues and discourse and open possibilities. It also precludes the notion of using research as authority.

Clearly, no research methodology or tradition can be done in isolation; the employment of the bricolage transcends unilateral commitments to a singular type of research. In the face of a wide variety of different knowledges and ways of seeing the universe, human beings' confidence in what they think they know collapses. In a countercolonial move, bricoleurs raise questions about any knowledges and ways of knowing that claim universal status. In this context, bricoleurs make use of this suspicion of universalism in combination with global knowledges to understand how they have been positioned in the world. Almost all of us from Western backgrounds or non-Western colonized backgrounds have been implicated in some way in the web of universalism (Scatamburlo D'Annibale & McLaren, 2009). The inevitable conflicts that arise from this implication do not have to be resolved immediately by bricoleurs. At the base of these conflicts rests the future of global culture as well as the future of multicultural research and pedagogy. Recognizing that these are generative issues that engage us in a productive process of analyzing self and world is in itself a powerful recognition. The value of both this recognition and the process of working through the complicated conceptual problems are treasured by bricoleurs. Indeed, bricoleurs avoid any notion of finality in the resolution of such dilemmas. Comfortable with the ambiguity, bricoleurs as critical researchers work to alleviate human suffering and injustice even though they possess no final blueprint alerting them as to how oppression takes place (Kincheloe & Berry, 2004; Steinberg, 2011).

Toward a Critical Research

Within the context of multiple critical theories and multiple critical pedagogies, a critical research bricolage serves to create an equitable research field and disallows a proclamation to correctness, validity, truth, and the tacit axis of Western power through traditional research. Employing a rigorous and tentative context with the notions presented through Marxist examinations of power, critical theory's location and indictment of power blocs vis-à-vis traditional noncritical research methodologies, a critical pedagogical notion of emancipatory research can be located within a research bricolage (Fiske, 1993; Roth & Tobin, 2010). Without proclaiming a canonical and singular method, the critical bricolage allows the researcher to become participant and the participant to become researcher. By eschewing positivist approaches to both qualitative and quantitative research (Cannella & Steinberg, 2011; Kincheloe & Tobin, 2009) and refusing to cocoon research within the pod of unimethodological approaches, we believe critical theory and critical pedagogy continues to challenge regularly employed and obsessive approaches to research.

▣ NOTE

1. Thanks to Dr. Michael Watts, a local, for his suggestions and critique of this chapter.

▣ REFERENCES

Agger, B. (1992). *The discourse of domination: From the Frankfurt School to postmodernism.* Evanston, IL: Northwestern University Press.

Agger, B. (1998). *Critical social theories: An introduction.* Boulder, CO: Westview.

Ashcroft, B., Griffiths, G., & Tiffin, H. (Eds.). (1995). *The post-colonial studies reader.* New York: Routledge.

Au, W. (2007). Epistemology of the oppressed: The dialectics of Paulo Freire's theory of knowledge. *Journal for Critical Education Policy Studies, 5*(2). Available at http://www.jceps.com/index.php?pageID=article&articleID=100

Austin, J., & Hickey, A. (2008). Critical pedagogical practice through cultural studies. *International Journal of the Humanities, 6*(1), 133–140. Available at http://eprints.usq.edu.au/4490/

Bourdieu, P., & Wacquant, L. (1992). *An invitation to reflexive sociology.* Chicago: University of Chicago Press.

Bresler, L., & Ardichvili, A. (Eds.). (2002). *Research in international education: Experience, theory, and practice.* New York: Peter Lang.

Britzman, D. (1991). *Practice makes practice: A critical study of learning to teach.* Albany: SUNY Press.

Brock-Utne, B. (1996). Reliability and validity in qualitative research within Africa. *International Review of Education, 42,* 605–621.

Cannella, G., & Steinberg, S. (2011). *Critical qualitative research: A reader.* New York: Peter Lang.

Carspecken, P. F. (1993). *Power, truth, and method: Outline for a critical methodology.* Unpublished manuscript, Indiana University.

Carspecken, P. F. (1999). *Four scenes for posing the question of meaning and other essays in critical philosophy and critical methodology.* New York: Peter Lang.

Chapman, D. E. (Ed.). (2010). *Examining social theory: Crossing borders/reflecting back.* New York: Peter Lang.

Clough, P. T. (1998). *The end(s) of ethnography: From realism to social criticism* (2nd ed.). New York: Peter Lang.

Collins, J. (1995). *Architectures of excess: Cultural life in the information age.* New York: Routledge.

Cucinelli, G. (2010). *Digital youth praxis and social justice.* Unpublished doctoral dissertation, McGill University, Montréal, Québec, Canada.

Dale, J., & Hyslop-Margison, E. J. (2010). *Paulo Freire: Teaching for freedom and transformation.* Dordrecht, the Netherlands: Springer.

Darder, A. (2010). Schooling bodies: Critical pedagogy and urban youth [Foreword]. In Steinberg, S. R. (Ed.), *19 urban questions: Teaching in the city* (pp. xiii–xxiii). New York: Peter Lang.

Dei, G. (Ed.). (2011). *Indigenous philosophies and critical education.* New York: Peter Lang.

De Lissovoy, N., & McLaren, P. (2003). Educational "accountability" and the violence of capital: A Marxian reading. *Journal of Education Policy, 18,* 131–143.

Denzin, N. K. (1994). The art and politics of interpretation. In N. K. Denzin & Y. S. Lincoln (Eds.), *Handbook of qualitative research.* Thousand Oaks, CA: Sage.

Denzin, N. K. (2003). *Performative ethnography: Critical pedagogy and the politics of culture.* Thousand Oaks, CA: Sage.

Denzin, N. K., & Lincoln, Y. S. (2000). Introduction: The discipline and practice of qualitative research. In N. K. Denzin & Y. S. Lincoln (Eds.), *Handbook of qualitative research* (2nd ed.). Thousand Oaks, CA: Sage.

DeVault, M. (1996). Talking back to sociology: Distinctive contributions of feminist methodology. *Annual Review of Sociology, 22,* 29–50.

Fiske, J. (1993). *Power works, power plays.* New York: Verso.

Flecha, R., Gomez, J., & Puigvert, L. (Eds.). (2003). *Contemporary sociological theory.* New York: Peter Lang.

Flossner, G., & Otto, H. (Eds.). (1998). *Towards more democracy in social services: Models of culture and welfare.* New York: Aldine.

Foster, R. (1997). Addressing epistemologic and practical issues in multimethod research: A procedure for conceptual triangulation. *Advances in Nursing Education, 20*(2), 1–12.

Freire, A. M. A. (2000). Foreword. In P. McLaren, *Che Guevara, Paulo Freire, and the pedagogy of revolution.* Boulder, CO: Rowman & Littlefield.

Freire, P. (1970). *Pedagogy of the oppressed.* New York: Herder and Herder.

Freire, P. (1972). *Research methods.* Paper presented at a seminar on Studies in Adult Education, Dar es Salaam, Tanzania.

Freire, P. (1978). *Education for critical consciousness.* New York: Seabury.

Freire, P. (1985). *The politics of education: Culture, power, and liberation.* South Hadley, MA: Bergin & Garvey.

Freire, P., & Faundez, A. (1989). *Learning to question: A pedagogy of liberation.* London: Continuum.

Friedman, S. (1998). (Inter)disciplinarity and the question of the women's studies Ph.D. *Feminist Studies, 24*(2), 301–326.

Fuchs, M. (1993). The reversal of the ethnological perspective: Attempts at objectifying one's own cultural horizon: Dumont, Foucault, Bourdieu? *Thesis Eleven, 34*(1), 104–125.

Gadamer, H.-G. (1989). *Truth and method* (2nd rev. ed., J. Weinsheimer & D. G. Marshall, Eds. & Trans.). New York: Crossroad.

Giroux, H. (1983). *Theory and resistance in education: A pedagogy for the opposition.* South Hadley, MA: Bergin & Garvey.

Giroux, H. (1988). Critical theory and the politics of culture and voice: Rethinking the discourse of educational research. In R. Sherman & R. Webb (Eds.), *Qualitative research in education: Focus and methods.* New York: Falmer.

Giroux, H. (1997). *Pedagogy and the politics of hope: Theory, culture, and schooling.* Boulder, CO: Westview.

Giroux, H. (2010). *Zombie politics and the age of casino capitalism.* New York: Peter Lang.

Goldie, T. (1995). The representation of the indigenous. In B. Ashcroft, G. Griffiths, & H. Tiffin (Eds.), *The post-colonial studies reader.* New York: Routledge.

Grande, S. (2004). *Red pedagogy: Native American social and political thought.* Lanham, MD: Rowman & Littlefield.

Gresson, A. D., III. (2006). Doing critical research in mainstream disciplines: Reflections on a study of Black female individuation. In K. Tobin & J. Kincheloe (Eds.), *Doing educational research.* Rotterdam, the Netherlands: Sense Publishers.

Grinberg, J. (2003). "Only the facts?" In D. Weil & J. L. Kincheloe (Eds.), *Critical thinking: An encyclopedia.* New York: Greenwood.

Hammer, R., & Kellner, D. (2009). *Media/cultural studies: Critical approaches.* New York: Peter Lang.

Harper, D. (1987). *Working knowledge: Skill and community in a small shop.* Chicago: University of Chicago Press.

Hickey, A., & Austin, J. (2009). Working visually in community identity ethnography. *International Journal of the Humanities, 7*(4), 1–14. Available at http://eprints.usq.edu.au/5800/

Hinchey, P. (2009). *Finding freedom in the classroom: A practical introduction to critical theory.* New York: Peter Lang.

Hooley, N. (2009). *Narrative life: Democratic curriculum and indigenous learning.* Dordrecht, the Netherlands: Springer.

Horkheimer, M. (1972). *Critical theory.* New York: Seabury.

Horn, R. (2004). *Standards.* New York: Peter Lang.

Horton, M., & Freire, P. (1990). *We make the road by walking: Conversations on education and social change.* Philadelphia: Temple University Press.

Hyslop-Margison, E. J. (2009). Scientific paradigms and falsification: Kuhn, Popper and problems in education research. *Educational Policy, 20*(10), 1–17.

Hyslop-Margison, E. J., & Naseem, A. (2007). *Scientism and education: Empirical research as neo-liberal ideology.* Dordrecht, the Netherlands: Springer.

Janesick, V. (2010). *Oral history for the qualitative researcher: Choreographing the story.* New York: Guilford.

Jardine, D. (1992). The fecundity of the individual case: Considerations of the pedagogic heart of interpretive work. *British Journal of Philosophy of Education. 26*(1), 51–61.

Jardine, D. (1998). *To dwell with a boundless heart: Essays in curriculum theory, hermeneutics, and the ecological imagination.* New York: Peter Lang.

Jardine, D. (2006a). On hermeneutics: "What happens to us over and above our wanting and doing." In K. Tobin & J. L. Kincheloe (Eds.), *Doing educational research* (pp. 269–288). Rotterdam, the Netherlands: Sense Publishers.

Jardine, D. (2006b). *Piaget and education.* New York: Peter Lang.

Keesing-Styles, L. (2003). The relationship between critical pedagogy and assessment in teacher education. *Radical Pedagogy, 5*(1). Available at http://radicalpedagogy.icaap.org/content/issue5_1/03_keesing-styles.html

Kellner, D. (1995). *Media culture: Cultural studies, identity, and politics between the modern and the postmodern.* New York: Routledge.

Kincheloe, J. L. (1998). Critical research in science education. In B. Fraser & K. Tobin (Eds.), International handbook of science education (Pt. 2). Boston: Kluwer.

Kincheloe, J. L. (2001a). Describing the bricolage: Conceptualizing a new rigour in qualitative research. *Qualitative Inquiry, 7*(6), 679–692.

Kincheloe, J. (2001b). *Getting beyond the facts: Teaching social studies/social sciences in the twenty-first century* (2nd ed.). New York: Peter Lang.

Kincheloe, J. (2003a). Critical ontology: Visions of selfhood and curriculum. *JCT: Journal of Curriculum Theorizing, 19*(1), 47–64.

Kincheloe, J. L. (2003b). Into the great wide open: Introducing critical thinking. In D. Weil & J. Kincheloe (Eds.), *Critical thinking: An encyclopedia.* Santa Barbara, CA: ABC-CLIO.

Kincheloe, J. L. (2003c). *Teachers as researchers: Qualitative paths to empowerment* (2nd ed.). London: Falmer.

Kincheloe, J. L. (2004). Iran and American miseducation: Coverups, distortions, and omissions. In J. Kincheloe & S. Steinberg (Eds.), *The miseducation of the West: Constructing Islam.* New York: Greenwood.

Kincheloe, J. L. (2007). *Teachers as researchers: Qualitative paths to empowerment.* London: Falmer.

Kincheloe, J. L. (2008a). *Critical pedagogy primer* (2nd ed.). New York: Peter Lang.

Kincheloe, J. L. (2008b). *Knowledge and critical pedagogy.* Dordrecht, the Netherlands: Springer.

Kincheloe, J. L., & Berry, K. (2004). *Rigour and complexity in educational research: Conceptualizing the bricolage.* London: Open University Press.

Kincheloe, J. L., & Steinberg, S. R. (1993). A tentative description of post-formal thinking: The critical confrontation with cognitive theory. *Harvard Educational Review, 63,* 296–320.

Kincheloe, J. L., & Steinberg, S. R. (1997). *Changing multiculturalism: New times, new curriculum.* London: Open University Press.

Kincheloe, J. L., & Steinberg, S. R. (2008). Indigenous knowledges in education: Complexities, dangers, and profound benefits. In N. K. Denzin, Y. S. Lincoln, & L. T. Smith, (Eds.), *Handbook of critical and indigenous methodologies.* Thousand Oaks, CA: Sage Publishing.

Kincheloe, J. L., Steinberg, S. R., & Hinchey, P. (Eds.). (1999). *The postformal reader: Cognition and education.* New York: Falmer.

Kincheloe, J. L., & Tobin, K. (2009). The much exaggerated death of positivism. *Cultural Studies of Science Education, 4,* 513–528.

Kirylo, J. (2011). Paulo Freire: *The man from Recife.* New York: Peter Lang.

Kress, T. (2010). Tilting the machine: A critique of one teacher's attempts at using art forms to create postformal, democratic learning environments. *The Journal of Educational Controversy, 5*(1).

Kress, T., & Silva, K. (2009). Using digital video for professional development and leadership: Understanding and initiating teacher learning communities. In I. Gibson et al. (Eds.), *Proceedings of Society for Information Technology & Teacher Education International Conference 2009* (pp. 2841–2847). Chesapeake, VA: Association for the Advancement of Computing in Education (AACE).

Leistyna, P., Woodrum, A., & Sherblom, S. (1996). *Breaking free: The transformative power of critical pedagogy.* Cambridge, MA: Harvard Educational Review.

Lemke, J. L. (1998). Analyzing verbal data: Principles, methods, and problems. In B. Fraser & K. Tobin (Eds.), *International handbook of science education* (Pt. 2). Boston: Kluwer.

Lévi-Strauss, C. (1968). *The savage mind.* Chicago: University of Chicago Press.

Lincoln, Y. (2001). An emerging new bricoleur: Promises and possibilities—a reaction to Joe Kincheloe's "Describing the bricoleur." *Qualitative Inquiry, 7*(6), 693–696.

Lund, D., & Carr, P. (Eds.). (2008). *Doing democracy: Striving for political literacy and social justice.* New York: Peter Lang.

Lutz, K., Jones, K. D., & Kendall, J. (1997). Expanding the praxis debate: Contributions to clinical inquiry. *Advances in Nursing Science, 20*(2), 23–31.

Macedo, D. (2006). *Literacies of power: What Americans are not allowed to know* (2nd ed.). Boulder, CO: Westview.

Macrine, S., Hill, D., & McLaren, P. (Eds.). (2009). *Critical pedagogy: Theory and praxis.* London: Routledge.

Macrine, S., McLaren, P., & Hill, D. (Eds.). (2009). *Revolutionizing pedagogy: Educating for social justice within and beyond global neo-liberalism.* London: Palgrave Macmillan.

Manganaro, M. (1990). Textual play, power, and cultural critique: An orientation to modernist anthropology. In M. Manganaro (Ed.), *Modernist anthropology: From fieldwork to text.* Princeton, NJ: Princeton University Press.

Marcus, G. E., & Fischer, M. M. J. (1986). *Anthropology as cultural critique: An experimental moment in the human sciences.* Chicago: University of Chicago Press.

Mayo, P. (2009). *Liberating praxis: Paulo Freire's legacy for radical education and politics.* Rotterdam, the Netherlands: Sense Publishing.

McLaren, P. (1992). Collisions with otherness: "Traveling" theory, postcolonial criticism, and the politics of ethnographic practice—the mission of the wounded ethnographer. *International Journal of Qualitative Studies in Education, 5,* 77–92.

McLaren, P. (1997). *Revolutionary multiculturalism: Pedagogies of dissent for the new millennium.* New York: Routledge.

McLaren, P. (1999). *Schooling as a ritual performance: Toward a political economy of educational symbols and gestures* (3rd ed.). Boulder, CO: Rowman & Littlefield.

McLaren, P. (2001). Bricklayers and bricoleurs: A Marxist addendum. *Qualitative Inquiry, 7*(6), 700–705.

McLaren, P. (2002). Marxist revolutionary praxis: A curriculum of transgression. *Journal of Curriculum Inquiry Into Curriculum and Instruction, 3*(3), 36–41.

McLaren, P. (2003a). Critical pedagogy in the age of neoliberal globalization: Notes from history's underside. *Democracy and Nature, 9*(1), 65–90.

McLaren, P. (2003b). The dialectics of terrorism: A Marxist response to September 11: Part Two. Unveiling the past, evading the present. *Cultural Studies <=> Critical Methodologies, 3*(1), 103–132.

McLaren, P. (2009). E. San Juan, Jr.: The return of the transformative intellectual. *Left Curve, 33,* 118–121.

McLaren, P., & Farahmandpur, R. (2003). Critical pedagogy at ground zero: Renewing the educational left after 9–11. In D. Gabbard & K. Saltman (Eds.), *Education as enforcement: The militarization and corporatization of schools.* New York: Routledge.

McLaren, P., & Farahmandpur, R. (2006). Who will educate the educators? Critical pedagogy in the age of globalization. In A. Dirlik (Ed.), *Pedagogies of the global: Knowledge in the human interest* (pp. 19–58). Boulder, CO: Paradigm.

McLaren, P., & Jaramillo, N. (2010). Not neo-Marxist, not post-Marxist, not Marxian, not autonomist Marxism: Reflections on a revolutionary (Marxist) critical pedagogy. *Cultural Studies <=> Critical Methodologies, 10*(3), 251–262.

McLaren, P., & Kincheloe, J. L. (2007). *Critical pedagogy: Where are we now?* New York: Peter Lang.

McLeod, J. (2000, June). *Qualitative research as bricolage.* Paper presented at the annual conference of the Society for Psychotherapy Research, Chicago.

Myrsiades, K., & Myrsiades, L. (Eds.). (1998). *Race-ing representation: Voice, history, and sexuality.* Lanham, MD: Rowman & Littlefield.

Nocella, A. J., II, Best, S., & McLaren, P. (2010). *Academic repression: Reflections from the academic industrial complex.* Oakland, CA: AK Press.

Park, J. (2005). *Writing at the edge: Narrative and writing process theory.* New York: Peter Lang.

Peters, M. (1993). *Against Finkielkraut's la défaite de la pensés culture, post-modernism and education.* Unpublished manuscript, University of Glasgow, Scotland.

Pieterse, J., & Parekh, B. (1995). Shifting imaginaries: Decolonization, internal decolonization, postcoloniality. In J. Pieterse & B. Parekh (Eds.), *The decolonization of imagination: Culture, knowledge, and power.* Atlantic Highlands, NJ: Zed.

Porfilio, B., & Carr, P. (Eds.). (2010). *Youth culture, education, and resistance: Subverting the commercial ordering of life.* Rotterdam, the Netherlands: Sense Publishing.

Prakash, M., & Esteva, G. (2008). *Escaping education: Living as learning within grassroots cultures.* New York: Peter Lang.

Pryse, M. (1998). Critical interdisciplinarity, women's studies, and cross-cultural insight. *National Women's Studies Association Journal, 10*(1), 1–11.

Quail, C. B., Razzano, K. A., & Skalli, L. H. (2004). *Tell me more: Rethinking daytime talk shows.* New York: Peter Lang.

Quantz, R. A. (1992). On critical ethnography (with some postmodern considerations). In M. D. LeCompte, W. L. Millroy, & J. Preissle (Eds.), *The handbook of qualitative research in education.* New York: Academic Press.

Quintero, E., & Rummel, M. K. (2003). *Becoming a teacher in the new society: Bringing communities and classrooms together.* New York: Peter Lang.

Rodriguez, N. M., & Villaverde, L. (2000). *Dismantling White privilege.* New York: Peter Lang.

Roman, L., & Eyre, L. (Eds.). (1997). *Dangerous territories: Struggles for difference and equality in education.* New York: Routledge.

Roth, W.-M., & Tobin, K. (2010). Solidarity and conflict: Prosody as a transactional resource in intra- and intercultural communication involving power differences. *Cultural Studies of Science Education, 5*(4), 807–847.

Ryoo, J. J., & McLaren, P. (2010). Aloha for sale: A class analysis of Hawaii. In D. E. Chapman (Ed.), *Examining social theory: Crossing borders/reflecting back* (pp. 3–18). New York: Peter Lang.

San Juan, E., Jr. (1992). *Articulations of power in ethnic and racial studies in the United States.* Atlantic Highlands, NJ: Humanities Press.

Scatamburlo-D'Annibale, V., & McLaren, P. (2009). The reign of capital: A pedagogy and praxis of class struggle. In M. Apple, W. Au, & L. Armando Gandin (Eds.), *The Routledge international handbook of critical education* (pp. 96–109). New York and London: Routledge.

Scheurich, J. J., & Young, M. (1997). Coloring epistemologies: Are our research epistemologies racially biased? *Educational Researcher, 26*(4), 4–16.

Selfe, C. L., & Selfe, R. J., Jr. (1994). The politics of the interface: Power and its exercise in electronic contact zones. *College Composition and Communication, 45*(4), 480–504.

Semali, L., & Kincheloe, J. L. (1999). *What is indigenous knowledge? Voices from the academy.* New York: Falmer.

Shor, I. (1996). *When students have power: Negotiating authority in a critical pedagogy.* Chicago: University of Chicago Press.

Sikes, P. (2008). Researching research cultures: The case of new universities. In P. Sikes & A. Potts (Eds.), *Researching education from the inside: Investigations from within.* Abingdon, UK: Routledge.

Silverman, E. K. (1990). Clifford Geertz: Towards a more "thick" understanding? In C. Tilley (Ed.), *Reading material culture.* Cambridge, MA: Blackwell.

Siry, C. A., & Lang, D. E. (2010). Creating participatory discourse for teaching and research in early childhood science. *Journal of Science Teacher Education, 21,* 149–160.

Skalli, L. (2004). Loving Muslim women with a vengeance: The West, women, and fundamentalism. In J. L. Kincheloe & S. R. Steinberg (Eds.), *The miseducation of the West: Constructing Islam.* New York: Greenwood.

Slee, R. (2011). *The irregular school: Schooling and inclusive education.* London: Routledge.

Soto, L. (Ed.). (2000). *The politics of early childhood education.* New York: Peter Lang.

Souto-Manning, M. (2009). *Freire, teaching, and learning: Culture circles across contexts.* New York: Peter Lang.

Steinberg, S. R. (2000). The nature of genius. In J. L. Kincheloe, S. R. Steinberg, & D. J. Tippins (Eds.), *The stigma of genius: Einstein, consciousness, and education.* New York: Peter Lang.

Steinberg, S. R. (Ed.). (2001). *Multi/intercultural conversations.* New York: Peter Lang.

Steinberg, S. R. (2004a). Desert minstrels: Hollywood's curriculum of Arabs and Muslims. In J. L. Kincheloe & S. R. Steinberg (Eds.), *The miseducation of the West: Constructing Islam.* New York: Greenwood.

Steinberg, S. R. (2004b). Kinderculture: The cultural studies of childhood. In N. Denzin (Ed.), *Cultural studies: A research volume.* Greenwich, CT: JAI.

Steinberg, S. R. (2007). Cutting class in a dangerous era: A critical pedagogy of class awareness. In J. Kincheloe & S. Steinberg (Eds.), *Cutting class: Socioeconomic status and education.* Lanham, MD: Rowman & Littlefield.

Steinberg, S. R. (2008). Reading media critically. In D. Macedo & S. Steinberg (Eds.), *Media literacy: A reader.* New York: Peter Lang.

Steinberg, S. R. (2009). *Diversity and multiculturalism: A reader.* New York: Peter Lang.

Steinberg, S. R. (2010). Power, emancipation, and complexity: Employing critical theory. *Journal of Power and Education, 2*(2), 140–151.

Steinberg, S. R. (2011). Critical cultural studies research: Bricolage in action. In K. Tobin & J. Kincheloe (Eds.), *Doing educational research* (2nd ed.). Rotterdam, the Netherlands: Sense Publishing.

Steinberg, S. R. (in press). *The bricolage.* New York: Peter Lang.

Steinberg, S. R., & Kincheloe, J. L. (Eds.). (1998). *Students as researchers: Creating classrooms that matter.* London: Taylor & Francis.

Steinberg, S. R., & Kincheloe, J. L. (2011). Employing the bricolage as critical research in science education. In B. J. Fraser, K. Tobin, & C. J. McRobbie (Eds.), *The international handbook of research in science education* (2nd ed.). Dordrecht, the Netherlands: Springer.

Thayer-Bacon, B. (2003). *Relational "(e)pistemologies."* New York: Peter Lang.

Tobin, K. (2009). Repetition, difference and rising up with research in education. In K. Ercikan & W.-M. Roth (Eds.), *Generalizing from educational research* (pp. 149–172). New York: Routledge.

Tobin, K. (2010). Global reproduction and transformation of science education. *Cultural Studies of Science Education, 5.*

Tobin, K., & Llena, R. (2010). Producing and maintaining culturally adaptive teaching and learning of science in urban schools. In C. Murphy & K. Scantlebury (Eds.), *Coteaching in international contexts: Research and practice* (pp. 79–104). Dordrecht, the Netherlands: Springer.

Trueba, E. T., & McLaren, P. (2000). Critical ethnography for the study of immigrants. In E. T. Trueba & L. I. Bartolomé (Eds.), *Immigrant voices: In search of educational equity.* Boulder, CO: Rowman & Littlefield.

Varenne, H. (with McDermott, R. P.). (1996). Culture, development, disability. In R. Jessor, A. Colby, & R. Shweder (Eds.), *Ethnography and human development.* Chicago: University of Chicago Press.

Vicars, M. (2008). Is it all about me? How Queer! In P. Sikes & A. Potts (Eds.), *Researching education from the inside: Investigations from within.* Abingdon, UK: Routledge.

Viergever, M. (1999). Indigenous knowledge: An interpretation of views from indigenous peoples. In L. Semali & J. L. Kincheloe (Eds.), *What is indigenous knowledge? Voices from the academy.* Bristol, PA: Falmer.

Villaverde, L. (2007). *Feminist theories and education primer.* New York: Peter Lang.

Villaverde, L., & Kincheloe, J. L. (1998). Engaging students as researchers: Researching and teaching Thanksgiving in the elementary classroom. In S. R. Steinberg & J. L. Kincheloe (Eds.), *Students as researchers: Creating classrooms that matter.* London: Falmer.

Watts, M. (2006). Disproportionate sacrifices: Ricoeur's theories of justice and the widening participation agenda for higher education in the UK. *Journal of Philosophy of Education, 40*(3), 301–312.

Watts, M. (2008). Narrative research, narrative capital, narrative capability. In J. Satterthwaite, M. Watts, & H. Piper (Eds.), *Talking truth, confronting power: Discourse, power, resistance* (Vol. 6). Stoke on Trent, UK: Trentham Books.

Watts, M. (2009a). Higher education and hyperreality. In P. Smeyers & M. Depaepe (Eds.), *Educational research: Educationalisation of social problems.* Dordrecht, the Netherlands: Springer.

Watts, M. (2009b). Sen and the art of motorcycle maintenance: Adaptive preferences and higher education in England. *Studies in Philosophy and Education, 28*(5), 425–436.

Weil, D., & Kincheloe, J. (Eds.). (2004). *Critical thinking and learning: An encyclopedia for parents and teachers.* Westport, CT: Greenwood.

Wesson, L., & Weaver, J. (2001). Administration-educational standards: Using the lens of postmodern thinking to examine the role of the school administrator. In J. Kincheloe & D. Weil (Eds.), *Standards and schooling in the United States: An encyclopedia* (3 vols.). Santa Barbara, CA: ABC-CLIO.

Wexler, P. (2008). *Social theory in education.* New York: Peter Lang.

Willinsky, J. (2001). Raising the standards for democratic education: Research and evaluation as public knowledge. In J. Kincheloe & D. Weil (Eds.), *Standards and schooling in the United States: An encyclopedia* (3 vols.). Santa Barbara, CA: ABC-CLIO.

Willis, P. E. (1977). *Learning to labour: How working class kids get working class jobs.* Farnborough, UK: Saxon House.

Willis, P. (2000). *The ethnographic imagination.* Cambridge, UK: Polity.

Wright, H. K. (2003a). An introduction to the history, methods, politics and selected traditions of qualitative research in education [Editorial]. *Tennessee Education, 32*(2), 5–7.

Wright, H. K. (Ed.). (2003b). Qualitative research in education. *Tennessee Education, 32*(2).

Wright, H. K., & Lather, P. (Eds.). (2006). Paradigm proliferation in educational research. *International Journal of Qualitative Studies in Education, 19*(1).

Zizek, S. (1990). Beyond discourse analysis. In E. Laclau, (Ed.), *New reflections on the revolution of our time.* London: Verso.

10

CULTURAL STUDIES

Performative Imperatives and Bodily Articulations[1]

Michael D. Giardina and Joshua I. Newman

Cultural studies has always been propelled by its desire to construct possibilities, both immediate and imaginary, out of its historical circumstances. It has no pretensions to totality or universality; it seeks only to give us a better understanding of where we are so that we can get somewhere else (some place, we hope, that is better—based on more just principles of equality and the distribution of wealth and power).

—Lawrence Grossberg, 1997, p. 415

I. PROEM

In the first edition of the *Handbook of Qualitative Research,* John Fiske (1994) began his chapter on cultural studies with the following statement: "Cultural studies is such a contested and currently trendy term that I must disclaim any attempt to either define or speak for it" (p. 189). Nearly 20 years later, Fiske's comment about the contested terrain of the field still holds true; today we find multiple, if not competing cultural studies projects at work.[2] In fact, and as if to underscore the multiplicity of formations, the three chapters on cultural studies to appear in earlier editions of the *Handbook* have each taken strikingly varied approaches to the deployment of cultural studies as theory and method: Fiske addressed production and consumption of media texts and audiences; John Frow and Meghan Morris (2000) outlined a largely multiperspectival approach to culture in an ever-evolving, globalizing world; and Paula Saukko (2005) offered an integrative methodological approach to contextual, dialogic, and self-reflexive validity.

The fourth edition continues this evolutionary trend, as our chapter moves in yet another direction, delving largely into the performative imperatives and bodily articulations of cultural studies in an age of increasing uncertainty characterized by, among other major developments: the post-9/11/01 militarization of culture; a war of aggression by imperial Western powers in Iraq and Afghanistan; the further destabilization of the Middle East as

brought forth by, among other things, the Israeli/Palestinian conflict; the meltdown of worldwide financial markets and institutions; drastically rising levels of unemployment and widening gaps between the rich and poor; the growing threats of religious fundamentalism and theocratic nationalism; a condition of neoliberal capitalism run wild; and so forth.

In a collectivity of these instances, we especially find "the body" under various forms of duress:

- Assailed by (global) capital's twin logics of overconsumption (think: genetically modified food in a child's Happy Meal) and overproduction (think: corporate accumulation through the exploitation of underprivileged bodies);
- Gradually stripped of its plurality and subjected to homogenizing strategies of the global popular;
- Discursively confined to the frames of heteronormative, patriarchal, xenophobic, White paranoia;
- Increasingly mediated as both an *immanent threat* (e.g., as carrier of influenza or of Jihadist intent) and as *under threat* (e.g., loss of human rights, etc.); and
- Forced to become an apparatus for—and collateral casualty of—war and genocide in such places as Afghanistan, Bosnia, Kashmir, Iraq, Sudan, Zimbabwe, and elsewhere.

Indeed, these are tough times for the body, deeply entangled as it is in the now-banal conditions of (social and material) production and accumulation; enwrapped in the hegemony of a fundamentalist assault on women's rights, equal rights, and social and economic justice; and "enfleshed" (McLaren, 1988) by the spectacle of the fetishized commodity. These developments have brought about an increasingly complex commixture of bodily contact and separation wherein overconsuming bodies of the developed world are perpetually engaged in the spectacles of late capitalism, and yet in ways that alienate the consumer from the bodies residing on the other end of these exploitative chains of interdependence. We need only glimpse at the front pages of our ever-disappearing newsprint to see headlines heralding the latest legal victory for antichoice advocates or giddily proclaiming increased bodily surveillance technologies being deployed at our airports (e.g., controversial full-body X-ray scanners) or advertisements for the latest reality television fare trading in body maintenance (e.g., *Project Runway*), modification (e.g., *The Biggest Loser*), or mastery (e.g., *Man v. Food*).

In light of these bodily discourses and the increasing importance placed on physical cultural forms and their attendant derivations more generally, this chapter offers both an overview of and a new methodological direction for what David Andrews (2008), Andrews and Michael Silk (2011), Jennifer Hargreaves and Patricia Vertinsky (2006), Alan Ingham (1985), and Pirkko Markula and Richard Pringle (2006) among others, have variously identified by the term *physical* cultural studies—that is, an antidisciplinary intellectual domain aimed at understanding "the complex and diverse practices and representations of active embodiment" and the "empirical and political import of cultural physicalities" (Silk & Andrews, in press) in the historical present.[3] To this end, we argue that such a *radically embodied* project can be found within (at least) three generative coordinates: (1) locating the body within cultural studies' articulative and radically contextual politics; (2) parting ways with the allusive embodiments that haunt most poststructuralist imaginings of the textualized corporeal in favor of a "bodily participative" research paradigm whereby active, agentive human bodies (and their flesh politics) are engaged through sometimes messy, sometimes difficult, sometimes dangerous points of corporeal contact; and (3) self-reflexively wrestling with the bodily politics of research performance, the very research act of physical cultural studies itself.

We begin by situating physical cultural studies within broader discussions related to cultural studies, articulation, and contextual analysis. We then move to directly engage with the embodied practice/s of the research act. In so doing, we draw on various contributions to this fledgling field (including some of our own work), especially as related to sporting physicalities, to discuss the promise of bodily participation. In this way, we move to take up the case of the critically reflexive body within cultural studies writ large and how it can best serve the interests of a public pedagogy—and ultimately "as a means for communicating important social understandings, social criticism, and powerful emotion-laden social science research" (Lincoln, 2004, p. 140). We conclude by offering a rejoinder as to the future of [physical] cultural studies.

▣ II. REVISITING THE POLITICS OF CULTURAL STUDIES

Before moving forward, some history is in order. Cultural studies, in many of the recognizable iterations, has been a mainstay in contemporary academic discourse from at least the mid-1950s to the present. Institutionally speaking, its (Western) beginnings are popularly though not unproblematically traced to the establishment of the Centre for Contemporary Cultural Studies (CCCS) at the University of Birmingham, United Kingdom, in 1964. Founded by Richard Hoggart and later reaching its apex during the leadership of Stuart Hall, the early work of (British) cultural studies at the CCCS is generally looked on, from a theoretical standpoint, as having used "the methods of literary criticism to understand popular and mass culture and to develop criteria for critically evaluating specific texts" (Dworkin, 1997, p. 116). Leading the way were figures such as Hoggart (*The Uses of Literacy*, 1957), Raymond Williams (*Culture and Society*, 1958), and E. P. Thompson (*Making of the English Working Class*, 1963), the three of whom authored what are considered by many to be among the seminal texts of early (British) cultural studies research. Cast as active interventions into understanding a particular crisis of "English" national identity following World War II—specifically, why large segments of the working class chose to align with a particular political ideology that was seemingly at odds with and did not appear to represent so-called traditional working-class values—their work was *inherently* political.[4]

However, and whereas the CCCS under Hoggart focused on everyday lived experiences and transformations of the English working class, the Hall years took (British) cultural studies in a new direction—what he called "Marxism without guarantees" (Hall, 1982)—as he sought to account for the emergence of the New Right within British politics in the second half of the 1970s. As such, Hall's approach focused on the contextual specificities of cultural meanings, relations, and identities within the temporal and spatial boundaries of the object in question. Marking this shift were numerous texts that have become hallmarks in the field, among them *Policing the Crisis: Mugging, the State, and Law and Order* (Hall, Critcher, Jefferson, Clarke, & Roberts, 1978), *The Empire Strikes Back: Race and Racism in 70s Britain* (CCCS, 1982), *There Ain't No Black in the Union Jack: The Cultural Politics of Race and Nation* (Gilroy, 1987), and *Learning to Labor: How Working-Class Kids Get Working-Class Jobs* (Willis, 1977).[5]

Somewhat lost among this historical record, however, as David Andrews and Jon Loy (1993), and later David Andrews and Michael Giardina (2008), remind us, is that cultural physicalities—most notably those related to the sporting body, but also the leisuring body, active body, healthy body, and so on—have long been embedded within the development of the (British) cultural studies tradition. In point of fact, such physicalities have been a recurrent focus of cultural studies research ever since it was discussed within Hoggart's (1957) critique of postwar British working-class culture; the subtitle of Hoggart's book, *The Uses of Literacy: Aspects of Working-Class Life, With Special Reference to Publications and Entertainments,* expresses its implicit focus on popular institutions such as sport. Indeed, Hoggart's literary humanistic approach viewed sporting culture as a popular practice that easily related to the material conditions and experiences of working-class existence:

> At work, sport vies with sex as the staple conversation. The popular Sunday newspapers are read as much for their full sports reports as for their accounts of the week's crimes. Sports conversations start from personalities, often spoken of by their Christian as well as by their surnames, as "Jim Motson," "Arthur Jones," and "Will Thompson": technical details of play are discussed, often to the accompaniment of extraordinary feats of memory as to the history of matches many seasons back. The men talk about individuals whom they know, at least as figures on the field, in situations eliciting qualities they can respect and admire. (p. 91)

In addition, in his landmark study, *The Making of the English Working Class,* E. P. Thompson (1963) would a few years after Hoggart identify numerous physical, leisuring, and sporting practices that contributed to the creation of a coherent English working-class culture within conditions of industrialization. (for more see Andrews & Giardina, 2008, p. 398).

However, as Andrews and Loy (1993) rightfully point out in their authoritative excavation of "sport" from the CCCS's Working Papers in Cultural Studies and Occasional Stenciled Papers series, it was not until cultural studies was institutionalized at the CCCS that "sport-focused cultural studies projects began to appear" (p. 30) in wide circulation and with anything resembling regularity. A sampling of these papers speaks to the location of sport and the sporting/physical body within both the Centre and English culture; for example, we find Chas Critcher's writings on football (1971) and women's sports (1974) as a cultural practice; Paul Willis's studies of motorbike clubs (1971) and women's sports (1974); Rod Watson's (1973) account of public announcements of motor-racing fatalities; John Clarke's writings on football hooliganism vis-à-vis skinheads (1973, 1975); and Roy Peters's (1975) treatise on the television coverage of sport.

The impact of these early inroads was not lost on Toby Miller. Writing in his introduction to *A Companion to Cultural Studies,* Miller (2001) states:

> I recall my excitement when I first saw the cover of the Birmingham Center's *Working Papers in Cultural Studies 4* of 1973 … the bottom center-left read like this:

LITERATURE~SOCIETY
MOTOR RACING

> It seemed natural to me for these topics to be together (as is the case in a newspaper). But of course that is not academically "normal." To make them syntagmatic was *utterly sensible* in terms of people's lives and mediated realities, and *utterly improbable* in terms of intellectual divisions of labor and hierarchies of discrimination. (pp. 12–13, emphasis in original)

From these interventions forward, the corpus of sport/physical cultural studies blossomed—slowly at first, but later with increasing regularity—beginning to shape the field and take it in multiple directions in the early 1980s and into the 1990s (see, e.g., Andrews, 1993; Clarke & Critcher, 1985; Cole & Hribar, 1995; Gruneau, 1985; Gruneau & Whitson, 1993; Jackson, 1992; Tomlinson; 1981; Whannel, 1983). A guiding principle to take from this early work can be summed up in the work of two of the more prolific scholars associated with cultural studies' early formation, Clarke and Critcher. As they posit in *The Devil Makes Work: Leisure in Capitalist Britain* (1985), which focuses on the problematic politics of leisure and popular culture, their primary interest "is not really in 'leisure' itself, it is in what leisure can tell us about the development, structure, and organization of the whole society" (p. xviii). Updating Clarke and Critcher, we would suggest that our primary interest in the *physical* in this physical cultural studies of ours is not really about physical corporealities in and of themselves, but about what they can tell us about the development, structure, and organization of the whole society in the historical present.[6]

▣ III. So What *Is* This *Physical* In Physical Cultural Studies?

In much the same way as the author from whom we liberally re-appropriate our title of this section asked in his article, "What Is This "Black" in Black Popular Culture?" (Hall, 1993), we start with a question: *What sort of moment is this in which to pose the question of* physical *culture?* Clearly, it is an epoch in which the imperatives of economic growth and corporate-infused democracy in many cases supersedes society's will to ensure the health of its young, its poor, and its suffering; the laboring and leisuring vestiges of the Keynesian welfare state are being slowly eradicated in the pursuit of a "pure" free-market condition; access to spaces of bodily play and health have been colonized and made exclusionary for the purposes of capital accumulation; women's bodies, queer bodies, and Othered bodies of difference are still largely marginalized in most realms of global physical culture; and now, more then ever, the body is subjected to infantilizing, sexualizing, and objectifying disciplinary regimes.

This is an apt description, for definitional purposes at least, but how are we to better understand the ground-level impact of such developments on real people? How are we to better understand—and communicate—a cultural register about, as Arundhati Roy (2001) has so eloquently written, "what it's like to lose your home, your land, your job, your dignity, your past, and your future to an invisible force. To someone or something you can't see" (p. 32)—stories about what it's like to hate and feel despair, anger, and alienation in a world bursting at the seams as it struggles to reinvent itself and its dominant mythologies (see Denzin & Giardina, 2006). How can we come to see more clearly how "understanding is constituted by the cultural experiences embedded in [our] research" (Berry & Warren, 2009, p. 601) acts made meaningful by and through the "dynamic and dialectical relation of the text and body" (Spry, 2001, p. 711)?

Our answer, we believe, is that the best qualitative inquiries of physical culture—those that intercede on antihumane structures, practices, and symbolic acts within cultures of the active body—make use of *both* physical and ideological praxis to, as Ernesto Laclau and Chantal Mouffe (1985) posit, *articulate* the human experience with these broader contextual forces. These connections are meant to highlight "any practice establishing a relation among elements such that their identity is modified as a result of the articulatory practice" (Laclau & Mouffe, 1985, p. 105). Most often situated within Hall's (1996) work, the idea of the metaphoric lorry in conceptualizing the dialectic theory and method of articulation is quite helpful in understanding such a practice:

> "Articulate" means to utter, to speak forth, to be articulate. It carries that sense of language-ing, of expressing, etc. But we also speak of an "articulated" lorry (truck): a lorry where the front (cab) and back (trailer) can, but need not necessarily, be connected to one another. The two parts are connected to each other, but through a specific linkage, that can be broken. An articulation is thus the form of the connections that can make a unity of two different elements, under certain conditions. It is a linkage which is not necessary, determined, absolute, and essential for all time. You have to ask under what circumstances *can* a connection be forged or made. (pp. 141–142; emphasis in original)

Or, as Jennifer Daryl Slack (1996) puts it, articulation is *both* that connection between broader contextual formations and the empirical transference we seek to establish and, at the same time, the methodological *episteme* under which we operate. On the articulation of context and practice, and with particular regard to the ways in which practice produces context, Slack writes: "The context is not something *out there, within which practices occur or which influence the development of practices. Rather, identities, practices, and effects generally constitute the very context in which they are practices, identities, or effects*" (p. 125, emphases in original).

Thus is our *physical* cultural studies project not simply an exercise in context mapping or abstracted corporeal cartography, but a method of using the political and politicized body to directly engage and interact with human activity; that is, an articulatory praxis that produces, and is produced by, social, political, and economic context/s. Furthermore, if we are to emerge from the tautological impasses of our structural Marxist forbearers, then we must break free from the determinism of early Marxist-inspired social thought, instead placing value on the idea that the cultures of the body are neither *necessarily correspondent* to the overdetermining structural realm (much like the economic base determining the superstructure) nor *necessarily noncorrespondent* (culture as autonomous from economic relations) (see Hall, 1985; Laclau & Mouffe, 1985). In other words, and rephrasing Andrews (2002), we might say that the structure and influence of the body in any given conjuncture is a product of intersecting, multidirectional lines of articulation between forces and practices that compose the social contexts. The very uniqueness of the historical moment or conjuncture means there is a condition of no necessary correspondence, or indeed noncorrespondence, between physical culture and particular forces (i.e., economic). Forces do determine *givenness* of physical practices; however, their determinacy cannot be guaranteed in advance (p. 116).

While there are no necessary guarantees that the body will be produced in predictable ways, this is not to suggest that the weight of social, political, and economic structures is not always already bearing down on the body. To rework Karl Marx, and later C. Wright Mills (1959), *we make our own cultural physicalities, but not under conditions of our choosing.* To ignore this fundamental dialectic is at once to abstract the body and to depoliticize its existence. Amid the tides of the academic-industrial complex, decontextualized or antidialectic analyses of the body are *made political.* To feign political neutrality is itself a political act, one that bolsters the hegemony of a natural, taken-for-grantedness of the formations of contemporary life—as the radical historian, Howard Zinn (1996), famously reminded us, "you can't be neutral on a moving train." Informed by Richard Johnson's (1987) formulation of (British) cultural studies, Andrews (2008) makes this point clear: "Physical Cultural Studies researchers must remain vigilant in their struggle against 'the disconnection' that will surely occur if we produc[e] studies in which physical cultural forms are divorced from contextual analyses of 'power and social possibilities'" (p. 58). In critically studying the cultures *of* the body, we seek to better understand context *through* bodily practice, as well as the oppressive and liberatory potential of the human body as constrained by contextual forces.

As such, we should strive to produce or elicit a public pedagogy that peculiarizes the banalities of political and politicized bodies. Indeed, by revealing the social constructedness of the historical contexts acting on cultures of the body, those working on/in physical culture should foster critical consciousness among both those individuals whose social, cultural, and economic status is inextricably linked to past cultures of alienation and exploitation and those individuals whose lives continue to be challenged as a result. Ben Carrington (2001) makes this point clear when he laments the depoliticized nature of too much of what passes for [physical] cultural studies today:

> Being able to deconstruct the dialogic processes within a Nike commercial is one thing; connecting them to the exploitative economic production of the shoes themselves in Southeast Asia, through to their consumption in the deprived inner-cities of the West, and the meanings this produces, is quite another, and a process too often not addressed. (p. 286)

Although such pseudo-political, relatively textual work of the kind Carrington critiques may well "teach us about consumer culture, late-capitalism, and identity politics therein" on some general level, as Joshua I. Newman (2008) notes, it nevertheless "fails to engage the dialectics of practice through which these discursive formations are made meaningful, consequential, and powerful" (p. 2), to move beyond a critique of texts to the relationships between texts, contexts, and interventions into the material realities of everyday operations of power.

In aiming to avoid such a pitfall, we actively follow and endorse the Brazilian critical educator Paulo Freire (1970/2006)—whose pedagogical method was a *mélange* of counteroppressive politics and emancipatory education, of classroom instruction and everyday encounters—in cultivating a form of popular education intended to share in the communal practice of raising individual consciousness (*conscientization*) of the political and oppressive regimes acting against the human condition. For Freire, this critical consciousness, or *conscientização,* comes about when individuals develop an epistemological awareness of the ways dialogic, political, and economic structures act on their everyday lives. Such awareness is nurtured through constant dialogue with, and consideration of, the oppressive elements of one's life, and actively imagining and working to make real alternative, egalitarian social formations.

As Norman Denzin and Michael Giardina's (2010) Freirean-inspired volume on qualitative inquiry and human rights helps us to remember, the conduct of qualitative inquiry "is not *just* about 'method' or 'technique,'" but likewise *also* an inherently political project that works toward "making the world visible in ways that implement the goals of social justice and radical, progressive democracy" (p. 14, emphasis in original).[7] In practical terms, this means subscribing to a public pedagogy that is "never neutral, just as it is never free from the influence of language, social, and political forces" (Giroux, 2000, p. 8). The goal here is to foster an engaged social citizenship, in effect a version of what Peter McLaren (2000) refers to as a *revolutionary pedagogy* that

> creates a narrative space set against the naturalized flow of the everyday, against the daily poetics of agency, encounter, and conflict, in which subjectivity is constantly dissolved and reconstructed—that is, in which subjectivity turns-back-on-itself, giving rise to both the affirmation of the world through naming it, and an opposition to the world through unmasking and undoing the practices of concealment that are latent in the process of naming itself. (p. 185)[8]

To wit, the field's principal intermediaries have often professed that only through rigorous, empirical qualitative encounters can we begin to elucidate the complexities of contemporary physical culture (e.g., Andrews, 2008; Andrews & Silk, 2011; Hargreaves & Vertinsky, 2006; Ingham, 1985; Markula & Pringle, 2006; etc.). This much we agree on, and there is an ever-growing body of work along this general plane that speaks to the promise and potential of physical cultural studies.

Jacqueline Reich's (2010) excellent work on the mediated dimensions of early 20th-century American physical culture as embodied by famed fitness guru Charles Atlas (born Angelo Siciliano in Calabria, Italy) is one such example, as she makes meaningful the textual discourses of "body building photography and the creation and marketing of his iconic fitness plan," which allowed for his public transformation from "Italian immigrant to pillar of American masculinity" (p. 450). As a work of cultural history and media analyses, Reich's article provides a lively and revealing critical interrogation of a popular historical figure whose success and celebrity was predicated on his physicality and the meanings ascribed to it through the armatures of fledgling marketing and advertising industries of the day, especially as related to the complex racial politics of immigration. In a related fashion, Shari Dworkin and Faye Linda Wachs (2004) investigated the gendered discourses germane to postindustrial motherhood in their textual analysis of *Shape Fit Pregnancy* (a magazine aimed at "young, intelligent, affluent, and professional" middle-class women); it is a tour de force commentary on the popular and political forces working to shape normative ideals of femininity, success, and healthy bodies narratives at the height of the U.S. health and fitness empowerment boom. Of course, these are just two of a myriad number of works published in recent years (e.g., Aalten, 2004; Atkinson, 2008; Brace-Govan, 2002; Butryn & Masucci, 2009; Chase, 2008; Cole, 2007; Evers, 2006; Francombe, 2010; Fusco, 2006; Grindstaff & West, 2006; Helstein, 2007; Markula, 1995; Metz, 2008; Miah, 2004; Schultz, 2004; Scott, 2010; Thorpe, 2009; van Ingen, 2004; Wedgewood, 2004; Wheatley, 2005; etc.) that have worked to establish a provisional canon for physical cultural studies (see Andrews, 2008, for more on this).

We have elsewhere engaged in a lengthy debate over several of the core tenets of this fledgling field (see Giardina & Newman, in press), breaking in some regard with our contemporaries about both the definitional and historical legacies contained therein. For our purposes here, and more widely applicable to the broader cultural studies universe, let it be sufficient to say that we want to push beyond the mediated, the textual, the corporeally disembodied to a more actively engaged research act—whatever the focus of empirical inquiry—one that does not reduce itself to textual patterns, media representations, and/or grand corporeal narratives or erase the researcher's own body and politics from any empirical discussion (or do some combination of both).

What we are arguing against then, is the abstraction of politically enfleshed bodies, the disappearance of authorial bodies, and the empirical dialectics of the self, which have given way to rhetorical bravado and, in some cases, what reads like educated guesswork. If we want [physical] cultural studies to matter as an intellectual domain—*or at the least, to push the field(s) forward within the pages of this handbook*—we believe that it has to be more than an empty metaphor, a bland descriptor of *any* study focused on *any* object residing in the cultural realm: We do not think it is enough to write and report on bodies and physicality alone *as if* we were in the field of body studies writ large (or sociology of the body, etc.), or simply apply some form of cultural studies inheritance to sites and artifacts of physical culture (which we have seen as a growing trend the last decade). Ours, then, is a project that seeks to *move beyond writing and researching* about *bodies to writing and researching* through *bodies as a principal force of the research act.*[9]

Put differently, we cannot allow [physical] cultural studies to become a discipline of professional *convenience:* It has to mean more than simply critically "reading" an object of culture from a *distance* (e.g., from our couch or in front of our computers; on ESPN, in *Sports Illustrated,* or on *The New York Times* website, etc.).[10] We cannot allow it to suffer from the same ill fate of (sadly, much of American) cultural studies, which—as Michael Bérubé (2009) recently reported in a somewhat disheartening but nevertheless accurate report on its perception within the corporate university—"now means everything and nothing; it has effectively been conflated with 'cultural criticism' in general, and associated with a cheery 'Pop culture is fun!' approach" (n.p.).[11] We cannot stand to see the field abandon the shared promise of a politically engaged, performative field of inquiry in favor of producing academic and economic capital for the sake of doing so (we readily admit that we are not immune to this charge, either). And we cannot ignore the extent to which the fact of the physical, and the study thereof and *through*, is *consequential;* that spatialized and temporalized empirical physicality is still besieged to ever-increasing ends by normative, hegemonic, and sometimes dangerous forces.

Rather, as Denzin (2007b) reminds us, the best of [physical] cultural studies should be conceived of as an emphatically political, *activist-minded* project, a public intellectualism on the order of the kind Noam Chomsky advanced in his 1967 article "The Responsibility of Intellectuals," where he argued that intellectuals (i.e., you) have a moral, ethical, and professional obligation to speak the truth, to expose lies, and to see events in their historical perspective (see also Denzin & Giardina, 2006).[12] To do so, we must think through the "radical implications of cultural politics, the role of academics and cultural workers as oppositional public intellectuals, and the centrality of cultural pedagogy as moral and political practice" (Giroux, 2001, pp. 5–6)—indeed, thinking through the very political dimensions of our research acts. We must seek critical methodologies that "protest, resist, and help us represent, imagine, and perform radically free utopian spaces" (Denzin, 2007b, p. 40)—methodologies that aid us in creating the very worlds we are embedded within, one that

is always already performative, ideological, and pedagogical. We must see our pedagogy as a "kind of transformative intellectual practice that can encompass the variegated works of artists and critics, as well as researchers and educators" (Dimitriadis & Carlson, 2003, p. 3). We must, following Roy (2004), "never let a little hunk of expertise carry us off to our lair and guard against the unauthorized curiosity of passers-by" (p. 120), doing instead the opposite: "We must create links. Join the dots. Tell politics like a story. Communicate it. Make it real . . . And refuse to create barriers that prevent ordinary people from understanding what is happening to them" (p. 10).[13] And we must be committed, as Judith Butler (2004) writes in her moral polemic *Precarious Life,* to creating "a sense of the public in which oppositional voices are not feared, degraded, or dismissed, but valued for the instigation to a sensate democracy they occasionally perform" (p. 151).[14]

This is *not* to say, however, that our answers are or should be "determin[ed] in advance," that our politics should serve "as an excuse for not doing the work of coming to a better understanding of the context of struggle and possibility" (Grossberg, quoted in Cho, 2008, p. 121). Rather, and for example, as Freire (1999) expressed in *Pedagogy of Freedom:*

> My abhorrence of neoliberalism helps to explain my legitimate anger when I speak of the injustices to which the ragpickers among humanity are condemned. It also explains my total lack of interest in any pretension of impartiality. I am not impartial or objective; not a fixed observer of facts and happenings. I never was able to be an adherent of the traits that falsely claim impartiality or objectivity. That did not prevent me, however, from holding always a rigorously ethical position. Whoever really observes, does so from a given point of view. And this does not necessarily mean that the observer's position is erroneous. It is an error when one becomes dogmatic about one's point of view and ignores the fact that, even if one is certain about his or her point of view, it does not mean that one's position is always ethically grounded. (p. 22)

In the sections that remain, and keeping the above in mind, we outline one such theoretical and methodological progression for a radically embodied cultural studies project.

▣   IV. PHYSICAL CULTURAL STUDIES AND "BODILY PARTICIPATION"

If we allow that our (specific) project must be about more than just a topical focus on bodies and physicality (in whatever varied forms they may take, from the active to the inactive), *and* that we should endeavor to adhere in some manner to the best politics and practices of the (British) cultural studies tradition (i.e., antidisciplinary, self-reflective, political, theoretical, and radically contextual; Grossberg, 1997)—then, first and foremost, we would do well to begin thinking of the research *act* of [physical] cultural studies as necessarily being "an embodied activity" (Coffey, 1999, p., 59). For as Amanda Coffey (1999) writes in *The Ethnographic Self,* we must acknowledge the critical extent to which "our body and the bodies of others are central to the practical accomplishment of fieldwork. We locate our physical being alongside those of others, as we negotiate the spatial context of the field" (p. 59). This would seem an obvious point. And yet, as Michele K. Donnelly (2009) reminds us, the very *embodied* practice of one's research act is, paradoxically, most often overlooked, owing perhaps to the seeming "inevitability of the body's role" in conducting qualitative inquiry. Paul Atkinson, Sara Delamont and William Housley (2008) have written at length about avoiding this oversight in ethnography, explaining that

> the very idea of participant observation implies not merely "observation" but also the embodied presence of the ethnographer. Clearly we cannot engage with the social actors we work with without physical copresence; the element of "participation" is a bodily one. (p. 140)

As such, argues Donnelly (2009), "Participation involves making and maintaining space for the researcher in the field: socially, culturally, *and physically*" (p. 4, emphasis added).

To engender significant heuristic pedagogies of the subjugated and transformative body, we need to wield theories, strategies, and epistemologies that account for the minutiae, the variations, and the complex formations of physical culture (*especially those impacting our own bodies within the research act*). Tami L. Spry (2010), drawing on Elyse Pineau's (2000) notion of *performative embodiment,* describes the epistemological necessity of such an engaged research act in this manner:

> We live in our bodies and learn about self, others, and culture through analyzing the performances of our bodies in the world. The performing body is at once a pool of data, a collector of data, and then the interpreter of data in knowledge creation, in the process of epistemology. (p. 160)

The nature of knowing the body is—and always has been— both a politically entangled and dialectically meaningful enterprise. Because of this dynamic, adherents to the demands of a radically embodied project such as ours might do well, after D. Soyini Madison (2009), to

> embrace the body not only as the feeling/sensing home of our being—the harbor of our breath—but the vulnerability of how our body must move through the space and time of another—transporting our very being and breath—for the purpose of knowledge, for the purpose of realization and discovery. Body knowledge, knowledge through the body, is evidence of the present. . . . This is intersubjective vulnerability in existential and ontological order, because bodies rub against one another flesh to flesh in a marked present and where we live on and between the extremes of life and death. (p. 191)

But such an engaged, interventionist, reflexive, reciprocal, and practiced method can sometimes get messy, if not conflicted. For example, Loïc Wacquant (2004), in his widely hailed treatise on boxing culture on the South Side of Chicago during the late 1980s and early 1990s, *Body and Soul: Ethnographic Notebooks of an Apprentice Boxer,* presents his readership with a "carnal sociology" that finds his body coordinating three points of intersection: his flesh-and-blood bodily actions, his internal struggles with training, and his interactions with his trainer and fellow boxers. While serving as a notable early entrant into the pantheon of research *on* physical culture—and most assuredly using his own researcher body as the primary source of knowledge production—Wacquant ignores the politics of representation governing his enfleshed body and the context of his research act/s. In a stinging critique of Wacquant's text, Denzin (2007a) reminds us of this failing when he states,

> His method presumes a reality that is not shaped by cultural mythology, or self-aggrandizing statements. He wants his embodied method to go directly to the real, actual world of the boxer. [But his] carnal sociology *stays* at the level of the body, the white/black male body in the dying throes of a violent sport. This is a sociology that is outside of time, a sociology that some say time has passed by. (pp. 429–430, emphasis added)

By staying wholly at the level of the body, Wacquant's narrative lacks a "reflexive language of critique, and praxis, a way of looking into and beyond the repressive cultural categories of neoliberal capitalism" (p. 430).

Radically embodied cultural studies research, following Gretchen Rossman and Sharon F. Rallis (2003), is necessarily a complicated mélange of the "recursive, iterative, messy, tedious, challenging, full of ambiguity, and exciting" (p. 4). And as we strive to "forge the micro and the macro in a way that does not reduce the local experiences to props of social theories" (Saukko, 2005, p. 345), we must be sensitive to the ways in which our *own* bodies, and our *own* performances, shape the research encounter. But we must *also* be sensitive to the ways our research acts "reproduce a particular order of things that is shaped by the racial and cultural politics of neoliberalism" (Denzin, 2007a, p. 430) and remain cognizant of the wider conjunctural forces impacting the body (from the mediated to the political); it is not an either/or choice.

Scholars of this new generation have, to varying degrees, found themselves aligned with the very philosophical imperatives on offer by Madison and her contemporaries. Take the following three avatars of this direction.

Ashley Mears's (2008) ethnographic account of the New York fashion industry—in which she actively worked as a model during the course of her research—quite explicitly implicates her *researcher body* in the process of constructing a critical interrogation of gender, power, and cultural production. Her "bodily copresence" is necessarily pronounced in the following extract, in which she describes a model line-up and the degree to which the ever-present, critical gaze works "between and within models as they compare their bodies to one another":

> In the runway rehearsal line, the model before me comments as another model walks past, "Her waist is so small!" Standing backstage in our first look—little bikinis—the models scan each other from head to toe, *myself included*. After standing in line for a while, most of us end up folding our arms across our stomachs. Perhaps the others are tired, or bored, *but I do it to cover up*. (p. 438, emphases ours)

Research, for Mears, is not solely about the mechanical process of methodological expression: It is a personal[ized] and internal[ized] journey—a complicated, self-inhabiting one aimed at intervening directly into the conditions of production and consumption that shape, govern, and exploit women's bodies. This is not the work of a casual or detached ethnographer, carefully taking notes so as to catalog and "reproduce" the social world (as if this can be done!). Rather, through active bodily investment in working within and against the hegemonic spaces of the fashion industry, Mears works through her body to better understand the physical and psychic demands of emotional labor on others in the profession and challenge the chilling effects it can have on its most vulnerable purveyors (i.e., young women). As Ron Pelias (2005) makes clear, there is a considered recognition in such an approach "that individual bodies provide a potent database for understanding the political and hegemonic systems written on individual bodies" (p. 420).

In a related manner, Michael's (see Giardina, 2005, 2009) engaged bodily interrogations of transnational movement, power, and politics serve as a useful vivisection of the complex, conflictual, and continually shifting identity performances revealed in and through our fleeting global experiences with one another. Whether brushing up against the hyphenated spatial histories of British colonialism and Asian diaspora in London and Manchester or witnessing rampant expressions of xenophobic nationalism pervading the U.S. popular-public sphere in Yankee Stadium in New York, Michael actively sutured himself and his critique into and through the landscape of global social relations, including his own interpretive bodily interactions of disconnection and reconnection with place, home, and nation. As he reflexively stated at length while writing a stone's throw from the Baltic Sea:

> Yet, strangely, here in a country whose native tongue I can barely speak clearly without mistakenly ordering the wrong food off of a menu [I really thought I said something that came close to sounding like '*Jag skulle lik en hamburgaren och en soda, behaga*'], I ultimately feel oddly connected to the world. Although writing about someone *else's* flexible or hybrid performances of culture and identity has tended to be easy for me, writing about my *own* (dis) articulation with—nee implication in?—such frames generally proves a far different task. But the moment/s are there, in the text, behind the screen, in the performative acts, of coming face to face

with my own (trans)nationally unbounded, floating identity in the performances of others who have been materially and representationally Othered (often revealed unintentionally through some form of white privilege, such as passing through an airport security checkpoint unassailed by watchful eyes): On my way "home" from London once, I caught my reflection in a duty-free store mirror. I was wearing a dark blue fleece pullover, jeans, and Swiss-made Bally shoes. No one would have mistaken me for being "an American" unless we spoke to each other . . . and I didn't go on advertising that fact . . . But with a few hours to kill before my flight was scheduled to depart, I ate a bacon cheeseburger and drank a Coors Light at the T.G.I. Friday's restaurant in Heathrow's Terminal 3. "What could be more American than that?" (I thought at the time). (Giardina, 2009, p. 174)[15]

Here we see Michael presenting his life "as mutually and reciprocally co-articulated to the world and the participants in that world" (McCarthy et al., 2007, p. xx). Not only is his own body implicated in the performance of his research act itself, but *through it,* he exposes the "inviolable link between researchers' identities, experiences in the field, and substantive findings" (Joseph & Donnelly, in press). This is very much an explicit attempt to "reinforce the lost arts of humility, self-questioning, deep reflexivity and conversation in research, re-connecting ourselves to the fractured and divided worlds in which we live" (McCarthy et al., pp. xx). It is an attempt, as Carrington (2008) would say, to

> develop a reflexive account of the Self that opens up to critical interrogation of both the researcher's own biography in relation to those studied and the very act of inscribing or narrating that ethnography, the turning of the analytical gaze back on the researcher in an attempt to dissolve or at least problematize subject/object relations within the research process and even that we have a unified, fixed, singular Self. (p. 426)

Invoking Trinh T. Minh-ha (1991), it is an effort to "interrogate the realities our writing represents, to invoke the teller's story in the history that is being written and performed" (p. 188).

A third example is drawn from Donnelly's (2009) work on "women's onlyness" in roller derby, which explicitly ties the notion of placing our bodies within and among the empirical uncertainties of spatial and corporeal practice(s) discussed above all together:

> When beginning my roller derby research, I was keenly aware of *my* body with respect to impression management. I thought carefully about how to dress, how to speak, how to move, where to be, what to be, etc. . . . More problematically, I was initially apprehensive that my own *performing* body would not hold up to the demands of a physically-demanding, if not dangerous, sport. Yet the explicit physical experience of my research act had the unintended consequence of visibly transforming my body into that of someone who skates for several hours every week for a year . . . Importantly,

Coffey (1999) identifies that "In certain places taking part in the physicality of the setting may well be part of gaining insight or understanding into that setting." As I began to notice changes to *my* body, I came to notice—and to better understand—comments I had been hearing all along from my research participants about a "derby body," and, specifically, a "derby butt"—it was only in and through *my* body that I was able to make sense of those bodies performing around me. (p. 8, emphases ours)

Donnelly expresses the thorough-going use of knowledge produced by and through *her* researching body to better understand the aperçutive bodily interactions, feelings, and physicalities experienced by her research participants. Acting as what Cornel West (1991) would term a critical moral agent—one who "understands that the consequences of his or her interventions into the world are exclusively political, judged always in terms of their contributions to a politics of liberation, love, caring and freedom" (Denzin & Giardina, 2006)—Donnelly (and Mears, Giardina, and others doing similar work) is not merely presenting an engaging yet anecdotal look at body politics observed during her accounts of derby life. Rather, she illustrates how, for critical agents and provocateurs of cultural studies, "the body is implicated in the roles and relationships of fieldwork both in terms of how our body becomes part of our experience of the field and in the necessity (albeit often implicit) . . . to learn the skills and rules of embodiment in the particular social setting" (Coffey, 1999, p. 73).

By necessarily situating the researcher's physical body in and among bodies—sharing experiences of the physical ways in which we experience fieldwork—we are better able, as the examples above make clear, to elucidate the politics of gender, exclusion/inclusion, and corporeality acting upon and within these spaces of physical culture. In so doing, as Elin Diamond (1996) notes, we enable the incisive critique and reflexive re-evaluation of cultural contexts through one's own subjectivity (a subjectivity that, Kakali Bhattacharya, 2009, notes is "full of contradictions, inconsistencies, tensions, voices, and silences . . . [of] fractured shifts, border crossings, and negotiations between spaces" [p. 1065]). *But to do so ultimately means that the researcher's body (and self-perceptions thereof) is made vulnerable to, and by, the politically iniquitous circumstances into which the body has been thrust.* This we address in the following section.

▣ V. CRITICAL REFLECTIONS ON THE PHYSICAL (CULTURAL STUDIES) BODY

As we put forth in the section above, we believe the best critical analyses of the corporeal are those that envisage the body through both dialectically imaginative techniques and a conscientious, often stifling, self-awareness of researcher and research

act (see Langellier, 1999). As such, to convolute our simple social worlds—to excavate the plural dimensions of social life—we need to both make use of *and also reflect on* how our own bodies frame and are framed by the critical cultural analyses we undertake. In other words, we need to locate our vulnerable bodies within spatial praxes and be insatiably reflexive in how that (re)location produces new dimensions, complex relations, and new bodily epistemologies.

Carrington's (2008) work on racialized performativity, reflexivity, and identity is especially instructive of this position, as he interrogates (his own) black masculinity and the differently arrayed and performed iterations of black bodiedness he experienced during his research on and with a "black" cricket team in Leeds, England (e.g., as a black south Londoner being "read" by his older West Yorkshire teammates as "black British" rather than the "authentic" Caribbean-based identity they saw themselves as holding). In moving to problematize the signification of blackness itself, revealed to us through deeply personal and self-reflexive accounts of his position "in, but not fully of" the particular black cultural space within which he was located during his time as a participant-observer of the cricket club, he acknowledges that the crux of the matter was that:

I was coming to terms with my own black Britishness. . . . I started to engage those "most personal" aspects of my self; that is, I began to think about what it meant for me to be "black." . . . [M]y experiences in the field were proving difficult as I negotiated field relations in which my blackness was being questioned. The personal diary began to take the form of self-reflexive questions: How black *am* I? Am I *black* enough? What does such a question even mean? (pp. 434–435, emphases in original)

Susanne Gannon (2006), invoking the work of Roland Barthes, might say of Carrington's weighty confessional that his work reveals how "the lived body is a discursive and multiple but very present space where we do not go looking for any 'sacred originary' but for traces and unreliable fragments" (p. 483) through which to "foreground the dialogic relationship between the self and his or her tenuous and particular social/cultural/historical locations" (p. 477). Or, as Coffey (1999) would say,

[He is] engaged in a practice of writing and *rewriting* the body. This does not only include the writing of *other* bodies, as performers and physical entities of the social world. We are also engaged in responding to and writing our *own* bodies—as well or sick or fit or hurting or exposed or performing. (p. 131, our emphases)

Carrington is not alone in publicly confronting his intersubjective bodily tensions as he works through its embodied politics. Exposing us to a similar dilemma, Silk (2010) unmasks—if not openly questions—his research act in relation to his consuming identity within spectacularized space in Baltimore, Maryland, and the troubles this caused for his self-reflexive, political self:

> How could I live, work, as a supposed academic committed to social justice and overcoming social inequality, in Redwood Towers? How could I produce work that offered my narrative accounts from a position of comfort, and, of a city whose regeneration favored civic image over improved citizen welfare? Was writing about this spectacularized consumption space as I had previously done . . . only serving to glamorize, if not reify, the space, and conveniently ignoring the most pressing public health and social issues? Questions abounded. Social justice? For whom? For what purpose? Whose city was this? Who belonged? Who did not belong? Where did I fit? Did I fit? Did that even matter? . . . To understand, to expose, to intervene, I needed to shift—literally, by moving out of Redwood Towers to a rowhouse in Baltimore's historic Pigtown community. To do so, however, involved crossing a street that "friends" and colleagues had warned me never to go west of— Martin Luther King Expressway. (pp. 5–6)[16]

Jennifer L. Metz (2008) similarly conveys a sense of internal conflict at the use of her body within her research act. Her inner dialogue, contemplating an upcoming interview as part of her project on race and motherhood among professional athletes in the Women's National Basketball Association—and her ultimate re-visioning of identity through the performance of her "role" as a member of the working media, which allowed her to gain access to her participants—is illustrative:

> Arriving at the stadium tonight, members of the media and I enter through the side gate door, its plain, black steel doors missing the nostalgia-trimmed ambience of the consuming public's main entrance. Suddenly, I have become one of the cogs of the pro sport mechanism, and I am always taken aback by this change. And yet it should not be a surprise. After all, the radio station I'm stringing for grants me access to [do] the interviews, I have an audience that calls in to comment on our words, I record trailers for the interviews, and yet suddenly—at this moment—the vision of myself as "MEDIA" becomes a reality. How does one *act* as M-E-D-I-A, that corporate organism a great majority of my project seeks to rebuke for its marginalization of my participants as welfare queens and ghetto sisters? (p. 250)

Here Metz places herself not only on the side of her research participants but, paradoxically, acts on the side of the corporate media to gain access in the first place. To borrow liberally from Julianne Cheek's (2007) original argument regarding research strategies for surviving in the neoliberal moment, simply "understanding the spaces is not enough"; rather, we must ask, "What actions are we going to take? What positions are we going to adopt? . . . [and] [W]hat positions are there that we might adopt?" (p. 1054). Or, as Cheek puts it, "Unless we better understand how we are positioned in these spaces, and how we may, in turn, position ourselves, then there is the very real possibility that we will be worked over by the spaces rather

than working in them and, *importantly, on them*" (p. 1057, emphasis in original).

Turning again to our own work, we have each endeavored to account for, in sometimes painstaking (and introspectively painful) detail, how our own situated [researching] bodies have forged new cultural dialectics and conjunctures. Joshua's (Newman, in press) self-reflexive account of the rediscovery of his whiteness, his Southern-ness, and his masculinity through (auto)ethnographic engagement with the "New Sporting South," for example, serves as much as an analysis of the seemingly banal nature of Southern sporting fixtures like college football and stock car auto racing as it does as a critical inspection (and introspection) of the performative politics of engaged cultural studies research on the body. As he reflected on his ethnographic fieldwork on the New Sporting South, Joshua became increasingly concerned about how his own Southern, white body—*against his best intentions*—was becoming a site of identity-based power within these spaces.

Consider his fieldnotes addressing the power and politics of whiteness experienced in the early morning hours prior to a University of Memphis/Ole Miss game:

> The tailgating party to the immediate south of where I was located had begun to fraternize with a group I had joined, telling stories and offering predictions on the upcoming game. On his way back from the "pisser," one of the neighboring tailgaters, a middle-aged white man, stopped by our area to speak with us. He said, in a soft, almost timid voice: "Ya'll mind if I tell ya'll a nigger joke?" While I wanted to answer in the negative, I held my tongue and the all-white members of my group agreed that they did indeed want to hear the joke. (Fieldnotes, September 4, 2004)[17]

This is but one example of a number of overt racist offerings Joshua noted during his time at Ole Miss researching Southern whiteness. In this instance, power was productive in the sense that he was able to use it through research outcomes to create new pedagogies of sporting whiteness. But to do this, he had to make himself *visibly invisible*—using his body to gain access to research sites and moments but not forcing his new "self" onto the lived experiences he encountered. It was through encounters such as these that Joshua came to surmise that there was a "visible center" of identity politics at work within these empirical spaces, one that celebrated hetero-patriarchal Southern whiteness as the dominant cultural corporeality. In addition, and despite his own apprehensions toward these dominant cultural politics, he found that he *himself* was becoming part of that visible center. In short, *Joshua was blending his white, Southern, masculine self in with the crowd*. In large part due to his choice of research sites—two sporting spheres most deeply saturated by neo-Confederate forms of unchallenged whiteness (college football at Ole Miss, see Newman, 2010; and later NASCAR, see Newman & Giardina, 2008), and dialoging with the "white reign" that exists within those spaces (Kincheloe, Steinberg, Rodriguez, & Chennault, 1998)—his body became a symbol of conformity amongst thousands of other similarly white bodies. Like most spectators at these events, Joshua did not wear a Confederate flag T-shirt or less subtle race-based signifiers, yet his white skin was cloaked by the "ideological blanket" (Baudrillard, 1983) that always already covers these Southern sporting spaces.

To put it as explicitly as possible, Joshua's white-skinned body—and all of its ideological and phenotypical entanglements made meaningful in the contemporary South—*is inextricably (and inevitably) bound to the conduct of his research acts.* And it is because of these entanglements that those of us seeking to do radically contextual, politically engaged, rigorously empirical, physical cultural studies must remain cognizant of how our bodies articulate with the formations of power that exist within the research space. Although this particular encounter is one of the more problematic of Joshua's ethnographic experiences within that cultural field, the politics of inquiry beg questions such as, if Joshua was not identified within the boundaries of a white, Southern, masculine researcher (i.e., an 'insider'), how would these and other interactions have been different?

* * *

The above examples call for and embrace a heightened, re-engaged sense of what Merleau-Ponty refers to as *corporeal reflexivity:* a self-awareness of the researcher as *embodied subject* (see Vasterling, 2003), both a discursive property in the physical world and an agent subjected to the existential structures acting on those discourses. In "reading for the best" of the phenomenologists' work,[18] as Stuart Hall (1986) would put it, we can surmise that Merleau-Ponty's model of *intercorporeality* illuminates the meaning-making processes active within and between bodies and the power-knowledge relations produced within the bodily encounters we seek to better understand (Kelly, 2002). Rosalyn Diprose's (2002) synthetic interpretations of Merleau-Ponty's imperative for corporeal reflexivity are worth quoting here at length:

> And while it may seem as if my corporeal reflexivity is already in place before the world or the other, which would allow the imaginary in my body to dominate, it is also the case that it is the other's body entering my field that "multiplies it from within," and it is through this multiplication, this decentering, that "as a body, I am exposed to the world" (Merleau-Ponty, 1973, p. 138). This exposure to the world through the disturbance of the other's body "is not an accident intruding from outside upon a pure cognitive subject . . . or a content of experience among many others but our first insertion into the world and into truth" (ibid. p. 139). (pp. 183–184)

Much like Deleuze's (1988) notion of "the double," discourses of the body produce embodiments: meaningful texts

projected out of similarity and difference. We are thus reminded of his famous dictum that identification is the "interiorization of the outside" (p. 98), the connection between the external discourses of identity and the internal definitions of the self. And by suturing our researcher-bodies into cultural fields of bodily texts (through adornment, gesticulation, physicality, musculature, deportment, etc.), we must not only remain aware of how our bodies are intruding on the bodies of others, but also of how we are engaging and producing various "differential processes."

It is at such a moment that we become all the more aware of the dual nature of subjectivity; at once *a subject* with some agency in shaping various experience (such as those in the research field), and yet *subjected to* the power imbedded in our own bodies and our own performances (of past and present). So we do not offer any answers on this front, but only use these reflections on the self, the body, and the politics of reflexivity and articulation to call for a messier, bottom-up qualitative engagement with the body; one that seeks to counter the nomothetic tendencies and "objective" mythologies of modern (sociology's) scientific paradigms with a contemplative method of articulation. In this regard, we defer to Alan Ingham (1997), who, in laying the groundwork for physical cultural studies and attending to its embodied-ethnographic imperative, sharply postulated:

> In "physical culture," all of us share genetically endowed bodies, but to talk about physical culture requires that we try to understand how the genetically endowed is socially constituted or socially constructed, as well as socially constituting and constructing. In this regard, we need to know how social structures and cultures impact our social presentation of our "em-bodied" selves and how our embodied selves reproduce and transform structures and cultures; how our attitudes towards our bodies relate to our self- and social identities. (p. 176)

Make no mistake: While this soulfully naked positionality might bring about risk, discomfort, and uncertainty, that sense of vulnerability and doubt can be empowering (see Stewart, Hess, Tracy, & Goodall, 2009). These uncertainties are produced out of a sense of "belonging" and in this way demand that the researcher reflect on what constitutes the self; what aesthetic, embodied, performative, (auto) biographical discourses have come to be intertwined within the research act. Vulnerability provides a lens through which to understand the tenuous body and conditions that make it unsettled. Furthermore, the vulnerable body gives the (auto) ethnographer perspective—reminding us at once that we inhabit a political body and that we are in the same instance responsible for, and answerable to, our interpretations and representations of others' corporealities (Butler, 2001). Hence, in studying the complex relations of the body, the self, and pedagogy—and representing the self and the Other in just and reflexive ways—we must be aware of, and limit the violence created by, our em-bodied selves along the way.

◨ VI. CODA: AN INCONVENIENT [PHYSICAL] CULTURAL STUDIES

A deeply articulative physical cultural studies, to rephrase Carol Rambo Ronai (1992), should engage in a "continuous dialectic of experience" (p. 396); experience that is both constituent and constitutive of context and through which we frame our discursive-constituted selves. Just as we critically interpret the texts produced by various cultural intermediaries, and even though it could get a bit messy, we ourselves must endeavor to locate our *selves* and *our* bodies in the scholarship we produce. Again invoking the work of Ingham (e.g., 1985), we must therefore make use of our bodies to understand how power operates on the body. Furthermore, we must avoid the temptation to mobilize a progressive aesthetic without fully realizing the potential for interceding through constructivist learning and emancipatory narratives. These are not bodies of society (as some sociologists would lead us to believe) but rather bodies *about* society.

If we do indeed agree that a truly articulative physical cultural studies necessarily looks to illuminate—rather than nomothetically generalize or reduce—the messiness of the human experience (and the corporealities thereof), then we need to cultivate investigations of the body *from the ground up*. This, we have suggested, can be done by developing carefully crafted critical dialogue with the authors of embodiment and empirical praxis and by cultivating those performative representations of practice that come through human interaction—through sharing knowledge and experience with other human beings. We need to understand how and why the body is meaningful, as well as "the conditions of emergence," as Butler (2009) terms it, which have made the body meaningful. In short, we must "act as participants and performers in the meanings that we seek to elicit from the subordinated worlds that we try to understand and intervene into, worlds in which we are densely implicated as meaning makers, cultural citizens, and fellow travelers" (McCarthy et al., p. xx).

◨ NOTES

1. The authors thank Norman K. Denzin, Yvonna S. Lincoln, David L. Andrews, Michael Silk, Paula Saukko, and Della Pollock for crucial feedback on earlier iterations of this chapter. The authors also thank Michele K. Donnelly, Ryan King-White, Steven J. Jackson, Jennifer L. Metz, Adam Beissel, C. L. Cole, and Mark Falcous for conversations related to this project. Special thanks to Doug Booth, Dean of the School of Physical Education at the University of Otago, New Zealand, for his generous support of this project at a key moment in its development. This chapter draws on and revisits arguments related to Giardina (2005) and Newman (2008). Portions of this chapter, especially Section IV on physical cultural studies and bodily participation, are reprinted from Giardina & Newman (in press) with kind permission of Human Kinetics.

2. Among them, the Birmingham School, Black British cultural studies, Latin American revisions of Birmingham, Australian cultural studies, Black feminist cultural studies, African cultural studies, Canadian cultural studies, and various forms in between that have developed both *de*pendently and *in*dependently of the British tradition (for more see McCarthy et al., 2007).

3. However, as Stuart Hall's (1985) re-reading of Louis Althusser made clear some time ago, these structural formations hold no guaranteed sway over our "knowing bodies" (see Lattimer, 2009). Rather, they can—and most certainly *will*—be contested. If our task within the *Handbook* is to push [physical] cultural studies forward to new frontiers, then our contribution seeks to consider the body logics, "body pedagogics" (Shilling, 2007), and, ultimately, the "new body ontologies" (as Butler, 2009, would have it) of such a project while still in its formative stages.

4. Carrington (2001) likewise reminds us that such a project was philosophically "aimed at popular education for working-class adults . . . in the hope that a genuinely socialist democratic society could be created . . . as a form of political struggle" (pp. 277–278), rather than a solely academic pursuit.

5. Hall (1996) is quick to point out that the various intellectual shifts occurring at the Centre did not come easily and without struggle, especially concerning feminism and the politics of race.

6. Despite the lofty canonical position ascribed to these early texts—as well as the generative movement(s) surrounding the institutionalization and international expansion of cultural studies—it is important to note that they do operate within a specific historical context (i.e., post–WWII Britain) and that a wholesale appropriation of so-called "British" cultural studies has always been problematic because of its explicit sensitivity to that context. As Jon Stratton and Ien Ang (1996) remind us: "We have to recognize that the intellectual practices which we now bring together under the category of 'cultural studies' were developed in many different (but not random) places in the world, and that there were local conditions of existence for these practices which determined their emergence and evolution. It is undeniable that 'Birmingham' has played a crucial role in the growth of the international cultural studies network as we know it today. But there was never just a one-sided and straightforward expansion of British cultural studies to other locations" (p. 374).

7. They continue, pointing to the demands for such a project within the current moment: "This is a historical present that cries out for emancipatory visions that inspire transformative inquiries, and for inquiries that can provide the moral authority to move people to struggle and resist oppression" (Denzin & Giardina, 2010, p. 15).

8. Critical pedagogues maintain that every dimension of schooling and every form of educational practice (from the classroom to the television screen to the sporting arena) are politically contested sites (see Kincheloe & McLaren, 2000). Through its focus on grappling with issues of ethical responsibility and the enactment of democratic ideals of equality, freedom, and justice in the pursuit of positively altering the

material conditions of everyday life, physical cultural studies as a form of critical pedagogy therefore acts as a transformative practice that "seeks to connect with the corporeal and the emotional in a way that understands at multiple levels and seeks to assuage human suffering" (Kincheloe, 2004, p. 2) at every level of injustice.

9. This is a play on Clifton Evers's (2006) phrase.

10. Poststructuralism has served us well, but along the way discourse and textuality became the end, rather than the means, toward understanding the human condition.

11. Bérubé (2009) goes on to summarize one prominent view of the current landscape as such: "Anybody writing about *The Bachelor* or *American Idol* is generally understood to be 'doing' cultural studies, especially by his or her colleagues elsewhere in the university. In a recent interview, Stuart Hall … gave a weary response to this development, one that speaks for itself: 'I really cannot read another cultural-studies analysis of Madonna or *The Sopranos*.'"

12. As Roy (2003) explains: "The starting premise of Chomsky's method is not ideological, but it *is* intensely political. He embarks on his course of inquiry with an anarchist's instinctive mistrust of power. He takes us on a tour through the bog of the U.S. establishment, and leads us through the dizzying maze of corridors that connects the government, big business, and the business of managing public opinion" (p. 83).

13. Roy (2004) continues, in unsparing terms: "The language of the Left must become more accessible, must reach more people. We must acknowledge that if we don't reach people, it's our failure. Every success of Fox News is a failure for us. Every success of major corporate propaganda is our failure. It is not enough to moan about it. We have to do something about it. Reach ordinary people, break the stranglehold of mainstream propaganda. It's not enough to be intellectually pristine and self-righteous" (p. 147).

14. In the current era of methodological conservatism with respect to growing hostility toward qualitative inquiry in general, and critical scholarship in particular, Lauren Berlant's (1997) words are worth noting: "The backlash against cultural studies is frequently a euphemism for discomfort with work on contemporary culture around race, sexuality, class, and gender. It is sometimes a way of talking about the fear of losing what little standing intellectual work has gained through its studied irrelevance (and superiority) to capitalist culture. It expresses a fear of popular culture and popularized criticism. It expresses a kind of antielitism made in defense of narrow notions of what proper intellectual objects and intellectual postures should be" (p. 265).

15. Miguel A. Malagreca (2007), writing about the borders of geography and desire, strikes a similar chord: "Though I am writing this paragraph almost two years after I started this essay at a café located at the center of Piazza di Spagna, I can still feel the frantic pace of life that inundates that square at any time during the day. … Everybody brings to this piazza a share of cultural capital. Tourist guides are meaningless in their anodyne descriptions of this Roman epicenter of wealth and tourism. I let myself be captured by the intoxicating coexistence of parallel cultures, and anti-cultures formed by the lower-class bargaining in small shops, North African and Latin American immigrants selling imitations of fashion items in the streets, and the working girls streetwalking in daylight. I was at home; oh, yes I was" (pp. 92–93).

16. Redwood Towers, located in the heart of downtown Baltimore near Oriole Park at Camden Yards and the Inner Harbor, advertises itself thusly: "Nestled atop eight floors of above-ground parking, The Redwood offers residents breathtaking views, serene surroundings, and exclusive amenities. Whatever you imagined about Baltimore City living, THIS is even better: a 24-hour fitness center, broadband internet access in every apartment, a short walk to the Light Rail, MARC train, Camden Yards, Hippodrome and Inner Harbor, plus our relaxing imaginative skydeck, The Park on Nine. Does it get any better than this? We don't think so either" (http://www.ar-cityliving.com/redwood/).

17. Repeating the "joke" itself does not add anything to the retelling of the situation, which is why we have omitted it from the chapter. Suffice to say, it is not so much that the joke was explicitly racist (which it was) *but that it was viewed as normative and normalized by those telling/listening to it in the space of the ethnography* and that the researcher/researcher body became implicated in its telling by remaining silent.

18. For a more detailed reconciliation of Jean Paul Sartre's idea of being-for-itself and Merleau-Ponty's phenomenological conceptions of self-discovery of fundamental meaning, see Kujundzic and Buschert's (1994) article, "Instruments of the Body." For our purposes here, let it suffice to oversimplify: the role of the body in each theorist's work is complex, but each acknowledges various relational interdependencies between the body, conceptions of the body, and the physical and ideological worlds.

◼ REFERENCES

Aalten, A. (2004). "The moment when it all comes together": Embodied experience in ballet. *European Journal of Women's Studies, 11*(4), 263–276.

Andrews, D. L. (2002). Coming to terms with cultural studies. *Journal of Sport & Social Issues, 26*(1), 110–117.

Andrews, D. L. (1993). *Deconstructing Michael Jordan: Popular culture, politics, and postmodern America.* Unpublished doctoral dissertation, University of Illinois at Urbana-Champaign.

Andrews, D. L. (2008). Kinesiology's inconvenient truth and the physical cultural studies imperative. *Quest, 60*(1), 45–60.

Andrews, D. L., & Giardina, M. D. (2008). Sport without guarantees: Toward a cultural studies that matters. *Cultural Studies ⇔ Critical Methodologies, 8*(4), 395–422.

Andrews, D. L., & Loy, J. W. (1993). British cultural studies and sport: Past encounters and future possibilities. *Quest, 45*(2), 255–276.

Andrews, D. L., & Silk, M. (Eds.). (2011). *Physical cultural studies: An anthology.* Philadelphia, PA: Temple University Press.

Atkinson, M. (2008). Exploring male femininity in the crisis: Men and cosmetic surgery. *Body & Society, 14*(1), 67–87.

Atkinson, P., Delamont, S., & Housley, W. (2008). *Contours of culture: Complex ethnography and the ethnography of complexity.* Walnut Creek, CA: AltaMira Press.

Baudrillard, J. (1983). *Simulations.* New York: Semiotext[e].

Berlant, L. (1997). *The queen of America goes to Washington City: Essays on sex and citizenship.* Durham, NC: Duke University Press.

Berry, K., & Warren, J. T. (2009). Cultural studies and the politics of representation: Experience⇔subjectivity⇔research. *Cultural Studies⇔Critical Methodologies, 9*(5), 597–607.

Bérubé, M. (2009, 14 September). What's the matter with cultural studies? The popular discipline has lost its bearings. *The Chronicle of Higher Education.* Available at http://chronicle.com/article/Whats-the-Matter-With/48334/

Bhattacharya, K. (2009). Negotiating shuttling between transnational experiences: A de/colonizing approach to performance ethnography. *Qualitative Inquiry, 15*(6), 1061–1083.

Brace-Govan, J. (2002). Looking at bodywork: Women and three physical activities. *Journal of Sport and Social Issues, 24*(4), 404–421.

Butler, J. (2001). *Giving an account of oneself.* New York: Fordham University Press.

Butler, J. (2004). *Precarious life: The power of mourning and violence.* London: Verso.

Butler, J. (2009). *Frames of war: When is life grievable?* London: Verso.

Butryn, T., & Masucci, M. (2009). Traversing the matrix: Cyborg athletes, technology, and the environment. *Journal of Sport & Social Issues, 33*(3), 285–307.

Carrington, B. (2001). Decentering the centre: Cultural studies in Britain and its legacy. In T. Miller (Ed.), *A companion to cultural studies* (pp. 275–297). Oxford, UK: Blackwell.

Carrington, B. (2008). "What's the footballer doing here?" Racialized performativity, reflexivity, and identity. *Cultural Studies⇔Critical Methodologies, 8*(4), 423–452.

Centre for Contemporary Cultural Studies. (1982). *The empire strikes back: Race and racism in 70s Britain.* London: Routledge.

Chase, L. (2008). Running big: Clydesdale runners and technologies of the body. *Sociology of Sport Journal, 27*(2), 130–147.

Cheek, J. (2007). Qualitative inquiry, ethics, and the politics of evidence: Working within these spaces rather than being worked over by them. In N. K. Denzin & M. D. Giardina (Eds.), *Ethical futures in qualitative research: Decolonizing the politics of evidence* (pp. 99–108). Walnut Creek, CA: Left Coast Press.

Cho, Y. (2008). We know where we're going, but we don't know where we are: An interview with Lawrence Grossberg. *Journal of Communication Inquiry, 32*(2), 102–122.

Clarke, J. (1973). *Football hooliganism and the skinheads* (Centre for Contemporary Cultural Studies Stenciled Occasional Paper Series, No. 42, pp. 38–53). Birmingham, UK: University of Birmingham.

Clarke, J. (1975, Summer). *Skinheads and the magical recovery of community* (Working Papers in Cultural Studies, Nos. 7–8, pp. 99–105). Birmingham, UK: University of Birmingham.

Clarke, J., & Critcher, C. (1985). *The devil makes work: Leisure in capitalist Britain.* London: Macmillan.

Coffey, A. (1999). *The ethnographic self: Fieldwork and the representation of identity.* London: Sage.

Cole, C. L. (2007). Bounding American democracy: Sport, sex, and race. In N. K. Denzin & M. D. Giardina (Eds.), *Contesting empire/globalizing dissent: Cultural studies after 9/11* (pp. 152–166). Boulder, CO: Paradigm.

Cole, C., & Hribar, A. (1995). Celebrity feminism: Nike-style post-Fordism, transcendence, and consumer power. *Sociology of Sport Journal, 12*(4), 347–369.

Critcher, C. (1971). *Football and cultural values* (Working Papers in Cultural Studies, No. 1, pp. 103–119). Birmingham, UK: University of Birmingham.

Critcher, C. (1974). *Women in sport* (Working Papers in Cultural Studies, No. 5, pp. 77–91). Birmingham, UK: University of Birmingham.

Deleuze, G. (1988). *Foucault.* Minneapolis, MN: University of Minnesota Press.

Denzin, N. K. (2003). *Performance ethnography: Critical pedagogy and the politics of culture.* Thousand Oaks, CA: Sage.

Denzin, N. K. (2007a). Book review: Loïc Wacquant *Body & Soul: Notebooks of an Apprentice Boxer. Cultural Sociology, 1*(3), 429–430.

Denzin, N. K. (2007b). *Flags in the window: Dispatches from the American war zone.* New York: Peter Lang.

Denzin, N. K., & Giardina, M. D. (Eds.). (2006). *Contesting empire/globalizing dissent: Cultural studies after 9/11.* Boulder, CO: Paradigm.

Denzin, N. K., & Giardina, M. D. (2010). *Qualitative inquiry and human rights.* Walnut Creek, CA: Left Coast Press.

Diamond, E. (1996). Introduction. In E. Diamond (Ed.), *Performances and cultural politics* (pp. 1–12). New York: Routledge.

Dimitriadis, G., & Carlson, D. (2003). Introduction: Aesthetics, popular representation, and democratic public pedagogy. *Cultural Studies⇔Critical Methodologies, 3*(1), 3–7.

Diprose, R. (2002). *Corporeal generosity: On giving with Nietzsche, Merleau-Ponty, and Levinas.* New York: State University of New York Press.

Donnelly, M. K. (2009, November). *Women-only leisure activities: Physicality, inevitability, and possibility in embodied ethnography.* Paper presented at the annual conference of the North American Society for the Sociology of Sport, Ottawa, Ontario, Canada.

Dworkin, D. (1997). *Cultural Marxism in postwar Britain: History, the New Left, and the origins of cultural studies.* Durham, NC: Duke University Press.

Dworkin, S., & Wachs, F. L. (2004). "Getting your body back": Postindustrial fit motherhood in *Shape Fit Pregnancy* magazine. *Gender & Society, 18*(5), 610–624.

Evers, C. (2006). How to surf. *Journal of Sport & Social Issues, 30*(3), 229–243.

Fiske, J. (1994). Cultural practice and cultural studies. In N. K. Denzin & Y. S. Lincoln (Eds.), *Handbook of qualitative research.* Thousand Oaks, CA: Sage.

Francombe, J. (2010). "I cheer, you cheer, we cheer": Physical technologies and the normalized body. *Television & New Media, 11*(5), 350–366.

Freire, P. (1970/2006). *Pedagogy of the oppressed.* New York: Continuum.

Freire, P. (1999). *Pedagogy of freedom: Ethics, democracy, and civic courage.* Lanham, MD: Rowman & Littlefield.

Frow, J., & Morris, M. (2000). Cultural studies. In N. K. Denzin & Y. S. Lincoln (Eds.), *Handbook of qualitative inquiry* (2nd ed., pp. 315–346). Thousand Oaks, CA: Sage.

Fusco, C. (2006). Spatializing the (im)proper: The geographies of abjection in sport and physical activity space. *Journal of Sport & Social Issues, 30*(1), 5–28.

Gannon, S. (2006). The (im)possibilities of writing the self-writing: French poststructural theory and autoethnography. *Cultural Studies⇔Critical Methodologies, 6*(4), 474–495.

Giardina, M. D. (2005). *Sporting pedagogies: Performing culture & identity in the global arena.* New York: Peter Lang.

Giardina, M. D. (2009). Flexibly global? Performing culture and identity in an age of uncertainty. *Policy Futures in Education, 7*(2), 172–184.

Giardina, M. D., & Newman, J. I. (in press). What is the 'physical' in physical cultural studies? *Sociology of Sport Journal, 28*(1).

Gilroy, P. (1987). *"There ain't no black in the Union Jack": The cultural politics of race and nation.* London: Hutchinson.

Giroux, H. A. (2000). *Impure acts: The practical politics of cultural studies.* New York: Routledge.

Giroux, H. A. (2001). Cultural studies as performative practice. *Cultural Studies ⇔ Critical Methodologies, 1*(1), 5–23.

Grindstaff, L., & West, E. (2006). Cheerleading and the gendered politics of sport. *Social Problems, 53*(4), 500–518.

Grossberg, L. (1997). *Bringing it all back home. Essays in cultural studies.* Durham, NC: Duke University Press.

Gruneau, R. S. (1985). *Class, sports, and social development.* Amherst: University of Massachusetts Press.

Gruneau, R. S., & Whitson, D. (1993). *Hockey night in Canada: Sport, identities, and cultural politics.* Toronto, ON: Garamond Press.

Hall, S. (1982). The problem of ideology: Marxism without guarantees. In B. Matthews (Ed.), *Marx 100 years on* (pp. 57–86). London: Lawrence & Wishart.

Hall, S. (1985). Signification, representation, ideology: Althusser and the post-structuralist debates. *Critical Studies in Mass Communication, 2,* 91–114.

Hall, S. (1986). Gramsci's relevance for the study of race and ethnicity. *Journal of Communication Inquiry, 10*(2), 5–27.

Hall, S. (1993). What is this "black" in black popular culture? *Social Justice, 20*(1–2), 104–115.

Hall, S. (1996). On postmodernism and articulation. In D. Morley & K. Chen (Eds.), *Stuart Hall: Critical dialogues in cultural studies* (pp. 131-150). London: Routledge.

Hall, S., Critcher, C., Jefferson, T., Clarke, J., & Roberts, B. (1978). *Policing the crisis: Mugging, the state, and law and order.* London: Macmillan.

Hargreaves, J., & Vertinsky, P. (Eds.). (2006). *Physical culture, power, and the body.* London: Routledge.

Helstein, M. (2007). Seeing your sporting body: Identity, subjectivity, and misrecognition. *Sociology of Sport Journal, 24*(1), 78–103.

Hoggart, R. (1957). *The uses of literacy: Aspects of working-class life, with special reference to publications and entertainments.* London: Chatto & Windus.

Ingham, A. (1985). From public sociology to personal trouble: Well-being and the fiscal crisis of the state. *Sociology of Sport Journal, 2*(1), 43–55.

Ingham, A. G. (1997). Toward a department of physical cultural studies and an end to tribal warfare. In J. Fernandez-Balboa (Ed.), *Critical postmodernism in human movement, physical education, and sport* (pp. 157–182). Albany: State University of New York Press.

Jackson, S. J. (1992). *Sport, crisis, and Canadian identity in 1988: A cultural analysis.* Unpublished doctoral dissertation, University of Illinois at Urbana-Champaign.

Johnson, R. (1987). What is cultural studies anyway? *Social Text, 6*(1), 38–79.

Joseph, J., & Donnelly, M. K. (in press). Drinking on the job: The problems and pleasure of ethnography and alcohol. *International Review of Qualitative Research.*

Kelly, S. D. (2002). Merleau-Ponty on the body. *Ratio, 15*(4), 376–391.

Kincheloe, J. L. (2004). *Critical pedagogy.* New York: Peter Lang.

Kincheloe, J. L., & McLaren, P. (2000). Rethinking critical theory and qualitative research. In N. K. Denzin & Y. S. Lincoln (Eds.), *Handbook of qualitative research* (2nd ed., pp. 279–313).

Kincheloe, J. L., Steinberg, S. R., Rodriguez, N. M., & Chennault, R. E. (Eds.). (1998). *White reign: Deploying whiteness in America.* New York: St. Martin's Griffin.

Kujundzic, N., & Buschert, W. (1994). Instruments and the body: Sartre and Merleau-Ponty. *Research in Phenomenology, 24*(2), 206–215.

Laclau, E., & Mouffe, C. (1985). *Hegemony and socialist strategy: Towards a radical democratic politics.* London: Verso.

Langellier, K. (1999). Personal narrative, performance, performativity: Two or three things I know for sure. *Text and Performance Quarterly, 19*(1), 125–144.

Lattimer, J. (2009). Introduction: Body, knowledge, words. In J. Lattimer & M. Schillmeier (Eds.), *Un/knowing bodies* (pp. 1–22). Malden, MA: Blackwell.

Lincoln, Y. S. (2004). Perfoming 9/11: Teaching in a terrorized world. *Qualitative Inquiry, 10*(1), 140–159.

Madison, D. S. (2009). Dangerous ethnography. In N. K. Denzin & M. D. Giardina (Eds.), *Qualitative inquiry and social justice* (pp. 187–197). Walnut Creek, CA: Left Coast Press.

Malagreca, M. (2007). Writing queer across the borders of geography and desire. In C. McCarthy, A. Durham, L. Engel, A. Filmer, M. D. Giardina, & M. Malagreca (Eds.), *Globalizing cultural studies: Ethnographic interventions in theory, method, and politics* (pp. 79-100). New York: Peter Lang.

Markula, P. (1995). Firm but shapely, fit but sexy, strong but thin: The postmodern aerobicizing female bodies. *Sociology of Sport Journal, 12*(4), 424–453.

Markula, P., & Pringle, R. (2006). *Foucault, sport, and exercise.* London: Routledge.

McCarthy, C., Durham, A., Engel, L., Filmer, A., Giardina, M. D., & Malagreca, M. (2007). Confronting cultural studies in globalizing times. In C. McCarthy, A. Durham, L. Engel, A. Filmer, M. D. Giardina, & M. Malagreca (Eds.), *Globalizing cultural studies: Ethnographic interventions in theory, method, and policy* (pp. xvii-xxxiv). New York: Peter Lang.

McLaren, P. (1988). Schooling the postmodern body: Critical pedagogy and the politics of enfleshment. *Journal of Education, 170*(3), 53–83.

McLaren, P. (2000). *Che Guevara, Paulo Freire, and the pedagogy of revolution.* Lanham, MA: Rowman & Littlefield.

Mears, A. (2008). Discipline of the catwalk: Gender, power, and uncertainty in fashion modeling. *Ethnography, 9*(4), 429–456.

Metz, J. L. (2008). An interview on motherhood: Racial politics and motherhood in late capitalist sport. *Cultural Studies <=> Critical Methodologies, 8*(2), 248–275.

Miah, A. (2004). *Genetically modified athletes: Biomedical ethics, gene doping, and sport.* London: Routledge.

Miller, T. (2001). Introduction. In T. Miller (Ed.), *A companion to cultural studies* (pp. 1–20). London: Blackwell.

Mills, C. W. (1959). *The sociological imagination.* London: Oxford University Press.

Minh-ha, T. T. (1991). *When the moon waxes red: Representation, gender, and cultural politics.* London: Routledge.

Newman, J. I. (2008). *Notes on physical cultural studies.* Paper presented at the annual conference of the North American Society for the Sociology of Sport, Denver, CO.

Newman, J. I. (2010). *Embodying Dixie: Studies in the body pedagogics of Southern whiteness.* Melbourne, Australia: Common Ground Press.

Newman, J. I. (in press). [Un]comfortable in my own skin: Articulation, reflexivity, and the duality of self. *International Review of Qualitative Research.*

Newman, J. I., & Giardina, M. D. (2008). NASCAR and the "Southernization" of America: Spectactorship, subjectivity, and the confederation of identity. *Cultural Studies⇔Critical Methodologies, 8*(4), 497–506.

Pelias, R. J. (2005). Performative writing as scholarship: An apology, an argument, an anecdote. *Cultural Studies⇔Critical Methodologies, 5*(4), 415–424.

Peters, R. J. (1975). *Television coverage of sport* (Center for Contemporary Cultural Studies Stenciled Occasional Paper Series 48). Birmingham, UK: University of Birmingham.

Pineau, E. L. (2000). "Nursing mother" and articulating absence. *Text and Performance Quarterly, 20*(1), 1–19.

Reich, J. (2010). "The world's most perfectly developed man": Charles Atlas, physical culture, and the inscription of American masculinity. *Men and Masculinities, 12*(4), 444–461.

Ronai, C. R. (1992). The reflexive self through narrative: A night in the life of an erotic dancer/researcher. In C. Ellis & M. G. Flaherty (Eds.), *Investigating subjectivity: Research on live experience* (pp. 102–124). Thousand Oaks, CA: Sage.

Rossman, G. B., & Rallis, S. F. (2003). *Learning in the field: An introduction to qualitative research* (2nd ed.). Thousand Oaks, CA: Sage.

Roy, A. (2001). *Power politics.* Cambridge, MA: South End Press.

Roy, A. (2003). *War talk.* Cambridge, MA: South End Press.

Roy, A. (2004). *An ordinary person's guide to empire.* Cambridge, MA: South End Press.

Saukko, P. (2005). Methodologies for cultural studies: An integrative approach. In N. K. Denzin & Y. S. Lincoln (Eds.), *The SAGE handbook of qualitative research* (3rd ed., pp. 343–357).

Schultz, J. (2004). Discipline and push-up: Female bodies, femininity, and sexuality in popular representations of sports bras. *Sociology of Sport Journal, 21*(2).

Scott, S. (2010). How to look good (nearly) naked: The performative regulation of the swimmer's body. *Body & Society, 16*(2), 143–168.

Shilling, C. (2007). Introduction: Sociology and the body. In C. Shilling (Ed.), *Embodying sociology* (pp. 2–18). Oxford, UK: Blackwell.

Silk, M. (2010). Postcards from Pigtown. *Cultural Studies⇔Critical Methodologies, 10*(2), 143-156.

Silk, M., & Andrews, D. L. (in press). Physical cultural studies. *Sociology of Sport Journal, 28*(1).

Slack, J. D. (1996). The theory and method of articulation in cultural studies. In D. Morley and K. H. Chen (Eds.), *Stuart Hall: Critical dialogues in cultural studies* (pp. 112–127). London: Routledge.

Spry, T. (2001). Performing autoethnography: An embodied methodological praxis. *Qualitative Inquiry, 7*(6), 706–732.

Spry, T. (2010). Some ethical considerations in preparing students for performative autoethnography. In N. K. Denzin & M. D. Giardina (Eds.), *Qualitative inquiry and human rights* (pp. 158–170). Walnut Creek, CA: Left Coast Press.

Stewart, K. A., Hess, A., Tracy, S. J., & Goodall, H. L., Jr. (2009). Risky research: Investigating the "perils" of ethnography. In N. K. Denzin & M. D. Giardina (Eds.), *Qualitative inquiry and social justice.* Walnut Creek, CA: Left Coast Press.

Stratton, J., & Ang, I. (1996). On the impossibility of a global cultural studies: "British" cultural studies in an "international" frame. In D. Morley & K. Chen (Eds.), *Stuart Hall: Critical dialogues* (pp. 360–392). London: Routledge.

Thompson, E. P. (1963). *The making of the English working class.* New York: Vintage.

Thorpe, H. (2009). Bourdieu, feminism, and female physical culture: Gender reflexivity and the habitus-field complex. *Sociology of Sport Journal, 26*(4), 491–516.

Tomlinson, A. (Ed.). (1981). The sociological study of sport: Configuration and interpretive studies. *Proceedings of Workshop of British Sociological Association and Leisure Studies Association.* Brighton, UK: Brighton Polytechnic.

van Ingen, C. (2004). Therapeutic landscapes and the regulated body in Toronto Front Runners. *Sociology of Sport Journal, 21*(3).

Vasterling, V. (2003). Body and language: Butler, Merleau-Ponty and Lyotard on the speaking embodied subject. *International Journal of Philosophical Studies, 11*(2), 205–223.

Wacquant, L. (2004). *Body and soul: Ethnographic notebooks of an apprentice-boxer.* New York: Oxford University Press.

Watson, R. (1973). *The public announcement of fatality* (Working Papers in Cultural Studies, 4). Birmingham, UK: University of Birmingham.

Wedgewood, (2004). Kicking like a boy: Schoolgirl Australian Rules Football and bi-gendered 'female embodiment. *Sociology of Sport Journal, 21*(2), 140–162.

West, C. (1991). Theory, pragmatisms, and politics. In J. Arac & B. Johnson (Eds.), *Consequences of theory.* Baltimore, MD: Johns Hopkins University Press.

Whannel, G. (1983). *Blowing the whistle: Culture, politics, and sport.* London: Routledge.

Wheatley, E. E. (2005). Disciplining bodies at risk: Cardiac rehabilitation and the medicalization of fitness. *Journal of Sport and Social Issues, 29*(2), 198–221.

Williams, R. (1958). *Culture and society.* London: Chatto & Windus.

Willis, P. (1971). *The motorbike club within a subcultural group* (Working Papers in Cultural Studies, No. 2, pp. 53–70). Birmingham, UK: University of Birmingham.

Willis, P. (1977). *Learning to labor: How working-class kids get working-class jobs.* Farnborough, England: Saxon House.

Zinn, H. (1996). *You can't be neutral on a moving traing.* Boston, MA: South End Press.

11

CRITICAL HUMANISM AND QUEER THEORY

Living With the Tensions

Ken Plummer

Failure to examine the conceptual structures and frames of reference which are unconsciously implicated in even the seemingly most innocent factual inquiries is the single greatest defect that can be found in any field of inquiry.

—John Dewey (1938, p. 505)

Most people in and outside of the academy are still puzzled about what queerness means, exactly, so the concept still has the potential to disturb or complicate ways of seeing gender and sexuality, as well as the related areas of race, ethnicity and class.

—Alexander Doty (2000, p. 7)

Research—like life—is a contradictory, messy affair. Only on the pages of "how-to-do-it" research methods texts or in the classrooms of research methods courses can it be sorted out into linear stages, clear protocols, and firm principles. My concern in this chapter lies with some of the multiple, often contradictory assumptions of inquiries. Taking my interest in sexualities/gay/queer research as a starting point and as a tension, I see "queer theory" and "critical humanism" as one of my own tensions. I have tried to depict each and to suggest some overlaps, but my aim has not been to reconcile the two. That is not possible and probably is not even desirable. We have to live with the tensions, and awareness of them is important background for the self-reflexive social researcher.

▣ SOCIAL CHANGE AND ZOMBIE RESEARCH

This discussion should be seen against a background of rapid social change. Although for many, research methods remain the same over time (they just get a bit more refined with each

generation), for others of us, changes in society are seen to bring parallel changes in research practices. To put it bluntly, many claim we are moving into a postmodern, late modern, globalizing, risk, liquid society. A new global order is in the making that is much more provisional and less authoritative than that of the past; it is a society of increasing self-reflexivity and individuation, a network society of flows and mobilities, a society of consumption and waste (Bauman, 2000, 2004; Beck, 2003; Giddens,1991; Urry, 2000).

As we tentatively move into these new worlds, our tools for theory and research need radical overhaul. German sociologist Ulrich Beck, for example, speaks of "zombie categories"; we move among the living dead! Zombie categories are categories from the past that we continue to use even though they have long outlived their usefulness and even though they mask a different reality. We probably go on using them because at present we have no better words to put in their place. Yet dead they are.

Beck cites the example of the concept of "the family" as an instance of a zombie category, a term that once had life and meaning but for many now means very little. I suggest that we could also cite most of our massive research methodology apparatus as partially zombified. I am not a major fan of television, but when I choose to watch a documentary, I often am impressed by how much more I get from it than from the standard sociological research tract. Yet the skills of a good documentary maker are rarely the topics of research methods courses, even though these skills—from scriptwriting and directing to camera movements and ethics—are the very stuff of good 21st-century research. And yes, some research seems to have entered the world of cyberspace, but much of it simply replicates the methods of quantitative research, making qualitative research disciplined, quantitative, and antihumanistic. Real innovation is lacking. Much research at the end of the 20th century—to borrow Beck's term again—truly was zombie research (Beck, 2003).

Table 11.1 suggests some links between social change and social research styles. The background is the authoritative scientific account with standard research protocols. As the social world changes, so we may start to sense new approaches to making inquiries. My concern in this chapter is largely with the arrival of queer theory.

◼ A REFLEXIVE INTRODUCTION

How research is done takes us into various language games—some rational, some more contradictory, some qualitative, some quantitative. The languages we use bring with them all manner of tensions. Although they sometimes help us chart the ways we do research, they often bring their own contradictions and problems. My goal here is to address some of the incoherencies I have found in my own research languages and inquiries and to

suggest ways of living with them. Although I will draw widely from a range of sources and hope to provide some paradigmatic instances, the chapter inevitably will be personal. Let me pose the key contradiction of my inquiries. (We all have our own.)

The bulk of my inquiries have focused on sexualities, especially lesbian and gay concerns, with an ultimate eye on some notion of sexual justice. In the early days, I used a relatively straightforward symbolic interactionism to guide me in relatively straightforward fieldwork and interviewing in and around London's gay scene of the late 1960s. At the same time, I engaged politically, initially with the Homosexual Law Reform Society and then with the Gay Liberation Front in its early years. I read my Becker, Blumer, Strauss, and Denzin! At the same time, I was coming out as a young gay man and finding my way in the very social world I was studying. More recently, such straightforwardness has come to be seen as increasingly problematic. Indeed, there was always a tension there: I just did not always see it (Plummer, 1995).

On one hand, I have found myself using a language that I increasingly call that of critical humanism, one allied to symbolic interactionism, pragmatism, democratic thinking, storytelling, moral progress, redistribution, justice, and good citizenship (Plummer, 2003). Inspirations range from Dewey to Rorty, Blumer to Becker. All of these are quite old and traditional ideas, and although I have sensed their postmodernized affinities (as have others), they still bring more orthodox claims around experience, truths, identities, belonging to groups, and a language of moral responsibilities that can be shared through dialogues (Plummer, 2003).

By contrast, I also have found myself at times using a much more radicalized language that nowadays circulates under the name of queer theory. The latter must usually be seen as at odds

Table 11.1 Shifting Research Styles Under Conditions of Late Modernity

Current Social Changes	Possible Changes in Research Style
Toward a late modern world	Toward a late modern research practice
Postmodern/fragmentation/pluralization	The 'polyphonic' turn
Mediazation	The new forms of media as both technique and data
Stories and the death of the grand narrative	The storytelling/narrative turn
Individualization/choices/unsettled identities	The self-reflexive turn
Globalization-glocalization hybridization/ diaspora	The hybridic turn: decolonizing methods (L. T. Smith, 1999)
High tech/mediated/cyborg/post-human	The high-tech turn
Knowledge as contested	The epistemological turn
Postmodern politics and ethics	The political/ethical turn
The network society	Researching flows, mobilities, and contingencies
Sexualities as problematic	The queer turn

with the former: Queer theory puts everything out of joint, out of order. "Queer," for me, is the postmodernization of sexual and gender studies. "Queer" brings with it a radical deconstruction of all conventional categories of sexuality and gender. It questions all the orthodox texts and tellings of the work of gender and sexuality in the modern world (and all worlds). It is a messy, anarchic affair—not much different from intellectual anarchists or political International Situationists. "Queer" would seem to be antihumanist, to view the world of normalization and normality as its enemy, and to refuse to be sucked into conventions and orthodoxy. If it is at all sociological (and it usually is not), it is gothic and romantic, not classical and canonical (Gouldner, 1973). It transgresses and subverts.

On one hand, then, I am quite happy about using the "new language of qualitative method" (Gubrium & Holstein, 1997); on the other, I am quite aware of a queer language that finds problems everywhere with orthodox social science methods (Kong, Mahoney, & Plummer, 2002). Again, these tensions are very much products of their time (queer theory did not exist before the late 1980s). Yet, retrospectively, it would seem that I have always walked tightropes between an academic interactionism, a political liberalism, a gay experience, and a radical critique.

But of course, as usual, there are more ironies here. Since the late 1980s, I have more or less considered myself "post-gay." So who was that young man from the past who studied the gay world? Likewise, those wild queer theorists have started to build their textbooks, their readers, and their courses, and they have proliferated their own esoteric cultlike worlds that often seem more academic than the most philosophical works of Dewey. Far from breaking boundaries, queer theorists often have erected them, for while they may not wish for closure, they nevertheless find it. Queer theories have their gurus, their followers, and their canonical texts. But likewise, humanists and new qualitative researchers—finding themselves under siege from postmodernists, queer theorists, some feminists, and multiculturalists and the like—have also fought back, rewriting their own histories and suggesting that many of the critiques laid at their door are simply false. Some, like Richard Rorty—the heir apparent to the modern pragmatism of Dewey and James—fall into curious traps: Himself labeled a postmodernist by others, he condemns postmodernists as "posties" (Rorty, 1999). Methodological positions often lead in directions different from those originally claimed.[1]

So here am I, like many others, a bit of a humanist, a bit post-gay, a sort of a feminist, a little queer, a kind of a liberal, and seeing that much that is queer has the potential for an important radical change. In the classic words of interactionism, Who am I? How can I live with these tensions?

This chapter is not meant to be an essay of overly indulgent self-analysis, but rather one in which, starting to reflect on such a worry, I am simply showing tensions that many must confront these days. Not only am I not alone in such worries, but I also am fairly sure that all reflective qualitative inquiries will face their own versions of them, just as most people face them in their daily lives. Ambivalence is the name of the game.

In this chapter, I plan to deal with three interconnected issues raised by qualitative research— all focused on just how far we can "push" the boundaries of qualitative research into new fields, strategies, and political/moral awareness—and how this has been happening continuously in my own work. New languages of qualitative method benefit from new ideas that at least initially may be seen as opposition. This is how they grow and how the whole field of qualitative research becomes more refined. In what follows, I will explore:

- What is critical humanism and how to do a critical humanist method
- What is queer and how to do a queer method
- How the contradictions can be lived through

▣ THE CRITICAL HUMANIST PROJECT

How different things would be . . . if the social sciences at the time of their systematic formation in the nineteenth century had taken the arts in the same degree they took the physical science as models.

—Robert Nisbet (1976, p. 16)

There is an illusive center to this contradictory, tension-ridden enterprise that seems to be moving further and further away from grand narratives and single, overarching ontological, epistemological, and methodological paradigms. This center lies in the humanistic commitment of the qualitative researcher to study the world always from the perspective of the interacting individual. From this simple commitment flow the liberal and radical politics of qualitative research. Action, feminist, clinical, constructivist, ethnic, critical, and cultural studies researchers all unite on this point. They all share the belief that a politics of liberation must always begin with the perspectives, desires, and dreams of those individuals and groups who have been oppressed by the larger ideological, economic, and political forces of a society or a historical moment.

—Denzin & Lincoln (1994, p. 575)

I use the term "critical humanism" these days to suggest orientations to inquiry that focus on human experience—that is, with the structure of experience and its daily lived nature—and that acknowledge the political and social role of all inquiry. It goes by many names—symbolic interactionism,[2] ethnography, qualitative inquiry, reflexivity, cultural anthropology, and life story research, among others—but they all have several concerns at heart. All these research orientations have a focus on human subjectivity, experience, and creativity: They start with people living their daily lives. They look at their talk, their feelings, their actions, and their bodies as they move around in social worlds and experience the constraints of history and a material world of inequalities and exclusions. They make methodological claims for a naturalistic "intimate familiarity" with these lives, recognizing their own part in such study. They make no claims for grand abstractions or universalism—assuming an inherent ambivalence and ambiguity in human life with no "final solutions," only damage limitations—while simultaneously sensing both their subjects' ethical and political concerns and their own in conducting such inquiries. They have pragmatic pedigrees, espousing an epistemology of radical, pragmatic empiricism that takes seriously the idea that knowing—always limited and partial—should be grounded in experience (Jackson, 1989). It is never neutral, value-free work, because the core of the inquiry must be human values. As John Dewey remarked long ago, "Any inquiry into what is deeply and inclusively (i.e., significantly) human enters perforce into the specific area of morals" (1920, p. xxvi). Impartiality may be suspect; but a rigorous sense of the ethical and political sphere is a necessity. Just why would one even bother to do research were it not for some wider concern or value?

What are these values? In the most general terms, critical humanism champions those values that give dignity to the person,[3] reduce human sufferings, and enhance human well-being. There are many such value systems, but at a minimum they probably must include the following:

1. A commitment to a whole cluster of *democratizing values* (as opposed to totalitarian ones) that aim to *reduce/remove human sufferings.* They take as a baseline *the value of the human being* and often provide a number of suggested *human rights*—freedom of movement, freedom of speech, freedom of association, freedom against arbitrary arrest, and so on. They nearly always include the *right to equality.* This commitment is strongly antisuffering and provides a major thrust toward both equality and freedom for all groups, including those with "differences" of all kinds (Felice, 1996).

2. An ethics of *care* and *compassion.* Significantly developed by feminists, this is a value whereby looking after the other takes on a prime role and whereby *sympathy*, *love*, and even *fidelity* become prime concerns (Tronto, 1993).

3. A politics of *recognition* and *respect.* Following the work of Axel Honneth (1995) and significantly shaped earlier by George Herbert Mead, this is a value whereby others are always acknowledged and a certain level of empathy is undertaken.

4. The importance of *trust.* This value recognizes that no social relationships (or society, for that matter) can function unless humans have at least some modicum of trust in each other (O'Neill, 2002).

Of course, many of these values bring their own tensions: We must work through them and live with them. A glaring potential contradiction, for example, may be to talk of humanistic values under capitalism, for many of the values of humanism must be seen as stressing nonmarket values. They are values that are not necessarily given a high ranking in a capitalist economy. Cornel West has put this well:

> In our own time it is becoming extremely difficult for non-market values to gain a foothold. Parenting is a non market activity; so much sacrifice and service go into it without any assurance that the providers will get anything back. Mercy, justice: they are non market. Care, service: non market. Solidarity, fidelity: non market. Sweetness and kindness and gentleness. All non market. Tragically, non market values are relatively scarce. . . . (West, 1999, p. 11)

The Methodologies of Humanism

These values strongly underpin critical humanism. In his classic book *The Human Perspective in Sociology*, T. S. Bruyn (1966) locates this humanistic perspective as strongly allied to the methods of participant observation. Elsewhere, I have suggested an array of life story strategies for getting at human experience. The task is a "fairly complete narrating of one's entire experience of life as a whole, highlighting the most important aspects" (R. Atkinson, 1998, p. 8). These may be long, short, reflexive, collective, genealogical, ethnographic, photographic, even auto/ethnographic (Plummer, 2000). Life stories are prime humanistic tools, but it is quite wrong to suggest that this means that the stories only have a concern with subjectivity and personal experience.[4]

Throughout all of this, there is a pronounced concern not only with the humanistic understanding of experience but also with ways of telling the stories of the research. Usually, the researcher is present in many ways in the text: The text rarely is neutral, with a passive observer. Chris Carrington's (1999) study of gay families, for example, makes it very clear from the outset his own location within a single-parent family: "I grew up in a working-poor, female-headed, single parent family. Throughout much of my childhood, in order to make ends meet, my mother worked nights as bar tender. There were periods where she could not get enough hours and our family had to turn to food stamps and welfare" (p. 7). Likewise, Peter Nardi's (1999) study of gay men's friendships is driven by his own passion for friends: "What follows is partly an attempt to make sense of my own experiences with friends" (p. 2). Humanistic inquiries usually reveal humanistic researchers.

Most commonly, as in Josh Gamson's *Freaks Talk Back* (1998) and Leila Rupp and Verta Taylor's *Drag Queens at the 801 Cabaret* (2003), the method employed will entail triangulation—a combination of cultural analysis tools.[5] Here, multiple sources of data pertaining to texts, production, and reception are collected and the intersections among them analyzed. In Rupp and Taylor's study of drag queens, they observed, tape recorded, and transcribed 50 drag performances, along with the dialogue, music, and audience interactions, including photographs and dressing up themselves. They collected data on the performances through weekly meetings of the drag artists and semistructured life histories, and they conducted focus groups on people who attended the performances. In addition, they looked at weekly newspapers (such as the gay paper *Celebrate*) and others to partially construct the history of the groups. Their research has a political aim, humanistic and sociological, and yet queer too, showing that combinations are possible. Enormous amounts of research have been written on all of this (e.g., Clifford & Marcus, 1986; Coffey, 1999; Coles, 1989; Ellis & Flaherty, 1992; Hertz, 1997; Reed-Danahay, 1997; Ronai, 1992).

A further recent example of such work is Harry Wolcott's (2002) account of Brad, the Sneaky Kid. Wolcott, an educational anthropologist, is well known for his methodological writings and books, especially in the field of education. This book started life in the early 1980s as a short journal article on the life story of Brad, a troubled 19-year-old. The story is aimed at getting at the human experience of educational failure, in particular, the lack of support for those who are not well served by our educational systems.

This would have been an interesting life story but an unexceptional one had it not been for all the developments that subsequently emerged around it. What are not told in the original story are the details of how Wolcott met Brad, how he had gay sex with him, and how he got him to tell his life story. Much follows after the original story, which later takes curious turns: Brad develops schizophrenia and returns one night to Wolcott's house to burn it down in an enraged attempt to kill him. This leads to the complete destruction of Wolcott's home and all his belongings (and those of his schoolteacher partner). A serious court case ensues in which Brad is tried and sent to prison. Despite Brad's guilt, Wolcott is himself scrutinized regarding his relationship, his homosexuality, and even his role as an anthropologist. Brad's family is especially unhappy about the relationship with Wolcott, but so are many academics. Ultimately, Brad is institutionalized. Eventually, the story is turned into an intriguing ethnographic play. I have only read the text of the play and not seen it performed. Judging by the text presented here, it comes across as a collage of 1980s pop music, sloganized slides, and a drama in two layers—one about Brad's relationship with Wolcott and another about Wolcott's ruminations, as a professor, on the plights of ethnography.

I mention this study because although it started out as a life story gloss—a simple relaying of Brad's story—because of the curious circumstances that it led to, a much richer and complex story was revealed that generated a host of questions and debates about the ethical, personal, and practical issues surrounding fieldwork. Sexuality and gender were pretty much at the core. It is a gripping tale of the kinds of issues highlighted by

all humanistic research. Indeed, within the book a second major narrative starts to appear—that of Harry Wolcott himself. He was always present, of course, but his story takes over as he reveals how he had regular sex with the young man, his partner's disapproval of Brad, and how one night he returns to his house to find a strong smell of oil and Brad screaming "You fucker. I'm going to kill you. I'm going to kill you. I'm going to tie you up and leave you in the house and set the house on fire" (p. 74). Luckily, Harry escapes, but unluckily, his house does not. It goes up entirely in flames, with all of his and his partner's belongings. This may be one of the core dramatic moments in life story telling— certainly an "epiphany"! After that, a major chapter follows that tells of the working of the court and how Wolcott himself is almost on trial.

When the story of Sneaky Kid was first published in 1983, it was a 30-page essay; it has grown into a book of more than 200 pages (Wolcott, 2002). The original article does not tell much about the relationship from which it grew or much of the other background; the book tells much more, but it raises sharply the issue of just how much remains left out. The book serves as a sharp reminder that all social science, including life stories, consists of only partial selections of realities. There is always much going on behind the scenes that is not told. Here we have the inevitable bias, the partiality, the limits, the selectivity of all stories told—but I will not take these issues further here.

▣ THE TROUBLES WITH HUMANISM

Although I think humanism has a lot to offer qualitative inquiry, it is an unfashionable view these days: Many social scientists seem to want to turn only to discourse and language. But this discourse is not incompatible with doing this, as it evokes the humanities (much more so than other traditions), widens communities of understanding by dialoguing with the voices of others, and takes a strong democratic impulse as the force behind its thinking and investigating. As a form of imagery to think about social life, this is all to the good. It brings with it the possibility for such inquiry to engage in poetry and poetics, drama and performance, philosophy and photography, video and film, narrative and stories.

Nevertheless, these days humanism remains a thoroughly controversial and contested term— and not least among queer theorists themselves. We know, of course, the long-standing attacks on humanism from theologies, from behavioral psychologies, and from certain kinds of philosophers: There is a notorious debate between the humanist Sartre's *Existentialism and Humanism* and Heidegger's *Letter on Humanism.* More recent attacks have denounced "humanism" as a form of white, male, Western, elite domination and colonization that is being imposed throughout the world and that brings with it too strong a sense of the unique individual. It is seen as contra postmodernism. In one telling statement, Foucault proclaims, "The modern individual— objectified, analyzed, fixed—is a historical achievement. There is no universal person on whom power has performed its operations and knowledge, its enquiries" (1979, pp. 159–160). The "Human Subject" becomes a Western invention. It is not a progress or a liberation, merely a trapping on the forces of power.

This loose but important cluster of positions critical of humanism—usually identified with a postmodern sensibility— would include queer theorists, multicultural theorists, postcolonialists, many feminists, and antiracists, as well as post-structural theorists. Although I have much sympathy with these projects and the critical methodologies they usually espouse (e.g., L. T. Smith, 1999), I also believe in the value of the pragmatic and humanist traditions. How can I live with this seeming contradiction?

Let me look briefly at what the critics say. They claim that Humanists propose some kind of common and hence universal "human being" or self: a common humanity that blinds us to wider differences and positions in the world. Often this is seen as a powerful, actualizing, and autonomous force in the world: The individual agent is at the center of the action and of the universe. This is said to result in overt individualism strongly connected to the Enlightenment project (Western, patriarchal, racist, colonialist, etc.) which turns itself into a series of moral and political claims about progress through a liberal and democratic society. Humanism is linked to a universal, unencumbered "self" and to the "modern" Western liberal project. Such ideas of the human subject are distinctly "Western" and bring with them a whole series of ideological assumptions about the centrality of the white, Western, male, middle-class/ bourgeois position; hence, they become the enemies of feminism (human has equaled male), ethnic movements (human has equaled white superiority), gays (human has equaled heterosexual), and all cultures outside the Western Enlightenment project (human here has equaled the middle-class West).

A More Complex Humanism?

Such claims made against "humanism" demean a complex, differentiated term into something far too simple. Humanism can, it is true, come to mean all of the above, but the term does not have to. Alfred McLung Lee (1978, pp. 44–45) and others have charted both a long history of humanism and many forms of it. Attacks usually are waged at a high level of generality, and specifics of what constitutes "the human" often are seriously overlooked. But, as I have suggested elsewhere, for me this "human being" is never a passive, helpless atom. Humans must be located in time and space: They are always stuffed full of their culture and history, and they must "nest" in a universe of contexts. Human beings are both embodied, feeling animals and creatures with great symbolic potential. They engage in symbolic communication and are dialogic and intersubjective:

There is no such thing as the solitary individual. Human lives are shaped by chance, fateful moments, epiphanies, and contingencies. There is also a continuous tension between the specificities and varieties of humanities at any time and place, and the universal potentials that are to be found in all humans. And there is a continuous engagement with moral, ethical, and political issues.

Curiously, it is also clear that many of the seeming opponents of humanism can be found wanting to hold onto some version of humanism after all. Indeed, it is odd that some of the strongest opponents lapse into a kind of humanism at different points in their argument. For instance, Edward Said—a leading postcolonial critic of Western-style humanism—actually urges another kind of humanism, "shorn of all its 'unpleasantly triumphalist weight,'" and in recent work he actually claims to be a humanist (Said, 1992, p. 230; 2003).

Indeed, at the start of the 21st century, there have been many signs that the critique of humanism that pervaded the previous century has started to be reinvigorated as a goal of inquiry. More and more contemporary commentators, well aware of the attacks above, go on to make some kinds of humanist claims. It would not be hard to find signs of humanism (even if the authors disclaimed them!) in major studies such as Nancy Scheper-Hughes's *Death Without Weeping* (1994), Stanley Cohen's *States of Denial* (1999), and Martha Nussbaum's *Sex and Social Justice* (1999). For me, they are clearly inspired by a version of humanism with the human being at the heart of the analysis, with care and justice as core values, and with the use of any methods at hand that will bring out the story.[6] So whatever the critiques, it does appear that a critical humanism still has its place in social science and qualitative inquiry. But before going too far, we should see what queer theory has to say on all this.

▣ A QUEER PROJECT

Queer articulates a radical questioning of social and cultural norms, notions of gender, reproductive sexuality and the family.

—Cherry Smyth (1992, p. 28)

Queer is by definition whatever is at odds with the normal, the legitimate, the dominant. There is nothing in particular to which it necessarily refers.

—David Halperin (1995, p. 62)

Queer theory emerged around the mid-to late 1980s in North America, largely as a humanities/ multicultural-based response to a more limited "lesbian and gay studies." While the ideas of Michel Foucault loom large (with his talks of "regimes of truth" and "discursive explosions"), the roots of queer theory (if not the term) usually are seen to lie in the work of Teresa de Lauretis (Halperin, 2003, p. 339) and Eve Kosofsky Sedgwick, who argued that

> many of the major nodes of thought and knowledge in twentieth century Western culture as a whole are structured—indeed fractured—by a chronic, now endemic crisis of homo/heterosexual definition, indicatively male, dating from the end of the nineteenth century. . . . an understanding of any aspect of modern Western culture must be, not merely incomplete, but damaged in its central substance to the degree that it does not incorporate a critical analysis of modern Homo/heterosexual definition. (1990, p. 1)

Judith Butler's work has been less concerned with the deconstruction of the homo/heterosexual binary divide and more interested in deconstructing the sex/gender divide. For her, there can be no kind of claim to any essential gender: It is all "performative," slippery, unfixed. If there is a heart to queer theory, then, it must be seen as a radical stance around sexuality and gender that denies any fixed categories and seeks to subvert any tendencies toward normality within its study (Sullivan, 2003).

Despite these opening suggestions, the term "queer theory" is very hard to pin down (some see this as a necessary virtue for a theory that refuses fixed identity). It has come to mean many things: Alexander Doty can suggest at least six different meanings, as follow. Sometimes it is used simply as a synonym for lesbian, gay, bisexual, transgender (LGBT). Sometimes it is an "umbrella term" that puts together a range of so-called "non-straight positions." Sometimes it simply describes any non-normative expression of gender (which could include straight). Sometimes it is used to describe "non-straight things" not clearly sign-posted as lesbian, gay, bisexual, or transgender but that bring with them a possibility for such a reading, even if incoherently. Sometimes it locates the "non-straight work, positions, pleasures, and readings of people who don't share the same sexual orientation as the text they are producing or responding to." Taking it even further, Doty suggests that "queer" may be a particular form of cultural readership and textual coding that creates spaces not contained within conventional categories such as gay, straight, and transgendered. Interestingly, what all his meanings have in common is that they are in some way descriptive of texts and they are in some way linked to (usually transgressing) categories of gender and sexuality (Doty, 2000, p. 6).

In general, "queer" may be seen as partially deconstructing our own discourses and creating a greater openness in the way we think through our categories. Queer theory must explicitly challenge any kind of closure or settlement, so any attempts at definition or codification must be nonstarters. Queer theory is, to quote Michael Warner, a stark attack on "normal business in the academy" (1992, p. 25). It poses the paradox of being inside the academy while wanting to be outside it. It suggests that a "sexual order overlaps with a wide range of institutions and social ideologies, to challenge the sexual order is sooner or later to encounter these institutions as a problem" (Warner, 1993, p. 5). Queer theory is really poststructuralism (and postmodernism) applied to sexualities and genders.

To a limited extent, queer theory may be seen as another specific version of what Nancy Harstock and Sandra Harding refer to as standpoint theory (though I have never seen it discussed in this way). Initially developed as a way to analyze a position of women's subordination and domination, it suggests that an "opposition consciousness" can emerge that transcends the more taken-for-granted knowledge. It is interesting that hardly any men have taken this position up, but other women—women of race and disability, for example—have done so. Men seem to ignore the stance, and so too do queer theorists, yet what we may well have in queer theory is really something akin to a "queer standpoint."

Certain key themes are worth highlighting. Queer theory is a stance in which

- both the heterosexual/homosexual binary and the sex/gender split are challenged.
- there is a de-centering of identity.
- all sexual categories are open, fluid, and non-fixed (which means that modern lesbian, gay, bisexual, and transgender identities are fractured, along with all heterosexual ones).
- it offers a critique of mainstream or "corporate" homosexuality.
- it sees power as being embodied discursively. Liberation and rights give way to transgression and carnival as a goal of political action, what has been called a "politics of provocation."
- all normalizing strategies are shunned.
- academic work may become ironic, is often comic and paradoxical, and is sometimes carnivalesque: "What a difference a gay makes," "On a queer day you can see forever" (cf. Gever, Greyson, & Parmar, 1993).
- versions of homosexual subject positions are inscribed everywhere, even in heterosexualities.
- the deviance paradigm is fully abandoned, and the interest lies in a logic of insiders/outsiders and transgression.
- its most common objects of study are textual—films, videos, novels, poetry, visual images.
- its most frequent interests include a variety of sexual fetishes, drag kings and drag queens, gender and sexual playfulness, cybersexualities, polyamoury, sadomasochism, and all the social worlds of the so-called radical sexual fringe.

▣   A QUEER METHODOLOGY?

What are the implications of queer theory for method (a word it rarely uses)? In its most general form, queer theory is a refusal of all orthodox methods—a certain disloyalty to conventional disciplinary methods (Halberstam, 1998, pp. 9–13). What, then, does queer method actually do? What does it look like? In summary, let me give a few examples of what a queer methodology can be seen to offer.

The Textual Turn: Rereadings of Cultural Artifacts. Queer methods overwhelmingly employ an interest in and analysis of texts—films, literature, television, opera, musicals. This is perhaps the most commonly preferred strategy of queer theory. Indeed, Michael Warner has remarked that "almost everything that would be called queer theory is about ways in which texts—either literature or mass culture of language—shape sexuality." More extremely, he continues, "you can't eliminate queerness . . . or screen it out. It's everywhere. There's no place to hide, hetero scum!" (Warner, 1992, p. 19). The locus classicus of this way of thinking usually is seen to be Sedgwick's *Between Men* (1985), in which she looked at a number of key literary works (from Dickens to Tennyson) and reread these texts as driven by homosexuality, homosociality, and homophobia. Whereas patriarchy might condemn the former, it positively valorizes the latter (Sedgwick, 1985). In her wake have come hosts of rereadings around such themes. In later works, she gives readings to work as diverse as Diderot's *The Nun*, Wilde's *The Importance of Being Earnest*, and authors such as James and Austen (Sedgwick, 1990, 1994). In her wake, Alexander Doty gives queer readings to mass culture products such as "the sitcom"—from lesbian readings of the sitcoms *I Love Lucy* or *The Golden Girls*, to the role of "feminine straight men" such as Jack Benny, to the bisexual meanings in *Gentlemen Prefer Blondes* (Doty, 1993, 2000). Indeed, almost no text can escape the eyes of the queer theorist.

Subversive Ethnographies: Fieldwork Revisited. These are often relatively straightforward ethnographies of specific sexual worlds that challenge assumptions. Sasho Lambevski (1999), for instance, attempted to write "an insider, critical and experiential ethnography of the multitude of social locations (class, gender, ethnicity, religion) from which 'gays' in Macedonia are positioned, governed, controlled and silenced as subaltern people" (p. 301). As a "gay" Macedonian (are the terms a problem in this context?) who had spent time studying HIV in Australia, he looks at the sexual conflicts generated between the gay Macedonians and gay Albanians (never mind the Australian connection). Lambevski looks at the old cruising scenes in Skopje, known to him from before, that now take on multiple and different meanings bound up with sexualities, ethnicities, gender playing, and clashing cultures. Cruising for sex here is no straightforward matter. He describes how, in approaching and recognizing a potential sex partner as an Albanian (in an old cruising haunt), he feels paralyzed. Both bodies are flooded with ethnic meaning, not simple sexual ones, and ethnicities reek of power. He writes: "I obeyed by putting the (discursive) mask of my Macedonicity over my body" (p. 398). In another time and place, he may have reacted very differently.

Lambevski is overtly critical of much ethnography and wishes to write a queer experiential ethnography, not a confessional one (1999, p. 298). He refuses to commit himself to what he calls "a textual lie," which "continues to persist in much of what is considered a real ethnographic text." Here bodies, feelings, sexualities, ethnicities, and religions all can be left out easily. Nor, he claims, can ethnography simply depend on site observation or one-off interviewing. There is a great chain of connection: "The gay scene is inextricably linked to the Macedonian school system, the structuring of Macedonian and Albanian families and kinship relations, the Macedonian state and its political history, the Macedonian medical system with its power to mark and segregate 'abnormality' (homosexuality)" (1999, p. 400). There is a chain of social sites, and at the same time his own life is an integral part of this (Macedonian, queer, Australian, gay). Few researchers have been so honest regarding the tensions that infuse their lives and the wider chains of connectedness that shape their work.

I find it hard to believe that this is not true for all research, but it is usually silenced. Laud Humphreys's classic *Tearoom Trade* (1970), for example—admittedly, written some 30 years earlier—cannot speak of Humphreys's own gayness, his own bodily presence (though there is a small footnote on the taste of semen!), his emotional worlds, his white middle-classness, or his role as a white married minister. To the contrary, although he does remind the reader of his religious background and his wife, this serves more as a distraction. As important as it was in its day, this is a very different kind of ethnography. The same is true of a host that followed it. They were less aware of the problematic nature of categories and the links to material worlds. They were, in a very real fashion, "naïve ethnographies"—somehow thinking "the story could be directly told as it was." We live in less innocent times, and queer theory is a marker for this.

Scavenger Methodologies: The Raiding of Multiple Texts to Assemble New Ones. A fine example of queer "method" is Judith Halberstam's work on "female masculinity" (1998). Suggesting that we have failed to develop ways of seeing that can grasp the different kinds of masculinities that women have revealed both in the past and the present, she wrote a study that documents the sheer range of such phenomena. In her own work, she "raids" literary textual methods, film theory, ethnographic field research, historical survey, archival records, and taxonomy to produce her original account of emerging forms of "female masculinity" (Halberstam, 1998, pp. 9–13). Here we have aristocratic European cross-dressing women of the 1920s, butch lesbians, dykes, drag kings, tomboys, black "butch in the hood" rappers, trans-butches, the tribade (a woman who practices "unnatural vices" with other women), the gender invert, the stone butch, the female-to-male transsexual (FTM), and the raging bull dyke! She also detects—through films as diverse as

Alien and *The Killing of Sister George*—at least six prototypes of the female masculine: tomboys, Predators, Fantasy Butches, Transvestites, Barely Butches, and Postmodern Butches (1998, chap. 6).

In introducing this motley collection, she uses a "scavenger methodology. . . [of] different methods to collect and produce information on subjects who have been deliberately or accidentally excluded from traditional studies of human behavior" (1998, p. 13). She borrows from Eve Kosofsky Sedgwick's "nonce taxonomy": "The making and unmaking and remaking and redissolution of hundreds of old and new categorical meanings concerning all the kinds it takes to make up a world" (Sedgwick, 1990, p. 23). This is the mode of "deconstruction," and in this world the very idea that types of people called homosexuals or gays or lesbians (or, more to the point, "men" and "women") can be simply called up for study becomes a key problem in itself. Instead, the researcher should become more and more open to start sensing new worlds of possibilities.

Many of these social worlds are not immediately transparent, whereas others are amorphously nascent and forming. All this research brings to the surface social worlds only dimly articulated hitherto—with, of course, the suggestion that there are more, many more, even more deeply hidden. In one sense, Halberstam captures rich fluidity and diversity—all this going on just beneath the surface structures of society. But in another sense, her very act of naming, innovating terms, and categorizing tends itself to create and assemble new differences.

Performing Gender and Ethnographic Performance. Often drawing upon the work of Judith Butler, who sees gender as never essential, always unfixed, not innate, never natural, but always constructed through performativity—as a "stylized repetition of acts" (1990, p. 141)—much of the work in queer theory has been playing around with gender. Initially fascinated by drag, trans-gender, and transsexualism, and with Divas, Drag Kings, and key cross-genderists such as Del LaGrace Volcano and Kate Bornstein (1995), some of it has functioned almost as a kind of subversive terrorist drag. It arouses curious, unknown queered desires emancipating people from the constraints of the gendered tyranny of the presumed "normal body" (Volcano & Halberstam, 1999). Others have moved out to consider a wide array of playing with genders—from "faeries" and "bears," to leather scenes and the Mardi Gras, and on to the more commercialized/normalized drag for mass consumption: RuPaul, Lily Savage, and Graham Norton.

Sometimes performance may be seen as even more direct. It appears in the work of alternative documentaries, in "video terrorism" and "street theater," across cable talk shows, experimental artworks, and activist tapes. By the late 1980s, there was a significant expansion of lesbian and gay video (as well as film and film festivals), and in the academy, posts were created to deal with this— along with creation of more informal groupings.

(See, for example, Jennie Livingston's film *Paris Is Burning* [1990], which looks at the "ball circuit" of poor gay men and transgenderists, usually black, in the late 1980s in New York City, or Ang Lee's *The Wedding Banquet* [1993], which reconfigured the dominant "rice queen" image).[7]

Exploring New/Queered Case Studies. Queer theory also examines new case studies. Michael Warner, for example, looks at a range of case studies of emergent publics. One stands out to me: It is the details of a queer cabaret (a counter-public?) that involves "erotic vomiting." Suggesting a kind of "national heterosexuality" that, along with "family values," saturates much public talk, he argues that multiple queer cultures work to subvert these. He investigates the queer counter public of a "garden variety leather bar" where the routines are "spanking, flagellation, shaving, branding, laceration, bondage, humiliation, wrestling—as they say, you know, the usual" (Warner, 2002, pp. 206–210). But suddenly this garden-variety S&M bar is subverted by the less than usual: a cabaret of what is called erotic vomiting.

The Reading of the Self. Most of the researches within queer theory play with the author's self: It is rarely absent. D. A. Miller's (1998) account, for example, of the Broadway musical and the role its plays in queer life is an intensely personal account of the musical, including snapshots of the author as child, with the albums played.

▣   WHAT'S NEW?

As interesting as many of these methods, theories, and studies most certainly are, I suggest that there is really very little that could be called truly new or striking here. Often, queer methodology means little more than literary theory rather belatedly coming to social science tools such as ethnography and reflexivity (although sometimes it is also a radical critique of orthodox social science—especially quantitative— methods). Sometimes it borrows some of the oldest of metaphors, such as drama. Queer theory does not seem to me to constitute any fundamental advance over recent ideas in qualitative inquiry—it borrows, refashions, and retells. What is more radical is its persistent concern with categories and gender/sexuality—although, in truth, this has long been questioned, too (cf. Plummer, 2002; Weston, 1998). What seems to be at stake, then, in any queering of qualitative research is not so much a methodological style as a political and substantive concern with gender, heteronormativity, and sexualities. Its challenge is to bring stabilized gender and sexuality to the forefront of analyses in ways they are not usually advanced and that put under threat any ordered world of gender and sexuality. This is just what is, indeed, often missing from much ethnographic or life story research.

◨ THE TROUBLES WITH QUEER

Responses to queer theory have been mixed. It would not be too unfair to say that outside the world of queer theorists—the world of "straight academia"—queer theory has been more or less ignored and has had minimal impact. This has had the unfortunate consequence of largely ghettoizing the whole approach. Ironically, those who may most need to understand the working of the heterosexual-homosexual binary divide in their work can hence ignore it (and they usually do), whereas those who least need to understand it actively work to deconstruct terms that really describe themselves. Thus, it is comparatively rare in mainstream literary analysis or sociological theory for queer to be taken seriously (indeed, it has taken three editions of this handbook to include something on it, and the so-called seventh moment of inquiry (see Lincoln & Denzin, Epilogue, this volume) has so far paid only lip service to it!). More than this, many gays, lesbians, and feminists themselves see no advance at all in a queer theory that, after all, would simply "deconstruct" them, along with all their political gains, out of existence. Queer theorists often write somewhat arrogantly, as if they have a monopoly on political validity, negating both the political and theoretical gains of the past. Let me reflect on some of the standard objections to queer theory.

First, for many, the term itself is provocative: a pejorative and stigmatizing phrase from the past is reclaimed by that very same stigmatized grouping and had its meaning renegotiated; as such, it has a distinct generational overtone. Younger academics love it; older ones hate it. It serves to write off the past worlds of research and create new divisions.

Second, it brings a category problem, what Josh Gamson (1995) has described as a Queer Dilemma. He claims that there is simultaneously a need for a public collective identity (around which activism can galvanize) and a need to take apart and blur boundaries. As he says, fixed identity categories are the basis for both oppression and political power. Although it is important to stress the "inessential, fluid and multiple sited" forms of identity emerging within the queer movement, he also can see that there are very many from within the lesbian, gay, bisexual, and transgender movement (LGBT, as it is currently clumsily called) who also reject its tendency to deconstruct the very idea of gay and lesbian identity—hence abolishing a field of study and politics when it has only just gotten going.

There are also many radical lesbians who view it with even more suspicion, as it tends to work to make the lesbian invisible and to reinscribe tacitly all kinds of male power (in disguise), bringing back well-worn arguments about S&M, porn, and transgender politics as anti-women. Radical lesbian feminist Sheila Jeffreys (2003) is particularly scathing, seeing the whole queer movement as a serious threat to the gains of radical lesbians in the late 20th century. With the loss of the categories of woman-identified-woman and radical lesbian in a fog of (largely masculinist) queer deconstruction, it becomes impossible to see the roots of women's subordination to men. She also accuses it of a major elitism: The languages of most of its proponents ape the language of male academic elites, and lose all the gains that were made by the earlier, more accessible writings of feminists who wrote for and spoke to women in the communities, not just other academics. Lilian Faderman claims it is "resolutely elitist" and puts this well:

> The language queer scholars deploy sometimes seems transparently aimed at what lesbian feminists once called the "big boys" at the academy. Lesbian-feminist writing, in contrast, had as primary values clarity and accessibility, since its purpose was to speak directly to the community and in so doing reflect change. (1997, pp. 225–226).[8]

There are many other critics. Tim Edwards (1998) worries about a politics that often collapses into some kind of fan worship, celebration of cult films, and weak cultural politics. Stephen O. Murray hates the word "queer" itself because it perpetuates binary divisions and cannot avoid being a tool of domination, and he worries about excessive preoccupation with linguistics and with textual representation (2002, pp. 245–247). Even some of queer theory's founders now worry if the whole radical impulse has gotten lost and queer theory has become normalized, institutionalized, even "lucrative" within the academy (Halperin, 2003).

From many sides, then, doubts are being expressed that all is not well in the house of queer. There are problems that come with the whole project, and in some ways I still find the language of the humanists more conducive to social inquiry.

◨ QUEER THEORY MEETS CRITICAL HUMANISM: THE CONFLICTUAL WORLDS OF RESEARCH

Conflict is the gadfly of thought . . . a sine qua non of reflection and ingenuity.

— John Dewey
(1922, p. 300)

And so we have two traditions seemingly at serious odds with each other. There is nothing unusual about this—all research positions are open to conflict from both within and without. Whereas humanism generally looks to experience, meaning, and human subjectivity, queer theory rejects this in favor of representations. Whereas humanism generally asks the researcher to get close to the worlds he or she is studying, queer theory almost pleads for distance—a world of texts, defamiliarization, and deconstruction. Whereas humanism brings a liberal democratic project with "justice for all," queer theory aims to prioritize the oppressions of sexuality and gender and urges a more radical change. Humanism usually favors a calmer conversation and dialogue, whereas queer is carnivalesque, parodic, rebellious, and playful. Humanism champions the voice of the public intellectual; queer theory is to be found mainly in the universities and its own self-generated social movement of aspiring academics.

Yet there are some commonalities. Both, for instance, would ask researchers to adopt a critically self-aware stance. Both would seek out a political and ethical background (even though, in a quite major way, they may differ on this—queer theory has a prime focus on radical gender change, and humanism is broader). And both assume the contradictory messiness of social life, such that no category system can ever do it justice.

On a closer look, several of the above differences overlap. Many critical humanisms can focus on representations (though fewer queer theorists are willing to focus on experience). Critical humanists often are seen as social constructionists, and this hardly can be seen as far removed from deconstructionists. There is no reason why critical humanism cannot take the value and political stances of queer theorists (I have and I do), but the moral baselines of humanism are wider and less specifically tied to gender. Indeed, contemporary humanistic method enters the social worlds of different "others" to work a catharsis of comprehension. It juxtaposes differences and complexities with similarities and harmonies. It recognizes the multiple possible worlds of social research—not necessarily the standard interviews or ethnographies, but the roles of photography, art, video, film, poetics, drama, narrative, autoethnography, music, introspection, fiction, audience participation, and street theater. It also finds multiple ways of presenting the "data," and it acknowledges that a social science of any consequences must be located in the political and moral dramas of its time. One of those political and moral dramas is "queer."

But there again, the histories, canons, and gurus of critical humanism and queer theory are indeed different, even though, in the end, they are not nearly as at odds with each other as one could be led to believe. Yes, they are not the same; and it is right that they should maintain some of their key differences. But no, they are not so very different either. It is no wonder, then, that I find that I can live with both. Contradiction, ambivalence, and tension reside in all critical inquiries.

▣ NOTES

1. As Dmitri Shalin noted more than a decade ago, "The issues that symbolic interactionism has highlighted since its inception and that assured its maverick status in American sociology bear some uncanny resemblance to the themes championed by postmodernist thinkers" (1993, p. 303). It investigates "the marginal, local, everyday, heterogeneous and indeterminate" alongside the "socially constructed, emergent and plural" (p. 304). Likewise, David Maines (2001) has

continued to sustain an earlier argument that symbolic interactionism, by virtue of its interpretive center, finds an easy affinity with much of postmodernism, but because of that same center, has no need for it (pp. 229–233). He finds valuable the resurgence of interest in interpretive work, the importance now given to writing "as intrinsic to method," the concern over multiple forms of presentation, and the reclaiming of value positions and "critical work" (Maines, 2001, p. 325). In addition, as is well known, Norman K. Denzin has been at the forefront in defending postmodernism within sociology/cultural studies and symbolic interactionism, in numerous books and papers (e.g., Denzin, 1989, 1997, 2003).

2. For some, "interactionism" has become almost synonymous with sociology; see Maines (2001) and P. Atkinson and Housley (2003).

3. The liberal, humanist feminist philosopher Martha Nussbaum (1999, p. 41) suggests a long list of "human capabilities" that need cultivating for a person to function as a human being. These include concerns such as "bodily health and integrity" senses, imagination, and thought; emotions; practical reason; affiliation; concern for other species; play; control over one's environment; and life itself. To this I might add the crucial self-reflexive process, a process of communication that is central to the way we function.

4. In Bob Connell's rich study of *Masculinities* (1995)—a study that is far from being either avowedly "humanist" or "queer"—he takes life stories as emblematic/symptomatic of "crisis tendencies in power relations (that) threaten hegemonic masculinity directly" (p. 89). He looks at four groups of men under crisis— radical environmentalists, gay and bisexual networks, young working-class men, and men of the new class. Connell implies that I do not take this seriously (1995, p. 89). However, even in the first edition of my book *Documents of Life* (Plummer, 1983), I make it quite clear that among the contributions of the life story, it can be seen as a "tool for history," as a perspective on totality, and as a key focus on social change! (pp. 68–69).

5. Or, as Rupp and Taylor call it, "the tripartite model of cultural investigation" (2003, p. 223).

6. Likewise, I can sense a humanism at work in the writings of Cornel West, Jeffrey Weeks, Seyla Benhabib, Anthony Giddens, Zygmunt Bauman, Agnes Heller, Jürgen Habermas, Michel Bakhtin, and many others. Never mind the naming game, in which they have to come out as humanists (though some clearly do); what matters are the goals that they see will produce adequate understanding and social change for the better. In this respect, a lot of them read like humanists manqué.

7. See, for example, *Jump Cut, Screen, The Celluloid Closet, Now You See It?, The Bad Object Choices* collective, and the work of Tom Waugh and Pratibha Parmar.

8. See also Simon Watney's critiques to be found in *Imagine Hope* (2000). Watney is far from sympathetic to radical lesbianism, but his account has distinct echoes. Queer theory has often let down AIDS activism.

▣ REFERENCES

Atkinson, P., & Housley, W. (2003). *Interactionism.* London: Sage.
Atkinson, R. (1998). *The life story interview.* Thousand Oaks, CA: Sage.

Bauman, Z. (2000). *Liquid society.* Cambridge, UK: Polity.

Bauman, Z. (2004). *Wasted lives: Modernity and its outcasts.* Cambridge, UK: Polity.

Beck, U. (2003). *Individualization.* London: Sage.

Bornstein, K. (1995). *Gender outlaw.* New York: Vintage.

Bruyn, T. S. (1966). *The human perspective in sociology.* Englewood Cliffs, NJ: Prentice Hall.

Butler, J. (1990). *Gender trouble.* London: Routledge.

Carrington, C. (1999). *No place like home: Relationships and family life among lesbians and gay men.* Chicago: University of Chicago Press.

Clifford, J., & Marcus, G. E. (Eds.). (1986). *Writing culture.* Berkeley: University of California Press.

Coffey, A. (1999). *The ethnographic self: Fieldwork and the representation of identity.* London: Sage.

Cohen, S. (1999). *States of denial.* Cambridge, UK: Polity.

Coles, R. (1989). *The call of stories: Teaching and the moral imagination.* Boston: Houghton Mifflin.

Connell, R. W. (1995). *Masculinities.* Cambridge, UK: Polity.

Denzin, N. K. (1989). *Interpretive biography.* London: Sage.

Denzin, N. K. (1997). *Interpretive ethnography: Ethnographic practices for the 21st century.* London: Sage.

Denzin, N. K. (2003). *Performance ethnography.* London: Sage.

Denzin, N., & Lincoln, Y. (Eds.). (1994). *Handbook of qualitative research.* London: Sage.

Dewey, J. (1920). *Reconstruction of philosophy.* New York: Henry Holt.

Dewey, J. (1922). *Human nature and conduct.* New York: Henry Holt.

Dewey, J. (1938). *Logic of inquiry.* New York: Henry Holt.

Doty, A. (1993). *Making things perfectly queer: Interpreting mass culture.* Minneapolis: University of Minnesota Press.

Doty, A. (2000). *Flaming classics: Queering the film canon.* London: Routledge.

Edwards, T. (1998). Queer fears: Against the cultural turn. *Sexualities, 1*(4), 471–484.

Ellis, C., & Flaherty, M. G. (Eds.). (1992). *Investigating subjectivity: Research on lived experience.* London: Sage.

Faderman, L. (1997). Afterword. In D. Heller (Ed.), *Cross purposes: Lesbians, feminists and the limits of alliance.* Bloomington: Indiana University Press.

Felice, W. F. (1996). *Taking suffering seriously.* Albany: State University of New York Press.

Foucault, M. (1979). *The history of sexuality.* Middlesex, UK: Harmondsworth.

Gamson, J. (1995). Must identity movements self-destruct?: A queer dilemma. *Social Problems, 42*(3), 390–407.

Gamson, J. (1998). *Freaks talk back: Tabloid talk shows and sexual nonconformity.* Chicago: University of Chicago Press.

Gever, M., Greyson, J., & Parmar, P. (Eds.). (1993). *Queer looks: Perspectives on lesbian and gay film and video.* New York: Routledge.

Giddens, A. (1991). *Modernity and self-identity.* Cambridge, UK: Polity.

Gouldner, A. (1973). *For sociology: Renewal and critique in sociology today.* London: Allen Lane.

Gubrium, J., & Holstein, J. (1997). *The new language of qualitative research.* Oxford, UK: Oxford University Press.

Halberstam, J. (1998). *Female masculinity.* Durham, NC: Duke University Press.

Halperin, D. (1995). *Saint Foucault: Towards a gay hagiography.* New York: Oxford University Press.

Halperin, D. (2003). The normalization of queer theory. *Journal of Homosexuality, 45*(2–4), 339–343.

Hertz, R. (Ed.). (1997). *Reflexivity and voice.* Thousand Oaks, CA: Sage.

Honneth, A. (1995). *The struggle for recognition: The moral grammar of social conflicts.* Cambridge, UK: Polity.

Humphreys, L. (1970). *Tearoom trade.* Chicago: Aldine.

Jackson, M. (1989). *Paths toward a clearing: Radical empiricism and ethnographic inquiry.* Bloomington: Indiana University Press.

Jeffreys, S. (2003). *Unpacking queer politics.* Oxford, UK: Polity.

Kong, T., Mahoney, D., & Plummer, K. (2002). Queering the interview. In J. F. Gubrium & J. A. Holstein (Eds.), *The handbook of interview research* (pp. 239–257). Thousand Oaks, CA: Sage.

Lambevski, S. A. (1999). Suck my nation: Masculinity, ethnicity and the politics of (homo)sex. *Sexualities, 2*(3), 397–420.

Lee, A. (Director). (1993). *The wedding banquet* [Motion picture]. Central Motion Pictures Corporation.

Lee, A. M. (1978). *Sociology for whom?* Oxford: Oxford University Press.

Lincoln, Y. S., & Denzin, N. K. (1994). The fifth moment. In N. K. Denzin & Y. S. Lincoln (Eds.), *Handbook of qualitative research* (pp. 575–586). Thousand Oaks, CA: Sage.

Livingston, J. (Director), & Livingston, J., & Swimar, B. (Producers). (1990). *Paris Is Burning* [Motion picture]. Off White Productions.

Maines, D. (2001). *The fault lines of consciousness: A view of interactionism in sociology.* New York: Aldine de Gruyter.

Miller, D. A. (1998). *Place for us: Essay on the Broadway musical.* Cambridge, MA: Harvard University Press.

Murray, S. O. (2002). Five reasons I don't take queer theory seriously. In K. Plummer (Ed.), *Sexualities: Critical concepts in sociology* (Vol. 3, pp. 245–247). London: Routledge.

Nardi, P. (1999). *Gay men's friendships: Invincible communities.* Chicago: University of Chicago Press.

Nisbet, R. (1976). *Sociology as an art form.* London: Heinemann.

Nussbaum, M. C. (1999). *Sex and social justice.* New York: Oxford University Press.

O'Neill, O. (2002). *A question of trust: The BBC Reith Lectures 2002.* Cambridge, UK: Cambridge University Press.

Plummer, K. (1983). *Documents of life.* London: Allen and Unwin.

Plummer, K. (1995). *Telling sexual stories.* London: Routledge.

Plummer, K. (2001). *Documents of life 2: An invitation to a critical humanism.* London: Sage.

Plummer, K. (Ed.). (2002). *Sexualities: Critical concepts in sociology* (4 vols.). London: Routledge.

Plummer, K. (2003). *Intimate citizenship.* Seattle: University of Washington Press.

Reed-Danahay, D. E. (Ed.). (1997). *Auto/ethnography: Rewriting the self and the social.* Oxford, UK: Berg.

Ronai, C. R. (1992). A reflexive self through narrative: A night in the life of an erotic dancer/researcher. In C. Ellis & M. G. Flaherty (Eds.), *Investigating subjectivity: Research on lived experience* (pp. 102–124). Newbury Park, CA: Sage.

Rorty, R. (1999). *Philosophy and social hope.* Middlesex, UK: Penguin.

Rupp, L., & Taylor, V. (2003). *Drag queens at the 801 Cabaret.* Chicago: University of Chicago Press.

Said, E. (2003). *Orientalism* (2nd ed.). New York: Cambridge.

Scheper-Hughes, N. (1994). *Death without weeping.* Berkeley: University of California Press.

Sedgwick, E. K. (1985). *Between men: English literature and male homosexual desire.* New York: Columbia University Press.

Sedgwick, E. K. (1990). *Epistemology of the closet.* Berkeley: University of California Press.

Sedgwick, E. K. (1994). *Tendencies.* London: Routledge.

Shalin, D. N. (1993). Modernity, postmodernism and pragmatic inquiry. *Symbolic Interaction, 16*(4), 303–332.

Smith, L. T. (1999). *Decolonizing methodologies: Research and indigenous peoples.* London: Zed Books.

Smyth, C. (1992). Lesbians talk queer notions. London: Scarlet Press.

Sullivan, N. (2003). *A critical introduction to queer theory.* Edinburgh: University of Edinburgh Press.

Tronto, J. (1993). *Moral boundaries: A political argument for an ethic of care.* London: Routledge.

Urry, J. (2000). *Sociology beyond societies: Mobilities for the twenty-first century.* London: Routledge.

Volcano, D. L., & Halberstam, J. (1999). *The drag king book.* London: Serpent's Tail.

Warner, M. (1991). *Fear of a queer planet: Queer politics and social theory.* Minneapolis: University of Minnesota.

Warner, M. (1992, June). From queer to eternity: An army of theorists cannot fail. *Voice Literary Supplement, 106,* pp. 18–26.

Warner, M. (1999). *The trouble with normal: Sex, politics, and the ethics of queer life.* Cambridge, MA: Harvard University Press.

Warner, M. (2002). *Public and counterpublics.* New York: Zone Books.

Watney, S. (2000). *Imagine hope: AIDS and gay identity.* London: Routledge.

West, C. (1999). The moral obligations of living in a democratic society. In D. B. Batstone & E. Mendieta (Eds.), *The good citizen* (pp. 5–12). London: Routledge.

Weston, K. (1998). *Longslowburn: Sexuality and social science.* London: Routledge.

Wolcott, H. F. (2002). *Sneaky kid and its aftermath.* Walnut Creek, CA: AltaMira.

POSTSCRIPT 2011 TO LIVING WITH THE CONTRADICTIONS

Moving on: Generations, Cultures and Methodological Cosmopolitanism

Ken Plummer

Stop worrying where you're going, move on. If you can know where you're going, you've gone. Just keep moving on.

Stephen Sondheim, *Sunday in the Park with George*

The contradictory tensions of life and research go on and on. This chapter was written about a decade ago, and life and the tensions it raises, while unresolved, have moved on. The tensions of life do not stop coming—and since I wrote this article, other tensions have become more prominent. Some of the puzzles of critical humanism and queer theory, which are now relatively well established as intellectual orthodoxies, have been supplanted for me by the dilemmas posed by two further issues: generation and culture. Moving on, I see the need for the development of generational and cosmopolitan methodologies. In a very brief afterword, let me hint at these ever-expanding tensions.

Transgressing Generations

At the heart of my current concerns and tensions is the awareness that I am growing old: All academics do. With this startling revelation has come clear awareness that all intellectual life is (partially, if not primarily) organized through the tensions of generational standpoints—even though this is rarely discussed. Having recently read Randall Collins's (1998) magisterial *The Sociology of Philosophies* and re-read Karl Mannheim's (1937/1952) classic account of generations, I can see that academic life—like all social life—is bound up with different ages functioning through their generational cohorts and their generational networks of *interaction ritual chains*. Different generations assemble ideas that reach out to that generation; the task of later intellectual generations is always to move beyond the wisdoms of their predecessors. In intellectual life, there is always a premium on "saying something new": This is the way to forge ahead in a new career. Repeating the old stories of earlier research programs will get you nowhere. The 45 years of my own academic tensions have been but a mere blip in the vast cycle of generations of people thinking about the world we live in from many standpoints and contrasting cultures. Our intellectual, political, and research agendas have restricted time-bound concerns. They come and go, and very few ideas survive more than a generation or two. A few—"the classics"—are singled out from the vast pantheons of knowledges and libraries to be celebrated; most are lost forever, often in their own time (and even more so these days, when there is such a flood of academic work). Intellectual generations are both *diachronic* (bound up with intellectual generations that will rarely last more than 30 years) and *synchronic:* They create tensions across the generations at any one moment in time—as the different trainings and world moments come into radical, often conflictual interplay. All this suggests the significance of memory and time: Older ideas will always be replaced by the latest fashion. We need memories of the past even as old ideas are supplanted by the new ones of a new generation. There is a premium on the new-fashioned rather than the old-fashioned. Social life and social research are always lived at the intersectional ties of generations (a fact often overlooked by those theorists who focus usually only on the class, gender, and ethnicity of intersectional theory) (see Plummer, 2010, on all this).

One quick example, from a multitude of possibilities, must suffice. The term *queer* was the word of my childhood and youthful homosexuality, the word of hostility and hatred run rampant. It was a word that my time in the Gay Liberation Front of the early 1970s struggled so hard to resist. I loathed the idea of queer. It was bad news. Twenty years later, it became the flag post of a new generation—decked out in full radical irony. I have had to live with the term and recognize the new position it supposedly represents. But this does not make me happy with it. Recently, in an issue of *GLQ,* the premier queer journal, a leading review article offered a clear sense of how the different appropriations of queer varied by different generations. In a telling interchange between the young Matt Houlbrook, author of the much acclaimed *Queer London* (2005), and Jeffrey Weeks, the well-established author of the path-breaking account, *Coming Out* (1977), Chris Waters recounts how their different positions in the world radically shaped their contrasting historical interpretations. Houlbrook was only 2 years old when Weeks published his classic work; now, 30 years on, Houlbrook writes his history from a very different stance. This is, of course, as it must inevitably be. Houlbrook describes the homosexual past in "elegiac terms," arguing that in "exploring the history of queer London in the first half of the twentieth century, we should lament possibilities long lost as much as we celebrate opportunities newly acquired" (p. 140). It is precisely this nostalgia for lost possibilities that does not sit well with Jeffrey Weeks, the young radical of the 1970s who is now himself retired. Pressed on his attitude toward this past at a BBC radio discussion, Houlbrook says, "I think I'm going to have to admit being very nostalgic for this lost world," to which Weeks quickly responds, "I think you can only be nostalgic if you didn't live it. . . . Those of us who had the misfortune to live that life until the 1970s don't feel nostalgic about it" (Waters, 2008, p. 141). It seems to me that they lived in different worlds—and this is precisely what the shaping of generations does. In one sense, all of our qualitative researches are the records of successive generational cohorts jostling with each other. And the message is: Always keep your eye on the next generation; things are bound to change again. The stories we tell of social life are bound up with the generations we live in. All social sciences—like all of social life—entail the telling of generational narratives, and with this comes the presence of continuing tensions and even contradiction.

Transgressing Culture

My second, linked concern here is one I have been aware of for a long time. It is the sheer limitations of my own cross-cultural knowledge—the tensions over the differences across cultures and languages we are born into which are the limiting horizons of our thinking and practices. The vast populations of the world in China, India, Muslim cultures, Latin America, and the rest—who speak languages different from mine and live with religions I do not understand—have long been outside of the provinces of much so-called Western social science. Of course, these days we are sophisticated in postcolonial shapings, standpoint theories, and intersectional knowledges. We are aware that there are disturbing differences in our continents of knowledge and a massive arrogance about the localized theories and methods of the West, especially with the limited worldview of the Anglo American west. We have been told off for ignoring "Southern Theory," as Raewyn Connell (1998) rightly puts it. There is so much politically located knowledge that is beyond that of the limited West, and especially the even more limited United States, and yet our books and our researches carry on more or less in linguistic, emotional, and political ignorance of it. My earlier "Western models" of being gay, for example, now look absurdly quaint in the world scheme—the tales of a little White English boy struggling to come out in the swinging 1960s—when read against the almost genocidal plights of gay and lesbian people in, say, Uganda or Iran today.

At the same time, I am so glad to sense a new generation in the making that is developing research and languages that take this seriously and who are working hard to get us beyond the Western fix. Queer theory and critical humanism are simultaneously highly privileged local Western disputes; but they also travel quite well, having become debates in different ways across a wide range of cultures. Still the question now has to be posed: Is a global critical humanism possible—and, even more, is it possible to generate a transnational queer studies? Here we are hurled straight into major debates about the continuously hybridic and emergent character of cultures. The challenge is on as to whether we seek an abstracted universalism from the North over a truly grounded understanding from the grounded analytics of the daily practices of the South? This is personally a massive challenge. I was born a little boy in England, and although I am a little travelled, I do not speak the dialects of other local cultures, nor do I have the idiosyncratic deep wisdoms that their cultures harbor. It is simply wrong for me to pronounce on them. Sadly, the new (mainly Western) languages of internationalism, globalization, and decolonization make me feel increasingly marginal. It is fair to say I found my own cultural and generational voice in the 1960s, and I was able to have my say; but now is either the time to be silent or to quietly assist a new generational process of those billions of voices in the world without privilege who now need to have their own indigenous global say. We westerners babble and babble on with our theories and methods—often extending our own privileges further and further. It may be time for the Western academic party to call it a day.

Again let me give some examples from many. Travis Kong is a Hong Kong-born man, immediately hybridized between the English colony and his Chineseness. He has lived his life at this juncture. But more: He was working class and gay. This did not make life easy for him or for his ability to advance in academic life. His important book, *Chinese Male Homosexualities* (2010), is written at the intersections of an emerging global mosaic of multiple global sexualities and their challenging diversities at the start of the 21st century. There is much to learn from it. His study shows "the complexity of globalization in the kaleidoscopic life of Chinese homosexualities in Hong Kong, London, and China" (p. 8). We are immediately in three different lands with their wildly different histories and symbolic meanings. At the heart of this book are three case studies of what might conventionally be called gay men in three cities: Hong Kong, London, and Guangdong. The men are all in some sense "Chinese" and "gay," but in these very terms lie definitional problems of the mysterious essence. At various times, we enter the worlds of the diasporic and feminized "golden boys," the *tongzhi*, the "money boy," the *memba,* and the "potato." These new worlds have meant the creation of different hybridized gay identities (Kong, 2010). Same-sex sexualities are never cut from the same cloth. A homosexual is not a homosexual is not a homosexual. We have to learn to live with (and love) the varieties, the differences, the hybridic.

His study finds three broad clusters of Chinese men in search of meaning in their sexual lives in worlds of rapid social change, postcolonialism, globalization—and confusion. It tells of emerging new sexual stories in which we find new sexually imagined communities. It is a study, too, of the plurality, multiplicities, and differences of sexualities in the contemporary world—a move right away from any sense that there is one true Chinese sexual way, or indeed any one true sexual way anywhere. Shaped as they may be by familistic Confucian ideals from the past, new sexual identities are under construction, and this book suggests a patchwork of such differences as new identities generate the hybridization of sexualities. One hybridic dimension Kong points out is the queering of the straights and the straightening the queers. Here, too, sexualities are always on the move—across time and space, we find sexscapes, sexual flows, and sexual mobilities: These men transform their lives as they move to spaces full of new possibilities and from the cities of the East to the cities of the West—from communism to capitalism, from rural to global city, from colonialism to postcolonialism.

There is, then, much to savor in Kong's book. It is a study of hybridic sexualities and their refashioning across the globe. And in this, Kong's book joins a newly emerging field of study, in Latin America and in Asia, which looks at the shifting internal and external borders of the sexualities of nations and countries. There has been a recent flourishing of new work by new scholars who reject the presumptions of much Western theorization

about queer and gay. The 2005 Asian Queer Studies Conference was emblematic of this when it brought together in Bangkok some 600 academics and activists. It marked a turning point for all this challenging new work. What we are charting here are hybridic and cosmopolitan sexualities in their lived political contexts. Often, as here, this research produces ethnographic work about the complexities and subtleties of grounded lives in specific locations, work that is always much more messy, contradictory, and ambiguous than wider theories or dogmatic positions allow.

There are many other examples, including the growing work on international sex work and trafficking, which suggests large movements of people across the globe for sexual purposes involving money and often coercion and violence. Queer theory and critical humanism have so far not had that much to say on this. There are instead the standard formal studies—usually condemnatory—which give the global statistics of this "horror" provided by global agencies. But there are also growing numbers of accounts that are more sensitive to the multiple meanings emergent within both the cultures of departure and the cultures of arrival—and the struggles here between actors' shifting definitions, their senses of agency, and the constraining and enabling social structures they negotiate. As women (and men, and children) move from their home cultures (and their ambivalent expectations of work, family, sex) to their new host cultures, they have to negotiate new hybridic sexualities. In the work of Kaoru Aoyama (2009), Laura Agustín (2007), Tiantian Zheng (2009), and others, we find their own lives and those of women in tension. Here we find that migrants often make personal choices to travel and work in the sex industry, even as they are part of a dynamic and exploitative global economy. They all are advocates of listening to the voices of the migrant women.

Aoyama, for example, who is from Japan but studied in the United Kingdom, became interested in the women who leave Thailand for the sex industry in Japan. Here she can depict a theme of much recent work: An international world of sex slavery, trafficking, and coercion runs parallel to the needs of migrating women, who find sex work to be a means to support themselves and their families. Her concern is with the multiple complex paths of human agency shaped within changing culture as sex workers struggle in different places and at different times of their lives with different material conditions and different meanings. Aoyama's (2009) study above all is concerned with studying women's human agency, and it is compelling in detailing its empirical complexity. There is no one pattern of agency but multiple routes and sites.

Likewise, Zheng (2009) focused on the ways in which sexuality and sex work are shifting in a post-Mao landscape. Becoming a field worker in the city of Dalian, she closely observed young Chinese women who become karaoke bar hostesses. She interrogated not only their business situation but also their family and earlier backgrounds. Unsurprisingly, her research was far from welcome in China. She discloses some of her many difficulties in conducting such research—not least as she becomes partially engaged in waiting on the men herself, the only acceptable role that would allow her to be present in such bars as a Chinese woman. All this, however, helped her get close to the data, and it becomes another finely observed ethnography, which investigates what goes on in the club, the kinds of roles that the girls have to play, and their lives outside. Patriarchy (or the gender order) and male dominance can be observed at work everywhere, as can the damage being done to the women. Zheng directly links this sex work to patriarchy and masculinity in China as it is in serious postsocialist change. But the stories are more complex than this. The girls are from poor backgrounds, and Zheng shows the sheer poverty and degradation of their village lives—"the wretched living conditions in the countryside" (p. 150): Ironically, damned as these hostesses' city experiences are, they seem a lot better off. At least, they can send money back to their families. So the stories they tell here start to get more nuanced. They rationalize that their work is for the benefit of their family. More: There is real change in their lives. Often they started their city work in the sweatshops but found themselves moving up and beyond—so that they are now furnished with the very clothes made in the sweatshops. Up to a point, the quality of their lives is better, although they remain full of problems. Yet, at the same time, they daily confront the downright objectionable behavior of many of the men with whom they have to deal. Zheng's work joins a growing body of well-researched feminist ethnographies of the international sex work scene. Sex worker sexualities confront and construct hybridities. Like Kong's queer sexualities, sex worker sexualities are not cut from the same cloth. A sex worker is not a sex worker is not a sex worker.

Toward Methodological Cosmopolitanism

A vast cosmos so stuffed with conundrum.

A muddle of life with its dialogic drift.

An infinity of lists and contradictory cuts.

The bordered boundaries we break beyond.

So here we here are, never to agree?

So let me conclude this short afterthought by suggesting that we need to become *methodological cosmopolitans*. One recent book by Robert Holton (2009) lays out more than 30 positions and debates within the theory of cosmopolitanism, and I cannot go into all this. I most certainly do not mean the elite and exclusive chatter of the sophisticated intelligentsia of urbane intellectual and university life. Rather, I suggest more straightforwardly *a down-to-earth methodological, epistemological, and political stance that dialogues openly with the now*

massive diversity and fragmentation of the vast enterprise of understanding the complexities of global human social life—its labyrinths, its "infinity of lists," its "incorrigible pluralities" (Eco, 2009). We need to develop ways to comprehend the truly radically different ways of speaking across cultures and generations and to set them into tolerant, empathetic dialogues with each other.

As I see it, a cosmopolitan methodology needs to dialogue across multiple disciplines, across multiple academic and everyday life conventions, across generations, and across multiple cultures. Methodological cosmopolitanism means both a respect and willingness to listen, learn, and dialogue across the vast array of tensions found in research methods, epistemological stances, theoretical concerns, and political actions across the world. Ultimately, methodological cosmopolitanism links to the wider project of intimate citizenship—it is a phrase to add to the tools we need to understand the diversities of a global flourishing life. Intimate citizenship (Plummer, 2003) has human rights and responsibilities at its core—the rights and responsibilities of citizens to have and lead a good personal life in all their differences and tensions. The challenge is to develop a methodology that can match this task. We will keep moving on.

◨ REFERENCES

Agustin, L. M. (2007). *Sex at the margins: Migration, labor markets and the rescue industry.* London: Zed.

Aoyama, K. (2009). *Thai migrant sexworkers: From modernization to globalization.* Hampshire, UK: Palgrave.

Collins, R. (1998). *The sociology of philosophies: A global theory of intellectual change.* Cambridge, MA: Harvard University Press.

Connell, R. (2007). *Southern theory: The global dynamics of knowledge in social science.* Cambridge, UK: Polity Press.

Eco, U. (2009). *The infinity of lists.* Bloomsbury, UK: MacLehose Press.

Holton, R. (2009). *Cosmopolitanisms.* Hampshire, UK: Palgrave.

Houlbrook, M. (2005). *Queer London: Perils and pleasure in the sexual metropolis 1918–1957.* Chicago: University of Chicago Press.

Kong, T. (2010). *Chinese male homosexualities.* London: Routledge.

Mannheim, K. (1952). The problem of generations. In *Collected works of Karl Mannheim* (Vol. 5, pp. 276–320). London: Routledge. (Original work published 1937)

Plummer, K. (2003). *Intimate citizenship.* Seattle: University of Washington Press.

Plummer, K. (2010). Generational sexualities, subterranean traditions, and the hauntings of the sexual world: Some preliminary remarks. *Symbolic Interaction, 33*(2), 163–190.

Waters, C. (2008). Distance and desire in the new British social history. *GLQ, 14*(1), 139–155.

Weeks, J. (1977). *Coming out: Homosexual politics in Britain from the nineteenth century to the present.* London: Quartet.

Zheng, T. (2009). *Red lights: The lives of sex workers in postsocialist China.* Minneapolis: University of Minnesota.

12

ASIAN EPISTEMOLOGIES AND CONTEMPORARY SOCIAL PSYCHOLOGICAL RESEARCH[1]

James H. Liu

Any analysis of Asian epistemologies and their influence on contemporary social psychological research in Asia should begin by situating itself within recent flows of world history where Western science, industry, and political, economic, and military power have dominated the globe. Global forms of both natural and social sciences have had their origins in Western epistemologies and social practices. Social sciences like anthropology, sociology, and psychology all emerged in European societies in the 19th century, which was perhaps coincidentally the peak of Western nationalism and imperialism. Not coincident to this, elements of racism were both implicitly and explicitly embedded within early theories and practices of social science (Smith, 1999). It took the global cataclysm of World War II and all its aftermaths for racism to be put to bed as a legitimate basis for social science theorizing (Cartwright, 1979).

Given this type of "societal anchoring" (Moscovici, 1961/2008) in a particular historical moment where one civilization had apparently achieved ascendancy above all others through a particular formula of success, it is not surprising that social scientists in Asia found themselves in the position of having to react to forces put into motion by Western societies. First, social sciences in Asia (as in Western societies) have been and continue to be poor cousins to natural and physical sciences in terms of funding and visibility concerning national priorities. Second, modernization has provided a master set of discourses and practices whereby importation of Western ideas and practices is taken for granted as necessary to increase national strength and autonomy (see, e.g., Pandey, 2004). Within these overarching frames, following Western universities by importing logical positivism (an epistemology itself borrowed from the natural sciences) as the basis for Asian social sciences occurred largely without debate. Not only epistemology, but the structure and content of Asian social sciences were borrowed wholesale from the West as Asian universities were established throughout the late 19th and 20th centuries. In most disciplines in most countries, the first textbooks were translations of standard texts from North America and Europe.

This was the historical situation, and given continuing disparities in power, prestige, and influence distributed between developed and developing societies, between Western and non-Western scholars, and between English and non-English speakers (Moghaddam, 1989; Moghaddam & Taylor, 1985, 1986), it is not surprising that in the main, Asian social sciences remain for the most part thoroughly situated within Anglo-empiricist global norms, positioning the social sciences within epistemologies, or theories of knowledge, and practices drawn from the natural sciences.

If historical differences in power and prestige between Asia and the West were responsible for the structural foundation and mainline development of Asian social sciences, then the subsequent rise of Asian societies such as Japan, China, India, Taiwan, Philippines, and South Korea as indispensable components of the global economy has served as the impetus for an important countermovement. This is the rise of Asian epistemologies and Asian forms of psychological knowledge that emphasize cultural differences with the West rather than imitation (Liu, Ng, Gastardo-Conaco, & Wong, 2008). While decidedly less central than the first movement, this countermovement contains potentiality for the future because the world is moving toward both economical integration *and* the distribution of political, military, and economic power across multiple cultural centers.

▣ A SURVEY OF RECENT DEVELOPMENTS IN PSYCHOLOGY

The necessarily simplified introduction provided here sets the platform to launch a focused discussion of how Asian epistemologies have and will influence the theory and practice of psychology and especially social and cross-cultural psychology. Different patterns may be prevalent in other social sciences, like sociology or anthropology. In psychology, cracks in the edifice of borrowing from the West became visible in the 1960s, with the emergence of the subdiscipline of cross-cultural psychology out of shared interest among scholars in both Western and non-Western societies (the latter who frequently began their careers by attaining doctorates in Western universities and then returned home). Its goals were to (1) empirically test the generality and transportability of theories of psychology and (2) develop theories and constructs better suited to explain and predict behavior, cognition, and emotion in non-Western societies (see Berry, Poortinga, Segall, & Dasen, 1992; Ward, 2007). Although initially situated across subdisciplines in psychology, a cross-cultural approach has been most influential in social and personality studies, perhaps as a result of empirical demonstrations of the limitations of the "transport and test" model (e.g., Amir & Sharon, 1987). As time went on, powerful theories began to emerge to account for cross-cultural differences. In the 1980s, the seminal *Culture's Consequences* by Geert Hofstede (1980/2001), with its statistical analysis of survey data from countries around the world, found dimensions of cultural variation that located Western societies' psychological profiles as not universal, but culture-bound syndromes most notably characterized by individualism and low power distance (see Smith & Bond, 1993, for an update of this literature).

This trend of making psychological phenomena contingent on culture through scientific arguments has continued to the present day. Hazel R. Markus and Shinobu Kitayama (1991) famously made virtually all theories in social and personality psychology contingent on the construal of self as independent or interdependent (this making an element of culture into a discrete variable amenable to experimental manipulation). Ongoing published dialogue between North Americans and East Asians, mainly Japanese and then Chinese, has become a major feature of cross-cultural psychology. One example of recent issues that have engaged attention is the question of whether the requirement for positive self-esteem is universal (see Heine, Lehman, Markus, & Kitayama, 1999, versus Brown & Kobayashi, 2003). Recently, the flagship *Journal of Cross-Cultural Psychology* achieved an impact rating of 2.0, marking an unprecedented level of influence according to such indicators,[2]

while the *International Journal of Intercultural Relations* has also been an important contributor to the profile of the subdiscipline (with an impact factor of 1.0 in recent years). Perhaps because of this success, social constructionist epistemologies have had little influence on cross-cultural psychology. The majority of its adherents appear content to operate within empiricist practices and scientific discourses that have become the norm in this growing field (for a brief discussion, see Liu et al., 2010; for a comprehensive overview, see Berry et al., 1997).

Following from cross-cultural psychology, the Asian Association of Social Psychology held its inaugural conference in 1995 and established the *Asian Journal of Social Psychology* in 1998. Recently, an influential former editor of the journal wrote that "In a nutshell, AJSP is able to promote research that addresses cultural issues, and the journal seems to have developed a reputation as a "cultural" journal" (Leung, 2007, p. 10). But on the downside, "No obvious theoretical framework comes to mind when one thinks of Asian social psychology. Except for the indigenous psychologists, most Asian social psychologists work on topics that are popular in the West." (Leung, 2007, p. 11). The term *indigenous* in debates in psychology is used to refer to an intellectual movement that arose in reaction to the Western mainstream and seeks to reflect the social, political, and cultural character of local peoples around the world (Allwood & Berry, 2006). This movement has been especially prominent in Asia, as part of an intellectual decolonization (or de-Westernization) of psychology. For the most part, it does *not* refer to a psychology of first peoples, that is, a psychology of aboriginal peoples positioned as minorities within a politically dominant Westernized majority. Linda Waimarie Nikora, Michelle Levy, Bridgette Masters, and Moana Waitoki decried psychology's use of the term indigenous in their vignette on indigenous psychology for Maori, who are first peoples of New Zealand. But they wrote further,

> Terminology aside, the objectives of an indigenous psychology are agreeable: That is, to develop psychologies that are not imposed or imported; that are influenced by the cultural contexts in which people live; that are developed from within the culture using a variety of methods; and that result in locally relevant psychological knowledge. (Nikora, Levy, Masters, & Waitoki, 2004, quoted in Allwood & Berry, 2006, p. 255)

Indigenous psychology movements sprang up in India, Taiwan, and the Philippines in the 1970s and in Korea in the 1980s under the leadership of charismatic leaders that strongly influenced social science agendas in these and other Asian societies (Sinha, 1997). Whereas cross-cultural psychology has been and continues to be strongly influenced by positivist forms of empiricism dedicated to testing the generalizability and applicability of psychological theories to different populations, indigenous psychologies have been more varied in terms of their philosophical, epistemological, and political stands concerning the production and use of social science knowledge. Several overlapping definitions of indigenous psychology have been offered by major Asian protagonists (see Kim, Yang, & Hwang, 2006, or Allwood & Berry, 2006, for overviews). Virgilio Enriquez (1990) described indigenous psychology as a system of psychological thought and practice rooted in a particular cultural tradition, while Uichol Kim and John W. Berry (1993), defined it as "the scientific study of human behavior (or mind) that is native, that is not transported from other regions, and that is designed for its people" (p. 2). Among its more epistemologically and philosophically oriented advocates, David Ho (1998) views indigenous psychology as "the study of human behavior and mental processes within a cultural context that relies on values, concepts, belief systems, methodologies, and other resources indigenous to the specific ethnic or cultural group under investigation" (p. 93). The most influential programmatic developer of indigenous psychology, Kuo-shu Yang (2000) defined it as "an evolving system of psychological knowledge based on scientific research that is sufficiently compatible with the studied phenomena and their ecological, economic, social, cultural, and historical contexts" (p. 245). All the major protagonists agree that indigenous psychology involves knowledge and practice native to or rooted in particular societies and their cultural traditions. They vary in their commitments to global science, on the one hand, and locally informed action on the other.

The differences between Taiwan and the Philippines, where indigenous psychology has been most prolific (each with large regular meetings attended by hundreds of scholars), are instructive as to variations in theory versus practice. Both emerged in the late 1970s under the auspices of a talented, energetic founder capable of mobilizing both people and funding toward the enterprise. While the tenor of their research aims was similar, Enriquez's (1990, 1992) vision differed substantially from Yang's (1999, 2000) with respect to focus of application.

Enriquez was not opposed to natural science epistemologies in principle, but in practice, he thought that they were often inappropriately applied. He wrote extensively about the process of the indigenization of psychological science (Enriquez, 1990), both by adapting Western scientific constructs to the local culture and by developing local systems of psychological knowledge on its own terms (indigenization from without and within). The Philippines has been and continues to be a developing nation, with a current gross national product (GNP) of less than $2,000 U.S. per capita and a transparency rating putting it in the bottom quartile, along with other countries in the world struggling with endemic corruption.

In this societal climate, Filipino indigenous psychology is highly engaged with communities on a myriad of issues, and it is published mainly in Tagalog (the Filipino national language, which is especially dominant in Luzon; see Enriquez, 1992). It has a thriving relationship with other academic disciplines, government ministries, and nongovernment organizations (NGOs),

which results in what could be described as participant action-oriented research (McTaggart, 1997) or community-based participatory research (Minkler & Wallerstein, 2003). As such, ethnographic (qualitative field-based) inquiry is its predominant method of choice. Enriquez (1992) refers to this as "indigenization from within." Its outputs are mainly in monographs (e.g., Aguiling-Dalisay, Yacat, & Navarro, 2004) and internal reports for the commissioning agencies, and it uses primarily qualitative methods developed indigenously (see Pe-Pua & Protacio-Marcelino, 2000, for a recent English-language overview). Publications in international journals are rare, but at least as frequent as for works from other developing nations in Southeast Asia.

Filipino indigenous psychology could be described as highly applied, with development focused on content and ethnographic methods (e.g., how to work with illiterate sex workers) without concurrent development of an epistemology grounded in indigenous philosophical traditions. This pattern of focusing on applied research using the local language without taking a strong position on epistemology would be characteristic of much of Southeast Asia, including Indonesia, Malaysia, and Vietnam, but these latter nations would be less coherent in terms of the use of indigenously compatible theory, practice, and methods compared to the Philippines. Most of this work flies under the radar of the international scholarly community as it is published mainly in monographs, funder-mandated reports, and local journals.

Enriquez, the charismatic founder of Filipino indigenous psychology, died in 1994 at age 52, leaving a huge void that has not been filled. Yang, in contrast, has been and continues to be active in shaping Chinese indigenous psychology from Taiwan for more than three decades. In contrast to the Philippines, Taiwanese indigenous psychology has been more consistent with the norms of research practice prevalent in cross-cultural psychology, which are highly empiricist and quantitative but use paper-and-pencil surveys rather than being based on laboratory experiments as in mainstream psychology. Taiwan is a newly industrialized economy, where GNP and living standards are comparable with the lower half of the Organization for Economic Cooperation and Development (OECD). It is also a newly democratizing society, having achieved significant advances in free elections, gender equality, and the development of civil society over the last two decades. This, together with a weak version of the "publish or perish" academic culture prevalent in North America, helps account for an alternative pathway for the development of indigenous psychology taken here compared to the Philippines (see Allwood & Berry, 2006, for a global perspective on this).

Chinese indigenous psychology is internationally one of the more visible among all the indigenous psychologies in the world (see, e.g., Yang, 1999, or Hwang, 2005a, for accounts; see Bond, 1996, for a more cross-cultural approach to Chinese psychology).

Chinese indigenous psychology has its own journal, which has been published regularly in Chinese from Taiwan for two decades, and regular conferences are attended by many hundreds of scholars, often involving Mainland China and Hong Kong. In his most ambitious statement, Yang (2000) offers a program of development in indigenous psychology capable of unifying cultural psychology (derived from anthropology; see Cole, 1995), with its commitment to qualitative methods and "human science" epistemologies, and cross-cultural psychology, with its focus on quantitative and "natural science" epistemologies. He views psychology as consisting of a hierarchically organized system of indigenous psychologies:

> Beyond the imperative of indigenization, no other restraints need to be imposed upon activities of indigenous research. . . . Psychologists in any society may legitimately strive to construct an indigenous psychology for their people that is as comprehensive in scope as the current indigenous American psychology. . . . For example, some indigenously-oriented Chinese psychologists have set their hearts on developing an indigenous Chinese psychology comparable to the North American one in scope and depth. (Yang, 2000, p. 246)

It is understandable, given their population size and time-honored philosophical traditions, that Chinese people might have higher expectations for their indigenous psychology than many other peoples.

Most Asian indigenous psychologists in practice prefer particular methods (e.g., Yang is survey oriented, Enriquez was ethnography oriented), but in principle, they do not regard their activities as constrained by methods warranted by a Western form of epistemology. For Yang (2000), who draws liberally from Enriquez's thinking, the key concept is *indigenous compatibility*, defined pragmatically in terms of "empirical study . . . conducted in a manner such that the researcher's concepts, theory, methods, tools, and results adequately represent, reflect, or reveal the natural elements, structure, mechanism, or process of the studied phenomenon embedded in its context" (p. 250). He offers several do's and don'ts rather than a philosophically oriented system to achieving indigenous compatibility. For example,

> Don't uncritically or habitually apply Western psychological concepts, theories, methods, and tools to your research before thoroughly understanding and immersing yourself in the phenomenon being studied.
>
> Don't overlook Western psychologists' important experiences in developing their own indigenous psychologies, which may be usefully transferred to the development of non-Western indigenous psychologies.
>
> Don't think in terms of English or any other foreign language during the various stages of the research process in order to prevent distortion or inhibition of the indigenous aspects of contemplation involved in doing research. (p. 251)

On the other hand,

> Do tolerate ambiguous or vague states and suspend decisions as long as possible in dealing with theoretical, methodological, and empirical problems until something indigenous emerges in your mind during the research process.
>
> Do be a typical native in the cultural sense when functioning as a researcher.
>
> Do take the studied psychological or behavioral phenomenon and its sociocultural context into careful consideration.
>
> Do give priority to culturally unique psychological and behavioral phenomena or characteristics of people in your society, especially during the early stages of the development of an indigenous psychology in a non-Western society.
>
> Do base your research on an intellectual tradition of your own culture. (p. 251)

This highly pragmatic approach, rooted in research practices rather than epistemology, can be said to characterize the modal Asian indigenous psychology response to issues involving the social construction of knowledge. Indigenously oriented East Asians in economically developed societies like Taiwan (or Korea and Japan) as a rule have not used theoretical race, gender, or ethics critiques to challenge prevailing empiricist norms for the practice of psychology. Rather, all of these issues have been examined within an overarching empiricist umbrella that favors quantitative but also makes use of qualitative methods.

Kashima (2005) has argued that this approach is deeply rooted in Asian traditions of knowledge, which may give them an advantage in examining questions that fundamentally involve complexity and multiplicity at their very root, like culture. He challenges Clifford Geertz's (2000, p. 197) assertion that "bringing so large and misshapen a camel as anthropology into psychology's tent is going to do more to toss things around than to arrange them in order." Although this is simplifying his argument considerably, Kashima (2005) locates contemporary epistemological struggles between hermeneutic and empiricist schools of thought within a Western dualist ontology that separates mind from matter, human nature from material nature. He claims,

> If we take a view that intentionality is materially realized, meaning is part of a causal chain, and social scientific investigation is also part of complex causal processes, we can adopt a monist ontology, in which human nature is not distinct from, but continuous with, material nature. (p. 35)

▣ IMPLICATIONS OF CHINESE EPISTEMOLOGIES FOR SOCIAL PSYCHOLOGICAL RESEARCH

Being understandably better versed in Western philosophy than contemporary Chinese philosophy (which until recently has only been available in Chinese), Kashima (2005) states further,

"What we need is a monist ontology that is not the materialist ontology of the Enlightenment. It is difficult to speculate what it looks like until some philosophical investigations clarify this" (p. 36). In fact, the great neo-Confucian philosopher Mou Tsung-san (or Zongshan) (1970) used Immanuel Kant, one of the Enlightenment philosophers who contributed to the emergence of Western dualism, as a starting point to develop an autonomous moral metaphysics (see S. H. Liu, 1989a, for an English-language review of neo-Confucianism). While epistemology was not a central concern for ancient Chinese, Mou's work is emblematic of contemporary Chinese philosophers carrying their intellectual inheritances forward into dialogue with Western thinking. Unlike most Western philosophers, Mou allows for the possibility of the "intuitive illumination" of the cognitive mind (i.e., enlightenment in the highest sense), whereas Kant allowed only sensible intuition.

Kant followed from and expanded upon René Descartes' mind-body dualism by formulating a dualism of phenomenon and *noumenon* (thing-in-itself). Kant was convinced that only God has intellectual intuition (noumenon, thing-in-itself), while humans have to rely on sensible intuition (or evidence from the senses). Pure reason can only construct knowledge of the phenomenal world. According to Kant, human beings cannot know things-in-themselves (noumena), and hence, it is impossible for us to have knowledge of metaphysics because this would end in antinomies.[3] Mou, by contrast, reinterprets "intellectual intuition" to mean "intuitive illumination" (following Eastern traditions such as Buddhism, Confucianism, and Taoism) and posits that humans are capable of this, no matter what their faith. He proposes a transcendental dialectic where the mind, while unable to produce acceptable proof of the metaphysical ultimate, nevertheless can realize the thing-in-itself as "thusness" or "suchness," the exact opposite of phenomenal knowledge constructed by the cognitive mind, bound in time and space. Because Mou's transcendental dialectic does not deal with empirically verifiable knowledge, it is similar to Soren Kierkegaard's position that "subjectivity is truth." However, it nonetheless describes a rational process that departs radically from Kierkegaard's irrational approach and hence avoids dualism.

Western Enlightenment thinkers, influenced by Christian traditions, saw the metaphysical ultimate as God and tended to view it (as Kant and Descartes did) as transcending the phenomenal world. For Kant, freedom of the will, immortality of the soul, and existence of God are postulates of practical reason. Following from these, an epistemology emerges consistent with a dualism between mind and matter and a division between natural and human phenomena because for Christians like Descartes and Kant, it was important to maintain their religion as a valid system of knowledge in the face of their own logic and rationality.

As culture's effects are largely implicit, Kashima's (2005) point is that without necessarily being aware of it, contemporary Western social scientists have maintained an unnecessarily sharp division between natural and human phenomena as part of their particular cultural program (see also Kim, 2000): Some carry on with a natural science paradigm in an Enlightenment vein, and others react against this as an affront to human agency and dignity. As most social scientists are not philosophically trained, they have a tendency to translate their cultural ontology into an almost religious commitment on methodological issues that might be described by philosophers as "methodolatry": the conflation of ontological issues with methodology. As Paul Tillich (1951) observes, value must have an ontological basis. The value of scientific observations formalizing sensible intuition compared to the phenomenology and hermeneutics of intuitive illumination cannot be reduced to any formula involving emotive responses or subjective utilities, and it cannot be deduced or induced by any form of logical or empirical proof. Hence, to privilege one set of research practices, which are derived from a particular value system associated with a particular ontology, as providing "the answer" to all the social sciences' contributions to the human condition is methodolatry.

In general terms, Asian philosophical traditions allow for human beings to have the ability to grasp ontological reality, although they may reach radically different conclusions about what this might be. This means that rather than seeing methodology as the solution to problems involving the privileging of different value systems in social science research (methodolatry), Asian implicit theory (or folk beliefs) is based on holism and perpetual change, where "a tolerance of contradiction, an acceptance of the unity of opposites, and an understanding of the coexistence of opposites as permanent, not conditional or transitory, are part of everyday lay perception and thought" (Spencer-Rodgers, Williams, & Peng, 2007, p. 265; see Nisbett, Peng, Choi, & Norenzayan, 2001, for an overview of East Asian holistic thinking). In practical terms, this means that Asian traditions do not privilege scientific methods of observation above the intuitive illumination of the original mind but rather see these as complementary forms of knowing.

Confucian traditions in particular tend to see the metaphysical ultimate as a creative principle functioning incessantly to guide the becoming of worldly phenomena. *Jen* or *Ren* (defined as humanity) is identified with *Shengsheng* (creative creativity) by Song-Ming neo-Confucian philosophers (see S. H. Liu, 1998a, for an extended treatment). It is thus a "moral principle" in the broadest sense of the term, from which continually changing aspects of being in time and space emerge. In the most powerful and complete statement of contemporary neo-Confucian philosophy by Mou Tsung-san, the Kantian dualism between phenomenon (perception of reality) and noumenon (the thing-in-itself) is not accepted. While Mou (1975) is sympathetic to Martin Heidegger's (1977) notion of human beings as *Dasein* (being there), a being in the world, and psychological states like anxiety and care as modes of existence, Mou argued that a phenomenological ontology is capable of giving only a description of human existence and not a value basis. Hence, according to Mou (1975), the best that Heidegger (1977) can achieve is an inner metaphysics and not the transcendent metaphysics that Asian intellectual traditions demand (see S. H. Liu, 1989a, for a more extended version of these arguments, and Bhawuk, 2008a, for contemporary social psychological work following on this theme from Indian philosophical traditions).

Chinese social scientists, like Western social scientists, might not explicitly reference philosophy as they conduct their research, but like Westerners, they have followed their own implicit cultural program, and many have proceeded to conduct research that frequently combines qualitative and quantitative methods and blurs the boundaries between empiricism and hermeneutics. For instance, many indigenous Chinese psychology papers combine quantitative and qualitative methods, and the warp and weft of their papers is the interweaving of Chinese tradition with contemporary mainstream psychology references. A "hot topic" at indigenous Chinese psychology conferences (most of it published in Chinese) has been the relationship between mother-in-law and daughter-in-law, which is fraught with the weight of contending cultural expectations between younger and older generations. Qualitative or quantitative approaches to analyzing surveys, interviews, and ethnographies have been used as acceptable forms of inquiry. Culture is most often *not* the explicit topic of inquiry but rather is embedded within the processes and objects of inquiry.

There are obvious exceptions to these rules. For example, Lee (2004) is a strong advocate of a hermeneutical phenomenological approach to indigenous research on ethnicity in Taiwan, following Heidegger, and feminist scholars in Asia as elsewhere tend to prefer qualitative to quantitative forms of inquiry. However, the most popular text advising social science graduate students on how to do thesis research (Bih, 2005) is completely ecumenical with respect to methodology, advising only that the research method should suit the research question. Having said this, the flagship *Chinese Journal of Psychology,* representing "the establishment" in Taiwan, still favors quantitative research, and as a whole, the university system privileges the contributions of the natural sciences above those of the humanities and social sciences.

For many Asian social scientists, the six statements that Kashima (2005) uses to describe a generic epistemic position on culture and its implications for psychology would be uncontroversial:

1. Culture is socially and historically constructed.

2. People construe themselves using concepts and other symbolic structures that are available.

3. People develop a theory of mind (i.e., a theory of how the mind works) to understand others.

4. People have beliefs about the world, and they act on those beliefs.

5. People engage in meaningful action.

6. Culture is constitutive of the mind. (p. 20)

For indigenous psychologists in particular, the division between human and natural phenomena and the polemics between advocates of different forms of knowledge construction would appear to be problematic. Kashima's (2005) summary seems more like a good starting point than a bone of contention: "To put it simply, the argument is that human agency and self-reflexivity make human society and culture dynamic (i.e., changing over time) and knowledge and human activities historically and culturally contingent" (p. 22). Gergen (1973), for example, treats the historical contingency of psychological phenomena as a call to revolution, whereas Liu and Hilton (2005) see it as grist for their mill. The former sees historical contingency as evidence requiring the overthrow of a methodological hegemony, whereas the latter see it as a description of the operation of human agency and cultural construction through time requiring empirical investigation (Liu & Hilton, 2005) and philosophical reflection (Liu & Liu, 1997).

As Leung (2007) noted, Asian social psychologists have not yet fully capitalized on the relative freedom from methodolatry that their philosophical traditions provide in terms of creating notable breakthroughs. He criticized Asians for their lack of ambition, citing a relative paucity of sustained programmatic research. As the current editor-in-chief of the *Asian Journal of Social Psychology,* I would have to concur: Most of the 200 or so papers submitted to the journal annually lack imagination, consisting to a significant extent of replications and minor variations on a theme established by quantitative research from the United States.

Although indigenous research in Asian social, personality, clinical, cultural, and cross-cultural psychology is still in its formative years, several characteristics would appear to be foundational. The first is the aforementioned lack of preoccupation for translating epistemological concerns into methodological boundaries. The second is an overwhelming concern with social relationships and social interconnectedness. The third is a naturalistic approach to culture as a relatively uncontested element of basic psychology. Liu and Liu (1997, 1999, 2003) describe this as a psychology of interconnectedness.

It remains to be seen whether Asian social psychologists will be able to fulfill the epistemological promise of their philosophical traditions (Ward, 2007). Asian universities, like universities all over the world, privilege the natural sciences and aspire to and internalize standards set by Western universities. They push their faculty to publish in prestigious journals, which are most often controlled by American universities and American or European academics. The Shanghai Jiaotong University's rankings of the best 500 universities in the world, which was constructed for the purpose of providing "objective standards" to aim for in developing a "world class" Chinese university ranked in the top 100 (Beijing University and Tsinghua are aspirants), completely favors the natural sciences and virtually disregards contributions from the humanities and social sciences. Given these circumstances (see Adair, Coelho, & Luna, 2002; Leung 2007; Ward, 2007), it is highly unlikely that Asian academics will be able to produce philosophically and epistemologically autonomous bodies of work. Rather in the near future, global psychology will emerge as a patchwork quilt of pluralistic practices connected to a still-dominant American center (see Liu et al., 2008; Moghaddam, 1989).

Japanese social psychology is a good example of both the variety and constraint in the patchwork quilt. Mainstream Japanese social psychology is thoroughly enmeshed in an empiricist dialogue with American social psychology on epistemic grounds set by North America. While there certainly is a small indigenous psychology movement in Japan (see Behrens, 2004), it does not have the scope or ambition of the movements in Taiwan or the Philippines. Perhaps in reaction to this, recently a dissident faction emerged in Japan challenging the mission of the mainstream on epistemic grounds, constructing arguments that pit quantitative versus qualitative methods and contrast human science with natural science in ways that would be very familiar to qualitative researchers in North America (see Atsumi, 2007; Sugiman, 2006).

Overall, the volume of increase in submissions to *Asian Journal of Social Psychology,* from 168 in 2007 to 182 in 2008 to 210 in 2009 (a 10% increase per annum), and the massive increase in Asian authors in social psychology over the last 10 years—even after controlling for the impact of *AJSP* (Haslam & Kashima, in press)—point to the potential inherent in the region and its peoples. As the methodological and theoretical skills, together with the cultural confidence of Asians, increase following the trend-setting success of their economies, one cannot help but be excited about possible breakthroughs coming out of Asia that pierce the dichotomy between natural science and human science in innovative ways. Hence, it is fitting to close this chapter by providing a brief introduction to a few of the more prominent epistemologically informed projects that have emerged in Asian psychology in recent years.

▣ THREE EPISTEMOLOGICALLY INFORMED ASIAN RESEARCH PROJECTS

The work of Hong Kong clinical psychologist David Y. F. Ho is unusual in that it is both informed by Chinese philosophy and written in English, making it accessible to international audiences. Two of his more imaginative pieces on Chinese indigenous psychology contain an interpretative analysis of classic Chinese culture stories (Ho, 1998a) and a humorous dialogue between a Confucianist and a clinical psychologist (Ho, 1989) around such culturally loaded terms as *propriety* and *impulse*

control. These excursions are underpinned by a serious commitment to what he and his colleagues call methodological relationalism (Ho, Peng, Lai, & Chan, 2001). This is a general conceptual framework for the analysis of thought and action that takes a person's embeddedness in a network of social relations as the fundamental unit of analysis:

> Actions of individuals must be considered in the context of interpersonal, individual-group, individual-society, and intergroup relations. . . . In particular, each interpersonal relationship is subject to the interactive forces of other interpersonal relationships. This consideration introduces the dialectical construct of *metarelation* or *relation of relations.* (Ho, Peng, et al., p. 937)

Two basic analytic units are used in Ho et al.'s (2001) approach to personality and social psychology, person-in-relations (focused on the target person in different relational contexts), and persons-in-relation (focused on different persons interacting within a relational context). The authors' quantitative work attempts to deconstruct the hegemony of person-versus-situation formulations of behavior as "consistent" versus "inconsistent" by introducing an intermediate layer, person-in-relations. They have argued cogently that relationships can transcend the person-versus-situation dichotomy because they are neither intrinsically part of the person nor intrinsically part of the situation; people are situated in a web of relations that help them to navigate through a situation in particular ways (see Ho & Chau, 2009, for an empirical demonstration).

They have also developed a qualitative approach called investigative research (Ho, Ho, & Ng, 2006), arguing that "neither a psychology predicated on methodological individualism nor a sociology based on methodological holism is fully equipped to account for the total complexities involved" (p. 19) in understanding the relationships between individuals in society.

> A social fact, though not reducible to facts about the individual, is nonetheless a fact about the social behaviour of, manifested by, individuals; and a psychological fact is a social fact wherever it refers to behaviour occurring in the presence of others, actual, imagined, or implied. Each contains and is contained by the other. A knowledge of one enhances, and a lack of knowledge of one diminishes, the understanding of the other. (pp. 19–20)

Ho et al. (2006) propose to base their research methodology on two metatheoretical propositions:

> 1. The conceptualization of psychological phenomena is, in itself, a psychological phenomenon. As a metalevel phenomenon, it requires further study. 2. The generation of psychological knowledge is culture dependent: Cultural values and presuppositions inform both the conceptualization of psychological phenomena and the methodology employed to study them. Accordingly, the role of the knowledge generator, given his or her cultural values and presuppositions, cannot be separated or eliminated from the

process of knowledge generation. These propositions do not necessarily negate positivism. Rather, they challenge positivism to have greater sensitivity to culture dependence and to broaden its scope of investigation. (p. 22)

At this point, rather than elaborating their epistemological position, they argue for *reflexivity* in applying three "intellectual attitudes germane to investigative research" based on the two presuppositions. These appear to be pragmatic and dialectical, but grounded in realist epistemology:

The first stresses the importance of critically examining the evidence in the truth-seeking process. The second confronts the inherently complex, even deceptive, nature of social phenomena; vigilance against deception is integral to seeking truth. The third sees the recognition of ignorance and knowledge generation as twin aspects of the same process. (p. 22)

Rather than offer any standard techniques or procedures, they describe investigative research as "disciplined, naturalistic, and in-depth," guaranteeing data quality and acting in the service of social conscience.

Ho et al. (2006) advocate *both* the use of reflexivity *and* moving from exploration to confirmation (e.g., from qualitative exposition to quantitative hypothesis testing) as research methods. They state their admiration for good investigative reporting produced by journalists (e.g., in their verification of source information and their dedication to truth seeking), but they do not state in a clear, programmatic way how such journalistic training could be applied to social science research. From the perspective of the Western-trained methodologist, Ho et al.'s (2006) program might not appear sufficiently compelling—it is lacking in details, and the thorny questions of confrontation between truth-value and desire to do good in the process of investigative research are not articulated. But the Asian ontological and epistemological systems described previously can help Western scholars to make sense of this desire and their pragmatic means of achieving it. For Ho, Ng, and Ho (2007), reifying a dividing line between qualitative forms of "human science" and quantitative forms of "natural science" just does not make sense, and they react against this with an almost moral sense of indignation.

For 30 years, Taiwanese social psychologist Kwang-guo Hwang has conducted a research program in indigenous Chinese psychology built on foundations of traditional Chinese theories of knowledge. Whereas Ho could be described as something of a lone wolf, working out an epistemologically sophisticated program of indigenous research in a Hong Kong social science thoroughly entrained by Western paradigms, Hwang has had the good fortune to have spent his career working within and contributing to a highly developed and collaborative indigenous psychology in Taiwan (Hwang, 2005c; Yang, 1999). Whereas Ho's primary dialogue partners are Westerners and

Westernized or bicultural Asians, the capstone of Hwang's (2009) work, *Confucian Relationalism: Philosophical Reflection, Theoretical Construction, and Empirical Research,* was written in Chinese and directed toward Chinese social scientists. Because Hwang's prodigious output consists mainly of books written in Chinese whose thrust is theoretical rather than empiricist, the work is almost impossible to do justice to in a few paragraphs. It is possible here only to give a flavor of the work. It should be noted that while Hwang is situated in social psychology, he has read widely in the philosophy of science and the sociology of science, and his writing is clearly directed toward social scientists and not just psychologists.

Hwang's (2005b, 2005c, 2006, 2009) mission is to realize a comprehensive epistemology of social sciences for Chinese (and by extension other non-Westerners) that provides the philosophical foundations for engaging in fruitful dialogue with one another and Westerners. Hwang's project is consistent with Mou's basic premise that different philosophical and cultural traditions provide alternative (and overlapping) ontological bases for constructing the phenomenology of subjective experience and the epistemology of its examination. The foundational statement of his work on Confucian relationalism was a model of face and favor (Hwang, 1987) that analyzed the inner structure of Confucianism for managing social relations and social exchange.

In his book on *Knowledge and Action,* Hwang (1995) argued that Western culture emphasizes the importance of philosophy for pursuing knowledge, whereas Chinese cultural traditions of Taoism, Confucianism, legalism, and martial school are concerned about wisdom for action. Consistent with an approach based on constructive realism (Wallner & Jandl, 2006), Hwang (1995) argues that since psychology's foundations and current practices are grounded in Western philosophy, genuine progress in indigenous psychology comes through constructing a scientific microworld consistent with Western philosophy, while maintaining a comprehensive understanding of the influence of Chinese cultural traditions on the daily life of Chinese people.

To familiarize Chinese social scientists with the major schools of Western philosophical thought influencing social science thinking, Hwang (2006) wrote *The Logic of Social Science* in Chinese. Here he recognizes that to construct a coherent scientific microworld, social scientists must not only be able to recognize themselves as fish swimming in a phenomenological sea of cultural constructions, but *they must be able to translate these insights into the systematic forms that scientific microworlds require.* Moreover, these microworlds often share much in common with one another, so that communication and translation of concepts between them is a critical feature of scientific *and* phenomenological insight.

Based on a philosophy of postpositivism, his summary work *Confucian Relationalism: Philosophical Reflection, Theoretical Construction, and Empirical Research* (Hwang, 2009) emphasizes that the epistemological goal of indigenous psychology is the construction of a scientific microworld constituted by a series of theories that reflect both universal human minds in general and the particular mentalities of a given culture. In view of the fact that most theories of Western social psychology have been constructed on the presumption of individualism, Hwang (2009) explained how he constructed the face and favor model (Hwang, 1987) with four elementary forms of social behavior and used it as a framework to analyze the deep structure of Confucianism (Hwang, 2001). Then he illustrated the nature of Confucian ethics in sharp contrast to Western ethics and constructed a series of theories based on relationalism to illuminate social exchange, the concept of face, achievement motivation, organizational behaviors, and conflict resolution in Confucian societies.

Hwang's (2009) project lays out the foundations and issues a call for programmatic development that could generate decades of research in indigenous social science, particularly if in the future mainland Chinese decide to pursue this avenue of research. Students and professors from various places attend his seminars at National Taiwan University. Hwang is primarily a theorist rather than an empiricist, and so this is a slow developing project that is focused on the big picture. We should not expect to see immediate results. Rather, as the Chinese Culture Connection (1987) noted, one of the most salient characteristics of Chinese people is long-term orientation, and in the 21st century, Chinese will need time on the order of decades to work out their response in the social sciences to the foundations and practices laid out by the West.

Recent work from Indian scholars, which draws from their great philosophical traditions to create a metaphysically oriented psychology, is another topic of international interest. According to a recent definitional statement by Ajit K. Dalal and Girishwar Misra (2010),

> More than materialistic-deterministic aspects of human existence, IP (Indian psychology) takes a more inclusive spiritual-growth perspective on human existence. In this sense no clear distinction is made between psychology, philosophy and spirituality, as conjointly they constitute a comprehensive and practical knowledge or wisdom about human life.

Hence, what appears consistent among Asian scholars drawing from their massive and distinct traditions is a questioning of Western ontology and reconsideration of whether discretely methodological forms of knowing should hold such a privileged position in generating and reifying social science knowledge (Bhawuk, 2008b; Paranjpe, 1984). But there are differences as well. If Chinese philosophical traditions have drawn Chinese social psychologists into thinking about social relatedness and holistic interconnectedness as fundamental ontological postulates, Indian philosophical traditions have a similar pull into the spiritual depths of the phenomenology, epistemology, and practice of Self (*atman*) (Bhawuk, 2003).

Sinha (1933), as cited by Dharm Bhawuk (2008a), described Indian psychology as based on metaphysics. Rather than beginning with Erik Erikson or Sigmund Freud, Indian scholars like Anand C. Paranjpe (1998) begin with Vedic traditions like the Upanishads, which includes such basic tenets as the "truth should be realized, rather than simply known intellectually" (Bhawuk, 2010). According to Bharati (1985, as cited in Bhawuk, 2008a), the self has been studied as "an ontological entity" in Indian philosophy from time immemorial, and far more "intensively and extensively" than in any of the other societies in the East (Confucian, Chinese, or Japanese) or the West (either secular thought or Judeo-Christian-Muslim traditions). The basic methodology is the practice of meditation, and the goal of meditative practices is to uncover the nature of the true self (*atman*), unencumbered by even such a fundamental phenomenological unity as time (Bhawuk, 2008a; Rao & Paranjpe, 2008). Even the basic dividing line between the knower and the known cannot be maintained if the meditative practices of Indian philosophy are accepted as an important and valid form of knowledge. In marked contrast to Chinese philosophers' tendency to maintain distinctions (*li-i-fen-shu*, one principle, many manifestations) while seeking for unity (*tien-ren-ho-i*, heaven and humanity in union; see S. H. Liu, 1989b, or Liu & Liu, 1999), Indian philosophers have plumbed the very depths of knowing to collapse even basic polarities such as good and evil or being and nonbeing under the glare of intuitive illumination.

The contemporary Indian psychology movement appears to be in the process of constructing a psychology of self that is simultaneously a practice of self-realization. Bhawuk has gone so far as to propose a general methodology for translating classic Indian scriptures into psychological models of theory and practice:

> For example, in the second canto of the *Bhagavad-Gita* a process of how desire and anger cause one's downfall is presented. The sixty-second verse delineates this process by stating that when a person thinks about sense objects, he or she develops an attachment to it. Attachment leads to desire, and from desire anger is manifested. The sixty-third verse further develops this causal link by stating that anger leads to confusion (*sammoha*) or clouding of discretion about what is right or wrong, confusion to bewilderment, to loss of memory or what one has learned in the past, to destruction of *buddhi* (i.e., intellect or wisdom) to the downfall of the person or his or her destruction. (Bhawuk, 2010)

Even a cursory reader of Bhawuk's work will recognize that the phenomenological layer of concepts described in Indian psychology are not only distinct from comparable Western concepts, but also systematic and compelling, once their internal logic is discerned. Bhawuk does not appear to privilege any Western forms of empiricism or phenomenology in terms of validating or providing an understanding of this system. Similarly, advocates of transcendental meditation were quite happy for Western scientists to measure them during meditation and find that their oxygen consumption and heart rate decreased, skin resistance increased, and electroencephalographs showed changes in frequencies suggesting low stress (Bhawuk, 2008b; see Rao & Paranjpe, 2008, for a more detailed review). This scientific knowledge did not change the subjective practice and goals of transcendental meditation one bit.

Bhawuk's (2008a, 2010) models are entirely theoretical at this point: In Hwang's (2006, 2009) terms, they represent a translation of the philosophical microworld of Indian philosophy into a psychological microworld of relationships between variables: How this translation will then impact on the cultural macroworld of the practice of Indian religions by lay people or inspire qualitative or quantitative investigation is anyone's guess. It is mind-boggling to realize that one of the most profound statements on the consequences of anger and desire on the human condition, known and practiced for millennia as part of the root philosophy of one of the world's great cultures, has only recently made its entry into the psychological literature (see Bhawuk, 1999, 2008a). Bhawuk (2010) carefully situates his construction of psychological models within the context of his daily meditative practices and as part of his family life. The dualism of qualitative versus quantitative methodologies never comes up as an issue in his writing. His quest is to expand the boundaries of science, not divide it into analytical portions circumscribed by methodological differences that seem almost quaint beside the monumental questions of being and nonbeing, time and permanence, probed by Indian philosophy (Bhawuk, 1999).

By comparing Indian culture with the culture of science, Bhawuk (2008b) argued that science itself has a culture, which is characterized by tenets like objectivity, impersonalness, reductionism, and rejection of the indeterminate. He cautioned that, as cultural or cross-cultural researchers, we need to be sensitive to the fact that science also has a culture, and our research might benefit from adopting worldviews, models, questions, and methods that are characteristic of indigenous cultures, especially those of non-Western origin. Bhawuk stressed the need for crossing disciplinary boundaries and recommended that we go beyond multiple-method and use multiparadigmatic research strategies to understand various worldviews in their own contexts. A team of researchers from various academic disciplines can help us find linkages across disciplines and paradigms.

◨ CONCLUSION

While each of the three programs reviewed above is exceptional, they should be situated within the greater flows of history and institutional practices that characterize social science research in Asia. Ho, Hwang, and Bhawuk mobilize the intellectual capital inherent in their cultures to innovate original solutions to perennial problems in social science. The first two are senior scholars toward the final phase of their careers, whereas the third is a senior scholar in his prime; none of them is under survival pressures in terms of career development. The far more typical submission to *Asian Journal of Social Psychology* or other major culture-oriented journals in psychology is a replication of a Western model with minor variations, the primary justification for which is "no data from (fill in the country) has to their knowledge been collected to test this model." Social psychological research in Asia can be characterized by tension between scholars living within a phenomenological layer of cultural constructions as a visible part of their everyday life, and producing English-language publications that are devoid of such meaningful content and dedicated toward the pragmatics of career advancement according to top-down standards imported from the "best" (read Western) universities.[4]

The pockets of innovation cited here have not changed the overall institutional climate favoring natural science research and practices in Asia, nor have they touched the publications prestige gradient, where English language (JCI/SSCI) journals are valued above local language outputs.[5] Asian social psychologists appear highly pragmatic in carving out their careers amid a disjuncture between their subjective experiences and dominant institutional practices (see Adair et al., 2002, for bibliometric evidence of the massive extent of Western dominance of the published literature in psychology, and Haslam & Kashima, in press, for challenges to this trend).

Some researchers working with qualitative paradigms in Western institutions have made highly conscious, sometimes ethical choices in working with particular methodologies. At best, their work reflects the polish and cohesiveness of intellectual rigor. At worst, it dissolves into hair-splitting methodolatry and promotion of group interests using methodology as a means of academic combat. Researchers in Asian institutions seem typically to be more pragmatic, at worst, sublimating their phenomenological experiences into whatever methodological paradigm is dominant and can be used to promote self-advancement and, at best, developing late in their careers an ecumenical and innovative orientation toward methodology. To change the shape of this gradient, it would be necessary for there to be more collaboration and communication between open-minded Westerners with influence on international journals and Asian scholars with a passion for probing the depths of their cultural resources and expanding the breadth of their disciplinary practices (see Liu et al., 2008). In this process, bicultural members of the Asian academic diaspora have and will continue to play a major role (Liu & Ng, 2007).

In the interim, Asian scholars working pragmatically in disciplines not of their cultural making can operate at the margins to adopt an alternative system of meaning for reconciling the disjuncture between their phenomenological experience of the world as a cultural construction and their professional judgment of how to best further their careers. To illustrate with personal experience, early in my career, I would sometimes write quantitative descriptive papers without hypotheses and be forced into a hypothetico-deductive model by international journal editors. At this point in mid-career, I have internalized mainstream psychological discourses to such an extent that writing in such a mode requires little effort and has significant benefits. But in terms of meaning, I regard the hypothetico-deductive model in psychology as a post-hoc explanatory model rather than as a universal model of prediction and control. This is not to deny there might be a deep structure that underlies human psychology, but even where it exists, this deep structure can find expression only through interactions with the phenomenological layer of subjective experiences, which is mediated by culture's concepts and an institutional layer of societal governance. Therefore, I treat all statements of causality in psychology as contingent on the phenomenological and institutional layer operating in the situation at the time of survey or experiment administration. In my own papers, I articulate this symbolic layer of meaning with great detail, and I tend to be cautious about other more careless statements of universality. In some sense, I do treat social psychology, in Kenneth Gergen's (1973) famous phrase, as history (Liu & Hilton, 2005; Liu & Liu, 1997), but I do not see how privileging qualitative methodology over quantitative methodology or vice versa solves the problem of the historical and cultural contingency of human behavior. When I write for quantitative journals, I follow their dominant discourses and practices for communicating how the phenomenological layer of culture conditions individual behavior, emotion, and cognition; when I write for qualitative journals, I do the same thing (Liu & Mills, 2006). I view methodologies as no more and no less than different prisms through which the objects of inquiry are refracted and communicated. In terms of their relative strengths, qualitative research is useful for telling us the *what* of a phenomenon, and quantitative research the *how much, how prevalent,* and *under what conditions is it causal.*

Putting these together, Liu and Sibley (2009) advocate four steps in the interweaving of qualitative and quantitative methods to describe and prescribe national political cultures.

1. Ascertaining the symbolic landscape through open-ended survey methods that give an overview of the major historically warranted symbols prevalent in a society; this may include quantitative analysis techniques and representative samples but must involve open-ended inquiry.

2. Describing the discursive repertoires that make use of political symbols in everyday talk through various institutionally mediated channels; this may involve archival analysis, interview, or focus group methods; the key is to examine naturalistic discourse for thematic content.

3. Operationalizing symbolic representations such as legitimizing myths or ideologies by converting naturally occurring talk to quantitative scale measures or experimental conditions; making maximum use of empiricist techniques to make causal inferences.

4. Moving from representation to action using both empirical findings and personal reflexivity as resources; applying findings from social science with a full awareness of their conditional and contingent nature when giving policy advice.

As the editor of a journal, I am open to either or both modes of communication (Liu, 2008), but I believe that a researcher must understand the internal logic of each prism to be able to blend and transcend their influences. I see the ultimate arbitrator of methodology as value, and I see value as having an ontological status that precedes rather than being derived from epistemology. The two research values I subscribe to besides truth value (see Liu & Liu, 1999) are (1) indigenous compatibility (Yang, 2000)—to what extent does the research reflect the phenomenology of cultural and institutional systems from which observations were derived, and (2) practical value—to what extent does the research provide subjective utility to academic *and* lay communities in which the researcher resides. I believe the future for Asian social psychology, to paraphrase Tomohide Atsumi (2007), is to fly with the two wings of scientific inquiry and practical utility. In the latter area, I hope that an Asian social psychology unencumbered by dualism will be able to make substantial breakthroughs in the future (Liu, Ng, et al., 2008; Liu & Liu, 2003).

In conclusion, S. H. Liu (1989a) summarizes the methodological advice of Hsiung Shih-Li, another eminent neo-Confucian philosopher (and a teacher of Mou Tsung-san) as follows:

> The scientific way of thinking has to posit an external physical world as having an independent existence of its own. From a pragmatic point of view this procedure is perfectly justifiable. But it has the danger of hypostatizing functions into ontological substances and hence committing a metaphysical fallacy. In order to guard against the natural tendency of man to fall into such a naïve attitude, philosophy has to adopt two important methodological procedures. In the first place, we have to appeal to a specific analytical method which purports to destroy all attempts to identify phenomenal functions with ontological principle itself by finding out all the contradiction or absurdities involved in such untenable metaphysical conjectures. (p. 25)

On this first point, qualitative researchers have done well, constructing a phalanx of *posts* to deconstruct naïve attempts to reify natural science models in human science. But Hsiung's second point is more radical and cuts right to the heart of what it means to be a social scientist and a human being:

> This is exactly what the Buddhist philosophy has done in attempting to sweep away phenomena in order to realize the ontological depth of all beings. However, in adopting these negative procedures of the Buddhist philosophy one is tempted to emphasize only the silent aspect of the ontological principle and neglect its creative aspect. In the second place, therefore, we have to appeal to a specific method of inner illumination. It is only through such illumination that we are able to realize the infinite creative power of the ontological principle. (p. 25)

The great Asian philosophical traditions of both China and India recognize the possibility of both sensible intuition and intuitive illumination. They converge in both providing theories of not only knowledge but also practice. Indian philosophers have delved most deeply into the nature of self, and contemporary Indian psychologists have developed this into a body of knowledge with both theoretical and practical implications. Chinese philosophers have synthesized Indian philosophical insights (particularly those of the Buddhists) to develop a moral metaphysics that leads directly to a psychology of ethical social relations. While these are early days in the development of indigenous psychologies, there is hope for the future as the 21st century unfolds. The twin scourges of the end of cheap oil and the continuation of global warming will likely require a more practically oriented social science (see Liu et al., 2009), particularly in developing societies where these challenges will be felt most keenly. In the critical transition period between a fossil fuel-driven global economy and a mixed energy economy, there is the possibility that the epistemological breakthroughs in Asian philosophy may be translated into concrete practices of social science, where quantitative and qualitative methods are used hand in glove to assist societal development and creating global consciousness. In their summary of open-ended survey data from 24 societies on representations of world history, Liu et al. (2009) commented, "if there is a lay narrative of history, it might be that out of suffering comes great things" (p. 678). Conversely, even if the sky does not fall down, it may still be useful to work on the margins to connect the centers of an increasingly interconnected world of parallel and distributed cultural values (Liu, 2007/2008).

◼ NOTES

1. The author wishes to acknowledge the generous comments made by Dharm Bhawuk, K. K. Hwang, Isamu Ito, and Shu-hsien and An-yuan Liu on previous drafts of this chapter; thanks also go to Norman Denzin and Alison Jones for incisive and helpful reviews.

2. While there is considerable debate about the value of such indices, an impact rating in the 2s is comparable to top journals in anthropology and approaches those for sociology, whereas an impact factor of 1 or higher is very respectable (impact factors less than 1 are less prestigious).

3. *Antimonies* are fundamental contradictions between two sets of laws, each of which are reasonable given their premises.

4. Asian education systems emphasize rote learning, and Asian-educated scholars often do not realize that *pure* rote learning (i.e., replication without innovation) is not valued in most international journals.

5. It is more, not less difficult for Asians who have learned English as a second language to publish in international *qualitative* compared to *quantitative* journals. A bibliometric study on international social psychology publications by Haslam and Kashima (in press) reported that the *Journal of Cross Cultural Psychology, International Journal of*

Intercultural Relations, Asian Journal of Social Psychology, and *Journal of Social Psychology* were the most popular outlets for Asian authors: All of these are predominantly quantitative in orientation, though AJSP also publishes qualitative papers and the other two culture-oriented journals have had recent special issues on qualitative forms of inquiry.

◼ REFERENCES

Adair, J. G., Coelho, A. E. L., & Luna, J. R. (2002). How international is psychology? *International Journal of Psychology, 37*(3), 160–170.

Aguiling-Dalisay, G. H., Yacat, J. A., & Navarro, A. M. (2004). *Extending the self: Volunteering as Pakikipagkapwa.* Quezon City: University of the Philippines, Center for Leadership.

Allwood, C. M., & Berry, J. W. (2006). Origins and development of indigenous psychologies: An international analysis. *International Journal of Psychology, 41*(4), 243–268.

Amir, Y., & Sharon, I. (1987). Are social psychological laws cross-culturally valid? *Journal of Cross-Cultural Psychology, 18*(4), 383–470.

Atsumi, T. (2007). Aviation with fraternal wings over the Asian context: Using nomothetic epistemic and narrative design paradigms in social psychology. *Asian Journal of Social Psychology, 10,* 32–40.

Behrens, K. Y. (2004). A multifaceted view of the concept of *Amae:* Reconsidering the indigenous Japanese concept of relatedness. *Human Development, 47*(1), 1–27.

Berry, J. W., Poortinga, Y. H., Pandey, J., Dasen, P. R., Saraswathi, T. S., & Kagitcibasi, C. (1997). *Handbook of cross-cultural psychology* (2nd ed.). Boston: Allyn & Bacon.

Berry, J. W., Poortinga, Y. H., Segall, M. H., & Dasen, P. R. (1992). *Cross-cultural psychology: Research and applications.* Newbury Park, CA: Sage.

Bharati, A. (1985). The self in Hindu thought and action. In A. H. Marsella, G. DeVos, & F. L. K. Hsu (Eds.), *Culture and self: Asian and Western perspectives.* New York: Tavistock.

Bhawuk, D. P. S. (1999). Who attains peace: An Indian model of personal harmony. *Indian Psychological Review, 52*(2/3), 40–48.

Bhawuk, D. P. S. (2003). Culture's influence on creativity: The case of Indian spirituality. *International Journal of Intercultural Relations, 27*(1), 1–22.

Bhawuk, D. P. S. (2008a). Anchoring cognition, emotion, and behavior in desire: A model from the *Gita.* In K. R. Rao, A. C. Paranjpe, & A. K. Dalal (Eds.), *Handbook of Indian psychology* (pp. 390–413). New Delhi, India: Cambridge University Press.

Bhawuk, D. P. S. (2008b). Science of culture and culture of science: Worldview and choice of conceptual models and methodology. *The Social Engineer, 11*(2), 26–43.

Bhawuk, D. P. S. (2010). Methodology for building psychological models from scriptures: Contributions of Indian psychology to indigenous & global psychologies. *Psychology and Developing Societies.*

Bih, H. D. (2005). *Why didn't teacher tell me?* (in Chinese). Taipei: Shang-yi Publishers.

Bond, M. H. (Ed.). (1996). *The handbook of Chinese psychology.* Hong Kong: Oxford University Press.

Brown, J. D., & Kobayashi, C. (2003). Motivation and manifestation: Cross-cultural expression of the self-enhancement motive. *Asian Journal of Social Psychology, 6,* 85–88.

Cartwright, D. (1979). Contemporary social psychology in historical perspective. *Social Psychology Quarterly, 42*(1), 82–93.

Chinese Culture Connection. (1987). Chinese values and the search for culture-free dimensions of culture. *Journal of Cross-Cultural Psychology, 18,* 143–164.

Cole, M. (1995). Socio-cultural-historical psychology: Some general remarks and a proposal about a meso-genetic methodology. In J. V. Wertsch, P. del Rio, & A. Alvarez (Eds.), *Sociocultural studies of mind* (pp. 187–214). Cambridge, UK: Cambridge University Press.

Dalal, A., & Misra, G. (2010). The core and context of Indian psychology, *Psychology and Developing Societies.*

Enriquez, V. (1990). *Indigenous psychologies.* Quezon City, Philippines: Psychological Research & Training House.

Enriquez, V. G. (1992). *From colonial to liberation psychology: The Philippine experience.* Manila, Philippines: De La Salle University Press.

Geertz, C. (2000). Imbalancing act: Jerome Bruner's cultural psychology. In C. Geertz (Ed.), *Available light* (pp. 187–202). Princeton, NJ: Princeton University Press.

Gergen, K. J. (1973). Social psychology as history. *Journal of Personality and Social Psychology, 26,* 309–320.

Haslam, N., & Kashima, Y. (in press). The rise and rise of social psychology in Asia: A biobliometric analysis. *Asian Journal of Social Psychology.*

Heidegger, M. (1977). *Basic writings from* Being and Time *(1927) to* The Task of Thinking *(1964)* (D. F. Kell, Ed. & Trans.). New York: Harper and Row.

Heine, S. H., Lehman, D. R., Markus, H. R., & Kitayama, S. (1999). Is there a universal need for positive self-regard? *Psychological Review, 106,* 766–794.

Ho, D. Y. F. (1989). Propriety, sincerity, and self-cultivation: A dialogue between a Confucian and a psychologist. *International Psychologist, 30*(1), 16–17.

Ho, D. Y. F. (1998). Filial piety and filicide in Chinese family relationships: The legend of Shun and other stories. In U. P. Gielen & A. L. Comunian (Eds.), *The family and family therapy in international perspective* (pp. 134–149). Trieste, Italy: Edizioni LINT.

Ho, D. Y. F., & Chau, A. W. L. (2009). Interpersonal perceptions and metaperceptions of relationship closeness, satisfaction, and popularity: A relational and directional analysis. *Asian Journal of Social Psychology, 12,* 173–184.

Ho, D. Y. F., Ho, R. T. H. H., & Ng, S. M. (2006). Investigative research as a knowledge-generation method: Discovering and uncovering. *Journal for the Theory of Social Behavior, 36*(1), 17–38.

Ho, D. Y. F., Peng, S. Q., Lai, A. C., & Chan, S. F. F. (2001). Indigenization and beyond: Methodological relationism in the study of personality across cultural traditions. *Journal of Personality, 69*(6), 925–953.

Ho, R. T. H, Ng, S. M., & Ho, D. Y. F. (2007). Responding to criticisms of quality research: How shall quality be enhanced? *Asian Journal of Social Psychology, 10*(3), 277–279.

Hofstede, G. (2001). *Culture's consequences* (2nd ed.). Thousand Oaks, CA: Sage. (Original work published 1980)

Hwang, K. K. (1987). Face and favor: The Chinese power game. *American Journal of Sociology, 92,* 944–974.

Hwang, K. K. (1995). *Knowledge and action* (in Chinese). Taipei: Psychological Publisher.

Hwang, K. K. (2001). The deep structure of Confucianism: A social psychological approach. *Asian Philosophy, 11*(3), 179–204.

Hwang, K. K. (2005a). From anticolonialism to postcolonialism: The emergence of Chinese indigenous psychology in Taiwan. *International Journal of Psychology, 40*(4), 228–238.

Hwang, K. K. (2005b).The necessity of indigenous psychology: The perspective of constructive realism. In M. J. Jandl & K. Greiner (Eds.), *Science, medicine, and culture* (pp. 284–294). New York: Peter Lang.

Hwang, K. K. (2005c). A philosophical reflection on the epistemology and methodology of indigenous psychologies. *Asian Journal of Social Psychology, 8*(1), 5–17.

Hwang, K. K. (2006). Constructive realism and Confucian relationism: An epistemological strategy for the development of indigenous psychology. In U. Kim, K. S. Yang, & K. K. Hwang (Eds.), *Indigenous and cultural psychology: Understanding people in context* (pp. 73–108). New York: Springer.

Hwang, K. K. (2009). *Confucian relationalism: Philosophical reflection, theoretical construction, and empirical research* (in Chinese). Taipei: Psychological Publisher.

Kashima, Y. (2005). Is culture a problem for social psychology? *Asian Journal of Social Psychology, 8,* 19–38.

Kim, U. (2000). Indigenous, cultural, and cross-cultural psychology: A theoretical, conceptual, and epistemological analysis. *Asian Journal of Social Psychology, 3,* 265–288.

Kim, U., & Berry, J. W. (Eds.). (1993). *Indigenous psychologies: Research and experience in cultural context.* Newbury Park, CA: Sage.

Kim, U., Yang, K. S., & Hwang, K. K. (Eds.). (2006). *Indigenous and cultural psychology: Understanding people in context.* New York: Springer.

Lee, W. L. (2004). Situatedness as a goal marker of psychological research and its related methodology. *Applied Psychological Research, 22,* 157-200. (in Chinese, published in Taiwan).

Leung, K. (2007). Asian social psychology: Achievements, threats, and opportunities. *Asian Journal of Social Psychology, 10,* 8–15.

Liu, J. H. (2007/2008). The sum of my margins may be greater than your center: Journey and prospects of a marginal man in the global economy. *New Zealand Population Review, 33/34,* 49–67.

Liu, J. H. (2008). Editorial statement from the new editor. *Asian Journal of Social Psychology, 11,* 103–104.

Liu, J. H., & Hilton, D. (2005). How the past weighs on the present: Social representations of history and their role in identity politics. *British Journal of Social Psychology, 44,* 537–556.

Liu, J. H., & Liu, S. H. (1997). Modernism, postmodernism, and neo-Confucian thinking: A critical history of paradigm shifts and values in academic psychology. *New Ideas in Psychology, 15*(2), 159–177.

Liu, J. H., & Liu, S. H. (1999). Interconnectedness and Asian social psychology. In T. Sugiman, M. Karasawa, J. H. Liu, & C. Ward (Eds.), *Progress in Asian social psychology* (Vol. 2, pp. 9–31). Seoul, Korea: Kyoyook Kwahaksa.

Liu, J. H., & Liu, S. H. (2003). The role of the social psychologist in the "Benevolent Authority" and "Plurality of Powers" systems of historical affordance for authority. In K. S. Yang, K. K. Hwang, P. B. Pedersen, & I. Daibo (Eds.), *Progress in Asian social psychology: Conceptual and empirical contributions* (Vol. 3, pp. 43–66). Westport, CT: Praeger.

Liu, J. H., & Mills, D. (2006). Modern racism and market fundamentalism: The discourses of plausible deniability and their multiple functions. *Journal of Community and Applied Social Psychology, 16*, 83–99.

Liu, J. H., & Ng, S. H. (2007). Connecting Asians in global perspective: Special issue on past contributions, current status, and future prospects of Asian social psychology. *Asian Journal of Social Psychology, 10*(1), 1–7.

Liu, J. H., Ng, S. H., Gastardo-Conaco, C., & Wong, D. S. W. (2008). Action research: A missing component in the emergence of social and cross-cultural psychology as a fully inter-connected global enterprise. *Social & Personality Psychology Compass, 2*(3), 1162–1181.

Liu, J. H., Paez, D., Slawuta, P., Cabecinhas, R., Techio, E., Kokdemir, D., et al. (2009). Representing world history in the 21st century: The impact of 9–11, the Iraq War, and the nation-state on the dynamics of collective remembering. *Journal of Cross-Cultural Psychology, 40*, 667–692.

Liu, J. H., Paez, D., Techio, E., Slawuta, P., Zlobina, A., & Cabecinhas, R. (2010). From gist of a wink to structural equivalence of meaning: Towards a cross-cultural psychology of the collective remembering of world history. *Journal of Cross-Cultural Psychology, 41*(3), 451–456.

Liu, J. H., & Sibley, C. G. (2009). Culture, social representations, and peacemaking: A symbolic theory of history and identity. In C. Montiel & N. Noor (Eds.), *Peace psychology in Asia.* Heidelberg, Germany: Springer.

Liu, S. H. (1989a). Postwar neo-Confucian philosophy: Its development and issues. In C. W. H. Fu & G. E. Spiegler (Eds.), *Religious issues and interreligious dialogues.* New York: Greenwood Press.

Liu, S. H. (1989b). Toward a new relation between humanity and nature: Reconstructing t'ien-jen-ho-i. *Zygon, 24,* 457–468.

Liu, S. H. (1998). *Understanding Confucian philosophy: Classical and Sung-Ming.* Westport, CT: Greenwood.

Markus, H. R., & Kitayama, S. (1991). Culture and self: Implications for cognition, emotion and motivation. *Psychological Review, 98,* 224–253.

McTaggart, E. (1997). *Participatory action research: International contexts and consequences.* Albany: State University of New York Press.

Minkler, M., & Wallerstein, N. (Eds.). (2003). *Community-based participatory research for health.* San Francisco: Jossey-Bass.

Moghaddam, F. M. (1989). Specialization and despecialization in psychology: Divergent processes in the three worlds. *International Journal of Psychology, 24,* 103–116.

Moghaddam, F. M., & Taylor, D. M. (1985). Psychology in the developing world: An evaluation through the concepts of "Dual Perception" and "Parallel Growth." *American Psychologist, 40,* 1144–1146.

Moghaddam, F. M., & Taylor, D. M. (1986). What constitutes an "appropriate psychology" for the developing world? *International Journal of Psychology, 21,* 253–267.

Moscovici, S. (2008). *Psychoanalysis: Its image and its public.* Cambridge, UK: Polity. (Original work published 1961)

Mou, T. S. (1970). Hsin-t'i yu hsing-t'i [Mind and human nature] (in Chinese). In *Philosophy East and West* (Vol. 20). Taipei: Cheng Chung Press.

Mou, T. S. (1975). *Phenomenon and the thing-in-itself* (in Chinese). Taipei: Student Book.

Nikora, L. W., Levy, M., Masters, B., & Waitoki, M. (2004). Indigenous psychologies globally—A perspective from Aotearoa/New Zealand. Hamilton, New Zealand: University of Waikato, Maori & Psychology Research Unit.

Nisbett, R. E., Peng, K., Choi, I., & Norenzayan, A. (2001). Culture and systems of thought: Holistic versus analytic cognition. *Psychological Review, 108*(2), 291–310.

Pandey, J. (2004). *Psychology in India revisited: Development in the discipline: Vol. 4. Applied social and organizational psychology.* New Delhi: Sage.

Paranjpe, A. C. (1984). *Theoretical psychology: The meeting of East and West.* New York: Plenum Press.

Paranjpe, A. C. (1998). *Self and identity in modern psychology and Indian thought.* New York: Plenum Press.

Pe-Pua, R., & Protacio-Marcelino, E. (2000). Sikolohiyang Pilipino [Filipino psychology]: A legacy of Virgilio G. Enriquez. *Asian Journal of Social Psychology, 3,* 49–71.

Rao, K. R., & Paranjpe, A. C. (2008). Yoga psychology: Theory and application. In K. R. Rao, A. C. Paranjpe, & A. K. Dalal (Eds.), *Handbook of Indian psychology* (pp. 163–185). New Delhi: Cambridge University Press.

Sinha, D. (1997). Indigenizing psychology. In J. W. Berry, Y. H. Poortinga, & J. Pandey (Eds.), *Handbook of cross-cultural psychology* (pp. 130–169). Boston: Allyn & Bacon.

Smith, L. T. (1999). *Decolonizing methodologies.* London: Zed.

Smith, P. B., & Bond, M. H. (1993). *Social psychology across cultures* (2nd ed.). New York: Harvester Wheatsheaf.

Spencer-Rodgers, J., Williams, M. J., & Peng, K. P. (2007). How Asian folk beliefs of knowing affect the psychological investigation of cultural differences. In J. H. Liu, C. Ward, A. Bernardo, M. Karasawa, & R. Fischer (Eds.), *Progress in Asian social psychology: Casting the individual in societal and cultural contexts.* Seoul, Korea: Kyoyook Kwahasaka.

Sugiman, T. (2006). Theory in the context of collaborative inquiry. *Theory and Psychology, 16,* 311–325.

Tillich, P. (1951). *Systematic theology* (Vols. 1–3). Chicago: University of Chicago Press.

Wallner, F., & Jandl, M. J. (2006). The importance of constructive realism for the indigenous psychologies approach. In U. Kim, K. S. Yang, & K. K. Hwang (Eds.), *Indigenous and cultural psychology: Understanding people in context* (pp. 49–72). New York: Springer.

Ward, C. (2007). Asian social psychology: Looking in and looking out. *Asian Journal of Social Psychology, 10,* 22–31.

Yang, K. S. (1999). Toward an indigenous Chinese psychology: A selective review of methodological, theoretical, and empirical accomplishments. *Chinese Journal of Psychology, 4,* 181–211.

Yang, K. S. (2000). Monocultural and cross-cultural indigenous approaches: The royal road to the development of a balanced global psychology. *Asian Journal of Social Psychology, 3,* 241–264.

13

DISABILITY COMMUNITIES

Transformative Research for Social Justice

Donna M. Mertens, Martin Sullivan, and Hilary Stace

The purpose of the present Convention is to promote, protect and ensure the full and equal enjoyment of all human rights and fundamental freedoms by all persons with disabilities, and to promote respect for their inherent dignity.

—Article 1, UN Convention on Rights of Persons With Disabilities, 2006

If all the disability research conducted from the late 19th century had been carried out with the above article as its guiding principle, then, we would suggest that the lot of disabled people would be qualitatively and quantitatively so much better today. It was not until the advent of the disability rights movement (DRM) on both sides of the Atlantic in the late 1960s that the basic humanity of "the disabled" began to be recognized and they became either "disabled people" or "people with disability."

As "the disabled," terrible things have been done to this group in the name of science; they were imprisoned in nursing homes, surgically mutilated, sterilized, lobotomized, euthanized, shocked into passivity, placed in chemical and physical straitjackets, denied education, denied employment, and denied meaningful lives (see Braddock & Parrish, 2001; Linton, 1998; Morris, 1991; Sobsey, 1994). Most of these procedures were done with the best of intention; they were the product of cutting-edge scientific research of the day into the cause and cure of disability. As the objects of this research, "the disabled" became dehumanized conditions, categories, and examples to be cured, ameliorated, or cared for in institutions, rather than human beings to be loved, cherished, and nurtured in their families and communities.

Following the examples set by antiracist, feminist, and gay rights movements of the 1960s, disabled people throughout the Western world began to organize to challenge the oppressive stereotypes that blighted their lives. In this chapter, we illustrate how social research has been complicit in fostering oppressive stereotypes of disabled people and suggest ways to do nonoppressive disability research.

RECONCEPTUALIZING DISABILITY, RETHINKING METHODOLOGY

In both the United States and the United Kingdom, the intellectual underpinnings of the DRM were provided by what became known globally as "the social model of disability." In the United States, Hahn (1982) described this as a socio-political approach in which disability is regarded as a product of interactions between individual and environment. Later, Hahn (1988) went on to say that this perspective could guide research viewing disabled citizens as an oppressed minority. Indeed, in the United States the minority model of disability (see Fine & Asch, 1988) is more commonly referred to than the social model of disability. Albrecht (2002) explains this with reference to the tradition of American pragmatism, which promoted the linking of arguments for civil rights for disabled people with a minority group approach rather than providing a comprehensive theoretical explanation for disability as in the United Kingdom. For Albrecht, American pragmatism can be summarized as "the practical consequences of believing in a particular concept or social policy" (2002, p. 21); other American citizens who experienced prejudice and discrimination on the basis of visible differences—age, ethnicity/race, gender—had successfully fought for their civil rights by adopting a minority group model that had allowed them to present a united front in their struggle. If it worked for these minorities, why would it not work for disabled American citizens of whom, by 1986, many felt constituted a minority group in the same sense as "Blacks" or "Hispanics" (Hahn, 2002, p. 173)?

In the United Kingdom, the social model was largely developed by Mike Oliver (1982, 1990), following the *Fundamental Principles of Disability* booklet published by the Union of the Physically Impaired Against Segregation (UPIAS) in 1976. Here, the separation of impairment and disability lies at the heart of UPIAS's analysis of disability and, subsequently, the social model. For UPIAS, people have impairments and disability is the negative social response to impairment in terms of the exclusion of impaired people from the political, economic, and social organizations of their communities. According to Shakespeare and Watson (2001), this redefinition of disability changed the consciousness of people with disability. Disability no longer resided in their bodies or minds; it was framed as a problem of social oppression and disabled people described themselves as an oppressed minority. For example, Groce (1985/2003) conducted a retrospective anthropological study of a community on Martha's Vineyard, Massachusetts, that had a high level of genetic deafness. The deaf people who lived there at that time did not live in a disabling society because everyone learned to use sign language. A person in a wheelchair is only disabled if there is no cut in the sidewalk or no elevator in a multistory building.

When discussing complex concepts such as disability, it is not unusual for language-based challenges to surface. In disability studies, the terms impairment and disability are subject to such challenges. Members of disability communities seek to make clear the difference between an inherent characteristic of a person and the response of society to that characteristic. Hence, in the New Zealand Ministry of Health's (2001) governmental documents and the United Nations' (2006) Convention on the Rights of People With Disabilities, we find the following definitions that provide an explanation of this separation.

> Disability is not something individuals have. What individuals have are impairments. They may be physical, sensory, neurological, psychiatric, intellectual or other impairments. Disability is the process which happens when one group of people create barriers by designing a world only for their way of living, taking no account of the impairments other people have. . . . Our society is built in a way that assumes that we can all move quickly from one side of the road to the other; that we can all see signs, read directions, hear announcements, reach buttons, have the strength to open heavy doors and have stable moods and perceptions. (New Zealand Ministry of Health, 2001, p. 1)

In the United Nations' Convention on the Rights of People With Disabilities, people with disabilities are defined as "those who have long-term physical, mental, intellectual or sensory impairments which in interaction with various barriers may hinder their full and effective participation in society on an equal basis with others (United Nations, 2006, p. 1).

Disability Rights Movement advocates in the United States agree with the importance of making a distinction between an inherent characteristic and the societal response to that characteristic (Gill, Kewman, & Brannon, 2003). In the United States, however, there is more emphasis on persons with disabilities as representing dimensions of diversity in society. People with disabilities are "hindered primarily not by their intrinsic differences but by society's response to those differences" (Gill et al., p. 306). Rather than saying they have "an impairment," disability rights advocates self-define as "disabled and proud" (Triano, 2006).

With this radically new self-perception came disability activism as disabled people began to mobilize for equal citizenship in the United States (Anspach, 1979), the United Kingdom (Campbell & Oliver, 1996), and Canada, Europe, Australia, and New Zealand (Barnes, Oliver, & Barton, 2002), and to demand their right to be economically and socially productive rather than exist in a state of forced dependency on welfare, charity, and good will (Scotch, 2001).[1]

The social model was the polar opposite to the hegemonic individual medical model of disability that had dominated conventional wisdom and created a state of dependency for people with disability. The medical model characterized disability as a personal problem for which individuals ought to seek medical intervention in the hope of cure, amelioration, or care. Such medicalized perceptions are underpinned by personal tragedy theory (Oliver, 1990), which casts disabled people as the tragic victims of some terrible circumstance or event and who, rightfully, need to be pitied. Moreover, it created a climate of "aesthetic and existential anxiety" (Hahn, 1988), in which disability and disabled people were to be avoided at all costs.

In Britain, the first disability studies course, The Handicapped Person in the Community, was offered by the Open University in 1975. Vic Finkelstein, a founding member of UPIAS, was key in its development. Each year more and more disabled people were involved in the production of course materials and by the time it closed in 1994, it had been renamed The Disabling Society to reflect its wider content (Barnes et al., 2002). In the United States, a similar link between disability activism, the academy, and the emergence of disability studies exists. Following the 1977 White House Conference on Handicapped Individuals, which brought disability rights activists and academics together, the first disability studies course was offered in the United States (Pfeiffer & Yoshida, 1995). It was in the area of medical sociology, focused on living with a disability, and its main tutor was a disabled person. In 1981, disabled sociologist Irving K. Zola founded the *Disability Studies Quarterly* and cofounded the U.S.-based Society for Disability Studies. Twelve disability studies courses were taught in U.S. institutions at the time; within five years, the number had almost doubled to 23 (Pfeiffer & Yoshida, 1995, p. 482).

With this blossoming of disability studies programs across the United States and the United Kingdom, it was not long before disabled academics and students began asking questions about "the what, how, and who" of disability research (Sullivan, 2009, p. 72). Oliver (1992) argued that up to that point most disability research was dominated by the individual medical model of disability, which couched disability in terms of functional limitations and individual deficits that needed to be explored in the hope of finding a cure or an explanation as to why people with disability were not participating in their communities. Oliver (1992, p. 104) exemplifies this positivist research with a question taken from a survey of disabled adults that asks, "What complaint causes your difficulty in holding, gripping or turning things?" He contrasts this with an alternative question—"What defects in design of everyday equipment like jars, bottles and tins cause you difficulty in holding, gripping or turning them?" (p. 104)—which paints an entirely different picture.

Oliver (1992) described the challenge to positivist approaches presented by the interpretive paradigm, which sees all knowledge as socially constructed and a product of a particular time and place. From this perspective, disability is a social problem requiring education, attitude change, and social adjustment on the part of both abled and disabled people. While Oliver concedes that the interpretive or constructivist paradigm is an advance on positivism, it does not go far enough as it does not change the relationship between the researcher and the researched/disabled people. He says that in another context this kind of research has been called

> "the rape model of research" (Reinharz, 1985) in that researchers have benefitted by taking the experience of disability, rendering a faithful account of it and then moving on to better things while the disabled subjects remain in exactly the same social situation they did before the research began. (Oliver, 1992, p. 109)

According to Oliver, such research is very alienating for the disabled subjects. He proposed a new paradigm for disability research, the emancipatory paradigm, in which the social relations of research production would be fundamentally changed. The main features of the emancipatory paradigm are similar to the participatory action model of research; they include:

- research is political in nature, rooted in the social model and its focus is on exposing and changing disabling structures to extend the control disabled people have over their own lives;
- the lopsided power relationship between the researcher and the researched is to be redressed by researchers placing their skills at the disposal of disabled people;

- control is now in the hands of people with disability, who can determine the what, the how, and the when of the research process (see Goodley, 2000);
- focus on the strengths and coping skills of people with disabilities, rather than on deficits;
- conduct research that examines the contextual and environmental factors that either facilitate or impede a person with a disability's integration in society (Wright, 1983).

In Doing Disability Research, an edited collection by Colin Barnes and Geof Mercer (1997), a number of researchers reflect upon reconciling the theoretical purity of the emancipatory research paradigm with the practicalities of maintaining intellectual and ethical integrity in their research. Some of the difficulties listed include:

- the majority of funding goes to research based on the individual medical model; hence, little funding is available for participatory and action research;
- while interpretive forms of research may be about improving the existing social relations of research production, they are not about changing them (Oliver, 1992);
- participatory research might increase the input of research subjects into the research process, but control, ultimately, rests with the researcher who is also the one likely to derive the most benefit from the research (Oliver, 1997);
- difficulties arise when applying the emancipatory model to groups with other than physical and sensory impairment, such as psychiatric system survivors (Beresford & Wallcraft, 1997) or people with learning difficulties (Booth & Booth, 1997);
- questions surface about the role of nondisabled researchers (Lunt & Thornton, 1997; Priestly, 1997; Stone & Priestly, 1996) and who owns the research (Priestly, 1997; Shakespeare, 1997).

Other challenges include how to ensure that disabled people are the leaders in research and have maximum involvement in research, and making certain the research is accountable to disabled people (Beazley, Moore, & Benzie, 1997; Frank, 2000). In addition, researchers struggle with the question of integrity of the research. For example, Tom Shakespeare, prominent academic and activist in the British disability movement, contends that he needs to retain choice and control over the process and to "follow my own intellectual and ethical standards, rather than trying to conform to an orthodoxy" (Shakespeare, 1997, p. 185).

In making this statement, Shakespeare reflects the position of many of the contributors to *Doing Disability Research* when trying to deploy the emancipatory paradigm in the field in its purest sense. They had reported having to make some compromises in order to complete their research. These compromises did not mean that the expertise research subjects brought to the project was not acknowledged or drawn upon; rather it meant researchers were guided by the spirit of the emancipatory paradigm rather than adhering rigidly to its principle. Researchers justified, for example, not handing complete control of the process over to their research subjects, by insisting that the theoretical and practical motivation for their research was the social model of disability and, therefore, ethical. This justification, however, rests upon the uncritical acceptance of the social model of disability within disability studies.

In the past decade, critiques of the social model of disability have gathered momentum within disability studies. These critiques are based mainly on feminist and poststructural thought and appear to be creating a self-titled space for their play, critical disability studies (CDS). Helen Meekosha and Russell Shuttleworth (2009) ask, "What's So 'Critical' About Critical Disability Studies?" In this article, they trace the evolution of the term "critical disability studies" in various scholarly articles over the past decade, and observe that "the influx of humanities and cultural studies scholars with their postmodern leanings and decentering of subjectivity during the 1990s, especially in the US, enabled a more self-conscious focus on critical theorizing to take hold in disability studies" (p. 4). This means moving beyond the social model and the binary way of thinking about disability that emerges from the conceptual distinction between impairment and disability. Drawing upon feminism, critical race theory, queer theory, postmodern thought, and postcolonial theory, CDS has set new "terms of engagement" in disability studies:

> The struggle for social justice and diversity continues but on another plane of development: one that is not simply social, economic and political, but also psychological, cultural, discursive and carnal. (Meekosha & Shuttleworth, 2009, p. 4)

This means that CDS will deploy critical theory in a more nuanced and complex exploration of disability in terms of intersectionality; that is, the intersection of disability with the social divisions of gender, race, class, and sexuality. Or, in other words, an account of disability that is embodied, gendered, raced, classed, and sexed.

In a rather devastating critique of the "so-called emancipatory approach" to disability research, Meekosha and Shuttleworth (2009) argue that while it might have democratized the research process,

> The researcher's methodological and theoretical expertise was considered technical skills . . . not critical-interpretive skills in the analysis of interaction, meaning, and the unmasking of ideologies and hierarchies. (p. 8)

So the internal inconsistencies outlined above by researchers deploying the emancipatory research model and the critique offered from CDS raises an important question about the possibility of doing nonalienating disability research: Is there a research paradigm that is able to capture disability as a complex, embodied relationship between people with impairments and their natural and social environments? Yes, there is one: the transformative paradigm.

◻ THE TRANSFORMATIVE PARADIGM
IN DISABILITY RESEARCH

The historical strands of disability research in terms of the disability minority group model and the social oppression model formed the groundwork for the emancipatory paradigm. As Sullivan (2009) notes, the emancipatory paradigm emerged to respond to these models of disability and changed approaches to research in the disability communities to place priority on participatory modes, place power in the hands of those with disabilities, and challenge oppressive research relations. However, several tensions arose in the implementation of the emancipatory paradigm that led to the recognition of the transformative paradigm by some as having potential to address the issues of power and privilege that sustain oppressive conditions for people with disability. The transformative paradigm is a framework of philosophical assumptions that directly engages members of culturally diverse groups with a focus on increased social justice (Mertens, 2009, 2010). The tensions that arose in the emancipatory paradigm are contrasted with the different stance of the transformative paradigm in Table 13.1.

The transformative paradigm provides a framework for research in the disability communities that is more attuned to handling diversity within communities, aims to build on strengths within communities, develops solidarity with other groups that are marginalized, and changes identity politics to a socio-cultural perspective. Critical disability theory is one theory that is commensurate with the transformative paradigm. However, the transformative paradigm provides a metaphysical umbrella that covers the intersection of discrimination based on disability and gender, race, ethnicity, age, religion, national origin, indigenous status, immigration, and other relevant dimensions of diversity used as a basis for discrimination and oppression. The encompassing features of the transformative paradigm are explored in subsequent sections of this chapter.

Transformative Paradigm: Basic Belief Systems

Building on the work of Egon Guba, Yvonna Lincoln, and Norman Denzin (Denzin & Lincoln, 2005; Guba & Lincoln, 2005), the transformative paradigm is defined by four basic assumptions: axiology (the nature of ethics), ontology (the nature of reality), epistemology (the nature of knowledge and the relationship between the knower and that-which-would-be known), and methodology (the nature of systematic inquiry). The transformative axiological assumption is given priority as the assumption that provides guidance for the full set of transformative assumptions. It is based on the belief that ethics is

Table 13.1 Contrasting the Emancipatory and Transformative Paradigms (Mertens, 2009; Sullivan, 2009)

	Emancipatory	*Transformative*
Focus	Focuses exclusively on disability as the central focus	Focuses on dimensions of diversity associated with differential access to power and privilege, including disability, gender, race/ethnicity, social class, sexual orientation, and other contextually important dimensions of diversity
Role of researcher/ participants	Assumes participants are "conscious of their situation and ready to take leadership" (Sullivan, 2009, p. 77)	Team approach; partnerships are formed; capacity building undertaken as necessary
Model of research	Participatory action research; interpretive approaches	Multiple and mixed methods; culturally respectful; supportive of diverse needs
Tone	Sets up an "us" against "them" tone	Acknowledges the need to work together to challenge oppressive structures

defined in terms of the furtherance of human rights, the pursuit of social justice, the importance of cultural respect, and the need for reciprocity in the researcher-participant relationship. Hence, transformative research seeks to challenge the status quo of an oppressive, hegemonic system in order to bring about a more equitable society.

For example, the differential identification of children with disabilities in the United States on the basis of gender and race/ethnicity is viewed through a human rights lens and is suggestive of the need to conduct research that addresses the wider social conditions that explain this situation. Males, especially those from racial and ethnic minority groups, are two times more likely to be identified as having a disability than females (Gravois & Rosenfeld, 2006; U.S. Department of Education, 2004).

The transformative axiological assumption stimulates the recognition of diversity within disability communities and leads researchers to ask such questions as, Are girls from all ethnic groups being underidentified? How are the indicators for various disabilities (e.g., learning disabilities or autism) differently manifested in girls and boys? In members of dominant and minority groups? What are the motivations for parents to have their children identified or not identified? How do these motivations differ based on race/ethnicity? What is the quality of services provided to children who are identified as having a disability and how does this differ on the basis of type of disability, gender, race, or ethnicity? How can diagnostic and referral strategies be adapted to be more culturally fair? How can the quality of special education services be improved to be more responsive to cultural differences among the students? What is the relationship between quality of special education services and success in later life? What is the meaning of using the term special education, which disability advocates have suggested is used to condone exclusion and segregation (Mertens, Holmes, &

Harris, 2009; Mertens, Wilson, & Mounty, 2007)? Other researchers who have made contributions at the intersection of gender and disability include Fine and Asch (1988; Asch & Fine, 1992), Doren and Benz (2001), and Rousso (2001).

Ontologically, the transformative paradigm questions the nature of reality as it is defined in the postpositivist and constructivist paradigms. Postpositivists hold that there is one reality that is waiting to be discovered and that can be captured within a specified level of probability. Constructivists' view of reality is that there are multiple socially constructed realities wherein the researcher and researched co-construct meaning. However, the transformative paradigm's ontological assumption suggests that there may be one reality about which there are many opinions and that differential access to power influences which version of reality is given privilege (Mertens, 2009, 2010). The researcher then has a responsibility to uncover the various versions of reality and to interrogate them to determine which version is most in accord with furthering social justice and human rights. This raises questions about how the researcher becomes competent in each cultural context in order to accurately reveal issues related to oppression and resilience.

This leads to the epistemological assumption that researchers need to have sufficient grounding in the culture of the communities in which they work, as well as recognize the limitations of their grounding, to conduct research in ways that are viewed as credible and useful to members of those communities. For researchers to serve as agents of prosocial change, as is recommended by the American Psychological Association (2002), they need sufficient understanding of the relevant dimensions of diversity to combat racism, prejudice, bias, and oppression in all their forms. The transformative ontological assumption leads to the epistemological stance in which researchers strive for a level of cultural competency by building rapport despite differences, gaining trust of community members, and reflecting upon and recognizing their own biases (Edno, Joh, & Yu, 2003; Symonette, 2009).

Given the focus of the transformative paradigm on the often-overlooked strengths that are found in communities, it comes as no surprise that theoretical lenses that also focus on social justice, human rights, and resiliency are commensurate with this stance, including, indigenous and postcolonial theories (Barnes, McCreanor, Edwards, & Borell, 2009; Bishop, 2008; Chilisa, 2009; Cram, 2009; Denzin, Lincoln, & Smith, 2008; LaFrance & Crazy Bull, 2009), queer theories (Dodd, 2009), critical race theories (Thomas, 2009), and feminist theories (Brabeck & Brabeck, 2009).

The benefits of combining these theoretical lenses with disability rights theories to research in disability communities are reflected in comments that one of the author's (Mertens) students made after reading about Māori indigenous theories. Heidi Holmes, a Deaf woman who is a member of the American Sign Language community, based her comment on interpretations she made from her readings of Māori academic literature (by nondisabled writers), which she viewed as presenting a homogenous picture of the Māori community because of a lack of mention of disability or deafness or other relevant dimensions of diversity (Cram, Ormond, & Carter, 2004; Smith, 2005). She reflects on the need to and benefits of recognizing multiple dimensions of diversity within the Māori community, including disability and deafness.

> What about diversity in the Māori community? Are they all the same? From a Deaf perspective, to what extent do Māori values/culture overlap with the Deaf community? There is a sharing of anger, frustration, discrimination, and oppression. . . . What about studying deaf people? Should we come up with different cultural values within the Deaf culture, such as hard of hearing, cochlear implants, oral, little hard of hearing, deaf of deaf, deaf of hearing, etc.? There is no one approach to the group of deaf people like the Māori approach. How's it work with Māori? Are they all the same? (Heidi Holmes, personal communication, February 5, 2006)[2,3]

Recognition of multiple dimensions of diversity is necessary for conducting research in the space that overlaps indigenous and disability and deaf communities.

A point of overlap between indigenous peoples and people with disabilities is evidenced in the adaptation of the Aboriginal Indigenous Terms of Reference (ITR)[4] (Osborne & McPhee, 2000) to research proposed within the American Sign Language community (Harris, Holmes, & Mertens, 2009). The ITR were developed to explicitly recognize

> what the community believes needs to be taken into consideration [when determining the focus of the research], how the issue will be dealt with and any special requirements the community puts on

> the issue. This way, the community is empowered to determine what the issue is, how it should be dealt with and what things need to be taken into account. The combination of the community's viewpoint and guidelines is what makes up the Indigenous Terms of Reference. (Osborne & McPhee, 2000, p. 4)

The ITR and the adapted terms of reference for the deaf and disability communities are problematic in the sense that they suggest that there is one homogeneous Māori community, disability community, or deaf community. As Harris, Holmes, and Mertens (2009) struggled to decide what to call the community for which they were proposing terms of reference, they were aware of the diversity within the deaf community in terms of such characteristics as race, ethnicity, gender, sexuality, national origin, and religion. However, they were also aware of the dimensions of diversity uniquely associated with the deaf community in terms of use of either an oral or visual communication system. Visual systems include American Sign Language, cued speech, exact signed English, or Pidgin Signed English. They were aware of the use of assistive listening devices such as hearing aids and cochlear implants that are used by some deaf people to enhance their ability to hear sounds. They were also aware of the contentious history between advocates for sign language and those who favor speech and speech reading. Thus, they decided to write the terms of reference, not for the deaf community, but for the American Sign Language community, referring to people whose primary experience and allegiance are with Sign Language,[5] as well as the community and culture of Deaf people. However, researchers who are interested in studying Sign Language communities should always be conscious of the complexity of deaf people and the Sign Language community.

We present here an adaptation of the American Sign Language Community's Terms of Reference (Harris, Holmes, & Mertens, 2009) more broadly defined for research in the disability community. This is offered with an understanding of the tension in discussing communities as homogeneous groups, but the necessity to recognize communities that have been pushed to the margins of society. The Disability Terms of Reference might include

1. The authority for the construction of meanings and knowledge within the disability community rests with the community's members.

2. Investigators should acknowledge that disability community members have the right to have those things that they value to be fully considered in all interactions.

3. Investigators should take into account the worldviews of the disability community in all negotiations or dealings that impact the community's members.

4. In the application of disability communities' terms of reference, investigators should recognize the diverse

experiences, understandings, and ways that reflect their contemporary cultures.

5. Investigators should ensure that the views and perceptions of the critical reference group are reflected in any process of validating and evaluating the extent to which Disability communities' terms of reference have been taken into account.

6. Investigators should negotiate within and among disability groups to establish appropriate processes to consider and determine the criteria for deciding how to meet cultural imperatives, social needs, and priorities. (Adapted from Harris, Holmes, & Mertens, 2009, p. 115)

These Disability Terms of Reference lead to methodological implications consistent with the transformative paradigm. The transformative methodological assumption emanates from the previously described assumptions and brings focus to the contextually relevant dimensions of diversity and engagement with members of the disability community. Research in the transformative paradigm is a site of multiple interpretive practices. It has no specific set of methods or practices of its own. This approach to research draws on several theories, methods, and techniques. Quantitative, qualitative, or mixed methods can be used; however, the inclusion of a qualitative dimension in methodology is critical in order to establish a dialogue between the researchers and the community members. Mixed methods designs can be considered in order to address the community's information needs. However, the methodological decisions are made with a conscious awareness of contextual and historical factors, especially as they relate to discrimination and oppression. Thus, the formation of partnerships with researchers and disability communities is an important step in addressing methodological questions in research.

Overall, the transformative paradigm provides a useful framework for addressing the role that researchers play when dealing with issues related to oppression, discrimination, and power differences. The transformative paradigm places central importance on the dynamics of power inequalities that have been the legacy of many members of disability communities with regard to whose version of reality is privileged. The transformative epistemological assumption raises questions about the nature of relationships among researchers in terms of who controls the investigation, especially when it is conducted by a team of members and nonmembers of disability communities. Transformative methodological assumptions encourage researchers who are interested in investigating a topic within a disability community to follow research guidelines developed by the community itself. The transformative axiological assumption puts issues of social justice and human rights at the forefront of decision making with regard to research in disability communities.

▣ EXAMPLES OF RESEARCH THAT TRANSFORMS THE LIVES OF PEOPLE WITH DISABILITIES

Many examples of transformative research reveal the potential to change lives of research participants, as well as to have a transformative effect on the researcher and the wider social context. In this chapter, we use detailed examples of four studies to illustrate research that transforms the lives of those who are pushed to society's margins. Other examples of such research include Abma's (2006) work with patient involvement in research about themselves; Cushing, Carter, Clark, Wallis, and Kennedy's (2009) transformative research with people with disabilities; Horvath, Kampfer-Bohach, and Farmer Kearns's (2005) study of involvement of people who are deaf and blind in research; Kroll, Barbour, and Harris's (2007) use of focus groups with people with disabilities; Rapaport, Manthorpe, Moriarty, Hussein, and Collins's (2005) work with advocacy research with people with disabilities; Ryan's (2007) work in Australia conducting qualitative research with people with disabilities; collaborative research by MacArthur, Gaffney, Kelly, and Sharp (2007) promoting the agency and voice of disabled children in the New Zealand mainstream school setting; and initiatives such as focus group work in Christchurch, New Zealand, led by young adults with Down syndrome, on transitions from school to post school including work and independent living (Gladstone, Dever, & Quick, 2009).

▣ COURT ACCESS WITH DEAF AND HARD OF HEARING PEOPLE

Mertens (2000) conducted a study of court access for deaf and hard of hearing people in the United States that was guided by an advisory board deliberately chosen to reflect the relevant dimensions of diversity in the deaf and hard of hearing communities in terms of hearing status, mode of communication (American Sign Language, speech reading), and use of assistive listening devices (cochlear implants, hearing aids), as well as position in the court (deaf judges, deaf and hearing attorneys who work with deaf clients, interpreters, judicial educators, and police officers who work on cases involving the deaf community).

Based on the transformative paradigm's assumptions, data collection began with discussions between the researcher (Mertens) and members of the advisory board in terms of how to get an accurate picture of the challenges faced by deaf and hard of hearing people in court. This led to a decision to use focus groups to collect data from deaf and hard of hearing people's experiences in court that reflected relevant dimensions of diversity in the deaf and hard of hearing communities, as well as the appropriate accommodations needed to support effective communication.

One focus group included hearing and deaf co-moderators, American Sign Language interpreters, and court reporters, who used real-time captioning to display what was being said (signed) during the focus group on big TV screens in the room. The group also included a deaf and blind woman with an interpreter who signed into her hands so she could feel the communication, as well as a deaf man with a low level of literacy who did not understand sign language as a language. In his case, a deaf interpreter used some signs combined with pantomime and gestures to convey the questions being asked by the moderator. In order to provide appropriate supports, the researcher needs to have a deep understanding of the diversity within the community and what is needed to provide support for meaningful and effective communication.

This level of support was necessary in order to ascertain the experiences of diverse deaf and hard of hearing individuals to serve as a basis for judicial training that would lead to plans to improve court access for the targeted population. The researcher used data from each stage of the study to inform decisions about research methods in subsequent stages. Hence, meetings with the advisory board constitute a qualitative data collection moment; focus group data were analyzed after each one to determine if gaps were still evident in terms of relevant dimensions of diversity. Subsequently, participant observation and quantitative surveys were used to collect data on the effectiveness of the training for judicial teams that included judges, judicial educators, and deaf or hard of hearing advocates from each state. Transformative aspects of this study include the conscious inclusion of the diverse voices of the deaf and hard of hearing people as the starting point, provision of appropriate supports for effective communication, critical reflection on missing voices throughout the process, involvement of the deaf and hard of hearing people in the development and implementation of the training, and formation of teams to prepare plans for increasing accessibility of their courtrooms that included deaf and hard of hearing persons.

As a point of contrast, when Mertens was invited to present about this study in the United Kingdom, a faculty member in the audience stated that they did not have problems like that in Britain. In fact, they had recently had a student do her dissertation on the same topic; she sent a survey form to all the courts to ask if they had interpreters. All the courts wrote back saying that they did have interpreters.

▣ TEACHER TRAINING FOR
  DEAF STUDENTS WITH DISABILITY

Mertens, Holmes, and Harris (2008) conducted an evaluation of a teacher training program that used a transformative cyclical approach. The teacher training program at Gallaudet University was designed to prepare teachers of color and/or who were deaf

or hard of hearing to teach deaf or hard of hearing students who had an additional disability. In keeping with the transformative paradigm, Mertens used her knowledge of the diversity within the deaf community to assemble a team of researchers who represented relevant dimensions of diversity, including individuals who were hearing, deaf, and hard of hearing (who used a cochlear implant). This team read through program documents such as the funding proposal and seven years of annual reports. They did not assume that they understood which issues would be of importance, rather, they asked for and received permission to conduct observations at a reflective seminar being held for program graduates.

These qualitative data were used as a basis for developing interview questions for use with the seminar participants. To accommodate differences in communication, the hearing and hard of hearing researchers interviewed the hearing and hard of hearing teachers using speech; the deaf researchers interviewed the deaf teachers using American Sign Language. The use of a transformative lens led to the identification of the participants' concerns about their preparation for oppressive conditions that they encountered in their early job placements and the adequacy of their preparation for students who reflected dimensions of diversity, such as choice of communication mode, home language, and type of additional disability. These data were used to develop a quantitative online survey to determine if program graduates who did not attend the seminar had similar experiences.

Being aware of the critical dimensions of diversity relevant in this context, the researchers identified challenges the new teachers encountered with their students, such as home languages other than English or children coming to school with no language at all. They also reported experiences of marginalization and other educators' low expectations for the deaf students. The data that follow are from fieldnotes, interviews, and the online survey. They illustrate the new teachers' challenges and serve as a basis for potential changes in teacher preparation programs that resulted from the use of a cyclical transformative approach.

Low expectations and marginalization. When I graduated, I thought I was ready to teach. Then the principal gave me my list of students and my classroom and just washed his hands of me. You're on your own. The principal did not require me to submit weekly plans like other teachers because he thought I'd only be teaching sign language. But I told him, I'm here to really teach. We (my students and I) were not invited to field day or assemblies. That first year really hit me—so I changed schools and this one is definitely better. Now I'm in a school where people believe that deaf students can learn. (Graduate student fieldnotes, May 2007)

Diversity in the student body. My students are under 5 years old and they come with zero language and their behavior is awful. They can't sit for even a minute. Kids come with temper tantrums and run out of the school building. I have to teach these kids language; I see them start to learn to behave and interact with others. My biggest challenge is seeing three kids run out of school at the same time. Which one do I run after? One kid got into the storm drain. I'm only one teacher and I have an assistant, but that means there is still one kid we can't chase after at the same time as the other two. (Program graduate interview, May 2007)

The quantitative data from the online survey revealed that only a few graduates reported having a mentor from Gallaudet (13%). About half reported having a mentor in their own school during their first year of teaching (47%). The interview questions combined quantitative data from the survey and qualitative data from the interviews and fieldnotes. The combined data reflect the new teachers feeling a lack of the support needed to address the challenges they encountered.

The data from the interviews, fieldnotes, and surveys were presented to the program faculty and staff at the cooperating schools that had served as sites for student teachers. University faculty also were asked to respond to the issues raised by their recent graduates. They expressed concern about the need to recognize the conditions into which their graduates move and include that as part of the teacher preparation curriculum. The university faculty also were moved to action to provide ongoing support for the new teachers.

As a result of reviewing the study's findings, the Gallaudet Department of Education established an online student teaching seminar during the Fall 2007 semester. That semester, the seminar was taught concurrently with the student teaching experience in an online format using the *Black Board* delivery system. The initial feedback from the student teachers was extremely positive. The Department of Education decided to offer the online seminar for all program graduates as well as student teachers from all concentration areas, not just to students in the multiple disabilities program. The goal of the online seminar is to offer ongoing support from not only a faculty member, but also from other new teachers and students located across the country. The transformative cycle was complete when the study's findings related to the need to better prepare new teachers to respond to diversity in students and work against oppression, resulted in the development of the online seminar that brought together students and faculty from different concentration areas in a way that enhances their early teaching experiences.

▣ SPINAL CORD INJURY RESEARCH IN NEW ZEALAND

Transformative research in other parts of the world includes that of Martin Sullivan and colleagues (Derrett, Crawford, Beaver, Fergusson, Paul, and Herbison), who are currently undertaking research in New Zealand on the first two years of spinal cord injury (SCI) and the transition from spinal unit to community. The aims of the research are (1) to explore how the interrelationship(s) of body, self, and society shape the life chances, life choices, and subjectivity of a cohort of people with SCI; and (2) to investigate how entitlement to rehabilitation and compensation through the Accident Compensation Commission (ACC)[6] affects socioeconomic and health outcomes. Most of the team have SCI and have been developing their own research skills through their participation in this project. The research utilizes a transformative parallel design using quantitative and qualitative methods concurrently. Quantitative material will be collected in three structured interviews with all participants recruited over a two-year period. The first interview is at 4 months following SCI. This is a face-to-face interview undertaken by the trained, onsite interviewers (who themselves have a SCI) prior to participants' discharge from the spinal unit. The second and third structured follow-up interviews will be undertaken by telephone at 12 and 24 months after SCI. The timing of these interviews complies with international recommendations for follow-up times in injury outcome studies (van Beeck et al., 2005).

These three structured interviews are designed to collect factual evidence on the "what" of the world SCI people inhabit: their life chances, attitudes, health status, support services, work and income, personal and social relationships, and life satisfaction, as this constitutes the framework in which they create their subjectivity. Two qualitative, face-to-face interviews with a subsample of 20 participants explore in greater depth the meanings participants now attach to these phenomena and how these phenomena and meanings are shaping their life choices and subjectivity. On the advice of advocacy groups made up of ex-patients of the spinal units, these qualitative interviews will be held at 6 and 18 months after discharge from the spinal units for specific reasons: At 6 months, the person ought to be settled in at home with alterations complete and the necessary personal supports in place to be thinking about venturing out into the broader world if she or he has not already done so. At 18 months, any neurological recovery that was going to occur would have occurred and the person will be thinking, "So this is it for the rest of my life." Inferences derived from facts and personal perspectives will be triangulated to enrich understandings of what actually happens to SCI people in the first two years postinjury.

At the time of writing (August 2009), recruitment is complete and the second round of quantitative interviews has begun. The questionnaire being used repeats a number of questions from the first interview and includes a new set of questions on issues such as what has changed for participants since the first interview, their use of and satisfaction with support services in general and with personal assistants and ACC in particular. These later questions are yielding some interesting data that will be followed up in depth in the second qualitative interview.

Preliminary analyses suggest that participants are so newly paralyzed that most of their energy is directed at their personal situation of learning to live with new bodies and new lives. Notwithstanding, we have had a high response rate to this study as most want to contribute their experience in order to improve or transform the lot of all people with SCI.

▣ DEINSTITUTIONALIZATION

In Melbourne, Australia, academics Patsie Frawley and Christine Bigby (Robertson, Frawley, & Bigby, 2008) challenged the research power dynamic by partnering with a person with intellectual disability, Alan Robertson, as a researcher on a project requiring lived expertise of institutionalization, which Robertson had. The evaluation of a large deinstitutionalization project into smaller residential community accommodation required a researcher with an astute understanding of what constituted "homeliness" in this context. Robertson and Frawley visited the homes and Robertson conducted interviews with the residents. Robertson instructed Frawley, as his research assistant, what to photograph, including indications of inappropriate furniture and fittings and lack of personal belongings. Frawley was responsible for transcribing the tapes and finding a suitable place in the community to use as their office as Robertson was uncomfortable in the university setting. Robertson has traveled to conferences around Australia to report on the research, which was published in easy-to-read "plain language" as "Making Life Good in the Community: When Is a House a Home?"

▣ PREDICTIONS

The Disability Rights Movement and the transformative paradigm intertwine to suggest new criteria for interpretive approaches to research with people with disability. First, a focus on social justice suggests a need for a redistribution of resources in terms of who is supported for research and under what conditions. Second, the United Nations declaration that opened this chapter illuminates the need for a focus on human rights in terms of both individual and group dignity. Third, research with disability communities needs to be based on appropriate cultural engagement that takes into account the diversity of cultural and ethnic settings in which these communities exist as well as the diversity found within those communities.

This shift in criteria leads to predictions of what the field of disability research will look like in 10 years. Research conducted using the transformative paradigm illustrates the benefits of the use of mixed methods; we predict that this approach will become the norm in transformative disability research and

disabled people will be increasingly rewritten in terms of resilience and purposive subjectivity.

Political struggles will increase between those who seek cure and those who identify as disabled, Aspie,[7] People First,[8] Deaf,[9] neuro-diverse[10]: This struggle has implications for the type and funding of research. For example, there is a debate raging between those who seek a cure for autism (such as Generation Rescue) and those who claim autism is nothing more than a variant of the human condition and are proud to claim their status as a cultural minority (such as the Autistic Self Advocacy Network [ASAN], 2011). While the former valorize neurotypicality and seek a cure for autism, the latter valorize neurodiversity, embrace it, and seek inclusion, respect, and equity.

There is bound to be increasing tension between these two groups. One is extremely powerful with high-profile celebrities fronting, while behind the scenes powerful interests fund their efforts. What the other lacks in resources, it makes up with passion, global Internet networks, and the righteousness and justice of their cause. In the meantime, ASAN (www.autisticadvocacy. org) is calling for quality-of-life-type research to be done with them as part of their struggle for inclusion. Significantly, it is linking more and more with other groups of disability activists. This means that over the next decade some hard decisions have to be made over what type of research is done and which is funded; this is likely to be intensely political. Will it be scientific research into the genetics of autism or even interventions that lack any scientific evidence base in search of a "cure"? Or will it be transformative research done with networks such as ASAN, seeking their expertise on research proposals and mentoring? What needs to change for this to happen? Currently, the research process itself is political and even the research application and ethical approval processes are inaccessible to many of those with impairments: for example, those with vision or intellectual impairment.

However, times are changing. The U.S. government's powerful Interagency Autism Advisory Committee now has autistic members to provide expertise and advice on research. In another victory in the battle of "who speaks for me?"—one of those invited to address an IAAC subcommittee was a nonverbal autistic woman who is also a prolific writer and poet. She learned to type, and, thereby, to communicate with words, using the augmentive communication method known as "facilitated communication" (FC), whereby a facilitator supports the hand or arm of a person using some kind of keyboard.

Although not the inventor of FC, Douglas Biklen's name has been associated with it over the last two decades and his 2005 book *Autism and the Myth of the Person Alone* includes the personal narratives of several autistic people who do not use the spoken word for communication. Through FC, they have found not only their written voice but have also challenged the neurotypical cultural assumptions of the "aloneness" of autism. Yet FC incites fury among many academics and behaviorist psychologists, who believe the facilitator produces the works and thus FC is fraudulent and exploitative of the autistic participants. The fact that many autistic individuals such as Australian Lucy Blackman have gone on from FC to type independently and dispute their lack of agency does not sway the critics of FC (Blackman, 1999).

Disability culture is also threatened by the eugenic potential of developments in genetic and other technology. Unless disabled people in all their diversity win transformative research power, we predict that by 2100 the human race will have become boringly neuro-typical and genetically homogenous after generations of genetic manipulations in pursuit of some mythical "norm" (Alper et al., 2002; Parens & Asch, 2000). We hope the sentiments of Article 1 of the UN Convention on Rights of Persons With Disabilities, with which we began this chapter, prevail in 2100 and that humanity is increasingly eccentrically, creatively, and interestingly neuro-diverse.

▣ PEDAGOGIC CHALLENGES

We need to rethink who are the experts on lived experience of disability, and therefore who should be the teachers and researchers. Christopher Newell (2006; Goggin & Newell, 2005), who died in his prime in 2008, was an Australian medical ethicist, priest, and disability activist who personified a transformative pedagogic approach. He began his working life in a sheltered workshop and ended up as an associate professor lecturing medical students in ethics. When he was unwell, he would often lecture his students from his hospital bed, which he would get wheeled into the lecture theater in the Hobart Hospital. He let his students see him at his most vulnerable and disabled so as to challenge them to respect the humanity of the disabled people they would meet in their professional lives. Christopher was an academic activist to the core. He was also an Anglican priest who chose not to take the high road but trod the back roads and byways where he befriended, counseled, and supported the abandoned, destitute, and lonely—many who were disabled people. His raison d'être was "moving disability from other to us"—a task he struggled to inculcate in the consciousness and public life of Australia (Newell, 2005).

As more people with impairments, including intellectual impairments, assert their rights and expectations to attend tertiary education, the academy will have to rise to the challenge. Trinity College Dublin enrolls students with intellectual impairment in a mainstream undergraduate course. Using a buddy system, the students also attend electives in any other Trinity College course of their choosing (O'Brien et al., 2008).

Such initiatives challenge the tenets of academia and lead to such questions as whether a PhD dissertation in plain language or sign language would be acceptable or even desirable? Should more value be placed on grey literature websites, blogs, and personal stories for those who may not use an academic paradigm?

What about language? Who will have power over acceptable usage? How will disabled people—including those with intellectual impairment—be represented on editorial boards of academic journals and other publications? What about access to research funding? As mentioned above, the application process and deliberation process will have to become much more attuned to the needs of disabled researchers.

Many nonverbal disabled students—some previously institutionalized—have successfully attained tertiary qualifications, but there are challenges to finding the appropriate support. Barriers—physical and attitudinal—will need to be removed by universities. Role models are required. But most significant for change will be transformative research led and mentored by disabled people themselves. Linton (2006) urges researchers in the field of disabilities to be cognizant of the value brought by inclusion of many types of people with disabilities. She notes that war veterans who return with disabling conditions have critically important perspectives that can inform research decisions. By combining forces, the issues of equity and discrimination as they relate to disability rights can be brought to greater visibility for researchers, policy makers, and legislators in the areas of employment, mobility, health care, education, transportation, and the arts.

▣ CONCLUSION

Thirty years ago, the institutionalization of disabled children and adults was the norm and debates were emerging about intelligence and whether disabled children required education. These arguments are still current in many countries. Thirty years on from now there could be two possible scenarios. We are at a crossroads whereby we can go down the path of technology, genetics, further politicization of research, and the domination of research agendas by treatment and cure rather than good living. Or disability could be welcomed in from the margins to the center where, say, the possibilities for the deployment of technology to support disabled people having a good life are researched from within a transformative paradigm in which disabled people, in all their diversity, act as the guides and mentors of the research process. If we achieve this, then the UN Convention on the Rights of Persons With Disabilities will, indeed, be a living document.

The pathway to full realization of human rights and social justice for people with disability is not smooth. A history of oppression and neglect has to be overcome. The transformative paradigm provides a way of thinking about the enduring consequences of this history, as well as a way forward through the conduct of research viewed as an instrument in the struggle. Heterogeneity within the disability communities leads to issues that will not be easily addressed. Challenges remain with regard to the role of nondisabled researchers and how they can work respectfully with disabled people to further the goal of social transformation. Partnerships can be difficult to begin and maintain; however, much strength is gained by bringing those with and without disabilities together. The transformative spirit can be viewed as a positive force to guide those who walk this path.

◧ NOTES

1. For additional information about the history of the disability rights movement, see the University of California, Berkeley, oral history website (2010) at http://bancroft.berkeley.edu/collections/drilm/index.html.

2. This quote first appeared in Mertens, 2009, p. 209.

3. The authors wish to thank Kirsten Smiler for bringing to our attention the work of Mason Durie and his colleagues at the Research Centre for Māori Health & Development at Massey University (http://www.massey.ac.nz/~wwwcphr/restph.htm). Since 1994, they have conducted Best Outcomes for Māori: Te Hoe Nuku Roa, a longitudinal study of Māori households (http://www.tehoe-nukuroa.org.nz/). This survey is designed to examine the diversity within the Māori community. Smiler's MA thesis focuses on issues of language and identity for Māori members of the New Zealand Deaf community (Smiler, 2004; Smiler & McPhee, 2007); she has an undergraduate degree in Māori studies. She is in the process of conducting her PhD dissertation research on successful interventions with deaf Māori and their families (http://www.victoria.ac.nz/hsrc/projects/maori.aspx).

Adrienne Wiley (2009) of Johns Hopkins University has written one of the few research papers on Māori experience of disability—an evaluation of Objective 11 of the New Zealand Disability Strategy, which promotes the participation of disabled Māori.

4. The Indigenous Terms of Reference (ITR) were developed by the faculty and staff of the Centre for Aboriginal Studies at Curtin University in Perth, Australia. Osborne and McPhee (2000) note that the original concept of the ITR "is attributed to Lilla Watson (1985), a Queensland Murri woman, who developed the concept Aboriginal (or Murri) Terms of Reference to among other things describe an Indigenous worldview and context for Indigenous cultural and social recognition" (p. 2). The Centre for Aboriginal Studies, Curtin University, adopted aspects of the concept and developed them into an area of competence within an undergraduate Community Development and Management course from 1990.

5. The capitalization of the term Sign Language signifies a cultural group similar to African American or Jewish.

6. In 1974, New Zealanders gave up their right to sue for compensation following an accident in exchange for a no-fault compensation system funded from levies on employers, wages, motor vehicle registration, and petrol. It is administered centrally by the Accident Compensation Corporation (ACC) and includes a cash lump sum, all costs associated with hospitalization, rehabilitation, aides, appliances, home and workplace alterations, personal care, continence and medical supplies, and home support; earnings-related compensation (80% of earnings at time of accident) are paid during the period of rehabilitation (or for the rest of a person's working life in the case of SCI) and a suitably modified motor vehicle. All people in New Zealand, including visitors, are covered by ACC. People who get their SCI through congenital or illness-related causes are covered by the far less generous and means-tested Disability Support Services through the Ministry of Health.

7. Many people with Asperger's syndrome refer to themselves as Aspies.

8. People First is the international name of the self-advocacy organization of people with intellectual disability.

9. People who identify as Deaf do not see themselves as disabled but as members of a linguistic minority. They use sign language and refer to themselves as culturally Deaf. Some of the people who are hard of hearing or hearing impaired and use hearing aids, cochlear implants, and try to talk, describe themselves as deaf and disabled.

10. In the literature, people with autism are referred to as having autism spectrum disorder while people without autism are described as neuro-typical. People on the spectrum who object to being defined as having a disorder, have adopted the term neuro-diverse (see Robertson, 2010) as a less offensive descriptor.

◧ REFERENCES

Abma, T. A. (2006). Patients as partners in a health research agenda setting: The feasibility of a participatory methodology. *Evaluation in the Health Professions, 29,* 424–439.

Albrecht, G. L. (2002). American pragmatism: Sociology and the development of disability studies. In C. Barnes, M. Oliver, & L. Barton (Eds.), *Disability studies today* (pp. 18–37). Cambridge, UK: Polity.

Alper, J. S., Ard, C., Asch, A., Beckwith, J., Conrad, P., & Geller, L. (Eds.). (2002). *The double-edged helix: Social implication of genetics in a diverse society.* Baltimore: Johns Hopkins University Press.

American Psychological Association. (2002). *Ethical principles of psychologists and code of conduct.* Washington, DC: Author. Available at http://www.apa.org/ethics/code/index.aspx

Anspach, R. (1979). From stigma to identity politics: Political activism among the physically disabled and former mental patients. *Social Science & Medicine, 13*(A), 765–773.

Asch, A., & Fine, M. (1992). Beyond pedestals: Revisiting the lives of women with disabilities. In M. Fine (Ed.), *Disruptive voices: The possibilities of feminist research* (pp. 139–172). Ann Arbor: University of Michigan Press.

Autistic Self Advocacy Network (ASAN). (2011). *About the Autistic Self Advocacy Network.* Available at www.autisticadvocacy.org

Barnes, C., & Mercer, G. (Eds.). (1997). *Doing disability research.* Leeds, UK: The Disability Press.

Barnes, C., Oliver, M., & Barton, L. (Eds.). (2002). *Disability studies today.* Cambridge, UK: Polity.

Barnes, H. W., McCreanor, T., Edwards, S., & Borell, B. (2009). Epistemological domination: Social science research ethics in Aotearoa. In D. M. Mertens & P. G. Ginsberg (Eds.), *Handbook of social research ethics* (pp. 442–457). Thousand Oaks, CA: Sage.

Beazley, S., Moore, M., & Benzie, D. (1997). Involving disabled people in research: A study of inclusion in environmental activities. In C. Barnes & G. Mercer (Eds.), *Doing disability research* (pp. 142–157). Leeds, UK: The Disability Press.

Beresford, P., & Wallcraft, J. (1997). Psychiatric system survivors and emancipatory research: Issues, overlaps and differences. In C. Barnes & G. Mercer (Eds.), *Doing disability research* (pp. 67–87). Leeds, UK: The Disability Press.

Biklen, D. (2005). *Autism and the myth of the person alone.* New York: New York University Press.

Bishop, R. (2008). Te Kotahitanga: Kaupapa Māori in mainstream classrooms. In N. K. Denzin, Y. S. Lincoln, & L. T. Smith (Eds.), *Handbook of critical & indigenous methodologies* (pp. 285–307). Thousand Oaks, CA: Sage.

Blackman, L. (1999). *Lucy's story: Autism and other adventures.* London: Jessica Kingsley Publishers.

Booth, T., & Booth, W. (1997). Making connections: A narrative study of adult children of parents with learning difficulties. In C. Barnes & G. Mercer (Eds.), *Doing disability research* (pp. 123–141). Leeds, UK: The Disability Press.

Brabeck, M. M., & Brabeck, K. M. (2009). Feminist perspectives on research ethics. In D. M. Mertens & P. G. Ginsberg (Eds.), *Handbook of social research ethics* (pp. 39–53). Thousand Oaks, CA: Sage.

Braddock, D. L., & Parish, S. L. (2001). An institutional history of disability. In G. L. Albrecht, K. D. Seelman, & M. Bury (Eds.), *Handbook of disability studies* (pp. 11–68). Thousand Oaks, CA: Sage.

Campbell, J., & Oliver, M. (1996). *Disability politics: Understanding our past, changing our future.* London: Routledge.

Chilisa, B. (2009). Indigenous African-centered ethics: Contesting and complementing dominant models. In D. M. Mertens & P. G. Ginsberg (Eds.), *Handbook of social research ethics* (pp. 407–425). Thousand Oaks, CA: Sage.

Cram, F. (2009). Maintaining indigenous voices. In D. M. Mertens & P. G. Ginsberg (Eds.), *Handbook of social research ethics* (pp. 308–322). Thousand Oaks, CA: Sage.

Cram, F., Ormond, A., & Carter, L. (2004). *Researching our relations: Reflections on ethics and marginalization.* Paper presented at the Kamehameha Schools 2004 Research Conference on Hawaiian Well-being, Kea'au, HI.

Cushing, L. S., Carter, E. W., Clark, N., Wallis, T., & Kennedy, C. H. (2009). Evaluating inclusive educational practices for students with severe disabilities using the program quality measurement tool. *Journal of Special Education, 42,* 195–208.

Denzin, N. K., & Lincoln, Y. S. (Eds.). (2005). *The SAGE handbook of qualitative research* (3rd ed.). Thousand Oaks, CA: Sage.

Denzin, N. K., Lincoln, Y. S., & Smith, L. T. (2008). *Handbook of critical & indigenous methodologies.* Thousand Oaks, CA: Sage.

Dodd, S.-J. (2009). LGBTQ: Protecting vulnerable subjects in *all* studies. In D. M. Mertens & P. G. Ginsberg (Eds.), *Handbook of social research ethics* (pp. 474–488). Thousand Oaks, CA: Sage.

Doren, B., & Benz, M. (2001). Gender equity issues in the vocational and transition services and employment outcomes experienced by young women with disabilities. In H. Rousso & M. Wehmeyer (Eds.), *Double jeopardy: Addressing gender equity in special education.* Albany: SUNY Press.

Edno, T., Joh, T., & Yu, H. C. (2003). *Voices from the field: Health and evaluation leaders on multicultural evaluation.* Oakland, CA: Social Policy Research Associates.

Fine, M., & Asch, A. (1988). Disability beyond stigma: Social interaction, discrimination, and activism. *Journal of Social Issues, 44*(1), 3–21.

Frank, G. (2000). *Venus on wheels.* Berkeley: University of California Press.

Gill, C. J., Kewman, D. G., & Brannon, R. W. (2003). Transforming psychological practice and society. *American Psychologist, 58*(4), 305–312.

Gladstone, C., Dever, A., & Quick, C. (2009, August 26–27). *"My Life When I Leave School" Transition Project: Self-determination and young intellectually disabled students in the transition from school to post school life.* Paper presented at the From Theory to Practice: Knowledge and Practices Conference, the 6th annual conference of the New Zealand Association for the Study of Intellectual Disability, Hamilton, New Zealand.

Goggin, G., & Newell, C. (2005). *Disability in Australia: Exposing a social apartheid.* Sydney, Australia: University of New South Wales Press.

Goodley, D. (2000). *Self-advocacy in the lives of people with learning difficulties.* Buckingham, UK: Open University Press.

Gravois, T. A., & Rosenfield, S. A. (2006). Impact of instructional consultation teams on the disproportionate referral and placement of minority students in special education. *Remedial and Special Education, 27,* 42–52.

Groce, N. E. (2003). *Everyone here spoke sign language.* Cambridge, MA: Harvard University Press. (Original work published 1985)

Guba, E. G., & Lincoln, Y. S. (2005). Paradigmatic controversies, contradictions, and emerging confluences. In N. K. Denzin & Y. S. Lincoln (Eds.), *The SAGE handbook of qualitative research* (3rd ed., pp. 191–216). Thousand Oaks, CA: Sage.

Hahn, H. (1982). Disability and rehabilitation policy: Is paternalistic neglect really benign? *Public Administration Review, 43,* 385–389.

Hahn, H. (1988). The politics of physical differences: Disability and discrimination. *Journal of Social Issues, 44*(1), 39–47.

Hahn, H. (2002). Academic debates and political advocacy: The US disability movement. In C. Barnes, M. Oliver, & L. Barton (Eds.), *Disability studies today* (pp. 162–190). Cambridge, UK: Polity.

Harris, R., Holmes, H., & Mertens, D. M. (2009). Research ethics in sign language communities. *Sign Language Studies, 9*(2), 104–131.

Horvath, L. S., Kampfer-Bohach, S., & Farmer Kearns, J. (2005). The use of accommodations among students with deaf-blindness in large-scale assessment systems. *Journal of Disability Policy Studies, 16,* 177–187.

Kroll, T., Barbour, R., & Harris, J. (2007). Using focus groups in disability research. *Qualitative Health Research, 17,* 690–698.

LaFrance, J., & Crazy Bull, C. (2009). Researching ourselves back to life: Taking control of the research agenda in Indian country. In D. M. Mertens & P. G. Ginsberg (Eds.), *Handbook of social research ethics* (pp. 153–149). Thousand Oaks, CA: Sage.

Linton, S. (1998). *Claiming disability: Knowledge and identity.* New York: New York University Press.

Linton, S. (2006). *My body politic.* Ann Arbor: University of Michigan Press.

Lunt, N., & Thornton, P. (1997). Researching disability employment policies. In C. Barnes & G. Mercer (Eds.), *Doing disability research* (pp. 108–122). Leeds, UK: The Disability Press.

MacArthur, J., Gaffney, M., Kelly, B., & Sharp, S. (2007). Disabled children negotiating school life: Agency, difference, teaching practice and education policy. *International Journal of Children's Rights, 15*(1), 99–120.

Meekosha, H., & Shuttleworth, R. (2009). What's so "critical" about critical disability studies? *Australian Journal of Human Rights, 15*(1), 47–75.

Mertens, D. M. (2000). Deaf and hard of hearing people in court: Using an emancipatory perspective to determine their needs. In C. Truman, D. M. Mertens, & B. Humphries (Eds.), *Research and inequality* (pp. 111–125). London: Taylor & Francis.

Mertens, D. M. (2009). *Transformative research and evaluation.* New York: Guilford.

Mertens, D. M. (2010). *Research and evaluation in education and psychology: Integrating diversity with quantitative, qualitative, and mixed methods* (3rd ed.). Thousand Oaks, CA: Sage.

Mertens, D. M., Holmes, H., & Harris, R. (2008, February). *Preparation of teachers for students who are deaf and have a disability.* Presentation at the annual meeting of the Association of College Educators of the Deaf and Hard of Hearing, Monterey, CA.

Mertens, D. M., Holmes, H., & Harris, R. (2009). Transformative research and ethics. In D. M. Mertens & P. G. Ginsberg (Eds.), *Handbook of social research ethics* (pp. 85–102). Thousand Oaks, CA: Sage.

Mertens, D. M., Wilson, A., & Mounty, J. (2007). Gender equity for people with disabilities. In S. Klein et al. (Eds.), *Handbook for achieving gender equity through education* (pp. 583–604). Mahwah, NJ: Lawrence Erlbaum.

Morris, J. (1991). *Pride against prejudice.* London: Women's Press.

Newell, C. (2005). Moving disability from other to us. In P. O'Brien & M. Sullivan (Eds.), *Allies in emancipation: Shifting from providing service to being of support.* Melbourne: Thomson/Dunmore.

Newell, C. (2006). Representation or abuse? Rhetorical dimensions of genetics and disability. *Interaction, 20*(1), 28–33.

New Zealand Ministry of Health. (2001). *New Zealand disability strategy.* Wellington, NZ: Author.

O'Brien, P., Shevlin, M., O'Keeffe, M., Kenny, M., Fitzgerald, S., Espiner, D., & Kubiack, J. (2008, November 24–26). *Opening up a whole new world: Students with intellectual disability being included within a university setting.* Paper presented at the 43rd Australian Society for the Scientific Study of Intellectual Disabilities Conference, University of Melbourne, Australia.

Oliver, M. (1982). A new model of the social work role in relation to disability. In J. Campling (Ed.), *The handicapped person: A new perspective for social workers?* London: RADAR.

Oliver, M. (1990). *The politics of disablement.* Basingstoke, UK: Macmillan.

Oliver, M. (1992). Changing the social relations of research production? *Disability, Handicap & Society, 7*(2), 101–114.

Oliver, M. (1997). Emancipatory research: Realistic goal or impossible dream? In C. Barnes & G. Mercer (Eds.), *Doing disability research* (pp. 15–31). Leeds, UK: The Disability Press.

Osborne, R., & McPhee, R. (2000, December 12–15). *Indigenous terms of reference* (ITR). Presentation at the 6th annual UNESCO-ACEID International Conference on Education, Bangkok, Thailand.

Parens, E., & Asch, A. (Eds.). (2000). *Prenatal testing and disability rights.* Washington, DC: Georgetown University Press.

Pfeiffer, D., & Yoshida, K. (1995). Teaching disability studies in Canada and the USA. *Disability & Society, 10*(4) 475–495.

Priestley, M. (1997). Who's research? A personal audit. In C. Barnes & G. Mercer (Eds.), *Doing disability research* (pp. 88–107). Leeds, UK: The Disability Press.

Rapaport, J., Manthorpe, J., Moriarty, J., Hussein, S., & Collins, J. (2005). Advocacy and people with learning disabilities in the UK: How can local funders find value for money? *Journal of Intellectual Disabilities, 9,* 299–319.

Reinharz, S. (1979). *On becoming a social scientist.* San Francisco: Jossey-Bass.

Robertson, A., Frawley, P., & Bigby, C. (2008). *Making life good in the community: When is a house a home? Looking at how homely community houses are for people with an intellectual disability who have moved out of an institution.* Melbourne, Australia: La Trobe University and the State Government of Victoria.

Robertson, S. M. (2010). Neurodiversity, quality of life, and autistic adults: Shifting research and professional focuses onto real-life challenges. *Disability Studies Quarterly, 30*(1).

Rousso, H. (2001). *Strong proud sisters: Girls and young women with disabilities.* Washington, DC: Center for Women Policy Studies.

Ryan, J. (2007). Learning disabilities in Australian universities: Hidden, ignored, and unwelcome. *Journal of Learning Disabilities, 40,* 436–442.

Scotch, R. K. (2001). *From good will to civil rights.* Philadelphia: Temple University Press.

Shakespeare, T. (1997). Researching disabled sexuality. In C. Barnes & G. Mercer (Eds.), *Doing disability research* (pp. 177–189). Leeds, UK: The Disability Press.

Shakespeare, T., & Watson, N. (2001). The social model of disability: An outdated ideology? In S. N. Barnarrt & B. M. Altman (Eds.), *Exploring theories and expanding methodologies: Where are we and where do we need to go?* (pp. 9–28). Greenwich, CT: JAI.

Smiler, K. (2004). *Maori deaf: Perceptions of cultural and linguistic identity of Maori members of the New Zealand deaf community.* Unpublished master's thesis, Victoria University of Wellington, New Zealand.

Smiler, K., & McKee, R. L. (2007). Perceptions of "Maori" deaf identity in New Zealand. *Journal of Deaf Studies and Deaf Education, 12*(1), 93–111.

Smith, L. T. (2005). On tricky ground: Researching the native in the age of uncertainty. In N. K. Denzin & Y. S. Lincoln (Eds.), *The SAGE handbook of qualitative research* (3rd ed., pp. 85–108). Thousand Oaks, CA: Sage.

Sobsey, D. (1994). *Violence and abuse in the lives of people with disabilities: The end of silent acceptance?* Baltimore: Brooks.

Stone, E., & Priestly, M. (1996). Parasites, pawns and partners: Disability research and the role of non-disabled researchers. *British Journal of Sociology, 47*(4), 699–716.

Sullivan, M. (2009). Philosophy, ethics and the disability community. In D. M. Mertens & P. G. Ginsberg (Eds.), *Handbook of social research ethics* (pp. 69–84). Thousand Oaks, CA: Sage.

Symonette, H. (2009). Cultivating self as a responsive instrument: Working the boundaries and borderlands for ethical border crossings. In D. M. Mertens & P. G. Ginsberg (Eds.), *Handbook of social research ethics* (pp. 279–294). Thousand Oaks, CA: Sage.

Thomas, V. G. (2009). Critical race theory: Ethics and dimensions of diversity in research. In D. M. Mertens & P. G. Ginsberg (Eds.), *Handbook of social research ethics* (pp. 54–68). Thousand Oaks, CA: Sage.

Triano, S. (2006). Disability pride. In G. Albrecht (Ed.), *Encyclopedia of disability* (Vol. 1, pp. 476–477). Thousand Oaks, CA: Sage.

Union of the Physically Impaired Against Segregation (UPIAS). (1976). *Fundamental principles of disability* [Booklet]. London: Author.

United Nations. (2006). *Convention on the rights of people with disabilities.* New York: Author.

United States Department of Education. (2004). *Twenty-sixth annual report to Congress on the implementation of the Individuals With Disabilities Act* (Vol. 2). Washington DC: Author.

University of California, Berkeley. (2010). *The disability rights and independent living movement.* Available at http://bancroft.berkeley.edu/collections/drilm/index.html

van Beeck, E., Larsen, C. F., Lyons, R., Meerding, W. J., Mulder, S., & Essink-Bot, M. L. (2005). *Draft guidelines for the conduction of empirical studies into injury-related disability.* Amsterdam: Eurosafe (European Association for Injury Prevention and Safety Promotion). Available at http://www.eurosafe.eu.com/csi/catalogus.nsf/c1af8df8ec2b154bc12570b500682709/62760ac72216e50ac12570d60036ccc9/$FILE/ER-285.pdf

Watson, L. (1985, July 8–11). *The establishment of aboriginal terms of reference in a tertiary institution.* Paper presented at the Aborigines and Islanders in Higher Education: The Need for Institutional Change National Conference, Townsville, Australia.

Wiley, A. (2009). At a cultural crossroads: Lessons on culture and policy from the New Zealand Disability Strategy. *Disability and Rehabilitation, 31*(14), 1205–1214.

Wright, B. A. (1983). *Physical disability: A psychosocial approach.* New York: Harper & Row.

Part III

STRATEGIES OF INQUIRY

The civic-minded qualitative researcher thinks historically, interactionally, and structurally. He or she attempts to identify the many persuasions, prejudices, injustices, and inequities that prevail in a given historical period (Mills, 1959, p. 7). Critical scholars seek to examine the major public and private issues and personal troubles that define a particular historical moment. In doing so, qualitative researchers self-consciously draw upon their own experience as a resource in such inquiries. They always think reflectively and historically, as well as biographically. They seek strategies of empirical inquiry that will allow them to make connections between lived experience, social injustices, larger social and cultural structures, and the here and now. These connections will be forged out of the interpretations and empirical materials that are generated in any given inquiry.

Empirical inquiry, of course, is shaped by paradigm commitments and by the recurring questions that any given paradigm or interpretive perspective asks about human experience, social structure, and culture. More deeply, however, the researcher always asks how the practices of qualitative inquiry can be used to help create a free, democratic society. Critical theorists, for example, examine the material conditions and systems of ideology that reproduce class and economic structures. Queer, constructivist, cultural studies, critical race, and feminist researchers examine the stereotypes, prejudices, and injustices connected to race, ethnicity, and gender. There is no such thing as value-free inquiry, although in qualitative inquiry this premise is presented with more clarity. Such clarity permits the value-commitments of researchers to be transparent.

The researcher-as-interpretive-bricoleur is always already in the material world of values and empirical experience. This world is confronted and constituted through the lens that the scholar's paradigm or interpretive perspective provides. The world so conceived ratifies the individual's commitment to the paradigm or perspective in question. This paradigm is connected at a higher ethical level to the values and politics of an emancipatory, civic social science.

As specific investigations are planned and carried out, two issues must be immediately confronted: research design and choice of strategy of inquiry. We take them up in order. Each devolves into a variety of related questions and issues that must also be addressed.

RESEARCH DESIGN[1]

The research design, as discussed in our Introduction and analyzed by Julianne Cheek in this part of the *Handbook,* situates the investigator in the world of experience. Five basic questions structure the issue of design:

1. How will the design connect to the paradigm or perspective being used?

That is, how will empirical materials be informed by and interact with the paradigm in question?

2. How will these materials allow the researcher to speak to the problems of praxis and change?

3. Who or what will be studied?

4. What strategies of inquiry will be used?

5. What methods or research tools for collecting and analyzing empirical materials will be utilized?

These questions are examined in detail in Part IV of the *Handbook.*

PARADIGM, PERSPECTIVE, AND METAPHOR

The positivist, postpositivist, constructionist, and critical paradigms dictate, with varying degrees of freedom, the design of a qualitative research investigation. This can be viewed as a continuum moving from rigorous design principles on one end to emergent, less well-structured directives on the other. Positivist research designs place a premium on the early identification and development of a research question, a set of hypotheses, a research site, and a statement concerning sampling strategies, as well as a specification of the research strategies and methods of analysis that will be employed. A research proposal laying out the stages and phases of the study may be written. In interpretive research, a priori design commitments may block the introduction of new understandings. Consequently, while qualitative researchers may design procedures beforehand, designs always have built-in flexibility, to account for new and unexpected empirical materials and growing sophistication.

The stages of a study can be conceptualized as involving reflection, planning, entry, data collection, withdrawal from the field, analysis, and write-up. Julianne Cheek (Chapter 14) observes that the degree of detail involved in the proposal will vary, depending on the funding agency. Funding agencies fall into at least six categories: local community funding units, special purpose, family-sponsored, corporate or national foundations, and federal governmental agencies. Depending on the requirements of the funder, proposals may also include a budget, a review of the relevant literature, a statement concerning human subjects protection, a copy of consent forms, interview schedules, and a timeline. Positivist designs attempt to anticipate all of the problems that may arise in a qualitative study (although interpretivist designs do not). Such designs provide rather well-defined road maps for the researcher. The scholar working in this tradition hopes to produce a work that finds its place in the literature on the topic being studied.

In contrast, much greater ambiguity and flexibility are associated with postpositivist and nonpositivist designs, those based, for example, on the constructivist or critical theory paradigms, or the critical race, feminist, queer, or cultural studies perspectives. In studies shaped by these paradigms and perspectives, there will be less emphasis on formal grant proposals, well-formulated hypotheses, tightly defined sampling frames, structured interview schedules, and predetermined research strategies and methods and forms of analysis. The researcher may follow a path of discovery, using as a model qualitative works that have achieved the status of classics in the field. Enchanted, perhaps, by the myth of the Lone Ethnographer, the scholar hopes to produce a work that has the characteristics of a study done by one of the giants from the past (Bronislaw Malinowski, Margaret Mead, Gregory Bateson, Erving Goffman, Ernest Becker, Claude Lévi-Strauss, Harry Wolcott). As a result, qualitative researchers often at least begin by undertaking studies that can be completed by one individual after prolonged engagement.

◨ THE POLITICS AND PRACTICES OF
FUNDING QUALITATIVE INQUIRY

Cheek's chapter complicates and deconstructs the relationship between money, ethics, and research markets. She examines the politics and practices involved in funding qualitative inquiry,

including seeking, gaining, and accepting funding. The politics of funding privileges certain forms of inquiry. A concern for the politics of evidence—what is evidence—leads to problems surrounding research design and sample size. Pressures to employ mixed methods procedures can complicate matters.

Cheek shows how qualitative research is a commodity that circulates and is exchanged in this political economy. Funding involves selling one's self to a funding agency. Such agencies may not understand the nuances of qualitative research practice. She discusses the problems associated with institutional review boards (IRBs) and ethics committees. In Australia, researchers cannot conduct research on human subjects until they have formal ethics approval from the University Research Ethics Committee. In the United States and the United Kingdom, as well as Australia, the original focus of IRBs and the context from which they emerged was medicine. Qualitative research is often treated unfairly by ethics committees. Such research, it may be charged, is unscientific. In effect, IRBs have become methodological review boards, institutionalizing only one brand, or version, of science. In the United Kingdom, the Royal College of Physicians' guidelines make the point that badly designed research is unethical. This means that judgment is being passed on the scientific as well as the ethical merits of research. Cheek observes that in too many instances "it seems that qualitative researchers have become the fall guys for ethical mistakes in medical research." Cheek notes that many times qualitative researchers are unable to answer in advance all of the questions that are raised by such committees. Issues of control over the research are also central. As she observes, "taking money from a sponsor [in order to conduct research] is not a neutral activity." This issue shades into another, namely, what happens when the researcher's findings do not please the funder?

There are problems in accepting external funding. Faculty are increasingly under pressure to secure external funding for their research. Such pressures turn research into a commodity that is bought and sold. Cheek observes that these are dangerous times. The conservative discourse of the marketplace has become preeminent. It is the market, not the judgment of stakeholders and peers, that now determines the worth of what we do. Are we writing for inquiry purposes, or for funding reasons?

Choreographing the Dance of Design

V. J. Janesick (2000, 2010) presents a fluid view of the design process. She observes that the essence of good qualitative research design requires the use of a set of procedures that are at once open-ended and rigorous. Influenced by Martha Graham, Merce Cunningham, Alvin Ailey, Elliot Eisner, and John Dewey, she approaches the problem of research design from an aesthetic, artistic, and metaphorical perspective. With Dewey and Eisner, she sees research design as a work of improvisational, rather than composed, art—as an event, a process, with phases connected to different forms of problematic experience, and their interpretation and representation. Art molds and fashions experience. In its dance form, art is a choreographed, emergent production with distinct phases: warming up, stretching exercises, and design decisions, cooling down, interpretation, and writing the narrative.

Who and What Will Be Studied?

The who and what of qualitative studies involve cases, or instances, of phenomena and/or social processes. Three generic approaches may be taken to the question of who or what will be studied. First, a single case, or single process, may be studied, what Robert Stake (2005) calls the intrinsic case study. Here, the researcher examines in detail a single case or instance of the phenomenon in question, for example, a classroom, an arts program, or a death in the family.

Second, the researcher may focus on a number of cases. Stake (2005) calls this the collective case approach. These cases are then analyzed in terms of specific and generic properties. Third, the researcher can examine multiple instances of a process as that process is displayed in a variety of different cases. Denzin's (1993) study of relapse in the careers of recovering alcoholics examined types of relapses across several different types of recovering careers. This process approach is then grounded or anchored in specific cases.

Research designs vary, of course, depending on the needs of multi-, or single-focused case and process inquiries. Different sampling issues arise in each situation. These needs and issues also vary by the paradigm that is being employed. Every instance of a case or process bears the stamp of the general class of phenomenon to which it belongs. However, any given instance is likely to be particular and unique. Thus, for example, any given classroom is like all classrooms, but no classroom is the same.

For these reasons, many postpositivist, constructionist, and critical theory qualitative researchers employ theoretical or purposive, and not random, sampling models. They seek out groups, settings, and individuals where (and for whom) the processes being studied are most likely to occur. At the same time, a process of constant comparison between groups, concepts, and observations is necessary, as the researcher seeks to develop an understanding that encompasses all instances of the process, or case, under investigation. A focus on negative cases is a key feature of this process.

These sampling and selection issues would be addressed differently by a postmodern ethnographer in the cultural studies tradition. This investigator would be likely to place greater stress on the intensive analysis of a small body of empirical materials (cases and processes), arguing, after Jean-Paul Sartre (1981, p. ix), that no individual or case is ever just an individual or a case. He or she must be studied as a single instance of more universal social experiences and social processes.

The individual, Sartre (1981, p. ix) states, is "summed up and for this reason universalized by his [her] epoch, he [she] in turn resumes it by reproducing him- [her-]self in it as a singularity." Thus to study the particular is to study the general. For this reason, any case will necessarily bear the traces of the universal, and, consequently, there is less interest in the traditional positivist and postpositivist concerns with negative cases, generalizations, and case selection. The researcher assumes that the reader will be able, as Robert Stake (2005) argues, to generalize subjectively from the case in question to his or her own personal experiences.

An expansion on this strategy is given in the method of instances (see Denzin, 1999; Psathas, 1995). Following George Psathas (1995, p. 50), the "method of instances" takes each instance of a phenomenon as an occurrence that evidences the operation of a set of cultural understandings currently available for use by cultural members.

An analogy may be useful. In discourse analysis, "no utterance is representative of other utterances, though of course it shares structural features with them; a discourse analyst studies utterances in order to understand how the potential of the linguistic system can be activated when it intersects at its moment of use with a social system" (Fiske, 1994, p. 195). This is the argument for the method of instances. The analyst examines those moments when an utterance intersects with another utterance, giving rise to an instance of the system in action.

Psathas clarifies the meaning of an instance: "An instance of something is an occurrence . . . an event whose features and structures can be examined to discover how it is organized" (1995, p. 50). An occurrence is evidence that "the machinery for its production is culturally available . . . [for example,] the machinery of turn-taking in conversation" (pp. 50–51).

The analyst's task is to understand how this instance and its intersections work, to show what rules of interpretation are operating, to map and illuminate the structure of the interpretive event itself. The analyst inspects the actual course of the interaction "by observing what happens first, second, next, etc., by noticing what preceded it; and by examining what is actually done and said by the participants" (Psathas, 1995, p. 51). Questions of meaning are referred back to the actual course of interaction, where it can be shown how a given utterance is acted upon and hence given meaning. The pragmatic maxim obtains here (Peirce, 1905). The meaning of an action is given in the consequences that are produced by it, including the ability to explain past experience, and predict future consequences.

Whether the particular utterance occurs again is irrelevant. The question of sampling from a population is also not an issue, for it is never possible to say in advance what an instance is a sample of (Psathas, 1995, p. 50). Indeed, collections of instances "cannot be assembled in advance of an analysis of at least one, because it cannot be known in advance what features delineate each case as a 'next one like the last'" (Psathas, 1995, p. 50).

This means there is little concern for empirical generalization. Psathas is clear on this point. The goal is not an abstract, or empirical, generalization; rather the aim is "concerned with providing analyses that meet the criteria of unique adequacy" (1995, p. 50). Each analysis must be fitted to the case at hand, each "must be studied to provide an analysis *uniquely adequate* for that particular phenomenon" (p. 51, italics in original).

Strategies of Inquiry

A strategy of inquiry describes the skills, assumptions, enactments, and material practices that researchers-as-methodological-bricoleurs use when they move from a paradigm and a research design to the collection of empirical materials. Strategies of inquiry connect researchers to specific approaches and methods for collecting and analyzing empirical materials. The case study, for example, relies on interviewing, observing, and document analysis. Research strategies locate researchers and paradigms in specific empirical, material sites, and in specific methodological practices, for example, making a case an object of study (see Bent Flyvbjerg, Chapter 17).

We turn now to a brief review of the strategies of inquiry discussed in this volume. Each is connected to a complex literature with its own history, its own exemplary works, and its own set of preferred ways for putting the strategy into motion. Each strategy also has its own set of problems involving the positivist, postpositivist, and postmodern legacies.

Mixed Methods Research

John W. Creswell (Chapter 15) and Charles Teddlie and Abbas Tashakkori (Chapter 16) examine controversies and issues in mixed methods research (MMR), or the third methodological moment. Although there is considerable debate over what constitutes mixed methods research, Creswell and Teddlie and Tashakkori suggest that it is inquiry that focuses on collecting, analyzing, and mixing both quantitative and qualitative empirical materials in a single study, or a series of studies. Creswell identifies 11 key controversies and questions being raised about mixed methods research. These issues include disagreements over definitions; just what is a mixed methods study; paradigm debates—that is, are there incommensurable and incompatible (and irresolvable) differences between paradigms?; how does the current conversation privilege postpositivism?; and what value is added by mixed methods? In giving voice to these controversies, Creswell creates the space for a reassessment of the mixed methods movement and where it is taking the interpretive community.

Teddlie and Tashakkori (and Creswell) offer a history of this field, noting overlaps with recent developments in emergent methods (Hess-Biber & Leavy, 2008), parallels with earlier arguments for triangulation (Denzin, 1970),[2] as well as discourse in the fields of evaluation, nursing, education, disability studies, and sociology. For these researchers, MMR is characterized by eclecticism, paradigm pluralism, a celebration of diversity, a rejection of dichotomies, an iterative approach to inquiry, an emphasis on the research question, and a focus on signature MMR design and analysis strategies (QUAL/QUAN): parallel, sequential, multilevel, sequential mixed, and so on. A typology of designs is reviewed.

Three dominant paradigms—pragmatism, transformative, and dialectical—are also reviewed, even as these authors discuss the arguments against a continued focus on paradigms. Some contend the term paradigm is outmoded. We disagree. Criticisms of MMR include the incompatibility thesis, a pervasive postpositivist bias, the tendency to subordinate QUAL to QUAN, cost, superficial methodological bilingualism, and an entanglement in superficial philosophical debate (e.g., forms of pragmatism). Teddlie and Tashakkori believe many of these issues will be resolved in the next decade.

▣ A PRAGMATIC ASIDE

As pragmatists trained in, or sympathetic to, the Chicago School, we are not so certain (Denzin, 2010; Lincoln, 2010). So we respectfully demur.

The MMR links to the pragmatism of John Dewey, William James, Margaret Mead, and Charles Peirce are problematic for us. Classic pragmatism is not a methodology per se. It is a doctrine of meaning, a theory of truth. It rests on the argument that the meaning of an event cannot be given in advance of experience. The focus is on the consequences and meanings of an action or event in a social situation. This concern goes beyond any given methodology—that is, the interpreter examines and inspects, and reflects upon an action and its consequences. Nor are they revealed by a given methodology. The MMR community does not seem to have a method for ascertaining meaning at this level.

Neopragmatists Richard Rorty, Jürgen Habermas, and Cornel West extend the classic doctrine. They endorse a thoroughly interpretive, hermeneutic pragmatism that is explicitly anti-positivist, antifoundational, and radically contextual. Basing an argument for mixed methods on this version of pragmatism seems misplaced.

The compatibility thesis for the MMR community asserts that combining qualitative and quantitative methods is a good thing; that is, there is no incompatibility between QUAN and QUAL at the practical or epistemological levels. Under this reading, pragmatism rejects paradigm conflicts between QUAN and QUAL epistemologies. Pragmatism is thus read as a practical and applied research philosophy that supports mixed or multiple methods of social science inquiry (Maxcy, 2003, p. 85).

An additional warrant for this is given by K. R. Howe (1988), who appeals to a "what works," or practical consequences, version of pragmatism. This is cash register pragmatism, not classic pragmatism. But this version of what works is not the point. The pragmatist focus is on the consequences of action, not on combining methodologies. And here, the MMR is of little help.

It is one thing to endorse pluralism, or multiple frameworks (Schwandt, 2007, p. 197), but it is quite another to build a social science on a cash register pragmatism. What works means two things, or has two consequences. First, it is a mistake to forget about paradigmatic, epistemological, or methodological differences between and within QUAN/QUAL frameworks. These are differences that matter, but they must not distract us from the second problem. As currently formulated, MMR offers few strategies for assessing the interpretive, contextual level of experience where meaning is created.

The Case Study

Bent Flyvbjerg (Chapter 17) employs a commonsense definition of the case study as the intensive analysis of an individual unit. He examines, and then refutes, five misunderstandings about this strategy of inquiry: (1) general rather than case knowledge is more valuable; (2) one cannot generalize from an individual case; (3) the case study is not suited to theory building; (4) the case study has a tendency to confirm the researcher's biases; and (5) it is difficult to develop generalizations based on specific case studies.

He demonstrates that concrete case knowledge is more valuable than the vain search for predictive theories and universals. It is possible to generalize from a single case (Charles Darwin, Isaac Newton, Albert Einstein), and it is useful for generating and testing hypotheses. It contains no greater bias toward verification of the researcher's preconceived notions than any other method of inquiry. Often it is not desirable to generalize from case studies. Flyvbjerg clarifies the methodological value of the case study and goes some distance in establishing its importance to the social sciences.

Robert Stake (2005) contends that the case study is not a methodological choice, but a choice of object to be studied, for example, a child, or a classroom. Ultimately, the researcher is interested in a process, or a population of cases, not an individual case per se. Stake identifies several types of case studies (intrinsic, instrumental, collective). Each case is a complex historical and contextual entity. Case studies have unique conceptual structures, uses, and problems (bias, theory, triangulation, telling the story, case selection, ethics). Researchers routinely provide information on such topics as the nature of the case, its historical background, its relation to its contexts and other cases, as well as to the informants who have provided information. In order to avoid ethical problems, the case study researcher needs constant input from conscience, from stakeholders, and from the research community.

Performance Ethnography

Judith Hamera (Chapter 18) offers a nuanced, and detailed, discussion of the complex relationship between performance studies, ethnography (and autoethnography), and critical pedagogy. She connects these formations to critical pedagogy theory. Performance ethnography is a way of inciting culture, a way of bringing culture alive, a way of fusing the pedagogical with the performative, with the political. Hamera's chapter addresses the key terms (reflexivity, performance, ethnography, performativity, aesthetics), the philosophical contingencies, the procedural pragmatics, and the pedagogical and political possibilities that exist in the spaces and practices of performance ethnography. Her arguments complement the Tedlock and Spry chapters (19 and 30, respectively) in this handbook.

Performance is an embodied act of interpretation, a way of knowing, a form of moral discourse. A politics of possibility organizes the project. Performance ethnography can be used politically, to incite others to moral action. Performance ethnography strengthens a commitment to a civic-minded discourse, a kind of performative citizenship advocated by Zora Neale Hurston, Dwight Conquergood, Soyini Madison, Bella Pollock, and others. Performance ethnography is a way for critical scholars to make sense of this historical movement, a form of action that helps us imagine radically free utopian spaces.

Narrative Ethnography

Barbara Tedlock (Chapter 19) reminds us that ethnography "involves an ongoing attempt to place specific encounters, events, and understandings into a fuller, more meaningful context." Tedlock shows how participant observation has become the observation of participation. As a consequence, the doing, framing, representation, and reading of ethnography have been dramatically changed in the last two decades. The fields of passionate, narrative, evocative, gonzo ethnography and autoethnography have emerged out of this discourse.

Tedlock observes that early anthropology in the United States included a tradition of social criticism and public engagement. Franz Boas, Ruth Benedict, and Margaret Mead shaped public opinion through their social criticisms and their calls for public and political action. By the mid 1960s, the term critical anthropology gained force in the context of the civil rights movement and growing opposition to the Vietnam War. Critical theory in anthropology was put into practice through the production of plays. An indigenous political theater based on the works of Bertolt Brecht, Augusto Boal, Paulo Freire, and others gained force in Latin America, Africa, and elsewhere.

Victor and Edith Turner, and Edward Bruner, developed performance ethnography in the 1980s. Culture was seen as a performance, and interpretation was performative. Ethnodrama and public ethnography emerged as vehicles for addressing social issues. Public ethnography is a discourse that engages with critical issues of the time. It is an extension of critical anthropology. In the late 1990s, under the editorship of Barbara and Dennis Tedlock, the *American Anthropologist* began to publish politically engaged essays. Tedlock observes that "within this politically engaged environment, social science projects serve the communities in which they are carried out, rather than serving external communities of educators, policymakers, military personnel and financiers." Thus does public ethnography take up issues of social justice.

Today, we inhabit a space of braided narrative, double consciousness, performance, creative nonfiction, history, drama, and magical realism, memories forgotten, recaptured, "overtake us as spiders weaving the dreamcatchers of our lives." Amen.

Analyzing Interpretive Practice

In Chapter 20, James Holstein and Jaber Gubrium extend their more than two-decade-long constructivist project offering a new language of qualitative research that builds on ethnomethodology, conversational analysis, institutional studies of local culture, and Foucault's critical approach to history and discourse analysis. Their chapter masterfully captures a developing consensus in the interpretive community. This consensus seeks to show how social constructionist approaches can be profitably combined with poststructuralist discourse analysis (Foucault), and the situated study of meaning and order as local, social accomplishments.

Holstein and Gubrium draw attention to the interpretive narrative procedures and practices that give structure and meaning to everyday life. These reflexive practices are both the topic of, and the resources for, qualitative inquiry. Knowledge is always local, situated in a local culture, and embedded in organizational and interactional sites. Everyday stereotypes and ideologies, including understandings about race, class, and gender, are enacted in these sites. The systems of power, what Dorothy Smith (1993) calls the ruling apparatuses, and relations of ruling in society, are played out in these sites. Holstein and Gubrium build on Smith's project, elaborating a critical theory of discourse and social structure. Holstein and Gubrium then show how reflexive discourse and discursive practices transform the processes of analytic and critical bracketing. Such practices make the foundations of local social order visible. This emphasis on constructivist analytics, interpretive resources, and local resources enlivens and dramatically extends the reflexive turn in qualitative research. With this apparatus, we can move to dismantle and contest oppressive realities that threaten to derail social justice initiatives.

Grounded Theory

Kathy Charmaz (Chapter 21) is a leading exponent of the constructivist approach to grounded theory. Grounded theory is a method of qualitative inquiry "in which data collection and analysis reciprocally inform each other through an emergent iterative process." The term "grounded theory" refers to a theory developed from successive conceptual analyses of empirical materials. Charmaz shows how grounded theory methods offer rich possibilities for advancing qualitative justice research in the 21st century. Grounded theorists have the tools to describe and go beyond situations of social justice. They can offer interpretations and analyses about the conditions under which injustice develops, changes, or is maintained. They can enact an explicit value stance and agenda for change. Some focus on a social justice issue because it illuminates a theoretical problem. Those who explicitly identify as social justice researchers use words like should and ought.

Charmaz suggests that grounded theory, in its essential form, consists of systematic inductive guidelines for collecting and analyzing empirical materials to build middle-range theoretical frameworks that explain collected empirical materials. Her chapter outlines the history of this approach, from the early work of Glaser and Strauss, to its transformations in more recent statements by Glaser, Strauss, and Corbin. She contrasts the positivist-objectivist positions of Glaser, Strauss, and Corbin with her own more interpretive constructivist approach, which stakes out a middle ground between postmodernism and positivism. Grounded theory may be the most widely employed interpretive strategy in the social sciences today. It gives the researcher a specific set of steps to follow that are closely aligned with the canons of "good science." But on this point Charmaz is clear: It is possible to use grounded theory without embracing earlier proponents' positivist leanings (a position long adopted by Guba and Lincoln [see Chapter 6 in this volume, and Lincoln & Guba, 1985]). She notes that grounded theory mixed methods projects are increasing in fields such as education and health.

Charmaz reviews the basic strategies used by grounded theorists. She moves these strategies into the space of social justice inquiry. She offers key criteria, basic questions that can be asked of any grounded theory study of social justice. Does a study exhibit credibility, and originality? Does it have resonance—is it connected to the worlds of lived experience? Is it useful? Can it be used by people in their everyday worlds? Does it contribute to a better society? With these criteria, she reclaims the social justice tradition of the early Chicago School, while moving grounded theory firmly into this new century. Her constructivist grounded theory is consistent with a symbolic interactionist pragmatism. Constructive grounded theory will be a method for the 21st century.

▣ IN THE NAME OF HUMAN RIGHTS

Leading South Afrikaner scholar and poet Antjie Krog (Chapter 22) conducted two years of radio interviews and reportage for the South African Truth and Reconciliation

Commission (TRC). Her essay—humorous, autobiographical, painful—opens with a 100-year-old account of a young widow's family story constructed on the basis of human footprints around a water hole. There are several voices in the story: She asks who is the scholar here? Who is raw material? Is it Bleek, the recorder of the original narration? Is it Liebenberg, the scholar of the tracking? Is it Krog, the author of this chapter? Is it the Bushman narrator? Is it the woman in the story? Who has the right to tell this story? Who has the right to enter into this discourse? How does the subaltern speak?

Krog playfully recounts her experience with an academic administrator who told her she was raw material, not a scholar. She then discusses the story of Mrs. Konile, whose TRC testimony was first read as incoherent raw material. It is not that the subaltern cannot speak. They cannot be heard by the privileged of either the first or third worlds. We have a duty to listen and to act, to hear *testimonios* as cries to be heard.

Riffs on Participatory Action Research

Mary Brydon-Miller, Michael Kral, Patricia Maguire, Susan Noffke, and Anu Sabhlok (Chapter 23) contend that participatory action research (PAR) combines theory and practice in a participatory way. It presumes that knowledge generation is a collaborative process. "Each participant's diverse experiences and skills are critical to the outcome of the work" (p. 387). The goal of PAR is to solve concrete community problems by engaging community participants in the inquiry process. PAR is like jazz improvisation—seemingly effortless and spontaneous, but in actuality the result of "rigorous training" (p. 387). PAR is also like the banyan tree—a gathering place of common people, a place of community discussion and decision making. PAR, like the banyan tree, provides a space and place where community members can come together, a place for discussion, dialogue.

Brydon-Miller and colleagues review the several different traditions and histories of PAR, noting that much of the early development of PAR took place outside of traditional academic settings in the "south," or third world. The history is dense, ranging from Paulo Freire's critical pedagogy project in Brazil, to Fals Borda's initiatives in Latin America, the Scandinavian folkschool movement, participatory action networks in Asia, and Australia (Stephen Kemmis and Robin McTaggart), the global young people's initiatives of Michele Fine and associates, to the struggles of feminist, literacy, social justice, labor, civil rights, and academic advocates. Traditionally, PAR challenges the distinction between theory and method. Strategies for collecting, analyzing, understanding, and distributing empirical materials cannot be separated from epistemology, social theory, or ethical stances.

The authors ground their reading of PAR in three sites: Sabhlok's dissertation research with the Self Employed Women's Association in Gujarat, a state in India; Alicia Fitzpatrick's work with American Indian nation youth in the southwestern

United States; and Michael Kral's Inuit suicide prevention and reclamation project in Nunavut, Canada. Each case study demonstrates the power of PAR to "challenge and unsettle existing structures of power and privilege, to provide opportunities for those least often heard to share their knowledge and wisdom, and for people to work together to bring about positive social change and to create more just and equitable political and social systems" (p. 396).

Qualitative Health Research

Janice Morse (Chapter 24) observes that because of its subject matter qualitative health care research is distinct from other forms of qualitative inquiry. Qualitative health researchers deal with serious quality-of-life issues, as well as life-and-death situations. Morse considers the origins, history, content, and scope of qualitative health research. She offers a content analysis of all articles published in 2009 in *Qualitative Health Research*. She shows how qualitative methods can be adapted for use in clinical settings, and ends by making the case that qualitative health research is an important disciplinary subfield in its own right. The classics in the field—*Boys in White*; *Asylums*; *Awareness of Dying*; *Good Days, Bad Days*—are foundational. Their influence extends far beyond the field of health care research.

In 1997, Morse established the International Institute of Qualitative Methodology (IIQM) and *Qualitative Health Research*, a Sage monthly international journal. IIQM soon established links to 115 universities, through hubs in eight international sites. Today, there is a global network of mentors, sponsor journals, and congresses supporting qualitative health research.

Morse examines the many quandaries involved in health care inquiry, ranging from issues with IRBs, to problems with informed consent, studying ill, dying, and diseased persons, and medical staff fears of evaluation. She proposes some research strategies that work to make clinical research possible, ranging from retrospective interviews, to involving caregivers as coinvestigators. She concludes, "Qualitative health research is a specialized form of qualitative research . . . [with] its own needs for education, training, methods, and dissemination of knowledge." We agree.

▣ CONCLUSION

Together, the chapters in Part III show how qualitative research can be used as a tool to create social change and advance social justice initiatives. Once the previously silenced are heard, they can then speak for themselves as agents of social change. Research is connected to political action, systems of language and meaning are changed, and paradigms are challenged. How to interpret these voices is the topic of Part IV of the *Handbook*. In the meantime, listen to the voices in Part III; these are calls to action.

▣ NOTES

1. Mitch Allen's comments have significantly shaped our treatment of the relationship between paradigms and research designs.

2. Denzin's call for triangulation involved combining multiple qualitative methodologies—life story, case study, interviewing, participant observation, ethnography. It did not include combining qualitative and quantitative methodologies.

▣ REFERENCES

Denzin, N. K. (1970). *The research act in sociology.* London: Butterworths.

Denzin, N. K. (1993). *The alcoholic society: Addiction and recovery of self.* New Brunswick, NJ: Transaction.

Denzin, N. K. (1999). Cybertalk and the method of instances. In S. Jones (Ed.), *Doing Internet research: Critical issues and methods for examining the net* (pp. 107–126). Thousand Oaks, CA: Sage.

Denzin, N. K. (2010). *The qualitative manifesto.* Walnut Creek, CA: Left Coast Press.

Fiske, J. (1994). Audiencing: Cultural practice and cultural studies. In N. K. Denzin & Y. S. Lincoln (Eds.), *Handbook of qualitative research* (pp. 189–198). Thousand Oaks, CA: Sage.

Hess-Biber, S. N., & Leavy, P. (Eds.). (2008). *Handbook of emergent methods.* New York: Guilford.

Howe, K. R. (1988). Against the quantitative-qualitative incompatibility thesis, or dogmas die hard. *Educational Researcher, 17*(8), 10–16.

Janesick, V. J. (2000). The choreography of qualitative research design. In N. K. Denzin & Y. S. Lincoln (Eds.), *Handbook of qualitative research* (2nd ed., pp. 379–399). Thousand Oaks, CA: Sage.

Janesick, V. J. (2010). *"Stretching" exercises for qualitative researchers* (3rd ed.). Thousand Oaks, CA: Sage.

Lincoln, Y. (2010). What a long, strange trip it's been . . . : Twenty-five years of qualitative and new paradigm research. *Qualitative Inquiry, 16*(1), 3–9.

Lincoln, Y. S., & Guba, E. G. (1985). *Naturalistic inquiry.* Beverly Hills, CA: Sage.

Maxcy, S. J. (2003). Pragmatic threads in mixed methods research in the social sciences: The search for multiple modes of inquiry and the end of the philosophy of formalism. In A. Tashakkori & C. Teddlie (Eds.), *Handbook of mixed methods in social & behavioral research* (pp. 51–90). Thousand Oaks, CA: Sage.

Mills, C. W. (1959). *The sociological imagination.* New York: Oxford University Press.

Peirce, C. S. (1905). What pragmatism is. *The Monist, 15*(2), 161–181.

Psathas, G. (1995). *Conversation analysis.* Thousand Oaks, CA: Sage.

Sartre, J.-P. (1981). *The family idiot: Gustave Flaubert, 1821–1857* (Vol. 1).Chicago: University of Chicago Press.

Schwandt, T. A. (2007). *The SAGE dictionary of qualitative inquiry* (3rd ed.). Thousand Oaks, CA: Sage.

Smith, D. E. (1993). High noon in textland: A critique of Clough. *Sociological Quarterly, 34,* 183–192.

Stake, R. E. (2005). Qualitative case studies. In N. K. Denzin & Y. S. Lincoln (Eds.), *The SAGE handbook of qualitative research* (3rd ed., pp. 443–466). Thousand Oaks, CA: Sage.

14

THE POLITICS AND PRACTICES OF FUNDING QUALITATIVE INQUIRY

Messages About Messages About Messages . . .

Julianne Cheek

Dear Reader,

This chapter is a series of reflections about how it is, and might be, possible to fund qualitative inquiry. Put another way, it is about how to sell our research ideas and research expertise in order to gain the resources we need to be able to do our research. For, no matter what type of qualitative inquiry we do, we are all selling, be it selling our time and/or our labor or/and our research projects. While speaking to issues that in some way touch and affect all qualitative inquirers, the discussion is primarily directed to readers who are perhaps a little unsure of how to begin funding their qualitative inquiry and/or navigate the research marketplace. My hope is that the chapter will provide this group of readers with both practical information that they can use to help them gain funding, as well the impetus to continue the ongoing intellectual work and reflection that thinking about the politics and practices of funding qualitative inquiry demands of us all.

At this point, some readers may be thinking that a discussion about funding qualitative inquiry is not relevant to them. If you are one of these readers, then perhaps you are thinking that the type of research you are interested in does not need large amounts of funding or interaction with funders, it really only needs your time. My response to this is that implicit within this type of thought is the assumption that funding for qualitative research is synonymous with, or limited to, support from sources removed from researchers themselves, most often thought of in terms of monetary support. Also implicit is the assumption that the researcher's time is somehow not funding for research. This overlooks the fact that funding for research comes in many different forms.

For example, funding can be in the form of money for salaries to employ research staff, bring on research students, or even release chief investigators from other duties in order to be able to conduct the research. Support can also come in the form of money for the purchase of equipment such as data recorders and computers, or it can involve in-kind (nonmonetary) support where the equipment is either supplied or loaned for the duration of the project. Monetary, or in some cases in-kind, funding can also be gained to enable necessary travel or access to specialist skills and services such as translation expertise. Other forms of funding for qualitative inquiry can come either directly or indirectly from institutional incentive schemes to "reward" researchers who attract research income for that institution through their grants, publications, and successful research student completions. Such rewards can take the form of, for example, infrastructure grants to establish research centers within that institution, an increased percentage of workload allocation for research, and/or allocation of modest discretionary funds for approved research-related purposes such as conference attendance or research assistant hours.

Morse (2002) points out that it is a myth that qualitative research is cheap, or cheaper than other types of research, to do. It seems to me that this myth has arisen and flourished throughout much of qualitative research's history both without, and also interestingly within, the qualitative research community because of limited notions of funding and what constitutes funding for research. All researchers, regardless of their methodological or substantive interests, require some way of funding their research. This is true of any study, even those qualitative studies largely done by researchers working in their "own time" and not requiring large pieces of equipment, specialized workplaces/laboratories, and/or expensive consumables such as chemicals. It is this allocation of paid work time and/or donation of discretionary private or unpaid time to the study that is in effect funding the research. Thus, I think that funding our qualitative inquiry is a question that concerns all of us.

There were many forms that a chapter about funding and qualitative inquiry could take. For example, one possibility was to focus attention on "how to write a winning proposal" for funding. But this runs the risk of limiting the discussion to one about writing for funders, and often very specific types of funders running large research-granting programs. Funding, as we have seen, can take many forms. It also runs the risk of reducing the discussion to an instrumental "how to" one. Another possibility was to focus on the political dimensions affecting funding for qualitative inquiry, looking, for example, at how we can position our research in relation to a politics of evidence that has been so pervasive for the past decade. But this has already been written about, and written about well, in ongoing and vigorous calls for qualitative inquiry to resist the excesses of a politics of evidence with normalizing, exclusionary, and positivist tendencies (Denzin & Giardina, 2008; Holmes, Murray, Perron, & Rail, 2006). Further, a possible danger of such a focus is limiting the discussion of funding and qualitative inquiry to one that is always in reaction or relation to this politics, a politics imposed on qualitative inquiry from without. This may actually have the effect of sustaining the centrality of that politics in defining the parameters of the discussions we both do have, and importantly might have, about funding and qualitative inquiry. In this way, the politics of evidence can become a distractor displacing qualitative inquiry, funding, and even evidence as the focus (Long, 2010; Morse, 2006a).

Given this, how have I decided to structure the discussion? The explorations and reflections to follow about how it is, and might be, possible to fund qualitative inquiry, focus on two different, yet interconnected areas. The first area is what I have decided to call the *practices* involved in the seeking, gaining, and accepting of funding. It is about the practicalities of funding qualitative inquiry such as where funding might come from, the forms that funding might take, and strategies in locating appropriate funding for a particular piece of research. It is also about crafting a proposal for funding, and what it involves. This includes decisions about whom and what will be studied, as well as when and how and why. But equally, the discussion to follow is about how these decisions can be affected by the politics of funding and the politics of research. It is this *politics* that is the second area of focus in the chapter. It has been my experience that practices associated with locating and obtaining funding for qualitative inquiry cannot be seen apart from the politics that affects them at every point in both the research and funding process. Where a qualitative inquirer chooses to seek funding from, how they go about locating and applying for funding, and

what they do when they receive funding, is as much a political discussion as it is a practical one.

What is this politics that I am talking about? The form of political thought that has dominated the thinking of many Western governments and administrators in the past decades, is one derived from neoliberal understandings. This is a politics promoting competition, efficiency, quality, notions of the marketplace, and audit-derived understandings of accountability (Cheek, 2005; Kvale, 2008; Torres, 2002). It is a politics in which one type of research evidence is privileged over another. Implicit within it are "contested notions of evidence. It is both about the way we do our research (the methods or procedures used to produce the evidence) and concomitant ways of thinking about what form(s) evidence must take in order for it to be considered valid, acceptable and therefore of use" (Cheek, 2008, p. 20). For as Denzin (2009, p. 142), drawing on Morse (2006a), points out, it "is not a question of evidence or no evidence." Instead, it is about who can say that something is evidence and something else is not. This shifts the emphasis from evidence per se to the *politics* of evidence, which is about the power that enables one type of research finding to be deemed as evidence and another not. This is a politics that has percolated down to the level of many funders of research, affecting the way that funding for qualitative inquiry is thought about and allocated (Cheek, 2005, 2006; Hammersley, 2005; Morse, 2006b; Stronach, 2006).

Thinking about it, few researchers have remained immune from the effects of the infiltration of this politics into the research arena. For example, at a macro political level qualitative researchers have seen the proliferation of endless forms of government-driven reviews and audits in the quest to ensure, and provide "evidence" of, research efficiency, output, and quality (Cannella & Lincoln, 2004; Denzin, 2009). At a more micro level of research practices, limited understandings of research and research evidence have, and continue to, affect the types of methods used, able to be used, and/or even called methods (Atkinson & Delamont, 2006; Morse, 2006b; Torrance, 2006). As a result, over the past decade, there has been ongoing tension, collision, and at times rupture, at the interfaces of qualitative inquiry and a politics of evidence (Denzin & Giardina, 2006, 2007a, 2008, 2009). These tensions, collisions, and ruptures have spilled over to, and impacted, ways we think about, and do, fund our qualitative inquiry. Given all this, I believe that exploring the interfaces between the politics and practices of funding our qualitative inquiry is absolutely crucial, not only for trying to better understand how best to fund our inquiry, but also because this interface has the potential to shape the form that our inquiry does, and even can, take.

How can this be? How can the interfaces between the politics and practices of funding qualitative inquiry shape the form that our inquiry takes, or might take? An example of what I am thinking about is if we were to choose a particular substantive area for study due to a perception that this will make it more likely to get funded. So, if, for example, a funder has a priority area to which funding will be allocated, then it may be that substantive area we tailor our research to and not another. Or, thinking about methodology, if a funder prefers one type of research approach over another, then we write our research design to fit that preferred approach. Of course, writing within and to the parameters set by funders may not necessarily be a "bad" thing. Indeed, some might call it just common sense. However, it does raise questions about what is driving what—is the funding driving the research, or the research the funding, or something in between? To some extent, this has the potential to turn the original intent of funding, which was as an enabler or support for research, on its head. The means of support (the funding) becomes as much, or even more, the goal as the research to be supported (the research problem). What the researcher might in fact be doing is buying the funding as much as they are selling the research project.

All of the above raises a number of complex issues for qualitative researchers to think about, and at times difficult choices to make, related to the practices and politics affecting the way they fund their qualitative inquiry. For me, chief among these is constantly reflecting on what is my prime motivation—enabling the research or gaining the funding? Am I selling research or buying funding or something in between? How far am I prepared to modify my research in order to get funding for it and, importantly, what is the reason for my decision making in this regard? How is it possible to find ways to fit the politics and funding to qualitative inquiry, and not qualitative inquiry to the politics and funding? What am I prepared to give up, change, or adapt in terms of my qualitative inquiry and what am I not? Many qualitative inquirers have had to make a number of tradeoffs in relation to funding their qualitative inquiry when faced with questions such as these. Talking about such tradeoffs can be confronting and difficult. Perhaps this is why there is not much writing about them. Yet what we *don't* say about what we do can be just as, if not more, instructive and important to consider as what we actually *do* say about what we do (Cheek, 2010a).

Thinking about all this, I came to the conclusion that what I wanted to explore was how it might be possible to be pragmatic and realistic about funding qualitative inquiry in a world where there are demands on most of us to produce research "products," yet at the same time feel like we have some control over what those products are and what they might be. It seemed crucial to me that this discussion must speak, and relate directly, to the everyday reality we live and work in. How can we as qualitative inquirers both thrive and survive (in that order) in this reality? This is a particularly important question for less experienced and beginning researchers, especially those who are seeking their first research or academic post, or who are yet to gain tenure. How do you grapple with, and even accommodate, demands

of administrators, tenure and search committees, research agencies, and the research marketplace without the feeling of selling out either in terms of oneself or the type of qualitative inquiry one employs? All of these questions are central and important dimensions of exploring the ideas about ideas about ideas, and messages about messages about messages (Bochner, 2009), related to funding qualitative research. A starting point for unpacking and exposing this cascade of ideas and messages pertains to locating funding for qualitative inquiry, and it is to this that our discussion now turns.

▣ LOCATING FUNDING: WHERE TO BEGIN AND SOME HARD QUESTIONS TO ASK YOURSELF WHEN DOING SO

How and where do you begin when trying to locate funding for your qualitative research? Of course, the answer to this depends to a large extent on the research problem you wish to investigate, the type of research you want to do, and what type of funding you need. But in all of this there is a key, immutable point. This point is that, regardless of what strategies you may employ when thinking about how to fund your qualitative inquiry, the first step toward attracting that funding is the development of a well-thought-out research problem and concomitant research design. You must have a very clear idea of what you want to gain funding for, or put another way, what you want to research, why, and how. There are already excellent discussions published pertaining to qualitative research design, one of which is the next chapter in this handbook (Creswell, Chapter 15, this volume). I will not repeat those discussions here. Instead, I want to emphasize that it is only after you are sure that you have spent the time and energy required on the plain hard thinking that Morse (2008) reminds us sits behind the deceptive simplicity of the design of an elegant qualitative inquiry, that you are ready to take the next step in locating funding for that research.

This next step involves not only ascertaining what types of funding are available, but also the way those funding sources understand and think about research. Are they going to be interested in qualitative projects? Do they have particular emphases or priorities in terms of the substantive issues or areas that they will fund? One problem faced by qualitative inquirers in trying to find out this type of information about potential funders and funding schemes is that qualitative inquiry has not been particularly active in, or good at, building up collective knowledge about either funding sources or the experience of interacting with them. There is no collective list of potential funders, for example, nor is there systematic collection of information about the experiences of qualitative researchers who either have attracted funding from a particular source, or who sit on review panels of funding agencies. As a result, much collective wisdom is lost.

Nevertheless, strategies do exist that can be employed in the quest to locate possible funding sources for qualitative inquiry.

A beginning strategy might be to use formal networks, such as meetings like the Congress of Qualitative Inquiry held annually at the University of Illinois, and/or informal networks of qualitative inquirers to find out from where funding has been gained for qualitative inquiry, in what form, and for what type of research project. Just as important, also try to find out where funding has not been awarded to qualitative inquiries and the reasons why. This will give you a clue as to possible "unfriendly" funders for qualitative inquiry and may save you much time and energy from the outset. Another strategy might be to access and use databases and/or publications that list potential funding sources, the frequency of their call for applications seeking funding, and the success rates of applications in the respective schemes. These are often country-specific such as the GrantSearch electronic database in Australia that is published in hard copy every two years as the GrantSearch Register of Australian Funding. It advertises that it has over 3,000 entries identifying funding sources for study, travel, research, business and professional development, the arts, sport and recreation, and community groups offered by all levels of government, universities, foundations, the private sector, and overseas sources specifying Australian applicants (GrantSearch Register of Australian Funding, 2010). Of course, the relative importance of, and opportunities for, funding for qualitative inquiry afforded by each of the above categories of potential funding sources will vary from country to country and even within countries.

It is most important to bear in mind when trying to locate funding for qualitative inquiry that accepting funding from a sponsor "links the researcher and research inexorably with the values of that funder" (Cheek, 2005, p. 400). Consequently, in the search for potential funders, at some point it will be necessary to reflect on whether or not funding should be sought from a particular funding source. For example, most researchers will hesitate if offered funding from tobacco companies. For some qualitative researchers working in health-related areas, whether to seek part of the large sums of funds often available from pharmaceutical drug companies is a vexing question creating fine lines to be traversed. And what of seeking support from companies whose profits are derived in some way from outsourcing labor offshore in developing countries where people work for very little and barely enough to survive? Does a funder's environmental record matter? In considering such questions, the politics of funding affecting qualitative inquirers comes from another position—the politics that comes with the funder. Taking up a position in respect to this politics is something that all qualitative inquirers will have to reflect on when working out their strategy to locate funding. Further, some places where qualitative inquirers are employed may have policies about where funding cannot be accepted from. It is important to find this out.

Once potential funding sources have been identified, the search for funding can be refined further with respect to the chances of having a particular qualitative inquiry supported by a particular funder. Sometimes funders have specific substantive areas that they will fund, or priority areas for funding that may change annually. Your project will have a much better chance of being funded if it is in keeping with these areas of interest identified by the funder. Thus, it is important to find out if a funder has priority or interest areas, and if they do, what these areas are. Of course, this can raise dilemmas for you as well in terms of what you are prepared to trade off with respect to your research problem and design in order to "fit" a priority area. You will have to make your own decisions about this. If I am completely honest, at times some of my research has been "fitted" to a particular funder's requirements as seeking and gaining the funding has been my prime motivation. Thinking about it, this has been for a number of reasons. One is the constant pressure that I have experienced over the course of my career, like many other researchers, to attract funding for my research. Another reason is the harsh reality that funds for research are not easy to get and there are not a lot of them. Given this, I have grappled with whether I needed to be pragmatic about all this, and if so how pragmatic was I prepared to be? Was it better to try to fund exactly the project I wanted to do, or could I live with researching something else because the funding was available to do that project and not the one I had originally intended to do? Was it better to be doing some research in an area or none at all, which may have been the outcome if I did not shape my research to fit with the funding call or the funder's agenda? There are no easy answers to these questions. Nor is it a matter of right or wrong. However, each potential position that can be adopted in this regard has a cost of some kind. How much modification we are prepared to make to our research problem and why we would do that are questions each of us as qualitative inquirers will need to think about.

Finding out what a particular funder has previously funded is another way of refining the search for potential funding and funders for qualitative inquiry. Have there been any qualitative projects funded and if so what type of qualitative inquiry has been supported? Try to get hold of any funding guidelines, application forms, or other relevant documents such as annual reports detailing funding or research activities published by the funder. This is because "shaping all application forms or guidelines provided by funders are assumptions, often unwritten and unspoken, about research and the way that research is understood" (Cheek, 2005, p. 394). If it is a struggle to fit the qualitative inquiry around, or into, the sections and headings that are required as part of the funding application and/or there is no record of funding qualitative inquiries, then this strongly suggests that qualitative inquiry may not be something that has been part of the thinking of the funder when constructing the funding scheme. At the very least, this indicates that more needs to be found out about this funder's position with respect to qualitative inquiry.

The composition of the panel that will review the applications and the process that they will follow can provide further useful information in this regard. Is there anyone on the review panel with expertise and experience in qualitative inquiry? And if the answer to this is a "yes," then what exactly does that mean? For "yes" can mean qualitative inquirers with extensive experience, or it might mean those with an "acquaintance" with qualitative research, having "attended a short session on qualitative methods at a conference—and they use these isolated 'facts' as gold standards" (Morse, 2003a, p. 740). If, on the other hand, the answer is a "no," then is there the possibility for the panel to seek advice from external reviewers for applications outside of the panel's collective expertise? For if there are no persons on the review board with qualitative expertise, and the standard procedure is either not to, or not be able to, seek and/or take into account opinion from outside the panel, then this does not augur well for the likely success of the proposal. If expert external opinion is sought, then how that expert advice is sought is equally important to consider. The way this expertise is sought can often be very ad hoc (Morse 2003a).

Being entirely honest about something that maybe we often do not say about what we do, I have at times weighed in my mind whether having external expert opinion, even if it is positive about the proposed research, really changes anything anyway. For example, I have sat on granting and review panels where 0.5 or less of a point in a score out of a possible 100 points has separated those applications that will be funded from those that will not. Every panel member's score awarded to a grant is therefore crucial. Even if only one member has concerns, spoken or unspoken, about supporting qualitative inquiry, that member's score can in effect make an application uncompetitive in the funding competition where success rates may be lower than 20%. Thus, the traditional practice of averaging the scores from panel members for a particular application often works against the chances of a qualitative inquiry being funded.

If, after considering the above facets of the practices and processes followed by a funder, it seems as though there is a real opportunity for qualitative inquiry to be funded, then a next step in locating funding can be to approach the funding agency in person. This can be done in different ways and for different purposes depending on a particular sponsor's way of organizing the allocation of funding. If the allocation of funding is organized on the basis of annual granting rounds, then there is little point to meet to ask for funding outside of those funding rounds. In addition, to do so may give the impression that you have not done your homework with respect to this funder's way of operating. In this instance, you may seek to talk to the office and/or manager with responsibility for the oversight of the funding rounds. This is not only to acquire information about that process, but also to introduce both the researcher(s) and the idea for the research to the people who will be dealing with the application on an administrative basis. This enables them to

put a face to a name and demonstrates that you are serious both about your idea, and having a working partnership with them with respect to this research. If a potential sponsor has no set dates for funding rounds but applications can be made at any time for funding, or it is not entirely clear how to apply for funding, then it is important to find the appropriate person in the organization to talk to about your research idea and you as a researcher. This can be difficult, so it is important to persist politely when trying to gain entry to the organization and navigate the various levels of contact that are usually gone through before eventually finding the right person to talk to.

If you are successful in getting a meeting time with that person, make sure that you are very well prepared thereby respecting the time that they have given you, as well as making the most of it in terms of putting across your research idea. Try to think about the research from their point of view. Why should they fund this? What do you offer them? The key is to be clear and precise in the message that you are trying to convey and the research that you are trying to sell. Offer to send a one-page concept paper to them before the meeting that outlines the research idea, the type of support being sought, and any preliminary work that has been done. The concept paper should also have information about you and any other researchers that constitute the research team for the project. This is because track record is an important consideration in a potential funder's mind. Avoid submitting a standard format, multipage, and multipurpose curriculum vitae (CV) that may or may not be read by the person. Rather, present a track record in the form of a modified CV with accompanying text that is written in such a way to draw attention to, and highlight, aspects of your research career that demonstrate that, based on your past experience, you have the expertise to do this research, on time, and within budget (Cheek, 2005). If you do not have a lot of research experience, then a starting point can be to join a research team where some members do have this experience. You can then build up your own track record while researching with, and learning from, them as part of the team. Another strategy for building up a track record is to look for seed funding opportunities directed to less-experienced, early-career researchers. In these schemes, small amounts of funding are made available to less-experienced researchers for research projects, thereby helping them overcome the eternal issue highlighted here: how to get a track record without having a track record!

▣ LOCATING FUNDING: CONTRACTING
TO DO A PIECE OF RESEARCH

A different strategy when trying to locate potential funding for qualitative inquiry is to look for calls or advertisements from government or other organizations seeking to pay researchers to do a specific piece of research. This specificity may pertain to any or all of the substantive focus of the research, the aspects of that substantive focus to be picked up on and how, as well as the time frame in which the research must be done, written up, and the results "delivered." To varying degrees, and not always, there may be some scope for the researcher to develop the focus and methods that will underpin the research. Advertisements for this type of research funding can often be found in the contract/ tender section of newspapers. They ask interested researchers and/or organizations to submit a proposal, on a commercial basis, for conducting the specific project. What the funder is looking for in tendering or contracting the research is to buy the expertise and time needed to do the research because they do not have it themselves. For example, tenders may seek researchers able to conduct a specific number of workshops, focus groups, and/or interviews in order to attain certain prespecified research aims. Here, the emphasis is on researchers fitting both themselves and their research to a specific, time-delimited funding opportunity.

An understandable part of the funder's thinking when devising and awarding a tender for research is the question of what is the least amount of money that they need to spend in order to get the expertise they need and the research done. This is not to suggest that the cheapest proposal will necessarily be successful as the expertise and experience of the researchers will be major considerations as will the overall research design. Rather, it is about getting value for money, which "means not only meeting high standards in the research; it also means considering how much, or little, money needs to be allocated to attain those standards" (Cheek, 2005, p. 394). What all of this means is that the budget will play a major part throughout the entire funding application process. It may also mean that the maximum amount of possible funding that has been set aside for this project makes it impossible to conduct a qualitative study that can meet the aims of the research required. It is easy to underestimate the amount of time needed for a qualitative project and associated project costs. It is also easy to fall into the trap of trying to be more competitive in the tender round by undercosting researcher expertise and project costs and/or overcommitting as to what it is possible to "deliver" in the time available to do the research. Help in costing projects and making sure that what is promised is realistic can be obtained from research and/ or business development offices increasingly found in universities and research organizations. These offices also have the remit to source and look out for calls for tendered research.

Tendered research is characterized by short time frames at every part of the research process. There may be only days to prepare an application, and just weeks or a few months to complete the research. Not all qualitative inquiry and approaches are suited to such short time frames. This is compounded if the funder requires interim reports throughout the research in that it is not always easy, or even possible, to "chunk" any interim findings into bits that can be delivered in a piecemeal fashion.

Time and energy put into such interim reporting can cause much frustration in that this diverts researcher time and focus from the bigger research picture. A danger of such emphasis on time delimited results is that it can encourage the rise of atheoretical, or at the very least shallow and thin, qualitative research designs and findings. Morse (2003b, p. 846) notes that not having enough time to do and think about our qualitative research "will kill a project or result in a project that has not become all that it could—and should—be." Her observation forces us to think carefully about what the short time frames for the delivery of the research findings might actually mean in practice. Stripped of the theoretical understandings that underpin them, such research designs can become little more than a collection of techniques or a series of steps to be followed.

For some qualitative researchers, gaining funding from contracted research has been part of everyday reality for many years. For other qualitative inquirers, tendered research has only relatively recently become an increasingly important income source as the contemporary research context has continued to see a decline in many countries in the relative amount of monetary and other support given by governments and other funders for researcher-generated projects. Qualitative inquirers thinking about tendering for research need to think deeply about how much of the control of the design they are prepared to give away or can live with giving away. How will they navigate the tensions arising from the entry of "fast capitalist texts" (Brennan, 2002, p. 2) into their qualitative inquiry with, for example, research products being transformed into "deliverables" and research timelines into "milestones"? This is not just a matter of semantics. It is an outworking of messages about messages within messages. One message is that research can be reduced to, or thought of, as a series of tasks. A message that then sits within this message is that once these tasks have been identified, the performance of the research team and the quality of the research can be assessed in relation to whether they have fulfilled these tasks.

For example, I was involved, as part of a team, in doing a contracted piece of research for the Australian government (Cheek et al., 2002). Part of the contract for the research outlined the tasks to be performed by the research team, as well as associated performance criteria that would be used as indicators of success in fulfilling those tasks. One of the tasks stipulated in the Standards and Best Practice section of the contract was to gain "informed consent of all participants." The associated performance indicator identified in the contract was "signed consent forms." Thus, gaining informed consent of all participants was in effect reduced to being synonymous with the production of signed consent forms. Our research team had a far wider view of the contested and complex issues surrounding what both informed consent, and the gaining of that consent, might mean in practice. Yet we signed this contract and attempted to navigate this potential tension by delivering what the funder required in terms of signed consent forms while also ensuring that consent and ethical considerations extended far beyond this. However, we never talked about this tension with the funder. Why was this? Perhaps, looking back, there was a pragmatic edge to what we traded off in terms of meeting the requirements imposed by the funding body while at the same time conducting the research in a way that we felt did not compromise the qualitative inquiry. This research addressed an important substantive area and the results were going to be used by government as part of initiatives being thought about in this area at the time. If we were not doing this research, then perhaps others would be who did not have the same problem with, for example, reducing gaining the informed consent of participants to a task and the production of signed consent forms.

There were, and are, no easy answers to these sorts of issues and tradeoffs. Nor are there right or wrong answers. Rather, it is a matter of thinking deeply about what we are prepared to give up or trade off and what we are not when selling ideas or time to funders. As I have mentioned before, as researchers we are all selling something all of the time. Thus the idea of selling is not an issue for me. However, what is an issue for me, and what I still constantly reflect on, is what I am trying to sell, how I am going about selling it, and most importantly why? Questions that all qualitative inquirers might usefully ask themselves in this regard include the following: Are we selling a research problem, our time, our expertise, or all of the above? How do we feel about selling our time and expertise to do what is, to some extent at least, someone else's research? What are we prepared to trade off in doing that research and what are we not? The importance of reflecting on these types of questions is one of the key messages in this chapter.

▣   Locating Funding: Yet More Considerations *Before* Submitting an Application for Funding

However, even having employed all, or some, of the above strategies in the quest to locate funding for qualitative inquiry, there is yet more homework to do before any application for funding should be made. In the excitement of the possibility of having found a potential funder, or the pressure of a deadline for research proposal or tender submissions, it is easy to overlook the question of what the funder's expectations will be if funding is gained. For example, to what extent will they want to be involved in the research process? What parts of it, if any, will they wish to have control of? Will they expect to have access to the data or parts of it? Will they want representatives on the research team, at team meetings, or to form advisory boards to give input to the form and direction of the research as it unfolds? How will this fit with the research design? How will it affect the researcher/funder/participant relationships in the research and does that matter?

It is just as crucial to consider these matters at the point of applying for funding as it is to focus on trying to write a winning proposal. Before rushing in, or being carried away by the tantalizing possibility of having located a funding source for your research, it is important to remember that applying for and accepting funding from any source is not a neutral activity (Cheek 2005). It involves entering into intellectual and contractual agreements with funding organizations and their agents, such as the manager assigned responsibility for ensuring that the research is completed and on time. Concentrating on winning the funding and worrying later about such details is a poor and dangerous strategy. There is little point in getting the funding if the research cannot proceed as planned because of different expectations on the part of the funder and the researcher as to their respective roles in the research process. All of this is not to paint a picture of doom and gloom. Rather, it is to caution against not thinking through the partnership and agreement you are about to enter into, and making sure that it is one that you want to be in.

The extent and formality of such agreements will vary from funder to funder. Not all funders will have formal expectations about their involvement in, or the way that, the research should be done. Nevertheless, no matter the degree of formality or even informality, it is essential that early on, and absolutely before funding is accepted, researchers and funders discuss, agree upon, and record any assumptions that they have about the research and how it will be done. This includes what can be said about the research and by whom. Not being able to write about and publish part of the research, or any of it at all, because of contractual issues is the uncomfortable place in which some qualitative researchers, including myself (Cheek, 2005), have found themselves. It is thus critical to negotiate how the findings of a study will, and can be, reported, when, and where; the use(s) to which the research data can be put and who owns that data; as well as what can be published in scholarly literature by the researcher(s). Making such agreements is central to building a positive workable relationship with funders built on mutual respect for each other's point of view. It is crucial that this open communication continues throughout the duration of the research project. This is particularly so if something goes wrong or if, for whatever reason, aspects of the research plan, such as the timelines for the research, have to be changed.

It is equally important that these issues have been discussed and negotiated among the research team itself so that there is an agreed position with respect to them *before* they are discussed with the funder. This discussion and negotiation is not always easy, as often there are competing views in the team that need to be navigated. However, one thing is for certain—these competing views will not go away by ignoring them. In addition, the team will need to discuss matters such as what support individual members will have access to; individual responsibilities in contributing to the research especially with respect to time frames and the final report; how often to meet; how disagreements among team members will be resolved; and, very importantly, who in the team will have contact with the funder and on what basis. This is critical to the smooth functioning of the team throughout the project. Having these things thought through, under control, and working well during the research will give the funder confidence in the team. It also makes the life of the funder much easier if they know whom to deal with on the team and the message that they are receiving about the research is clear and consistent.

There should also be agreement between funders and researchers as to what, and by whom, participants in the research will be told about the agreement between the funder and the researcher(s). Qualitative inquiry is built on building trust between participants in the research and those conducting the research. Central to this trust is being open and honest about who is funding the research and the agreements that have been entered into with that funder with respect to what can, and cannot, be said about that research and any findings. It is also about telling participants exactly who will have access to what data from the research, and how participant anonymity will be ensured. This is particularly important to allay any fears on the part of participants who may be in dependent positions in relation to the funder, such as, for example, the funder's tenants, suppliers, or employees. Such considerations are ethical ones as much as they are practical ones.

Thus, locating any type of funding for qualitative inquiry involves much more than simply locating funding sources. It involves thinking about how the research and researcher will be affected by seeking, and possibly gaining, that funding. In all funding agreements, to a greater or lesser extent, researchers will give away some of their freedom with respect to the project and the way it might be, and actually is, done. Making decisions about how much freedom they are prepared to give up in this regard is one of the fine lines (Cheek, 2008) that qualitative inquirers walk when navigating the funding and research marketplaces. Thinking about how to navigate these fine lines must be done as part of the search for potential funding sources, well before the research funding is gained, and absolutely before the funding has been accepted and the research is actually underway.

▣ WRITING TO ATTRACT FUNDING:
  CRAFTING A PROPOSAL FOR FUNDING

Having done all this, and having located a potential source or way of funding your research, the next step is crafting a proposal to seek that funding. The discussion that I want to have here is not about qualitative research design and/or how to write a proposal per se for a qualitative inquiry. Good examples of this type of discussion already exist (see, for example, Carey

& Swanson, 2003; Morse, 2003b; Penrod, 2003). Rather, the discussion is focused on exposing, in order to be better able to navigate, the tensions inherent in writing a funding proposal for multiple audiences sometimes with multiple understandings of what constitutes "good" research or even research. Or, to put it even more simply, how do you write a proposal for a number of different interested parties and still retain the project you want to do both in terms of intent and approach? Shaping and tailoring a research idea to a particular audience is a crucial part of the craft of writing for funding, a craft built on a set of skills that are largely learned and refined by practice.

When crafting the funding proposal, it will need to be congruent with the guidelines of a funder as well as meeting the requirements of other interested parties such as ethics committees that will need to approve the research before it can proceed. Thus, each application for funding, even for the same project, will vary depending on the characteristics and requirements of the audience being approached. It is not good strategy to work up a generic proposal that is then submitted to a number of funders for two reasons. First, because it may be that individuals serve on more than one review panel and therefore may have seen the proposal before, giving the impression of a nontargeted and recycled funding proposal. Second, when a proposal is written for a potential sponsor's consideration, it is written for a particular audience, whose members have assumptions and expectations of the form a proposal should take and the level of detail that it should contain. These will vary from funding scheme to funding scheme. Thus, as emphasized previously, it is important for researchers to know their audience(s) and its (their) expectations. Sometimes, it may even be possible to have access to research applications that have won funding thereby providing an excellent guide as to how much detail is expected in the proposal and the style of presentation expected by those reviewing the application.

An essential part of crafting the application involves following the guidelines. At the most basic of levels this means conforming to any word or page limits either for individual sections of the application, or the application as a whole. It also means addressing all sections or areas that the guidelines identify. For example, if the guidelines state that an application should comprise a 100-word summary, and then, in no more than 10 pages of text that uses 10-point font or larger, outline the research aims, outcomes, dissemination strategies, significance and novelty value in relation to existing work in the area, methods, ethical considerations, budget and budget justification, and provide a statement about the researcher(s) track record(s), then that is *exactly* what the application should do. A good strategy is to use each of the areas identified in the guidelines as needing to be addressed as headings for the different sections of the text that make up the application. This organizes and gives structure to your writing, as well as making it clear where and how you are addressing these areas. Often, when scoring applications, reviewers will have a sheet where they assign a score to each of these sections, so making it easy to see where and how you have addressed them makes good sense.

A good research proposal is one that is clear and concise. Use accessible language that can be read and understood by panel members who may not be working and/or researching in your exact area of expertise. Writing that is full of jargon will not impress. There must be a clear statement of the question/issue/problem that the research is addressing and everything that is written following that should be in terms of how it is related to addressing the question/issue/problem. Thus, for example, the description of the methods should be explicitly related back to how they enable the aims of the research to be met. Any literature review must move beyond description to be an argument of why the proposed research is needed in relation to both what has preceded it, and is happening currently, in the relevant substantive area. Show how the proposed research fits with the funder's priorities and funding remit. Make these connections throughout the discussion and document. Do not make the reader have to do the work, or expect that they will, as if they are a reviewer with many applications to read it is likely that they will not have the time to do this. Never underestimate the importance of parts of an application that may seem to be more routine or administrative, such as the budget.

It is important to bear in mind that when crafting your proposal, you are selling your research *and* your expertise to this funder. This is why the statement of the researcher's track record is so important. Publications are the hard currency of the track record needed for researchers to be able to compete for funding. Without a strong publishing track record, it will be much more difficult to win funding. Largely for matters of expediency in terms of ease of data collection and manipulation, publication track records are increasingly being assessed using quantitative metrics. The most pervasive of these is impact factor. Although related to impact of a specific journal (Cheek, Garnham, & Quan, 2006), increasingly the term "impact factor" is being used interchangeably with the term "impact" in the commonspeak of funders and university administrators. One effect of this is that in many funding schemes, impact factor is being used when assessing the quality and impact of the work of researchers. Thus, guidelines for applying for funding require the publication track record of researchers to list the impact factor of the journal that a particular journal article was published in. The erroneous assumption is that the higher the impact factor, the higher the scholarship and quality of the applicant's published work. Thus, a supposed measure of the impact of a *journal* is transformed into a measure of the impact and quality of a particular *article*. This in turn is transformed into measures of the impact and quality of a particular *researcher* (Cheek et al., 2006).

This poses considerable challenges for qualitative inquirers and creates messages about messages about messages about where, and why, to publish. Chief among them is, given the increasing emphasis placed on a metric-based, impact-factor-driven publishing track record when assessing applications for funding, should qualitative inquirers seek only to publish in "high-impact" journals? How, when presenting their publishing track record to funding bodies not familiar with qualitative inquiry, do qualitative inquirers demonstrate the impact of publishing in places such as this handbook, for example, which does not have an impact factor? What does opting in or out of trading in the currency developed around metric-based publication measures do to our chances of locating and gaining funding for our research? Where to publish their research and why is another of the fine lines all qualitative inquirers walk.

▣ WRITING OUR METHODS FOR FUNDING: WALKING FINE LINES AND MAKING TRADEOFFS WE CAN LIVE WITH

Part of writing for funding involves making decisions about how we will write our methods for particular audiences. This involves making decisions not only about what methods we might employ but also about what details we might, or might not, give about those methods. I am going to use three examples drawn from my own experience of writing my methods for funding to explore some of the tensions that this writing, and the decisions that underpinned it, created for me. The first of these is about fine lines I walked, and tradeoffs I made, when writing sample size into a qualitative funding proposal in response to needing to have a detailed and justified budget for the proposed project.

Example One: Deciding What to Say and/or Not to Say About Sample Size

The production of a well-thought-out and justified budget is an important part of crafting a research proposal for funding. It is quite reasonable for those who provide funding for research to ask whether or not the proposed project represents appropriate use of the funds for which they have responsibility. It is also reasonable for them to require a justification of the amount of funding being sought. Reasons need to be given for the amount and type of funding sought. Often this means a detailed level of precision and justification in the budget with respect to what exactly funding is being applied for and why. For example, here is a short excerpt from the budget justification section of a grant proposal I was part of developing:

A professional transcriber will be required to transcribe the audio taped interviews. It is estimated that every hour of taped data will take three hours to transcribe. We anticipate 50 interviews of each at least 1 hour, therefore 50 hours minimum of taped interviews plus allowance for 10 hours of extra time as experience has shown some interviews will go for longer. Thus 60 hours of tape x 3 transcribing hours per hour of tape recording = 180 hours of transcription @ $17.50 per hour (quoted by L. . . Supplies, a professional transcribing company). 180 hours x $17.50 = $3,150.

From this justification, it is clear that one of the costs associated with the interview component of the study is that of transcribing the interviews. Thus, this part of the budget is premised on multiplying the number of interviews to be done by the transcription costs associated with one interview to come up with an overall dollar figure.

While on one level this represents very clear crafting of a budget and its justification, on another it raised, and still raises, questions about the tradeoffs that were involved in deriving these numbers.

These questions center on whether it was possible to give precise details about the number of interviews that would be done before the research was underway. In keeping with the inductive approach of qualitative inquiry, we could not be sure of exactly how many participants we would interview in the study. However, costs associated with doing each interview, such as the transcription of the interview, were the major component of the funding being sought. How could we come up with a budget and justify it if a large part of the budget for the project reflected costs related to those interviews but we couldn't say how many we would do? After all, it was not unreasonable for the review committee to expect that we could provide justification for our budget.

After much thought, we decided that we would nominate a specific number of interviews to be conducted in the proposal even though we were not entirely comfortable in so doing. We nominated a sample size of 50 interviews. This number was based on the sample size used in a study reported in the research literature on which, in part, our study drew. We wrote in the proposal that our *anticipated* sample size of a total of 50 interviews (25 older people and 25 family members) was *guided* by that study. We were fortunate enough to gain funding for the project, and after we had completed the research we published the findings. In this article, we specified the sample size we had used (Cheek & Ballantyne, 2001). We then applied for funding for a related study that also involved interviewing older people and their families. When writing this proposal, we were able to refer to our published first study and state that drawing on our previous experience we anticipated that 25 interviews of older people and 25 interviews of family members would meet the needs of the study. We gained funding for this study, carried out the research, and published it. More studies followed and the results were also published (Cheek, Ballantyne, Byers, & Quan, 2007; Cheek, Ballantyne, Gilham, et al., 2006; Cheek, Ballantyne, Roder-Allen, & Jones, 2005). By this stage, we had provided ourselves with a way of overcoming the problem of how it is

possible to specify the number of interviews to be conducted when it is not possible to do so with absolute certainty. We could refer to our own work.

Throughout the entire process of doing this, we were employing tradeoffs. If we did not give a number with respect to the number of interviews to be conducted, then it was impossible to apply for funding. This was because we could not stipulate a budget, and also, if I am completely honest, because we knew that without a defined sample size it would be impossible to attract funding from the particular granting scheme we were applying to. Yet, if we did give a number then we may be complicit in giving the impression that it is possible to be certain about aspects of qualitative research design that it is not possible to be certain about. Using words in the proposal such as "it is anticipated" that 25 interviews will meet the needs of the study, and writing in our report of the research about why sometimes we did not end up having 25 interviews of older people and/or their family members but more or less (see Cheek & Ballantyne, 2001), were some of the ways we tried to navigate the tradeoffs we knew we were making. In addition, we found that interviewing 25 people from each group of participants (older people and family members) in these studies provided us with the richness of data and information that we sought. And we are optimistic that our results have influenced practice and attempted to address issues pertaining to social justice and human rights for these groups of people (Cheek, 2010a, 2010b; Cheek, Corlis, & Radoslovich, 2009). Nevertheless, there were very fine lines to walk when making these decisions, and no doubt there may be readers who feel as though we may have sold out a little (or a lot) in what we did.

Morse (2003a) also shares her reflections on the tradeoffs she made when needing to specify a sample size in order to be able to specify a budget. She does this using her experience of having had an application for funding rejected by a review committee. She explains that when crafting this proposal, in keeping with the tenets of qualitative inquiry underpinning her research design, she wrote, "sampling would continue until saturation is reached" (Morse, 2003a, p. 740). However, she then went on to specify an actual number for the sample largely because of the need to have a number in order to be able to develop a budget. The number chosen, and the reasons for providing it, drew on her long experience of doing qualitative inquiry. "Experience has taught me that you must calculate some number as the requested dollar amount; experience has also taught me that it is folly to minimize, rather than maximize, the sample size" (p. 740). However, in this instance, the tradeoff of giving of a number became a problem. Morse puts it this way: "It did not matter what the number was. My sin was to have produced an actual number: 'In qualitative research' they [the review committee] told me, 'the sample size cannot be predicted'" (p. 740). While there may be some merit in such a view, what in effect this did was reduce complex design issues to simplistic "rules" stripped of any theoretical understandings, while effectively dismissing years of experience of conducting qualitative inquiry. Further, in effect it made getting funding for qualitative inquiry nearly impossible, for if there could be no guide to the sample size in the study, then a budget for this research was impossible to develop. If no budget could be developed, then what would be applied for in terms of amount of funding sought?

Both Morse's and my own experiences highlight that when crafting an application qualitative inquirers face the dilemma of negotiating the pragmatics of trying to write a competitive proposal without selling out to expediency. Writing for funding is as much a political as it is a practical activity. Consequently, how and what to write, and equally what to leave out or not write, are much more than decisions about style or following sections in guidelines. Rather, they are decisions about what to defend and what to give up when crafting a proposal. What aspects of qualitative inquiry, both substantively and methodologically, can be/might be/must never be tailored to meet the understandings and requirements of audiences such as funders? These are questions that confront all qualitative inquirers. I wish to continue the conversation about these questions and the fine lines they involve in terms of the decisions made and paths chosen when writing for funding, using the example of writing for ethics committees.

Example Two: Writing for Ethics or "Getting Our Methods Through Ethics" to Get the Funding?

It may be a bit unusual or unexpected for some readers to think about writing for ethics committees (also referred to as Institutional Review Boards or IRBs in some countries), as part of a discussion about writing our methods for funding. But the fact is that, in some ways, ethics committees are often as much a part of the decision making about if a research project is able to attract funding as are formal research review panels. This is because most funders will require that research is approved by formally constituted ethics committees before funds can be given and/or released. Even if such approval is not a requirement of a particular funder, then, in Australia, for example, university-based researchers must obtain ethics approval from the relevant university ethics committee before their research can proceed. Thus, a critical part of the process of gaining funding for qualitative inquiry is the navigation of, and writing for, the formal review process of the ethics of the research.

On the surface, this seems something that all those committed to qualitative inquiry surely would support. Yet there has been an uneasy tension between the role and function of ethics committees on the one hand, and qualitative inquiry and inquirers on the other. Such tensions increased markedly in the late 1990s and continue as we enter the second decade of the 21st century. Many tensions emanate from disagreements about which parts of the research are ethical issues, and therefore part

of the remit of an ethics committee, and which are not. This has particularly pertained to questions of methods against the backdrop of the emergence of a politics of evidence embodying narrow views of what constitutes research evidence (see the opening section of this chapter). Lincoln and Cannella have explored aspects of these tensions with great insight (Cannella & Lincoln, Chapter 5, this volume; Lincoln & Cannella, 2007). One of their key concerns is whether ethics committees are making what are in fact methodological recommendations rather than ethical ones.

Ethics committees argue that part of their remit is to stop research proceeding that is unethical in terms of it being of poor research design and therefore unable to provide any meaningful or useful results (Lacey, 1998). Thus, the risks of the research far outweigh any possible benefits. While on one level this is a reasonable position for ethics committees to adopt, it is not reasonable if only certain understandings of research and evidence are in play. In the United States (Riesman, 2002), the United Kingdom (Ramcharan & Cutcliffe, 2001), and Australia (Cheek, 2005), ethics committees emerged from the traditions of medicine and medical science, and largely in response to issues that had arisen in relation to questionable medical practices. The methods and traditions of medicine and science are those more in keeping with conventional quantitative methods (Lacey, 1998) and understandings of research as deductively based (van den Hoonaard, 2001). Guidelines drawn up by ethics committees have thus tended to reflect this way of thinking about research.

So what does this mean for how we write our methods? Some qualitative inquirers have experienced having their research blocked by ethics committees, or even dismissed out of hand, on the basis that it is not the experimental, deductively based research that some members of the committee believe that all research should be (Denzin & Giardina, 2007b; Lincoln & Cannella, 2002; Lincoln & Tierney, 2002). The research is thus deemed unethical on the basis of research paradigm. Although, over time, there has been some mellowing of this situation with ethics committees recognizing that qualitative inquiry (or at least some forms of it) is a legitimate form of research, tensions remain. For example, qualitative researchers have found that some of the information ethics committees require is difficult to provide at the outset of an inductively based inquiry. This can pertain to, for example, the requirement to specify the exact wording of every question that participants will be asked during an interview. Or it might be in relation to specifying an exact number in regard to the sample size, thereby once again raising the issues we talked about previously in relation to specifying a sample size when constructing a budget for the research. Further, the practice of some ethics committees of requiring ethics approval every time questions are added in response to emergent themes, or when more participants are needed to be interviewed than first anticipated, creates much frustration for qualitative researchers.

Tensions also arise from the fact that there is sometimes inconsistency between ethics committees. If the qualitative inquiry is to be conducted on several sites, such as several hospitals, for example, and each site has its own ethics committee that will not accept the approval given by another ethics committee, then qualitative inquirers can find themselves in the position of having gained approval from some committees and not others for the same research. This creates many problems and can even jeopardize the funding received, as the research cannot proceed according to the application that was funded. The time that it can take to navigate the minefield of one or more ethics committees can mitigate against the relatively short time frames of tendered and contract research, thereby in effect making it uncompetitive in winning tenders. As noted previously (Cheek, 2005), what all this potentially means is that ethics committees can be more powerful than any funder in determining what research will and will not get funding or even go ahead. At times, the last and definitive word on method has belonged to ethics committee members, not members of peer-review panels constituted by the funder. And this includes national, so-called "gold star," peer-reviewed funding schemes!

So what to do about all this? Perhaps the key is to ensure the maintenance, in some form, of the two separate discussions that Lincoln and Cannella (2007) noted should be ongoing about ethics and qualitative inquiry. The first of these pertains to the effects of the increasing bureaucratization of ethics and concomitant tightening regulation of research. It is a conversation about how we might navigate the reality of having to submit our proposed research to ethics committees who may not necessarily be supportive of, or expert in, qualitative inquiry. In the section of this chapter about locating funding, I talked a lot about trying to understand the way that funders think so that we can try to work with, while not necessarily accepting, those understandings. The same applies to ethics committees. Be proactive. Discuss your project with the Chair of the committee. See if it is possible to get copies of previous applications to the committee to get an idea of their depth and the way they have been written. Do the hard reflective and reflexive work about what you are prepared to give up and what you are not in terms of your research and its methods in order to get it "through ethics."

However, only to focus on "getting ethics" or "getting through" ethics committees is to be complicit in reducing understandings of ethics to a focus on expediency and instrumentality. There is another discussion to be had, as Lincoln and Cannella (2007) remind us. This is a broader discussion. It is about what ethics is, and might be, in terms of research. It involves creating spaces in our teaching, the enactment of our research, and our research training where we model the ethics of our research and research practices (Lincoln & Cannella, 2007; Schwandt, 2007). In this discussion, there is less emphasis on the mechanics of navigating ethics processes and more on the reasons for talking about ethics in the first place. Rather than simply dwelling on the negative effects of the increasing interference of ethics committees in matters pertaining more to methodology than to ethics, this conversation can be a more positive and constructive one to have. For as Hurdley, while reflecting on the bureaucratization of ethics, reminds us, there is still the possibility of "constant active reflection within these confines" in that "limited horizons do not inhibit depth and richness of insight" (Hurdley, 2010, p. 518).

Example Three: Mixed Methods—Writing for Design or Writing for Funding?

The third example I wish to use to explore tensions around writing our methods for funding shifts the focus to a discussion about choosing what methods to use in our research, and importantly *how* we choose to use those methods. Mixed methods provide an interesting vehicle for this discussion. Recently, there has been an upsurge of interest in, and concomitant writing about, what is termed mixed methods research both without, and significantly within, the folds of qualitative inquiry. There are now whole journals, books (Creswell & Plano Clark, 2007; Greene, 2007; Hesse-Biber, 2010; Morse & Niehaus, 2009; Teddlie & Tashakkori, 2009), and handbooks (Tashakkori & Teddlie, 2003) devoted to it. Advocates of mixed methods argue that this type of research design can add a type of depth and multi-dimensionality to the analysis not otherwise possible. While this may be so, it also seems to me that mixed methods as a research approach is very vulnerable to being reduced to a technique in the way that it is thought about and used, and that writing for funding exacerbates this vulnerability. What do I mean by this and why do I think this? The rest of this section of the chapter unpacks these ideas.

Despite the rise in the popularity and prominence of mixed methods research design, it remains a difficult to define and somewhat contested concept. In large part, this is due to the different emphases given to what is to be mixed, and how, when mixed methods is written about and/or discussed. Most often, mixed methods refers to the use of qualitative and quantitative methods in the same study or program or sequence of studies (Creswell, 2003; Hesse-Biber, 2010; Morse & Niehaus, 2009). However, it can also refer to the use of two methods from the same paradigm, for example, two qualitative methods in the same study, although this way of "mixing" methods has not achieved the same prominence. Responding to the traditional primacy of quantitative approaches over qualitative ones in mixed methods discourses, Hesse-Biber identifies the centering of qualitative approaches to mixed methods research, and providing "researchers with a more detailed understanding of qualitative mixed methods perspectives and practices" (2010, p. 9) as two of her four aims when writing her recent book on mixed methods.

Morse and Niehaus (2009) provide the following definition of mixed methods that attempts to pick up on and incorporate the diversity of emphases in mixed method design. They write, "a mixed method design is a scientifically rigorous *research project*, driven by the inductive or deductive *theoretical drive*, and comprised of a qualitative or quantitative *core component* with qualitative or quantitative *supplementary component(s)*" (2009, p. 14, emphasis in the original). In this definition, the emphasis is on mixed methods in terms of an overall research design that is theoretically informed, rather than on the fact that methods have been "'mixed" in a study and/or the specific mix of methods that will be employed. It stands in stark contrast to a "methods-centric" conception of the field of mixed methods research with a focus "primarily on the construction of mixed methods designs, often to the detriment of how they interact with the research problem" (Hesse-Biber, 2010, p. vi). This is a significant point that we will return to later in this discussion as it has implications for the way that we think about writing an application for funding employing a mixed methods approach.

It is interesting in terms of the recent rise of interest in mixed methods that, in reality, this research approach is not actually new. It has been around in various forms for many years, including in qualitative inquiry. For example, think of ethnography involving observations, interviews, and sometimes other ways of collecting data (Morse & Niehaus, 2009). What is new is the degree of prominence this approach has assumed, especially in the past decade. In part, this is due to renewed scholarship in the area exemplified by the work of Morse and Niehaus (2009), Hesse-Biber (2010), Creswell and Plano Clark (2007), Mertens (2005), and others. But this in itself cannot fully account for the degree of heightened interest in mixed methods and the almost fervent urgency with which students and more experienced researchers take classes and workshops in it. Perhaps the politics and practices of the research marketplace can add another layer to the possible explanations for this elevated interest.

Interviewed in 2004 (the transcript of the interview can be found in Hesse-Biber & Leavy, 2006, p. 335), Morse made an interesting comment when asked about how she saw what was termed "multimethods" evolving. Her response was, "I think it is going to get in a terrible mess but will sort itself out in the end." When asked what she meant by this, part of her response was as follows: "I think that the pressure to do mixed methods, in order to get funding, overwhelms or overrides the goals of qualitative inquiry. I think the funding agencies say they fund qualitative inquiry, meaning that they really do fund mixed methods. This still places qualitative inquiry in an inferior position" (p. 335). Here, Morse points to the way that writing for funding can affect the methods chosen *and* the way we choose to write our methods. An example of this was when I was approached to be part of a large quantitatively focused funding bid on the basis that a (small) qualitative component would give an edge to the competitiveness of the grant in question. This perception arose from the fact that in the previous funding round more studies that had a mix of qualitative and quantitative methods were successful.

The assumption was that it was the fact that there were different methods mixed that gave the successful proposals their competitive edge, rather than the theoretical drive of the design and the congruity of the mixing of methods with that drive.

How to respond to approaches such as this creates dilemmas for qualitative inquirers. On the one hand, it may be that, even if for questionable reasons, the qualitative "bit" was added as an afterthought to a quantitatively driven project, doing that qualitative component of the research may enhance the findings of the research. Participating in the research as part of the team may offer the opportunity to heighten that team's understandings of qualitative inquiry. Joining a strong research team may also help build track record for the individual qualitative researcher. On the other hand, to accept such invitations runs the risk that methods that have been stripped of their theoretical underpinnings are considered to constitute qualitative inquiry. Similarly, perpetuating an instrumental use of mixed methods design puts the emphasis when thinking about that design on the use of different methods per se, rather than the use of different methods in an overall coherent and theoretically driven design. In so doing, it reduces mixed methods design *itself* to a technique rather than a theoretically grounded research approach. As Morse and Niehaus (2009) and Hesse-Biber (2010) emphasize, mixed methods is a theoretically driven study design, not simply the insertion of a mix of methods into a study.

All of this raises important points for qualitative inquirers to reflect on with respect to their motives in writing their methods in particular ways for funding. One starting point for such reflection is a somewhat unsettling and provocative question. When we write our application for funding, is the reason for choosing a mixed method design (or any other design for that matter) that the question will be better answered by this approach, or is it that by using this approach as opposed to a wholly qualitative design the chances of funding will be increased? Another starting point might be to reflect upon why qualitative inquiry within a mixed methods frame is seemingly more acceptable to some funders than research that is wholly qualitative. Perhaps this is because in some funder's minds mixed methods fits "well with the global economic imperative of the 1990s to do more with less and with the rising evidence-based practice movement" (Giddings, 2006, p. 196).

Perhaps, too, it is because descriptions of mixed methods read superficially might seem familiar to many funders and members of grant review panels. The QUAN → qual, QUAN + qual, QUAL → quan, and QUAL + quan symbols developed by Morse (1991) appear reassuringly similar to what science "should" look like. So does the language in the following excerpt from a description of some of the strategies within mixed methods: "The sample selection for the qualitative and quantitative components of sequential (QUAL → quan) or simultaneous (QUAL + quan) triangulation must be independent" (Morse, 1991, p. 122). Morse's incorporation of notations, symbols, and arrows into complicated figures that visually represent the mixed method strategy bears an uncanny resemblance to traditional ways of representing science and scientific "speak." While this in itself is not a problem, what is a problem is if that resemblance is what reassures funders and others that the research is sound and the methods reliable, rather than what those symbols are representing and conveying. Yet the tradeoff is that without this type of conceptualization and notation, the risk is that understandings of this approach will become muddled and unclear.

None of this is to suggest that mixed methods are not a valuable and important part of qualitative inquiry. Nor should it give the impression that a mixed methods research design is necessarily more inherently problematic than any other research approach with respect to the tensions and potential ruptures that exist at the interface of qualitative inquiry and funding. Rather, what it is to suggest is that mixed methods provides an excellent example of the fine lines we walk between selling and not selling out when we write our methods for funding. Achieving prominence in the repertoire of methods and study designs looked upon favorably by funders may not necessarily work in the interests of mixed methods design. As Hesse-Biber and Leavy point out, "more is not necessarily better" (2006, p. 334) with respect to methods; but it can be if, and it is *if* that is the crucial word here, it is the right design used for the right reasons to explore the problem or question that the research seeks to address.

◼ FUNDING: WHEELING AND DEALING IN A RESEARCH MARKETPLACE

Our conversation about the politics and practices of funding qualitative inquiry has ranged widely across issues such as how to locate funding, what that funding might be, how we might think about writing for that funding, and what this might mean in practice. At various points throughout this conversation, we have touched on the notion of a research marketplace where research is bought and sold. In this research marketplace, funding for qualitative inquiry is about much more than supporting a specific research project. Funding is a commodity, a unit of exchange. For example, individual researchers can trade their research for jobs, promotion, and tenure based on their research "performance." Institutions also trade in this marketplace, trading institutional research performance for rewards such as research infrastructure funding from governments, as well as a competitive ranking in the marketplace. The discussion to follow explores these ideas further using the metaphor of "wheeling and dealing" in a research marketplace as the organizing construct. As such, it is the most politically overt section of the chapter, shifting the focus of our discussion about the politics and practice of funding qualitative inquiry from the more micro level of specific projects, funders, or methods to a more macro, wider societal context.

Arising in the late 1990s, continuing throughout the first decade of the 2000s, and exacerbated by the global financial crisis of 2008 and 2009, there has been a period of economic restraint and funding cuts by governments in most Western countries. As a result, most universities and other research-intensive organizations are forced to run "mixed economies," where part of their funding comes from government operating grants and the other part is reliant on the generation of income from staff activities. This has created an imperative, and increasingly an expectation, for individual staff members to generate income for cash-strapped institutions. Gaining research funding is one way staff can do this, as funding for research can generate income for institutions in several ways. The first is from profit arising directly from the research undertaken. This is when contracted and tendered research can become very interesting and attractive. Normally, in this type of research funding it is possible to charge consultancy market rates both for the research team's time as well as research costs and overheads. This is in contrast to funding schemes that have a set limit on the amount of costs and overheads that they will pay, and which may not pay at all for the researchers' time if they are already employed by a university, for example. Given all this, it might seem that universities, research organizations, and researchers alike would focus more and more on contracted and tendered research funding. However, it is not this simple. There are messages about the messages.

These messages about messages are related to the fact that research income is also one of the variables in the schemes developed and used by governments to assess the research performance of higher education and research-oriented institutions. In many countries, a complex array of ever-changing formulas have been used in the quest to transform the individual research problems and projects of researchers employed in institutions into measures of collective institutional research performance (Cheek, 2006; Torrance, 2006). In Australia, for example, over the course of the last decade there have been several versions of proposed ways of measuring and ensuring quality and excellence in research. All have been premised on audit-driven notions of accountability designed to give government, industry, business, and the wider community assurance of the excellence of research conducted in Australia's higher education institutions. The current scheme in play at the time of writing this chapter is called Excellence in Research for Australia (ERA). ERA is a complex scheme that, like so many of its predecessors, as it unfolds becomes even more complex. There are 12 key 2010 documents alone listed on the Excellence in Research for Australia website (July 2010) including a 53-page booklet of evaluation guidelines and an 89-page booklet detailing submission guidelines! "Rewards" will flow from the Australian government to higher education institutions based on their research performance as calculated by the formulas embedded in the ERA scheme. Those institutions ranked highest will receive the largest share of the pool of dollars allocated for distribution.

However, that is not the end of the story. The results of the ERA assessment exercise will be publicly released and published at both institutional and national levels. This will enable comparisons to be made with respect to research performance not only between institutions, but also within institutions at the departmental level. League tables can be drawn up and published where the rankings of institutions in terms of this formulaic-determined research performance are displayed. Universities ranked highly in these tables can, and do, market themselves in the research marketplace as being "the top," a "top," or in the "top eight," and so on research universities in Australia. This marketing can then be used to attract sponsors of funding wishing to be associated with highly ranked and therefore prestigious research universities. It can also be used as part of the marketing to attract more research students. In turn, this generates more research outputs in the form of completed research student projects and publications, which in turn generates more income through assessment exercises that use research income, numbers of research student completions, and publications as part of their formula. In this way, not only do schemes such as ERA generate "rewards" for research performance, they also contribute to the possibility of this cycle of "rewards" continuing. Thus, layer upon complicated layer is deposited in a landscape known as the research marketplace.

But there are still more messages about the messages. In this landscape, not all research income is necessarily equal. In these assessment exercises, schemes, and/or formulas, often it is *where* the money comes from, not just the absolute amount of it, that matters. One effect of this has been to create a complicated system of classifying funding. For example, in Australia at the time of writing this chapter, funding received is divided into various categories depending on its source. Highly sought after and prized are grants that appear on the Australian Competitive Grant Register (2010). The number of Australian Competitive Grants (ACG), as well as the amount of dollars attached to them, held by a particular university or individual researcher have been, and are, used as a measure of quality and research activity for institutions and individual researchers alike.

Yet there are even more messages about these messages to unpack. For, in Australia, among all grants, National Health and Medical Research Council (NH&MRC) and Australian Research Council (ARC) grants are prized the most. Historically, these funding schemes have attracted more "rewards" from government both monetarily and prestige-wise. Indeed, institutions will sometimes draw up tables of their relative national ranking with respect to number of grants, and amount of dollars, received from the NH&MRC and ARC alone. This is particularly the case when doing so presents them in a better light than their relative ranking on the league table of national research performance. This selective reporting of performance is part of wheeling and dealing in national and, increasingly, international research marketplaces. At the time of writing this chapter, the website of the University of Sydney (2010) proudly boasts that "Sydney researchers scored the 'research double' topping both the National Health and Medical Research Council (NH&MRC) and the Australian Research Council (ARC) for the number and dollar value of new, peer-reviewed, researcher driven research projects commencing in 2009." It goes on to state that "Sydney not only leads the Nation, it achieved a personal best by attracting record amounts of funding from both the ARC and NH&MRC for research projects starting in 2009." Here, the language of a competitive marketplace such as "research double," "topping," and "personal best" is overt. In this research marketplace, it is the amount and type of funding that both counts and is counted!

The effect of such wheeling and dealing, and the translation of research funding into other forms of funding, cascades down to the level of the individual researcher. The amount of dollars received by that researcher, and where they come from, are used as indicators of individual research performance and impact. What those dollars are used for, or even if they are really needed, is not the point. For example, in some types and fields of research the grant might be for large pieces of equipment. In such cases, the amount of the funding is actually as much a measure of the expense of the equipment as it is of the quality or the intensity of the research in a given institution. Another effect at the level of the individual researcher of this increasing emphasis on funding as being able to provide additional income for institutions, is that researchers and/or areas without this research income are increasingly viewed as not being as research-productive or research-intensive as those researchers or departments attracting large amounts of dollars. Writing on one's CV that one has funded research by working on it for some 30 hours a week above paid workload allocation does not have the same currency in a market-driven research context as does writing that one has attracted some hundreds of thousands of dollars of funding in the past two years, even if that funding has been to buy an expensive piece of equipment or equip a laboratory. This works against many qualitative inquirers whose major form of support or funding for their research is their own time either as a percentage of their paid workload, or the time they use on weekends or evenings to do research over and above any such allocated percentage.

An effect of differentially rewarding types of funding for research is that the gaining of specific categories of funding has become an end in itself, rather than that funding being a means to support a specific piece of research. The danger in this is that the research design becomes subverted and shaped by the demands and expectations of highly prized funding sources. For example, if a so called "gold star" research scheme has historically been adverse to funding qualitative research in general, or some types of qualitative research specifically, how do qualitative researchers position themselves in relation to this? Does this mean we abandon this source of funding, or do we abandon our qualitative inquiry and put up approaches and techniques that we think the funder will support? If we choose not to abandon our qualitative inquiry altogether, do we try to make our qualitative inquiry look more like the type of research that these funders and their decision-making panels will expect and/or understand? How far are we prepared to go in terms of shaping our research to fit these schemes? Is it better to opt out of, or opt to be in, this research marketplace and the spaces it creates?

But it is not as simple as that. There is ambivalence in these categories of opting out and opting in and their possible effects. Opting in, and thereby gaining currency in this marketplace, may paradoxically provide opportunities to promote and advance qualitative inquiry. Perhaps there is the possibility that external funding can be used as a currency to gain institutional power and positions on panels awarding funds in granting schemes where qualitative inquirers can work from within to try to change practices that work against qualitative inquiry. Furthermore, in many institutions, attracting this type of "prized" funding is "rewarded" with infrastructure monies and in-kind support to build critical masses of researchers and develop research centers. In turn, this may provide the opportunity to fund forms of qualitative inquiry that may not otherwise be able to attract monetary or other forms of support, thereby helping resist the potential reduction and homogenization of qualitative inquiry into forms acceptable to external funding agencies. Thus, paradoxically, although participating in a research marketplace does not necessarily advance qualitative inquiry, it might. Yet even given this possibility, a question that lingers in my mind is whether this is an example of a way to resist the excesses of the research marketplace by turning it against itself, or a case of selling out—at least in part. Is there a danger in this that inadvertently a two-tiered structure of qualitative researcher is created, a new binary where those with the currency highly prized in the research marketplace, the "haves," are the primary and dominant category, with their poor cousins, the marketplace "have nots" dependent on them?

Denzin and Lincoln remind us that all of us are "stuck in the present working against the past as we move into a politically charged and challenging future" (2005, p. xv). At times, this sense of being stuck in a present working against the past has resulted in feelings of fatigue and frustration on my part. For example, I wonder about where the next research productivity reform or audit will come from and what form it will take. I think about how much time I have used, and will again use, to respond to relentless demands for demonstrating delimited forms of evidence about the impact/quality/products (or some other form of research marketspeak) of my research only to find that with a change of government it has all changed anyway. Thinking about the past decade, these audits and reviews congeal and blur into each other. They are new versions of the same old rhetoric, ideas about ideas about ideas, depositing layer upon layer on bedrock of positivistic-influenced assumptions about research and evidence.

In all of this, it has not always been easy or comfortable for me to navigate the tension between critiquing the spaces I research in, and surviving in the spaces of the research marketplace of which, whether I like it or not, my research has been part (Cheek, 2007). At times, I have questioned my motives in applying for certain types of external funding for my research. Was my motive to obtain the external funding to do a project I was burning to do, or was it that I needed the external funding to score points on some form of research marketplace scorecard? If the answer is both of the above, then which mattered most to me? In wheeling and dealing in this market, and enjoying the privileges and trappings that this can "buy," such as tenure, promotion, and research infrastructure, have I had to do things that I would rather not talk about—the "yes but" thinking? Have I been part of what Stronach and Torrance (1995) referred to close to two decades ago as preserving little private spaces to be radical in, while doing something else in various dimensions of my professional life? And, if so, could I be accused of saying a lot about what I do in this chapter and not enough about what I do not say about what I do (Cheek, 2010a)?

Like so many other qualitative inquirers, I am still searching for some sort of middle ground in all of this. I cannot tell you exactly what this middle ground might look like, as I suspect it will vary for each of us. However, I can tell you that the middle ground that I am looking for is a place from which to engage the politics and practices of funding qualitative inquiry and the research marketplace on my terms, not someone else's. It is a place from which any engagement with those politics and practices is because I choose to engage with them not because I have to, or because they constantly engage my time and energy destructively (Cheek, 2008). In this middle ground, I can choose how to engage with a politics not of my own making.

◨ SIGNING OFF FOR NOW: THE POLITICS AND PRACTICE OF FUNDING AS AN ONGOING CONVERSATION

So reader, how to conclude? It is customary to call the final section of an extended discussion such as the one in this chapter a conclusion. However, somehow at this point the word "conclusion" does not seem right or an easy fit with what has been, and remains to be, said. It seems too final, too much like having the last word when there is no last word to be had. At best, this chapter could only ever hope to be "a comment within an ongoing discussion" (Maxey, 1999, p. 206). This ongoing discussion is about moving away from reactionary positions be they in terms of a funder's requirements, or a politics of evidence, or a critique of qualitative inquiry from within qualitative inquiry itself. It is an ongoing discussion about being in places that are of our own making as qualitative inquirers and places that we choose to be in. From such places, we can apply for funding and take the hard decisions that this involves with respect to what is negotiable and what is not, and know that we and not others have made those decisions. From such places, we can be confident that the thinking and reflection that sits behind taking up such positions will contribute to the ongoing development and strength of qualitative inquiry. This is because when we are forced to try to articulate on what basis we have made these decisions we develop further our understandings about what enables something to be called both "qualitative" and "inquiry." And, importantly, such development comes out of qualitative inquiry itself rather than being imposed on it. Put another way, it enables us to choose the messages about the messages about the messages that we will take on board, and those that we will ignore.

▣ References

Atkinson, P., & Delamont, S. (2006). In the roiling smoke: Qualitative inquiry and contested fields. *International Journal of Qualitative Studies in Education, 19*(6), 747–755.

Australian Competitive Grants Register (ACGR). (2010). Available at http://www.innovation.gov.au/Section/Research/Pages/Australian CompetitiveGrantsRegister(ACGR).aspx

Bochner, A. P. (2009). Warm ideas and chilling consequences. *International Review of Qualitative Research, 2*(3), 357–370.

Brennan, M. (2002). *The politics and practicalities of grassroots research in education.* Available at http://www.staff.vu.edu.au/alnarc/forum/marie_brennan.html

Cannella, G. S., & Lincoln, Y. S. (2004). Dangerous discourses II: Comprehending and countering the redeployment of discourses (and resources) in the generation of liberatory inquiry. *Qualitative Inquiry, 10*(2), 165–174.

Carey, M. A., & Swanson, J. (2003). Pearls, pith, and provocation: Funding for qualitative research. *Qualitative Health Research, 13*(6), 852–856.

Cheek, J. (2005). The practice and politics of funded qualitative research. In N. K. Denzin & Y. S. Lincoln (Eds.), *The SAGE handbook of qualitative research* (3rd ed., pp. 387–409). Thousand Oaks, CA: Sage.

Cheek, J. (2006). The challenge of tailor-made research quality: The RQF in Australia. In N. K. Denzin & M. D. Giardina (Eds.), *Qualitative inquiry and the conservative challenge* (pp. 109–126). Walnut Creek, CA: Left Coast Press.

Cheek, J. (2007). Qualitative inquiry, ethics, and the politics of evidence: Working within these spaces rather than being worked over by them. In N. K. Denzin & M. D. Giardina (Eds.), *Ethical futures in qualitative research: Decolonizing the politics of knowledge* (pp. 9–43). Walnut Creek, CA: Left Coast Press.

Cheek, J. (2008). A fine line: Positioning qualitative inquiry in the wake of the politics of evidence. *International Review of Qualitative Research, 1*(1). Walnut Creek, CA: Left Coast Press.

Cheek, J. (2010a). Human rights, social justice, and qualitative research: Questions and hesitations about what we say about what we do.

In N. K. Denzin & M. D. Giardina (Eds.), *Qualitative inquiry and human rights* (pp. 100–111). Walnut Creek, CA: Left Coast Press.

Cheek, J. (2010b). A potent mix: Older people, transitions, practice development and research. *Journal of Research in Nursing, 15, 2.*

Cheek, J., & Ballantyne, A. (2001). Moving them on and in: The process of searching for and selecting an aged care facility. *Qualitative Health Research, 11*(2), 221–237.

Cheek, J., Ballantyne, A., Byers, L., & Quan, J. (2007). From retirement village to residential aged care: What older people and their families say. *Health and Social Care in the Community, 15*(1), 8–17.

Cheek, J., Ballantyne, A., Gilham, D., Mussared, J., Flett, P., Lewin, G., et al. (2006). Improving care transitions of older people: Challenges for today and tomorrow. *Quality in Aging, 7*(4), 18–25.

Cheek, J., Ballantyne, A., Roder-Allen, G., & Jones, J. (2005). Making choices: How older people living in independent living units decide to enter the acute care system. *International Journal of Nursing Practice, 11*(2), 52–57.

Cheek, J., Corlis, M., & Radoslovich, H. (2009). Connecting what we do with what we know: Building a community of research and practice. *International Journal of Older People Nursing, 4*(3), 233–238.

Cheek, J., Garnham, B., & Quan, J. (2006). "What's in a number": Issues in providing evidence of impact and quality of research(ers). *Qualitative Health Research, 16*(3), 423–435.

Cheek, J., Price, K., Dawson, A., Mott, K., Beilby, J., & Wilkinson, D. (2002). *Consumer perceptions of nursing and nurses in general practice.* Available at http://www.health.gov.au/internet/main/publishing.nsf/Content/work-pr-nigp-res-cons-rept

Creswell, J. W. (2003). *Research design: Qualitative, quantitative and mixed methods approaches* (2nd ed.). Thousand Oaks, CA: Sage.

Creswell, J. W., & Plano Clark, V. L. (2007). *Designing and conducting mixed methods research.* Thousand Oaks, CA: Sage.

Denzin, N. K. (2009). The elephant in the living room: Or extending the conversation about the politics of evidence. *Qualitative Research, 9,* 139–160.

Denzin, N. K., & Giardina, M. D. (2006). *Qualitative inquiry and the conservative challenge.* Walnut Creek, CA: Left Coast Press.

Denzin, N. K., & Giardina, M. D. (2007a). *Ethical futures in qualitative research: Decolonizing the politics of knowledge.* Walnut Creek, CA: Left Coast Press.

Denzin, N. K., & Giardina, M. D. (2007b). Introduction. In N. K. Denzin & M. D. Giardina (Eds.), *Ethical futures in qualitative research: Decolonizing the politics of knowledge* (pp. 9–43). Walnut Creek, CA: Left Coast Press.

Denzin, N. K., & Giardina, M. D. (2008). *Qualitative inquiry and the politics of evidence.* Walnut Creek, CA: Left Coast Press.

Denzin, N. K., & Giardina, M. D. (2009). *Qualitative inquiry and social justice.* Walnut Creek, CA: Left Coast Press.

Denzin, N. K., & Lincoln, Y. S. (2005). Preface. In N. K. Denzin & Y. S. Lincoln (Eds.), *The SAGE handbook of qualitative research* (3rd ed., pp. ix–xix). Thousand Oaks, CA: Sage.

Excellence in Research for Australia. (2010). *Key 2010 documents.* Available at http://www.arc.gov.au/era/key_docs10.htm

Giddings, L. S. (2006). Mixed-methods research: Positivism dressed in drag? *Journal of Research in Nursing, 11*(3), 195–203.

GrantSearch Register of Australian Funding. (2010). *Funding at your fingertips.* Available at http://www.grantsearch.com.au/

Greene, J. C. (2007). *Mixed methods in social inquiry.* San Francisco: Jossey-Bass.

Hammersley, M. (2005). Close encounters of a political kind: The threat from the evidence-based policy-making and practice movement. *Qualitative Researcher, 1,* 2–4.

Hesse-Biber, S. N. (2010). *Mixed methods research: Merging theory with practice.* New York: Guilford.

Hesse-Biber, S. N., & Leavy, P. (2006). *The practice of qualitative research.* Thousand Oaks, CA: Sage.

Holmes, D., Murray, S. J., Perron, A., & Rail, G. (2006). Deconstructing the evidence-based discourse in health sciences: Truth, power, and fascism. *International Journal of Evidence-Based Healthcare, 4*(3), 180–186.

Hurdley, R. (2010). In the picture or off the wall? Ethical regulation, research habitus, and unpeopled ethnography. *Qualitative Inquiry, 16*(6), 517–528.

Kvale, S. (2008). Qualitative inquiry between scientistic evidentialism, ethical subjectivism and the free market. *International Review of Qualitative Research, 1*(1), 5–18.

Lacey, E. A. (1998). Social and medical research ethics: Is there a difference? *Social Sciences in Health, 4*(4), 211–217.

Lincoln, Y. S., & Cannella, G. S. (2002, April). *Qualitative research and the radical right: Cats and dogs and other natural enemies.* Paper presented at the 66th annual meeting of the American Educational Research Association, New Orleans, LA.

Lincoln, Y. S., & Cannella, G. S. (2007). Ethics and the broader rethinking/reconceptualization of research as construct. In N. K. Denzin & M. D. Giardina (Eds.), *Ethical futures in qualitative research: Decolonizing the politics of knowledge* (pp. 67–84). Walnut Creek, CA: Left Coast Press.

Lincoln, Y. S., & Tierney, W. G. (2002, April). *"What we have here is a failure to communicate . . .": Qualitative research and institutional review boards (IRBs).* Paper presented at the 66th annual meeting of the American Educational Research Association, New Orleans, LA.

Long, B. (2010). [Review of the book *Qualitative inquiry and the politics of evidence*]. *Qualitative Health Research, 20*(3), 432–434.

Maxey, I. (1999). Beyond boundaries? Activism, academia, reflexivity and research. *Area, 31*(3), 199–208.

Mertens, D. (2005). *Research and evaluation in education and psychology: Integrating diversity with quantitative, qualitative, and mixed methods.* Thousand Oaks, CA: Sage.

Morse, J. M. (1991). Approaches to qualitative-quantitative methodological triangulation. *Methodology Corner, 40*(2), 120–123.

Morse, J. M. (2002). Myth #53: Qualitative research is cheap. *Qualitative Health Research, 12*(10), 1307–1308.

Morse, J. M. (2003a). The adjudication of qualitative proposals. *Qualitative Health Research, 13*(6), 739–742.

Morse, J. M. (2003b). A review committee's guide for evaluating qualitative proposals. *Qualitative Health Research, 13*(6), 833–851.

Morse, J. M. (2006a). The politics of evidence. *Qualitative Health Research, 16*(3), 395–404.

Morse, J. M. (2006b). Reconceptualizing qualitative inquiry. *Qualitative Health Research, 16*(3), 415–422.

Morse, J. M. (2008). Deceptive simplicity. *Qualitative Health Research, 18*(10), 1311.

Morse, J. M., & Niehaus, L. (2009). *Mixed method design: Principles and procedures.* Walnut Creek, CA: Left Coast Press.

Penrod, J. (2003). Getting funded: Writing a successful qualitative small-project proposal. *Qualitative Health Research, 13*(6), 821–832.

Ramcharan, P., & Cutcliffe, J. R. (2001). Judging the ethics of qualitative research: Considering the "ethics as process" model. *Health and Social Care in the Community, 9*(6), 358–366.

Riesman, D. (2002, November/December). Reviewing social research. *Change,* 9–10.

Schwandt, T. (2007). The pressing need for ethical education: A commentary on the growing IRB controversy. In N. K. Denzin & M. D. Giardina (Eds.), *Ethical futures in qualitative research: Decolonizing the politics of knowledge* (pp. 85–98). Walnut Creek, CA: Left Coast Press.

Stronach, I. (2006). Enlightenment and the "heart of darkness": (Neo) imperialism in the Congo, and elsewhere. *International Journal of Qualitative Studies in Education, 19*(6), 757–768.

Stronach, I., & Torrance, H. (1995). The future of evaluation: A retrospective. *Cambridge Journal of Education, 25*(3), 283–300.

Tashakkori, A., & Teddlie, C. (Eds.). (2003). *Handbook of mixed methods in social and behavioral research.* Thousand Oaks, CA: Sage.

Teddlie, C., & Tashakkori, A. (2009). *Foundations of mixed methods research: Integrating quantitative and qualitative approaches in the behavioral and social sciences.* Thousand Oaks, CA: Sage.

Torrance, H. (2006). Research quality and research governance in the United Kingdom. In N. K. Denzin & M. D. Giardina (Eds.), *Qualitative inquiry and the conservative challenge* (pp. 127–148). Walnut Creek, CA: Left Coast Press.

Torres, C. A. (2002). The state, privatization and educational policy: A critique of neo-liberalism in Latin America and some ethical and political implications. *Comparative Education, 38*(4), 365–385.

University of Sydney. (2010). *Research achievements.* Available at http://www.usyd.edu.au/research/about/major_achievements.shtml

van den Hoonaard, W. C. (2001). Is research-ethics review a moral panic? *Canadian Review of Sociology and Anthropology, 38*(1), 19–36.

15

CONTROVERSIES IN MIXED METHODS RESEARCH

John W. Creswell

Mixed methods has emerged in the last few years as a research approach popular in many disciplines and countries, and supported through diverse funding agencies. With such growth, it is not surprising that critical commentaries have surfaced through papers presented at conferences and in published journal articles. These critics have come from both within (e.g., Greene, 2008; Morse, 2005; Creswell, Plano Clark, & Garrett, 2008) and outside (Denzin & Lincoln, 2005; Howe, 2004) the mixed methods community. Although concerns have mounted, they have been largely ignored by social scientists and the mixed methods community.

This chapter gives voice and focus to these controversies. I discuss 11 far-ranging controversies from basic concerns about defining and describing mixed methods, to philosophical debates, and on into the procedures for conducting a study.

For each controversy, I present critical questions, diverse stances, and lingering questions. At the end of this chapter, I reflect on the implications of these controversies. I hope this discussion will help mixed methods researchers, students, and policy makers appreciate the still-unanswered questions, view the multiple perspectives that have emerged, and reflect on new commitments that the mixed methods field needs to make. For qualitative researchers, I hope that this reflection will encourage the continued discussion of the strong vital role that qualitative research has and continues to play in mixed methods research.

The thoughts to follow will reflect my own writings of the last 20 years and will include, at times, a self-reflective critique. My methodological development consisted of formal training as a postpositivist in the 1970s, self-education as a constructivist through teaching qualitative courses in the 1980s, and advocacy for mixed methods through my writings and teachings from the 1990s to the present. As one spokesperson for mixed methods, many controversies have come to my attention through scholarly

papers presented at conferences, articles published in the *Journal of Mixed Methods Research* (JMMR) while I served as founding coeditor for the last five years, and papers sent to me by authors who wanted me to keep abreast of emerging issues. As I look across these diverse materials, I hope to foster the ongoing conversation about the controversies and the many possible answers that scholars have offered to them.

SOME RECENT QUESTIONS

Some of the controversies that I will present figured prominently in a discussion in March 2009. I was attending and presenting at the University of Aberdeen in Scotland (Creswell, 2009d) at the Economic & Social Research Council (ESRC) Seminar Series sponsored by the Health Services Research Unit at the University of Aberdeen. I had finished my overview of mixed methods research to a gathering of 50 scholars primarily from the health sciences. They had assembled in historic Elphinstone Hall, an ancient venue with a high, vaulted, hammer-beam roof, banners hanging from the rafters, and oak-paneled walls lined with pictures of distinguished scholars dating back centuries. Much to my surprise, the conference organizer suddenly asked small groups to form and record their questions about both the advantages and the challenges of using mixed methods research. Not wanting to miss a key opportunity to capture their challenges and critical thoughts, I hastily began taking notes. They spoke about claims being made about the value of mixed methods research ("Is mixed methods seen as the answer to everything?" "Are there undue expectations raised by mixed methods that cannot be fulfilled?"), about philosophical and theoretical issues ("Is there opposition to mixed methods from those who hold strong worldview positions?" "Does a dominant paradigm prevail in mixed methods?" "Is qualitative research working on an even playing field with quantitative in mixed methods?"), and about the procedures and processes of research ("Is there a good fit between the research question and mixed methods?" "Do researchers have expertise and competence in both areas?").

The irony of "new" voices of concern about mixed methods arising in the "old," historic setting of Elphinstone Hall did not escape my attention. But, in retrospect, hearing the issues was not surprising. Concerns have been voiced in recent respected journal articles (Giddings, 2006; Howe, 2004), in the third edition of this handbook (Denzin & Lincoln, 2005), in conference presentations (Holmes, 2006), and in articles published in the *Journal of Mixed Methods Research*. In 2006, I had presented my views about unresolved issues in a journal article on the role of

qualitative research in mixed methods (Creswell, Shope, Plano Clark, & Green, 2006), and at a panel presentation made at the 2007 International Qualitative Inquiry Congress (Creswell, 2007). In light of these discussions, it is timely to address these controversies. In this chapter, I address 11 controversies and raise several questions, as outlined in Table 15.1. The controversies, as a group, reflect what Kuhn (1970) said years ago about the transition period in research:

The proliferation of competing articulations, the willingness to try anything, the expression of explicit discontent, the recourse to philosophy, and to debate over fundamentals, all these are symptoms of a transition from normal to extraordinary research. (p. 91)

Table 15.1 Eleven Key Controversies and Questions Being Raised in Mixed Methods Research

Controversies	Questions Being Raised
1. The changing and expanding definitions of mixed methods research	What is mixed methods research? How should it be defined? What shifts are being seen in its definition?
2. The questionable use of qualitative and quantitative descriptors	Are the terms "qualitative" and "quantitative" useful descriptors? What inferences are made when these terms are used? Is there a binary distinction being made that does not hold in practice?
3. Is mixed methods a "new" approach to research?	When did the conceptualization of mixed methods begin? Does mixed methods predate the period often associated with its beginning? What initiatives began prior to the late 1980s?
4. What drives the interest in mixed methods?	How has interest grown in mixed methods? What is the role of funding agencies in its development?
5. Is the paradigm debate still being discussed?	Can paradigms be mixed? What stances on paradigm use in mixed methods have developed? Should the paradigm for mixed methods be based on scholarly communities?
6. Does mixed methods privilege postpositivism?	In the privileging of postpositivism in mixed methods, does it marginalize qualitative, interpretive approaches and relegate them to secondary status?
7. Is there a fixed discourse in mixed methods?	Who controls the discourse about mixed methods? Is mixed methods nearing a "metanarrative?"
8. Should mixed methods adopt a bilingual language for its terms?	What is the language of mixed methods research? Should the language be bilingual or reflect quantitative and qualitative terms?
9. Are there too many confusing design possibilities for mixed methods procedures?	What designs should mixed methods researchers use? Are the present designs complex enough to reflect practice? Should entirely new ways of thinking about designs be adopted?
10. Is mixed methods research misappropriating designs and procedures from other approaches to research?	Are the claims of mixed methods overstated (because of misappropriation of other approaches to research)? Can mixed methods be seen as an approach lodged within a larger framework (e.g., ethnography)?
11. What value is added by mixed methods beyond the value gained through quantitative or qualitative research?	Does mixed methods provide a better understanding of a research problem than either quantitative or qualitative research alone? How can the value of mixed methods research be substantiated through scholarly inquiry?

◨   CHANGING AND EXPANDING DEFINITIONS

Heading the list of controversies would certainly be the funda-mental question: What is mixed methods research? How should it be defined? To answer these questions requires a brief historical review of shifts in the definition of mixed methods over the years. For example, an early definition of mixed methods came from writers in the field of evaluation, Greene, Caracelli, and Graham (1989). They emphasized the mixing of *methods* and the disentanglement of methods and paradigms when they said,

> In this study, we defined mixed-method designs as those that include at least one quantitative method (designed to collect num-bers) and one qualitative method (designed to collect words), where neither type of method is inherently linked to any particular inquiry paradigm. (p. 256)

Ten years later, the definition had shifted from mixing two methods to mixing in all phases of the research process, and mixed methods was being seen as a *methodology* (Tashakkori & Teddlie, 1998). Included within this process would be mixing from philosophical (i.e., worldview) positions, to final infer-ences, and to the interpretations of results. Thus, Tashakkori and Teddlie (1998) defined mixed methods as the combination of "qualitative and quantitative approaches in the methodology of a study" (p. ix). These authors reinforced this methodological orientation in their preface to the *Handbook of Mixed Methods in Social & Behavioral Research* by writing, "mixed methods research has evolved to the point where it is a separate method-ological orientation with its own worldview, vocabulary, and techniques" (Tashakkori & Teddlie, 2003, p. x).

A few years later, when Plano Clark and I (Creswell & Plano Clark, 2007) wrote a definition for mixed methods into our introductory book, we blended *both* a methods and a method-ological orientation along with a central assumption being made with this type of research. We said,

> Mixed methods research is a research design with philosophi-cal assumptions as well as methods of inquiry. As a methodol-ogy, it involves philosophical assumptions that guide the direc-tion of the collection and analysis and the mixture of qualita-tive and quantitative approaches in many phases of the research process. As a method, it focuses on collecting, analyzing, and mixing both quantitative and qualitative data in a single study or series of studies. Its central premise is that the use of quan-titative and qualitative approaches, in combination, provides a better understanding of research problems than either approach alone. (p. 5)

This definition was patterned on describing an approach using multiple meanings, such as found in Stake's (1995) definition of a case study. Our definition of mixed methods had both a philoso-phy and a method orientation, and it conveyed components of the *core characteristics* of mixed methods that I advance today in workshops and presentations (e.g., see Creswell, 2009a). In mixed methods, the researcher

- collects and analyzes persuasively and rigorously both qualita-tive and quantitative data (based on research questions);
- mixes (or integrates or links) the two forms of data concur-rently by combining them (or merging them), or sequentially by having one build on the other, and in a way that gives prior-ity to one or to both;
- uses these procedures in a single study or in multiple phases of a program of study;
- frames these procedures within philosophical worldviews and a theoretical lens; and
- combines the procedures into specific research designs that direct the plan for conducting the study.

These core characteristics have provided some common features for describing mixed methods research. They evolved from many years of reviewing mixed methods articles and determining how researchers use both qualitative and quantita-tive methods in their studies.

I am not alone in proposing some common features. In a highly cited JMMR article, Johnson, Onwuegbuzie, and Turner (2007) suggested a composite definition for mixed methods based on 19 definitions provided by 21 highly published mixed methods researchers. After sharing these definitions, they noted the variations in definitions, from what was being mixed (e.g., methods, methodologies, or types of research), the place in the research process in which mixing occurred (e.g., data collection, data analysis), the scope of the mixing (e.g., from data to worldviews), the purpose or rationale for mixing (e.g., breadth, corroboration), and the elements driving the research (e.g., bottom-up, top-down, the core component). Incorporat-ing these diverse perspectives, the authors end with a compos-ite definition:

> Mixed methods research is the type of research in which a researcher or team of researchers combines elements of qualitative and quantitative research approaches (e.g., use of qualitative and quantitative viewpoints, data collection, analysis, inference tech-niques) for the purposes of breadth and depth of understanding and corroboration. (p. 123)

In this definition, the authors do not view mixed methods simply as methods, but more as a methodology that spans from viewpoints to inferences. They do not view mixed methods as only data collection, but rather as the more general combination of qualitative and quantitative research. They incorporate diverse viewpoints, but do not specifically mention paradigms (as in the Greene et al., 1989, definition) or philosophy (as in the

Creswell & Plano Clark, 2007, definition). Their purposes for mixed methods—breadth and depth of understanding and corroboration—do not speak to how the research question may suggest mixed methods rather than force-fitting a line of inquiry into either a quantitative or qualitative approach. Perhaps most important, they suggest that there is a common definition that should be used.

Another definition has been advanced by Greene (2007), who stated that mixed methods was an orientation toward looking at the social world

> that actively invites us to participate in dialogue about multiple ways of seeing and hearing, multiple ways of making sense of the social world, and multiple standpoints on what is important and to be valued and cherished. (p. 20)

This definition has moved mixed methods into an entirely new realm of conceptualization, and perhaps a useful one. Defining mixed methods as "multiple ways of seeing" opens up broad applications beyond using it as only a research method. It can be used, for example, as an approach to think about designing documentaries (Creswell & McCoy, in press), or a means for "seeing" participatory approaches to HIV-infected populations in the Eastern Cape of South Africa (Olivier, de Lange, Creswell, & Wood, 2009). Lately, I have begun my workshops on mixed methods by indicating that we have many instances of mixed methods in our social world. I start with Al Gore's film-documentary, *An Inconvenient Truth*, about global warming and Gore's combined use of mixed method-like statistical trends and personal stories (David, Bender, Burns, & Guggenheim, 2006). Defining mixed methods as a way of seeing opens up applications for it in many aspects of social life.

However, I still have unresolved concerns after reviewing these diverse definitions. Do we need a common definition or common set of core characteristics? Will such common features limit what we see as mixed methods? Do we need multiple definitions? For those individuals new to mixed methods, do they need a commonly accepted definition to convey the purpose of their research and to convince others of the legitimacy of their approach?

▣ THE QUESTIONABLE USE OF QUALITATIVE AND QUANTITATIVE DESCRIPTORS

Researchers talk about mixed methods using descriptors such as "qualitative" and "quantitative." The use of statistics and stories in Gore's film reinforces a binary distinction between qualitative and quantitative research. Are the terms "qualitative" and "quantitative" useful descriptors to use? What are the inferences being made when these terms are used? This controversy has brought forward one group of writers who have found the terms "qualitative" and "quantitative" intermingled with designs and paradigms, rather than referring to methods of data collection and analysis. It also has brought forward another group of writers who feel that the use of these terms fosters an unacceptable binary or dichotomy that minimizes the diversity in methods.

Giddings (2006) felt that the terms "qualitative" and "quantitative" became normative descriptors for research paradigms in the 1970s and 1980s, and that the term "qualitative" gave nonpositivist researchers "a place to stand" (p. 199). When writers have used the term "qualitative paradigm," it has often been in the context of the qualitative-quantitative debates in evaluation and the social sciences (Greene, 2007). Greene pointed out that it was helpful to separate the research methods of "qualitative" and "quantitative" from broader philosophical issues, and to refrain from intermingling methods and philosophy. Another intermingling occurs at the design level. Vogt (2008) took the strong position: "To think in terms of quantitative and qualitative *designs* is a category mistake" (p. 1, emphasis added). He felt that all research designs—such as surveys, document analysis, experiments, and quasi-experiments—could accommodate data coded as numbers and words.

The use of "qualitative" and "quantitative" has been further discouraged because it creates a binary distinction that does not hold in practice. Often writers equate "qualitative" to text data and "quantitative" to numbers data. In a recent *JMMR* article, Sandelowski, Voils, and Knafl (2009) countered this binary thinking by pointing out that counting often involved qualitative judgments, and that numbers often related to context. Further, qualitative data are sometimes transformed in data analysis into categorical data, and a binary configuration overlooked both within-group (e.g., qualitative) and between-group similarities (e.g., qualitative and quantitative). Resonating with this thought, Giddings (2006) stated that binary positioning made methodological diversity invisible.

Adding confusion to the meaning of "qualitative" and "quantitative" have been those who felt that mixed methods should mean collecting mono-methods—*multiple* qualitative sources of data or quantitative sources of data (Shank, 2007; Vogt, 2008) instead of collecting *both* qualitative and quantitative data (mixed methods). Some writers have been clear that multiple sources of one kind of data (i.e., qualitative or quantitative data) should be called "multiple methods" (Morse & Niehaus, 2009, Appendix 1), not mixed methods. Again, regardless as to how mixed methods is viewed, both perspectives rely on a normative, binary distinction between "qualitative" and quantitative" to reinforce their positions.

A strong case can be made that "qualitative" and "quantitative" should refer to methods. A useful diagram is advanced by Crotty (1998), who provided a conceptual framework for sorting out these layers of research into epistemology, theoretical perspectives (e.g., feminist theory), methodology, and methods. But to throw out the terms "quantitative" and "qualitative" seems to disrupt a long-established pattern of communication that has been used in the social, behavior, and health sciences. Until we have replacement terms, a means of discourse across fields is helpful, but we need to be careful how we use the terms. On the issue of the binary distinction, writers in the mixed methods field have tended to dismiss the dichotomy in favor of a continuum for presenting qualitative and quantitative differences (Creswell, 2008; Tashakkori & Teddlie, 2003). Writers in mixed methods are also careful to distinguish "multi-method studies" in which multiple types of *qualitative* or *quantitative* data are collected (see Creswell & Plano Clark, 2007) from "mixed methods studies" that incorporate collecting *both* qualitative and quantitative data. In the health sciences, the term "multi-method" is typically used to convey studies in which both forms of data are gathered (e.g., see Stange, Crabtree, & Miller, 2006), although in a study of National Institutes of Health–funded projects, Plano Clark (2009) found that "multimethod" meant multiple methods of quantitative *or* qualitative data 64% of the time, and "mixed methods" 36% of the time.

In light of these discussions about intermingling and the binary distinction, should we refrain from using the terms "qualitative" and "quantitative?" Why do mixed methods writers not clearly distinguish among methods, designs, and paradigms? Should mixed methods involve multiple qualitative or quantitative methods or some combine of both?

▣   THE NEW VERSUS THE OLD

Historically, researchers have used both forms of methods in these studies. This leads to another controversy: Is mixed methods a "new" approach or is it simply pouring new ideas into old packaging? Emphasizing the "new," recent writers have called mixed methods the third methodological "movement" (following quantitative and qualitative) (Tashakkori & Teddlie, 2003, p. 5), the "third research paradigm" (Johnson & Onwuegbuzie, 2004, p. 15), and "a new star in the social science sky" (Mayring, 2007, p. 1). Claims such as these have left some critics to wonder "exactly what the new mixed methods movement is claiming. The major proponents insist that what they developed is a new way of doing research" (Holmes, 2006, p. 2).

I often date the beginnings of mixed methods back to the late 1980s and early 1990s with the coming together of several publications all focused on describing and defining what is now known as mixed methods. These writers worked independently and they came from sociology in the United States (Brewer & Hunter, 1989) and in the United Kingdom (Fielding & Fielding, 1986), from evaluation in the United States (Greene et al., 1989), from management in the United Kingdom (Bryman, 1988), from nursing in Canada (Morse, 1991), and from education in the United States (Creswell, 1994). A critical mass of writings came together within a short space of time, and all of these individuals were writing books, chapters, and articles on an

approach to research that moved from simply using quantitative and qualitative approaches as distinct, separate strands in a study to research that actually linked or combined them. At this time, qualitative inquiry had become largely accepted as a legitimate methodology in the social sciences and was moving into the "blurred genres" stage (Denzin & Lincoln, 2005). Philosophical debates between quantitative and qualitative researchers were still underway (Reichardt & Rallis, 1994) but beginning to soften, and new methodologies to address the complex problems of society were being encouraged.

In retrospect, I now wonder if these writers were truly the first individuals to talk about combining quantitative and qualitative data. Individuals in my workshops have for some time been saying that mixed methods is not "new." Holmes (2006) raised this question when he commented,

> The major proponents insist that what they have developed is a new way of doing research—an alternative to qualitative and quantitative research, but what's new about that? . . . ethnographers and other social researchers have been gathering data using mixed methods at least since the 1920s, and case study researchers and anyone using triangulation have also been using mixed methods. (p. 2)

To probe whether or not it is a "new" idea requires returning to historical documents in fields such as sociology, evaluation, and action research. How does the pre-late-1980s discussion fit with what is known about mixed methods today? Three threads of thinking prior to the late 1980s can give us insight: the use of multiple methods, the discussions about using qualitative research within a research world largely dominated by quantitative research, and the informal initiatives to combine methods.

In terms of multiple methods, in 1959 Campbell and Fiske advanced the use of multiple methods in convergent and discriminant validation of psychological traits using a multitrait-multimethod matrix. They felt that more than one trait as well as more than one method must be employed in the validation process. Their discussion, however, was limited to multiple *quantitative* sources of data. During the 1970s, Denzin (1978) identified several types of combinations of methodologies in the study of the same phenomena or programs through his idea of *data triangulation*—the use of various data sources in a study. He said, "I now offer as a final methodological rule the principle that multiple methods should be used in every investigation" (Denzin, 1978, p. 28).

Throughout the 1970s and on into the 1980s, several noted authors were calling for the use of qualitative research on equal footing with more quantitative-experimental methods (Patton, 1980). Campbell (1974) gave a noted presentation at the American Psychological Association meeting on "Qualitative Knowing in Action Research" for the Kurt Lewin Award address. He suggested that a true scientific approach was to eliminate the question of the position of ultimate authority between quantitative and qualitative research and to reestablish the importance of qualitative research. Cronbach (1975), in his well-known article "Beyond the Two Disciplines of Scientific Psychology," cast doubt on the idea that the social sciences could be modeled only on the natural sciences. Both Campbell and Cronbach started out as quantitative researchers and then embraced qualitative or naturalistic research through their writing.

Other authors began combining methods informally, and these writers were clearly the pioneers of mixed methods thinking today. In sociology, Sieber (1973) discussed the "interplay" of fieldwork and survey methods, and identified procedures for combining the two methods. Lamenting the fact that there were "too few examples to adduce general principles" (p. 1358), Sieber suggested the need for a "new style of research" (p. 1337). He further discussed the sequence of both methods with "concurrent scheduling" and "interweaving" the two methods (p. 1357). Equally important, he cited a number of studies that incorporated both interviews and surveys, and he discussed his own projects that included these forms of data collection (Sieber & Lazersfeld, 1966).

Another example of early mixed methods thinking comes from the field of evaluation in which Patton advanced "methodological mixes" (Patton, 1980, p. 108). He advocated for the use of anthropological naturalistic research in evaluation based on the "holistic-inductive paradigm" to complement the more traditional "hypothetical-deductive" approach. He recommended several models for program evaluation built on this combination. A design could be the pure hypothetical-deductive approach with an experimental design, quantitative data, and a statistical analysis, or a pure qualitative approach with naturalistic inquiry, qualitative measurement, and a content analysis. Then he suggested four "mixed form" models (p. 112) that varied from using experimental or naturalistic designs, qualitative or quantitative measurements, and often the transformation of qualitative data into counts. The diagram he sketched for the four models was remarkably similar to diagrams of mixed methods designs presented by recent authors (e.g., Johnson & Onwuegbuzie, 2004; Tashakkori & Teddlie, 1998).

Taking these readings as a whole, a good case can be made that mixed methods was underway much earlier than the late 1980s. These early writers focused on gathering multiple methods, including both quantitative (e.g., surveys) and qualitative (e.g., interviews) data. They initiated a language for mixed methods through such terms as the more general word "interplay" and more specific terms, such as "concurrent scheduling" (Sieber, 1973, pp. 1353, 1358). They provided examples of studies that employed multiple methods, and they took a process approach of thinking about the "interplay" through design, data collection, and data analysis. They conceptualized different types of mixed methods designs, such as those involving data transformation (Patton, 1980), and those including one form of method building on the other (Sieber, 1973).

On the other hand, although these early writers were interested in the "interplay" of quantitative and qualitative data, they did not specifically discuss how they would integrate the two data sources, or the reasons for integration as mixed methods is described today (e.g., see Bryman, 2006). They did not explicate the vast array of design possibilities in response to different purposes that is seen today (Creswell & Plano Clark, 2007, 2011). Although they started the discussion about names for the designs, they had a limited repertoire for designs (e.g., concurrent scheduling) as compared to the extensive list of design possibilities discussed recently (see Creswell & Plano Clark, 2011). They did not have a notation system (e.g., pluses and arrows) for providing a shorthand description of designs that would begin to emerge in 1991 (see Morse, 1991). Some of the detailed discussions about procedures (e.g., developing an instrument based on qualitative data), the use of mixed methods questions (Creswell & Plano Clark, 2011), or the larger philosophical issues (see Greene, 2007) were not present in their discussions.

The pre-late-1980s writers did, however, lay a foundation for mixed methods. As Tashakkori and Teddlie (2003) commented, these early writers "were mostly unaware that they were doing anything out of the ordinary" (p. 5). They used informal, commonsense ways of conducting research. A colleague recently remarked, "What is most amazing about mixed methods is that all of these (current) writers have taken ideas that have been around for a long time and spun them into a way of research, a methodology!" (Duane Shell, personal communication, August 17, 2009). Today, we have systematic, detailed, and defined ways of thinking about mixed methods research. But is a systematic approach better than the more intuitive early approach? Why do current mixed methods researchers (including myself) not give more credit to the early researchers who had the initial ideas that have now been embraced today as mixed methods?

▣ WHAT REALLY DRIVES MIXED METHODS?

The ideas of a "new movement" or a "new star" suggest that some trends in methodology are building. What has promoted the escalation of interest in mixed methods? As suggested at the Aberdeen, Scotland, seminar, is it simply a response to funding initiatives?

Interest in mixed methods has grown since the *Handbook of Mixed Methods in Social & Behavioral Research* (Tashakkori & Teddlie, 2003) was published eight years ago (Creswell, 2009b). This handbook, consisting of four sections covering 759 pages, addressed current and future issues, methodological issues, and analytical issues. Using the base year of 2003 as a rough benchmark, it has been documented how interest has developed in the use of the term "mixed methods," as reported in funded projects at the National Institutes of Health (Plano Clark, 2010). Journals exclusively devoted to reporting mixed methods empirical studies and methodological discussions have also been initiated, such as in 2007, the *Journal of Mixed Methods Research* (Sage); in 2008, the *International Journal of Multiple Research Approaches* (eContent Management Pty); and in 2009, the *International Journal of Mixed Methods in Applied Business & Policy Research* (Academic Global). To these journals, I can also add journals started much earlier, such as *Quality and Quantity* (1967, Springer), *Field Methods* (1989, Sage), and the *International Journal of Social Research Methodology* (1998, Routledge). In addition, a number of recent journals have published special issues focusing exclusively or largely on mixed methods, such as *Research in Schools* (2006), *Annals of Family Medicine* (2004), and the *Journal of Counseling Psychology* (2005). At least 16 major books have been written about mixed methods, including recent books by Creswell and Plano Clark (2011), Greene (2007), Plano Clark and Creswell (2008), Teddlie and Tashakkori (2009), and Morse and Niehaus (2009). Mixed methods books are being published that have a distinct discipline focus, such as for nursing and health researchers (Andrew & Halcomb, 2009) and psychologists (Mayring, Huber, Gurtler, & Kiegelmann, 2007; Todd, Nerlich, McKeown, & Clarke, 2004). Chapters can be found in methods books in discipline fields such as social work (Engel & Schutt, 2005) and family research (Greenstein, 2006). An international conference on mixed methods has been offered in the United Kingdom for the last five years, along with international publications on mixed methods around the globe: in psychology from Europe (Mayring et al., 2007), in nursing from Australia (Andrew & Halcomb, 2009), in linguistics from Japan (Heigham & Croker, 2009), in the social sciences from Switzerland (Bergman, 2008), and in education from South Africa (Creswell & Garrett, 2008).

In light of these developments, I must ask what has given impetus to this interest? It may well be that funding sources have encouraged mixed methods research with the global economic imperative—starting in the 1990s—to do more with less (Giddings, 2006). In a mixed methods study of family adoption practices, Miall and March (2005) wrote about how their funders forced them to change their questions and design from their initial plan of starting with quantitative questions that would be intentionally followed by qualitative questions. Holmes (2006) alleged that mixed methods reduced researchers to "depersonalized technicians," which tacitly supported funding agencies to seek projects with convergence on a single answer rather than differences in opinions and beliefs.

On the other side, certainly the legitimacy of qualitative research has encouraged researchers to think in a pluralistic way. Interdisciplinary research problems now call for addressing complex issues using skilled methodologists from both quantitative and qualitative research who bring diverse approaches to studies (Mayring et al., 2007). Still, questions linger about whether mixed methods is simply a response to funding interests and whether the research questions addressed by mixed

methods researchers truly merit a "mixed" methodology. Those coming from a philosophical, postmodern perspective have suggested that researchers are "accepting uncritically and undigested" mixed methods (Freshwater, 2007, p. 145).

▣ THE PARADIGM DEBATE CONTINUES

Philosophically oriented writers for years have debated whether mixed methods research is possible because it mixes worldviews or paradigms. They ask: Can paradigms (ontologies or realities) be mixed? Some writers adhere to the idea that paradigms or worldviews have rigid boundaries and cannot be mixed. Holmes (2006) asked: "Can we really have one part of the research which takes a certain view about reality nested alongside another which takes a contradictory view? How would we reconcile, or even work with, competing discourses within a single project?" (p. 5). The logic being used here was that mixed methods was untenable because methods were linked to paradigms, and therefore the researcher, in using mixed methods research, was mixing paradigms. This stance has been described as the purist stance (see Rossman & Wilson, 1985), and it has been called the "incompatibility thesis" (Howe, 2004) and discussed in the mixed methods literature as mixing viewpoints (Johnson et al., 2007, p. 123). Individuals that hold this position view paradigms as having discrete and impermeable boundaries, an idea reinforced by the clear-cut boxes and lines around the alternative inquiry paradigms in the literature (e.g., see Guba & Lincoln's tables, 2005; or Creswell's table of worldviews, 2009c). Granted, by 2005, Guba and Lincoln had taken down these artificial boundaries by declaring cautiously that elements of paradigms might be blended together in a study. Contributing to this perspective was certainly a "delinking" of paradigms and methods, such as conveying that many different research methods would be linked to certain paradigms, and that a paradigm justification did not dictate specific data collection and analysis methods (Johnson & Onwuegbuzie, 2004).

With the gate now opened to thinking about use of multiple paradigms, mixed methods writers have now taken varied stances on incorporating paradigms into mixed methods. For example, a dialectic stance by Greene and Caracelli (1997) suggested that multiple paradigms might be used in mixed methods studies, but that each paradigm needed to be honored and that their combined use contributed to healthy tensions and new insights. In my writings, I took a similar stance, but suggested that multiple paradigms related to different phases of a research design (Creswell & Plano Clark, 2007, 2011), thus linking paradigms to research designs. For example, a mixed methods study that begins with a quantitative survey phase reflects an initial postpositivist leaning, but, in the next qualitative phase of focus groups, the researcher shifts to a constructivist paradigm. Relinking paradigms and designs makes sense.

Still others advocated for one underlying paradigm that fits mixed methods, and some found their paradigm in pragmatism with historical roots back to Charles Peirce, William James, John Dewey, Richard Rorty, and others (Johnson & Onwuegbuzie, 2004; Tashakkori & Teddlie, 2003). Pragmatism emphasizes the importance of the research questions, the value of experiences, and practical consequences, action, and understanding of real-world phenomena. Advocates said that it is a "philosophical partner for mixed methods research" (Johnson & Onwuegbuzie, 2004, p. 16). A different paradigmatic stance, suggested by Mertens (2003, 2009), is found in the transformative-emancipatory framework that made explicit the goal for research to "serve the ends of creating a more just and democratic society that permeates the entire research process" (Mertens, 2003, p. 159). Mertens thus creatively relates this goal to different phases in designing a mixed methods study.

Whether the paradigm for mixed methods involves a single paradigm, multiple paradigms, or phased-in paradigms, Morgan (2007) recently reminded the mixed methods community of the importance of Kuhn's (1970) original description of a paradigm. Using the definition of a paradigm as "shared belief systems that influence the kinds of knowledge researchers seek and how they interpret the evidence they collect" (Morgan, 2007, p. 50), Morgan found paradigms to be (1) worldviews, an all-encompassing perspective on the world; (2) epistemologies, incorporating ideas from the philosophy of science such as ontology, methodology, and epistemology; (3) "best" or "typical" solutions to problems; and (4) shared beliefs of a "community of scholars" in a research field. It is this last perspective (embraced by Kuhn, 1970) that Morgan strongly endorses, and he discussed how researchers share a consensus in specialty areas about what questions are most meaningful and which procedures are most appropriate for answering their questions.

Another mixed methods writer, Denscombe (2008), agreed with this perspective and took it one step further. Denscombe outlined how "communities" may work using such ideas as sharing identity, researching common problems, forming networks, collaborating in pursuing knowledge, and developing informal groupings. This line of thinking has focused attention on the emerging fragmentation of the mixed methods field in which various disciplines adopt mixed methods in different ways, create unique practices, and cultivate their own specialized literatures. For example, at the Veterans Administration Research Center in Ann Arbor, Michigan, in the health sciences, colleagues have conceptualized mixed methods as formative and summative evaluation procedures (Forman & Damschroder, 2007). This conceptualization adapts mixed methods to the Veterans Administration health services context of intervention research. The rise of discipline-oriented mixed methods books is another instance of adapting mixed methods to scholarly communities. Still, I wonder if discipline fragmentation of mixed methods will lead to further philosophical differences among scholars in mixed methods. Will the scholarly community line of thinking continue or will the conversation return to the difficulty of mixing realities? Is the idea of mixing realities actually all about whether one paradigm takes precedence over another in mixed methods research?

◻ MIXED METHODS PRIVILEGES POSTPOSITIVISM

Critics make the allegation that mixed methods favors postpositivist thinking over more interpretive approaches. Does mixed methods privilege postpositivist thinking and marginalize interpretive approaches? Several authors have taken this position. The context for many of these concerns resides in what is seen as a conservative challenge to qualitative inquiry (Denzin & Giardina, 2006). Denzin and Giardina believe that conservative regimes enforce scientifically based models of research (SBR). For example, the 2001 No Child Left Behind Act (NCLB) in education emphasized accountability, high-stakes testing, and performance scores for students. The model for research being advanced was to "apply rigorous, systematic, and objective methodology to obtain reliable and valid knowledge" (Ryan & Hood, 2006, p. 58). Within this context, qualitative research is marginalized, and it minimizes complex and dynamic contexts, subtle social differences produced by gender, race, ethnicity, linguistics status, and class, and multiple kinds of knowledge (Lincoln & Canella, 2004). In 2002, one year after the No Child Left Behind Act was implemented, the National Research Council established guidelines in their report, *Scientific Research in Education,* that called for a quantitative approach to research through guiding principles asking for significant questions that could be empirically studied, relevant theory, methods closely tied to the research questions, explanations of findings using a logical chain of reasoning, replicated studies and generalizations, and disseminated research for critique by the professional scientific community (Ryan & Hood, 2006; Shavelson & Towne, 2002). Howe (2004) called the National Research Council's perspective "mixed-methods experimentalism" (p. 48) and felt that it assigned a prominent role to quantitative experimental research and a lesser role to qualitative, interpretive research. Further, this approach "elevates quantitative-experimental methods to the top of the methodological hierarchy and constrains qualitative methods to a largely auxiliary role in pursuit of the *technocratic* aim of accumulating knowledge of 'what works'" (Howe, 2004, pp. 53–54). He also stated, "It is not that qualitative methods can never be fruitfully and appropriately used in this way, but their natural home is within an interpretivist framework with the democractic aim of seeking to understand and give voice to the insider's perspective" (p. 54). This interpretivist aim values outcomes assessed by various stakeholders, includes all relevant voices in the dialogue, and engages in qualitative data collection procedures to promote dialogue, such as participant observation, interviews, and focus groups. This dialogue also needs to be *critical* with the views of participants subjected to rational scrutiny.

Howe's theme was echoed again in the following years' publication of the third edition of this handbook (Denzin & Lincoln, 2005). Denzin and Lincoln talked directly about the "mixed-methods movement" as taking qualitative methods out of their natural home, which is in the "critical, interpretive framework" (p. 9). Finally, in a provocative article by Giddings (2006), titled, "Mixed-methods Research: Positivism Dressed in Drag?" the issue of the hegemony of positivism and the marginalization of nonpositivist research methodologies in mixed methods was addressed. She conveyed the idea that certain "thinking" went on in research that was reflected in methodologies and "the 'thinking' of positivism continues in the 'thinking' of mixed methods" (p. 200). Giddings felt that this mixed methods "thinking" was expressed through analysis and prescriptive styles, structured approaches to research design and data collection, and the use qualitative aspects "fitted in" (p. 200).

There is little doubt that a good case can be made that, in certain approaches, mixed methods researchers have relegated qualitative inquiry to a secondary role. A good example would be the embedded research design (Creswell & Plano Clark, 2007), in which qualitative methods often provide a supportive role in experimental, intervention studies. Our feeling has long been that the use of qualitative approaches *whatever their role* in traditional quantitative experiments elevates qualitative research to a new status and opens the door for seeing qualitative research as a legitimate form of inquiry. Whether this will materialize can certainly be debated. The structured ways of designing mixed methods projects that we embrace in our text (Creswell & Plano Clark, 2007) also reinforces Giddings' idea of the structured "thinking" in our approach to mixed methods. In mixed methods data analysis, the use of "manifest effect sizes" by Onwuegbuzie and Teddlie (2003, p. 356) reinforces a postpositivist leaning of mixed methods.

On the other hand, many studies in mixed methods can be found that give priority to qualitative methods. Some designs subordinate quantitative methods to qualitative methods (see the exploratory sequential design mentioned by Creswell & Plano Clark, 2007). Also, the writings on applying the transformative-emancipatory framework to mixed methods emphasize qualitative research (Mertens, 2009). A close reading of the National Research Council's report on scientific research in education shows that the types of questions recommended for scholarly educational research were both quantitative (descriptive, experimental) as well as qualitative (exploratory), a point that Howe (2004) concedes. Although more critical, interpretivist articles are needed in the mixed methods field, some evidence exists that the number of articles is growing. A recent paper (Sweetman, Badiee, & Creswell, 2010) has identified mixed methods studies

that honor the inclusion and dialogue of communities of action within Mertens's transformative-emancipatory framework. This paper examined several mixed methods studies that addressed disability, ethnic, feminism, and social class as theoretical standpoints and advanced ways that researchers might incorporate these standpoints into their mixed methods projects. Further evidence of standpoint epistemology—typically found in qualitative research—is found in recently published articles in *JMMR* addressing women's social capital (Hodgkin, 2008) and African American women's interest in science (Buck, Cook, Quigley, Eastwood, & Lucas, 2009). Despite these studies, what is the evidence that mixed methods research marginalizes interpretive approaches? Do we need more mixed methods research that incorporates an interpretive perspective? Is the use of qualitative research in a supportive role in intervention studies marginalizing qualitative inquiry, or is it advancing it within fields that traditionally honored experimental methods? Do we need more articles that embrace "mixed methods interpretivism," in which quantitative research is relegated to a secondary role within qualitative research, as Howe (2004) would recommend?

▣ A FIXED DISCOURSE IN MIXED METHODS

Unquestionably, more interpretive, theoretical studies in mixed methods would broaden the audience and discourse of it. This raises another controversy about the discourse of mixed methods. Some critics are asking: Is there a dominant discourse in mixed methods? Is messiness allowed in? These questions speak to the issue of mixed methods privileging postpositivist thinking—a postmodern concern about the discourse in mixed methods. Who controls this discourse and the language that is being used in mixed methods research? Several authors have weighed in on this issue.

A recent important article takes up these concerns (Freshwater, 2007). Freshwater is an editor and leading researcher in nursing as well as a postmodernist. She was concerned about how mixed methods was being "read" and the discourse that followed. Discourse was defined as a set of rules or assumptions for organizing and interpreting the subject matter of an academic discipline or field of study in mixed methods. The uncritical acceptance of mixed methods as an emerging dominant discourse ("is nearing becoming a metanarrative" [Freshwater, 2007, p. 139]) impacts how it is located, positioned, presented, and perpetuated. She called on mixed methods writers to make explicit the internal power struggle between the mixed methods text as created by the researcher and the text as seen by the reader/audience. Mixed methods, she felt, was too "focused on fixing meaning" (p. 137). Expanding on this, she stated that mixed methods was mainly about doing away with "indeterminancy and moving toward incontestability" (p. 137), citing as key examples the objective third-person style of writing, the flatness, and the disallowance for competing interpretations to coexist. She requested that mixed methods researchers adopt a "sense of incompleteness" (p. 138) and recommended that reforms required the

> need to explore the possibility of hybridization in which a radical intertextuality of mixing forms, genres, conventions, and media is encouraged, where there are no clear rules of representation and where the researcher, who is in reality working with radical undecidability and circumscribed indeterminacy, is able to make this experience freely available to readers and writers. (p. 144)

These ideas were a positive criticism, and a call for mixed methods writers to insert questions into their discourses, to acknowledge the messiness of mixed methods, and to recognize that it is a field still in "adolescence" (Tashakkori & Teddlie, 2003, p. x).

Still, by providing a visual of the mixed methods research process that follows linear development, Johnson and Onwuegbuzie (2004) erase the "messiness." A certain tidiness is given when specific names are assigned to research designs (e.g., explanatory sequential designs—Creswell & Plano Clark, 2007), when researchers do not attend to the "messiness" in conducting the designs (e.g., see Creswell et al., 2008), and when writers look for a consensus in definitions (Johnson et al., 2007). These examples all point toward "fixing" or the field "being fixed." But these points open up further questions, such as how should mixed methods writers discuss its messiness, its blurred borders, and its problems? Will unstructured mixed methods serve well the beginning researcher as well as the more experienced researcher?

▣ TO BE BILINGUAL OR NOT

A related issue is whether any one ideological camp dominates the language of mixed methods research. Is there a dominant language or set of terms for mixed methods? Vygotsky and Cole (1978) propose that the sociocultural perspective of language shapes how individuals make sense of the world, and that the learning process consists of a gradual internalization of this language. What is the language of mixed methods? One issue being discussed is whether we need a "bilingual" language for mixed methods research so that it does not favor quantitative or qualitative research. Raising this question is reminiscent of concerns in qualitative research in the early 1980s around the topic of qualitative validity, and how terms such as trustworthiness and authenticity created a "new," distinct language to discuss validity (Lincoln & Guba, 1985).

As the language of mixed methods develops, a confusing picture has emerged about the nomenclature to use. For example, in writing about validity, Onwuegbuzie and Johnson (2006) intentionally called validity "legitimation" and thereby created a new

word in the mixed methods lexicon. In our specification of types of research designs, we created new names, such as the "exploratory sequential design," to provide a descriptive label signifying that the design would first fulfill the intent of exploring using qualitative data followed by explanation using quantitative data (Creswell & Plano Clark, 2007). Illustrating an example of a made-up bilingual term, writers in a recent psychology text used the term "qualiquantology" to express their discomforting hybridity of mixing qualitative and quantitative methods (Stenner & Rogers, 2004).

Other writers in the mixed methods field use a less bilingual vocabulary. Leaning toward a more quantitative language, Teddlie and Tashakkori (2009) use the term "inferences," or "meta-inferences," to denote when the results are incorporated into a coherent conceptual framework to provide an answer to the research question. Although "inferences" may relate to either qualitative or quantitative research, it seems to be employed frequently in drawing conclusions from a sample to a population in a quantitative study. Another example is the use of the term "construct validity" by Leech, Dellinger, Brannagan, and Tanaka (2010) as an overarching validity concept for mixed methods research. This term is drawn from quantitative measurement ideas. On the qualitative side, the idea of personal transformation advanced by Mertens (2009) clearly has qualitative roots. Unquestionably, the language that has emerged is both bilingual and oriented toward one form of inquiry (quantitative or qualitative). The use of glossaries in recent mixed methods books suggests the need for a common vocabulary (see Morse & Niehaus, 2009; Teddlie & Tashakkori, 2009). These examples, however, raise difficult questions about who controls the language of mixed methods, how it is conveyed, and what the language should be. It also introduces questions about how the writing up of mixed methods proposals and projects influences what gets approved, funded, and published.

▣ A BAFFLING (AND COMPLEX) ARRAY OF DESIGNS

It is not only the language that introduces confusion and controversy into the mixed methods discourse. In research designs—a topic that has filled the pages of mixed methods writings—researchers are confronted by a baffling array of names and types of ways to conduct mixed methods research. How might a mixed methods researcher conduct a mixed methods study? When my colleague, Vicki Plano Clark, and I wrote an introduction to the field for beginning mixed methods researchers (Creswell & Plano Clark, 2007), we presented 12 different classification systems of designs drawn from diverse fields of evaluation, nursing, public health, and education.

Not wanting to add to the confusion, we suggested a parsimonious set of designs. Triangulation (or now called convergent) designs involved one phase of qualitative and quantitative data collection gathered concurrently. Explanatory or exploratory designs required two phases of data collection, quantitative data collection followed sequentially by qualitative data collection (or vice versa). Embedded designs, in which one form of data was embedded within another, may be either a single- or a double-phase design with concurrent or sequential approaches. In all of these designs, we focused on the weight given to qualitative and quantitative data, the timing of both forms of data, and the mixing of the data in the research process. To present these designs, we used a modified notation system first developed by Morse (1991), and we sketched diagrams of procedures and advanced guidelines for constructing these diagrams found in the literature (Ivankova, Creswell, & Stick, 2006).

We now know that these designs are not complex enough to mirror actual practice, although our thinking at the time was to advance designs for the first-time mixed methods researcher. Also, we are more aware of the complex designs being used and reported in the literature. For example, Nastasi and colleagues wrote about a complex evaluation design with multiple stages and the combination of both sequential and concurrent phases (Nastasi et al., 2007). The designs reported in journals have incorporated "unusual blends" of methods, such as combinations of quantitative and qualitative longitudinal data, discourse analysis with survey data, secondary data sets with qualitative follow-ups, and the combination of qualitative themes with survey data to produce new variables (Creswell, 2011). The representation of designs has also advanced joint matrices for arraying both quantitative and qualitative data in the same table, an approach encouraged by the matrix feature of qualitative software products (see Kuckartz, 2009).

Our designs and the many classifications bring a typology approach to mixed methods design. Arguing that we need an alternative to typologies, Maxwell and Loomis (2003) conceptualized a systems approach of five interactive dimensions of the research process consisting of the purpose, the conceptual framework, the questions, the methods, and the issue of validity. With this approach, they provided a fuller, more expansive view of the way to conceptualize mixed methods designs. Another approach comes from the creative thinking of Hall and Howard (2008). They suggested a synergistic approach in which two or more options interacted so that their combined effect was greater than the sum of the individual parts. Instead of looking at mixed methods as a priority of one approach over the other, or a weighting of one approach, the researcher considered their value and representations equal. The researcher also viewed the two as equal from an ideology of multiple points of view, balancing objectivity with subjectivity. Collaboration consisted of the equal skill expertise about qualitative and quantitative methodologies on a research team.

The synergistic approach, along with other challenges to typological perspectives has contributed to a softening of the differences between qualitative and quantitative research, provided answers to questions about dominance of one method over the other (e.g., Denzin & Lincoln, 2005), and honored the formation of research teams with diverse expertise. In light of these discussions, are typologies of research designs outdated? Are newer, more free-flowing designs an improved way to think about designing a mixed methods study?

▣ MISAPPROPRIATING DESIGNS

Another procedural question about designs is whether mixed methods is misappropriating designs from other fields. As mixed methods continues to grow in popularity and use, is the field misappropriating traditional designs and calling them "mixed methods" (thereby overstating the value and claims of mixed methods)? Several examples stand out. Scale development (DeVellis, 1991) has been available to the researcher for many years in quantitative research. Early phases of scale development often call for an initial exploration, even though this may consist of reviewing the literature rather than conducting an extensive qualitative data collection procedure, such as the use of focus groups (Vogt, King, & King, 2004). One might argue that scale development should be a distinct procedure from mixed methods research, and yet, mixed methods designs with the purpose of developing an instrument are available in the journal literature (e.g., Myers & Oetzel, 2003).

Another example would be content analysis, a quantitative procedure involving the collection of qualitative data and its transformation and analysis by quantitative counts. In this approach, *both* qualitative and quantitative are not collected, but both qualitative research (in data collection) and quantitative research (in data analysis) are employed. If one views mixed methods as collecting both quantitative and qualitative data, then content analysis does not qualify as mixed methods research. Is content analysis a separate approach or is data transformation also a part of mixed methods designs as suggested by Sandelowski et al. (2009)? What are appropriate boundaries for mixed methods research?

Perhaps mixed methods is actually a subordinated set of procedures used within a large number of designs. I call this approach using a "framework" for conducting mixed methods procedures. It is basically the idea that some larger framework becomes a placeholder within which the researcher gathers quantitative and qualitative data (or conducts mixed methods procedures). This idea first surfaced when a participant at a workshop asked, "Is ethnography mixed methods research?" The sense of this question was that ethnographers have traditionally collected both quantitative and qualitative data and used both in their description and analysis of culture-sharing groups. Morse and Niehaus (2009) discussed this question, and concluded that many ethnographers do see their methodology as a distinct approach, and that ethnography needs to be viewed as independent of mixed methods.

But I wonder if seeing mixed methods as a subordinate procedure within ethnography is the most appropriate stance. Researchers seem to use mixed methods within larger frameworks of many types. Evidence for these frameworks comes from using mixed methods procedures within narrative studies (Elliot, 2005), experiments (Sandelowski, 1996), and case studies (Luck, Jackson, & Usher, 2006). Other frameworks can be seen as well, such as using mixed methods within a social network analysis (Quinlin, 2010), an overarching research question (Yin, 2006), a feminist lens (Hesse-Biber & Leavy, 2007), or in action research (Christ, 2009). If the mixed methods designs can be stretched to include these different frameworks, then the potential for extending use of mixed methods in many ways is possible. But where is the boundary between mixed methods and other designs? Is a boundary needed? If mixed methods researchers are claiming other designs for their own, can their claims be justified?

◨ **VALUE ADDED?**

Regardless of the design and whether it is appropriate, the utility of mixed methods research—from a pragmatic approach—is tied to whether it is a valuable approach. In our earlier definition (Creswell & Plano Clark, 2007), we end with the assumption that the combination of methods provides a better understanding than either quantitative method or qualitative method alone. Can this assumption be substantiated? In tracing the recent history of mixed methods, I referred to a question asked by the president of Sage Publications during a luncheon meeting. He asked me, "Does mixed methods provide a better understanding of a research question than either quantitative or qualitative research alone?" (Creswell, 2009b, p. 22). This difficult question is central to justifying mixed methods and giving it legitimacy. Unfortunately, it remains unanswered in the mixed methods community.

I can provide a hypothetical series of studies on how it *might* be addressed. One approach is to turn to research procedures used in early studies that compared participant observation with survey results (Vidich & Shapiro, 1955) or interviews with surveys (Sieber, 1973) and examine if the two databases converge or diverge in understanding a research problem. A second approach is to proceed with an experiment in which groups of readers examine a study divided into a qualitative, a quantitative, and a mixed methods part. In this experiment, outcomes are specified such as the quality of interpretation, the inclusion of more evidence, the rigor of the study, or the persuasiveness of the study, and the three groups could be compared experimentally. A third approach is to examine some outcomes suggested by authors of published mixed methods studies. One such outcome might be "yield," such as that advanced by O'Cathain, Murphy, and Nicholl (2007), in which they assess it by the number of publications and whether the authors of a mixed methods study actually integrate the data. Other outcomes could be

analyzed using qualitative document analysis approaches, and themes developed from statements of value posed by authors of mixed methods empirical articles and methodological studies. For example, authors from the field of communication studies suggested that the value of mixed methods lies in addressing limitations in the results learned from one method:

> To address more thoroughly this question, and account for some of the possible limitations of study-one, a broader based assessment of students' involvement in intercultural communication courses was pursued. (Corrigan, Pennington, & McCroskey, 2006, pp. 15–16)

Other options may also exist. The mixed methods community does not have an adequate answer to this controversy, and so I ask: When and how can we begin to answer this question? Does a mixed method better address the core research question being asked in a study than either quantitative or qualitative alone? What criteria should be used in assessing it? Why have mixed methods researchers not pursued this issue more vigorously?

◨ **CONCLUSION**

Striking at the heart of its existence, critical comments about mixed methods are being made about its meaning and definition (raising concerns about expectations, as I learned at Aberdeen). The form of this conversation has been to debate whether mixed methods is a "method," a "methodology," some combination, or a way of seeing. Related to this larger issue is whether it is a "new" way of researching, reinforces a slanted use of terms, and creates a false binary distinction between quantitative and qualitative data (and research).

Assuming that mixed methods researchers take paradigms (i.e., worldviews, beliefs, values) seriously (an assumption that several writers have questioned; see Holmes, 2006, and Sale, Lohfeld, & Brazil, 2002), I see the paradigm discussion as an important discussion in the mixed methods literature. Diverse stances have emerged from a single paradigm perspective, such as pragmatism or the transformational-emancipatory perspective, to multiple paradigm use in a dialectic approach, and to relating the paradigm to the design. Some discussion has moved away from which one paradigm, or how many to use, to a focus on paradigm use within communities of scholars. Still, critics are concerned about whether the current approaches to mixed methods privilege postpositivist thinking and create discourses that "fix" the otherwise messy content of mixed methods.

No subject has been so widely discussed in the mixed methods literature as its designs and its methods. This emphasis places importance on the methods, sometimes at the expense of minimizing the importance of the research question in directing scholarly inquiry (Gurtler, Huber, & Kiegelmann, 2007). At other times, critics of the mixed methods literature see a baffling list of different types of designs with unusual names, the potential of mixed methods claiming many more designs than it deserves, and having questionable outcomes.

The implications of these controversies are that many of them are interrelated and my sorting them out here is contrived—a heuristic. When authors talk about the controversies, I have found their discussion to cover many topics rather than an in-depth analysis of any one controversy. Also, the range of controversies is quite extensive, stretching from basic issues of the legitimacy and meaning of mixed methods to its philosophical underpinnings, and on to the pragmatics of conducting a mixed methods study. Fundamentally, my position is that the mixed methods community needs to squarely place these controversies on the table for discussion and honor their presence.

Some readers will say that I have overlooked critical controversies such as the relationship of research problems to methods, validity, and evaluation of mixed methods, the writing of a mixed methods study, and the common question of "who cares about methods?" Other readers will undoubtedly see my views as deliberately "transgressive " (Richardson, 1997): a turn to challenging mixed methods rather than advocating for it. Others will see my remarks as an attempt to open up the discourse about mixed methods, much like I have advocated in authored and coauthored editorials for the *Journal of Mixed Methods Research*. Still others might consider my justifications both for and against the issues as evidence of postpositivist leanings (or even worse the creation of new metanarratives). All of these renderings may be both right and wrong. As a pragmatist, I can confidently say that I am interested in the consequences of this discussion of controversies, the seeds of which were sprouted at Aberdeen. Perhaps rather than finding irony in the space of Elphinstone Hall in Scotland, I should have seen instead the long shadows that the walls were casting. In the end, I advise those interested in mixed methods to reassess their commitment to controversies now being raised. As Kuhn (1970) said, "A revolution is for me a special sort of change involving a certain sort of reconstruction of group commitments" (p. 181).

▣ **REFERENCES**

Andrew, S., & Halcomb, E. J. (Eds.). (2009). *Mixed methods research for nursing and the health sciences.* Chichester, UK: Blackwell.

Bergman, M. M. (2008). *Advances in mixed methods research.* London: Sage.

Brewer, J., & Hunter, A. (1989). *Multimethod research: A synthesis of styles.* Newbury Park, CA: Sage.

Bryman, A. (1988). *Quantity and quality in social research.* London and New York: Routledge.

Bryman, A. (2006). Integrating quantitative and qualitative research: How is it done? *Qualitative Research, 6*(1), 97–113.

Buck, G., Cook, K., Quigley, C., Eastwood, J., & Lucas, Y. (2009). Profiles of urban, low SES, African-American girls' attitudes toward science: A sequential explanatory mixed methods study. *Journal of Mixed Methods Research, 3*(4), 386–410.

Campbell, D. T. (1974). *Qualitative knowing in action research.* Paper presented at the annual meeting of the American Psychological Association, New Orleans, LA.

Campbell, D. T., & Fiske, D. W. (1959). Convergent and discriminant validation by the multitrait-multimethod matrix. *Psychological Bulletin, 56,* 81–105.

Christ, T. (2009). Designing, teaching, and evaluating two complementary mixed methods research courses. *Journal of Mixed Methods Research, 3*(4), 292–325.

Corrigan, M. W., Pennington, B., & McCroskey, J. C. (2006). Are we making a difference? A mixed methods assessment of the impact of intercultural communication instruction on American students. *Ohio Communication Journal, 44,* 1–32.

Creswell, J. W. (1994). *Research design: Qualitative, quantitative, and mixed methods approaches.* Thousand Oaks, CA: Sage.

Creswell, J. W. (2007, May). *Concerns voiced about mixed methods research.* Paper presented at the International Qualitative Inquiry Congress, University of Illinois, Champaign.

Creswell, J. W. (2008). *Educational research: Planning, conducting, and evaluating quantitative and qualitative research* (3rd ed.). Upper Saddle River, NJ: Pearson Education.

Creswell, J. W. (2009a, October). *The design of mixed methods research in occupational therapy.* Presentation to the Society for the Study of Occupation, New Haven, CT.

Creswell, J. W. (2009b). *How SAGE has shaped research methods: A 40-year history.* London: Sage.

Creswell, J. W. (2009c). *Research design: Qualitative, quantitative, and mixed methods approaches* (3rd ed.). Thousand Oaks, CA: Sage.

Creswell, J. W. (2009d, March). *What qualitative evidence means for mixed methods intervention trials in the health sciences.* Paper presented at the Economic & Social Research Council (ESRC) Research Seminar hosted by the Health Services Research Unit, Kings College, University of Aberdeen, Scotland.

Creswell, J. W. (2010). Mapping the developing landscape of mixed methods research. In A. Tashakkori & C. Teddlie (Eds.), *SAGE handbook of mixed methods in social & behavioral research* (2nd ed., pp. 45–68). Thousand Oaks, CA: Sage.

Creswell, J. W., & Garrett, A. L. (2008). The "movement" of mixed methods research and the role of educators. *South African Journal of Education, 28,* 321–333.

Creswell, J. W., & McCoy, B. R. (in press). The use of mixed methods thinking in documentary development. In S. N. Hesse-Biber (Ed.), *The handbook of emergent technologies in social research.* Oxford, UK: Oxford University Press.

Creswell, J. W., & Plano Clark, V. L. (2007). *Designing and conducting mixed methods research.* Thousand Oaks, CA: Sage.

Creswell, J. W., & Plano Clark, V. L. (2011). *Designing and conducting mixed methods research* (2nd ed.). Thousand Oaks, CA: Sage.

Creswell, J. W., Plano Clark, V. L., & Garrett, A. L. (2008). Methodological issues in conducting mixed methods research designs. In M. M. Bergman (Ed.), *Advances in mixed methods research* (pp. 66–83). London: Sage.

Creswell, J. W., Shope, R., Plano Clark, V. L., & Green, D. O. (2006). How interpretive qualitative research extends mixed methods research. *Research in the Schools, 13,* 1–11.

Cronbach, L. J. (1975). Beyond the two disciplines of scientific psychology. *American Psychologist, 30,* 116–127.

Crotty, M. (1998). *The foundations of social research: Meaning and perspective in the research process.* London: Sage.

David, L., Bender, L., Burns, S. (Producers), & Guggenheim, D. (Director). (2006). *An inconvenient truth* [Motion picture]. United States: Paramount Classics.

Denscombe, M. (2008). Communities of practice: A research paradigm for the mixed methods approach. *Journal of Mixed Methods Research, 2,* 270–283.

Denzin, N. K. (1978). *The research act: A theoretical introduction to sociological methods.* New York: McGraw-Hill.

Denzin, N. K., & Giardina, M. D. (2006). Introduction: Qualitative inquiry and the conservative challenge. In N. K. Denzin & M. D. Giardina (Eds.), *Qualitative inquiry and the conservative challenges* (pp. ix–xxxi). Walnut Creek, CA: Left Coast Press.

Denzin, N. K., & Lincoln, Y. S. (Eds.). (2005). *The SAGE handbook of qualitative research* (3rd ed.). Thousand Oaks, CA: Sage.

DeVellis, R. F. (1991). *Scale development: Theory and application.* Newbury Park, CA: Sage.

Elliot, J. (2005). *Using narrative in social research: Qualitative and quantitative approaches.* London: Sage.

Engel, R. J., & Schutt, R. K. (2005). *The practice of research in social work.* Thousand Oaks, CA: Sage.

Fielding, N. G., & Fielding, J. L. (1986). *Linking data.* Beverly Hills, CA: Sage.

Forman, J., & Damschroder, L. (2007, February). *Using mixed methods in evaluating intervention studies.* Presentation at the Mixed Methodology Workshop at the national meeting of the Veterans Administration Health Services Research & Development, Arlington, VA.

Freshwater, D. (2007). Reading mixed methods research: Contexts for criticism. *Journal of Mixed Methods Research, 1*(2), 134–145.

Giddings, L. S. (2006). Mixed-methods research: Positivism dressed in drag? *Journal of Research in Nursing, 11*(3), 195–203.

Greene, J. C. (2007). *Mixed methods in social inquiry.* San Francisco, CA: John Wiley.

Greene, J. C. (2008). Is mixed methods social inquiry a distinctive methodology? *Journal of Mixed Methods Research, 2*(1), 7–22.

Greene, J. C., & Caracelli, V. J. (Eds.). (1997). Advances in mixed-method evaluation: The challenges and benefits of integrating diverse paradigms. *New Directions for Evaluation, 74.* San Francisco: Jossey-Bass.

Greene, J. C., Caracelli, V. J., & Graham, W. F. (1989). Toward a conceptual framework for mixed-method evaluation designs. *Educational Evaluation and Policy Analysis, 11*(3), 255–274.

Greenstein, T. N. (2006). *Methods of family research* (2nd ed.). Thousand Oaks, CA: Sage.

Guba, E. G., & Lincoln, Y. S. (2005). Paradigmatic controversies, contradictions, and emerging confluences. In N. K. Denzin & Y. S. Lincoln (Eds.), *The SAGE handbook of qualitative research* (3rd ed., pp. 191–215). Thousand Oaks, CA: Sage.

Gurtler, L., Huber, L., & Kiegelmann, M. (2007). Conclusions: The reflective use of combined methods—a vision of mixed methodology. In P. Mayring, G. L. Huber, L. Gurtler, & M. Kiegelmann (Eds.), *Mixed methodology in psychological research* (pp. 243–245). Rotterdam/Taipei: Sense Publishers.

Hall, B., & Howard, K. (2008). A synergistic approach: Conducting mixed methods research with typological and systemic design considerations. *Journal of Mixed Methods Research, 2*(3), 248–269.

Heigham, J., & Croker, R. A. (2009). *Qualitative research in applied linguistics: A practical introduction.* London: Palgrave Macmillan.

Hesse-Biber, S. N., & Leavy, P. L. (2007). *Feminist research practice: A primer.* Thousand Oaks, CA: Sage.

Hodgkin, S. (2008). Telling it all: A story of women's social capital using a mixed methods approach. *Journal of Mixed Methods Research, 2*(4), 296–316.

Holmes, C. A. (2006, July). Mixed (up) methods, methodology and interpretive frameworks. Paper presented at the Mixed Methods Conference, Cambridge, UK.

Howe, K. R. (2004). A critique of experimentalism. *Qualitative Inquiry, 10,* 42–61.

Ivankova, N. V., Creswell, J. W., & Stick, S. L. (2006). Using mixed methods sequential explanatory design: From theory to practice. *Field Methods, 18*(1), 3–20.

Johnson, R. B., & Onwuegbuzie, A. J. (2004). Mixed methods research: A research paradigm whose time has come. *Educational Researcher, 33,* 14–26.

Johnson, R. B., Onwuegbuzie, A. J., & Turner, L. A. (2007). Toward a definition of mixed methods research. *Journal of Mixed Methods Research, 1*(2), 112–133.

Kuckartz, U. (2009). *Realizing mixed-methods approaches with MAX-QDA.* Unpublished manuscript, Department of Education, Phillipps-Universitaet, Marburg, Germany. Available at http://maxqda .com/download/MixMethMAXQDA-Nov01-2010.pdf

Kuhn, T. S. (1970). *The structure of scientific revolutions* (2nd ed.). Chicago: University of Chicago Press.

Leech, N. L, Dellinger, A. B., Brannagan, K. B., & Tanaka, H. (2010). Evaluating mixed research studies: A mixed methods approach. *Journal of Mixed Methods Research, 4*(1), 17–31.

Lincoln, Y. S., & Cannella, G. S. (2004). Qualitative research, power, and the radical right. *Qualitative Inquiry, 10*(2), 175–201.

Lincoln, Y. S., & Guba, E. G. (1985). *Naturalistic inquiry.* Beverly Hills, CA: Sage.

Luck, L., Jackson, D., & Usher, K. (2006). Case study: A bridge across the paradigms. *Nursing Inquiry, 13*(2), 103–109.

Maxwell, J., & Loomis, D. (2003). Mixed methods design: An alternative approach. In A. Tashakkori & C. Teddlie (Eds.), *Handbook of mixed methods in social & behavioral research* (pp. 241–272). Thousand Oaks, CA: Sage.

Mayring, P. (2007). Introduction: Arguments for mixed methodology. In P. Mayring, G. L. Huber, L. Gurtler, & M. Kiegelmann (Eds.), *Mixed methodology in psychological research* (pp. 1–4). Rotterdam/Taipei: Sense Publishers.

Mayring, P., Huber, G. L., Gurtler, L., & Kiegelmann, M. (Eds.). (2007). *Mixed methodology in psychological research.* Rotterdam/Taipei: Sense Publishers.

Mertens, D. M. (2003). Mixed methods and the politics of human research: The transformative-emancipatory perspective. In A. Tashakkori & C. Teddlie (Eds.), *Handbook of mixed methods in social & behavioral research* (pp. 135–164). Thousand Oaks, CA: Sage.

Mertens, D. M. (2009). *Transformative research and evaluation.* New York: Guilford.

Miall, C. E., & March, K. (2005). Community attitudes toward birth fathers' motives for adoption placement and single parenting. *Journal of Family Issues, 26,* 380–410.

Morgan, D. L. (2007). Paradigms lost and pragmatism regained: Methodological implications of combining qualitative and quantitative methods. *Journal of Mixed Methods Research, 1*(1), 48–76.

Morse, J. M. (1991). Approaches to qualitative-quantitative methodological triangulation. *Nursing Research, 40,* 120–123.

Morse, J. M. (2005). Evolving trends in qualitative research: Advances in mixed methods designs. *Qualitative Health Research, 15,* 583–585.

Morse, J. M., & Niehaus, L. (2009). *Mixed method design: Principles and procedures.* Walnut Creek, CA: Left Coast Press.

Myers, K. K., & Oetzel, J. G. (2003). Exploring the dimensions of organizational assimilation: Creating and validating a measure. *Communication Quarterly, 51*(4), 438–457.

Nastasi, B. K., Hitchcock, J., Sarkar, S., Burkholder, G., Varjas, K., & Jayasena, A. (2007). Mixed methods in intervention research: Theory to adaptation. *Journal of Mixed Methods Research, 1*(2), 164–182.

No Child Left Behind Act of 2001, Pub. L. No. 107–110, 115 Stat. 1425 (2002).

O'Cathain, A., Murphy, E., & Nicholl, J. (2007). Integration and publications as indicators of "yield" from mixed methods studies. *Journal of Mixed Methods Research, 1*(2), 147–163.

Olivier, T., de Lange, N., Creswell, J. W., & Wood, L. (2009, July). *Teachers as video producers and agents of change: A transformative mixed methods approach.* Paper presented at the fifth annual Mixed Methods Conference, Harrogate, UK.

Onwuegbuzie, A. J., & Johnson, R. B. (2006). Types of legitimation (validity) in mixed methods research. *Research in the Schools, 13*(1), 48–63.

Onwuegbuzie, A. J., & Teddlie, C. (2003). A framework for analyzing data in mixed methods research. In A. Tashakkori & C. Teddlie (Eds.), *Handbook of mixed methods in social & behavioral research* (pp. 351–383). Thousand Oaks, CA: Sage.

Patton, M. Q. (1980). *Qualitative evaluation methods.* Beverly Hills, CA: Sage.

Plano Clark, V. L. (2010). The adoption and practice of mixed methods: U.S. trends in federally funded health-related research. *Qualitative Inquiry, 16*(6), 428–440.

Plano Clark, V. L., & Creswell, J. W. (2008). *The mixed methods reader.* Thousand Oaks, CA: Sage.

Quinlin, E. (2010). Representations of rape: Transcending methodological divides. *Journal of Mixed Methods Research, 4*(2), 127–143.

Reichardt, C. S., & Rallis, S. F. (Eds.). (1994). *The qualitative-quantitative debate: New perspectives.* San Francisco: Jossey-Bass.

Richardson, L. (1997). *Fields of play: Constructing an academic life.* New Brunswick, NJ: Rutgers University Press.

Rossman, G. B., & Wilson, B. L. (1985). Numbers and words: Combining quantitative and qualitative methods in a single large-scale evaluation study. *Evaluation Review, 9*(5), 627–643.

Ryan, K. E., & Hood, L. K. (2006). Guarding the castle and opening the gates. In N. K. Denzin & M. D. Giardina (Eds.), *Qualitative inquiry and the conservative challenge* (pp. 57–77). Walnut Creek, CA: Left Coast Press.

Sale, J. E. M., Lohfeld, L. H., & Brazil, K. (2002). Revisiting the quantitative-qualitative debate: Implications for mixed-methods research. *Quality and Quantity, 36,* 43–53.

Sandelowski, M. (1996). Using qualitative methods in intervention studies. *Research in Nursing & Health, 19*(4), 359–364.

Sandelowski, M., Voils, C. I., & Knafl, G. (2009). On quantitizing. *Journal of Mixed Methods Research, 3*(3), 208–222.

Shank, G. (2007). How to tap the full potential of qualitative research by applying qualitative methods. In P. Mayring, G. L. Huber, L. Gurtler, & M. Kiegelmann (Eds.), *Mixed methodology in psychological research* (pp. 7–13). Rotterdam/Taipei: Sense Publishers.

Shavelson, R. J., & Towne, L. (Eds.) (2002). *Scientific research in education.* Washington, DC: National Research Council, National Academy Press.

Sieber, S. D. (1973). The integration of fieldwork and survey methods. *American Journal of Sociology, 78,* 1335–1359.

Sieber, S. D., & Lazarsfeld, P. F. (1966). *The organization of educational research* (USOE Cooperative Research Project No. 1974). New York: Columbia University, Bureau of Applied Social Research.

Stake, R. (1995). *The art of case study research.* Thousand Oaks, CA: Sage.

Stange, K. C., Crabtree, B. F., & Miller, W. L. (2006). Publishing multimethod research. *Annals of Family Medicine, 4,* 292–294.

Stenner, P., & Rogers, R. S. (2004). Q methodology and qualiquantology. In Z. Todd, B. Nerlich, S. McKeown, & D. D. Clarke (Eds.), *Mixing methods in psychology: The integration of qualitative and quantitative methods in theory and practice* (pp. 101–120). Hove and New York: Psychology Press.

Sweetman, D., Badiee, M., & Creswell, J. W. (2010). Use of the transformative framework in mixed methods studies. *Qualitative Inquiry, 16*(6), 441–454.

Tashakkori, A., & Teddlie, C. (1998). *Mixed methodology: Combining qualitative and quantitative approaches.* Thousand Oaks, CA: Sage.

Tashakkori, A., & Teddlie, C. (Eds.). (2003). *Handbook of mixed methods in social & behavioral research.* Thousand Oaks, CA: Sage.

Tashakkori, A., & Teddlie, C. (2003). The past and future of mixed methods research: From data triangulation to mixed model designs. In A. Tashakkori & C. Teddlie (Eds.), *Handbook of mixed methods in social & behavioral research* (pp. 671–701). Thousand Oaks, CA: Sage.

Teddlie, C., & Tashakkori, A. (2009). *Foundations of mixed methods research: Integrating quantitative and qualitative approaches in the social and behavioral sciences.* Thousand Oaks, CA: Sage.

Todd, Z., Nerlich, B., McKeown, S., & Clarke, D. D. (2004). *Mixing methods in psychology: The integration of qualitative and quantitative methods in theory and practice.* Hove and New York: Psychology Press.

Vidich, A. J., & Shapiro, G. (1955). A comparison of participant observation and survey data. *American Sociological Review, 20*(1), 28–33.

Vogt, D. S., King, D. W., & King, L. A. (2004). Focus groups in psychological assessment: Enhancing content validity by consulting members of the target population. *Psychological Assessment, 16,* 231–243.

Vogt, P. W. (2008). Quantitative versus qualitative is a distraction: Variations on a theme by Brewer & Hunter (2006). *Methodological Innovations Online, 3,* 1–10.

Vygotsky, L. S., & Cole, M. (1978). *Mind in society the development of higher psychological processes.* Cambridge, MA: Harvard University Press.

Yin, R. K. (2006). Mixed methods research: Are the methods genuinely integrated or merely parallel? *Research in the Schools, 13*(1), 41–47.

16

MIXED METHODS RESEARCH

Contemporary Issues in an Emerging Field

Charles Teddlie and Abbas Tashakkori[1]

The field of mixed methods research (MMR), which we have called the "third methodological movement," has evolved as a result of discussions about methods and paradigms in the social and behavioral sciences that have been ongoing for at least three decades. The "paradigm debate" between quantitatively oriented and qualitatively oriented researchers was based on sets of interlocking epistemological, ontological, and methodological assumptions. MMR offers a third alternative based on pragmatism, which argues that the two methodological approaches are compatible and can be fruitfully used in conjunction with one another (e.g., Howe, 1988; Tashakkori & Teddlie, 1998).

This chapter briefly presents several important issues in contemporary MMR, including a definition of MMR, theoretical and conceptual issues, issues in conducting MMR, and criticisms of the third methodological movement. We advise the reader to consider this to be a "sampler" of some of the contemporary issues relevant to MMR and, if you are interested, to continue your exploration of the field by reading some of the numerous cited references.

DEFINITIONS AND ORIGINS OF MIXED METHODS RESEARCH

Definition of Mixed Methods Research

As writing in the field of MMR has become more sophisticated, several authors have labored to identify and define exactly *what mixed methods research is* (e.g., Creswell, 2010; Greene, 2007, 2008; Johnson, Onwuegbuzie, & Turner, 2007; Tashakkori & Teddlie, 1998, 2003a). There is even continued debate over what the field should be called with variants including, but certainly not limited to, MMR, multimethod research, mixed methods, mixed methodology, mixed research, integrated research, and so forth.

Fortunately, there appears to be some consensus around "mixed methods research" as the de facto term due to common usage (e.g., the names of the leading journal in the field and of a handbook now in its second edition). We suspect that this term will endure, since it now has the trappings of a "brand name" that has been widely disseminated throughout the social and behavioral sciences.

As for the definition of MMR, Johnson et al. (2007) presented 19 alternative meanings from leaders in the field. While these meanings had varying levels of specificity, the authors of this analysis settled upon the following "composite" definition:

> Mixed methods research is the type of research in which a researcher or team of researchers combines elements of qualitative and quantitative research approaches (e.g., use of qualitative and quantitative viewpoints, data collection, analysis, inference techniques) for the broad purposes of breadth and depth of understanding and corroboration. (Johnson et al., 2007, p. 123)

From our perspective, this definition works because it includes what we believe is an essential characteristic of MMR: *methodological eclecticism,* a term that has only occasionally been used (e.g., Hammersley, 1996; Yanchar & Williams, 2006). Hammersley originally described this characteristic as follows:

> What is being implied here is a form of methodological eclecticism; indeed, the *combination* of quantitative and qualitative methods is often proposed, on the ground that this promises to cancel out the respective weaknesses of each method. (Hammersley, 1996, p. 167, italics in original)

Our definition of methodological eclecticism goes beyond simply combining qualitative (QUAL) and quantitative (QUAN) methods to cancel out respective weaknesses. *Eclectic,* the root word of eclecticism, means "choosing what appears to be the best from diverse sources, systems, or styles."[2] For us, *methodological eclecticism* involves *selecting and then synergistically integrating the most appropriate techniques from a myriad of QUAL, QUAN, and mixed methods* in order to more thoroughly investigate a phenomenon of interest. A researcher employing methodological eclecticism is a *connoisseur*[3] *of methods* who knowledgeably (and often intuitively) selects the best techniques available to answer research questions that frequently evolve during the course of an investigation.[4]

Origins of Mixed Methods Research, With an Emphasis on Qualitative Methods

MMR emerged as a distinct orientation in the late 1970s from applied fields in the social and behavioral sciences, such as evaluation, nursing, and education (e.g., Greene, Caracelli, & Graham, 1989; Miles & Huberman, 1984, 1994; Morse, 1991; Patton, 1980, 1990, 2002; Reichardt & Cook, 1979; Rossman & Wilson, 1985). Its origin in the applied, rather than pure, human sciences was not coincidental, since those disciplines often require a pragmatic, wide-angle lens utilizing all data sources available to answer practical questions.

Numerous studies from this early MMR involved researchers adding a QUAL component to a study that was initially a QUAN-only project in order to make greater sense out of the numerical findings. In evaluation research, this involved adding a formative component (how or why did the program succeed or fail) to the summative component (did the program work). In the human sciences, this distinction relates to causal effects (i.e., *whether* X causes Y) as opposed to causal mechanisms (i.e., *how* did X cause Y) (e.g., Shadish, Cook, & Campbell, 2002).[5]

In our own research, we have found that information gleaned from narratives generated by participants and investigators often proves to be the most valuable source in understanding complex phenomena. For example, qualitatively oriented case studies of differentially effective schools express the complexity of evolving contextual and behavioral patterns in those institutions much more thoroughly than statistical summaries of numeric indicators (Teddlie & Stringfield, 1993). A simple way of saying this is that narratives (stories) are intrinsically more interesting (and often more enlightening) than numbers to many researchers, the participants in their studies, and their audiences. It is no coincidence that several MMR pioneers (e.g., Creswell, Miles and Huberman, Morse, Patton) have also written QUAL methods texts.

We want to unambiguously express our regard for the powerful contributions of QUAL methods in this fourth edition of *The SAGE Handbook of Qualitative Research* due to the concern that some scholars have expressed about MMR subordinating QUAL methods to a secondary role behind QUAN methods (e.g., Denzin & Lincoln, 2005; Howe, 2004). This is not how we interpret the MMR literature we have reviewed from the past 30-plus years. In fact, QUAL + quan studies emphasizing the detailed, impressionistic perceptions of human "data-gathering instruments" and their interpretations of their outcomes are among the most valuable of all the extant MMR literature.

We also believe that MMR can add an important dimension to QUAN research. There has been much discussion (e.g., Mosteller & Boruch, 2002; Shavelson & Towne, 2002) about the importance of randomized controlled trials (RCTs) in the social and behavioral sciences. While RCTs may represent the "gold standard" for the identification of causal effects, the addition of a QUAL component (e.g., case studies) to the design allows researchers to discuss causal mechanisms as well. There are several examples of this in the MMR literature, including (1) the mixed methods intervention program of research in the health sciences (Song, Sandelowski, & Happ, 2010) and (2) the group-case method (also known as experimental ethnography) in several disciplines (e.g., Teddlie, Tashakkori, & Johnson, 2008). MMR enables researchers to examine issues in those fields in ways that traditional QUAN methods cannot alone.

◧ SOME CONTEMPORARY CHARACTERISTICS OF MIXED METHODS RESEARCH

We begin this section by acknowledging that there are others writing in the MMR field who will disagree with the inclusion of some or all of the characteristics described below, or with our interpretation of certain of those characteristics. Such is the nature of most emerging fields in academia, as new ideas are put forth and contested by those highly interested in the topic. This is especially the case with regard to MMR, whose development has been enhanced greatly by the juxtaposition of diverse perspectives. (Table 16.1 presents eight contemporary characteristics of MMR.)

We described the first contemporary characteristic of MMR in the previous section of this chapter: *methodological eclecticism.* This characteristic stems from rejection of the incompatibility (of methods) thesis, which stated that it is inappropriate to mix QUAL and QUAN methods in the same study due to epistemological differences between the paradigms that are purportedly related to them. Howe (1988) countered this point of view with his *compatibility thesis,* which contends that "combining quantitative and qualitative methods is a good thing" and "denies that such a wedding of methods is epistemologically incoherent" (p. 10). Howe proposed pragmatism as an alternative paradigm, a suggestion that has been endorsed by

Table 16.1 Eight Contemporary Characteristics of Mixed Methods Research

	Description of Characteristic
1.	Methodological eclecticism
2.	Paradigm pluralism
3.	Emphasis on diversity at all levels of the research enterprise
4.	Emphasis on continua rather than a set of dichotomies
5.	Iterative, cyclical approach to research
6.	Focus on the research question (or research problem) in determining the methods employed within any given study
7.	Set of basic "signature" research designs and analytical processes
8.	Tendency toward balance and compromise that is implicit within the "third methodological community"

many others (e.g., Biesta, 2010; Johnson & Onwuegbuzie, 2004; Maxcy, 2003; Tashakkori & Teddlie, 1998).

Methodological eclecticism not only means that we are free to combine methods, but that we do so by choosing what we believe to be the best tools for answering our questions. We have called this choice of "best" methods for answering research questions, "design quality"[6] and have included it as an essential part of our framework for determining the inference quality of MMR (Tashakkori & Teddlie, 2008). Furthermore, we believe that the best method for any given study in the human sciences may be purely QUAL, purely QUAN, or (in many cases) mixed.

Schulenberg (2007) presented a complex example of methodological eclecticism in a mixed methods study of the processes that occur in police decision making. Her data sources included interviews administered to individual officers, documents provided by the interviewees, QUAL data gathered from police department websites, documents obtained from provincial governments, census data, and tabulations of statistical data on the proportion of apprehended youth actually charged with crimes. The interview data gathered from police officers were originally QUAL in nature (from semistructured protocols), but were also converted into numbers (quantitized).

Schulenberg (2007) used these diverse data sources to generate five separate databases that addressed her research questions and hypotheses. She employed eight types of QUAL techniques and six types of statistical QUAN techniques including *t* tests, chi-squares, multiple regression, analysis of variance, manifest and latent content analysis, the constant comparative method, and grounded theory techniques. The *methodological*

eclecticism (connoisseurship) of this criminologist/sociologist is apparent.

The second contemporary characteristic of MMR is *paradigm pluralism,* or the belief that a variety of paradigms may serve as the underlying philosophy for the use of mixed methods. This characteristic is, of course, a function of the rejection of the incommensurability (of paradigms) thesis, which is widely accepted within the MMR community.

We believe that contemporary MMR is a kind of "big tent" in that researchers who currently use mixed methods come from a variety of philosophical orientations (e.g., pragmatism, critical theory, the dialectic stance). We believe that it is both unwise, and unnecessary, at this time to exclude individuals from the MMR community because their conceptual frameworks are different. We agree with Denzin's (2008, p. 322) paraphrase of a theme originally stated by Guba (1990): "A change in paradigmatic postures involves a personal odyssey; that is we each have a personal history with our preferred paradigm and this needs to be honored."

While paradigm pluralism is widely endorsed by many mixed methods scholars, theoretical and conceptual dialogues related to MMR have been, and will continue to be, of great importance. Recent developments and controversies in this area are summarized later.

The third characteristic of contemporary MMR is *a celebration of diversity at all levels of the research enterprise* from the broader more conceptual dimensions to the narrower more empirical ones. This is demonstrated in methodological eclecticism and paradigm pluralism, but also extends to other issues. For example, MMR can simultaneously address a diverse range of confirmatory and exploratory questions,[7] while single approach studies often address only one or the other. Additionally, MMR provides the opportunity for an assortment of divergent views in conclusions and inferences due to the complexity of the data sources and analyses.

MMR emerged partially out of the literature on triangulation (e.g., Campbell & Fiske, 1959; Denzin, 1978; Patton, 2002) and has commonly been associated with the *convergence* of results from different sources. Nevertheless, there is a growing awareness (e.g., Erzberger & Kelle, 2003; Greene, 2007; Johnson & Onwuegbuzie, 2004; Tashakkori & Teddlie, 2008) that an equally important result of combining information from different sources is divergence or dissimilarity, which can then provide greater insight into complex aspects of the same phenomenon and/or to the design of a new study or phase for further investigation.

The fourth characteristic of contemporary MMR is *an emphasis on continua rather than a set of dichotomies.* A hallmark of MMR is its replacement of the "either-or" with continua that describe a range of options (e.g., Newman, Ridenour, Newman, & DeMarco, 2003; Patton, 1980, 1990, 2002; Ridenour

& Newman, 2008; Tashakkori & Teddlie, 2003c). For example, we have applied what we called the QUAL-MIXED-QUAN multidimensional continuum to a variety of research issues including statement of purpose, research questions, designs, sampling, data collection and analysis, and validity or inference quality (Teddlie & Tashakkori, 2009). The either-or dichotomies (e.g., explanatory or exploratory questions, statistical or thematic analyses) have been replaced with a range of options (including integrated questions and innovative methods for mixed data analysis).

The fifth characteristic of contemporary MMR *is an iterative, cyclical approach to research.* MMR is characterized by the cycle of research, which includes both deductive and inductive logic[8] in the same study (e.g., Krathwohl, 2004; Tashakkori & Teddlie, 1998). The cycle may be seen as moving from grounded results (facts, observations) through inductive logic to general inferences (abstract generalizations or theory), then from those general inferences (or theory) through deductive logic to tentative hypotheses or predictions of particular events/outcomes. Research may start at any point in the cycle: Some researchers start from theories or abstract generalizations, while others start from observations or other data points. This cycle may be repeated iteratively as researchers seek deeper levels of a phenomenon. We believe that all research projects go through a full cycle at least once, regardless of their starting point (e.g., Teddlie & Tashakkori, 2009).

This cyclical approach to research may also be conceptualized in terms of the distinction between

- the *context or logic of justification*—the process associated with the testing of predictions, theories, and hypotheses, and
- the *context or logic of discovery*—the process associated with understanding a phenomenon in more depth, the generation of theories and hypotheses.

While several authors writing in MMR acknowledge the logic of justification as a key part of their research, they also emphasize the importance of the *context of discovery,* which involves creative insight possibly leading to new knowledge (e.g., Hesse-Biber, 2010; Johnson & Gray, 2010; Teddlie & Johnson, 2009). This discovery component of MMR often, but not always, comes from the emergent themes associated with QUAL data analysis.

We also conceptualize the cyclical nature of research as a kind of "ebb and flow" that characterizes some of the *signature MMR processes,* such as sequential research designs. More details on these signature MMR processes are presented later in this section.

The sixth characteristic endorsed by many writing in MMR is *a focus on the research question (or research problem) in determining the methods/approaches employed within any given study* (e.g., Bryman, 2006; Johnson & Onwuegbuzie, 2004; Tashakkori

& Teddlie, 1998). This *centrality of the research question* was initially intended to move researchers (particularly novice researchers) beyond intractable philosophical issues associated with the paradigms debate and toward the selection of methods that were best suited for their investigations.

Much has been written about the starting point for research; that is, do researchers start with a worldview or conceptual problem, a general purpose for conducting research, a research question, or some combination thereof? Newman et al. (2003) have argued convincingly that during the past four decades the research purpose has gained in importance relative to the research question. We maintain, however, that once a researcher has decided what she is interested in studying (e.g., what motivates the study, purpose, personal/political agenda, etc.), the specifics of her research questions will determine the choice of the best tools to use and how to use them. Experienced researchers are well aware of the fact that research questions undergo (often small) modifications and refocusing during the course of a study. Nevertheless, research questions generally direct the path of a research project.

MMR questions are usually broad, calling for both in-depth, emergent QUAL data *and* focused and preplanned QUAN data. These broad "umbrella" questions are often followed by more specific subquestions. In some (sequential) MMR projects, however, mixed questions emerge after the data are collected and analyzed, rather than being stated as initial "umbrella" questions. For example, a broad and emergent question may be asked and answered by collecting and analyzing QUAL data, followed by a question regarding the pervasiveness of the findings in a broader context or with regard to generalizablity to a population. Despite the emergent (or sometimes preplanned) sequence in these MMR studies, both groups of findings must be incorporated toward broader understandings (i.e., meta-inferences).

The seventh characteristic of contemporary MMR is a *set of basic research designs and analytical processes,* most of which are agreed upon, although they go by different names and diagrammatic illustrations. For example, we refer to *parallel mixed designs* (Teddlie & Tashakkori, 2009, p. 341, italics in original) as,

> a family of MM designs in which mixing occurs in an independent manner either simultaneously or with some time lapse. The QUAL and QUAN *strands* are planned and implemented in order to answer related aspects of the same questions.

These designs have also been referred to as concurrent, simultaneous, and *triangulation designs* (Creswell & Plano Clark, 2007, p. 85), but there is much commonality across their definitions.

Earlier in this section, we referred to *signature MMR design and analysis processes,* such as sequential mixed designs or conversion procedures. We call these design and analysis processes "signature" terms because they help to define MMR in relation to QUAN or QUAL methods; that is, they are unique to MMR and help set this approach apart from the other two. These signature design and analysis processes include the following:

- *Sequential mixed designs* "are a family of MM designs in which mixing occurs across chronological phases (QUAL, QUAN) of the study; questions or procedures of one *strand* emerge from or are dependent on the previous strand; *research question*s are built upon one another and may evolve as the study unfolds" (Teddlie & Tashakkori, 2009, p. 345, italics in original).
- *Quantitizing* refers to the process of converting qualitative data to numerical codes that can be statistically analyzed (e.g., Miles & Huberman, 1994; Sandelowski, Voils, & Knafl, 2009).
- *Qualitizing*[9] refers to the process by which quantitative data are transformed into data that can be analyzed qualitatively (e.g., Tashakkori & Teddlie, 1998).

More signature designs and analytical procedures indigenous to MMR are discussed later. While there is general agreement about the existence of these unique MMR design and analytical processes, there is considerable disagreement about terminology and definitions, and these disagreements widen as more complex typologies are generated. For example, many believe that a complete typology of MMR designs is not possible due to the emergent nature of the QUAL component of the research and the ability of MMR designs to mutate, while others seek agreement on a set number of basic designs for the sake of simplicity and pedagogy.

The eighth contemporary characteristic of MMR is *a tendency toward balance and compromise that is implicit within the "third methodological community."* MMR is based on rejecting the either-or of the incompatibility thesis; therefore, we as a community are inclined toward generating a balance between the excesses of both the QUAL and QUAN orientations, while forging a unique MMR identity. This balance is in keeping with Johnson and Onwuegbuzie's (2004) depiction of pragmatism as seeking a middle ground between philosophical dualisms and finding workable solutions for seemingly insoluble conceptual disputes.

In this context, we refer again to Denzin's (2008) paraphrase of three of Guba's (1990) themes regarding paradigms:

- "There needs to be decline in confrontationalism by alternative paradigm proponents."
- "Paths for fruitful dialog between and across paradigms need to be explored."
- "The three main interpretive communities . . . must learn how to cooperate and work with one another" (Denzin, 2008, p. 322).

We believe that most mixed methods researchers are in agreement with these themes that call for compromise in dialogues among the three methodological communities.

▣ THEORETICAL AND CONCEPTUAL ISSUES IN MIXED METHODS RESEARCH

While there is agreement on some broad characteristics of MMR, there are several ongoing dialogues regarding basic theoretical and conceptual issues within MMR. We concentrate on two: (1) issues related to the paradigms, which are also referred to by several other terms such as stances, approaches, frameworks, perspectives, mental models, and so forth and (2) issues related to the language of MMR.

Issues Related to the Use of Paradigms (or Conceptual Frameworks or Mental Models)

In this section, we first provide more details on the concept of paradigm pluralism. Then we present three alternative paradigmatic positions for MMR, followed by a discussion of some arguments against the continued focus on paradigm issues.

We presented paradigm pluralism as one of the contemporary characteristics of MMR earlier. The belief that multiple paradigms may serve as the underlying conceptual framework for MMR is a practical solution to some thorny philosophical and conceptual issues: Researchers simply use the philosophical framework that best fits their particular "intellectual odyssey."

Most MMR scholars can agree with paradigm pluralism as a starting point, but then they have to (1) consider the alternative paradigmatic positions and (2) *ascertain which of those positions is most closely related to their own perspective.* The following three paradigmatic positions[10] are the most widely accepted in contemporary MMR:

- pragmatism and its interpretations,
- frameworks associated with the axiological assumption (Mertens, 2007), and
- the dialectical stance, which involves using multiple assumptive frameworks within the same study (e.g., Greene, 2007; Greene & Caracelli, 2003).

Before examining these positions further, we need to briefly reconsider the ramifications of paradigm pluralism, which was posited in opposition to the single paradigm-single method thesis (e.g., postpositivism and QUAN methods; constructivism and QUAL methods). Denzin (2008, p. 317) considers the rejection of the single paradigm-single method thesis to be historical:

When the field went from one to multiple epistemological paradigms, many asserted that there was incompatibility between and across paradigms, not just incompatibility between positivism and its major critic, constructivism. . . . Ironically, as this discourse evolved, the complementary strengths thesis emerged, and is now accepted by many in the mixed-methods community. Here is where history starts to be rewritten. That is multiple paradigms can be used in the same mixed-methods inquiry. . . . Thus the demise of the single theoretical and /or methodological paradigm was celebrated.

It is important to realize that Denzin's analysis emphasizes not only paradigm pluralism, but also that researchers *may use multiple frameworks in the same study,* which is supported by only one of the contemporary positions noted above (the dialectical stance). Researchers who prescribe to pragmatism or a framework based on the axiological assumption typically use only that perspective in their research.

Pragmatism and Its Interpretations

There is an affinity for pragmatism as the paradigm of choice for many mixed methodologists (e.g., Tashakkori & Teddlie, 1998). This affinity is a historical one going back to Howe's (1988) postulation of the compatibility thesis based on pragmatism. The pragmatic approach to philosophical issues is appealing to many applied scientists who utilize a kind of "everyday pragmatism" (Biesta, 2010) in their solution of research and evaluation problems.

A more *philosophically nuanced pragmatism* has emerged recently (e.g., Biesta, 2010; Greene & Hall, 2010; Johnson & Onwuegbuzie, 2004; Maxcy, 2003; Teddlie & Tashakkori, 2009). This pragmatism asks, "Apart from the rejection of the either-or, what does pragmatism mean for MMR?" We briefly describe three recent interpretations of pragmatism (Johnson and colleagues; Biesta, 2010; Greene, 2007) that have advanced the conversation.

Johnson and colleagues have ventured into a kind of "paradigm or systems building" with regard to *philosophical pragmatism.* Johnson and Onwuegbuzie (2004) presented 21 characteristics of pragmatism in an effort to more completely delineate the tenets of this philosophy and how they relate to MMR.

Johnson et al. (2007) defined three pragmatisms: of the right, of the left, and of the center (*classical pragmatism*). Johnson (2009, p. 456) further defined *dialectical pragmatism* as a "supportive philosophy for mixed methods research" that combines classical pragmatism with Greene's (2007) dialectical approach. The cumulative contribution of Johnson and colleagues' work is that we now have a clearly articulated and detailed account of pragmatism as it relates to MMR.

In contrast, Biesta (2010, p. 97) contends that "pragmatism should not be understood as a philosophical position among others, but rather as a set of philosophical tools that can be used to address problems." Biesta emphasizes that John Dewey warned against philosophical system building. Biesta concludes that Deweyan pragmatism contributes to the dismantling of the epistemological dualism of objectivity/subjectivity.

The major contribution of Dewey is that he engages with this discussion from a different starting point so that the either/or of objectivism and subjectivism loses its meaning. . . . This is tremendously important for the field of mixed methods research as it does away with alleged hierarchies between different approaches and rather helps to make the case that different approaches generate *different* outcomes, *different* connections between doing and undergoing, between actions and consequences, so that we always need to judge our knowledge claims pragmatically, that is in relation to the processes and procedures through which the knowledge has been generated. (Biesta, 2010, p. 113, italics in original)

Biesta concludes that philosophical pragmatism leads us to understand that no methodological approach is intrinsically better than another in knowledge generation. We have to evaluate the results from our research studies in terms of how good a job we did in selecting, utilizing, and integrating all the available methodological tools. Did we succeed in our efforts at *methodological eclecticism?*

Greene (2007) referred to pragmatism as the *alternative paradigm* (to the dominant traditional ones) that promotes the active mixing of methods and integration of research findings. Greene and Hall (2010) further described how thinking pragmatically affects the manner in which mixed researchers conduct their research. For Greene and Hall and others (e.g., Biesta, 2010; Johnson & Onwuegbuzie, 2004), pragmatism results in a problem-solving, action-oriented inquiry process based on a commitment to democratic values and progress.

Frameworks Associated With the Axiological Assumption

Mertens (2007) identified four basic assumptions associated with paradigms that were previously delineated by Guba and Lincoln (2005): axiological, epistemological, ontological, and methodological. Mertens, Bledsoe, Sullivan, & Wilson (2010, p. 195) further described the *axiological assumption* that "takes precedence and serves as a basis for articulating the other three belief systems because the transformative paradigm emerged from the need to be more explicit about how researchers can address issues of social justice." The axiological assumption is based on "power differences and ethical implications that derive from those differences" between marginalized and other groups (Mertens et al., p. 195).

In discussions of pragmatism, the philosophical issues that are emphasized are epistemological in nature concerning issues such as what is knowledge, how is it acquired, and the relationship between the knower and "known." On the other hand, scholars working within transformative or critical frameworks (e.g., feminism) give precedence to axiological considerations, which center on the nature of value judgments. This *axiological assumption* means that scholars working within transformative/critical frameworks have a different perspective on research methods. For these scholars, mixed methods are tools that are used in the service of value systems that are always foremost.

The Dialectic Stance or Way of Thinking

The dialectic stance assumes that all paradigms have something to offer, and that employing multiple paradigms contributes to greater understanding of phenomena under study. Pragmatism and axiologically oriented frameworks utilize one perspective exclusively, while the dialectical stance calls for the juxtaposition of multiple assumptive frameworks within the same study. Greene (2007, p. 114) expresses it thus:

> I have adopted the stance that method cannot be divorced from the inquirer's assumptions about the world and about knowledge, the inquirer's theoretical predispositions, professional experience, and so forth.... So when one mixes methods, one may also mix paradigmatic and mental model assumptions as well as broad features of inquiry methodology.

Greene's dialectical stance directs attention away from the so-called incommensurable attributes of paradigms and toward different and distinctive (but not inherently incompatible) attributes such as distance-closeness, outside-insider, emic and etic, particularity and generality, and so forth. Greene and Hall's (2010) dialectical stance agrees with Biesta's (2010) pragmatism in that these philosophical systems are *not* "paradigm packages" with interlocking philosophical assumptions or beliefs.

Arguments Against the Continued Focus on "Paradigms"

The term "paradigm" has played a crucial role in the development of the three methodological communities since the initial publication of Kuhn's (1962) *The Structure of Scientific Revolutions.* Recently, authors have expressed increasing doubt about the utility of the continued focus on paradigm issues in MMR. For instance, Bazeley (2009, p. 203) concluded that "Although the epistemological arguments of the 'paradigm wars' sharpened our thinking about issues related to mixed methodology, their lingering legacy has been to slow the progress of integration of methods."

Morgan (2007) deconstructed the term "paradigm" into four possible (and not mutually exclusive) positions:

- paradigms as worldviews (ways of perceiving and experiencing the world),
- paradigms as epistemological stances (which Morgan called the *metaphysical paradigm*),
- paradigms as model examples (i.e., "exemplars" demonstrating how research is conducted), and
- paradigms as shared beliefs about types of questions, methods of study (and so on) among a community of scholars or within a field of study.

Morgan argued that Guba and Lincoln (e.g., Lincoln & Guba, 1985; Guba & Lincoln, 1994, 2005) used the metaphysical paradigm to draw attention to QUAL research as an alternative to QUAN research. This metaphysical version focused on the basic assumptions or beliefs noted above, with a special emphasis on epistemological considerations, drawing essential, incommensurable differences between the QUAL and QUAN perspectives thereby leading to the paradigm wars.

Morgan further argued that now is the time to move from what he considers the outmoded concept of *metaphysical paradigms*[11] to paradigms as *shared beliefs in a research field* due to conceptual problems with the former position (e.g., a *strong* stand on incommensurability) and to the fact that the latter position is a more accurate interpretation of Kuhn's use of the term.[12] Morgan's focus on shared beliefs in a research field has contributed to an increasing emphasis on the *community of scholars* perspective (e.g., Creswell, 2010; Tashakkori & Creswell, 2008), which is a position that has been reinforced by Denscombe's (2008) discussion of the nature that such a community might take.

The Language of Mixed Methods Research

We previously identified the language of mixed methods as one of the major issues in MMR (Teddlie & Tashakkori, 2003). At that time, we distinguished between MMR using a *bilingual language* that combined QUAL and QUAN terms or generating a *new language* with terms unique to the field itself. Since that time, we have seen manifestations of both tendencies.

For instance, we recently (Teddlie & Tashakkori, 2009, p. 282) generated a list of common analytical processes used in both QUAL and QUAN research that are examples of a bilingual language. These processes are cognitively interchangeable, although one uses numbers and the other employs words as data. For example, a bilingual mixed methods researcher knows that cluster analysis employs the same *modus operandi* as the categorizing process of the constant comparative method: that is, maximizing between-group variation and minimizing within-group variation. Other examples include comparing analyses from one part of a sample with analyses from another part of the sample; comparison of actual results with expected results; and contrasting components of research design or elements to find differences.

Recognition of these common processes is a step in the direction of developing a language that crosses methodological lines. On the other hand, Box 16.1 presents a partial list of unique terms related to mixed methods data analysis that have emerged since the 1990s. The emergence of new analytical processes constitutes one of the most creative areas in MMR and often comes from researchers working on practical solutions for answering their research questions using available QUAL and QUAN data. Using mixed data analysis as an example, it appears that the language used in MMR will involve both bilingual terms and unique mixed terms (e.g., Box 16.1).

Box 16.1 Partial List of Data Analysis Terms Indigenous to Mixed Methods Research

A partial list of MMR data analysis terms includes

- crossover track analysis
- data conversion or transformation
- data importation
- fully integrated mixed data analysis
- fused data analysis
- inherently mixed data analysis
- integrated data display
- integrated data reduction
- iterative sequential mixed analysis
- morphed data analysis
- multilevel mixed data analysis
- narrative profile formation
- parallel mixed data analysis
- parallel track analysis
- qualitizing
- quantitizing
- sequential mixed data analysis
- single track analysis
- typology development
- warranted assertion analysis

These terms were generated or employed by several authors, including Bazeley, 2003; Caracelli & Greene, 1993; Greene, 2007; Greene, Caracelli, & Graham, 1989; Li, Marquart, & Zercher, 2000; Onwuegbuzie & Combs, 2010; Onwuegbuzie, Johnson, & Collins, 2007; Onwuegbuzie & Teddlie, 2003; Tashakkori & Teddlie, 1998; Teddlie & Tashakkori, 2003, 2009.

Box 16.2 How Worldviews Affect MMR Praxis

There is general agreement that a researcher's worldview affects the manner in which that person conducts his or her research, yet there have been few explicit discussions of how that occurs in MMR. Several of the chapters in the recently published second edition of the *SAGE Handbook of Mixed Methods in Social & Behavioral Research* presented detailed versions of actual or hypothetical researchers and how they conducted MMR within a particular worldview (or assumptive framework or mental mode). These chapters included

- Greene and Hall (2010) described Michelle (a hypothetical researcher), who is conducting research on the interactions among middle school children as they go through their daily routines. Michelle's perspective is that of the *dialectic inquirer* who is attuned to the values underlying the multiple philosophical frameworks (constructivist epistemology, feminist ideology) that guide her research.
- Greene and Hall (2010) described Juan (another hypothetical researcher), whose perspective is that of a *pragmatic inquirer* who is studying schools that are struggling to simultaneously serve the needs of their diverse study bodies and to meet the accountability mandates of NCLB (No Child Left Behind).
- Hesse-Biber's (2010) description of research conducted within the *feminist tradition,* including studies as diverse as forestland usage in Nepal and sex work in Tijuana.
- Mertens and colleagues' (2010) description of research conducted within the *transformative paradigm tradition,* including studies on inclusive education for disabled people in New Zealand and poverty reduction in Rwanda.

▣ ISSUES IN CONDUCTING
MIXED METHODS RESEARCH

Issues related to how to conduct MMR appear to have gained in importance relative to discussions of theoretical and conceptual issues recently. This trend is probably a reflection of the growing acceptance of MMR as a distinct methodological orientation and increased curiosity regarding the specifics of exactly how such research is conducted, disseminated, and utilized.

Chapters in the second edition of the *SAGE Handbook of Mixed Methods in Social & Behavioral Research* not only describe how to do MMR, but also illustrate how researchers' worldviews affect the manner in which they conduct their research. Box 16.2 presents information on how the different paradigmatic orientations summarized in the previous section of this chapter affect MMR praxis.

While there are several broad issues in conducting MMR (from generating research questions through making inferences from integrated data analyses), we can only discuss a couple here. We have selected research design, because there has been substantial work done in this area, and data analysis, because this is an area where considerable creative energy is currently being expended.

The Design of Mixed Methods Studies: A Diversity of Options

Design typologies have long been an important feature of MMR starting with Greene et al. (1989) writing in the field of evaluation and Morse (1991) in nursing. The reasons for the importance of MMR design typologies include their role in establishing a common language for the field, providing possible blueprints for researchers who want to employ MMR designs, legitimizing MMR by introducing designs that are clearly distinct from those in QUAN or QUAL research, and providing useful tools for pedagogical purposes.

Recently, some authors have contended that there is an overemphasis on research design typologies (e.g., Adamson, 2004; Bazeley, 2009), arguing that other areas (e.g., data analysis) should be stressed more. While such design typologies may not be featured as extensively in future writing in the field, they will continue to be an essential element of MMR. This is partly due to the fact that many of the proposed data analysis procedures in MMR are actually design-bound; that is, they are related to a specific type (or family) of designs (e.g., sequential data analysis in sequential mixed methods designs).

While some authors argue for a set number of prespecified designs, others contend that MMR design typologies can never be exhaustive, due to the iterative nature of MMR research projects (i.e., new components or strands might be added during the course of a project). This is an important point, since many inexperienced researchers want a design "menu" from which to select the "correct" one, similar to those provided in QUAN research (e.g., Shadish et al., 2002). In contrast, researchers using mixed methods are encouraged to continuously reexamine the results from one strand of a study compared to the results from another, and make changes both in the design and data collection procedures accordingly.

Researchers seeking their own *MMR design family* have a variety of viable options in the current "marketplace" (e.g., Creswell & Plano Clark, 2007; Greene, 2007; Leech & Onwuegbuzie, 2009; Maxwell & Loomis, 2003; Morse, 1991, 2003; Teddlie & Tashakkori, 2009). Nastasi, Hitchcock, and Brown (2010) recently examined various design typologies,[13] divided them into basic and complex categories, and determined that they differed with regard to nine distinct criteria or dimensions.

While some find the lack of consensus regarding the specific number and types of designs disconcerting, we believe that this is a healthy sign and that the most useful of the typologies will survive. The ultimate value of these typologies lies in their ability to provide researchers with viable design options to choose from and build upon (i.e., modify, expand, combine) when they are planning or implementing their MMR studies. The diversity in design typologies can be best exemplified by briefly examining two points of view that are distinct and have continued to evolve since first introduced: those of Jennifer Greene and our own. Other perspectives are equally valuable, but we chose these two because they make particularly interesting contrasts.

Greene contends that researchers cannot divorce method from "assumptive frameworks" when designing MMR studies; therefore, she encourages mixing those frameworks in single research studies. Her designs are anchored in mixing methods for five basic purposes that emerged from Greene et al. (1989): triangulation, complementarity, development, initiation, and expansion. Caracelli and Greene (1993) distinguished between *component designs* in which the methods are connected or mixed only at the level of inference and *integrated designs* in which the methods are integrated throughout the course of the study.

Greene (2007) presented two examples of component designs (convergence, extension) and four examples of integrated designs (iteration, blending, nesting or embedding, mixing for reasons of substance or values). These six examples of MMR designs map onto the five basic purposes for mixing with each example aligned with one or two of the original purposes. Greene (2007, p. 129) concludes that designing a MMR study does *not* involve following a formula or set of prescriptions, but rather is "an artful crafting of the kind of mix that will best fulfill the intended purposes for mixing within the practical resources and contexts at hand."

In our approach to MMR, we have always treated design as separable from research purpose. That is not to deny the importance of purpose; obviously, if you did not have a purpose for doing a study, you would not be doing it. We think purpose is a complex, psycho-socio-political concept and we believe each individual has a multiplicity of purposes for doing research ranging from "advancing your career" to "understanding complex phenomena" to "improving society." These purposes are intertwined and often change over time.

Our design typology has evolved as MMR has developed over the past decade (Tashakkori & Teddlie, 1998, 2003c; Teddlie & Tashakkori, 2009). The base of our system is a three-stage model of the research process that evolved from Patton's (2002, p. 252) "pure and mixed strategies" for conducting research. These three stages are conceptualization (formulation of questions specific to the research study), experiential (methodological operations, data generation, analysis, and so forth), and inferential (emerging theories, explanations, inferences, and so forth). Mixed designs are those in which the QUAL and QUAN approaches are integrated across the three stages. There are (currently) four families of mixed methods designs in our typology: parallel, sequential, conversion, and fully integrated. These families are based on what we call "type of implementation process"; that is, how does the integration of the QUAL and QUAN strands actually occur when conducting a study. Increasingly, MMR studies seem to use a combination of the basic configurations, often leading to fully integrated designs with multiple types/sources of data.

Similar to Greene's perspective, we distinguish between whether integration occurs at only one stage of the process (for us, the experiential stage) or throughout the study. Our latest solution to this thorny issue is the distinction between mixed and *quasi-mixed designs;* the former was defined in the previous paragraph, while we define the latter as designs in which two types of data are collected, but there is little or no integration of findings and *inferences* from the study.

Both of these perspectives regarding MMR designs

- reflect coherent and internally consistent perspectives,
- are currently viable as they continue to evolve in interesting ways related to changes in the field,
- are heuristic in terms of informing MMR dissertations and other projects, and
- have advanced the MMR designs literature over time (and have, themselves, evolved as a result).

In comparing our position regarding MMR design with hers, Greene (2007, p. 117) concluded,

> my own thinking about mixed methods design shares considerable intellectual space with those of Tashakkori and Teddlie, but also contains some differences.... There is certainly ample space in the contemporary mixed methods conversation for these complementary yet distinct sets of ideas.

Mixed Methods Data Analysis

Mixed methods data analyses are the processes whereby QUAN and QUAL data analysis strategies are combined, connected, or integrated in research studies (Teddlie & Tashakkori, 2009). Much creative energy is currently being expended on topics related to MMR data analysis, especially that involving integrated computer-generated applications (e.g., Bazeley, 2010). Bazeley (2009, p. 206) recently concluded that an indicator of the maturation of MMR would come when it moves from "a literature dominated by foundations and design typologies" toward a field "in which there are advances in conceptualization and breakthroughs derived from analytical techniques that support integration."

We limit our discussion of analysis issues to two topics foreshadowed in the previous section on the language of mixed methods: (1) the identification of analogous analytical processes in QUAL and QUAN research, and (2) the generation of a unique lexicon of MMR analysis procedures indigenous to the area. The analogous processes represent what Greene (2007, p. 155) called, "using aspects of the analytic framework of one methodological tradition within the analysis of data from another tradition."

In an early demonstration of this process, Miles and Huberman (1984/1994) took matrices from the QUAN tradition (e.g., contingency tables filled with numbers or percentages generated from chi-square analysis) and applied that framework to the QUAL tradition by crossing two dimensions and then completing the cells with narrative information. In one example, Miles and Huberman (1994) illustrated the implementation of a longitudinal school improvement project by using columns that represented years and rows that represented levels of intervention. Cross-case comparisons between schools demonstrated

where there were differences in reform implementation between more and less successful schools.

Similarly, Onwuegbuzie (2003) applied the QUAN concept of effect sizes to generate an analogous QUAL typology including three broad categories (manifest, adjusted, and latent QUAL effect sizes). Effect sizes in QUAN research refer to the strength of the relationship between two numeric variables calculated by statistical indices. The generation of effect sizes in QUAL research is an analytical process in which the strength of the relationship between narrative variables is calculated after these variables have been quantitized.

In the future, we believe that MMR researchers will increasingly apply the analytical frameworks used in either the QUAL or QUAN tradition in developing analogous techniques within the other tradition. This requires both appropriate training in the QUAN and QUAL approaches and the ability to creatively see analogous processes from the mixed methods perspective.

Similarly, creative insight on the part of a variety of researchers has resulted in the lengthy list of data analysis terms indigenous to MMR in Box 16.1. These terms refer to general analytical processes (e.g., data conversion); specific techniques within more general analytical processes (e.g., crossover track analysis within parallel mixed data analysis); and complex iterative mixed data analyses utilizing multiple computer programs. Bazeley (2003, p. 385, italics added) has called the latter process *fused data analysis*, which she describes as follows:

> Software programs . . . offer . . . the capacity of qualitative data analysis (QDA) software to incorporate quantitative data into a qualitative analysis, and to transform qualitative coding and matrices developed from qualitative coding into a format which allows statistical analysis. . . . *The "fusing" of analysis then takes the researcher beyond blending of different sources to the place where the same sources are used in different but interdependent ways in order to more fully understand the topic at hand.*

Another noteworthy trend in mixed methods data analysis was discussed in the section on the third characteristic of contemporary MMR: the celebration of diversity at all levels of the research enterprise. This characteristic is exemplified in mixed methods data analysis by the growing awareness that divergence of findings and inferences across the QUAL and QUAN strands is equally as informative as convergence (or even more so), because that divergence leads researchers to more complex understandings and toward further research studies.

◻ CRITIQUES OF MIXED METHODS RESEARCH

Several criticisms of MMR have been voiced, especially as the field has become more visible since the turn of the 21st century. In this section, we briefly review some of the most salient of those criticisms.

From a historical perspective, the most common criticism of MMR is the incompatibility thesis, which stated that it is inappropriate to mix QUAL and QUAN methods in the same study due to epistemological differences between the paradigms that are purportedly related to them (e.g., Howe, 1988). This issue was addressed in the discussion regarding the first contemporary characteristic of MMR, *methodological eclecticism,* which contends that we are free to combine the best methodological tools in answering our research questions. While the philosophical justification for methodological eclecticism is important, the historical argument against the incompatibility thesis is probably more compelling: Researchers have been fruitfully combining QUAL and QUAN methods throughout the history of the social and behavioral sciences resulting in multilayered research that is distinct from either QUAL or QUAN research alone.

Criticisms of MMR from the QUAL research and postmodern communities (e.g., Denzin & Lincoln, 2005; Howe, 2004; Sale, Lohfeld, & Brazil, 2002) have involved several issues, which have in turn been addressed by the MMR community (e.g., Creswell, Shope, Plano-Clark, & Green, 2006; Teddlie et al., 2008). Perhaps the most salient of these issues is the concern that MMR subordinates QUAL methods to a secondary position to QUAN methods. As noted in the first section of this chapter, we unequivocally express our regard for the powerful contributions of QUAL methods and interpret the overwhelming majority of truly mixed research as involving a thorough integration of both methods. Fortunately, recent literature (e.g., Creswell et al., 2006; Denzin, 2008) indicates that the QUAL and MMR communities can be involved in a productive discourse respectful of diverse viewpoints and cognizant of our many points of agreement.

Valuable criticisms of MMR include logistical ones (i.e., its implementation in actual research studies), including concerns about the costs of such research and about who does the research (e.g., teams of researchers, solo investigators). We believe that the employment of QUAL, QUAN, or MMR approaches in any given study depends on the research questions that are being addressed and that many issues are best and most efficiently answered using either the QUAL-only or QUAN-only approach. MMR techniques should be used only when necessary to adequately answer the research questions, because the mixed approach is inherently more expensive than the QUAL or QUAN alone orientations. Mixed studies take longer to conduct, which is a major issue for doctoral students, as well as researchers operating under stringent timelines to complete contracted work. Researchers bidding for contracts using MMR should be especially careful to provide accurate budgets for what it would take to do the work comprehensively, especially the QUAL component, which may involve time-consuming ethnographies. MMR projects that underestimate the time and money required to complete all components of the design will likely result in "QUAL-light" research that does not deliver what was promised.

As for who does the research, there is concern that a "minimal competence model or methodological" bilingualism is "superficial, perhaps even unworkable" (Denzin, 2008, p. 322). Issues of mixed methods pedagogy are beyond the scope of this chapter, but there is an active literature developing in this area (e.g., Christ, 2009, 2010; Creswell, Tashakkori, Jensen, & Shapley, 2003; Tashakkori & Teddlie, 2003b) that includes details on current MMR courses being taught and how they have evolved over time (Christ, 2010). The collaborative approach to MMR has been described by Shulha and Wilson (2003) and successful examples of it are found in the literature (e.g., Day, Sammons, & Qu, 2008).

In our discussion of "methodological connoisseurship," we indicated that mixed methodologists knowledgeably (and often intuitively) select the best techniques available to answer research questions that may evolve during the course of a research project. The question arises: How is such experience and judgment developed across diverse methods, especially in the QUAL area? There is no simple answer to this question, but we believe that a combination of coursework and field experiences is necessary to begin the journey toward "methodological connoisseurship." The field experiences are crucial and we advocate an active mentorship between professors who are mixed methodologists and their graduate students. Preferably, this mentorship would include field experiences in research projects where the professor is the principal investigator and/or dissertations in which the student is required to conduct extensive QUAL and QUAN research to answer different parts of the research questions being investigated. We have served on several dissertation committees where students have completed successful MMR projects and have begun their journey toward becoming "methodological connoisseurs." (See Schulenburg [2007] and Ivankova, Creswell, & Stick [2007], for examples of research articles based on mixed methods dissertations.)

Another criticism of MMR concerns the quality of the writing of many articles and chapters in the field. Leech (2010) conducted interviews with early developers of the field who concluded that authors need to do a better job of (1) expressing where their research fits within the current MMR literature; (2) presenting their own definition of MMR; (3) explaining where and how the mixing of methods occurred in their research; and (4) explicitly describing their philosophical orientation. Creswell (2009) has recently presented a preliminary "map" delineating subareas of MMR that should help authors in "locating" themselves within the field. The multiple definitions of MMR presented by Johnson et al. (2007) should help authors in describing their own perspectives, while the various design typologies offer options with regard to how authors can describe the mixing of methods in their research projects. Furthermore, the explicit delineation of at least three philosophical orientations in the field (pragmatism, frameworks associated with the axiological assumption, the dialectic stance) with other emerging alternatives (e.g., critical realism) provides authors with alternative philosophical orientations from which to choose and then make explicit in their writings.

Finally, Freshwater (2007), Greene (2007), Greene and Hall (2010), and others have expressed a concern that MMR is prematurely headed toward some "fixed" unity or consensus for social inquiry that will preclude the consideration of and respect for multiple approaches. For example, Freshwater (2007, p. 141) criticizes the "idolatry of integration and coherence," which she sees as "rife throughout nursing and the healthcare literature." This concern is akin to the apprehension that Smith and Heshusius (1986) voiced about "closing down the conversation" with regard to the quantitative-qualitative debate. We can understand this concern intellectually, since one of the characteristics of MMR is a "tendency toward balance and compromise," but we do not see MMR as becoming a static unified approach toward social inquiry that will stifle diverse viewpoints.

Perhaps our confidence that MMR leads toward a "celebration of diversity at all levels of the research enterprise" comes from our experiences in editing two volumes of the *SAGE Handbook of Mixed Methods in Social & Behavioral Research,* which have presented

- a wide variety of philosophical and conceptual models for MMR,
- an increasingly diverse set of methodological tools that can be employed in all aspects of conducting integrated research, especially those related to data analysis and the inferential process, and
- a diversity of applications of MMR across disciplinary boundaries and within specific lines of research.

Closely linked to this perspective regarding the inherent diversity of MMR is our perception of it as an extension of everyday sense making. Everyday problem solvers (naïve researchers) use multiple approaches concurrently or in sequence, examine a variety of evidence in decision making (or even in forming impressions), and question the credibility of their impressions, conclusions, and decisions. Although using a different type of data, more sophisticated methods of analysis, and more stringent standards of evidence and inference, a mixed methods researcher (the *methodological connoisseur* described earlier) follows the same general path that is characterized by a reliance on diverse sources of evidence.

◫ WHERE WILL WE BE IN 10 YEARS?

It is always difficult to predict the future, especially for a field that has only formally emerged in the past 15 to 20 years. The following comments are, therefore, our best guesses based on what we see as the trajectory of the field and are presented with the acknowledgment that future historic events could radically change the course of MMR.

1. There will be a gradual acceptance of pragmatism as the primary philosophical orientation associated with MMR, just as constructivism is associated with QUAL research and postpositivism with QUAN methods. Philosophical pragmatism as it relates to MMR will be defined more precisely. Other philosophical points of view will exist along with pragmatism as a basis for MMR, and this will be acceptable due to the belief of most mixed methodologists in paradigm pluralism. There will be relatively less emphasis on discussion of theoretical and conceptual issues.

2. A generic set of MMR designs will emerge over time and will be popularized in textbooks. These designs will include "signature" designs plus others that will emerge. Debates about which typology (among the half dozen or so most well-known ones) will subside as this generic set of prototypical designs is popularized. There will be relatively less emphasis on discussion of design issues in MMR.

3. Analysis issues will become more important, fueled by advances in the computer analysis of mixed methods data (e.g., Bazeley, 2009, 2010). Within MMR, data will be conceptualized "less in terms of words or numbers and more in terms of transferable units of information" (Teddlie & Tashakkori, 2009, p. 283). Mixed methodologists will develop widely accepted principles of mixed methods data analysis that will supersede the typologies that currently exist. The development of these principles of mixed methods data analysis is crucial to the continuation of MMR as a separate methodological movement.

4. MMR will continue to be adopted throughout the social and behavioral sciences. The form that it takes within any particular discipline will depend on the existing conceptual and methodological orientations within those fields. A challenge for mixed methodologists will be to develop and maintain a "core identity" (e.g., a set of commonly understood methodological principles) that cuts across disciplinary lines.

5. An alternative future is for MMR to continue to pave the way for human sciences research to be more inclusive (eclectic) and research question oriented. This will result in fewer projects being identified as purely QUAL or QUAN, and more that are simply called "research projects" (not labeled specifically as MMR). Unless mixed methodologists develop a core identity of commonly understood methodological principles, it may simply be absorbed into this eclectic blend of research methodologies.

◫ NOTES

1. We wish to express our gratitude to Norman Denzin, Yvonna Lincoln, and Harry Torrance for their very helpful comments and suggestions on earlier versions of this chapter.

2. This definition was taken from *The American Heritage Dictionary of the English Language* (1969, p. 412).

3. Denzin and Lincoln (2005, p. 4) similarly refer to QUAL researchers as *bricoleurs,* who creatively use a variety of QUAL methodological practices.

4. We do not want readers to confuse our use of the term "connoisseur of methods" with the well-known "educational connoisseurship" of Eisner (1998), which involves the art of appreciation and is a "qualitative, artistically grounded approach to educational evaluation" (Eisner, 1979, p. 11).

5. We are not implying that causal effects are examined exclusively by QUAN research or causal mechanisms solely by QUAL research. There are many examples of QUAN results being used descriptively and of QUAL results employed in examining the causes of phenomena (e.g., Maxwell, 2004; Yin, 2003).

6. *Design quality* is the degree to which the investigator has utilized the most appropriate procedures for answering the research question(s) and implemented them effectively. It consists of *design suitability, fidelity, within-design consistency, and analytic adequacy* (Tashakkori & Teddlie, 2008).

7. We do not believe in the dichotomy of QUAL and QUAN approaches on the basis of type of questions. Both exploratory and confirmatory questions may be found in QUAN and in QUAL research.

8. Abductive logic is a third type of logic that occurs when a researcher observes a surprising event and then tries to determine what might have caused it (e.g., Erzberger & Kelle, 2003; Peirce, 1974). It is the process whereby a hypothesis is generated, so that the surprising event may be explained.

9. Quantitizing and qualitizing refer to techniques that convert a QUAN-only or QUAL-only into a MMR study. Some researchers within the QUAL community (e.g., poststructuralists) are unlikely to utilize these techniques.

10. Critical realism (Maxwell & Mittapalli, 2010) has recently been proposed as another framework for the use of mixed methods, but its inclusion is beyond the scope of this chapter.

11. In his critique of the metaphysical paradigm, Morgan (2007, p. 68) acknowledged the valuable contribution that it had made in shifting discussions from mechanical concerns about methods only to larger philosophical and conceptual issues.

12. These arguments then lead Morgan (2007, p. 68) to an alternative position, which he called the *pragmatic approach* that concentrates on "methodology as an area that connects issues at the abstract level of epistemology and the mechanical level of actual methods." Morgan's approach emphasizes issues such as abduction, intersubjectivity, and transferability that supersede the traditional dichotomies (e.g., induction/deduction).

13. Maxwell and Loomis (2003) presented a systemic perspective on research design in MMR that was *non-typological* in nature: the interactive model of design, which consisted of five components (i.e., purposes, conceptual framework, research questions, methods, validity).

□ REFERENCES

Adamson, J. (2004). [Review of the book *Handbook of mixed methods in social & behavioral research*]. *International Journal of Epidemiology, 33*(6), 1414–1415.

Bazeley, P. (2003). Computerized data analysis for mixed methods research. In A. Tashakkori & C. Teddlie (Eds.), *Handbook of mixed methods in social & behavioral research* (pp. 385–422). Thousand Oaks, CA: Sage.

Bazeley, P. (2009). Integrating data analyses in mixed methods research. *Journal of Mixed Methods Research, 3*(3), 203–207.

Bazeley, P. (2010). Computer assisted integration of mixed methods data sources and analysis. In A. Tashakkori & C. Teddlie (Eds.), *SAGE handbook of mixed methods in social & behavioral research* (2nd ed., pp. 431–467). Thousand Oaks, CA: Sage.

Biesta, G. (2010). Pragmatism and the philosophical foundations of mixed methods research. In A. Tashakkori & C. Teddlie (Eds.), *SAGE handbook of mixed methods in social & behavioral research* (2nd ed., pp. 95–117). Thousand Oaks, CA: Sage.

Bryman, A. (2006). Paradigm peace and the implications for quality. *International Journal of Social Research Methodology Theory and Practice, 9*(2), 111–126.

Campbell, D. T., & Fiske, D. W. (1959). Convergent and discriminant validation by the multitrait-multimethod matrix. *Psychological Bulletin, 56,* 81–105.

Caracelli, V. J., & Greene, J. C. (1993). Data analysis strategies for mixed-method evaluation designs. *Educational Evaluation and Policy Analysis, 15*(2), 195–207.

Christ, T. W. (2009). Designing, teaching, and evaluating two complementary mixed methods research courses. *Journal of Mixed Methods Research, 3*(4), 292–325.

Christ, T. W. (2010). Teaching mixed methods and action research: Pedagogical, practical, and evaluative considerations. In A. Tashakkori & C. Teddlie (Eds.), *SAGE handbook of mixed methods in social & behavioral research* (2nd ed., pp. 643–676). Thousand Oaks, CA: Sage.

Creswell, J. W. (2009). Mapping the field of mixed methods research. *Journal of Mixed Methods Research, 3*(2), 95–108.

Creswell, J. W. (2010). Mapping the developing landscape of mixed methods research. In A. Tashakkori & C. Teddlie (Eds.), *SAGE handbook of mixed methods in social & behavioral research* (2nd ed., pp. 45–68). Thousand Oaks, CA: Sage.

Creswell, J. W., & Plano Clark, V. L. (2007). *Designing and conducting mixed methods research.* Thousand Oaks, CA: Sage.

Creswell, J., Shope, R., Plano-Clark, V., & Green, D. (2006). How interpretive qualitative research extends mixed methods research. *Research in the Schools, 13*(1), 1–11.

Creswell, J., Tashakkori, A., Jensen, K., & Shapley, K. (2003). Teaching mixed methods research: Practice, dilemmas and challenges. In A. Tashakkori & C. Teddlie (Eds.), *Handbook of mixed methods in social & behavioral research* (pp. 619–638). Thousand Oaks, CA: Sage.

Day, C., Sammons, P., & Qu, Q. (2008). Combining qualitative and quantitative methodologies in research on teachers' lives, work, and effectiveness: From integration to synergy. *Educational Researcher, 37*(6), 330–342.

Denscombe, M. (2008). Communities of practice: A research paradigm for the mixed methods approach. *Journal of Mixed Methods Research, 2,* 270–283.

Denzin, N. K. (1978). *The research act: A theoretical introduction to sociological method* (2nd ed.). New York: McGraw-Hill.

Denzin, N. K. (2008). The new paradigm dialogs and qualitative inquiry. *International Journal of Qualitative Studies in Education, 21,* 315–325.

Denzin, N. K., & Lincoln, Y. S. (2005). Introduction: The discipline and practice of qualitative research. In N. K. Denzin & Y. S. Lincoln

(Eds.), *The SAGE handbook of qualitative research* (3rd ed., pp. 1–32). Thousand Oaks, CA: Sage.

Eisner, E. W. (1979). The use of qualitative forms of evaluation for improving educational practice. *Educational Evaluation and Policy Analysis, 1*(6), 11–19.

Eisner, E. W. (1998). *The enlightened eye: Qualitative inquiry and the enhancement of educational practice.* Upper Saddle River, NJ: Merrill.

Erzberger, C., & Kelle, U. (2003). Making inferences in mixed methods: The rules of integration. In A. Tashakkori & C. Teddlie (Eds.), *Handbook of mixed methods in social & behavioral research* (pp. 457–490). Thousand Oaks, CA: Sage.

Freshwater, D. (2007). Reading mixed methods research: Contexts for criticism. *Journal of Mixed Methods Research, 1*(2), 134–146.

Greene, J. C. (2007). *Mixing methods in social inquiry.* San Francisco: Jossey-Bass.

Greene, J. C. (2008). Is mixed methods social inquiry a distinctive methodology? *Journal of Mixed Methods Research, 2*(1), 7–22.

Greene, J. C., & Caracelli, V. J. (2003). Making paradigmatic sense of mixed-method practice. In A. Tashakkori & C. Teddlie (Eds.), *Handbook of mixed methods in social & behavioral research* (pp. 91–110). Thousand Oaks, CA: Sage.

Greene, J. C., Caracelli, V. J., & Graham, W. F. (1989). Toward a conceptual framework for mixed-method evaluation designs. *Educational Evaluation and Policy Analysis, 11*, 255–274.

Greene, J. C., & Hall, J. (2010). Dialectics and pragmatism: Being of consequence. In A. Tashakkori & C. Teddlie (Eds.), *SAGE handbook of mixed methods in social & behavioral research* (2nd ed., pp. 119–143). Thousand Oaks, CA: Sage.

Guba, E. G. (1990). Carrying on the dialog. In E. G. Guba (Ed.), *The paradigm dialog* (pp. 368–378). Thousand Oaks, CA: Sage.

Guba, E. G., & Lincoln, Y. S. (1994). Competing paradigms in qualitative research. In N. K. Denzin & Y. S. Lincoln (Eds.), *Handbook of qualitative research* (pp. 105–117). Thousand Oaks, CA: Sage.

Guba, E. G., & Lincoln, Y. S. (2005). Paradigmatic controversies, contradictions, and emerging confluences. In N. K. Denzin & Y. S. Lincoln (Eds.), *The SAGE handbook of qualitative research* (3rd ed., pp. 191–215). Thousand Oaks, CA: Sage.

Hammersley, M. (1996). The relationship between qualitative and quantitative research: Paradigm loyalty versus methodological eclecticism. In J. T. E. Richardson (Ed.), *Handbook of qualitative research methods for psychology and the social sciences* (pp. 159–174). Leicester, UK: BPS Books.

Hesse-Biber, S. (2010). Feminist approaches to mixed methods research: Linking theory and praxis. In A. Tashakkori & C. Teddlie (Eds.), *SAGE handbook of mixed methods in social & behavioral research* (2nd ed., pp. 169–192). Thousand Oaks, CA: Sage.

Howe, K. R. (1988). Against the quantitative-qualitative incompatibility thesis or dogmas die hard. *Educational Researcher, 17*, 10–16.

Howe, K. R. (2004). A critique of experimentalism. *Qualitative Inquiry, 10*(1), 42–61.

Ivankova, N. V., Creswell, J. W., & Stick, S. (2006). Using mixed methods sequential explanatory design: From theory to practice. *Field Methods, 18*(1), 3–20.

Johnson, R. B. (2009). Comments on Howe: Toward a more inclusive "Scientific Research in Education." *Educational Researcher, 38*, 449–457.

Johnson, R. B., & Gray, R. (2010). A history of philosophical and theoretical issues for mixed methods research. In A. Tashakkori & C. Teddlie (Eds.), *SAGE handbook of mixed methods in social & behavioral research* (2nd ed., pp. 69–94). Thousand Oaks, CA: Sage.

Johnson, R. B., & Onwuegbuzie, A. (2004). Mixed methods research: A research paradigm whose time has come. *Educational Researcher, 33*(7), 14–26.

Johnson, R. B., Onwuegbuzie, A. J., & Turner, L. A. (2007). Toward a definition of mixed methods research. *Journal of Mixed Methods Research, 1*(2), 112–133.

Krathwohl, D. R. (2004). *Methods of educational and social science research: An integrated approach* (2nd ed.). Long Grove, IL: Waveland Press.

Kuhn, T. S. (1962). *The structure of scientific revolutions.* Chicago: University of Chicago Press.

Leech, N. L. (2010). Interviews with the early developers of mixed methods research. In A. Tashakkori & C. Teddlie (Eds.), *SAGE handbook of mixed methods in social & behavioral research* (2nd ed., pp. 253–272). Thousand Oaks, CA: Sage.

Leech, N. L., & Onwuegbuzie, A. J. (2009). A typology of mixed methods research designs. *Quality and Quantity, 43*, 265–275.

Li, S., Marquart, J. M., & Zercher, C. (2000). Conceptual issues and analytic strategies in mixed-method studies of preschool inclusion. *Journal of Early Intervention, 23*, 116–132.

Lincoln, Y. S., & Guba, E. G. (1985). *Naturalistic inquiry.* Beverly Hills, CA: Sage.

Maxcy, S. (2003). Pragmatic threads in mixed methods research in the social sciences: The search for multiple modes of inquiry and the end of the philosophy of formalism. In A. Tashakkori & C. Teddlie (Eds.), *Handbook of mixed methods in social & behavioral research* (pp. 51–90). Thousand Oaks, CA: Sage.

Maxwell, J. A. (2004). Causal explanation, qualitative research, and scientific inquiry in education. *Educational Researcher, 33*(2), 3–11.

Maxwell, J. A., & Loomis, D. (2003). Mixed methods design: An alternative approach. In A. Tashakkori & C. Teddlie (Eds.), *Handbook of mixed methods in social & behavioral research* (pp. 241–272). Thousand Oaks, CA: Sage.

Maxwell, J. A., & Mittapalli, K. (2010). Realism as a stance for mixed method research. In A. Tashakkori & C. Teddlie (Eds.), *SAGE handbook of mixed methods in social & behavioral research* (2nd ed., pp. 145–167). Thousand Oaks, CA: Sage.

Mertens, D. M. (2007). Transformative paradigm: Mixed methods and social justice. *Journal of Mixed Methods Research, (1)*3, 212–225.

Mertens, D. M., Bledsoe, K. L., Sullivan, M., & Wilson, A. (2010). Utilization of mixed methods for transformative purposes. In A. Tashakkori & C. Teddlie (Eds.), *SAGE handbook of mixed methods in social & behavioral research* (2nd ed., pp. 193–214). Thousand Oaks, CA: Sage.

Miles, M. B., & Huberman, M. A. (1984). *Qualitative data analysis: A sourcebook for new methods.* Thousand Oaks, CA: Sage.

Miles, M. B., & Huberman, M. A. (1994). *Qualitative data analysis: An expanded sourcebook.* (2nd ed.). Thousand Oaks, CA: Sage.

Morgan, D. (2007). Paradigms lost and pragmatism regained: Methodological implications of combining qualitative and quantitative methods. *Journal of Mixed Methods Research, (1)*1, 48–76.

Morse, J. M. (1991). Approaches to qualitative-quantitative methodological triangulation. *Nursing Research, 40*(2), 120–123.

Morse, J. M. (2003). Principles of mixed methods and multimethod research design. In A. Tashakkori & C. Teddlie (Eds.), *Handbook of mixed methods in social & behavioral research* (pp. 189–208). Thousand Oaks, CA: Sage.

Mosteller, F., & Boruch, R. (Eds.). (2002). *Evidence matters: Randomized trials in education research.* Washington, DC: Brookings Institution Press.

Nastasi, B. K., Hitchcock, J. H., & Brown, L. (2010). An inclusive framework for conceptualizing mixed methods design typologies: Moving toward fully integrated synergistic research models. In A. Tashakkori & C. Teddlie (Eds.), *SAGE handbook of mixed methods in social & behavioral research* (2nd ed., pp. 305–338). Thousand Oaks, CA: Sage.

Newman, I., Ridenour, C., Newman, C., & DeMarco, G. M. P., Jr. (2003). A typology of research purposes and its relationship to mixed methods research. In A. Tashakkori & C. Teddlie (Eds.), *Handbook of mixed methods in social & behavioral research* (pp. 167–188). Thousand Oaks, CA: Sage.

Onwuegbuzie, A. J. (2003). Effect sizes in qualitative research: A prolegomenon. *Quality & Quantity: International Journal of Methodology, 37,* 393–409.

Onwuegbuzie, A., & Combs, J. (2010). Emergent data analysis techniques in mixed methods research: A synthesis. In A. Tashakkori & C. Teddlie (Eds.), *SAGE handbook of mixed methods in social & behavioral research* (2nd ed., pp. 397–430). Thousand Oaks, CA: Sage.

Onwuegbuzie, A. J., Johnson, R. B., & Collins, K. M. T. (2007). Conducting mixed analysis: A general typology. *International Journal of Multiple Research Approaches, 1*(1), 4–17.

Onwuegbuzie, A. J., & Teddlie, C. (2003). A framework for analyzing data in mixed methods research. In A. Tashakkori & C. Teddlie (Eds.), *Handbook of mixed methods in social & behavioral research* (pp. 351–384). Thousand Oaks, CA: Sage.

Patton, M. Q. (1980). *Qualitative evaluation methods.* Thousand Oaks, CA: Sage.

Patton, M. Q. (1990). *Qualitative research and evaluation methods* (2nd ed.). Thousand Oaks, CA: Sage.

Patton, M. Q. (2002). *Qualitative research and evaluation methods* (3rd ed.). Thousand Oaks, CA: Sage.

Peirce, C. S. (1974). *Collected papers* (C. Hartshore, P. Weiss, & A. Burks, Eds.). Cambridge, MA: Harvard University Press.

Reichardt, C. S., & Cook, T. D. (1979). Beyond qualitative *versus* quantitative methods. In T. D. Cook & C. S. Reichardt (Eds.), *Qualitative and quantitative methods in program evaluation* (pp. 7–32). Thousand Oaks CA: Sage.

Ridenour, C. S., & Newman, I. (2008). *Mixed methods research: Exploring the interactive continuum.* Carbondale: Southern Illinois University Press.

Rossman, G., & Wilson, B. (1985). Numbers and words: Combining quantitative and qualitative methods in a single large scale evaluation study. *Evaluation Review, 9,* 627–643.

Sale, J., Lohfeld, L., & Brazil, K. (2002). Revisiting the qualitative-quantitative debate: Implications for mixed-methods research. *Quality and Quantity, 36,* 43–53.

Sandelowski, M., Voils, C. I., & Knafl, G. (2009). On quantitizing. *Journal of Mixed Methods Research, 3*(3), 208–222.

Schulenberg, J. L. (2007). Analyzing police decision-making: Assessing the application of a mixed-method/mixed-model research design. *International Journal of Social Research Methodology, 10,* 99–119.

Shadish, W., Cook, T., & Campbell, D. (2002). *Experimental and quasi-experimental designs for general causal inference.* Boston: Houghton Mifflin.

Shavelson, R. J., & Towne, L. (Eds.). (2002). *Scientific research in education.* Washington, DC: National Research Council, National Academy Press.

Shulha, L., & Wilson, R. (2003). Collaborative mixed methods research. In A. Tashakkori & C. Teddlie (Eds.), *Handbook of mixed methods in social & behavioral research* (pp. 639–670). Thousand Oaks, CA: Sage.

Smith, J. K., & Heshusius, L. (1986). Closing down the conversation: The end of the quantitative-qualitative debate among educational researchers. *Educational Researcher, 15,* 4–12.

Song, M., Sandelowski, M., & Happ, M. B. (2010). Current practices and emerging trends in conducting mixed methods intervention studies. In A. Tashakkori & C. Teddlie (Eds.), *SAGE handbook of mixed methods in social & behavioral research* (2nd ed., pp. 725–747). Thousand Oaks, CA: Sage.

Tashakkori, A., & Creswell, J. (2008). Envisioning the future stewards of the social-behavioral research enterprise. *Journal of Mixed Methods Research, 2*(4), 291–295.

Tashakkori, A., & Teddlie, C. (1998). *Mixed methodology: Combining the qualitative and quantitative approaches.* Thousand Oaks, CA: Sage.

Tashakkori, A., & Teddlie, C. (Eds.). (2003a). *Handbook of mixed methods in social & behavioral research.* Thousand Oaks, CA: Sage.

Tashakkori, A., & Teddlie, C. (2003b). Issues and dilemmas in teaching research methods courses in social and behavioral sciences: US perspective. *International Journal of Social Research Methodology, 6,* 61–77.

Tashakkori, A., & Teddlie, C. (2003c). The past and future of mixed methods research: From data triangulation to mixed model designs. In A. Tashakkori & C. Teddlie (Eds.), *Handbook of mixed methods in social & behavioral research* (pp. 671–702). Thousand Oaks, CA: Sage.

Tashakkori, A., & Teddlie, C. (2008). Quality of inference in mixed methods research: Calling for an integrative framework. In M. M. Bergman (Ed.), *Advances in mixed methods research: Theories and applications* (pp. 101–119). London: Sage.

Teddlie, C., & Johnson, B. (2009). Methodological thought before the twentieth century. In C. Teddlie & A. Tashakkori (Eds.), *The foundations of mixed methods research: Integrating quantitative and qualitative techniques in the social and behavioral sciences* (pp. 40–61). Thousand Oaks, CA: Sage.

Teddlie, C., & Stringfield, S. (1993). *Schools make a difference: Lessons learned from a 10-year study of school effects.* New York: Teachers College Press.

Teddlie, C., & Tashakkori, A. (2003). Major issues and controversies in the use of mixed methods in the social and behavioral sciences. In A. Tashakkori & C. Teddlie (Eds.), *Handbook of mixed methods in social & behavioral research* (pp. 3–50). Thousand Oaks, CA: Sage.

Teddlie, C., & Tashakkori, A. (2009). *The foundations of mixed methods research: Integrating quantitative and qualitative techniques in the social and behavioral sciences.* Thousand Oaks, CA: Sage.

Teddlie, C., Tashakkori, A., & Johnson, B. (2008). Emergent techniques in the gathering and analysis of mixed methods data. In S. Hesse-Biber & P. Leavy (Eds.), *Handbook of emergent methods in social research* (pp. 389–413). New York: Guilford.

Yanchar, S. C., & Williams, D. D. (2006). Reconsidering the compatibility thesis and eclecticism: Five proposed guidelines for method use. *Educational Researcher, 35*(9), 3–12.

Yin, R. K. (2003). *Case study research: Design and methods* (3rd ed.). Thousand Oaks, CA: Sage.

17

CASE STUDY

Bent Flyvbjerg[1]

[C]onduct has its sphere in particular circumstances. That is why some people who do not possess theoretical knowledge are more effective in action (especially if they are experienced) than others who do possess it. For example, suppose that someone knows that light flesh foods are digestible and wholesome, but does not know what kinds are light; he will be less likely to produce health than one who knows that chicken is wholesome.

—Aristotle

◨ What Is a Case Study?

Definitions of "case study" abound. Some are useful, others not. Merriam-Webster's dictionary (2009) defines a case study straightforwardly as follows:

> *Case Study.* An intensive analysis of an individual unit (as a person or community) stressing developmental factors in relation to environment.

According to this definition, case studies focus on an "individual unit," what Robert Stake (2008, pp. 119–120) calls a "functioning specific" or "bounded system." The decisive factor in defining a study as a case study is the choice of the individual unit of study and the setting of its boundaries, its "casing" to use Charles Ragin's (1992, p. 217) felicitous term. If you choose to do a case study, you are therefore not so much making a methodological choice as a choice of what is to be studied. The individual unit may be studied in a number of ways, for instance qualitatively or quantitatively, analytically or hermeneutically, or by mixed methods. This is not decisive for whether it is a case study or not; the demarcation of the unit's boundaries is. Second, the definition stipulates that case studies are "intensive." Thus, case studies comprise more detail, richness, completeness, and variance—that is, depth—for the unit of study than does cross-unit analysis. Third, case studies stress "developmental factors," meaning that a case typically evolves in time, often as a string of concrete and interrelated events that occur "at such a time, in

such a place" and that constitute the case when seen as a whole. Finally, case studies focus on "relation to environment," that is, context. The drawing of boundaries for the individual unit of study decides what gets to count as case and what becomes context to the case.

Against Webster's commonsensical definition of case study, the Penguin *Dictionary of Sociology* (Abercrombie, Hill, & Turner, 1984, p. 34; and verbatim in the 1994 and 2006 editions) has for decades contained the following highly problematic, but unfortunately quite common, definition of case study:

> *Case Study.* The detailed examination of a single example of a class of phenomena, a case study cannot provide reliable information about the broader class, but it may be useful in the preliminary stages of an investigation since it provides hypotheses, which may be tested systematically with a larger number of cases.

This definition is indicative of much conventional wisdom about case study research, which, if not directly wrong, is so oversimplified as to be grossly misleading. The definition promotes the mistaken view that the case study is hardly a methodology in its own right, but is best seen as subordinate to investigations of larger samples. Whereas it is correct that the case study is a "detailed examination of a single example," it is wrong that a case study "cannot provide reliable information about the broader class." It is also correct that a case study *can* be used "in the preliminary stages of an investigation" to generate hypotheses, but it is wrong to see the case study as a pilot method to be used only in preparing the real study's larger surveys, systematic hypotheses testing, and theory building. The Penguin definition juxtaposes case studies with large-sample, statistical research in an unfortunate manner that blocks, instead of brings out, the productive complementarity that exists between the two types of methodology, as we will see below.

John Gerring (2004, p. 342) has correctly pointed out that the many academic attempts to clarify what "case study" means has resulted in a definitional morass, and each time someone attempts to clear up the mess of definitions it just gets worse. If we need a definition of what a case study is, we are therefore better off staying with commonsensical definitions like that from Webster's above than with more loaded academic definitions like that from the Penguin *Dictionary of Sociology*.

◨ THE CASE STUDY PARADOX

Case studies have been around as long as recorded history and today they account for a large proportion of books and articles in psychology, anthropology, sociology, history, political science, education, economics, management, biology, and medical science. For instance, in recent years roughly half of all articles in the top political science journals have used case studies, according to

Alexander George and Andrew Bennett (2005, pp. 4–5). Much of what we know about the empirical world has been produced by case study research, and many of the most treasured classics in each discipline are case studies.

But there is a paradox here. At the same time that case studies are widely used and have produced canonical texts, it may be observed that the case study as a methodology is generally held in low regard, or is simply ignored, within the academy. For example, only 2 of the 30 top-ranked U.S. graduate programs in political science require a dedicated graduate course in case study or qualitative methods, and a full third of these programs do not even offer such a course. In contrast, all of the top 30 programs offer courses in quantitative methods and almost all of them require training in such methods, often several courses (George & Bennett, 2005, p. 10). In identifying this paradox of the case study's wide use and low regard, Gerring (2004, p. 341) rightly remarks that the case study survives in a "curious methodological limbo," and that the reason is that the method is poorly understood.

In what follows, we will try to resolve Gerring's paradox and help case study research gain wider use and acceptance by identifying five misunderstandings about the case study that systematically undermine the credibility and use of the method. The five misunderstandings can be summarized as follows:

Misunderstanding No. 1	General, theoretical knowledge is more valuable than concrete case knowledge.
Misunderstanding No. 2	One cannot generalize on the basis of an individual case; therefore, the case study cannot contribute to scientific development.
Misunderstanding No. 3	The case study is most useful for generating hypotheses; that is, in the first stage of a total research process, while other methods are more suitable for hypotheses testing and theory building.
Misunderstanding No. 4	The case study contains a bias toward verification, that is, a tendency to confirm the researcher's preconceived notions.
Misunderstanding No. 5	It is often difficult to summarize and develop general propositions and theories on the basis of specific case studies.

The five misunderstandings may be said to constitute the conventional view, or orthodoxy, of the case study. We see that theory, reliability, and validity are at issue; in other words, the very status of the case study as a scientific method. In what follows, we will correct the five misunderstandings one by one and thereby clear the ground for a use of case study research in the social sciences that is based on understanding instead of misunderstanding.

◨ MISUNDERSTANDING No. 1

General, theoretical knowledge is more valuable than concrete case knowledge.

In order to understand why the conventional view of case study research is problematic, we need to grasp the role of cases and theory in human learning. Here, two points can be made. First, the case study produces the type of concrete, context-dependent knowledge that research on learning shows to be necessary to allow people to develop from rule-based beginners to virtuoso experts. Second, in the study of human affairs, there appears to exist only context-dependent knowledge, which thus presently rules out the possibility for social science to emulate natural science in developing epistemic theory, that is, theory that is explanatory and predictive. The full argument behind these two points can be found in Flyvbjerg (2001, Chaps. 2–4). For reasons of space, I can only give an outline of the argument here. At the outset, however, we can assert that if the two points are correct, it will have radical consequences for the conventional view of the case study in research and teaching. This view would then be problematic.

Phenomenological studies of human learning indicate that for adults there exists a qualitative leap in their learning process from the rule-governed use of analytical rationality in beginners to the fluid performance of tacit skills in what Pierre Bourdieu (1977) calls virtuosos and Hubert and Stuart Dreyfus (1986), true human experts. Here we may note that most people are experts in a number of everyday social, technical, and intellectual skills like giving a gift, riding a bicycle, or interpreting images on a television screen, while only few reach the level of true expertise for more specialized skills like playing chess, composing a symphony, or flying an airplane.

Common to all experts, however, is that they operate on the basis of intimate knowledge of several thousand concrete cases in their areas of expertise. Context-dependent knowledge and experience are at the very heart of expert activity. Such knowledge and expertise also lie at the center of the case study as a research and teaching method; or to put it more generally yet—as a method of learning. Phenomenological studies of the learning process therefore emphasize the importance of this and similar methods; it is only because of experience with cases that one can at all move from being a beginner to being an expert. If people were exclusively trained in context-independent knowledge and rules, that is, the kind of knowledge that forms the basis of textbooks, they would remain at the beginner's level in the learning process. This is the limitation of analytical rationality; it is inadequate for the best results in the exercise of a profession, as student, researcher, or practitioner.

In teaching situations, well-chosen case studies can help students achieve competence, while context-independent facts and rules will bring students just to the beginner's level. Only few institutions of higher learning have taken the consequence of this. Harvard University is one of them. Here both teaching and research in the professional schools are modeled to a wide extent on the understanding that case knowledge is central to human learning (Christensen & Hansen, 1987; Cragg, 1940).

It is not that rule-based knowledge should be discounted; such knowledge is important in every area and especially to novices. But to make rule-based knowledge the highest goal of learning is topsy-turvy. There is a need for both approaches. The highest levels in the learning process, that is, virtuosity and true expertise, are reached only via a person's own experiences as practitioner of the relevant skills. Therefore, beyond using the case method and other experiential methods for teaching, the best that teachers can do for students in professional programs is to help them achieve real practical experience, for example, via placement arrangements, internships, summer jobs, and the like.

For researchers, the closeness of the case study to real-life situations and its multiple wealth of details are important in two respects. First, it is important for the development of a nuanced view of reality, including the view that human behavior cannot be meaningfully understood as simply the rule-governed acts found at the lowest levels of the learning process, and in much theory. Second, cases are important for researchers' own learning processes in developing the skills needed to do good research. If researchers wish to develop their own skills to a high level, then concrete, context-dependent experience is just as central for them as to professionals learning any other specific skills. Concrete experiences can be achieved via continued proximity to the studied reality and via feedback from those under study. Great distance from the object of study and lack of feedback easily lead to a stultified learning process, which in research can lead to ritual academic blind alleys, where the effect and usefulness of research becomes unclear and untested. As a research method, the case study can be an effective remedy against this tendency.

The second main point in connection with the learning process is that there does not and probably cannot exist predictive theory in social science. Social science has not succeeded in producing general, context-independent theory and has thus in the final instance nothing else to offer than concrete, context-dependent knowledge. And the case study is especially well suited to produce this knowledge. In his later work, Donald Campbell (1975, p. 179) arrives at a similar conclusion. Earlier, he (Campbell and Stanley, 1966, pp. 6–7) had been a fierce critic of the case study, stating that "such studies have such a total absence of control as to be of almost no scientific value." Now he explained that his work had undergone "an extreme oscillation away from my earlier dogmatic disparagement of case studies." Using logic that in many ways resembles that of the phenomenology of human learning, Campbell explains,

> After all, man is, in his ordinary way, a very competent knower, and qualitative common-sense knowing is not replaced by quantitative knowing.... This is not to say that such common sense naturalistic observation is objective, dependable, or unbiased. But it is all that we have. It is the only route to knowledge—noisy, fallible, and biased though it be. (1975, pp. 179, 191)

Campbell is not the only example of a researcher who has altered his views about the value of the case study. Hans Eysenck (1976, p. 9), who originally saw the case study as nothing more than a method of producing anecdotes, later realized that "sometimes we simply have to keep our eyes open and look carefully at individual cases—not in the hope of proving anything, but rather in the hope of learning something!" Final proof is hard to come by in social science because of the absence of "hard" theory, whereas learning is certainly possible. More recently, similar views have been expressed by Charles Ragin, Howard Becker, and their colleagues in explorations of what the case study is and can be in social inquiry (Ragin & Becker, 1992).

As for predictive theory, universals, and scientism, so far social science has failed to deliver. In essence, we have only specific cases and context-dependent knowledge in social science.

The first of the five misunderstandings about the case study—that general theoretical (context-independent) knowledge is more valuable than concrete (context-dependent) case knowledge—can therefore be revised as follows:

Predictive theories and universals cannot be found in the study of human affairs. Concrete case knowledge is therefore more valuable than the vain search for predictive theories and universals.

▣ MISUNDERSTANDING No. 2

> One cannot generalize on the basis of an individual case; therefore, the case study cannot contribute to scientific development.

The view that one cannot generalize on the basis of a single case is usually considered to be devastating to the case study as a scientific method. This second misunderstanding about the case study is typical among proponents of the natural science ideal within the social sciences. Yet even researchers who are not normally associated with this ideal may be found to have this viewpoint. According to Anthony Giddens, for example,

Research which is geared primarily to hermeneutic problems may be of generalized importance in so far as it serves to elucidate the nature of agents' knowledgeability, and thereby their reasons for action, across a wide range of action-contexts. Pieces of ethnographic research like . . . say, the traditional small-scale community research of fieldwork anthropology—are not in themselves generalizing studies. But they can easily become so if carried out in some numbers, so that judgements of their typicality can justifiably be made. (1984, p. 328)

It is correct that one can generalize in the ways Giddens describes, and that often this is both appropriate and valuable. But it would be incorrect to assert that this is the only way to work, just as it is incorrect to conclude that one cannot generalize from a single case. It depends upon the case one is speaking of, and how it is chosen. This applies to the natural sciences as well as to the study of human affairs (Platt, 1992; Ragin & Becker, 1992).

For example, Galileo's rejection of Aristotle's law of gravity was not based upon observations "across a wide range," and the observations were not "carried out in some numbers." The rejection consisted primarily of a conceptual experiment and later of a practical one. These experiments, with the benefit of hindsight, are self-evident. Nevertheless, Aristotle's view of gravity dominated scientific inquiry for nearly 2,000 years before it was falsified. In his experimental thinking, Galileo reasoned as follows: If two objects with the same weight are released from the same height at the same time, they will hit the ground simultaneously, having fallen at the same speed. If the two objects are then stuck together into one, this object will have double the weight and will according to the Aristotelian view therefore fall faster than the two individual objects. This conclusion ran counter to common sense, Galileo found. The only way to avoid the contradiction was to eliminate weight as a determinant factor for acceleration in free fall. And that was what Galileo did. Historians of science continue to discuss whether Galileo actually conducted the famous experiment from the leaning tower of Pisa, or whether this experiment is a myth. In any event, Galileo's experimentalism did not involve a large random sample of trials of objects falling from a wide range of randomly selected heights under varying wind conditions, and so on, as would be demanded by the thinking of the early Campbell and Giddens. Rather, it was a matter of a single experiment, that is, a case study, if any experiment was conducted at all. (On the relation between case studies, experiments, and generalization, see Bailey, 1992; Griffin, Botsko, Wahl, & Isaac, 1991; Lee, 1989; Wilson, 1987.) Galileo's view continued to be subjected to doubt, however, and the Aristotelian view was not finally rejected until half a century later, with the invention of the air pump. The air pump made it possible to conduct the ultimate experiment, known by every pupil, whereby a coin or a piece of lead inside a vacuum tube falls with the same speed as a feather. After this experiment, Aristotle's view could be maintained no longer. What is especially worth noting in our discussion, however, is that the matter was settled by an individual case due to the clever choice of the extremes of metal and feather. One might call it a critical case: For if Galileo's thesis held for these materials, it could be expected to be valid for all or a large range of materials. Random and large samples were at no time part of the picture. Most creative scientists simply do not work this way with this type of problem.

Carefully chosen experiments, cases, and experience were also critical to the development of the physics of Isaac Newton, Albert Einstein, and Niels Bohr, just as the case study occupied a central place in the works of Charles Darwin. In social science, too, the strategic choice of case may greatly add to the generalizability of a case study. In their classical study of the "affluent worker," John Goldthorpe, David Lockwood, Frank Beckhofer, and Jennifer Platt (1968–1969) deliberately looked for a case that was as favorable as possible to the thesis that the working class, having reached middle-class status, was dissolving into a society without class identity and related conflict (see also Wieviorka, 1992). If the thesis could be proved false in the favorable case, then it would most likely be false for intermediate cases. Luton, then a prosperous industrial center outside of London with companies known for high wages and social stability—fertile ground for middle-class identity—was selected as a case, and through intensive fieldwork the researchers discovered that even here an autonomous working-class culture prevailed, lending general credence to the thesis of the persistence of class identity. Below we will discuss more systematically this type of strategic sampling.

As regards the relationship between case studies, large samples, and discoveries, William Beveridge (1951; here quoted from Kuper & Kuper, 1985) observed immediately prior to the breakthrough of the quantitative revolution in the social sciences, "[M]ore discoveries have arisen from intense observation [of individual cases] than from statistics applied to large groups." This does not mean that the case study is always appropriate or relevant as a research method, or that large random samples are without value. The choice of method should clearly depend on the problem under study and its circumstances.

Finally, it should be mentioned that formal generalization, be it on the basis of large samples or single cases, is considerably overrated as the main source of scientific progress. Economist Mark Blaug (1980)—a self-declared adherent to the hypothetico-deductive model of science—has demonstrated that while economists typically pay lip service to the hypothetico-deductive model and to generalization, they rarely practice what they preach in actual research. More generally, Thomas Kuhn has shown that the most important precondition for science is that researchers possess a wide range of practical skills for carrying out scientific work. Generalization is just one of these. In Germanic languages, the term "science" (*Wissenschaft*) means literally "to gain knowledge." And formal generalization is only one of many ways by which people gain and accumulate knowledge. That knowledge cannot be formally generalized does not mean that it cannot enter into the collective process of knowledge accumulation in a given field or in a society. Knowledge may be transferable even where it is not formally generalizable. A purely descriptive, phenomenological case study without any attempt to generalize can certainly be of value in this process and has often helped cut a path toward scientific innovation. This is not to criticize attempts at formal generalization, for such attempts are essential and effective means of scientific development. It is only to emphasize the limitations, which follows when formal generalization becomes the only legitimate method of scientific inquiry.

The balanced view of the role of the case study in attempting to generalize by testing hypotheses has been formulated by Harry Eckstein:

> [C]omparative and case studies are alternative means to the end of testing theories, choices between which must be largely governed by arbitrary or practical, rather than logical, considerations . . . [I]t is impossible to take seriously the position that case study is suspect because problem-prone and comparative study deserving of benefit of doubt because problem-free. (1975, pp. 116, 131, emphasis in original; see also Barzelay, 1993)

Eckstein here uses the term "theory" in its "hard" sense, that is, comprising explanation and prediction. This makes Eckstein's dismissal of the view that case studies cannot be used for testing theories or for generalization stronger than my own view, which is here restricted to the testing of "theory" in the "soft" sense, that is, testing propositions or hypotheses. Eckstein shows that if predictive theories would exist in social science, then the case study could be used to test these theories just as well as other methods.

More recently, George and Bennett (2005) have demonstrated the strong links between case studies and theory development, especially through the study of deviant cases, and John Walton (1992, p. 129) has similarly observed that "case studies are likely to produce the best theory." Already, Eckstein noted, however, the striking lack of genuine theories within his own field, political science, but apparently failed to see why this is so:

> Aiming at the disciplined application of theories to cases forces one to state theories more rigorously than might otherwise be done—provided that the application is truly "disciplined," i.e., designed to show that valid theory compels a particular case interpretation and rules out others. As already stated, this, unfortunately, is rare (if it occurs at all) in political study. One reason is the lack of compelling theories. (1975, pp. 103–104)

The case study is ideal for generalizing using the type of test that Karl Popper called "falsification," which in social science forms part of critical reflexivity. Falsification is one of the most rigorous tests to which a scientific proposition can be subjected: If just one observation does not fit with the proposition, it is considered not valid generally and must therefore be either revised or rejected. Popper himself used the now famous example of "All swans are white," and proposed that just one observation of a single black swan, that is, one deviant case, would falsify this proposition and in this way have general significance and stimulate further investigations and theory building. The case study is well suited for identifying "black swans" because of its in-depth approach: What appears to be "white" often turns out on closer examination to be "black." Deviant cases and the falsifications they entail are main sources of theory development, because they point to the development of new concepts, variables, and causal mechanisms, necessary in order to account for the deviant case and other cases like it.

We will return to falsification in discussing the fourth misunderstanding of the case study below. For the present, however, we can correct the second misunderstanding—that one cannot generalize on the basis of a single case and that the case study cannot contribute to scientific development—so that it now reads:

> One can often generalize on the basis of a single case, and the case study may be central to scientific development via generalization as supplement or alternative to other methods. But formal generalization is overvalued as a source of scientific development, whereas "the force of example" and transferability are underestimated.

▣ MISUNDERSTANDING NO. 3

> The case study is most useful for generating hypotheses, while other methods are more suitable for hypotheses testing and theory building.

The third misunderstanding about the case study is that the case method is claimed to be most useful for generating hypotheses in the first steps of a total research process, while hypothesis-testing and theory-building is best carried out by other methods later in the process, as stipulated by the Penguin definition of case study at the beginning of this chapter. This misunderstanding derives from the previous misunderstanding that one cannot generalize on the basis of individual cases. And since this misunderstanding has been revised as above, we can now correct the third misunderstanding as follows:

The case study is useful for both generating and testing of hypotheses but is not limited to these research activities alone.

Eckstein—contravening the conventional wisdom in this area—goes so far as to argue that case studies are better for testing hypotheses than for producing them. Case studies, Eckstein (1975, p. 80) asserts, "are valuable at all stages of the theory-building process, but most valuable at that stage of theory-building where least value is generally attached to them: the stage at which candidate theories are tested." George and Bennett (2005, pp. 6–9) later confirmed and expanded Eckstein's position, when they found that case studies are especially well suited for theory development because they tackle the following tasks in the research process better than other methods:

- Process tracing that links causes and outcomes (see Box 17.1)
- Detailed exploration of hypothesized causal mechanisms
- Development and testing of historical explanations
- Understanding the sensitivity of concepts to context
- Formation of new hypotheses and new questions to study, sparked by deviant cases

Even rational choice theorists have begun to use case study methods to test their theories and hypotheses, which, if anything, should help deflate the decades-old antagonism between quants and qualts over case study research (Bates, Greif, Levi, Rosenthal, & Weingast, 1998; Flyvbjerg, 2006).

Box 17.1   Falsifying Nobel Prize Theories Through Process Tracing

Some years ago, the editor of *Harvard Business Review* contacted me and asked for a comment on an article he was printing by Princeton psychologist Daniel Kahneman. The editor was puzzled by the fact that Kahneman's Nobel Prize–winning theories on decision making under uncertainty explained failure in executive decisions in terms of inherent optimism (Lovallo & Kahneman, 2003), whereas my group and I explained similar phenomena in terms of strategic misrepresentation, that is, lying as part of principal-agent behavior (Flyvbjerg, Holm, & Buhl, 2002). Who was right, the editor asked? Optimism is unintentional self-deception, whereas lying is intentional deception of others. The question therefore boiled down to whether deception, which caused failure—that much we agreed upon—was intentional or not. The statistical methods that both Kahneman and I had relied upon in our studies of deception could not answer this question. It was now necessary to process trace all the way into people's heads in order to understand whether intention was present or not. Through a number of case studies and interviews, my group and I established that deception is in fact often intentional, especially for very large and expensive decisions taken under political and organizational pressure. We thus falsified optimism as a global explanation of executive failure and developed a new and more nuanced theory that combines optimism and strategic misrepresentation in accounting for failure (Flyvbjerg, 2007).

Testing of hypotheses relates directly to the question of "generalizability," and this in turn relates to the question of case selection. Here, generalizability of case studies can be increased by the strategic selection of cases (for more on the selection of cases, see Ragin, 1992; Rosch, 1978). When the objective is to achieve the greatest possible amount of information on a given problem or phenomenon, a representative case or a random sample may not be the most appropriate strategy. This is because the typical or average case is often not the richest in information. Atypical or extreme cases often reveal more information because they activate more actors and more basic mechanisms in the situation studied. In addition, from both an understanding-oriented and an action-oriented perspective, it is often more important to clarify the deeper causes behind a given problem and its consequences than to describe the symptoms of the problem and how frequently they occur. Random samples emphasizing representativeness will seldom be able to produce this kind of insight; it is more appropriate to select some few cases chosen for their validity.

Table 17.1 summarizes various forms of sampling. The *extreme, or deviant, case* can be well suited for getting a point across in an especially dramatic way, which often occurs for well-known case studies such as Sigmund Freud's "Wolf-Man" and Michel Foucault's "Panopticon." The deviant case is also

Table 17.1 Strategies for the Selection of Samples and Cases

Type of Selection	Purpose
A. Random selection	To avoid systematic biases in the sample. The sample's size is decisive for generalization.
1. Random sample	To achieve a representative sample that allows for generalization for the entire population.
2. Stratified sample	To generalize for specially selected subgroups within the population.
B. Information-oriented selection	To maximize the utility of information from small samples and single cases. Cases are selected on the basis of expectations about their information content.
1. Extreme/deviant cases	To obtain information on unusual cases, which can be especially problematic or especially good in a more closely defined sense. To understand the limits of existing theories and to develop new concepts, variables, and theories that are able to account for deviant cases.
2. Maximum variation cases	To obtain information about the significance of various circumstances for case process and outcome; e.g., three to four cases that are very different on one dimension: size, form of organization, location, budget, etc.
3. Critical cases	To achieve information that permits logical deductions of the type, "If this is (not) valid for this case, then it applies to all (no) cases."
4. Paradigmatic cases	To develop a metaphor or establish a school for the domain that the case concerns.

particularly well suited for theory development, because it helps researchers understand the limits of existing theories and to develop the new concepts, variables, and theories that will be able to account for what were previously considered outliers.

In contrast, a *critical case* can be defined as having strategic importance in relation to the general problem. The above-mentioned strategic selection of lead and feather for the test of

whether different objects fall with equal velocity is an example of critical case selection. This particular selection of materials provided the possibility to formulate a type of generalization that is characteristic of critical cases, a generalization of the sort, "If it is valid for this case, it is valid for all (or many) cases." In its negative form, the generalization would be, "If it is not valid for this case, then it is not valid for any (or only few) cases" (see also Box 17.2).

Box 17.2 Critical Case for Brain Damage

An occupational medicine clinic wanted to investigate whether people working with organic solvents suffered brain damage. Instead of choosing a representative sample among all those enterprises in the clinic's area that used organic solvents, the clinic strategically located a single workplace where all safety regulations on cleanliness, air quality, and the like, had been fulfilled. This model enterprise became a critical case: If brain damage related to organic solvents could be found at this particular facility, then it was likely that the same problem would exist at other enterprises that were less careful with safety regulations for organic solvents. Via this type of strategic choice, one can save both time and money in researching a given problem, and one may generalize in the following manner from a critical case: "If it is valid for this case, it is valid for all (or many) cases." In its negative form, the generalization would be, "If it is not valid for this case, then it is not valid for any (or only few) cases." In this instance, the occupational medicine clinic found brain damage related to organic solvents in the model enterprise and concluded that the problem needed to be dealt with in all enterprises in its jurisdiction.

How does one identify critical cases? This question is more difficult to answer than the question of what constitutes a critical case. Locating a critical case requires experience, and no universal methodological principles exist by which one can with certainty identify a critical case. The only general advice that can be given is that when looking for critical cases, it is a good idea to look for either "most likely" or "least likely" cases, that is, cases that are likely to either clearly confirm or irrefutably

falsify propositions and hypotheses. A model example of a "least likely" case is Robert Michels's (1962) classic study of oligarchy in organizations. By choosing a horizontally structured grass-roots organization with strong democratic ideals—that is, a type of organization with an especially low probability of being oligarchic—Michels could test the universality of the oligarchy thesis, that is, "If this organization is oligarchic, so are most others." A corresponding model example of a "most likely" case is

W. F. Whyte's (1943) study of a Boston slum neighborhood, which according to existing theory should have exhibited social disorganization, but in fact showed quite the opposite (see also the articles on Whyte's study in the April 1992 issue of the *Journal of Contemporary Ethnography*).

Cases of the "most likely" type are especially well suited to falsification of propositions, while "least likely" cases are most appropriate for tests of verification. It should be remarked that a most likely case for one proposition is the least likely for its negation. For example, Whyte's slum neighborhood could be seen as a least likely case for a hypothesis concerning the universality of social organization. Hence, the identification of a case as most or least likely is linked to the design of the study, as well as to the specific properties of the actual case.

A final strategy for the selection of cases is choice of the *paradigmatic case.* Thomas Kuhn has shown that the basic skills, or background practices, of natural scientists are organized in terms of "exemplars," the role of which can be studied by historians of science. Similarly, scholars like Clifford Geertz and Michel Foucault have often organized their research around specific cultural paradigms: A paradigm for Geertz lay for instance in the "deep play" of the Balinese cockfight, while for Foucault, European prisons and the "Panopticon" are examples. Both instances are examples of paradigmatic cases, that is, cases that highlight more general characteristics of the societies in question. Kuhn has shown that scientific paradigms cannot be expressed as rules or theories. There exists no predictive theory for how predictive theory comes about. A scientific activity is acknowledged or rejected as good science by how close it is to one or more exemplars, that is, practical prototypes of good scientific work. A paradigmatic case of how scientists do science is precisely such a prototype. It operates as a reference point and may function as a focus for the founding of schools of thought.

As with the critical case, we may ask, "How does one identify a paradigmatic case?" How does one determine whether a given case has metaphorical and prototypical value? These questions are even more difficult to answer than for the critical case, precisely because the paradigmatic case transcends any sort of rule-based criteria. No standard exists for the paradigmatic case because it sets the standard. Hubert and Stuart Dreyfus see paradigmatic cases and case studies as central to human learning. In an interview with Hubert Dreyfus (author's files), I therefore asked what constitutes a paradigmatic case and how it can be identified. Dreyfus replied,

> Heidegger says, you recognize a paradigm case because it shines, but I'm afraid that is not much help. You just have to be intuitive. We all can tell what is a better or worse case—of a Cézanne painting, for instance. But I can't think there could be any rules for deciding what makes Cézanne a paradigmatic modern painter. . . . [I]t is a big problem in a democratic society where people are supposed to

justify what their intuitions are. In fact, nobody really can justify what their intuition is. So you have to make up reasons, but it won't be the real reasons.

One may agree with Dreyfus that intuition is central to identifying paradigmatic cases, but one may disagree it is a problem to have to justify one's intuitions. Ethnomethodological studies of scientific practice have demonstrated that all variety of such practice relies on taken-for-granted procedures that feel largely intuitive. However, those intuitive decisions are accountable, in the sense of being sensible to other practitioners or often explicable if not immediately sensible. That would frequently seem to be the case with the selection of paradigmatic cases. We may select such a case on the basis of taken-for-granted, intuitive procedures but are often called upon to account for that selection. That account must be sensible to other members of the scholarly communities of which we are part. This may even be argued to be a general characteristic of scholarship, scientific or otherwise, and not unique to the selection of paradigmatic social scientific case studies. For instance, it is usually insufficient to justify an application for research funds by stating that one's intuition says that a particular research should be carried out. A research council ideally operates as society's test of whether the researcher can account, in collectively acceptable ways, for his or her intuitive choice, even though intuition may be the real, or most important, reason why the researcher wants to execute the project.

It is not possible consistently, or even frequently, to determine in advance whether or not a given case—Geertz's cockfights in Bali, for instance—is paradigmatic. Besides the strategic choice of case, the execution of the case study will certainly play a role, as will the reactions to the study by the research community, the group studied, and, possibly, a broader public. The value of the case study will depend on the validity claims that researchers can place on their study, and the status these claims obtain in dialogue with other validity claims in the discourse to which the study is a contribution. Like other good craftspeople, all that researchers can do is use their experience and intuition to assess whether they believe a given case is interesting in a paradigmatic context, and whether they can provide collectively acceptable reasons for the choice of case.

Concerning considerations of strategy in the choice of cases, it should also be mentioned that the various strategies of selection are not necessarily mutually exclusive. For example, a case can be simultaneously extreme, critical, and paradigmatic. The interpretation of such a case can provide a unique wealth of information, because one obtains various perspectives on and conclusions about the case according to whether it is viewed and interpreted as one or another type of case. Finally, a case that the researcher initially thought was one type may turn out to be another, upon closer study (see Box 17.3).

Box 17.3   From Critical Case to Extreme Case, Unwittingly

When I was planning a case study of rationality and power in urban policy and planning in Aalborg, Denmark, reported in Flyvbjerg (1998a), I tried to design the study as a "most likely" critical case in the following manner: If rationality in urban policy and planning were weak in the face of power in Aalborg, then, most likely, they would be weak anywhere, at least in Denmark, because in Aalborg the rational paradigm of policy and planning stood stronger than anywhere else. Eventually, I realized that this logic was flawed, because my research of local relations of power showed that one of the most influential "faces of power" in Aalborg, the Chamber of Industry and Commerce, was substantially stronger than its equivalents elsewhere. This had not been clear at the outset because much less research existed on local power relations than research on local planning. Therefore, instead of a critical case, unwittingly I ended up with an extreme case in the sense that both rationality and power were unusually strong in Aalborg. My study thus became one of what happens when strong rationality meets strong power in the arena of urban policy and planning. But this selection of Aalborg as an extreme case happened to me; I did not deliberately choose it. It was a frustrating experience, especially during those several months after I realized I did not have a critical case until it became clear that all was not lost because I had something else. As a case researcher charting new terrain, one must be prepared for such incidents, I believe.

◻ MISUNDERSTANDING NO. 4

> The case study contains a bias toward verification, that is, a tendency to confirm the researcher's preconceived notions.

The fourth of the five misunderstandings about case study research is that the method maintains a bias toward verification, understood as a tendency to confirm the researcher's preconceived notions, so that the study therefore becomes of doubtful scientific value. Jared Diamond (1996, p. 6), for example, holds this view. He observes that the case study suffers from what he calls a "crippling drawback," because it does not apply "scientific methods," which Diamond understands as methods useful for "curbing one's tendencies to stamp one's preexisting interpretations on data as they accumulate."

Francis Bacon (1853, p. xlvi) saw this bias toward verification not simply as a phenomenon related to the case study in particular, but as a fundamental human characteristic. Bacon expressed it like this:

> The human understanding from its peculiar nature, easily supposes a greater degree of order and equality in things than it really finds. When any proposition has been laid down, the human understanding forces everything else to add fresh support and confirmation. It is the peculiar and perpetual error of the human understanding to be more moved and excited by affirmatives than negatives.

Bacon certainly touches upon a fundamental problem here, a problem that all researchers must deal with in some way. Charles Darwin (Barlow, 1958, p. 123), in his autobiography, describes the method he developed in order to avoid the bias toward verification:

I had . . . during many years followed a golden rule, namely, that whenever a published fact, a new observation or thought came across me, which was opposed to my general results, to make a memorandum of it without fail and at once; for I had found by experience that such facts and thoughts were far more apt to escape from the memory than favorable ones. Owing to this habit, very few objections were raised against my views, which I had not at least noticed and attempted to answer.

The bias toward verification is general, but the alleged deficiency of the case study and other qualitative methods is that they ostensibly allow more room for the researcher's subjective and arbitrary judgment than other methods: They are often seen as less rigorous than are quantitative, hypothetico-deductive methods. Even if such criticism is useful, because it sensitizes us to an important issue, experienced case researchers cannot help but see the critique as demonstrating a lack of knowledge of what is involved in case study research. Donald Campbell and others have shown that the critique is fallacious, because the case study has its own rigor, different to be sure, but no less strict than the rigor of quantitative methods. The advantage of the case study is that it can "close in" on real-life situations and test views directly in relation to phenomena as they unfold in practice.

According to Campbell, Ragin, Geertz, Wieviorka, Flyvbjerg, and others, researchers who have conducted intensive, in-depth case studies, typically report that their preconceived views, assumptions, concepts, and hypotheses were wrong and that the case material has compelled them to revise their hypotheses on essential points. The case study forces upon the researcher the type of falsifications described above. Ragin (1992, p. 225) calls this a "special feature of small-N research," and goes on to explain that criticizing single case studies for being inferior to multiple case studies is misguided, since even single case studies "are multiple in most research efforts because ideas and evidence may be linked in many different ways."

Geertz (1995, p. 119) says about the fieldwork involved in most in-depth case studies that "The Field" itself is a "powerful disciplinary force: assertive, demanding, even coercive." Like any such force, it can be underestimated, but it cannot be evaded. "It is too insistent for that," says Geertz. That he is speaking of a general phenomenon can be seen by simply examining case studies, such as those by Eckstein (1975), Campbell (1975), and Wieviorka (1992). Campbell (1975, pp. 181–182) discusses the causes of this phenomenon in the following passage:

> In a case study done by an alert social scientist who has thorough local acquaintance, the theory he uses to explain the focal difference also generates prediction or expectations on dozens of other aspects of the culture, and he does not retain the theory unless most of these are also confirmed.... Experiences of social scientists confirm this. Even in a single qualitative case study, the conscientious social scientist often finds no explanation that seems satisfactory. Such an outcome would be impossible if the caricature of the single case study ... were correct—there would instead be a surfeit of subjectively compelling explanations.

According to the experiences cited above, it is falsification and not verification that characterizes the case study. Moreover, the question of subjectivism and bias toward verification applies to all methods, not just to the case study and other qualitative methods. For example, the element of arbitrary subjectivism will be significant in the choice of categories and variables for a quantitative or structural investigation, such as a structured questionnaire to be used across a large sample of cases. And the probability is high that (1) this subjectivism survives without being thoroughly corrected during the study, and (2) that it may affect the results, quite simply because the quantitative/structural researcher does not get as close to those under study as does the case study researcher and therefore is less likely to be corrected by the study objects "talking back." George and Bennett (2005, p. 20) describe this all-important feature of case study research like this:

> When a case study researcher asks a participant "were you thinking X when you did Y," and gets the answer, "No, I was thinking Z," then if the researcher had not thought of Z as a causally relevant variable, she may have a new variable demanding to be heard.

Statistical methods may identify deviant cases that can lead to new hypotheses, but in isolation these methods lack any clear means of actually identifying new hypotheses. This is true of all studies that use existing databases or that collect survey data based on questionnaires with predefined standard questions. Unless statistical researchers do their own archival work, interviews, or face-to-face surveys with open-ended questions—like case study researchers—they have no means of identifying

left-out variables (George & Bennett, 2005, p. 21). According to Ragin (1992, p. 225; see also Ragin, 1987, pp. 164–171):

> This feature explains why small-N qualitative research is most often at the forefront of theoretical development. When N's are large, there are few opportunities for revising a casing [that is, the delimitation of a case]. At the start of the analysis, cases are decomposed into variables, and almost the entire dialogue of ideas and evidence occurs through variables. One implication of this discussion is that to the extent that large-N research can be sensitized to the diversity and potential heterogeneity of the cases included in an analysis, large-N research may play a more important part in the advancement of social science theory.

Here, too, this difference between large samples and single cases can be understood in terms of the phenomenology for human learning discussed above. If one thus assumes that the goal of the researcher's work is to understand and learn about the phenomena being studied, then research is simply a form of learning. If one assumes that research, like other learning processes, can be described by the phenomenology for human learning, it then becomes clear that the most advanced form of understanding is achieved when researchers place themselves within the context being studied. Only in this way can researchers understand the viewpoints and the behavior that characterizes social actors. Relevant to this point, Giddens states that valid descriptions of social activities presume that researchers possess those skills necessary to participate in the activities described:

> I have accepted that it is right to say that the condition of generating descriptions of social activity is being able in principle to participate in it. It involves "mutual knowledge," shared by observer and participants whose action constitutes and reconstitutes the social world. (1982, p. 15)

From this point of view, the proximity to reality, which the case study entails, and the learning process that it generates for the researcher will often constitute a prerequisite for advanced understanding. In this context, one begins to understand Beveridge's conclusion that there are more discoveries stemming from intense observation of individual cases than from statistics applied to large groups. With the point of departure in the learning process, we understand why the researcher who conducts a case study often ends up by casting off preconceived notions and theories. Such activity is quite simply a central element in learning and in the achievement of new insight. More simple forms of understanding must yield to more complex ones as one moves from beginner to expert.

On this basis, the fourth misunderstanding—that the case study supposedly contains a bias toward verification, understood as a tendency to confirm the researcher's preconceived ideas—is revised as follows:

The case study contains no greater bias toward verification of the researcher's preconceived notions than other methods of inquiry. On the contrary, experience indicates that the case study contains a greater bias toward falsification of preconceived notions than toward verification.

▣ MISUNDERSTANDING NO. 5

> It is often difficult to summarize and develop general propositions and theories on the basis of specific case studies.

Case studies often contain a substantial element of narrative and one can get into a terrible quicksand today talking about the matter of narrative in social science (for a good overview of narrative inquiry, see Chapter 25 in this volume by Susan Chase; Todd Landman, in press). After certain strands of discourse theory have defined any text as narrative and everything as text, it seems that narrative is everything. But if something is everything, maybe it is nothing, and we are back to square one. It is difficult to avoid the subject of narrative completely, however, when considering the case study and qualitative research. In my own work, when I think about narrative, I do not think of discourse theory but of Miles Davis, the jazz icon. When asked how he kept writing classics through a four-decades-long career, he answered, "I first write a beginning, then a middle, and finally the ending." Narrative suggests questions about plot, that is, a sequence of events and how they are related, and Davis set out the naked minimum. Obviously, plots and narratives may be hatched in many ways. But if you write the kind of classic narrative that Davis talks about, with a beginning, a middle, and an end, you typically first try to get the attention of the reader, often by means of a hook, that is, a particularly captivating event or problematic that leads into the main story. You then present the issues and who are involved, including their relationships. Gradually, you reel in the reader to a point of no return, from where the main character—who in a case study need not be a person but could be, say, a community, a program, or a company—has no choice but to deal with the issues at hand, and in this sense is tested. At this stage, typically, there is conflict and the conflict escalates. Finally, harmony is restored by the conflict being resolved, or at least explained, as may be the appropriate achievement in a social science narrative.

To Alasdair MacIntyre (1984, pp. 214, 216), the human being is a "story-telling animal," and the notion of a history is as fundamental a human notion as the notion of an action. Other observers have noted that narrative seems to exist in all human societies, modern and ancient, and that it is perhaps our most fundamental form for making sense of experience (Mattingly, 1991, p. 237; Novak, 1975, p. 175; see also Abbott, 1992; Arendt, 1958; Bal, 1997; Carr, 1986; Fehn, Hoestery, & Tatar, 1992; Rasmussen, 1995;

Ricoeur, 1984). Narrative thus seems not only to be the creation of the storyteller, but seems also to be an expression of innate relationships in the human mind, which we use to make sense of the world by constructing it as narrative.

The human propensity for narrative involves a danger, however, of what has been called the narrative fallacy. The fallacy consists of a human inclination to simplify data and information through overinterpretation and through a preference for compact stories over complex data sets (Taleb, 2010, p. 63). It is easier to remember and make decisions on the basis of "meaningful" stories than to remember strings of "meaningless" data. Thus, we read meaning into data and make up stories, even where this is unwarranted. As a case in point, consider the inspirational accounts of how the Internet led to a "new economy" where productivity had been disconnected from share prices; or the fairy tale that increasing real estate prices are enough to sustain economic growth in a nation. Such stories are easy to understand and act on—for citizens, policy makers, and scholars—but they are fallacies and as such they are treacherous. In social science, the means to avoid the narrative fallacy is no different from the means to avoid other error: the usual systematic checks for validity and reliability in how data are collected and used.

Dense narratives based on thick description will provide some protection against the narrative fallacy. Such narratives typically approach the complexities and contradictions of real life. Accordingly, they may be difficult or impossible to summarize into neat formulas, general propositions, and theories (Benhabib, 1990; Mitchell & Charmaz, 1996; Roth, 1989; Rouse, 1990; White, 1990). This tends to be seen by critics of the case study as a drawback. To the case study researcher, however, a particularly "thick" and hard-to-summarize narrative is not a problem. Rather, it is often a sign that the study has uncovered a particularly rich problematic. The question, therefore, is whether the summarizing and generalization, which the critics see as an ideal, is always desirable. Friedrich Nietzsche (1974, p. 335, para. 373) is clear in his answer to this question. "Above all," he says about doing science, "one should not wish to divest existence of its *rich ambiguity*" (emphasis in original).

Lisa Peattie (2001, p. 260) explicitly warns against summarizing dense case studies: "It is simply that the very value of the case study, the contextual and interpenetrating nature of forces, is lost when one tries to sum up in large and mutually exclusive concepts." The dense case study, according to Peattie, is more useful for the practitioner and more interesting for social theory than either factual "findings" or the high-level generalizations of theory.

The opposite of summing up and "closing" a case study is to keep it open. Two strategies work particularly well in ensuring openness. First, when writing up their case studies, authors may demur from the role of omniscient narrator and summarizer. Instead, they may choose to tell the story in its diversity, allowing the story to unfold from the many-sided, complex, and sometimes-conflicting stories that the actors in the case have told researchers. Second, authors of case studies may avoid linking their study with the theories of any one academic specialization. Instead, they may choose to relate the case to broader philosophical positions that cut across specializations. In this way, authors leave scope for readers of different backgrounds to make different interpretations and draw diverse conclusions regarding the question of what the case is a case of. The goal is not to make the case study be all things to all people. The goal is to allow the study to be different things to different people. Here it is useful to describe the case with so many facets—like life itself—that different readers may be attracted, or repelled, by different things in the case. Readers are not pointed down any one theoretical path or given the impression that truth might lie at the end of such a path. Readers will have to discover their own path and truth inside the case. Thus, in addition to the interpretations of case actors and case narrators, readers are invited to decide the meaning of the case and to interrogate actors' and narrators' interpretations in order to answer that categorical question of any case study: "What is this case a case of?"

Case stories written like this can neither be briefly recounted nor summarized in a few main results. The case story is itself the result. It is a "virtual reality," so to speak. For the reader willing to enter this reality and explore it inside and out, the payback is meant to be a sensitivity to the issues at hand that cannot be obtained from theory. Students can safely be let loose in this kind of reality, which provides a useful training ground with insights into real-life practices that academic teaching often does not provide.

If we return again briefly to the phenomenology for human learning, we may understand why summarizing case studies is not always useful and may sometimes be counterproductive. Knowledge at the beginner's level consists precisely in the reduced formulas that characterize theories, while true expertise is based on intimate experience with thousands of individual cases and on the ability to discriminate between situations, with all their nuances of difference, without distilling them into formulas or standard cases. The problem is analogous to the inability of heuristic, computer-based expert systems to approach the level of virtuoso human experts, even when the systems are compared with the experts who have conceived the rules upon which these systems operate. This is because the experts do not use rules but operate on the basis of detailed case experience. This is *real* expertise. The rules for expert systems are formulated only because the systems require it; rules are characteristic of expert *systems*, but not of real human *experts*.

In the same way, one might say that the rule formulation that takes place when researchers summarize their work into theories is characteristic of the culture of research, of researchers, and of theoretical activity, but such rules are not necessarily part of the studied reality constituted by Bourdieu's (1977, pp. 8, 15) "virtuoso social actors." Something essential may be lost by this summarizing—namely the possibility to understand virtuoso social acting, which, as Bourdieu has shown, cannot be distilled into theoretical formulas—and it is precisely their fear of losing this "something"

that makes case researchers cautious about summarizing their studies. Case researchers thus tend to be skeptical about erasing phenomenological detail in favor of conceptual closure.

Ludwig Wittgenstein shared this skepticism. According to Gasking and Jackson, Wittgenstein used the following metaphor when he described his use of the case study approach in philosophy:

> In teaching you philosophy I'm like a guide showing you how to find your way round London. I have to take you through the city from north to south, from east to west, from Euston to the embankment and from Piccadilly to the Marble Arch. After I have taken you many journeys through the city, in all sorts of directions, we shall have passed through any given street a number of times—each time traversing the street as part of a different journey. At the end of this you will know London; you will be able to find your way about like a born Londoner. Of course, a good guide will take you through the more important streets more often than he takes you down side streets; a bad guide will do the opposite. In philosophy I'm a rather bad guide. (1967, p. 51)

This approach implies exploring phenomena firsthand instead of reading maps of them. Actual practices are studied before their rules, and one is not satisfied by learning only about those parts of practices that are open to public scrutiny; what Erving Goffman (1963) calls the "backstage" of social phenomena must be investigated, too, like the side streets that Wittgenstein talks about.

With respect to intervention in social and political affairs, Andrew Abbott (1992, p. 79) has rightly observed that a social science expressed in terms of typical case narratives would provide "far better access for policy intervention than the present social science of variables." Alasdair MacIntyre (1984, p. 216) similarly says, "I can only answer the question 'What am I to do?' if I can answer the prior question 'Of what story or stories do I find myself a part?'" In a similar vein, Cheryl Mattingly (1991, p. 237) points out that narratives not only give meaningful form to experiences we have already lived through, they also provide us a forward glance, helping us to anticipate situations even before we encounter them, allowing us to envision alternative futures. Narrative inquiries do not—indeed, cannot—start from explicit theoretical assumptions. Instead, they begin with an interest in a particular phenomenon that is best understood narratively. Narrative inquiries then develop descriptions and interpretations of the phenomenon from the perspective of participants, researchers, and others.

William Labov and Joshua Waletzky (1966, pp. 37–39) write that when a good narrative is over, "it should be unthinkable for a bystander to say, 'So what?'" Every good narrator is continually warding off this question. A narrative that lacks a moral that can be independently and briefly stated, is not necessarily pointless. And a narrative is not successful just because it allows a brief moral. A successful narrative does not allow the question to be raised at all. The narrative has already supplied the answer before the question is asked. The narrative itself is the answer (Nehamas, 1985, pp. 163–164).

A reformulation of the fifth misunderstanding, which states that it is often difficult to summarize specific case studies into general propositions and theories, thus reads as follows:

It is correct that summarizing case studies is often difficult, especially as concerns case process. It is less correct as regards case outcomes. The problems in summarizing case studies, however, are due more often to the properties of the reality studied than to the case study as a research method. Often it is not desirable to summarize and generalize case studies. Good studies should be read as narratives in their entirety.

It must again be emphasized that despite the difficulty or undesirability in summarizing certain case studies, the case study as such can certainly contribute to the cumulative development of knowledge, for example, in using the principles to test propositions described above under the second and third misunderstandings.

CURRENT TRENDS IN CASE STUDY RESEARCH

This chapter began by pointing out a paradox in case study research, namely, that even as case studies are widely used in social science and have produced many of the classic texts here, it may be observed that the case study as a methodology is generally held in low regard, or is simply ignored, within large and dominant parts of the academy. This state of affairs has proved remarkably long-lived.

However, as pointed out by George and Bennett (2005, pp. 4–5), recently a certain loosening of positions has taken place. A more collaborative approach is gaining ground, where scholars begin to see that different methodological approaches have different strengths and weaknesses and are essentially complementary. The old and often antagonistic division between quants and quals is losing ground as a new generation of scholars trained in both quantitative and qualitative methods is emerging. For these scholars, research is problem-driven and not methodology-driven, meaning that those methods are employed that for a given problematic best help answer the research questions at hand. More often than not, a combination of qualitative and quantitative methods will do the task best. Finally, some of the most ambitious claims regarding how the quantitative revolution would make possible a social science on a par with natural science in its ability to explain and predict have been scaled back, making room for the emergence of a more realistic and balanced attitude to what social science can and cannot do. The chapters in this volume on mixed methods, by John Creswell (Chapter 15), and Charles Teddlie and Abbas Tashakkori (Chapter 16), are good examples of this loosening of positions and more balanced attitude.

If the moment of the quantitative revolution in social science is called positivistic, as is often the case, then today we are in a postpositivist and possibly post-paradigmatic moment (Schram, 2006). My own efforts at developing a social science suited for this particular moment have been concentrated on what I call "phronetic social science," named after the ancient Greek term for practical wisdom, or common sense, *phronesis* (Flyvbjerg, 2001; Schram & Caterino, 2006). And this is what the new social science is: commonsensical. It is common sense to give up wars that cannot be won, like the methods war over quantitative versus qualitative methods, or the science wars, which pit social science against natural science. It is also common sense to finally acknowledge that case studies and statistical methods are not conflicting but complementary (see Box 17.4).

Box 17.4   Complementarity in Action: From Case Studies to Statistical Methods, and Back

My current research on megaprojects was originally sparked by events at the Channel tunnel, which links the United Kingdom and France, and the Danish Great Belt tunnel, linking Scandinavia with continental Europe. These are the two longest underwater rail tunnels in Europe, each costing several billion dollars. Soon after construction of the Channel tunnel began, costs started escalating, and at the opening of the tunnel, in 1994, costs had doubled in real terms leaving the project in serious financial trouble. But maybe the British and French had just been unlucky? Perhaps the Danes would do better on the Great Belt tunnel? Not so. Here the cost overrun was larger still, at 120% in real terms, and the tunnel proved financially nonviable even before it opened to traffic in 1997, several years behind schedule. I did a case study of these two hugely expensive projects in order to document and understand the apparent incompetence in their planning and execution (Flyvbjerg, Bruzelius, & Rothengatter, 2003). The study raised the inevitable question of whether the Channel and Great Belt tunnels were outliers regarding cost overrun and viability or whether such extreme lack of ability to build on budget was common for large-scale infrastructure projects. Searching the world's libraries and asking colleagues, I found that no study existed that answered these questions in a statistically valid manner. I therefore decided to do such a study and my group and I now turned from case studies to statistical methods. To our amazement, our studies showed, with a very high level of statistical significance, that the Channel and Great Belt projects were not outliers, they were normal; nine out of ten projects have cost overrun. Even more surprisingly, when we extended our data back in time we

(Continued)

(Continued)

found that for the 70 years for which we were able to find data there had been no improvement in performance regarding getting cost estimates right and staying on budget. The same apparent error of cost underestimation and overrun was being repeated decade after decade. We now began debating among ourselves whether an error that is being repeated over and over by highly trained professionals is really an error, or whether something else was going on. To answer this question, we went back to case studies and process tracing (see Box 17.1). We found that cost overrun and lack of viability were not best explained by simple error but by something more sinister and Machiavellian, namely strategic misrepresentation of costs and benefits by promoters during appraisal in order to get projects funded and built. From my initial case-based curiosity with the outcomes at the Channel and Great Belt tunnels—and by going from case studies to statistical methods and back—my group and I had uncovered a deep-rooted culture of deception in the planning and management of large-scale infrastructure projects (Flyvbjerg, 2007). As a recent spin-off from this research, my group and I are now investigating whether the success of one in ten projects in staying on budget—documented in our statistical studies—may be replicated or is due to luck. Here, again, we are back to case study research, now studying success as a deviant case.

The complementarity between case studies and statistical methods may be summarized as in Table 17.2. The main strength of the case study is depth—detail, richness, completeness, and within-case variance—whereas for statistical methods it is breadth. If you want to understand a phenomenon in any degree of thoroughness—say, child neglect in the family or cost overrun in urban regeneration—what causes it, how to prevent it, and so on, you need to do case studies. If you want to understand how widespread the phenomenon is, how it correlates with other phenomena and varies across different populations, and at what level of statistical significance, then you have to do statistical studies. If you want to understand both, which is advisable if you would like to speak with weight about the phenomenon at hand, then you need to do both case studies and

statistical analyses. The complementarity of the two methods is that simple, and that beautiful.

When you think about it, it is amazing that the separation and antagonism between qualitative and quantitative methods often seen in the literature, and in university departments, have lasted as long as they have. This is what happens when tribalism and power, instead of reason, rules the halls of academia. As such, it is testimony to the fact that academics, too, are all too human, and not testimony to much else. The separation is not a logical consequence of what graduates and scholars need to know in order to do their studies and do them well; quite the opposite. Good social science is opposed to an either/or and stands for a both/and on the question of qualitative versus quantitative methods. The *International Encyclopedia of the Social & Behavioral Sciences*

Table 17.2 Complementarity of Case Studies and Statistical Methods

	Case Studies	*Statistical Methods*
Strengths	■ Depth	■ Breadth
	■ High conceptual validity	■ Understanding how widespread a phenomenon is across a population
	■ Understanding of context and process	■ Measures of correlation for populations of cases
	■ Understanding of what causes a phenomenon, linking causes and outcomes	■ Establishment of probabilistic levels of confidence
	■ Fostering new hypotheses and new research questions	
Weaknesses	■ Selection bias may overstate or understate relationships	■ Conceptual stretching, by grouping together dissimilar cases to get larger samples
	■ Weak understanding of occurrence in population of phenomena under study	■ Weak understanding of context, process, and causal mechanisms
	■ Statistical significance often unknown or unclear	■ Correlation does not imply causation
		■ Weak mechanisms for fostering new hypotheses

(Smelser & Baltes, 2001, p. 1513) is certainly right when it points out that the case study and statistical methods can "achieve far more scientific progress together than either could alone."

This being said, it should nevertheless be added that the balance between case studies and statistical methods is still biased in favor of the latter in social science, so much so that it puts case studies at a disadvantage within most disciplines. For the time being, it is therefore necessary to continue to work on clarifying methodologically the case study and its relations to other social science methods in order to dispel the methodological limbo in which the method has existed for too long. This chapter is intended as such clarification.

◨ NOTE

1. The author wishes to thank Maria Flyvbjerg Bo for her help in improving an earlier version of this chapter.

◨ REFERENCES

Abbott, A. (1992). What do cases do? Some notes on activity in sociological analysis. In C. C. Ragin & H. S. Becker (Eds.), *What is a case? Exploring the foundations of social inquiry* (pp. 53–82). Cambridge, UK: Cambridge University Press.

Abercrombie, N., Hill, S., & Turner, B. S. (1984). *Dictionary of sociology*. Harmondsworth, UK: Penguin.

Arendt, H. (1958). *The human condition*. Chicago: University of Chicago Press.

Bacon, F. (1853). Novum organum. In *Physical and metaphysical works of Lord Bacon* (Vol. 1). London: H. G. Bohn.

Bailey, M. T. (1992). Do physicists use case studies? Thoughts on public administration research. *Public Administration Review, 52*(1), 47–54.

Bal, M. (1997). *Narratology: Introduction to the theory of narrative* (2nd ed.). Toronto: University of Toronto Press.

Barlow, N. (Ed.). (1958). *The autobiography of Charles Darwin*. New York: Norton.

Barzelay, M. (1993). The single case study as intellectually ambitious inquiry. *Journal of Public Administration Research and Theory, 3*(3), 305–318.

Bates, R., Greif, A., Levi, M., Rosenthal, J.-L., & Weingast, B. (1998). *Analytic narratives*. Princeton, NJ: Princeton University Press.

Benhabib, S. (1990). Hannah Arendt and the redemptive power of narrative. *Social Research, 57*(1), 167–196.

Beveridge, W. I. B. (1951). *The art of scientific investigation*. London: Heinemann.

Blaug, M. (1980). *The methodology of economics: Or how economists explain*. Cambridge, UK: Cambridge University Press.

Bourdieu, P. (1977). *Outline of a theory of practice*. Cambridge, UK: Cambridge University Press.

Campbell, D. T. (1975). Degrees of freedom and the case study. *Comparative Political Studies, 8*(1), 178–191.

Campbell, D. T., & Stanley, J. C. (1966). *Experimental and quasi-experimental designs for research*. Chicago: Rand McNally.

Carr, D. (1986). *Time, narrative, and history*. Bloomington: Indiana University Press.

Christensen, C. R., & Hansen, A. J. (Eds.). (1987). *Teaching and the case method*. Boston, MA: Harvard Business School Press.

Cragg, C. I. (1940). Because wisdom can't be told (Harvard Business School Reprint 451–005). *Harvard Alumni Bulletin,* 1–6.

Diamond, J. (1996, November 14). The roots of radicalism. *The New York Review of Books,* pp. 4–6.

Dreyfus, H., & Dreyfus, S. (with Athanasiou, T.). (1986). *Mind over machine: The power of human intuition and expertise in the era of the computer*. New York: Free Press.

Eckstein, H. (1975). Case study and theory in political science. In F. J. Greenstein & N. W. Polsby (Eds.), *Handbook of political science* (Vol. 7, pp. 79–137). Reading, MA: Addison-Wesley.

Eysenck, H. J. (1976). Introduction. In H. J. Eysenck (Ed.), *Case studies in behaviour therapy*. London: Routledge and Kegan Paul.

Fehn, A., Hoestery, I., & Tatar, M. (Eds.). (1992). *Neverending stories: Toward a critical narratology*. Princeton, NJ: Princeton University Press.

Flyvbjerg, B. (2001). *Making social science matter: Why social inquiry fails and how it can succeed again*. Cambridge, UK: Cambridge University Press.

Flyvbjerg, B. (2006). A perestroikan straw man answers back: David Laitin and phronetic political science. In S. F. Schram & B. Caterino (Eds.), *Making political science matter: Debating knowledge, research, and method* (pp. 56–85). New York and London: New York University Press.

Flyvbjerg, B. (2007). Policy and planning for large-infrastructure projects: Problems, causes, cures. *Environment and Planning B: Planning and Design, 34*(4), 578–597.

Flyvbjerg, B., Bruzelius, N., & Rothengatter, W. (2003). *Megaprojects and risk: An anatomy of ambition*. Cambridge, UK: Cambridge University Press.

Flyvbjerg, B., Holm, M. K. S., & Buhl, S. L. (2002). Underestimating costs in public works projects: Error or lie? *Journal of the American Planning Association, 68*(3), 279–295.

Gasking, D. A. T., & Jackson, A. C. (1967). Wittgenstein as a teacher. In K. T. Fann (Ed.), *Ludwig Wittgenstein: The man and his philosophy* (pp. 49–55). Sussex, UK: Harvester Press.

Geertz, C. (1995). *After the fact: Two countries, four decades, one anthropologist*. Cambridge, MA: Harvard University Press.

George, A. L., & Bennett, A. (2005). *Case studies and theory development in the social sciences*. Cambridge, MA: MIT Press.

Gerring, J. (2004). What is a case study and what is it good for? *The American Political Science Review, 98*(2), 341–354.

Giddens, A. (1982). *Profiles and critiques in social theory*. Berkeley: University of California Press.

Giddens, A. (1984). *The constitution of society: Outline of the theory of structuration*. Cambridge, UK: Polity Press.

Goffman, E. (1963). *Behavior in public places: Notes on the social organization of gatherings*. New York: Free Press.

Goldthorpe, J. H., Lockwood, D., Beckhofer, F., & Platt, J. (1968–1969). *The affluent worker* (Vols. 1–3). Cambridge, UK: Cambridge University Press.

Griffin, L. J., Botsko, C., Wahl, A.-M., & Isaac, L. W. (1991). Theoretical generality, case particularity: Qualitative comparative analysis of trade union growth and decline. In C. C. Ragin (Ed.), *Issues and alternatives in comparative social research* (pp. 110–136). Leiden, The Netherlands: E. J. Brill.

Kuper, A., & Kuper, J. (Eds.). (1985). *The social science encyclopedia.* London: Routledge and Kegan Paul.

Labov, W., & Waletzky, J. (1966). Narrative analysis: Oral versions of personal experience. In *Essays on the verbal and visual arts: Proceedings of the American Ethnological Society* (pp. 12–44). Seattle, WA: American Ethnological Society.

Landman, T. (in press). Phronesis and narrative analysis. In B. Flyvbjerg, T. Landman, & S. Schram (Eds.), *Real social science: Applied phronesis.* Cambridge, UK: Cambridge University Press.

Lee, A. S. (1989). Case studies as natural experiments. *Human Relations, 42*(2), 117–137.

Lovallo, D., & Kahneman, D. (2003, July). Delusions of success: How optimism undermines executives' decisions. *Harvard Business Review,* 56–63.

MacIntyre, A. (1984). *After virtue: A study in moral theory* (2nd ed.). Notre Dame, IN: University of Notre Dame Press.

Mattingly, C. (1991). Narrative reflections on practical actions: Two learning experiments in reflective storytelling. In D. A. Schön (Ed.), *The reflective turn: Case studies in and on educational practice* (pp. 235–257). New York: Teachers College Press.

Merriam-Webster Online Dictionary. (2009). *Case study.* Available at http://www.merriam-webster.com/dictionary/case%20study

Michels, R. (1962). *Political parties: A study of the oligarchical tendencies of modern democracy.* New York: Collier.

Mitchell, R. G., Jr., & Charmaz, K. (1996). Telling tales, writing stories: Postmodernist visions and realist images in ethnographic writing. *Journal of Contemporary Ethnography, 25*(1), 144–166.

Nehamas, A. (1985). *Nietzsche: Life as literature.* Cambridge, MA: Harvard University Press.

Nietzsche, F. (1974). *The gay science.* New York: Vintage.

Novak, M. (1975). "Story" and experience. In J. B. Wiggins (Ed.), *Religion as story.* Lanham, MD: University Press of America.

Peattie, L. (2001). Theorizing planning: Some comments on Flyvbjerg's *Rationality and power. International Planning Studies, 6*(3), 257–262.

Platt, F. (1992). "Case study" in American methodological thought. *Current Sociology, 40*(1), 17–48.

Ragin, C. C. (1987). *The comparative method: Moving beyond qualitative and quantitative strategies.* Berkeley: University of California Press.

Ragin, C. C. (1992). "Casing" and the process of social inquiry. In C. C. Ragin & H. S. Becker (Eds.), *What is a case? Exploring the foundations of social inquiry* (pp. 217–226). Cambridge, UK: Cambridge University Press.

Ragin, C. C., & Becker, H. S. (Eds.). (1992). *What is a case? Exploring the foundations of social inquiry.* Cambridge, UK: Cambridge University Press.

Rasmussen, D. (1995). Rethinking subjectivity: Narrative identity and the self. *Philosophy and Social Criticism, 21*(5–6), 159–172.

Ricoeur, P. (1984). *Time and narrative.* Chicago: University of Chicago Press.

Rosch, E. (1978). Principles of categorization. In E. Rosch & B. B. Lloyd (Eds.), *Cognition and categorization* (pp. 27–48). Hillsdale, NJ: Lawrence Erlbaum.

Roth, P. A. (1989). How narratives explain. *Social Research, 56*(2), 449–478.

Rouse, J. (1990). The narrative reconstruction of science. *Inquiry, 33*(2), 179–196.

Schram, S. F. (2006). Return to politics: Perestroika, phronesis, and post-paradigmatic political science. In S. F. Schram & B. Caterino (Eds.), *Making political science matter: Debating knowledge, research, and method* (pp. 17–32). New York and London: New York University Press.

Schram, S. F., & Caterino, B. (Eds.). (2006). *Making political science matter: Debating knowledge, research, and method.* New York and London: New York University Press.

Smelser, N. J., & Baltes, P. B. (Eds.). (2001). *International encyclopedia of social & behavioral sciences.* Elmsford, NY: Pergamon.

Stake, R. E. (2008). Qualitative case studies. In N. K. Denzin & Y. S. Lincoln (Eds.), *Strategies of qualitative inquiry* (3rd ed., pp. 119–150). Thousand Oaks, CA: Sage.

Taleb, N. N. (2007). *The black swan: The impact of the highly improbable* (2nd ed.). London and New York: Penguin.

Walton, J. (1992). Making the theoretical case. In C. C. Ragin & H. S. Becker (Eds.), *What is a case? Exploring the foundations of social inquiry* (pp. 121–137). Cambridge, UK: Cambridge University Press.

White, H. (1990). *The content of the form: Narrative discourse and historical representation.* Baltimore: Johns Hopkins University Press.

Whyte, W. F. (1943). *Street corner society: The social structure of an Italian slum.* Chicago: University of Chicago Press.

Wieviorka, M. (1992). Case studies: History or sociology? In C. C. Ragin & H. S. Becker (Eds.), *What is a case? Exploring the foundations of social inquiry* (pp. 159–172). Cambridge, UK: Cambridge University Press.

Wilson, B. (1987). Single-case experimental designs in neuro-psychological rehabilitation. *Journal of Clinical and Experimental Neuropsychology, 9*(5), 527–544.

18

PERFORMANCE ETHNOGRAPHY

Judith Hamera

My students did not understand "Sandy Sem's"[1] response to her parents' traumatic negotiation of their survivor status, and the relationship between that status and a class assignment she was given by her teacher. "Sandy's" mother and father were victims and survivors of the Khmer Rouge autohomeogenocide in Cambodia between 1975 and 1979, and their experiences of these atrocities haunted them in their new lives as refugees in Long Beach, California. As I observe in my analysis of the family's use of Khmer classical dance (Hamera, 2007, pp. 138–171), her parents would not share details of their ordeal, or even much about their lives, and "Sandy" did not press them. Raised in a cultural moment celebrating memoir and self-disclosure; with understandable pride in their own cultures of origin and family traditions; and deeply, if perhaps unreflectively, inheritors to the idea of testimony as both personally and socially redemptive, my class could not easily assimilate her logic, articulated in a fieldnote of mine that I shared with them:

> "Sandy Sem": We had some project for school, to talk about our culture and our families—like grandmothers and grandfathers and stuff. But you can't ask them [her parents] about that, him [her father] especially because he gets mad and the teacher— right—she's going to believe that. And I'm going to go in, okay, and say: "My family's from Cambodia and everybody's dead from the war or over here somewhere but nobody says," okay? So I just made it up.

How could she just "make it up"? How could she not want to push her parents to "tell the truth"? Don't they, and didn't she, have an obligation to know and share everything about where she came from? How could others learn from what was "just made up," and enabling others to learn was important, right? Why was I, their professor, who proclaimed commitment to

rigorous inquiry, not pushing back at the family to "speak the truth to power"? Isn't that what good critical scholars do?

Try as we might, we couldn't come to a collective understanding of why "Sandy's" response might be useful, or necessary, or "right." What if we translate her situation into space, I asked? Where would she sit? Would she look at her audience? Who is her audience? Does she have one? Where are her parents? Who else is around—and where are they? Desks were moved and space, literal and conceptual, opened up. "Sandy's" position was embodied by, not one, but two students. One was sitting facing the audience, looking down at a blank page in an open notebook. "The ethnographer" stood on one side with her own notebook and "the teacher," holding a grade book, stood on other. Another "Sandy" sat with her back to the first, looking in the opposite direction. From that direction, receding in a diagonal as if toward a vanishing point, were "the parents," silent except for occasional sounds (sighs of exhaustion, sharp intakes of breath as if in pain), their backs to one another, and further still were "the others" and "the ancestors." The "others" and "the ancestors" were moving, sometimes in tight circles around one another, sometimes randomly across the space. They murmured, barely audible.

Here, between the murmurs, the paralinguistic articulations of pain and resignation, the inadequacy of notebooks and grade book: Here was the logic of "Sandy's" response. The "others" and "ancestors" were too far away to hear, the "ethnographer" and "teacher" too removed in other ways to understand. "Sandy's" logic was born of a nuanced reading of context—verbal and even more important, extra-verbal: the circulation of affective energy in her home, in her parents' lives, and in their histories. She had negotiated the collapse of time and space ("here and now," "there and then") in her personal, familial, and cultural pasts, and made a decision that my students could only grasp by engaging and embodying that circulation in a charged environment in the best way they knew how.

They also came to understand her tactical resistance to the teacher's imperative to narrate her family for a "show and tell," however well intended, within a larger politics and commerce of testimony. "Truths," particularly those of the disenfranchised, could and were so easily co-opted by an all-pervasive corporate media culture pedaling disclosure as a commodity: whether simple sensationalism, "ethnic color" as discursive décor, or alibi for an easy sentimental but apolitical empathy. This was a form of what Jon McKenzie (2001b) has so aptly identified as the imperative to "perform or else." In this context, silence and subterfuge were personally prudent and socially productive, useful, even empowering for "Sandy" in ways that demanded respect: a respect generated by a critical performance intervention that was both hers and theirs, one that was, in some important way, shared, however imperfectly and asymmetrically. They understood the simplicity of their own earlier interpretations. Oh, one student observed, *this* is what Victor Turner meant by performance as making, not faking.

◨ PERFORMANCE ETHNOGRAPHY
   AS A STRATEGY OF INQUIRY

This example demonstrates the utility of performance ethnography as pedagogy, but the method is more than a pedagogical technique. In fact, performance ethnography is vitally important as a pedagogical tool for precisely the same reasons it is a potent conceptual and methodological one (see also Alexander, 2006; Denzin, 2003, 2006). It exposes the dynamic interactions between "power, politics, and poetics" (Madison, 2008, p. 392), and challenges researchers to represent these interactions to make meaningful interventions: those that produce new understanding and insist that this understanding generate more just circumstances.[2]

Performance ethnography offers the researcher a vocabulary for exploring the expressive elements of culture, a focus on embodiment as a crucial component of cultural analysis and a tool for representing scholarly engagement, and a critical, interventionist commitment to theory in/as practice. In some cases, performance ethnography takes performance *per se* as an object of study. In others, it uses the idea of performance to tease apart phenomena not normally thought of in these terms. Some performance ethnographers stage their research as a form of interpretation and/or publication, as my class did with "Sandy Sem." Some use performative writing techniques to enact research dynamics on the page. These options are mutually reinforcing, not mutually exclusive, as illustrated by the examples in this discussion, and particularly the case study: D. Soyini Madison's *Water Rites* (2006c).

This chapter offers some of the basic epistemological, historical, and methodological infrastructure of performance ethnography and examines provocative new possibilities. It is inevitably selective and partial. The method itself is suspicious of the putatively "finished," preferring instead the Bakhtinian (1984) notion of the "unfinalizeable": the idea that there can never be a last definitive word, only penultimate ones. In addition to this theoretical commitment, the sheer number of intellectual turns contributing to performance ethnography speaks to its institutional unfinalizability as well as its vitality. This orientation to research is both interdisciplinary and polydisciplinary: interdisciplinary because it relies on and forges connections between a variety of fields—communication and theater studies, for example, or music and folklore. It is polydisciplinary because so many areas claim and contribute to it: anthropology, communication, dance, ethnomusicology, folklore, performance studies, theater studies. A definitive list would include nearly every academic formation in the humanities and qualitative social sciences, and many of these are themselves interdisciplinary. Performance ethnography is inter- and polydisciplinary because performance itself demands it. Plato considered this one of performance's great weaknesses: that it could not enclose a discrete field of knowledge to claim as its own private preserve. Theater and performance artists, on the other hand, appreciate that *poiesis* requires integrating knowledge from multiple areas of expertise (specialized knowledge), the full scope of the senses (embodied knowledge), critique (politically engaged conceptual knowledge), and pragmatic knowledge (know-how).

The institutional situation of performance ethnography is relational—betwixt and between the disciplines—as are its practices; the idea of relationality binds method and meta-methodology together. The relationality of performance ethnography also requires teasing out complex exchanges between specific practices and the larger context, which must be construed broadly. It includes the standard "when, where, and how" of a field site, as well as the power and privilege differentials that permeate it, the historical relationships that organize it, and the tropes that emerge to shape what can and cannot be said, enacted, and understood about it. The performance ethnographer explores the interanimating relationships that produce context: precisely the oscillation between "here/now" and "there/then" that so permeated the "Sems'" lives. There is no "now" innocent of history, and no "local" fully exempt from global flows of people, resources, and capital (see, for example, Alexander, 2008).

In keeping with Dwight Conquergood's (2006b) call for rhetorical reflexivity, performance ethnography generally, and my perspective here, are explicitly critical. That is, performance ethnography is inherently committed to what D. Soyini Madison calls "the doing or 'performance' of critical theory" (2005, p. 15) as its strategy of inquiry.[3] First, it assumes congruence, not division, between theory and method. Methodology is infused with theoretical commitments and theory is incarnated through methodology. Madison's emphasis on "doing" critical theory underscores the action-oriented nature of critical theory in practice. "Doing" makes a move and, coupled with "critical," that move is one of activation and activism, of unsettling and challenging conventional meaning and advocating for change (see Denzin, 2003, 2006; Madison, 2008). Framing performance ethnography as the "doing of critical theory" honors the tradition of, and ongoing commitments to, "intellectual rebellion" that define this research: investments in interrogating what often passes for the conventional wisdom (Madison, 2005, p. 13). "Doing" critical theory means investigating our research sites, our own methods and motives, our tactics of scholarly representation, and the structures of our own privilege. It means repeatedly and explicitly asking, Who benefits? Who decides? Who decides who decides? Does it have to be this way? What are the alternatives? As Jill Dolan (2005), Raymond Williams (1981), and others have observed, there is a utopian element to performance, one shared by performance ethnography's critical project: not proscriptive, not "pie in the sky," but "processual, as an index to the possible, to the 'what if.'" (Dolan, p. 13). The subjunctive dimension of performance enables ethnographers to investigate what is, and imagine, inspire, and initiate what could be: justice, engaged citizenship, generative public discourse, and transformative political *poiesis*.

▣ METHODOLOGICAL INFRASTRUCTURE

We typically think of infrastructure as the nearly invisible but indispensable support that makes viable communities possible: roads, phone and data networks, utilities. These are basic public goods. When infrastructure crumbles through neglect, or when it is privatized for the profit of the few, possibilities for social exchange diminish. Performance ethnography has intellectual infrastructure: keywords, formative figures, and key questions that also make community possible.[4] Scholars may draw upon some of them or all of them, entering and exiting at different points depending on their research trajectories. In inter- and polydisciplinary practices like performance ethnography, conceptual infrastructure provides a pluralist, contested, yet shared terrain: continually in flux but nevertheless a common intellectual inheritance on which we depend and to which we contribute as we define or refine our own research.

Keywords

Raymond Williams (1983) famously used the idea of "keywords" to examine shifting social, historical, and political values adhering to terms like "culture," "industry," and "democracy." These "historical semantics" (p. 23) expose the mutability and political utility of such words, as well as attempts to arrest their meaning. Performance ethnography has its own set of keywords; "critical," discussed above, is certainly one of them. Boundaries between definitions and ethics blur in performance ethnography; definitions point to necessary ethical clarification and ethics shape definitions. Definitions of keywords enable the researcher to operationalize responsibilities for ethical and rigorous engagement. A complete survey of all important keywords in performance ethnography is beyond the scope of this chapter, but four in particular are essential to understanding the method's conceptual infrastructure and the interpretive criteria that characterize ethical, generative research: performance, ethnography, performativity, and aesthetics.

Performance is a productively elastic term "on the move" (Conquergood, 1995). In *Opening Acts* (Hamera, 1991b, 2006b), I define it as both an event and a heuristic tool that illuminates presentational and representational elements of culture (p. 5). Performance makes and does things: materially, affectively, imaginatively. To use performance as a method of inquiry, the researcher gives focused attention to the denotative, sensory elements of the event: how it looks, sounds, smells, shifts over time. This also includes accounting for the event's affective dynamics: which emotions seem "authorized" and encouraged, which silenced, how they can be expressed and contained, how emotions and behaviors intersect to produce meaning. Performance as a strategy of inquiry also demands that the researcher place her site of inquiry within larger sets of ongoing historical, political, intellectual, and aesthetic conversations. It requires approaching cultural work—both that of the researcher and that of the researched—as imaginative in its most precise sense: as co-created within and between communities, as expressive and meaningful, and as embedded in the specifics of time and place, even as it may create its own unique visions of both.

Performance ethnographers view "performance" expansively by focusing on the expressive dimensions of culture, and then tracing the social, rhetorical force of particular expressions, including those characterizing the research act itself. From this perspective, both live and mediated events are performances. Both theatrical expressions that "key" audiences by signaling acts to be regarded with heightened awareness, and banal, nearly invisible practices of everyday life are performances (Bauman, 1977; Berger & del Negro, 2004; Hamera, 2006b, pp. 12–21; Hamera, 2007). Silence is a performance, as "Sandy Sem" illustrated. Rituals of state power—executions, civil defense drills, deployments of folk practices—and resistance to that power—urban rebellions—are performances (Afary, 2009; Alexander Craft, 2008; Conquergood, 2002a; Davis, 2007). Interpersonal conversations are performances (Hawes, 2006).

Ethnography, "participant observation," meets "performance" on the terrain of expression. Where traditional ethnography asks, "How and why do my research interlocutors express what they do?" performance ethnography takes a more layered and critical approach, examining expression *about* the site as well as *within* it. It demands explicit attention to the politics of representing that expression, not just to conventions of accurately recording and interpreting it. Performance ethnography lifts up the "graph," the always already taken-for-grantedness of writing. As the braided genealogy below demonstrates, this is far more complex than an imagined dichotomy between text and performance. Rather, "performance" reminds "ethnography" that embodiment and the politics of positionality are as central to representing the field-work encounter as they are to participating in it. "Performance" makes a claim on ethnography, as do other modifiers like "critical," "feminist," and "indigenous." This claim concerns both the *subject* of inquiry—expressive culture as constitutive of social life—and the *practices* of inquiry on the page, the stage, or both.

Performativity is one way that performance makes and does something. Performative utterances make interventions in the world as they are spoken. Through their repetition, these utterances stabilize the power of words and, by extension, the authorities and conventions undergirding that power. Judith Butler (1993) redeployed J. L. Austin's formulation to describe the apparently stable character of identity. This stability, she argues, does not result from a set of essential, unchanging, innate characteristics. Rather, it is an effect produced through repetition. Performance theorists have used performativity to theorize multiple dimensions of identity, and the material and ideological exigencies that constrain or enable particular kinds of repetitions. Elin Diamond (1996) describes the methodological utility in this move from a theoretical notion of performativity to analysis of a specific enactment: "[a]s soon as performativity comes to rest on *a* performance, questions of embodiment, of social relations, of ideological interpellations, of emotional and political effects, all become discussable" and interruptible (p. 5). Scholars examine and instigate these interruptions as they interrogate the rhetorical force of performatives, along with their roles in forging communal coherence or inserting relational or even intrapersonal instability (see Alexander, 2006; Dolan, 2005; Hamera, 2007; Johnson, 2003; Muñoz, 2006; Pollock, 2006, 2007).

Aesthetics are the criteria and implicit social contracts that shape how performance and performative repetitions are perceived and understood. As the genealogies discussed below demonstrate, performance ethnography's deep roots in the creative arts and criticism mean that aesthetics are a crucial component of its conceptual infrastructure. Aesthetics are never exempt from context. They always require a modifier: "feminist," "Black," "butoh," "White Eurocentric," "queer," "15th-century," and so on.

Commonly reduced to the study of formal properties in the so-called "fine arts," aesthetics are in fact deeply and profoundly communal and political, and by no means only elite matters. The properties and presumptions intrinsic to the production and consumption of culture are expressive currency, binding members of communities together. Politics suffuses aesthetic judgments, including what counts as "beautiful" or "creative," and what institutions are authorized to make and enforce these views. Aesthetics support decision making among our research interlocutors in the field as well as our own on stage, and on the page.

Performance ethnographers do not see aesthetics as the unique property of "the arts." Rather, they are inseparable from lived experience, and the imaginative work of meaning making. The research process itself, whether qualitative or quantitative, is organized by aesthetic conventions. Both physicists and performance ethnographers talk about "beautiful theories," demonstrating that aesthetics are important intellectual criteria, even if what "beautiful" means varies with context. In performance ethnography, it is useful to think of aesthetics as sets of interpretive and expressive strategies to be interrogated, deployed, or resisted. The researcher must be mindful of the history and specific ideological freight each strategy carries. She needs to know the unique conventions, standards of taste, genres, and techniques circulating, however implicitly, within her site. This demands precision and, for this reason, awareness of aesthetics serves as an important interpretive criterion of rigorous performance ethnography. Consider Harris Berger's (1999) study of heavy metal, jazz, and rock musicians in America's so-called "rust belt" cities. He writes,

> Observing that a piece of music is infused with a quality of aggressiveness, for example, is only the starting point of our description of the participant's experience; merely adding contextual and bodily dimensions to the account does not suffice.... The righteous rage of an American Christian metal band and the disgusted rage of an English hardcore outfit are not the same ... (pp. 251–252)

Berger notes the detailed genre distinctions that are deeply meaningful to these musicians, and provides painstaking accounts of their musical syntax. Aesthetics matter deeply here. They frame the communication and perception of communal identity for both musicians and the researcher. Aesthetics drives analysis in this ethnography, and finds expression in extensive thick description of the technical elements in these groups' music, sustained immersion in the field, and multisite comparisons across communities.

Aesthetics also organize how performance ethnographers stage their work. Do they strive for audience empathy with their interlocutors, or for an alienation that activates the audience, turning them into spect-actors.[5] As the analysis of *Water Rites* (Madison, 2006c) below illustrates, careful shaping of research in/as performance is as crucial to its social force as the dynamics, rhythms, and textures of metal, jazz, and rock are to Berger's musicians.

Genealogies

Performance ethnography's strengths and complexities as an orientation to research can be productively examined through select examples from its complex genealogy. In some cases, this means retrospectively reclaiming works that implemented the core commitments and practices of performance ethnography without using a specific disciplinary affiliation. In some cases, it involves recognizing the centripetal pull of performance across multiple disciplines: anthropology, folklore, the oral interpretation of literature, speech communication, and theater, among others. The genealogy below should not be read as strictly linear. It is not a list of who "begat" whom. Anthropologically informed negotiations of performance, ethnography, and aesthetics did not birth oral interpretation and communication scholars' explorations of these same terms. This is a braided genealogy, one in which relationships between keywords and strategies are, and continue to be, in conversation with one another. "From" in the subheadings does not indicate an eventual convergence on a methodological consensus but rather a disciplinary starting point for ongoing conversations.

From Anthropology

Zora Neale Hurston's (1990) *Mules and Men* is one particularly rich example of performance ethnography. In her introduction to the work, Hurston describes Black Southern folk culture that "was fitting me like a tight chemise. I couldn't see for wearing it" (p. 1). She credits the "spy-glass of Anthropology" for giving her the ability to navigate the challenges of participant observation but, in fact, her work stands in contrast to the ocularcentric metaphor with which it begins. Instead, Hurston presents the "telling and the told" (Madison, 1998; Pollock, 1990) through what we now call "orature." As articulated by Ngugi wa Thiong'o (1998, 2007), "orature" describes the interpenetration of speech, writing, music, dance, even the cinematic, so as to resist simplistic dichotomies between text and performance. As a director/choreographer, novelist and playwright as well as an anthropologist, Hurston was keenly attuned to the theatricality of the lore she collected. She focused careful attention on the contexts and exchanges that inspired "breakthroughs into performance" (Hymes, 1981), noting not just the clearly bounded "folk tales" but their elasticity as they stretched to accommodate a wide range of social performances: teasing, joke telling, and "big old lies." Moreover, using literary and dramatic devices from the "fourth wall" to free indirect discourse, she shares the explicit theatricality of her research practice with her readers so they can feel the grains of the voices, the pacing, the overall flow of events. Even her footnotes ventriloquate voices from the field, blurring the positions of researcher/writing and researched/speaking into orature in this most "textual" of devices.[6]

Victor Turner (1982) was also interested in this "both-and" quality of performance ethnography. He described performance itself as a "liminal" experience: betwixt and between consensual reality and fantasy, neither simply here, nor simply there. For Turner, performance is constitutive; in a profound challenge to the antitheatrical bias that has constricted Western epistemology since Plato, Turner asserts that performance makes, not fakes, social life. Working closely with director and performance theorist Richard Schechner (1985), Turner applied the performance paradigm to the ethnographic enterprise. Schechner explicitly positions performance ethnography "between theatre and anthropology." Central to this task was his identification of the shared liminality of the ethnographer and the performer using psychoanalyst D. W. Winnicott's idea of the transitional object. The ethnographer was not a "native" just as a performer was not the character. Yet she was "not-not" the native/character either. This liminality—this threshold status—is intellectually productive; it encourages self-reflexivity with the recognition that identity is not immutable but fluid, social, and contextual. And it opens up conceptual spaces betwixt and between identities for an imaginative, even poetic theorizing of cultural processes.

From Oral Interpretation and Communication

The oral interpretation of literature, rooted in the elocutionary movement, is based on the premise that performance is an embodied hermeneutic tool: a way of "doing" analysis that moves beyond inscription to enactment.[7] Central to this commitment to performance is Wallace Bacon's (1979) "sense of the other," the idea that embodiment in performance constructed through detailed analysis could generate critical insight into multiple dimensions of difference in literary texts. Dwight Conquergood drew from and radicalized oral interpretation to fashion a performance ethnography that demanded "body-to-bodyness" (Olomo/Jones, 2006, p. 341) beyond the boundaries of the field encounter and the margins on the page.[8] In his classic essay "Rethinking Ethnography," Conquergood (1991/2006b) clearly articulates the importance of a return to the body. His own fieldwork in a Hmong refugee camp (1988), and among Chicago street gangs (1997), foregrounded the corporeality of culture: its processuality as an ensemble of behaviors, and dances with history and politics. Research methods are not a separate category of experience in this view. They are also enactments. Conquergood made the move from ethnographic inscription to ethnographic enactment, from writing to performing culture. Performance-based research held out the promise of truly "radical research" (2002b). Conquergood (1991/2006b) argued for, and skillfully demonstrated, rhetorical reflexivity by asking bracing questions about interrelationships between culture and power, expanding these questions beyond the preserve of "the field" to include genres of academic production. At its most profound, Conquergood's commitment to performance as a tool of knowledge production challenges the scriptocentric academy, and text-based knowledges that often disenfranchise those outside its own economies. He argued for performance-based methods that "revitalize the connection between practical knowledge (knowing how), propositional knowledge (knowing that), and political savvy (knowing who, when, and where)" (Conquergood, 2002b, p. 153). D. Soyini Madison summarizes all of these contributions in her generative recasting of performance ethnography as coperformance:

> Coperformance as dialogical performance means you not only do what subjects do, but you are intellectually and relationally invested in their symbol-making practices as you experience them with a range of yearnings and desires. Coperformance, for Conquergood, . . . is a "doing with" that is a deep commitment. (2005, p. 168)

Moral Maps

Conquergood (1982) offered what is, for many, a definitive way to examine the ethical pitfalls of performance ethnography in his essay, "Performing as a Moral Act." The goal of the ethnographer is coperformance, achieved dialogically through the persistent posing of unsettling questions like the "Key Questions" below. These questions; a disciplined grasp and thick description of aesthetics in the field, on the stage, and on the page; self-reflexivity; and the commitment to "doing" critical theory help the researchers avoid four fundamental ethical errors. The *curator's exhibitionism* is an error of aesthetics: confusing a prurient desire to showcase the "exotic" with a rigorous understanding of how expressive behaviors actually work. The "show and tell" impulse "Sandy Sem" subverted with her silence might fall here. The *custodian's rip-off* fails the fundamental relationality of coperformance. Here, field sites and interlocutors are raw materials to be recoded as products of the researcher's putatively autonomous "genius," a form of intellectual piracy. The *enthusiast's infatuation* marks a failure to rigorously "do" critical theory. Where the first two positions see research interlocutors as objects for display or raw materials for self-fashioning and self-promotion, this position absorbs all differences into a romantic celebration of a simple difference from, or similarity to, the self. Irreducible difference is ignored, difficult issues superficially glossed over or excused. The *cynic's cop-out* insists on the unintelligibility of difference and the inability to overcome distances inevitably encountered in ethnographic research. This is an alienated and alienating stand, ultimately impotent, bereft of the utopian sense of performance so crucial to sustained efforts to do critical theory.

Drawing on the anthropological and oral interpretation traditions, performance studies scholars continue to grapple with the ways "performance," "ethnography," "performativity," and "aesthetics" inform one another. They continue to raise questions about how performance works, challenging assumptions that corporeality and textuality are mutually exclusive representational modes in the field or in scholarly inquiry. Dance practices are especially fruitful sites: Limit cases, because they are so often reduced to untranslatable embodiment. In my studies of dancing communities in Los Angeles (Hamera, 2007), I argue that dance is enmeshed in language: in the stories and demonstrations that train the next generation, in the productive imprecision of metaphor that describes how a move looks or feels, in the institutional prose (laws, syllabi, press kits, word of mouth) that enables or circumscribes it. For me, analyzing the ways dance constructs diverse communities means dancing with my interlocutors as much as listening to and writing about them. The commitment to "dance with" as well as "write about" also opens up opportunities for challenging hegemonic assumptions about genres of performance. Ballet and modern dance both carry the imprimatur of elite "high art," at odds with the material circumstances of most of the artists who create it. Training with amateur and professional dancers showed me these techniques' other lives: as homeplaces for a wide range of performers to come together, bound in solidarity and in difference, sometimes briefly, sometimes over decades, by the rigors of their shared rituals.

Key Questions

There are no prescriptions for operationalizing performance ethnography. The complexities of each site, each researcher's embodied particularity, each location in place and history demands its own unique negotiations. But this does not mean blind or naïve reinvention of good research practices. A set of key questions for performance ethnographers raised throughout the research process reminds us of our aesthetic, ethical, and intellectual responsibilities. Madison has marked the popularity of performance as a mode of research by wryly observing, "everyone I know and don't know is thinking, speaking, and writing in the language of performance, or trying to" (2006b, p. 243). This plurality of disciplines and discourses enables creative and generative play with epistemological and methodological conventions, but performance ethnography is not a playground without accountability and innocent of history. On the contrary, the productive pliability of performance and its multiple disciplinary locations require the researcher to articulate her own conceptual commitments by answering basic questions about her research design. This is the methodological and ethical equivalent of the site survey: the meticulous accounting for how performance opens up a specific research site in demographic and discursive detail. Like the site survey, these questions orient the researcher, pointing her to ever more nuanced understanding of what it means to "profess performance" from her unique disciplinary or interdisciplinary orientation (Jackson 2004).[9] Answering these questions exposes performance ethnography as relational in yet one more sense: it positions the researcher vis-à-vis other individual methodologists using similar vocabularies.

1. How does performance emerge in my research site? Because the term refers to both events and a heuristic tool, its use in specific contexts of research demands critical reflection and precision. Does it announce itself through its self-conscious theatricality? Is "performance" a term I use to explain expressive force, expressive techniques, both or neither? Do my research interlocutors think of what they are doing as performance, or is this a term I am using to communicate something powerful about their actions to the audiences for my research? What conceptual permissions does "performance" offer me as a researcher? What dangers does it hold? What preconceptions do I bring to the term? Am I assuming performance is inherently creative, derivative, live, resistant, reactionary?

2. Where is my performance located in time and place, and how do these times and places intersect with history, with other places, other institutions? What global matrices construct the "local" in my site? Which historical ones undergird the "here and now"?

3. When I use "performance" and reflect on my own assumptions underlying this use, which scholarly conversations am I participating in, however implicitly? What obligations does participation in these conversations impose? Do I need to understand specific techniques, vocabulary, bibliography? How does my use of "performance" contribute to, challenge, or subvert turns in these conversations?

4. How do I conceptualize the act of research itself as a performance, beyond the simple idea of demonstrating "competent execution": the techno-bureaucratic definition of the term? How have I engaged my interlocutors? As coperformers? As "extras" or props? How do I represent exchanges with my interlocutors, in all of their sensory and social complexities? To what extent am I translating performances, not only in the sense of moving between languages, or between verbal and nonverbal modes of communication, but also between modes of representation, especially corporeality and textual fixity? How do I understand and communicate the entire research endeavor as a set of aesthetic, ethical, political, and rhetorical elements, decisions, and responsibilities?

5. How and where does my research make meaningful interventions? What changes as a result of my work? What good does it do, what is it good for, and what does "good" mean in this research context? Who does it serve? How do I share my research with my interlocutors who are represented in my work? How are they affected? What do I want my audiences to do as a result of exposure to my research?

These five sets of questions capture the processual dynamics at the heart of the method. Further, they serve as interpretive cautions: reminders of our responsibilities to our research communities broadly construed.[10] They invite the researcher to reflect on the ways "performance" circulates in her scholarship.

▣ WATER RITES: A CASE STUDY

Water Rites, conceived and directed by D. Soyini Madison (2006c) and realized in performance by students at the University of North Carolina at Chapel Hill, is an exemplary illustration of how a deeply ethical and coperformative representational strategy, a critically engaged commitment to intervene in the politics of privatization, and a disciplined deployment of aesthetics actually work in performance ethnography. This multimedia production relies on fluidity of form as well as content. Like the vision of water as a public good to which it is committed, the work flows between genres: memoir and personal narrative, movement, ethnographic field notes, sound, projections (both the techno-managerial PowerPoint slide and documentary photographs), and *actos:* short, highly politicized and often highly satirical sketches.[11] The result is a model of engaged performance ethnography. It is itself a water rite, turning on the phonic relation of "rite/right": a ritual that reinforces shared humanity and an entitlement arising from this same nonnegotiable status. It demands that its viewers pay attention to the politics of water, pay attention to the human costs and institutional profits, pay attention to their own memories, consumption, and taken-for grantedness of water, pay attention to what must be different.[12] A full analysis of *Water Rites* exceeds the scope of this essay, but a brief discussion of three key moments demonstrates the aesthetic potency of performance ethnography as a critical method.

Water Rites opens by establishing the intimate coupling of free flow and restriction that characterize the binaries of water politics, casting this coupling as both a personal and global exigency.

RECORDER 1

Dear Journal: October 12, 1998, University of Ghana, Legon—Accra, Ghana, West Africa. There is no water in my house—the pipes are dry. There's no water left in my storage containers. There's no water anywhere here in Legon. I can't find water and it scares me. They warned us about the pipes drying up, but I never thought it would go on this *long.* How could there not be water?

RECORDER 2

Dear Journal: January 2006, London, England. These are the facts: More than 1 billion people lack access to clean and affordable water and about 2 billion lack access to sanitation . . .

RECORDER 1

Kweku, said he will come and we will search for water . . . he told me he knows where we can get enough to fill the containers. I just want him to hurry up and get here. It's just too scary not having water . . . too weird and scary. I worry how the students here are managing?

RECORDER 2

In the urban areas of Ghana, only 40 percent of the population has a water tap that is flowing; 78 percent of the poor in urban areas do not have piped water.

Though the Recorders give voice to Madison's research and experience, they are not simple figures of ethnographic authority. They record the interpenetration of personal affect, demographic context, and the larger realities of global water politics. Note how the Recorders' statements are themselves a flow as discourse moves from medium to medium and place to place. The "Dear Journal" indicator of fixed, written affect, dissolves into speaking, which in turn struggles to stay afloat in a rising tide of near primal anxiety: "How could there not be water?" Ellipses and repetition underscore this anxiety and the failure of language to fully capture it: "It's just too scary not having water . . . too weird and scary." Likewise, speaking dissolves the distance between University of Ghana, Legon—Accra, Ghana, West Africa, and London, England. The first location shows us the consequences of policies forged, in part, in the second; the second, free of the policy-inflicted exigencies of the first, is a source of "facts" to illuminate and enlarge those experiences. Writing and information can flow freely across borders and genres for a privileged Western subject. For much of the world, life-sustaining water does not.

The personal and the factual are interanimating registers of discourse in *Water Rites,* but they are not the only ones. *Water Rites* shows as well as tells. One of the most compelling examples is the use of empty plastic water bottles. Dozens of them form rivers of bottles, moats of bottles: aggregate yet highly individual. They provide visual continuity throughout the performance and/because they continually remind the audience of the social costs of private water. The sounds of the empty bottles hitting the floor, their transparency, the way they roll—all concretize both flow and restriction acoustically, visually, tactilely. One of the dramatic punctums organizing the piece demonstrates the polyvalence of water as it circulates in Madison's ethnographic work, through individual performers' memories of water, and through global networks of privatization and profit.

Sounds of water rise and the "Donkey and Fetching Water Scene" is projected on both screens. As the Fetching Water scenes are projected, sounds of water rise to a high pitch as actors rise from their islands as if they are moving through water. They leave their boxes on the island—feeling the opposing force of the water—the actors rise and begin to search among the water bottles for the special one that they want—they read the various brands and inspect the size

and shape of some of the bottles until they find the one they want. When each actor finds the "right" water bottle, they reach to the floor against the force of the water and lay down holding the bottle in various semi-fetal positions with their backs to the audience. *The water sounds fade but they can still be heard.*

This seemingly small nonverbal moment is itself a water rite. The force of water is registered in multiple ways: scenes documenting the unrelenting, body-breaking labor of fetching water; the roar of rushing water; and the performers' kinesthetic struggles against the current. That force is juxtaposed against the triviality of brand choice: a privileged way to "fetch water" where one can afford to have a "special" (clean, safe) kind. Even when reduced to accessories held in all those bottles, the sounds of water can still be heard. What to make of the semi-fetal positions of the performers? Perhaps they are allegories for the way ideologies of privatization and environmental devastation have infantilized consumers who choose not to ask, "Who decides who gets water? Who decides who decides?" Perhaps they remind the audience that we all come from water, that our fetal and evolutionary homes were water worlds. Perhaps they are exhausted, unable to swim any longer against the riptides of global capital and the institutions channeling it.

Water Rites shares narratives of both exhaustion and activation: accounts from West and South Africa, India, and Bolivia. These accounts are affect-ing in a double sense. They are *affective* in Sara Ahmed's (2004) definition: a form of cultural politics that is social and rhetorical rather than individual and interior. They also demand an *effect* from the audience, one activated not by pathos or solely by personal empathy. Local and international water activists and corporate stand-ins affirm what must be done, sometimes by negation. But one particular provocation pushes the audience beyond an instrumental view of performance and change, challenging them to deploy their own privileged access to facts and global mobility, as established in the work's opening moments discussed above.

MADELINE

And every once in a while one of you people . . . will whine or someone will yell at us, WHY WON'T YOU LISTEN? And I reply the same way that I always reply. You're either a beggar or a chooser. And if you have such a problem with it, get out of the street, get out of your hemp clothes and your teeshirts with defiant phrases and your classrooms where you discuss over and over again what's *wrong* with the international system. Stop throwing around your buzzwords and get out of your idea that *you* are going to change anything by being small. *Especially* you, who was born big, was born with privilege and money and the stamp *American* that won't come off no matter how hard you rub it or how many tattoos you put over it. You accomplish nothing by celebrating your smallness. And the only thing I have to say for myself is a piece of advice for you. Become a chooser—maybe you'll be a better listener than me. . . . Maybe you'll rewrite the manual. Until you do, I'm afraid I can't help you.

Who is Madeline? A person of authority at the World Bank or the International Monetary Fund? A generic person of privilege and influence feeling so-called "compassion fatigue"? An internal voice members of the audience hear but would like to disown? Whoever she is, she demands that the audience enact their commitments. This is a call to move beyond Gayatri Spivak's (1990) mandate to "unlearn your privilege" (p. 42). It is a demand to acknowledge our own complicity, look within, deploy it in critical interventions, and be accountable in the attempt (see Alexander Craft, McNeal, Mwangola, & Zabriskie, 2007, p. 56).

Water Rites demonstrates how performance ethnography does more than represent the problematics of water privatization; it intervenes in them. In so doing, it offers tactics, themes, and commitments central to emerging research: novelizing ethnographic discourse, exploring the performative potential of objects, and probing the inextricable links binding the present and the past, the local and the global.

▣ EMERGING PARADIGMS AND NEW DIRECTIONS IN PERFORMANCE ETHNOGRAPHY

Performance ethnography gives a lot of permission. Its potency as an analytical, political, and representational tool has attracted scholars alert to new opportunities to explore expressive culture, embodiment, and aesthetics and to do critical theory. This innovative work looks both forward and backward. It seeks out new genealogies, new modes of performance and the performative, and new forms of scholarly representation.

One emerging trajectory in performance ethnography is historical and involves intersections of performance and the archive. Scholars interested in this intersection draw on the insights of Diana Taylor (2003), and her useful formulations "the archive" and "the repertoire." The archive "exists as documents, maps, literary texts, letters, archeological remains, bones, videos films, CDs, all those items supposedly resistant to change" (p. 19), while the repertoire "enacts embodied memory: performances, gestures, orality, dance, singing—in short, all those acts usually thought of as ephemeral, nonreproducible knowledge" (p. 20). Taylor is not interested preserving the conventional false binary that separates these two spheres, but instead shows that they thoroughly interpenetrate one another: both are mediated, highly selective, and citational. Both are "mnemonic resources" (Roach, 1996, p. 26). Scholars are actively investigating the interanimation of the archive and the repertoire in/as performance ethnography, and dance studies offers a compelling example.

In *Choreographing the Folk,* Anthea Kraut (2008) examines Zora Neale Hurston's work as director and choreographer, offering close analyses of her concert *The Great Day,* and particularly the Bahamanian fire dance that was a central feature of the production. Kraut discusses Hurston's deployment of folk idioms in

performance, and in materials that supported it, including promotional literature and correspondence with patrons and colleagues. Of special interest are the ways Hurston's theories of the folk in motion on stage emerge as distinct from those in her work on the page. Kraut analyzes the sometimes-fraught negotiations between Hurston, her patrons, and collaborators, her attempts to delineate distinct genres of African American vernacular dance, and her own assertions of aesthetic/ethnographic authority. Especially important in this analysis is the highly racialized entertainment market for products of performance ethnography, and particularly those framed as "folk," in the 1930s. Consistent with the inter- and polydisciplinary nature of performance ethnography, Kraut's book contributes to African American and American studies as well as performance and dance studies. Among many other contributions, it recovers dance for the history of performance ethnography and charts the contested, commercial path one ethnographer took to stage the results of her inquiry.

Performance ethnography can also illuminate new sites using the idea of performance in ways that may seem counterintuitive. For example, the process of commodification, the circulation of objects, and the imagined communities constructed by them can be productively viewed as performances. Examining objects and social processes through performance does not dispense with embodiment as a crucial concern. Rather, it expands focus to include the ways embodiment is invoked, ventriloquated, or staged through specific markets and desires. Ngugi wa Thiong'o (1999) observes, "There is a performance to space, to architecture, to sculpture." When spaces and objects are infused with exoticism, difference, and marketability, the idea of performance can illuminate the flows of power and pleasure that define commodity situations.

Genres of Native American art are productive examples. Here, the performance ethnographer examines the circulation of these objects in specific contexts, teasing out the complex pleasures and fantasies undergirding their consumption. I have argued (Hamera, 2006a, 2006c) that, in the case of Navajo folk art, the invisible, putatively "vanishing" native is both brought to life and frozen in forms that could be superficially viewed as naïve, politically innocent, and timeless—products of a homogenous "folk," yet appreciated for their seeming idiosyncrasy. Further, the art object functions as a perpetual performance of inclusion and appreciation for the collector, one that offers absolution from his/her position of relative privilege vis-à-vis the artist; exemption from the often sordid history of non-Indian desires for, and designs on, Indian objects; and recognition for "a good eye" replete with multicultural aesthetic sophistication. At the same time, consuming the objects, especially those that are characterized in terms of "authenticity," offers the collector vicarious immersion in native culture.

Finally, and perhaps most controversially, performance ethnographers may weave performance and aesthetics into "novelized" accounts. "Novelizing" comes from Mikhail Bakhtin's view of the novel's social situation. Michael Bowman (1995) operationalizes novelizing as

> a willingness to engage in a kind of verbal-textual-semiotic "misrule" which carries with it an "indeterminacy," as Bakhtin would say, "a certain semantic open-endedness," which has the *potential* to destabilize canonical notions of performance/text relations, of performance process, as well as of performance/audience relations. Although a novelistic production may have its preferred meanings, values, or political-cultural agenda, it also contains voices and values that contradict the ones it prefers. (p. 15)

As *Water Rites* demonstrates, performance ethnographers novelize the stage by including both multiple, contrary voices (indigenous activists, recorders, "Madeline," dismissive yuppies), and multiple media (image, sound, movement, *actos,* personal narrative). Sometimes these voices and media reinforce each other, but often they contradict and problematize, shifting the interpretive burden to the audience with the hope that they "will be better listeners than me" (see "Madeline" above). They even go beyond to "ethnographize" the process of novelizing the stage, detailing the politics of adaptation and performance itself (see Goldman, 2006).

Novelizing can be applied to the entire ethnographic project itself, challenging norms of solitary authorship as an extension of the historical anthropologist "hero." From this perspective, we can conceptualize novelizing authorship and methodologies as a quilt: not the type that subsumes all difference into a unified whole but one that stitches together sometimes contradictory aesthetics and commitments. As Renee Alexander Craft et al. (2007) observes, collaborative interventions are not always seamless meetings of the minds.

> One of my sister-quilters picks up a piece of cloth to add to her quilt-pattern. I grimace. With so many prettier pieces in her pile, I wonder why she has chosen that one. I look up to ask, when I see her eyes fixed on the fabric in my hand, her eyebrows knitting and un-knitting like mine. We meet each other's gaze, laugh, tease, and continue working. (p. 78)

Alexander Craft's "sister quilters" are performance ethnographers coming together across a wide range of boundaries to offer a manifesto for Black feminist performance ethnography, one committed to novelizing conventional understandings of gender and Blackness in/as cultural practices by examining "modalities of blackness within discourses of Africa, modalities of Africa within discourses of blackness, and all of the messiness in between" (Alexander Craft et al., 2007, p. 62) and by "minding the gaps" (p. 70).

Novelizing can go even further: Fieldwork can be communicated through actual novels. This is not as fraught an operation as it might first appear, as Kamala Visweswaran (1994) reminds us. Fiction and ethnography are never fully discrete discourses; each hinges on devices, tropes, and claims that define the other. Indeed, even my use of "Sandy Sem" here can be read as inserting a fiction into ethnography, with the quotation marks around her name designating the pseudonym—a reminder that "truth" is never simple, or even fully knowable. Yet novelizing the ethnographic text does not propel the researcher out of the realm of politics, ethics, and rigor. She must still address the key questions above, still commit to "doing" critical theory. Perhaps no one meets this burden as well as Martiniquen novelist Patrick Chamoiseau, best known for his prize-winning work *Texaco.* His earlier novel, *Solibo Magnificent* (1997), offers a fully novelized ethnography remarkable for its sensitivity to orature, to what performance can and cannot change, and to the complexities of postcolonial politics on his island nation. This beautiful and poignant novel includes a caution given to the ethnographer/narrator/writer. Chamoiseau, playfully reinscribed by master storyteller Solibo as "Oiseau de Cham," (an allusion to the biblical Shem and, literally, "bird of the field") is reminded of the ultimate limits of the ethnographic enterprise, particularly one committed to "the word," the rich and irreducible corporeality of cultural performance.

> (Solibo Magnificent used to tell me: "Oiseau de Cham, you write. Very nice. I, Solibo, I speak. You see the distance? . . . you want to capture the word in your writing. I see the rhythm you try to put into it, how you want to grab words so they ring in the mouth. You say to me: Am I doing the right thing, Papa? Me, I say: One writes but words, not the word, you should have spoken. To write is to take the conch out of the sea to shout: here's the conch! The word replies: where's the sea? But that's not the most important thing. I'm going and you're staying. I spoke but you, you're writing, announcing that you come from the word. You give me your hand over the distance. It's all very nice, but you just touch the distance") (pp. 28–29)

Chamoiseau reminds us that, as performance ethnographers, we all reach across the distances separating the linearity of language—written or spoken—from the flux of experience. Sometimes just touching that distance by novelizing ethnography is the best we can do, whether we speak, write, dance, or paint the performances we encounter.

These examples of new directions in performance ethnography share common themes that have characterized the method from its inception. They are deeply concerned with the transnational: the interpenetration of locales across nation-state boundaries or within them (the day-to-day politics of neocolonialism; fantasies of engaging native others through Navajo folk art). They examine structures of community formation, whether as an imagined community of "the folk," a sorority of ethnographers, or the solidarity or atomization of the audience-performer relationship (Alexander Craft et al., 2007; Chamoiseau, 1997; Hamera, 2006a, 2006c; Kraut, 2008). Multiple dimensions of difference, and the intersectionality of difference in/as performance are explicit elements of each.

▣ CONCLUSION

The themes outlined above, particularly difference, connection, and transnationalism, coupled with the critical commitment to interrogate structures of oppression, have taken on new urgency in a post-9/11 context. As Norman Denzin and Michael Giardina (2007) argue, this context impels artists and scholars "to try to make sense of what is happening, to seek nonviolent regimes of truth that honor culture, universal human rights, and the sacred; and to seek critical methodologies that protect, resist, and help us represent and imagine radically free utopian spaces" (p. 10). Performance ethnography is such a method, incarnating critical interventions so they live in the flesh as well as on the page or the screen, though we must continually resist the temptation to conflate all performance with utopian space. As Jon McKenzie (2001a) reminds us, performance itself is an agent of globalization and its discontents. In the spirit of Denzin and Giardina, "Jenny Sem" and Solibo Magnificent challenge us to interpret and represent what can, must, cannot and will not be said in our research sites. We take up the challenge because the power of performance, as paradigm and shared corporeality, gives us the radical hope that acts of poiesis will productively intervene in our understanding of the world, and in the world itself.

▣ NOTES

1. Both "Sem" and "Sandy" are pseudonyms. As the example indicates, the Sem family's experiences led them to impose thresholds of secrecy that I was never able to fully cross. For a full discussion of these dynamics, see Hamera (2007).

2. Madison's formulation resonates with Dwight Conquergood's (2002b) characterization of performance studies itself as composed of creativity (artistry), communication (analysis), and citizenship (activism).

3. Critical theory approaches social formations, embedded in their specific histories, with the goal of teasing out the intricate workings of power. In so doing, it seeks a more just and emancipatory order. "Critical theory" in the generic sense includes critical race theory, disability studies, feminism, indigenous knowledges, Marxism, poststructuralism, psychoanalysis, and other methods that interrogate structures and practices of domination. "Critical Theory" as a specific body of literature was defined by members of the Frankfurt School as a more radical hermeneutic form of Marxism. For examples of how critical theory broadly construed enters performance studies, see Madison and Hamera (2006, pp. 1–64).

4. My view of infrastructure here resonates strongly with Shannon Jackson's (2005) "infrastructural memory," a construct that

links aesthetics, discussed later in this section, and materiality in productive ways. "Infrastructural memory" is especially useful in understanding emerging relationships between performance and the archive, as noted in the final section of this essay.

5. "Spect-actors" is Augusto Boal's (1979) term for activated spectators: those driven to intervene in the theatrical experience to address injustice rather than simply passively consume the event and, by extension, the status quo. Activating spect-actors is a crucial component in his *Theatre of the Oppressed.*

6. See, for example, Hurston, 1990, p. 94.

7. For a history of oral interpretation within the academic construction of "performance," see Jackson (2004). For a history of the move from elocution to oral interpretation, see Edwards (1999). For a critique of elocution as the performance of whiteness naturalized, see Conquergood (2006a).

8. See Jackson (2009) for a deft theorizing of the relationship between oral interpretation and ethnography.

9. Jackson's book provides a valuable history of the institutionalization of performance, important background for those who want to fully understand its circulation across disciplines, and its disciplinary debts, presumptions, and vocabularies.

10. See Pollock (2006) for a complementary set of qualities to these questions: international, immersive, incorporative, integrative, and interventionist.

11. Luis Valdez developed the *acto* as part of his work with El Teatro Campesino and the United Farm Workers Union. More information on the form is available from his *Actos* (1971), and Eugène van Erven's *Radical People's Theatre* (1988), pp. 43–53.

12. Madison articulates an ethnographic ethic of "paying attention" in her article, "The Dialogic Performative in Performance Ethnography" (2006a).

▣ REFERENCES

Afary, K. (2009). *Performance and activism: Grassroots discourse after the Los Angeles rebellion of 1992.* Lanham, MD: Lexington Books.

Ahmed, S. (2004). *The cultural politics of emotion.* New York: Routledge.

Alexander, B. K. (2006). *Performing Black masculinity: Race, culture, and queer identity.* Lanham, MD: AltaMira Press.

Alexander, B. K. (2008). Queer(y)ing the postcolonial through the West(ern). In N. K. Denzin, Y. S. Lincoln, & L. T. Smith (Eds.), *Handbook of critical and indigenous methodologies* (pp. 101–131). Thousand Oaks, CA: Sage.

Alexander Craft, R. (2008). "Una raza, dos etnias": The politics of be(com)ing/performing "Afropanameño." *Latin American and Caribbean Ethnic Studies, 3*(2), 123–149.

Alexander Craft, R., McNeal, M., Mwangola, M., & Zabriskie, Q. M. (2007). The quilt: Towards a twenty-first-century Black feminist ethnography. *Performance Research, 12*(3), 54–83.

Bacon, W. (1979). *The art of interpretation* (3rd ed.). New York: Holt, Rinehart & Winston.

Bakhtin, M. (1984). *Problems of Dostoevsky's poetics* (C. Emerson, Ed. & Trans.). Minneapolis: University of Minnesota Press.

Bauman, R. (1977). *Verbal art as performance.* Rowley, MA: Newbury House.

Berger, H. M. (1999). *Metal, rock, and jazz: Perception and the phenomenology of musical experience.* Hanover, NH: Wesleyan University Press.

Berger, H. M., & del Negro, G. (2004). *Identity and everyday life: Essays in the study of folklore, music, and popular culture.* Middletown, CT: Wesleyan University Press.

Boal, A. (1979). *Theatre of the oppressed* (C. A. McBride & M. O. L. McBride, Trans.). New York: Theatre Communications Group.

Bowman, M. S. (1995). "Novelizing" the stage: Chamber theatre after Breen and Bakhtin. *Text and Performance Quarterly, 15*(1), 1–23.

Butler, J. (1993). *Bodies that matter: On the discursive limits of sex.* New York: Routledge.

Chamoiseau, P. (1997). *Solibo magnificent* (R. M. Réjouis & V. Vinkurov, Trans.). New York: Vintage.

Conquergood, D. (1982). Performing as a moral act: Ethical dimensions of the ethnography of performance. *Literature in Performance, 5*(2), 1–13.

Conquergood, D. (1988). Health theatre in a Hmong refugee camp. *TDR: The Drama Review, 32*(3), 174–208.

Conquergood, D. (1995). Of caravans and carnivals: Performance studies in motion. *TDR: The Drama Review, 39*(4), 137–141.

Conquergood, D. (1997). Street literacy. In J. Flood, S. B. Heath, & D. Lapp (Eds.), *Handbook of research on teaching literacy through the communicative and visual arts* (pp. 334–375). New York: Macmillan.

Conquergood, D. (2002a). Lethal theatre: Performance, punishment, and the death penalty. *Theatre Journal, 54*(3), 339–367.

Conquergood, D. (2002b). Performance studies: Interventions and radical research. *TDR: The Drama Review, 46*(2), 145–156.

Conquergood, D. (2006a). Rethinking elocution: The trope of the talking book and other figures of speech. In J. Hamera (Ed.), *Opening acts: Performance in/as communication and cultural studies* (pp. 141–160). Thousand Oaks, CA: Sage.

Conquergood, D. (2006b). Rethinking ethnography. In D. S. Madison & J. Hamera (Eds.), *Handbook of performance studies* (pp. 351–365). Thousand Oaks, CA: Sage. (Original work published 1991)

Davis, T. (2007). *Stages of emergency: Cold war nuclear civil defense.* Durham, NC: Duke University Press.

Denzin, N. K. (2003). *Performance ethnography: Critical pedagogy and the politics of culture.* Thousand Oaks, CA: Sage.

Denzin, N. K. (2006). The politics and ethics of performance pedagogy: Toward a pedagogy of hope. In D. S. Madison & J. Hamera (Eds.), *Handbook of performance studies* (pp. 325–338). Thousand Oaks, CA: Sage.

Denzin, N. K., & Giardina, M. D. (2007). Introduction: Cultural studies after 9/11. In N. K. Denzin & M. D. Giardina (Eds.), *Contesting empire, globalizing dissent: Cultural studies after 9/11* (pp. 1–19). Boulder, CO: Paradigm.

Diamond, E. (1996). *Performance and cultural politics.* New York: Routledge.

Dolan, J. (2005). *Utopia in performance: Finding hope at the theatre.* Ann Arbor: University of Michigan Press.

Edwards, P. (1999). Unstoried: Teaching literature in the age of performance studies. *Theatre Annual, 52,* 1–147.

Goldman, D. (2006). Ethnography and the politics of adaptation: Leon Forrest's *Divine Days.* In D. S. Madison & J. Hamera (Eds.), *Handbook of performance studies* (pp. 366–384). Thousand Oaks, CA: Sage.

Hamera, J. (2006a). Disruption, continuity, and the social lives of things: Navajo folk art and/as performance. *TDR: The Drama Review, 46*(4), 146–160.

Hamera, J. (Ed.). (2006b). *Opening acts: Performance in/as communication and cultural studies.* Thousand Oaks, CA: Sage.

Hamera, J. (2006c). Performance, performativity, and cultural poiesis in practices of everyday life. In D. S. Madison & J. Hamera (Eds.), *Handbook of performance studies* (pp. 46–64). Thousand Oaks, CA: Sage.

Hamera, J. (2007). *Dancing communities: Performance, difference and connection in the global city.* Basingstoke, UK: Palgrave Macmillan.

Hawes, L. C. (2006). Becoming other-wise: Conversational performance and the politics of experience. In J. Hamera (Ed.), *Opening acts: Performance in/as communication and cultural studies* (pp. 23–48). Thousand Oaks, CA: Sage.

Hurston, Z. N. (1990). *Mules and men.* New York: Harper & Row.

Hymes, D. (1981). *"In vain I tried to tell you": Essays in Native American ethnopoetics.* Philadelphia: University of Pennsylvania Press.

Jackson, S. (2004). *Professing performance: Theatre in the academy from philology to performativity.* Cambridge, UK: Cambridge University Press.

Jackson, S. (2005). *Touchable stories* and the performance of infrastructural memory. In D. Pollock (Ed.), *Remembering: Oral history performance* (pp. 46–66). New York: Palgrave Macmillan.

Jackson, S. (2009). Rhetoric in ruins: Performing literature and performance studies. *Performance Research, 14*(1), 4–15.

Johnson, E. P. (2003). *Appropriating blackness: Performance and the politics of authenticity.* Durham, NC: Duke University Press.

Kraut, A. (2008). *Choreographing the folk: The dance stagings of Zora Neale Hurston.* Minneapolis: University of Minnesota Press.

Madison, D. S. (1998). That was my occupation: Oral narrative, performance, and Black feminist thought. In D. Pollock (Ed.), *Exceptional spaces: Essays in performance and history* (pp. 319–342). Chapel Hill: University of North Carolina Press.

Madison, D. S. (2005). *Critical ethnography: Methods, ethics, and performance.* Thousand Oaks, CA: Sage.

Madison, D. S. (2006a). The dialogic performative in performance ethnography. *Text and Performance Quarterly, 26*(4), 320–324.

Madison, D. S. (2006b). Performing theory/embodied writing. In J. Hamera (Ed.), *Opening acts: Performance in/as communication and cultural studies* (pp. 243–265). Thousand Oaks, CA: Sage.

Madison, D. S. (2006c, March 2–6). *Water rites* [Multimedia performance]. Chapel Hill: University of North Carolina.

Madison, D. S. (2008). Narrative poetics and performative interventions. In N. K. Denzin, Y. S. Lincoln, & L. T. Smith (Eds.), *Handbook of critical and indigenous methodologies* (pp. 391–405). Thousand Oaks, CA: Sage.

Madison, D. S., & Hamera, J. (Eds.). (2006). *Handbook of performance studies.* Thousand Oaks, CA: Sage.

McKenzie, J. (2001a). Performance and global transference. *TDR: The Drama Review, 45*(3), 5–7.

McKenzie, J. (2001b). *Perform or else: From discipline to performance.* New York: Routledge.

Muñoz, J. E. (2006). Stages: Queers, punks, and the utopian performative. In D. S. Madison & J. Hamera (Eds.), *Handbook of performance studies* (pp. 9–20). Thousand Oaks, CA: Sage.

Olomo, O. O. O./Jones, J. L. (2006). Performance and ethnography, performing ethnography, performance ethnography. In D. S. Madison & J. Hamera (Eds.), *Handbook of performance studies* (pp. 339–345). Thousand Oaks, CA: Sage.

Pollock, D. (1990). Telling the told: Performing like a family. *The Oral History Review, 18*(2), 1–35.

Pollock, D. (2006). Marking new directions in performance ethnography. *Text and Performance Quarterly, 26*(4), 325–320.

Pollock, D. (2007). The performative "I." *Cultural Studies <=> Critical Methodologies, 7*(3), 239–255.

Roach, J. (1996). *Cities of the dead: Circum-Atlantic performance.* New York: Columbia University Press.

Schechner, R. (1985). *Between theatre and anthropology.* Philadelphia: University of Pennsylvania Press.

Spivak, G. C. (1990). *The post-colonial critic: Interviews, strategies, dialogues.* New York: Routledge.

Taylor, D. (2003). *The archive and the repertoire: Performing cultural memory in the Americas.* Durham, NC: Duke University Press.

Turner, V. (1982). *From ritual to theatre.* New York: PAJ.

Visweswaran, K. (1994). *Fictions of feminist ethnography.* Minneapolis: University of Minnesota Press.

wa Thiong'o, N. (1998). Oral power and Europhone glory: Orature, literature, and stolen legacies. In *Penpoints, gunpoints, and dreams: Towards a critical theory of the arts and the state in Africa* (pp. 103–128). Oxford, UK: Clarendon.

wa Thiong'o, N. (1999). Penpoints, gunpoints, and dreams: An interview by Charles Cantalupo. *Left Curve, 23.* Available at http://www.leftcurve.org/LC23webPages/ngugu.html

wa Thiong'o, N. (2007). Notes toward a performance theory of orature. *Performance Research, 12*(3), 4–7.

Valdez, L. (1971). *Actos.* San Juan Bautista, CA, Cucuracha Press.

van Erven, E. (1988). *Radical people's theatre.* Bloomington: Indiana University Press.

Williams, R. (1981). *Politics and letters: Interviews with* New Left Review. London: Verso.

Williams, R. (1983). *Keywords: A vocabulary of culture and society.* New York: Oxford University Press.

19

BRAIDING NARRATIVE ETHNOGRAPHY WITH MEMOIR AND CREATIVE NONFICTION

Barbara Tedlock

Being there seeing, hearing, and meditating; being here dreaming, remembering, and inscribing. For years, I have recorded stories lurking inside my conversations with Mayan women returning from market with baskets of squawking turkeys; stories bursting forth during the sharing of a pink kola nut with a Yoruba woman on a 707 lazily circling the island of Manhattan; stories bubbling up in five gallons of red-chili deer meat on top a woodstove at the Pueblo of Zuni; stories swelling inside a Mongolian Ger filled with red-and-gold lacquered wooden chests, Chinese bronze mirrors, reindeer-hide tambourine drums, and wispy spirit placements. How does one enact such strange realities? Tapes, videos, notes, sketches, maps, and photos tell of an overanxious urge to preserve. But far more obsesses me since I have spent my time not so much in walking a particular path, but rather in spiraling along multiple alternative paths.

Writing evokes other writing and mirrors reflect other selves. The Velázquez painting *Las Meninas,* or "The Ladies in Waiting," captures a suspended moment with members of the royal court including the child Margarita, heir to the Spanish throne, staring outward implicating us as both observers and the observed. Behind and above Margarita's right shoulder hangs a painted mirror on the back wall reflecting her parents, as the king and queen in each of us. An actual mirror set up in the small room devoted to the painting at the Museo del Prado enhances the illusion; we see ourselves reflected in the vacuous center of the canvas. This creates an anomalous third space between self and other, interior and exterior, thought and emotion, truth and illusion. By creating an enchanted sacred spot, we encourage interactions in which each moment becomes two moments, history and memory, suspended in our consciousness. Such double consciousness negates the control of lineal history with its regime of cool curiosity, impersonal self-confidence, cultural completeness, ethnic purity, rational essentialism, and exoticism.

Holders of brushes and holders of cameras cannot trace a really Real reality outside the self, but instead mirror reality. As the South Indian novelist and social anthropologist Amitav Ghosh suggests, real life can only be grasped as a performance within the theater of writing that produces the presence it describes. So, why not admit that we are busy generating written mutterings? Our brush-and-camera reality creates contact zones where people meet, hauntings happen, and horizons fuse. Given our postmodern sensibility, we celebrate pop stars like Madonna and Britney with their Arabic henna hand designs and Hindu forehead *bindis* over their six chakras, seats of concealed wisdom. These cultural icons cut loose from their moorings create profound strangeness.

Field ethnographers, like street photographers, seek the magical in the quotidian: lemon-yellow flowers framed in gray-and-purple thunderstorms. Raghubir Singh, one of India's foremost ethnographic photographers, evokes the surge of life during his ongoing act of living it. In *River of Colour* (1998), he arranges his photographs tenderly yet starkly, revealing his engagement with his subjects. His rich documentation offers cultural immersion in the ongoing rush of experiencing common lifeways: cow-dung cakes drying in the morning sun, people gathering at the village well, a ragged peacock pecking at grains of millet, children shooting marbles while their fathers push carts and label shipping crates. Unlike the colonial photographers, who documented the intensely wounded life in the slums of Calcutta, his photos playfully capture the reverberating color and poetry of rural life. Walter Benjamin, if he had seen Singh's photographs, might have noticed that he had captured "a child's view of color" (Benjamin, 1996, p. 50), both as a magical substance and as an animal. Adults and children, others and ourselves, do not live in different worlds but rather live differently in the same world, tasting other ways of life in cultural co-participation, solidarity, and friendship.

▣ POSTMODERN GONZO JOURNALISM AND ETHNOGRAPHY

Gonzo is South Boston Irish American slang for the last person left standing after an all-night drinking marathon. Gonzo is also the title of a 1960 hit song written by James Booker, a flamboyant New Orleans rhythm and blues keyboardist famous for his raw-wired musical arrangements and heroin addiction. Bill Cardoso, a *Boston Globe* editor, invented the concept of "gonzo journalism" and applied it to Hunter Thompson's remarkable essay, "The Kentucky Derby Is Decadent and Depraved" (1970/1979). Rather than describing and honoring that year's winners of the derby, Thompson focused on himself, how bored yet frightened he felt trapped inside the huge drunk-and-disorderly crowd.

Gonzo ethnography, like gonzo journalism, is a postmodern documentary style that encourages a blend of observation with participation and rationality with altered states of consciousness. In so doing, they inscribe the Real while evoking solidarity with participants inside an exuberant unmapped performance space. An example is the cultural anthropologist Bruce Grindal's evocation of an African ritual. During his fieldwork in Ghana, he witnessed a death divination in which the corpse, sitting cross-legged on a cowhide, was propped up against the wall of his compound. Then, a praise singer danced and sang around him, until

> I began to see the *goka* [praise singer] and the corpse tied together in the undulating rhythms of the singing, the beating of the iron hoes, and the movement of feet and bodies. Then I saw the corpse jolt and occasionally pulsate, in a counterpoint to the motions of the goka. At first I thought that my mind was playing tricks with my eyes, so I cannot say when the experience first occurred; but it began with moments of anticipation and terror, as though I knew something unthinkable was about to happen. The anticipation left me breathless, gasping for air. In the pit of my stomach I felt a jolting and tightening sensation, which corresponded to moments of heightened visual awareness.
>
> What I saw in those moments was outside the realm of normal perception. From both the corpse and the goka came flashes of light so fleeting that I cannot say exactly where they originated. The hand of the goka would beat down on the iron hoe, the spit would fly from his mouth, and suddenly flashes of light flew like sparks from a fire.
>
> Then I felt my body become rigid. My jaws tightened and at the base of my skull I felt a jolt as though my head had been snapped

off my spinal column. A terrible and beautiful sight burst upon me. Stretching from the amazingly delicate fingers and mouths of the *goka* strands of fibrous light played upon the head, fingers, and toes of the dead man. The corpse, shaken by spasms, then rose to its feet, spinning and dancing in frenzy. As I watched, convulsions in the pit of my stomach tied not only my eyes but also my whole being into this vortex of power. It seemed that the very floor and walls of the compound had come to life, radiating light and power, drawing the dancers in one direction and then another. Then a most wonderful thing happened. The talking drums on the roof of the dead man's house began to glow with a light so strong that it drew the dancers to the rooftop. The corpse picked up the drumsticks and began to play. (Grindal, 1983, p. 68)

Now is the time for passionate ethnographic memoir, a blend of magical realism and a hard-driving narrative line in which a performer "is telling it like it is." Here the author is the active part of the story, a person so enthralled by hearing his own voice and listening to others telling the tale that he cannot remove himself from the narrative. The closest parallels to these memoirs are docudramas with unscripted humorous situations, POV radio, and Japanese *gakino tsukai*, or "crazy television." These and other genres create a contact zone between performers and audiences as a grittily realistic yet sacred performance space opening outward to an enchanted way of knowing and being in the world.

In the past, under the regime of colonialism, fieldwork produced two independent things: reportable nonparticipatory observation and nonreportable total participation. When ethnographers agreed to such a split, they cultivated rapport not friendship, compassion not sympathy, respect not belief, understanding not solidarity, and admiration not love. We did this, I fear, because we thought that if we cultivated friendship, sympathy, belief, solidarity, and love, we might lose it all—join history with memory and solidarity with objectivity—and "go native." Or so our tribal elders scared us into believing.

One way out of this impasse was to take the gamble and, as Australians like to say, "Go troppo!" by which they mean, "Go crazy." George Harrison, lead guitarist of the Beatles, released his album *Gone Troppo* in 1982, but it flopped. His son Dhani remastered and reissued it in 2004, and since then it has built a large international audience. Apparently, it was only a matter of timing between failure and success. This may also hold true for ethnography. Here I'm thinking about my classmate, Timothy Knab, who during the 1980s undertook linguistic research on the Nahuatl language spoken in Cuetzalan, Mexico. During his research, Tim became ensnared and ended up apprenticing himself to a group of shamans. In the early 1980s, after he wrote his doctoral dissertation, he was unable to find a publisher for his book until Harper San Francisco took the risk and released it as a work of creative nonfiction titled *The War of the Witches: A Journey Into the Underworld of the Contemporary Aztecs* (1995).

In his deeply evocative ethnography, Tim Knab unveiled how he learned to hear and tell stories and dreams in culturally recognizable ways. Later, he republished much of the same information in the genre of narrative ethnography with the University of Arizona Press, *The Dialogue of Earth and Sky: Dreams, Souls, Curing, and the Modern Aztec Underworld* (2004). These books show how his initial research as a linguist to a dying language gradually evolved into the work of apprentice to a living culture. They also reveal the strange blend of performativity and sovereignty of a nomad who learned to both live in and write about other cultural settings.

▣   Nomadic Thought and Becoming

Undertaking documentary fieldwork in a location far from home creates radically new experiences producing a blend of wonder and shock that may result in an epiphany, or sudden reperception of reality. This leads to the understanding that one cannot simply impose one's worldview on others. If one avoids either an ethnocentric rejection or a facile assimilation of the strange, then one may reconceptualize both within a third in-between space. This space can accommodate multiple individuals with various cultural and ethnic identities who interact and in so doing change while maintaining certain of their unique qualities. When an ethnographer refuses to either occupy or conquer the third space, then nomadic thought, which does not separate differences into oppositional dualities, arises creating an overlapping dialogue based on *becoming.*

Becoming refers to a process of ongoing transformation based on multiple dynamic interactions of the type one experiences during an extended sojourn abroad. The Lithuanian-French philosopher Emmanuel Levínas (1969) envisioned travel as a return to the self in such a way that experiences with otherness did not provoke a substantive transmutation in the attitude of the traveler. A traveling ethnographer's project hinges on translating otherness without sacrificing difference to the logic of the same. Levínas's teacher, Edmund Husserl, theorized that consciousness is characterized by *intentionality,* a tendency toward owning external objects as well as internal and external psychic systems. Levínas rejected this notion of intentionality as a form of violence and pointed out that consciousness desires to conquer the world by objectifying it. He, like Jacques Derrida, rejected the notion that the Other must become the Same; instead there is a metaphysical element that remains totally strange and although it wants desperately to be heard it can never be understood. Michel Foucault (1977) admonished us to prefer difference to uniformity, flows to unities, and mobile arrangements to fixed systems.

Later, in *A Thousand Plateaus* (1987) Gilles Deleuze and Félix Guattari argued that what is real is the *becoming* that is central to the development of rhizomatic theory. In their philosophy, a rhizome, or rootlike plant stem forming an entwined spherical mass, is a metaphor for an epistemology that spreads in all directions at once. A rhizome is reducible to neither the one nor the many; it has neither a beginning nor an end, but always a middle from which it grows. The development of rhizomic thought without hierarchies produces *nomadic space,* a place where individuals are shaped by new experiences and identities that may lead to the development of double consciousness. This nomadic state of being moves beyond unified identities and affirms unique differences between people (Deleuze & Guattari, 1986).

▣   Double Consciousness

First introduced into European philosophy by Friedrich Hegel (1807/1952), double consciousness entered American intellectual life by way of the writings of W. E. B. Du Bois. In *The Souls of Black Folk* (1903/1989), he described both the curse and the gift of African Americans who live between contradictory identities; that of "an American, a Negro; two souls, two thoughts, two unreconciled strivings; two warring ideals in one dark body, whose dogged strength alone keeps it from being torn asunder" (Du Bois, 1903, p. 215). More recently, double consciousness has been explored so as to include the worldviews of Whites and Browns. Whites live a double racial life, one colorblind and one race conscious, while Browns live suspended within a combination of whiteness and otherness (Bonilla-Silva, 2003).

As I conceptualize double consciousness, it is an equilibristic construction of identity that stresses the performativity of a nomadic subject. By endlessly citing the conventions of the social world around us, we produce our own reality through speech acts that combine language and gesture. My analysis rests on the experiential ethnographic approach pioneered by Victor and Edith Turner (1982) and practiced by a number of other ethnographers. The Turners pointed out that feeling and will, as well as thought, constitute the structure of cultural experience. To aid their students in understanding how people the world over experience the richness of their local lives, they experimented with rendering ethnography in a form of instructional theater. At the Universities of Chicago and Virginia, and New York University, they set up workshops in which members worked to acquire a kinetic understanding of other cultures. They experimented with the social dramas from their own Central African fieldwork and encouraged other ethnographers to perform dramas from their fieldwork. Stanley Walens, an ethnographer among Northwest coast Native Americans, scripted, narrated, and performed a set of rituals from his memoir *Feasting With Cannibals* (2001).

Experiential ethnographers acquire entrance into and partial enculturation within the worlds they study. During fieldwork, they may become actors and weave themselves into local cultures. Deborah Wong (2008), a Japanese American ethnographer as well as a *tako* drumming ethnomusicologist, reports that the field is simultaneously everywhere and nowhere, and thus everyone is in some sense an insider. While ethnomusicologists seem especially well suited to a performance approach, other areas of culture are also available. As the French ethnographer Jeanne Favret-Saada observed in her memoir *Deadly Words: Witchcraft in the Bocage,* "to understand the meaning of this discourse [witchcraft] there is no other solution but to practice it oneself, to become one's own informant, to penetrate one's own amnesia, and to try and make explicit what one finds unstateable in oneself" (Favret-Saada, 1980, p. 22).

◾ PERFORMING ETHNOGRAPHY

Ethnography as an enterprise consists of the examination, reflection, and shaping of human experience. Experiencing other ways of life while working and speaking with others in vulnerability and solidarity is central to the human sciences today (Tedlock, 2009). Combining participatory experience with memory and embodied performance is a rapidly emerging social practice. Performing ethnography encourages alternative strategies for the exploration, narration, celebration, writing, and rewriting of personal identities and social realities. Milton Singer's (1972) cultural performance, Victor and Edith Turner's (1982) performance ethnography, and Richard Schechner's (1989) intercultural performance merged into what we now call the "performance turn" in the social sciences (Conquergood, 1989).

Beginning in the 1980s and continuing into the early years of the 21st century, the Turners and Dwight Conquergood helped to shift ethnographers from interpretation studies toward performance studies. Dwight Conquergood performed his ethnographic work in refugee camps in Thailand and the Gaza Strip as well as among Hmong refugees in Chicago and during state executions in Texas and Indiana (Conquergood, 1985, 1992, 1998, 2002). He and others argued that social rituals draw their meaning and affective resonance from the traditions they reenact and that they never simply repeat but rather reverberate within these traditions (Schechner, 1985, pp. 36–37). These scholars advocate for performance as a "border discipline" expanding the meaning of texts by privileging embodied ethnographic research.

Performing ethnography produces a mimetic parallel or alternate instance through which the subjective is envisioned and made available to witnesses. In so doing, it creates a paradoxical location in which new possibilities for "the observation of participation" (Tedlock, 1991), or the living in while representing the world, emerges. Several recent ethnographers have centered their research and practice on the critical pedagogy and progressive politics of performative cultural studies (Alexander, 1999, 2002;

Allen & Garner, 1995; Denzin, 2003; Kondo, 1997; Laughlin, 1995; Madison, 2005). Such work uses dialogue, performative writing, kinesis, and staging that directly involves the arrangement of scenery, performers, and audience members (Garoian, 1999; Schutz, 2001).

◾ PERFORMATIVITY AND CULTURAL MEMORY

Performativity describes the reiterative power of discourse to create and produce the phenomena it regulates and constrains. The concept was initially developed in speech-act theory by John Austin (1962, 1970). Utterances such as "I promise," "I swear," and "I do" not only describe something but they also make it happen. In feminist studies, the concept was extended by Judith Butler (1990, 1997), who theorized gender, heterosexuality, and homosexuality as acts one performs; thus, something one *does* rather than expressions of what one *is.*

During the height of Vietnam antiwar protests, in the California of the 1960s, popular theater groups, such as Bread and Puppet and El Teatro Campesino (or "The Farmworkers' Theater"), performed all over the state. These progressive collectives produced free street theater for the masses. After each show, Bread-and-Puppet performers served fresh homemade bread with strong garlic aioli to the audience as a way of creating community. Members of El Teatro Campesino stood on the flatbeds of trucks parked in the grape fields outside Delano, California. There, these predominantly Mexican migrant laborers enacted events from their own lives and those of their audiences. Luis Valdez, a member of the San Francisco Mime Troupe, supported the United Farm Workers' strike against Gallo Vineyards by producing skits for the striking workers during which they showcased their Chicano identity (Montejano, 1999).

Chicano performance culture blends the theatricality of popular performances with the performativity of historical events such as Reies Lopez Tijerina's 1967 raid on the courthouse in Tierra Amarilla, New Mexico. Like Pancho Villa's 1916 raid on Columbus, New Mexico, Reies Tijerina reasserted Mexican American ownership of the American Southwest. Villa's cunning ability to elude North American forces became part of the folklore that was rhetorically reiterated in Tijerina's later flight from U.S. authorities.

> Immediately [Tijerina] and a small band of followers became targets of the largest manhunt in New Mexico history. National Guard convoys, state police from all northern counties, local sheriffs and unofficial posses, Jicarilla Apache police and cattle inspectors, all joined the search. Equipped with two ammunition-less tanks, clattering helicopters, droning spotter planes, a hospital van, and patrolling jeeps, these forces combed every hamlet, gully, and pasture for the insurrectionists who had staged the "bold daylight raid." (Nabokov, 1970, p. 12)

Here we see Reies Tijerina *performing* Pancho Villa.

This style of performance uses a strategy that the Mexican performance artist Guillermo Gómez-Peña calls "reverse anthropology." In an interview with the philosopher Eduardo Mendieta, Gómez-Peña explained that anthropology uses the power and knowledge of the dominant culture to study marginalized others, while in reverse anthropology, "we [the marginalized others] occupy a fictional space" in order "to push the dominant culture to the margins, treat it as exotic and unfamiliar." (Mendieta & Gómez-Peña, 2001, p. 543)

Another striking example of the power of grassroots participatory performance is the work of Sistren, a Jamaican theater group that collectively wrote and produced *Lionheart Gal: Life Stories of Jamaican Women* (1987). The dramas of women's oppression they scripted and enacted were their own, including those of their director, Honor Ford-Smith, who served as a working member of the group rather than an outside researcher and director. Sistren recorded, transcribed, and edited, as a collective, dozens of life stories and enacted them publicly in theater workshops with farmworkers and slum dwellers (Sistren, 1987, pp. 14–16).

In North America, there is a long history during which native peoples were disenfranchised by means of violence, laws, and treaties. To confront this, dance-dramas based on indigenous mythology were created and performed by survivors. As Leslie Marmon Silko wrote in her novel *Almanac of the Dead,*

> The Ghost Dance has never ended, it has continued, and the people have never stopped dancing; they may call it by other names, but when they dance, their hearts are reunited with the spirits of beloved ancestors and loved ones recently lost in the struggle. Throughout the Americas, from Chile to Canada, the people have never stopped dancing; as the living dance, they are joined again with all our ancestors before them, who cry out, who demand justice, and who call the people to take back the Americas! (1991, p. 1)

When Rosalie Jones (Daystar), a Chippewa-Cree dancer, joined the faculty of the Institute of American Indian Arts in Santa Fe, New Mexico, she began choreographing dances based on animal stories. In 1980, she formed a modern-dance company called Daystar: Classical Dance-drama of Indian America to perform and explore the spirituality behind Native American dance culture (Magill, 1998). Her dances provided the place where she connected with and communicated American Indian spiritual practices. In a masked shamanic dance she called "Wolf: A Transformation," she choreographed the *Anishinaabe* creation story in which Wolf was a companion to First Man. During the performance, a young male dancer crouched before the audience, wearing a wolf head and fur. By slowly turning his head side-to-side he connected wolfishness with humanness. Then he shed his wolf head, only to quickly reinhabit Wolf.

Non-native audience members reported that as they shifted their awareness, they became active witnesses rather than passive tourists. This response is similar to Native Americans during sacred ceremonies.

◨ NARRATIVE ETHNOGRAPHY
AND CREATIVE NONFICTION

Narrative is a fundamental means of imposing order on otherwise random and disconnected events and experiences. Since narratives are embedded within discourse and give shape to experience, storytelling and the self are closely linked. Narrative identity encourages a subjective sense of self-continuity while we symbolically integrate the events of our lived experience into the plot of our life stories. The pleasure of narrative is that it seamlessly translates *knowing* into *telling about* the way things really happened.

There are many narrative forms: history, drama, biography, autobiography, creative nonfiction, and narrative ethnography. Both narrative ethnography and creative nonfiction have characters, action, and shifting points of view. They follow a story-like narrative arc with a beginning, middle, and end, as well as high and low points of dramatic development including moments of tension and revelation. They also have an emotional arc consisting of inner conflict that meshes with the narrative arc. In a successful narrative ethnography, as the heroine is confronted with major decisions, dangerous threats, and emotionally powerful critiques from her family and society, we learn indirectly of her inner emotional life.

Before continuing with laying out the characteristics of narrative ethnography and creative nonfiction, I note that another, rather different, understanding of "narrative ethnography" has recently emerged in social science (Gubrium & Holstein, 2008). Here a set of methodological concepts including narrative resources, environment, embeddedness, and control are used primarily to prompt new research questions. To accomplish this, the ethnographic act and end product are collapsed into a single, highly abstract rhetorical field and reified as "an emergent method," combining epistemological, methodological, and analytical sensibilities. In so doing, the written genre is nearly erased.

The roots of the written genre of narrative ethnography lie at the crossroads between life-history and memoir. Vincent Crapanzano, in *Tuhami: Portrait of a Moroccan* (1980), documents both the life of his subject and his own responses to working with him. Over time, they evolved into reciprocal objects of transference to one another. While Tuhami was initially the main character, Crapanzano emerged in the writing process as a secondary character. The result is a psychologically rich double portrait. A similar intertwining of a biography with the story of the ethnographic encounter structured Laurel Kendall's *The Life and Hard Times of a Korean Shaman* (1988). Here, in a series of exchanges reproduced from memory and captured on tape, Kendall represents herself and her field assistant as sympathetic students of a Korean woman shaman. With the addition of personal and theoretical interludes (in typographically marked sections), we witness a female shaman actively engaging with a female ethnographer, her field assistant, and her readers.

An overlap between biography and personal memoir also structures Ruth Behar's *Translated Woman* (1993). Here she confessed how worried, yet relieved, she was when she realized that after nearly three years of studying what colonial women had said to their inquisitors and developing relationships with a number of townswomen she had let one of her subjects take over her research. Throughout the text she portrays her inner feelings by using an italic font: *"I am remembering the hurt I had felt several days before. While I was sitting in the half-open doorway reading, a boy had run past, gotten a peek at me, and yelled out with what to me sounded like venom in his voice, 'Gringa!'"* (Behar, 1993, p. 250). Since she is Cuban American, this insult from a fellow Hispanic was not only totally unexpected but also deeply painful.

What these psychologically rich intersubjective documents contribute is an unsettling of the boundaries that were once central to the notion of a self studying another. Instead, this form of border-zone cultural coproduction emerged as a new direction of ethnographic interchange and cultural inscription as a form of creative nonfiction. Creative nonfiction, like narrative ethnography, is factually accurate, and written with attention to literary style: However, the story is polyphonic with the author's voice and those of other people woven together. In creative nonfiction, the story is told using scenes rather than exposition and, as in narrative ethnography, the author-as-character is either the central figure or the central consciousness, or both. This type of artful emotional documentary discourse has emerged as a powerful literary genre infused with the rhetoric, metaphors, and other tropes that are commonly used in lyrical poetry and narrative fiction. Its sheer literariness distinguishes it from narrative ethnography.

Narrative ethnographers privilege traditional narrative techniques and include the main principles of expository writing, augments, and citing appropriate sources. Only some creative nonfiction writers use either narrative techniques or citation. Others deemphasize narrative in favor of deep reflection on experience and lyric or collage forms. An example of this tradition in creative nonfiction is *The Mirror Dance* (Krieger, 1983), a highly literary composite story told by means of a multiple-person stream of consciousness. To accomplish this, Susan Krieger constructed the account by paraphrasing her interview and documentary evidence without allowing herself any analytical commentary or even citation, as she might have if she had chosen to cast the work as a narrative ethnography. Other authors wrote creative nonfiction as a way to simultaneously refuse anonymity and authority (Eber, 1995; Tedlock, 1992). Instead, their work sought connection, intimacy, and passion. More recently, creative nonfiction has been used as a way to explore the lives of real people working in extra-legal worlds as a way of not revealing their locations and blowing their covers (Nordstrom, 2004, 2007).

▣ TERRE HUMAINE OR HUMAN EARTH

Fifty-five years ago, Jean Malaurie, now professor of Arctic anthropology and ecology at the *École des Hautes Études en Sciences Socials* in Paris, initiated *Terre Humaine* (literally "human earth") as a literary collection. In responding to the utopian appeal of the French revolution—liberty, equality, fraternity—he encouraged authors to write directly from personal experience and commitment. He convinced Editions Plon, at that time the second-largest publishing house in France, to accept the books he selected as a series (Balandier, 1987). Today, there are more than 85 titles that have sold over 11 million copies worldwide. The bestseller so far is *Le cheval d'orgueil: Mémoires d'un Breton du pays bigouden* by Pierre Jakez Hélias (1975). The author initially wrote in Breton, the Celtic language spoken in Brittany, then translated it himself into French for publication in the series.

The writing featured in Terre Humaine falls mainly into the area of creative nonfiction, which today is taught as "the fourth genre" alongside poetry, fiction, and drama in many writing programs worldwide. These literary works center on the human condition and bear witness to what each author saw, experienced, and understood. As one of the early authors, the ethnographer and folklorist Bruce Jackson, observed: "The great vision of Terre Humaine is that understanding is always a collaborative venture between those who are seen and those who are seeing, between those who speak and those who write, between those who write and those who read" (Jackson, 1999, p. 141).

The earliest books in the series were Malaurie's own Arctic travelogue *Les Derniers Rois de Thulé* (1955), and Claude Lévi-Strauss's Amazonian travelogue *Tristes Tropiques* (1955). After these successful launches, Malaurie sought out, translated, and reprinted many other examples of what he described as *la littérature du reel,* or "the literature of reality," which includes travelogues, life histories, memoirs, and autobiographies. In 1956, he found and republished Victor Segalen's remarkable documentary novel *Les immémoriaux* (1907/1956). This French naval doctor, explorer, and ethnographer of Breton origin expressed concern about the extinction of tribal civilizations in Oceania. While he presented his work as a set of harmless folkloristic recitations from ancient indigenous oral lore, it functions as an indictment of French imperialism and missionary Christianity, which nearly destroyed native Tahitian culture by a combination of mismanagement, syphilis, and drugs.

Jean Malaurie revealed his own emotional commitment to the dignity, complexity, and humanity of indigenous peoples in his five editions of *Derniers Rois*. The book steadily grew in length and complexity over the years from 328 pages of text, illustrations, and maps in the 1955 first edition to 854 pages by the final edition of 1989. He revealed his ethical stance again when he considered translating *Sun Chief: The Autobiography of a Hopi Indian* (1942). Although this remarkable life story was initially published by the American ethnographer Leo Simmons under his own name, Malaurie removed the name of Leo Simmons from the title page, returning the rightful authorship and royalties to the Hopi Indian whose life story it was, Don Talayesva (1959).

Other popular books in the series include Pierre Clastres's *Chronique des Indiens Guayaki* (1972), based on fieldwork in South America during the mid-1960s. Clastres lived among a recently contacted indigenous group in Paraguay where, although he could understand their language (since he spoke a neighboring dialect), they refused to converse with him. He hauntingly describes the situation, "they were still green," "hardly touched, hardly contaminated by the breezes of our civilization," "a society so healthy that it could not enter into a dialogue with me, with another world" (Clastres, 1972, pp. 96–97). His translator into English, the poet and prize-winning novelist Paul Auster, noted that the book is not only the true story of one man's experiences but that it is a portrait of him and that he writes with "the cunning of a novelist" (Auster, 1998, pp. 7–9).

Malaurie also selected and translated into French James Agee and Walker Evans's famous book *Let Us Now Praise Famous Men* (1941), as *Trois familles de métayers en 1936 en Alabama* (1972). At the time of their research, the writer Agee and the photographer Evans were employees of the Farm Security Administration who visited Hale County, Alabama, and became intimately acquainted with three White sharecropper families. Over a period of eight weeks, they recorded these families' struggle for survival in the aftermath of the Great Depression. The resulting book is partly documentary and partly literary, evoking the dark shacks and depleted fields of the American South.

Among the many other contributors to the series were Georges Balandier (1957), Margaret Mead (1963), Theodora Kroeber (1968), Guwa Baba and Mary Smith (1969), Bruce Jackson (1975), Alexander Alland (1984), Eric Rosny (1981), Colin Turnbull (1987), Robert Murphy (1990), Philippe Descola (1994), Roger Bastide (2000), Darcy Ribeiro (2002), Barbara Glowczewski (2004), and Barbara Tedlock (2004). Key elements in these works are firsthand experience, thick description, character development, point of view, and voice. The authors refrain from using the passive voice of a laboratory report ("it was concluded that . . ."); instead their voices are active, in the first person, passionate, and even theatrical. They portray themselves reflexively as bearing witness to both themselves and to history. Since they play important roles—be it hero, victim, or witness—they attribute motives to themselves as well as to others. Their choice of linguistic forms—including word order, tense, pronouns, and

evidentials—vividly convey their points of view and cast their narrators, protagonists, and listeners in an ethically engaged performative manner.

As Bruce Jackson noted, these authors step into other worlds, stay a short while, then return to our world to bear witness. "They document their passage in ways that become for us not simply a report of experience, but an experience in itself. Their work is, in a phrase Malaurie wrote to me a letter, *plus un document qu'un documentaire*" (Jackson, 2005, p. 15). In other words, each of these works is more of a literary document than a documentary account. Each is a complex, stand-alone, three-dimensional work of art within the theater of writing rather than a simple chronological diary entry.

▣   ▣   ▣

As a child, I spent most summers and holidays in my grandmother's log home on the prairie of northern Saskatchewan. Skipping behind her on riverside trails, she pointed out dozens of living rocks and edible plants: blackberries, bearberries, deer berries, violets, mints, fiddleheads, chickweed, and wild mushrooms. Sitting together on boulders nibbling violets and mints, she told me stories of a world filled with people, only some of whom were human beings. My favorite stories were about rock persons and cumulus clouds who gave advice, and deer, badger, and bear persons who healed.

To keep her language alive in me, a half-blood child, Nokomis explained key words in her *Anishinaabe* (Ojibwe) language; rocks are *asin,* in the singular and *asiniig,* in the plural. And since the *-iig* suffix, is used only for animate possessions, this means that rocks are alive. She was certain about this since she herself had seen rocks move and heard them speak and sing. In time, she said, I might also hear and speak with rocks. She warned me though that it could only happen if I spent time in the North all alone so that my schooling could not erase the magic of the natural world. As an Anglican lay preacher and traditional Ojibwe herbalist, midwife, and storyteller, she explained to me the differences and similarities between these spiritualities—pointing out that while Christians *talked about* guardian angels, Indians *talked to* guardian spirits. "These are our brothers and sisters, the animals," she insisted. For her, the two ideas were nearly the same and she admonished me not to choose one over the other. Instead, I should walk in balance along the edges of these worlds. "There is beauty and strength in being both: a double calling, a double love."

Becoming an ethnographer, a highly suspect enterprise within most Native North American communities, has ironically enabled me to fulfill my grandmother's expectations. Today, while telling my own story alongside and entangled within the telling of others' stories, I have realized that many narrative bits are mirages, seductively real phenomena that I photograph and describe only to discover they depend upon the theater of my imagination for life. Other scraps, like rainbow spokes and wheels in air, evaporate since the shadows we cast, the ones other people see, are not accurate reflections of who we really are, were, or ever will be. The memories we hide from eventually catch us; overtake us as spiders weaving the dreamcatchers of our lives.

▣ REFERENCES

Agee, J., & Evans, W. (1972). *Louons maintenant les grands hommes: Trois familles de métayers en 1936 en Alabama.* Paris: Editions Plon, Collection Terre Humaine. (Translated into French from *Let Us Now Praise Famous Men* [1941])

Alexander, B. K. (1999). Performing culture in the classroom: An instructional (auto)ethnography. *Text and Performance Quarterly, 19,* 307–331.

Alexander, B. K. (2002). Performing culture and cultural performance in Japan: A critical (auto)ethnographic travelogue. *Theatre Annual: A Journal of Performance Studies, 55,* 1–28.

Alland, A. (1984). *La danse de l'araignée: Un ethnologue Américain chez les Abron (Côte-d'Ivoire).* Paris: Editions Plon, Collection Terre Humaine. (Translated into French from *When the Spider Danced: Notes From an African Village* [1975])

Allen, C. J., & Garner, N. (1995). Condor qatay: Anthropology in performance. *American Anthropologist, 97*(1), 69–82.

Auster, P. (1998). Translator's note. In *Chronicle of the Guayaki Indians* (pp. 7–13). New York: Zone Books.

Austin, J. L. (1962). *How to do things with words.* London: Oxford University Press.

Austin, J. L. (1970). Performative utterances. In *Philosophical papers* (pp. 233–252). London: Oxford University Press.

Baba, G., & Smith, M. F. (1969). *Baba de Karo: L'autobiographie d'une musulmane haoussa du Nigeria.* Paris: Editions Plon, Collection Terre Humaine. (Translated from the English version *Baba of Karo: A Woman of Muslim Hausa* [1954])

Balandier, G. (1957). *L'Afrique ambiquë.* Paris: Editions Plon, Collection Terre Humaine. (Translated into English as *Ambiguous Africa: Cultures in Collision* [1966])

Balandier, G. (1987). "Terre Humaine" as a literary movement. *Anthropology Today, 3,* 1–2.

Bastide, R. (2000). *Le condomblé de Bahia (Brésil).* Paris: Editions Plon, Collection Terre Humaine.

Behar, R. (1993). *Translated woman.* Boston: Beacon.

Benjamin, W. (1996). A child's view of color (1913). In M. Bullock & M. W. Jennings (Eds.), *Walter Benjamin selected writings: Vol. 1. 1913–1926.* Cambridge, MA: Harvard University Press.

Bonilla-Silva, E. (2003). *The double consciousness of Black, White, and Brown folks in the 21st century.* Paper presented at the meeting of the American Sociological Association, Atlanta, GA.

Butler, J. (1990). *Gender trouble.* New York: Routledge.

Butler, J. (1997). *Excitable speech: A politics of the performative.* London: Routledge.

Clastres, P. (1972). *Chronique des Indiens Guayaki: Ce que savent les Aché, chasseurs nomads du Paraguay.* Paris: Editions Plon, Collection Terre Humaine. (Translated into English as *Chronicle of the Guayaki Indians* [1998])

Conquergood, D. (1985). Performing as a moral act: Ethical dimensions of the ethnography of performance. *Literature in Performance, 5,* 1–13.

Conquergood, D. (1989). Poetics, play, process and power: The performance turn in anthropology. *Text and Performance Quarterly, 9,* 81–88.

Conquergood, D. (1992). Fabricating culture: The textile art of Hmong refugee women. In E. C. Fine & J. H. Speer (Eds.), *Performance, culture, and identity* (pp. 206–248). Westport, CT: Praeger.

Conquergood, D. (1998). Beyond the text: Toward a performative cultural politics. In S. J. Dailey (Ed.), *The future of performance studies: Visions and revisions* (pp. 25–36). Annandale, VA: National Communication Association.

Conquergood, D. (2002). Lethal theatre: Performance, punishment, and the death penalty. *Theatre Journal, 54,* 339–367.

Crapanzano, V. (1980). *Tuhami: Portrait of a Moroccan.* Chicago: University of Chicago Press.

Deleuze, G., & Guattari, F. (1986). *Nomadology: The war machine.* New York: Semiotext(e).

Deleuze, G., & Guattari, F. (1987). *A thousand plateaus* (B. Massumi, Trans.). Minneapolis: University of Minnesota.

Denzin, N. K. (2003). *Performance ethnography: Critical pedagogy and the politics of culture.* Thousand Oaks, CA: Sage.

Descola, P. (1994). *Les lances du crépuscule: Relations Jivaros, Haute-Amazonie.* Paris: Editions Plon, Collection Terre Humaine. (Translated into English as *The Spears of Twilight: Life and Death With the Last Free Tribe of the Amazon* [1996])

Du Bois, W. E. B. (1989). *The souls of Black folk: Essays and sketches.* New York: Penguin. (Original work published 1903)

Eber, C. (1995). *Women and alcohol in a highland Maya town.* Austin: University of Texas Press.

Favret-Saada, J. (1980). *Deadly words: Witchcraft in the Bocage.* Cambridge, UK: Cambridge University Press.

Foucault, M. (1977). Preface (R. Hurley, M. Seem & H. Lane, Trans.). In G. Deleuze & F. Guattari, *Anti-Oedipus: capitalism and schizophrenia.* New York: Viking.

Garoian, C. R. (1999). *Performing pedagogy: Toward an art of politics.* Albany: State University of New York Press.

Glowczewski, B. (2004). *Rêves en colère: La pensée en réseau des aborigènes d'Australie.* Paris: Editions Plon, Collection Terre Humaine. (Translated into English and published by Editions Plon as *Dreams in Anger* [2004])

Grindal, B. (1983). Into the heart of Sisala experience: Witnessing death divination. *Journal of Anthropological Research, 39*(1), 60–80.

Gubrium, J. F., & Holstein, J. A. (2008). Narrative ethnography. In N. Hesse-Biber & P. Leavy (Eds.), *Handbook of emergent methods* (pp. 241–264). New York: Guilford.

Hegel, G. W. F. (1952). *Phenomenology of the spirit* (A. V. Miller, Trans.). Oxford, UK: Oxford University Press. (Original work published 1807)

Hélias, P. J. (1975). *Le cheval d'orgueil: Mémoires d'un Breton du pays bigouden.* Paris: Editions Plon, Collection Terre Humaine. (Translated into English as *Horse of Pride: Life in a Breton Village* [1978])

Jackson, B. (1975). *Leurs prisons: Autobiographies de prisonniers et d'ex-détenus Américains.* Paris: Editions Plon, Collection Terre

Humaine. (Translated into French from *In the Life: Versions of the Criminal Experience* [1972])

Jackson, B. (1999, October). The ethnographic voice. *Il Polo,* 139–141. Available at http://www.acsu.buffalo.edu/~bjackson/ETHNOG-RAPHY.HTM

Jackson, B. (2005). "Plus un document qu'un documentaire": The voices of Terre Humaine. In M. Berne & J.-M. Terrace (Eds.), *Terre humaine: Cinquante ans d'une collection* (pp. 14–23). Paris: Bibliothèque Nationale de France.

Kendall, L. (1988). *The life and hard times of a Korean shaman: Of tales and the telling of tales.* Honolulu: University of Hawaii Press.

Knab, T. J. (1995). *The war of the witches: A journey into the underworld of the contemporary Aztecs.* San Francisco: Harper.

Knab, T. J. (2004). *The dialogue of earth and sky: Dreams, souls, curing, and the modern Aztec underworld.* Tucson: University of Arizona Press.

Kondo, D. K. (1997). *About face: Performing race in fashion and theater.* New York: Routledge.

Krieger, S. (1983). *The mirror dance: Identity in a women's community.* Philadelphia: Temple University Press.

Kroeber, T. (1968). *Ishi: Testament du dernier Indien sauvage de l'Amérique du Nord.* Paris: Editions Plon, Collection Terre Humaine. (Translated from *The Last Testament of a Wild Indian of North America* [1961])

Laughlin, R. M. (1995). "From all for all": A Tzotzil-Tzeltal tragicomedy. *American Anthropologist, 97,* 528–542.

Levínas, E. (1969). *Totality and infinity: An essay on exteriority* (A. Lingis, Trans.). Pittsburgh: Duquesne University Press.

Lévi-Strauss, C. (1955). *Tristes tropiques.* Paris: Editions Plon, Collection Terre Humaine. (Translated into English with the same title [1973])

Madison, D. S. (2005). Critical ethnography as street performance: Reflections of home, race, murder, and justice. In N. K. Denzin & Y. S. Lincoln (Eds.), *The SAGE handbook of qualitative research* (3rd ed., pp. 537–546). Thousand Oaks, CA: Sage.

Magill, G. L. (1998, August). Rosalie Jones: Guiding light of Daystar—Native American choreographer. *Dance Magazine,* 1–3.

Malaurie, J. (1955). *Les derniers rois de Thulé.* Paris: Editions Plon, Collection Terre Humaine. (Translated into English as *The Last Kings of Thule: With the Polar Eskimos, as They Face Their Destiny* [1982])

Mead, M. (1963). *Moeurs et sexualité en Océanie.* Paris: Editions Plon, Collection Terre Humaine. (Translated from the English *Manners and Sexuality in Oceania,* combining materials from her earlier books *Coming of Age in Samoa* [1928] and *Sex and Temperament in Three Primitive Societies* [1935])

Mendieta, E., & Gómez-Peña, G. (2001). A Latino philosopher interviews a Chicano performance artist. *Napantla: Views from South, 2*(3), 539–554.

Montejano, D. (1999). On the question of inclusion. In D. Montejano (Ed.), *Chicano politics and society in the late twentieth century* (pp. xi–xxvi). Austin: University of Texas Press.

Murphy, R. F. (1990). *Vivre à corps perdu: Le témoignage et le combat d'un anthropologue paralysé.* Paris: Editions Plon, Collection Terrie Humaine. (Translated into French from *The Body Silent* [1987])

Nabokov, P. (1970). *Tijerina and the courthouse raid.* Berkeley, CA: Ramparts.

Nordstrom, C. (2004). *Shadows of war: Violence, power, and international profiteering in the twenty-first century.* Berkeley: University of California Press.

Nordstrom, C. (2007). *Global outlaws: Crime, money, and power in the contemporary world.* Berkeley: University of California Press.

Ribeiro, D. (2002). *Carnets indiens: Avec les Indiens Urubus-Kaapor, Brésil.* Paris: Editions Plon, Collection Terre Humaine. (Translated from the Portuguese version, *Diarios Indios—os Urubus-Kaapor* [1996])

Rosny, E. (1981). *Les yeux de ma chèvre: Sur les pas des maîtres de la nuit en pays Douala (Cameroun).* Paris: Editions Plon, Collection Terre Humaine. (Translated into English as *Healers in the night* [1985])

Schechner, R. (1985). *Between theater and anthropology.* Philadelphia: University of Pennsylvania Press.

Schechner, R. (1989). Intercultural themes. *Performing Arts Journal, 33/34,* 151–162.

Schutz, A. (2001). Theory as performative pedagogy: Three masks of Hannah Arendt. *Educational Theory, 51,* 127–150.

Segalen, V. (1956). *Les immémoriaux.* Paris: Editions Plon, Collection Terre Humaine. (Translated into English as *A Lapse of Memory* [1995])

Silko, L. M. (1991). *Almanac of the dead.* New York: Simon & Schuster.

Singer, M. (1972). *When a great tradition modernizes.* New York: Praeger.

Singh, R. (1998). *River of colour: The India of Raghubir Singh.* London: Phaidon.

Sistren (with Ford-Smith, H.). (1987). *Lionheart gal: Life stories of Jamaican women.* Toronto: Sister Vision.

Talayesva, D. (1959). *Soleil Hopi: L'autobiographie d'un Indien Hopi.* Paris: Editions Plon, Collection Terre Humaine. (Translated into French from *Sun Chief: The Autobiography of a Hopi Indian* [1942])

Tedlock, B. (1991). From participant observation to the observation of participation: The emergence of narrative ethnography. *Journal of Anthropological Research, 47,* 69–94.

Tedlock, B. (1992). *The beautiful and the dangerous: Encounters with the Zuni Indians.* New York: Viking.

Tedlock, B. (2004). *Rituels et pouvoirs: Les Indiens Zuñis Nouveau-Mexique.* Paris: Editions Plon, Collection Terre Humaine. (Translated into French from *The Beautiful and the Dangerous: Encounters with the Zuni Indians* [1992])

Tedlock, B. (2009). Writing a storied life: Nomadism and double consciousness in transcultural ethnography. *Etnofoor, 21*(1), 21–38.

Thompson, H. S. (1970/1979). The Kentucky Derby is decadent and depraved. In *The great shark hunt: Gonzo papers: Vol. 1. Strange tales from a strange time.* New York: Summit Books.

Turnbull, C. M. (1987). *Les Iks: Survivre par la cruauté: Nord-Ouganda.* Paris: Editions Plon, Collection Terre Humaine. (Translated into French from *The Mountain People* [1972])

Turner, V., & Turner, E. (1982). Performing ethnography. *The Drama Review, 26*(2), 33–50.

Walens, S. (2001). *Feasting with cannibals: An essay on Kwakiutl cosmology.* Princeton, NJ: Princeton University Press.

Wong, D. (2008). Moving: From performance to performative ethnography and back again. In G. Barz & T. J. Cooley (Eds.), *Shadows in the field: New perspectives for fieldwork in ethnomusicology* (pp. 76–89). New York: Oxford University Press.

20

THE CONSTRUCTIONIST ANALYTICS OF INTERPRETIVE PRACTICE

James A. Holstein and Jaber F. Gubrium

For the last half century, qualitative inquiry has focused increasingly on the socially constructed character of lived realities (see Denzin & Lincoln, 2005; Holstein & Gubrium, 2008). Much of this has centered on the interactional constitution of meaning in everyday life, the leading principle being that the world we live in and our place in it are not simply and evidently "there," but rather variably brought into being. Everyday realities are actively constructed in and through forms of social action. The principle supplies the basis for a constructionist perspective on qualitative inquiry that is both an intellectual movement and an empirical research perspective that transcends particular disciplines.

With its growing popularity, however, the constructionist approach has become particularly expansive and amorphous. Often it seems that the term "constructionism" can be applied to virtually every research approach imaginable. James Jasper and Jeff Goodwin (2005), for example, have wryly noted, "We are all social constructionists, almost" (p. 3). But there is a drawback to this popularity, because, as Michael Lynch (2008) suggests, the perspective may have become too diverse and diffuse to adequately define or assess. In the process, constructionism sometimes loses its conceptual bearings.

Elsewhere (Holstein & Gubrium, 2008), we have argued that constructionism resists a single portrait but is better understood as a *mosaic* of research efforts, with diverse (but also shared) philosophical, theoretical, methodological, and empirical underpinnings. This does not mean, however, that just anything goes under the constructionist rubric. We should resist the temptation to conflate constructionism with other contemporary or postmodern modes of qualitative inquiry; it is not synonymous with symbolic interactionism, social phenomenology, or ethnomethodology, for example, even as it shares their abiding concerns

with the dynamics of social interaction. Nor should we equate all variants of constructionism.

Darin Weinberg (2008) has argued that two important threads weave throughout the mosaic of constructionist thought: antifoundationalist sensibilities and a resistance to reification. These threads, of course, also wend through early statements of analytic philosophy, critical theory, pragmatism, and the hermeneutic tradition (see Weinberg, 2008). Joel Best (2008) traces the origins of the term "social constructionism" within sociology as far back as the early-20th century. He notes numerous appearances of the term in disciplines as varied as anthropology, history, and political science in the earlier parts of that century. At the same time, proto-constructionist sensibilities were evident in the work of a variety of scholars including W. I. Thomas (1931), George Herbert Mead (1934), Alfred Schutz (1962, 1964, 1967, 1970), and Herbert Blumer (1969), among many others. Best, however, suggests that the expansive popularity of the perspective, or perhaps the term, burst forth in the wake of the 1966 publication of Peter Berger and Thomas Luckmann's *The Social Construction of Reality: A Treatise in the Sociology of Knowledge.*

This chapter outlines the development of a constructionist analytics of interpretive practice, a particular variant of constructionist inquiry. In our view, the approach unites enough common elements to constitute a recognizable, vibrant research program. The program centers on the interactional constitution of lived realities within discernible contexts of social interaction. We use the term "analytics" because the approach and its variants produce understandings of the construction process by way of distinctive analytic vocabularies, what Blumer (1969) might have called a systematically linked set of "sensitizing concepts" spare enough not to overshadow the empirical, yet robust enough to reveal its constructionist distinctive contours. Our analytics of interpretive practice is decidedly theoretical, not just descriptive, but concertedly minimalist in its conceptual thrust. The chapter's aim is neither historic nor comprehensive. Rather, it looks more narrowly at the development of a particular strain of constructionist studies that borrows liberally, if somewhat promiscuously, from the traditions of social phenomenology, ethnomethodology, ordinary language philosophy, and Foucauldian discourse analysis.

▣ CONCEPTUAL SOURCES

The constructionist analytics of interpretive practice has diverse sources. For decades, constructionist researchers have attempted to document the agentic processes—the *hows*—by which social reality is constructed, managed, and sustained. Alfred Schutz's (1962, 1964, 1967, 1970) social phenomenology, Berger and Luckmann's (1966) social constructionism, and process-oriented strains of symbolic interactionism (e.g., Blumer, 1969; Hewitt, 1997; Weigert, 1981) have offered key elements to this constructionist project. More recently, ethnomethodology and conversation analysis (CA) have arguably supplied a more communicatively detailed dimension by specifying the interactive procedures through which social order is accomplished (see Buckholdt & Gubrium, 1979; Garfinkel, 1967, 2002, 2006; Heritage, 1984; Holstein, 1993; Lynch, 1993; Maynard & Clayman, 1991; Mehan & Wood, 1975; Pollner, 1987, 1991).[1] Discursive constructionism (see Potter & Hepburn, 2008)—a variant of discourse analysis bearing strong resemblances to CA—also has emerged to examine everyday descriptions, claims, reports, assertions, and allegations as they contribute to the construction and maintenance of social order.

A related set of concerns has emerged along with ethnomethodology's traditional interest in how social action and order are accomplished, reflecting a heretofore suspended interest in *what* is being accomplished, under *what* conditions, and out of *what* resources. Such traditionally naturalistic questions have been revived, with greater analytic sophistication and with a view toward the rich, varied, and consequential contexts of social construction. Analyses of reality construction are now re-engaging questions concerning the broad cultural and the institutional contexts of meaning making and social order. The empirical horizons, while still centered on processes of social accomplishment, are increasingly viewed in terms of what we have called "interpretive practice"—the constellation of procedures, conditions, and resources through which reality is apprehended, understood, organized, and conveyed in everyday life (Gubrium & Holstein, 1997; Holstein, 1993; Holstein & Gubrium, 2000b). The idea of interpretive practice turns us to both the *hows* and the *whats* of social reality; its empirical purview relates to both how people methodically construct their experiences and their

worlds and the contextual configurations of meaning and institutional life that inform and shape reality-constituting activity. This attention to both the *hows* and the *whats* of the social construction process echoes Karl Marx's (1956) maxim that people actively construct their worlds but not completely on, or in, their own terms.

This concern for constructive action-in-context not only makes it possible to understand more fully the construction process, but also foregrounds the realities themselves that enter into and are reflexively produced by the process. Attending closely to the *hows* of the construction process informs us of the mechanisms by which social forms are brought into being in everyday life, but it may shortchange the shape and distribution of these realities in their own right. The *whats* of social reality tend to be deemphasized in research that attends exclusively to the *hows* of its construction. We lose track of consequential *whats, whens,* and *wheres* that locate the concrete, yet constructed, realities that emerge.

Ethnomethodological Sensibilities

Ethnomethodology is perhaps the quintessential *how* analytic enterprise in qualitative inquiry. While indebted to Edmund Husserl's (1970) philosophical phenomenology and Schutz's social phenomenology (see Holstein & Gubrium, 1994), ethnomethodology struck a new course, addressing the problem of order by combining a "phenomenological sensibility" (Maynard & Clayman, 1991) with a paramount research concern for the mechanisms of practical action (Garfinkel, 1967; Lynch, 2008). From an ethnomethodological standpoint, the social world's facticity is accomplished by way of members' discernible interactional work, the mechanics of which produces and maintains the accountable circumstances of their lives.[2] Ethnomethodologists focus on how members "do" social life, aiming in particular to document the distinct processes by which they concretely construct and sustain the objects and appearances of the life world. The central phenomenon of interest is the in situ *embodied* activity and the practical production of accounts (Maynard, 2003). This leads to inquiries into how mundane practices are actually carried out, such as doing gender (Garfinkel, 1967), counting people and things (see Martin & Lynch, 2009), or delivering good or bad news (see Maynard, 2003).

The policy of "ethnomethodological indifference" (Garfinkel & Sacks, 1970) prompts ethnomethodologists to temporarily suspend all commitments to a priori or privileged versions of the social world. This turns the researcher's attention to how members accomplish a sense of social order. Social realities such as crime or mental illness are not taken for granted; instead, belief in them is temporarily suspended in order to make visible how they become realities for those concerned. This brings into view the ordinary constitutive work that produces the locally unchallenged appearance of stable realities.

This policy vigorously resists judgmental characterizations of the correctness of members' activities (see Lynch, 2008). Contrary to the common sociological tendency to ironicize and criticize commonsense formulations from the standpoint of ostensibly correct sociological understanding, ethnomethodology takes members' practical reasoning for what it is—circumstantially adequate ways of interpersonally constituting the world at hand. The abiding guideline is succinctly conveyed by Melvin Pollner's "Don't argue with the members!" (personal communication; see Gubrium & Holstein, 2011).

Ethnomethodological research is keenly attuned to naturally occurring talk and social interaction, orienting to them as constitutive elements of the settings studied (see Atkinson & Drew, 1979; Maynard, 1984, 1989, 2003; Mehan & Wood, 1975; Sacks, 1972). This has taken different empirical directions, in part depending upon whether the occasioned dynamics of social action and practical reasoning or the structure of talk is emphasized. Ethnographic studies tend to focus on locally accountable social action and the settings within which social interaction constitutes the practical realities in question. Such studies consider the situated content of talk in relation to local meaning structures (see Gubrium, 1992; Holstein, 1993; Lynch & Bogen, 1996; Miller, 1991; Pollner, 1987; Wieder, 1988). They combine attention to how social action and order is built up in everyday communication with detailed descriptions of place settings as those settings and their local understandings and perspectives serve to mediate the meaning of what is said in the course of social interaction. The texts produced from such analytics are highly descriptive of everyday life, with both conversational extracts from the settings and ethnographic accounts of interaction being used to convey the methodical production of the subject matter in question. To the extent the analysis of talk in relation to social interaction and setting is undertaken, this tends to take the form of (non-Foucauldian) discourse analysis, which more or less critically orients to how talk, conversation, and other communicative processes are used to organize social action. Variations on this analytic have also emerged in a form of discursive constructionism that resonates strongly with ethnomethodology and CA, but orients more to epistemics and knowledge construction (Potter & Hepburn, 2008; also see Nikander, 2008; Potter, 1996, 1997; Potter & Wetherell, 1987; Wodak, 2004; Wooffitt, 2005).

Studies that emphasize the structure of talk itself focus on the conversational "machinery" through which social action emerges. The focus here is on the sequential, utterance-by-utterance, socially structuring features of talk or "talk-in-interaction," a familiar term of reference in conversation analysis (see Heritage, 1984; Sacks, Schegloff, & Jefferson, 1974; Silverman, 1998; Zimmerman, 1988). The analyses produced from such studies are detailed explications of the communicative processes by which speakers methodically and sequentially construct their concerns in conversational practice. Often bereft of ethnographic detail

except for brief lead-ins that describe place settings, the analytic sense conveyed is that biographical and social particulars can be understood as artifacts of the unfolding conversational machinery, although the analysis of what is called "institutional talk" or "talk at work" has struck a greater balance with place settings in this regard (see, for example, Drew & Heritage, 1992). While some contend that CA's connection to ethnomethodology is tenuous because of this lack of concern with ethnographic detail (Atkinson, 1988; Lynch, 1993; Lynch & Bogen, 1994; for counterarguments see Maynard & Clayman, 1991 and ten Have, 1990), CA clearly shares ethnomethodology's interest in the local and methodical construction of social action (Maynard & Clayman, 1991).

Recently, Garfinkel, Lynch, and others have elaborated what they refer to as a "postanalytic" ethnomethodology that is less inclined to universalistic generalizations regarding the enduring structures or machinery of social interaction (see Garfinkel, 2002, 2006; Lynch, 1993; Lynch & Bogen, 1996). This program of research centers on the highly localized competencies that constitute specific domains of everyday "work," especially the (bench)work of astronomers (Garfinkel, Lynch, & Livingston, 1981), biologists and neurologists (Lynch, 1985), forensic scientists (Lynch, Cole, McNally, & Jenkins, 2008) and mathematicians (Livingston, 1986), among many others. The aim is to document the "haecceity"—the "just thisness"—of social practices within circumscribed domains of knowledge and activity (Lynch, 1993). The practical details of the real-time work of these activities are viewed as an *incarnate* feature of the knowledges they produce. It is impossible to separate the knowledges from the highly particularized occasions of their production. The approach is theoretically minimalist in that it resists a priori conceptualization or categorization, especially historical time, while advocating detailed descriptive studies of the specific, local practices that manifest order and render it accountable (Bogen & Lynch, 1993).

Despite their success at displaying a panoply of social production practices, CA and postanalytic ethnomethodology in their separate ways tend to disregard an important balance in the conceptualizations of talk, setting, and social interaction that was evident in Garfinkel's early work and Harvey Sacks's (1992) pioneering lectures on conversational practice (see Silverman, 1998). Neither Garfinkel nor Sacks envisioned the machinery of conversation as productive of recognizable social forms in its own right. Attention to the constitutive *hows* of social realities was balanced with an eye to the meaningful *whats*. Settings, cultural understandings, and their everyday mediations were viewed as reflexively interwoven with talk and social interaction. Sacks, in particular, understood culture to be a matter of practice, something that served as a resource for discerning the possible linkages of utterances and exchanges. Whether they wrote of (Garfinkel's) "good organizational reasons" or (Sacks's) "membership categorization devices," both initially avoided the reduction of social practice to highly local-ized or momentary haecceities of any kind.

Some of the original promise of ethnomethodology may have been short-circuited as CA and postanalytic ethnometh-odology have increasingly restricted their investigations to the relation between social practices and the immediate accounts of those practices (see Pollner 2011a, 2011b, 2011c). A broader constructionist analytics aims to retain ethnomethodology's interactional sensibilities while extending its scope to both the constitutive and constituted *whats* of everyday life. Michel Fou-cault, among others, is a valuable resource for such a project.

Foucauldian Inspirations

If ethnomethodology documents the accomplishment of everyday life at the interactional level, Foucault undertook a parallel project in a different empirical register. Appearing on the analytic stage at about the same time as ethnomethodology in the early 1960s, Foucault considers how historically and cul-turally located systems of power/knowledge construct subjects and their worlds. Foucauldians refer to these systems as "dis-courses," emphasizing that they are not merely bodies of ideas, ideologies, or other symbolic formulations, but are also working attitudes, modes of address, terms of reference, and courses of action suffused into social practices. Foucault (1972, p. 48) him-self explains that discourses are not "a mere intersection of things and words: an obscure web of things, and a manifest, visible, colored chain of words." Rather, they are "practices that systematically form the objects [and subjects] of which they speak" (p. 49). Even the design of buildings such as prisons reveals the social logic that specifies ways of interpreting persons and the physical and social landscapes they occupy (Foucault, 1979).

Similar to the ethnomethodological view of the reflexivity of social interaction, Foucault views discourse as operating reflex-ively, at once both constituting and meaningfully describing the world and its subjects. But, for Foucault, the accent is as much on the constructive *whats* that discourse constitutes as it is on the *hows* of discursive technology. While this implies an ana-lytic emphasis on the culturally "natural," Foucault's treatment of discourse as social practice suggests, in particular, the importance of understanding the practices of subjectivity. If he offers a vision of subjects and objects constituted through dis-course, he also allows for an unwittingly active subject who simultaneously shapes and puts discourse to work in construct-ing our inner lives and social worlds (Best & Kellner, 1991; Foucault, 1988).

Foucault is particularly concerned with social locations or institutional sites—the asylum, the hospital, and the prison, for example—that specify the practical operation of discourses, linking the discourse of particular subjectivities with the con-struction of lived experience. Like ethnomethodology, there is

an interest in the constitutive quality of systems of discourse; it is an orientation to practice that views lived experience and subjectivities as always already embedded and embodied in their discursive conventions.

Several commentators have pointed to the parallel between what Foucault (1980) refers to as systems of "power/knowledge" (or discourses) and ethnomethodology's formulation of the constitutive power of language use (Atkinson, 1995; Gubrium & Holstein, 1997; Heritage, 1997; Miller, 1997b; Potter, 1996; Prior, 1997; Silverman, 1993). The correspondence suggests that what Foucault's analytics documents historically as "discourses-in-practice" in varied institutional or cultural sites may have a counterpart in what ethnomethodology's analytics traces as "discursive practice" in varied forms of social interaction (Holstein & Gubrium, 2000b, 2003).[3] We use these terms—discourses-in-practice and discursive practice—throughout the chapter to flag the parallel concerns.

While ethnomethodologists and Foucauldians draw upon different intellectual traditions and work in distinct empirical registers, their similar concerns for social practice are evident; they both attend to the constitutive reflexivity of discourse. Neither discursive practice nor discourse-in-practice is viewed as being caused or explained by external social forces or inter-nal motives. Rather, they are taken to be the operating mecha-nism of social life itself, as actually known or performed in real time and in concrete places. For both, "power" lies in the articulation of distinctive forms of social life as such, not in the application of particular resources by some to affect the lives of others. While discourses-in-practice are represented by "regimens/regimes" or lived patterns of action that broadly (historically and institutionally) "discipline" and "govern" adherents' worlds, and discursive practice is manifest in the dynamics of talk and interaction that constitute everyday life, the practices refer in common to the lived "doing" or ongoing accomplishment of society.

If ethnomethodologists emphasize *how* members use every-day methods to account for their activities and their worlds, Foucault (1979) makes us aware of the related conditions of possibility for *what* the results are likely to be. For example, in a Western postindustrial society, to seriously think of medicine and voodoo as equally viable paradigms for understanding sickness and healing would seem idiosyncratic, if not preposter-ous, in most conventional situations. The power of medical discourse partially lies in its ability to be "seen but unnoticed," in its ability to appear as *the* only possibility while other possi-bilities are outside the plausible realm.

It bears repeating that both ethnomethodological and Foucauldian approaches to empirical material are analytics, not explanatory theories in the causal sense. Conventionally understood, theory purports to explain the state of matters in question. It responds to *why* concerns, such as why the sui-cide rate is rising or why individuals are suffering depression.

Ethnomethodology and the Foucauldian project, in contrast, aim to answer how it is that individual experience is understood in particular terms such as these. They are pretheoretical in this sense, respectively seeking to arrive at an understanding of how the subject matter of theory comes into existence in the first place, and of what the subject of theory might possibly become. The parallel lies in the common goal of documenting the practiced stuff of such realities.

Still, this remains a parallel—not a shared—scheme. Because Foucault's project (and most Foucauldian projects) operates in a historical register, real-time talk and social interaction are understandably missing from empirical materials under examination (but see Kendall & Wickham, 1999, for example). While Foucault himself points to sharp turns in the discursive formations that both shape and inform the shifting realities of varied institutional spheres, contrasting extant social forms with the "birth" of new ones, he provides little or no sense of the everyday interactional technology by which this is achieved (see Atkinson, 1995, Holstein & Gubrium, 2000b). Certainly, he elaborates the broad birth of new technologies, such as the emergence of new regimes of surveillance in medicine and modern criminal justice systems (Foucault, 1975, 1979), but he does not provide us with a view of how these operate on the ground. The everyday *hows,* in other words, are largely missing from Foucauldian analyses.

Conversely, ethnomethodology's commitment to documenting the real-time, interactive processes by which social action and order are rendered visible and accountable precludes a broad substantive perspective on constitutive resources, possibilities, and limitations. Such *whats* are largely absent in ethnomethodological work. It is one thing to show in interactive detail that our everyday encounters with reality are ongoing accomplishments, but it is quite another to derive an understanding of what the general parameters of those everyday encounters might be. The machinery of talk-in-interaction tells us little about the massive resources that are taken up in, and that guide, the operation of conversation, or about the consequences of producing particular results and not others, each of which is an important ingredient of practice. Members speak their worlds and their subjectivities, but they also articulate particular forms of life as they do so. Foucauldian considerations offer ethnomethodology an analytic sensitivity to the discursive opportunities and possibilities at work in talk and social interaction, without casting them as external templates for the everyday production of social order.

🔲 DIMENSIONS OF CONSTRUCTIONIST ANALYTICS

The constructionist analytics of interpretive practice reflects both ethnomethodological and Foucauldian impulses. It capitalizes on key sensibilities from their parallel projects, but it is

not simply another attempt at bridging the so-called macro-micro divide. That debate usually centers on the question of how to conceptualize the relationship between preexisting larger and smaller social forms, the assumption being that these are categorically distinct and separately discernible. Issues raised in the debate perpetuate the distinction between, say, social systems on the one hand, and social interaction, on the other.

In contrast, those who consider the ethnomethodological and Foucauldian projects to be parallel operations focus their attention instead on the interactional, institutional, and cultural variabilities of socially constituting discursive practice or discourses-in-practice, as the case might be. They aim to document how the social construction process is shaped across various domains of everyday life, not in how separate theories of macro and micro domains can be linked together for a fuller account of social organization. Doctrinaire accounts of Garfinkel, Sacks, Foucault, and others may continue to sustain a variety of distinct projects, but these projects are not likely to inform one another; nor will they lead to profitable dialogue between dogmatic practitioners who insist on viewing themselves as speaking different analytic languages. In our view, what we need is an openness to new, perhaps hybridized, analytics of reality construction at the crossroads of institutions, culture, and social interaction.

Beyond Ethnomethodology

Some ethnomethodologically informed varieties of CA have turned in this direction by analyzing the sequential machinery of talk-in-interaction as it is patterned by institutional context, bringing a greater concern for the *whats* of social life into the picture. Some field-based studies with ethnomethodological sensibilities have extended their concerns beyond the narrow *hows* of social interaction to include a wider interest in *what* is produced through interaction, in response to *what* social conditions. Still other forms of discourse analysis have similarly focused on the discursive resources brought to bear in situated social interaction or the kinds of objects and subjects constituted though interaction (see Wooffitt, 2005). These trends have broadened the empirical and analytic purview.

CA studies of "talk at work," for example, aim to specify how the "simplest systematics" of ordinary conversation (Sacks, Schegloff, & Jefferson, 1974) is shaped in various ways by the reflexively constructed speech environments of particular interactional regimes (see Boden & Zimmerman, 1991; Drew & Heritage, 1992). Ethnomethodologically oriented ethnographers approach the problem from another direction by asking how institutions and their respective subjectivities are brought into being, managed, and sustained in and through members' social interaction (or "reality work") (see Atkinson, 1995; Dingwall, Eekelaar, & Murray, 1983; Emerson, 1969; Emerson

& Messinger, 1977; Gubrium, 1992; Holstein, 1993, Mehan, 1979; Miller, 1991, 1997a). Foucault has even been inserted explicitly into the discussion, as researchers have drawn links between everyday discursive practice and discourses-in-practice to document in local detail how the formulation of everyday texts such as psychiatric case records or coroners' reports reproduce institutional discourses (see Prior, 1997). Others taking related paths have noted how culturally and institutionally situated discourses are interactionally brought to bear, to produce social objects and institutionalized interpersonal practices (see Hepburn, 1997, and Gubrium & Holstein, 2001).

In their own fashions, these efforts consider both the *hows* and the *whats* of reality construction. But this is analytically risky business. Asking *how* questions without having an integral way of getting an analytic handle on *what* questions renders concerns with the *whats* rather arbitrary. While talk-in-interaction is locally "artful," as Garfinkel (1967) puts it, not just anything goes. On the other hand, if we swing too far analytically in the direction of contextual or cultural imperatives, we end up with the cultural, institutional, or judgmental "dopes" that Garfinkel (1967) decried.

Accenting Analytic Interplay

To broaden and enrich ethnomethodology's analytic scope and repertoire, researchers have extended its purview to the institutional and cultural *whats* that come into play in social interaction. This has not been a historical extension, such as Foucault might pursue, although that certainly is not ruled out. In our own constructionist analytics, we have resurrected a kind of "cautious" (self-conscious) naturalism that addresses the practical and sited production of everyday life (Gubrium, 1993a). More decidedly constructionist in its concern for taken-for-granted realities, this balances *how* and *what* concerns, enriching the analytic impulses of each. Such an analytics focuses on the *interplay,* not the synthesis, of discursive practice and discourses-in-practice, the tandem projects of ethnomethodology and Foucauldian discourse analysis. In doing so, the analytics assiduously avoids theorizing social forms, lest the discursive practices associated with the construction of these forms be taken for granted. By the same token, it concertedly keeps institutional or cultural discourses in view, lest they be dissolved into localized displays of practical reasoning or forms of sequential organization for talk-in-interaction. First and foremost, a constructionist analytics of interpretive practice has taken us, in real time, to the "going concerns" of everyday life, as Everett Hughes (1984) liked to call social institutions. This approach focuses attention on how members artfully put distinct discourses to work as they constitute their social worlds.

Interplay connotes the acceptance of a dynamic relationship, not a to-be-resolved tension, between the *hows* and *whats* of interpretive practice. We have intentionally avoided analytically

privileging either discursive practice or discourses-in-practice. Putting it in ethnomethodological terms, in our view the aim of a constructionist analytics is to document the interplay between the practical reasoning and interactive machinery entailed in constructing a sense of everyday reality, on the one hand, and the institutional conditions, resources, and related discourses that substantively nourish and interpretively mediate interaction on the other. Putting it in Foucauldian terms, the goal is to describe the interplay between institutional discourses and the "dividing practices" that constitute local subjectivities and their domains of experience (Foucault 1965). The symmetry of real-world practice has encouraged us to give equal treatment to both its articulative and substantive engagements.

Constructionist researchers have increasingly emphasized the interplay between the two sides of interpretive practice. They are scrutinizing both the artful processes and the substantive conditions of meaning making and social order, even if the commitment to a multifaceted analytics sometimes remains implicit. Douglas Maynard (1989), for example, notes that most ethnographers have traditionally asked, "How do participants see things?" while ethnomethodologically informed discourse studies have asked, "How do participants do things?" While his own work typically begins with the later question, Maynard cautions us not to ignore the former. He explains that, in the interest of studying how members *do* things, ethnomethodological studies have tended to deemphasize factors that condition their actions. Recognizing that "external social structure is used as a resource for social interaction at the same time as it is constituted within it" (p. 139), Maynard suggests that ethnographic and discourse studies can be mutually informative, allowing researchers to better document the ways in which the "structure of interaction, while being a local production, simultaneously enacts matters whose origins are externally initiated" (p. 139). "In addition to knowing how people 'see' their workaday worlds," writes Maynard (p. 144), researchers should try to understand how people "discover and exhibit features of these worlds so that they can be 'seen.'"

Maynard (2003) goes on to note significant differences in the way talk and interaction typically are treated in conversation analytic versus more naturalistic, ethnographic approaches to social process. His own work, like many similarly grounded CA studies, exploits what Maynard terms the "limited affinity" between CA concerns and methods and more field-based ethnographic techniques and sensibilities (see Maynard, 2003, chapter 3). While a broad-based constructionist analytics would argue for a deeper, more "mutual affinity" (Maynard, 2003) between attempts to describe the *hows* and *whats* of social practice, there is clearly common ground, with much of the difference a matter of emphasis or analytic point of departure.

Expressing similar interests and concerns, Hugh Mehan has developed a discourse-oriented program of "constitutive ethnography" that puts "structure and structuring activities on an equal footing by showing *how* the social facts of the world emerge from structuring work to become external and constraining" (1979, p. 18, emphasis in the original). Mehan examines "contrastive" instances of interpretation in order to describe both the "distal" and "proximate" features of the reality-constituting work people do "within institutional, cultural, and historical contexts" (1991, pp. 73, 81).

Beginning from similar ethnomethodological and discourse analytic footings, David Silverman (1993) likewise attends to the institutional venues of talk and social construction (Silverman, 1985, 1997). Seeking a mode of qualitative inquiry that exhibits both constitutive and contextual impulses, he suggests that discourse studies that consider the varied institutional contexts of talk bring a new perspective to qualitative inquiry. Working in the same vein, Gale Miller (1994, 1997b) has proposed "ethnographies of institutional discourse" that serve to document "the ways in which setting members use discursive resources in organizing their practical actions, and how members' actions are constrained by the resources available in the settings" (Miller, 1994, p. 280). This approach makes explicit overtures to both conversation analysis and Foucauldian discourse analysis (see Miller, 1997a, and Weinberg, 2005) for rigorous empirical demonstrations of analytic interplay.

Dorothy Smith (1987, 1990a, 1990b) has been similarly explicit in addressing a version of the interplay between the *whats* and *hows* of social life from a feminist point of view, pointing to the critical consciousness made possible by the perspective. Hers has been an analytics initially informed by ethnomethodological and, increasingly, Foucauldian sensibilities. Moving beyond ethnomethodology, she calls for what she refers to as a "dialectics of discourse and the everyday" (Smith, 1990a, p. 202).

A concern for interplay, however, should not result in integrating an analytics of discursive practice with an analytics of discourse-in-practice. To integrate one with the other is to reduce the empirical purview of a parallel enterprise. Reducing the analytics of discourse-in-practice into discursive practice risks losing the lessons of attending to institutional differences and cultural configurations as they mediate, and are not "just talked into being" through, social interaction. Conversely, figuring discursive practice as the mere residue of institutional discourse risks a totalized marginalization of local artfulness.

Analytic Bracketing

A constructionist analytics that eschews synthesis or integration requires procedural flexibility and dexterity that cannot be captured in mechanical scriptures or formulas. Rather, the analytic process is more like a skilled juggling act, alternately concentrating on the myriad *hows* and *whats* of everyday life. This requires a new form of bracketing to capture the interplay

between discursive practice and discourses-in-practice. We refer to this technique of oscillating indifference to the construction and realities of everyday life as "analytic bracketing" (see Gubrium & Holstein, 1997). While we have given it a name, it resonates anonymously in other constructionist analytics.

Recall that ethnomethodology's interest in the *hows* by which realities are produced requires a studied, temporary indifference to those realities. Ethnomethodologists typically begin their analysis by setting aside belief in the objectively real in order to bring into view the everyday practices by which subjects, objects, and events come to have an accountable sense of being observable, rational, and orderly. The ethnomethodological project moves forward from there, documenting how discursive practice constitutes social action and order by identifying the particular interactional mechanisms at play. Ludwig Wittgenstein (1953, p. 19) is instructive as he advocates taking language "off holiday" in order to make visible how language works to produce the objects it is otherwise viewed as principally describing.

Analytic bracketing works somewhat differently. It is employed throughout analysis, not just at the start. As analysis proceeds, the researcher intermittently orients to everyday realities as both the *products* of members reality-constructing procedures and as *resources* from which realities are reflexively constituted. At one moment, the researcher may be indifferent to the structures of everyday life in order to document their production through discursive practice. In the next analytic move, he or she brackets discursive practice in order to assess the local availability, distribution, and/or regulation of resources for reality construction. In Wittgensteinian terms, this translates into attending to both language-at-work and language-on-holiday, alternating considerations of how languages games, in particular institutional discourses, operate in everyday life and what games are likely to come into play at particular times and places. In Foucauldian terms, it leads to alternating considerations of discourses-in-practice on the one hand and the locally fine-grained documentation of related discursive practices on the other.

Analytic bracketing amounts to an orienting procedure for alternately focusing on the *whats* then the *hows* of interpretive practice (or vice versa) in order to assemble both a contextually scenic and a contextually constitutive picture of everyday language-in-use. The objective is to move back and forth between discursive practice and discourses-in-practice, documenting each in turn, and making informative references to the other in the process. Either discursive machinery or available discourses and/or constraints becomes the provisional phenomenon, while interest in the other is temporarily deferred, but not forgotten. The analysis of the constant interplay between the *hows* and *whats* of interpretive practice mirrors the lived interplay between social interaction and its immediate surroundings, resources, restraints, and going concerns.

Because discursive practice and discourses-in-practice are *mutually* constitutive, one cannot argue definitively that analysis should begin or end with either one, although there are predilections in this regard. Smith (1987, 1990a, 1990b), for example, advocates beginning "where people are"; we take her to mean the places where people are concretely located in the institutional landscape of everyday life. Conversely, conversation analysts insist on beginning with discursive practice (i.e., everyday conversation), even while a variety of unanalyzed *whats* typically inform their efforts.

Wherever one starts, neither the cultural and institutional details of discourse nor its real-time engagement in social interaction predetermines the other. If we set aside the need for an indisputable resolution to the question of which comes first, last, or has priority, we can designate a suitable point of departure and proceed from there, so long as we keep firmly in mind that the interplay within interpretive practice requires that we move back and forth analytically between its facets. In the service of not reifying the components, researchers continuously remind themselves that the analytic task centers on the *dialectics* of two fields of play, not the reproduction of one by the other.

While we advocate no rule for where to begin, we need not fret that the overall task is impossible or logically incoherent. Maynard (1998, p. 344), for example, compares analytic bracketing to "wanting to ride trains that are going in different directions, initially hopping on one and then somehow jumping to the other." He asks, "How do you jump from one train to another when they are going in different directions?" The question is, in fact, merely an elaboration of the issue of how one brackets in the first place, which is, of course, the basis for Maynard's and other ethnomethodologists' and conversation analysts' own projects. The answer is simple: knowledge of the *principle* of bracketing makes it possible. Those who bracket the lifeworld or treat it indifferently, as the case might be, readily set aside aspects of social reality every time they get to work on their respective corpuses of empirical material. It becomes as routine as rising in the morning, having breakfast, and going to the workplace.[4] On the other hand, the desire to operationalize bracketing of any kind, analytic bracketing included, into explicitly codified and sequenced procedural moves would turn bracketing into a set of recipe-like, analytic directives, something surely to be avoided. We would assume that no one, except the most recalcitrant operationalist, would want to substitute a recipe book for an analytics.[5]

The alternating focus on discursive practice and discourses-in-practice reminds us not to appropriate either one naïvely into our analysis. It helps sustain ethnomethodology's important aim of distinguishing between members' resources and our own. Analytic bracketing is always substantively temporary. It resists full-blown attention to discourses as systems of power/knowledge, separate from how these operate in lived experience. It also is enduringly empirical in that it does not take the everyday operation of discourses for granted as the truths of a setting *tout court*.[6]

Resisting Totalization

Located at the crossroads of discursive practice and discourses-in-practice, a constructionist analytics works against analytic totalization or reduction. It accommodates the empirical realities of choice and action, allowing the analytic flexibility to capture the interplay of structure and process. It restrains the propensity of a Foucauldian analytics to view all interpretations as artifacts of particular regimes of power/knowledge. Writing in relation to the broad sweep of his "histories of the present," Foucault was inclined to overemphasize the predominance of discourses in constructing the horizons of meaning at particular times or places, conveying the sense that discourses fully detail the nuances of everyday life. A more interactionally sensitive analytics of discourse—one operating in tandem with a view to discursive practice—resists this tendency.

Because interpretive practice is mediated by discourse through institutional objectives and functioning, the operation of power/knowledge can be discerned in the myriad going concerns of everyday life. Yet, those matters that one institutional site brings to bear are not necessarily what another puts into practice. Institutions constitute distinct, yet sometimes overlapping, realities. While an organized setting may deploy a gaze that confers agency or subjectivity upon individuals, for example, another may constitute subjectivity along different lines (see, for example, Gubrium, 1992; Miller, 1997a; Weinberg, 2005).

If interpretive practice is complex and fluid, it is not socially arbitrary. In the practice of everyday life, discourse is articulated in myriad sites and is socially variegated; actors methodically build up their intersubjective realities in diverse, locally nuanced and biographically informed terms. This allows for considerable slippage in how discourses do their work; it is far removed from the apparently uniform, hegemonic regimes of power/knowledge in some Foucauldian readings. Discernible social organization nonetheless is evident in the going concerns referenced by participants, to which they hold their talk and interaction accountable.

Accordingly, a constructionist analytics deals with the perennial question of what realities and/or subjectivities are being constructed in the myriad sites of everyday life (see Hacking, 1999). In practice, diverse articulations of discourse intersect, collide, and work against the construction of common or uniform subjects, agents, and social realities. Interpretations shift in relation to the institutional and cultural markers they reference, which, in turn, fluctuate with respect to the varied settings in which social interaction unfolds. Discourses-in-practice refract one another as they are methodically adapted to practical exigencies. Local discursive practice makes totalization impossible, instead serving up innovation, diversification, and variation (see Abu-Lughod, 1991, 1993; Chase, 1995; Narayan & George, 2002).

▣   DIVERSE DIRECTIONS

Considering and emphasizing diverse analytic dimensions, variations on the constructionist analytics of interpretive practice continue to develop in innovative directions. Some are now "maturing," such as the "institutional ethnography" (IE) that Dorothy Smith and her colleagues have pioneered, and continue to expand. Others are of more recent vintage, such as the growth of discursive constructionism or Gubrium and Holstein's (2009) development of a constructionist analytics for narrative practice. Old or new, in their own fashions all take up the interplay of discursive practice and discourses-in-practice, variously emphasizing the *hows* and the *whats* of everyday life.

Ethnography of Narrative Practice

Let us begin with a recent development centered on how to analyze the interpretive practices associated with narrative and storytelling. Narrative analysis has become a popular mode of qualitative inquiry over the past two decades. If (almost) everyone is a constructionist, today nearly everyone also seems to be doing what they call narrative analysis. As sophisticated and insightful as the new wave of narrative analysis has become, most of this research is focused closely on texts of talk (e.g., Riessman, 1993). Researchers collect stories in interviews about myriad aspects of social life, then the stories are transcribed and analyzed for the way they emplot, thematize, and otherwise construct what they are about.

While attempts at narrative analysis have evinced constructionist sensibilities from the start, the socially situated, unfolding activeness of the narrative process has been shortchanged. The emphasis on the transcribed texts of stories tends to strip narratives of their social organization and interactional dynamics, casting narrative as a social product, not as social process. Emphasis is more on the text-based *whats* of the story and how that is organized, than on the *hows* of narrative production. Paul Atkinson (1997), among others, promotes a shift in focus:

> The ubiquity of the narrative and its centrality . . . are not license simply to privilege those forms. It is the work of anthropologists and sociologists to examine those narratives and to subject them to the same analysis as any other forms. We need to pay due attention to their construction in use: how actors improvise their personal narratives. . . . We need to attend to how socially shared resources of rhetoric and narrative are deployed to generate recognizable, plausible, and culturally well-informed accounts. (p. 341)

This reorientation encourages researchers to consider the circumstances, conditions, and goals of narratives—how storytellers work up and accomplish things with the accounts they produce. Adapting once more from Wittgenstein (1953, 1958), storytellers not only tell stories, they *do* things with them.

Capitalizing on Atkinson's and others suggestion, we have recently turned our brand of constructionist analytics to issues of narrative production (see Gubrium & Holstein, 2009). The challenge is to capture narrative's active, socially situated dimensions by moving outside of story texts to the occasions and practical activities of story construction and storytelling. By venturing into the domain of *narrative practice*, we gain access to the content of accounts and their internal organization, to the communicative conditions and resources surrounding how narratives are assembled, conveyed, and received, and to storytelling's everyday consequences.

The focus on practice highlights the reflexive interplay between discursive practice and discourse-in-practice. The narrative analysis of story transcripts may be perfectly adequate for capturing the internal dynamics and organization of stories, but it isolates those stories from their interactional and institutional moorings. For example, a transcript may not reveal a setting's discursive conventions, such as what is usually talked about, avoided, or discouraged under the circumstances. It may not reveal the consequences of a particular narrative told in a specific way. In order to understand how narrative operates in everyday life, we need to know the details and mediating conditions of narrative occasions. These details can only be discerned from direct consideration of the mutually constitutive interplay between what we have called "narrative work" and "narrative environments."

Narrative work refers to the interactional activity through which narratives are constructed, communicated, sustained, or reconfigured. The leading questions here are, "How can the process of constructing accounts be conceptualized?" and "How can the empirical process be analyzed?" Some of this is visible in story transcripts, but typically, narrative analysts tend to strip these transcripts of their interactional and institutional contexts and conversational character. This commonly results in the transcribed narrative appearing as a more-or-less finished, self-contained product. The in situ work of producing the narrative within the flow of conversational interaction disappears.

To recapture some of this narrative activity, we examine narrative practice for some of the ways in which narratives are activated or incited (see Holstein & Gubrium, 1995, 2000b). Working by way of analytic bracketing, these studies concentrate on conversational dynamics, machinery, and emerging sequential environments (many traditional CA concerns), while retaining sensitivity to broader contextual issues. Other studies focus on narrative linkages and composition, the ways in which horizons of meaning are narratively constructed (see Gubrium, 1993b; Gubrium & Holstein, 2009). Studies of narrative performativity document the ways in which narratives are produced and conveyed in and for particular circumstances and audiences (see Bauman, 1986; Abu-Lughod, 1993; Ochs & Capps, 2001). Collaboration and control are additional key concerns in analyzing narrative practice (see Holstein & Gubrium, 1995, 2000b; Norrick, 2000; Young, 1995). Because they are interactionally produced, narratives are eminently social accomplishments.

The other side of our analytics of narrative practice centers on narrative environments—contexts within which the work of narrative construction gets done. Narratives are assembled and told to someone, somewhere, at some time, with a variety of consequences for those concerned. (In contrast to CA, we do not limit narrative environments to the machinery of speech exchanges.) All of this has a discernible impact on how stories emerge, what is communicated, and to what ends. The environments of storytelling shape the content and internal organization of accounts, just as internal matters can have an impact on one's role as a storyteller. In turning to narrative environments, the analytic emphasis is more on the *whats* of narrative reality than on its *hows,* although, once again, analytic bracketing makes this a matter of temporary emphasis, not exclusive focus. One key question here is, "How is the meaning of a narrative influenced by the particular setting in which it is produced, with the setting's distinctive understandings, concerns, and resources, rather than in another setting, with different circumstances?" A second question is, "What are the purposes and consequences of narrating experience in particular ways?" A turn to the narrative environments of storytelling is critical for understanding what is at stake for storytellers and listeners in presenting accounts or responding to them in distinctive ways.

A growing body of work addresses such questions in relation to formal and informal settings and organizations, from families, to friendship networks, professions, and occupations (see Gubrium & Holstein, 2009). The comparative ethnographies of therapeutic organizations conducted by Miller (1997a) and Weinberg (2005) are exemplary in this regard. The influence of narrative environments is portrayed even more strikingly in *Out of Control: Family Therapy and Domestic Disorder* (Gubrium, 1992), which describes the narrative production of domestic troubles in distinctly different family therapy agencies. Susan Chase's (1995) *Ambiguous Empowerment: The Work Narratives of Women School Superintendents* and Amir Marvasti's (2003) *Being Homeless: Textual and Narrative Constructions* offer nuanced examinations of the accounts of some of society's most and least successful members, accenting the environmentally sensitive narrative work that is done to construct vastly different accounts of life and its challenges.

To move beyond transcribed texts, narrative analysis requires a methodology that captures the broad and variegated landscape of narrative practice. In essence, the researcher must be willing to move outside stories themselves and into the interactional, cultural, and institutional fields of narrative production, bringing on board a narrative ethnography of storytelling (see Gubrium &

Holstein, 2008, 2009).[7] Applied to storytelling, this ethnographic approach is attuned to the discursive dynamics and contours of narrative practice. It provides opportunities for the close scrutiny of narrative circumstances, their actors, and actions in the process of constructing accounts. This clearly resonates with contextually rich work done in the ethnography of communication (Hymes, 1964), the study of orally performed narratives (Bauman, 1986; Briggs & Bauman, 1992; Ochs & Capps, 2001), and ethnographically grounded studies of folk narratives (Glassie, 1995, 2006).

Concern with the production, distribution, and circulation of stories in society requires that we step outside of narrative texts and consider questions such as who produces particular kinds of stories, where are they likely to be encountered, what are their purposes and consequences, who are the listeners, under what circumstances are particular narratives more or less accountable, how do they gain acceptance, and how are they challenged? Ethnographic fieldwork helps supply the answers. In systematically observing the construction, use, and reception of narratives, we have found that their internal organization, while important to understand in its own right, does not tell us much about how stories operate in society. This does not diminish the explanatory value of text-based narrative analysis, but instead highlights what might be added to that approach if we attended to narrative practice.

Institutional Ethnography

Another approach relating discursive practice and discourse-in-practice is Smith's "institutional ethnography" (IE) research program.[8] IE emerged out of Smith's (1987, 1990a, 1990b, 1999, 2005) feminist work that explored the ruptures between women's everyday experience and dominant forms of knowledge that, while seemingly neutral and general, concealed particular standpoints grounded in gender, race, and class (McCoy, 2008). The approach takes the everyday world as both its point of departure and its problematic. Inquiry begins with ongoing activities of actual people in the world, "starting where people are," as Smith characteristically puts it. The aim is to map the translocal processes of administration and governance that shape lives and circumstances by way of the linkages of ruling relations. Recognizing that such connections are accomplished primarily through what is often called textually mediated social organization, IE focuses on texts-in-use in multiple settings. Across a range of locations—embodying people's everyday concerns, professional, administrative and management practices, and policy making—IE studies examine the actual activities that coordinate these interconnected sites (see DeVault & McCoy, 2002).

The dominant form of coordination is what Smith calls "ruling relations"—a mode of knowledge that involves the "continual transcription of the local and particular activities of our lives into abstracted and generalized forms . . . and the creation of a world in texts as a site of action" (Smith, 1987, p. 3). In IE, "text" orients the analyst to forms of representation (written spoken, visual, digital, or numeric) that exist materially separate from embodied consciousness. Such texts provide mediating linkages between people across time and place, making it possible to generate knowledge separate from individuals who possess such knowledge. Modern governance and large-scale coordination occur through rapidly proliferating, generalized, and generalizing, text-based forms of knowledge. These texts promote the "ruling relations [that generate] forms of consciousness and organization that are objectified in the sense that they are constituted externally to particular people and places" (Smith, 2005, p. 13). But to appreciate how texts do their coordinative work, the researcher must view them "in action" as they are produced, used, and oriented to by particular people in ongoing, institutional courses of action (see DeVault & McCoy, 2002; McCoy, 2008).

Therein lie the institutional and ethnographic dimensions of the approach. In IE, "institution" refers to coordinated and intersecting work processes and courses of action. "Ethnography" invokes concrete modes of inquiry used to discover and describe these activities. The IE researcher's goal is not to generalize about the people under study, but to identify and explain social processes that have generalizing effects. Practitioners of IE characteristically have critical or liberatory goals, an aim that we will address shortly. They pursue inquiry to elucidate the ideological and social processes that produce the experience of domination and subordination. As Smith and colleagues often point out, institutional ethnography offers a sociology *for* people not just about them (see DeVault & McCoy, 2002; McCoy, 2008; Smith, 2005).

While IE is not typically categorized as a variant of constructionism (McCoy, 2008), its conceptual antecedents and empirical interests often converge with the general constructionist project, especially with respect to the ways in which discursive resources and constraints affect social life and social forms. Centering on textually (discursively) mediated social relations, IE studies examine how forms of consciousness and organization are objectified or constituted as if they were external to particular people and places. IE analysis, however, strives to show that, at the same time, seemingly obdurate forms of social life are realized in concerted actions—produced, used, and oriented to by actual persons in ongoing, institutional courses of action (McCoy, 2008; Smith, 2005). From the standpoint of IE, the interplay between structures and agency is key to the social organization of lived experience.

As an alternative "sociology for people," IE has been adopted by researchers working in a wide variety of disciplines and settings: in education, social work, nursing and other health sciences, as well as sociology (see McCoy 2008; Smith, 2006). In a general sense, IE addresses the socially organized and organizing "work" done in varied domains of everyday life.

Work is construed in a very broad sense—activities that involve conscious intent and acquired skill; including emotional and thought work as well as physical labor or communicative action. It is not confined to occupational employment, although this form of work is also ripe for analysis. Marjorie DeVault (1991), for example, has examined the work of feeding a family, while several IE studies have investigated various aspects of mothers' experience and the deeply consequential mothering work done by women in diverse domestic and organizational settings (see Brown, 2006; Griffith & Smith, 2004; Weigt, 2006). Other studies have examined the situated experience of living with HIV infection (Mykhalovskiy & McCoy, 2002), child rearing and housing (Luken & Vaughan, 2006), nursing home care (Diamond, 1992), and job training and immigrant labor (Grahame, 1998). IE investigations conducted in more formal (occupational) work settings include studies of the work performed by teachers (Manicom, 1995), security guards (Walby, 2005), social workers (De Montigny, 1995), nurses (Campbell & Jackson, 1992; Rankin & Campbell, 2006), and policing in the gay community (G. Smith, 1988). Across these IE studies, the goal is to discover how lives are socially organized and coordinated. The analytic basis for all these projects is to display the interplay between institutional practices and individual actions. If IE resists a constructionist designation, it nonetheless shares many of the sensibilities embodied in a constructionist analytics.

Discursive Constructionism

Another innovative approach has been grouped loosely under the banners of discursive constructionism, or DC, and discourse analysis, or DA (see Potter, 1996; Potter & Hepburn, 2008). Its constructionist analytics also centers on the interplay of interpretive practice. As Jonathan Potter and Alexa Hepburn (2008) note, the DC label is itself a construction that supplies a particular sense of coherence to a body of more-or-less related work. If it is not singularly programmatic, it nevertheless represents a cogent analytic perspective that addresses the reflexive complexity of social interaction.

Centering attention on everyday conversations, arguments, talk-at-work, and other occasions where people are interacting, DC focuses on action and practice rather than linguistic structure. The approach emerged from the discourse analytic tradition in the sociology of scientific knowledge (e.g., Gilbert & Mulkay, 1984) and within a broader perspective developed within social psychology (e.g., Potter & Wetherell, 1987; see Hepburn, 2003). It is indebted in many ways to ethnomethodology (especially work by Harvey Sacks), and draws heavily on CA methods and findings. DC differs from CA, however, because it explicitly brings substantive issues of social construction to the fore; it is more concerned with the *whats* of social interaction than CA generally has been. While there are many other subtle distinctions, areas of overlap are substantial (Wooffitt, 2005), and in recent years DC and CA have found increasing areas of convergence (Potter & Hepburn, 2008).

DC approaches social construction in two fashions. In one, investigation aims to describe how discourse is constructed in the sense that it is assembled from a range of different resources with different degrees of structural organization. At the most basic level, these resources are words and grammatical structures, but they also include broader elements such as categories, metaphors, idioms, rhetorical conventions, and interpretive repertoires. The second approach emphasizes the constructive aspects of discourse in the sense that assemblages of words, repertoires, categories, and the like assemble and produce stabilized versions of the world and its actions and events. Central to DC is the notion that discourse does far more than describe objective states of affairs; it is used to construct versions of the world that are organized for particular purposes (Potter & Hepburn, 2008).

Following this commitment, DC treats all discourse as situated. At one level, it is located in the sequential environment of conversation (see Sacks, Schegloff, & Jefferson, 1974) and other forms of mediated interaction (e.g., turn allocation in legal or medical proceedings, screen prompts on computer displays). On another level, discourse is institutionally embedded. That is, it is generated within, and gives sense and structure to routine, ongoing practices such as family conversations, shopping transactions, and twelve-step meetings, for example. On a third level, discourse is situated rhetorically, in that discursive constructions are produced to advocate a particular version and counters possible alternatives (Potter & Hepburn, 2008). In this regard, analysis of the interest-related and consequential *whats* of discursive constructions is imperative.

While DC incorporates a view of discourse-in-practice, it stops short of the extended notion of discourse used in some of Foucault's work. DC's view of discourse is more restricted, emphasizing its use in everyday practice. Nevertheless, DC is dynamic and flexible enough to potentially address phenomena that Foucauldian analysis might also contemplate, or to conscript some of Foucault's insights about institutions, practice, and the nature of subjectivity into its own service (Potter & Hepburn, 2008). For example, Margret Wetherell (1998) argues that social identities cannot be understood apart from consideration of the discourses that provide the subject positions through which those identities are produced.

DC is not a "coherent and sealed system" (Potter & Hepburn, 2008, p. 291). Its field of interest is extremely broad, including but not restricted to, studies in discursive psychology and social psychology (e.g., Edwards, 2005; Edwards & Potter, 1992; Hepburn, 2003; Potter, 2003; Potter & Wetherell, 1987), cognition (e.g., Potter & te Molder, 2005), race and racism (e.g., Wetherell & Potter, 1992), gender (see Speer & Stokoe, in press), age (e.g., Nikander, 2002), facts (e.g., Wooffitt, 1992), and emotion (Edwards, 1999).

DC is not without its analytic tensions. For example, the issue of social structure and context remains a subject of debate. There is considerable contention regarding how the researcher might analyze utterances within conversation with an eye to identifying transcending discourses, subject positions, or repertoires. As in Foucauldian or critical discourse analysis (see, for example, Fairclough, 1995; Van Dijk, 1993; Wodak & Meyer, 2009), the issue is how critically to address the substantive *whats* of social construction while attending to the interactional dynamics and circumstances that construct them (Wooffitt, 2005). The danger in turning too fully to the study of transcendent discourse (writ large) is that it can shortchange the artful human conduct and agency involved in discursive practice (Wooffitt, 2005).

▣ SUSTAINING A CRITICAL CONSCIOUSNESS

This brings us to the concluding issue of how to maintain a critical consciousness in constructionist research while upholding a commitment to the neutral stance of bracketing. We have just noted that this is a desire shared by both DC and IE. But it does pose competing aims: documenting the social construction of reality, on the one hand, and critically attending to dominant and marginalized discourses and their effects on our lives, on the other. Exclusive attention to the constructive *hows* of interpretive practice cannot by itself sustain a critical consciousness.

Our way of addressing the issue comes by way of analytic bracketing. Our constructionist analytics sustains a critical consciousness by exploiting the critical potential of the analytic interplay of discourse-in-practice and discursive practice. Attending to both the constitutive *hows* and substantive *whats* of interpretive practice provides two different platforms for critique. The continuing enterprise of analytic bracketing does not keep us comfortably ensconced throughout the research process in a domain of indifference to the lived realities of experience, as phenomenological bracketing does. Nor does analytic bracketing keep us engaged in the unrepentant naturalism of documenting the world of everyday life as if it were fully objective and obdurate. Rather, it continuously rescues us from the analytic lethargies of both endeavors.

When questions of discourse-in-practice take the stage, there are grounds for problematizing or politicizing what otherwise might be too facilely viewed as socially or individualistically constructed, managed, and sustained. The persistent urgency of *what* questions cautions us not to assume that agency, artfulness, or the machinery of social interaction is the whole story. The urgency prompts us to inquire into broader environments and contingencies that are built up across time and circumstance in discursive practice. These are the contemporaneous conditions that inform and shape the construction process, and the personal and interpersonal consequences of having constituted the world in a particular way. While a constructionist view toward interpretive practice does not orient naturalistically to the "real world" as such, neither does it take everyday life as built from the ground up in talk-in-interaction on each and every communicative occasion. This allows for distinctly political observations since the analytics can point us toward matters of social organization and control that implicate matters beyond immediate interaction. It turns us to wider contexts (as constructed as they may be) in search of sources of action, control, change, or stability.

When discursive practice commands the analytic spotlight, there are grounds for critically challenging the representational hegemony of taken-for-granted realities. Researchers unsettle or deconstruct taken-for-granted realities in search of their construction to reveal the constitutive processes that produce and sustain them. Critically framed, persistent *how* questions remind us to bear in mind that the everyday realities of our lives—whether they are being normal, abnormal, law-abiding, criminal, male, female, young, or old—are realities we *do*. Having done them, they can be undone. We can move on to dismantle and reassemble realities, producing and reproducing, time and again, the world we inhabit. Politically, this recognizes that, in the world we inhabit, we could enact alternate possibilities or alternative directions, even if commonsense understandings make this seem impossible. If we make visible the constructive fluidity and malleability of social forms, we also reveal a potential for change (see Gubrium & Holstein, 1990, 1995; Holstein & Gubrium, 1994, 2000b, 2004, 2008).

The critical consciousness of a constructionist analytics deploys the continuous imperative to take issue with discourse or discursive practice when either one is foregrounded in research or seemingly obdurate in everyday life, thus turning the analytics on itself as it pursues its goals. In this sense, analytic bracketing is its own form of critical consciousness. Politically framed, the interplay of discourse-in-practice and discursive practice transforms analytic bracketing into critical bracketing, offering a basis not only for documenting interpretive practice, but also for critically commenting on its own constructions.

◫   NOTES

1. Some self-proclaimed ethnomethodologists, however, might reject the notion that ethnomethodology is in any sense a "constructionist" or "constructivist" enterprise (see Lynch, 1993, 2008). Some reviews of the ethnomethodological canon also clearly imply that constructionism is anathema to the ethnomethodological project (see Maynard, 1998; Maynard & Clayman, 1991).

2. While clearly reflecting Garfinkel's pioneering contributions, this characterization of the ethnomethodological project is perhaps closer to the version conveyed in the work of Melvin Pollner (1987, 1991) and D. Lawrence Wieder (1988) than some of the more recent "postanalytic" or conversation analytic forms of ethnomethodology. Indeed, Garfinkel (1988, 2002), Lynch (1993), and others might object to how we ourselves portray ethnomethodology. We would contend, however, that there is much to be gained from a studied "misreading" of the ethnomethodological "classics," a practice that Garfinkel himself advocates for the sociological classics more generally (see Lynch, 1993). With the figurative "death of the author" (Barthes, 1977), those attached to doctrinaire readings of the canon should have little ground for argument.

3. Other ethnomethodologists have drawn upon Foucault, but without necessarily endorsing these affinities or parallels. Lynch (1993), for example, writes that Foucault's studies can be relevant to ethnomethodological investigations in a "restricted and 'literal' way" (p. 131), and resists the generalization of discursive regimes across highly occasioned "language games." See McHoul (1986) and Lynch and Bogen (1996) for exemplary ethnomethodological appropriations of Foucauldian insights.

4. There are other useful metaphors for describing how analytic bracketing changes the focus from discourse-in-practice to discursive practice. One can liken the operation to shifting gears while driving a motor vehicle equipped with a manual transmission. One mode of analysis may prove quite productive, but it will eventually strain against the resistance engendered by its own temporary analytic orientation. When the researcher notes that the analytic engine is laboring under, or being constrained by, the restraints of what it is currently geared to accomplish, she can decide to virtually shift analytic gears in order to gain further purchase on the aspects of interpretive interplay that were previously bracketed. Just as there can be no prescription for shifting gears while driving (i.e., one can never specify in advance at what speed one should shift up or down), changing analytic brackets always remains an artful enterprise, awaiting the empirical circumstances it encounters. Its timing cannot be prespecified. Like shifting gears while driving, changes are not arbitrary or undisciplined. Rather they respond to the analytic challenges at hand in a principled, if not predetermined, fashion.

5. This may be the very thing Lynch (1993) decries with respect to conversation analysts who attempt to formalize and professionalize CA as a "scientific" discipline.

6. Some critics (see Denzin, 1998) have worried that analytic bracketing represents a selective objectivism, a form of "ontological gerrymandering." These, of course, have become fighting words among constructionists. But we should soberly recall that Steve Woolgar and Dorothy Pawluch (1985) have suggested that carving out some sort of analytic footing may be a pervasive and unavoidable feature of any sociological commentary. Our own constant attention to the interplay between discourse-in-practice and discursive practice continually reminds us of their reflexive relationship. Gerrymanderers stand their separate ground and unreflexively deconstruct; analytic bracketing, in contrast, encourages a continual and methodical deconstruction of empirical groundings themselves. This may produce a less-than-tidy picture, but it also is designed to keep reification at bay and ungrounded signification under control.

7. The term "narrative ethnography," which is an apt designation for an ethnographic approach to narrative, is also associated with

another approach to qualitative inquiry. Some researchers have applied the term to the critical analysis of representational practices in ethnography. Their aim is to work against the objectifying practices of ethnographic description. Practitioners of this form of narrative ethnography use the term to highlight researchers' narrative practices as they craft ethnographic accounts. They feature the interplay between the ethnographer's own subjectivity and the subjectivities of those whose lives and worlds are in view. Their ethnographic texts are typically derived from participant observation, but are distinctive because they take special notice of the researcher's own participation, perspective, voice, and especially his or her emotional experience as these operate in relation to the field of experience in view. Anthropologists Barbara Tedlock (1991, 1992, 2004), Ruth Behar (1993, 1996), and Kirin Narayan (1989), and sociologists Carolyn Ellis (1991), Laurel Richardson (1990a, 1990b), and others (Ellis & Flaherty, 1992; Ellis & Bochner, 1996) are important proponents of this genre. The reflexive, representational engagements of field encounters are discussed at length in H. L. Goodall's (2000) book *Writing the New Ethnography,* while Carolyn Ellis (2004) offers a description of the autoethnographic approach to narratives.

8. According to McCoy (2008), institutional ethnographers generally resist the tendency to be subsumed under the constructionist umbrella. By not affiliating with constructionism, she argues, IE has been free to participate in constructionist conversations, but on its own terms. This independent positioning is important for the IE project that aims to begin, not from theoretical vantage points, but from the actualities of people's lives.

◘ REFERENCES

Abu-Lughod, L. (1991). Writing against culture. In R. Fox (Ed.), *Recapturing anthropology* (pp. 137–162). Santa Fe, NM: SAR Press.

Abu-Lughod, L. (1993). *Writing women's worlds: Bedouin stories.* Berkeley: University of California Press.

Atkinson, J. M., & Drew, P. (1979). *Order in court.* Atlantic Highlands, NJ: Humanities Press.

Atkinson, P. (1988). Ethnomethodology: A critical review. *Annual Review of Sociology, 14,* 441–465.

Atkinson, P. (1995). *Medical talk and medical work.* London: Sage.

Atkinson, P. (1997). Narrative turn or blind alley? *Qualitative Health Research, 7,* 325–344.

Barthes, R. (1977). *Image, music, text.* New York: Hill & Wang.

Bauman, R. (1986). *Story, performance, and event: Contextual studies of oral narrative.* Cambridge, UK: Cambridge University Press.

Behar, R. (1993). *Translated woman: Crossing the border with Esperanza's story.* Boston: Beacon.

Behar, R. (1996). *The vulnerable observer: Anthropology that breaks your heart.* Boston: Beacon.

Berger, P. L., & Luckmann, T. (1966). *The social construction of reality.* New York: Doubleday.

Best, J. (2008). Historical development and defining issues of constructionist inquiry. In J. Holstein & J. Gubrium (Eds.), *Handbook of constructionist research* (pp. 41–64). New York: Guilford.

Best, S., & Kellner, D. (1991). *Postmodern theory: Critical interrogations.* New York: Guilford.

Blumer, H. (1969). *Symbolic interactionism.* Englewood Cliffs, NJ: Prentice Hall.

Boden, D., & Zimmerman, D. (Eds.). (1991). *Talk and social structure.* Cambridge, UK: Polity.

Bogen, D., & Lynch, M. (1993). Do we need a general theory of social problems? In J. Holstein & G. Miller (Eds.), *Reconsidering social constructionism: Debates in social problems theory* (pp. 213–237). Hawthorne, NY: Aldine de Gruyter.

Briggs, C. L., & Bauman, R. (1992). Genre, intertextuality, and social power. *Journal of Linguistic Anthropology, 2,* 131–172.

Brown, D. (2006). Working the system: Re-thinking the role of mothers and the reduction of "risk" in child protection work. *Social Problems, 53,* 352–370.

Buckholdt, D. R., & Gubrium, J. F. (1979). *Caretakers: Treating emotionally disturbed children.* Beverly Hills, CA: Sage.

Campbell, M., & Jackson, N. (1992). Learning to nurse: Plans, accounts, and actions. *Qualitative Health Research, 2,* 475–496.

Chase, S. E. (1995). *Ambiguous empowerment: The work narratives of women school superintendents.* Amherst: University of Massachusetts Press.

De Montigny, G. A. J. (1995). *Social working: An ethnography of frontline practice.* Toronto: University of Toronto Press.

Denzin, N. K. (1998). The new ethnography. *Journal of Contemporary Ethnography, 27,* 405–415.

Denzin, N. K., & Lincoln, Y. S. (Eds.). (2005). *The SAGE handbook of qualitative research* (3rd ed.). Thousand Oaks, CA: Sage.

DeVault, M. L. (1991). *Feeding the family: The social organization of caring as gendered work.* Chicago: University of Chicago Press.

DeVault, M. L., & McCoy, L. (2002). Institutional ethnography: Using interviews to investigate ruling relations. In J. F. Gubrium & J. A. Holstein (Eds.), *Handbook of interview research: Context and method* (pp. 751–776). Thousand Oaks, CA: Sage.

Diamond, T. (1992). *Making gray gold: Narratives of nursing home care.* Chicago: University of Chicago Press.

Dingwall, R., Eekelaar, J., & Murray, T. (1983). *The protection of children: State intervention and family life.* Oxford, UK: Blackwell.

Drew, P., & Heritage, J. (Eds.). (1992). *Talk at work.* Cambridge, UK: Cambridge University Press.

Edwards, D. (1999). Shared knowledge as a performative and rhetorical category. In J. Verschueren (Ed.), *Pragmatics in 1998: Selected papers from the 6th International Pragmatics Conference* (Vol. 2, pp. 130–141). Antwerp, Belgium: International Pragmatics Association.

Edwards, D. (2005). Discursive psychology. In K. L. Fitch & R. E. Sanders (Eds.), *Handbook of language and social interaction* (pp. 257–273). Hillsdale, NJ: Lawrence Erlbaum.

Edwards, D., & Potter, J. (1992). *Discursive psychology.* London: Sage.

Ellis, C. (1991). Sociological introspection and emotional experience. *Symbolic Interaction, 14,* 23–50.

Ellis, C. (2004). *The ethnographic I: A methodological novel about autoethnography.* Walnut Creek, CA: AltaMira.

Ellis, C., & Bochner A. P. (Eds.). (1996). *Composing ethnography: Alternative forms of qualitative writing.* Walnut Creek, CA: AltaMira.

Ellis, C., & Flaherty, M. (Eds.). (1992). *Investigating subjectivity.* Newbury Park, CA: Sage.

Emerson, R. M. (1969). *Judging delinquents.* Chicago: Aldine de Gruyter.

Emerson, R. M., & Messinger, S. (1977). The micro-politics of trouble. *Social Problems, 25*, 121–134.

Fairclough, N. (1995). *Critical discourse analysis.* London: Longman.

Foucault, M. (1965). *Madness and civilization.* New York: Random House.

Foucault, M. (1972). *The archaeology of knowledge.* New York: Pantheon.

Foucault, M. (1975). *The birth of the clinic.* New York: Vintage.

Foucault, M. (1979). *Discipline and punish.* New York: Vintage.

Foucault, M. (1980). *Power/knowledge.* New York. Pantheon.

Foucault, M. (1988). The ethic of care for the self as a practice of freedom. In J. Bernauer & G. Rasmussen (Eds.), *The final Foucault* (pp. 1–20). Cambridge: MIT Press.

Garfinkel, H. (1967). *Studies in ethnomethodology.* Englewood Cliffs, NJ: Prentice Hall.

Garfinkel, H. (1988). Evidence for locally produced, naturally accountable phenomena of order, logic, reason, meaning, method, etc. in and as of the essential quiddity of immortal ordinary society: Vol. 1. An announcement of studies. *Sociological Theory, 6*, 103–109.

Garfinkel, H. (2002). *Ethnomethodology's program: Working out Durkheim's aphorism.* Lanham, MD: Rowman & Littlefield.

Garfinkel, H. (2006). *Seeing sociologically: The routine grounds of social action.* Boulder, CO: Paradigm Publishers.

Garfinkel, H., Lynch, M., & Livingston, E. (1981). The work of a discovering science construed with materials from the optically discovered pulsar. *Philosophy of the Social Sciences, 11*, 131–158.

Garfinkel, H., & Sacks, H. (1970). On the formal structures of practical actions. In J. C. McKinney & E. A. Tiryakian (Eds.), *Theoretical sociology* (pp. 338–366). New York: Appleton-Century-Crofts.

Gilbert, G. N., & Mulkay, M. (1984). *Opening Pandora's box: A sociological analysis of scientists' discourse.* Cambridge, UK: Cambridge University Press.

Glassie, H. H. (1995). *Passing the time in Ballymenone: Culture and history of an Ulster community.* Bloomington: Indiana University Press.

Glassie, H. H. (2006). *The stars of Ballymenone.* Bloomington: Indiana University Press.

Goodall, H. L., Jr. (2000). *Writing the new ethnography.* Walnut Creek, CA: AltaMira.

Grahame, K. M. (1998). Asian women, job training, and the social organization of immigrant labor markets. *Qualitative Sociology, 53*, 75–90.

Griffith, A. I., & Smith, D. E. (2004). *Mothering for schooling.* New York: Routledge Falmer.

Gubrium, J. F. (1992). *Out of control: Family therapy and domestic disorder.* Newbury Park, CA: Sage.

Gubrium, J. F. (1993a). For a cautious naturalism. In J. Holstein & G. Miller (Eds.), *Reconsidering social constructionism* (pp. 89–101). New York: Aldine de Gruyter.

Gubrium, J. F. (1993b). *Speaking of life: Horizons of meaning for nursing home residents.* Hawthorne, NY: Aldine de Gruyter.

Gubrium, J. F., & Holstein, J. A. (1990). *What is family?* Mountain View, CA: Mayfield.

Gubrium, J. F., & Holstein, J. A. (1995). Life course malleability: Biographical work and deprivatization. *Sociological Inquiry, 53*, 207–223.

Gubrium, J. F., & Holstein, J. A. (1997). *The new language of qualitative method.* New York: Oxford University Press.

Gubrium, J. F., & Holstein, J. A. (Eds.). (2001). *Institutional selves: Troubled identities in a postmodern world.* New York: Oxford University Press.

Gubrium, J. F., & Holstein, J. A. (2008). Narrative ethnography. In S. Hesse-Biber & P. Leavy (Eds.), *Handbook of emergent methods* (pp. 241–264). New York: Guilford.

Gubrium, J. F., & Holstein, J. A. (2009). *Analyzing narrative reality.* Thousand Oaks, CA: Sage.

Gubrium, J. F., & Holstein, J. A. (2011). "Don't argue with the members." *The American Sociologist, 42*.

Hacking, I. (1999). *The social construction of what?* Cambridge, MA: Harvard University Press.

Hepburn, A. (1997). Teachers and secondary school bullying: A postmodern discourse analysis. *Discourse and Society, 8*, 27–48.

Hepburn, A. (2003). *An introduction to critical social psychology.* London: Sage.

Heritage, J. (1984). *Garfinkel and ethnomethodology.* Cambridge, UK: Polity.

Heritage, J. (1997). Conversation analysis and institutional talk: Analyzing data. In D. Silverman (Ed.), *Qualitative research: Theory, method and practice* (pp. 161–182). London: Sage.

Hewitt, J. P. (1997). *Self and society.* Boston: Allyn & Bacon.

Holstein, J. A. (1993). *Court-ordered insanity: Interpretive practice and involuntary commitment.* Hawthorne, NY: Aldine de Gruyter.

Holstein, J. A., & Gubrium, J. F. (1994). Phenomenology, ethnomethodology, and interpretive practice. In N. K. Denzin & Y. S. Lincoln (Eds.), *Handbook of qualitative research* (pp. 262–272). Newbury Park, CA: Sage.

Holstein, J. A., & Gubrium, J. F. (1995). *The active interview.* Thousand Oaks, CA: Sage.

Holstein, J. A., & Gubrium, J. F. (2000a). *Constructing the life course* (2nd ed.). Dix Hills, NY: General Hall.

Holstein, J. A., & Gubrium, J. F. (2000b). *The self we live by: Narrative identity in a postmodern world.* New York: Oxford University Press.

Holstein, J. A., & Gubrium, J. F. (2003). A constructionist analytics for social problems. In *Challenges and choices: Constructionist perspectives on social problems* (pp. 187–208). Hawthorne, NY: Aldine de Gruyter.

Holstein, J. A., & Gubrium, J. F. (2004). Context: Working it up, down, and across. In C. Seale, G. Gobo, J. F. Gubrium, & D. Silverman (Eds.), *Qualitative research practice* (pp. 297–343. London: Sage.

Holstein, J. A., & Gubrium, J. F (Eds.). (2008). *Handbook of constructionist research.* New York: Guilford.

Hughes, E. C. (1984). Going concerns: The study of American institutions. In D. Riesman & H. Becker (Eds.), *The sociological eye* (pp. 52–64). New Brunswick, NJ: Transaction Books.

Husserl, E. (1970). *Logical investigations.* New York: Humanities Press.

Hymes, D. (1964). The ethnography of communication. *American Anthropologist, 66*, 6–56.

Jasper, J. M., & Goodwin, J. (2005). From the editors. *Contexts, 4*(3), 3.

Kendall, G., & Wickham, G. (1999). *Using Foucault's methods.* London: Sage.

Livingston, E. (1986). *The ethnomethodological foundations of mathematics.* London: Routledge & Kegan Paul.

Luken, P. C., & Vaughan, S. (2006). Standardizing childrearing through housing. *Social Problems, 53*, 299–331.

Lynch, M. (1985). *Art and artifact in laboratory science.* London: Routledge & Kegan Paul.

Lynch, M. (1993). *Scientific practice and ordinary action.* Cambridge, UK: Cambridge University Press.

Lynch, M. (2008). Ethnomethodology as a provocation to constructionism. In J. A. Holstein & J. F. Gubrium (Eds.), *Handbook of constructionist research* (pp. 715–733). New York: Guilford.

Lynch, M., & Bogen, D. (1994). Harvey Sacks' primitive natural science. *Theory, Culture, and Society, 11,* 65–104.

Lynch, M., & Bogen, D. (1996). *The spectacle of history.* Durham, NC: Duke University Press.

Lynch, M., Cole, S., McNally, R., & Jenkins, K. (2008). *Truth machine: The contentious history of DNA fingerprinting.* Chicago: University of Chicago Press.

Manicom, A. (1995). What's class got to do with it? Class, gender, and teachers' work. In M. Campbell & A. Manicom (Eds.), *Knowledge, experience, and ruling relations: Studies in the social organization of knowledge* (pp. 135–148). Toronto: University of Toronto Press.

Martin, A., & Lynch, M. (2009). Counting things and people: The practices and politics of counting. *Social Problems, 56,* 243–266.

Marvasti, A. (2003). *Being homeless: Textual and narrative constructions.* Lanham, MD: Lexington Books.

Marx, K. (1956). *Selected writings in sociology and social philosophy* (T. Bottomore, Ed.). New York: McGraw-Hill.

Maynard, D. W. (1984). *Inside plea bargaining.* New York: Plenum.

Maynard, D. W. (1989). On the ethnography and analysis of discourse in institutional settings. In J. Holstein & G. Miller (Eds.), *Perspectives on social problems* (Vol. 1, pp. 127–146). Greenwich, CT: JAI.

Maynard, D. W. (1998). On qualitative inquiry and extramodernity. *Contemporary Sociology, 27,* 343–345.

Maynard, D. W. (2003). *Bad news, good news: Conversational order in everyday talk and clinical settings.* Chicago: University of Chicago Press.

Maynard, D. W., & Clayman, S. E. (1991). The diversity of ethnomethodology. *Annual Review of Sociology, 17,* 385–418.

McCoy, L. (2008). Institutional ethnography and constructionism. In J. A. Holstein & J. F. Gubrium (Eds.), *Handbook of constructionist research* (pp. 701–714). New York: Guilford.

McHoul, A. (1986). The getting of sexuality: Foucault, Garfinkel, and the analysis of sexual discourse. *Theory, Culture, and Society, 3,* 65–79.

Mead, G. H. (1934). *Mind, self, and society.* Chicago: University of Chicago Press.

Mehan, H. (1979). *Learning lessons: Social organization in the classroom.* Cambridge, MA: Harvard University Press.

Mehan, H. (1991). The school's work of sorting students. In D. Zimmerman & D. Boden (Eds.), *Talk and social structure* (pp. 71–90). Cambridge, UK: Polity.

Mehan, H., & Wood, H. (1975). *The reality of ethnomethodology.* New York: Wiley.

Miller, G. (1991). *Enforcing the work ethic.* Albany: SUNY Press.

Miller, G. (1994). Toward ethnographies of institutional discourse. *Journal of Contemporary Ethnography, 23,* 280–306.

Miller, G. (1997a). *Becoming miracle workers: Language and meaning in brief therapy.* Hawthorne, NY: Aldine de Gruyter.

Miller, G. (1997b). Building bridges: The possibility of analytic dialogue between ethnography, conversation analysis, and Foucault. In D. Silverman (Ed.), *Qualitative research: Theory, method and practice* (pp. 24–44). London: Sage.

Mykhalovskiy, E., & McCoy, L. (2002). Troubling ruling discourses of health: Using institutional ethnography in community-based research. *Critical Public Health, 12,* 17–37.

Narayan, K. (1989). *Storytellers, saints, and scoundrels: Folk narrative in Hindu religious teaching.* Philadelphia: University of Pennsylvania Press.

Narayan, K., & George, K. N. (2002). Personal and folk narrative as culture representation. In J. F. Gubrium & J. A. Holstein (Eds.), *Handbook of interview research* (pp. 815–832). Thousand Oaks, CA: Sage.

Nikander, P. (2002). *Age in action: Membership work and stages of life categories in talk.* Helsinki, Finland: Academia Scientarum Fennica.

Nikander, P. (2008). Constructionism and discourse analysis. In J. A. Holstein & J. F. Gubrium (Eds.), *Handbook of constructionist research* (pp. 413–428). New York: Guilford.

Norrick, N. R. (2000). *Conversational narrative: Storytelling in everyday talk.* Amsterdam: John Benjamins Publishing.

Ochs, E., & Capps, L. (2001). *Living narrative: Creating lives in everyday storytelling.* Cambridge, MA: Harvard University Press.

Pollner, M. (1987). *Mundane reason.* Cambridge, UK: Cambridge University Press.

Pollner, M. (1991). Left of ethnomethodology: The rise and decline of radical reflexivity. *American Sociological Review, 56,* 370–380.

Pollner, M. (2011a). The end(s) of ethnomethodology. *The American Sociologist, 42.*

Pollner, M. (2011b). Ethnomethodology from/as/to business. *The American Sociologist, 42.*

Pollner, M. (2011c). Reflections on Garfinkel and ethnomethodology's program. *The American Sociologist, 42.*

Potter, J. (1996). *Representing reality: Discourse, rhetoric, and social construction.* London: Sage.

Potter, J. (1997). Discourse analysis as a way of analyzing naturally-occurring talk. In D. Silverman (Ed.), *Qualitative research* (pp. 144–160). London: Sage.

Potter, J. (2003). Discursive psychology: Between method and paradigm. *Discourse & Society, 14,* 783–794.

Potter, J., & Hepburn, A. (2008). Discursive constructionism. In J. A. Holstein & J. F. Gubrium (Eds.), *Handbook of constructionist research* (pp. 275–294). New York: Guilford.

Potter, J., & te Molder, H. (2005). Talking cognition: Mapping and making the terrain. In H. te Molder & J. Potter (Eds.), *Conversation and cognition* (pp. 1–54). Cambridge, MA: Cambridge University Press.

Potter, J., & Wetherell, M. (1987). *Discourse and social psychology.* London: Sage.

Prior, L. (1997). Following in Foucault's footsteps: Text and context in qualitative research. In D. Silverman (Ed.), *Qualitative research: Theory, method and practice* (pp. 63–79). London: Sage.

Rankin, J. M., & Campbell, M. L. (2006.) *Managing to nurse: Inside Canada's health care reform.* Toronto: University of Toronto Press.

Richardson, L. (1990a). Narrative and sociology. *Journal of Contemporary Ethnography, 9,* 116–136.

Richardson, L. (1990b). *Writing strategies: Reaching diverse audiences.* Newbury Park, CA: Sage.

Riessman, C. K. (1993). *Narrative analysis.* Thousand Oaks, CA: Sage.

Sacks, H. (1972). An initial investigation of the usability of conversational data for doing sociology. In D. Sudnow (Ed.), *Studies in social interaction* (pp. 31–74). New York: Free Press.

Sacks, H. (1992). *Lectures on conversation* (Vols. 1 and 2). Oxford, UK: Blackwell.

Sacks, H., Schegloff, E., & Jefferson, G. (1974). A simplest systematics for the organization of turn-taking for conversation. *Language, 50,* 696–735.

Schutz, A. (1962). *The problem of social reality.* The Hague, the Netherlands: Martinus Nijhoff.

Schutz, A. (1964). *Studies in social theory.* The Hague, the Netherlands: Martinus Nijhoff.

Schutz, A. (1967). *The phenomenology of the social world.* Evanston, IL: Northwestern University Press.

Schutz, A. (1970). *On phenomenology and social relations.* Chicago: University of Chicago Press.

Silverman, D. (1985). *Qualitative methodology and sociology.* Aldershot, UK: Grower.

Silverman, D. (1993). *Interpretive qualitative data.* London: Sage.

Silverman, D. (Ed.). (1997). *Qualitative research.* London: Sage.

Silverman, D. (1998). *Harvey Sacks: Conversation analysis and social science.* New York: Oxford University Press.

Smith, D. E. (1987). *The everyday world as problematic.* Boston: Northeastern University Press.

Smith, D. E. (1990a). *The conceptual practices of power: A feminist sociology of knowledge.* Toronto: University of Toronto Press.

Smith, D. E. (1990b). *Texts, facts, and femininity.* London: Routledge.

Smith, D. E. (1999). *Writing the social: Critique, theory, and investigations.* Toronto: University of Toronto Press.

Smith, D. E. (2005). *Institutional ethnography: A sociology for people.* Lanham, MD: AltaMira.

Smith, D. E. (Ed.). (2006). *Institutional ethnography as practice.* Lanham, MD: AltaMira.

Smith, G. W. (1988). Policing the gay community: An inquiry into textually mediated relations. *International Journal of Sociology and the Law, 16,* 163–183.

Speer, S. A., & Stokoe, E. (Eds.). (in press). *Conversation and gender.* Cambridge, UK: Cambridge University Press.

Tedlock, B. (1991). From participant observation to the observation of participation: The emergence of narrative ethnography. *Journal of Anthropological Research, 47,* 69–94.

Tedlock, B. (1992). *The beautiful and the dangerous: Encounters with the Zuni Indians.* New York: Viking.

Tedlock, B. (2004). Narrative ethnography as social science discourse. *Studies in Symbolic Interaction, 27,* 23–31.

ten Have, P. (1990). Methodological issues in conversation analysis. *Bulletin de Methodolgie Sociologique, 27,* 23–51.

Thomas, W. I. (1931). *The unadjusted girl.* Boston: Little, Brown.

Van Djik, T. A. (1993). Principles of critical discourse analysis. *Discourse and Society, 4,* 249–283.

Walby, K. (2005). How closed-circuit television surveillance organizes the social: An institutional ethnography. *Canadian Journal of Sociology, 30,* 189–214.

Weigert, A. J. (1981). *Sociology of everyday life.* New York: Longman.

Weigt, J. (2006). Compromises to carework: The social organization of mothers' experiences in the low-wage labor market after welfare reform. *Social Problems, 53,* 332–351.

Weinberg, D. (2005). *Of others inside: Insanity, addiction, and belonging in America.* Philadelphia: Temple University Press.

Weinberg, D. (2008). The philosophical foundations of constructionist research. In J. A. Holstein & J. F. Gubrium (Eds.), *Handbook of constructionist research* (pp. 13–39). New York: Guilford.

Wetherell, M. (1998). Positioning and interpretive repertoires: Conversation analysis and post-structuralism in dialogue. *Discourse and Society, 9,* 387–412.

Wetherell, M., & Potter, J. (1992) *Mapping the language of racism: Discourse and the legitimation of exploitation.* New York: Columbia University Press.

Wieder, D. L. (1988). *Language and social reality.* Washington, DC: University Press of America.

Wittgenstein, L. (1953). *Philosophical investigations.* New York: Macmillan.

Wittgenstein, L. (1958). *Philosophical investigations.* New York: Macmillan.

Wodak, R. (2004). Critical discourse analysis. In C. Seale, G. Gobo, J. F. Gubrium, & D. Silverman (Eds.), *Qualitative research practice.* London: Sage.

Wodak, R., & Meyer, M. (Eds.). (2009). *Methods of critical discourse analysis.* London: Sage.

Wooffitt, R. (1992). *Telling tales of the unexpected: The organization of factual discourse.* London: Harvester/Wheatsheaf.

Wooffitt, R. (2005). *Conversation analysis and discourse analysis: A comparative and critical introduction.* London: Sage.

Woolgar, S., & Pawluch, D. (1985). Ontological gerrymandering. *Social Problems, 32,* 214–227.

Young, A. (1995). *The harmony of illusions.* Princeton, NJ: Princeton University Press.

Zimmerman, D. H. (1988). On conversation: The conversation analytic perspective. In J. A. Anderson (Ed.), *Communication yearbook* (Vol. 2, pp. 406–432). Newbury Park, CA: Sage.

21

GROUNDED THEORY METHODS IN SOCIAL JUSTICE RESEARCH

Kathy Charmaz[1]

Qualitative research has long attracted researchers who hope that their studies will matter in the public arena as well as in their disciplines. Yet many qualitative studies have been conducted that posed intriguing intellectual questions, addressed an interesting population, or explored an understudied phenomenon without raising explicit questions concerning social justice or policies that result in inequities. Such studies could often be taken a step or two further to explicate and explore social justice issues and subsequently reframe discussion of the studied phenomenon. What does social justice research entail? How can qualitative researchers move in this direction? What tools do they need?

When I speak of social justice inquiry, I mean studies that attend to inequities and equality, barriers and access, poverty and privilege, individual rights and the collective good, and their implications for suffering. Social justice inquiry also includes taking a critical stance toward social structures and processes that shape individual and collective life. I cast a wide net here across areas and levels of analysis in which questions about social justice arise. I include micro, meso, and macro levels of analysis, the local and global, as well as relationships between these levels. In the past, many social justice researchers have assumed that they must focus on macro structural relationships, but issues concerning social justice occur in micro situations and meso contexts, as well as in macro worlds and processes. Social scientists can study how the macro affects the micro and how micro processes also influence larger social entities. Global, national, and local social and economic conditions shape and are shaped by collective and individual meanings and actions. Yet when, how, and to what extent these conditions affect specific groups and individuals may not be fully recognized.

This chapter builds on my argument in the third edition of *The SAGE Handbook of Qualitative Research* (Denzin & Lincoln, 2005): Qualitative researchers can use grounded theory methods to advance social justice inquiry. These methods begin with inductive logic, use emergent strategies, rely on comparative inquiry, and are explicitly analytic. All these attributes give social justice researchers tools to sharpen and specify their analyses that will increase the analytic power and influence of their work while simultaneously expediting the research process. Grounded theory methods not only offer social justice researchers tools for developing innovative analyses but also for examining established concepts afresh. To mine the largely untapped potential of grounded theory methods for social justice inquiry, social justice researchers need to understand the logic of the method, the development of its different versions, their epistemological roots, and how they might use it.

Research in the area of social justice addresses differential power, prestige, resources, and suffering among peoples and individuals. It focuses on and furthers equitable resources, fairness, and eradication of oppression (Feagin, 1999).[2] Some reports in social justice inquiry begin with an explicit value stance and an agenda for change (see, e.g., these grounded theory studies: Karabanow, 2008; Nack, 2008; Sakamoto, Chin, Chapra, & Ricciar, 2009; Ullman & Townsend, 2008).[3] Other research reports often convey a taken-for-granted concern with social justice (e.g., Dumit, 2006; Foote-Ardah, 2003; Frohmann, 1991, 1998; Gagné, 1996; Hyde & Kammerer, 2009; Jiménez, 2008; Lio, Melzer, & Reese, 2008; Lutgen-Sandvik, 2008; Mevorach, 2008; Moore, 2005; Swahnberg, Thapar-Björkert, & Bertero, 2007; Tuason, 2008; Veale & Stavrou, 2007). Still other authors indicate that they chose a controversial topic that has social justice implications because it could illuminate a theoretical problem (Einwohner & Spencer, 2005; Ogle, Eckman & Leslie, 2003; Spencer & Triche, 1994).[4] Researchers may, however, begin their studies with an interest in a social issue rather than an impassioned commitment to changing it (Wasserman & Clair, 2010). Yet, the very process of witnessing their participants' lives and analyzing their data may elicit concerns about social justice that they had not understood earlier or anticipated.

Many researchers hold ideals of creating a good society and a better world and thus pursue empirical studies to further their ideas. For those who identify themselves as social justice researchers, "shoulds" and "oughts" are part of the research process and product. Claiming an explicit value position and studying controversial topics can result in having one's work contested. Hence, some researchers remain silent about their value commitments and instead choose to frame their studies in conceptual terms, rather than social justice concerns.

Grounded theory is a method of qualitative inquiry[5] in which data collection and analysis reciprocally inform and shape each other through an emergent iterative process. The term, "grounded theory," refers to this method and its product, a theory developed from successive conceptual analysis of data. Researchers may adopt grounded theory strategies while using a variety of data collection methods. Grounded theory studies have frequently been interview studies, and some studies have used documents (Clarke, 1998; Einwohner & Spencer, 2005; Mulcahy, 1995; Star, 1989) or ethnographic data (e.g., Casper, 1998; Thornberg, 2007; Wasserman & Clair, 2010; Wolkomir, 2001, 2006). It is often difficult, however, to discern the extent to which researchers have engaged grounded theory strategies (Charmaz, 2007, 2010; Timmermans & Tavory, 2007).

The strategies of grounded theory provide a useful toolkit for social justice researchers to employ. Grounded theory practice consists of emergent research decisions and actions that particularly fit social justice studies. The grounded theory emphases on empirical scrutiny and analytic precision fosters creating nuanced analyses of how social and economic conditions work in specific situations, whether or not researchers take their work into explicit theory construction (see, e.g., Ball, Perkins, Hollingsworth, Whittington, & King, 2009; Dixon, 2007; Jackson-Jacobs, 2004; Lazzari, Ford, & Haughey, 1996; Sixsmith, 1999; Speed & Luker, 2006). Such analyses not only contribute to knowledge, but also can inform those practices and policies that social justice researchers seek to change.

Researchers can learn how to use grounded theory guidelines and put them to use for diverse research objectives, including interrogating social justice questions. To date, few grounded theory studies in social justice inquiry demonstrate theory construction. Many, however, show how grounded theory guidelines have sharpened thematic analyses. This chapter aims to clarify the method and its evolution, illuminate how researchers have used specific grounded theory guidelines,

and demonstrate how this method complements social justice inquiry.

The constructivist revision of Glaser and Strauss's (1967) classic statement of grounded theory assumes that people construct both the studied phenomenon and the research process through their actions. This approach recognizes the constraints that historical, social, and situational conditions exert on these actions and acknowledges the researcher's active role in shaping the data and analysis. The constructivist version is particularly useful in social justice inquiry because it (1) rejects claims of objectivity, (2) locates researchers' generalizations, (3) considers researchers' and participants' relative positions and standpoints, (4) emphasizes reflexivity, (5) adopts sensitizing concepts such as power, privilege, equity, and oppression, and (6) remains alert to variation and difference (see Bryant & Charmaz, 2007; Charmaz, 2006, 2009b; Clarke, 2005; Clarke & Friese, 2007).

Nonetheless, adopting strategies common to all versions of grounded theory will advance social justice studies and, therefore, I discuss works that use all versions of grounded theory.

My discussion relies on a selective review of grounded theory studies in social justice inquiry. The burgeoning number and range of relevant works across disciplines and professions precludes a comprehensive review. The selected studies (1) show connections between social justice inquiry and grounded theory, (2) reveal debates concerning grounded theory, and (3) demonstrate ways to use this method. Like other qualitative research projects, many social justice studies indicate having used grounded theory strategies only for coding and confuse developing thematic topics for theoretical categories. One purpose of this chapter is to help researchers to make informed choices about when, how, and to what extent they adopt grounded theory logic and strategies. Grounded theory strategies can help scholars with diverse pursuits without necessarily developing a grounded theory.[6] The point is to make clear decisions and to be aware of their implications.

▣ THE LOGIC OF GROUNDED THEORY

Grounded theory is a method of social scientific theory construction. As Glaser and Strauss (1967) first stated, the grounded theory method consists of flexible analytic guidelines that enable researchers to focus their data collection and to build middle-range theories. These guidelines emphasize studying processes in the field setting(s), engaging in simultaneous data collection and analysis, adopting comparative methods, and checking and elaborating our tentative categories. We grounded theorists begin with a systematic inductive approach to inquiry but do not stop with induction as we subject our findings and tentative categories to rigorous tests.

Fundamentally, grounded theory is an iterative, comparative, interactive, and abductive method (Bryant & Charmaz, 2007; Charmaz, 2006, 2007, 2008e; Charmaz & Henwood, 2008). The grounded theory method leads researchers to go back and forth between analysis and data collection because each informs and advances the other. By asking analytic questions during each step in the iterative process, the researcher raises the abstract level of the analysis and intensifies its power. Using comparative methods throughout the analytic—and writing—processes sharpens a researcher's emerging analysis. Moreover, using a comparative approach in an iterative process keeps grounded theorists interacting with their data by asking analytic questions of these data and emerging analyses. Thus the strength of grounded theory not only resides in its comparative methodology but moreover, in its *interactive* essence (Charmaz, 2006, 2007, 2008a, 2008e, 2009b).

The method encourages researchers to become active, engaged analysts. Abductive reasoning keeps researchers involved. As grounded theorists, we engage in abductive reasoning when we come across a surprising finding during inductive data collection. Then we consider all possible theoretical accounts for this finding, form hypotheses or questions about them, and subsequently test these explanations with new data (Peirce, 1958; Reichert, 2007; Rosenthal, 2004). Abductive reasoning advances theory construction.

What does using the method involve? Grounded theory prompts us to study and interact with our data by moving through comparative levels of analysis. First, we compare data with data as we develop codes; next, we compare data with codes; after that, we compare codes and raise significant codes to tentative categories; then, we compare data and codes with these categories; subsequently, we treat our major category(ies) as a concept(s), and last, we compare concept with concept, which may include comparing our concept with disciplinary concepts. The analytic comparisons we make during our current phase of inquiry shape what we will do in the next phase and cannot be ascertained beforehand. The method prompts us to interact with our participants, data, codes, and tentative categories. Through these interactions, our nascent analyses emerge and take form (Charmaz, 2006, 2007, 2008b, 2008c, 2008e). This comparative, interactive process of inquiry leads us to move back and forth between data collection and analysis as each informs the other (Charmaz & Henwood, 2008). The grounded theory emphasis on theory construction influences how we interact with our participants and the questions we bring to the empirical world (see Charmaz, 2009a, 2009b).

The comparisons sharpen our analyses and the iterative data collection allows us to test our ideas and to check our emerging theoretical concepts. Grounded theorizing involves imaginative interpretations and rigorous examination of our data and nascent analyses (Charmaz, 2006; Kearney, 2007; Locke, 2007). Our systematic scrutiny not only increases analytic precision but also keeps us close to the data and, thus, strengthens our claims about it. Such an approach helps social justice researchers make their work visible and their voices heard.

In short, the logic of grounded theory involves fragmenting empirical data through coding and working with resultant codes to construct abstract categories that fit these data and offer a conceptual analysis of them (Charmaz, 2006; Glaser, 1978, 1998). Grounded theorists start with empirical specifics to move toward general statements about their emergent categories and the relationships between them. This approach allows social justice researchers to address problems in specific empirical worlds and to theorize how their categories may apply to other situations and iniquities (Dixon, 2007; Lutgen-Sandvik, 2008; Rivera, 2008; Shelley, 2001; Wolkomir, 2001).

▣ GROUNDED THEORY STRATEGIES IN SOCIAL JUSTICE INQUIRY

The analytic power of grounded theory offers qualitative researchers distinct advantages in pursuing social justice inquiry. Five grounded theory strengths make it a particularly useful toolkit for social justice researchers. First, this method contains tools for analyzing and situating processes. Thus, the logic of grounded theory leads to (1) defining relevant processes, (2) demonstrating their contexts, (3) specifying the conditions in which these processes occur, (4) conceptualizing their phases, (5) explicating what contributes to their stability and/or change, and (6) outlining their consequences. Adopting this logic can help social justice researchers attend to the construction of inequities and how people act toward them. Thus, grounded theory logic can lead researchers to make explicit interpretations of what is happening in the empirical world and to offer an analysis that depicts how and why it happens.

Second, grounded theory can aid researchers in explicating their participants' implicit meanings and actions (see, for example, McPhail & DiNitto, 2005). A task for social justice researchers is to see beyond the obvious. The most significant meanings and actions in a field setting are often implicit. Successive, meticulous grounded theory analysis can help researchers to define implicit meanings and actions and to theorize tentative but plausible accounts of them. Subsequently, the grounded theory guideline of checking hunches and conjectures encourages researchers to subject their tentative ideas to rigorous scrutiny and to develop more robust analyses.

Third, the purpose of grounded theory is to construct middle-range theory from data. Hence, grounded theory can aid social justice researchers to increase the abstract level of conceptualization of their analyses. Social justice researchers can then identify the conditions under which their categories emerge, specify relationships between these categories, and define the consequences. Thus, they can build complexity into their analyses that challenges conventional explanations of the studied phenomenon.

Fourth, the constructivist version of grounded theory attends to context, positions, discourses, and meanings and actions and thus can be used to advance understandings of how power, oppression, and inequities differentially affect individuals, groups, and categories of people. Last, but extremely significant, grounded theory methods provide tools to reveal links between concrete experiences of suffering and social structure, culture, and social practices or policies (Charmaz, 2007; Choi & Holroyd, 2007; Einwohner & Spencer, 2005; Rier, 2007; Sandstrom, 1990, 1998).

To date, few researchers who adopt grounded theory methods have explicitly framed their work as contributions to social justice inquiry or made social justice issues a central focus (but see Mitchell & McCusker, 2008; Sakamoto et al., 2009; Tuason, 2008). Implicit concerns about justice, however, form a silent frame in numerous grounded theory studies. Many researchers' studies assume the significance of social justice goals (see, e.g., Carter, 2003; Ciambrone, 2007; Hyde & Kammerer, 2009; Jones, 2003; Karabanow, 2008; Mcintyre, 2002; Roxas, 2008; Scott, 2005; Scott, London, & Gross, 2007; Wasserman & Clair, 2010), and other studies advance these goals through the content of their analyses (Frohmann, 1998; Quint, 1965; Sakamoto et al., 2009; Sixsmith, 1999; Swahnberg et al., 2007; Ullman & Townsend, 2008; Valdez & Flores, 2005; Veale & Stavrou, 2007). In keeping with grounded theory logic, social justice issues may arise through grappling with data *analysis* as well as through learning what is happening during data collection or starting from an explicit standpoint of pursuing social justice.

Researchers who pursue social justice goals enrich the contributions of development of the grounded theory method. Their attentiveness to context, constraint, power, and inequality advances attending to structural, temporal, and situational contexts in qualitative research generally and in grounded theory studies specifically. Social justice researchers are attuned to the silent workings of structure and power. They can offer grounded theorists important reminders of how historical conditions and larger social conditions shape current situations.

The critical stance of social justice inquiry combined with its structural focus can aid grounded theorists to locate subjective and collective experience in larger structures and increase understanding of how these structures work (Charmaz, 2005; Clarke, 2003, 2005; Maines, 2001; Rivera, 2008). The narrow focus and small size of many grounded theory studies have militated against the authors finding variation in their data, much less seeing how structure and historical process affect both the data and analysis. Like most qualitative researchers over the past 50 years, grounded theorists have often concentrated on overt processes and overt statements. A social justice standpoint brings critical inquiry to covert processes and invisible structures. Thus, we can discover contradictions between rhetoric and realities, ends and means, and goals and outcomes. This stance furthers understandings of the tacit, the liminal, and the marginal that otherwise might remain unseen and ignored, such as latent sources of conflict. The critical edge of social justice inquiry can help us subject our data to new tests and create new connections in our theories (Charmaz, 2005).

Recent grounded theory studies show increased engagement with social justice issues. To varying degrees, these studies address power, agency, structural constraints, resources, and analyze a wide range of questions including specific problems of impoverished, oppressed, stigmatized, and disenfranchised people (Choi & Holroyd, 2007; Ciambrone, 2007; Hyde & Kammerer, 2009; Mevorach, 2008; Ryder, 2007; Scott et al., 2007; Sixsmith, 1999; Tuason, 2008; Ullman & Townsend, 2008; Veale & Stavrou, 2007; Wilson & Luker, 2006; Wolkomir, 2001) as well as those that interrogate relationships between a social justice issue and social structure (Gunter, 2005; McDermott, 2007; Mitchell & McCusker, 2008). To date, the latter frequently emerge in the implications of studying a pressing issue or small group of people who suffer multiple and cumulative effects of iniquities (e.g., Dixon, 2007; Jiménez, 2008; Valadez, 2008; Wasserman & Clair, 2010; Wolkomir, 2001, 2006; Zieghan & Hinchman, 1999). Researchers in diverse disciplines and professions have primarily used grounded theory methods for small studies of individual behavior, as have researchers using other qualitative approaches. That does not, however, preclude adopting grounded theory to develop organizational and structural studies, as studies in organizations (O'Connor, Rice, Peters, & Veryzer, 2003; Scott, 2005; Vandenburgh, 2001) and the sociology of science have already shown (Casper, 1998; Clarke, 1998; Star, 1989). Grounded theory methods have gained a foothold in participatory action research (PAR) (Dick, 2007; Foster-Fishman, Nowell, Deacon, Nievar, & McCann, 2005; Kemmis & McTaggart, 2005; McIntyre, 2002; Poonamallee, 2009; Sakamoto et al., 2009; Teram, Schachter, & Stalker, 2005), a method that holds powerful potential for re-envisioning life and thus for advancing emancipatory change.

◨ RECONSTRUCTING GROUNDED THEORY

Grounded Theory as a Specific, General, and Generalized Method

Grounded theory is simultaneously a method that invokes specific strategies, a general method with guidelines that has informed qualitative inquiry, and a method whose strategies have become generalized, reconstructed, and contested. Strauss and Corbin (1994) observed over 15 years ago that grounded theory has become a general qualitative method. Grounded theory methodological strategies of simultaneous data collection and analysis, inductive coding, and memo writing have permeated qualitative research. Authors who claim to use grounded theory may, however, be conducting a more general form of qualitative research. Some authors' claims to be using grounded theory are attempts to legitimize inductive qualitative research; others result from naïve readings of the method. The abstract guidelines and dense writing in the early grounded theory texts led to misunderstandings of the method and confused readers (Piantanida, Tananis, & Grubs, 2004).

As grounded theory has become a general method, researchers may only adopt one or two grounded theory strategies (Foster-Fishman et al., 2005; Mitakidou, Tressou, & Karagianni, 2008). Other researchers may adopt more strategies but misunderstand them. And consistent with Virginia Olesen's (2007) statement about her work, some researchers may understand grounded theory strategies but their research questions and objectives lead them to combine grounded theory strategies with other qualitative approaches. Researchers frequently combine grounded theory strategies, especially coding, with narrative and thematic analyses (see, e.g., Cohn, Dyson, & Wessley, 2008; Hansen, Walters, & Baker, 2007; Harry, Sturges, & Klingner, 2005; Mathieson & Stam, 1995; Moreno, 2008; Salander, 2002; Sakamoto et al., 2009; Somerville, Featherstone, Hemingway, Timmis, & Feder, 2008; Tuason, 2008; Williamson, 2006; Wilson & Luker, 2006).

Naïve misunderstandings of grounded theory can prevent researchers from realizing its analytic power. In brief, misunderstandings about grounded theory arise in three main areas: coding, theoretical sampling,[7] and theory construction. I outline these misunderstandings and describe principles of grounded theory coding here but discuss them more thoroughly elsewhere (Charmaz, 2006, 2007, 2008b, 2008c). Cathy Urquhart (2003) questions whether grounded theory is, in essence, a coding technique. Coding is crucial but grounded theory is much more than a coding technique. Many researchers, however, use it for just that and appear to rely on CAQDAS (Computer Assisted Qualitative Data Analysis Software) to do their coding (see Bong, 2007, for problems of grounded theory coding with CAQDAS).

Grounded theory coding strategies include sorting, synthesizing, and summarizing data but, moreover, surpass these forms of data management. Rather, the fundamental characteristic of grounded theory coding involves taking data apart and defining how they are constituted. By asking what is happening in small segments of data and questioning what theoretical category each segment indicates, grounded theorists can take a fresh look at their data and create codes that lead to innovative analyses. By simultaneously raising questions about power and connections with larger social units, social justice researchers can show how data are constituted in ways that elude most grounded theorists.

Early grounded theory works (Glaser, 1978; Glaser & Strauss, 1967) lack clarity on what theoretical sampling means and how to conduct it. This lack of clarity combined with researchers' preconceptions of the term "sampling" created frequent misunderstandings. Theoretical sampling occurs *after* the initial data collection and analysis. It means sampling data to fill out the properties of an emergent conceptual category (Charmaz, 2006; Glaser, 1978, 1998; Morse, 2007). This strategy also helps a researcher to discover variation in the category and differences between categories. Thus, grounded theorists conduct theoretical sampling *after* they have developed tentative categories of data, *not* before they begin to collect data.

The objective of theoretical sampling is theory construction. Jane Hood (2007) contends that textbook authors often mistake theoretical sampling for purposive sampling, which sets criteria for representation of key attributes when planning initial data collection. Sharon Nepstad's (2007) methodological statement assumes this common misunderstanding: "Then I contacted staff members at these solidarity organizations and with their input I constructed a purposive theoretical sample (Glaser & Strauss, 1967) to ensure a diverse representation of geographic regions, age range, gender, and levels of participation in the movement" (p. 474).

In another area of common misunderstanding, many grounded theorists claim to construct theory but neglect to explicate what they assume theory encompasses. As I (Charmaz, 2006, p. 133) have argued, their assumptions about what constitutes theory suggest a range of meanings that include (1) a description, (2) an empirical generalization, (3) relationships between variables, and (4) an abstract understanding of relationships between concepts. If we define theory as either explaining the relationships between concepts or offering an abstract understanding of them, most studies that purport to have produced theory actually do not. Their authors assert that they construct theory but their analyses attest to their efforts to synthesize data and condense themes. Despite lofty claims to the contrary, most grounded theorists do not produce theory, although some move toward theory construction. And numerous authors produce mundane descriptions under the guise of doing grounded theory. The potential of grounded theory for theory construction has yet to be fully explored and exploited.

Grounded Theory as a Specific Method

We can discern convergent approaches between recognized grounded theorists that undergird grounded theory as a specific method. *How* grounded theorists use their methodological strategies differs from other qualitative researchers who study topics and structures instead of actions and processes. How we collect data and what we do with it matters. Research *actions* distinguish grounded theory from other types of qualitative inquiry (Charmaz, 2010). Grounded theorists representing each version engage in the following actions:

1. Conduct data collection and analysis simultaneously in an iterative process

2. Analyze actions and processes rather than themes and structure

3. Use comparative methods

4. Draw on data (e.g. narratives and descriptions) in service of developing new conceptual categories

5. Develop inductive categories through systematic data analysis

6. Emphasize theory construction rather than description or application of current theories

7. Engage in theoretical sampling

8. Search for variation in the studied categories or process

9. Pursue developing a category rather than covering a specific empirical topic (Charmaz, 2010)

Researchers who engage in the first five actions give their studies a distinctive analytic cast that differs from that of other qualitative works, particularly those that remain descriptive. Studies by grounded theorists reach across individuals and events to reveal a collective analytic story. Detailing conceptual categories takes precedence over participants' accounts and summarized data. A grounded theorist presents excerpts and summaries of data to demonstrate the connection between data and category and to offer evidence for the robustness of the category. A much smaller number of researchers engage in the remaining actions but they move their analyses into theory construction (Charmaz, 2010).

Despite agreement about these nine research actions, what stands as a bona fide grounded theory study may remain ambiguous (Charmaz, 2008e, 2010; Timmermans & Tavory, 2007). Methods statements in published works seldom address analytic strategies, much less detail them.[8] Certain studies such as Qin and Lykes (2006), Roschelle and Kaufman (2004), and Wolkomir (2001) illustrate distinctive grounded theory logic because they conceptualize a problematic process, construct analytic categories from inductive, comparative coding of data, define the properties of the categories, specify the relationships between categories, and outline the consequences of the processes.

If readers cannot discern distinctive grounded theory logic in the analysis, then it becomes difficult to determine whether authors' claims to using grounded theory methods are mistaken or are aimed to legitimize inductive qualitative research. Nonetheless, some authors' analyses may not indicate a grounded theory approach but their methodological descriptions reveal a sophisticated understanding of the method. Consider Henry Vandenburgh's (2001) statement in his study of organizational deviance:

> I followed the stages suggested by Turner (1981) in interpreting Strauss, first developing categories that used available data to suggest nominal classifications fitting these data closely. I then saturated these categories by accumulating all of the examples I could from my interview data that fit each category. Next, I then abstracted a definition for each category by stating the criteria for putting further instances of this specific type of phenomena into the category. I continued to use the categories by making follow-up calls based upon some of the questions raised. I then further exploited the categories by inspecting them to see if they suggested additional categories, suggested more general or specific instances, or suggested their opposites. I noted and developed links between categories by becoming aware of the patterned relationships between them, and by developing [a] hypothesis about these links. Finally, I considered the conditions under which the links held by theorizing about these relationships and the contexts that conditioned them. I then made conditions to existing theory. (2001, p. 62)

Like Vandenburgh, other grounded theorists may reveal their use of the method in their methodological discussions rather than in their analyses. Monica Casper (1997, 2007) and Robert Thornberg (Thornberg & Charmaz, in press) each have telling discussions that illuminate their studies (see Casper, 1998; Thornberg, 2007, 2009).

Grounded Theory as Contested From Within

Grounded theory is a contested method from both within and without (see Boychuk, Duchscher, & Morgan, 2004; Charmaz, 2006, 2009a; Kelle, 2005).[9] The contested status of the method further complicates what stands as a grounded theory study today. Since its inception in 1967, the grounded theory method has undergone both clarification and change by all of its major proponents. Grounded theory has become an evolving *general* qualitative method with three versions: constructivist, objectivist, and postpositivist. Major texts that teach readers how to use grounded theory represent each version of grounded theory (Bryant & Charmaz, 2007; Charmaz, 2006; Corbin & Strauss, 2008; Glaser, 1978, 1998; Strauss & Corbin, 1990, 1998).

Constructivist grounded theory adopts the methodological strategies of Glaser and Strauss's classic statement but integrates relativity and reflexivity throughout the research process.

As such, this approach loosens grounded theory from its positivist, objectivist roots and brings the researcher's roles and actions into view. Constructivist grounded theory uses methodological strategies developed by Barney Glaser, the spokesperson for objectivist grounded theory, yet builds on the social constructionism inherent in Anselm Strauss's symbolic interactionist perspective (Charmaz, 2006; 2007, 2008). Constructivist grounded theory views knowledge as located in time, space, and situation and takes into account the researcher's construction of emergent concepts.

Objectivist grounded theory shares an emphasis on constructing emergent concepts but emphasizes positivist empiricism with researcher neutrality while aiming for abstract generalizations independent of time, place, and specific people (Glaser, 1978, 1998, 2001). Unlike many positivists of the past, however, Glaser evinces little concern for establishing criteria for data collection or for evaluating its quality. He maintains that "all is data" (2001, p. 145) but leaves unexamined what researchers may define as "all." For Glaser, a concern with data reflects the "worrisome accuracy" (Glaser, 2002, para. 2) characterizing the conventional qualitative research that he argues against. Phyllis Noerager Stern (2007), a major proponent of Glaserian grounded theory, finds that a small number of cases is sufficient to saturate the researcher's emerging analytic categories. Glaser contends that examining many cases through the comparative process renders data objective. His previous view that research participants will tell researchers their main concern about what is happening in their setting (Glaser, 1992) likely contributed to the notion of discovering theory in the data, as though it simply resided there.[10] I have long argued that we cannot assume that participants' overt statements represent the most significant data (Charmaz, 1990, 1995, 2000). Instead their statements may take for granted fundamental processes that shape their lives or provide a strategic rhetoric to manage an impression (Charmaz, 1990, 2000, 2008f). Constructivist grounded theory contrasts with its objectivist predecessor in several fundamental ways, as I indicate above and summarize below. Postpositivist grounded theory (Corbin & Strauss, 2008; Strauss & Corbin, 1990, 1998) takes a middle ground between the two versions. It places less emphasis on emergence than the objectivist and constructivist approaches, as it provides preconceived coding and analytic frameworks to apply to data. Yet postpositivist grounded theory views reality as fluid, evolving, and open to change. Strauss and Corbin's early books made grounded theory a method of application rather than innovation (Charmaz, 2007).

In her recent reflection, however, Juliet Corbin (2009) outlines how her approach to research has changed. She describes having been imbued with methodological prescriptions of earlier decades that shaped writing the first two editions of *Basics of Qualitative Research* (Strauss & Corbin, 1990, 1998). These prescriptions led qualitative researchers to (1) study data to find the theory embedded in them; (2) maintain objectivity; (3) avoid "going native"; and (4) capture a semblance of "reality" in data and present them as "theoretical findings," while simultaneously believing that no one truth existed (Corbin, 2009, pp. 36–37). Corbin's list combined with the technical procedures in Strauss and Corbin's (1990, 1998) *Basics of Qualitative Research* confirms my earlier contentions (Charmaz, 2000, 2002) that their earlier editions contain objectivist threads. Corbin (2009) now, however, endorses engaging in reflexivity, takes a value stance that furthers social justice, believes in multiple realities, and disavows rigid application of technical procedures. These changes mark the updated third edition of *Basics of Qualitative Research* (Corbin & Strauss, 2008) and bring it closer to constructivist grounded theory.

The three versions of grounded theory share commitments to conceptualizing qualitative data through analyzing these data, constructing theoretical analyses, and adopting key grounded theory strategies. Each version emphasizes systematic inquiry using transparent strategies, begins with an inductive logic, emphasizes constructing theory, and aims to construct useful analyses for research participants, policy makers, and relevant practitioners (Charmaz, 2009b). Which strategies each version adopts, creates, or discards often differ in crucial ways that reflect more than favored or disfavored techniques. Differences in epistemology and ontology come into play.

Epistemological Differences in Versions of Grounded Theory

Grounded theory contained the seeds of divergence from its beginnings. Glaser's Columbia University positivism and theoretical background in structural-functionalism[11] and Strauss's University of Chicago pragmatism[12] drew on conflicting philosophical and methodological presuppositions about the nature of reality, objectives of inquiry, and the research process and practice. The legacy of Anselm Strauss rests on pragmatism and its development in symbolic interactionism (Charmaz, 2008d). Differences between a grounded theory informed by positivism and one informed by pragmatism appear most starkly in the contrast between objectivist grounded theory and constructivist grounded theory articulated in the second edition of this handbook (Charmaz, 2000; see also Bryant, 2002; Bryant & Charmaz, 2007, in press; Charmaz, 2002, 2006, 2007, 2008e, 2009b; Charmaz & Bryant, 2011; Charmaz & Henwood, 2008).

Objectivist grounded theory assumes that a neutral observer discovers data in a unitary external world. In this view, researchers can separate their values from "facts" residing in this world and suggests what Kelle (2007) calls "epistemological fundamentalism" (p. 205). In this approach, data gathering does not raise questions about researchers' tacit assumptions, privileged statuses, or the particular locations from which they view studied life. The researcher stands outside the studied phenomenon. Data are "there" rather than constructed. Researchers can add reflexivity about data collection and their roles, if they wish. Ordinarily, however, the neutral but passive observer simply gathers data to analyze as the authoritative expert and active analyst. In the objectivist logic, the number of cases corrects the researcher's possible biases. This approach gives priority to the researcher's voice and analysis and treats the researcher's representation of participants as straightforward, not as inherently problematic. A hazard is that researchers may import their unacknowledged presuppositions into the research process and product. Objectivist grounded theory aims for parsimonious abstract generalizations about relationships between variables that explain empirical phenomena. These generalizations constitute a middle-range theory explaining the studied phenomenon.

Constructivist grounded theory adopts a contrasting relativist approach that shifts its ontological and epistemological grounds (Charmaz, 2009b) and aligns them with the pragmatist tradition of Anselm Strauss (see Charmaz, 2008a, 2008d, 2009b; Reichert, 2007; Strübing, 2007). Here, realities are multiple and the viewer is part of what is viewed. Subjectivities matter. Values shape what stands as fact. To the extent possible, constructivist grounded theorists enter the studied phenomenon and attempt to see it from the inside. Researchers and participants co-construct the data through interaction. Data reflect their historical, social, and situational locations, including those of the researcher (Charmaz, 2009a, 2009b). Representations of the data are inherently problematic and partial. These concerns involve constructivist grounded theorists in reflexivity throughout inquiry as an integral part of the research process (see also Mruck & Mey, 2007; Neill, 2006). Rather than aiming for theoretical generalizations, constructivist grounded theory aims for interpretive understanding. The quest for generalizations erases difference and obscures variation (see also Clarke, 2003, 2005, 2006; Clarke & Friese, 2007). For constructivists, generalizations remain partial, conditional, and situated. Moreover, generalizations are not neutral. As Norman Denzin (2007) avows, interpretation is inherently political.

All these contrasts alter the processes and products of inquiry, as do differences in grounded theory practice, such as the contested place of the literature review. Glaser (1978, 1998, 2003) advocates conducting the literature review after developing an independent analysis to avoid forcing the data into preconceived categories and theories. However, few doctoral students and professional researchers begin their studies without knowledge of their fields (Charmaz, 2006; Lempert, 2007). They must include thorough literature reviews in dissertation proposals, grant applications, and today even in some human

subjects IRB (institutional review board) applications. Karen Henwood and Nick Pidgeon's (2003) concept of theoretical agnosticism makes more sense than theoretical innocence. They argue that researchers need to subject all possible theoretical explanations of a phenomenon to rigorous scrutiny—whether from the literature or their own analysis. Perhaps most significantly, constructivist grounded theorists contend that researchers' starting points and standpoints, including those occurring throughout inquiry, influence the research process and product.

▣ GROUNDED THEORY IN MIXED METHODS SOCIAL JUSTICE INQUIRY

Researchers have identified grounded theory as a useful qualitative method to adopt in mixed methods research. Despite the growing number of studies that purport to use grounded theory studies in mixed methods to increase knowledge of the studied topic, few of these studies have a clear focus on social justice. The place of grounded theory in mixed methods social justice inquiry has yet to be developed. Thus, I offer brief concerns about mixed methods here that grounded theory social justice researchers might consider.

Mixed methods research usually means using both quantitative and qualitative methods to gain more, and a more nuanced, analysis of the research problem. Definitions of mixed methods and what they mean for inquiry are, however, contested and multiple. For mixed methods specialists (see, e.g., Cameron, 2009; Creswell, 2003; Morgan, 2007), the rapid rise of mixed methods is a movement heralding a paradigm shift analogous to the qualitative revolution that Denzin and Lincoln proclaimed in 1994. For many researchers, mixed methods are tools that produce findings, regardless of their reasons for adopting these tools, analyzing the findings, and deciding whether and to what extent to use each of the subsequent analyses. For a few researchers, mixed methods simply means using more than one method, whether or not these methods mix quantitative and qualitative research.[13] R. Burke Johnson, Anthony J. Onwuegbuzie, and Lisa A. Turner (2007) view mixed methods as combining qualitative and quantitative approaches including their respective perspectives, analyses, and forms of inference. They point out that mixed methods mean combining elements of methods for "breadth and depth of understanding and corroboration" (p. 123).

The discussion of mixed methods takes into account Norman Denzin's (1970) early call for triangulation (Greene, 2006; Morse, 1991; Tashakkori & Teddlie, 2003). Numerous researchers advocate methodological pluralism. Others take a more skeptical view and see the quantitative data and analysis as not only dominating mixed methods projects, but also "quantitizing" qualitative data by transforming them into numbers (Sande-lowski, Voils, & Knafl, 2009).[14] However, Creswell, Shope, Plano Clark, and Green (2006) and Creswell and Plano Clark (2007) argue that qualitative methods extend mixed methods practice and may be given priority in mixed methods projects. In practice, researchers use mixed methods for varied purposes including to (1) construct instruments, (2) corroborate findings, (3) reduce cultural and investigator biases, (4) improve clinical trials, (5) address research participants' experience, (6) demonstrate credibility, (7) increase generalizability, and (8) inform professional practice and/or public policy.

Questions arise about integrating the results and analyses in mixed methods studies. To what extent should integration of quantitative and qualitative findings be a major methodological goal? What should be done when the qualitative and quantitative data have conflicting results? Bryman (2007) finds that mixed methods researchers often dismiss the qualitative analyses. He contends that "the key issue is whether in a mixed methods project, the end product is more than the sum of the individual quantitative and qualitative parts" (p. 8). In practice, that may not occur, nor may it have been the researchers' intent, as Bryman observes.

Mixed methods research designs often consist of complicated procedures and hence require team efforts. Grounded theory mixed methods projects are steadily increasing in fields such as education and health in which funded team research is common. Social justice research, especially in its explicit forms, is less likely to be a funded team project staffed by an array of methodological specialists having different but complementary skills. Few researchers are equally skilled in both quantitative and qualitative methods. Social justice research is likely to be an unfunded individual pursuit or a participatory action research project in which the researchers are members of and responsible to local communities.

Thomas W. Christ (2009) correctly observes that the goals of mixed methods research typically contrast with those in transformative inquiry in which social justice goals dominate. As he points out, researchers conduct critical and transformative research "to improve communities or reduce oppression, not to generalize results from a non-representative sample to a larger population" (p. 293). However, Donna Mertens (2007, 2010) argues eloquently for using mixed methods in a transformative paradigm to further social justice and Deborah K. Padgett (2009) states, "Social justice values do not have to be sidelined" (p. 101). Their purposes are explicit rather than hidden under a bland—and exclusive—term like "public sociology" (Burawoy, 2004).

Because social justice researchers may face skeptical audiences, presenting multiple forms of data in an integrated analysis may buttress their reports. The emerging philosophical foundations for mixed methods would support their efforts. Discussions

are occurring that position mixed methods in pragmatism, and thus fit grounded theory research in social justice (see, e.g., Duemer & Zebidi, 2009; Feilzer, 2010; Morgan, 2007).

Researchers in education are among the most attuned to using grounded theory in mixed methods studies for social justice goals. As is evident in other studies, however, other researchers may assume, rather than state, social justice goals and use grounded theory in limited or extensive ways. To cite one interesting example, Sahin-Hodoglugil et al. (2009) used mixed methods in a randomized controlled clinical trial to study the effect of a low-cost HIV prevention method, the diaphragm. This method gave women control because they could use it without their male partners' knowledge. The authors invoked an iterative process in which both quantitative data and qualitative findings informed each other. Sahin-Hodoglugil et al. used insights about covert diaphragm use from the qualitative analysis to inform the analytic framework for the quantitative data and then subsequently explored some findings in the quantitative portion by gathering qualitative data. These researchers discovered that covert use was more complicated than they had anticipated and occurred along a continuum with disclosure.

In short, social justice researchers who can bring multiple types of solid data to their analyses make their reports less easy to dismiss. The test of mixed methods studies resides in doing credible work in all adopted methods to answer the research questions, fulfill the research goals, and convince relevant audiences of the significance of the reports.

▣ USING GROUNDED THEORY STRATEGIES IN SOCIAL JUSTICE INQUIRY

In this section, I offer several specific examples of grounded theory in practice. Coding, memo writing, theoretical sampling and saturation, sorting memos are all part of the process. These grounded theory strategies have been described elsewhere in detail (Charmaz, 1990, 2001, 2002, 2005, 2006; Corbin & Strauss, 2008; Glaser, 1978, 1998, 2001, 2003; Strauss, 1987; Strauss & Corbin, 1990, 1998), so I merely introduce several examples of how constructing grounded theory analyses animate social justice inquiry.

Coding for Processes

By using gerunds to code for actions, grounded theorists make individual or collective action and process visible and tangible. Social justice researchers can use grounded theory coding strategies to show how people enact injustice and inequity.[15] Gerunds define actions and enable grounded theorists to envision implicit actions and to identify how they are linked (see, e.g., Schwalbe, 2005; Schwalbe, Godwin, Schrock, Thompson, & Wolkomir, 2000).

Coding data for actions and mining the theoretical potential of both data and codes make grounded theory distinctive (Charmaz, 2006, 2008b, 2008c). Coding with gerunds pinpoints actions and thus helps grounded theorists to define what is happening in a fragment of data or a description of an incident. Gerunds enable grounded theorists to see implicit processes, to make connections between codes, and to keep their analyses active and emergent. In contrast, coding for topics and themes helps the researcher to sort and synthesize the data but neither breaks them apart as readily as grounded theory coding for actions nor fosters seeing implicit relationships between topics and themes.

Line-by-line coding, the initial grounded theory coding with gerunds, is a heuristic device to bring the researcher into the data, interact with it, and study each fragment of it (see Box 21.1). This type of coding helps to define implicit meanings and actions, gives researchers directions to explore, spurs making comparisons between data, and suggests emergent links between processes in the data to pursue and check. The data excerpts in the textboxes tell the story of a middle-aged woman with lupus erythematosus whose friends rushed her to a doctor to reassess her

medications during a medical crisis. These medications often cause multiple side effects including confusion, depression, blurred vision, and inappropriate emotional responses. After two hospital transfers, this woman's medical crisis became redefined as a psychiatric crisis. Subsequently, her claims of having physical symptoms were unacknowledged and her requests for lupus medications went unheeded. She aroused the doctor's ire when he discovered another patient helping her complete the detailed confidential intake survey. Getting help with the survey broke hospital rules but this woman's vision problems meant that she could not read the survey and so just filled circles randomly after her doctor forbade her from having help. Her actions in one incident after another made sense given her situation, but fit neither hospital protocol nor her treatment program. Nevertheless, she attempted to present her views and to become her own advocate while her illness worsened. A psychiatrist who was unaware of her medical history or ignored it could invoke the same incidents as justifying his treatment approach. In this case, the grounded theory codes chronicle the woman's progressive loss of control over her life and her illness. Thus, the initial codes in the excerpt become the details substantiating a more general code, "resisting spiraling powerlessness."

Box 21.1a Initial Coding for Topics and Themes

Examples of Codes	Narrative Data to Be Coded
Friends' support Hospitalization Conflict with doctor	P: They called the clinic to see if they could see me, if they would reevaluate some of my meds and stuff, and they said, "Oh yeah." When I got there they decided that they were going to put me in, put me away or whatever. And I ended up with a really bad doctor. Really bad. I even brought charges against him, but I lost.
	I: What did he do?
Hospital transfer Loss of choice of doctor Conflict with doctor Physician control Threats Powerlessness Lack of physical care	P: They put me in this one place, then the next day they sent me over to West Valley [hospital 60 miles away], and they didn't have any female doctors there, they only had male, so you didn't have a choice, and you get one and that's who you get the whole time you're there. For some reason he just took a disliking, I guess, and I tried to tell him about some of the problems I had with my Lupus and stuff, and angered him. [He had ordered her to take off her dark glasses.] And I wore [dark] glasses all the time and I tried to tell him, you know, that if he would turn off the fluorescent lights, I would take off the glasses. And he felt I was just being stubborn. I gave him the name and number of my doctor that makes the glasses and he just ripped it up in front of me and threw it away. And you have to go to group sessions all the time while you're there. I went, I just didn't speak to anybody. But I went. I went to everything they said I had to. And everyday he'd say that he was going to lock me up again. After ten days he did. He called me in to this little room and there were these two big guys there and they grabbed me and put me on a gurney and tied me down. And he sent me to a lockup ward, wouldn't let me make a call or do anything. And my potassium was down really bad and, I mean my whole–I was *sick*. They weren't giving me the pills that I needed and he just wouldn't even acknowledge that I had Lupus. It was bad.

Box 21.1b   Initial Grounded Theory Coding

Examples of Codes	Initial Narrative Data to Be Coded
Receiving friends' help in seeking care Requesting regimen reevaluation Gaining medical access Being admitted to hospital Getting a "bad" doctor Taking action against MD Being sent away Preferring a female MD Losing choice; dwindling control Getting stuck with MD Accounting for MD's behavior Trying to gain a voice—explaining symptoms Remaining unheard Asserting self Attempting to bargain Being misjudged Countering the judgment Offering evidence, being discounted Facing forced attendance Maintaining silence "Following" orders Receiving daily threats MD acting on threat Being overpowered Experiencing physical constraints Experiencing immediate loss of control Witnessing one's deterioration Being denied medication MD rejecting illness claims	P: They [her friends] called the clinic to see if they could see me, if they would reevaluate some of my meds and stuff, and they said, "Oh yeah." When I got there they decided that they were going to put me in, put me away or whatever. And I ended up with a really bad doctor. Really bad. I even brought charges against him, but I lost. I: What did he do? P: They put me in this one place, then the next day they sent me over to West Valley [hospital 60 miles away], and they didn't have any female doctors there, they only had male, so you didn't have a choice, and you get one and that's who you get the whole time you're there. For some reason he just took a disliking, I guess, and I tried to tell him about some of the problems I had with my Lupus and stuff, and angered him. [He had ordered her to take off her dark glasses.] And I wore [dark] glasses all the time [because of her photosensitivity] and I tried to tell him, you know, that if he would turn off the fluorescent lights, I would take off the glasses. And he felt I was just being stubborn. I gave him the name and number of my doctor that makes the glasses and he just ripped it up in front of me and threw it away. And you have to go to group sessions all the time while you're there. I went, I just didn't speak to anybody. But I went. I went to everything they said I had to. And everyday he'd say that he was going to lock me up again. After ten days he did. He called me in to this little room and there were these two big guys there and they grabbed me and put me on a gurney and tied me down. And he sent me to a lockup ward, wouldn't let me make a call or do anything. And my potassium was down really bad and, I mean my whole—I was *sick*. They weren't giving me the pills that I needed and he just wouldn't even acknowledge that I had Lupus. It was bad.

Comparing the codes in the boxes demonstrates that grounded theory codes preserve the character of the data, provide a precise handle on the material, and point to places that need further elucidation. Coding gives the researcher leads to pursue in subsequent data collection. In vivo codes use research participants' terms as codes to uncover their meanings and understand their emergent actions. Zieghan and Hinchman's (1999) in vivo codes: "breaking the ice," "figuring out how to help," "trying to understand" gave form to their study of college students who tutored adult learners. Note that they code in gerunds and thus portray the tutors' actions as they wrestled with dealing with their situations. Despite the student tutors increased

awareness of poverty and lack of opportunity, the authors learned "that the border between campus life and the adult literacy community is a site of reproduction rather than transformation" (p. 99).

If coding in gerunds is so fruitful, why don't more researchers use them? In my view, the English language favors thinking in structures, topics, and themes rather than thinking in actions and processes. In addition, Strauss and Corbin's (1990, 1998) books have instructed thousands of researchers but emphasize gerunds less than Charmaz's (1990, 2006, 2008c) and Glaser's (1978, 1998) works. Many researchers report beginning with open coding, the initial coding in which the researcher examines and categorizes the data. Some turn next to axial coding, a type of coding to relate categories to subcategories, or to thematic coding but do not build fresh conceptual categories. Those following Strauss and Corbin (1990, 1998) often adopt complicated coding procedures to generate themes (Ball et al., 2009; Morrow & Smith, 1995; Sakamoto et al., 2009; Ullman & Townsend, 2008). Ullman and Townsend's (2008) coding procedures generated themes such as "Definitions of Feminist/Empowerment Approaches," "Importance of Control," "Techniques for Empowerment," and "Advocate Versus Agency Orientations," rather than a theory or conceptualized process. Although many authors found Strauss and Corbin's coding procedures to be helpful, some like Judy Kendall (1999) did not. She states, "I became so distracted by working the model to its natural conclusion that I stopped thinking about what the data were telling me in regard to the research question" (p. 753).

Grounded theory coding need not be complex. By engaging in thorough coding early in the research process and comparing data and codes, the researcher can identify which codes to explore as tentative categories. In turn, selecting categories expedites inquiry because the researcher then uses these categories to sort large batches of data. This approach is particularly useful in social justice research projects that address pressing social issues and policies. Grounded theory coding preserves empirical detail and simultaneously moves the project toward completion.

Considering CAQDAS

Increasingly, grounded theorists turn to one of the CAQDAS programs for coding the data. Several CAQDAS programs aim to be compatible with grounded theory logic and treat qualitative inquiry as interchangeable with grounded theory, or their conception of grounded theory. Software developers may have been criticized for their reliance on grounded theory. Ironically, however, their products may fit general qualitative coding for topics and themes more than coding for processes and engaging in comparative analysis. Grounded theory coding involves more than merely applying labels, identifying topics, and labeling themes, although many researchers do not realize it.

Depending on the grounded theorist's skills and objectives, advantages of using CAQDAS may include (1) relative ease of searching, retrieving, sorting, separating, and categorizing data and codes, (2) the ability to work at multiple levels of analysis simultaneously, (3) visibility of both the data and analytic processes, (4) document-sharing capacities for team research, and (5) management and organization of the data and emerging analysis. Since the advent of CAQDAS, numerous researchers (see, e.g., Fielding & Lee, 1998; Glaser, 2003; Weitzman, 2000) have raised varied questions concerning the conceivable advantages of using CAQDAS. Their concerns included users becoming too close or too distanced from their data, software design driving the analysis, and users being able to produce results without understanding the analytic process or range of analytic approaches. CAQDAS has gained much wider audiences since the early years and the software supports more intricate functions (Fielding & Lee, 2002). As the software becomes more sophisticated, its effects on knowledge production may change. Bringer, Johnston, and Brackenridge (2006) state that effective use of the grounded theory method holds greater significance than whether or not a software package is adopted. Konopásek (2008) avers that "The software . . . extends the researcher's mental capabilities to organise, to remember, and to be systematic. But while doing so it essentially remains a stupid instrument" (para. 2). Yet for Konopásek, the software not only consists of tools but also a virtual environment in which a set of mediations and embodied and practice-based knowledge production occurs.

Udo Kelle (2004) argues that CAQDAS requires its users to explicate their data management strategies and subsequently to think about their methodological and epistemological significance (p. 473). Does it? To what extent? Kelle acknowledges that CAQDAS could be viewed as a step in the rationalization and mechanization of qualitative research with the production of trivial results. The plethora of simplistic CAQDAS papers written under the guise of grounded theory confirms that this problem exists. However, Kelle contends that CAQDAS helps users to gain clarity about the processes involved in theory construction. Yet many researchers use CAQDAS—and grounded theory—primarily for coding without venturing into theory construction. Nonetheless, Kelle presents a worthy goal. But first, researchers need to learn to use the method.

Defining Extant Concepts Through Data Analysis

Social justice researchers use concepts that reflect structural arrangements and collective forces. They could adopt grounded theory strategies to refine or redefine these concepts by their empirical properties. In this sense, they can subject a sensitizing concept to rigorous *empirical* analysis. In the following example, I examined my data about the experience of illness and used marginalization as a sensitizing concept (Blumer, 1969; van den Hoonaard, 1997) from which to begin analysis. A 46-year old woman I call Marilyn (Charmaz, 2008f) recounted becoming ill and disabled with chronic fatigue syndrome and environmental illness.

Marilyn compares and contrasts her now unending saga of illness with the story of her past successes.

> I did a lot of things that were very challenging, and, you know, I used to work 50, 60 hours a week, and made good money and had great benefits, and had a life and all of that stuff changed one year—really abruptly in one year. And since then, everything is gone—from the financial to memory to—and everything in between. (p. 7)

Marilyn planned to testify at a city council meeting to restrict wood burning because of its devastating effects on people with lung disease, asthma, and chemical sensitivities. Because she anticipated possible discomfort from odors in a packed, closed room, she arrived 15 minutes before the time for community members to speak, but she had a long wait. Marilyn recounted:

I had to wear my mask and by two hours the charcoal [filter in the mask] was shot and by the time I got up to speak, you know, my voice was going, my brain was starting to go, I was having problems formulating words, so it's, and then of course when you wear a mask, people, you know, you've seen mothers kind of pull their kids close to them. (pp. 8–9)

In her first statement, Marilyn is making identity claims about who she had been. Her second statement reveals the combined effects of appearance and of losing control over timing. The subsequent increasing visibility of Marilyn's difference led to her being discredited, devalued—othered. As I coded data and compared incidents, it was apparent that visible difference marginalized people with illnesses and disabilities. But what did marginalization mean? Through coding data, I identified properties of marginalization that linked its social origins with subjective experience and thus stated, "Marginalization means boundaries or barriers, distance or separation, and division or difference. Disconnection, devaluation, discrimination, and deprivation exemplify experiences of marginalization" (Charmaz, 2008f, p. 9). Moreover, I showed how people enacted marginalization.

Using grounded theory strategies in similar ways can assist social justice researchers to infuse taken-for-granted concepts with specific meanings.[16] Furthermore, it can help them avoid reifying, objectifying, and universalizing ideas without putting them to the test.

Developing Categories and Discovering Meanings

Grounded theory has long been touted as a method of discovery—of data and of theory. The constructivist critique argues that such "discoveries" are constructions located in space, time, and situation. Yet grounded theory can give us tools for constructing new understandings. Learning how participants define their situations, attempting to grasp what they assume, and understanding the problems that confront them become major sources of our "discoveries" and of the categories through which we conceptualize these discoveries. In his study of street youth, Jeff Karabanow (2008) discovered what leaving the street meant to these young people. He explains a crucial category and part of the process of leaving the streets as "cutting street ties."

Cutting street ties meant leaving friends, surrogate families, and a culture associated with the downtown core. For many young people, friends and surrogate families were forged as a result of, or during, very stressful survival situations. (p. 781)

Karabanow's category speaks to meanings of the past as well as of the present and future. Cutting street ties meant more than merely leaving the streets. It occurred in a complex context in which the streets often held greater appeal to the young people than having shelter. Karabanow's analysis has resonance and power not only because of the clarity of his analysis but also because of the strength of his data: 128 interviews with street youth, 50 interviews with service providers, lengthy experience with the topic, and the help of two research assistants who lived on the streets.

Note that Karabanow's category, cutting street ties is a precondition for the larger process of leaving the streets. Grounded theory analyses gain this kind of specificity when researchers scrutinize the data to explicate the conditions that produce the studied process or phenomenon. In their study of women who had suffered childhood sexual abuse, Susan L. Morrow and Mary Lee Smith (1995) looked for causal conditions that led to the abuse. They identify two strategies that their research participants had used to cope with it: "keeping from being overwhelmed by threatening or dangerous feelings," and "managing helplessness, powerlessness, and lack of control" (pp. 27–28). Morrow and Smith find that these women had had few resources available for help and thus had adopted psychological strategies that focused inward on self and emotions such as reducing the intensity of troubling feelings, avoiding or escaping feelings, or using self-induced physical pain to override emotional pain.

As Morrow and Smith learned what these women had done to cope with their situations, they also learned the meanings the women had held. Social justice inquiry often focuses on people who experience horrendous coercion and oppression. Not surprisingly their categories reflect the untenable situations they observe. Angela Veale and Aki Stavrou (2007) studied the reintegration of Ugandan child abductees who had been forced to fight against the Uganda People's Defense Force (UPDF), their own people. Veale and Stavrou state that the child abductees are part rebel soldier, part a child of his or her village but yet identified as "external to it—as the aggressor" (p. 284). Veale and Stavrou's category depicting this conflicted identity is "Managing Contradictions." Veale and Stavrou state,

The UPDF is Ugandan, and fighting the UPDF is a source of sadness, because they fought to kill the enemy in order to survive. Victor expressed the conflict as follows:

Victor: When fighting against the Ugandan Army, I felt partly as army, partly as civilian.

Interviewer: You used to steal food from Ugandan families. What did you feel when you steal food?

Victor: I feel very bad 'cause the food I am going to steal is [from] my father or guardian, my brothers' or sisters' guardian.

For these youth, this dual role as soldier–abductee
could not be resolved. (p. 285)

The irony of Veale and Stavrou's category resides in the impossibility of resolution: The contradictions exceeded the confronted reality and yet these very contradictions were the reality.

Each of the categories discussed above remain close to the studied experience and address what is happening in the data. Grounded theorists construct theoretical categories, in contrast, as they ask what theoretical questions and concepts the data indicate and Michelle Wolkomir's (2001) analysis suggests below.

Conceptualizing a Process

The emphasis on coding in action terms enables grounded theorists to discern processes that might otherwise remain invisible. Scrutinizing these processes can help social justice researchers refine their concepts, form nuanced analyses, see how powerful cultural scripts are acted upon, and become attuned to possibilities for change. In her study of gay and ex-gay Christian men in a support group, Michelle Wolkomir (2001) outlines how the men engaged in "ideological maneuvering" (p. 407) to evade and subvert Christian ideology that condemned their sexuality and viewed them as "egregious sinners" (p. 408). She argues that such ideological revision requires sustained effort, particularly when conducted by marginalized groups without power.

Consistent with a grounded theory emphasis on analyzing social and social psychological processes, Wolkomir's major conceptual category, ideological maneuvering, is a process. She developed her analysis of this process through studying the men's actions and observing the tensions they faced from their perspectives. How could they avoid stigma and claim moral Christian identities? Wolkomir's (2001) guiding analytic question for her article asks, "Under what conditions is such change [ideological change] likely to occur, and how is it accomplished?" (p. 407). By raising such questions and defining these conditions, Wolkomir brings analytic precision to her analysis. Moreover, her work provides a theoretical concept that can be transported and tested in other empirical studies.

Wolkomir's article reveals the underpinnings of her grounded theory analysis while simultaneously providing an insightful analysis of the overall process and major conceptual category. Wolkomir states that the process of ideological maneuvering entails three subprocesses: (1) "selective dismantling of existing ideology to open new interpretive space; (2) constructing a new affirming ideology; and (3) authenticating new self-meanings" (p. 408). She treats these subprocesses as analytic categories and then demonstrates the actions constituting each one. Note that Wolkomir's categories are active, specific, and rooted in the data. Her categories depict how the men dealt with the Christian ideology that condemned and excluded them. Wolkomir found that for one support group, dismantling the existing ideology explicitly included "redefining sin" (p. 413). These men discovered new scriptural reasons to believe that the significance of homosexual sin had been exaggerated and "concluded that their homosexual sin was no worse than selfishness or gossip" (p. 414).

Not only does Wolkomir show how these men challenged and shifted reigning ideas and hierarchical relationships but she also specifies the conditions under which changes occur. Wolkomir's analysis does not end with successful ideological maneuvering. Instead, she positions her analysis in relation to its larger implications of her study. Wolkomir concludes that inequalities limit such ideological revision and, in turn, ideological maneuvering reproduces inequality because it allows the larger oppressive ideology to remain intact. In short, Wolkomir's grounded theory analysis advances our understanding of how ideological change can occur while simultaneously specifying its limits.

Wolkomir's processual analysis demonstrates grounded theory in practice. Her approach reveals how people confer meaning on their situation and enact ideological stances. Yet Wolkomir's analysis does more. It contains strong links between detailed ethnographic description, substantive processual categories, and development of a theoretical concept, ideological maneuvering. Wolkomir then situates her concept and frames her article in the larger theoretical discourse on ideology, and by doing so offers a dynamic analysis of relationships between agency and structure. Wolkomir's nuanced theoretical account contributes to knowledge in a substantive area, theoretical ideas in her discipline, and useful understandings for social justice scholars and activists.

Defining Variation

Defining variation in a process or phenomenon is an important grounded theory goal, particularly of the postpositivist and constructivist versions. Researchers who conduct thorough research may discover variation within their findings and subsequent analyses. Learning how to handle variation analytically and how to write about it strengthens the analytic precision and usefulness of the grounded theory.

Grounded theorists compare their analyses with the extant literature, which can serve as data to illuminate the properties of emergent categories. Ordinarily, grounded theorists develop their analyses first and then use the relevant literature for comparative analysis. Researchers who gain intimate knowledge of their participants and settings may, however, define sharp differences with the literature early in their research. They may then construct their analyses from this position. Both Wasserman and Clair (2010) and Roschelle and Kaufman (2004) discovered that homelessness was not monolithic and sought to demonstrate the variation they found. Roschelle and Kaufman focused on homeless kids in their ethnographic study of an organization that served and sheltered homeless families. They offered new representations of these children that challenged earlier conclusions of homeless children having developmental and psychiatric problems. In a similar logic, Wasserman and Clair argued that homeless men who had networks on the streets were safer than those who used shelters.

The features of the setting and context of the studied phenomenon thus shape behavior and events. Curtis Jackson-Jacobs (2004) realized that he had found a strategic site that challenged earlier knowledge about crack cocaine uses and their worlds. He analyzed the settings and context of crack cocaine use among four college students as a strategic site and made systematic comparisons between it and previously reported sites of crack use.

Following Glaser and Strauss's (1967) guidelines, Jackson-Jacobs (2004) treated the literature as data to analyze the variation between his strategic site to arrive at causal generalizations and thus build theory. Jackson-Jacobs revealed that two conditions in his study contrasted with prior research on crack cocaine use and altered our knowledge of it. First, college student crack cocaine users could keep their drug use bounded because they (1) had resources, (2) wanted to avoid being identified as crack users, (3) treated smoking crack as a leisure pursuit, (4) purchased crack from friends, and (5) gave higher priority to their conventional involvements.[17] Second, these students had substantial residential mobility within a "safe" area where college students lived, not where drug dealers and users hung out. The residential location gave the men a benign environment that camouflaged hard drug use. Mobility allowed them to move if tensions arose or when they feared being identified.

By following the men over time, Jackson-Jacobs witnessed the explanatory power of the two conditions he had specified. One man lost control of his drug use and suffered the stigma of his friends' viewing him as a failure and of his mother discovering his crack habit. Another man's situation changed upon moving home. Location matters. This man no longer had the safety and relative anonymity afforded by his former neighborhood. He and his suburban friends now had to buy rock cocaine in the nearby urban ghetto, which changed the conditions, meanings, and consequences of drug use. They got into trouble with the dealers and police and experienced violence at the hands of both.

◙   COMPARATIVE ANALYSIS WITH EXTANT LITERATURES

In the above ethnographies, the authors report that they were struck by the difference between the portrayals of their studied phenomenon in the literature and what they later observed. In conventional grounded theory practice, researchers develop their analyses first and then return to the literature, whether to position their studies or to use the literature as data. Roz Dixon (2007) delineates how she developed her analysis in her exploratory study of the early school experiences and peer relationships of 35 deaf adults. Early coding revealed that deaf children's classmates subjected them to physical and psychological attacks. Dixon's early codes consisted of incidents such as "pulled my hearing aids out," "damaged hearing aids," and "banged my ears." The children not only rejected a deaf classmate, but also colluded so that this child broke group norms, as the following account indicates:

Bryony: If there was a lot of noise, I didn't stand a hope of hearing anything . . . and frequently when things were going on and (the teachers) were trying to tell things, people would start drumming desks (*she demonstrates making the sound of very gentle knocking on the table top*).

Interviewer: So that you couldn't hear?

Bryony: So that I couldn't hear. . . . I wouldn't know what (had been said) . . . and the teacher would get pissed off with having to keep repeating, you know, "How many times do I have to tell you" sort of thing, and you'd say "Sorry, I didn't hear" . . . "Well, it's funny you heard for the last half hour." . . . and, you know, and I think the teachers were very suspicious of me." (Dixon, 2007, p. 12)

Apparently, Bryony had had no troubles with friendships at school until classmates excluded her after she became deaf. Dixon does not specify how the teacher's exasperated response contributed to the simultaneous processes of experiencing ostracism and harassment. The teacher likely gave her students license to break classroom deportment rules and further harass Bryony.[18] Dixon's detailed grounded theory coding of her interview data revealed signs of ostracism such as in Bryony's interview. Subsequently, Dixon used the literature as data to identify general properties of ostracism, the properties of temporary coercive ostracism, and the properties of actual exclusion. She states:

It was during the process of subsequently organizing the codes that some codes seemed particularly suggestive of ostracism. To test this hypothesis, a literature review was conducted to clarify the nature, function and parameters of ostracism. A set of codes was developed describing behaviours and contextual factors which might be seen in the interview data if ostracism had been at work. All data was reanalyzed. (2007, p. 9)

Consistent with conventional grounded theory practice, Dixon first coded her data and studied her codes. Subsequently, she examined and coded how other authors treated ostracism.

This coding enabled her to define types of ostracism as well as the "generic features of ostracism" (pp. 13–14), which she tested with her data. While acknowledging inherent limitations of using retrospective accounts, Dixon distinguishes conditions that link ostracism and bullying and specifies conditions under which ostracism may be desirable.[19] Dixon's grounded theory analysis led her to define both problems related to ostracism by children and creative interventions for handling it.

▣ SUMMARY AND CONCLUSIONS

The implications of the above discussion are fivefold. First, the review of grounded theory clarifies strategies and approaches that grounded theorists share. These strategies and approaches distinguish the method from other types of qualitative research. Simultaneously, the influence of grounded theory on the development of qualitative methods becomes more apparent.

Second, delineating the similarities between versions of grounded theory and juxtaposing their differences creates a space for methodological explication of foundational assumptions and of research practice. Grounded theory studies range between objectivist and constructionist approaches and often contain elements of both. Yet attending to foundational assumptions and to research practice constitute pivotal turns toward engaging in reflexivity. And that may heighten awareness of our research choices and actions and, moreover, deeper understandings of our research participants' situations because we see them in new light.

Third, the constructivist version reclaims grounded theory from being a method of application to a method of innovation. Wolkomir's (2001) analysis exemplifies the difference between application and innovation. She uses the method to learn about her participants' views and concerns, not to apply a set of rules to her data. Under these conditions, grounded theory remains an emergent method. Both the form and specific content of the method arise as the researcher grapples with the problems at hand. Thus, the emergent character of the method contributes to its flexibility. This flexibility gives social justice researchers a mutable frame for their studies that they can adapt to fit their research problems and budding analyses.

Fourth, constructivist grounded theory acknowledges the foundations of its production, calls for reproduction of the method on new grounds, and moves inquiry beyond what is overt and obvious. In these ways, constructivist grounded theory answers earlier criticisms of objectivist grounded theory that emanate from feminist scholarship (Olesen, 2007), postmodernism, performance, and interpretive ethnography (Denzin, 2007), as well as critiques by qualitative methodologists.[20] Constructivist grounded theory challenges positivist elements that ignore reflexivity, overlook ethical issues, disregard issues of

representation, and do not attend to researchers' agency in constructing and interpreting data (Olesen, 2007).

Fifth, constructivist grounded theory acknowledges the dual roots of the method in positivism and pragmatism and seeks to develop the emphasis on pragmatism. Consistent with pragmatism, constructivist grounded theory acknowledges multiple perspectives and multiple forms of knowledge. Its practitioners become attuned to nuances in empirical worlds that elude researchers who assume a unitary method and unitary knowledge and thus are ill-quipped to grasp such nuances as the unheard voice of dissent and the silence of suffering. Both may remain imperceptible when researchers use objectivist grounded theory or, for that matter, conventional research methods. Classic grounded theory set forth tools for developing theoretical sensitivity. Constructivist grounded theory adds tools for increasing critical sensitivity and thus holds considerable potential for social justice inquiry.

The constructivist turn in grounded theory has clarified the strategies of the classic statements and generated a resurgence of interest in the method. It complements new developments such as Adele E. Clarke's (2005) situational analysis and incorporates methodological developments. It offers mixed methods researchers a set of useful tools and holds promise of informing software development. Constructivist grounded theory is and will be a method for the 21st century.

▣ NOTES

1. I thank Adele E. Clarke, Norman K. Denzin, and the following members of the Sonoma State University Faculty Writing Program—Sheila Katz, Lena McQuade, Suzanne Rivoire, Tom Rosin, and Richard Senghas—for their helpful comments on an earlier version of this chapter.

2. Such emphases often start with pressing social problems, collective concerns, and impassioned voices. In contrast, Rawls's (1971) emphasis on fairness begins from a distanced position of theorizing individual rights and risks from the standpoint of the rational actor under hypothetical conditions. Conceptions of social justice must take into account both collective goods and individual rights and recognize that definitions of rationality as well as of "rational" actors are situated in time, space, and culture—and both can change. To foster justice, Nussbaum (2000, p. 234) argues that promoting a collective good must not subordinate the ends of some individuals over others. She observes that women suffer when a collective good is promoted without taking into account the internal power and opportunity hierarchies within a group.

3. Throughout my discussion, I focus on studies that identify grounded theory as their method of inquiry.

4. Their approach may follow conventions of framing research for academic consumption rather than indicating a stance on social justice.

5. One of the originators of grounded theory, Barney Glaser, has consistently argued that researchers can use the method for quantitative

as well as qualitative research, and his recent book (2008) reaffirms this argument. To date, few researchers have acted on it.

6. But see earlier contentious arguments about purism in methods (Glaser, 1992; Greckhamer & Koro-Ljungberg, 2005; May, 1996; Stern, 1994; Wilson & Hutchinson, 1996).

7. Theoretical sampling means sampling to develop the properties of a theoretical category, not to sample for representation of a population.

8. The task of offering methodological discussions has been taken up in other venues, such as this handbook. Publishers' qualitative methods lists, journals such as *Qualitative Inquiry* and *International Journal of Social Research Methods,* as well as methodological articles in substantive journals bring methodological proclivities and practices into view. Where once methodological confessional tales focused on what happened in the research site, now authors such as Wasserman and Clair (2010), Suddaby and Greenwood (2005), and Harry, Sturges, and Klingner (2005) reveal remarkable candor in making their analytic strategies transparent. These authors invert the backstage of analytic work and bring it to the front stage of discussion. Although I view each discussion as only partly grounded theory (but see Suddaby's [2006] astute depiction of what is not grounded theory), I admire their candor and willingness to enter the methodological fray.

9. I have addressed criticisms of grounded theory in the third edition of this handbook in some detail, so I will only present those from within grounded theory here.

10. Over the years, Glaser (2001, 2003) has changed his view and now states that the researcher conceptualizes participants' main concern.

11. Structural-functionalism was the reigning theory of the 1950s. It invokes a biological metaphor, addresses the structure of social institutions, and evaluates how well they accomplished key societal tasks such as socializing children and controlling crime. Structural-functionalism assumes consensus between individuals and segments of society, studies social order, and emphasizes social roles within institutions (see Merton, 1957; Parsons, 1951).

12. Pragmatism not only informed Strauss's work, but also he stayed within and developed the pragmatist tradition through symbolic interactionism. This perspective emphasizes interaction, language, and culture as shaping the construction of meaning and action. It assumes a dynamic relationship between agentic, reflective actors and society and thus sees social institutions and society as constructed, not as given (see Blumer, 1969; Reynolds & Herman, 2003; Strauss, 1959/1969, 1993).

13. Fielding and Cisneros-Puebla (2010) integrate CAQDAS and GIS (geographic information system) methods in an innovative mixed methods approach.

14. These authors affirm that "qualitizing" quantitative data also occurs.

15. Schwalbe et al.'s (2000) article and Harris's (2001, 2006a, 2006b) studies exemplify how inequalities are enacted.

16. At the time I worked on this analysis, the literature contained many studies that used marginalization as a significant concept. However, authors left its meanings implicit and understood rather than taking them as problematic.

17. This last point resonates with Patrick Biernacki's (1986) grounded theory of natural recovery from heroin use. Based on his findings, Biernacki constructs an analysis of identity. Relinquishing

heroin use without treatment turned on the significance of having and maintaining conventional identities.

18. For a grounded theory of how school rules are enacted, see Thornberg (2007).

19. The reciprocal effects of interactional dynamics come into play here. Not all children who were ostracized were bullied. Dixon found that temporary ostracism by peers kept some children's angry outbursts in check and manageable.

20. See Charmaz (2005) for these critiques and my responses.

▣ References

Ball, M. M., Perkins, M. M., Hollingsworth, C., Whittington, F. J., & King, S. V. (2009). Pathways to assisted living: The influence of race and class. *Journal of Applied Gerontology, 28,* 81–108.

Biernacki, P. L. (1986). *Pathways from heroin addition: Recovery without treatment.* Philadelphia: Temple University Press.

Blumer, H. (1969). *Symbolic interactionism.* Englewood Cliffs, NJ: Prentice Hall.

Bong, S. A. (2007). Debunking myths in CAQDAS use and coding in qualitative data analysis: Experiences with and reflections on grounded theory methodology. *Historical Social Research, 32*(Suppl. 19), 258–275.

Boychuk Duchscher, J. E., & Morgan, D. (2004). Grounded theory: Reflections on the emergence vs. forcing debate. *Journal of Advanced Nursing, 48*(6), 605–612.

Bringer, J. D., Johnston, L. H., & Brackenridge, C. H. (2006). Using computer-assisted qualitative data analysis software to develop a grounded theory project. *Field Methods, 18*(3), 245–266.

Bryant, A. (2002). Re-grounding grounded theory. *The Journal of Information Technology Theory and Application, 4,* 25–42.

Bryant, A., & Charmaz, K. (2007). Grounded theory in historical perspective: An epistemological account. In A. Bryant & K. Charmaz (Eds.), *The SAGE handbook of grounded theory* (pp. 31–57). London: Sage.

Bryant, A., & Charmaz, K. (in press). Grounded theory. In P. Vogt & M. Williams (Eds.), *The SAGE handbook of methodological innovations in the social sciences.* London: Sage.

Bryman, A. (2007). Barriers to integrating quantitative and qualitative research. *Journal of Mixed Methods Research, 1*(1), 8–22.

Burawoy, M. (2004). For public sociology. *American Sociological Review, 70*(1), 4–28.

Cameron, R. (2009). A sequential mixed model research design: Design, analytical and display issues. *International Journal of Multiple Research Approaches, 3*(2), 140–152.

Carter, P. L. (2003). Black cultural capital, status positioning, and schooling conflicts for low-income African-American youth. *Social Problems, 50*(1), 136–155.

Casper, M. J. (1997). Feminist politics and fetal surgery: Adventures of a research cowgirl on the reproductive frontier. *Feminist Studies, 23*(2), 232–262.

Casper, M. J. (1998). *The making of the unborn patient: A social anatomy of fetal surgery.* New Brunswick, NJ: Rutgers University Press.

Casper, M. J. (2007). Fetal surgery then and now. *Conscience 28*(3), 24–28.

Charmaz, K. (1990). "Discovering" chronic illness: Using grounded theory. *Social Science and Medicine, 30*(11), 1161–1172.

Charmaz, K. (1995). Between positivism and postmodernism: Implications for methods. In N. K. Denzin (Ed.), *Studies in symbolic interaction, 17,* 43–72.

Charmaz, K. (2000). Constructivist and objectivist grounded theory. In N. K. Denzin & Y. S. Lincoln (Eds.), *The SAGE handbook of qualitative research* (2nd ed., pp. 509–535). Thousand Oaks CA: Sage.

Charmaz, K. (2002). Grounded theory analysis. In J. F. Gubrium & J. A. Holstein (Eds.), *The SAGE handbook of interview research* (pp. 675–694). Thousand Oaks, CA: Sage.

Charmaz, K. (2005). Grounded theory in the 21st century: Applications for advancing social justice studies. In N. K. Denzin & Y. S. Lincoln (Eds.), *The SAGE handbook of qualitative research* (3rd ed., pp. 507–535). Thousand Oaks, CA: Sage.

Charmaz, K. (2006). *Constructing grounded theory: A practical guide through qualitative analysis.* London: Sage.

Charmaz, K. (2007). Constructionism and grounded theory. In J. A. Holstein & J. F. Gubrium (Eds.), *Handbook of constructionist research* (pp. 319–412). New York: Guilford.

Charmaz, K. (2008a). A future for symbolic interactionism. In N. K. Denzin (Ed.), *Studies in symbolic interaction, 32,* 51–59.

Charmaz, K. (2008b). Grounded theory. In J. A. Smith (Ed.), *Qualitative psychology: A practical guide to research methods* (2nd ed., pp. 81–110). London: Sage. (Revised and updated version of the 2003 chapter)

Charmaz, K. (2008c). Grounded theory as an emergent method. In S. N. Hesse-Biber & P. Leavy (Eds.), *The handbook of emergent methods* (pp. 155–170). New York: Guilford.

Charmaz, K. (2008d). The legacy of Anselm Strauss for constructivist grounded theory. In N. K. Denzin (ed.), *Studies in symbolic interaction, 32,* 127–141.

Charmaz, K. (2008e). Reconstructing grounded theory. In L. Bickman, P. Alasuutari, & J. Brannen (Eds.), *The SAGE handbook of social research methods* (pp. 461–478). London: Sage.

Charmaz, K. (2008f). Views from the margins: Voices, silences, and suffering. *Qualitative Research in Psychology, 5*(1), 7–18.

Charmaz, K. (2009a). Recollecting good and bad days. In W. Shaffir, A. Puddephatt, & S. Kleinknecht (Eds.), *Ethnographies revisited: The stories behind the story.* New York: Routledge.

Charmaz, K. (2009b). Shifting the grounds: Constructivist grounded theory methods for the twenty-first century. In J. M. Morse, P. N. Stern, J. Corbin, B. Bowers, K. Charmaz, & A. E. Clarke, *Developing grounded theory: The second generation* (pp. 127–154). Walnut Creek, CA: Left Coast Press.

Charmaz, K. (2010). Studying the experience of chronic illness through grounded theory. In G. Scambler & S. Scambler (Eds.), *New directions in the sociology of chronic and disabling conditions: Assaults on the lifeworld* (pp. 8–36). London: Palgrave.

Charmaz, K., & Bryant, A. (2010). Grounded theory. In B. McGaw, E. Baker, & P. P. Peterson (Eds.), *The international encyclopedia of education* (pp. 401–406). Oxford, UK, Elsevier.

Charmaz, K., & Henwood, K. (2008). Grounded theory in psychology. In C. Willig & W. Stainton-Rogers (Eds.), *The SAGE handbook of qualitative research in psychology* (pp. 240–260). London: Sage.

Choi, S. Y. P., & Holroyd, E. (2007). The influence of power, poverty and agency in the negotiation of condom use for female sex workers in Mainland China. *Culture, Health and Sexuality, 9*(5), 489–503.

Christ, T. W. (2009). Designing, teaching, and evaluating two complementary mixed methods research courses. *Journal of Mixed Methods Research, 3*(4), 292–325.

Ciambrone, D. (2007). Illness and other assaults on self: The relative impact of HIV/AIDS on women's lives. *Sociology of Health & Illness, 23*(4), 517–540.

Clarke, A. E. (1998). *Disciplining reproduction: Modernity, American life sciences and the "problem of sex."* Berkeley: University of California Press.

Clarke, A. E. (2003). Situational analyses: Grounded theory mapping after the postmodern turn. *Symbolic Interaction 26,* 553–576.

Clarke, A. E. (2005). *Situational analysis: Grounded theory after the postmodern turn.* Thousand Oaks, CA: Sage.

Clarke, A. E. (2006). Feminisms, grounded theory, and situational analysis. In S. Hess-Biber & D. Leckenby (Eds.), *The SAGE handbook of feminist research methods* (pp. 345–370). Thousand Oaks, CA: Sage.

Clarke, A. E., & Friese, C. (2007). Situational analysis: Going beyond traditional grounded theory. In K. Charmaz & A. Bryant (Eds.), *The SAGE handbook of grounded theory* (pp. 694–743). London: Sage.

Cohn, S., Dyson, C., & Wessley, S. (2008). Early accounts of Gulf War illness and the construction of narratives in UK service personnel. *Social Science & Medicine, 67,* 1641–1649.

Corbin, J. (2009). Taking an analytic journey. In J. M. Morse, P. N. Stern, J. Corbin, B. Bowers, K. Charmaz, & A. E. Clarke, *Developing grounded theory: The second generation* (pp. 35–53). Walnut Creek, CA: Left Coast Press.

Corbin, J., & Strauss, A. (2008). *Basics of qualitative research* (3rd ed.). Thousand Oaks, CA: Sage.

Creswell, J. W. (2003). *Research design: Qualitative, quantitative, and mixed methods design* (2nd ed.). Thousand Oaks, CA: Sage.

Creswell, J. W., & Plano Clark, V. L. (2007). *Designing and conducting mixed methods research.* Thousand Oaks, CA: Sage.

Creswell, J. W., Shope, R., Plano Clark, V. L., & Green, D. O. (2006). How interpretive qualitative research extends mixed methods research. *Research in the Schools, 13*(1), 1–11.

Denzin, N. K. (1970). *The research act: A theoretical introduction to sociological methods.* Chicago: Aldine.

Denzin, N. K. (2007). Grounded theory and the politics of interpretation. In A. Bryant & K. Charmaz (Eds.), *The SAGE handbook of grounded theory* (pp. 454–471). London: Sage.

Denzin, N. K., & Lincoln, Y. S. (1994). Preface. In N. K. Denzin & Y. S. Lincoln (Eds.), *Handbook of qualitative research* (pp. ix–xii). Thousand Oaks, CA: Sage.

Denzin, N. K., & Lincoln, Y. S. (2005). *The SAGE handbook of qualitative research* (3rd ed.). Thousand Parks, CA: Sage.

Dick, B. (2007). What can grounded theorists and action researchers learn from each other? In A. Bryant & K. Charmaz (Eds.), *The SAGE handbook of grounded theory* (pp. 398–416). London: Sage.

Dixon, R. (2007). Ostracism: One of the many causes of bullying in groups? *Journal of School Violence, 6*(3), 3–26.

Duemer, L. S., & Zebidi, A. (2009). The pragmatic paradigm: An epistemological framework for mixed methods research. *Journal of Philosophy and History of Education, 59,* 164–168.

Dumit, Joseph. (2006). Illnesses you have to fight to get: Facts as forces in uncertain, emergent illnesses. *Social Science & Medicine, 62*(3), 577–590.

Einwohner, R. L., & Spencer, J. W. (2005). That's how we do things here: The construction of sweatshops and anti-sweatshop activism in two campus communities. *Sociological Inquiry, 75*(2), 249–272.

Feagin, J. R. (1999). Social justice and sociology: Agendas for the twenty-first century. *American Sociological Review, 66*(1), 1–20.

Feilzer, M. V. (2010). Doing mixed methods research pragmatically: Implications for the rediscovery of pragmatism as a research paradigm. *Journal of Mixed Methods Research, 4*(4), 6–16.

Fielding, N., & Cisneros-Puebla, C. (2010). CAQDAS-GIS convergence: Toward a new integrated mixed method research practice. *Journal of Mixed Methods Research, 3*(4), 349–370.

Fielding, N., & Lee, R. M. (1998). *Computer analysis and qualitative field research.* London: Sage.

Fielding, N., & Lee, R. M. (2002). New patterns in the adoption and use of qualitative software. *Field Methods, 14*(2), 197–216.

Foote-Ardah, C. E. (2003). The meaning of complementary and alternative medicine practices among people with HIV in the United States: Strategies for managing everyday life. *Sociology of Health & Illness, 25*(5), 481–500.

Foster-Fishman, P., Nowell, B., Deacon, Z., Nievar, M. A., & McCann, P. (2005). Using methods that matter: The impact of reflection, dialogue, and voice. *American Journal of Community Psychology, 36*(3/4), 275–291.

Frohmann, L. (1991). Discrediting victims' allegations of sexual assault: Prosecutorial accounts of case rejections. *Social Problems, 38*(2), 213–226.

Frohmann, L. (1998). Constituting power in sexual assault cases: Prosecutorial strategies for victim management. *Social Problems, 45*(3), 393–407.

Gagné. P. (1996). Identity, strategy and feminist politics: Clemency for women who kill. *Social Problems, 43*(1), 77–93.

Glaser, B. G. (1978). *Theoretical sensitivity.* Mill Valley, CA: Sociology Press.

Glaser, B. G. (1992). *Basics of grounded theory analysis.* Mill Valley, CA: Sociology Press.

Glaser, B. G. (1998). *Doing grounded theory: Issues and discussions.* Mill Valley, CA: Sociology Press.

Glaser, B. G. (2001). *The grounded theory perspective: Conceptualization contrasted with description.* Mill Valley, CA: Sociology Press.

Glaser, B. G. (2002). Constructivist grounded theory? *Forum: Qualitative Sozialforschung/Qualitative Social Research, 3*(3). Available at http://www.qualitative-research.net/index.php/fqs/article/view/825

Glaser, B. G. (2003). *The grounded theory perspective II: Description's remodeling of grounded theory methodology.* Mill Valley, CA: Sociology Press.

Glaser, B. G. (2008). *Doing quantitative grounded theory.* Mill Valley, CA: Sociology Press.

Glaser, B. G., & Strauss, A. L. (1967). *The discovery of grounded theory.* Chicago: Aldine.

Greckhamer, T., & Koro-Ljungberg, M. (2005). The erosion of a method: Examples from grounded theory. *International Journal of Qualitative Studies in Education, 18*(6), 729–750.

Greene, J. C. (2006). Toward a methodology of mixed methods social inquiry. *Research in the Schools, 13*(1), 94–99.

Gunter, V. J. (2005). News media and technological risks: The case of pesticides after *Silent Spring. The Sociological Quarterly, 46*(4), 671–698.

Hansen, E. C., Walters, J., & Baker, R. W. (2007). Explaining chronic obstructive pulmonary disease (COPD): Perceptions of the role played by smoking. *Sociology of Health & Illness, 29*(5), 730–749.

Harris, S. R. (2001). What can interactionism contribute to the study of inequality? The case of marriage and beyond. *Symbolic Interaction, 24*(4), 455–480.

Harris, S. R. (2006a). *The meanings of marital equality.* Albany: State University of New York Press.

Harris, S. R. (2006b). Social constructionism and social inequality: An introduction to a special issue of JCE. *Journal of Contemporary Ethnography, 35*(3), 223–235.

Harry, B., Sturges, K. M., & Klingner, J. K. (2005). Mapping the process: An exemplar of process and challenge in grounded theory analysis. *Educational Researcher, 34*(2), 3–13.

Henwood, K., & Pidgeon, N. (2003). Grounded theory in psychological research. In P. M. Camic, J. E. Rhodes, & L. Yardley (Eds.), *Qualitative research in psychology: Expanding perspectives in methodology and design* (pp. 131–155). Washington, DC: American Psychological Association.

Hood, J. (2007). Orthodoxy vs. power: The defining traits of grounded theory. In A. Bryant & K. Charmaz (Eds.), *The SAGE handbook of grounded theory* (pp. 151–164). London: Sage.

Hyde, J., & Kammerer, N. (2009). Adolescents' perspectives on placement moves and congregate settings: Complex and cumulative instabilities in out-of-home care. *Children and Youth Services Review, 31,* 265–273.

Jackson-Jacobs, C. (2004). Hard drugs in a soft context: Managing trouble and crack use on a college campus. *Sociological Quarterly, 45*(4), 835–856.

Jiménez, T. R. (2008). Mexican immigrant replenishment and the continuing significance of ethnicity and race. *American Journal of Sociology, 113*(6), 1527–1567.

Johnson, R. B., Onwuegbuzie, A. J., & Turner, L. A. (2007). Toward a definition of mixed methods research. *Journal of Mixed Methods Research, 1,* 112–133.

Jones, S. J. (2003). Complex subjectivities: Class, ethnicity, and race in women's narratives of upward mobility. *Journal of Social Issues, 50*(4), 804–820.

Karabanow, J. (2008). Getting off the street: Exploring the processes of young people's street exits. *American Behavioral Scientist, 51*(6), 772–788.

Kearney, M. H. (2007). From the sublime to the meticulous: The continuing evolution of grounded formal theory. In A. Bryant & K. Charmaz (Eds.), *The SAGE handbook of grounded theory* (pp. 127–150). London: Sage.

Kelle, U. (2004). Computer-assisted qualitative data analysis. In C. Seale, G. Gobo, J. F. Gubrium, & D. Silverman (Eds.), *Qualitative research practice* (pp. 473–489). London: Sage.

Kelle, U. (2005, May). Emergence vs. forcing: A crucial problem of "grounded theory" reconsidered. *Forum: Qualitative Sozialforsung/Qualitative Sociology, 6*(2). Available at http://www.qualitative research.net/index.php/fqs/article/view/467

Kelle, U. (2007). The development of categories: Different approaches of grounded theory. In A. Bryant & K. Charmaz (Eds.), *The SAGE handbook of grounded theory* (pp. 191–213). London: Sage.

Kemmis, S., & McTaggart, R. (2005). Participatory action research: Communicative action and the public sphere. In N. K. Denzin & Y. S. Lincoln (Eds.), *The SAGE handbook of qualitative research* (3rd ed., pp. 559–603). Thousand Oaks, CA: Sage.

Kendall, J. (1999). Axial coding and the grounded theory controversy. *Western Journal of Nursing Research, 21*(6), 743–757.

Konopásek, Z. (2008). Making thinking visible with Atlas.ti: Computer assisted qualitative analysis as textual practices. *Forum: Qualitative Sozialforschung/Qualitative Social Research, 9*(2). Available at http://nbn-resolving.de/urn:nbn:de:0114-fqs0802124

Lazzari, M. M., Ford, H. R., & Haughey, K. J. (1996). Making a difference: Women of action in the community. *Social Work, 41*(2), 197–205.

Lempert, L. B. (2007). Asking questions of the data: Memo writing in the grounded theory tradition. In A. Bryant & K. Charmaz (Eds.), *The SAGE handbook of grounded theory* (pp. 245–264). London: Sage.

Lio, S., Melzer, S., & Reese, E. (2008). Constructing threat and appropriating "civil rights": Rhetorical strategies of gun rights and English only leaders. *Symbolic Interaction, 31*(1), 5–31.

Locke, K. (2007). Rational control and irrational free-play: Dual-thinking modes as necessary tension in grounded theorizing. In A. Bryant & K. Charmaz (Eds.), *The SAGE handbook of grounded theory* (pp. 565–579). London: Sage.

Lutgen-Sandvik, P. (2008). Intensive remedial identity work: Responses to workplace bullying trauma and stigmatization. *Organization, 15*(1) 97–119.

Maines, D. R. (2001). *The faultline of consciousness: A view of interactionism in sociology.* New York: Aldine.

Mathieson, C., & Stam, H. (1995). Renegotiating identity: Cancer narratives. *Sociology of Health & Illness, 17*(3): 283–306.

May, K. (1996). Diffusion, dilution or distillation? The case of grounded theory method. *Qualitative Health Research, 6*(3), 309–311.

McDermott, K. A. (2007). "Expanding the moral community" or "blaming the victim"? *American Education Research Association Journal, 44*(1), 77–111.

Mcintyre, A. (2002). Women researching their lives: Exploring violence and identity in Belfast, the North of Ireland. *Qualitative Research, 2*(3), 387–409.

McPhail, B. A., & DiNitto, D. M. (2005). Prosecutorial perspectives on gender-bias hate crimes. *Violence Against Women, 11*(9), 1162–1185.

Mertens, D. M. (2007). Transformative paradigm: Mixed methods and social justice. *Journal of Mixed Methods Research, 1*(3), 212–235.

Mertens, D. M. (2010). *Research and evaluation in education and psychology: Integrating diversity with quantitative, qualitative, and mixed methods.* Thousand Oaks, CA: Sage.

Merton, R. K. (1957). *Social theory and social structure.* Glencoe, IL: Free Press.

Mevorach, M. (2008). Do preschool teachers perceive young children from immigrant families differently? *Journal of Early Childhood Teacher Education, 29*, 146–156.

Mitakidou, S., Tressou, E., & Karagianni, P. (2008). Students' reflections on social exclusion. *The International Journal of Diversity in Organisations, Communities and Nations, 8*(5), 191–198.

Mitchell, R. C., & McCusker, S. (2008). Theorising the UN convention on the rights of the child within Canadian post-secondary education: A grounded theory approach. *International Journal of Children's Rights, 16*, 159–176.

Moore, D. L. (2005). Expanding the view: The lives of women with severe work disabilities in context. *Journal of Counseling and Development, 83*(3), 343–348.

Moreno, M. (2008). Lessons of belonging and citizenship among hijas/os de inmigrantes Mexicanos. *Social Justice, 35*(1), 50–75.

Morgan, D. (2007). Paradigms lost and pragmatism regained: Methodological implications of combining qualitative and quantitative research. *Journal of Mixed Methods Research 1*(1): 48-76.

Morrow, S. L., & Smith, M. L. (1995). Constructions of survival and coping by women who have survived childhood sexual abuse. *Journal of Counseling Psychology, 42*(1), 24–33.

Morse, J. M. (1991). Approaches to qualitative-quantitative methodological triangulation. *Nursing Research, 40*(2), 120–123.

Morse, J. M. (2007). Sampling in grounded theory. In A. Bryant & K. Charmaz (Eds.), *The SAGE handbook of grounded theory* (pp. 229–254). London: Sage.

Mruck, K., & Mey, G. (2007). Grounded theory and reflexivity. In A. Bryant & K. Charmaz (Eds.), *The SAGE handbook of grounded theory* (pp. 515–538). London: Sage.

Mulcahy, A. (1995). Claims-making and the construction of legitimacy: Press coverage of the 1981 Northern Irish hunger strike. *Social Problems, 42*(4), 449–467.

Nack, A. (2008). *Damaged goods? Women living with incurable sexually transmitted diseases.* Philadelphia: Temple University Press.

Neill, S. J. (2006). Grounded theory sampling: The contribution of reflexivity. *Journal of Research in Nursing, 11*(3), 253–260.

Nepstad, S. E. (2007). Oppositional consciousness among the privileged: Remaking religion in the Central America solidarity movement. *Critical Sociology, 33*(4), 661–688.

Nussbaum, M. C. (2000). Women's capabilities and social justice. *Journal of Human Development, 1*, 219–247.

O'Connor, G. C., Rice, M. P., Peters, L., & Veryzer, R. W. (2003). Managing interdisciplinary, longitudinal research teams: Extending grounded theory-building methodologies. *Organization Science, 14*(4), 353–373.

Ogle, J. P., Eckman, M., & Leslie, C. A. (2003). Appearance cues and the shootings at Columbine High: Construction of a social problem in the print media. *Sociological Inquiry, 73*(1), 1–27.

Olesen, V. (2007). Feminist qualitative research and grounded theory. In A. Bryant & K. Charmaz (Eds.), *The SAGE handbook of grounded theory* (pp. 417–435). London: Sage.

Padgett, D. K. (2009). Qualitative and mixed methods in social work knowledge development. *Social Work, 54*(1), 101–105.

Parsons, T. (1951). *The social system.* Glencoe, IL: Free Press.

Peirce, C. S. (1958). *Collected papers.* Cambridge, MA: Harvard University Press.

Piantanida, M., Tananis, C. A., & Grubs, R. E. (2004). Generating grounded theory of/for educational practice: The journey of three epistemorphs. *International Journal of Qualitative Studies in Education, 17*(3), 325–346.

Poonamallee, L. (2009). Building grounded theory in action research through the interplay of subjective ontology and objective epistemology. *Action Research, 7*(1), 69–83.

Qin, D., & Lykes, M. B. (2006). Reweaving a fragmented self: A grounded theory of self-understanding among Chinese women students in the United States of America. *International Journal of Qualitative Studies in Education, 19*(2), 177–200.

Quint, J. C. (1965). Institutionalized practices of information control. *Psychiatry, 28*(May), 119–132.

Rawls, J. (1971). *A theory of justice.* Cambridge, MA: Belknap.

Reichert, J. (2007). Abduction: The logic of discovery in grounded theory. In A. Bryant & K. Charmaz (Eds.), *The SAGE handbook of grounded theory* (pp. 214–228). London: Sage.

Reynolds, L. T., & Herman, N. J. (Eds.). (2003). *Handbook of symbolic interaction.* Walnut Creek, CA: AltaMira.

Rier, D. (2007). Internet social support groups as moral agents: The ethical dynamics of HIV+ status disclosure. *Sociology of Health & Illness, 29*(7), 1–16.

Rivera, L. A. (2008). Managing "spoiled" national identity: War, tourism, and memory in Croatia. *American Sociological Review, 73*(4), 613–634.

Roschelle, A. R., & Kaufman, P. (2004). Fitting in and fighting back: Stigma management strategies among homeless kids. *Symbolic Interaction, 27*(1), 23–46.

Rosenthal, G. (2004). Biographical research. In C. Seale, G. Gobo, J. F. Gubrium, & D. Silverman (Eds.), *Qualitative research practice* (pp. 48–64). London: Sage.

Roxas, K. (2008). Who dares to dream the American dream? *Multicultural Education, 16*(2), 2–9.

Ryder, J. A. (2007). "I wasn't really bonded with my family": Attachment, loss and violence among adolescent female offenders. *Critical Criminology, 15*(1), 19–40.

Sahin-Hodoglugil, N. N., vander Straten, A., Cheng, H., Montgomery, E. T., Kcanek, D., Mtetewa, S., et al. (2009). A study of women's covert use of the diaphragm in an HIV prevention trial in sub-Saharan Africa. *Social Science & Medicine, 69,* 1547–1555.

Sakamoto, I., Chin, M., Chapra, A., & Ricciar, J. (2009). A "normative" homeless woman? Marginalisation, emotional injury and social support of transwomen experiencing homelessness. *Gay and Lesbian Issues and Psychology Review, 5*(1), 2–19.

Salander, P. (2002). Bad news from the patient's perspective: An analysis of the written narratives of newly diagnosed cancer patients. *Social Science & Medicine, 55,* 721–732.

Sandelowski, M., Voils, C. I., & Knafl, G. (2009). On quantitizing. *Journal of Mixed Methods Research, 3,* 208–222.

Sandstrom, K. L. (1990). Confronting deadly disease: The drama of identity construction among gay men with AIDS. *Journal of Contemporary Ethnography, 19,* 271–294.

Sandstrom, K. L. (1998). Preserving a vital and valued self in the face of AIDS. *Sociological Inquiry, 68*(3), 354–371.

Schwalbe, M. (2005). Identity stakes, manhood acts, and the dynamics of accountability. In N. K. Denzin (Ed.), *Studies in symbolic interaction 28,* 65–81. Bingley, UK: Emerald Publishing Group.

Schwalbe, M., Godwin, S., Holden, D., Schrock, D., Thompson, S., & Wolkomir, M. (2000). Generic processes in the reproduction of inequality: An interactionist analysis. *Social Forces, 79,* 419–452.

Scott, E. K. (2005). Beyond tokenism: The making of racially diverse feminist organizations. *Social Problems, 52*(2), 232–254.

Scott, E. K., London, A. S., & Gross, G. (2007). "I try not to depend on anyone but me": Welfare-reliant women's perspectives on self-sufficiency, work, and marriage. *Sociological Inquiry, 77*(4), 601–625.

Shelley, N. M. (2001). Building community from "scratch": Forces at work among urban Vietnamese refugees in Milwaukee. *Sociological Inquiry, 71*(4), 473–492.

Sixsmith, J. A. (1999). Working in the hidden economy: The experience of unemployed men in the UK. *Community, Work and Family, 2*(3), 257–277.

Somerville, C., Featherstone, K., Hemingway, H., Timmis, A., & Feder, G. S. (2008). Performing stable angina pectoris: An ethnographic study. *Social Science & Medicine, 66*(7), 1497–1508.

Speed, S., & Luker, K. A. (2006). Getting a visit: How district nurses and general practitioners "organise" each other in primary care. *Sociology of Health & Illness, 28*(7), 883–902.

Spencer, J. W., & Triche, E. (1994). Media constructions of risk and safety: Differential framings of hazard events. *Sociological Inquiry, 64*(2), 199–213.

Star, S. L. (1989). *Regions of the mind: Brain research and the quest for scientific certainty.* Stanford, CA: Stanford University Press.

Stern, P. N. (1994). Eroding grounded theory. In J. Morse (Ed.), *Critical issues in qualitative research methods* (pp. 212–223). Thousand Oaks, CA: Sage.

Stern, P. N. (2007). On solid ground: Essential properties for growing grounded theory. In A. Bryant & K. Charmaz (Eds.), *The SAGE handbook of grounded theory* (pp. 114–126). London: Sage.

Stern, P. N. (2009). Glaserian grounded theory. In J. M. Morse, P. N. Stern, J. Corbin, B. Bowers, K. Charmaz, & A. E. Clarke, *Developing grounded theory: The second generation* (pp. 23–29). Walnut Creek, CA: Left Coast Press.

Strauss, A. L. (1969). *Mirrors and masks: The search for identity.* Mill Valley, CA: Sociology Press. (Original work published 1959)

Strauss, A. L. (1987). *Qualitative analysis for social scientists.* New York: Cambridge University Press.

Strauss, A. L. (1993). *Continual permutations of action.* New York: Aldine.

Strauss, A., & Corbin, J. (1990). *Basics of qualitative research: Grounded theory procedures and techniques.* Newbury Park, CA: Sage.

Strauss, A., & Corbin, J. (1994). Grounded theory methodology: An overview. In N. K. Denzin & Y. S. Lincoln (Eds.), *Handbook of qualitative research* (pp. 273–285). Thousand Oaks, CA: Sage.

Strauss, A., & Corbin, J. (1998). *Basics of qualitative research: Grounded theory procedures and techniques* (2nd ed.). Thousand Oaks, CA: Sage.

Strübing, J. (2007). Research as pragmatic problem-solving: The pragmatist roots of empirically grounded theorizing. In A. Bryant & K. Charmaz (Eds.), *The SAGE handbook of grounded theory* (pp. 580–601). London: Sage.

Suddaby, R. (2006). From the editors: What grounded theory is not. *Academy of Management Journal, 49*(4), 633–642.

Suddaby, R., & Greenwood, R. (2005). Rhetorical strategies of legitimacy. *Administrative Science Quarterly, 50*(1), 35–67.

Swahnberg, K., Thapar-Björkert, S., & Berterö, C. (2007). Nullified: Women's perceptions of being abused in health care. *Journal of Psychosomatic Obstetrics and Gynecology, 28*(3), 161–167.

Tashakkori, A., & Teddlie, C. (2003). The past and future of mixed methods research: From data triangulation to mixed model designs. In A. Tashakkori & C. Teddlie (Eds.), *Handbook of mixed methods in social & behavioral research* (pp. 671–701). Thousand Oaks, CA: Sage.

Teram, E., Schachter, C. L., & Stalker, C. A. (2005). The case for integrating grounded theory and participatory action research: Empowering clients to inform professional practice. *Qualitative Health Research, 15*(8), 1129–1140.

Thornberg, R. (2007). Inconsistencies in everyday patterns of school rules. *Ethnography and Education, 2*(3), 401–416.

Thornberg, R. (2009). The moral construction of the good pupil embedded in school rules. *Education, Citizenship and Social Justice, 4*(3), 245–261.

Thornberg, R., & Charmaz, K. (in press). Grounded theory. In S. Lapan, M. Quartaroli, & F. Riemer (Eds.), *Qualitative research: An introduction to methods and designs.* San Francisco: Jossey-Bass.

Timmermans, S., & Tavory, I. (2007). Advancing ethnographic research through grounded theory practice. In A. Bryant & K. Charmaz (Eds.), *The SAGE handbook of grounded theory* (pp. 493–512). London: Sage.

Tuason, M. T. G. (2008). Those who were born poor: A qualitative study of Philippine poverty. *Journal of Counseling Psychology, 55*(2), 158–171.

Turner, B. A. (1981). Some practical aspects of qualitative data analysis: One way of organizing the cognitive processes associated with the generation of grounded theory. *Quantity and Quality, 15,* 225–247.

Ullman, S. E., & Townsend, S. M. (2008). What is an empowerment approach to working with sexual assault survivors? *Journal of Community Psychology, 36*(3), 299–312.

Urquhart C. (2003). Re-grounding grounded theory-or reinforcing old prejudices? A brief response to Bryant. *Journal of Information Technology Theory and Application, 4*(3), 43–54.

Valadez, J. R. (2008). Shaping the educational decisions of Mexican immigrant high school students. *American Educational Research Journal, 45*(4), 834–860.

Valdez, A., & Flores, R. (2005). A situational analysis of dating violence among Mexican American females associated with street gangs. *Sociological Focus, 38*(2), 95–114.

Vandenburgh, H. (2001). Physician stipends as organizational deviance in for-profit psychiatric hospitals. *Critical Sociology, 27*(1), 56–76.

van den Hoonaard, W. C. (1997). *Working with sensitizing concepts: Analytical field research.* Thousand Oaks, CA: Sage.

Veale, A., & Stavrou, A. (2007). Former Lord's Resistance Army child soldier abductees: Explorations of identity in reintegration and reconciliation. *Peace and Conflict: Journal of Peace Psychology, 13*(3): 273–292.

Wasserman, J. A., & Claire, J. M. (2010). *At home on the street: People, poverty, and a hidden culture of homelessness.* New York: Lynne Rienner.

Weitzman, E. A. (2000). Software and qualitative research. In N. K. Denzin & Y. S. Lincoln (Eds.), *The SAGE handbook of qualitative research* (2nd ed., pp. 803–820). Thousand Oaks, CA: Sage.

Williamson, K. (2006). Research in constructivist frameworks using ethnographic techniques. *Library Trends, 55*(1), 83–101.

Wilson, H. S., & Hutchinson, S. A. (1996). Methodologic mistakes in grounded theory. *Nursing Research, 45*(2), 122–124.

Wilson, K., & Luker, K. A. (2006). At home in hospital? Interaction and stigma in people affected by cancer. *Social Science & Medicine, 62,* 1616–1627.

Wolkomir, M. (2001). Wrestling with the angels of meaning: The revisionist ideological work of gay and ex-gay Christian men. *Symbolic Interaction, 24*(4), 407–424.

Wolkomir. M. (2006). *Be not deceived: The sacred and sexual struggles of gay and ex-gay Christian men.* New Brunswick, NJ: Rutgers University Press.

Ziegahn, L., & Hinchman, K. A. (1999). Liberation or reproduction: Exploring meaning in college students' adult literacy tutoring. *Qualitative Studies in Education, 12*(1), 85–101.

22

IN THE NAME OF HUMAN RIGHTS

I Say (How) You (Should) Speak (Before I Listen)[1]

Antjie Krog

It is the year 1872. A Bushman shaman called //Kabbo narrates an incident to a German philologist Wilhelm Bleek in Cape Town, South Africa. In the narration, which took Bleek from April 13 to September 19 to record and translate from /Xam into English, the following two paragraphs appear, describing how a young woman tracks down her nomadic family:

> She [the young widow] arrives with her children at the water hole. There she sees her younger brother's footprints by the water. She sees her mother's footprint by the water. She sees her brother's wife's spoor by the water.
>
> She tells her children: "Grandfather's people's footprints are here; they had been carrying dead springbok to the water so that people can drink on their way back with the game. The house is near. We shall follow the footprints because the footprints are new. We must look for the house. We must follow the footprints. For the people's footprints were made today; the people fetched water shortly before we came." (Lewis-Williams, 2002, p. 61)

For more than a hundred years, these words seemed like just another interesting detail in an old Bushmen story, until researcher Louis Liebenberg went to live among modern Bushmen. In his book, *The Art of Tracking: The Origin of Science* (1990), Liebenberg insists that what seems to be an instinctive capacity to track a spoor, is actually the Bushmen using intricate decoding, contextual sign analysis to create hypotheses.

Liebenberg distinguishes three levels of tracking among the Bushmen: first, simple tracking that just follows footprints. Second, systematic tracking involving the gathering of information from signs until a detailed indication is built up of the action. Third, speculative tracking that involves the creation of a working hypothesis on the basis of (1) the initial interpretation of signs, (2) a knowledge of behavior, and (3) a knowledge of the

terrain. According to Liebenberg, these skills of tracking are akin to those of Western intellectual analysis, and he suggests that all science actually started with tracking (Brown, 2006, p. 25).

Returning to the opening two paragraphs, one sees that the young widow effortlessly does all three kinds of tracking identified by Liebenberg. She identifies the makers of the footprints, their coming and going, that they were carrying something heavy and/or bleeding, that they were thirsty, that they drank water on the way back from hunting, she identifies the game as a springbok, she establishes when the tracks were made and then puts forward a hypothesis of what they were doing and where and how she will find her family that very day.

The question I want to pose here is: Is it justified to regard Wilhelm Bleek (as the recorder of the narration), Louis Liebenberg (as a scholar of tracking), and myself (for applying the tracking theory to the narration) as the scholars/academics, while //Kabbo (Bushman narrator) and the woman in the story (reading the tracks) are "raw material"?

How does this division respect Article 19 of the Universal Declaration of Human Rights of the United Nations?

> Everyone has the right to freedom of opinion and expression; this right includes freedom to hold opinions without interference *and to seek, receive and impart information and ideas through any media and regardless of frontiers.* (emphasis added; available at http://www.un.org/en/documents/udhr/)

Who May Enter the Discourse?

The rights of two groups will be discussed in this essay: First, the rights of those living in marginalized areas but who produce virtually on a daily basis intricate knowledge systems of survival. Second, the rights of scholars coming from those marginalized places, but who can only enter the world of acknowledged knowledge in languages not their own and within discourses based on foreign and estrang-*ing* structures.

Although Gayatri Spivak describes the one group as subaltern, she deals with both of these groups in her famous essay, "Can the Subaltern Speak?" suggesting that the moment that the subaltern finds herself in conditions in which she can be heard, "her status as a subaltern would be changed utterly; she would cease to be subaltern" (Williams & Chrisman, 1994, p. 190).

"Mrs. Khonele" as Subaltern

During the two years of hearings conducted by the South African Truth and Reconciliation Commission (TRC), 2,000 testimonies were given in public. Instead of listening to the impressive stories of well-known activists, the commission went out of its way to provide a forum for the most marginalized narratives from rural areas given in indigenous languages. In this way, these lives and previously unacknowledged narratives were made audible and could be listened to through translation to become the first entry into the South African psyche of what Spivak so aptly calls in her piece, *Subaltern Studies—Deconstructing Historiography,* "news of the consciousness of the subaltern" (Williams & Chrisman, 1994, p. 203).

Covering the hearings of the truth commission for national radio, one woman's testimony stayed with me as the most incoherent testimony I had to report on. I considered the possibility that one needed special tools to make sense of it and wondered whether clarification could be found in the original Xhosa, or was the woman actually mentally disturbed, or were there vestiges of "cultural supremacy" in me that prevented me from hearing her?

Trying to find her testimony later on the truth commission's website proved fruitless. There was no trace of her name in the index. Under the heading of the Gugulethu Seven incident, her surname was given incorrectly as "Khonele," and she was the only mother in this group to be presented without a first name. Her real name was Notrose Nobomvu Konile, but I later found that even in her official identity document her second name was given incorrectly as "Nobovu." (Notrose Konile's TRC testimony is available online at http://www.justice.gov.za/trc/hrvtrans/heide/ct00100.htm.)

One might well ask: Is it at all possible to hear this unmentioned, incorrectly identified, misspelled, incoherently testifying, translated, and carelessly transcribed woman from the deep rural areas of South Africa?

I asked two colleagues at the University of the Western Cape—Nosisi Mpolweni from the Xhosa department, and Professor Kopano Ratele from the psychology department and women and gender studies—to join me in a reading of the testimony. Mpolweni and Ratele immediately became interested. Using the original Xhosa recording, we started off by

transcribing and retranslating. Then we applied different theoretical frameworks (Elaine Scarry, Cathy Garuth, Soshana Felman, Dori Laub, G. Bennington, etc.) to interpret the text; and, finally, we visited and reinterviewed Konile. What started out as a casual teatime discussion became a project of two and a half years and finally a book: *There Was This Goat—Investigating the Truth Commission Testimony of Notrose Nobomvu Konile* (Krog, Mpolweni, & Ratele, 2009).

But first, some concepts need to be introduced that play a role the moment that the voice of the subaltern becomes audible.

The Fluke of "Raw Material"

I was proud to be appointed by a university that, during apartheid, deliberately ignored the demands of privileged White academia and focused unabashedly on the oppressed communities surrounding the campus. The university prided itself, and rightly so, on being the University of the Left and threw all its resources behind the poor.

Since the first democratic election in 1994, South Africa has been trying to become part of what is sometimes called "a normal dispensation." Some months after my appointment at the university five years ago, I was asked to send a list of what I had published that year. Fortunately, or so I thought, I was quite active: a nonfiction book, poetry, controversial newspaper pieces, and more. So imagine my surprise to receive an e-mail saying that none of the listed writings "counted."

I went to see the dean of research. The conversation went like this:

"Why do my publications not count?

"It's not peer reviewed."

"It was reviewed in all the newspapers!"

"But not by peers."

Wondering why the professors teaching literature would not be regarded as my peers I asked, "So who are my peers?"

"Of course you are peerless," this was said somewhat snottily, "but I mean the people in your field."

"So what is my field?"

"The people working . . . ," and his hands fluttered, "in the areas about which you write."

"Well," I said, "when I look at their work I see that they all quote me."

His face suddenly beamed: "So you see! You are raw material!"

Initially, I thought nothing of the remark, but gradually came to realize how contentious, judgmental, and excluding the term "raw material" was. Who decides who is raw material? Are

Konile and //Kabbo and the Bushman woman "raw material"? Looking back on our project, I found myself asking, why did we three colleagues so easily assume that Konile was "raw material" and not a cowriter of our text? Why are her two testimonies and one interview in which she constructs and analyzes, deduces and concludes, less of an academic endeavor than our contribution? Her survival skills after the devastating loss of her son were not perchance remarks, but careful calculations and tested experiences from her side. During our interview, we even asked her to interpret her text. Why should she enter our book and the academic domain as raw material? Should she not be properly credited as a cotext producer on the cover like the three of us?

I began wondering: What would be the questions another Gugulethu mother would ask Konile? Or to move to another realm: How would one cattle herder interview another cattle herder? How would one cattle herder analyze and appraise the words of a fellow cattle herder? How would such an interview differ from me interviewing that cattle herder? And, finally, how can these experiences enter the academic discourse *without* the conduit of a well-meaning scholar? How shall we ever enter any new realm if we insist that all information must be processed by ourselves for ourselves?

The Fluke of Discipline

After being downgraded to "raw material," I duly applied to attend a workshop on how to write "un-raw" material in order to meet one's peers through unread but accredited journals. The workshop had been organized by the university after it became clear that our new democratic government wanted universities to come up with fundable research. We were obliged to compete with the established and excellently resourced former White universities and their impressive research histories.

I walked into this organized workshop. There were about 40 of us. I was the only White person. During smoke breaks, the stories poured out. The professor in math told the following:

One Sunday a member of the congregation told me that he was installing science laboratories in the schools of the new South Africa, that it was very interesting because every school was different. So this went on every Sunday until I said to him that he should write it down. So after I had completely forgotten about it, he pitched up [arrived] with a manuscript this thick [about four inches] and joked: Is this not a MA thesis? I looked and indeed it was new, it was methodically researched and systematically set out and riveting to read. So where to now? I said it was not math so he should take it to the science department. Science said it was more history than science. History said no . . . and so forth.

The group that attended the workshop was by no means subaltern, but first-generation educated men and women from formerly disadvantaged communities in apartheid South Africa. As we attended subsequent workshops in writing academic papers, one became aware of how the quality of "on-the-ground experience" was being crushed into a dispirited nothingness through weak English and the specific format of academic papers. We learned how easily an important story died within the corset of an academic paper, how a crucial observation was nothing without a theory, and how a valuable experience dissolved outside a discipline.

The Fluke of Theory

The last story is about a seminar I attended on the Black body. Opening the seminar, the professor said that when he was invited, he thought that the paper he was preparing would already have been accepted by an accredited journal and the discussion could then have taken place together with the peer reviews. The journal had, however, rejected the piece, so . . . maybe the discussion should start from scratch.

The paper he presented was indeed weak. As he was speaking, one had the distinct feeling of seeing a little boat rowing with all its might past waves and fish and flotillas and big ships and fluttering sails to a little island called Hegel. The oar was kept aloft until, until. . . . At last, the oar touched Hegel. Then the rowing continued desperately until the oar could just-just touch the island called Freud or Foucault. In the meantime, you want to say, forget these islands, show us what is in your boat, point out the fish that you know, how did you sidestep that big ship, where did you get these remarkable sails?

The discussion afterward was extraordinary. Suddenly, the professor was released from his paper and the Black students and lecturers found their tongues and it became a fantastic South African analysis. Afterward, I asked the professor: "Why didn't you write what you have just said?" He answered, "Because I can't find a link between what I know and existing literature. It's a Catch-22 situation: I cannot analyze my rural mother if it is assumed that there is no difference between her mind and the average North American or Swedish mind. On the other hand, my analysis of my rural mother will only be heard and understood if it is presented on the basis of the North American and Swedish mind."

▣   ACADEMICS FROM MARGINALIZED COMMUNITIES

Both of my colleagues, Nosisi Mpolweni and Kopano Ratele, were the first in their families to be tertiary educated, while I was the fourth generation of university-educated women. Right through our collective interpretative analysis on the testimony of Konile, the power relations among us changed. The project started with my initiative, but I quickly became the one who knew the least. Ratele was the best educated of us three, having already published academically. Nosisi made an invaluable input with her translations and knowledge about Xhosa culture. I could write well, but not academically well. English was our language, but only Ratele could speak it properly. During our field trip to interview Konile, the power swung completely to Nosisi, while I, not understanding Xhosa, had no clout during our fieldwork excursions.

However, during our discussions, I became aware that while we were talking my colleagues had these moments of perfect formulation—a sort of spinning toward that sentence that finally says it all. We would stop and realize: Yes, this was it. This was the grasp we were working toward, but when we returned with written texts, these core sentences were nowhere to be seen in the work.

For one of our sessions, I brought a tape recorder. We were discussing why Konile so obsessively used the word "I" within her rural collective worldview. I transcribed the conversation, sent everybody chunks, and here is the text returned by Ratele:

> Mrs. Konile dreamt about the goat the night before she heard that her son was killed. The TRC however was not a forum for dreams, but for the truth about human rights abuses. I suggest that through telling about the dream, Mrs. Konile was signaling to the TRC her connection to the ancestral worlds.
>
> The dream revealed that she was still whole, that she was in contact with the living and the dead and she clearly experienced little existential loneliness.... Her son's death is what introduced her to a loneliness, a being an "I." She had become an individual through the death of her son—selected, cut off, as it were to become an individual. She was saying: "I am suffering, because I had been forced to become an individual." The word "I" was not talking about her real psychological individuality. Mrs. Konile was using "I" as a form of complaint. She was saying: "I don't want to be I. I want to be us, but the killing of my son, made me into an 'I.'" (Krog et al., 2009, pp. 61–62)

As a White person steeped in individuality, I initially did not even notice the frequency of the word "I," but when I did it merely confirmed to me that the notion of African collective-*ness* was overrated, despite the emphasis it receives from people like Nelson Mandela and Archbishop Desmond Tutu. The conclusion Ratele reached, however, was the opposite, and it was a conclusion I could not have reached, and, up until now, also one that no other White TRC analyst had reached.

For me, this was the big breakthrough not only for our book, not only in TRC analysis, but also in our method of working. The confidence of the spoken tone, a confidence originating from the fact that somebody was talking from within and out of a world he knows intimately, had been successfully carried over onto paper. Ratele was crossing "frontiers" to get past all the barriers lodged in education, race, background, structure, language,

and academic discipline to interpret his own world from out of its postcolonial, postmodern past and racial awarenesses with a valid confidence that speaks into and even beyond exclusive and prescriptive frameworks.

My guess is that my colleague would never have been able to write this particular formulation without first talking it, and talking it to us—a Black woman who understood him and a White woman who did not.

We wrote an essay about Konile's dream in our three different voices, but the piece was rejected by a South African journal for allowing contradictory viewpoints to "be" in the essay, for having a tone that seemed oral, for not producing any theory that could prove that Konile was somehow different from other human beings, and so on. The piece was, however, I am glad to say, accepted by Norman Denzin, Yvonna Lincoln, and Linda Smith for their book on indigenous methodologies.

▣ CONCLUSION: RESEARCH AS RECONCILIATORY CHANGE

These examples, ranging from a Bushman shaman to a Black professor of psychology, expose the complexities of doing research in a country emerging from divided histories and cultures. It also poses ethical questions about the conditions we set for people to enter academic discourse. Spivak indeed stresses that ethics is not a problem of knowledge but a call of relationship (Williams & Chrisman, 1994, p. 190). When she claims that the subaltern "cannot speak," she means that the subaltern as such cannot be heard by the privileged of either the first or third worlds. If the subaltern were able to make herself heard, then her status as a subaltern would be changed utterly; she would cease to be subaltern. But is that not the goal of our research, "that the subaltern, the most oppressed and invisible constituencies, as such might cease to exist" (Williams & Chrisman, 1994, p. 5)?

French philosopher Deleuze rightly remarks that the power of minorities "is not measured by their capacity to enter into and make themselves felt within the majority system" (Deleuze & Quattari, 1987, p. 520). At the same time, Deleuze points out that it is precisely these different forms of minority-becoming that provide the impulse for change, but change can only occur to the extent that there is adaptation and incorporation on the side of the standard or the majority.

We have to find ways in which the marginalized can enter our discourses in their own genres and their own terms so that we can learn to hear them. They have a universal right to *impart information and ideas through any media and regardless of frontiers,* and we have a duty to listen and understand them through engaging in new acts of becoming.

▣ NOTE

1. This chapter extends and inserts itself into the discussion of *testimonio,* as given in John Beverley's article, "*Testimonio,* Subalternity, and Narrative Authority" (Denzin & Lincoln, 2005, pp. 547–558).

▣ REFERENCES

Brown, D. (2006). *To speak of this land—Identity and belonging in South Africa and beyond.* Scottsville, South Africa: University of KwaZulu-Natal Press.

A thousand., & Quattari, F. (1987). *Thousand plateaus: Capitalism and schizophrenia* (B. Massumi, Trans.). Minneapolis: University of Minnesota Press.

Denzin, N. K., & Lincoln, Y. S. (Eds.). (2005). *The SAGE handbook of qualitative research* (3rd ed.). Thousand Oaks, CA: Sage.

Denzin, N. K., Lincoln, Y. S., & Smith, L. T. (Eds.). (2008). *Handbook of critical and indigenous methodologies.* Thousand Oaks, CA: Sage.

Krog, A., Mpolweni, N., & Ratele, K. (2009). *There was this goat—Investigating the truth commission testimony of Notrose Nobomvu Konile.* Scottsville, South Africa: University of KwaZulu-Natal Press.

Lewis-Williams, J. D. (Ed.). (2000). *Stories that float from afar—Ancestral folklore of the San of Southern Africa.* Cape Town, South Africa: David Philip.

Liebenberg, L. (1990). *The art of tracking: The origin of science.* Cape Town, South Africa: David Philip.

Spivak, G. C. (1988). Can the subaltern speak? In C. Nelson & L. Grossberg (Eds.), *Marxism and the interpretation of culture.* New York: Macmillan.

Williams, P., & Chrisman, L. (Eds.). (1994). *Colonial discourse and postcolonial theory: A reader.* New York: Harvester Wheatsheaf.

23

JAZZ AND THE BANYAN TREE

Roots and Riffs on Participatory Action Research

Mary Brydon-Miller, Michael Kral,
Patricia Maguire, Susan Noffke, and Anu Sabhlok[1]

When Charles Mingus, Charlie Parker, Dizzy Gillespie, Max Roach, and Bud Powell took the stage at Massey Hall in 1953 playing *Perdido,* it seemed effortless, as if this incredible music just exploded from the stage to engulf the audience. But the truth is that this apparently spontaneous eruption of perfectly crafted music was the result of rigorous training—an integration of music theory and years of practice. Genuine improvisation is only possible when the players understand that each voice contributes a vital component to the overall structure of the piece. The willingness to innovate and explore that is the heart of great jazz music is made possible by the individual expertise of the musicians and their respect for one another's differing contributions.

Participatory action research (PAR) is like jazz.[2] It is built upon the notion that knowledge generation is a collaborative process in which each participant's diverse experiences and skills are critical to the outcome of the work. PAR combines theory and practice in cycles of action and reflection that are aimed toward solving concrete community problems while deepening understanding of the broader social, economic, and political forces that shape these issues. And PAR is responsive to changing circumstances, adapting its methods, and drawing on the resources of all participants to address the needs of the community.

Participatory action research is also like the banyan tree. The great Indian poet Rabindranath Tagore immortalized the "shaggy-headed banyan tree," a symbol of learning, meditation, reflection, and enlightenment in both Hindu and Buddhist traditions. But the banyan tree is also a gathering place of common people, a place of community discussion and decision making. By spreading out its branches and putting down deep roots, the banyan tree extends its reach and creates new spaces for living

and learning. Similarly, the participatory action research process provides a space within which community partners can come together and a process by which they can critically examine the issues facing them, generating knowledge and taking action to address these concerns.

In this chapter, we draw upon the power of metaphor and narrative to provide readers new to participatory action research a framework for understanding and appreciating this practice, while at the same time, we hope, providing those already familiar with PAR some unexpected insights into this approach. We begin by defining participatory action research and by locating this approach within the broader context of action research practice. We then provide a brief historical overview of participatory action research and the major contributors to the development of this practice. Following this, we consider the relationship between some of the theoretical frameworks that have informed participatory action research and the ways in which methodological choices both reflect and deepen our understanding of theory, including a discussion of new frameworks for conceptualizing research ethics in the context of PAR. We then provide three exemplars from recent projects to illustrate how PAR works in practice and to identify common themes and concerns rising out of the process. We highlight these research narratives because we believe that these descriptions best demonstrate the reach of participatory action research, the diversity of issues that are being addressed, and the wealth of approaches that have been developed to create new knowledge and to bring about meaningful change in communities. And because these narratives bring to life the most important element of PAR— the people involved in doing it. We close with a discussion of pedagogical strategies for deepening understanding and improving practice in both classroom and community settings and a reconsideration of our metaphors and their implications for the future.

◼ DEFINING PARTICIPATORY ACTION RESEARCH

Participatory action research is the sum of its individual terms, which have had and continue to have multiple combinations and meanings, as well as a particular set of assumptions and processes. Most important to this chapter are the intentions behind the terms, as well as the particular historical and contemporary meanings these convey. Participation is a major characteristic of this work, not only in the sense of collaboration, but in the claim that all people in a particular context (for both epistemological and, with it, political reasons) need to be involved in the whole of the project undertaken. Action is interwoven into the process because change, from a situation of injustice toward envisioning and enacting a "better" life (as understood from those in the situation) is a primary goal of the work. Research as a social process of gathering knowledge and asserting wisdom belongs to all people, and has always been part of the struggle toward greater social and economic justice locally and globally. While participatory action research falls under the broader framework of action research approaches, all of which share a belief that knowledge is generated through reflection on actions designed to create change (Reason & Bradbury, 2008), PAR is distinct in its focus on collaboration, political engagement, and an explicit commitment to social justice.

◼ EXPLORING THE ROOTS OF PARTICIPATORY ACTION RESEARCH

It is always problematic to construct a history of PAR, whose roots, like those of the banyan tree, are deep and wide. Distinguished by a collaborative ethos, any origins narrative of PAR involves the tension between giving credit to seminal individuals' work while avoiding the "one great expert" trap. As McDermott (2007) notes, the theoretical underpinnings and the practical methodologies of PAR "resisted traditional models of knowledge construction which privilege expert knowledge" (p. 405), so we endeavor here to place the history of PAR within the global social and political contexts both within and outside the academy that framed its creation, while acknowledging some of the key contributors to this development.

All approaches to knowledge production or inquiry are shaped by the historical contexts in which they emerge. The seeds of PAR were sown in the early 20th century as the social sciences emerged as new disciplines. African American and feminist voices (DuBois, 1973; Lengerman & Niebrugge-Brantley, 1998; Reinharz, 1992) were part of a broad effort to create new forms of inquiry deeply connected to collective action for social justice (Greenwood, personal communication; see also Messer-Davidow, 2002; Price, 2004). Meanwhile critiques of positivist social science research exposed claims to values-free knowledge production as untenable and challenged the basic tenets of objectivity and generalizability (Fay, 1975; Kuhn, 1970; Mills, 1961), laying the groundwork for more politically informed and socially engaged forms of knowledge creation. PAR has also been deeply influenced by feminist critiques of the social construction of knowledge (Calloway, 1981; Maguire, 1987, 2001a; Mies, 1982; Reid & Frisby, 2007; Reinharz, 1992; Smith, 1989), acknowledging that the identities and positionalities of those involved in knowledge creation affect its processes and outcomes.

Affirming the notion that ordinary people can understand and change their own lives through research, education, and action, PAR emerged with other challenges to existing structures of power by fostering opportunities for the development of more participatory and democratic solutions to social problems. PAR's development has been informed by numerous social struggles and movements, including workers' movements (Adams, 1975), women's movements (Maguire, 2001b), and human rights and peace movements (Tandon, 1996). One such influence was the postcolonial reconceptualization of international development assistance in the 1960s and 1970s in response to a quarter century of failed development policies (Frank, 1973; Furtado, 1973).

Much of the early development of PAR took place outside of traditional academic settings, particularly in the "south," or third world, away from Western European and North American contexts. In the early 1970s, Marja-Liis Swantz, a Finnish social scientist with the Tanzanian Bureau of Land Use and Productivity, and her students at the University of Dar es Salaam used the term "participant research" to describe their participatory development and research work with Tanzanian villagers (Swantz, 1974, 2008; see also Hall, 1993). During the same period, Orlando Fals Borda and other Latin American sociologists were using the term *investigation yaccion*—participatory action research—to describe "research that involved investigating reality in order to transform it" (quoted in de Souza, 1988, p. 35; see also Fals Borda, 1977, 1979).

Another key contribution to the development of PAR was the reframing of adult education as an empowering alternative to traditional and colonial education to unsettle relationships based on dominance (Freire, 1970; Horton & Freire, 1990; Kindervatter, 1979; Nyerere, 1969). Confronting "contradictions between their philosophy of adult education and their practice of research methodology" (Tandon, 1988, p. 5), adult educators sought to develop research approaches that reflected the same democratic and collaborative values that informed their pedagogy. Paulo Freire's literacy work in Brazil (1970), the Scandinavian folkschool movement, the contributions of Myles Horton and the other founders of the Highlander Research and Education Center (Horton & Freire, 1990; Lewis, 2001), and regional participatory research networks, such as the Society for Participatory Research in Asia (see Tandon, 2005), which grew out of the International Council for Adult Education funded Participatory Research Project (PRP) under the directorship of Budd Hall (2001), all contributed to the further growth of critical pedagogy and participatory action research.

These international connections and networks, formal and informal, provided an important opportunity for links between the long-standing tradition of action research in education and other social research areas. For example, in the action research emerging in Australia in the mid 1980s, Kemmis and McTaggart, through the suggestion of Giovanna di Chiro, formed connections to the work of Fals Borda, describing their work as participatory action research. This explicitly socially critical perspective on action research also built on participatory norms developed in the literacy work of Marie Brennan and Lynton Brown (Kemmis & McTaggart, personal correspondence; see also Kemmis & McTaggart, 2005).

Following the United Nations' Convention on the Rights of the Child (1989), which recognized children's right to participate in projects affecting them, there has been considerable growth of PAR involving young people as agents rather than objects of research (Fine & Torre, 2005). Whether in community-based organizations or school-related projects, young people often frame problems—and solutions—differently from adults in those settings (Cammarota & Fine, 2008; Fernández, 2002; Groundwater-Smith & Downes, 1999; Guishard, 2009; Hutzel, 2007; Lewis, 2007; McIntyre, 2000; Morgan et al., 2004; Tuck, 2009).

Taken together, these varied "origins" of PAR show distinct characteristics. One is that of continued challenges to forms of knowledge generation that position nondominant groups as outsiders. Popular knowledge generation is a crucial element to PAR. But perhaps most important is that PAR has grown from and with social movements (Mies, 1982). PAR's emergence within the academy came from discipline-specific activist scholarship. The social struggles of literacy and development workers, feminists, labor activists, civil rights workers, and activist academics all informed the foundation of PAR, and are an essential part of considering its use in contemporary social research.

Theoretical and Methodological Frameworks

In participatory action research, the distinction between theory and method is challenged by the assumption that theory is informed by practice and practice a reflection on theory. Methods for collecting, analyzing, understanding, and distributing data cannot be separated from the epistemologies, social theories, and ethical stances that shape our understanding of the issues we seek to address.

While the theories, methods, and the methodologies in PAR are varied and evolve differently within every context, it is the belief in collaboration and respect for local knowledge, the commitment to social justice, and trust in the ability of democratic processes to lead to positive personal, organizational, and community transformation that provide the common set of principles that guide this work (Brydon-Miller, Greenwood, & Maguire, 2003). PAR stems from the understanding that knowledge(s) are plural and that those who have been systematically excluded from knowledge generation need to be active participants in the research process, especially when it is about them. The nature of the data collected, the methods of analysis, and the resulting reflections and actions emerge out of a collaborative engagement within a community and often, but not always, with an academic researcher as a partner in the process. In order to reflect the principles of PAR, this academic researcher must be willing to embrace the hard work of examining how his or her multiple identities shape and inform engagement with community members.

This critical examination of issues of identity requires an analysis of the dynamics of power and privilege. From the outset, reflection on these aspects of the research relationship has informed the practice of participatory action research. The early Gramscian and feminist-informed PAR work were developed as responses to the political agendas and social contexts of their time. The critiques of colonialism embedded in these works (Fals Borda & Mora-Osejo, 2003) and the role of women's groups in the efforts toward social reconstruction (Mies, 1982) must be seen as efforts toward theorizing power relations. In the contemporary context, these would be recognized as part of postcolonial theorizing, and PAR workers in that era were aware of the emergent writings of Freire and his analysis of systems of oppression (1970). While other social theories have been employed in relation to PAR, the integral connection between social theory and social action has been an essential part of participatory action research strategies. Theoretical frames are seen as integrally connected to politics, meaning not only in the examination of the workings of power, but also the nature of participation, and in activism as a contributor to the fight for social justice. One example of current resonances between social struggle and social research is reflected in the use of narratives within PAR and the prominence of counter narrative within critical race theory (Brydon-Miller, 2004; Ladson-Billings & Tate, 1995). As Delgado notes, "stories can shatter complacency and challenge the status quo" (2000, p. 61), key objectives of the participatory action research process as well. Another emergent intersection between an existing theoretical framework and PAR is in the exploration of the ways in which the notion of borderlands scholarship (Anzaldúa, 2007; Torre & Ayala, 2009a) can deepen the practice of PAR. There is also an important component of reciprocity in the relationship with both critical race theory and borderlands scholarship, in that

PAR offers concrete strategies for making manifest the critical perspectives and demands for social justice embodied in these frameworks (Brydon-Miller, 2004; Torre & Ayala, 2009b).

Recent social theory also emphasizes agency, subjectivity, and pragmatism. Toulmin (1988) described the more recent philosophical turn from universal, general, and timeless thinking to oral, particular, local, and timely investigations. Writers from the later Frankfurt School challenged positivist inquiry by looking at local and contextualized meanings, as well as moral, historical, and political realities needing to be tied to research methodology (Rabinow & Sullivan, 1985). Ortner (2006) argues for the need to incorporate subjectivity, agency, and power into social research, what Burke (2005) calls the return of the actor. In PAR, the research subject becomes this actor, whose contexts and communities are woven into the research tapestry. And PAR argues for including diverse actors in the process. It is an indigenizing of research practice, and indigenous organizations such as the National Aboriginal Health Organization (NAHO) of Canada have developed ethical principles calling for deep community participation in, and shared ownership of, all aspects of research projects (NAHO, 2007; Royal Commission on Aboriginal Peoples, 1993; Smith, 1999). In many indigenous communities, researchers have made a bad name for themselves. Linda Tuhiwai Smith (1999) writes that research "is probably one of the dirtiest words in the indigenous world's vocabulary" (p. 1). This is because, like the colonizers, researchers have, too many times, come into indigenous communities to collect their stories to disappear without a word coming back or any benefit returning to the community. This has been experienced by many indigenous communities as another form of dispossession. Some researchers have come as ethnographers who were later seen as spies working with "informants" (see Deloria, 1997). The natives are now talking back, in the contexts of decolonization and reclamation of control over their lives, part of self-representation and what is now being called indigenism (Niezen, 2003).

Asad (1993) referred to such practices as the reconstitution of colonized subjectivities. Increasingly, the addition of collaboration, reciprocity, consultation, and public engagement are entering the social research agenda. A public anthropology is on the rise, which includes public participation and empowerment in research directed toward solving problems in the world (Lamphere, 2004; Rylko-Bauer, Singer, & Van Willigen, 2006). This practical or pragmatic emphasis is also appearing in methodological texts (Creswell & Plano Clark, 2007; Maxwell, 2005).

The openness and willingness to allow for multiple knowledge(s) within the PAR process, as well as the rejection of the assumptions embedded in a positivist worldview and the emphasis on critical reflection, open the doors to an eclectic approach to using existing methods as well as the development of methodological innovations. PAR draws upon both quantitative (Krzywkowski-Mohn, 2008; Merrifield, 1993; Schulz et al., 1998) and qualitative approaches, adhering to the belief that the issues facing the community and the research questions they generate to address these issues should drive the method. But, for the most part, the focus is on creating dialogue and generating knowledge through interaction. Often, PAR researchers use methods and communication strategies that have a "hands-on" nature (Kindon, Pain, & Kesby, 2007), especially when working with marginalized communities.

One area of particular note is in the use of arts as a way of both generating and recording the PAR process. PAR's emphasis on multiple ways of knowing creates fertile ground for experimentation and innovation with arts-based methodologies. Art offers both a means of expression and the potential to challenge and change. Arts-based methods employ a wide repertoire including storytelling (Sangtin Writers & Nagar, 2006), visual arts (Bastos & Brydon-Miller, 2004; Hutzel, 2007), photography (McIntyre & Lykes, 2004; Wang, 1999), performance (Boal, 1985; Guhathakurta, 2008; Pratt, 2004), fiber arts, indigenous arts (Lykes, 2001), and newer forms of media art. In addition, new technologies such as geographic information systems (GIS), which allow users to gather, synthesize, and represent information relating phenomena such as environmental hazards, health outcomes, income and educational disparities, or the incidence of crime to a location (Mapedza, Wright, & Fawcett, 2003), afford opportunities for innovative PAR methods that can help communities to re-envision change. These innovative data collection techniques must be seen in light of the epistemological shifts ongoing in research strategies. The use of the arts, of graphic organizers, and other techniques were not developed as ways to gain greater access to information for the "outside" researchers to analyze and interpret. Rather, the "methods" within PAR are determined not only to gain a richer understanding, but to generate new ways to consider actions within the socio-political sphere (Cammarota & Fine, 2008). Importantly, the "action" part of PAR has been a way to develop both the strategies for change in oppressive social situations as well as a sense of hope and agency among participants (Mies, 1996).

Clearly, there is a wealth of approaches available to researchers and community partners to carry out a PAR process, but whichever specific methods are selected, they emerge from the context, the interactions between the "outsider" and community knowledge forms, the questions to be addressed, the nature of actions, and the way the project evolves. The process is as critical as the product; and while the research results are important, PAR researchers pay careful attention to developing the skills and capacities of the community participants through the research process (Kesby, Kindon, & Pain, 2005; Maguire, 1987).

The theory and methods of PAR require a fundamental reconsideration of research ethics as well. The current system relies upon a contractual model of research ethics with an emphasis on informed consent and the academic researcher's ownership and control of data. This model reinforces the power and authority of the academic researcher and abets what Newkirk (1996) has referred to as the seduction and betrayal of research subjects lured into revealing intimate aspects of their lives only to have those details made public through the interpretations and representations of the researcher. The current model of research ethics also calls into question the principles of caring and commitment that are at the heart of participatory action research by recasting these relationships as potential sites of coercion.

An alternative model builds instead upon the notion of covenantal ethics, "an ethical stance enacted through relationship and commitment to working for the good of others" (Brydon-Miller, 2009, p. 244; see also Hilsen, 2006; May, 2000). The ethical grounding of participatory action research might be best conceived of as a system of community covenantal ethics framed by relationships of reciprocal responsibility, collaborative decision making, and power sharing (Brydon-Miller, 2007). This framework draws upon feminist research ethics (Brabeck, 2000) and is consistent with the current efforts within many indigenous communities to establish community-based systems of research ethics guidelines (Battiste, 2007). In Canada, the Royal Commission on Aboriginal Peoples (1993) published ethical guidelines for research where participation is key: "Researchers shall establish collaborative procedures to enable community representatives to participate in the planning, execution, and evaluation of research results" (p. 39). The National Aboriginal Health Organization (NAHO, 2007) also has a document on ethical research practices in indigenous communities arguing, "all partners are involved in the entire scope of the research project through the planning, implementation, data analysis and reporting stages" (p. 2). The challenge now before us is to synthesize these various efforts to reconceptualize research ethics and to clearly articulate criteria by which the ethics of participatory action research might be examined.

▣ VARIATIONS ON THE THEME OF PARTICIPATORY ACTION RESEARCH

The following exemplars present PAR projects in very different communities around the world, dealing with issues as diverse as youth suicide, response to and recovery from natural disasters, and the reclamation of local knowledge to enhance health outcomes for school-aged children. The methods used in these studies also highlight the range of ways in which PAR can be practiced. But all three demonstrate the commitment of the researchers to developing close and trusting relationships with community partners, their shared dedication to working for positive social change, and their willingness to reflect upon the challenges and contradictions of participatory action research.

▣ WELCOMING THE UNINVITED GUEST: PARTICIPATORY ACTION RESEARCH AND THE SELF EMPLOYED WOMEN'S ASSOCIATION

Anu Sabhlok

My perceived vulnerability, especially the fact that I was pregnant, created unforeseen advantages for me as a researcher. I was welcomed, well almost forced into homes. Immediately groups of women gathered around me. Ah! I thought I didn't have to work very hard towards getting respondents. But again, I was in for surprises. Before I could ask any of my questions I was fielding theirs, "what makes you come here in a pregnant state? Where is your husband, how did your family allow you to travel? Why is your nose not pierced?"–it was almost as if the gaze was reversed. I was the subject and they were the researchers. Sharing with them my stories, however, did help form a bond and created a space where research became a dialogue rather than a one-way interview.

—Anu Sabhlok, excerpts from my fieldnotes, 2001

Dissertations usually are very individualistic ventures—it is your question, your research, and eventually your degree. However, as my experience during my dissertation work in Gujarat shows, the dynamics of more participatory, community-centered research are difficult to predict and often unfold in surprising ways during the process. Gujarat, a state in western India, witnessed a massive earthquake in 2001, the year I started my doctoral work on earthquake rehabilitation. In 2002, violent Hindu-Muslim riots broke out resulting in more than 50,000 people having to flee to relief camps, primarily from the minority Muslim communities.

My initial agenda as I embarked upon this project as an architecture and geography student was to work on post-earthquake structures and temporary relief housing. However, many discussions with the local populations revealed that it was not the number of houses constructed or amount of money received in aid that was important; rather there were questions of access, corruption, economic liberalization (pulling out of the welfare state), religious ideology, and power that needed to be addressed. Thus, the research focus shifted to understanding the multiple meanings of relief work from the perspective of those that perform relief and those that receive it.

Sometimes "experts" are invited by development agencies to conduct PAR, other times an organic collaboration develops between the academics and the community. In my case, I wanted to do research that was participatory, that had a social justice agenda, and that dismantled the researcher-researched hierarchy. This desire stemmed from my feminist epistemological position and training. The collaboration was not organic and I was not invited—how then does one become part of a community and engage with them collaboratively toward a common agenda?

It is also important to address the question of distance—both geographic and social. How does engaged, participatory action research happen across 10,000 miles? I established e-mail contact with an organization that I admired, SEWA: Self Employed Women's Association. With roots in Marxist and Gandhian struggles, SEWA is the world's largest trade union of poor, informal-sector women workers. SEWA women had adopted five relief camps and had initiated *Shantipath Kendras* (centers of peace) to cultivate dialogue among the diverse religious communities. Through my work, I wanted to express solidarity with SEWA. Ironically, SEWA women, busy in their grassroots work, did not welcome me in easily. Bridging the gap with the working-class women in SEWA seemed almost impossible. However, I think our shared commitment toward building a more just, more peaceful, and a more democratic country (and Gujarat State in particular) paved the way for a collaborative process.

Eventually, after two months of visiting SEWA offices, I was accepted as a volunteer in SEWA academy. SEWA academy is the research section of SEWA and has been conducting grassroots research for the past 30 years. SEWA academy trains poor, often illiterate, women in the tools of research and helps them form a team of what they call "barefoot researchers." These teams go into urban and rural homes to identify and document issues relevant to informal-sector poor women. SEWA women also produce participatory videos to make their voices heard in the public domain. After a few "sharing sessions," we decided to document the multiple meanings of relief for those at the grass-roots. Initially, we organized focus groups where collective and individual experiences of relief work were shared. Then, accompanied by one or two SEWA women, I started conducting in-depth interviews and life-story sharing sessions with individual SEWA women. At the end of every week, we would again have a "sharing session" at the SEWA academy, where I would present an analysis of the interviews and the SEWA team would share results from the survey research they were conducting. A day before I returned to the United States, SEWA academy conducted an "experience sharing" session wherein we reflected upon the collaborative experience and attempted to make sense of the data. This process of collecting and exchanging understandings of relief efforts deepened and complicated our collective understanding of relief on both the individual and community levels. Through discussions, there emerged a collectively generated analysis of the relief process that not only outlined the numbers of those affected and resources distributed but also revealed how power (at the international, national, and local levels) and issues of identity (particularly religious, gender, class, and caste) play out during the relief process.

Back in the United States in 2005, I transcribed, translated, and re-analyzed the interviews and discussions and wrote out a 300-page document that was "my" dissertation. While we retained intermittent contact during the writing phase, there was very little collaboration with SEWA. I graduated in 2007 with a doctoral degree in geography and women's studies from Pennsylvania State University (Sabhlok, 2007). I struggle with numerous questions as I write this section: To what extent was this process PAR? While the research questions emerged out of a dialogue and we collaborated and shared the data collection with a collective agenda of social justice through research, yet the written product was my analysis and my writing. I struggle with questions of belonging—I was sometimes accepted as an insider (we shared similar commitments) and at other times my outsider position (differences in class and educational background) was made apparent. While the theoretical analysis in my dissertation might be of little use to SEWA women, I think it was the process of collaborative research, sharing, and reflection that enriched their and my perspectives (and therefore action) on disaster relief among other seemingly mundane things. This example brings to light some of the challenges that surface when academics engage with organizations that are independently deeply immersed in PAR processes.

SEWA's website shows a banyan tree. The top part of the tree shows thick foliage as its members, and the numerous roots stemming from branches represent the cooperatives, the social security organizations, the SEWA academy, and the various unions in SEWA. The interconnectedness of the roots gives the tree its strength and the new branches and roots that emerge create new spaces for SEWA's collective engagement and actions. Looking back, I see myself as part of these new roots that grow independently and yet are always connected to the tree. Looking ahead, I visualize that connection growing stronger. The dissertation as I see it was just the starting point for a long-term engagement with women in SEWA.

▣ "Don't Worry—Our Community's Been Gardening for Centuries": Youth Participatory Action Research in a School Setting

Patricia Maguire, from the work
of Alicia Fitzpatrick and her students

In this exemplar, Alicia Fitzpatrick, a high school teacher, and 15 students used participatory action research to create new spaces in school for learning and living (Eriacho et al., 2007; Fitzpatrick et al., 2007). A former Peace Corps volunteer, Alicia taught science for three years in an alternative high school on an American Indian nation in the southwestern United States. She was also a graduate student in a teacher education master's program that promoted critical reflection and action research with transformative intentions. During the program, teachers engage in action research (AR) to improve their own practices and to more deeply understand—and attempt to unsettle—the inequitable power arrangements and relationships that shape U.S. schooling (Maguire & Horwitz, 2005).

It was in the context of taking my teacher action research course, that this PAR project emerged from Alicia's conversation with a student. Reflecting on changes she had already made to her teaching practices, Alicia noted of her classroom lab activities, "I was still uncomfortable with the fact that the students were finding solutions to problems that had already been solved" (Eriacho et al., 2007, p. 6). While grappling with teaching practices she might improve through action research, one of her students, Alex, talked with her about the quality of school lunches. Noting widespread problems of diabetes and obesity, he thought school lunches should include fresh produce and vegetarian options. Reporting on this conversation, Alex noted two things. First, Ms. Fitzpatrick actually listened; and second, she asked him, "How do you see a solution to this problem?" (Eriacho et al., 2007, p. 7). His reply, that the school should grow its own food, framed the problem to be addressed by students and teacher through PAR.

Building on Alex's suggestion to develop an agricultural management elective, 15 students (with Ms. Fitzpatrick's facilitation and extensive inclusion of community members as resources) used a series of research-action cycles to codevelop the agricultural course curriculum through which they planned, built, and operated a school greenhouse and traditional garden. The core group of nine male and five female students included teen parents, students who had been expelled from other schools, and students identified for special education services. As they began, Richelle, one of the students, noted that the community viewed students from the alternative school as "bad influences" who would never graduate (Eriacho et al., 2007, p. 14). The project slowly unsettled this perception as students demonstrated leadership qualities while interacting with community members.

Alicia and these students studied what was happening for them as they codeveloped the agricultural class, school greenhouse, and garden. They maintained reflection journals and analyzed photographs of their work using Photovoice protocols (Wang, 1999). A group of students and Alicia cowrote the final paper for her graduate AR course (Eriacho et al., 2007). They were the first student-teacher team to ever copresent at the New Mexico Center for Teaching Excellence's annual Teacher Action Research conference (Fitzpatrick et al., 2007).

While PAR has increasingly been utilized with youth in community organizations, development projects, and after-school programs (Fernández, 2002; Hutzel, 2007; McIntyre, 2000; Nairn, Higgins, & Sligo, 2007), school-based teacher-student PAR is less common. Indeed, Groundwater-Smith and Downes (1999) have criticized the teacher-as-researcher movement, in which teachers examine their own practices, for positioning students as objects of teachers' studies rather than collaborators in classroom AR. In the school greenhouse and garden project, Fitzpatrick intentionally created processes for meaningful student leadership and decision making. This was challenging as the structures needed to support students as coresearchers

are not well represented in teacher action research literature (Brydon-Miller & Maguire, 2009). Throughout the AR course, Alicia continuously talked with other teachers—and her students—about how to create processes and spaces for students' control of the project and agricultural class.

Fitzpatrick and her students worked to move up Hart's Ladder of Child Participation (1997; see also Arnstein, 1969), from the bottom rungs of manipulated or token student participation to more meaningful youth-initiated and -directed participation. This movement required the teacher to share power and control with the students and community members in ways quite uncommon in the wake of No Child Left Behind mandates. Initially, Alicia was reluctant to even take on creation of the agricultural course, noting that she did not know much about gardening. Alex reassured her, "Don't worry—our community's been gardening for centuries." The students knew that given a framework, they could tap into the vast and long-term, community-held knowledge of traditional gardening practices in the arid Southwest. The teens were likewise required to stretch out of their comfort zones to develop curriculum, initiate contact with community resources, speak before public officials, and keep regular school attendance.

Teacher and students moved into another uncomfortable zone when they confronted the inequitable power arrangements of gender at play in the project. Alicia noticed that when students initially divided up the work, the boys took on greenhouse construction while the girls worked on the project website. She was also aware of the dominance of male voices in the classroom. "I instantly started to brainstorm ways in which females could speak up . . . not . . . strategies that would help the males become active listeners." She realized her teacher silence condoned male dominance and was surprised that her first impulse had been "to fix the girls" (personal communication, February 5, 2007). Despite her intellectual understanding of schools as raced, classed, and gendered spaces (Maguire & Berge, 2009), Alicia, a community outsider, was reluctant to start a conversation about project gender issues. But given space and structures, students did it themselves through Photovoice. For example, Farrah wrote about the time when analyzing photographs, a male student pointed out that there were only boys in one photo and only girls in another. "Once that was brought up, Ms. Fitzpatrick started asking questions about why that was. I said I would have liked to join the guys but then I would have been the only girl and I would have felt out of place" (Eriacho, et al., 2007, p. 23). Alicia continued the narrative: "After Farrah shared her thoughts . . . a male student sitting next to her said, 'What? You felt like this?' She looked at him and said, 'Yes.'" Only the girls had been aware of the very visible gender divisions. This is not unusual in PAR projects when gender is a quality assigned to females but not males and the gendering mechanisms at work are ignored (Maguire, 2001b). After the discussion, two girls joined the previously all-male construction group.

The greenhouse and garden project is an example of combining PAR and a greening project to change school landscapes into "learnscapes" (Lewis, 2007). As Lewis (2004) noted, the school grounds offered a rare uncolonized *third space* in schools where participatory research and education projects could take root.

▣ "I'm Doing This From My Heart": Inuit Suicide Prevention and Reclamation in Nunavut, Canada

Michael Kral

Suicide among indigenous youth is an epidemic in the circumpolar north. Canadian Inuit have among the highest suicide rates in the world. Why suicide? A brief history puts this in perspective. Nunavut is an Inuit political territory established in 1999 in the central and eastern Canadian Arctic. It is about the size of India, with 26 communities and a population of about 27,000. Almost all community members are Inuit, speaking Inuktitut and English. Significant contact by outsiders or *Qallunaat* began with Scottish and American whalers in the late 19th and early 20th centuries, and continued with the trinity of the fur trade, missionaries, and police. The most colossal change in Inuit history began in the 1950s, when the Canadian government tried to help Inuit during a tuberculosis epidemic and time of hunger. Inuit were moved from their family camps to aggregated settlements, children were placed in day or residential/boarding schools where they were not allowed to speak their language. *Qallunaat* Northern Service Officers ran the settlements and established a foreign electoral system, gender roles changed particularly for men, the new welfare state created poverty, and intergenerational segregation began for a people for whom kinship was at the center of social organization. Many of those children grew up without proper parenting skills and traditional ways, and their children began killing themselves in the 1980s in numbers that have only continued to grow (see Brody, 2000; Condon, 1988; Kral & Idlout, 2009; Wenzel, 1991).

In 1994, I attended a national conference on suicide prevention in Iqaluit, now the capital of Nunavut, Canada. This was my first visit to the Arctic. I was asked to chair a panel on suicide research based on my involvement with the Canadian Association for Suicide Prevention, which was hosting the conference. I was the only *Qallunaat* on that panel. Most of those attending were Inuit, and discussions were in Inuktitut and English with simultaneous translation. At this session, Inuit from many communities spoke about what they needed to know to prevent suicide among their youth. One older woman stood up and began a discussion in Inuktitut by saying, "In my community we have many suicides. My relative lives in another community and they have no suicides. We need to learn from that community. We need

to ask about wellness." Inuit also spoke about how to gather this information. This gathering of knowledge would need to take an oral and collective angle, in keeping with Inuit culture. Sharing was emphasized. Elders and youth should be involved with other community members. On the last day of the conference as a number of us sat around a table, I suggested we put these ideas into a project.

I was nervous as an outsider, not wanting to be seen as yet another colonial researcher taking advantage. I asked an Inuit elder with a translator what she thought of my being involved with them in such a project. Her response was that if my heart is in it, this is good. We developed a project over the next three years. An Inuit steering committee was organized by Eva Adams, an Inuk (Inuit) living in Nunavut, and I put together an academic research team. Our work together was based on the research questions and methodology offered by Inuit at the conference. We applied for and received a grant, and conducted our first study with two Nunavut communities, primarily with members of their youth committees, who helped shape the questions and how the research would take place. Elders finalized the open-ended interview, with Anthony Qrunnut volunteering as a test interviewee. I was greatly relieved when at the end, after an anxious (for me) pause, he smiled and said those were the questions he would ask. The Inuit steering committee provided the most direction during planning and data collection. We completed our report in 2003 with our grant money having expired years earlier, but the time taken was worth it. This work cannot be rushed. This is where I first learned about PAR, which emerged as an indigenous methodology.

Many Inuit have been involved in our participatory research, and we have a core group of four Inuit and myself. I am still the only *Qallunaat* in our Nunavut research group, and am honored to be so accepted. Not everyone in the Inuit communities accepts me, however, and this is a reality I live with. Yet my Inuit partners acknowledge that I can provide a helpful outsider's perspective. My main partner has been Lori Idlout, executive director of the Embrace Life Council, an Inuit organization devoted to suicide prevention and community wellness in Nunavut. Lori believes in PAR and cannot see us working any other way, telling me, "PAR is the most culturally sensitive form of academic research in social sciences when working with Inuit. The methodology is such that decision making is collaborative between the researcher and the community involved." (Idlout, personal communication, June 2010). The participatory process has been directed by Inuit in our research through their local knowledge and expertise. It is the process that has been the product, the most important factor.

Natar Ungalaq, our coresearcher in Igloolik, Nunavut, has long been involved in helping youth in his community. He said that money was not his priority when we were discussing people getting paid from our grant. "I'm doing this from my heart." This is the enduring spirit of our collaborative research on a difficult topic, which has now turned into individual and community success stories. Inuit have been decolonizing, moving to reclaim control over their lives, and our research aligns with this (Kral, Wiebe, Nisbet, Dallas, Okalik, Enuaraq, & Cinotta, 2009). Our research is Inuit-driven yet a true collaboration of minds, hearts, and cultures with love of life, people and community, and Inuit pride, as our motivating forces.

▣ LOCAL RIFFS ON COMMON THEMES

These exemplars highlight a number of issues related to the practice of participatory action research. The question of the roles and relationships of the researcher and the community members involved in the process are often cast in terms of insider/outsider dynamics, but this kind of dichotomizing overlooks the complex nature of these relationships and the possibility that the researcher might occupy multiple positions simultaneously. Anu's position as a young Indian woman brings her into the homes of the people in the community, but her caste, class, and educational status set her apart. Alicia's status as outsider necessitates a shift in her role as teacher to that of learner challenging the traditional systems of power and authority embedded in our educational practices. And while Michael is clearly identified as an outsider, a *Qallunaat,* which leads some members of the community to question his participation, his long-term commitment to working to address critical community concerns has led to the development of effective interventions as well as warm friendships (see also Humphrey, 2007; Johns, 2008).

The importance of community knowledge and expertise and the question of who owns and controls that knowledge and who benefits from the research are also central concerns in PAR. Alex's reassurances to Alicia that the understanding of effective agricultural techniques already resides in his community and her willingness to respect that knowledge and to put her own skills and resources to work on behalf of the community represents the kinds of genuine partnerships that are at the heart of participatory action research. Likewise, in Michael's exemplar, his research partner Lori notes that PAR is more resonant with local culture and mores than other forms of social research. PAR is in keeping with indigenous cosmologies where relationships are at the center, a form of research that is "evaluated by participant-driven criteria" (Denzin & Lincoln, 2008, p. 11). It is a decolonizing of methods and of academia, a political stance in the redistribution of power with a focus on sharing and mutual respect. Fine, Tuck, and Zeller-Berkman (2008) show that "participatory methods respond to these crises in politics by deliberately inverting who constructs research questions, designs, methods, interpretations, and products, as well as who engages in surveillance" (pp. 160–161).

The action side of PAR is to make the findings useful, and there is clear benefit to the communities participating in these projects. The impact of community-based programs in reducing youth suicide in Inuit communities is testament to the ability of participatory action research to create positive change. Likewise, Alicia's students benefit not only by having access to healthier food, but in developing stronger academic skills and deepening their sense of pride and their acceptance as contributing members of their communities. And even as Anu questions her effectiveness in bringing about change it is clear that her partners have acquired new skills and new understandings and that her on-going commitment to working for positive social change will continue to benefit the community.

At the same time it is important to acknowledge broader implications of PAR and the importance of transferability of knowledge to other researchers, other communities, and other settings. In each case the research being conducted by these scholars—and this includes both the academic and community-based researchers—deepens our collective understanding of important issues and provides strategies for others to draw upon in working to address similar concerns. Michael's work with his partners is already being expanded to other communities in the circumpolar region while Alicia's project provides an opportunity for other teachers to consider ways in which they might bring the knowledge already extant in their communities into their classrooms to enrich their students'—and their own—learning experiences. Anu's work enables an alternative narrative to government and international non-governmental organizational studies on disaster relief, one that is constructed in partnership with those that perform the hard on-the-ground labor of providing relief.

Each of these exemplars also reflects the criteria of the system of community covenantal ethics described earlier: reciprocal responsibility, collaborative decision making, and power sharing. Michael's project is explicitly founded in systems of indigenous research ethics that inform this understanding of research ethics. And by partnering with SEWA and responding to the questions and concerns of that community Anu also reflects these values. Sharing power with her students and working together to reach decisions regarding the process, Alicia allows them to create a project that is both personally and culturally meaningful to them. Drawing upon such work provides clear models for the further development of this notion of community covenantal ethics and compelling demonstrations of how they might be applied in practice.

▣ PRACTICAL PEDAGOGICAL ISSUES OF
IMPLEMENTATION

Jazz challenges assumptions about what music is, how it works, and what makes it beautiful. In teaching participatory action research, as with jazz, the first step is in challenging the dominant aesthetic of positivist research and traditional teaching methods in order to find beauty and productivity in the unpredictable and polyvocal processes of collaboration and group work. As a former student, Beverly Eby, noted after completing a year-long participatory action research course, "it took a whole quarter to learn that it was okay to do this."

Participatory action research provides spaces and specific processes where people, whatever their "instrument" or level of skill, can work together to "make music." The participants, whether expert or novice, must do the work of critical self-reflection, examining their own identities, positions of power and privilege, and interaction styles, as well as how these continuously impact the research process. The pedagogy of PAR must be congruent with these underlying values, proactively inviting and facilitating the contributions of and dialogue among diverse learners (Maguire, 2001b). This is as potentially scary as it is exciting because it means relinquishing control in the classroom to allow students to take the lead.

Participatory action research has been described as the intersection of popular education, community-based research, and collaborative action aimed at achieving positive social change (Brydon-Miller, 2001; Hall, 1993). PAR draws from non-formal and experiential education and community participatory appraisal techniques (Chambers, 1980; Kindervatter, 1979). Teaching PAR, whether this takes place in a university classroom or in a community center or other informal educational setting, must also combine these three elements. Traditional forms of "banking education" (Freire, 1970) must give way to more generative forms of exploration, discovery, and play while at the same time remaining attentive to the well-articulated theoretical and methodological foundations of this approach. Challenging accepted pedagogical practices in this way creates a "constructive disruption" (Cochran-Smith & Lytle, 2009, p. 86) of the basic structures of education by shifting power from the instructor to a system that actively encourages participation and ownership by all participants, whether they are students in a classroom or community partners.

▣ CONCLUSION

Using metaphors, if done uncritically, can be tricky. In closing, let us consider some of the limitations of the jazz and banyan tree metaphors and the way they might provide a more nuanced understanding of some of the issues of power and privilege that are at the heart of participatory action research (PAR). The Bania, a subcaste of Indian traders who sat under banyan trees to strategize and plan, have been accused of making heavy profits at the cost of the peasantry (Kumar, 1983). While not the fault of the banyan tree, which merely offers its shade, digging just below the surface of the metaphor raises issues of cooptation and expropriation. Who benefits from the strategizing taking place in the shade of the banyan, and for what and whose purposes? In a similar vein, we are concerned about the potential for participatory action research methods to be coopted by those using the language of collaboration and community control to encourage participation in processes that ultimately serve the interests of those in power in political, economic, as well as academic spheres.

From its inception, jazz has been criticized as being a male-dominated musical form, in which cultural and gender constraints colluded to keep many women out and to keep those women instrumentalists who made it in, under-recognized. Race solidarity did not trump gender discrimination. Tucker (n.d.) asked, "If women have played jazz all along, why don't we know more about them?" The reasons for our limited knowledge of African American women's and other women's contributions to the world of jazz are similar to reasons that at one time we knew little of the extensive contributions of women to the early development of PAR. The historical records that become the canon were often written by men, about their male colleagues, citing their male colleagues' publications (Maguire, Brydon-Miller, & McIntyre, 2004). There was little professional expectation that male social scientists or project directors would be well-versed in evolving feminist scholarship or women working in projects.

For great jazz music and banyan trees to endure, they have had to grow and change. The banyan extends its reach by sending down shoots that take root and become new trees. And musicians such as Charles Mingus and the others had to welcome on stage new generations of jazz musicians and even entirely new musical forms. So, too, PAR grows, changes, and expands with each new group of practitioners. Anu, Michael, Alicia, and the other students and younger scholars working in the field represent those new forms and new voices in participatory action research. The commitment of these individuals to their community partners and their dedication to tackling tough issues and to working to achieve meaningful social change reflect the influence and values of their predecessors. At the same time, their theoretical and methodological innovations, including an increased access to technology and the opportunity for increased international dialogue and the exchange of experiences and practices this makes possible, create fertile ground for further growth and development of the field. What comes next is therefore integrally connected to the social struggles evident in the global and local contexts and how those at the local level use the methods of research to understand and change the world.

We believe passionately in the power of participatory action research to push us to challenge and unsettle existing structures of power and privilege, to provide opportunities for those least often heard to share their knowledge and wisdom, and for people to work together to bring about positive social change and to create more just and equitable political and social systems.

▣ NOTES

1. The authors are listed in alphabetical order and have contributed equally to this work.

2. We wish to thank Mary's former student, Warren Foster, for first suggesting the jazz metaphor. It is also worth noting here that we chose this moment in jazz history because, to quote Rick VanMatre, director of Jazz Studies at the College-Conservatory of Music at the University of Cincinnati, "most modern jazz musicians still consider Mingus, Parker, and the others on this Massey Hall concert to be at the highest level of artistry and collaboration." VanMatre goes on to note that "the principles of working your butt off, studying, analyzing, and achieving a personal best, then allowing your more intuitive, right brain, collaborative nature to help you interact with others to create a whole greater than the sum of its parts, is common to all great jazz, and all great collaborative research" (personal communication). While jazz forms, like participatory action research methods, evolve, this dedication to artistry and collaboration remain as the most crucial elements of both practices. We thank Norman Denzin and Rick VanMatre for their insights and reflections on the history of jazz and the connections between jazz and PAR.

▣ REFERENCES

Adams, F. (1975). *Unearthing seeds of fire: The idea of Highlander.* Winston-Salem, NC: John F. Blair.

Anzaldúa, G. (2007). *Borderlands/La Frontera: The new Mestiza* (3rd ed.). San Francisco: Aunt Lute Books.

Arnstein, S. R. (1969). A ladder of citizen participation. *Journal of the American Planning Association, 35*(4), 216–224.

Asad, T. (1993). Afterword: From the history of colonial anthropology to the anthropology of Western hegemony. In G. W. Stocking (Ed.), *Colonial situations: Essays on the contextualization of ethnographic knowledge* (pp. 314–324). Madison: University of Wisconsin Press.

Bastos, F., & Brydon-Miller, M. (2004). Speaking through art: Subalternity and refugee women artists. In B. M. Lucas & A. B. Lopez (Eds.), *Global neo-imperialism and national resistance: Approaches from postcolonial studies* (pp. 107–118). Vigo, Spain: Universidade de Vigo.

Battiste, M. (2007). Research ethics for protecting indigenous knowledge and heritage: Institutional and researcher responsibilities. In N. K. Denzin & M. D. Giardina (Eds.), *Ethical futures in qualitative research: Decolonizing the politics of knowledge* (pp. 111–132). Walnut Creek, CA: Left Coast Press.

Boal, A. (1985). *Theatre of the oppressed.* London: Pluto.

Brabeck, M. M. (Ed.). (2000). *Practicing feminist ethics in psychology.* Washington, DC: American Psychological Association.

Brody, H. (2000). *The other side of Eden: Hunters, farmers, and the shaping of the world.* New York: North Point Press/Farrar, Straus & Giroux.

Brydon-Miller, M. (2001). Education, research, and action: Theory and methods of participatory action research. In D. L. Tolman & M. Brydon-Miller (Eds.), *From subjects to subjectivities: A handbook of interpretive and participatory methods* (pp. 76–89). New York: New York University Press.

Brydon-Miller, M. (2004). The terrifying truth: Interrogating systems of power and privilege and choosing to act. In M. Brydon-Miller, P. Maguire, & A. McIntyre (Eds.), *Traveling companions: Feminism, teaching, and action research* (pp. 3–19). Westport, CT: Praeger.

Brydon-Miller, M. (2007, September). *The community covenant: Understanding the ethical challenges of participatory action research.* Paper presented at Arbeidsforskningsinstituttet/Work Research Institute, Oslo, Norway.

Brydon-Miller, M. (2009). Covenantal ethics and action research: Exploring a common foundation for social research. In D. Mertens & P. Ginsberg (Eds.), *Handbook of social research ethics* (pp. 243–258). Thousand Oaks, CA: Sage.

Brydon-Miller, M., Greenwood, D., & Maguire, P. (2003). Why action research? *Action Research, 1*(1), 9–28.

Brydon-Miller, M., & Maguire, P. (2009). Participatory action research: Contributions to the development of practitioner inquiry in education. *Educational Action Research, 17*(1), 79–93.

Burke, P. (2005). *History and social theory.* Ithaca, NY: Cornell University Press.

Calloway, H. (1981). Women's perspective: Research as re-vision. In P. Reason & J. Rowan (Eds.), *Human Inquiry* (pp. 457–472). New York: John Wiley.

Cammarota, J., & Fine, M. (Eds.). (2008). *Revolutionizing education: Youth participatory action research in motion.* New York: Routledge.

Chambers, R. (1980). *Rapid rural appraisal: Rationale and repertoire* (IDS Discussion Paper No. 155). Brighton, UK: University of Sussex, Institute of Development Studies.

Cochran-Smith, M., & Lytle, S. (2009). *Inquiry as stance: Practitioner research for the next generation.* New York: Teachers College Press.

Condon, R. G. (1988). *Inuit youth: Growth and change in the Canadian Arctic.* New Brunswick, NJ: Rutgers University Press.

Creswell, J. W., & Plano Clark, V. L. (2007). *Designing and conducting mixed methods research.* Thousand Oaks, CA: Sage.

Delgado, R. (2000). Storytelling for oppositionists and others: A plea for narrative. In R. Delgado & J. Stefancic (Eds.), *Critical race theory: The cutting edge* (pp. 60–70). Philadelphia: Temple University Press.

Deloria, V., Jr. (1997). Anthros, Indians, and planetary reality. In T. Biolsi & L. J. Zimmerman (Eds.), *Indians and anthropologists: Vine Deloria Jr. and the critique of anthropology* (pp. 209–221). Tucson: University of Arizona Press.

Denzin, N. K., & Lincoln, Y. S. (2008). Introduction: Critical methodologies and indigenous inquiry. In N. K. Denzin, Y. S. Lincoln, & L. T. Smith (Eds.), *Handbook of critical & indigenous methodologies* (pp. 1–20). Thousand Oaks, CA: Sage.

de Souza, J. F. (1988). A perspective of participatory research in Latin America. *Convergence, 21*(2/3), 29–38.

DuBois, W. E. B. (1973). *The education of Black people: Ten critiques.* Amherst: University of Massachusetts Press.

Eriacho, R., Fitzpatrick, A., Jamon, A., Lahaleon, T., LaRue, F., Lewis, K., Poncho, G., Quam, R., & Tsethlikai, G., & Tsethlikai, S. (2007).

A student initiative to improve school lunch by practicing traditional agricultural and modern greenhouse management practices. Unpublished manuscript, Western New Mexico University, Gallup Graduate Studies Center, Silver City, NM.

Fals Borda, O. (1977). *For praxis: The problem of how to investigate reality in order to transform it.* Paper presented at the Cartagena Symposium on Action Research and Scientific Analysis, Cartagena, Colombia.

Fals Borda, O. (1979). Investigating reality in order to transform it. *Dialectical Anthropology, 4*(1), 33–55.

Fals Borda, O., & Mora-Osejo, L. (2003) Context and diffusion of knowledge: A critique of Eurocentrism. *Action Research, 1*(1), 20–37.

Fay, B. (1975). *Social theory and political practice.* London: George Allen and Unwin.

Fernández, M. (2002). *Creating community change: Challenges and tensions in community youth research* (JGC Issues Brief). Stanford, CA: John W. Gardner Center for Youth and Their Communities.

Fine, M., & Torre, M. (2005). Resisting and researching: Youth participatory action research. In S. Ginwright, J. Cammarota, & P. Noguera (Eds.), *Social justice, youth, and their communities* (pp. 269–285). New York: Routledge.

Fine, M., Tuck, E., & Zeller-Berkman, S. (2008). Do you believe in Geneva? Methods and ethics at the global-local nexus. In N. K. Denzin, Y. S. Lincoln, & L. T. Smith (Eds.), *Handbook of critical & indigenous methodologies* (pp. 157–180). Thousand Oaks, CA: Sage.

Fitzpatrick, A., Concho, G., Jamon, A., Lahaleon, T., LaRue, F., Tsethlikai, G., & Tsethlikai, S. (2007, June 8). *Youth action research for sustainable agriculture in a rural Southwest USA schoolyard.* Paper presented at the Center for Teaching Excellence Fifteenth Annual Action Research Conference at Eastern New Mexico University, Portales, NM.

Frank, A. (1973). The development of underdevelopment. In C. K. Wilber (Ed.), *The political economy of development and underdevelopment* (pp. 94–103). New York: Random House.

Freire, P. (1970). *Pedagogy of the oppressed.* New York: Seabury.

Furtado, C. (1973). The concept of external dependence. In C. K. Wilber (Ed.), *The political economy of development and underdevelopment* (pp. 118–123). New York: Random House.

Groundwater-Smith, S., & Downes, T. (1999). *Students: From informants to co-researchers.* Paper presented at the Australian Association of Research in Education Annual Conference, Melbourne, Australia. Available at http://www.aare.edu.au/99pap/gro99031.htm

Guhathakurta, M. (2008). Theatre in participatory action research: Experiences from Bangladesh. In P. Reason & H. Bradbury (Eds.), *The SAGE handbook of action research: Participative inquiry and practice* (2nd ed., pp. 510–521). London: Sage.

Guishard, M. (2009). The false paths, the endless labors, the turns now this way and now that: Participatory action research, mutual vulnerability, and the politics of inquiry. *Urban Review, 41,* 85–105.

Hall, B. (1993). Introduction. In P. Park, M. Brydon-Miller, B. Hall, & T. Jackson (Eds.), *Voices of change: Participatory research in the United States and Canada* (pp. xiii–xxii). Westport, CT: Bergin and Garvey.

Hall, B. (2001, December 5). *In from the cold? Reflections on participatory research from 1970–2002.* Inaugural Professorial Lecture, University of Victoria, Victoria, British Columbia, Canada.

Hart, R. (1997). *Children's participation: The theory and practice of involving young citizens in community development and environmental care.* New York: UNICEF/Earthscan.

Hilsen, A. I. (2006). And they shall be known by their deeds: Ethics and politics in action research. *Action Research, 4*(1), 23–36.

Horton, M., & Freire, P. (1990). *We make the road by walking: Conversations on education and social change.* Philadelphia: Temple University Press.

Humphrey, C. (2007). Insider-outsider: Activating the hyphen. *Action Research, 5*(1), 11–26.

Hutzel, K. (2007). Reconstructing a community, reclaiming a playground: A participatory action research study. *Studies in Art Education, 48*(3), 299–320.

Johns, T. (2008). Learning to love our black selves: Healing from internalized oppressions. In P. Reason & H. Bradbury (Eds.), *The SAGE handbook of action research: Participative inquiry and practice* (2nd ed., pp. 473–486). London: Sage.

Kemmis, S., & McTaggart, R. (2005). Participatory action research: Communicative action and the public sphere. In N. K. Denzin & Y. S. Lincoln (Eds.), *The SAGE handbook of qualitative research* (3rd ed., pp. 559–603). Thousand Oaks, CA: Sage.

Kesby, M., Kindon, S., & Pain, R. (2005). "Participatory" approaches and diagramming techniques. In R. Flowerdew & D. Martin (Eds.), *Methods in human geography: A guide for students doing a research project* (pp. 144–166). London: Pearson Prentice Hall.

Kindervatter, S. (1979). *Nonformal education as an empowering process.* Amherst: University of Massachusetts, Center for International Education.

Kindon, S., Pain, R., & Kesby, M. (Eds.). (2007). *Participatory action research approaches and methods.* London: Routledge.

Kral, M. J., & Idlout, L. (2009). Community wellness and social action in the Canadian Arctic: Collective agency as subjective well-being. In L. J. Kirmayer & G. G. Valaskakis (Eds.), *Healing traditions: The mental health of aboriginal peoples in Canada* (pp. 315–334). Vancouver: University of British Columbia Press.

Kral, M. J., Wiebe, P., Nisbet, K., Dallas, C., Okalik, L., Enuaraq, N., & Cinotta, J. (2009). Canadian Inuit community engagement in suicide prevention. *International Journal of Circumpolar Health, 68,* 91–107.

Krzywkowski-Mohn, S. (2008). *Diabetic control and patient perception of the Scheduled in Group Medical Appointment at the Cincinnati Veterans Administration Medical Center.* Unpublished doctoral dissertation, University of Cincinnati, OH.

Kuhn, T. (1970). *The structures of scientific revolutions* (2nd ed.). Chicago: University of Chicago Press.

Kumar, K. (1983). Peasants' perception of Gandhi and his program: Oudh, 1920–1922. *Social Scientist, 11*(5), 16–30.

Ladson-Billings, G., & Tate, W. F. (1995). Towards a critical race theory of education. *Teachers College Record, 97*(1), 47–69.

Lamphere, L. (2004). The convergence of applied, practicing, and public anthropology in the 21st century. *Human Organization, 63,* 431–443.

Lengermann, P., & Niebrugge-Brantley, J. (1998). *The women founders: Sociology and social theory: 1830–1930.* Boston: McGraw-Hill.

Lewis, H. (2001). Participatory research and education for social change: Highlander Research and Education Center. In P. Reason & H. Bradbury (Eds.), *Handbook of action research: Participative inquiry and practice* (pp. 356–362). London: Sage.

Lewis, M. E. (2004). A teacher's schoolyard tale: Illuminating the vagaries of practicing participatory action research (PAR) pedagogy. *Environmental Education Research, 10*(1), 89–115.

Lewis, M. E. (2007, April). *Developing and practicing participatory action research (PAR) pedagogy in a NYC high school greenhouse project: An insider's narrative inquiry.* Presentation at the annual meeting of the American Educational Research Association, Chicago, IL.

Lykes, M. B. (in collaboration with the Association of Maya Ixil women—New Dawn, Chajul, Guatemala). (2001). *Creative arts and photography in participatory action research in Guatemala.* In P. Reason & H. Bradbury (Eds.), *Handbook of action research: Participative inquiry and practice* (pp. 363–371). Thousand Oaks, CA: Sage.

Maguire, P. (1987). *Doing participatory research: A feminist approach.* Amherst: University of Massachusetts, Center for International Education.

Maguire, P. (2001a). The congruency thing: Transforming psychological research and pedagogy. In D. Tolman & M. Brydon-Miller (Eds.), *From subjects to subjectivities: A handbook of interpretive and participatory methods* (pp. 276–289). New York: New York University Press.

Maguire, P. (2001b). Uneven ground: Feminisms and action research. In P. Reason & H. Bradbury (Eds.), *Handbook of action research: Participative inquiry and practice* (pp. 59–69). London: Sage.

Maguire, P., & Berge, B.-M. (2009). Elbows out, arms linked: Claiming spaces for feminisms and gender equity in educational action research. In B. Somekh & S. Noffke (Eds.), *Handbook of educational action research* (pp. 398–408). London: Sage.

Maguire, P., Brydon-Miller, M., & McIntyre, A. (2004). Introduction. In M. Brydon-Miller, P. Maguire, & A. McIntyre (Eds.), *Traveling companions: Feminism, teaching, and action research* (pp. ix–xix). Westport, CT: Praeger.

Maguire, P., & Horwitz, J. (2005, April 11). *Nurturing transformative teacher action research in a teacher education program: Possibilities and tension.* Paper presented at the annual meeting of the American Education Research Association, Montreal, Canada.

Mapedza, E., Wright, J., & Fawcett, R. (2003). An investigation of land cover change in Mafungautsi Forest, Zimbabwe, using GIS and participatory mapping. *Applied Geography, 23*(1), 1–21.

Maxwell, J. A. (2005). *Qualitative research design: An interactive approach* (2nd ed.). Thousand Oaks, CA: Sage.

May, W. F. (2000). *The physician's covenant: Images of the healer in medical ethics* (2nd ed.). Louisville, KY: Westminster John Knox Press.

McDermott, C. (2007, June 6–9). Teaching to be radical: The women activist educators of Highlander. In L. Servage & T. Fenwick (Eds.), *Proceedings of the joint international conference of the Adult Education Research Conference and the Canadian Association for the Study of Adult Education* (pp. 403–408), Mount Saint Vincent University, Halifax, Nova Scotia, Canada.

McIntyre, A. (2000). *Inner-city kids: Adolescents confront life and violence in an urban community.* New York: New York University Press.

McIntyre, A., & Lykes, M. B. (2004). Weaving words and pictures in/through feminist participatory action research. In M. Brydon-Miller, P. Maguire, & A. McIntyre (Eds.), *Traveling companions: Feminism, teaching, and action research* (pp. 57–77). Westport, CT: Praeger.

Merrifield, J. (1993). Putting scientists in their place: Participatory research in environmental and occupational health. In P. Park, M. Brydon-Miller, B. Hall, & T. Jackson (Eds.), *Voices of change: Participatory research in the United States and Canada* (pp. 65–84). Westport, CT: Bergin and Garvey.

Messer-Davidow, E. (2002). *Feminism: From social activism to academic discourse.* Durham, NC: Duke University Press.

Mies, M. (1982). *Fighting on two fronts: Women's struggles and research.* The Hague, the Netherlands: Institute of Social Studies.

Mies, M. (1996). Liberating women, liberating knowledge: Reflections on two decades of feminist action research. *Atlantis, 21*(6), 10–24.

Mills, C. W. (1961). *The sociological imagination.* New York: Grove.

Morgan, D., Pacheco, V., Rodriguez, C., Vazquez, E., Berg, M., & Schensul, J. (2004). Youth participatory action research on hustling and its consequences: A report from the field. *Children, Youth, and Environments, 14*(2), 201–228. Available at http://www.colorado.edu/journals/cye

Nairn, K., Higgins, J., & Sligo, J. (2007, June 9). Youth researching youth: "Trading on" subcultural capital in peer research methodologies. *Teachers College Record.* Available at http://www.tcrecord.org/content.asp?contentid=14515

National Aboriginal Health Organization (NAHO). (2007). *Considerations and templates for ethical research practices.* Ottawa, Ontario, Canada: Author.

Newkirk, T. (1996). Seduction and betrayal in qualitative research. In P. Mortensen & G. Kirsch (Eds.), *Ethics and representation in qualitative studies of literacy* (pp. 3–16). Urbana, IL: National Council of Teachers of English.

Niezen, R. (2003). *The origins of indigenism.* Berkeley: University of California Press.

Nyerere, J. (1969). Education for self-reliance. *Convergence, 3*(1), 3–7.

Ortner, S. B. (2006). *Anthropology and social theory: Culture, power, and the acting subject.* Durham, NC: Duke University Press.

Pratt, G. (2004). *Working feminism.* Philadelphia: Temple University Press.

Price, D. (2004). *Threatening anthropology.* Durham, NC: Duke University Press.

Rabinow, P., & Sullivan, W. M. (1985). *Interpretive social science: A second look.* Berkeley: University of California Press.

Reason, P., & Bradbury, H. (2008). *The SAGE handbook of action research: Participative inquiry and practice* (2nd ed.). Thousand Oaks, CA: Sage.

Reid, C., & Frisby, W. (2007). Continuing the journey: Articulating dimensions of feminist participatory action research (FPAR). In P. Reason & H. Bradbury (Eds.), *The SAGE handbook of action research: Participative inquiry and practice* (2nd ed., pp. 93–105). London: Sage.

Reinharz, S. (1992). *Feminist methods in social research.* New York: Oxford University Press.

Royal Commission on Aboriginal Peoples. (1993). *Integrated research plan: Appendix B. Ethical guidelines for research.* Ottawa, Ontario, Canada: Office of the Solicitor General.

Rylko-Bauer, B., Singer, M., & Van Willigen, J. (2006). Reclaiming applied anthropology: Its past, present, and future. *American Anthropologist, 108,* 178–190.

Sabhlok, A. (2007). *SEWA in relief: Gendered geographies of disaster relief in Gujarat, India.* Unpublished doctoral dissertation, Pennsylvania State University, State College.

Sangtin Writers, & Nagar, R. (2006). *Playing with fire: Feminist thought and activism through seven lives in India.* Minneapolis: University of Minnesota Press.

Schulz, A. J., Parker, E. A., Israel, B. A., Becker, A. B., Maciak, B. J., & Hollis, R. (1998). Conducting a participatory community-based survey for a community health intervention on Detroit's East Side. *Journal of Public Health Management and Practice, 4*(2), 10–24.

Smith, D. E. (1989). *The everyday world as problematic: A feminist sociology.* Boston: Northeastern University Press.

Smith, L. T. (1999). *Decolonizing methodologies: Research and indigenous people.* London: Zed.

Swantz, M. L. (1974). *Participant role of research in development.* Unpublished manuscript, Bureau of Resource Assessment and Land Use Planning, University of Dar es Salaam, Tanzania.

Swantz, M. L. (2008). Participatory action research as practice. In P. Reason & H. Bradbury (Eds.), *The SAGE handbook of action research: Participative inquiry and practice* (2nd ed., pp. 31–48). London: Sage.

Tandon, R. (1988). Social transformation and participatory research. *Convergence, 21*(2/3), 5–18.

Tandon, R. (1996). The historical roots and contemporary tendencies in participatory research. In K. de Koning & M. Martin (Eds.), *Participatory research in health* (pp. 19–26). Johannesburg, South Africa: Zed.

Tandon, R. (2005). *Participatory research: Revisiting the roots.* New Delhi, India: Mosaic Books.

Torre, M., & Ayala, J. (2009a). Envisioning participatory action research entremundos. *Feminism and Psychology, 19*(3), 387–393.

Torre, M., & Ayala, J. (2009b, August). *Participatory echoes of Chataway: A symposium reflecting on the social change legacy of Cynthia Joy Chataway.* Paper presented at the annual meeting of the American Psychological Association, Toronto, Ontario, Canada.

Toulmin, S. (1988). The recovery of practical philosophy. *The American Scholar, 57,* 337–352.

Tuck, E. (2009). Re-visioning action: Participatory action research and indigenous theories of change. *Urban Review, 41,* 47–65.

Tucker, S. (n.d.). *Women in jazz.* Available at http://www.pbs.org/jazz/time/time_women.htm

United Nations. (1989). *Convention on the rights of the child.* New York: Author.

Wang, C. (1999). Photovoice: A participatory action research strategy applied to women's health. *Journal of Women's Health, 8,* 185–192.

Wenzel, G. W. (1991). *Animal rights, human rights: Ecology, economy and ideology in the Canadian Arctic.* Toronto, Ontario, Canada: University of Toronto Press.

24

WHAT IS QUALITATIVE HEALTH RESEARCH?

Janice M. Morse

W hat is qualitative *health* research? And why do qualitative health researchers need their own courses, insist on their own journals, and require specialized methodological texts? Surely knowing qualitative inquiry and being adept in qualitative methods is all that is required to be a good qualitative health researcher? No?

In this chapter, I argue that the complexities encountered by qualitative researchers in the context of health care, the seriousness of the conditions of their participants, the life-and-death nature of the topics that they study, and the clinical significance of their findings, make the qualitative health researcher distinct from researchers who do other forms of qualitative inquiry, and the product, qualitative health research, distinct. Here, I consider the origins, content, and scope of qualitative health research. Then, from a content analysis of all articles published in 2009 in *Qualitative Health Research* (Volume 19, $N = 142$ articles), I discuss the current areas of emphases, clinical application, and the contribution of these articles to the medical and allied health care fields. In the final section, I discuss why and how qualitative methods must be adapted for use with the ill and in the clinical setting, and close by making an argument that qualitative health research is developing as an important disciplinary subfield in its own right.

▣ CLASSICAL FOUNDATIONS

Qualitative inquiry has had a presence in hospitals and health care institutions and clinics since the 1950s, when ethnographies (conducted mainly by sociologists) began to appear. Some of this work endures as classics. It includes the study of the socialization of medical students, *Boys in White* (Becker, Geer,

Hughes, & Strauss, 1961); Erving Goffman's *Asylums* (1961), a study of "mental patients and other inmates" in Washington, D.C.; Barney Glaser and Anselm Strauss's *Awareness of Dying* (1965); Jeanne Quint's (1967) *The Nurse and the Dying Patient;* and Talcott Parsons's *The Sick Role and the Role of the Physician Reconsidered* (1975). At the University of California, San Francisco, in the 1970s, the student collaborators of Glaser and Strauss—Julie Corbin, Shizuko Fagerhaugh, David Maines, Barbara Suczek, and Carolyn Weiner—published *Chronic Illness and the Quality of Life* (Strauss et al., 1975), building an important foundation for qualitative health research. In nursing, qualitative inquiry was dependent on methodological texts from other disciplines such as anthropology and sociology. From the mid 1980s, they commenced writing their own methods texts, and moved away from other disciplines to be more centrally established into nursing. They also conducted significant qualitative inquiry situated in health. Of this work, Carole Germain's *The Cancer Unit* (1979) and Patricia Benner's *From Novice to Expert* (1984) have been influential. In medicine, the early work of Arthur Kleinman (*Patients and Healers in the Context of Culture* [1980]) was groundbreaking. Most of these studies used ethnography, with the exception of the work at the University of California, San Francisco, where Barney Glaser and Anselm Strauss (1967) were developing grounded theory, and this work continues development through the "second generation" of their students, Phyllis Stern, Julie Corbin, Barbara Stevens, Kathy Charmaz, and Adele Clarke (Morse et al., 2009).

In phenomenology, Jan van den Berg's *The Psychology of the Sick Bed* (1960) was significant. Then, mainly through the work of Max van Manen—his methodological text (van Manen, 1990), the Human Science Conferences, and the journal, *Phenomenology and Pedagogy* (Vols. 1–10, 1983–1992)—phenomenology gained importance in North America. Van Manen, primarily with his students from nursing, developed the phenomenology of illness as an important area, investigating *meaning* in the embodiment of illness.

Ben Crabtree and Will Miller's *Doing Qualitative Research* (1992) and the Family Practice Qualitative Interest Group annual conferences and publications of edited books were important in moving qualitative inquiry into medicine. Narrative inquiry in medicine began with Arthur Kleinman's *The Illness Narratives* (1986) and Howard Brody's *Stories of Sickness* (1987), bringing the patients' experience into medicine (see Engel, Zarconi, Pethtel, & Missimi, 2008) for the evaluation of care. Since the late 1990s, focus groups as a method and now mixed method design have gained significance, particularly in psychiatry and family medicine.

Qualitative inquiry, as a method and a way to approach research, moved into health care first through the work of a group of nurse anthropologists, most notably Madeleine Leininger, Pamela Brink, Margarita Kay, Eleanor Bowens, and Noel Chrisman, who, through the American Anthropological Association's Council of Nurse-Anthropologists (CONAA), provided a supportive forum for qualitative research. In the mid 1980s, qualitative texts began to appear (Field & Morse, 1985; Leininger 1985), courses were offered in graduate programs, and the National Institutes of Health's National Center for Nursing Research (now NINR, National Institute for Nursing Research) was urged to fund qualitative research. Publication of qualitative health research was scattered among several journals, such as *Social Science in Medicine,* or in disciplinary specialist journals, such as *Medical Anthropology* or *Symbolic Interaction.* In 1991, *Qualitative Health Research* (QHR) was launched as a quarterly journal, and it has continued to expand to its current 12 issues annually.

Gaining this acceptance has not been easy for qualitative researchers. Health care research was, and is still, dominated by medical research with an agenda of treatment and cure, using experimental clinical trials, rather than the more subjective, experiential agenda of qualitative inquiry. Medical researchers tend to focus on disease, rather than the person, and with their limited contact with the patient in practice, they have backstaged "the patient's experience." Further, medical research funding agencies historically have been heavily—if not solely—populated by quantitative researchers who had little or no understanding of or appreciation for the principles of qualitative inquiry, so that gaining recognition was slow.

Even today, in medicine, qualitative research continues to be devalued. In Britain, the Cochrane criteria (Cochrane, 1972/1989; Sackett, 1993) are used to evaluate evidence in medicine. This system establishes a hierarchy of evidence that assigns the most credibility to the clinical trial, and the lowest to "mere opinion." While I do not believe that Cochrane intended to include qualitative inquiry in this hierarchy, researchers tended to assign qualitative inquiry to the lowest rank, categorizing it as "mere opinion." The result of this classification was that qualitative inquiry was denigrated as invalid and of little worth. For instance, in the 1990s in Australia, where the Cochrane system was used by the National Health and Medical Research Council (NHMRC) to make funding decisions, qualitative inquiry was ranked as noncompetitive, and this took considerable energy to successfully reverse. Today, the NHMRC does fund qualitative inquiry, supporting workshops and was commissioning qualitative research. In Britain in the 1990s, likewise, qualitative inquiry was not considered generalizable, and was inappropriate for Cochrane reviews. It took a committee (led by Jennie Popay) to have qualitative inquiry recognized as an adequately rigorous method to be included in reviews to determine evidence. The *British Medical Journal* (*BMJ*) now publishes a regular column on qualitative methods, and in the United States, the occasional qualitative article appears in the *New England Journal of Medicine* and in *JAMA,* the *Journal of the American Medical Association.*

To support the development of qualitative inquiry, the International Institute of Qualitative Methodology (IIQM) was established in 1997 at the University of Alberta, Canada, and funded by a grant from the Alberta Heritage Foundation for Medical Research.[1] In an effort to establish qualitative inquiry as a discipline, the IIQM sponsored the annual Qualitative Health and Advances in Qualitative Methods research conferences in North America and internationally, held lecture and workshop series (called *Thinking Qualitatively*), and consequently facilitated training for a generation of qualitative health researchers. The IIQM developed and supported a multilingual, open-access, online journal called the *International Journal of Qualitative Methods* (*IJQM*), supported *Qualitative Health Research* (a Sage monthly international journal), initiated the *Qual Press* (publishing monographs), and a postdoc and predoc training program (EQUIPP, Enhancing Qualitative Understanding of Illness Process and Prevention), which provided an internship for international trainees. National and international outreach included links with 115 universities through hubs of eight international sites, which in turn spawned additional research, centers, conferences, and organizations internationally.

Where are we today, in 2011? Qualitative health research is now published in the major medical journals, although such articles still remain in the minority. Qualitative inquiry is considered an essential part of graduate research training in many universities, although it is not on an equal basis with quantitative inquiry. We still have some distance to go, and as recently as 2008, qualitative research was excluded from a Canadian urology conference (Morse, 2008). An earlier crisis, that is, the lack of mentors and supervisors to support new qualitative researchers, is easing as competence is gained and more researchers become experienced in qualitative inquiry, and issues of quality and standards are resolved. There is no doubt that the field is maturing. It simply takes time.

🔲 WHAT DO QUALITATIVE HEALTH RESEARCHERS STUDY?

Presently, qualitative health research is taking place in all areas where health care is administered: in the hospitals and nursing homes; in the clinics, schools, and workplaces; and in the community—on the streets, in the parks, and at people's homes. It focuses on the experiences of illness, on patient states and behaviors, on the healthy and the sick. It includes accidents, acute onset, and chronicity. The caregivers' experiences, including both lay and professional caregivers and their interactions with the sick person, is an important area. Qualitative inquiry examined cultural perception of disease and responses to illness. The focus was generally on the individual, the family, or groups. Qualitative health research also encompasses the education of health care professionals, and health care information provided the patients and their families.

To illustrate, using some concrete examples of the nature of the scope of qualitative health research, in the next section I provide a broad overview from *Qualitative Health Research,* Volume 19, Issues 1–12, 2009.[2]

🔲 THE CONTRIBUTIONS OF *QUALITATIVE HEALTH RESEARCH*

Research focused on health using qualitative methods can be broadly classified several ways. One may, for instance, classify participants by age group, ethnicity, medical specialty, or disease—or by behavioral concept or type of research method. For the purposes of this chapter, however, I will classify studies by the broadest system, that is, by *use*—categories used in clinical application, educational topics, and groups of subsequent research. These categories are examining the identification of health care needs, barriers and access to health care, processes of seeking health care, responses to illness, adjusting to being ill, living with illness, responses to treatment, behaviors and experiences of professional care providers, experiences of lay caregivers, perspectives of professional caregivers, and experiences of both lay and professional caregivers. There is also considerable research emphasis on studying support systems in illness and experiences of recovering from illness. Research that contributes to medicine, nursing, and other allied health professions is also important, assisting with delineating symptoms. Finally, I address evaluating health care, and aspects of teaching health care to patients as well as the education of members of the health professions.

The processes of recognizing that one is ill (or the responses to an acute episode or accident), the processes of seeking and receiving care, and the processes of recovering are presented in Table 24.1.

1. *The identification of health care needs*

Research in this category identifies silent or emerging health care problems that are not presently addressed, or poorly addressed, within the system. These problems may be sorted into subcategories with certain common factors that are jeopardizing health; they may be factors that the health care system overlooks; they may be factors in the environment that impair health. In these studies, authors point out inequities or health problems, usually using ethnographic or interview methods. For instance, one of De Marco, Thorburn, and Kue's (2009) malnourished participants said, "in a country as affluent as

Table 24.1 Categories and Examples of Articles in *Qualitative Health Research* in 2009

Article Title (abbreviated)	Author(s)	Citation
The identification of health care needs		
Perceptions of food insecurity among rural and urban Oregonians	De Marco et al.	*19,* 1010–1024
Menopause: Mapping the complexities of coping strategies	Kafanelis et al.	*19,* 30–41
Identifying patterns of seeking health care		
Cancer patients' accounts of negotiating a plurality of therapeutic options	Broom	*19,* 1050–1059
Treatment seeking by Samoan people in Samoa and New Zealand	Norris et al.	*19,* 1466–1475
The experience of involuntarily childless Turkish immigrants in the Netherlands	Van Rooij et al.	*19,* 621-632

(Continued)

Table 24.1 (Continued)

Article Title (abbreviated)	Author(s)	Citation
Describing the illness experience		
Disclosing a cancer diagnosis to friends and family	Hilton et al.	19, 744–754
Dignity violation in health care	Jacobson	19, 1536–1547
Illness meanings of AIDs among women with HIV: Merging immunology and life experience	Scott	19, 454-465
The culture of pediatric palliative care	Davies et al.	19, 5–16
Adjusting to illness/Living with illness		
Adjusting to life after esophagectomy	McCorry et al.	19,1485–1494
Making sense of living under the shadow of death	Kenne Sarenmalm et al.	19, 1116–1130
Social support and unsolicited advice in a bipolar disorder online forum	Vayreda & Antaki	19, 931–942
Pill taking from the perspective of HIV-infected women who are vulnerable to antiretroviral failure	Stevens & Hildebrandt	19, 593–604
Disclosure outcomes, coping strategies, and life changes among women living with HIV in Uganda	Medley et al.	19, 1744–1754
Breast cancer patients' experiences of external-beam radiotherapy	Schnur et al.	19, 668–676
Experiences of kidney graft failure	Ouellette et al.	19, 1131–1138
Experiences and practices of professional care providers		
Birth talk in second stage labor	Bergstrom et al.	19, 954–964
Positions in doctors' questions during psychiatric interview	Ziółkowska	19, 1621–1631
Experiences and practices of lay caregivers		
Hope experience of older bereaved women who cared for a spouse with terminal cancer	Holtslander & Duggleby	19, 388–400
Perspectives of both lay and professional caregivers		
Encounter between informal and professional care at the end of life	James et al.	19, 258–271
Empathy and empowerment in general practitioners who have been patients	Fox et al.	19, 1580–1588
Identification and analysis of support systems		
Being there for another with serious mental illness	Champlin	19, 1525–1535
Family presence during resuscitation and invasive procedures: Perceptions of nurses	Miller & Stiles	19, 1431–1442
Reflections on the Illness Experience		
The paradox of childhood cancer survivorship	Cantrell & Conte	19, 312–322
Normalization strategies of children with asthma	Protudjer et al.	19, 94–104
Dreams of my daughter: An ectopic pregnancy	Lahman	19, 272–278

America, people should be eating." Not surprisingly, poor health, low income, and unemployment were contributors to poor nutrition, but these studies documented the importance of nutrition assistance programs, alternative food sources, and social support ("godsends") to supplement diets. The authors recommend that policy should focus on increasing human capital for the prevention of malnutrition.

Does this recommendation sound obvious? By way of contrast, let us examine the descriptions of the complexities of coping with menopause in Australian women (Kafanelis, Kostanski, Komesaroff, & Stojanovska, 2009). These authors wrote that women's responses "surged and ebbed in intensity, leaving women to stumble, collapse, shift, settle, and meander from one episode or event to the next.... This fluidity allowed women to create, uncover and reinterpret their experiences ... which enabled women to maintain, consolidate and stabilize various coping strategies" (p. 39). Kafanelis et al. identify three styles of responses from such confusion: inventive (able to respond in an effective manner and work through their experiences), troubled (women who responded with anxiety, negativity, and increased conflict), and reactive (women who were determined, active, well-informed, and optimistic). This study illustrates the power of qualitative inquiry in its ability to mirror life, to reveal the implicit, and to be of use for those providing care. It is apparent that in good qualitative inquiry, mundane questions can produce significant findings.

2. *Identifying patterns of seeking health care*

The Western model of providing care is simple: The patient suspects that he/she may have a health problem, and comes to the physician for care. The physician diagnoses the problem, gives treatment recommendations to the patient, who then complies with the treatment and recovers. But this model collapses with some patients, when barriers to care impede access.

Different processes and choices acted upon when ill, the interpretation of the illness symptoms, and the effects of the input of others from the family were described by a team of New Zealand and Samoan researchers, explaining how Samoans choose to use the Western (*palagi*) or the traditional Samoan health care and healing resources—or both. The fact that the individual may select one course of action, and still be overruled by the family, has important implications for all cultures in which the family is the unit for treatment decisions (Norris, Fa'alau, Va'ai, Churchward, & Arroll, 2009). The model can break down even when the patient gets to the physician. In the United States, the patient is asked his or her preferences for therapy, and the advantages, effectiveness, and side effects of each alternative is explained. The multiple options for treatment can be overwhelming, yet the Western system expects patients to be involved in decisions about treatment options. Compounding the patient's choice among these therapeutic options are the

number of complementary medicines available. Broom (2009) describes patients "piecing together" therapies from both the Western and the complementary systems, using "intuitive" and "objective" scientific knowledge, and making their own decisions to assess therapeutic effectiveness of their treatments.

3. *The illness experience*

This category forms the main component of qualitative health research as a subdiscipline (and accounts for the majority of articles published in *QHR*). It includes response to illness, adjusting to illness, and living with illness.

Researchers are fascinated with the worlds of the acutely and chronically ill—the dramatic change in one's self, to one's lifestyle, and to one's being. Illness impacts and alters the core of the self (Jacobson, 2009). Becoming ill involves breaking the bad news to loved ones (Hilton, Emslie, Hunt, Chapple, & Ziebland, 2009), and their distress in turn compounds the suffering of the ill person.

In an article from Uganda, Medley, Kennedy, Lynyolo, and Sweat (2009) examined how women with HIV cope with the changes after learning their diagnosis. Stevens and Hildebrandt (2009) provide interesting perspectives of women who "had difficulty" taking their antiretroviral medications. Taking this medication "produced existential angst, interfered with functioning or caused a loss of self." Whenever they took their medications, they were reminded that they had HIV—in fact *not* taking their medications was viewed as a "positive thing—it enabled them to feel fully human, dignified" and "pleased that they managed to take at least some of their medications some of the time" (p. 601).

The complex emotions associated with being told that cancer has recurred were vividly described by Kenne Sarenmalm, Thorén-Jönsson, Gaston-Johansson, and Öhlén (2009). The feelings of "shock, fear, anxiety, sadness and depression, apathy and listlessness" were overwhelming for these women. The losses were acute—women felt "like a nothing" with the loss of femininity, loss of their physical appearance, and their attractiveness. "I usually feel like an 'it' on legs." They experienced a loss of self, a loss of control, loss of power to influence the situation, loss of independence, and of being dependent on others (p. 1121).

Schnur, Ouellette, Bovbjerg, and Montgomery (2009) described the experiences of women with breast cancer receiving external beam radiotherapy. The side effects experienced were not "just" side effects. In addition to being considered as a "problem with treatment," side effects were evaluated as an "omen of symptoms to come" or as an "indicator of personal unworthiness." For instance, fatigue may be "expected and considered normal to the physician," but is a sign of "weakness" to the women. Skin toxicity may be seen as "sunburn" to others, but to the women it would be considered "hideous." Women were self-critical ("self-downing"), and had difficulty in giving themselves permission to "take it easy." They were "merciless" and the self-criticism went beyond their treatment and pervaded their "self-identity as a conscious worker, a good parent, a good patient, and an attractive woman" (p. 673).

Treatments are not always "successful," as Ouelette, Achille, and Pâquet (2009) note. Kidney patients whose grafts had failed and who were forced to return to dialysis appeared to redefine their "representation of reality." How well they achieved that depended on their comparison with others, and how they sought normalcy. How well they were able to regain a sense of control of their lives depended on individual characteristics and the time since graft failure.

These articles uncover nuances of the illness experience, reinvestigating needs and opportunities for interaction and interventions.

4. *Experiences and practices of professional care providers*

Illness brings dependency, a reliance on others for assistance with the most intimate and private functions. Professional caregiving is easier for the patient to accept than intimate care provided by, perhaps, one's children. But dependency on the professional caregiver involves more than just being a recipient of physical care. Bergstrom, Richards, Proctor, Bohrer, Morse, & Roberts (2009) describe how "talking through"—providing "comfort talk" to women in second stage labor—can assist women to overcome the terror of the second stage of labor, to regain control, and cooperate with caregivers. A caregiver's therapeutic use of self is crucial for the women to have a safe labor experience.

The impact of the ill person's experiences on the caregiver is profound and extends after the patient's death. Holtslander and Duggleby (2009) explored the concept of *hope* in older bereaved women who had cared for their spouse with terminal cancer. When one's spouse is diagnosed with cancer, the loss of confidence, loss of security in the future, and overwhelming losses accompanied the loss of hope. Without hope, "one doesn't want to go on, this reality doesn't matter." "They don't care." The trick is to "search for a new hope" to fill the void, and this will enable them to "go on."

Interviews with both lay *and* professional caregivers provide us with two types of perspectives: (1) those comparing and contrasting lay and professional caregivers and (2) those studying professionals who had also been patients. The experience of being a patient changes the way a physician subsequently interacts with patients—in the realms of empathy, self-disclosure, and recognition of the "disempowering" role of being a patient. Fox et al. (2009) recommend that this content be introduced into medical students' curricula.

Family caregivers have a specialized knowledge. They are experts and protectors, possessing practical knowledge about "what care is best" (or "least harmful") for their loved one. Yet, when the loved one is transferred into the hospital, they are forced back to the sidelines, hovering and observing care, waiting for death (James, Andershed, & Ternestedt, 2009).

5. *Recovering from illness*

The course of rehabilitation has various patterns: rapid and complete, partial or prolonged. In the last, the person lives in the community as a disabled person. Champlin (2009, p. 1525) describes the process of the continuing care for a mentally ill person as one of both accepting and grieving for the changed person, accepting the challenge of caregiving, and recognizing it as never-ending and unpredictable, feeling isolated and ambiguous about the caregiving charge, yet knowing the other well and accepting the responsibility.

Is recovery ever complete? Cantrell and Conte (2009) describes the experiences of childhood cancer, in which the adolescents have the paradoxical experiences of their current state of functioning and, at the same time, being off treatment and having survived. They had to learn to refocus from the uncertainty of illness and treatment, to the fact that they could now dream, hope, and plan, and to reorient from "losses and missed opportunity to what they could accomplish and experience in the future" (p. 320).

6. *Research that contributes to the examination of nursing and other professions*

The original compendium of medical signs and symptoms in medicine came from careful observations of the single patient—qualitative observations. This contribution formed the very basis of medicine, and this role of symptom identification still continues. Qualitative inquiry assists with delineating symptoms, and also identifying new syndromes.

But the conundrum of diagnoses—particularly psychiatric diagnoses—continues. In an article examining beliefs, sensations, and symptoms of PTSD (post-traumatic stress disorder), Spoont and her colleagues (2009) use narrative to explore post-trauma suffering of veterans, and the process by which they themselves came to label their suffering as PTSD. While the experiences of some veterans were clear and validating (and treatment options available), the experiences of others were more ambiguous, and the veterans were uncertain about both the diagnostic label and treatment options. The lack of clarity with the diagnostic label left veterans unable to validate their own suffering as PTSD, unsure how they should proceed with seeking care, and "even contributed to a denial of their suffering." The authors conclude that the diagnostic label of PTSD serves to "provide meaning and validation for those who experience trauma-related suffering" and yet concerns about "possible malingering (i.e., feigning illness for personal gain)" exist (p. 1463).

7. *"Knowing the patient"*

This is an important category, with qualitative research targeting specific information that will be used in patient assessment and subsequent interventions. As shown in some of the studies mentioned in Table 24.2, these assessments are not

Table 24.2   Examples of Articles Making Substantive Contributions to Improving Medicine and Nursing in *Qualitative Health Research* in 2009

Article Title (abbreviated)	Author(s)	Citation
Children's pain assessment in Northern Thailand	Forgeron et al.	*19*, 71–81
Development of a health screening questionnaire for women in welfare transition programs	Lutz et al.	*19*, 105–115
Ecological validity of neuropsychological testing	Gioia	*19*, 1495–1503
Mind and body strategies for chronic pain and rheumatoid arthritis	Shariff et al.	*19*, 1037–1049
PTSD: Beliefs about sensations, symptoms, and mental illness	Spoont et al.	*19*, 1456–1465
Considering culture in physician–patient communication during colorectal screening	Gao et al.	*19*, 778–789
Exploring tuberculosis patients' adherence to treatment regimens and prevention programs	Naidoo et al.	*19*, 55–70
Helping direct and indirect victims of national terror: Experiences of Israeli social workers	Shamai & Ron	*19*, 42–54

included in routine assessment protocol, but rather they are investigated for further understanding of signs and symptoms (see, for instance, Spoont et al., 2009, for PTSD), as groundwork for developing assessment tools (see, for instance, Forgeron et al., 2009, for work with children's pain in Northern Thailand), or to develop models, such as Shariff et al.'s (2009) research in managing the pain of arthritis.

These articles promote important applicable information that may be transferred to the clinical arena or contribute directly to the development of quantitative tools. For instance, medicine used careful observation to establish a compendium of signs and symptoms as early as the 18th century; this effort continues today. Spoont et al.'s (2009) article on post-traumatic stress disorder jumps into the debate about the disorder (Is it a natural response to extreme trauma, or a psychiatric disorder?) with interviews with veterans who had submitted disability claims. Approximately half of the 40 participants were currently in mental health treatment, and all had had a period of military service. The sample included males and females. All were extremely distressed. Determining if their symptoms were consistent with

the prototypical description of PTSD, or if they had an ambiguous presentation with symptoms varying in intensity, resulted in self-recognition of the symptoms and difficulty in diagnosis. When considered in context, veterans struggled with the normalcy of the symptoms versus the psychiatric-illness presentation. The ambiguous nature of the syndrome both helped and hindered veterans. Spoont et al. (2009) noted that on one hand it validated their suffering while on the other hand the lack of clarity regarding the diagnostic label left them uncertain whether or not they should seek treatment.

8. *Health care evaluation*

Qualitative inquiry is excellent for evaluation, and evaluations range from health in the community, to the evaluations of health care institutions, to patient self-evaluations of care (see Table 24.3). These articles pertain to the health care system, rather than patient symptoms, as in the previous category. An example of using ethnographic methods to determine local explanatory models of how mothers perceive diarrheal disease in young children in Northern Thailand was published by

Table 24.3   Examples of Articles on Health Care Evaluation in *Qualitative Health Research* in 2009

Article Title (abbreviated)	Author(s)	Citation
Gaps between patients, media, and academic medicine in discourses on gender and depression	Johansson et al.	*19*, 633–644
Creating a quality improvement dialogue: Utilizing knowledge from frontline staff, managers, and experts to foster health care quality improvement	Parker et al.	*19*, 229–242
Elder authority and situational diagnosis of diarrheal diseases as normal infant development in Northeast Thailand	Pylypa	*19*, 965–975
From trauma to PTSD: Beliefs about sensations, symptoms, and mental illness	Spoont et al.	*19*, 1456–1465

Pylypa (2009). Mothers believed that diarrhea *(thai su)* was necessary to "lighten the body" so that infants could achieve a new developmental stage, such as sitting, standing, or walking. Therefore, she concluded, "mothers do not direct much attention to prevention, nor manage diarrheal cases in a manner consistent with biomedical recommendations" (p. 965), and it is "unlikely that health education will eliminate this strongly held ethnomedical belief" (p. 974). Importantly, the author notes that this information was "only accessible as a result of the use of an in-depth, qualitative approach to interviewing that has been largely absent from Thai studies of diarrheal disease" (p. 974).

Parker et al. (2009) combined the approaches of the local and the "expert" quality improvement personnel to provide a comprehensive perspective on quality improvement. These researchers noted that as dialogue must be established with the frontline managers and staff members, the best method for such conversations to take place is face to face, and the focus of these discussions sometimes differed (for instance, workload versus costs). Participants in this study felt the project was worth the time and effort. Further, health care organizations must be prepared not only to pay staff to care for patients, but also to work toward improvement in that care.

9. *Teaching health care to patients and the education of health professionals*

Articles in this category (see Table 24.4) address teaching techniques and strategies used by the health professions, including unique programs (Oman, Moulds, & Usher, 2009) and the development of models for teaching (Jenkins, Mabbett, Surridge, Warring, & Gwynn, 2009). Teaching in medicine and nursing has some unique characteristics. Rounds are used, and this necessarily involves patients. Care to an individual patient is provided by different practitioners and different specialties, and these professionals are responsible for different aspects of care. Qualitative inquiry reveals the strengths and weaknesses inherent in the utilization, evaluation, and improvement of such unique teaching structures.

From the above variety of contexts, perspectives, and conditions that were the focus of the research, we will now explore the question of necessary adaptations to qualitative methods to be used in health care research.

▣ DOING CLINICAL QUALITATIVE HEALTH RESEARCH

The process of trying to "get inside" the world of the ill, the disabled, and the dying evokes special ethical and methodological problems: ethically because the impact of the illness leaves little room for participation, and methodologically because of the restrictions of the illness process on the participants. Each of these two aspects will be discussed.

IRB Quandaries With QHR

Qualitative research may be considered invasive and intrusive in participant's lives. As IRBs (institutional review boards) consider these factors during review, sometimes they seek to

Table 24.4 Examples of Articles on Education and Models in *Qualitative Health Research* in 2009

Article Title (abbreviated)	Author(s)	Citation
Medical education		
Patients' involvement in hospital bedside teaching encounters	Monrouxe et al.	*19,* 918–930
Medical programs		
Professional satisfaction and dissatisfaction among Fiji specialist trainees: What are the implications for preventing migration?	Oman et al.	*19,* 1246–1258
Communication channels in general internal medicine: Improved interprofessional collaboration	Gotlib Conn et al.	*19,* 943–953
Health education		
Health materials and strategies for the prevention of immigrants' weight-related problems	Ferrari et al.	*19,* 1259–1272
Models for nursing		
Cooperative inquiry into action learning and praxis development in a community nursing module	Jenkins et al.	*19,* 1303–1320

protect participants by arguing that patients do not have the energy, inclination, or the time to participate in qualitative research (McIntosh & Morse, 2009). For example, if a patient is in pain, or struggling for breath, IRBs sometimes consider that they should not be "disturbed" for research, for the consent procedures, to be interviewed, or even have their privacy interrupted with the presence of an observer. If a patient is dying, qualitative research takes precious time that the patient should be spending with his or her family, or used productively to complete tasks they must accomplish while they are able. The research is considered to have limited benefit for the participant and their family, and the "benefit" of qualitative inquiry for subsequent patients is also considered limited if the generalizability of findings is questioned. Thus, IRBs are frequently reticent to approve requests for access, even when the interviewer is a nurse accustomed to working with the ill (Morse, 2002).

However, researchers report that these perceptions about patients not wanting to participate are incorrect, and participants appreciate the support, the opportunity to talk and to be listened to, and accept the presence of an observer. Patient's report they can see the benefit of research to others, and do not consider the research process as an invasion of their privacy. If the IRBs' concerns are correct and some patients do not wish to be involved in research, certainly some do wish to participate, and the concern regarding patients' time should not be the consideration for nonapproval. Patients themselves should be given the opportunity to choose if they wish to participate or not.

One area that has not been well studied is emergency care, in which treatment takes precedence over consent procedures. Medical research involving drug trials and other treatments has special regulations to allow the research to continue without consent, with the patient's consent replaced by the consent of two physicians. Such dispensation has not been obtained for qualitative inquiry, even though it does not generally involve treatments or impact care.

The cast of thousands. The complexity of care and the number of people involved directly or indirectly with patient care, even in one 8- or 12-hour shift, make the obtaining of consents for the conduct of observational research using videos very difficult. If the focus of the research is the patient, consent must be also obtained from all of those who interact with the patient, who come in contact with the patient, from the physician to the cleaning staff and the person who delivers the newspaper, as they will be on the tape. In one project involving videotaping trauma care, we obtained as many consents as possible before the project started—this involved EMTs (emergency medical technicians), a number of medical specialties, all staff who wished to participate, and the auxiliary staff, such as radiologists and the cleaning staff. Then, when we were actually taping, we posted signs on the door stating the we were taping for research purposes, and stood by the door, reminding people as they rushed in that we were collecting research data. Any person,

including the patient, could ask for the video recorder to be switched off at any time. The tapes were secured until the patient was able to give consent or declined to be in the study, and in the latter case the tapes were immediately erased. This demonstrates that such research can be conducted, but certainly not easily.

Vulnerable participants. Research conducted with institutionalized patients is considered a risk, as patients are *vulnerable.* Therefore, all research that is to be conducted in an institution is subject to IRB approval, and a review of the clinical areas may be conducted with respect to research burden—that is, are too many projects being conducted in the same area, resulting in too much "work" for the patient participants or too much time involvement for staff?

Participant vulnerability is clear when the patients are cognitively impaired (and unable to provide consent, but may be required to *assent,* and the guardian *consent*), or are minors (and the child is expected to provide *assent* to the research, and the parents to provide *consent*).

Consent becomes an issue with institutionalized patients. When the person obtaining consent (or conducting the research) is the same as the one providing care, the perceived coercive effects are considered a risk. The potential participant may ask, "What if care will be withheld if I do not participate?" Therefore, an explanation of the study, or the consent itself, is usually obtained by someone who is not involved with direct patient care, and the fact that the patient may withdraw from the project at any time without penalty is stressed. Staff must know who is participating in research, and usually a copy of the consent form is placed in the patient's file. Many institutions employ an ombudsman to oversee research projects and to manage questions or complaints, if any.

Privacy laws. Once the researcher has permission to conduct research in a hospital, locating patients with the necessary characteristics may be an issue. In the United States, federal privacy legislation HIPPA (Health Insurance Portability and Accountability Act of 1966, available at http://www.hhs.gov/ocr/privacy/), may prohibit contact. Canada has two federal privacy laws, the Privacy Act and the Personal Information Protection and Electronic Documents Act (PIPEDA; available at http://www.priv.gc.ca/fs-fi/02_05_d_15_e.cfm#contenttop), and many provinces have their own privacy laws. For instance, Alberta has FOIP (Freedom of Information and Protection of Privacy Act Canada; available at http://foip.alberta.ca/legislation/index.cfm), which prevents staff from releasing names and any other information to nonstaff without obtaining the patient's permission. This means that those who are eligible to participate in a study must be first contacted by a staff member, or a letter sent from the hospital to the patients, asking if they would be interested in hearing about and possibly participating in the study.

Access/gatekeepers/"getting in." Unfortunately, having administrative approval does not ensure that the researcher has access to patients. Some institutions require "administrative approval" from the physician and the head of the department. First, the physician: Even if your research does not involve treatments, it is a courtesy to inform the physician about the project. If the research does involve treatments—or perhaps involves identifying former patients—try to get the physician interested in your research, and perhaps involve him as a collaborator. The physician's permission will be essential before you invite patients to participate in your research.

Getting permission of the charge nurse is important. The nurses will be helping to ease you into the setting, to "learn the ropes," and if necessary, to assist you with patient identification and in gaining the cooperation of the staff. She (or he) will be setting up meetings for you with the staff to explain your study and to gain their interest, because whether or not they are involved, they must know who you are and why you are there, and when you are (or are not) on the unit. Be certain to learn the names of all staff members, including support staff. Take donuts to staff meetings—they are the currency of the hospital—and show your appreciation on special occasions and when withdrawing by taking a special cake for the staff (Kayser-Jones, 2003).

Empowered patients. Occasionally, you will be in a long-term care unit in which patient governance is significant and strong. In this case, it is important to have the patients' governing body's permission before you proceed. This may involve a meeting with the executive committee or even the council president to explain your study and get them interested. In turn, they will invite you to the next general meeting to explain your study, answer questions, and to get their permission to proceed. Again, this group has the power to provide you with permission for entry or to block your access.

The fear of documentation. The greatest fear of documenting clinical practice for staff is the fear of evaluation. This ranges from an informal discomfort (staff have asked, "Please hire non-nurses to monitor the cameras" so they would not feel their care was being evaluated), to a worry that a staff member may be recognized in your data, and embarrassed, or reprimanded for "doing something wrong," to a fear that the data may subpoenaed and used against the staff member. This makes clinical data "highly sensitive," and issues of trust, confidentiality, and anonymity very important.

This brings another issue to the fore: What happens if the researcher observes suboptimal care practices? Are they obligated to become a whistleblower? At what point does the researcher intervene?

The costs of intervening are always high—the researcher may, very likely, lose the research site. Of course, it depends on what the problem is; remember that the researcher's first obligation is always to the patient. An excellent model for discussion of such a problem is the fieldwork of Kayser-Jones (2002). When studying the hydration and nutrition of Alzheimer's patients in a nursing home, she and her research team meticulously documented the food intake of residents, weighing patients and recording the timing of their feedings. But at the same time, the researchers did not hesitate to give patients food and water if they so requested, but documented these actions, making it a part of the data. They also discussed the problems of thirst and hunger with the nursing home administrators in an attempt to elicit change—and, in addition to publishing their findings, gave testimony to government.

Ownership of data. If you are conducting research in a clinical setting, to whom does your research data belong? If the research is funded by an external agency, the research data belongs to the university (to whom the grant was awarded); in obtaining permission to conduct the research, the institution in which the research is being conducted relinquishes rights to those data. However, the researcher has the responsibility to protect these data. Data are always sensitive—names must be removed from transcripts, and code numbers assigned as pseudonyms. Data should not be reported case by case, as the more identifying tags are listed, the easier it is to identify each participant. Data must be kept in a locked cabinet, in a locked space. Similarly, the identity of the research site should not be revealed, and perhaps not even the city in which the study was conducted. Some institutions will ask to see articles for approval, prior to publication—this is a very awkward situation in which to find yourself—do not agree to such a requirement, for you may be asked to alter or adjust your report, which is a type of censorship.

If your data are sensitive, they may be subpoenaed—patients may wish to use information to sue caregivers, or a third party may want to use it as evidence. Researchers can protect their data (and therefore their participants) by applying for a NIH Certificate of Confidentiality (available at http://grants1.nih .gov/grants/policy/coc/background.htm). These certificates are available even to researchers who are not funded by the National Institutes of Health.

Difficulty in Conducting Research With the Acutely Ill and Dying

Qualitative inquiry is often delimited by the domain in which care is provided: in the hospital, sometimes limited to one department, patient unit, physician's office, or patient's room; in the community clinic; school office; or the patient's home. These studies are usually ethnographic or participant observation. While the research may center on the patient, it often includes the patient's family, the caregiver, or the interaction between the patient and the caregiver. Because of the way hospitals are organized, with patients with a similar condition sharing the same space—such as patients in a cardiac unit, undergoing renal dialysis, or chemotherapy—it is often difficult to get a private space to conduct an interview.

Silenced by disease. Patients are silenced by disease. They may be unconscious or on a respirator and unable to talk, have had oral surgery, or a dry mouth and unable to articulate. They may be confused, or mentally ill, and unable to express themselves coherently. They may have too much pain to be able to focus on the interview. They may be too fatigued or too sleepy to participate. Treatment also mutes patients—the patient may have an oral brace, may have gone through surgery, or may lack dentures and be unable to speak. They may have medications to make them drowsy, be very breathless, or lack the will to talk. Sometimes patients are stunned by shock and pain, and unable to comprehend anything but the overwhelming agony.

Confusing their physical state, patients are psychologically trying to grapple with their injuries or illness. Patients are too bewildered to make sense of whatever is going on with them, let alone report it to a researcher.

Instability/rapid change/inaccessibility. Often, if the patient's condition is unstable, it may rapidly deteriorate, so that nurses and physicians must try to resuscitate the person. While the researcher may interview the patient about the resuscitation afterward, drugs administered may erase the patient's memory. Again, patients grapple to make sense of what has happened to them, and often are unable to report the changes to another.

In this setting, the urgency/pace of care and the number of people involved in the care means that the researcher may not even be able to see the patient, let alone collect data. Participants may die, be transferred, be in a condition that suddenly deteriorates, or be discharged. When one can start data collection, interviews may be difficult. Patients lack time, have no privacy, and no quiet, uninterrupted space to participate in an interview. Finally, the bodily smells, the sounds of agony, and the sight of injuries, may combine to make the researcher feel faint, or nauseated, and unable to continue collecting data. If such feelings occur, leave the setting before you become an additional person for the staff to care for.

▣ DOES QUALITATIVE HEALTH RESEARCH REQUIRE MODIFICATION OF METHODS?

Qualitative inquiry traditionally used orthodox qualitative methods, and in the 1970s and 1980s, qualitative health researchers used the same texts as researchers in anthropology, sociology, and education. The only exception was grounded theory, developed at the University of California, San Francisco, School of Nursing, and later adopted by other disciplines. Thus, until the mid 1980s, there was some methodological consistency between disciplines.

Given that patients and institutions by their very nature interfere with data collection, it is surprising that any qualitative research is conducted in the hospital. Certainly, interview research is difficult. Some types of observational research are easier, provided one can actually see the patient. But when using observational methods alone, the researcher may have difficulty in interpreting the patient's experience. What did that waving of her hand in the air actually *mean*?

Given these limitations and difficulties, there are some strategies that work to make clinical qualitative research possible, or almost possible.

1. Do retrospective interviews: Interestingly, patients never forget (or forget very slowly) about significant events. Illness and hospitalization is one of those events that is hard to forget. In fact, asking patients to recall after the fact is often a more effective design than repeating interviews throughout the event. If patients have had time to reflect on the illness and learn what effect it has on their lives, to move from the suppression of emotions in enduring to emotional reflection and suffering, then the interviews will be of much better quality. The patients will have the emotional expression needed to do good qualitative description. Remember that the emotions felt in recalling the event mirror the emotions felt when the event was originally experienced, thus the interviews retain their validity. Furthermore, the "retrospective" methods work well with grounded theory, placing events in sequential order, an organization that simplifies the ordering of the process.

2. If one has difficulty in accessing the patient due to privacy or treatment constraints, collect data for shorter, more delineated periods of real time, modifying the topic accordingly. For instance, the topic of "breaking bad news" may be easily accomplished by audio- or videotaping consultations. Such recordings may be analyzed using conversational analysis, and later be combined with interviews to obtain a more rounded perspective.

3. Another perspective may be to involve the caregivers as coinvestigators, as some type of participatory action research (PAR). The caregiver may supplement ongoing data collection with stories of other cases, of other incidents, thus developing your database. Similarly, a patient's relative or caregiver may be an excellent source of observational data.

4. The most important piece of advice is to "take your time." Make the effort to spend time in the clinical setting, and to get to know the staff and the patients.

▣ CONCLUSION

I began this chapter by asking, Is qualitative health research a subdiscipline within the area of qualitative inquiry? And what are those features that make qualitative health research different?

The nature of illness itself gives people a different perspective on life. The ill may experience pain, immobility, a changed *self*. Socially, they may experience a loss of employment, changes in financial status, dependency on others (sometimes even for intimate functions), restrictions on day-to-day living, and perhaps an inability to eat or breathe without assistance. The time that must be allotted to therapies and medical appointments, and the threat of limited life remaining, demands that personal time be prioritized before an external researcher's request for interviews and/or observation. For the researcher, illness produces a different world, with a very different atmosphere from the day-to-day world "outside." This changed environment contains the intensity of birth and death, of pain, of suffering, and the joy of surviving. The researcher does not have priority in this world, and the researcher must take a back seat to the individual's treatment/healing agenda. Furthermore, to do such research requires additional skills and specialized knowledge: The researcher must know the formal rules and codes of conduct in hospitals, yet be empathetic toward the ill, and not become overwhelmed and upset by stories of dying or sights and smells in hospitals. The researcher must have the knowledge and skills to understand the acutely ill, in order, for instance, to break the interview into manageable chunks to prevent fatigue, or to manage the emotions of acute distress.

Julianne Cheek (personal communication, 2010) asked an interesting question: Are the features of the illness experience and health care environment the same for qualitative or quantitative research? If so, does that give qualitative and quantitative health research more in common than when both are compared to another discipline's research, such as educational research? In other words, does the environment and topic differentiate research more than differences in the research method itself?

I conclude "not entirely," for we must also consider the fact that quantitative researchers can more easily—and are expected to—remain detached from their subjects and "objective" in their approach to the topic. This means that they are not emotionally involved with their participants and their topic, and in this regard, quantitative methods, therefore, cannot be grouped with qualitative methods.

Where is the discipline likely to go in the future? As the numbers of students using qualitative health research methods increase and these students graduate, so does the present problem of inadequate numbers of supervisors and mentors dissipate, and the subdiscipline itself become stronger. The numbers of texts, courses, and workshops increase and qualitative heath research becomes mainstream. This push will be helped by unexpected sources—for instance, the push for evaluation and for mixed methods will help qualitative inquiry in places that the influx of qualitative inquiry per se may be slower.

Thus, I conclude by maintaining that qualitative health research is a specialized form of qualitative research. The emotional components of qualitative health research are different enough, and the ethical issues separate enough, that qualitative inquiry has its own needs for education, training, methods, and dissemination of knowledge. The conditions in the institutions and the features of ill participants are different enough substantively for qualitative health research to be considered a "specialty," requiring specialized knowledge, research design, and modification of methods.

▣ **NOTES**

1. AHFMR, as an establishment grant to Janice Morse.
2. *QHR* is published monthly, with approximately 12 to 14 articles per issue. The content analysis here was derived from the 142 articles published in Volume 19 in 2009.

▣ **REFERENCES**

Becker, H. S., Geer, B., Hughes, E. C., & Strauss, A. L. (1961). *Boys in white: Student culture in medical school.* Chicago: University of Chicago Press.

Benner, P. (1984). *From novice to expert: Excellence and power in clinical nursing practice.* Englewood Cliffs, NJ: Prentice Hall.

Bergstrom, L., Richards, L., Proctor, A., Bohrer Avila, L., Morse, J. M., & Roberts, J. E. (2009). Birth talk in second stage labor. *Qualitative Health Research, 19,* 954–964.

Brody, H. (1987). *Stories of sickness.* New York: Oxford University Press.

Broom, A. (2009). Intuition, subjectivity, and le bricoleur: Cancer patients' accounts of negotiating a plurality of therapeutic options. *Qualitative Health Research, 19,* 1050–1059.

Cantrell, M. A., & Conte, T. M. (2009). Between being cured and being healed: The paradox of childhood cancer survivorship. *Qualitative Health Research, 19,* 312–322.

Champlin, B. E. (2009). Being there for another with a serious mental illness. *Qualitative Health Research, 19,* 1525–1535.

Cochrane, A. L. (1972/1989). *Effectiveness and efficiency: Random reflections on health services.* London: British Medical Journal. [Original publication London: Nuffield Provincial Hospitals Trust, 1972]

Crabtree, B. F., & Miller, W. L. (1992). *Doing qualitative research.* Thousand Oaks, CA: Sage.

Davies, B., Larson, J., Contro, N., Reyes-Hailey, C., Ablin, A. R., Chesla, C. A., et al. (2009). Conducting a qualitative culture study of pediatric palliative care. *Qualitative Health Research, 19,* 5–16.

De Marco, M., Thorburn, S., & Kue, J. (2009). "In a country as affluent as America, people should be eating": Experiences with and perceptions of food insecurity among rural and urban Oregonians. *Qualitative Health Research, 19*(7), 1010–1024.

Engel, J. D., Zarconi, J., Pethtel, L. L., & Missimi, S. A. (2008). *Narrative in health care: Healing patients, practitioners, profession, and community.* Oxford, UK: Radcliffe.

Ferrari, M., Tweed, S., Rummens, J. A., Skinner, H. A., & McVey, G. (2009). Health materials and strategies for the prevention of immigrants' weight-related problems. *Qualitative Health Research, 19,* 1259–1272.

Field, P. A., & Morse, J. M. (1985). *Nursing research: The application of qualitative approaches.* London: Croom Helm.

Forgeron, P. A., Jongudomkarn, D., Evans, J., Finley, G. A., Thienthong, S., Siripul, P., et al. (2009). Children's pain assessment in Northeastern Thailand: Perspectives of health professionals. *Qualitative Health Research, 19,* 71–81.

Fox, F. E., Rodham, K. J., Harris, M. F., Taylor, G. J., Sutton, J., Scott, J., et al. (2009). Experiencing "The other side": A study of empathy and empowerment in general practitioners who have been patients. *Qualitative Health Research, 19,* 1580–1588.

Gao, G., Burke, N., Somkin, C. P., & Pasick, R. (2009). Considering culture in physician-patient communication during colorectal cancer screening. *Qualitative Health Research, 19,* 778–789.

Germain, C. P. (1979). *The cancer unit: An ethnography.* Wakefield, MA: Nursing Resources.

Gioia, D. (2009). Understanding the ecological validity of neuropsychological testing using an ethnographic approach. *Qualitative Health Research, 19,* 1495–1503.

Glaser, B. G., & Strauss, A. (1965). *Awareness of dying.* Chicago: Aldine.

Glaser, B. G., & Strauss, A. (1967). *Discovery of grounded theory.* Chicago: Aldine.

Goffman, E. (1961). *Asylums: Essays on the social situation of mental patients and other inmates.* New York: Anchor Books, Doubleday.

Gotlib Conn, L., Lingard, L., Reeves, S., Miller, K., Russell, A., & Zwarenstein, M. (2009). Communication channels in general internal medicine: A description of baseline patterns for improved interprofessional collaboration. *Qualitative Health Research, 19,* 943–953.

Hilton, S., Emslie, C., Hunt, K., Chapple, A., & Ziebland, S. (2009). Disclosing a cancer diagnosis to friends and family: A gendered analysis of young men's and women's experiences. *Qualitative Health Research, 19,* 744–754.

Holtslander, L. F., & Duggleby, W. D. (2009). The hope experience of older bereaved women who cared for a spouse with terminal cancer. *Qualitative Health Research, 19,* 388–400.

Jacobson, N. (2009). Dignity violation in health care. *Qualitative Health Research, 19,* 1536–1547.

James, I., Andershed, B., & Ternestedt, B.-M. (2009). The encounter between informal and professional care at the end of life. *Qualitative Health Research, 19,* 258–271.

Jenkins, E. R., Mabbett, G. M., Surridge, A. G., Warring, J., & Gwynn, E. D. (2009). A cooperative inquiry into action learning and praxis development in a community nursing module. *Qualitative Health Research, 19,* 1303–1320.

Johansson, E. E., Bengs, C., Danielsson, U., Lehti, A., & Hammarström, A. (2009). Gaps between patients, media, and academic medicine in discourses on gender and depression: A metasynthesis. *Qualitative Health Research, 19,* 633–644.

Kafanelis, B. E., Kostanski, M., Komesaroff, P. A., & Stojanovska, L. (2009). Being in the script of menopause: Mapping the complexities of coping strategies. *Qualitative Health Research, 19,* 30–41.

Kayser-Jones, J. (2002). Malnutrition, dehydration, and starvation in the midst of plenty: The political impact of qualitative inquiry. *Qualitative Health Research, 12,* 1391–1405.

Kayser-Jones, J. (2003). Continuing to conduct research in nursing homes despite controversial findings: Reflections by a research scientist. *Qualitative Health Research, 13,* 114–128.

Kenne Sarenmalm, E., Thorén-Jönsson, A.-L., Gaston-Johansson, F., & Öhlén, J. (2009). Making sense of living under the shadow of death: Adjusting to a recurrent breast cancer illness. *Qualitative Health Research, 19,* 1116–1130.

Kleinman, A. (1980). *Patients and healers in the context of culture.* Berkeley: University of California Press.

Kleinman, A. (1986). *The illness narratives.* New York: Basic Books.

Lahman, M. K. E. (2009). Dreams of my daughter: An ectopic pregnancy. *Qualitative Health Research, 19,* 272–278.

Leininger, M. M. (1985). *Qualitative research methods in nursing.* New York: Grune & Stratton.

Lutz, B. J., Kneipp, S., & Means, D. (2009). Development of a health screening questionnaire for women in welfare transition programs in the United States. *Qualitative Health Research, 19,* 105–115.

McCorry, N. K., Dempster, M., Clarke, C., & Doyle, R. (2009). Adjusting to life after esophagectomy: The experience of survivors and carers. *Qualitative Health Research, 19,* 1485–1494.

McIntosh, M., & Morse, J. M. (2009). Institutional review boards and the ethics of emotion. In N. K. Denzin & M. D. Gardina (Eds.), *Qualitative inquiry and social justice* (pp. 81–107). Walnut Creek, CA: Left Coast Press.

Medley, A. M., Kennedy, C. E., Lynyolo, S., & Sweat, M. D. (2009). Disclosure outcomes, coping strategies, and life changes among women living with HIV in Uganda. *Qualitative Health Research, 19,* 1744–1754.

Miller, J. H., & Stiles, A. (2009). Family presence during resuscitation and invasive procedures: The nurse experience. *Qualitative Health Research, 19,* 1431–1442.

Monrouxe, L. V., Rees, C. E., & Bradley, P. (2009). The construction of patients' involvement in hospital bedside teaching encounters. *Qualitative Health Research, 19,* 918–930.

Morse, J. M. (2002). Interviewing the ill. In J. Gubrium & J. Holstein (Eds.), *Handbook of interview research* (pp. 317–330). Thousand Oaks, CA: Sage.

Morse, J. M. (2008). Excluding qualitative inquiry: An open letter to the Canadian Urological Association [Editorial]. *Qualitative Health Research, 18*(6), 583.

Morse, J. M., Stern, P. N., Corbin, J., Bowers, B., Charmaz, K., & Clarke, A. (2009) *Grounded theory: The second generation.* Walnut Creek, CA: Left Coast Press.

Naidoo, P., Dick, J., & Cooper, D. (2009). Exploring tuberculosis patients' adherence to treatment regimens and prevention programs at a public health site. *Qualitative Health Research, 19,* 55–70.

Norris, P., Fa'alau, F., Va'ai, C., Churchward, M., & Arroll, B. (2009). Navigating between illness paradigms: Treatment seeking by Samoan people in Samoa and New Zealand. *Qualitative Health Research, 19,* 1466–1475.

Oman, K. M., Moulds, R., & Usher, K. (2009). Professional satisfaction and dissatisfaction among Fiji specialist trainees: What are the implications for preventing migration? *Qualitative Health Research, 19,* 1246–1258.

Ouellette, A., Achille, M., & Pâquet, M. (2009). The experience of kidney graft failure: Patients' perspectives. *Qualitative Health Research, 19,* 1131–1138.

Parker, L. E., Kirchner, J. E., Bonner, L., Fickel, J. J., Ritchie, M. J., Simons, C. E., et al. (2009). Creating a quality improvement dialogue: Utilizing knowledge from frontline staff, managers, and experts to foster health care quality improvement. *Qualitative Health Research, 19,* 229–242.

Parsons, T. (1975). The sick role and the role of the physician reconsidered. *The Millbank Memorial Fund Quarterly. Health and Society, 53*(3), 257–278.

Protudjer, J. L. P., Kozyrskyj, A. L., Becker, A. B., & Marchessault, G. (2009). Normalization strategies of children with asthma. *Qualitative Health Research, 19,* 94–104.

Pylypa, J. (2009). Elder authority and situational diagnosis of diarrheal diseases as normal infant development in Northeast Thailand. *Qualitative Health Research, 19,* 965–975.

Quint, J. (1967). *The nurse and the dying patient.* New York: Macmillan.

Sackett, D. L. (1993). Rules of evidence and clinical recommendations. *Canadian Journal of Cardiology, 9*(6), 487–489.

Schnur, J. B., Ouellette, S. C., Bovbjerg, D. H., & Montgomery, G. H. (2009). Breast cancer patients' experiences of external-beam radiotherapy. *Qualitative Health Research, 19,* 668–676.

Scott, A. (2009). Illness meanings of AIDS among women with HIV: Merging immunology and life experience. *Qualitative Health Research, 19,* 454–465.

Shamai, M., & Ron, P. (2009). Helping direct and indirect victims of national terror: Experiences of Israeli social workers. *Qualitative Health Research, 19,* 42–54.

Shariff, F., Carter, J., Dow, C., Polley, M., Salinas, M., & Ridge, D. (2009). Mind and body management strategies for chronic pain and rheumatoid arthritis. *Qualitative Health Research, 19,* 1037–1049.

Spoont, M. R., Sayer, N., Friedemann-Sanchez, G., Parker, L. E., Murdoch, M., & Chiros, C. (2009). From trauma to PTSD: Beliefs about sensations, symptoms, and mental illness. *Qualitative Health Research, 19,* 1456–1465.

Stevens, P. E., & Hildebrandt, E. (2009). Pill taking from the perspective of HIV-infected women who are vulnerable to antiretroviral treatment failure. *Qualitative Health Research, 19,* 593–604.

Strauss, A., Corbin, J. S., Fagerhaugh, S., Glaser, B., Maines, D., Suczek, B., & Weiner, C. (1975). *Chronic illness and the quality of life.* St. Louis, MO: Mosby.

van den Berg, J. H. (1960). *The psychology of the sick bed.* Pittsburgh, PA: Duquesne University Press.

van Manen, M. (1990). *Researching the lived experience.* London, Ontario, Canada: Althouse Press.

Van Rooij, F. B., van Balen, F., & Hermanns, J. M. A. (2009). The experiences of involuntarily childless Turkish immigrants in the Netherlands. *Qualitative Health Research, 19,* 621–632.

Vayreda, A., & Antaki, C. (2009). Social support and unsolicited advice in a bipolar disorder online forum. *Qualitative Health Research, 19,* 931–942.

Ziółkowska, J. (2009). Positions in doctors' questions during psychiatric interviews. *Qualitative Health Research, 19,* 1621–1631.

Part IV

METHODS OF COLLECTING AND ANALYZING EMPIRICAL MATERIALS

Nothing stands outside representation. Research involves a complex politics of representation. This world can never be captured directly; we only study representations of it. We study the way people represent their experiences to themselves and to others. Experience can be represented in multiple ways, including rituals, myth, stories, performances, films, songs, memoirs, and autobiography, writing stories, autoethnography. We are all storytellers, statisticians, and ethnographers alike.

The socially situated researcher creates through interaction and material practices those realities and representations that are the subject matter of inquiry. In such sites, the interpretive practices of qualitative research are implemented. These methodological practices represent different ways of generating and representing empirical materials grounded in the everyday world. Part IV examines the multiple practices and methods of analysis that qualitative researchers-as-methodological-bricoleurs now employ.

▣ NARRATIVE INQUIRY

Today narrative inquiry is flourishing; it is everywhere. We know the world through the stories that are told about it. Even so, as Susan Chase reminds us, narrative inquiry as particular type of qualitative inquiry is a field in the making. Modifying her earlier formulation of narrative that focused on retrospective meaning making, Chase now defines narrative, after Jaber Gubrium and James Holstein, as "meaning making through the shaping or ordering of experience." She provides an excellent overview of this field, discussing the multiple approaches to narrative, storytelling as lived experience, narrative practices and narrative environments, the researcher and the story, autoethnography, performance narratives, methodological and ethical issues, big and small stories content analysis, going beyond written and oral texts, narrative and social change, Latin American *testimonios,* collective stories, public dialogue, the need for meta-analysis of the vast array of narrative studies.

Narratives are socially constrained forms of action, socially situated performances, ways of acting in and making sense of the world. Narrative researchers often write in the first person, thus "emphasizing their own narrative action."Narrative inquiry can advance a social change agenda. Wounded storytellers can empower others to tell their stories. *Testimonios,* as emergency narratives, can mobilize a nation against social injustice, repression, and violence. Collective stories can form the basis of a social movement. Telling the stories of marginalized people can help create a public space requiring others to hear what they don't want to hear.

▣ CRITICAL ARTS-BASED INQUIRY

Critical arts-based inquiry situates the artist-as-researcher in a research paradigm committed to democratic, ethical agendas. Like participatory action research (PAR), critical arts-based inquiry demonstrates an activist approach to inquiry. Arts-based inquiry uses the aesthetics, methods, and practices of the literary, performance, and visual arts, as well as dance, theater, drama, film, collage, video, and photography. Arts-based inquiry is intertextual. It crosses the borders of art and research. Susan Finley writes a history of this methodology, locating it in the postcolonial, postmodern context. As the same time, she critiques neoliberal trends that are critical of social-justice based projects. She takes up the performative turn in qualitative inquiry, moving to a people's pedagogy that performs a radical ethical aesthetic. She shows how activist art can be used to address issues of political significance, including engaging community participants in acts of political, self-expression.

When grounded in a critical performance pedagogy, arts-based work can be used to advance a progressive political agenda that addresses issues of social inequity. Thus do researchers take up their "pens, cameras, paintbrushes, bodies" and voices in the name of social justice projects. Such work exposes oppression, targets sites of resistance, and outlines a transformative praxis that performs resistance texts. Finley ends with a rubric for evaluating critical-arts based research, asking whether the research demonstrates indigenous skills, openly resists structures of domination, performs useful public service, gives a voice to the oppressed, critiques neoconservative discourse, and brings passion to its performances, moving persons to positive social action.

◪ ORAL HISTORY

Linda Shopes discusses moral history as a way of collecting and interpreting human memories to foster knowledge and human dignity. Because they interview persons, oral historians implement the open-ended interview as a form of social inquiry.

As Chase observed, we live in a narrative, storytelling, interview society, in a society whose members seem to believe that interviews (and stories) generate useful information about lived experience and its meanings. The interview and the life-story narrative have become taken-for-granted features of our mediated, mass culture. But the life story, the oral history, and the personal narrative are negotiated texts, sites where power, gender, race, and class intersect.

Andrea Fontana and James H. Frey (2005) review the history of the interview in the social sciences, noting its three major forms—structured, unstructured, and open-ended—while showing how the tool is modified and changed during use. They also oral history interviews, creative interviewing, gendered, feminist and postmodern, or multi-voiced interviewing. Shopes takes up where Fontana and Frey ended, with oral history.

The oral history is a recorded interview, preserved for the record and made accessible to others. The oral history interview is historical in intent—it seeks new knowledge about the past through an individual biography. Oral history is understood as both an act of memory and an inherently subjective account of the past. The oral history interview elicits information that requites interpretation. The oral history interview is an inquiry in depth.

Oral historians are closely aligned with the projects of interpretive sociologists and anthropologists such as Chase, Holstein, and Gubrium. Shopes reviews the history of oral history, discussing its different meanings, and interpretations, from the 19th century to the present, from slave narratives to elite interviews and the oral histories of underrepresented groups. Thus, the method has helped to democratize history, as have recent developments in digital media.

Oral histories, as noted in our Introduction to Part I, have taken the lead in confronting the legal and ethical issues involved in qualitative inquiry. The ethical initiatives by oral historians have created a space within current institutional review board (IRB) structures for truthful inquiry, for commitments to a "utopian striving to know how things really are, and of how things may be." Shopes has been the leader in this discourse.

◪ RECONTEXTUALIZING OBSERVATIONAL METHODS

Going into a social situation and looking is another important way of gathering materials about the social world. Drawing on previous arguments (Angrosino & Mays de Pérez, 2000), Michael Angrosino and Judith Rosenberg fundamentally rewrite the methods and practices of naturalistic observation. All observation involves participation in the world being studied. There is no pure, objective, detached observation; the effects of the observer's presence can never be erased. Further, the colonial concept of the subject (the object of the observer's gaze) is no longer appropriate. Observers now function as collaborative participants in action inquiry settings. Angrosino and Rosenberg argue that observational interaction is a tentative, situational process. It is shaped by shifts in gendered identity, as well as by existing structures of power. As relationships unfold, participants validate the cues generated by others in the sitting. Finally, during the observational process people assume situational identities, which may not be socially or culturally normative.

Like Clifford Christians (Chapter 4, this volume) and Linda Shopes, Angrosino and Rosenberg offer compelling criticisms of institutional review boards (IRBs), noting that positivistic social scientists seldom recognize the needs of observational ethnographers. In many universities, the official IRB is tied to the experimental, hypothesis-testing, so-called scientific paradigm. This paradigm creates problems for the postmodern observer, for the scholar who becomes part of the world that is being studied. To get approval for their research, scholars may have to engage in deception (in this instance of the IRB). This leads some ethnographers to claim that their research will not be intrusive, and hence will not cause harm. Yet interactive observers are by definition intrusive. When collaborative inquiry is undertaken subjects become stakeholders, persons who shape the inquiry itself. What this means for consent forms—and forms of participatory inquiry more broadly—is not clear. Alternative forms of ethnographic writing, including the use of fictionalized stories, represent one avenue for addressing this ethical quandary.

An ethic of "proportionate reason" is offered. This utilitarian ethic attempts to balance the benefits, costs, and consequences of actions in the field, asking if the means to an end are justified by the importance and value of the goals attained. This ethic can then be translated into a progressive social agenda. This agenda stresses social, not commutative, distributive, or legal justice. A social justice ethic asks the researcher to become directly involved with the poor and the marginalized, to become an advocate, to facilitate empowerment in communities. Inquirers, seeking utopian visions, and progressive agendas, act as advocates, enacting pedagogies of service.

The worlds of observation are changing, new audio and visual technologies (see Prosser, Chapter 29, this volume) are now available, the stage has gone global, the Internet creates virtual worlds and virtual ethnographies (Gaston, Chapter 31, this volume) are commonplace. And the search for social justice continues. This chapter demystifies the observation method. Observation is no longer the key to some grand analysis of culture or society. Instead, observational research now becomes a method that focuses on differences, on the lives of particular people in concrete, but constantly changing, human relationships. The relevance and need for a radical ethics of care and commitment becomes even more apparent.

▣ VISUAL METHODOLOGY: A MORE SEEING RESEARCH

Jon Prosser's chapter outlines the key facets of contemporary visual research, concluding with future challenges. Visual researchers use the word *visual* to refer to phenomena that can be seen, can be given meaning. Since the 1960s, researchers have used visual images for one of two purposes: empirically, to document reality, or symbolically, to study the meaning of images produced by visual culture. Today a visual fluency for qualitative researchers is presumed.

Today visual sociologists and anthropologists use digital photography, motion pictures, the World Wide Web, interactive CDs, CD-ROMs, and virtual reality as ways of forging connections between human existence and visual perception. These forms of visual representation represent different ways of recording and documenting what passes as social life. Often called the mirror with a memory, photography takes the researcher into the everyday world, where the issues of observer identity, the subject's point of view, and what to photograph become problematic.

Prosser discusses four current trends and issues in the evolving field of visual methodology: (1) the representations of visual research, (2) technology and visual methods, (3) participatory visual methods, and (4) training in visual methodology. We are moving into a space where data (empirical materials) can be better and more effectively represented visually. The digital camera, software for storing large volumes of imagery, and visual compliant software—ATLAS.ti, NVivo, Transana, Observer XT—enable researchers to store, analyze, map, measure, and represent complex human interactions and communication structures. New participatory visual methods use photo-elicitation methods, photovoice, video diaries, photo-narratives, and various other hypermedia techniques. Training in visual method is burgeoning. In the United Kingdom, the Economic and Social Research Council (ESRC) has sponsored a nationwide training program aimed at teaching visual methods to a cross-section of qualitative researchers.

Prosser predicts that in the next decade there will be a greater alignment between visual methodologies and arts-based research. This alignment will lead to innovations in visual sociology, visual ethnography, and disability studies. (Prosser offers moving visual examples from his current study, which explores the perceptions of disabled people with limited communication skills.) He concludes with observations on the threats to visual research from ethical review boards. He endorses an ethics of care model. A preoccupation with biomedical regulatory ethics will slow down the development of visual methods. IRBs insist that confidentiality be maintained, that subjects remain anonymous. But in many cases, subjects are pleased and willing to be identified, and further, their very identifiability is critical to the research project. In such situations, the researcher is urged to develop an ethical covenant with those being studied so that only mutually agreed upon materials will be published.

We need to learn how to experiment with visual (and nonvisual) ways of thinking. We need to develop a critical, visual sensibility, a sensibility that will allow us to bring the gendered material world into play in critically different ways. We need to interrogate critically the logics of cyberspace and its virtual realities. The rules and methods for establishing truth that hold these worlds together must also be better understood.

▣ AUTHOETHNOGRAPHY: MAKING THE PERSONAL POLITICAL

Personal experience reflects the flow of thoughts and meanings persons have in their immediate situations. These experiences can be routine or problematic. They occur within the life of a person. When they are talked about, they assume the shape of a story or a narrative. Lived experience cannot be studied directly, because language, speech, and systems of discourse mediate and define the very experience one attempts to describe. We study the representations of experience, not experience itself. We examine the stories people tell one another about the experiences they have had. These stories may be personal experience narratives, or self-stories, interpretations made up as the person goes along

Many now argue that we can only study our own experiences. The researcher becomes the research subject. This is the topic of autoethnography. Tami Spry's chapter (Chapter 30, this volume) reflexively presents the arguments for writing reflexive, personal narratives. Indeed her multivoiced text is an example of such writing; it performs its own narrative reflexivity. She masterfully reviews the arguments for studying personal experience narratives, anchoring her text in the discourses of critical performance studies.

She reviews the history of and arguments for this writing form, the challenge to create texts that unfold in the life of the writer, while embodying tactics that enact a progressive politics of resistance. Such texts, when performed (and writing is a form of performance), enact a politics of possibility. They shape a critical awareness, disturb the status quo, and probe questions of identity. Spry writes out of her own history with this method and in so doing takes the reader to Judith Hamera's (Chapter 18, this volume) and Barbara Tedlock's (Chapter 19, this volume) chapters on performance and narrative ethnography.

Spry shows how performative autoethnography, as a critical reflexive methodology, provides a framework for making the personal political in the spaces of a post–September 11, 2001, world. She writes and performs from the spaces of hurting, healing, and grieving bodies; her own grief at the loss of her son at childbirth, the death of her father, and the bombings of 9/11. She offers a pedagogy of hope, a critical and indigenous ethnography. Her essay is about autoethnography as a radical resistant democratic practice, a political practice intended to create a space for dialogue and debate about issues of injustice. Her chapter tells by showing performance fragments, absent histories, embodied possibilities, the storytelling, performative I. Personal biography collides with culture and structure, turning historical discourse back on itself. Her performative I is embodied, liminal, accountable, wild, free, moral.

Spry ends with writing exercises, performative practices, a call for collaborative performances grounded in the belief that together we can create a local and global respect, love, and care for one another.

▣ ONLINE ETHNOGRAPHY

Sarah Gatson (Chapter 31, this volume) discusses two main versions of online ethnography: as an extension of traditional collaborative and multisited/extended-case ethnography, on the one hand, and autoethnographic inquiry, on the other hand. Under this form, the inquirer grounds an online map upon herself. Gatson reviews the classic works in each of these genres, noting that the online site is already inscribed and performative. She suggests that computer-mediated construction of self, other, and social structure constitutes a unique phenomenon of study. Offline, the body is present, and can be responded to by others. Identity construction is a situated, face-to-face process. By contrast, online, the body is absent, and interaction is mediated by computer technology and the production of written discourse. Gaston examines many of the issues that can arise in the qualitative study of Internet-mediated situations. These are issues connected to definitions of what constitutes the field or boundaries of a text, as well as what counts as text or empirical material. How the other is interpreted and given a textual presence is also problematic, as are ethical issues that are complex.

Ethical guidelines for Internet research vary sharply across disciplines and nations. She acknowledges with Judith Davidson and Silvana di Gregorio (Chapter 38, this volume) the ethical complexities when virtual worlds are the inquiry site. Gatson troubles the issue of informed consent, asking who gives permission to who, and for what, when one's research site is a public

venue "with not even an unlocked door whose opening announces a certain basic level of entry and at most slightly opaque windows that block certain kinds of participation." Of course, under a communitarian, feminist ethical model, researchers enter into a collaborative relationship with a moral community of online interactants. Attempts are made to establish agreed upon understandings concerning privacy, ownership of materials, the use of personal names, and the meaning of such broad principles as justice and beneficence.

Gaston draws material from several online multilocal ethnographies, arguing that we are in a new space—Ethnography 2.0—where online subjects talk back, interact with us, read our research, criticize our work, all while eroding the walls we build around ourselves as objective outsiders studying the virtual worlds of others. We have become the subject. In this space, it is essential to reflect carefully on the ethical issues framing our studies.

▣ ANALYZING TALK AND TEXT

Qualitative researchers study spoken and written records of human experience, including transcribed talk, films, novels, and photographs. Interviews give the researcher accounts about the issues being studied. The topic of the research is not the interview itself. Research using naturally occurring empirical materials—tape recordings of mundane interaction—constitute topics of inquiry in their own right. This is the topic of Anssi Peräkylä and Johanna Ruusuvuori's chapter (Chapter 32).

With Chase, Shopes, and Gubrium and Holstein, Peräkylä and Ruusuvuori treat interview materials as narrative accounts, rather than pictures of reality. Texts are based on transcriptions of interviews, and other forms of talk. These texts are social facts; they are produced, shared, and used in socially organized ways. Peräkylä and Ruusuvuori discuss semiotics, discourse analysis (DA), critical discourse analysis (CDA), and historical critical discourse analysis (HAD), after Michel Foucault. They review instances of each of these types of discourse analysis.

Peräkylä and Ruusuvuori also discuss membership categorization analysis (MCA), which is a less familiar form of narrative analysis. Drawing on the work of Harvey Sacks (see Silverman, 1998), they illustrate the logic of MCA. With this method, the researcher asks how persons use everyday terms and categories in their interactions with others.

Peräkylä and Ruusuvuori then turn to the analysis of talk. Two main social science traditions inform the analysis of transcripts, conversation analysis (CA) and DA. Peräkylä and Ruusuvuori review and offer examples of both traditions, arguing that talk is socially organized action. It is structurally organized, and as such it creates and maintains its own version of intersubjective reality. They show how this work has direct relevance for political and social justice concerns. Many CA studies have shown, for example, how specific interactional practices contribute to the maintenance or change of the gender system.

To summarize: Text-based documents of experience are complex. But if talk constitutes much of what we have, then the forms of analysis outlined by Peräkylä and Ruusuvuori represent significant ways of making the world and its words more visible.

◪ FOCUS GROUPS, PEDAGOGY, POLITICS, AND INQUIRY

George Kamberelis and Greg Dimitriadis (Chapter 33) continue to significantly advance the discourse on focus group methodology. Building on their previous treatments of this topic in earlier editions of the *Handbook*, they show how focus groups have been used in market and military research, in emancipatory pedagogy, and in first-, second-, and third-generation feminist inquiry. Kamberelis and Dimitriadis place these three genealogies in dialogue with one another while exploring new dangers faced by focus group research in the current political climate. They reimagine focus group work as performative, and as almost always involving multiple functions that are pedagogical, political, and empirical. The performative turn shapes a politics of evidence—that is how do we enact strong or weak evidence.

Kamberelis and Dimitriadis contrast the dialogical, critical theory approach to focus groups with their use in propaganda and market research. In the marketing context, focus groups are used to extract information from people on a given topic. This information is then used to manipulate people more effectively. Critical pedagogy theorists, such as Paulo Freire and Jonathan Kozol use focus groups for imagining and enacting the "emancipatory political possibilities of collective work."

They contrast these two approaches to the, history of focus groups in feminist inquiry, noting its use in first-, second-, and third-wave feminist formations for consciousness-raising purposes (CRP). They draw on Esther Madriz (2000), who offers a model of focus group interviewing that emphasizes a feminist ethic of empowerment, moral community, emotional engagement, and the development of long-term, trusting relationships. This method gives a voice to women of color, who have long been silenced. Focus groups facilitate women writing culture together. As a Latina feminist, Madriz places focus groups within the context of collective testimonies and group resistance narratives. Focus groups reduce the distance between the researcher and the researched. The multivocality of the participants limits the control of the researcher over the research process.

Within this history, focus groups have been used to elicit and validate collective testimonies, to give a voice to the previously silenced by creating a safe space for sharing one's life experiences. The critical insights and practices of consciousness-raising groups have helped us move more deeply into the praxis-oriented commitments of the seventh and eighth moments. In these spaces, as the work of Janice Radway, Patricia Lather, and Chris Smithies documents, focus groups can become the vehicle for allowing participants to take over and own the research. In these ways, focus groups become the sites where pedagogy, politics, and interpretive inquiry intersect and inform one another.

When this happens, inquiry becomes directly involved in the complexities of political activism and policy making. Often this clashes with local IRB offices. They offer examples from several sites showing how this can happen.

Virginia Olesen (Chapter 7, this volume) reminds us that women of color experience a triple subjugation based on class, race, and gender oppression. Critical focus groups, as discussed by Kamberelis and Dimitriadis, create the conditions for the emergence of a critical race consciousness, a consciousness focused on social change. It seems that with critical focus groups, critical race theory and progressive politics have found their methodology.

◪ CONCLUSION

The researcher-as-methodological bricoleur should have a working familiarity with each of the methods of collecting and analyzing empirical materials that are presented in this part of this handbook. This familiarity includes understanding the history of each method and technique, as well as hands-on experience with each. Only in this way can the limitations and strengths of each be fully appreciated. At the same time, the investigator will more clearly see how each, as a set of material, interpretive practices, creates its own subject matter.

In addition, it must be understood that each paradigm and perspective, as presented in Part II, has a distinct history with these methods of research. Although methods-as-tools are somewhat universal in application, they are not uniformly used by researchers from all paradigms, and when used, they are fitted and adapted to the particularities of the paradigm in question. However, researchers from all paradigms and perspectives can profitably make use of each of these methods of collecting and analyzing empirical materials.

◪ REFERENCES

Angrosino, M. V., & Mays de Pérez, K. A. (2000). Rethinking observation: From method to context. In N. K. Denzin & Y. S. Lincoln (Eds.), *Handbook of qualitative research* (2nd ed., pp. 673–702). Thousand Oaks: Sage.

Fontana, A., & Frey, J. H. (2005). The interview: From neutral stance to political involvement. In N. K. Denzin & Y. S. Lincoln (Eds.), *The SAGE handbook of qualitative research* (3rd ed., pp. 695–728). Thousand Oaks: Sage.

Madriz, E. (2000). Focus groups in feminist research. In N. K. Denzin & Y. S. Lincoln (Eds.), *Handbook of qualitative research* (2nd ed., pp. 835–850). Thousand Oaks: Sage.

Silverman, D. (1998). *Harvey Sacks: Social science & conversation analysis*. Oxford, UK: Polity Press.

25

NARRATIVE INQUIRY

Still a Field in the Making

Susan E. Chase

uch has happened in narrative inquiry since the third edition of this handbook. Many books have been published, including Michael Bamberg's *Narrative—State of the Art* (2007); D. Jean Clandinin's *Handbook of Narrative Inquiry* (2007); Jaber Gubrium and James Holstein's *Analyzing Narrative Reality* (2009); Mary Jo Maynes, Jennifer Pierce, and Barbara Laslett's *Telling Stories: The Use of Personal Narratives in the Social Sciences and History* (2008); Dan McAdams, Ruthellen Josselson, and Amia Lieblich's *Identity and Story: Creating Self in Narrative* (2006); and Catherine Kohler Riessman's *Narrative Methods for the Human Sciences* (2008). The journal *Narrative Inquiry* continues to thrive. So do research centers, such as the Life Story Center at the University of Southern Maine; the Center for Myth and Ritual in American Life at Emory University; the Narrative Therapy Centre of Toronto; the Centre for Narrative Practice in Sheffield, United Kingdom; and the Dulwich Centre in Adelaide, Australia. Digital collections of written, audio, and video narratives are expanding, including National Public Radio's StoryCorps project, the September 11 Digital Archive, and the Voices of the Holocaust Project.

Clearly, narrative inquiry is still flourishing. It is also still evolving. In this update of my chapter in the third edition of this handbook, I focus on recent contributions as I present multiple approaches to narrative research, address methodological issues, and explore how narratives and narrative research make personal and social change possible. I also sketch some ideas about the future of narrative inquiry.

▣ MULTIPLE APPROACHES

Narrative inquiry is a particular type—a subtype—of qualitative inquiry.[1] What distinguishes narrative inquiry is that it begins with the biographical aspect of C. Wright Mills' (1959)

famous trilogy—biography, history, and society. Narrative inquiry revolves around an interest in life experiences as narrated by those who live them. Narrative theorists define narrative as a distinct form of discourse: as meaning making through the shaping or ordering of experience, a way of understanding one's own or others' actions, of organizing events and objects into a meaningful whole, of connecting and seeing the consequences of actions and events over time. Narrative researchers highlight what we can learn about anything—history and society as well as lived experience—by maintaining a focus on narrated lives.

Within this framework, however, researchers' interests differ substantially. Without claiming to be comprehensive or exhaustive in my categories, I outline several approaches within contemporary narrative inquiry.[2]

The Story and the Life

Some researchers focus on the relationship between people's life stories and the quality of their life experiences. These researchers usually emphasize *what* people's stories are about—their plots, characters, and sometimes the structure or sequencing of their content.[3] In explaining this approach, D. Jean Clandinin and Jerry Rosiek (2007) argue that everyday experience itself—that taken for granted, immediate, and engrossing daily reality in which we are all continuously immersed—is where narrative inquiry should begin and end. They implore researchers to listen to people's stories about everyday experience "with an eye to identifying new possibilities within that experience" (p. 55). Beginning and ending with experience means tempering the academic impulse to generalize from specific stories to broader concepts, or to impose theoretical concepts (such as false consciousness) on people's stories. Rather, the goal of this approach is to work collaboratively with research participants to improve the quality of their everyday experiences.[4] This approach can be thought of as pragmatic or applied.

Along similar lines, psychologists who conduct narrative research focus on the relationship between people's stories and their identity development or personal well-being.[5] In *Identity and Story: Creating Self in Narrative*, editors McAdams, Josselson, and Lieblich (2006) summarize Erik Erikson's classic theory of identity development and then demonstrate narrative inquiry's contribution to an understanding of identity. They define *narrative identity* as "internalized and evolving life stories" (p. 5), and they present research focused on three questions: whether people's identity constructions through storytelling reveal the self's unity, multiplicity, or both; how self and society contribute to people's constructions of narrative identity; and how people's stories display stability, growth, or both, in their identities.[6]

The question of how narrative makes personal growth possible grounds the field of narrative therapy (Adler & McAdams, 2007; Baddeley & Singer, 2007; Cohler, 2008; Josselson, 1996; McAdams, 2006; White & Epston, 1990). While acknowledging that biographical, social, cultural, and historical circumstances condition the stories people tell about themselves, narrative therapists propose that the stories people tell affect how they live their lives. The aim of narrative therapy is to "help people resolve problems by discovering new ways of storying their situation" (Lock, Epston, & Maisel, 2004, p. 278).[7]

Storytelling as Lived Experience

Some researchers study narrative *as* lived experience, as itself social action. These researchers are as interested in *how* people narrate their experiences as in what their stories are about. These researchers treat an understanding of storytelling practices as essential to grasping what narrators are communicating. In this approach, narration is the practice of constructing meaningful selves, identities, and realities.

Many of these researchers use in-depth interviewing as their method of gathering narrative data. Some produce detailed transcripts of their interviews to pay close attention to the narrator's linguistic practices (such as word choice, repetition, hesitation, laughter, use of personal pronouns) and to how storytelling is embedded in the interaction between researcher and narrator (Bell, 2009; Chase, 1995, 2010; Riessman, 1990, 2002a, 2002b, 2008). Whether or not they produce detailed transcripts, however, these researchers are interested in how narrators make sense of personal experience in relation to cultural discourses.[8] In this approach, researchers treat narratives as a window to the contradictory and shifting nature of hegemonic discourses, which we tend to take for granted as stable monolithic forces. Unlike Clandinin and Rosiek, whose pragmatic approach resists theoretical abstraction, these researchers view identifying oppressive discourses—and the ways in which narrators disrupt

them—as a worthy goal of narrative inquiry. These researchers show that people create a range of narrative strategies in relation to cultural discourses, and that individuals' stories are constrained but not determined by those discourses.

This approach to narrative inquiry has been used to explore a broad range of topics. Rachelle Hole (2007) examines how Deaf women construct their identities through incorporating and resisting cultural narratives about difference, normalcy, passing, and Deaf culture. Helena Austin and Lorelei Carpenter (2008) explore how mothers of children diagnosed with ADHD resist dominant cultural assumptions about mothering—which are expressed in friends' and professionals' judgments about them as troublesome and troubled mothers. Sunil Bhatia (2008) shows that first-generation Indian immigrants' stories about their lives after September 11, 2001, embody disruptions in their sense of race, place, and safety in the United States, thus challenging mainstream psychology's concept of acculturation as a linear process. Alexandra Adame and Roger Knudson (2007) argue that people in the psychiatric survivors' movement resist the dominant psychiatric discourse of chemical imbalance, "broken brains," and individual normalcy, and offer an alternative discourse. That alternative is about striving to live a good life through "a collective journey of peer-support and political activism" (p. 175).

Narrative Practices and Narrative Environments

Some researchers focus specifically on the relationship between people's narrative practices and their local narrative environments. Gubrium and Holstein (2009) describe that relationship as a *reflexive interplay*, which means that people's narrative practices are shaped by and shape their narrative environments. These researchers are more interested in understanding *narrative reality* in any local context—what does and doesn't get said, about what, why, how, and to whom—than they are in understanding individuals' stories per se. They argue that understanding narrative reality in any context requires substantial attention both to narrative environments and narrative practices. Thus, this approach depends on "ethnographic sensibilities," that is, systematic consideration of "the communicative mechanisms, circumstances, purposes, strategies, and resources that shape narrative production" (pp. vii–viii). Gubrium and Holstein do not dismiss the use of in-depth interviews or a focus on broad cultural discourses, but they propose that understanding what gets said requires an ethnographic understanding of local contexts and interactional circumstances.

Gubrium and Holstein suggest that narrative environments include such diverse entities as intimate relationships, local cultures, occupations, and organizations. Each of these environments provides myriad circumstances and resources that condition but don't determine the stories people tell (and don't tell). Ethnographic sensibilities are needed for understanding narrative environments, but they are also needed for understanding narrative practices: the mechanics of how stories are activated, how storytellers create and develop meaning through interaction with each other, how speakers collaborate with each other or struggle for control over narrative meanings, and how narrators perform their identities for specific audiences and with specific (but not always intended) consequences. Gubrium and Holstein define a "good story" not in terms of linguistic criteria, but as *any* communication—even a word or a nod—that people treat as "narratively adequate in the circumstances, functioning to smoothly facilitate casual yet consequential interaction" (2009, p. 201).

Comparative ethnography lends itself to this approach. For example, in his study of addiction and mental illness, Darin Weinberg (2005) conducted fieldwork at two residential centers based in the same treatment model. Both centers aimed to "empower clients as agents of their own recoveries," but each center developed a distinct therapeutic orientation. One program addressed insanities and addictions as resources for understanding clients' past problems, but the other program addressed insanities and addictions in terms of clients' plans for their immediate futures (pp. 13–14). Comparative ethnography makes it possible to explore how narrative realities differ from place to place or shift over time.[9]

The Researcher and the Story

Some researchers treat *their* stories about life experience (including research itself as a life experience) as a significant and necessary focus of narrative inquiry. Sometimes their aim is to create a more equitable relationship between the researcher and those she or he studies by subjecting the researched *and* the researcher to an analytic lens. And sometimes researchers' aim is to explore a topic or research question more fully by including the researcher's experience of it.

Barbara Myerhoff pioneered this approach in *Number Our Days* (1979/1994), an ethnographic study of a community of elderly immigrant Jews in California. Since Myerhoff's groundbreaking study, many researchers have become more explicit about their experiences as they work to understand the other's voice, life, and culture. In her study of Esperanza's (a Mexican woman's) life story, Ruth Behar (1993/2003) writes about her comadre relationship with Esperanza and dilemmas she encountered as she became an anthropologist. In her portrait of Jewish communities in Cuba, Behar (2007) discusses her roots in Jewish Cuban culture and her search for home. As she explores women's struggles with anorexia and the discourses that govern treatment, Paula Saukko (2008) describes her battle with the disease. C. J. Pascoe (2007) discusses her self-presentation—her "least-gendered" identity—as a young woman doing ethnographic research on teenagers' sexual and gender identities in high school. Kris Paap (2006) uses the journal she kept while

working as a carpenter's apprentice as the basis for her cultural analyses of interactions at construction sites. Her experiences with coworkers, bosses, and the work itself feature heavily in her argument that structural insecurity in construction work creates classed, racial, and gendered labor practices that harm even the white male workers who engage in them.

Autoethnographers develop another version of this approach. They turn the analytic lens fully and specifically on themselves as they write, interpret, or perform narratives about their own culturally significant experiences. In autoethnography, also called interpretive biography (Denzin, 2008), the researcher and the researched are one and the same (Ellis, 2004, 2009; Jones, 2005). Recent examples include stories of childhood (Denzin, 2008); stories about September 11, 2001 (Denzin, 2008; Schneider, 2006); and stories about learning about autoethnography (Scott-Hoy and Ellis, 2008). Autoethnographers sometimes present or perform their narratives as plays, as poems, or as novels (Denzin, 1997, 2000, 2003, 2008; Ellis, 2004, 2009; Madison, 2006; Richardson, 2002; Saldaña, 2008). Scott-Hoy and Ellis (2008) experiment with painting as an autoethnographic presentation. The goal of autoethnography, and of many performance narratives, is to *show* rather than to *tell* (Denzin, 2003, p. 203; Saldaña, 2008, p. 201) and thus to disrupt the politics of traditional research relationships, traditional forms of representation, and traditional social science orientations to audiences (Langellier & Peterson, 2006; Miller & Taylor, 2006).

▣ METHODOLOGICAL ISSUES

No matter what approach they take, narrative researchers work closely with individuals and their stories. As a result, narrative inquiry involves a particular set of issues concerning the research relationship, ethics, interpretation, and validity. After discussing these briefly, I address two topics that have come to the fore in recent years: the limits of interviews as a source of narrative data, and the use of visual narratives as data and forms of presenting research.

When narrative researchers gather data through in-depth interviews, they work at transforming the interviewee-interviewer relationship into one of narrator and listener. This requires a shift from the conventional practice of asking research participants to generalize about their experiences (as qualitative researchers often do), to inviting narrators' specific stories (Chase, 2005). It also requires a shift from the conventional practice of treating the interview schedule as structuring or even semi-structuring the interview to treating it as a guide that may or may not be useful when one follows the narrator's story. Amia Lieblich (in Clandinin & Murphy, 2007) suggests that narrative interviewing requires emotional maturity, sensitivity, and life experience, all of which may take years to develop (p. 642). Similarly, Don Polkinghorne (in Clandinin & Murphy, 2007) suggests that narrative interviewing involves an intensive inter-action with the narrator and the patience to encourage narra-tors to explore memories and deeper understandings of their experiences (p. 644). In my undergraduate course on qualitative research methods, as students prepare for interviewing, I ask them what they will do if the interviewee cries. Sometimes a student says that she or he will change the subject, a response that lets me know the student is not ready for narrative inter-viewing. The latter requires the researcher to be a witness to a wide range of emotions.

Specific ethical issues arise in narrative research. Unlike qualitative researchers in general, who usually present short excerpts from interviews or fieldwork in their published work, narrative researchers often publish or perform longer stories from individuals' narratives. This increases the risk that narra-tors will feel vulnerable or exposed by narrative work. Lieblich (in Clandinin & Murphy, 2007) suggests that because narrative researchers do not know in advance exactly how they will use the narratives they collect, they should return to narrators to inform them—and ask again for permission to use their stories—when they *do* know how they plan to present, publish, or perform the work.

Josselson's (2007b) article, "The Ethical Attitude in Narrative Research" may be most comprehensive discussion of ethical issues in narrative work. She writes about the need to explain narrative research to participants, the particular problems raised by informed consent forms (which usually assume a researcher can say in advance everything the narrator needs to know), how to work with institutional review boards (IRBs), and writing research reports. Rather than listing specific rules for ethical practice, she implores researchers to develop an "ethical attitude," which must be carefully developed in each research situation.

When narrative researchers interpret narratives heard dur-ing interviews, they begin with narrators' voices and stories, thereby extending the narrator–listener relationship and the active work of listening into the interpretive process (Chase, 2005). This is a move away from a traditional theme-oriented method of analyzing qualitative material. Rather than locating distinct themes *across* interviews, narrative researchers listen first to the voices *within* each narrative (Riessman 2008, p. 12). For Polkinghorne, this is what distinguishes narrative inquiry from qualitative inquiry generally (in Clandinin & Murphy, 2007, pp. 633–634).

Martyn Hammersley (2008) notes that all qualitative research needs to be assessed in terms of validity, which means evaluating whether researchers' claims are sufficiently sup-ported by evidence (2008, pp. 162–163). But issues of validity also take particular forms in narrative research. Polkinghorne (2007) points out that narrative research "issues claims about the meaning life events hold for people. It makes claims about how people understand situations, others, and themselves"

(p. 476). The researchers' primary aim is not to discover whether narrators' accounts are accurate reflections of actual events, but to understand the meanings people attach to those events (p. 479). Nonetheless, he reminds us that words are not always sufficient to communicate meaning, that narrators are selective in the meanings they narrate, and that context and audience (e.g., an interview situation) shape what meanings get expressed. Narrative researchers do not need to claim that their interpretation is the only possibility, but they do "need to cogently argue that theirs is a viable interpretation grounded in the assembled texts" (p. 484).

In discussing the validity of *narrators'* stories—or the trust-worthiness of their stories, as she prefers to call it—Riessman (2008) argues that stories that "diverge from established 'truth' can sometimes be the most interesting, indicating silenced voices and subjugated knowledge" (p. 186). Similarly, Josselson (2007a) points out that narrative research allows for the study of "people's lives as lived, people whose life experience ha[s] been lost in the search for central tendencies" (p. 8). Because much narrative research reveals experiences and meanings that have not previously been exposed by other types of research, narra-tive researchers must present careful evidence for their claims from narrators' accounts (Riessman, 2008, p. 186). In addition, narrative researchers can strengthen their arguments by dis-cussing cases that don't fit their claims and by considering alternative interpretations (p. 191). They should also document their procedures for collecting and interpreting data (p. 193).

Beyond Interviews

Although narrative researchers have used many sources of data—diaries, letters, autobiographies, and field notes of natu-rally occurring conversations—in-depth interviews continue to be the most common source of narrative data (Bell, 2009, p. 171; Riessman 2008, p. 26; Hammersley, 2008, p. 89). In recent years, this privileging of interviews has been a topic of discussion and debate.

Big Stories and Small Stories

Mark Freeman (2006) calls the narrative material gathered from interviews *big stories*. He argues that their particular value as data is that they allow the narrator distance from and thus the opportunity to reflect on significant life events. Narrative researchers also value interviews for the window—a frequently used metaphor—they offer to the narrative environment exter-nal to the interview. Through close attention to both the content of narrators' stories and how they speak—for example, unself-consciously, hesitantly, or defensively—a researcher can hear the influence of narrative environments on narrative practice. Anal-ysis of patterns across interviews with similarly situated people contributes to a stronger understanding of those environments and their impact on individual narratives. But the metaphor of the window also indicates its limits. Looking out at narrative environments from inside the narrative, the narrative as a window limits how and how much of the narrative environment can be seen.

With this limit in mind, Riessman (2008) argues that ethnographic study of participants' settings facilitates stronger understanding of their stories (p. 26), including stories told during interviews. She describes this as the *dialogic/ performance* approach, which highlights "'who' an utterance may be directed to, 'when,' and 'why,' that is, for what purposes?" (p. 105). Here, "Attention expands from detailed attention to a narrator's speech—what is said and/or how it is said—to the dialogic environment in all its complexity. Historical and cultural context, audiences for the narrative, and shifts in the interpreter's positioning over time are brought into interpretation" (pp. 136–137).

Some narrative theorists resist the privileged status of big stories produced during interviews by arguing for greater attention to *small stories*. Alexandra Georgakopoulou (2007) defines small stories as a constant and natural feature of everyday life; they include talk about very recent events, such as what happened this morning, as well talk about what might happen in the near future (2007, p. 150). Moreover, "with a small stories perspective in mind, it is not just tellings or retellings that form part of the analysis: refusals to tell or deferrals of telling are equally important in terms of how the participants orient to what is appropriate . . . in a specific environment, what the norms for telling and tellability are" (p. 151). This resonates with Gubrium and Holstein's (2009) focus on the reflexive interplay between narrative environments and narrative practices. An interest in how stories are produced and received in society "requires that we step outside of narrative texts" to ask "who produces particular kinds of stories, where are they likely to be encountered, what are their purposes and consequences, who are the listeners, under what circumstances are particular narratives more or less accountable, how do they gain acceptance, and how are they challenged?" (p. 23)

Content Analyses

My own narrative work has relied heavily on in-depth interviews (for example, Chase, 1995). Recently though, I have been influenced by arguments about the limits of interviews and have sought ways to move beyond sole reliance on them. My book, *Learning to Speak, Learning to Listen: How Diversity Works on Campus* (Chase, 2010) offers an example of how it is possible to do this when long-term ethnographic study—and thus sustained attention to small stories—is not an option.

I conducted a case study of how students engage issues of race, class, gender, ability, and sexual orientation at City University (a pseudonym), a predominantly white private university. These days, most U.S. colleges and universities proclaim commitments to diversity (for example, in their mission statements), but this institutionalization of diversity does not always translate into serious engagement with diversity issues on campus. What interested me about City University (CU) is that a critical mass of students, faculty, and administrators has succeeded over the years in making organizational and cultural changes that strongly support students of color, women students, and gay, lesbian, bisexual, and transgender (GLBT) students. This critical mass of people has succeeded in making diversity issues an integral—if contentious—part of the narrative environment. One consequence is that CU students of color and GLBT students (among others) feel entitled to—and *do*— speak out when they perceive injustice on campus. Another consequence is that some CU students have learned to listen to those whose social identities and social locations differ from their own. I argue that students' speaking and listening across social differences are at once shaped by CU's narrative environment and contribute to it.

In my study I focused on events leading up to and culminating in a public protest by students of color who were frustrated by what they perceived as the university's lack of serious attention to racial issues. My major source of data was in-depth interviews with a wide range of individual students, groups of students in many different campus organizations, as well as with faculty, staff, and administrators. By interviewing people and groups that are differently situated on campus, I was able to get different views (or windows) on CU's narrative environment.

But I also wanted more direct access to that narrative environment. I did a limited amount of ethnographic observation, but long-term ethnography was not a practical option for me. Instead I conducted extensive quantitative and qualitative content analyses of key documents, most notably the student newspaper and the student government minutes, but also CU's curriculum, calendar of events, and website.

The broader understanding of CU's narrative environment that I gained from the content analyses allowed me to interpret puzzling aspects of the interviews. For example, I noticed a certain silence during my interview with Rachelle, one of the students who led the campus protest. As Rachelle told me about her personal development during college, she spoke about how she had become more open to GLBT people. As an African American raised in a Pentecostal tradition, this was a major change for her. She explained that while she was growing up, even *talking* about sexual orientation was taboo, never mind interacting with gays or lesbians. She said that at CU she had become more open-minded with the help of African American friends who showed her that it was possible to interact with—and even become friends with—GLBT people, without losing her faith.

Although Rachelle could tell me *this* story about becoming more open and tolerant, she said little about a related topic. When I asked about her current religious beliefs, she stated simply, "[homosexuality] is just something I don't feel is right but that's just for my own personal belief." If my study were *only* based on interviews, I would have noticed that Rachelle did not expand on this story of continuing to embrace her religious perspective that homosexuality is wrong, but I would not have understood *why* she had nothing more to say about this. The content analyses allowed me to see that Rachelle's relative silence about her belief that homosexuality is wrong was shaped by CU's narrative environment. The content analyses demonstrated that at CU an unquestioned acceptance of GLBT people and GLBT rights constitutes the *preferred story* (Gubrium and Holstein, 2009; Riessman, 2008) about sexual orientation. This preferred story is expressed routinely in articles, editorials, and letters in the student newspaper and in the student government's noncontroversial passage of resolutions in support of GLBT students and GLBT rights. Given that preferred story in CU's narrative environment, Rachelle's story about how she has become more open-minded about GLBT issues *and* her relative silence about how she still believes that homosexuality is wrong make sense. Both aspects of her personal narrative reflect the influence of that preferred story in CU's narrative environment.

The content analyses also showed that in CU's environment racial issues are much more contentious than are issues related to sexual orientation. The student newspaper, student government, and the administration were proactive and supportive in response to anonymous homophobic incidents on campus. By comparison, these same entities' responses to racial issues were interpreted by students of color as slow and unsupportive. The difference between the uncontentiousness of sexual orientation and the contentiousness of race in CU's narrative landscape helped me to understand students of color's frustration and thus their decision to stage a protest. The content analyses, in conjunction with the interviews, helped me to demonstrate the reflexive interplay between CU's narrative environment and students' narrative practices—such as Rachelle's relative silence about her religious belief and the student of color's public protest.

Beyond Written and Oral Texts

Even when narrative researchers move beyond interviews, their data sources are usually oral or written texts—such as field notes about naturally occurring talk or the documents I used in my study. Some narrative researchers, however, challenge the assumption that narratives are found only or primarily in spoken or written formats. Riessman (2008) contends that visual images are so central to our everyday lives that social scientists must attend to them if they are to understand more fully how people communicate meaning (see also Bach, 2007; Harper, 2005; and Weber, 2008). Narrative researchers who study visual images treat them as socially situated narrative texts that demand interpretation.[10]

Some narrative researchers focus on visual images that others have already made—such as photographs, films, or paintings (Riessman, 2008, p. 141; Weber, 2008, p. 48). For instance, Susan Bell (2002) analyzes the photographs of Jo Spence, a British feminist photographer who was diagnosed with breast cancer before the emergence of the women's breast cancer movement. Bell chose three photographs out of hundreds Spence had taken to interpret what Spence was communicating about her illness experience. One photograph is of Spence getting a mammogram, another of her breast the day before surgery, and the third of herself in bed shortly before she died. Bell interprets the three photographs in detail, concentrating on Spence's face, posture, and body; the rooms in which she was located, the objects in the rooms (including technological devices and medical equipment); and the way Spence framed the images as photographs and described them in accompanying texts. Through her interpretations, Bell demonstrates how Spence resisted having her illness experience defined by the medical world (Bell, 2002, 2006; Riessman, 2008, pp. 153–159).

Other narrative researchers collaborate with research subjects in the construction of visual images (Riessman, 2008, p. 141; Weber, 2008, p. 47). For example, physician and filmmaker Gretchen Berland gave video equipment to three adults with physical disabilities who use wheelchairs for mobility. For 2 years, these adults recorded and commented on their everyday lives. Berland produced the film, *Rolling,* and appears in it from time to time, but the film foregrounds the three adults' stories (see http://www.thirteen.org/rolling/thefilm/ and Riessman, 2008, p. 143).

In her ethnographic study of a school program for pregnant teenagers, Wendy Luttrell (2003) discovered that the teens were uninterested in talking about their experiences in an in-depth interview format. So she suggested that they create self-portraits and collages, media that the teens found conducive to self-expression. As they worked on their projects, Luttrell listened to the girls converse with each other about the images they were constructing. When their work was complete, the girls presented the images to each other and engaged in further discussion about them. At the end of the year, Luttrell collected the images in a book format so that each girl could have a copy. Luttrell's data include the visual images, the words that each girl attached to the images she had made, the group's conversations about the images, as well as Luttrell's broader ethnographic observations about interactions in the classroom and the program's place in the school and community. She demonstrates that each girl, in her own way, struggles against demeaning portrayals of pregnant teens within narrative environments (the school and American culture broadly) that make that struggle painful and difficult (see also Riessman, 2008, pp. 164–172).

In her study of Muslim women's experiences in the United States after the terrorist attacks of September 11, 2001, Mei-Po

Kwan (2008) uses visual narrative in yet another way. She asked 37 women in Columbus, Ohio, to carefully record their activities and trips outside the home on one particular day. Then she conducted interviews with the women about how their activities and sense of safety have changed since September 11th. She also asked them to indicate on a map where they go during their everyday lives, which areas they considered safe or unsafe before September 11th, and how their feelings about those areas have changed since then. Kwan uses these multiple sources of data to construct a visual narrative that shows changes in time and space in Muslim women's feelings of fear and safety pre– and post–September 11th. Kwan uses three-dimensional (3-D) geographical information systems (GIS) to "illuminate the impact of the fear of anti-Muslim hate violence on the daily lives of Muslim women and to help articulate their emotional geographies in the post–September 11 period" (p. 653). In Kwan's study, the visual narrative is not the data, but a powerful means of presenting Muslim women's narratives in the post–September 11th world.

▣ NARRATIVE INQUIRY, PERSONAL CHANGE, AND SOCIAL CHANGE

Like other qualitative researchers, narrative researchers continue to be compelled by the relationship between their work and possibilities for change and social justice. Some study how narratives make change happen, and some collect and present narratives to make change happen. In either case, there is a sense of *urgency,* of the need for personal and social change. In the following, I characterize that urgency in several ways: the urgency of speaking, the urgency of being heard, the urgency of collective stories, and the urgency of public dialogue. When narrative inquiry focuses on personal or social change, the relation between narrator and audience becomes central.

The Urgency of Speaking

Sometimes the act itself of narrating a significant life event facilitates positive change. In discussing a breast cancer survivor's narrative, Kristen Langellier (2001) writes, "The wounded storyteller reclaims the capacity to tell, and hold on to, her own story, resisting narrative surrender to the medical chart as the official story of the illness" (p. 146; see also Bell, 2002, 2009; Capps & Ochs, 1995; Frank, 1995; Lieblich, McAdams, & Josselson, 2004). Along similar lines, George Rosenwald and Richard Ochberg (1992) claim that self-narration can lead to personal emancipation—to "better" stories of life difficulties or traumas. In these cases, the narrator is his or her own audience, the one who needs to hear alternative versions of his or her identity or life events, and the one for whom changes in the narrative can "stir up changes" in the life (p. 8).

Researchers and practitioners in narrative therapy point out that creating alternative narratives of one's self or life can be extraordinarily difficult. For example, in their discussion of a woman who has been hospitalized multiple times for self-starvation, Andrew Lock, David Epston, and Richard Maisel (2004) show that it was only when the woman learned to treat "anorexia" as separate from herself that she began to develop a voice that confidently resisted the "voice of anorexia." In this case, the woman's externalization of anorexia was accomplished through therapeutic sessions in which one therapist literally spoke the punitive voice of "anorexia," the woman repudiated that voice in her own words, and a second therapist supported her in doing so.

The question of whether, how, and under what conditions the telling of traumatic experiences facilitates healing and emotional well-being is an important topic for narrative researchers (Naples, 2003). For instance, shortly after World War II, adult survivors of the Holocaust worked relentlessly to gather and publish child survivors' testimonies. At that time, some claimed that children's testimonies had therapeutic value for the children, but they did not always provide evidence for that claim (Cohen, 2007). Taking a life course perspective, Bertram Cohler (2008) suggests that *when* in the course of their lives survivors tell their stories shapes their meaning for survivors and the role the stories play in survivors' lives. He analyzes the memoirs of two Holocaust survivors, one of whom wrote her memoir right after the war. She described atrocities she had witnessed, but largely excluded her own experiences, as if she could not integrate them into her life story. The other woman wrote her memoir half a century after World War II as an émigré to the United States. She recounts her experiences before, during, and after the war, forming them into a "characteristic American redemptive account of successfully overcoming adversity" (Cohler, 2008, p. 1).

Some narrative therapists who work with Holocaust survivors find that successful therapy consists not in integrating the trauma into one's life story, but rather putting the traumatic narrative "into a capsule separated from other parts of the life story" (Shamai & Levin-Megged, 2006, p. 692). By defining successful therapy as separation of the traumatic story from the life story, these researchers counter traditional notions of therapeutic success. The difference lies, at least in part, in the severity of the trauma.

The Urgency of Being Heard

For some individuals and groups, the urgency of storytelling arises from the need and desire to have *others* hear one's story. Citing René Jara, John Beverly (2005) describes Latin American *testimonios* as "emergency" narratives that involve "a problem of repression, poverty, marginality, exploitation, or simply survival.... The voice that speaks to the reader through the text . . . [takes] the form of an 'I' that demands to be recognized, that wants or needs to stake a claim on our attention" (p. 548).

Of course, more than Latin American *testimonios* are narrated with this urgent voice. The stories of many marginalized groups and oppressed people shape the contemporary narrative landscape. To name just a few: transgendered people (Girshick, 2008); parents of children with disabilities (Goodley & Tregaskis, 2006); Hmong immigrants to the United States after the Vietnam War (Faderman, 2005); Latino and Asian American college students on predominantly white campuses (Garrod & Kilkenny, 2007; Garrod, Kilkenny, & Gómez, 2007); and the victims and survivors of gendered, racial, ethnic, and sexual violence (Bales & Trodd, 2008; Deer, Clairmont, Martel, & While Eagle, 2008). Indeed, "naming silenced lives" and "giving voice" to marginalized people—or in Riessman's (2008) more collaborative term, "amplifying" others' voices (p. 223)—have been primary goals of narrative research for several decades (McLaughlin & Tierney, 1993; Personal Narratives Group, 1989).

The urgency of speaking and being heard drives the ongoing collection and publication of narratives about many forms of social injustice. Examples include the personal narratives of refugees of the war in Bosnia and Croatia (Mertus , Tesanovic, Metikos, & Boric, 1997); the stories of September 11, 2001, survivors (www.911digitalarchive.org); testimonies about genocide in Rwanda (http://www.voicesofrwanda.org/); and the stories of survivors and witnesses of the Holocaust (the Voices of the Holocaust Project [www.iit.edu]; the Fortunoff Video Archive for Holocaust Testimonies at Yale University [http://www.library.yale.edu/testimonies]; Voice/Vision: Holocaust Survivor Oral History Archive [http://holocaust.umd.umich.edu/interviews.php]; and the University of Southern California's Shoah Foundation Institute's archive [http://college.usc.edu/vhi/]).[11]

The same urgency to get survivors' voices heard drives the work of Father Patrick Desbois (2008), a French Catholic priest, who has traveled through the Ukraine to find the unmarked mass graves of a million and a half Jews killed by Nazi mobile death squads during World War II. In addition to honoring the victims with proper burials, he films the testimonies of eyewitnesses who were children at the time. These acts of genocide are not well known, and many of the eyewitnesses have never before spoken publicly about their experiences.[12]

The act of speaking to be heard references an "other" who needs to hear, to listen, to pay attention. Mary Gergen and Kenneth Gergen (2007) state, "Audiences who listen to a story from a witness become themselves second-order witnesses. They create for themselves the visual images, sounds, and visceral responses of the witness. One might say that they engage in empathetic listening, in which they come to feel with the storyteller" (p. 139). When the story is about pain, trauma, and injustice, listening itself can be painful. Listening requires the willingness to put the other's story at the center of one's attention, to resist defensive reactions, and to acknowledge the limits of one's ability to put oneself in another's shoes (Chase, 2010).

The Urgency of Collective Stories

Stories about injustice are often more than individual stories. Laurel Richardson defines collective stories as those that connect an individual's story to the broader story of a marginalized social group (Richardson, 1990). In discussing the collective stories of sexual abuse survivors and gays and lesbians, Kenneth Plummer (1995) writes, "For narratives to flourish, there must be a community to hear. . . . For communities to hear, there must be stories which weave together their history, their identity, their politics. The one—community—feeds upon and into the other—story" (p. 87).

When survivors or marginalized or oppressed groups tell their collective stories, they demand social change. It may be a demand that people never forget the atrocities of the past. It may be a demand that educational curricula be transformed so that young people learn how to prevent what previous generations have suffered. It may be a demand that people who hold legal, cultural, or other forms of power take action to bring about justice. Thus, collective stories become integral to social movements (see also Davis, 2002).

Along these lines, Bell (2009) shows how women's personal narratives played a role in successfully challenging conventional medicine's treatment of women who were exposed prenatally to DES (a drug linked to a rare vaginal cancer and poor reproductive results). Bell gathered DES daughters' narratives through in-depth interviews as well as letters to the editor they had written in various media (which were collected and published in *DES Action Voice*). Through close attention to how DES daughters narrated their experiences, Bell shows how some of them became activists who created a feminist, embodied health movement. And through close examination of the proceedings of a National Institutes of Health–sponsored workshop, Bell demonstrates how DES daughters and biomedical scientists collaboratively "destabilized the discourse of science as usual" (p. 10).

In *To Plead Our Own Cause*, Kevin Bales and Zoe Trodd (2008) present verbatim the oral and written narratives of men, women, and children who have been enslaved in countries across the globe as soldiers, in prison camps, in workplaces, and as sexual objects. Some of these narrators have become activists, working with various organizations to get their stories heard. Many of the narratives in the book were originally elicited by or written for abolitionist and human rights organizations, public awareness campaigns, and congressional sessions on slavery-related bills. Indeed these testimonies were instrumental in getting the Victims of Trafficking and Violence Protection Act passed into U.S. law in 2000.

The Urgency of Public Dialogue

William Gamson (2002) writes, "Deliberation and dialogue in a narrative mode . . . lends itself more easily [than abstract argument] to the expression of moral complexity." In this sense, "storytelling facilitates a healthy, democratic, public life" (p. 197).[13] Many narrative researchers hope their work will stimulate dialogue about complex moral matters and about the need for social change. And they look for creative ways to present their work to the public (Barone, 2007; Knowles & Cole, 2008; Madison & Hamera, 2006; Mattingly, 2007).

Some researchers use ethnotheater—turning narrative data into theater performances—as a means of accomplishing this goal. Anna Deavere Smith (1993, 1994, 2004) has been a leader in this regard. In her solo stage performances, based on the words of people she interviewed, Smith has explored events such as the riots in Los Angeles after the acquittal of the police officers who beat Rodney King, and the riots in Crown Heights, Brooklyn, after a car carrying a Hasidic spiritual leader killed a black 7-year-old child from Guyana. In her most recent one-woman show, *Let Me Down Easy*, Smith performs the narratives of many different people, all of whom address "the fragility and resilience of the human body" (Isherwood, 2008). In this show, she presents people's stories about "the steroid scandal in sports, cancer therapies, African folk healing, the genocide in Rwanda, the tragedy of Katrina and the ailing American healthcare system" (Isherwood, 2008). In such performances, Smith presents a wide array of voices, and in so doing, attempts to create public dialogue about emotionally and politically charged issues and events.

During three years of ethnographic fieldwork in Ghana, Soyini Madison (2006) studied the debate surrounding a tradition of sending girls to a village shrine for years or even a lifetime to atone for a crime or violation committed by a (usually male) family member. Local human rights activists view that practice as tantamount to the girls' enslavement. Traditionalists view it as a matter of moral and cultural education and as protecting the girls from the shame of their family member's action. Traditionalists point out that the girls "are esteemed as 'queens' with special powers" (p. 398). On the basis of her fieldwork, Madison (2006) wrote, "*Is It a Human Being or a Girl?*," a play revolving around three major themes: the debate between the human rights activists and the traditionalists; critiques of corporate globalization, which produces the poverty underlying the traditional practice; and Madison's social location as an African American academic and how that shaped her ethnographic interests. In the play, five performers acted and spoke the various voices, based largely on narratives Madison collected during in-depth interviews. Madison's aim in presenting her fieldwork in the form of ethnotheater was to stimulate public dialogue in the local community about the moral issues involved.

Moisés Kaufman, a leading proponent and practitioner of ethnotheater, writes,

There are moments in history when a particular event brings the various ideologies and beliefs prevailing in a culture into sharp focus. At these junctures the event becomes a lightning rod of sorts, attracting and distilling the essence of these philosophies and convictions. By paying careful attention in moments like this to people's words, one is able to hear the way these prevailing ideas affect not only individual lives but also the culture at large. (2001, p. v)

Kaufman's play, *The Laramie Project*, is based on one such moment in history: the murder of Matthew Shepard in 1998. Four weeks after the murder, members of Kaufman's Tectonic Theater Project traveled to Laramie, Wyoming, to interview many of the people involved as well as other townspeople. A year and a half and 200 interviews later, the Tectonic Theater Project performed the first of many productions of the play, in which actors played the people they had interviewed, speaking their words verbatim.

Ten years later, Kaufman and his colleagues returned to Laramie and interviewed many of the same people again. One interviewed Aaron McKinney, who is currently serving two life sentences for the murder of Matthew Shepard. Kaufman and his colleagues turned these interviews into another play, which was staged in dozens of theaters across the United States and in other countries on October 12, 2009, the 11th anniversary of Shepard's death. The Tectonic Theater's goal in returning to Laramie was to find out what had and hadn't changed in the community during those 10 years. The new play shows that many Laramie residents have wanted to "move on," because "Laramie is not *that* kind of community."[14] Interestingly, Wyoming's hate crime legislation still excludes sexual orientation, but on October 28, 2009, President Barack Obama signed into law the Matthew Shepard and James Byrd, Jr. Hate Crimes Prevention Act. According to the Human Rights Campaign,

[This law] gives the Department of Justice (DOJ) the power to investigate and prosecute bias-motivated violence by providing the DOJ with jurisdiction over crimes of violence where a perpetrator has selected a victim because of the person's actual or perceived race, color, religion, national origin, gender, sexual orientation, gender identity or disability.[15]

Whether or not *The Laramie Project* had a direct impact on the federal legislation, it certainly created public dialogue in theater venues across the country and the globe.

▣ STILL A FIELD IN THE MAKING

As I worked on this update to my original chapter in the third edition of this handbook, I found, as before, that it is easier to identify complexities and multiplicities in the field of narrative inquiry than it is to identify commonalities. As I think about the future of this field, I suspect that this will continue to be the case.

One small but poignant indicator of increasing complexity and multiplicity lies in how I changed the definition of narrative.

In the original chapter, I wrote, "Narrative is *retrospective* meaning making—the shaping or ordering of *past* experience" [emphasis added]. In this update, I wrote that narrative is "meaning making through the shaping or ordering of experience." I deleted "retrospective" and "past" because of recent developments in this field. As noted earlier, in the last few years, some researchers have focused on "small stories" in everyday situated interaction (Bamberg, 2006; Georgakopoulou, 2007) and on the need for ethnographic sensibilities for understanding how small stories are produced in and organize social interaction (Gubrium & Holstein, 2009). These researchers demonstrate that the definition of narrative as *retrospective* meaning making is partial. They influenced me to broaden the definition.

I suspect that complexity and multiplicity will also persist in ideas about what narrative *inquiry* is. A number of researchers present overviews of and distinctions within narrative inquiry concerning its interests, goals, and methods (Bamberg, 2007; Clandinin & Rosiek, 2007; Gubrium & Holstein, 2009; Polkinghorne, in Clandinin & Murphy, 2007; Riessman, 2008). I have been influenced by their categories and yet I still came up with my own. Summarizing their conversation with Elliot Mishler about the future of narrative inquiry, D. Jean Clandinin and M. Shaun Murphy (2007) write, "Elliot notes that we cannot police the boundaries of narrative inquiry. For him, the field . . . will be defined from within the different communities of narrative inquirers with researchers picking up on each others' work that helps them address issues salient to their own research problems" (p. 636). This emphasis on different narrative research communities strikes me as an accurate description of what is happening and will continue to happen.

Yet the boundaries of narrative research communities are also fluid. Lieblich (in Clandinin & Murphy, 2007, pp. 640–641) notes that graduate students interested in narrative research still have trouble finding support for work they want to do. Mishler (in Clandinin & Murphy, 2007, p. 641) notes that even established narrative researchers often feel alone in their departments. As narrative researchers look outside their departments, disciplines, and across national boundaries to find colleagues with whom they share interests, narrative research communities will change and evolve.

Furthermore, it is not always clear which communities we should belong to, which colleagues we need to converse with, which conversations we need to cultivate. I still wonder sometimes who *my* colleagues are. Colleagues who use the same narrative methods but whose research covers different topics? Colleagues who work on the same topics, but who don't use narrative methods, or even qualitative methods? Practitioners in the fields for whom my research might provide useful insights? Ideally, of course, I would converse with all of these colleagues, but sometimes one has to choose. Because narrative inquiry is

still a field in the making, I suspect that narrative researchers will continue to ask these questions about colleagues, conversations, and communities.

As discussed earlier, the last few years have seen an expansion in the kinds of data narrative researchers use in their studies. This will probably continue as well. Examples presented in this chapter speak to the value of combining interviews and ethnographic observation (Riessman, 2008); photographs and autobiographical writings (Bell, 2002, 2006; Behar, 2007); interviews, letters to the editor, autobiographical film, and workshop proceedings (Bell 2009); ethnography and participants' collages and self-portraits (Luttrell, 2003); activity diaries, in-depth interviews, maps, and geographical information systems (GIS) (Kwan, 2008); and interviews and content analyses of documents (Chase, 2010). Using multiple sources of data underscores that any view is partial and that narrative environments are multiple and layered. Given the explosion of new technologies, narrative researchers are likely to seek new data sources, adding to the complexity and multiplicity of narrative research. And with these new data sources, new ethical issues will arise. Mishler points out that the increased use of visual narratives raises questions about how to protect the rights of those whose images we use (in Clandinin & Murphy, 2007, p. 649). For example, presenting or publishing photos that include people's faces makes it impossible to conceal their identities.

Another issue that is close to my heart has to do with the generally critical character of narrative research. Like qualitative research generally, narrative research often critiques cultural discourses, institutions, organizations, and interactions that produce social inequalities. Narrative researchers frequently look for the collusive or resistant strategies that narrators develop in relation to the constraints of their narrative environments. As Plummer (1995) demonstrates, social movements research reveals resistant narratives that develop in activist communities. Narrative researchers note that those resistant narratives change others' beliefs, attitudes, and actions. But narrative researchers are less likely to study the *audience* side of this narrative process. The urgency to speak, to get heard, to develop collective narratives, and to create public dialogue—all of these are about the need to influence an audience. What does it look like when the audience *is* influenced. What does an audience's *listening* look like? A focus on these questions would encourage a hopeful aspect in narrative research.

Along similar lines, I suggest that we need to know more about narrative environments that make possible and even encourage creative explorations of self, identity, community, and reality. In this vein, some researchers study the intimate environment of therapy and some study the macro environment of social movements. I would also like to see studies of the mundane environments of everyday life. Even as they constrain, some families, friendships, classrooms, workplaces, and organizations *also* provide members with narrative resources for creating strong relationships and vibrant communities.

In other words, I suggest that we have as much to learn from narrative inquiry into environments where something is working as we do from inquiry into environments where injustice reigns. And I don't believe we have to give up intellectual skepticism to ask these questions. When something is working—when individuals, groups, or communities marshal ordinary resources in their everyday lives to strengthen their relationships and their communities—what is going on narratively in those environments? Karen Gallas (1994) offers an example. As a teacher, she did ethnographic research on sharing time in her own first-grade classroom. She discovered that certain sharing time (narrative) practices hindered and others supported a homeless student's social and language development. (See Riessman's [2008, pp. 125–136] analysis of Gallas's research.) Such research makes both a theoretical contribution to social science and a practical contribution to the field of education. Listening to and observing this child in interaction with her peers allowed Gallas to figure out what facilitated the child's effective speaking practices and the other students' listening. An interest in what works requires a focus on the urgency of speaking and the urgency of being heard, as well as on what it means to *listen.* Here, attention includes the recipients of stories, the audiences for performances.

The complexities and multiplicities in contemporary narrative inquiry offer novice and seasoned researchers a great deal of freedom in the topics and interests they pursue and the methods and approaches they use. At the same time, it is impossible for anyone to keep up with the field as a whole. Josselson (2007a) notes the proliferation of narrative studies, the "array of fascinating, richly-detailed expositions of life as lived, well-interpreted studies full of nuance and insight that befit the complexity of human lives" (p. 8). She also points out that it is impossible to read them all, an observation I share, having attempted to follow developments in this field over the years. Given this situation, Josselson (2007a) suggests that one important issue for contemporary narrative inquiry is "the challenge of accumulating knowledge." She argues that we need a meta-analysis of the vast array of narratives studies (pp. 7–8).

According to Josselson (2007a), a meta-analysis would include the comparison of narrators' language structures across studies, the search for patterns and differences across studies on the same topic, the creation of criteria for determining what constitutes a pattern or a difference, the assessment of whether similar findings across studies of the same phenomenon give us confidence in those findings, the search for patterns across studies of empirically different phenomena, and the articulation of "the frontiers of ignorance," what researchers do not yet understand. Finally, a meta-analysis

would attend to the practical implications of narrative studies, what the findings of our studies tell us about how people act in the social world and about the kind of social world we all are creating (pp. 13–14).

I especially like Josselson's idea that this meta-analysis requires conversation among narrative researchers. It requires new colleagues and communities.

▣ NOTES

1. In my chapter in the third edition of this handbook, I covered important terms, historical background, and the analytic lenses that ground contemporary narrative inquiry.

2. Bamberg (2007), Mishler (1995), Polkinghorne (1995), and Riessman (2008) also make distinctions among types of narrative research in the social sciences, but because they exclude some kinds of work that I want to include (and include some kinds that I want to exclude), I construct my own categories here.

3. Riessman (2008) calls this focus on *what* questions a thematic approach to narrative analysis.

4. In outlining their approach to narrative inquiry, Clandinin and Rosiek (2007) draw heavily on the Deweyan definition of experience, which they explain in detail. They also describe the borders between their approach and several major theoretical paradigms (postpositivism, Marxism or critical theory, and poststructuralism).

5. Because quantitative modes of inquiry are so dominant in psychology, some psychologists treat narrative inquiry as synonymous with qualitative inquiry. But the psychologists I cite here carve out a distinctly narrative approach.

6. Josselson, Lieblich, and McAdams have edited four other books in the American Psychological Association's series, *The Narrative Study of Lives:* Josselson, Lieblich, and McAdams (2003; 2007); Lieblich, McAdams, and Josselson (2004); and McAdams, Josselson, and Lieblich (2001).

7. Several research centers focus specifically on narrative therapy: Narrative Therapy Centre of Toronto; Centre for Narrative Practice in Sheffield, United Kingdom; and the Dulwich Centre in Adelaide, Australia.

8. In Riessman's (2008) terms, this approach includes aspects of structural narrative analysis as well as aspects of dialogic/performative analysis.

9. For more examples of comparative ethnographies, see Gubrium and Holstein (2001) and Holstein and Gubrium (2000).

10. In the next three paragraphs, I rely heavily on Riessman's (2008) Chapter 6, "Visual Analysis."

11. These narrative collections are online, which means they are available to the general public as well as to researchers. In some cases, the websites make it possible for people to add their own stories.

12. See also a report about Father Patrick Desbois's work on the website of the U.S. Holocaust Memorial Museum (http://www.ushmm.org/museum/exhibit/focus/desbois/).

13. Gamson is writing specifically about media discourse about abortion, but his argument is useful for other topics and contexts.

14. Quotes in http://austinist.com/2009/10/14/the_laramie_ project_ten_years_later.php. See also http://community.laramieproject. org/content/About/ and http://artsbeat.blogs.nytimes.com/2009/10/15/ the-laramie-project-10-years-later-draws-50000-theatergoers/.

15. http://www.hrc.org/laws_and_elections/5660.htm

▣ REFERENCES

Adame, A. L., & Knudson, R. M. (2007). Beyond the counter-narrative: Exploring alternative narratives of recovery from the psychiatric survivor movement. *Narrative Inquiry, 17,* 157–178.

Adler, J. M., & McAdams, D. P. (2007). The narrative reconstruction of psychotherapy. *Narrative Inquiry, 17,* 179–202.

Austin, H., & Carpenter, L. (2008). Troubled, troublesome, troubling mothers: The dilemma of difference in women's personal motherhood narratives. *Narrative Inquiry, 18,* 378–392.

Bach, H. (2007). Composing a visual narrative inquiry. In D. J. Clandinin (Ed.), *Handbook of narrative inquiry: Mapping a methodology* (pp. 280–307). Thousand Oaks, CA: Sage.

Baddeley, J., & Singer, J. A. (2007). Charting the life story's path: Narrative identity across the life span. In D. J. Clandinin (Ed.), *Handbook of narrative inquiry: Mapping a methodology* (pp. 177–202). Thousand Oaks, CA: Sage.

Bales, K., & Trodd, Z. (Eds.). (2008). *To plead our own cause: Personal stories by today's slaves.* Ithaca, NY: Cornell University Press.

Bamberg, M. (2006). *Stories: Big or small: Why do we care? Narrative Inquiry, 16,* 139–147.

Bamberg, M. (Ed.). (2007). *Narrative—State of the art.* Philadelphia: John Benjamins.

Barone, T. (2007). A return to the gold standard? Questioning the future of narrative construction as educational research. *Qualitative Inquiry, 13,* 454–470.

Behar, R. (2003). *Translated woman: Crossing the border with Esperanza's story.* Boston: Beacon. (Original work published in 1993)

Behar, R. (2007). *An island called home: Returning to Jewish Cuba.* New Brunswick, NJ: Rutgers University Press.

Bell, S. E. (2002). Photo images: Jo Spence's narratives of living with illness. *Health: An Interdisciplinary Journal for the Social Study of Health, Illness and Medicine, 6,* 5–30.

Bell, S. E. (2006). Living with breast cancer in text and image: Making art to make sense. *Qualitative Research in Psychology, 3,* 31–44.

Bell, S. E. (2009). *DES daughters: Embodied knowledge and the transformation of women's health politics.* Philadelphia: Temple University Press.

Beverly, J. (2005). *Testimonio,* subalternity, and narrative authority. In N. K. Denzin & Y. S. Lincoln (Eds.), *The SAGE handbook of qualitative research* (3rd ed., pp. 547–557). Thousand Oaks, CA: Sage.

Bhatia, S. (2008). 9/11 and the Indian diaspora: Narratives of race, place, and immigrant identity. *Journal of Intercultural Studies, 29,* 21–39.

Capps, L., & Ochs, E. (1995). *Constructing panic: The discourse of agoraphobia.* Cambridge, MA: Harvard University Press.

Chase, S. E. (1995). *Ambiguous empowerment: The work narratives of women school superintendents.* Amherst: University of Massachusetts Press.

Chase, S. (2005). Narrative inquiry: Multiple lenses, approaches, voices. In N. K. Denzin & Y. S. Lincoln (Eds.), *The SAGE handbook of qualitative research* (3rd ed., pp. 651–679). Thousand Oaks, CA: Sage.

Chase, S. E. (2010). *Learning to speak, learning to listen: How diversity works on campus.* Ithaca, NY: Cornell University Press.

Clandinin, D. J. (Ed.). (2007). *Handbook of narrative inquiry: Mapping a methodology.* Thousand Oaks, CA: Sage.

Clandinin, D. J., & Murphy, M. S. (2007). Looking ahead: Conversations with Elliot Mishler, Don Polkinghorne, and Amia Lieblich. In D. J. Clandinin (Ed.), *Handbook of narrative inquiry: Mapping a methodology* (pp. 632–650). Thousand Oaks, CA: Sage.

Clandinin, D. J., & Rosiek, J. (2007). Mapping a landscape of narrative inquiry: Borderland spaces and tensions. In D. J. Clandinin (Ed.), *Handbook of narrative inquiry: Mapping a methodology* (pp. 35–75). Thousand Oaks, CA: Sage.

Cohen, B. (2007). The children's voice: Postwar collection of testimonies from child survivors of the Holocaust. *Holocaust and Genocide Studies, 21,* 73–95.

Cohler, B. J. (2008). Two lives, two times: Life-writing after Shoah. *Narrative Inquiry, 18,* 1–28.

Davis, J. E. (Ed.). (2002). *Stories of change: Narrative and social movements.* Albany: SUNY Press.

Deer, S., Clairmont, B., Martel, C. A., & White Eagle, M. L. (Eds.). (2008). *Sharing our stories of survival: Native women surviving violence.* Lanham, MD: AltaMira Press.

Denzin, N. K. (1997). *Interpretive ethnography: Ethnographic practices for the 21st century.* Thousand Oaks, CA: Sage.

Denzin, N. K. (2000). The practices and politics of interpretation. In N. K. Denzin & Y. S. Lincoln (Eds.), *Handbook of qualitative research* (2nd ed., pp. 897–922). Thousand Oaks, CA: Sage.

Denzin, N. K. (2003). The call to performance. *Symbolic Interaction, 26,* 187–207.

Denzin, N. K. (2008). Interpretive biography. In J. G. Knowles & A. L. Cole (Eds.), *Handbook of the arts in qualitative research* (pp. 117–125). Thousand Oaks, CA: Sage.

Desbois, Father P. (2008). *The Holocaust by bullets: A priest's journey to uncover the truth behind the murder of 1.5 million Jews.* New York: Palgrave Macmillan.

Ellis, C. (2004). *The ethnographic I: A methodological novel about autoethnography.* Walnut Creek, CA: AltaMira Press.

Ellis, C. (2009). *Revision: Autoethnographic reflections on life and work.* Walnut Creek, CA: Left Coast Press.

Faderman, L., with Xiong, G. (2005). *I begin my life all over: The Hmong and the American immigrant experience.* Boston: Beacon.

Frank, A. W. (1995). *The wounded storyteller: Body, illness, and ethics.* Chicago: University of Chicago Press.

Freeman, M. (2006). Life "on holiday"? In defense of big stories. *Narrative Inquiry, 16,* 131–138.

Gallas, K. (1994). *The languages of learning: How children talk, write, dance, draw, and sing their understanding of the world.* New York: Teachers College Press.

Gamson, W. A. (2002). How storytelling can be empowering. In K. A. Cerulo (Ed.), *Culture in mind: Toward a sociology of culture and cognition* (pp. 187–198). New York: Routledge.

Garrod, A., & Kilkenny, R. (Eds.). (2007). *Balancing two worlds: Asian American college students tell their life stories.* Ithaca, NY: Cornell University Press.

Garrod, A., Kilkenny, R., & Gómez, C. (Eds.). (2007). *Mi voz, mi vida: Latino college students tell their life stories.* Ithaca, NY: Cornell University Press.

Georgakopoulou, A. (2007). Thinking big with small stories in narrative and identity analysis. In M. Bamberg (Ed.), *Narrative—State of the art* (pp. 145–154). Philadelphia: John Benjamins.

Gergen, M. M., & Gergen, K. J. (2007). Narratives in action. In M. Bamberg (Ed.), *Narrative—State of the art* (pp. 133–143). Philadelphia: John Benjamins.

Girshick, L. B. (2008). *Transgender voices: Beyond men and women.* Hanover, NH: University Press of New England.

Goodley, D., & Tregaskis, C. (2006). Storying disability and impairment: Retrospective accounts of disabled family life. *Qualitative Health Research, 16,* 630–646.

Gubrium, J. F., & Holstein, J. A. (Eds.). (2001). *Institutional selves: Troubled identities in a postmodern world.* New York: Oxford University Press.

Gubrium, J. F., & Holstein, J. A. (2009). *Analyzing narrative reality.* Thousand Oaks, CA: Sage.

Hammersley, M. (2008). *Questioning qualitative inquiry: Critical essays.* Thousand Oaks, CA: Sage.

Harper, D. (2005). What's new visually? In N. K. Denzin & Y. S. Lincoln (Eds.), *The SAGE handbook of qualitative research* (3rd ed., pp. 747–762). Thousand Oaks, CA: Sage.

Hole, R. (2007). Narratives of identity: A poststructural analysis of three Deaf women's life stories. *Narrative Inquiry, 17,* 259–278.

Holstein, J. A., & Gubrium, J. F. (2000). *The self we live by: Narrative identity in a postmodern world.* New York: Oxford University Press.

Isherwood, C. (2008). The body of her work: Hearing questions of life and death. *The New York Times,* January 22. Available at http://www.nytimes.com/2008/01/22/theater/reviews/22easy.html

Jones, S. H. (2005). Autoethnography: Making the personal political. In N. K. Denzin & Y. S. Lincoln (Eds.), *The SAGE handbook of qualitative research* (3rd ed., pp. 763–791). Thousand Oaks, CA: Sage.

Josselson, R. (1996). *Revising herself: The story of women's identity from college to midlife.* New York: Oxford University Press.

Josselson, R. (2007a). Narrative research and the challenge of accumulating knowledge. In M. Bamberg (Ed.), *Narrative—State of the art* (pp. 7–15). Philadelphia: John Benjamins.

Josselson, R. (2007b). The ethical attitude in narrative research: Principles and practicalities. In D. J. Clandinin (Ed.), *Handbook of narrative inquiry: Mapping a methodology* (pp. 537–566). Thousand Oaks, CA: Sage.

Josselson, R., Lieblich, A., & McAdams, D. P. (Eds.). (2003). *Up close and personal: The teaching and learning of narrative research.* Washington, DC: American Psychological Association.

Josselson, R., Lieblich, A., & McAdams, D. P. (Eds.). (2007). *The meaning of others: Narrative studies of relationships.* Washington, DC: American Psychological Association.

Kaufman, M., & the members of the Tectonic Theater Project. (2001). *The Laramie project.* New York: Vintage.

Knowles, J. G., & Cole, A. L. (Eds.). (2008). *Handbook of the arts in qualitative research: Perspectives, methodologies, examples, and issues.* Thousand Oaks, CA: Sage.

Kwan, M-P. (2008). From oral histories to visual narratives: Re-presenting the post–September 11 experiences of the Muslim women in the USA. *Social & Cultural Geography, 9,* 653–669.

Langellier, K. M. (2001). You're marked: Breast cancer, tattoo, and the narrative performance of identity. In J. Brockmeier & D. Carbaugh (Eds.), *Narrative and identity: Studies in autobiography, self, and culture* (pp. 145–184). Amsterdam: John Benjamins.

Langellier, K. M., & Peterson, E. E. (2006). Shifting contexts in personal narrative performance. In D. S. Madison & J. Hamera (Eds.), *The SAGE handbook of performance studies* (pp. 151–168). Thousand Oaks, CA: Sage.

Lieblich, A., McAdams, D. P., & Josselson, R. (Eds.). (2004). *Healing plots: The narrative basis of psychotherapy.* Washington, DC: American Psychological Association.

Lock, A., Epston, D., & Maisel, R. (2004). Countering that which is called anorexia. *Narrative Inquiry, 14,* 275–301.

Luttrell. W. (2003). *Pregnant bodies, fertile minds: Gender, race, and the schooling of pregnant teens.* New York: Routledge.

Madison, D. S. (2006). Staging fieldwork/performing human rights. In D. S. Madison & J. Hamera (Eds.), *The SAGE handbook of performance studies* (pp. 397–418). Thousand Oaks, CA: Sage.

Madison, D. S., & Hamera, J. (Eds.). (2006). *The SAGE handbook of performance studies.* Thousand Oaks, CA: Sage.

Mattingly, C. F. (2007). Acted narratives: From storytelling to emergent dramas. In D. J. Clandinin (Ed.), *Handbook of narrative inquiry: Mapping a methodology* (pp. 405–425). Thousand Oaks, CA: Sage.

Maynes, M. J., Pierce, J. L., & Laslett, B. (2008). *Telling stories: The use of personal narratives in the social sciences and history.* Ithaca, NY: Cornell University Press.

McAdams, D. P. (2006). *The redemptive self: Stories Americans live by.* New York: Oxford University Press.

McAdams, D. P., Josselson, R., & Lieblich, A. (Eds.). (2001). *Turns in the road: Narrative studies of lives in transition.* Washington, DC: American Psychological Association.

McAdams, D. P., Josselson, R., & Lieblich, A. (Eds.). (2006). *Identity and story: Creating self in narrative.* Washington, DC: American Psychological Association.

McLaughlin, D., & Tierney, W. G. (Eds.). (1993). *Naming silenced lives: Personal narratives and processes of educational change.* New York: Routledge.

Mertus, J., Tesanovic, J., Metikos, H., & Boric, R. (Eds.). (1997). *The suitcase: Refugee voices from Bosnia and Croatia.* Berkeley: University of California Press.

Miller, L. C., & Taylor, J. (2006). The constructed self: Strategic and aesthetic choices in autobiographical performance. In D. S. Madison & J. Hamera (Eds.), *The SAGE handbook of performance studies* (pp. 169–187). Thousand Oaks, CA: Sage.

Mills, C. W. (1959). *The sociological imagination.* London: Oxford University Press.

Mishler, E. G. (1995). Models of narrative analysis: A typology. *Journal of Narrative and Life History, 5,* 87–123.

Myerhoff, B. (1994). *Number our days: Culture and community among elderly Jews in an American ghetto.* New York: Meridian/Penguin. (Original work published in 1979)

Naples, N. (2003). Deconstructing and locating survivor discourse: Dynamics of narrative, empowerment, and resistance for survivors of childhood sexual abuse. *Signs: Journal of Women in Culture and Society, 28,* 1151–1185.

Paap, K. (2006). *Working construction: Why white working-class men put themselves—and the labor movement—in harm's way.* Ithaca, NY: ILR/Cornell University Press.

Pascoe, C. J. (2007). *Dude, you're a fag: Masculinity and sexuality in high school.* Berkeley: University of California Press.

Personal Narratives Group. (Eds.). (1989). *Interpreting women's lives: Feminist theory and personal narratives.* Bloomington: Indiana University Press.

Plummer, K. (1995). *Telling sexual stories: Power, change, and social worlds.* London: Routledge.

Polkinghorne, D. E. (1995). Narrative configuration in qualitative analysis. In J. A. Hatch & R. Wisniewski (Eds.), *Life history and narrative* (pp. 5–23). London: Falmer.

Polkinghorne, D. E. (2007). Validity issues in narrative research. *Qualitative Inquiry, 13,* 471–486.

Richardson, L. (1990). Narrative and sociology. *Journal of Contemporary Ethnography, 19,* 116–135.

Richardson, L. (2002). Poetic representation of interviews. In J. F. Gubrium & J. A. Holstein (Eds.), *Handbook of interview research: Context and method* (pp. 877–892). Thousand Oaks, CA: Sage.

Riessman, C. K. (1990). *Divorce talk: Women and men make sense of personal relationships.* New Brunswick, NJ: Rutgers University Press.

Riessman, C. K. (2002a). Analysis of personal narratives. In J. F. Gubrium & J. A. Holstein (Eds.), *Handbook of interview research: Context and method* (pp. 695–710). Thousand Oaks, CA: Sage.

Riessman, C. K. (2002b). Positioning gender identity in narratives of infertility: South Indian women's lives in context. In M. C. Inhorn & F. van Balen (Eds.), *Infertility around the globe: New thinking on childlessness, gender, and reproductive technologies* (pp. 152–170). Berkeley: University of California Press.

Riessman, C. K. (2008). *Narrative methods for the human sciences.* Thousand Oaks, CA: Sage.

Rosenwald, G. C., & Ochberg, R. L. (Eds.). (1992). *Storied lives: The cultural politics of self-understanding.* New Haven, CT: Yale University Press.

Saldaña, J. (2008). Ethnodrama and ethnotheatre. In J. G. Knowles & A. L. Cole (Eds.), *Handbook of the arts in qualitative research* (pp. 195–207). Thousand Oaks, CA: Sage.

Saukko, P. (2008). *The anorexic self: A personal, political analysis of a diagnostic discourse.* Albany: SUNY Press.

Schneider, R. (2006). Never, again. In D. S. Madison & J. Hamera (Eds.), *The SAGE handbook of performance studies* (pp. 21–32). Thousand Oaks, CA: Sage.

Scott-Hoy, K., & Ellis, C. (2008). Wording pictures: Discovering he*art*ful autoethnography. In J. G. Knowles & A. L. Cole (Eds.), *Handbook of the arts in qualitative research* (pp. 127–140). Thousand Oaks, CA: Sage.

Shamai, M., & Levin-Megged, O. (2006). The myth of creating an integrative story: The therapeutic experience of Holocaust survivors. *Qualitative Health Research, 16,* 692–712.

Smith, A. D. (1993). *Fires in the mirror: Crown Heights, Brooklyn and other identities.* New York: Anchor.

Smith, A. D. (1994). *Twilight—Los Angeles, 1992 on the road: A search for American character.* New York: Anchor.

Smith, A. D. (2004). *House arrest: A search for American character in and around the White House, past and present.* New York: Anchor.

Weber, S. (2008). Visual images in research. In J. G. Knowles & A. L. Cole (Eds.), *Handbook of the arts in qualitative research* (pp. 41–53). Thousand Oaks, CA: Sage.

Weinberg, D. (2005). *Of others inside: Insanity, addiction and belonging in America.* Philadelphia: Temple University Press.

White, M., & Epston, D. (1990). *Narrative means to therapeutic ends.* New York: W. W. Norton.

26

CRITICAL ARTS-BASED INQUIRY

The Pedagogy and Performance of a Radical Ethical Aesthetic

Susan Finley

Critical arts-based inquiry situates the artist-as-researcher (or researcher-as-artist) in the new research paradigm of qualitative practitioners committed to democratic, ethical, and just research methodologies. It also demonstrates an activist approach to research in which the ultimate value of research derives from its usefulness to the community in which the research occurs.

Two recent events set the stage for my discussion of arts-based research. In the first example, I was invited to participate in an "Imagination Committee" to contemplate the future of education in a particular urban community. Participants were conscientiously drawn to honor cultural pluralism, with the intent of mobilizing culturally diverse communities within a single school district. We convened in one large group and then restructured into smaller learning pods that explored particular topics (e.g., student identity, technology, curriculum, etc.), with the instruction that each group would report back to the larger group in a closing exercise. The breakout groups were scheduled for about three hours work time.

In our smaller groups, we were encouraged to produce visual representations of our conversations, as well as written records. While we attended to our discussions in breakout groups, an artist traveled the room, talking with each work group, listening in on the various conversations, while sketching in a large pad. Our artist-researcher maintained his role as listener-observer throughout the morning and the lunch break, until we joined another as one large group to recap our visionary insights about how to shape the future of education in the context of this community. As the large group meeting convened, the artist-as-researcher exhibited his portrayal of our day of reflection on the "big screen." He had produced a comic strip-like series of panels

I associate with graphic novels. In the course of his drawings, he produced a sense of the complexities involved in developing a collaborative piece, which he conveyed through the nuance of his artistry—representations of movement, caricature portraits of individual speakers, and a Greek chorus that appeared in his cartoons. As he exhibited his work (sometimes collaged with drawings from the small group sessions), and with the assistance of a discussion moderator, the room buzzed with enthusiastic conversation, public responses to the art and its messages, and disagreements as to some of the representations (to which the artist-researcher responded by adding quick supplemental sketches).

The second example was a year later and with a different cast of characters—this time a group of individuals representing organizations that work with volunteers, refer volunteers, or who are committed to service learning as a pedagogical approach. In this setting, a visual artist was stationed in front of a large mural-like outline of buildings, parks, lakes, and the river that is central to the geography of the area. As large and small group discussions took place, the artist slowly completed the visual terrain using colored pencils until, at the end of the day, there existed a dynamic visual depiction of the future of service learning for the community.

These are examples of critical, qualitative arts-based research. It is a genre of research in which methodologies are emergent and egalitarian, local, and based in communal, reflective dialogue. These were performances of knowledge creation, taking shape in the context of complex conditions and in which art provided mechanisms and forms with which to see and hear each other's views on local socioeconomic systems, racial and cultural divides, and potential to develop common meeting spaces.

Each of these visual artists recorded images that emerged in dynamic conversations. Their visions were grounded in the particular place and the individual people who had committed themselves to the performance of a reflective dialogue with the other participants. Arts-based inquiry demystified the process of storytelling and facilitated participants' shared articulation of the experiences of living together, in harmony and in conflict. By engaging art as an emergent living practice, these research events brought audience and researcher to a place of possibility to experience the reciprocity of dialogue and representation, reflection and speaking, and speaking and listening. In each instance, art increased or rejuvenated participation and dialogue. The audience connected with the subject matter emotionally, simultaneously confirming the authenticity of their emotive, reflective dialogues that characterized their dialogues. These artists made no claims of truths, but clearly worked to represent reflective dialogue and explorations of futuristic possibility.

These examples drawn from my own experience represent a new approach to facilitating community discourse. They also speak to the ephemeral quality of many arts-based research methodologies. The art and the research are so localized as to be "in the moment." The intent in their creation is not to be replicated and distributed; in fact, in each of the given examples, the visual representations created in these localized discussions would be little appreciated outside the very local communities in which they were created. On a personal level, these instances of arts-based inquiry in community settings gave me a basis for optimism that arts-based research is one of the tools a community can use in the performance of community-based activism.

▣ CRITICAL ARTS-BASED INQUIRY

In the third edition of the *SAGE Handbook of Qualitative Research,* my discussion of arts-based inquiry focused on "performing revolutionary pedagogy" and explicated the usefulness of critical arts-based research in "doing qualitative inquiry when political activism is the goal" (S. Finley, 2005, p. 681). Arts-based inquiry first developed in the historical moment of a "crisis of representation" among qualitative researchers in anthropology and sociology who struggled with ways to represent new wave research that was local in nature and based in an ethics of care (Denzin & Lincoln, 2000; Geertz, 1988; Guba, 1967; Hammersley, 1992). Research in the genres of arts and humanities has since proliferated to the point of "post-experimental" status (Denzin & Lincoln, 2000) and the proliferation of poetic forms of research in the social sciences has led Zali Gurevitch (2002) to declare a "poetic moment" (p. 403) in qualitative inquiry. More recently, a "performative turn" in arts-based research has shifted focus away from the written text to performance as a "form of

research publication" (Denzin, 2003, p. 13; see also, Conquergood, 1988). Critical inquiry as performance art is particularly well suited to researchers who anticipate experiences of cultural resistance (Garoian, 1999) and positive social change through inclusive and emotional understandings created among communities of learner/participant/researchers.

In reworking my earlier handbook chapter for this newer edition, I have structured my discussion around three interrelated issues for critical arts-based researchers:

- First, I review neoliberal and neoconservative trends in research and curriculum that further institutionalize and reaffirm social divisions, deny access to creative participation and expression to particular groups of people, and stand in the way of postcolonizing research strategies (Cannella & Lincoln, 2004a, 2004b, Lincoln, 2005; Lincoln & Cannella, 2004a, 2004b).
- Second, I discuss the performative turn in qualitative inquiry and how it reinforces the potential of critical arts-based research as a revolutionary, activist, and aesthetic pedagogy (Alexander, 2005; Denzin, 2000; S. Finley, 2003a).
- Finally, the chapter moves into discussion (with examples) of arts-based inquiry at the heart of a people's pedagogy in which performances of critical arts-based research enact "a radical ethical aesthetic" (Denzin, 2000, p. 261) and attempt resistance to "the regressive structures of our every day lives" (Denzin, 1999, pp. 568, 572).

I believe that fulfillment of a resistance politics in research requires new urgency, requires renewed commitment, and calls for continuing development of research methodologies to support interpretive studies that extend democracy, freedom, and political voice into the everyday lives of politically oppressed people. What is called for in these times is political resistance that is intentional in its purpose of reversing efforts of neoliberal and neoconservative political forces that systematically counter progress that had been made to improve the human condition by "new wave" researchers whose work is based in an ethics of care (Lincoln, 1995).

In critical arts-based inquiry, arts are both a mode of inquiry and a methodology for performing social activism. Although many qualitative researchers draw on the arts and humanities as an epistemological construct that is useful as a communicative force, not all of these practitioners of arts-based research will join me in promoting a radical and revolutionary aesthetic. Instead, arts-based researchers have found "many and varied roles for the arts in social science research" (Knowles & Cole, 2008, p. xiii), not all of which consider social change the primary objective of their efforts.

In the wake of 9/11, neoconservative politics have taken hold in the United States. Coupled with economic collapse on a global scale, the face of new conservatism calls out qualitative researchers to political resistance. It is now more than ever the right for political activists in the academy to further develop and employ methodologies that inspire and facilitate progressive social action. Even those arts-based researchers who refuse the call to activate research for the purpose of cultural revolution (even in small, local efforts) cannot convincingly deny they have already entered the political fray. "It is an act of political defiance for arts-based researchers to say, 'I am doing art' and to mean, 'I am doing research.' . . . To hold that art and research can be synonymous is a charged political statement" (S. Finley, 2003a, p. 290; see also S. Finley, 2005, p. 685). The incidence of making political waves with research methodologies is not limited to qualitative or even arts-based researchers. All of research is political. Kenneth Howe (2009) declares the neoconservative regression in research methods to be "the new scientific orthodoxy" and further argues, "Whatever the methods employed, decisions about what factors to fix in the design and conduct of social research are unavoidable—and are unavoidably political" (p. 428). Norman Denzin and Yvonna Lincoln (2005; see also Smith, 1999, cited in Denzin & Lincoln) write, "Sadly, qualitative research, in many if not all of its forms (observation, participation, interviewing, ethnography), serves as a metaphor for colonial knowledge, for power, and for truth" (p. 1). Thus, a key question for arts-based researchers is, "How do we break through the complex barriers of colonial social conformity to an inclusive, pluralist aesthetic situated in the lives of others?"

▣ THE THREAT TO CRITICAL ARTS-BASED INQUIRY IMPOSED BY REINVIGORATED PALEOCON, NEOCON, AND NEOLIBERAL POLITICAL TRADITIONS IN RESEARCH AND EDUCATION

It stands to reason that social conditioning under what John Leaños and Anthony Villarreal (2007) have described as the "Judeo-Christian White Supremacist Heteronormative Capitalist Patriarchal Military Industrial Entertainment Prison University Complex" (p. 1) by definition limits possibility in an emerging tradition of an ethical and socially engaged arts-based research. Roadblocks to arts-based inquiry are rooted in early education curricular issues that follow all the way through higher education and include public policy for the conduct and funding of human studies. Although arts inquiry holds promise for an emerging research tradition that is postcolonial, pluralistic, ethical, and transformative in positive ways, the forces of neoconservative political agendas jeopardize its implementation.

Liberal arts education that builds a particular skills-base encourages active imagination, and the ability to engage in critical critique and dialogue are central features of curriculum and pedagogy to prepare researchers for critical arts-based

inquiry (Seidel, 2001). A profound shift toward capitalistic, business-strategies for educational organization and delivery, coupled with conservative schooling practices given emphasis in No Child Left Behind legislation, have depressed the value of arts and humanities education as they are pitted in competition for dollars against the profit-value assigned to sciences. The test-based, standardized accountability system in place in U.S. education holds little reward for arts-based teaching and learning. Richard Siegesmund observes "in an era of narrowly conceived outcomes for education, art is not taught" (1998, p. 199). As the character Shane Botin quips, kids go to school to "learn how to pass the weekly standardized tests to get the school more funding" (Benabib & Salsberg, 2009).

Also missing from the creative studies curriculum are opportunities and skills needed for critical analysis and social action. Leaños and Villarreal (2007) agree: "Critical pedagogy in general, and critical arts education in particular, have all but disappeared from school curriculum across the U.S." (p.1).[1] Even more alarming is the long-term influence of conservative educational policy and its institutionalization as sustained social practice. "The educational force of the culture actually works pedagogically to reproduce neoliberal ideology, values, identifications and consent," write Henry Giroux and Susan Giroux (2008). Elliot Eisner (2001b) had it right: "Education in our schools should look more like the arts, rather than the arts looking more like our schools" (p. 9).

In an unseemly paradox, standardization of arts education has been offered as a solution to remand the trend toward severe cutbacks in public school-based arts education. Despite the potential truth to an argument that uniform art standards could ensure the continuation of arts education in public schools, arts-education standards threaten cultural pluralism and more likely promise cultural reproduction (Eisner, 2001a). Standardizing arts-education would reaffirm neoliberal educational policies and further force arts-education into a market model, whereas culturally responsive arts education is directly linked to academic achievement and its absence with educational disempowerment (Hanley & Noblit, 2009).

Standardized arts education draws its design from an aesthetic education tradition with the goal of teaching school children strategies for art appreciation and aesthetic engagement—the primary learning goal in this model is teaching children how to be a good audience for the arts, rather than a critic or producer of arts. Defending standardization in arts-education, Laura Zakaras and Julia Lowell (2008) claim the virtue of standardized arts curriculum in its universal applicability to students, "regardless of their artistic talent, to enable them to have more satisfying encounters with works of art, now and in the future" (p. 20). Consistent with neoliberal design, the experience of art could be standardized for all students—but not in furtherance of a democratic goal of inclusive pluralism. That which is identified as "quality" art suited for teaching all children would draw from a limited classification of arts created by "masters." Zakaras and Lowell write,

> It is generally agreed that these perceptual skills are best learned in encounters with masterpieces, exemplary works of art that reward close attention and bring the entire range of aesthetic skills into play. Ideally these works represent a variety of historical periods, regions of the world, and genres of the art form, including folk, popular, classical, and ethnic cultures. (2008, p. 21)

It cannot go unnoted that identification of the artist, regions, techniques, forms of arts, and so forth in such a curriculum can be reduced to multiple-choice testing. Furthermore, the selection of masterpieces will be politically grounded. Zakaras and Lowell actually move beyond student-as-audience to consider that immersion in the masters can be transferred to skills-based arts education as well as appreciation: "In music, for example, learning to perform a challenging work of art requires the kind of attention to the work's components that often develops aesthetic perception and appreciation" (Zakaras & Lowell, 2008, p. 22). I refer to artist and arts teacher Twila Tharp to counter the masters' argument, as Tharp begs room for students to engage in original, creative explorations of art making:

> Repetition is a problem if it forces us to cling to our past successes. Constant reminders of the things that worked inhibit us from trying something bold and new. We lose sight of the fact that we weren't searching for a formula when we first did something great; we were in unexplored territory, following our instincts and passions wherever they might lead us. (2003, p. 217)

Instincts and passions reach beyond the possibilities of standardization and formulaic art making. A standardized curriculum is unlikely to include the passionate artist whose work is culturally and historically situated but uses techniques inconsistent with those that have been designated masterpieces worthy of study and reproduction. Standardization in arts education as appreciation and duplication of strokes of master artists leaves begging the notion of cultural pluralism.

"Whose art?" becomes a pivotal question. It leaves to reason that indigenous, counterculture, and outsider art forms will be little represented by standardized arts-appreciation curricula. In place of pluralism, predefined value definitions reify social worlds revered by convention. Will the masters of graffiti arts, graphic artists who create computer games, and works of unschooled artists be included as masterpieces? Will pluralism give way to eroticism and exploitation in "pornomiseria" (Faguet, 2009)?[2] Will the arts canon be standardized to censor radical political expressions? "The censorship of arts is intimately linked to the censorship of political dissent," Leaños and Villarreal (2007, p. 1) agree. "Being *pissed* is one of the artist's most valuable conditions," wrote Tharp. "Creativity is an act of

defiance. You're challenging the status quo. You're questioning accepted truths and principles" (Tharp, 2003, p. 133). Being pissed is what Denzin had in mind when he called qualitative researchers to engage "guerilla warfare" against the status quo in research. Leaños and Villarreal observe,

> In sum, the full range of art activity occurs within a "cultural arbitrary" (Bourdieu and Passeron, 1977) that establishes elite aesthetics as the norm, a norm further enacted and enforced by the everyday practices of arts education that privileges "beauty," "form," and "genius." As this high art aesthetic is central to the cultural arbitrary of traditional arts education in the U.S., artistic contributions that fall outside the narrow parameters of these norms, such as critical and politically engaged art practices, have been marginalized historically, and are practically absent from official school curriculum. (2007, p. 2)

Talent, genius, and *quality* are code words for cultural conservatism in arts. The narrow field of legitimacy introduced by these terms is the very definition of "high arts," the promotion of which further institutionalizes divisions along socioeconomic, racial, and gender lines. Expertism (and genius and natural talent) likewise can be traced to educational elitism. Tharp debunks the notion of natural genius through the example of Mozart: "Nobody worked harder than Mozart," Tharp observes, "By the time he was twenty-eight years old, his hands were deformed because of the hours he had spent practicing, performing, and gripping a quill pen to compose" (2003, p. 7). Yet, Mozart was born into the role of prodigy. "His first good fortune was to have a father who was a composer and a virtuoso on the violin, who could approach keyboard instruments with skill" (p. 7). Tharp concludes,

> Mozart was his father's son. Leopold Mozart had gone through an arduous education, not just in music, but also in philosophy and religion; he was a sophisticated, broad-thinking man, famous throughout Europe as a composer and pedagogue. . . . Leopold taught the young Wolfgang everything about music, including counterpoint and harmony. He saw to it that the boy was exposed to everyone in Europe who was writing good music or could be of use in Wolfgang's musical development. Destiny, quite often, is a determined parent. Mozart was hardly some naïve prodigy who sat down at the keyboard and, with God whispering in his ears, let the music flow from his fingertips. (2003, pp. 7–8)

Mozart had the advantages of education in both musical theory and technique. Pierre Bourdieu, for one, emphasized that legitimacy among artists has long depended on expertise with a particular set of craft skills ensconced within a defined aesthetics. Mozart learned well the craft skills needed to master the language of music. Tharp continues,

> Skill gives you the wherewithal to execute whatever occurs to you. Without it, you are just a font of unfulfilled ideas. Skill is how you

close the gap between what you can see in your mind's eye and what you can produce; the more skill you have, the more sophisticated and accomplished your ideas can be. With absolute skill comes absolute confidence, allowing you to dare to be simple. Picasso once said, while examining an exhibition of children's art, "When I was their age I could draw like Raphael, but it has taken me a whole lifetime to learn to draw like them." (2003, p. 163)

Teach children the skills to play, to paint, to create with confidence—let their hands form to the shape of pen, brush, or keyboard. If children don't have opportunities to learn skills, there will be no new masters—and maybe that is the point— cultural reproduction will not happen concurrently with social change.

In addition to financial restrictions that limit skills education in arts, also left on the cutting room floor is the teaching of critical arts education. To know about art or understand certain classical forms of art is not sufficient—as practitioners of a moral, ethical aesthetic, the goal is to move from knowledge and understanding to action. When the value of art depends on standardization of social conventions and traditions, the arts exist in a void of political and ethical involvement by artists (performers) and audiences (performers in the experience). Moreover, isolating convention from critical question and interpretation limits creative constructions of dynamic, evolutionary human activity. Simply, public school students of the 21st century for the most part are not being educated to approach the world as active citizens. In the same way that neoconservative educational policies sorely limit students' learning of craft skills, they displace critical discourse, performances of hope, and voices of public citizens sharing their creations of images of a better tomorrow. Giroux and Giroux (2006) write, "As the prevailing discourse of neoliberalism seizes the public imagination, there is no vocabulary for progressive social change, democratically inspired visions, critical notions of social agency, or the kinds of institutions that expand the meaning and purpose of democratic public life" (p. 25).

The conservative swell that threatens arts education repeats in form and content in the fields of human research. Forces of political conservatism reverberate in the call to claim human studies as "science" and to attach the notion of scientific rigor to understanding social phenomena. New life has returned to conservatism. Alternative ways of learning and knowing are cast aside as a traditionalist, capitalistic pedagogy (based in proof, truth, individualization, and competition) repeats its tired beat in the community of human researchers. This ideologically formed miseducation of arts and inquiry "prevents social art practices from becoming valued as a vital form of democratic engagement" (Leaños & Villarreal, 2007, p. 2).

Howe (2009) critiques positivist dogma "codified" in the National Research Council's (NRC) manifesto, *Scientific Research in Education* (2002) and "reinforced" in the American Educational

Research Association's *Standards for Reporting on Empirical Social Science Research in AERA Publications* (2006). Howe is disturbed by the definition of research in technocratic parlance that is largely silent to the "relationship between education science and democratic politics." This is a poignant silence that sets up a dichotomous relationship as if science and politics exist as "separate domains" (p. x). Working his critique from within the conceptual framework of "deliberative democracy" (and with references to Gutmann & Thompson, 2004, and Young, 2004, on this point), Howe says, "The possibility—and desirability—of culling political values from education research depends on moribund positivist principles" (p. 432).

Following the lead of the *SRE* report, the AERA task force reflected similar positivistic dogma in their publication of "Standards for Reporting on Empirical Social Science Research in AERA Publications." Like the SRE guidelines, AERA narrowed the focus of its report to "empirical social science research" (AERA, 2006). AERA guidelines differentiate "research methods" from "other forms of scholarship." Approaches to be excluded from the category of research include reviews and critiques of research traditions and practices; theoretical, conceptual, and methodological studies; and historical types of work. Further, the AERA guidelines expand on the NRC specific exclusion of history and philosophy as outside the realm of research because they are not based in empirical experimentations. AERA specifies excluded methodologies should include "scholarship more grounded in the humanities (e.g., history, philosophy, literary analysis, *arts-based inquiry* [emphasis added])" and the authors reason that these approaches are "beyond the scope of this document."

Humanities-oriented research standards were the subject of a second AERA committee report, "Standards for Reporting on Humanities-Oriented Research in AERA Publications" (2009). Howe (2009) was a member of the humanities research committee and argues vociferously against the AERA's reinforcement of

> a reductionist conception of empirical social science research" that dichotomizes educational science and the humanities. He defies the conclusion by the AERA task force authors who concluded that humanities-grounded research is not like scientific research because science is empirical and humanities is not. Howe argues that both methodologies are empirical but their differences are a matter of degree on a continuum of "blurred boundaries . . . overlap and complementarity." (Howe, 2009, p. 432)

The AERA Humanities-Research Task Force and authors of the "Standards for Reporting on Humanities-Oriented Research in AERA Publications" (2009) echoed Howe's analysis of overlap and complementarity in their report. Howe further argues against the perception that scientific study is apolitical and objective "Characterizing science as rhetorical in the sense that, I suggest, applies to Kuhn, Harding, Code, and Hacking does not require denying that science is 'a profoundly powerful form of inquiry' (Lessl, 2005, p. 2)" (Howe, 2009, p. 437).

> It only requires denying that scientific claims are above and beyond persuasive argumentation, that scientists are above and beyond weaknesses such as a blinkered perspective, ego involvement, resistance to novelty, an interest in the size of their paychecks, and the like, in the conduct of their research. (Howe, 2009, p. 437)

In much the same way that arts education has been compromised by neoconservative, capitalistic, and corporate ideology, in the current climate of the first decade of the 21st century, so has the culture of university research. Giroux (2009) argues that this cultural shift (a reactionary shift that takes social science out of the realm of arts and humanities) removes faculty and students from democratic language, values, and work. Consequently, Giroux says, when educational priorities labor to serve the "warfare state or the corporate state" (Giroux, p. 671), higher education falls short of its purpose of educating students in the performance of democratic governance. In such circumstances, higher education ceases to exist as "a crucial public sphere, responsible for both educating students for the workplace and providing them with the modes of critical discourse, interpretation, judgment, imagination, and experiences that deepen and expand a democracy" (Giroux, p. 671).

◨ Roadblocks, Gatekeepers, and Other Issues for Critical Research Formed in Arts

Conditions that interfere with the acceptance of arts-based research as a forum for performance pedagogy and political emancipation come from both within the community of artist-researchers and from the dominating structures of positivism that privilege science over other forms of knowledge production. Cathy Coulter and Mary Lee Smith (2009) observe that purpose, methods, ethics, and validity form the contested terrain of narrative inquiry based in fiction. In my view, an even more divisive discussion revolves around the issues of standards for assessing quality in arts-based research. This discussion encompasses fictional narrative, poetry, dance, film, and all other art disciplines that have made their way into usage as forms of inquiry and expressions of inquiry through arts. Eisner (2008) notes,

> One of the most formidable obstacles to arts-informed research is the paucity of highly skilled, artistically grounded practitioners, people who know how to use image, language, movement, in artistically refined ways. Schools of education, for example, seldom provide courses or even workshops for doctoral students to develop such skills. As a result, it is not uncommon to find this type of research appearing amateurish to those who know what the potentialities of the medium are. (p. 9)

Donald Blumenfeld-Jones (2008) (a classically trained dancer and arts-based researcher) concurs. He holds that a "vigorous education in dance" (p. 183) is required preparation for arts-based researchers who choreograph or perform their research. "Dance is, first and foremost, an art form" (p. 183). Blumenfeld-Jones writes,

> The art needs to be practiced. . . . Insights discovered through the practice of dance as an art form are only available through that practice, and the practice focuses on making art, not on coming to understand. To consider using dance as a primary mode of research, persons must first develop themselves as artists, understanding that the practice of art is, in many ways, no different than the practice of research (Blumenfeld-Jones, 2002, 2004a, 2004b). There are not many social scientists who are also well-educated dance artists, and without such grounding, the concern is that the emerging art will be poor and nothing significant can be gained from it. (2008, p. 184)

Johnny Saldaña (2008) takes a slightly different stance with regard to artistic expertism. He observes quality issues such as the over-inclusion of didactic content in ethnotheater scripts constructed by scholars who do not have theater training. He further notes that the director, designers, and performers need to be good at their craft if ethnotheater is to work as a research representation that is engaging and effective. Even so, he acquiesces that he would not want to discourage scholars who are inspired to playwriting. He asks only that those scholars, as with any playwright, would seek open critical feedback about the work.

Eisner (2008) suggests one solution to the dilemma of artistic expertism that I have tried. He encourages the formation of teams of social science researchers who work with practitioners of the arts. "It could be the case," Eisner wrote, "that such collaboration might provide a way to combine both theoretically sophisticated understandings and artistically inspired images" (2008, p. 9). For example, Macklin Finley and I worked with Saldaña (a theater professor and ethnodramatist) to condense a readers' theater script (S. Finley & M. Finley, 1998), a book of poetry (M. Finley, 2000) and other poems (e.g., S. Finley, 2000), as well as a short story (S. Finley & M. Finley, 1999) and several other artistic representations of our research with street youths into a single script. Elsewhere, Saldaña has opined that the goal of this type of shared work of adaptation has the goal of making the work "even better than its original source" (2008, p. 197). I disagree to some extent when this adage is applied to the *Street Rat* effort—in the instance of the readers' theater script, the adaptation was a vast improvement. In comparison with the short story and poetry formats, however, the forms of representation were different, but not improved. All of the dialogue of the play is excerpted from the previously published research poetry of S. Finley and M. Finley. Although I embrace the artfulness of the ethnodramatic staging of *Street Rat* and commend

Saldaña for creating with us a work that maintained the integrity of the original representations, I question the accessibility of the new script for local audiences. Our (Finley and Finley) practice had been to interact with audiences during poetry readings as a way to provoke communication about the social issues of poverty, school leaving, the place of arts in street life, and so on that came about in audience responses to the work (M. Finley, 2003, p. 604). Likewise, the readers' theater was produced in several settings and was sufficiently provocative of conversation with diverse audiences that included academics, social workers, in-school youth, and street youths (including some who were present in the text). When rehearsals, directors, stage props, and even finding a stage are requisite to the performance, it is much more difficult to arrange.

It might also be that greater sophistication with regard to understanding the conventions of theater and performance are required of audiences of a fully developed script, constructed by an expert in theater. If a sound educational base in an art is requisite for social science research, and given the neoconservative backlash against arts education (as discussed in the early pages of this chapter), then the future of arts-based research is in serious doubt.

Fear that traditional members of the academy will demean our work (e.g., Mayer, 2000) stokes arts-based researchers' concerns for quality control. Instead, we should keep at the forefront that we serve the dual purpose of unveiling oppression and transforming praxis. Despite the movement to fuse arts and research that is taking place in many social science disciplines (e.g., education, nursing, social science), many researchers do not identify as "arts-based" their work that draws on arts for either inspiration or form. In response, I contest Melissa Cahnmann-Taylor's (Cahnmann-Taylor & Siegesmund, 2008) statement, "There are still more researchers writing *about* arts-based research criteria than those producing examples of what it looks like in each area of the literary, visual, and performing arts" (p. 12). First, I see examples of the arts in qualitative research in a wide variety of forums, but I puzzle over why so many researchers who clearly experience art as qualitative research do not choose to describe their work as "arts-based." A cynical possibility is that some arts-informed qualitative research is not identified as such because work of this type has a special place on what Tom Barone (2008) has referred to as the "blacklist of research methodologies (indeed, one that disparages all non-experimental forms)" (p. 34) of qualitative research.

An alternative explanation that intrigues me is that the many names for arts in qualitative research are confusing—is it arts-based research, arts-based inquiry, arts-informed research, or A/R/Tography (for history and definition, see Sinner et al., 2006)? Similarly, with the use of arts to reach a broader, community-centered audience for research, do the multiple terms and their fine-tuned differences counter the goal of accessibility? Are the terms obtuse in their verbiage, and, therefore, more accessible to an audience of scholars, researchers, and policy wonks than for local audiences situated in communities where the research took place? Instead, could it be that the clarity of the goal of audience-participant driven research results in work that is, indeed local, and does not make its way into publications and other professional forums accessed by scholars looking for examples of arts-based research? Much of the work is ephemeral and can only be captured as description and in analytic discussions of that which must go unseen and unexperienced by an academic audience. Although multimedia technology offers some possibilities for expanding audiences to performances of arts-based inquiry, Carl Bagley (2008) correctly observes such re-presentations "would still constitute a (re) reading and (re)presentation of what was performed" (p. 54).

New constructions of what is possible in the realm of human studies motivated Eisner, Barone, Denzin, and others to call arts-based researchers to action when the field was being formed. If, however, the roles of researchers and standards for assessing quality in research are hidebound by tradition and inalienable definitions of "research" and "art," this call to transformative inquiry through art will be a difficult act to perform. Breaking with tradition would feature transformations in pedagogy and praxis—art would of necessity be taught as a method for accessing multiple ways of understanding human conditions and experiences.

To prepare arts-based researchers, Eisner (e.g., 1991/1998) proposed a graduate school curriculum that enhances students' skills of imagination, perception, and interpretation of the qualities of things, as well as teaching mastery of skills of artistic representation. "Art, music, dance, prose, and poetry are some of the forms that have been invented to perform this function," Eisner wrote (p. 235). Clive Seale (1999) visualized a studio apprenticeship model to teach research skills "in much the same way as artists learn to paint, draw, or sculpt" (p. 476). "Working knowledge" (Harper, 1987) requires deep understanding of materials, the skills to manipulate them, and intuitive, imaginative, and reflective thinking (S. Finley, 2001). In working knowledge, there is "kinesthetic correctness . . . [an] interplay of the theoretical and the empirical, the marriage of hand and mind in solving practical problems" (Harper, 1987, pp. 117–118). The product crafted from working knowledge is secondary to the mental and cultural experiences of the work. "Work in this instance is both a noun and a verb—it performs the dual purpose [of being and acting]" (S. Finley, 2001, p. 20; see also Sullivan, 2005, p. 241). Creative work is the site for dialogue and the source of further action.

"Discourse is the power which is to be seized," wrote Michel Foucault (1984, p. 110). "The great challenge to neoliberalism can only come through the reclaiming of a language of power, social movements, politics, and ethics that is capable of examining the effects of the neoliberal order" (Giroux & Polychroniou, 2008, p.1). "Higher education needs to be reclaimed as an ethical and political response to the demise of democratic public life," the authors conclude. In this contested terrain, art wields power to engender dialogue and can be a catalyst for reclaiming language and reviving the social imaginary to visions of hope. Leaños and Villarreal (2007) assert, "art's greatest potential to foster change is at the level of the micro-social, through tactical interruptions of bio-power, revealing the ways that we are complicit in the "normalizing" operations of power, opening up spaces for new forms of knowledge production, and spreading decolonial discourse" (p. 2).

Kathleen Casey (1995) explained that methodological shifts in research approaches are tied to political or theoretical interests charged by social and historical circumstances. She notes that, in the example of narrative research, the new paradigm researcher takes a stance that "deliberately defies the forces of alienation, anomie, annihilation, authoritarianism, fragmentation, commodification, deprecation, and dispossession" (p. 213). If it was important to defy such forces in the decades just past—during which academia saw the rise of arts-based inquiry in many (if not all) fields of human studies, it is even more important today, in the face of retrogressive conservatism, the loss of basic rights to free communication, and the commodification of the educational enterprise to capitalism.

▣ THE PERFORMATIVE TURN: ARTS-BASED RESEARCH AS A REVOLUTIONARY, CRITICAL, AND AESTHETIC PEDAGOGY

"Postmodern democracy cannot succeed," Denzin (2008a) argues, "unless critical qualitative scholars are able to adopt methodologies that transcend the limitations and constraints of a lingering, politically and racially conservative postpositivism" (p. x). He implores researchers (educational researchers in particular) to break the links that chain critical, qualitative inquiry to No Child Left Behind and similar politically and racially conservative postpositive frameworks for curriculum and pedagogy (Denzin, 2008a). For even longer, Denzin (1999) has urged a new movement in qualitative inquiry in which researchers "take up their pens" (and their cameras, paintbrushes, bodies, and voices) "so that we might conduct our own ground-level guerilla warfare against the oppressive structures of our everyday lives" (pp. 568, 572). This is the *performative turn* in qualitative research.

Resistance is a kind of performance that holds up for critique hegemonic texts that have become privileged stories told and retold. Performances "critique dominant cultural assumptions, to construct identity, and to attain political agency" (Garoian, 1999, p. 2). *Performativity* is the writing and rewriting of meanings to create a dynamic and open dialogue that continually disrupts the authority of meta-narratives. As Eisner (2001a)

writes, artists "invent fresh ways to show us aspects of the world we had not noticed; they release us from the stupor of the familiar" (p. 136). With reference to performance artist Suzanne Lacy, Charles Garoian (1999) observed that performance art opens a liminal and ephemeral space in which a community can engage in critical discourse. The community aspects of Lacy's work are accomplished by the involvement of diverse communities of participants as experts and actors examining their own oppression, where expertise is defined by participants' lives in the community. The participants in her work are co-researchers, critiquing and challenging themselves to understand their community and to overcome cultural oppressions that occur there. Thus, art, politics, pedagogy, and inquiry are brought together in performance.

Denzin (2008a), Charles Garoian and Yvonne Gaudelius (2008), and others defend the potential for *performance pedagogy* in a post–9/11 world to transform "everyday lives" by exposing and critiquing neoconservative/neoliberal constraints on human dignity and social justice (Denzin, 2008a, p. x). In performance, the emphasis is on *doing* (see Dewey, 1934/1958; also Giroux, 2001; Grossberg, 1996). As Denzin says, within a performance studies paradigm "inquiry is a form of activism . . . that inspires and empowers persons to act on their utopian impulses" (2008a, p. x). Resistance performances are creative constructions that "can strengthen the capacity of research groups to implement qualitative research as a solution to public health, social welfare, and education problems" (Denzin, 2008a, p. x).

Contextualized by the September 11 attacks and war in Iraq, Garoian and Gaudelius (2008) claim mass-media visual images can be characterized as "spectacle pedagogy" (p. 24) and can open spaces for participatory democracy. Specifically, strategies of collage, montage, assemblage, installation, and performance art are based in reflection and critique and thereby present pedagogical means by which artists/researchers/teachers/students can involve themselves meaningfully in understanding and responding to the complex political, social, and ethical ideologies conveyed in mass media (p. 37). As involved learners, spectacle critics (Garoian and Gaudelius focus on "students") act as cultural citizens and participants in the political processes of democratic social justice. Garoian and Gaudelius write,

> We characterize the spectacle pedagogy of visual culture in two opposing ways: First, as a ubiquitous form of representation, which constitutes the pedagogical objectives of mass mediated culture and corporate capitalism to manufacture our desires and determine our choices: the second, as a democratic form of practice that enables a critical examination of visual cultural codes and ideologies to resist social injustice. As the former spectacle pedagogy functions as an insidious, ever-present form of propaganda in the service of cultural imperialism, the latter represents critical citizenship, which aspires to cultural democracy. (2008, p. 24)

One of the creations of spectacle pedagogy of the first order—propaganda—is the myth that visual renditions in news, advertising, and other forms of mass media convey "truths." The historical context in which Garoian and Gaudelius situate their analysis of spectacle pedagogy accentuates the pervasiveness of visual cultural codes entrenched in capitalistic and colonial neocon political agendas. Thus, "critical performance pedagogy reflexively critiques those cultural practices that reproduce oppression" (Denzin, 2008a). Denzin writes, "Critical performance pedagogy moves from the global to the local, the political to the personal, the pedagogical to the performative" (2008b, p. 62).

Denzin writes, "In ethnodrama and radical pedagogy audience is transformed out of a consumer/consumption/entertainment space to a dialogical structure, a collaborative pedagogical assemblage, to use Garorian's term—a part of 'spectacle pedagogy'" (Denzin, personal correspondence, October 29, 2009). The performance itself creates an open text in which "meanings emerge within the sociology of space and are connected within the reciprocal relationships that exist between people and the political, dynamic qualities of place" (S. Finley, 2003a, p. 288; 2005, p. 689). Thus, performance creates specialized (open and dialogic) space that is simultaneously asserted for inquiry and expression. In this liminal space, distinctions are made between private and public spheres, thereby rendering personal identity, culture, and social order unstable, indeterminate, inchoate, and amenable to change. Giroux (1995) notes, "It is within the tensions between what might be called the trauma of identity formation and the demands of public life that cultural work is both theorized and made performative" (p. 5, cited in Garoian, 1999, pp. 40–41).

From within the liminal openings that are created by the performance/practice of arts-based inquiry, ordinary people, researchers as participants and as audiences can imagine new visions of dignity, care, democracy, and other decolonizing ways of being in the world. Once it has been imagined, it can be acted upon, or performed. In tracing the evolution of performance as a primary site for revolutionary research methodology, Denzin (2003) explained,

> Ethnography had to be taken out of a purely methodological framework and located first within a performative arena and then within the spaces of pedagogy, where it was understood that the pedagogical is always political. We can now see that interpretive ethnography's subject matter is set by a dialectical pedagogy. This pedagogy connects oppressors and the oppressed in capital's liminal, epiphanic spaces. (p. 31; for a more comprehensive discussion of the "dramaturgical turn," see Denzin, 1997, 2003; Garoian, 1999)

Gregory Ulmer (1994) similarly argued for a revolutionary pedagogy that makes its task the transformation of institutions by using the formalizing structures of the institution itself to experimentally rearrange reality for critical effect. He cited Umberto Eco (1984, p. 409) to make his case for engaging in "revolutionary" interventionist works that entertain the possibility, as in an ideal "guerilla" semiotics of "changing the circumstances by virtue of which the receivers choose their own codes of reading . . . This pragmatic energy of semiotic consciousness shows how a descriptive discipline can also be an active project" (Ulmer, 1994, p. 86; see also, Ulmer, 1989).

▣ Arts-Based Inquiry at
the Heart of a People's Pedagogy

In arts-based research within the paradigm of revolutionary pedagogy, the artfulness to be found in everyday living composes the aesthetic (Barone, 2001a; Barone & Eisner, 1997; Dewey, 1934/1958; Tolstoy, 1946/1996). For research to act locally, in its use of everyday, localized, and personal language, and in its reliance on texts that are ambiguous and open to interpretation, arts-based research draws audiences into dialogue and opens the possibility for critical critique of social structures (Barone, 2001a, 2001b). Denzin (2000) and others have encouraged artist-researchers to focus on the vernacular and to capture the visceral ephemeral moments in daily life in their representations of research. Communicating the "ordinary extraordinary" (Dissanayake, 1997) through vernacular expressions in the context of mass media popular culture—radio, television, film—does more than introduce dialogues that "automatically contain, constrain, or even liberate us," writes Joli Jensen (2002, p. 198). "Instead these cultural forms are part of an ongoing, humanly constructed conversation about the reality we are shaping as we participate in it" (p. 198). Thus, vernacular, expressive, and contextualized language forms open narratives that promote empathy and care (Barone, 2001b), and entreaties to the vernacular are encouraged as a means to more inclusive audience/participant voices in research representations.

Education in the arts is wrapped up in social privilege. I have previously noted my willingness to "hold open the possibility that the unschooled minds of untrained artists can construct and express ideas through the media of the arts. . . . I believe there is every possibility that the vernacular street performances of poetry, tagging, and fire dance are potentially meaningful" experiences of inquiry (S. Finley, 2003a, p. 292). If we define arts education through the informal venues of streets and communities, rather than through institutional delivery systems, arts-based researchers can perform a people's pedagogy.

Barone (2001a) observed that in arts-based inquiry representational media are "selected for their usefulness in recasting the contents of experience into a form with the potential for challenging (sometimes deeply held) beliefs and values" (p. 26). Connectivity among the forces of political resistance, pedagogy, interpretive performance, and arts-based methodological approaches crystallizes a way of understanding that is at once aesthetic and conducive to interpreting social structures and inspiring transformational action.

Arts-based research makes use of affective experiences, senses, and emotions. Its practitioners explore the bounds of space and place where the human body is a tool for gathering and exploring meaning in experience. Carl Bagley and Mary Beth Cancienne (2002) created a salon experience for artist-researchers to engage as a community of learners in the exploration of ways to use emotive, affective experiences, multiple senses, and bodies in coordination with intellect as ways of responding to the world. *Dancing the Data* (Bagley & Cancienne, 2002) is a compilation of performance research, accompanied by a CD-ROM where the researchers perform their interpretive work in a community setting. Denzin demonstrates the power of performance texts with his challenge to mythic, hegemonic texts and idealized views of Native American women and men (see *Searching for Yellowstone,* 2008a). Using family photographs, reminiscences from his own childhood, and descriptions of his family vacation retreat, he situates personal experiences within the complexities of political, social, and ethical ideologies conveyed in media and art. "Finding myself embedded in these representations" (p. 16), Denzin locates these stories in the current historical circumstances of race and gender "in a search for more realistic utopias, more just and more radically democratic social worlds for the twenty-first century" (p. 17). In constructing this text, Denzin uses strategies of assemblage that Garoian associates with spectacle pedagogy. Photo montage, personal journaling, script, poetry, and art criticism fold into each other and meld to create an experimental text that resists expectations for academic work. As a research text, it transforms the way that we perceive, read, write, and perform data and, in turn, challenges racial stereotypes with newly constructed representations of whiteness and of race. This movement from personal to global-political epitomizes performance texts.

Further examples of the merger of arts-based inquiry with the field of performance pedagogy include, for example, reader's theater—Robert Donmoyer and June Yennie-Donmoyer (2008; see also 1995, 1998), ethno-drama—Jim Mienczakowski (2000; also see, Mienczakowski, Smith, & Sinclair, 1996) and Saldaña (2008; also see Saldaña, Finley, & Finley, 2005).

Examples that follow of arts-based research oriented toward educational praxis or "a people's pedagogy" come largely from my own work. As an arts-based researcher, I seek opportunities to locate work within local communities and subscribe to participants' everyday language and vernacular in discourse. Arts-based inquiry that is locally situated facilitates individual and communal reflection by diverse communities of participants.

In general, its practitioners seek to understand and take action against the oppressive forces of politically conservative post-positivism that dominate and constrain the lives of ordinary citizens. In my work, I intend to create aesthetic spaces in which to experience transformational performances at a visceral level. I am, like Marcelo Diversi and Cláudio Moreira (2009), "reassured by the Rortyan notion that we are all stuck in a perpetual discussion about what the reality of oppression means to each and all of us" (p. 184). The potential exists for arts-based research to enact inquiry in the social world as one feature of a people's pedagogy (S. Finley, 2003b, 2005). Emancipation from colonizing human research that objectifies its participants (casting them as subjects) is not possible unless research is democratized and brought under the control of people in their daily lives.

Academic institutions—university classrooms, academic journals, and professional conferences—are one contested site for a people's pedagogy to be enacted. Other work is more appropriately taken to the streets and gathering places that bring people together in everyday life. In this genre of arts-based research, the researcher attempts to involve people as experts of their own lives and to create forums for outreach to venues outside the academy (Woo, 2008).

Journals such as *Qualitative Inquiry*[3] and *Cultural Studies<=>Critical Methodologies* initiated unique spaces for arts-based inquiry that challenges the academic status quo (and several other journals have stretched to be inclusive of arts-based work, despite rather fierce opposition in academic circles). A recent special issue of *Cultural Studies<=>Critical Methodologies* (Diversi & S. Finley, 2010) devoted to "critical homelessness" incorporates prose and poetry variously written by street youth, homeless activists, and residents of Dignity Village—a tent city in Portland, Oregon. Two of the included manuscripts are e-mail messages from homeless individuals reprinted in the pages of the journal.

Taken holistically, the special issue challenges the norms of academic research publication by including street authors— and crediting them for their own work as any journal would credit the work of an academic contributor. Poetry contributors were enlisted from my arts-based street research projects with homeless communities. To magnify the contrast between public citizens being included as authors of their own work in their own voices and being quoted as representative voices to illuminate points of view articulated by an academic voice, the journal also includes an article I have written about the experiences of a group of female street youth who participated in a poetry cooperative (S. Finley, 2010).

Through the structure of the co-op, female street youth shared their poetry and reflections on street life with each other and with me. Most of the exchanges of writing were accomplished through e-mail, which accommodated the transience of the youth participants. The group emerged from relationships

I had developed with the participants while otherwise investigating the lives of street youth. The idea behind this emergent phase of the overall project was to shape a version of "street education" in which all participants were, simultaneously, teachers and students—I would teach literacy skills and the youth would teach me about street life. I also had a larger purpose in mind to use the experience of arts-based inquiry to challenge these female youths to exercise their leadership potential and restructure their street life to confront male domination among street youths. In this way, "experiences are being investigated narratively, including inquiry experiences. They become curricular experiences for the inquirer—and possibly also for the audience, if the experiential narratives are read or listened to by others" (Conle, 2003, p. 4).

Poetry is included here too, but it is inserted in the pages of *my* article, framed by me and in service to my subjective purposes. In this more traditional form of academic discourse, my voice dominates, even in an article that is inclusive of youth commentary. In each of these two examples from the special issue on critical homelessness, the purpose for doing the research and the purpose behind the assemblage of the contributions to form a holistic artful representation was to create aesthetic spaces for "resistance performances" to challenge beliefs and values and to encourage transformational action.

Performances of arts-based research for audiences outside the community (of participants in the research or others who share the community experience) often draw on empathetic understanding to "move" the audience to action—or at least to reflective contemplation of the roles of oppressors and the oppressed. At one level, the audience to critical research performances responds to sensory stimuli—sound, color, movement, and their composition. Empathy provides another inroad to understanding through artful performances. "We seek out the arts in order to take a ride on the wings that arts provide," Eisner wrote (2008, p. 3). Planning for the audience who will experience the aesthetic will further shape the researcher's approach to representation (Woo, 2008).

Street Rat (M. Finley, 2000; adapted to theater by Saldaña, Finley, & Finley, 2005) was constructed from research recordings and transcripts and then staged as performances on Bourbon Street in New Orleans for a varied audience of street youths, tourists, business people—anyone who would stop to listen (M. Finley, 2003, p. 603). "The performances generated continuing dialogues with the youths who were featured in the poems, and they opened up new dialogues about homelessness and street life with tourists, business people, and other observers," Finley wrote, "What developed was a cyclic process of dialogue, poetic responses to dialogic performance, and [continuing] dialogue" (M. Finley, 2003, p. 603). This was performance pedagogy for the audiences of participants and (casual) observers of the participants in their daily lives—with the potential for challenging some of the derogatory stereotypes the observers

brought to their interactions with and understandings of the lives of street youths.

As part of the *At Home At School* (AHAS)[4] program, I designed a research project in which 20 youths (whose experiences were formed by poverty, homelessness, and living in foster care, and included several Deaf participants) were to engage in a new form of dialogic performance with *Street Rat*. As I proposed the project, the youth would read the poetry text (with the guidance of their theater teacher) and through their discussions of the text, participants would begin to substitute their own storied poems for those in the text to create a new and updated version of the poem/play that would include snippets of the original juxtaposed with new writings. Instead, the participants read the poetry together, discussed their responses to it, and put it aside to write their own script for a play entitled *All I Ask: A Look Into the Hardships Modern Teenagers Face* (AHAS, Not At-Risk Theatre Company, 2009). As Carola Conle (2003) observes, "Arts-based researchers use artistic means both to prompt inquiry and to represent their findings" (p. 10). Julia Colyar (2009) similarly discusses qualitative research writing as being "product, process, form of invention, and instrument of self-discovery" (p. 421). "Writing is product and process, noun and verb" (p. 423). Colyar (p. 424) quotes Denzin and Lincoln (2000): "'Fieldwork and writing blur into one another. There is, in the final analysis, no difference between writing and fieldwork'" (Denzin & Lincoln, 2000, p. 16). "Writing *is* inquiry. Writing is a kind of data collection" (Colyar, p. 424; also see Furman, 2006). For the AHAS Theatre Company, the duality of process and product, of inquiry and representation, manifested as self-reflective inquiry, reflection in community (and rewrite) and, finally, (re)presentation to a larger audience for the purpose of initiating discussion (the troupe performed their original work for a group of 150 youths enrolled in one of the AHAS summer programs).

Street Rat demonstrated to the theater group the potential of arts-based research as an approach to inquiry. Reading the poetry (often aloud to one another) served as a catalyst for the youths to initiate their own artistic processes. Their dialogue about their reading empowered them to action and encouraged them to take poetic "license" and to command their own arts-based research project. The theater troupe members then wrote and performed their own play—with themes that they chose to communicate their life experiences. They also changed their group name from the "AHAS Theatre Troupe" to the "Not at Risk Theatre Company." Their chosen purpose was to demonstrate to others that each individual in their group had faced tremendous systemic barriers, but that they individually and collectively regarded themselves as "not at risk." Their performance of their collective story ended with a group discussion with the actor/authors and the larger group of program participants (the audience comprised youths who were engaged in a variety of arts-inquiry projects that included mural painting and building a "green" and aesthetic outdoor classroom).

With their audience, the writer-performers explored the concepts of "at risk" and other labels that demean individual students and reinforce the systemic barriers experienced in educational settings. Through writing, they "named their reality" and created a product to communicate meaning about themselves (see Osterman & Kottkamp, 1993; see also Colyar, 2009). In the ensuing dialogue with the audience they further problematized the "at risk" label and engaged in a community dialogue about how they would live a future "not at risk." "It is the narrative repertoire of our imagination that helps us distinguish the world we live in from the world we want to live in" says Conle (2003, p. 4). Performance pedagogy enacted through the methods of writing and theater transformed the "everyday lives" of these youth through a process of exposing and critiquing insults to their personal dignity. In performance, they sought social justice. Inquiry took the form of activism and empowered the writers to act on their utopian impulses to change the way they were being defined by society. They created a resistance performance that implemented qualitative research as a solution to their real–world educational problems. For these youth, arts-based research was an active means by which to surpass oppressive social structures and rewrite their futures to include their personal dreams, desires, and goals as active citizens in a participatory democracy.

A good and productive discussion emanating from an artistic experience is only one step toward social change and breakthrough challenges to stereotyping. More powerful yet is that the youths have all continued with their educational pursuits (including reenrolling in school and seeking alternative educational options) and that they have continued to work as a group through the school year to improve their acting skills by engaging in workshops on body movement, expression, and so on, and one of their members created a new script performed by the Not at Risk Theatre Company.

A new kind of research pedagogy is needed in the context of diminishing democracy. Through arts-based research practices, citizen-scholars can employ the skills of the artist in creative roles such as shaping civic life so as to expand its democratic possibilities for all groups. The skills of the arts-based researcher include but are not limited to the manipulations of media or even the exercise of imagination, but also include preparedness to "directly confront the threat from fundamentalists of all varieties. . . . " To contest workplace inequalities, imagine democratically organized forms of work, and identify and challenge those injustices that contradict and undercut the most fundamental principles of freedom, equality, and respect for all people who make up the global public sphere (Giroux & Giroux, 2009, p. 29).

It is time to reaffirm the socially responsible political purpose of arts-based inquiry. Arts-based inquiry is a strategic means for political resistance to neo-cultural politics. It is a form of cultural resistance and a way to create a critical and dialogic space in which to engage in a struggle over the control of knowledge and the domination of discourse. The time is here to "perform revolutionary pedagogy" through an arts-based approach to inquiry that is socially responsible, locally useful, engaged in public criticism, and resistant to neoconservative discourses that threaten social justice and close down efforts toward a performative research ethics that facilitates critical race, indigenous, queer, feminist, and border studies.

Passion for a political cause or for individual people may be a better guide to creating quality arts-based inquiry than is preparatory education in the arts. For instance, Woo (2008) suggests important qualities of an artist-researcher include "an open mind, a tolerance for criticism, and willingness to learn [art-making techniques]" (p. 326). When the purpose of research is to provoke, to motivate, or to make meaning from experiences, it can be used to advance a progressive political agenda.

> One objective the arts-based researcher can serve is to provide tools and opportunity for participants to perform inquiry, reflect on their performances, and preserve, create, and rewrite culture in dynamic indigenous spaces. Thus in critical arts-based inquiry, the location of research changes from the isolated sanctuaries of the laboratory and constructed and bounded environments to places where people meet, including schools, homeless shelters, and neighborhoods. Socially responsible research for and by "the people" cannot reside inside the lonely walls of academic institutions. (S. Finley, 2008, p. 73–74)

In practice of a people's pedagogy research can become a tool for advancing critical race theory and opening space for an aesthetic of artist-researchers and participant-observers belonging to oppressed groups and individuals traditionally excluded from research locations.

In other contexts, I have hashed through the expert-quality issue for arts-based researchers in some detail. I am not ready to embrace a requisite of expert training in art as a condition of producing quality arts-based inquiry (as defined by its potential for audience engagement and response, Knowles & Cole, 2008, p. 67). I wouldn't want it to be taken that I do not favor intensive education in the arts. (The AHAS program features a curriculum that includes arts as separate disciplines as well as arts integration, and my 13-year-old daughter is currently a student at the public Vancouver School for Arts and Academics.) Indeed, I would strongly prefer that all children engage in deep aesthetic experiences of education. With reference to the youth-authored *All I Ask* script, I believe the participants benefited greatly from the theater exercises, reading of scripts, and discussions led by their theater teacher, Anne Averre. Although their script was a dynamic force for portraying lives and generating discussion in a local audience of peers, I am quite sure it would not receive awards for writing or quality of performance. It is not a "masterpiece" but it is an imaginative and visceral performance of critical discourse that pushes against tradition, hegemony, and oppression.

◨ CONCLUSION

The *Program for the Sixth International Congress of Qualitative Inquiry* (2010) continues the discussion about quality in arts-based research with a plenary session "On Rigor in Arts-Based Research." The question is not closed: How do we determine standards for both the processes and the products of arts-based inquiry? Are the expectations for critical arts-based inquiry different from other approaches to arts-and-research?

I have asked (S. Finley, 2003a): How do I assign grades to arts-based research? How do I determine which articles and proposals to recommend when I review my peers for publications and presentations of their arts-based research? Writing in *Qualitative Inquiry* (S. Finley, 2003a), I proposed a rubric to use in assessing arts-based inquiry; in the SAGE *Handbook of Qualitative Research,* third edition (S. Finley, 2005), I juxtaposed six traits of activist art with seven foundations that define arts-based research and seven bullets that form the framework of revolutionary pedagogy. Each of these movements is foundational to critical arts-based research, and each has its own expectations and standards. But I have to ask, Is "rigor" an appropriate term for contemplating arts-based research? Is the search for rigor in arts-based research another indication of neoconservative pedagogy working to reproduce the standards-laden language of education since 9/11?

Definitions and synonyms for *rigor* stand in stark contradiction with methodologies that are emergent, inclusive, and culturally responsive. Instead, rigor is likened to qualities of being unyielding or inflexible, austere and rigid, leading through fatality to the extremes of *rigor mortis* (S. Finley, 2007). For me, then, rigor is precluded by the key epistemological and ethical basis for using arts and research. Methodological rigor doesn't bring me any closer to understanding, How do we break through the complex barriers of colonial social conformity to an inclusive, pluralist aesthetic situated in the lives of others?

A people's pedagogy to replace neoconservative pedagogy and its constructs should define what is "good" in critical arts-based research. So I return to my initial rubric and refine it within the sociohistorical construct of neoconservatism.

What follows is a rubric for evaluating critical-arts based research in furtherance of a people's pedagogy and in opposition to post–9/11 neoconservative values.

- Does the research demonstrate indigenous or culturally relevant skills and practices?
- Does the research openly resist cultural dominance and demonstrations of meta-narratives of race, history, politics, and power?

- Are the researchers performing a useful, local, community service? Could the research be harmful in any way to the community of its participants?
- Who speaks? Are participants engaged in a process that uses the advantages of pluralism such as cacophony, bricolage, collage, and performance?
- Does the research defy limitations set by the hegemony of neoconservative research discourse?
- Is the research a performance of passionate and visceral communion?
- How likely are readers/viewers and participants to be moved to some kind of positive social action?

◨ NOTES

1. See Zakaras and Lowell (2008). This RAND report commissioned by the Wallace Foundation describes the fading optimism about public arts education of the 1960s and 1970s as the 1990s and beyond have ushered deep and continuing spending reductions in arts education (p. xiii). The focus of the report is determining conditions by which to "cultivate the capacity of individuals to have engaging experiences with works of art" (p. 14).

2. *Pornomiseria* describes voyeuristic, exploitive documentary-style films that depict poverty and human suffering in Latin American countries as a form of entertainment. See Michele Faguet, "Pornomiseria: or How Not to Make a Documentary Film," in *Afterall, 21,* Summer 2009, pp. 5–15.

3. For a review of the role *Qualitative Inquiry* has played in advancing arts-based inquiry, see S. Finley, 2003a).

4. For information about At Home At School programs, see http://AtHomeAtSchool.org

◨ REFERENCES

Alexander, B. K. (2005). Performance ethnography: The reenacting and inciting of culture. In N. K. Denzin & Y. S. Lincoln (Eds.), *The SAGE handbook of qualitative research* (3rd. ed., pp. 411–441). Thousand Oaks, CA: Sage.

American Educational Research Association. (2006). Standards for reporting on empirical social science research in AERA publications. *Educational Researcher, 35*(6), 33–40.

American Educational Research Association. (2009). Standards for reporting on humanities-oriented research in AERA publications. *Educational Researcher, 38*(6), 481–486.

At Home At School (AHAS) Not-At-Risk Theatre Company. (2009, August). *All I ask: A look into the hardships modern teenagers face.* [drama]. Unpublished script, At Home At School Program, Washington State University, Vancouver, WA.

Bagley, C. (2008). Educational ethnography as performance art: Towards a sensuous feeling and knowing. *Qualitative Research, 8,* 53–72.

Bagley, C., & Cancienne, M. B. (Eds.) (2002). *Dancing the data.* New York: Peter Lang.

Barone, T. (2001a). Science, art, and the predispositions of educational researchers. *Educational Researcher, 30*(7), 24–28.

Barone, T. (2001b). *Teaching eternity: The enduring outcomes of teaching.* New York: Columbia University, Teachers College Press.

Barone, T. (2008). How arts-based research can change minds. In M. Cahnmann-Taylor & R. Siegesmund (Eds.), *Arts-based research in education* (pp. 28–49). New York: Routledge.

Barone, T., & Eisner, E. (1997). Arts-based educational research. Section II of *Complementary Methods for Research in Education* (pp. 75–116, 2nd ed., R. M. Jaeger, Ed.). Washington, DC: American Educational Research Association.

Benabib, R., & Salsberg, M. (2009, July 20). *Weeds: Where the sidewalk ends.* [Television broadcast]. Showtime.

Blumenfeld-Jones, D. (2008). Dance, choreography, and social science research. In J. G. Knowles & A. L. Cole (Eds.), *Handbook of arts in qualitative research* (pp. 175–184). Thousand Oaks, CA: Sage.

Cahnmann-Taylor, M., & Siegesmund, R. (2008). *Arts-based research in education: Foundations for practice.* New York: Routledge.

Cannella, G. S., & Lincoln, Y. S. (2004a). Dangerous discourses II: Comprehending and countering the redeployment discourses (and resources) in the generation of liberatory inquiry. *Qualitative Inquiry, 10,* 165–174.

Cannella, G. S., & Lincoln, Y. S. (2004b). Epilogue: Claiming a critical public social science—Reconceptualizing and redeploying research. *Qualitative Inquiry, 10,* 298–309.

Casey, K. (1995). The new narrative research in education. *Review of Research in Education, 21,* 211–253.

Colyar, J. (2009). Becoming writing, becoming writers. *Qualitative Inquiry, 15*(2), 421–436.

Conle, C. (2003). An anatomy of narrative curricula. *Educational Researcher, 32*(3), 3–15.

Conquergood, D. (1988). Beyond the text: Toward a performance cultural politics. In S. J. Dailey (Ed.), *The future of performance studies: Visions and revisions* (pp. 25–36). Washington, DC: National Communication Association.

Coulter, C. A., & Smith, M. L. (2009). The construction zone: Literary elements in narrative research. *Educational Researcher, 38*(8), 577–590.

Denzin, N. K. (1997). Performance texts. In W. G. Tierney & Y. S. Lincoln (Eds.), *Representation and the text: Re-framing the narrative voice* (pp. 179–217). Albany: SUNY Press.

Denzin, N. K. (1999). Two-stepping in the 90s. *Qualitative Inquiry, 5,* 568–572.

Denzin, N. K. (2000). Aesthetics and the practices of qualitative inquiry. *Qualitative Inquiry, 6,* 256–265.

Denzin, N. K. (2003). *Performance ethnography: Critical pedagogy and the politics of culture.* Thousand Oaks, CA: Sage.

Denzin, N. K. (2004). *The First International Congress of Qualitative Inquiry.* Available at http://www.icqi.org/

Denzin, N. K. (2008a). *Searching for Yellowstone: Race, gender, family and memory in the postmodern west.* Walnut Creek, CA: Left Coast Press.

Denzin, N. K. (2008b). A critical performance pedagogy that matters. In J. A. Sandlin, B. D. Schultz, & J. Burdick. *Handbook of public pedagogy* (pp. 56–70). Thousand Oaks, CA: Sage.

Denzin, N. K., & Lincoln, Y. S. (2000). The discipline and practice of qualitative research. In N. K. Denzin & Y. S. Lincoln (Eds.), *Handbook of qualitative research* (2nd ed., pp. 1–28). Thousand Oaks, CA: Sage.

Denzin, N. K., & Lincoln, Y. S. (Eds.). (2005). *The SAGE handbook of qualitative research* (3rd ed.). Thousand Oaks, CA: Sage.

Dewey, J. (1958). *Art as experience.* New York: Capricorn. (Original work published in 1934)

Dissanayake, E. (1988). *What is art for?* Seattle: University of Washington Press.

Diversi, M., & Finley, S. (2010). Special issue on critical homelessness. *Cultural Studies<=>Critical Methodologies, 10*(1).

Diversi, M., & Moreira, C. (2009). *Betweener talk: Decolonizing knowledge production, pedagogy, and praxis.* Walnut Creek, CA: Left Coast Press.

Donmoyer, R., & Yennie-Donmoyer, J. (1995). Data as drama: Reflections on the use of readers' theater as a mode of qualitative data display. *Qualitative Inquiry, 20*(1), 74–83.

Donmoyer, R., & Yennie-Donmoyer, J. (1998). Reader's theater and educational research—Give me a for-instance: A commentary on Womentalkin'. *Qualitative Studies in Education, 11*(3), 397–402.

Donmoyer, R., & Yennie-Donmoyer, J. (2008). Readers' theater as a data display strategy. In J. G. Knowles & A. L. Cole (Eds.), *Handbook of arts in qualitative research* (pp. 209–224). Thousand Oaks, CA: Sage.

Eco, U. (1984). *La structure absente: Introduction a la reserche' semioteque* (U. Esposito-Torrigiani, Trans.). Paris: Mercured de France.

Eisner, E. (1998). *The enlightened eye: Qualitative inquiry and the enhancement of educational practice.* Upper Saddle River, NJ: Prentice Hall. (Original work published in 1991)

Eisner, E. (2001a). Concerns and aspirations for qualitative research in the new millennium. *Qualitative Research, 1,* 135–145.

Eisner, E. (2001b). Should we create new aims for art education? *National Art Education Association, 54*(5), 6–10.

Eisner, E. (2008). Arts and knowledge. In J. G. Knowles & A. L. Cole (Eds.), *Handbook of arts in qualitative research* (pp. 3–12). Thousand Oaks, CA: Sage.

Faguet, M. (2009). Pornomiseria: Or how not to make a documentary film. *Afterall, 21*(Summer), 5–15.

Finley, M. (2000). *Street rat.* Detroit: University of Detroit Press.

Finley, M. (2003). Fugue of the street rat: Writing research poetry. *Qualitative Studies in Education, 16*(4), 603–604.

Finley, S. (2000). "Dream child": The role of poetic dialogue in homeless research. *Qualitative Inquiry, 6,* 432–434.

Finley, S. (2001). Painting life histories. *Journal of Curriculum Theorizing, 17*(2), 13–26.

Finley S. (2003a). Arts-based inquiry in QI: Seven years from crisis to guerrilla warfare. *Qualitative Inquiry, 9,* 281–296.

Finley S. (2003b). The faces of dignity: Rethinking the politics of homelessness and poverty in America. *Qualitative Studies in Education, 16,* 509–531.

Finley, S. (2005). Arts-based inquiry: Performing revolutionary pedagogy. In N. K. Denzin & Y. S. Lincoln (Eds.), *The SAGE handbook of qualitative research* (3rd ed., pp. 681–694). Thousand Oaks, CA: Sage.

Finley, S. (2007). *Methodological rigor: Intellectual rigor mortis?* Paper presented at International Congress of Qualitative Inquiry, Urbana-Champaign, IL.

Finley, S. (2008). Arts-based research. In J. G. Knowles & A. L. Cole (Eds.), *Handbook of arts in qualitative research* (pp. 71–81). Thousand Oaks, CA: Sage.

Finley, S. (2010). "Freedom's just another word for nothin' left to lose": The power of poetry for young, nomadic women of the streets. *Cultural Studies<=>Critical Methodologies, 10,* 58–63.

Finley, S., & Finley, M. (1998). *Traveling through the cracks: Homeless youth speak out.* Paper presented at the American Educational Research Association, San Diego, CA.

Finley, S., & Finley, M. (1999). Sp'ange: A research story. *Qualitative Inquiry, 5,* 313–337.

Foucault, M. (1984). The order of discourse. In M. Shapiro (Ed.), *Language and politics* (pp. 108–138). London: Blackwell.

Furman, R. (2006). Poetic forms and structures in qualitative health research. *Qualitative Health Research, 16*(4), 560–566.

Garoian, C. R. (1999). *Performing pedagogy: Toward an art of politics.* Albany: SUNY Press.

Garoian, C. R., & Gaudelius, Y. M. (2008). *Spectacle pedagogy: Arts, politics, and visual culture.* Albany: SUNY Press.

Geertz, C. (1988). *Works and lives.* Cambridge, UK: Polity Press.

Giroux, H. A. (1995). Borderline artists, cultural workers, and the crisis of democracy. In C. Becker (Ed.), *The artist in society: Rights, rules, and responsibilities* (pp. 4–14). Chicago: New Art Examiner.

Giroux, H. A. (2001). Cultural studies as performative politics. *Cultural Studies<=>Critical Methodologies, 1,* 5–23.

Giroux, H. A. (2009). Democracy's nemesis: The rise of the corporate university. *Cultural Studies<=>Critical Methodologies, 9,* 669–695.

Giroux, H. A., & Giroux, S. S. (2006). Challenging neoliberalism's new world order: The promise of critical pedagogy. *Cultural Studies<=>Critical Methodologies, 6,* 21–32.

Giroux, H. A., & Giroux, S. S. (2008, December). Beyond bailouts: On the politics of education after neoliberalism. *Truthout.* Retrieved February 11, 2010, from http://www.truthout.org/123108A

Giroux, H. A., & Polychroniou, C. (2008, February). The scourge of global neoliberalism and the need to reclaim democracy. Retrieved January 19, 2011, from http://onlinejournal.com/artman/publish/article_2959.shtml

Grossberg, L. (1996). Toward a genealogy of the state of cultural studies. In C. Nelson & D. P. Gaonkar (Eds.), *Disciplinarity and dissent in cultural studies* (pp. 87–107). New York: Routledge.

Guba, E. (1967). The expanding concept of research. *Theory Into Practice, 6*(2), 57–65.

Gurevitch, Z. (2002). Writing through: The poetics of transfiguration. *Cultural Studies<=>Critical Methodologies, 2*(3), 403–413.

Gutmann, A., & Thompson, D. (2004). *Why deliberative democracy?* Princeton, NJ: Princeton University Press.

Hammersley, M. (1992). *What's wrong with ethnography?* London: Routledge.

Hanley, M. S., & Noblit, G. W. (2009). *Cultural responsiveness, racial identity and academic success: A review of the literature.* Pittsburgh, PA: Heinz Endowments.

Harper, D. (1987). *Working knowledge: Skill and community in a small shop.* Berkeley: University of California Press.

Howe, K. R. (2009). Positivist dogmas, rhetoric, and the education science question. *Educational Researcher, 38*(6), 428–440.

Jensen, J. (2002). *Is art good for us? Beliefs about high culture in American life.* Lanham, MD: Rowman & Littlefield.

Knowles, J. G., & Cole, A. L. (Eds.). (2008). *Handbook of the arts in qualitative research: Perspectives, methodologies, examples, and issues.* Thousand Oaks, CA: Sage.

Leaños, J. J., & Villarreal, A. J. (2007). Art education. In D. Gabbard (Ed.), *Knowledge and power in the global economy: The effects of school reform in a neoliberal/neoconservative age.* Available at http://www.leanos.net/Arts%20Education.html

Lincoln, Y. S. (1995). Emerging criteria for quality in qualitative and interpretive research. *Qualitative Inquiry, 1,* 275–289.

Lincoln, Y. S. (2005). Institutional review boards and methodological conservatism: The challenge to and from phenomenological paradigms. In N. K. Denzin & Y. S. Lincoln (Eds.), *The SAGE handbook of qualitative research* (3rd ed., pp. 165–181). Thousand Oaks, CA: Sage.

Lincoln, Y. S., & Cannella, G. S. (2004a). Dangerous discourses: Methodological conservatism and governmental regimes of truth. *Qualitative Inquiry, 10,* 5–14.

Lincoln, Y. S., & Cannella, G. S. (2004b). Qualitative research, power, and the radical right. *Qualitative Inquiry, 10,* 175–201.

Mayer, R. E. (2000). What is the place of science in education research? *Educational Researcher, 29*(6), 38–39.

Mienczakowski, J. (2000). Ethnodrama: Performed research—limitations and potential. In P. Atkinson, S. Delamont, & A. Coffey (Eds.), *Handbook of ethnography* (pp. 468–476). Thousand Oaks, CA: Sage.

Mienczakowski, J., Smith, R., & Sinclair, M. (1996). On the road to catharsis: A theoretical framework for change. *Qualitative Inquiry, 2*(4), 439–462.

National Research Council. (2002). *Scientific research in education.* Washington, DC: National Academy Press.

Osterman, K. F., & Kottkamp, R. B. (1993). *Reflective practice for educators: Improving schooling through professional development.* Newbury Park, CA: Corwin Press.

Program of the Sixth International Congress of Qualitative Inquiry (2010). University of Illinois at Urbana-Champaign. Retrieved January 20, 2011, from http://www.icqi.org/

Saldaña, J. (2008). Ethnodrama and ethnotheatre. In J. G. Knowles & A. L. Cole (Eds.), *Handbook of arts in qualitative research* (pp. 195–207). Thousand Oaks, CA: Sage.

Saldaña, J., Finley, S., & Finley, M. (2005). Street rat. In J. Saldaña (Ed.), *Ethnodrama: An anthology of reality theatre* (pp. 139–179). Walnut Creek, CA: AltaMira Press.

Seale, C. (1999). Quality in arts-based research. *Qualitative Inquiry, 5,* 465–478.

Seidel, K. (2001). Many issues, few answers—The role of research in K–12 arts education. *Arts Education Policy Review, 103*(2), 19–22.

Siegesmund, R. (1998). Why do we teach art today? Conceptions of art education and their justification. *Studies in Art Education, 39*(3), 197–214.

Sinner, A., Leggo, C., Irwin, R. L., Gouzouasis, P., & Grauer, K. (2006). Arts-based educational research dissertations: Reviewing the practices of new scholars. *Canadian Journal of Education, 29*(4), 1223–1270.

Smith, L. T. (1999). *Decolonizing methodologies: Research and indigenous peoples.* Dunedin, New Zealand: University of Otago Press.

Sullivan, G. (2005). *Art practice as research: Inquiry in the visual arts.* Thousand Oaks, CA: Sage.

Tharp, T. (2003). *The creative habit: Learn it and use it for life.* New York: Simon & Schuster.

Tolstoy, L. (1996). *What is art?* (A. Maude, Trans.). New York: Penguin. (Original work published in 1946)

Ulmer, G. (1989). *Teletheory.* New York: Routledge.

Ulmer, G. (1994). The heretics of deconstruction. In P. Brunette & D. Wills (Eds.), *Deconstruction and the visual arts: Art, media, architecture* (pp. 80–96). New York: Cambridge University Press.

Woo, Y. Y. J. (2008). Engaging new audiences: Translating research into popular media. *Educational Researcher, 37*(6), 321–329.

Young, I. M. (2004). *Inclusion and democracy.* New York: Oxford University Press.

Zakaras, L., & Lowell, J. F. (2008). *Cultivating demand for the arts: Arts learning, arts engagement, and state arts policy.* Santa Monica, CA: RAND Corporation.

27

ORAL HISTORY

Linda Shopes

▣ WHAT ORAL HISTORY IS, AND ISN'T

Oral history is a protean term: Within common parlance, it can refer to recorded speech of any kind or to talking about the past in ways ranging from casual reminiscing among family members, neighbors, or coworkers to ritualized accounts presented in formal settings by culturally sanctioned tradition-bearers. Most typically, the term refers to what folklorists call personal experience narratives—that is, orally transmitted, autobiographical stories crafted to communicate meaning or what is valued to others (Dolby, 1989). Oral history in this mode is exemplified most notably by the work of Studs Terkel (1967, 1970, 1974, 1984), whose multiple volumes have done much to popularize the term, and more recently, of David Isay (2007), whose StoryCorps project has been rekindling interest in the storied quality of everyday life. Typically, the term registers a certain democratic or populist meaning; *oral history* implies a recognition of the heroics of everyday life, a celebration of the quotidian, an appeal to the visceral.[1]

Among practitioners, however, oral history has a more precise meaning. The Oral History Association (2010) defines oral history as "a way of collecting and interpreting human memories to foster knowledge and human dignity." Donald Ritchie (2003), in his guide, *Doing Oral History,* describes it as "collect[ing] memories and personal commentaries of historical significance through recorded interviews." He continues, "An oral history interview generally consists of a well-prepared interviewer questioning an interviewee and recording their exchange in audio or video format. Recordings of the interview are transcribed, summarized, or indexed and then placed in a library or archives. These interviews may be used for research or excerpted in a publication, radio or video documentary, museum exhibition, dramatization, or other form of public presentation" (p. 19). Valerie Yow (2005), in her *Recording Oral History,* states, "Oral history is the recording of personal testimony delivered in

oral form." Distinguishing this practice from memoir, she notes that in oral history, "There is someone else involved who frames the topic and inspires the narrator to begin the act of remembering, jogs memory, and records and presents the narrator's words." Recognizing that various terms are used to describe this same activity, she concludes, "*Oral history* seems to be the [term] most frequently used to refer to the recorded in–depth interview" (pp. 3–4).

These definitions suggest six characteristics of oral history as a professional, disciplined practice. It is, *first,* an interview, an exchange between someone who asks questions, that is, the interviewer, and someone who answers them, referred to as the interviewee, narrator, or informant. It is not simply someone telling a story; it is someone telling a story in response to the queries of another; it is this dialogue that shapes the interview. Moreover, oral history generally involves only these two people. Although oral historians will occasionally conduct group interviews, these are generally done as preparation for or follow up to an individual interview. Oral historians value the intimacy of a one-on-one exchange. *Second,* oral history is recorded, preserved for the record, and made accessible to others for a variety of uses. Ritchie (2003) goes so far as to say, "An interview becomes an oral history only when it has been recorded, processed in some way, made available in an archive, library, or other repository, or reproduced in relatively verbatim form for publication. Availability for general research, reinterpretation, and verification defines oral history" (p. 24). These two primary characteristics of oral history suggest that it is properly understood as both process (that is the act of interviewing) and product (that is, the record that results from that interview).

Third, oral history interviewing is historical in intent; that is, it seeks new knowledge about and insights into the past through an individual biography. Although it always represents an interplay between past and present, the individual and the social, oral history is grounded in historical questions and hence requires that the interviewer has knowledge of both the subject at hand and the interviewee's relationship to that subject. *Fourth,* oral history is understood as both an act of memory and an inherently subjective account of the past. Interviews record what an interviewer draws out, what the interviewee remembers, what he or she chooses to tell, and how he or she understands what happened, not the unmediated "facts" of what happened in the past. An interview, therefore, renders an interpretation of the past that itself requires interpretation. *Fifth,* an oral history interview is an inquiry in depth. It is not a casual or serendipitous conversation but a planned and scheduled, serious and searching exchange, one that seeks a detailed, expansive, and reflective account of the past. Although framed by a broad set of questions or areas of inquiry, an oral history interview admits a high degree of flexibility, allowing the narrator to speak about what he or she wishes, as he or she wishes. Finally, oral history is fundamentally oral, reflecting both the conventions and dynamics of the spoken word. This may seem self–evident, but decades of relying on transcripts, which can never fully represent what was said, have obscured this fact. Only with the widespread adoption of digital technology are oral historians beginning to engage seriously with the orality of oral history.

Oral history generally distinguishes between life history and topical interviews: Life history interviews, often undertaken within local or community settings, record a narrator's biography, addressing topics such as family life; educational and work experiences; social, political, and religious involvements; and, at their best, the relationship of personal history to broader historical events and social themes. Typically, life history interviews aim at recording everyday life within a particular setting. Topical interviews, often done as part of a larger research project, focus on specific elements of an individual's biography, for example, participation in the U.S. civil rights movement, a topic well documented by oral history. In practice, many interviews include both life history and topical elements; lives, after all, are not easily compartmentalized.

Whether a life history or topical interview or some combination of the two, the best interviews have a measured, thinking-out-loud quality, as perceptive questions work and rework a particular topic, encouraging the narrator to remember details, seeking to clarify what is muddled, making connections among seemingly disparate recollections, challenging contradictions, evoking assessments. The best interviewers listen carefully between the lines of what is being said to discern what the narrator is trying to get at and have the confidence to ask the hard questions. Yet all interviews are shaped by the context within which they are conducted, as well as the particular interpersonal dynamic between narrator and interviewer: An interview can be a history lecture, a confessional, a verbal sparring match, an exercise in nostalgia, a moral tale, or any other of the ways people talk about their experiences.

Although the act of interviewing lies at the heart of oral history, best practices define the oral history process as considerably more extensive (Larson, 2006; MacKay, 2007; Ritchie, 2003;

Yow, 2005). The interview is preceded by careful preparation, including defining the focus of the inquiry, conducing background research in secondary and primary sources, developing skills in interview methods and in using recording technology, identifying and making contact with the narrator, cultivating rapport, conducting a preinterview, and developing an interview outline. An interview is then followed by a number of steps designed to facilitate preservation and access, including securing permission for others to use the interview by means of what is termed a *legal release;* making one or more copies of the original recording; placing these in a secure, publicly accessible repository; cataloguing or developing a finding aid for the interview; and developing a means of accessing what has been recorded without listening to the entire interview, by either transcribing or summarizing it, or, more recently, developing online search methods. If the interview is part of a larger project or program, additional considerations come into play, including project planning and design, management and staffing, office space, work flow, budget and funding, and the development of products or outcomes.

Oral history is thus distinguished from other kinds of interviewing. Its open-ended, subjective, historically inflected approach is quite unlike the highly structured opinion polls and surveys of current attitudes and behaviors conducted by sociologists, political scientists, and market researchers. Similarly, it is unlike interviews conducted by many journalists and documentary workers, who seek quotations to fit the story they are developing today rather than let the narrator define the plot of his or her own story for the historical record. (This is not to deny, however, that the line can be blurry; some journalists and documentarians are excellent oral historians, though they may not refer to themselves as such [Coles, 1997].) Oral history also differs from interviews done in a clinical or therapeutic setting. Although both are conducted in depth and recognize intersubjectivity and personal biography—and notwithstanding that an oral history interview often has a salutary effect on both narrator and interviewer—clinical interviewing posits dysfunction and seeks to help a person resolve personal problems, sometimes, as in narrative psychology (Bruner, 1990; Polkinghorne, 1988; Spence, 1982), by reframing the person's story. Oral history, however, does not seek to change the narrator; it proceeds from the assumption that the narrator has been an active agent in fashioning his or her life and life story.

Oral historians are perhaps most closely allied with anthropologists and qualitative sociologists in their approach to interviewing; all, in Clifford Geertz's (1974) resonate phrase, seek "the native's point of view." Oral historians, especially those interviewing individuals who share a particular social setting, will often engage in the anthropologist's practice of participant observation; and anthropologists and sociologists, though generally focusing on the ethnographic present, do recognize at times a historical dimension to the topic at hand (Atkinson, Coffey, & Delamont, 2003; di Leonardo, 1987; Mintz, 1979;

Silverman, 1997; Vansina, 1985). Oral historians also share certain approaches and practices with folklorists: Although folklorists focus on the formal and aesthetic qualities of traditional narratives, they and oral historians record firsthand accounts as part of the collective record of a culture; and within contemporary practice, both approach oral materials as subjective texts, constructions of language and mind, whose meaning demands a level of decoding (Abrahams, 1981; Davis, 1988; Jackson, 2007; Joyner, 1979).

Although oral history differs from the methods and purposes of other kinds of interviewing and allies most comfortably with the assumptions and intentions of history, it can be deeply interdisciplinary in the ways it seeks to understand interviews. Oral historians have looked to psychology for understanding the emotional undercurrents of an interview; to communications for the structure and dynamics of the interview exchange; to folklore and literary studies for the storied quality of interviews; to anthropology for the culture clash that often occurs as two different *mentalities* collide within the narrative; to cultural studies and critical race and gender studies for ways the social position of both narrator and interviewer underlies what is—and is not—said; to performance studies for the presentational quality of interviews; and to gerontology for understanding the way the imperatives of aging shape an interview. Indeed, much of the most creative thinking about oral history comes from practitioners trained and working in fields other than history.

This essay has thus far presented oral history in its own ethnographic present, for it has advanced a broad description of a practice that in fact has not been static but has evolved over several decades. Useful for laying out some generally—although not universally—agreed upon characteristics of oral history, for setting some boundaries, and for helping fix this indeed very protean term, this discussion has nonetheless stripped oral history of its own historical development. Thus, subsequent sections will discuss the development of oral history over time as both a method of research and mode of understanding the past (Gluck, 1999; Grele, 2007; Thomson, 2007). Collectively, they will address changes in practice, linking them to broader changes in the academy and within society, consider the politics of oral history as an intellectual and social practice, and outline oral history's institutional development. The chapter will conclude with a discussion of legal and ethical issues in oral history, including its problematic relationship with institutional review boards.

▣ EARLY DEVELOPMENTS: ORAL HISTORY
AS AN ARCHIVAL PRACTICE

Historians have long used oral sources for their work, either conducting interviews of their own or drawing on firsthand accounts recorded and preserved by others (Sharpless, 2006). No less than the ancient historian Thucydides interviewed participants for his

history of the Peloponnesian War, observing that "different eyewitnesses give different accounts of the same events, speaking out of partiality for one side or the other or else from imperfect memories" (Ritchie, 2003, p. 20). Accounts of Aztec and Inca life recorded by Spanish chroniclers in the 16th century and of Mexican and American settlers in California recorded by Hubert Howe Bancroft and his assistants remain valuable sources for historians today. Similarly, Henry Mayhew's inquiry into the living and working conditions of London's working classes in the mid-19th century is only the first in a long line of investigations that have relied heavily on evidence obtained by talking with the subjects of the inquiry; these social studies have both goaded reform and informed scholarly history.

Nonetheless, reliance on oral sources fell into disfavor during the late 19th and much of the 20th centuries, as the practice of history became increasingly professionalized and as positivism became the reigning academic paradigm. The German historian Leopold von Ranke's dictum that the goal of history was to recount "how it really was" (*wie es eigentlich gewesen)* described a form of scholarship that increasingly relied on the (paper) documentary record, or as C.-V. Langlois and Charles Seignobos, two French historians, put it, "There is no substitute for documents: no documents, no history" (Thompson, 1988, p. 51). Reliance on what had often been an informal practice of talking with people thus became suspect. Indeed, early efforts to record firsthand accounts of the past were often idiosyncratic or extemporaneous affairs, conducted according to methods that were more or less rigorous in any given case and with no intention of developing a permanent archival collection. Furthermore, the absence of mechanical—or digital—recording devices necessitated reliance on human note-takers, raising questions about accuracy and reliability.

Dating oral history's beginnings in the United States is a quixotic exercise at best. Some reckon its origins in the Depression-era Federal Writers Project (FWP), which recorded thousands of life histories with individuals from various regional, occupational, and ethnic groups during the late 1930s and early 1940s (Hirsch, 2006, 2007). The best known of the FWP interviews are the slave narratives, accounts by elderly men and women who had experienced slavery firsthand. Rediscovered by scholars in the 1970s, these narratives have become important sources for a reorientation of the historiography of American slavery from one that views slaves primarily as victims to one that recognizes the active agency of enslaved persons within a system of bondage (Blassingame, 1972; Genovese, 1974; Rawick, 1972). But what about the interviews James McGregor conducted in 1940 with survivors of the 1890 Wounded Knee Massacre? Or the interviews done by Bancroft and his associates?

Nonetheless, the Oral History Research Office (OHRO) at Columbia University, established by Columbia historian Allan Nevins in the late 1940s, is generally acknowledged as the first oral history program in the United States, a distinction likely related to OHRO's prominence in institutionalizing and professionalizing oral history in its modern incarnation (Starr, 1984). Recognizing that the bureaucratization of public affairs was tending to standardize the paper trail and that the telephone was replacing personal correspondence, Nevins came up then with the idea of conducting interviews with participants in recent history to supplement the written record. He wrote of the need "for obtaining a little of the immense mass of information about the more recent American past—the past of the last half century—which might come fresh and direct from men once prominent in politics, in business, in the professions, and in other fields; information that every obituary column shows to be perishing" (Starr, 1984, p. 8). It took a decade for this idea to reach fruition: Nevins and his amanuensis—for these early interviews were recorded in longhand—conducted their first interview in 1948 with New York civic leader George McAneny.

Several universities soon followed Columbia's lead and established their own oral history programs: the University of Texas in 1952, the University of California at Berkeley in 1954, and the University of California at Los Angeles and the University of Michigan in 1959. The Harry S. Truman Library and Museum inaugurated its oral history project in 1961, interviewing Truman's family, friends, and associates, thus initiating the practice of oral history at presidential libraries. Columbia's 1965 annual report listed some 89 projects nationwide, fostered partly by the development of recording technologies.[2] By the mid-1960s, oral history was well enough established to form the Oral History Association (OHA), founded in 1967. After publishing its annual proceedings for five years, in 1973 OHA began publishing an annual journal, the *Oral History Review* ; in 1987 the *Review* became a biannual publication. Recognizing the need to codify standards for oral history, it developed the first iteration of the current *Principles and Best Practices for Oral History* (2009) in 1968. The document is generally regarded as defining the parameters of best practice.[3]

Unlike previous interviewing initiatives, these early oral history programs were distinguished by both their permanence and their systematic and disciplined approach to interviewing. Staff and affiliates developed projects that included a number of interviews on a single topic and were designed to fill in gaps in the extant record. These were explicitly archival: The point was to record on tape, preserve, and make available for future research recollections deemed of historical significance. Archival exigencies have thus defined what have been generally understood as fundamental features of oral history and have been codified in established best practices. Two merit particular attention. First is the matter of releases: Because an interview is understood as a creative work, it is subject to the laws of copyright; and these laws deem the interviewee, as "author" of the interview, to be the owner of the copyright. It is by means of the legal release form that the interviewee signs over or "releases" to the sponsoring institution—or individual researcher or the repository that accepts completed interviews—rights to the interview; and, if the interviewee chooses,

sets certain limits to access. This is analogous to the deed of gift form by which archives typically acquire materials from donors, and indeed, some oral history interviews are transferred to an archive by a deed of gift (Neuenschwander, 2009). The legal status of the interviewer is unclear, but in practice he or she is often considered a cocreator of the interviewer and hence cosigner of the release.

Second is the matter of transcription: Transcribing interviews, that is rendering recorded speech in writing, has long been accepted as an essential part of the oral history process, on the assumption that a transcript will increase access considerably. Given that archives and the scholars who use them have historically been document driven, this assumption is understandable: Paper, unlike audio or visual media, is a familiar and comfortable form; it's easier and faster to scan a paper document than listen to or view an interview; words fixed on paper ensure accuracy of quotation in print; and they confer a certain intellectual authority on what could be construed as an ephemeral form. For years, oral historians accepted the transcript as the primary document of an oral history interview, despite its inevitable distortions, and early on, some programs destroyed or reused audiotapes (Allen, 1982; Baum, 1977; Mazé, 2006; Samuel, 1971). Only recently has the general consensus shifted away from the transcript and toward the recorded interview—the *oral* narrative—as the primary document; with the development of digital media has come a growing interest in supplementing—or supplanting—the transcript with digital access, topics that will be taken up later in this chapter.

Best practice also dictates that transcribed interviews be returned to the narrator for correction, amplification, and emendation, to obtain the fullest, most accurate account. This practice, coupled with the need for releases, can pit the rights and privileges of the narrator against the imperatives of scholarship. Law and custom give the narrator enormous control over the presentation of his or her story; and when, as is often the case, the narrator is someone who otherwise has little control over the circumstances of his or her life, this is certainly just. Still, a narrator can place restrictions on the interview by means of the release and can delete significant but unflattering, embarrassing, even incriminating information from the transcript, to the impoverishment of the historical record.

Because these early oral historians had been schooled in the Rankian document- and fact-based historiography of the times, they considered interviews to be a means of creating new facts that would lead to a more complete account of the past. The interviewee was viewed as a storehouse of information about "what actually happened"; the interviewer, a neutral presence who simply recorded these facts; and the interview, a document to be assessed like any other source for its reliability and verifiability. Michael Frisch (1990a, p. 160) has referred to this as the "more history" approach to oral history, "reducing [it] to simply another kind of evidence to be pushed through the historian's controlling mill." Because oral history was something of a maverick practice, dismissed by most historians as unreliable hearsay, a source of anecdote or color but little else, one finds a certain defensiveness among early practitioners and a strenuous effort to articulate systematic means of assuring and assessing the validity, reliability, and representativeness of interviews (Moss, 1977).

▣ SOCIAL HISTORY AND THE DEMOCRATIZATION OF ORAL HISTORY

The social movements and intellectual upheavals of the 1960s, 1970s, and beyond had enormous impact on oral history and, more modestly, vice versa. Who was interviewed, who interviewed them, what they were interviewed about, and the purpose of interviews all experienced significant shifts in these decades, not so much replacing the earlier, archival approach as building on it or occurring on a parallel track. Whereas early oral history programs, in line with the dominant historiography of the postwar era, had tended to interview the "elite"—that is, leaders in business, industry, and politics as well as distinguished individuals in the professions, the arts, and related fields—by the 1970s, oral history's scope widened considerably, in response to scholars' growing interest in the experiences of non-elites, ordinary people—anonymous Americans, as they were sometimes termed. As social history—that is, the history of social relationships among generally unequal and often competing groups—became the dominant historiographic paradigm, oral history became an essential tool for recovering the experiences of those to whom historians were now turning their attention. As Ronald Grele (2007, p. 12) has written, "The objective was to document the lives and past actions of classes of people heretofore ignored by historians; in particular the working class, but also racial and ethnic minorities, women, and sexual and political minorities. These are people whose lives were traditionally ignored or purposefully forgotten: people whose history [had been] . . . understood by examining documents provided by those who were outsiders to the communities under study, upper class commentators for the most part, but also journalists, social and other service workers, or anyone who had left a written record." Interviewers thus began asking about everyday life in working-class communities, about the differing experiences of women and men within families, about ways minorities created purposeful lives within deeply constraining circumstances. Interviewers began to ask not only "What happened?" but also "What was it like?" "What did you do about . . . ?" "How do you understand . . . ?"

At the same time, oral history became a practice carried out less exclusively for archival purposes and increasingly by individual scholars conducting interviews for their own research projects. In some cases, scholarly interests catalyzed the development of ongoing, multifaceted oral history programs, such as the Southern Oral History Program at the University of North Carolina at Chapel Hill, where scholarly research and archival collection building have gone hand in hand since 1973. Yet some scholar-interviewers, operating outside of an institutional oral history archive, did not always adhere to established standards with the same rigor as the pioneering oral history projects. Some were unwilling to pursue topics that lay outside their immediate interests, thereby limiting interviews' usefulness to others; some, less concerned about the future use of interviews and often with fewer resources than ongoing projects, failed to secure release forms, or to transcribe interviews, or even to place them in public archives.[4] The last especially is of concern, for it violates historians' professional commitment to open access to sources.

These shortcomings notwithstanding, oral history has played an important role in democratizing our collective understanding of the past; interviews have added new knowledge about previously excluded or underdocumented groups and have restored voice and agency to those whom the extant record has often objectified. To cite only one example, John Bodnar's *The Transplanted* (1985), its title deliberately playing off Oscar Handlin's *The Uprooted* (1951) and its interpretation deeply informed by the biographical narratives of dozens of interviewees, represented Eastern and Southern European immigrants to the United States not as disoriented and "uprooted," anomic individuals, unable to gain a footing in the new world, but as men and women actively deploying creative strategies to fashion a new life as transplants to that world. Collectively, interviews conducted within the social history paradigm have challenged dominant, top-down narratives of the past and addressed the relationship between subordination and agency. More recently, this kind of oral history has occurred within the context of ethnic and queer studies, as scholars probe notions of identity, break long-held silences, and broaden our understanding of "who counts" in history.

Oral history has democratized not only the historical record but also the practice of history, involving people outside the academy and established archives as producers and interpreters of their own history in addition to serving as interviewees. Increasingly during the 1970s and on into the present, local organizations and groups—historical societies, museums, and libraries and also churches, unions, senior centers, and other grassroots groups—have carried out oral history projects to document their own history, often developing performances, exhibitions, media productions, and other creative work to extend the reach of the interviews. It is probably accurate to state that since the mid-1970s, at least as many oral history projects have been located outside the academy as within; as early as 1973, a directory listing oral history centers in the United States located half of them outside of a college or university setting (Starr, 1984, p. 12).

Still, scholars have often become involved in these projects as organizers, workshop leaders, consultants, collaborators, and interviewers in a self-conscious effort to engage with communities, often those they themselves are studying. In recent years, academic involvement in community oral history projects has taken place under the rubric of "civic engagement" or "public history," as students and faculty work with community collaborators to document and present aspects of the local past. Frisch (1990b) has written about oral and public history's capacity for sharing authority between interviewer and interviewee, creating opportunities for a "profound sharing of knowledges, an implicit and sometimes explicit dialogue from very different vantages about the shape, meaning, and implications of history" (p. xxii). Academically trained oral historians working "in public" and their local partners frequently struggle with the implications of such sharing, as they confront differences between scholarly understandings of history as an interpretive activity and vernacular notions of storytelling; between a scholar's interest in a critical approach to the past and a community's self-interest in promoting a positive image or in avoiding unsavory aspects of its past; between academic languages and styles of work and less formalized practices. These differences, difficult as they are to negotiate, point to larger social differences, in class and race, in generation, in education, in social and political views; and in that context are often resolved only uneasily, or partially (Diaz & Russell, 1999; Lewis, Waller, & Hinsdale, 1995; Shopes, 1986, 2002a).

In a practice that is less about community history and more about advocacy, some have connected oral history with broader humanitarian and civic concerns. Oral historians have, for example, been involved in reminiscence work with older adults, engaging in the integrative process of life review (Bornat, 1993; Butler, 1963) and in "truth telling projects," designed to reconcile former antagonists (Lundy & McGovern, 2006; Minkley & Rassool, 1998). They have developed projects that both document and support redress for human rights abuses such as the internment of Japanese Americans during World War II (Densho, 2010; Dubrow, 2008) or more recently, the lengthy detention without trial of suspected Muslim terrorists in the United States (Shiekh, 2010). Oral historians have conducted interviews with displaced or homeless individuals as a means of stimulating and informing a broader activist agenda (Kerr, 2008), used the local knowledge gained in interviews to inform development projects around the globe (Cross & Barker, 2006; Slim & Thompson, 1993), and connected oral history with work for social change in numerous other ways.

Whether occurring inside or outside the academy or somewhere in between, oral history in this democratic mode has often been grounded not only in an interest in a more expansive sense of what counts as history and who counts as historians, but also in a progressive politics, an interest in using history to inform and at times to intervene in movements for equality and justice. As in other fields, the most forceful voice for a politically engaged oral history has often been feminist scholar/activists. Buoyed by the energy of the women's movement, they argued early on that "women's oral history was not merely *about* women.

It was *by* and *for* women, as well" (Gluck, 2006, p. 360). Recognizing that the personal is political, interviewers were eager to discover the female experience. Given the nature of that experience, women, it was argued, brought an especial empathy to interviewing "their sisters." At times, the goal was as much about using oral history as a means of consciousness-raising, empowerment, and change as it was about generating new knowledge; indeed, flushed with the excitement of opening the new field of women's history, interviewers tended to take what narrators said at face value—not, it must be said, an entirely inappropriate response for people who had too often been historically silent, or silenced (Anderson & Jack, 1991; Bloom, 1977; Gluck, 1977; Oakley, 1981).

Practice and reflection challenged these rather naïve formulations (Armitage, 1983; Armitage & Gluck, 1998; Gluck & Patai, 1991): Documentation of "the female experience" often failed to account for social and ideological differences (Geiger, 1990). Empathy could be a manipulative ploy, resulting in unguarded revelations that were later regretted and unrealistic expectations for a continuing relationship with the interviewer (Stacey, 1991). Efforts to raise someone's consciousness could become a patronizing refusal to hear another's point of view. And failure to subject narrator accounts to critical analysis reflected what Frisch (1990a) has referred to as "antihistory"—a counterpoint to the "more history" approach noted earlier—by which he means viewing "oral historical evidence because of its immediacy and emotional resonance, as something almost beyond interpretation or accountability, as a direct window on the feelings and ... [hence] on the meaning of past experience" (p. 160).

Recognizing that oral historians often share a broad sympathy with the people they interview and the intimacy that can develop in an interview, Yow (1997) has cautioned against "liking interviewees too much." Doing so can create an interview that is collusive, rather than searching: Hesitations, contradictions, and silences are not probed. Deeply painful memories are dismissed with ameliorative words. Cues to information that might shake the interviewer's positive view of a narrator are ignored. Challenging questions are not asked out of deference or to avoid an uncomfortable breach in mutual regard. To counter this tendency, Yow advocates a critical reflexivity, managing one's own emotional reactions to the narrator, challenging one's interests and ideological biases, thinking beyond the questions one intends to ask and developing alternative lines of inquiry.

Although not a concern of feminists exclusively, feminist oral historians were among the first to consider power relationships within the practice of oral history and, with colleagues addressing issues of race and ethnicity, have remained among the most sharply attentive to them (Coles, 1997; Gluck & Patai, 1991). How, they have asked, do knowledge production in oral history and the uses to which that knowledge is put reproduce unequal social relationship? Although doubts about "studying down" have somewhat attenuated over the years, it is still relevant to ask how the assumptions, questions, language, nonverbal cues, and modes of presentation of a relatively privileged interviewer can constrain a less privileged narrator. If scholars build careers based on recording and presenting life stories of others, however carefully and conscientiously they have pursued their work, it is still appropriate to ask how they might share some of the tangible rewards of that work with narrators and their communities. If oral history's much vaunted capacity to "give voice" assumes, naively, that narrators need the oral historian to find their voices, it is still useful to consider how, in the words of oral historian Alessandro Portelli (1997a, p. 69), we might responsibly "amplify their voices" within the public arena. If, as practice often demonstrates, a rough equality can exist within the bounded space of an interview, if indeed interviewees can retain the balance of power by deciding what to say and what to withhold, it is still useful to consider how one deploys power in presenting and interpreting the lives of those who have freely and in good faith shared them with us. Oral historians (James, 2000a; Kerr, 2003; Rouverol, 2003; Sitzia, 2003) continue to confront these questions, negotiating uneasy compromises with narrators, alternating their own interpretive voice with narrators' voices; sometimes letting narrators have their say with little comment sometimes deploying interviews within the context of their own narrative.

As a result of the range and depth of work in oral history since the 1960s, it has attained broad academic acceptance and the credibility it lacked in earlier years. Of course, the evidentiary value of oral history still had its critics. Historian Louise Tilly (1985, p. 41), with a bias toward quantitative evidence, referred to personal testimony, with its emphasis on the individual, as "ahistorical and unscientific."[5] Oral history has also shared in the critique of social history as overly concerned with the quotidian details of everyday life, celebratory of individual agency, and insufficiently attentive to ways the structures of power and relations of inequality constrain action. And local oral history work can be parochial and laced with nostalgia (Shopes, 1986, 2002a). Nonetheless, the dominance of social history during the 1970s and 1980s muted much of the earlier criticism, and we might summarize the two outstanding achievements of oral history in the democratic mode as restoring to our collective record of the past the voices of the historiographically—if not historically—silent and as providing a medium of exchange between the academy and communities.

Although the intent and topics of oral history interviews had shifted by the 1970s and the venues for practice broadened, as a source they were generally viewed much as they had always been, as transparent documents in the positivist tradition, purveyors of facts that were adjudged to be either true or false. Some oral historians, however, were gradually beginning to understand that something more was going on in an interview: that what a narrator said had something to do with the questions posed, the mental set of both narrator and interviewer, and the relationship between them; that narrators were telling stories, compressing years of living into a form that was shaped by language and culturally defined narrative conventions; that memory was not so much about the accuracy of an individual's recall and about how and why people remembered what they did; and that an interview was in many ways a performance, one that demanded our moral attention.

▣ FROM DOCUMENT TO TEXT: ORAL HISTORY'S MOVE TO INTERPRETIVE COMPLEXITY

Identifying a single turning point in the way oral history is understood is impossible; change has come from practitioners operating in a variety of intellectual contexts and has reflected broader theoretical currents. In the United States, Frisch (1990c) was perhaps the first to raise questions about the particular kind of historical evidence oral history provides in a review of Studs Terkel's *Hard Times appearing* in 1972. Unlike many reviewers, who lionized the book as the pure voice of the people, Frisch found the stories of individual failure and collective survival troubling, leading him to ask, "At what distance, in what ways, for what reasons, and in what patterns do people generalize, explain, and interpret experience? What cultural and historical categories do individuals use to help understand and present a view of experience?" (p. 11). By opening up these sorts of questions, Frisch suggested "oral history . . . encourages us to stand somewhat outside of cultural forms in order to observe their workings. Thus it permits us to track the elusive beats of consciousness and culture in way impossible to do within" (p. 13). Grele (1991a) brought a similar sort of reflexivity to oral history in a number of essays first published in the 1970s. Among his many insights is the especially fruitful one that an interview is a conversational narrative that incorporates three sets of structures—linguistic, performative, and cognitive—and that an analysis of these structures tells us a good deal about what, in addition to the obvious communication of information, is going on in an interview. In what is perhaps the most cited article in the oral history literature, Portelli (1991) analyzed why oral accounts of the death of Italian steel worker Luigi Trastulli, who had been shot during a workers' rally protesting NATO in 1949, routinely got the date, place, and reason for his death wrong, placing it instead in 1953, during street fights following announcement of the firing of more than two thousand steel workers. He argued that narrators manipulated the facts of Trastulli's death to render it less senseless, more comprehensible, and politically meaningful to them, concluding that "errors, inventions, and myths lead us through and beyond facts to their meaning" (p. 2).

These three seminal works and others (Passerini, 1980, 1987; Tonkin, 1992) initiated a gradual shift in the way oral historians think about their work. They did not change the methods of oral history—who is interviewed, what they are interviewed about, or how interviews are used; these elements of oral history have remained broadly democratic. Rather, these works have led practitioners into more theoretical territory, to focus less on the content of what a narrator has said and more on the meaning embedded in or lying underneath the words. This approach to interviews—as opposed to interviewing—arose from close attention to the dialogic exchange that lies at the heart of an interview, as well as sustained engagement with the narratives generated. This approach also reflected broader intellectual trends of the last decades of the 20th century, including what has been termed "the linguistic turn" in scholarship, in which attention to the semiotic and the symbolic have come to challenge the positivist paradigm. An interpretive approach to oral history has been further stimulated by the growing internationalization of oral history, bringing U.S. oral historians into closer contact with the work of their more theoretically inclined continental colleagues. Beginning in 1979, oral historians from around the world have been meeting biennially under the aegis of what became formalized in 1989 as the International Oral History Association. Beginning in 1980, work presented at these meetings has been published in a series of journals and annual publications, including the influential *International Journal of Oral History* published from 1980 through 1990.

It is difficult to summarize what is a diverse, complex, sometimes dense literature, but at bottom is the notion that interviews are hermeneutic acts, situated in time. Meaning is conveyed through language, which in turn is shaped by memory, myth, and ideology and through nonverbal expression and gesture, which give both immediacy and emotional depth to the exchange and further command the listener's attention. Interviews thus offer clues into narrators' subjectivities, or more accurately, the play of subjectivities—the intersubjectivity—between narrator and interviewer. Understood in this way, interviews are not documents in the traditional sense, to be mined for facts, but texts, to be interpreted for ways narrators understand—and want others to understand—their lives, their place in history, the way history works.

These more theoretical approaches to oral history can perhaps be approached by considering several examples from both published work and actual interviews. Consider the dialogic nature of an interview, the way it is the product—or expression—of two people talking. Historian Thomas Dublin (1998) came to understand this quite pointedly as he was reviewing family photographs with a husband and wife he had interviewed previously about the decline of the anthracite coal mining industry in Pennsylvania: "I expressed surprise at seeing so many pictures taken on [Tommy's] hunting trips with his buddies. When I commented that I had not realized how important hunting had been in Tommy's life, he responded good-naturedly, 'Well, you never asked'" (p. 21). Eva McMahan (1989, 2006) spins out the meaning of "asking"—and answering—by proposing a Conversational Analytic Framework for understanding the way meaning is actively negotiated within the interview exchange. By looking closely at the way the conversation moves and the rules that govern it, McMahon argues, we are able to see how "the oral history interview interaction is constitutive" (2006, p. 348) of meaning, not simply a recording of facts. Her work opens up rich possibilities for a rigorous analysis of interview dynamics. More practically, it has informed a more self-conscious, disciplined approach to interviewing.

Because an interview is a communicative event, communication sometimes becomes difficult or breaks down, pointing to issues of cognitive and social dissonance. Julie Cruikshank (1990) and David Neufeld (2008) describe how their interviews with Native Alaskan and First Nations Canadian elders resulted in life stories that did not conform to Western notions of autobiography as a chronological, ego-centered narrative, but rather mixed personal history with mythic, highly metaphorical stories. For Cruikshank, the challenge was to negotiate cultural differences about what properly constitutes a life history; for Neufeld, it was integrating these parallel narratives into historical programs for Parks Canada. Daniel James (2000b) describes the way incongruent expectations, an aggrieved narrator seeking to tell his version of Peronism, and James's own instance on penetrating the narrator's obfuscations while withholding his own views, resulted in a frustrating exchange and what he viewed as an act of symbolic violence.

Sometimes, meaning can be construed from what is not said, from silences in an interview. Luisa Passerini (1980, 1987) demonstrated the way the absence of talk can be not the result of "never asking," but of broad cultural significance. Recording life histories of members of the Turin, Italy, working class, she found that narrators frequently made no mention of Fascism, whose repressive regime nonetheless inevitably affected their lives. Even when questioned directly, narrators tended to jump directly from fascism's rise in the 1920s to its demise in World War II, avoiding any discussion of the years of fascism's reign. Passerini interprets this as evidence on the one hand "of a scar, a violent annihilation of many years in human lives, a profound wound in daily experience" among a broad swath of the population and, on the other, of people's preoccupation with the events of everyday life—"jobs, marriage, children"—even in deeply disruptive circumstances (1980, p. 9).

Addressing the narrative qualities of oral history, Mary Chamberlain (2006) has assessed ways that an oral history interviewee (or narrator) represents experience through language, drawing on a vast and diverse cultural repertoire to describe, structure, and make sense of his or her lived experience in ways that are, of necessity, highly selective. Chronology (first this, then this) and causality (this → this), for example, often are used to structure oral history interviews in Western societies. Similarly, narrators frequently make themselves the hero (or antihero) of their own stories, which can be partly attributed to the fundamentally ego-centered nature of oral history, but also represents the modern valorization of the individual, of living purposefully, overcoming odds (or not), progressing (or not) through life, achieving resolution (or not).

Interpreting oral history as narrative means looking for underlying patterns of meaning within the interview. Karen Fields (1994), reflecting on interviews she conducted with her grandmother, has argued that what "Gram" was trying to communicate to her, through anecdotes, stories, and commentaries, was not so much knowledge about her life as a black woman in the Jim Crow South, but wisdom and counsel for how Fields herself could live honorably in the present. Similarly, Linda Shopes (2002b) has developed the notion of iconic stories—concrete, specific stories embedded within the interview that "stand for" or sum up something the narrator reckons of particular importance. Often these are presented as unique or totemic events in the person's history, even as they include tropes common in folklore or popular culture. Grele (1991b) has analyzed closely the contrasting "structures of consciousness" present in interviews with two Jewish immigrants to New York, identifying "the particular vision of history articulated in [each] interview" (p. 213). It's not the content of the interview that interests him, but the way what's said conveys the broader ideological bases of the narrators' understanding of their personal pasts.

Oral history narratives thus connect the individual and the social, drawing on culturally agreed upon (or disputed) mental sets and modes of expression to tell one's story. These sets and modes are themselves deeply embedded in the culture: Portelli (1997b) notes how men tell war stories, women hospital stories, in both cases connecting their lives with gendered social experiences. Writing about the 1921 race riot in Tulsa, Oklahoma, Scott Ellsworth (1982) coined the phrase "segregation of memory" to describe the opposing ways Blacks and Whites remembered this gruesome event, the result of their own racialized experience of it. Alistair Thomson (1990) uses the ambiguous term "composure" to suggest a more complicated relationship between self and society is articulated in an interview: "In one sense we 'compose' or construct memories using the public language and meaning of our culture. In another sense we 'compose' memories, . . . which gives us a feeling of composure. . . . an alignment of our past, present and future lives. . . . The link between the two senses of composure is that the apparently private process of composing safe memories is in fact very public. . . . We compose our memories so that they will fit with what is publicly acceptable or if we have been excluded from general public acceptance, we seek out particular publics which affirm our identities and the way we want to remember our lives" (p. 25).

Thomson's use of the term "memories" to refer to the construction of narrative in oral history suggests how deeply implicated memory is in oral history, in both an organic and social sense. Oral history records accounts about the past, but the recording takes place in the present; memory is the bridge between the two. In line with the interpretive turn, oral historians have become less defensive about the evidentiary value of these memories and, drawing on the work of psychologists, have come to recognize that narrators do misremember: They collapse events, skew chronology, forget, and get details wrong; they "remember" as firsthand experiences what others, in fact, have told them and recall as true that which is false because they wish it to be so. Moreover, concerns about the reliability and validity of individual memories have become less important in recent years as, following Frisch's and Portelli's work, oral historians have turned concerns about accuracy on their head, recognizing that memories, like narrative, are highly social, expressive of ways the present mediates a narrator's recollection of the past and are generated within ideological, often politically charged contexts. Kim Lacy Rogers's (2006) interviews with veterans of the U.S. civil rights movement decades after their years of peak activism reveal a pervasive sense of disappointment and grief, as the movement's promise of equality has been only partially realized and economic and social distress continues to plague their communities. Portelli's (2003) study of a Nazi massacre in Rome and Susana Kaiser's (2005) work on postmemories of the military dictatorship in Argentina demonstrate ways oral history serves as a counter-memory of events that official histories have erased, distorted, or manipulated in service of a false consensus.

Theories of performance, drawn from both folklore and communications studies, have also informed the interpretation of oral history narratives. An interview is, most obviously, a performance for the interviewer, in which the narrator presents himself or herself as much through embodied movements—gesture, facial expressions, and the like—as through actual words, a fact that supports the use of video in interviewing. As Samuel Schrager (1983) has argued, an interview is also a cultural performance that looks both backward as a narrator relates well-rehearsed accounts of the past, told and retold to create a certain version of events, and forward, as he or she self-consciously speaks through the interviewer to "history," to the audience of future users whom he or she wants to inform, persuade, inspire. Jeff Friedman (2003), Della Pollock (2005) and others (Bauman, 1986; Denzin, 2003) have further theorized oral history as a doubly charged performance: the narrative encounter itself operates in the charged or liminal space between two people focusing their careful attention on each other to create something of value. It also charges the listener, first the interviewer, but then all who receive the interview, to pay attention, to witness, and also to act in ways that respond to the teller's story, sometimes through actual acting, that is via a dramatic production scripted from interviews and acted before an audience, and sometimes through acting in the world, with a moral vision inspired by the stories one has heard.

Although the interpretive approach to interview texts has dominated the discussion of oral history in recent years, it must be acknowledged that it has not been fully embraced by all who conduct or use interviews. In fact, most continue to consider oral history in the traditional documentary sense as one source among many or to highlight voices that have previously been muted in our collective understanding of the past. Some are concerned that a focus on the subjective, textual nature of interviews will obviate the need to triangulate them with other sources and assess their veracity; others that oral history will become more self-referential rather than remain the intellectually and socially expansive practice it has become. Still others are uneasy that critical analyses of interview texts create scholarly products that objectify narrators, distancing them from their own words. These are among the many questions in the field that remain open.

◧ THE DIGITAL REVOLUTION

Like the invention of movable type in the 15th century, digital media are transforming the culture, changing the ways we record and receive information; our scholarly practices, patterns of social interaction, and leisure pursuits; and, as some argue, the very way our brains work. And, like those living in the 15th century, we don't know where the digital revolution will lead, how the changes it is setting in motion will affect everyday practices and modes of thought, as well as the global economic, social, and political landscape. Oral historians share in both the transformations and uncertainty of the digital revolution.

Undoubtedly, digital media are transforming the way interviews are recorded, preserved, and accessed. Digital is now the preferred format for recording audio interviews, and some consensus exists on recording and preservation standards. However, given the rapid development of relatively easy to use, inexpensive video recording devices, video interviews are rapidly becoming the norm, though standards are less well established. It has been said that widespread access to the tools of digital media is making everyone a documentarian, creating a certain elasticity in what properly constitutes oral history, raising legitimate questions about quality, and requiring archivists to think anew about what sorts of materials have the potential for lasting significance to warrant acceptance from donors. Migration of analog recordings to digital format and the lack of metadata standards for cataloguing digital interviews are also concerns for those overseeing archival collections.

Perhaps the most significant impact of new technology on oral history to date is the remarkable access the Internet provides to interview collections (Grele, 2007; Thomson, 2007). Interviews once languishing in archives, used only by the occasional researcher, now are widely accessed by students and the interested public as well as by the scholarly community. Though generally heralded as a positive development, opportunities for misuse, violation of copyright, and unwelcome exposure to a vast audience, always present to a degree in oral history, have increased exponentially with Internet access. Likewise, the ethics of placing online interviews recorded in the pre-digital era for which the narrator gave no explicit permission for such "future use" is a continuing concern.

However, new media's impact on oral history extends to matters that are far more than technical; it can be argued that digital technology is shifting the terrain on which oral history has been practiced for the past six decades. Fundamentally, by allowing direct access to the primary document—the recorded interview—new media offer opportunities to restore the oral and the kinesthetic to oral history, and hence the layers of meaning communicated by tone, volume, velocity, pauses, and other nonverbal elements of oral communication, as well as the performative elements of the speaking body. Although oral historians continue to look forward to the development of sophisticated voice recognition software that will automate transcribing, some are also suggesting that direct access to digital audio and video recordings and the continuing growth of online publication may obviate the need for transcribing at all. Furthermore, the sonic and visual qualities of oral history interviews are grounded in the senses of hearing and seeing, which in turn are linked to neurophysiological receptors that trigger emotion. In short, hearing and seeing oral history interviews create a more emotional response in the user than does reading transcripts—a fact that enhances oral history's cultural power, connects it more deeply to the imaginative realm of the humanities, and challenges traditional notions of history as rational, critical inquiry.

Currently, some of the more creative work in oral history lies in the development of digital tools by means of which "the audio-video materials themselves—not the transcribed text version—can be searched, browsed, accessed, studied, and selected for use at a high level of specificity" (Frisch, 2006, p. 103). The implication is not simply greater access within and across interviews, but a "post-documentary sensibility," that is, the displacement of "the authority of the mediating intelligence or documentary authorship . . . by a sharable, dialogic capacity to explore, select, order, and interpret" interview materials in an "ongoing, contextually contingent, fluid construction of meaning," (p. 113); in other words, a radical democratization of the ways oral history can be used. This nonauthoritative approach to using interviews can be further enhanced by emerging modes of user-driven indexing via tagging. Steven High and David Sworn (2009, pp. 2–3), themselves advocates of digital oral history, nonetheless recognize "indexing can also conflict with the basic ethos of oral historical research: far from giving voice to interviewees, indexing risks sundering and de-contextualizing their life stories, . . . occluding the anomalous and specific in favour of the cross-referentiality afforded by topics and themes that are common to all interviews." Again, this sort of misuse of oral history is not new to the digital era; digital tools simply magnify enormously possibilities for doing so.

In these ways, new media are transforming oral history from an archival and research practice to a presentational one. Increasingly interviews are being conducted not to create a formal archive or to inform a research project, but to form the basis of a website devoted to a specific topic. Often this involves collaboration among diverse partners across disciplines and institutions; equally often, allied practices such as digital storytelling include active citizen participation. Although these shifts can further democratize an already democratic practice, they also threaten the depth, range, and especially the critical cast of archival oral history, as interviewers interview with an ear to the sound bite and interviewees speak more guardedly, mindful that their words no longer enjoy the protection of archival gatekeepers. These concerns too are neither new nor unique to oral history; still, new media place them front and center of the craft, even as oral historians share in larger debates about democracy and authority in a digital environment.

▣   LEGAL AND ETHICAL ISSUES IN ORAL HISTORY

Whereas legal issues in oral history can be understood as state-sanctioned rules for specific elements of practice, ethics refers to a higher standard governing the right conduct of relationships within the broad context of an interview or project. In his definitive *A Guide to Oral History and the Law,* historian and attorney John A. Neuenschwander (2009) outlines key legal issues in oral history: release agreements, related to ownership of the interview and copyright; subpoenas and Freedom of Information Act requests compelling the release of interviews; defamation; and privacy. All have important implications for oral history, but the two issues most commonly encountered are copyright, discussed earlier as a *sine qua non* of archival oral history, and defamation, defined as "a false statement of fact printed or broadcast about a person which tends to injure that person's interest" (p. 32). Insofar as "one who repeats or otherwise republishes defamatory matter is subject to the liability as if he had originally published it" (p. 33), any oral history project or program that makes available an interview that includes defamatory material is equally liable as the party making the original statement. Defamation is thus a serious issue for oral history, but it is also subject to several constraints and difficult to prove. For one, the injured party must be living—one cannot defame the dead; for another, statements construed as opinion, "nothing more than conjecture and rumor," are not considered defamatory. Confronted with a potentially defamatory statement, the oral historian has several courses of action: consult other sources in an attempt to determine if the statement, however extreme, is, in fact, true—if it's true, it's not defamatory;

close the defamatory portion of the interview until the defamed person has died; carefully edit the statement to excise the defamatory material while not significantly distorting the record; and delete the defamatory material—a problematic action that violates norms of academic freedom.

Legal issues, though at times complicated, are relatively straightforward when compared with ethical issues, which often require the exercise of judgment and involve matters over which conscientious practitioners may reasonably disagree. While there is a lively ethical narrative within oral history (Blee, 1993; K'Meyer & Crothers, 2007; Shopes, 2006), perhaps the best place to start to understand both fundamental ethical principles and some of the nuances is the Oral History Association's *Principles and Best Practices for Oral History* (2009). To generalize, these *Principles* define standards governing the oral historian's relationship to the narrator, to standards of scholarship for history and related disciplines, and to both current and future users of the interview. The first two of these relationships concern us here. Fundamental to the interviewee-interviewer relationship is the notion of informed consent—that is, that the interviewee is fully informed about the purpose, scope, and value of the interview; how it will proceed; its final disposition and the uses to which it will or may be put; and issues of copyright—in other words, everything the interviewee needs to know to make an informed decision about whether to consent to the interview—or not. Recognizing the rights of scholarship, that is, the second set of relationships, the *Principles* also state, "Interviewers must take care to avoid making promises that cannot be met, such as guarantees of control over interpretation and presentation of the interviews" (n.p.).

The *Principles* also recognize the dialectic quality of these dual allegiances and at least imply the potential for conflict: "Oral historians respect the narrators as well as the integrity of the research. Interviewers are obliged to ask historically significant questions ... [and] must also respect the narrators' equal authority in the interviews and honor their right to respond to questions in their own style and language. In the use of interviews, oral historians strive for intellectual honesty and the best application of the skills of their discipline, while avoiding stereotypes, misrepresentations, or manipulations of the narrators' words" (2009, n.p.).

The problem arises when responsibility to the narrator conflicts with the claims of scholarship and the broader public good. One might easily imagine lines of inquiry that discomfit a narrator or that lead to revelations, intended or not, that might be construed as damaging to the narrator or to others. One might just as easily imagine a narrator who deliberately misrepresents the facts of a situation, for whom intellectual honesty is not a value. Or consider the example of filmmaker Claude Lanzmann, who exposed perpetrators of the Nazi Holocaust by filming them with a hidden camera—verboten in oral history and other field-based practices—and then included their testimony in his epic film *Shoah.* Does the public's right to hold war criminals accountable trump Lanzmann's failure to secure informed consent? Or not?

Standard professional practice, privileging the rights of the individual narrator, would claim that Lanzmann acted unethically; broader civic or moral claims would suggest otherwise.

Although extreme, the example of Lanzmann points to what many oral historians believe to be a fundamental incongruity between their practice and federal regulations governing the ethics of research involving human subjects, codified as Title 45 Public Welfare, Part 46 Protection of Human Subjects (referred to as 45 CFR 46 or the Common Rule), with authority for implementation residing in the Office for Human Research Protections at the U.S. Department of Health and Human Services and delegated to local, often campus-based institutional review boards or IRBs (Schrag, 2010; Shopes, 2009) In brief, 45 CFR 46 includes "interaction" with human subjects as one of the research modes subject to ethical review by an IRB review and hence has been applied to oral history. Although the terms of 45 CFR 46 also "exempt" most interviewing from IRB review, only an IRB can confer an exemption, in effect requiring a researcher to submit his or her research for review.

Most problematic, however, is language in the Common Rule, which does not exempt—and hence raises concern about—interviews for which "disclosure of the human subjects' responses outside the research could reasonably place the subjects at risk of criminal or civil liability, or be damaging to the subjects' financial standing, employability, or reputation" (46.101 [b] [2]). IRBs have used this language to ask oral historians to submit detailed questionnaires in advance of any interview; to avoid sensitive, embarrassing, or potentially incriminating topics; to maintain narrator anonymity despite an interviewee's willingness to be identified; and to retain or destroy interviews and transcripts after the research project is completed—all of which violate fundamental practices and principles of oral history. At times information in an interview, if made public, can indeed place a person at risk of criminal or civil liability, or be damaging to one's financial standing, employability, or reputation. To constrain such inquiry *a priori,* many oral historians argue, undercuts the "integrity of the research" and impinges on academic freedom. The Oral History Association, in concert with the American Historical Association, has attempted to negotiate a broader exclusion from IRB review of oral history, but efforts to date have been largely unsuccessful. At best, they have alerted college- and university-based oral historians and their IRBs—which enjoy considerable autonomy—to potential conflicts and encouraged informed dialogue and mutual accommodation.

It is perhaps appropriate to conclude this section, and this essay, with Portelli's (1997a, p. 55) observation about law and ethics, which aptly summarizes the impulses, simultaneously humanistic, scholarly, and political, underlying much work in oral history: "Ultimately, in fact ethical and legal guidelines only make sense if they are the outward manifestation of a broader and deeper truth sense of personal and political commitment to honesty and to truth. . . . By commitment to honesty I mean personal respect for the people we work with and intellectual respect for the material we receive. By commitment to truth, I mean a utopian striving and urge to know 'how things really are' balanced by openness to the many variants of 'how things may be.'"

◫ NOTES

* In developing this article, the author has drawn in part on the background paper on oral history she wrote for the Mellon Project on Folklore, Ethnomusicology, and Oral History in the Academy; and gratefully acknowledges the American Folklore Society, copyright holder of the report, for permission to draw on it. The full text of the report is available at http://www.oralhistory.org/about/association-business/

1. First use of the term *oral history* to describe the practice of interviewing participants in past events is generally attributed to Allen Nevins, founder of Columbia University's Oral History Research Office. Oral historians find the term maddeningly imprecise and debated its utility during the early years of the Oral History Association's existence. Nevins's successor Louis Starr wrote in 1974, "Heaven knows, oral history is bad enough, but it has the sanction of a quarter century's usage, whereas presumably more beguiling substitutes like *living history* and *oral documentation* and sundry other variants have gone by the boards. Oral history is a misnomer to be sure. Let us cheerfully accept that fact that, like *social security* or the *Holy Roman Empire,* it is now hopelessly embedded in the language: one encounters it on every hand" (Morrissey, 1980, p. 40).

2. Wire recorders, based on German Magnetophones captured during World War II, first became available in 1948; Columbia began using them to record interviews in 1949. They were supplanted by reel-to-reel recorders, then in the mid-1960s by cassette tape recorders, which became standard for oral history until the digital revolution at the end of the 20th century.

3. The 1968 document, titled *Goals and Guidelines* (Oral History Association, 1969), was considerably amplified as a checklist of "evaluation guidelines" in 1979, and revised in 1990 and again in 1998 to take into account new issues and concerns, including new technologies and increasingly diverse uses of oral history. A thorough revision was undertaken in 2008–2009 to abbreviate and consolidate what had become a rather cumbersome document developed by accretion.

4. Editors of two oral history book series have estimated that releases had not been secured for perhaps one half of the interviews used in manuscripts they have reviewed.

5. For the full debate, see Tilly, "People's History and Social Science History" (1983); Thompson et al., "Between Social Scientists: Responses to Tilly" (1985); and Tilly, "Louise Tilly's Response to Thompson, Passerini, Bertaux-Wiame, and Portelli" (1985).

◫ REFERENCES

Abrahams, R. D. (1981). Story and history: A folklorist's view. *Oral History Review, 9,* 1–11.

Allen, S. E. (1982). Resisting the editorial ego: Editing oral history. *Oral History Review, 10,* 33–45.

Anderson, K., & Jack, D. C. (1991). Learning to listen: Interview techniques and analyses. In S. B. Gluck & D. Patai (Eds.), *Women's words: The feminist practice of oral history* (pp. 11–26). New York: Routledge.

Armitage, S. H. (1983). The next step. *Frontiers: Journal of Women's Studies, 7*(1), 3–8.

Armitage, S. H., & Gluck, S. B. (1998). Reflections on women's oral history: An exchange. *Frontiers: Journal of Women's Studies, 19*(3), 1–11.

Atkinson, P., Coffey, A., & Delamont, S. (2003). *Key themes in qualitative research: Continuities and change.* Walnut Creek, CA: AltaMira Press.

Baum, W. K. (1977). *Transcribing and editing oral history.* Nashville, TN: American Association for State and Local History.

Bauman, R. (1986). *Story, performance, and event: Contextual studies of oral narratives.* Cambridge, UK: Cambridge University Press.

Blassingame, J. (1972). *The slave community: Plantation life in the antebellum South.* New York: Oxford University Press.

Blee, K. M. (1993). Evidence, empathy, and ethics: Lessons from oral histories of the Klan. *Journal of American History, 80,* 596–606.

Bloom, L. Z. (1977). Listen! Women speaking. *Frontiers: Journal of Women's Studies, 2*(1), 1–3.

Bodnar, J. (1985). *The transplanted: A history of immigrants in urban America.* Bloomington: Indiana University Press.

Bornat, J. (Ed.). (1993). *Reminiscence reviewed: Perspectives, evaluations, achievements.* Buckingham, UK: Open University Press.

Bruner, J. (1990). *Acts of meaning.* Cambridge, MA: Harvard University Press.

Butler, R. N. (1963). The life review: An interpretation of reminiscence in the aged. *Psychiatry, 26,* 65–76.

Chamberlain, M. (2006). Narrative theory. In T. L. Charlton, L. E. Myers, & R. Sharpless (Eds.), *Handbook of oral history* (pp. 384–407). Lanham, MD: AltaMira Press.

Coles, R. (1997). *Doing documentary work.* New York: Oxford University Press.

Cross, N., & Barker, R. (2006). The Sahel Oral History Project. In R. Perks & A. Thomson (Eds.), *The oral history reader* (2nd ed., pp. 538–548). London: Routledge.

Cruikshank, J. (1990). *Life lived like a story: Life stories of three Yukon native elders.* Lincoln: University of Nebraska Press.

Davis, S. G. (1988). Review essay: Storytelling rights. *Oral History Review, 16,* 109–116.

Densho: The Japanese American Legacy Project. (2010). Available from http://densho.org

Denzin, N. K. (2003). The call to performance. *Symbolic Interaction, 26,* 187–208.

Diaz, R. T., & Russell, A. B. (1999). Oral historians: Community oral history and the cooperative ideal. In J. B. Gardner & P. S. LaPaglia (Eds.), *Public history: Essays from the field* (pp. 203–216). Malabar, FL: Kreiger Publishing.

di Leonardo, M. (1987). Oral history as ethnographic encounter. *Oral History Review, 15,* 1–20.

Dolby, S. S. (1989). *Literary folkloristics and the personal narrative.* Bloomington: Indiana University Press.

Dublin, T. With photographs by G. Harvan. (1998). *When the mines closed: Stories of struggles in hard times.* Ithaca, NY: Cornell University Press.

Dubrow, G. L. (2008). Contested places in public memory: Reflections on personal testimony and oral history in Japanese American heritage. In P. Hamilton & L. Shopes (Eds.), *Oral history and public memories* (pp. 125–143). Philadelphia: Temple University Press.

Ellsworth, S. (1982). *Death in a promised land: The Tulsa race riot of 1921.* Baton Rouge: Louisiana State University Press.

Fields, K. E. (1994). What one cannot remember mistakenly. In J. Jeffrey & G. Edwall (Eds.), *Memory and history: Essays on recalling and interpreting experience* (pp. 89–104). Lanham, MD: University Press of America.

Friedman, J. (2003). Muscle memory: Performing embodied knowledge. In R. C. Smith (Ed.), *The art and performance of memory: Sounds and gestures of recollection* (pp. 156–80). London: Routledge.

Frisch, M. (1990a). Oral history, documentary, and the mystification of power: A critique of *Vietnam: A Television History.* In M. Frisch, *A shared authority: Essays on the craft and meaning of oral and public history* (pp. 159–178). Albany: SUNY Press.

Frisch, M. (1990b). *A shared authority: Essays on the craft and meaning of oral and public history.* Albany: SUNY Press.

Frisch, M. (1990c). Oral history and *Hard Times:* A review essay. In M. Frisch, *A shared authority: Essays on the craft and meaning of oral and public history* (pp. 5–13). Albany: SUNY Press.

Frisch, M. (2006). Oral history and the digital revolution: Toward a post-documentary sensibility. In R. Perks & A. Thomson (Eds.), *The oral history reader* (2nd ed., pp. 102–114). London: Routledge.

Geertz, C. (1974). "From the native's point of view": On the nature of anthropological understanding. *Bulletin of the American Academy of Arts and Sciences, 28*(1), 26–45.

Geiger, S. (1990). What's so feminist about women's oral history? *Journal of Women's History, 2*(1), 169–182.

Genovese, E. (1974). *Roll, Jordon, roll: The world the slaves made.* New York: Pantheon.

Gluck, S. B. (1977). What's so special about women? Women's oral history. *Frontiers: Journal of Women's Studies, 2*(1), 3–13.

Gluck, S. B. (1999). From first generation oral historians to fourth and beyond. *Oral History Review, 26*(2), 1–9. Printed as part of Gluck, S. B., Ritchie, D. A., & Eynon, B. (1999). Reflections on oral history in the new millennium. *Oral History Review, 26*(2), 1–27.

Gluck, S. B. (2006). Women's oral history. Is it so special? In T. L. Charlton, L. E. Myers, & R. Sharpless (Eds.), *Handbook of oral history* (pp. 357–383). Lanham, MD: AltaMira Press.

Gluck, S. B., & Patai, D. (Eds.). (1991). Introduction. In S. B. Gluck & D. Patai (Eds.), *Women's words: The feminist practice of oral history* (pp. 1–5). New York: Routledge.

Grele, R. J. (1991a). *Envelopes of sound: The art of oral history* (2nd ed.). New York: Praeger.

Grele, R. J. (1991b). Listen to their voices: Two case studies in the interpretation of oral history interviews. In R. J. Grele, *Envelopes of sound: The art of oral history* (2nd ed., pp. 212–241). New York: Praeger.

Grele, R. J. (2007). Reflections on the practice of oral history: Retrieving what we can from an earlier critique. *Suomen Antropologi, 4,* 11–23.

Handlin, O. (1951). *The uprooted: The epic story of the great migrations that made the American people.* Boston: Little, Brown.

High, S. & Sworn, D. (2009). After the interview: The interpretive challenges of oral history video indexing. *Digital Studies Le champ numerique, 1*(2), 1–24.

Hirsch, J. (2006). *Portrait of America: A cultural history of the Federal Writers' Project.* Chapel Hill: University of North Carolina Press.

Hirsch, J. (2007). Before Columbia: The FWP and American oral history research. *Oral History Review, 34,* 1–16.

Isay, D. (2007). *Listening is an act of love.* New York: Penguin.

Jackson, B. (2007). *The story is true: The art and meaning of telling stories.* Philadelphia: Temple University Press.

James, D. (2000a). *Doña Maria's story: Life history, memory, and political identity.* Durham, NC: Duke University Press.

James, D. (2000b). Listening in the cold: The practice of oral history in an Argentine meatpacking community. In D. James, *Doña Maria's story: Life history, memory, and political identity* (pp. 119–156). Durham, NC: Duke University Press.

Joyner, C. W. (1979). Oral history as communicative event: A folkloristic perspective. *Oral History Review, 7,* 47–52.

Kaiser. S. (2005). *Postmemories of terror: A new generation copes with the legacy of the "dirty war."* New York: Palgrave Macmillan.

Kerr, D. (2003). "We know what the problem is": Using oral history to develop a collaborative analysis of homelessness from the bottom up. *Oral History Review, 30*(1), 27–46.

Kerr, D. (2008). Countering corporate narratives from the streets: The Cleveland Homeless Oral History Project. In P. Hamilton & L. Shopes (Eds.), *Oral history and public memories* (pp. 231–251). Philadelphia: Temple University Press.

K'Meyer, T. E., & Crothers, A. G. (2007). "If I see some of this in writing, I'm going to shoot you": Reluctant narrators, taboo topics, and the ethical dilemmas of the oral historian. *Oral History Review, 34,* 71–93.

Larson, M. A. (2006). Research design and strategies. In T. L. Charlton, L. E. Myers, & R. Sharpless (Eds.), *Handbook of oral history* (pp. 105–134). Lanham, MD: AltaMira Press.

Lewis, H. M., Waller, S. M., & Hinsdale, M. A. (1995). *It comes from the people: Community development and local theology.* Philadelphia: Temple University Press.

Lundy, P., & McGovern, M. (2006). "You understand again": Testimony and post-conflict transition in the North of Ireland. In R. Perks & A. Thomson (Eds.), *The oral history reader* (2nd ed., pp. 531–537). London: Routledge.

MacKay, N. (2007). *Curating oral histories: From interview to archive.* Walnut Creek, CA: Left Coast Press.

Mazé, E. A. (2006). The uneasy page: Transcribing and editing oral history. In T. L. Charlton, L. E. Myers, & R. Sharpless (Eds.), *Handbook of oral history* (pp. 237–271). Lanham, MD: AltaMira Press.

McMahan, E. M. (1989). *Elite oral history discourse: A study of cooperation and coherence.* Tuscaloosa: University of Alabama Press.

McMahan, E. M. (2006). A conversation analytic approach to oral history interviewing. In T. L. Charlton, L. E. Myers, & R. Sharpless (Eds.), *Handbook of oral history* (pp. 336–356). Lanham, MD: AltaMira Press.

Minkley, G., & Rassool, C. (1998). Orality, memory, and social history in South Africa. In S. Nuttall & C. Coetzee (Eds.), *Negotiating the past: The making of memory in South Africa* (pp. 89–99). Oxford, UK: Oxford University Press.

Mintz, S. W. (1979). The anthropological interview and the life history. *Oral History Review, 7,* 18–26.

Morrissey, C. T. (1980). Why call it "oral history"? Searching for early usage of a generic term. *Oral History Review, 8,* 20–48.

Moss, W. (1977). Oral history: An appreciation. *American Archivist, 40*(4), 429–439.

Neuenschwander, J. A. (2009). *A guide to oral history and the law.* New York: Oxford University Press.

Neufeld, D. (2008). Parks Canada, the commemoration of Canada, and northern Aboriginal oral history. In P. Hamilton & L. Shopes (Eds.), *Oral history and public memories* (pp. 3–29). Philadelphia: Temple University Press.

Oakley, A. (1981). Interviewing women: A contradiction in terms. In H. Roberts (Ed.), *Doing feminist research* (pp. 30–61). London: Routledge & Kegan Paul.

Oral History Association. (1969). Oral History Association adopts statement about goals and guidelines during Nebraska Colloquium. *Oral History Association Newsletter, 3*(1), 4.

Oral History Association. (2009). *Principles and best practices for oral history.* Available at http://www.oralhistory.org/do-oral-history/principles-and-practices

Oral History Association. (2010). Available at www.oralhistory.org

Passerini, L. (1980). Italian working-class culture between the wars: Consensus for fascism and work ideology. *International Journal of oral History, 1,* 1–27.

Passerini, L. (1987). *Fascism in popular memory: The cultural experience of the Turin working class.* (R. Lumley & J. Bloomfield, Trans.). Cambridge, UK: Cambridge University Press.

Polkinghorne, D. (1988). *Narrative knowing and the human sciences.* Albany: SUNY Press.

Pollock, D. (2005). Introduction: Remembering. In D. Pollock (Ed.), *Remembering: Oral history performance* (pp. 1–17). New York: Palgrave Macmillan.

Portelli, A. (1991). The death of Luigi Trastulli: Memory and the event. In A. Portelli, *The death of Luigi Trastulli and other stories: Form and meaning in oral history* (pp. 1–26). Albany: SUNY Press.

Portelli, A. (1997a). Tryin' to gather a little knowledge: Some thoughts on the ethics of oral history. In A. Portelli, *The battle of Valle Giulia: Oral history and the art of dialogue* (pp. 55–71). Madison: University of Wisconsin Press.

Portelli, A. (1997b). Oral history as genre. In A. Portelli, *The battle of Valle Giulia: Oral history and the art of dialogue* (pp. 3–23). Madison: University of Wisconsin Press.

Portelli, A. (2003). *The order has been carried out: History, memory, and meaning of a Nazi massacre in Rome.* New York: Palgrave Macmillan.

Rawick, G. P. (1972). *From sundown to sunup: The making of the Black community.* Westport, CT: Greenwood Press.

Ritchie, D. A. (2003). *Doing oral history: A practical guide* (2nd ed.). New York: Oxford University Press.

Rogers, K. L. (2006). *Life and death in the Delta: African American narratives of violence, resilience, and social change.* New York: Palgrave Macmillan.

Rouverol, A. J. (2003). Collaborative oral history in a correctional setting: Promise and pitfalls. *Oral History Review, 30*(1), 61–86.

Samuel, R. (1971). Perils of the transcript. *Oral History: Journal of the Oral History Society, 1*(2), 19–22.

Schrag, Z. (2010). *Ethical imperialism: Institutional review boards and the social sciences, 1965–2009.* Baltimore: Johns Hopkins University Press.

Schrager, S. (1983). What is social in oral history? *International Journal of Oral History, 4*(2), 76–98.

Sharpless, R. (2006). The history of oral history. In T. L. Charlton, L. E. Myers, & R. Sharpless (Eds.), *Handbook of oral history* (pp. 19–42). Lanham, MD: AltaMira Press.

Shiekh, I. (2010). *Being Muslim in America.* New York: Palgrave Macmillan.

Shopes, L. (1986). The Baltimore Neighborhood Heritage Project: Oral history and community involvement. In S. Benson, S. Brier, & R. Rosenzweig (Eds*.), Presenting the past: Critical perspectives on history and the public* (pp. 249–263). Philadelphia: Temple University Press.

Shopes, L. (2002a). Oral history and the study of communities: Problems, paradoxes, and possibilities. *Journal of American History, 69*(2), 588–598.

Shopes, L. (2002b). Making sense of oral history. *History matters: The U.S. survey course on the web.* Available from http://historymatters.gmu.edu/mse/oral/

Shopes, L. (2006). Legal and ethical issues in oral history. T. L. Charlton, L. E. Myers, & R. Sharpless (Eds.), *Handbook of oral history* (pp. 135–169). Lanham, MD: AltaMira Press.

Shopes, L. (2009). Human subjects and IRB review (2009). Available at http://www.oralhistory.org/do-oral-history/oral-history-and-irb-review

Silverman, D. (Ed.). (1997). *Qualitative research: Theory, method and practice.* London: Sage.

Sitzia, L. (2003). Shared authority: An impossible goal? *Oral History Review, 30*(1), 87–102.

Slim, H., & Thompson, P. (Eds.) (1993). *Listening for a change: Oral history and development.* London: Panos.

Spence, D. (1982). *Narrative truth and historical truth: Meaning and interpretation in psychoanalysis.* New York: W. W. Norton.

Stacey, J. (1991). Can there be a feminist ethnography? In S. B. Gluck & D. Patai (Eds.), *Women's words: The feminist practice of oral history* (pp. 111–119). New York: Routledge.

Starr, L. (1984). Oral history. In D. K. Dunn & W. K. Baum (Eds.), *Oral history: An interdisciplinary anthology* (pp. 3–26). Nashville, TN: American Association of State and Local History.

Terkel, S. (1967). *Division Street: America.* New York: Pantheon Books.

Terkel, S. (1970). *Hard times: An oral history of the Great Depression.* New York: New Press.

Terkel, S. (1974). *Working: People talk about what they do all day and how they feel about what they do.* New York: New Press.

Terkel, S. (1984). *The good war.* New York: Pantheon Books.

Thompson, P. (1988). *The voice of the past: Oral history.* New York: Oxford University Press.

Thompson, P., Passerini, L., Bertaux-Wiame, I., & Portelli, A. (1985). Between social scientists: Reponses to Tilly. *International Journal of Oral History, 6*(1), 19–40.

Thomson, A. (1990). Anzac memories: Putting popular memory theory into practice in Australia. *Oral History, 18*(2), 25–31.

Thomson, A. (2007). Four paradigm transformations in oral history. *Oral History Review, 34*(1), 49–70.

Tilly, L. (1983). People's history and social science history. *Social Science History, 7*(4), 457–474.

Tilly, L. (1985). Louise Tilly's response to Thompson, Passerini, Bertaux-Wiame, and Portelli with a concluding comment by Ronald J. Grele. *International Journal of Oral History, 6*(1), 40–47.

Tonkin, E. (1992). *Narrating our pasts: The social construction of oral history.* Cambridge, UK: Cambridge University Press.

Vansina, J. (1985). *Oral tradition as history.* Madison: University of Wisconsin Press.

Yow, V. R. (1997). "Do I like them too much?" Effects of the oral history interview on the interviewer and vice versa. *Oral History Review, 24,* 55–79.

Yow, V. R. (2005). *Recording oral history: A guide for the humanities and social sciences* (2nd ed.). Walnut Creek, CA: AltaMira Press.

28

OBSERVATIONS ON OBSERVATION

Continuities and Challenges

Michael Angrosino and Judith Rosenberg

Observation has been characterized as "the fundamental base of all research methods" in the social and behavioral sciences (Adler & Adler, 1994, p. 389) and as "the mainstay of the ethnographic enterprise" (Werner & Schoepfle, 1987, p. 257). Qualitative social scientists are observers both of human activities and of the physical settings in which such activities take place. In qualitative research, observations typically take place in settings that are the natural loci of activity. Such naturalistic observation is therefore an integral part of ethnographic fieldwork.

Naturalistic observation is a technique for the collection of data that are, in the ideal at least, as unobtrusive as possible. Even fieldworkers who think of themselves as participant observers usually strive to make the process as objective as possible despite their quasi-insider status. The notion of unobtrusive, objective observation has not, however, gone uncontested. The prescient, discipline-spanning scholar Gregory Bateson (1972) developed a "cybernetic" theory in which the observer is inevitably tied to what is observed. More recently, postmodernists in various disciplines have emphasized the importance of understanding researchers' "situations" (e.g., their gender, social class, ethnicity) as part of interpreting the products of their research.

The potency and pervasiveness of the postmodernist critique of traditional assumptions about objectivity have led some qualitative researchers to rethink and revise their approaches to observational methods. In a very important sense, we now function in a context of collaborative research in which the researcher no longer operates at a distance from those being observed. The latter are no longer referred to as "subjects" of research but as active partners who understand the goals of research and who help the researcher formulate and carry out the research plan. Judith Friedenberg (1998, p. 169), for example, has advocated

the solicitation of feedback on ethnographic constructions from study populations "using techniques that minimize the researcher's control of the interview situation and enhance intellectual dialog." Valerie Matsumoto (1996) sent a prepared set of questions for the people she was interested in interviewing for an oral history project. She assured them that any questions to which they objected would be eliminated. The potential respondents reacted favorably to this invitation to participate in the formulation of the research plan. As such situations have become the norm, Michael Angrosino and Kimberly Mays de Pérez (2000) advocated a shift away from thinking of observation strictly as a data collection technique; rather, it should also be seen as a context in which those involved in the research collaboration can interact.

To clarify that shift, it may be helpful to briefly review both the classic tradition of naturalistic observation and the more contextualized analysis of the research collaboration as it has developed in response to current challenges both academic (e.g., the postmodernist critique) and in the society that we aim to study.

◩ OBSERVATION-BASED RESEARCH: THE CLASSIC TRADITION

The creed of the classic tradition of observational research was explained by R. L. Gold (1997, p. 397), who noted that researchers believed it was both possible and desirable to develop standardized procedures that could "maximize observational efficacy, minimize investigator bias, and allow for replication and/or verification to check out the degree to which these procedures have enabled the investigator to produce valid, reliable data that, when incorporated into his or her published reports, will be regarded by peers as objective findings." Ethnographers were supposed to adhere to a "self-correcting investigative process" that included adequate and appropriate sampling procedures, systematic techniques for gathering and analyzing data, validation of data, avoidance of observer bias, and documentation of findings (Clifford, 1983, p. 129; Gold, 1997, p. 399).

According to Gold (1958), the sociological ethnographers of the first half of the 20th century often made implicit reference to a typology of roles that might characterize naturalistic research: the complete participant (a highly subjective stance whose scientific validity was automatically suspect), the participant-as-observer (only slightly less problematic), the observer-as-participant (more typically associated with anthropologists), and the complete (unobtrusive) observer. The purity of the latter type—difficult to attain even under controlled laboratory conditions, let alone in the field—was, as we can now see, compromised by the tendency of unobtrusive researchers to go about their business without informed consent, an ethical lapse that can no longer be tolerated. It is now very clear that the ethical imperative to provide informed consent paved the way toward the model of collaborative research that is now the norm because the process of obtaining such consent inevitably involves the people being studied in activity of research from the very beginning. In any case, the canons of observational research were modified long before the advent of the postmodernist critique as an awareness of relative degrees of researcher "membership" in a community under study entered the discussion (Adler & Adler, 1987). Nevertheless, even researchers who were active "members" were still enjoined to be careful "not to alter the flow of interaction unnaturally" (Adler & Adler, 1994, p. 380). The underlying assumption remained: A "natural" flow of social life could exist independent of the efforts of researchers to study it.

Anthropological ethnographers were less concerned than were their sociological cousins with obtrusiveness and its attendant delict, observer bias. Anthropological ethnographers were, however, still encouraged to seek objectivity in the midst of their acknowledged subjective immersion in a study community, and they did so by engaging in a three-step process of observation. First, there was "descriptive observation," which meant, to all intents and purposes, the observation of every conceivable aspect of the situation. Anthropologists at this point were supposed to be "childlike," assuming that they knew nothing and could take nothing for granted. There was, in effect, to be no sorting out of the important from the trivial based on assumptions carried into the field setting. As researchers became more familiar with the setting, however, they could move to the second step, "focused observation," at which point they could with some confidence discern the relevant from the irrelevant. Focused observation almost always involved interviewing because researchers could not rely on their own

intuition to make such discernments. Focused observations usually concentrated on well-defined types of group activities (e.g., religious rituals, classroom instruction, political campaigns). The third and final step was the most systematic—"selective observation"—at which point ethnographers could concentrate on the elements of social action that are most salient, presumably from the "native" point of view (Werner & Schoepfle, 1987, pp. 262–264).

▣ OBSERVATION-BASED RESEARCH
 IN LIGHT OF CURRENT CONCERNS

Contemporary fieldwork has three major attributes: (1) the increasing willingness of ethnographers to affirm or develop a more than peripheral membership role in communities they study; (2) the recognition of the possibility that it may be neither feasible nor possible to harmonize observer and insider perspectives to achieve an objective consensus about "ethnographic truth"; (3) the transformation of the erstwhile subjects of research into collaborative partners in research (e.g., Angrosino, 2007a; Creswell, 2007). The goal of contemporary observational research is not to replace the classic ideal of pure objectivity with one of total, membership-driven empathy. Both of these approaches remain as constituent elements in the process of observation-based research; they represent, however, extreme points at opposite ends of a continuum of research practice. The problem with both extremes is that they assume that it is both feasible and desirable to describe or interpret cultures and societies as if those depictions could exist without ethnographers being part of the action. Observation-based research nowadays must certainly consider the attributes and activities of ethnographers themselves; it is therefore considerably more subjective than those of the classic tradition would have countenanced. But it cannot become so utterly subjective that it loses the rigor of carefully conducted, clearly recorded, and intelligently interpreted observations; ethnography is more than casually observed opinion.

Angrosino and Mays de Pérez (2000, pp. 678–690) discuss the ways in which these factors have come to be established in the current ethnographic literature and the implications of these changes for both the conduct and the interpretation of observation-based research. They note that whereas classic ethnographic fieldworkers insisted on their objectivity and adopted limited participatory roles only as the ethics of "pure observation" were questioned, latter-day researchers consciously seek out and adopt situational identities that give them defined membership roles in the communities they study. As their membership roles deepen, ethnographers must become attuned to life as it is actually lived, which means that they must pay increasing attention to the ways in which their potential collaborators in study communities want to be studied. The older notion of imposing a predetermined "scientific" agenda (itself now often seen as a product of a Western, elite bias), which was so integral to the objective aims of the classic period, has been set aside. Although rarely acknowledged at the time, the classic approach was based on a model of interaction in which power resided in the ethnographer (who set the research agenda and implicitly represented the more generalized power of elite institutions); power is now clearly shared. In the case of certain applied or advocate social scientists, power is actually ceded to the study community; researchers of this orientation may well see themselves as agents of those communities in the same way that they once thought of themselves as extensions of their academic institutions or granting organizations.

The imperative to acknowledge the shift in power to study communities takes on particular importance when, as is now so often the case, ethnography is conducted "without the ethnos" (Gupta & Ferguson, 1996b, p. 2). In other words, few ethnographers function within the circumscribed communities that lent coherence to the cultures or societies that figured so prominently in the conceptual frameworks of the classic period of observational research. It is no longer possible to assume that "the cultural object of study is fully accessible within a particular site" (Marcus, 1997, p. 96). Much of the current "field" in which "fieldwork" is conducted consists of people who inhabit the "borders between culture areas," of localities that demonstrate a diversity of behavioral and attitudinal patterns, of "postcolonial hybrid cultures," and of the social change and cultural transformations that are typically found "within interconnected spaces" (Gupta & Ferguson, 1996a, p. 35). In the classic period, it was assumed that because people lived in a common space they therefore came to share social institutions and cultural assumptions. Nowadays, "it is the communities that are accidental, not the happenings" (Malkki, 1996, p. 92), particularly in the case of "virtual" communities that spring up, flourish, and then vanish—seemingly overnight—on the Internet. Ethnographers therefore no longer enjoy the luxury of assuming that the local scenes they observe are somehow typical or representative of any single culture or society. Rather, any observed community is more likely to be understood as a "nexus of interactions defined by interstitiality and hybridity" (Gupta & Ferguson, 1996a, p. 48). Researchers who depend on observation must therefore be increasingly careful not to confuse the shifting interactions of people with multiple affiliations in both real and virtual space as if they were the bounded communities of old. To take a clearly articulated membership role in such diffuse settings, a researcher must be willing to be explicit about his or her own gender, sexual identity, age, class, and ethnicity because such factors form the basis of his or her affinity with potential study collaborators, rather than the simple fact of hanging around in a defined space. These situational factors are likely to shift from one research project to the next, so ethnographers are in a position of having to "reinvent

themselves in diverse sites" (Giroux, 1995, p. 197). Norman Denzin (1997, p. 46) discusses the "mobile consciousness" of ethnographers who are aware of their "relationship to an ever-changing external world."

Much of the recent literature bearing on the creation, maintenance, and evolution of observers' identities has dealt with issues particular to women and lesbians and gay men. (This literature is vast, but a few representative studies that demonstrate the blending of observational sociocultural detail with analysis of personal "situations" include Behar, 1993; Blackwood, 1995; Lang, 1996; Walters, 1996; D. Wolf, 1996.) It is worth mentioning, however, that there are other identity issues that are of concern to researchers who study situations of political unrest and who come to be identified with politically proscribed groups (Hammond, 1996; Mahmood, 1996; Sluka, 1990), or who work with groups that are defined by their need for deceptive concealment, such as illegal immigrants (Chavez, Flores, & Lopez-Garza, 1990; Stepick & Stepick, 1990), or those involved in criminal activities (Agar & Feldman, 1980; Brewer, 1992; Dembo et al., 1993; Koester, 1994; van Gelder & Kaplan, 1992). In the post–9/11 era, it has been increasingly difficult to conduct ethnographic research in Muslim American communities, particularly if the research deals specifically with young men (Sirin & Fine, 2008).

▣ THE QUESTION OF CONTEXT: THE OVERLAPPING ROLES OF OBSERVATIONAL RESEARCHERS

In the classic period, ethnographers had to be concerned with only one audience—the academic/scientific community. Although that community could hardly be said to have spoken with one voice, it certainly did share a set of assumptions about what a proper research report looked and sounded like. In our own time, however, there seem to be as many different formats in which a report can be disseminated as there are constituencies to which ethnographers are now responsible. Researchers must therefore be concerned with the ways in which their observations come to be translated into the different voices suitable for multiple audiences. Traditional research reports favored the supposedly objective third-person voice, emanating from the "omniscient narrator" (Tierney, 1997, p. 27). The shift to collaborative research allows ethnographers to acknowledge their own presence; the once-banned "I" is now much more common as subjective experience comes to the fore, a trend apparently encouraged by feminist scholars who often felt marginalized by the academic world and its objectifying tendencies (M. Wolf, 1992, p. 52). This shift is no mere matter of stylistic preference; it reflects evolving self-images of ethnographers, changing relations between observers and those they observe (with explicit permission), and new perceptions about the diverse, and possibly even contradictory, audiences to whom ethnographic research is now addressed.

Ethnographers can no longer claim to be the sole arbiters of knowledge about the societies and cultures they study because they are in a position to have their representations read and contested by those for whom they presume to speak (Bell & Jankowiak, 1992). In effect, objective truth about a society or culture cannot be established because there are inevitably going to be conflicting versions of what happened. Researchers can no longer claim the privilege of authoritative knowledge when there are all too many other collaborators ready and able to challenge them. Margery Wolf (1992, p. 5) notes that as a fledgling ethnographer she was "satisfied to describe what I thought I saw and heard as accurately as possible, to the point of trying to resolve differences of opinion among my informants." She eventually came to realize "the importance of retaining those 'contested meanings.'" She wryly concludes that any member of the study community is likely to "show up on your doorstep with an Oxford degree and your book in hand" (1992, p. 137). In sum, the results of observational research can never be "reducible to a form of knowledge that can be packaged in the monologic voice of the ethnographer alone" (Marcus, 1997, p. 92). To be sure, given the complexities of publication and other genres for the dissemination of ethnographic research, it is still almost always the case that the researcher is the visible "author" of a report. Attempts to get the actual voices of all collaborators onto the public record have been spotty at best.

Observation-based research is not simply a data-collection technique; it forms the context in which ethnographic fieldworkers assume membership roles in communities they want to study. They do so in a process of negotiation with those who are already members and who might act as collaborators in the research process. They bring to that negotiation their own "situations" (i.e., gender, sexual orientation, age, social class, ethnicity), all of which must necessarily figure in the kinds of roles they might assume and the ways in which they will be allowed to interact with those already involved in the setting. For these reasons, naturalistic observation can only be understood in light of the results of specific interactive negotiations in specific contexts representing (perhaps temporary) loci of interests. The old notion that cultures or social institutions have an independent existence has been set aside. By the same token, neither cultures nor social institutions are reducible to the experiences of those who observe them. Observation, if it is to be useful to the research process, must be as rigorously conducted as it was in the classic period; our social scientific powers of observation must, however, be turned on ourselves and the ways in which our experiences interface with those of others in the same context if we are to come to a full understanding of sociocultural processes. Former generations of researchers were certainly not unaware of these experiential factors, but they were taught always to be aware of them so as to minimize them and hold them constant against the ethnographic truth.

The autonomous, enduring culture that embodies its own timeless truth may, however, no longer be an operative concept. After all, a researcher "never observes the behavioral event which 'would have taken place' in his or her absence, nor hears an account identical with that which the same narrator would have given to another person" (Behar, 1996, p. 6), so how could we ever be sure of that disembodied cultural reality? But the ways in which we as researchers negotiate the shifting sands of interaction, if we are careful to observe and analyze them, are important clues to the ways in which societies and culture form, maintain themselves, and eventually dissolve. In other words, the contexts may be evanescent, but the ways in which those contexts come to be may well represent enduring processes of human interaction.

▣ CURRENT CHALLENGES FOR OBSERVATION-BASED RESEARCHERS

The context of contemporary observation-based research is shaped by the situational characteristics of the researchers themselves and their potential collaborators, and by several important changes in the general intellectual climate, in academic culture, and in the nature of an increasingly globalized, seemingly borderless society. These issues have been dealt with elsewhere by Angrosino (2007a, 2007b). Only a few of these trends, those with perhaps the most direct bearing on the conduct of observation-based research, will be summarized here.

Ethical/Regulatory Constraints

It has been noted that the old ideal of purely objective observation ultimately ran afoul of a new ethical climate that privileged informed consent and confidentiality to the extent that these principles were encoded in guidelines and institutional structures governing the conduct of research funded by public moneys—which, in the contemporary context, is just about everything. The early history of research ethics is covered by Murray Lionel Wax and Joan Cassell (1979). The contemporary scene is covered by Carolyn Fluehr-Lobban (2003). A few of the highlights are discussed here.

Virtually all social research in our time is governed by the structure of institutional review boards (IRBs), which grew out of federal regulations beginning in the 1960s that mandated informed consent for all those participating in federally funded research. The perceived threat was from "intrusive" research (usually biomedical), participation in which was to be under the control of the "subjects," who had a right to know what was going to happen to them and to agree formally to all provisions of the research. They must be fully apprised of both direct benefits and potential risks (including risks to their privacy) entailed in the research. (See the National Commission for the

Protection of Human Subjects of Biomedical and Behavioral Research, 1979.) The right of informed consent, and the review boards that were eventually created to enforce it at each institution receiving federal moneys, radically altered the power relationship between researcher and "subject," allowing both parties to have a say in the conduct and character of research.

Ethnographic researchers, however, were initially uncomfortable with this situation—not, of course, because they wanted to conduct covert, harmful research, but because they did not believe that their research was "intrusive." Such a claim was of a piece with the assumptions typical of the observer-as-participant role. As ethnographers became more comfortable with more engaged participatory roles, they came to agree that their very presence was an occasion of change, although they continued to resist the notion that their "intrusion" was by definition harmful. Ethnographers were also concerned that the proposals sent to IRBs had to be fairly complete, so that all possibilities for doing harm might be adequately assessed. Their research, they argued, often grew and changed as it went along and could not always be set out with the kind of predetermined specificity that the legal experts seemed to expect (and that has always been appropriate in biomedical and other forms of clinical/experimental research).

In the 1980s, social scientists won from the federal Department of Health and Human Services an exemption from review for all social research except that dealing with children, people with disabilities, and others defined as members of "vulnerable" populations. Nevertheless, legal advisers at many universities (including the University of South Florida [USF] where both the authors have been based) have opted for caution and have been very reluctant to allow this near-blanket exemption to be applied. As a result, at USF it is possible for a proposal to undergo "expedited" (or "partial") review if it seems to meet the federal criteria for exemption, but a formal proposal must still be filed. This practice is required under guidelines promulgated by the U.S. Department of Health and Human Services in 2005 (Code of Federal Regulations, Title 45, Part 46) (see Office of the Federal Register, 2009).

USF now has two IRBs—one for biomedical research and one for "behavioral research." Because the latter is dominated by psychologists (by far the largest department in the social science division of the College of Arts and Sciences), this separate status rarely works to the satisfaction of qualitative researchers. Psychologists, used to dealing with hypothesis-testing, experimental, clinical/lab-based research, have been reluctant to recognize a subcategory of "observational" research design. As a result, the proposal format currently required by the behavioral research IRB is couched in terms of the individual subject rather than in terms of populations or communities, and it mandates the statement of a hypothesis to be tested and a "protocol for the experiment." Formats more congenial to the particular needs of qualitative researchers have not been fully explored or adopted at USF. It is perhaps plausible to maintain that qualitative research is really a species of humanistic scholarship and not "science" at all, social or otherwise. If qualitative inquiry is not "research" in the scientific sense, it must be automatically exempt from IRB oversight. This point of view, however, has not gained much traction. For one thing, qualitative researchers are, on the whole, unwilling to give up their scientific status. Moreover, they are unwilling to reinforce the suspicion that they are simply trying to evade their ethical responsibilities. The trick is to comply with currently accepted ethical standards without compromising the very premises of qualitative inquiry. But this form of inquiry as currently practiced really does confound traditional definitions of scientific research. For one thing, the kind of "collaborative" research currently in favor among qualitative researchers further militates against strict compliance with the guidelines for informed consent. In collaborative research, the ethnographer must discuss research plans with members of the prospective study community, so must these preliminary discussions also conform to norms of informed consent, or do the latter only apply to the formal research plans that ultimately emerge from the collaborative consultations?

Given the now widespread ethical suspicions about "pure" observational research, it is ironic that the only kind of social research that is explicitly mentioned and routinely placed in the "exempt" category at USF is that of observations of behavior in public spaces. But it was just this sort of "unobtrusive" observation that led to questions about the propriety of conducting research in the absence of informed consent in the first place. Having largely abandoned this genre of "public" research because of its ethical problems, will ethnographers return to it simply to avoid the philosophical and legal entanglements raised by the IRB structure for their kind of research?

A recent report from the Institute of Medicine (IOM, 2002), a body that one would think represents an old, established paradigm of research ethics, challenged researchers in all disciplines to rethink the fundamentals of research ethics. Its report pointed out that we have become used to asking basically negative questions (e.g., what is misconduct? how can it be prevented?). It might be preferable to consider the positive and ask, What is integrity? How do we find out whether we have it? How can we encourage it? The promotion of researcher integrity has both individual and institutional components, and those in charge of monitoring professional ethics should be in the business of "encouraging individuals to be intellectually honest in their work and to act responsibly, and encouraging research institutions to provide an environment in which that behavior can thrive" (Grinnell, 2002, p. B15). One possible way to accomplish this aim, constructed on a philosophy of "proportionate reason," was explored by Angrosino and Mays de Pérez (2000). The IOM went so far as to suggest that qualitative social researchers have a central role to play in the evolution of the structures of research ethics because they are particularly well equipped to conduct studies that could identify and assess the factors influencing integrity in research in both individuals and large social institutions.

The Changing Research Context: Technology

Participant observation once implied a lone researcher working in a self-contained community, armed only with a notebook and pen, and perhaps a sketch pad and a simple camera. The mechanics of observation-based research were revitalized by the introduction of audiotape recorders, movie cameras, and later video recorders. Note-taking has been transformed by the advent of laptop computers and software programs for the analysis of narrative data. But as our technological sophistication has increased, ethnographers have begun to realize that the technology helps us capture and fix "reality" in ways that are somewhat at variance with our lived experience as fieldworkers. The great value of naturalistic observation has always been that we have immersed ourselves in the ebb and flow, in the ambiguities of life as it is lived by real people in real circumstances. To that traditional perception we have now become increasingly aware of our own part in that ever-changing interactive context. But the more we fix this or that snapshot of that life and the more we have the capacity to disseminate this or that image globally and instantaneously, the more we risk violating our sense of what makes real life so particular and therefore so endlessly fascinating. Video recording (still or moving) poses definite challenges to the ethical norms of the protection of privacy and the maintenance of confidentiality, issues explored in detail by Lauren Clark and Oswald Werner (1997) and Werner and Clark (1998).

It may, perhaps, become necessary to turn our observational powers on the very process of observation, to understand ourselves as users of technology. Technological change is never merely additive; it is never simply an aid to doing what has always been done. It is, rather, *ecological* in the sense that a change in one aspect of behavior has ramifications throughout the entire system of which that behavior is a part. So the more sophisticated our technology, the more we change the way we do business. We need to begin to understand not only what happens when "we" encounter "them," but when "we" do so with a particular kind of powerful technology. That we possess this technology (and the means to use it) while many of our likely research collaborators do not means that the power differential—which the shift to collaborative research was supposed to ameliorate—has only been exacerbated. (See Nardi & O'Day, 1999, for an elaboration of these points.)

The Changing Research Context: Globalization

Globalization is the process by which capital, goods, services, labor, ideas, and other cultural forms move freely across international borders. In our own time, communities that once existed in some degree of isolation have been drawn into interdependent relationships that extend around the globe. Globalization has been facilitated by the growth of information technology. News from all corners of the world is instantaneously available. Although once we could assume that the behaviors and ideas we observed or asked about in a particular community were somehow indigenous to that community, now we must ask literally where in the world they might have come from. Aihwa Ong and Stephen Collier (2005) provide an extended treatment of the implications of globalization on social research in general, and ethnographic research in particular. A few highlights are summarized here.

Communities are no longer necessarily place-bound, and the traditional influences of geography, topography, and climate are much less fixed than in days past. Increasing numbers of people are now explicitly "transnational" in their orientation, migrating from homeland to other places for work or study, but maintaining their ties to home. Such constant movement was difficult for earlier generations of migrants to achieve, as the high cost and relative inefficiency of earlier modes of transportation and communication were prohibitive for all but the most affluent. Doing observation-based research in a "transnational" community presents obvious challenges. We could, of course, contrive to follow people around the globe, but doing so hardly seems practical in most cases. More often than not, we will continue to be place-bound researchers, but we will have to keep reminding ourselves that the "place" we are participating in and observing may no longer be the total social or cultural reality for all the people who are in some way or another affiliated with the community.

We can discern several aspects of the modern world that may help us take observational research beyond the small, traditional communities in which it developed. For one thing, we can now speak in terms of a world in which nations are economically and politically interdependent. The relationships of units within this global system are shaped in large measure by the global capitalist economy, which is committed to the maximization of profits rather than to the satisfaction of domestic needs. Some settings and events that might be studied by observational methods to contribute to our understanding of the global system include the nature of labor migration (Zuniga and Hernandez-Léon, 2001); the emergence of "outsourcing" and its impact on the traditional societies that are thus brought into the world of the dominant powers (Saltzinger, 2003); the transformation of the old Soviet sphere of influence (Wedel, 2002); and the dynamics of cultural diversity, multiculturalism, and culture contact (Maybury-Lewis, 2002). In the modern world, people are less defined by traditions of "high culture" and more likely to be influenced (and to be drawn together as a global "community") by popular culture. The study of popular culture has been a staple of "cultural studies" for some time, and it is now well established in the mainstream disciplines as well (Bird, 2003; Fiske & Hartley, 2003).

The Changing Research Context: Virtual Worlds

If they so choose, ethnographers can free themselves of "place" by means of the Internet—the "location" for so many of the most interesting communities on the contemporary scene. Virtual communities are characterized not by geographic proximity or long-established ties of heritage, but by computer-mediated communication and online interaction. They are "communities of interest" rather than communities of residence. Although some can last a while, they are mostly ephemeral in nature, and sometimes even by design.

Ethnography has demonstrably been carried out online (Jordan, 2009) although the nature of observation is necessarily somewhat altered. Living online is a 21st-century commonplace, and ethnography can certainly move into cyberspace along with the technology. Some cautions, however, are in order. First, electronic communication is based almost exclusively on the written word, or on deliberately chosen images. The ethnographer who is used to "reading" behavior through the nuances of gesture, facial expression, and tone of voice is therefore at something of a disadvantage. Moreover, it is very easy for people online to disguise their identities; sometimes the whole purpose of participating in an online group is to assume a new identity. This is not to suggest that all individuals who may be found in virtual communities are engaging in deception. Indeed, members of such communities are all members of nonvirtual human organizations as well. Brigitte Jordan (2009) advances the paradigm of hybrid spaces, wherein the real and virtual personae of the members of online communities are considered.

A potential advantage to the use of the Internet as a vehicle for qualitative research is the potential for access to individuals who are reluctant to communicate directly. Russel Ayling and Avril Mewes (2009) describe Internet interviewing with gay men as one example of this process, although this advantage might also apply to other groups requiring or preferring concealment.

Using online interactions as a source of "observation" does, however, present certain challenges. For example, online conversations may well have deeply nuanced subtexts that depart markedly from the superficial meaning of the typed words. In the case of face-to-face conversations, a researcher can observe gestures, body language, use of space, and intonation patterns to go beneath the surface of the discourse. Other cues are almost certainly available to online in-groups, so it is imperative that researchers develop an understanding of the full range of communicative strategies available to members of a virtual community so that they do not rely on the words alone. Angela Garcia et al. (2009) cite an example of an online study of "skinheads" in which it was observed that the participants had

established techniques for the conveyance of physicality, emotion, and feelings.

But are virtual communities really all that similar to traditional communities or social networks? How does electronic communication bring new communities into existence even as it enhances the ways in which older, established communities, now geographically dispersed, can keep in touch? Such questions lead us to the possibilities of research about specific people and their lives, as well as about the larger processes by which people define their lives.

Virtual ethnography also poses some ethical challenges that are similar to, but not exactly the same as, those that confront the fieldworker in traditional communities. The accepted norms of informed consent and protection of privacy and confidentiality continue to be important, even though we are dealing with people we do not see face-to-face. Although the Internet is a kind of public space (which means it might, in theory at least, be exempt from IRB rules), the people who "inhabit" it are still individuals entitled to the same rights as people in more conventional places. There are as yet no comprehensive ethical guidelines applicable to online research, but a few principles seem to be emerging by consensus. First, research based on content analysis of a public website need not pose an ethical problem, and it is *probably* acceptable to quote passages posted on public message boards, as long as they are not attributed to identifiable correspondents. Second, members of an online community should be informed if an ethnographer is also online "observing" their activities for research purposes. If at all possible, the researcher should obtain "signed" informed consent forms from the members before continuing to be an observing presence on the site. Doing so might be impossible if a site attracts transient users; it remains to be decided whether informing the webmaster alone is sufficient. Members of a virtual community under observation should be assured that the researcher will not use real names (or identifiable made-up names), e-mail addresses, or any other identifying markers in any publications based on the research. If the online group has posted its rules for entering and participating, those norms should be honored by the researcher, just as he or she would respect the values and expectations of any other community in which he or she intended to act as a participant observer. By conforming to those posted rules, the researcher is, in effect, drawing the members into a collaborative circle; the research is as much a result of the community's practices as it is of the researcher's agenda. Some online ethnographers have also decided to share drafts of research reports for comment by members of the virtual community. By allowing members to help decide how their comments are to be used, the researcher furthers the goals of collaborative research. Because researchers working in cyberspace are operating with social formations that are much potential as existing in current real time (that is, they are perpetually "under construction"), an ethical posture that is active and anticipatory is needed, in contrast to the essentially reactive ethics of prior forms of research (Hakken, 2003).[1]

◫ THE SEARCH FOR SOCIAL JUSTICE

The new contexts and challenges for observational research in our time as discussed earlier take on particular significance when researchers aim to move beyond academic discourse and use the fruits of their research to make an appreciable change in the world. As Norman Denzin and Michael Giardina (2009, p. 11) insist, this is a "historical present that cries out for emancipator visions, for visions that inspire transformative inquiries, and for inquiries that can provide the moral authority to move people to struggle and resist oppression." Observation-based research can certainly play a role in the pursuit of an agenda of human rights-oriented social justice, if only by producing vivid, evocative descriptive analyses of situations (such as those reviewed in the previous section of this article) that can serve a consciousness-raising function. To the extent that observational research is conducted in a participatory/collaborative mode, it can empower formerly "voiceless" people and communities. To the extent that the fruits of such research are widely disseminated through multiple media (not just the traditional academic outlets) in ways that express the multivocalic nature of the research process, those formerly voiceless communities are able to participate in a variety of public forums in which their non-mainstream positions can be effectively aired.

Angrosino (2005, p. 739) has defined *social justice* as the obligation of all people to apply moral principles to the systems and institutions of society; individuals and groups who seek social justice should take an active interest in necessary social and economic reforms. To that end, I have suggested three ways in which researchers can make a contribution to the pursuit of social justice.

First, the researcher should be directly connected to those marginalized by mainstream society; that is, the researchers should feel some sort of kinship (be it political or emotional) with those being studied and not treat them solely as depersonalized objects of research. There may certainly be communities of people who are deservedly marginalized, and social justice is certainly not served by having ethnographers directly connected to, say, White supremacists or purveyors of child pornography. (It is certainly possible to argue that we cannot tell researchers which groups they can or cannot empathize with. But since the codes of professional ethics associated with the various social science disciplines all emphasize an adherence to standards of human rights, it seems fair to conclude that if researchers choose to affiliate with groups that exist explicitly to violate the rights of others, then they do so outside the limits of accepted ethical professional practice.) There is, however, no shortage of communities of people marginalized because of the structures of oppression built into the current economic and political world system. Helping them might well involve intensive study of power elites, but a progressive agenda goes by the boards if the researcher comes to identify with those elites and sees the marginalized simply as "target populations" for policies and programs formulated on high. Direct connection necessarily involves becoming part of the everyday life of a marginalized community. Research in service to a progressive agenda flows from a degree of empathy (not simply "rapport," as traditional ethnographers might have defined it) that is not available to those who strive to maintain an objective distance.[2]

Second, the researcher should ask questions and search for answers. This might seem like such an obvious piece of advice that it hardly seems worth discussing. But we are in the habit of asking questions based primarily of our scholarly (i.e., distanced) knowledge of the situation at hand. We move in a more productive direction if we begin to ask questions based on our experience of life as it is actually experienced in the community under study. By the same token, we must avoid the sentimental conclusion that "the people" have all the answers, as if poverty and oppression automatically conferred wisdom and foresight. Asking the relevant questions might lead us to look within the community for answers drawing on its own untapped (and perhaps unrecognized) resources, or it might lead us to explore options beyond the community. One very effective role for the committed collaborative researcher might be that of culture broker, putting people in the study community in touch with other circles of interest to which they might not otherwise have had access.

Third, the researcher should become an advocate, which might mean becoming a spokesperson for causes and issues already defined by the community. It might also mean helping the people discern and articulate issues that may have been unstated or unresolved to that point. Advocacy often means engaging in some sort of conflict (either among factions within the community or between the community and the powers-that-be), but it can also mean finding ways to achieve consensus in support of an issue that has the potential to unite. In either case, one ends up working *with* the community, rather than working *for* the community, which implies a more distanced stance.

The overall goal of this process is to empower the community to take charge of its own destiny—to use research for its own ends and to assert its own position relative to the power elite. A researcher may well retain a personal agenda (e.g., collecting data to complete a dissertation), but his or her main aim should be to work with the community to achieve shared goals that move it toward a more just situation. Such a philosophy can be difficult to convey to students or other apprentice researchers, and so it might be instructive to consider a form of pedagogy that, although not specifically designed for this purpose, certainly serves these ends.

"Service learning" is basically a way of integrating volunteer community service with active guided reflection. Although encouraging students to volunteer is certainly praiseworthy in and of itself, service learning programs give students the opportunity to study social issues from social scientific perspectives so they can understand what is going on in the agencies in which they are working. The combination of theory and action is sometimes referred to as *praxis,* and it is one way in which an engaged, committed, advocacy-oriented form of social science is carried out. Students do not simply carry out a set of tasks set by the agency—tasks that in and of themselves may not seem particularly meaningful but that take on very clear meaning when the students carefully observe the setting, the people, the interactions—in short, the total context in which those tasks are conducted. By combining academic learning with community service, students experience praxis—the linkage of theory and practice—firsthand (Roschelle, Turpin, & Elias, 2000, p. 840). Service learning was designed explicitly to reinvigorate the spirit of activism that energized campuses in the 1960s. Institutions that accepted this challenge formed a support network (Campus Compact) to develop and promote service learning as a pedagogical strategy. Service learning is now a national movement that has received recent additional public exposure in the form of the Clinton Global Initiative University. This nonpartisan project, founded by former President Bill Clinton, is designed to reach across college campuses and stimulate students to confront the challenges of pressing global issues (Clinton, 2008). It is the responsibility of concerned faculty to see that students have a service learning experience that expresses the three aspects of a social justice agenda as discussed earlier.

The philosophical antecedent and academic parent of service learning is experiential learning (e.g., cooperative education, internships, field placements), which was based on the direct engagement of the learner in the phenomenon being studied. The critical distinguishing characteristic of service learning is its emphasis on enriching student learning while revitalizing the community. To that end, service learning involves students in course-relevant activities that address real community needs. Community agencies are encouraged to take the initiative in defining their own needs and approaching the campus representatives to see if a group of students under faculty mentorship might be interested in helping them achieve their goals. Course materials (e.g., textbooks, lectures, discussions, reflections) inform students' service, and the service experience is brought back to the classroom to inform the academic dialogue and the quest for knowledge. This reciprocal process is based on the logical continuity between experience and knowledge. Anne Roschelle, Jennifer Turpin, and Robert Elias (2000) point out the critical importance of after-the-fact evaluation to ensure both a productive learning experience for the students and value to the community served. Elizabeth Paul (2006) argues for the inclusion of critical evaluation (or, perhaps more specifically, needs assessment) before the community-based effort to ensure that limited resources are most effectively applied.

The pedagogy of service learning reflects research indicating that we retain 60% of what we do, 80% of what we do with active guided reflection, and 90% of what we teach or give to others. The pedagogy is also based on the teaching of information processing skills rather than on the mere accumulation of information. In a complex society, it is nearly impossible to determine what information will be necessary to solve particular problems, especially those of intractable social inequalities. All too often, the content that students learn in class is obsolete by the time they obtain their degrees. Service learning advocates promote the importance of "lighting the fire" (i.e., teaching students how to think for themselves). Learning is not a predictable linear process. It may begin at any point during a cycle, and students might have to apply their limited knowledge in a service situation before consciously setting out to gain or comprehend a body of facts or the evolutionary development of a personal theory for future application. To ensure that this kind of learning takes place, however, skilled guidance in reflection on the experience must occur. By providing students with the opportunity to have a concrete experience and then assisting them in the intellectual processing of that experience, service learning takes advantage of a natural learning cycle and allows students to provide a meaningful contribution to the community (Marullo & Edwards, 2000).

It is important to emphasize that the projects that form the basis of the students' experience are generated by agencies or groups in the community, not by faculty researchers. These projects can be either specific one-time efforts (e.g., a Habitat for Humanity home-building effort) or longer-term initiatives (e.g., the development of an after-school recreation and tutoring program based at an inner-city community center). All such activities build on the fundamentals of observational research. Student volunteers gradually adopt membership identities in the community and must nurture their skills as observers of unfamiliar interactions to carry out the specific mandates of the chosen projects and to act as effective change agents. In this way, even service learning projects affiliated with courses outside the social sciences require students to become practitioners of observational research methods. At USF, service learning has been a key feature of a diverse set of courses, including an anthropology seminar on community development, a sociology course on the effects of globalization, an interdisciplinary social science course on farm-worker and other rural issues, a psychology course on responses to the HIV/AIDS epidemic, a social work course on racial and ethnic relations, and a business seminar on workplace communication and cultural diversity issues.

In sum, service learning affects the professional educator as well as the novice or student. Service learning is more than traditional "applied social science," which often had the character of "doing for" a community. Service learning begins with the careful observation of a community by a committed student adopting a membership identity; he or she goes on to an active engagement in and with the community in ways that foster the goals of a social justice-oriented progressive political and social agenda.

◘ PROSPECTS FOR OBSERVATION-BASED RESEARCH: ARE WE IN A POST-POSTMODERN PERIOD?

Patricia Adler and Peter Adler (1994, p. 389) observed, "Forecasting the wax and wane of social science research methods is always uncertain." Nevertheless, it is probably safe to say that observation-based research is going to be increasingly committed to what Lila Abu-Lughod (1991, p. 154) called "the ethnography of the particular." Rather than attempting to describe the composite culture of a group or to analyze the full range of institutions that supposedly constitute the society, the observation-based researcher will be able to provide a rounded account of the lives of particular people, focusing on the lived experience of specific people and their ever-changing relationships. Angrosino (2005, p. 741) has expressed some doubt about the stability of the marriage between observation-based research and more positivistic forms of social science, but we are no longer so certain that a divorce is imminent, at least to the extent that there is an emerging consensus around a social justice agenda such that the disagreements are more about means rather than about the ends of research.

It also seems safe to predict that observation-based research, no less than any other genre of social research, will be influenced by changing technology and the inescapable presence of the online parallel world. Whether in the virtual world or the real world, observation-based researchers will continue to grapple with the ethical demands of their work. Those who seek "exemption" from the guidelines seem to be very much in the minority. On the other hand, the rise of a committed, social justice-oriented agenda means that ethical questions of an increasingly complex and vexing nature will continue to arise. A renewed framework for understanding research ethics, such as the one proposed by the IOM, may be one way to deal with this issue.

It seems clear that the once unquestioned hegemony of positivistic epistemology that encompassed even so apparently humanistic a research technique as observation has now been shaken to its roots by the postmodernist critique among other factors. But what lies beyond that critique? Postmodernists often seem to suggest that because absolute truth is an impossibility, any effort to take action is bound to be compromised by the situational biases of researchers and would-be reformers. But it is certainly possible to base sound reformist action on the foundation of the provisional truth that results in the negotiated contexts created by researchers and their collaborators in study communities, as the service learning experiment seems to demonstrate. The IOM-style reform of research ethics (see IOM, 2002, and the discussion thereof in an earlier section of this chapter) is also based on such provisional, negotiated, collaborative arrangements (rather than absolutist edicts from on high). It is clear in hindsight that we needed the postmodernist critique to help us rethink the assumptions of our traditions of research; it is equally clear that we now have the means to go forward with the fruits of our rethinking—if we but have the political will to do so.

◘ NOTES

1. The Association of Internet Researchers has produced a document on ethical practice (Ess & the AoIR Ethics Working Group, 2002), which may be a useful guide for those pursuing this type of research. See also Bruckman (2002).

2. As of this writing, the term *empathy* is at the center of a complex, but eye-opening political debate. We use the term here in a much more restricted sense, mainly to refer to the development of a researcher's primary commitment to the agenda of the community under study.

◘ REFERENCES

Abu-Lughod, L. (1991). Writing against culture. In R. G. Fox (Ed.), *Recapturing anthropology: Working in the present* (pp. 137–162). Santa Fe, NM: School of American Research.

Adler, P. A., & Adler, P. (1987). *Membership roles in field research.* Newbury Park, CA: Sage.

Adler, P. A., & Adler, P. (1994). Observational techniques. In N. K. Denzin & Y. S. Lincoln (Eds.), *Handbook of qualitative research* (pp. 377–392). Thousand Oaks, CA: Sage.

Agar, M., & Feldman, H. (1980). A four-city study of PCP users: Methodology and findings. In C. Akins & G. Beschner (Eds.), *Ethnography: A research tool for policymakers in the drug and alcohol fields* (pp. 80–146). Rockville, MD: National Institute on Drug Abuse.

Angrosino, M. V. (2005). Recontextualizing observation: Ethnography, pedagogy, and the prospects for a progressive political agenda. In N. K. Denzin & Y. S. Lincoln (Eds.), *The SAGE handbook of qualitative research* (3rd ed., pp. 729–745). Thousand Oaks, CA: Sage.

Angrosino, M. V. (2007a). *Doing ethnographic and observational research.* Thousand Oaks, CA: Sage.

Angrosino, M. V. (2007b). *Naturalistic observation.* Walnut Creek, CA: Left Coast Press.

Angrosino, M. V., & Mays de Pérez, K. A. (2000). Rethinking observation: From method to context. In N. K. Denzin & Y. S. Lincoln (Eds.), *Handbook of qualitative research* (2nd ed., pp. 673–702). Thousand Oaks, CA: Sage.

Ayling, R., & Mewes, A. J. (2009). Evaluating Internet interviews with gay men. *Qualitative Health Research, 19,* 566–576.

Bateson, G. (1972). *Steps to an ecology of mind: Collected essays in anthropology, psychiatry, evolution, and epistemology.* San Francisco: Chandler.

Behar, R. (1993). *Translated woman: Crossing the border with Esperanza's story.* Boston: Beacon Press.

Behar, R. (1996). *The vulnerable observer: Anthropology that breaks your heart.* Boston: Beacon Press.

Bell, J., & Jankowiak, W. R. (1992). The ethnographer vs. the folk expert: Pitfalls of contract ethnography. *Human Organization, 51,* 412–417.

Bird, S. E. (2003). *The audience in everyday life: Living in a media world.* New York: Routledge.

Blackwood. E. (1995). Falling in love with an-Other lesbian: Reflections on identity in fieldwork. In D. Kulick & M. Willson (Eds.), *Taboo: Sex, identity and erotic subjectivity in anthropological fieldwork* (pp. 51–75). London: Routledge.

Brewer, D. D. (1992). Hip hop graffiti writers' evaluations of strategies to control illegal graffiti. *Human Organization, 51,* 188–196.

Bruckman, A. (2002). *Ethical guidelines for research online.* Available at http://www.cc.gatech.edu/~asb/ethics

Chavez, L. R., Flores, E. T., & Lopez-Garza, M. (1990). Here today, gone tomorrow? Undocumented settlers and immigration reform. *Human Organization, 49,* 193–205.

Clark, L., & Werner, O. (1997). Protection of human subjects and ethnographic photography. *Cultural Anthropology Methods, 9,* 18–20.

Clifford, J. (1983). Power and dialogue in ethnography: Marcel Griaule's initiation. In G. W. Stocking, Jr. (Ed.), *Observers observed: Essays on ethnographic fieldwork* (pp. 121–156). Madison: University of Wisconsin Press.

Clinton, W. J. (2008). A new way for students and colleges to bring about global change. *Chronicle of Higher Education, 54*(25), A40.

Creswell, J. W. (2007). *Qualitative inquiry and research design: Choosing among five approaches* (2nd ed.). Thousand Oaks, CA: Sage.

Dembo, R., Hughes, P., Jackson, L., & Mieczkowski, T. (1993). Crack cocaine dealing by adolescents in two public housing projects: A pilot study. *Human Organization, 52,* 89–96.

Denzin, N. K. (1997). *Interpretive ethnography: Ethnographic practices for the 21st century.* Thousand Oaks, CA: Sage.

Denzin, N. K., & Giardina, M. D. (2009). Qualitative inquiry and social justice: Toward a politics of hope. In N. K. Denzin & M. D. Giardina (Eds.), *Qualitative inquiry and social justice* (pp. 11–52). Walnut Creek, CA: Left Coast Press.

Ess, C., & the Association of Internet Researchers (AoIR) Ethics Working Group. (2002). *Ethical decision-making and Internet research: Recommendations from the AoIR Ethics Working Group.* Available at http://www.aoir.org/reports/ethics.pdf

Fiske, J., & Hartley, J. (2003). *Reading television* (2nd ed.). New York: Routledge.

Fluehr-Lobban, C. (2003). Informed consent in anthropological research: We are not exempt. In C. Fluehr-Lobban (Ed.), *Ethics and the profession of anthropology* (2nd ed., pp. 159–178). Walnut Creek, CA: AltaMira Press.

Friedenberg, J. (1998). The social construction and reconstruction of the other: Fieldwork in El Barrio. *Anthropological Quarterly, 71,* 169–185.

Garcia, A. C., Standlee, A. J., Bechkoff, J., & Cui, Y. (2009). Ethnographic approaches to the Internet and computer-mediated communication. *Journal of Contemporary Ethnography, 38,* 52–84.

Giroux, H. A. (1995). Writing the space of the public intellectual. In G. A. Olson & E. Hirsh (Eds.), *Women writing culture* (pp. 195–198). Albany: SUNY Press.

Gold, R. L. (1958). Roles in sociological field observation. *Social Forces, 36,* 217–223.

Gold, R. L. (1997). The ethnographic method in sociology. *Qualitative Inquiry, 3,* 388–402.

Grinnell, F. (2002). *The impact of ethics on research.* Washington, DC: Institute of Medicine.

Gupta, A., & Ferguson, J. (1996a). Beyond "culture": Space, identity, and the politics of difference. In A. Gupta & J. Ferguson (Eds.), *Culture, power, place: Explorations in critical anthropology* (pp. 33–52). Durham, NC: Duke University Press.

Gupta, A., & Ferguson, J. (1996b). Culture, power, place: Ethnography at the end of an era. In A. Gupta & J. Ferguson (Eds.), *Culture, power, place: Explorations in critical anthropology* (pp. 1–32). Durham, NC: Duke University Press.

Hakken, D. (2003). An ethics for an anthropology in and of cyberspace. In C. Fluehr-Lobban (Ed.), *Ethics and the profession of anthropology* (2nd ed., pp. 179–195). Walnut Creek, CA: AltaMira Press.

Hammond, J. L. (1996). Popular education in the Salvadoran guerilla army. *Human Organization, 55,* 436–445.

Institute of Medicine. (2002). *Responsible research: A systems approach to protecting research participants.* Washington, DC: Institute of Medicine.

Jordan, B. (2009). Blurring boundaries: The "real" and the "virtual" in hybrid spaces. *Human Organization, 68,* 181–193.

Koester, S. K. (1994). Copping, running, and paraphernalia laws: Contextual variables and needle risk behavior among injection drug users in Denver. *Human Organization, 53,* 287–295.

Lang, S. (1996). Traveling woman: Conducting a fieldwork project on gender variance and homosexuality among North American Indians. In E. Lewin & W. L. Leap (Eds.), *Out in the field: Reflections on lesbian and gay anthropologists* (pp. 86–110). Urbana: University of Illinois Press.

Mahmood, C. K. (1996). Why Sikhs fight. In A. Wolfe & H. Yang (Eds.), *Anthropological contributions to conflict resolution* (pp. 7–30). Athens: University of Georgia Press.

Malkki, L. H. (1996). News and culture: Transitory phenomena and the fieldwork tradition. In A. Gupta & J. Ferguson (Eds.), *Anthropological locations: Boundaries and grounds of a field science* (pp. 86–101). Berkeley: University of California Press.

Marcus, G. E. (1997). The uses of complicity in the changing mise-en-scene of anthropological fieldwork. *Reflections, 59,* 85–108.

Marullo, S., & Edwards, B. (2000). The potential of university-community collaborations for social change. *American Behavioral Scientist, 43,* 895–912.

Matsumoto, V. (1996). Reflections on oral history: Research in a Japanese-American community. In D. L. Wolf (Ed.), *Feminist dilemmas in fieldwork* (pp. 160–169). Boulder, CO: Westview Press.

Maybury-Lewis, D. (2002). *Indigenous people, ethnic groups, and the state* (2nd ed.). Boston: Allyn & Bacon.

Nardi, B., & O'Day, V. (1999). *Information ecologies: Using technology with heart.* Cambridge: MIT Press.

National Commission for the Protection of Human Subjects of Biomedical and Behavioral Research (1979). *The Belmont report: Ethical principles and guidelines for the protection of human subjects of research.* Washington, DC: U.S. Department of Health, Education, and Welfare.

Office of the Federal Register, National Archives and Records Administration, and U.S. Government Printing Office (2009). *Code of federal regulations.* Available from http:www.gpoaccess.gov/cfr

Ong, A., & Collier, S. J. (2005). *Global assemblages: Technology, politics and ethics as anthropological problems.* Malden, MA: Blackwell.

Paul, E. L. (2006). Community-based research as scientific and civic pedagogy. *Peer Review, 8,* 12–16.

Roschelle, A. R., Turpin, J., & Elias, R. (2000). Who learns from social learning? *American Behavioral Scientist, 43,* 839–847.

Saltzinger, L. (2003). *Genders in production: Making workers in Mexico's global factories.* Berkeley: University of California Press.

Sirin, S. R., & Fine, M. (2008). *Muslim American youth: Understanding hyphenated identities through multiple methods.* New York: New York University Press.

Sluka, J. A. (1990). Participant observation in violent social contexts. *Human Organization, 49,* 114–126.

Stepick, A., & Stepick, C. D. (1990). People in the shadows: Survey research among Haitians in Miami. *Human Organization, 49,* 64–77.

Tierney, W. G. (1997). Lost in translation: Time and voice in qualitative research. In W. G. Tierney & Y. S. Lincoln (Eds.), *Representation and the text: Re-framing the narrative voice* (pp. 23–36). Albany: SUNY Press.

van Gelder, P. J., & Kaplan, C. D. (1992). The finishing moment: Temporal and spatial features of sexual interactions between streetwalkers and car clients. *Human Organization, 51,* 253–263.

Walters, D. M. (1996). Cast among outcastes: Interpreting sexual orientation, racial, and gender identity in the Yemen Arab Republic. In E. Lewin & W. L. Leap (Eds.), *Out in the field: Reflections of lesbian and gay anthropologists* (pp. 58–69). Urbana: University of Illinois Press.

Wax, M. L., & Cassell, J. (1979). *Federal regulations: Ethical issues and social research.* Boulder, CO: Westview Press.

Wedel, J. (2002). *Blurring the boundaries of the state-private divide: Implications for corruption.* Available at http://www.anthrobase.com/Txt/W/Wedel_J_01.htm

Werner, O., & Clark, L. (1998). Ethnographic photographs converted to line drawings. *Cultural Anthropology Methods, 10,* 54–56.

Werner, O., & Schoepfle, G. M. (1987). *Systematic fieldwork: Vol. 1. Foundations of ethnography and interviewing.* Newbury Park, CA: Sage.

Wolf, D. L. (1996). Situating feminist dilemmas in fieldwork. In D. L. Wolf (Ed.), *Feminist dilemmas in fieldwork* (pp. 1–55). Boulder, CO: Westview Press.

Wolf, M. A. (1992). *A thrice-told tale: Feminism, postmodernism, and ethnographic responsibility.* Palo Alto, CA: Stanford University Press.

Zuniga, V., & Hernandez-Léon, R. (2001). A new destination for an old migration: Origins, trajectories, and labor market incorporation of Latinos in Dalton, Georgia. In A. D. Murphy, C. Blanchard, & J. A. Hill (Eds.), *Latino workers in the contemporary South* (pp. 126–146). Athens: University of Georgia Press.

29

VISUAL METHODOLOGY

Toward a More Seeing Research

Jon Prosser

A striking phenomenon of visual research a decade ago was its apparent invisibility. The malaise for things visual was replaced by positive engagement following a general awakening to the significance and ubiquity of imagery in contemporary lives. Visuals are pervasive in public, work, and private space, and we have no choice but to look. Qualitative researchers are taking up the challenge to understand a society increasingly dominated by visual rather than verbal and textual culture.

Visual research focuses on what can be seen. How humans "see" is part nature part nurture being governed by perception that, like other sensory modes, is mediated by physiology, culture, and history. Visual researchers use the term *visible* ontologically in referring to imagery and naturally occurring phenomena that can be seen, emphasizing the physiological dimension and disregarding their meaning or significance. *Visual*, however, is not about an image or object in of itself but more concerned with the perception and the meanings attributed to them. The terms *to visualize* and *visualization* refer to researchers' sense-making attributes that are epistemologically grounded and include concept formation, analytical processes, and modes of representation (Grady 1996; Wagner, 2006).

Current issues are best understood by reflecting on recent debates that shape contemporary visual research. Since the 1960s, there has been broad agreement that the type of media, mode of production, and context in which visual data are set are important in determining the meaning ascribed to imagery. In short, how researchers and others construct imagery and the kinds of technology used to produce them, are considered intrinsic to the interpretations of the phenomena they are intended to represent. Between 1970 and 2000, a dual paradigmatic disparity existed (what Harper in 1998, termed *the two-headed beast*) between researchers using images generated for empirical purposes and those who studied meanings of images produced by visual culture. The terms *empirical* and *symbolic* were used to denote the relative differences in perspectives during this period. Empirically orientated researchers stressed the importance of theory building and image creation and addressed the relationship between visual data, trustworthiness, and context, whereas symbolically inclined researchers focused on critical analysis of everyday popular visual culture. Hence, during this period an intellectual tension existed between those who read symbolic imagery and social scientists who created images for research purposes. By 2000, visual methods achieved normative status in sociology, anthropology, geography, health studies, history, the arts, and even traditionally quantitative disciplines such as psychology and medicine. Increasingly, we live in a visual world and currently, no topic, field of study, or discipline is immune to the influences of researchers adopting a visual perspective. The most important competency in societies around the world in the 21st century is visual fluency, and qualitative researchers are developing visual methodologies to study that phenomenon.

This chapter outlines key facets of contemporary visual research. Emphasis is placed on fieldwork undertaken in the qualitative tradition exemplifying insightful approaches, areas of concern, and future possibilities. I will begin by outlining *current trends* and conclude with a discussion of *future challenges*.

CURRENT TRENDS

In this section, under four subheadings, I consider an eclectic mix of methods and studies to illustrate how the science and art of conducting visual research is currently evolving: *Representation of visual research,* a long-standing and contentious issue of considerable importance to visual researchers yet mostly ignored by nonvisual qualitative research; *technology and visual methods* are in the ascendancy because they provide powerful strategies for answering complex global research questions involving analysis of metadata; *participatory visual methods* are well established and included because they represent the most popular genre in visual methodology; and *training in visual methods* because it is not a luxury but an imperative following the burgeoning growth in visual research around the world.

Representation of Visual Research

Tim Berners-Lee created the World Wide Web in the early 1990s, and his graphical point-and-click browser, *WorldWide-Web,* was the precursor to providing access to multiple audiences worldwide. The subsequent standard graphics packages (Excel and Adobe) for creating tables, bar charts, graphs, and pie charts are adequate for most quantitative data but limit qualitative researchers' capacity to represent data effectively by their "one size fits all" approach and limited representational range. Scientists and social scientists are finding this new world of representation challenging, and Luc Pauwels (2006, p. x) points out why:

> While there seems to be an implicit but persistent belief that the rapid spread of visual technologies in almost every sector of society automatically will result in an increased visual literacy or competency, there is at least as much reason to believe that the already vulnerable link between the referent, its visual representation, and the functions it needs to serve will come under even greater pressure.

Currently there is a slow shift toward data visualization for summation of data, displaying information, and providing an opportunity for analysis. In their basic form, interactive graphics provide additional data as the cursor is moved around the screen. The *New York Times* has excellent examples including findings from the 2008 American Time Use Survey, which asked thousands of Americans to recall how they spent every minute of their day (do an Internet search for "How different groups spend their day interactive graphic"). Statistics, especially in the form of large numbers, are difficult to relate to a human scale. A project called *Running the Numbers*[1] (developed by Chris Jordan) adopts creative approach to making large numbers accessible and meaningful. The images, usually sets of photographs, portray specific quantities: 15 million sheets of office paper (5 minutes of paper use in the United States), 106,000 aluminum cans (30 seconds of can consumption in the United States), or the 32,000 breast augmentation surgeries that take place in the United States every month. But these are early days, and critical reflection is needed to determine what are advances in social science and visual representation and what are eye candy.

Representation of visual research is in a depressingly stagnant state because mainstream dissemination in academia remains hard copy text–based and conservative. The passion for the printed page, the "thingness" of books as a sensual experience, continues to dominate, and only slowly the screen is emerging as a site for presenting findings of visual research. Qualitative visual researchers currently struggle to present their work outside of the traditional word/print format despite the potential that digital delivery systems have to change the way visual research is represented (Banks, 2007, Pink, 2008; Ruby, 2005). Some multimedia presentations on the Web look little different from the usual sequential text on a printed page, whereas others work sympathetically with different media and connect to other online representation via digital video, clickable hypermedia links, and podcast or sound essays. Visual researchers' screen-based authoring and blogs are increasingly seen as places to publish and disseminate research and methods (see, for example, David Gauntlett's site: http://www.artlab.org.uk/). The combination of text, images, blogs, twitters, vlogging (video blogging), and digital hypermedia are now part of visual research vocabulary signposting a future direction of communicating visual research.

An unexpected twist in representational forms is the reemergence of photographs as visual presentations rather than representations of research—unexpected given long-standing arguments concerning "the myth of photographic truth." Barry Goldstein's "All Photos Lie" (2007) is yet another restatement of this argument. None of this is new to photographers or visual researchers. Photographers have always known that their photographs represent a highly selected sample of the "real" world about which their images convey some subjective and empirical truth (Becker, 1986; Fox Talbot,[2] 1844; Ruby, 2005) in the same way that wordsmiths know that structure, vocabulary, and tone of their texts contain misrepresentations.

Arguably, social scientists are too preoccupied by their own epistemology to reflect on how to take account of and apply other forms of inquiry. Jon Wagner (2007), in his essay *Observing Culture and Social Life: Documentary Photography, Fieldwork and Social Research,* provides a robust case for a more seeing research by revealing similarities in epistemology across the word-image interface. Central to visual studies and the representation of findings where imagery is involved is the relationship between words and images: Is the space taken up with mostly words or images; which is most influential; is it necessary to translate images into words; and is it necessary to provide captions for images? Marcus Banks (2007) opts for photographs as a quote within contextualizing text as though it were an interview quote. Elizabeth Chaplin (2006), in "The Convention of Captioning," being part artist and part visual sociologist, balances the arguments for and against captioning images before opting for the latter and prescribing that readers should work harder to interpret by relating text to image.

Figure 29.1 "Inspector, 2005," *In the Kitchen* (2009a, p. 143, courtesy Dona Schwartz)

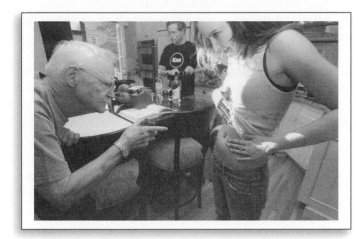

The next and most important step is for someone working within the epistemology of visual ethnography to represent a case for a photographic rather than word emphasis. Dona Schwartz's (2009a) domestic case study *In the Kitchen* is an exemplar of how this might look (see Figure 29.1). In her Keynote Photographer address at the 1st International Visual Methods Conference in 2009[3] and the challengingly titled "If a Picture Is Worth a Thousand Words, Why Are You Reading this Essay" paper (2007), she accepts that photographs are interpretations of the things to which they refer—they are not the thing itself. Schwartz acknowledges they are abstractions from ongoing time and space, a single arrested moment in time but, as Chaplin suggests, Schwartz challenges her audience to work harder at interpretation. More importantly, Schwartz makes strategic decisions and acts as a social scientist would but with important provisos:

> In making the case for making pictures I am suggesting that pictures can offer us ideas and an irreducible experience that cannot be restated or translated into linguistic terms. Articulations produced through photographs can offer us insights based on spatial and compositional arrangements, they can convey moods and emotions. They can generate novel ideas and inferences..... In the sciences, the idea of "productive ambiguity" with multiple readings giving rise to innovations that would have been unimagined had not a plurality of readings been possible. (2009b)

What Schwartz does is reverse the Banks and Chaplin call to place visual quotes in a contextualizing sea of words: "I am working to make pictures prime, to invest them with the thousand words I might otherwise write, and to present them in such a way as to insist that viewers read what the images have to say" (Schwartz, 2007, p. 320). A key element of this counterargument is the rejection of the widely held belief that visual literacy is somehow automatically improved because visual culture is so pervasive. Indeed, she argues that viewing images is a learned skill-based activity that differs from language, and being a fundamentally different communication system does not lend itself to context–free semiotic analysis. *In the Kitchen* has an erudite combination of word authors. The preface is a situating review of documentary photography of family settings by Alison Nordstrom (curator of photographs at George Eastman House in New York, the largest museum of photography in the world), followed by Schwartz's own rationale and methodological insights that are recognizable and acceptable to ethnographic and visual communication scholars. The following 170+ photographs are interspersed with poetry (by poet Marion Winik) that conjure up powerful mind imagery, inviting readers to search their own memory banks and examine their own values and beliefs about everyday family life. The history, ethnography, and humanities combination in this book illustrates how compelling interdisciplinary studies can be when research teams, working from a disciplinary base, are flexible enough to recognize the insights of others.

Technology and Visual Methods

Tools and techniques for seeing more and differently are key factors contributing to step changes in visual research. Advances such as the telescope, microscope, X-ray, ultrasound, MRI scanner, photography, and computers reflect our innate capacity to see, store, organize, and represent knowledge. Current documentary and participatory visual methods owe much to Steven Sasson's invention of the digital camera in 1975. Often, changes in research reflect changes in technology and vice versa, and qualitative visual researchers are well aware that technologies change what and how they study. With the advent of new technology for storing, organizing, analyzing, communicating, and presenting research, the qualitative-quantitative interface is being broadened, refined, and morphed. The cheapness and variety of image-making technology makes picture making and sharing a common activity. Still photography remains favored by participatory visual researchers, and video was widely used in the past by educationalists, anthropologists, ethnographers, and ethnomethodologists because it supports fine-grained analysis of complex social interaction (Goldman et al., 2007; Heath, Hindmarsh, & Luff, 2010).

Software for organizing and interpreting large volumes of imagery is becoming increasingly sophisticated. In the United Kingdom, the Economic and Social Research Council (ESRC) established the National Centre for e-Social Science (NCeSS), which increased interest in Computer Aided Qualitative Data Analysis Software (CAQDAS).[4] Consequently, there was an increased uptake of visual compliant software such as ATLAS.ti, NVivo, Transana (whose primary focus is audio/video analysis), and Observer XT, and the ESRC National Centre for Research Methods (NCRM) and Researcher Development Initiative (RDI) programs developed further software for analyzing digital data and provided nationwide training courses in their application.

The combination of improved software and training capacity enabled researchers with metadata to store, analyze, map, measure, and represent, for example, complex human communication with other interconnected entities.

The use of innovative digital technology and software opens new methodological possibilities for visual researchers. With colleagues, I am currently interested in an education paradox—although visual culture has increasingly come to dominate many areas of the social and personal lives of students and teachers, relatively little is known about the significance of visual culture to learning and teaching. The objective of our study is to generate a theoretical model of the dynamic relationship between societal visual culture and the visual culture of classrooms, with a view to understanding pedagogic consequences. As a starting point, given the paucity of theoretical models and the partial and fragmented nature of current knowledge, we will identify elements central to understanding visual culture of classrooms and examine how those elements are dynamically interrelated using the tentative framework illustrated in Figure 29.2 as a starting point.

We (eight researchers, each knowledgeable in different aspects of the visual culture of schooling) will employ a qualitatively driven, visually orientated, mixed method, interdisciplinary, participatory approach to developing a model. Data will be collected using orthodox visual methods and different media, for example:

■ Participatory visual methods, for example, graphical elicitation, photo elicitation, video diary, video elicitation, photovoice, timelines, and arts-based methods

Figure 29.2 A Starting Point for Modeling the Visual Culture of Classrooms (from Prosser, 2007)

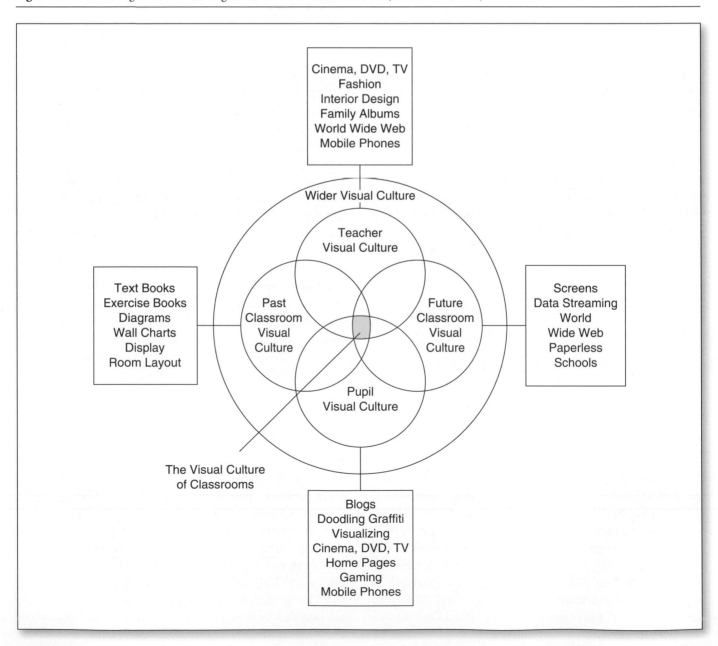

- Data collected via dual camera video to document versions of talk, bodily communication (e.g., gesture and facial expression, proxemic and kinesthetic data) and material transitional resources such as computer screens, textbooks, interactive whiteboards, classroom walls, and the learning space

- Data collected by eye tracking software (Duchowski, 2003; Tai , Loehr, & Brigham, 2006) to aid understanding how screens and textbooks in classrooms are scanned, and employ new collaboration-friendly geographical information systems) software that enables coding, annotating, and analysis of the use of classroom physical space with CAQDAS functionality

We are faced with two major problems: The first is analyzing, contrasting, and combining a diverse range of data and media; the second is creating a positive climate where ideas can be exchanged among the eight researchers and participants. The digitized data from the case study schools will be archived on and displayed via the Digital Replay System (DRS) to help resolve both problems. The DRS (see Figure 29.3 for an example but not of school data) is a next generation CAQDAS tool developed through the ESRC e-Social Science program. Like other CAQDAS tools, DRS enables the synchronization, replay, and analysis of audio and video recordings. Distinctively, DRS also enables these conventional forms of recording to be combined with system logs, or data "born digital," which record interaction within computational environments, or are the product of using computational techniques, to analyze data. These novel forms of digital data may include global positioning system (GPS) or WiFi logs, systems management server (SMS) logs, logs of network traffic, or logs generated by vision recognition software. Systems logs may be visualized, synchronized, and replayed alongside video and transcripts, or other conventional data, through the construction of log viewers.

The capacity of DRS to store and compare multiple forms of digitized data that consist of interrelated elements of societal and classroom visual culture is clearly important. However, so is the capacity of DRS to draw down and display information on screens or a large white wall for simultaneous research team and participant viewing. The synergistic possibilities for evolving multilayered models are considerable. It is possible, for example, to compare different digitized media across time (see the horizontal overlapping circles in Figure 29.2) and to contrast pupil and teacher visual subcultures (see the vertical overlapping circles and boxes in Figure 29.2) and compare them. Exciting possibilities for theory generation occur when the researchers and participants collaborate. We intend drawing on psychodynamic theory and applying the concept of using the DRS screen as a transitional space to stimulate creative play throughout the study with the ultimate aim of improving the quality of learning

Figure 29.3   Digital Replay System Example (courtesy Andrew Crabtree, Warwick University)

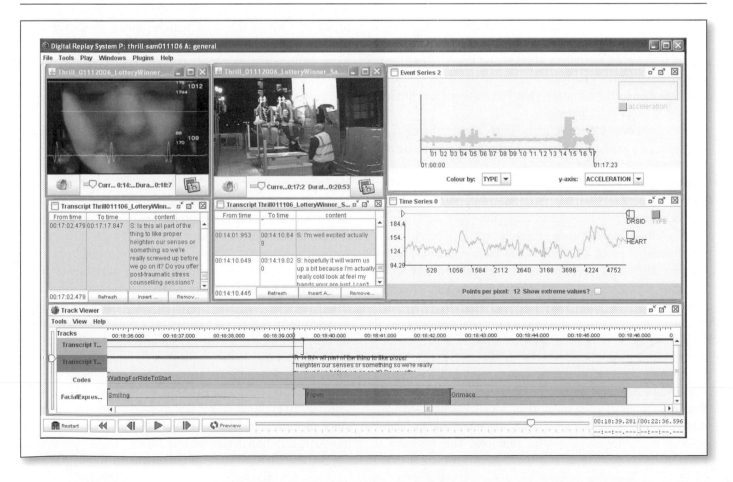

and teaching in classrooms. In addition, establishing an internationally searchable DRS archive opens other future synergistic opportunities. We are able to share different forms of digital data between our team members who might be located in different geographical locations in the United Kingdom and allow other research teams in different countries to access our archive and load their own data, enabling international comparisons of models of visual culture of classrooms to be made. When funding for research is constrained through a global fiscal downturn and costs of collecting metadata for analysis and synthesis are high, archiving with DRS for secondary analysis by others makes economic sense.

Participatory Visual Methods

Throughout the evolution of visual research, the researcher has been the instigator, designer, collector, interpreter, and producer in the research process. Post-1960 attention broadened to include the external narratives and combine researcher and participant insights. The earliest documented examples of this genre in visual research were by John Collier in mid-1950s (Collier & Collier, 1967/1986) who used photo elicitation as a way of stimulating interviewees' thinking during repeat interviews, and Sol Worth, John Adair, and Richard Chalfen's project (1972/1997) with Navajo in the mid-1960s. They were the forerunners of what is the most influential and abused methodological genre in contemporary visual research. Within the social sciences, broadly conceived participatory methodologies are evolving that are diverse and contested. Currently the role of technology, the potential gains and losses of participatory dissemination, and consequent ethical issues provide a major challenge. Participatory visual researchers are currently striving to meet that challenge.

Visual elicitation involves using photographs, drawings, or diagrams in a research interview to stimulate a response and remain the most popular and common method in participatory visual research. Photo elicitation is used as an "ice-breaker" or neutral third party when the power differential between researcher-researched is significant. Like most participatory visual methods, agreed upon protocols are rare. The method is not researcher-proof (Packard, 2008; Warren, 2005), and the biggest danger to democratization is when researchers come to the table with too many preconceptions in focus, process, or direction. After 50 years of application, even experienced practitioners think very carefully before exploring the meaning of images or objects with the interviewee (Harper, 2002).

Participants feel less pressured when discussing sensitive topics through intermediary artifacts. Because they do not speak directly about a topic on which they feel vulnerable but work through a material go-between (e.g., a doll, toy, line drawings, mobile phone, or memorabilia), they are more able to express difficult memories and powerful emotions. This approach is gaining in popularity because researchers believe that transitional objects have the capacity to be the locus of corporeal embodied memories. However, used injudiciously, without sensitivity, and under certain conditions, apparently innocuous visual stimuli and material culture can evoke inaccurate, distorted, unexpected, and even painful memories.

A long-standing strength of image/artifact elicitation is its capacity to evoke as well as create collective and personal memory. Chalfen's (1998) classic studies of how families in different cultures produce and use photographic albums and visual media to construct familial memory recognized that considering their temporal and contextual dimensions was of pivotal importance. The shift from looking in private through family snaps in plastic albums to semi-public browsing of web-based digitized multimedia family records illustrates changes in visual culture and the importance of considering lifestyle in any memory work. Making sense of web-based family photography is a relatively new but fast-growing area of study. Pauwels (2008) provides an insightful overview of this body of work by combining historical review, fieldwork, and identifying methodological problems. Current studies of the relationship between photography and memory draw on family collections, public archives, museums, newspapers, and art galleries for source material (see Kuhn & McAllister, 2006).

Marisol Clark-Ibanez (2007) uses participants' photographs in photo elicitation (she uses the term *auto-driving*) in her work with inner-city children to avoid overt voyeurism and to ensure topics relevant to the child remain central to the research agenda. Clark-Ibanez makes a strong case for using photo elicitation because she recognizes participants are expert in their own lives and able to define or refine the research, the agenda, and process. Obvious ground rules such as being sensitive to the participant's values, beliefs, lifestyle, and culture are emerging slowly, but some visual researchers use this method uncritically and without thinking through why or how it is to be used.

Photovoice and its variants (participative video, video diaries, photo-narratives, and photo-novella), which entail providing participants and collaborators with digital video or still cameras, remains the most commonly used visual method in social science research. An overtly political form entails giving participants cameras as an act of empowerment to generate changes in personal or community life or to influence policy directly. Here resultant photographs are circulated on the Internet; become local, regional, or national exhibitions; or are sent to government agencies as persuasive evidence. Such work tackles the power inequalities that create crises and sustain poverty and injustices through the suppression of marginalized voices. Researchers adopting this strategy are deeply committed to improving the lives of people with whom they work. Claudia Mitchell, for example, uses visual methods, particularly photovoice, to address problems arising from the rising tide of HIV and AIDS infection in sub-Saharan Africa. Mitchell has worked and lived in South Africa, Zambia, Malawi, Swaziland, Ethiopia, and Rwanda, serving community needs and acting as an adviser to various ministries. She draws together young people, researchers, and nongovernmental organizations and uses participatory visual and other arts-based methods to address issues of invisibility and marginalization. The work is important given that HIV/AIDS poses a major threat to development, poverty alleviation, and low life expectancy in sub-Saharan Africa. Five young people are being infected with HIV every minute, 7,000 every day, and 2.6 million every year. Worldwide, young people between the ages of 15 and 25 make up more than 40% of all new HIV and AIDS infections, with young women being disproportionately affected. Visual action research does not require publication to be life affirming and life saving.

Participatory video aims to create a narrative that conveys what respondents want to communicate in the manner they wish to communicate. Ruth Holliday (2007, p. 257) provides an insightful reflexive account of her use of video diaries to explore how "queer performances of identity are constructed as texts on the surface of bodies." She wanted to use a method in which participants were active in representing their own identities. The 15 diarists were given explicit instructions to explore their queer identities in three different spaces—work, rest (home), and play (the gay scene)—and to chart the similarities and differences in identity performances. They were also given specific instructions to film themselves in their typical choice of outfit for each setting and to comment on these settings, explaining to the camera what clothes, hairstyles, jewelry, and their other bodily arrangements were designed to portray. However, because of the constraints posed by the largely heterosexist culture in the United Kingdom, structured by "the closet," using a camera to film at a gay club (at play) proved impossible. Filming at work was equally difficult, and where it was done, it was carried out after hours so the spaces were strangely devoid of people. Hence much of the filming was carried out in the home.

Interestingly, Holliday considered and rejected feminist theory, the critique of objectivism, and masculine tendencies in research to arrive at distinctively queer methodology. She draws on the work of Judith Halberstam as partial justification of her choice of video diaries:

> A queer methodology, in a way, is a scavenger methodology that uses different methods to collect and produce information on subjects who have been deliberately or accidentally excluded from traditional studies of human behavior. (Holliday, 2007, p. 260)

In terms of generating theory, Holliday took the stance that given "self-filming" had an affinity with queer methodologies in their visual representations, encoding and decoding of queer (bodily) texts, a one-way (i.e., researcher) socio-semiotic reading of the videos was acceptable. The video diaries did, through the processes of watching, recording, and editing diaries before submission, afford participants the potential for a greater degree of reflection and control than other methods. However, clearly there were times when Holliday wanted to explore with the diarists, but did not, some of the points they had made, by follow-up video elicitation interviews.

It is interesting that beyond the initial specific instructions, the participants required no directions beyond working the video cameras. They knew how to behave in front of the camera and what was expected of them because they learned the technique from television. This is a clear demonstration of how visual methods and visual culture are inextricably linked. How so? In 1993, the British Broadcasting Corporation launched an audience inter-activity project aimed at encouraging people to record their lives on video to be aired nationwide. The resultant television program, *Video Nation*,[5] ran until 2001, preceding vlogging (video blogging) by 8 years. The program comprised quirky 2-minute shorts featuring everything from the profound to the gloriously trivial (my favorite was Scottish clan chief filming a flower-filled toilet in Finland). The program was highly popular in the United Kingdom and helped make video diaries in visual research fashionable and a genre that many, as in Holliday's study, are keen to emulate.

Holliday's diarists operated at a sophisticated level thanks to *Big Brother,* another highly popular television program in the United Kingdom. At the heart of Holliday's discussion is the idea of the confessional that is now commonplace across postmodern media and reality TV. In *Big Brother,* the show's pervasive filming of housemates makes for a variety of more or less self-conscience performances, but the soundproof "diary room" is set aside as a private space to talk directly to a voice that is "Big Brother." The viewing audiences expect brazen "truths" to be revealed in such diary confessions, although they know these truths are performances for the camera and public at large. Participants in Holliday's study are not naive but, rather, media savvy and well able to draw on available cultural codes to know how to perform the kind of reflexive mediated video diary required. Holliday is very much aware of this:

> Self-storying in the video diaries suggest that mediated confession is a performance in which participants knowingly, reflexively, and willingly engage; in a media-saturated and confession-saturated culture, this confirms the value of this method and suggests that need to more fully understand the culture work of the confession as a site of local mediated meaning making. (Holliday, 2007, p 278)

An unusual but important methodological twist in Holliday's study is that the participants, knowing how to behave in front of the camera and what was expected of them, are in control of the method.

Holliday ends with the familiar visual researchers' lament concerning representation of data and findings:

> In spite of the visual nature of the data that inform this chapter, I am left to present it using only text and a few still images. To capture the

flavour of the diaries in the text is extremely difficult and takes up an enormous amount of writing space . . . The nuances available in an audiovisual text are such that many simultaneous interpretations are possible . . . The diarists and their views are foregrounded in presentations, and the audience is similarly skilled at reading video diaries due to their near-constant use in lifestyle and reality TV (and indeed beyond). (Holliday, 2007, p. 276)

She bemoans the lack of acceptance in the wider academic community for representation beyond paper-based renditions. Having seen the video diaries many times, I can understand this position because they are highly informative, data-rich representations of participants' understanding of their own queer performance. Holliday's comments echo Peter Biella's view back in 1993 in "Beyond Ethnographic Film: Hypermedia and Scholarship" that mixed media studies are severely limited by text-based representations. Missing from major methodological texts are colors, movements, and sounds that are central representational forms in visual research.

Training in Visual Methods

Visual research in North America, Australasia, Scandinavia, Italy, and the United Kingdom is long-standing, distinctive, and above all, burgeoning. The benefits accrued from the rapid uptake include increased vitality, diversity, and a firmament that invites intellectual exchange. However, with increased popularity comes the need to support those new to visual methodology.

Newcomers extemporize by drawing knowledge from journals such as *Visual Studies,* accessing online resources such as *Visualanthropology.net,* joining organizations such as the International Visual Sociology Association (IVSA), and attending conferences and workshops. Many turn to books for insight. The best of the current bunch is Gregory Stanczak's (2007) edited *Visual Research Methods: Image, Society, and Representation.* This critically reflexive text covers different visual methodologies and analytic approaches from photography to virtual research, and ranges from sociology to religion and political science. The *Handbook of the Arts in Qualitative Research* edited by J. Gary Knowles and Ardra Cole (2008) is a 54-chapter celebration of the concepts, processes, and representational forms from the arts and a must-buy book for aspiring visual researchers with a creative turn. Another concise and erudite text is Banks' *Using Visual Data in Qualitative Research* (2007), which draws on exemplars from an anthropological tradition. Gauntlett's (2007) playful *Creative Explorations: New Approaches to Identities and Audiences* draws usefully on an eclectic array of disciplines including neuroscience and philosophy to explore how creative methods can be employed to understand identity and individuals' connections with the wider world. Pauwel's (2006) edited book *Visual Cultures of Science* brings together some first-rate essays that plug a gap in the literature by rethinking representational and visualization practices in the

communication of science. Finally, Paula Reavey's (2011) edited book *Visual Methods in Psychology: Using and Interpreting Images in Qualitative Research* will be welcomed by qualitatively orientated psychologists.

In the United Kingdom, as in many countries, the training provision before 2005 was mostly ad hoc. The ESRC began its involvement in visual methods training with the international seminar series Visual Evidence: the Use of Images in Social and Cultural Research (2000–2002), which brought together empirical and symbolic researchers from around the world. However, the Building Capacity in Visual Methods program (2006–2009), part of the ESRC Researcher Development Initiative, was the first nationwide integrated program aimed at teaching visual methods to a cross-section of qualitative researchers. The program provided a strategic capacity building plan to meet the training needs of contemporary and future visual researchers in the social sciences in the United Kingdom. The objectives were threefold:

- To provide methodology trainers, users of research, and active researchers new to visual research with core skills and resources to enable them to build a deep understanding of visual research methods
- To provide an ongoing, visual methods resource for researchers with experience in visual methods at intermediate level that is stimulating, challenging and grounded in "best practice"
- To establish a national infrastructure that was self-sufficient and developmental that meets the ongoing needs of the research community

There were two progressive and interrelated levels: The first level targeted new visual researchers inside and outside academia and from a broad spectrum of social science disciplines. This training program spanned 3 years and comprised two 2-day Introduction to Visual Methods workshops held at six monthly intervals each year at different locations around the United Kingdom. The second level was aimed at more experienced visual researchers in need of an intermediate level of visual methods training and built on the level one workshops. Stress was placed on approaches that were generic and transferable across the social sciences rather than on unique or specialized methods or fields. The second level of the training program spanned 3 years and comprised two 1-day Visual Methods Symposiums held at six monthly intervals each year. The culminating event of the program was the staging of the *1st International Visual Methods Conference* at the University of Leeds, September 2009, attended by 300 delegates from around the world.

The pedagogic dimension of Building Capacity in Visual Methods is worth considering as a model of delivery for anyone envisaging a national scheme. The first level—An Introduction to Visual Methods—presented visual research in a workshop format not as a stand-alone strategy taking one particular form but as theoretically and methodologically varied and compliant with the needs of a broad cross-section of qualitative researchers. Special consideration was given to ensuring that specific methods were related to the research process as a whole rather than to a technique applied in isolation. Hence, visual methods were discussed relative to different visual paradigms, visual media, analytical perspectives, and applied the study of a wide range of topics.

The level one 2-day workshops included overviews of visual frameworks (for example, visual ethnography), researcher-generated data (for example, object/photo elicitation and record/documentary techniques), participant-created data (for example, video diaries and arts-based methods), technology-based capture and analysis (for example, CAQDAS software-based analysis and eye-tracker technology), modes of analysis (for example, visual socio-semiotics, the internal and external narrative of an image), forms of representation (for example, graphics, mixed media, photography, and film), and visual ethics (for example, legal requirements, institutional ethics committee requirements, and issues of anonymity and confidentiality). Generic skills and knowledge were viewed as transferable, and workshop participants, being an interdisciplinary mix, were encouraged to learn from each other. The workshops were designed to be intensive experiences, and total engagement and commitment were expected. The team of 14 tutors (who also designed the program) was drawn from the most able visual methodologists in the United Kingdom and included Marcus Banks, Sarah Pink, Steve Higgins, Rob Walker, Gillian Rose, David Gauntlet, and me. Each 2-day workshop was repeated around the United Kingdom and usually comprised 12 short talks, 8 discussion sessions, and 4 hands-on practical activities. In addition, each participant was given a pedagogically informed hard-copy material to take away and digest, and ongoing support provided through an e-learning resource.

Practical activities typically consisted of short exercises involving photo elicitation, visual analysis, documenting complex events, and research design, designed to reinforce theoretical understanding and application of first principles. This element was included to give students a hands-on experience so they could gain confidence of techniques in action and gain an understanding of their potential pitfalls (Wiles et al., 2005).

The *hard-copy material* and supplementary basic texts (for example, Prosser & Loxley, 2008) were designed to complement face-to-face activities. The material was designed to meet the educative rather than strictly academic needs of participants. Hence, the papers were shaped by the authors (the tutors) asking themselves the question, "If I were new to visual research what sort of 'take-away' document would I need to help me apply visual methods to my own the field?" Each paper contained background information—for example, brief history, visual framework, and origin of the method—a protocol of how the method is applied in practice, analytical approaches that are appropriate with the method, and exemplars of good practice. The hard-copy material avoided jargon, explained technical words, and provided a short bibliography of the best literature available.

Face-to-face activities and hard-copy materials were further supported by a purposely designed *e-learning resource*. The tutors created a dedicated, dynamic, and evolving online learning environment based on open-source software that complied with interoperability standards. The environment included resources such as interactive exercises, documents, banks of still and moving images, and social software to enable discussion and allowed students individually or collaboratively to share their research stories. In this way, students continued to construct and benefit from the e-learning resources thus becoming part of a community of practice that supports learning across the life course. Furthermore, the flexibility of e-learning media and resources allowed a high level of personalization and enabled the inclusion of students who, for reasons of location, disability, or life constraints, might not otherwise be able to access visual research methods training. The project used Ning (www.ning.com) to build a space in which to enable an online community to develop. The Ning site tools including image and video uploading, video playback, e-mail to all users, discussion boards, blogs, and individual member profiles. After each of the training workshops, participants were invited to join the Visual Methods community on Ning. In addition to providing an informal space for online communication, the site hosted several online events during the life of the project. At each of these events, one of the tutors hosted a themed expert seminar—for example, "visual ethics"—usually for 1 day. During that time, members of the community used the discussion boards to ask questions about the theme of the session and to discuss these questions with the expert visual researcher hosting the seminar.

The rapid growth of interest in visual methodology caught many institutions and training facilities by surprise. Although those new to visual research or with limited experience are able to draw from journals, access visual online resources, join organizations, attend conferences and workshops, or read visual methods books, there is no substitute for pedagogically led training. An effective and efficient way to educate new visual researchers is for countries to train the trainers via a nationwide program designed specifically for that purpose.

▣   FUTURE CHALLENGES

In his summary for the visual chapter in the third edition of this book, Douglas Harper (2005, p. 760) looked to the future, partly in trepidation and partly in aspiration:

> My hope is that visual methods will become ever more important in the various research traditions where it already has a foothold . . . I hope that during the next decade, visual studies will become a world movement . . .

Harper's hopes have been fulfilled, for visual methods have permeated all disciplines and topics. My hope for the next decade is that visual methodology will have a greater impact on social science globally through refining its strengths, developing creative elements, and resolving emergent problematic aspects. Hence, this final section will consider potential growth and a limitation:

- A strength of visual methods: the capacity to create and innovate
- An opportunity for visual methods: postcards from the edge
- A threat to the visual methods: ethics regulation

A Strength of Visual Methods— The Capacity to Create and Innovate

A limiting factor of interviews conducted verbally is the narrow parameters of responses and that they favor the articulate. One of the strengths of visual research is the wide range of response possibilities and their capacity to harness the creative abilities of researchers and participants. More thought and imagination is needed in academic debate and that is why art is so important in visual research. Art can comprise complexity and contradiction, and unlike the arguments in an academic paper, art need not be linear. Art can describe, reflect, and evoke emotion, which dry facts or figures and cool logic rarely do. Art is often about stories, of lives and characters with whom an audience can identify. Above all, art can help us (researchers, participants, and interested communities) imagine what it might be like to live that life. It may not be obvious what the art is saying and maybe the artists do not know themselves, or do not know in a way that is communicated by words. Art is a tool for thinking and a very powerful means of expression and promoting discussion. Arts-based approaches invoke beyond-text sensations employed to access sensory phenomena that are highly meaningful in ways that are ineffable and invisible using conventional text-based methods. Arts-based research is vibrant, evolving rapidly, and defined by Shaun McNiff (2008, p. 29) as

> The systematic use of artistic process, the actual making of artistic expressions in all of the different forms of the arts, as a primary way of understanding and examining experience by both researchers and the people that they involve in their studies.

Elliot Eisner (2008, p. 7) claims a special provenance for including arts-based approaches to qualitative research:

> Langer (1957) claims that discursive language is the most useful scientific device humans have created but that the arts provide access to qualities of life that literal language has no great power to disclose.

Of the emergent visual paradigms, arts-based has the greatest potential to be innovative and insightful in terms of imagery, and Sandra Weber (2008, pp. 44–46) provides examples of why this might be the case:

> Images make us pay attention to things in new ways . . . images are likely to be memorable . . . images can be used to communicate more holistically, incorporating multiple layers, and evoking stories or questions; images can enhance empathic understanding and generalizability . . . through metaphor and symbol, artistic images can carry theory elegantly and eloquently . . . images encourage embodied knowledge . . . images can be more accessible than most forms of academic discourse . . . images provoke action for social justice.

Of course, there are different theoretical perspectives within arts-based research. Graeme Sullivan (2005) for example, critiques visual sociology and visual ethnography in particular as being too close to traditional sociological theory and methodological thinking that stress collaboration between "researchers and researched." Sullivan's hierarchical framework sees the artist occupying a position that is pivotal and supreme, where practice-based research within the artist's studio central, and hence the artist-as-researcher plays the sole role. It is somewhat ironic that Pink, a visual ethnographer who champions a variety of forms of arts-based participatory research, is criticized for limited reflexivity by Sullivan's version of art-based inquiry:

> Pink's text follows a strategy common to most research in critical and visual cultural inquiry in that it emphasizes the critique and analysis of phenomena, but has very little to say about the creation of new knowledge using visual means that might be taken within a research perspective. (Sullivan, 2005, p. xv)

Sullivan's provocative and narrow evocation of "artist as researcher" perspective underestimates Pink's capacity to work with diverse audiences and encourage agency in many. An indicator of the innovative possibilities, the energy and insights possible of an arts-based approach is revealed in Knowles and Cole's *Handbook of the Arts in Qualitative Research* (2008), mentioned earlier. It covers a gamut of arts-informed and arts-based research, each with a distinctive methodology and promoting the notion that art should be regarded as a form of knowledge and not merely an ornamental product of human experience. The 54 chapters cover mixed modes, media, methodologies, and representations, for example, "Collage as Inquiry" (Butler-Kisber, 2008), "From Research to Performance" (Cancienne, 2008), "Exhibiting as Inquiry" (Church, 2008), "Installation Art-as-Research" (Cole & McIntyre, 2008), "Psychology: Knowing the Self Through the Arts" (Higgs, 2008), and "Ethnodrama and Ethnotheatre" (Saldaña, 2008).

An Example of the Creative Possibilities of Visual Methods

A distinctive capacity of visual methods is to improve the quality and trustworthiness of data and findings by drawing on participants' own resourcefulness and ingenuity. But what does an innovative visual enquiry look like? In *Creative Explorations* (2007), Gauntlett explains how, by adopting the methodological middle ground between method-sparse postmodern and cultural studies thinking and method-limiting word-dominated approaches, fresh insights about people's experiences are gained. He claims his "new creative methods" draw on people's needs to engage with the world. Participants are invited to spend time creatively making something metaphorical or symbolic about their lives and then reflect on their creation. Participants are introduced to the notion of "creative explorations" via simple experiments—for example, "build a creature" and then "in the next two minutes, turn the creature into how you feel on Monday morning or Friday afternoon." Participants play with this idea, and, for example, a walrus-type creature is turned into a Friday afternoon feeling by adding a wagging tail, a zingy hairstyle, or a set of wheels, to appear obviously excited and looking forward to the weekend (Figure 29.4).

Figure 29.4   Build a creature and turn it into a Friday afternoon feeling (courtesy David Gauntlett).

Then the real work begins, and participants are asked to construct their identity and include key influences. In his current work, Gauntlett asks diverse groups of people, including architects, unemployed people, and social care workers, to construct models of their identity using LEGO blocks. He believes that a more hands-on, minds-on, approach by participants makes for more truthful results (see Figure 29.5 and www.artlab.org.uk).

Gauntlett draws from an eclectic array of theories and disciplines ranging from neuroscience to philosophy to provide the theoretical underpinning and rationale for his method. His approach is far removed from the approaches of psychotherapists and art therapists of a past era who asked their subjects to construct or make something and then referred to a diagnostic manual to give them the expert insight into what a patient's artwork "actually" meant. Gauntlet draws on phenomenologist Maurice Merleau-Ponty to support his premise, that it is the process, particularly the embodied experience, of creating identity that is crucial to participants' capacity to provide a more realistic response than that obtained by word-only interviews.

In addition, he cites Paul Ricoeur's *The Rule of the Metaphor* (2003) and his work on the concept of metaphor at the level of discourse and, more importantly, his work on narrative and

Figure 29.5   LEGO serious play—each part represents something meaningful (courtesy David Gauntlett).

identity. Gauntlett employs the human capacity to use meta-phors to advantage by providing participants with full LEGO kits so a tiger could be used in a model to represent ambition, pride, a driving force, and so forth, and a bridge could represent variously challenges, connecting people, or the chance to reach higher places. He makes a strong claim for his creative explora-tions approach to building knowledge of society:

> Pictures or objects enable us to present information, ideas or feel-ings simultaneously, without the material being forced into an order or a hierarchy. Language may be needed to *explain* the visual, but the image remains primary and shows the relationships between parts most effectively. (Gauntlett, 2007, p. 15)

Gauntlett's critique of orthodox verbal interviews rests on the belief that interviewers have unreasonable expectations of interviewees. He feels that people's brains do not usually con-tain ready-made lists of "what I think" about topics such as identity. This gives rise, he believes, to participants generating "instant answers" that are imprecise and inaccurate. LEGO maps of identity help participants to form words and ideas at the speed and in the way they feel is their expression in the time that's best for them. The serious LEGO approach is one example of an innovative visual researcher using the creative capacities of participants.

An Opportunity for Visual Methods—Postcards From the Edge

Because of the ubiquity of the semi-structured interview and the sample survey that favor the articulate (Mason,[6] 2002, 2008), participants with communication difficulties, learning difficulties, and other disabilities are inhibited from taking part in research. Despite the long-standing trend toward inclusive research and working "with" rather than "on" participants, the voice and agency of the least able in society is often missing. It is important to respect people with disability and to accept they can be powerful, beautiful, and sexy. Visual methodologists can make a major contribution here by adopting an egalitarian stance and by working alongside the most vulnerable, under-represented, and least researched and understood members of society. Ontological and epistemological difficulties, coupled with the current global fiscal downturn and emergent enthusi-asm for cost-effective research, will probably deflect the moral compass of research. It is relatively easy to obtain a research grant to fund social science curiosity but a lot tougher to make a difference to people's lives or advance social justice.

Currently in the United Kingdom, there are 850,000 people with dementia, 500,000 people with autistic spectrum disorders, and many others with intellectual disabilities who are habitually and systematically excluded from research data because of the underlying assumption that they are insufficiently articulate to contribute through interviews or sample surveys. Hence, little is known about the quality of life of autistic people after they leave school, research with dementia sufferers is limited, nothing is known about the quality of life of disabled people with a dual diagnosis (e.g., people with Down syndrome in their 50s who experience dementia 30 years ahead of the norm), and very little is known about the subjective well-being of disabled people with learning and communication difficulties.

Many in the disabled community communicate insights about themselves and their lives despite limited skills in speech and writing. Lester Magoogan,[7] for example, is a young man with Down syndrome who expresses his bubbly personality and unique perspectives on life through simple but evocative line drawings (see Figure 29.6). His images convey his cognitive understanding and emotional insight into "two-faced people," "coming out of nowhere," and "windy" in a way that language has no great power to reveal.

Figure 29.6   The Drawings of Lester Magoogan (copyright Lester Magoogan, www.lestermagoogan.co.uk)

Here I will provide three examples of using sensory methods with disabled people, each with an unadorned but important methodological message. The first is the simple strategy of asking two people with different disabilities to work together, the second illustrates the dangers of ocular centrism and participants "speaking" and the researcher hearing but not listening, and the third demonstrates the potential of technology as a collaborative tool of visual research to communicate with those previously considered noncommunicators. "Postcards from the edge" is an apt metaphor for this section because it comprises superficial messages on one side and concise, pithy visual statements from individuals on the other, and always the sender is a long way off—at the edge of research possibilities.

1. To be a child and disabled is to be doubly disadvantaged in terms of voice. A well-known "draw and write" method was used to help Jane, a young girl with a fire phobia who was experiencing recurrent nightmares with a fire theme. She is autistic, dyspraxic, aphasic with learning difficulties, experiences problems relating to others, and was unable to speak or write expressively. I asked Jane to draw her nightmare (Figure 29.7) and a disabled but slightly more able peer, her only friend, who through a language they had developed between them, helped her to write a prayer to

Figure 29.7 "Dear Lord Happ me because cos I am ily sad" (published with permission)

accompany the image. It became apparent from the drawing and words and later gesticulations that Jane had seen TV footage of the New York 9/11 disaster. She was deeply disturbed by the experience particularly by the sight of people leaping from the building to avoid the fire (see bodies at the base). Later counseling based on the visual information enabled Jane to escape the nightly replay of the film loop in her mind that caused the nightmares to recur.

The draw and write technique and other visual methods (see Thomson, 2008) are often used with "normal" children. Jane and her friend were able to combine their strengths to communicate their views to me. This is not always the case, and the underlying assumption is that children with a mental disability or learning difficulty are neither well informed nor sufficiently able to contribute. This legitimizes not seeking or accepting their views and justifies researchers too quickly seeking professionals, parents, or guardians to speak for them. It is an undeniable right of all human beings to be heard and participate in studies in which they are expert.

I am currently undertaking a study to explore the perceptions of disabled people with limited communication skills of their subjective well-being through a "sensescape" approach.

2.   Andrew is a participant in the project, and I have known him for 8 years. He is 42 years of age and suffered a brain hemorrhage when he was 12. He has significant physical and mental impairment. Despite very poor eyesight and motor skills, his paintings of his emotional state are skilful and insightful. However, for a long time I was unable to use them as part of an image-elicitation approach because of his limited interpersonal communication. When showing me his paintings, Andrew mostly sang apparently unrelated songs that were popular when he was 12 years old. I listened and tried to be patient, but it became very clear that he was becoming increasingly impatient with me. Then I realized that he was singing answers to my questions. When I asked him how he was feeling at the start of an interview, he would sing a Beatles song "It's been a hard day's night . . ." However, when asked about a drawing that to me denoted "sadness," he would come back with a Shakespearean quote "To be, or **NOT** to be: that is the question" (Andrew's emphasis), which I interpreted as a comment on his drawing and his state of mind (Is it better to live or to die?). In subsequent meetings, he continued to take songs memorized before his illness and by changing intonation or emphasis or by playing subtle word games with lyrics, he communicated his views. Now we conduct interviews by singing, and my job is to rotate his choice of song or turn of phrase until it catches the light of my reasoning.

When someone suffers a brain trauma as in Andrew's case, or has a congenital mental disability, elements of cognition are lost or depleted. The residual components reconfigure to make use of the remaining cerebral capacity, and this may entail increased capacity in one or more senses. Andrew communicates his views in two sensory forms—drawings and songs and

verse—drawn from his memory stores. Like other visual researchers, I believed that photography (i.e., imagery) is *the* central mnemonic device (for example family photo albums), but Andrew accessed his memory banks not for images but for tunes, lyrics, and prose that he could manipulate to his own ends. Only when I understood the combined visual and auditory/oracy senses and their respective roles, did I and Andrew make sense of each other. Visual researchers should not assume that the visual has primacy over other senses but, rather, other senses are part, but not all, of our engagement with the world and therefore important to our understanding of society (Mason & Davies, 2009).

The capacity to communicate where communication was once thought impossible is changing. Improvements in intensive care have led to an increase in the number of patients who survive severe brain injury. Some of these patients go on to have a good recovery but others awaken from the acute comatose state and do not show any signs of awareness. Those who yield no evidence of purposeful sensory response are given a diagnosis of a vegetative state.

3.   Martin Monti et al. (2010) reported a study in *The New England Journal of Medicine* that gives credibility to the possibility that some patients classified as vegetative are actually conscious, and a few may be able to communicate. The researchers used functional magnetic resonance imaging (fMRI) to scan 54 patients' brains to record any activity generated following verbal prompts and questions from the doctors. They found signs of awareness in five patients who demonstrated the ability to generate willful, neuroanatomically specific responses during two established mental-imagery tasks, one of whom was able to answer basic "yes" or "no to questions by activating different parts of his brain after being instructed to do so. In conclusion, Monti et al. stated,

> These results show that a small proportion of patients in a vegetative or minimally conscious state have brain activation reflecting some awareness and cognition. Careful clinical examination will result in reclassification of the state of consciousness in some of these patients. This technique may be useful in establishing basic communication with patients who appear to be unresponsive. (Monti et al., 2010, p. 579)

The results indicate how much we still have to learn about visual methods and sensory consciousness. Researchers can potentially communicate with people diagnosed as in a vegetative state through auditory or other sensory stimuli, record responses visually (fMRI), and ask simple questions requiring a "yes" or "no" response. Ethical problems will arise if doctors ask bigger questions—for example, whether the patient wants to live or die, and the answer is "die." This development should stimulate sensory qualitative researchers to rethink communicating with participants who are sensorially challenged. The study also illustrates the power of applied technology to question our assumptions of what is possible in participatory research.

The sensorial experiences of people with disabilities are important because sensory relationships are pivotal domains of cultural expression and the medium through which well-being is enacted and depicted. Sensory relations are also social relations. Understanding how people with intellectual disabilities navigate changing social and cultural landscapes through their senses is fundamental to understanding how micro and macro contexts affect, for example, happiness and well-being. Disabled people make sense of their lives through the interplay of sensory relations not accessible through discourse; words are mere proxies for their direct experiences. Text- and verbal-based approaches are limited because they fail to move beyond inherent psychophysical characteristics to reveal taken-for-granted, embodied, sensorial lives.

A Threat to Visual Research—Ethics Regulation

In framing my argument for a more "seeing" research, I stressed the importance of normalizing visual methods within mainstream disciplines, paradigms, and practices. I conclude this chapter on a somber note by describing a significant threat to the collaborative and developmental possibilities of visual methods.

Increased regulation and bureaucratization of the ethical review process is particularly noticeable in Europe, the United States, Canada, Australia, New Zealand, and the developing world (van den Hoonaard, 2002). Researchers, institutions, and funding bodies are bound together by a web of ethical regulation that depends on limited value notions. Ethical regulations in the United States are viewed by Norman Denzin and Yvonna Lincoln (2005, p. 1123) as "out of date for the purposes of qualitative research and entirely useless for the development of culturally, racially, and ethically sensitive methods." Unfortunately, the U.S. regulatory/medical model of research ethics was widely adopted in the United Kingdom (Tinker & Coomber, 2004) and in mainland Europe. An empirical study of researchers' experience and understanding of visual ethics in the United Kingdom by Rose Wiles and colleagues (2010, p. 2) concluded,

> Critiques of the process of ethical approval for social scientific research include the following concerns: (1) the capacity of ethics committees to make ethical judgments across a wide range of research approaches and contexts, some of which might require particular and specialist knowledge; (2) the model of regulation which is perceived to be based on that of biomedical research where there are actual physical risks to participants and disproportionate for social science research relative to the level of potential risk; (3) the consequences of ethical regulation on social research practice including making the use of some research approaches or topics difficult or impossible, impeding positive relations between researchers and participants, encouraging researchers to tell "half truths" (Atkinson, 2009) on forms for

ethical approval and encouraging researchers to think of ethics as a one-off event rather than as a series of issues that need consideration through the lifetime of a research project. Thus paradoxically a case has been made that the new regimes of ethical approval might actually be discouraging ethical thinking among researchers.

Visual researchers working within the qualitative paradigm shoulder an additional burden because visual methods comprise a wide array of approaches and types of media and hence raise particular challenges for ethics committees adopting a conservative, regulatory/medical approach to ethics. Confidentiality, legal issues (such as copyright), and dissemination of visual data are problematic for visual researchers faced with ethics committees with limited knowledge of visual methods. Because, for example, anonymity is considered central to ethical research, ethics committees may adopt a restrictive "safety first" stance when faced with a visual study in which plans to use or publish images that make people identifiable. An important claim made by visual researchers is that visual methods can reveal important information that text or word-based methods cannot. Hence, any attempt to disguise people without careful reasoning and due cause can remove the very point of the data and the moral rights of participants who wish to have their voices heard.

The primary danger of the current system of bureaucratic regulation and review process is the preoccupation with gaining ethical approval. There is a real possibility that genuine discussions of sensitive visual ethical dilemmas will be put to one side in the effort to demonstrate that a study is ethically sound and acceptable to the legal, regulatory conscious members who constitute ethics committees. In encouraging making visual research appear to be respectable in the eyes of ethics committees, the potential for raising genuine ethical concerns will be discouraged and diminished to the detriment of visually located ethical debate and future good practice. Qualitative visual researchers should know enough about the culture, society, or community through their research to make sound sensitive moral decisions. The overarching aim of any ethical regulatory system should be to develop visual researchers' integrity and knowledge base:

> Ethical reflexivity is a matter of awareness and sensitivity and is reflected in the degree of honesty and truthfulness in their dealings with others. These values are a measure of researchers' integrity and professionalism and are increasingly a requirement of research institutions and funding bodies aspiring to excellence. To act ethically is to value integrity, inclusiveness, personal security, privacy and dignity. For visual researchers, ethics guidelines and codes of practice cover important principles, but "visual" research brings additional, potentially distinct, ethical conundrums. (Clark, Prosser, & Wiles, 2010, p. 90)

For qualitative visual researchers, an "ethics of care" approach, which is an important but less common model that challenges the deontological framework underpinning biomedical ethics, is much preferred. Here, ethical decisions are made on the basis of care, compassion, and a desire to act in ways that benefit the individual or group that is the focus of research rather than following universalist principles or absolute norms and rules that may govern ethical decision making. Experienced qualitative visual researchers (e.g., Banks, 2007; Harper, 1998; Pink, 2008; Rose, 2007) seek to implement collaborative relationships in their research relationships that have some commonality with an ethics of care approach. The current trend toward a biomedical regulatory ethics may slow the development visual methods but nothing will stop visual methods' becoming one of the most important qualitative research methodologies in the 21st century.

▣ NOTES

1. Do an Internet search for "Running the Numbers" to find Chris Jordan's work. http://www.chrisjordan.com/current_set2.php

2. Stanczak (2007, p. 8) points out that William Fox Talbot, founder of paper-process photography, wrote, "It frequently happens, moreover—and this is one of the charms of photography—that the operator himself [sic] discovers on examination, perhaps long afterwards, that he has depicted many things he had no notion of at the time."

3. The 1st International Visual Methods Conference, convened by Jon Prosser, was held at Leeds University, United Kingdom, September 15–17, and was part of an Economic and Social Research Council (ESRC) Researcher Development Initiative program "Building Capacity in Visual Methods." The second conference will take place at the Open University, United Kingdom, in 2011.

4. CAQDAS Networking Project, University of Surrey: http://caqdas .soc.surrey.ac.uk/. The project is in receipt of its seventh consecutive term of funding by the UK ESRC. The research project Qualitative Innovations in CAQDAS (QUIC) is currently funded by the ESRC National Centre for Research Methods (NCRM).

5. See ongoing BBC *Video Nation* clips at http://www.bbc.co.uk/ videonation/

6. Real Life Methods and Realities projects led by Jennifer Mason at the University of Manchester are both ESRC, National Centre for Research Methods projects that are at the cutting edge of creative, qualitatively driven research methods. The Morgan Centre is currently the main center in the United Kingdom for innovative qualitative research methods. See http://www.socialsciences.manchester.ac.uk/ realities/ or search on the Internet for "Realities Manchester, UK."

7. Lester Magoogan has exhibited at Tate Modern and Lowry galleries in London and appeared on television in the United Kingdom and abroad. Lester Magoogan's art can be found at www.lester magoogan.co.uk

▣ REFERENCES

Atkinson, P. (2009). Ethics and ethnography. *21st Century Society: Journal of the Academy of Social Sciences, 4*(1), 17–30.

Banks, M. (2007). *Using visual data in qualitative research.* London: Sage.

Becker, H. (1986). *Doing things together.* Evanston, IL: Northwestern University Press.

Biella, P. (1993). Beyond ethnographic film: Hypermedia and scholarship. In J. Rollwagen (Ed.), *Anthropological film and video in the 1990s.* New York: Institute Press.

Butler-Kisber, L. (2008). Collage as inquiry. In J. G. Knowles & A. L. Cole (Eds.), *Handbook of the arts in qualitative research* (pp. 265–276). Thousand Oaks, CA: Sage.

Cancienne, M. B. (2008). From research to performance. In J. G. Knowles & A. L. Cole (Eds.), *Handbook of the arts in qualitative research* (pp. 397–406). Thousand Oaks, CA: Sage.

Chalfen, R. (1998). Interpreting family photography as a pictorial communication. In J. Prosser (Ed.), *Image-based research: A sourcebook for qualitative researchers* (pp. 214–234). London: Falmer Press.

Chaplin, E. (2006). The convention of captioning: W. G. Sebald and the release of the captive image. *Visual Sociology, 21*(1), 42–54.

Church, K. (2008). Exhibiting as inquiry. In J. G. Knowles & A. L. Cole (Eds.), *Handbook of the arts in qualitative research* (pp. 421–434). Thousand Oaks, CA: Sage.

Clark, A., Prosser, J., & Wiles, R. (2010). Ethical issues in image-based research. *Arts & Health, 2*(1), 81–93. Available at http://dx.doi .org/10.1080/17533010903495298

Clark-Ibanez, M. (2007). Inner-city children in sharper focus: Sociology of childhood and photo-elicitation interviews. In G. Stanczak (Ed.), *Visual research methods: Image, society, and representation* (pp. 167–196). Thousand Oaks, CA: Sage.

Cole, A. L., & McIntyre, M. (2008). Installation art-as-research. In J. G. Knowles & A. L. Cole (Eds.), *Handbook of the arts in qualitative research* (pp. 287–298). Thousand Oaks, CA: Sage.

Collier, J., & Collier, M. (1986). *Visual anthropology: Photography as a research method.* Albuquerque: University of New Mexico Press. (Original work by J. Collier published in 1967)

Denzin, N. K., & Lincoln, Y. S. (Eds.). (2005). Epilogue. The eighth and ninth moments—Qualitative research in/and the fractured future. In *The SAGE handbook of qualitative research* (3rd ed., pp. 1115–1126). Thousand Oaks, CA: Sage.

Duchowski, A. T. (2003). *Eye tracking methodology: Theory and practice.* London: Springer.

Eisner, E. (2008). Art and knowledge. In J. G. Knowles & A. L. Cole (Eds.), *Handbook of the arts in qualitative research* (pp. 3–12). Thousand Oaks, CA: Sage.

Fox Talbot, W. H. (1844). *The pencil of nature.* Cambridge, MA: Capo Press. (Original published in a series 1844–1846)

Gauntlett, D. (2007). *Creative explorations: New approaches to identities and audiences.* New York: Routledge.

Goldman, R., Pea, R., Barron, B., & Derry, S. (Eds.). (2007). *Video research in the learning sciences.* Mahwah, NJ: Lawrence Erlbaum.

Goldstein, B. M. (2007). All photos lie: Images as data. In G. Stanczak (Ed.), *Visual research methods: Image, society, and representation.* Thousand Oaks, CA: Sage.

Grady, J. (1996). The scope of visual sociology. *Visual Sociology, 11*(1), 10–24.

Harper, D. (1998). An argument for visual sociology. In J. Prosser (Ed.), *Image-based research: A sourcebook for qualitative researchers* (pp. 20–35). London: Falmer Press.

Harper, D. (2002). Talking about pictures: A case for photo-elicitation. *Visual Studies, 17*(1), 13–26.

Harper, D. (2005). What's new visually? In N. K. Denzin & Y. S. Lincoln (Eds.), *The SAGE handbook of qualitative research* (3rd ed., pp. 747–762). Thousand Oaks, CA: Sage.

Heath, C., Hindmarsh, J., & Luff, P. (2010). *Video in qualitative research.* London: Sage.

Higgs, G. E. (2008). Psychology: Knowing the self through arts. In J. G. Knowles & A. L. Cole (Eds.), *Handbook of the arts in qualitative research* (pp. 545–556). Thousand Oaks, CA: Sage.

Holliday, R. (2007). Performances, confessions, and identities: Using video diaries to research sexualities. In G. Stanczak (Ed.), *Visual research methods: Image, society, and representation* (pp. 255–280). Thousand Oaks, CA: Sage.

Knowles, J. G., & Cole, A. L. (Eds.). (2008). *Handbook of the arts in qualitative research.* Thousand Oaks, CA: Sage.

Kuhn, A., & McAllister, K. E. (2006). *Locating memory: Photographic acts.* New York: Berghahn Books.

Langer, S. K. (1957). *Problems of art: Ten philosophical lectures.* New York: Scribner.

Mason, J. (2002). *Qualitative researching.* Thousand Oaks, CA: Sage.

Mason, J. (2008). Tangible affinities and the real life fascination of kinship. *Sociology, 42*(1), 29–45.

Mason, J., & Davies, K. (2009). Coming to our senses? A critical approach to sensory methodology. *Qualitative Research, 9*(5), 587–603.

McNiff, S. (2008). Art-based research. In J. G. Knowles & A. L. Cole (Eds.), *Handbook of the arts in qualitative research* (pp. 29–40). Thousand Oaks, CA: Sage.

Monti, M. M., Vanhaudenhuyse, A., Coleman, M. R., Boly, M., Pickard, J., D., Tshibanda, L., et al. (2010). Willful modulation of brain activity in disorders of consciousness. *New England Medical Journal, 362*(7), 579–589.

Packard, J. (2008). "I'm gonna show you what it's really like out here": The power and limitation of participatory visual methods. *Visual Studies, 23*(1, April), 63–77.

Pauwels, L. (Ed.). (2006). *Visual cultures of science.* Hanover, NH: Dartmouth College Press.

Pauwels, L. (2008). A private practice going public? Social functions and sociological research opportunities of web-based family photography. *Visual Studies, 23*(1, April), 34–49.

Pink, S. (2008). *Doing visual ethnography* (2nd ed.). Thousand Oaks, CA: Sage.

Prosser, J. (2007). Visual methods and the visual culture of schools. *Visual Studies, 22*(1), 13–30.

Prosser, J., & Loxley. A. (2008). *Introducing visual methods.* ESRC National Centre for Research Methods. NCRM/010 Review papers. Available at http://eprints.ncrm.ac.uk/420/

Reavey, P. (2011). *Visual methods in psychology: Using and interpreting images in qualitative research.* London: Routledge.

Rose, G. (2007). *Visual methodologies* (2nd ed.). London: Sage.

Ricoeur, P. (2003). *The rule of the metaphor: Multi-disciplinary studies of the creation of meaning in language.* Toronto, ON: University of Toronto. Toronto.

Ruby, J. (2005). The last 20 years of visual anthropology. *Visual Studies, 20*(2), 159–170.

Saldaña, J. (2008). Ethnodrama and ethnotheatre. In J. G. Knowles & A. L. Cole (Eds.), *Handbook of the arts in qualitative research* (pp. 195–208). Thousand Oaks, CA: Sage.

Schwartz, D. (2007). If a picture is worth a thousand words, why are you reading this essay? *Social Psychology Quarterly, 70*(4), 319–321.

Schwartz, D. (2009a). *In the kitchen.* Heidelberg, Germany: Kehrer Verlag.

Schwartz, D. (2009b). *Visual art meets visual methods: Making a case for making pictures.* Keynote photographer's address, 1st International Visual Methods Conference, Clothworkers' Hall, University of Leeds, Leeds, UK.

Stanczak, G. (Ed.). (2007). *Visual research methods: Image, society, and representation.* Thousand Oaks, CA: Sage.

Sullivan, G. (2005). *Art practice as research: Inquiry in the visual arts.* Thousand Oaks, CA: Sage.

Tai, R. H., Loehr, J. F., & Brigham, F. J. (2006). An exploration of the use of eye-gaze tracking to study problem-solving on standardized science assessments. *International Journal of Research and Method in Education, 29*(2), 185–208.

Thomson, P. (Ed.). (2008). *Doing visual research with children and young people.* New York: Routledge.

Tinker, A., & Coomber, V. (2004). *University research ethics committees: Their role, remit and conduct.* London: King's College.

Van den Hoonaard, W. C. (2002). *Walking the tightrope: Ethical issues for qualitative researchers.* Toronto, ON: University of Toronto Press.

Wagner, J. (2006). Visible materials, visualised theory and images of social research. *Visual Studies, 21*(1), 55–69.

Wagner. J. (2007). Observing culture and social life: Documentary photography, fieldwork and social research. In G. Stanczak (Ed.), *Visual research methods: Image, society, and representation.* Thousand Oaks, CA: Sage.

Warren, S. (2005). Photography and voice in critical qualitative management research. *Accounting, Auditing and Accountability Journal, 18*, 861–882.

Weber, S. (2008). Visual images in research. In J. G. Knowles & A. L. Cole (Eds.), *Handbook of the arts in qualitative research* (pp. 44–45). Thousand Oaks, CA: Sage.

Wiles, R., Coffey, A., Robison, J., & Prosser, J. (2010). Ethical regulation and visual methods: Making visual research impossible or developing good practice? *Sociological Research Online.* Retrieved February 2011 from http://www.socresonline.org.uk/

Wiles, R., Durrant, G., De Broe, S., & Powell. J., (2005). *Assessment of the needs for training in research methods in the UK social science community.* ESRC National Centre for Research Methods. Available at http://www.ncrm.ac.uk/research/outputs/publications/reports.php

Worth, S., Adair, J., & Chalfen, R. (1997). *Through Navajo eyes: An exploration in film communication and anthropology.* Albuquerque: University of New Mexico Press. (Original work published in 1972)

30

PERFORMATIVE AUTOETHNOGRAPHY

Critical Embodiments and Possibilities

Tami Spry

When the "I" seeks to give an account of itself, an account that must include the conditions of its own emergence, it must, as a matter of necessity, become a social theorist.

—Judith Butler (2005, p. 8)

S/he who writes, writes. In uncertainty, in necessity. And does not ask whether s/he is given the permission to do so or not.

—Trinh T. Minh-ha (1989, p. 8)

Performance sometimes resists, exceeds, and overwhelms the constraints and strictures of writing.

—Dwight Conquergood (1991, p. 193)

This chapter is an ensemble piece. . . . It asks for a performance, one in which we might discover that our autoethnographic texts are not alone.

—Stacy Holman Jones (2005, p. 764)

This chapter seeks engagement body to body, with hurting and healing bodies, with "articulate bodies" (Pineau, 2000), and with necessarily uncertain bodies; it seeks co-performance in the entanglements of accounting for "I" and in the rupture and rapture of performance that may well exceed the constraints of its (and this) writing. It seeks co-presence with others in the "chorus of discordant voices" (Denzin, 2008) in qualitative research. It is a bid for autoethnographic ensemble, for social theorizing with those laboring for disruptive dialogue and transformative pedagogies on the page and in the stages of our lives together post–9/11, postmodernity, postcolonially, post-political monologism.

Critical reflection on loss and the development of hope will comprise the autoethnographic bones of the chapter. Fragmentation, dismemberment, delivery of body/story will be implemented as metaphors to interrupt colonizing narratives of 9/11

and other personal/political and local/global issues of loss toward a performative pedagogy of hope and possibility. In performative autoethnographic fashion, the essay must *do* autoethnography as it articulates autoethnographic theory/methodology praxis, pushing and pulling between the "constraints and strictures of writing" (Conquergood, 1991) and the sometimes messy, resistant, and epistemologically overwhelming performing body.

Autoethnography Lost and Found

Autoethnography is body and verse.

It is self and other and one and many.

It is ensemble, a cappella, and accompaniment.

Autoethnography is place and space and time.

It is personal, political, and palpable.

It is art and craft. It is jazz and blues.

It is messy, bloody, and unruly.

It is agency, rendition, and dialogue.

It is danger, trouble, and pain.

It is critical, reflexive, performative, and often forgiving.

It is the string theories of pain and privilege

forever woven into fabrics of power/lessness.

It is skin/flints of melanin and bodies

in the gendered hues of sanctuary and violence.

It is a subaltern narrative revealing the understory of hegemonic systems.

It is skeptical and restorative.

It is an interpreted body

of evidence.

It is personally accountable.

It is wholly none of these, but fragments of each.

It is a performance of possibilities.

Performative autoethnography is a critically reflexive methodology resulting in a narrative of the researcher's engagement with others in particular sociocultural contexts. Performative autoethnography views the personal as inherently political, focuses on bodies-in-context as co-performative agents in interpreting knowledge, and holds aesthetic crafting of research as an ethical imperative of representation. At least, this is what I have come to know, because autoethnography, for me, has been about performing what I had thought impossible, about moving in and out of trauma with words and blood and bone. It has been about dropping down out of the personal and individual to find painful and comforting connection with others in sociocultural contexts of loss and hope.

Autoethnographically inhabiting the process of losing a son in childbirth felt like the identification of body parts, as if each described piece of the experience were a cumbersome limb that I could snap off my body and lay on the ground. There was a bizarre and profound comfort in admitting to and describing this feeling of dismemberment and fragmentation. I had been trying to glue myself together with dominant cultural narratives about grief, performed on its "five stages," and tried time and time again to stifle the waves of wails coming from an empty body and from Twin Towers of ashes. An 18-month period brought the loss of a child, the bombings of 9/11, the death of a close colleague from cancer, the loss of our beloved Minnesota Senator Paul Wellstone, and the death of my father.

I am thankful for the disciplinary wisdom to view lived experience through theories of embodiment because it was only in trusting the embodied knowledge that "I am an un/learning body in the process of feeling," that I began to heal (Madison, 2006, p. 245). Reinhabiting the only space I ever lived with my son motivated a deeply embodied theorizing about the narrative disposition of this grief (Spry, 2006). I felt a deep somatic connection to that fractured self and space, like I was moving back into my body, a body that I had abandoned with the birth and subsequent death of our son. Like bell hooks, "I came to theory desperate, wanting to comprehend—to grasp what was happening in and around me. Most importantly, I wanted the hurt to go away. I saw in theory then a location for healing" (hooks, 1994, p. 59). After years of moving through pain with pen and paper, asking the nurse for these tools the morning after losing our child was the only thing I could make my body do. Though it felt as if my arms had bounced stiff and clumsily about my ankles that day, the language of bodies came pouring out.

However, the resulting performative autoethnography, "Paper and Skin: Bodies of Loss and Life" (Spry, 2004), was certainly not written at the bedside of my grieving body. To believe so would be to romanticize the processes of grief and of performative autoethnography. Articulating my own personal pain is *not* autoethnography. Novelist David Foster Wallace suggests, "All the attention and engagement and work you need to get from the reader can't be for your benefit; it's got to be for hers" (quoted in Max, 2009, p. 48). It is the intentional and critically reflexive connection of this narrative to larger social issues, to the politics, pleasure, and pain of other people, that distinguishes performative autoethnography as a methodology grounded in forging knowledge with others to dismantle and transform the inequities of power structures.

In "Autoethnography: Making the Personal Political," Stacy Homan Jones describes autoethnography as "a performance that asks how our personal accounts count" (2005, p. 764). Like Jones, I, too, come to autoethnography and research from performance studies, particularly from the disciplinary turns of performance studies toward ethnography and ethnography toward performance studies; performative autoethnography emerges from this academic and artistic space. The "crisis of representation," which I have contended elsewhere, was not so much a crisis for performance studies artist/scholars as it was a recognition of a familiar (Spry, 2006). Our disciplinary roots are grounded in interpretation, a process wrought with the crisis and complexity of representation. The "performative turn" in ethnography (Turner, 1986) has expanded the scope and recognition of the cultural/political implications of performance studies (Spry, 2006). Similarly, performance studies theories of embodiment and textual interpretation inform ethnographic methods of ethics, researcher positionality, cultural performances, and fieldwork (Conquergood, 1985, 1991; Schechner,

1985). Mary Strine mapped the "cultural turn" in performance studies, asserting that the "cultural-performance matrix" refocused perspectives on how performative forms and practices have served to "produce, sustain, and transform" systems of power and dominance and directed us toward less traditional texts (personal narratives, oral histories, performance art) (1998, pp. 6–7). Strine, with Conquergood, argued that this cultural-performance matrix signals a paradigmatic shift "from performance as a distinctive *act* of culture to performance as an integrated *agency* of culture" (Strine, 1998, p. 7).

Performance and ethnography continually turn back on themselves emerging as praxes of participatory civic social action (Alexander, 2006; Denzin, 2006, 2008; Jones, 2005; Madison & Hamera, 2006: Pelias, 2004). Nowhere is the complexity, utility, and ethical implications of these praxes more evident than in D. Soyini Madison's work, *Critical Ethnography: Method, Ethics, and Performance* (2005), which remains at this point in the development of critical/performative/pedagogical autoethnography the pivotal work informing auto/ethnographic performance scholarship. Performative autoethnography continues its epistemological development through insistence on the critique of our historical and sociocultural emergence. From this scholarly heritage, I find hope through loss with others in autoethnography.

Performative autoethnography then, as a personal, political, and social praxis, and as a critically reflexive methodology, can provide the framework to critically reflect on the ways in which our personal lives intersect, collide, and commune with others in the body politic in ways alternate to hegemonic cultural constructions. Autoethnography provides an apparatus to pose and engage the questions of our global lives.

This chapter is a textual "performance of possibilities" (Denzin, 2006b; Madison 2005) in a time when loss and hardship are found in our foreclosing homes, in our millions of No Children Left Behind, in our sisters and brothers recovering from and simultaneously shipping out to war(s). I hear Norman Denzin and Michael Giardina:

> Never before has there been a greater need for a militant utopianism that can help us imagine a world free of conflict, terror, and death; a world that is caring, loving, and truly compassionate: a world that honors healing. Postmodern democracy cannot succeed unless critical cultural scholars adopt methodologies that transcend the limitations and constraints of a lingering politically and racially conservative postpositivism. (2007, p. 12)

In a time of backlash with the rise in violent militia spurred on by media hate mongers (Keller, 2009) performative autoethnography is a methodology that can "transcend the limitations and constraints" of a lingering—and possibly growing—neoconservative postpositivism. It is a personal/political praxis, an aesthetic/epistemic performance, and a critical/indigenous/advocational ethnography that operates from a compassionate and lionhearted will to usurp and resist injustice.

What follows in this chapter then, is a response to Jones's 2005 call to "disrupt, produce, and imagine" co-performatively (p. 763). The essay will track representative histories ("Fragmented Histories Absent and Present"), offer interpretive criteria ("Performative Fragments and Embodied Possibilities"), discuss pedagogical developments ("Critical Fragments of Craft"), and posit future directions of autoethnography ("Concluding Fragments"). Throughout each section, the concept of the performative-I researcher disposition will be further developed as the nexus of performative autoethnography (Spry, 2006).

And so I have come to autoethnography yearning to know what is possible after absence, to embody the afterlife of loss, after 9/11, after an un/American politics of ignorance and bullying. In my/our body/politic are dark and empty and furtive places where things did not come to fruition, places from which there were no fruits of the labor. But writing performatively and autoethnographically with others in these darknesses reveals hope in the hollow, potential in the filling, and cleansing in the emptying.

▣ FRAGMENTED HISTORIES ABSENT AND PRESENT

> What we call *history* is strictly woven into each cultural environment. The making of historicizing therefore depends on "local knowledge."
>
> —Antonis Liakos (2008, p. 139)

> The past, home, is not a perfect memory—it will not save us.
>
> —Elizabeth Adams St.Pierre (2008, p. 122)

In considering a history of autoethnography, it becomes clear that the contestation of history itself has been a catalyst of autoethnography as a methodology, revealing autoethnography's potential to break and remake canons of history through localized subaltern knowledges. Much autoethnographic work has been about the business of recognizing and articulating the multiplicity of histories that exist within any past event or historical epoch, and examining the ways in which dominant narratives of History with a capital "H" perpetuate and maintain the racism, classism, and sexism that in/forms dominant thought, memory, and imagination (Grande, 2008; Liakos, 2008; St.Pierre, 2008). In *Searching for Yellowstone: Race, Gender, Family, and Memory in the Postmodern West* (2008), Norman K. Denzin employs critical autoethnographic reflexivity concerning his childhood at Yellowstone, popular representations, and scholarly discourses to "create a new version of the past, a new history. I want to create a chorus of discordant voices (and images) concerning Native Americans and their place in Yellowstone Park as well as in our collective imagination" (p. 18). Here, critical reflection on how one's personal experiences collide with hegemonic

History puts bodies in motion that break and remake historical memory and imagination, and foreground the power of performative autoethnography to empower critical collaborative meaning making to challenge and change, in this case, the romantic nostalgia reifying racist performances of Native Americans. Autoethnography can democratize historicizing by critically reflecting on the inherent collaboration and collisions between selves and others in the performativity of race, class, and other politicized identities. In critiquing the "minstrel shows that replay the Wild West," Denzin seeks to "replace old stereotypes with new understandings. I want to show how historical discourse can in fact turn back on itself, revise its stance toward the past, and perform new, progressive representations of cultural difference" (p. 23). Performative autoethnography makes present the absence of diverse, indigenous, and subaltern histories providing a multiplicity of representations that may challenge, argue, and disagree with one another, but, by their very presence, dispel the racial myths and minstrel shows of monologic history written by the politically powerful.

More specifically, the work of indigenous scholars critiques "history" as a hegemonic Westernized product of modernity in need of decolonization (Anzuldúa, 2007; Grande, 2008; Mutua-Kombo, 2009; Smith, 1999; Swadener & Mutua, 2008). In examining the process of canon and exclusion in historicizing, Liakos writes, "Since the eighteenth century, the tradition of history writing in Europe involves not only a description of the past, but also the imposition of a hierarchical view of the world, with Europe perched at the top" (2008, p. 143). And in the fashion of ethnography before the "crisis of representation," historians defined all that was non-Western and non-European as abject, abnormal, exotic, and "uncivilized." But here, for "historical discourse [to] in fact turn back upon itself" it must also view this "crisis" as a crisis only to colonizers; being represented as a subcategory to all that is moral and good was, and still is, merely a state of affairs to those colonized by History and the History of research. Research produced by privileged Whites sometimes reads as if they "discovered" the tragedy and inequity of racism, placing people of color as witnesses to their "findings" rather than as intellectual interlocutors, as evidence rather than as agents. In discussing decolonizing research schemata, Beth Blue Swadener and Kagendo Mutua argue,

> Non-Western knowledge forms are excluded from or marginalized in normative researched paradigms, and therefore non-Western/indigenous voices and epistemologies are silenced and subjects lack agency within such representations. Furthermore, decolonizing research recognizes the role of colonizing in the scripting and encrypting of a silent, inarticulate, and inconsequential indigenous subject and how such encryptions legitimize oppression. (2008, pp. 33–34)

Though only one in a large and efficacious body of methods underpinned by indigenous methodologies, performative autoethnography breaks the colonized encrypted code of what counts as knowledge redefining silence as a form of agency and positioning local knowledge at the heart of epistemology and ontology (Spry, 2008; Visweswaran, 2006). Performative autoethnography interrupts and opens constructs of history, thereby reframing what a "crisis" is, and to whom, and who has the discursive power to define the crisis and its antecedents and antidotes in the first place.

History, Time, and the Other

The process of performative autoethnography starts with a body, in a place, and in a time. In an earlier work, I considered autoethnography as an embodied praxis by employing Clifford Geertz's notion of "being here" and "being there" (Spry, 2001a). Though Geertz's construct recognizes time as an efficacious element in ethnography, its significance seems a taken–for–granted element, which, Johannes Fabian argues, can lead one to "disregard the many ways in which time is used to construct otherness" (2007, p. 49). How we represent the autoethnographic body "now" and "then" is of equal importance as "here" and "there." "What we know about the politics of time," writes Fabian, "should have epistemological consequences" (p. 49). Concepts of time and history go hand in hand in their ability to hegemonize and to proliferate representations of otherness as "inarticulate and inconsequential" (Swadener & Mutua, 2008). The Westernized terms *Third World, developing, underdeveloped,* and *primitive* exemplify attempts by dominant power systems to assign temporal constructs that situate the Other in the past, reifying Others as always and already "behind the times," as a quaint romanticized anecdote. Like the "Noble Savage," the Other is not represented as a contemporary, but as a hapless morally and intellectually inferior exotic. Fabian argues that the representation of time through language becomes a tool allowing the (auto)ethnographer to distance self from others, again positioning others as objects of study rather than as co-performers of knowledge, *co-present* in the time and process of knowledge production.

This is especially relevant to indigenous research where "natives" have been studied for the "richness of their past," rather than engaged as contemporary agents of epistemology (Denzin, 2008; Fabian, 1983; Grande, 2008; Liakos, 2008; Smith, 1999). Linda Tuhiwai Smith articulates the tradition of imperialism and colonialism in research, explaining in an understandably oft-quoted observation that the word *research* itself "is probably one of the dirtiest words in indigenous world's vocabulary" (1999, p. 1). Additionally, cultures view time in radically different ways, that is, circular, past and present as one, history as a space of still living experience, among other temporal constructs, which subsequently changes epistemological forms and content.

Living with others in time(s) and understanding the potentially colonizing apparatus of time is tantamount to autoethnographic research. In her discussion of a *Red* research construct of *Indianismo,* Sandy Grande explains that the "notion of *Indianismo* stands outside the polarizing debates of essentialism and post-modernism, recognizing that *both the timeless and temporal are essential for theorizing the complexity of indigenous realities*" (2008, p. 241, emphasis mine). In performative autoethnography, time is viewed as politically contested and contingent, as is cultural place, space, and identity—all of which effect and are effected by varying notions of temporality and history.

For a performative autoethnographer, the critical stance of the performing body constitutes a praxis of evidence and analysis through time and place. We offer our performing body as raw data of a critical cultural story. The performance work of Guillermo Gómez-Peña illustrates in stark clarity the exoticized and sexualized "Mexican" male body. Gómez-Peña embodies, and then flips these images by placing himself in a cage for audiences to observe "the noble savage." "The way in which society sees itself," writes Liakos, "determines both the historical view and vice versa: culture is historically determined not only because of its formation in time but also on account of the perceptions over time that constitute part of the warp and weft of culture" (2008, p. 139). Autoethnographic constructions and embodiments of history can trouble our perceptions of, "How far have we come, or not?" when considering cultural oppressions.

Performative autoethnography can interrupt master narratives that become "stuck in time" through its continual re/creation of knowledge by critically reflecting back on who we are, and where, and when. This kind of reflection constitutes a continual opening to the natures of temporality and its sociocultural representations, placing all of us in a coeval relationship with each other over time.

Déjà Vu by Any Other Name

In a 2006 special issue of the *Journal of Contemporary Ethnography,* editors Scott Hunt and Natalia Junco presented essays that responded to Leon Anderson's work on "analytic autoethnography" where the ethnographic researcher is "a full member in the research group or setting" and is "committed to an analytic research agenda focused on improving theoretical understandings of broader social phenomenon" (Anderson, 2006, p. 375). I mention this special issue here because it presents significantly varying conceptions of the historical development of autoethnography. In his essay, Anderson decides that Carolyn Ellis and Arthur Bochner's (2006) work on "evocative or emotional ethnography" (Anderson, 2006, p. 374) constitutes a rejection of "realist and analytic ethnographic epistemological assumptions," and may "eclipse other versions of what autoethnography can be" (pp. 377, 374). Glaringly absent from Anderson's essay is a large body of autoethnographic work constituting many versions of what performative "autoethnography can be" and already is, including works by Bryant Alexander (2006), Ken Gale and Jonathan Wyatt (2008), Craig Gingrich-Philbrook

(2001), H. Lloyd Goodall (2008), Stacy Holman Jones (2005), Ronald Pelias (2004), Elyse Lamm Pineau (2000), Chris Poulos (2009), Larry Russell (2004), Jonathan Wyatt (2008), and many others. In his response, "Analytic Autoethnography, or Déjà Vu All Over Again," Denzin writes, "Like others before him, Anderson does not want to review the debates that have gone on between the analytic and evocative schools of (auto)ethnography" (2006a, pp. 420–421). This want in literary review makes substantive comparative analysis difficult.

In their response, Ellis and Bochner, in concert with other essays in the series, argue that Anderson's defense of "realist" ethnography constitutes a call to stultify and bring under control the radicalizing, unruly, and creative elements of autoethnography that grew as a response to hegemonic reason/logic/analysis master narratives in anthropology and in academic discourse generally. In an essay outside of the special issue, "'Real Anthropology' and Other Nostalgias," Kath Weston argues that the term *real anthropology* encodes a certain historical consciousness when "the authority of the anthropologist remained intact, his or her identity secure . . . Its invocation implies that in recent times, when anthropology stumbled from grace, some policing of the boundaries of the discipline is necessary to separate acceptable from unacceptable topics or methods of study" (2008, pp. 128, 129). Surely, collaborative vigorous critique advances the development of theory and method; however, as Weston suggests, a policing of what is acceptable can draw us away from heuristic and pedagogical processes and possibilities.

Anderson is well advised by Denzin to engage the vast array of performance studies scholarship in autoethnography and critical ethnography. Performative autoethnography is forged within the ontological tension between its epistemological potential and its aesthetic imperative, the aesthetic/epistemic double-bind (Gingrich-Philbrook, 2005). To Anderson's plea "that other scholars will join me in reclaiming and refining autoethnography as a part of the analytic ethnographic tradition" (2006, p. 392), authors in performative autoethnography are merely representative of the myriad scholars whose work is evidence that good autoethnography is, of course, theoretically grounded at its outset and methodologically heuristic in process and product, advancing itself as a praxis of inquiry as it performatively *does* analysis.

Ultimately, however discordant our voices may be in the articulation of autoethnography, as evidenced in the special issue of the *Journal of Contemporary Ethnography* (Hunt & Junco, 2006), our history, our historical present as researchers, is enriched by the constant conversation of what we are doing and why we are doing it. In their introduction to *Ethnographica Moralia: Experiments in Interpretive Anthropology,* Neni Panourgia and George Marcus caution us "not to rest comfortably in our assumptions, in our disciplinary boundaries . . . but to interrogate their certainty and interrupt their narratives" (2008, p. 3).

I rely on the works of James Clifford, Dwight Conquergood, Craig Gingrich-Philbrook, D. Soyini Madison, George Marcus, Della Pollock, Mary Strine, Victor Turner, and others who view ethnography as performative, who see, as does Turner, performance as "the explanation and explication of life itself" (1986, p. 21). In the *co-performativity of meaning with others* I find myself, as autoethnographic researcher, in the constant negotiation of representation with others in always emergent, contingent, and power-laden historical contexts. I find scholarly desire and pedagogical purpose in Denzin's response in the special issue:

> Ethnography is not an innocent practice. Our research practices are performative, pedagogical, and political. Through our writing and our talk, we enact the worlds we study. These performances are messy and pedagogical. They instruct our readers about this world and how we see it. The pedagogical is always moral and political; by enacting a way of seeing and being, it challenges, contexts, or endorses the official, hegemonic ways of seeing and representing the other. (2006a, p. 422)

If autoethnography is knowledge forged collaboratively without a policing of disciplinary boundaries, then it may, in turn, become our history as well, always moving us forward, backward, in circle, or all at once. Rather than a historical modality of "civilizing" or "developing," performative autoethnography radicalizes scholarship through operating under the idea that, as Fabian argues, "In the real world *theory happens*" (2001, p. 5), in a present that includes the political inequities of privilege where researchers cannot displace otherness into a patronage of "native tradition," thus defining who and what is and is not "developed." "Theory," writes Fabian, "has no place unless it has time" (p. 5). In a post-9/11 America, autoethnography's local knowledge situates time and history as inherently critical tools.

▣ PERFORMATIVE FRAGMENTS
AND EMBODIED POSSIBILITIES

I could tell a White band from a Black band. I could just tell, it wouldn't go into my body.

—Miles Davis (2001)

Experiencing language as a transformative force was not an awareness that I arrived at through writing. I discovered it through performance.

—bell hooks (1999, p. 35)

Davis' words illustrate the corporeal embeddedness of knowledge. They reveal the inherency, the seamlessness, the materiality of the personal and political, in a manner where we cannot

tell where one ends and the other begins. His words speak a theory of embodiment, a theorizing of the embodied knowledge of, among other things, race. He could just "tell," his body telling, his telling body, that a particular composition wouldn't "go into my [his] body," not because he didn't *know* or understand the sound, but perhaps because he knew it too well, was *required* to know it, knew it as a racially compulsory verse in the soundscapes of power. In the agency of *telling*, of the telling body, in the critical assignment of language to experience, performative autoethnography is constructed. This is the basic foundation of autoethnography, the pulse of this methodology of the heart (Pelias, 2004), the peril of this anthropology that will break your heart (Behar, 1997). Embodied knowledge is the somatic (the body's interaction with culture) represented through the semantic (language), a linguistic articulation, a telling, of what does and does not go into the body, and why (Spry, 1998).

Performative-I Embodiment:
Loss, Co-Presence, and (Re)location

But whose body? Whose words? Where or who does the telling come from? What is the sociocultural and temporal location and implication of the autoethnographer? What is the relationship between autoethnographer and others when considering and employing embodied theory and methodologies (embodied praxis)? The effect of theories of embodiment on autoethnography are deftly articulated by Jones; she addresses "how body and voice are inseparable from mind and thought as well as how bodies and voices move and are privileged (and are restricted and marked) in very particular and political ways" (2005, p. 767). The politics of/in the body are central in performance ethnography; Madison writes,

> In performance studies we do a lot of talking about the body. For performance ethnographers, this means we must embrace the body not only as the feeling/sensing home of our being—the harbor of our breath—but the vulnerability of how our body must move through the space and time of another—transporting our very being and breath—for the purpose of knowledge, for the purpose of realization and discovery. (2009, p. 191)

Embodied knowledge is the research home, the methodological toolbox, the "breath" of the performative autoethnographer. It allows the researcher to reflect on the myriad ways in which, for example, Miles' statement is packed with the politics of race, the politics of his body as the home of his being. It is, among other things, his social, cultural, and temporal embodied location as a distinguished musician, as an African American man, and as a person raised in financial privilege that embodies the ways in which he, and we as readers, make meaning of his words.

The consideration of *researcher location* in relation to others seems tantamount in the present development of autoethnography. As such, I am interested in continuing to develop a researcher location that I have termed a "performative-I" as a plural and performative researcher embodiment (Spry, 2006). The performative-I disposition is rooted in critical ethnography, critical cultural theory, politics of identity, and performance studies where the researcher seeks to develop a critique of her or his political standpoint and sociocultural situatedness to disarm the power structures restricting individual and group access and ability for social change and social justice. "Representation" writes Madison, "happens at different points along power's spectrum—we are all 'vehicles and targets' of power's contagion and omnipresence" (2009, p. 193). The performative-I disposition encourages the researcher to locate self in relation to others in the "both/and" of "power's contagion" seeking to understand how we—as both vehicles and targets—can effectively negotiate and transform power's contagion.

A performative-I location of autoethnography constitutes a focal shift away from identities as *constructed* socioculturally/politically/historically and more toward the *negotiation* of these subjectivities in meaning making. This process of negotiating—in this case, masculine subjectivities—is deeply salient in Gale and Wyatt's book, *Between the Two: A Nomadic Inquiry Into Collaborative Writing and Subjectivity* (2009). Further, in *Performing Black Masculinity* (2006), Bryant Alexander writes, "identity politics are not situated on the body, but serve as a constellation of resources in the cultural negotiation of the ideal—in the form of nostalgia, remembrance, and remorse" (p. xiv); this "constellation of [sociocultural] resources" is central to the performative-I disposition, inviting us toward an examination of how we *co-performatively function* within a particular sociocultural, political, or historical context to (re)make meaning that illustrates the complex negotiations between selves and others in contexts, and reflects a multiplicity of cultural narratives. The performative-I research location embodies a plural sense of self collaboratively co-performing meaning in sociocultural contexts. The autoethnographer seeks to articulate with others the co-construction of culture, history, and power-provoking critical reflection on differences in power and privilege for the purpose of community efficacy.

Surely, the term "ethnographic-I" resonates here for many readers as it has been offered up and fruitfully developed in autoethnographic studies (Ellis, 2004). The conceptualization of "ethnographic-I" has served as a useful definitional guidepost for the conceptualization and composition of autoethnography (Ellis, 2009; Goodall, 2000; Richardson, 2007). In an intentional move away from realist ethnography, Ellis and Bochner (2006) advocate autoethnography that explicates emotional dimensions of self-reflection on lived experience. Goodall's work in "new ethnography" establishes a grounded rhetorical and literary stance while embracing the messy unpredictability of deep communicative engagements with others in cultural contexts (2000, 2008).

A performative-I location or disposition of embodiment, however, offers a different researcher positionality and perspective in

autoethnography because of its foundational conflation of performativity, performance studies, and ethnography with established autoethnographic research. However, in the methodological fashion of autoethnography, the emergence of a performative-I was the result of a deep and all-consuming loss and the necessity to save my life after death.

All understandings of research and life broke apart for me into pieces large and small, sharp edged, inchoate, and seemingly irretrievable after the loss of our child. Writing was the only thing I could make myself do. My arms literally ached from the absence. I felt dismembered mentally and physically, phantom limbs holding a baby. My subject position went from a destabilized "me" to a chaotic but oddly comforting "we." And because, as hooks suggests, theory heals, Sidonie Smith's words acted as balm:

> And so the cultural injunction to be a deep, unified, coherent, autonomous "self" produces necessary failure, for the autobiographical subject is amnesiac, incoherent, heterogeneous, interactive. In that very failure lies the fascination of autobiographical storytelling as performativity. (1998, p. 108)

"Paper and Skin" is a "necessary failure" of a coherent unified self. In Smith's concept of autobiographical performativity, the autobiographical subject is not an intact coherent self waiting within the body to be recorded through language; rather, she is a conflation of effects, a "constellation of resources" created through a performative process of critical narration that resists notions of individual coherency; the performative-I disposition is a coupling of this sense of subjective incoherency with critical ethnographic reflexivity.

After much frustration with dominant cultural narratives of grief as well as my own writing process at the time, I finally gave in to the rupture. In their call for a "deconstructive autoethnography," Alecia Jackson and Lisa Mazzei deftly critique a hegemonic methodological trend of the "I" in autoethnography where "the goal is coherence, comfort, and continuity through mediated truth" (2008, p. 300). Rather, they argue for an "I" that confronts "experience as questionable, as problematic, and as incomplete—rather than as a foundation for truth" (p. 304). In the comfort of incoherency and incompleteness, I began to find relief. I began to experience rupture and fragmentation as a form and function of performative ethnographic representation. In her deeply moving book, *Telling Bodies Performing Birth*, Della Pollock speaks with women whose stories of birth suggest such rupture. She writes,

> The stories that emerge in each case rise up against the norms that deny their integrity, that prefer silence, conformity, and invisibility. In the corporealities of performance, they break through normative reiteration into the time-space of terrifying exhilarating possibility. They bend to the breaking point the comic-hero norm of birth storytelling, making story answer to performance, performance to

difference, and difference to its origins in absence, in silence, in the blank expanse of not knowing and unknowing that remains impenetrably unknown. (1999, pp. 27–28)

As I let myself fall apart, I let myself see the pieces. I let myself fall into the presence of absence, into "the blank expanse of not knowing." My own experience of the multiplicity and partiality of knowledge became deeply embodied; I understood this as an autoethnographic stance, as a construction of self that seemed to navigate, to negotiate the interrelations between self/other/bodies/language/culture/history in ways that were markedly different from any previous "I" positionality in my research. Butler writes,

I speak as an "I," but do not make the mistake of thinking that I know precisely all that I am doing when I speak in that way. I find that my very formation implicates the other in me, that my own foreignness to myself is, paradoxically, the source of my ethical connection with others. (2005, p. 84)

Butler (2005) suggests a decentering of epistemological authority in the "I" of autoethnographic writing. This decentering of authority made great sense walking in the fragments and rubble of loss personally in our family and nationally in 9/11. Pollock describes a performative self that "is not merely multiple," it moves itself "forward . . . and between selves/structures" (1998, p. 87). Any sense of knowledge located firmly within the boundaries of my own body fell away, and I began to feel within the concept of performativity and co-presence, an engagement with others in culture that was no longer centered in "I." As Conquergood believed, performance is about struggle to push through what seems fixed, static, or hopeless. My emptied body needed to speak of absence, of the incoherence of here/now/who, to embody "the indecidability of meaning, of self, of narrative—without requiring self-identification or mastery" (Jackson & Mazzei, 2008, p. 305). Performative writing can "make absence present and yet recover presence from structural, realist mimesis for poesies" (Pollock 1998, p. 81). Moving from mimetic imitation of "deep autonomous selfhood" into the deconstructive motile vision of empty arms falling, I began and structured the piecemeal form of "Paper and Skin: Bodies of Loss and Life."

Embodying this performative turn on grief activated an intervention on dominant cultural performances of grief, and I could feel a methodological shift in my positionality within this and other fields of study from participant-observer actor to co-performative agent within the research context; from within this experience, I understood more fully "the vulnerability of how our body must move through the space and time of another—transporting our very being and breath—for the purpose of knowledge, for the purpose of realization and discovery" (Madison, 2009, p. 191). From this embodied negotiation with others in loss emerged a performative-I disposition. The hearts and minds of words and lives of the works I had

been reading began to make a more deeply embodied and more problematic "felt-sense."

The work of Dwight Conquergood, Norman Denzin, Kristen Langellier, Soyini Madison, Della Pollock, and others articulate performativity as having the capability of resisting and interrupting sedimented social meanings and normative performances that become oppressive, hegemonic, silencing. "Performativity," write Madison and Hamera, "is the interconnected triad of identity, experience, and social relations. . . . performativities are the many markings substantiating that all of us are subjects in a world of power relations" (2006, p. xix). Postcolonial critic Homi Bhabha (1993) enacts performativity to disrupt, dislodge, and dislocate hegemonic constructs of race and imperialism. Performative autoethnography presents alternate versions and options of reacting to and experiencing sociocultural expectations, thereby resisting and intervening on normative constructs of human being and reified structures of power. In this way performativity, a performative-I disposition, functions as agency of culture rather than as an act of culture (Strine, 1998).

Based in performativity, the performative-I location is a negotiation of critical agency and personal/political accountability.

Performative-I Critical Agency

With performativity and dialogue, the performative-I disposition requires collaborative co-performative meaning making and critical agency at once. Here, autoethnography is a collection of representational fragments of knowledge assembled with others from a plural sense of self, a dialectic of co-presence with others in the field of study concerning how bodies are read in various contexts of culture and power. In her work on Red pedagogy, Grande articulates a collaborative critical agency in research, "[Red pedagogy] is a space of engagement. It is the liminal and intellectual borderlands where indigenous and nonindigenous scholars encounter one another, working to remember, redefine, and reverse the devastation of the original colonialist 'encounter'" (2008, p. 234). The colonializing historicities of these encounters are critiqued and transformed, reassembled from a plural sense of self, a dialectic of co-presence where selves and others challenge and recognize their "overlapping cultural identifications" (p. 234) that may be in communion or in conflict with social and power relations. Vershawn Ashanti Young offers clarity of such overlaps in a dialectic of co-presence in *Your Average Nigga: Performing Race, Literacy, and Masculinity* where he seeks to "illustrate the intersection between what I call the burden of racial performance and the problems that I and other Blacks face in the ghetto and in school, particularly in college" (2007, p. 12). This kind of performative-I dialectic identification does not romanticize collaboration or assume that autoethnography is a manifestation of agreement or consensus or solution or emotional connection between selves and others. Within critical agency is the ethical awareness of representing one's own reflection on the complex interaction and negotiation between selves and others in complex sociocultural power-laden contexts.

For example, at a recent meeting of the Congress of Qualitative Inquiry, a groundbreaking conference that provides an international and interdisciplinary communal space for deep critical qualitative inquiry, I attended a panel on performance and metaphor. The presentations were provocative and heuristic, generating discussion that continued to open the epistemological efficacy of performance methodologies. The conversation turned to the uses of metaphor in autoethnographic research, and a participant asked how the autoethnographer gets others to understand his use of metaphor for the purposes of critical reflection. This was an iconic moment for me in relation to a performative-I research disposition. Though the young man seemed sincere and well-intentioned, his question illustrated, it seems to me, the imperialist impulse to "make others understand" rather than doing understanding with others. In "Rethinking Collaboration: Working the Indigene-Colonizer Hyphen," Alison Jones and Kuni Jenkins write, "I do not argue for a rejection of collaboration. Rather, I unpack its difficulties to suggest a less dialogical and more uneasy, unsettled relationship, based on learning (about difference) from the Other rather than learning about the Other" (2008, p. 471). Though I conceptualize dialogic engagement as inherently uneasy and unsettled, Jones and Jenkin's point is fundamental to the imperializing potential of research when the researcher's own power dynamics are left unconsidered and uncritiqued.

The performative-I autoethnographic positionality would encourage the researcher to seek understanding of others' uses of metaphor, and to critically reflect on the nuances and difficulties in the ontological situation of one's own in/ability to understand, or as Jackson and Mazzei write, "To interrogate the process of constituting a performative 'I' is to put experience under erasure, to expose the uncertainty of 'who' that 'I' could become, and to open up what can be known about the 'I'" (2008, p. 305). Through the performative-I re/positioning of the autoethnographer, a reader may learn about the complexity of cultural interaction, how the researcher's uses of dominant discourse effect these interactions, how we are all, as Madison suggests, "'vehicles and targets' of power's contagion and omnipresence" (2009, p. 193). The performative-I disposition calls forth dialogue between the personal and the politicized body.

Performative-I Accountability

Jazz great Wynton Marsalis writes of deep jazz swing, "It's a matter of understanding what a thing means to you, and being dedicated to playing that even if its meaning casts a cold eye on you yourself" (2005, p. 59). Marsalis captures the praxis of the performative-I in articulating the necessity for *agency* as well as

accountability. The autoethnographer, who may certainly carry privilege into the research context, must be acutely aware of the power dynamics involved in representation; the autoethnographer must be able to engage in reflexive critique of her or his own social positioning, must be "dedicated to playing that," to recognize the necessity to be what Butler calls a "social theorist." She writes, "When the 'I' seeks to give an account of itself, it can start with itself, but will find that this self is already implicated in a social temporality that exceeds its own capacities for narration . . . an account that must include the conditions of its own emergence" (2005, pp. 8–9). The performative-I researcher disposition assumes the inherency of accountability in autoethnography. "There is no 'I,'" writes Butler, "that can stand apart from the social conditions of its emergence" (2005, p. 7). Performativity requires the perpetual critique of cultural performance, as well as intervening on sediment dominant patterns of oppression (Denzin 2003, 2005; Madison & Hamera, 2006).

In discussing the postcolonial notion of the *mestizaje,* an indigenous research positioning of critically reflexive multiple subjectivities, Grande writes, "Unlike liberal notions of subjectivity, it [mestizaje] also roots identity in the discourses of power" (2008, p. 239). Woven into a subversive performativity and Butler's accounting of "I" is the idea that one must critique the always and already power-laden structures of everyday life experience, that one cannot "stand apart from the social conditions" of one's emergence. The performative autoethnographer assumes that the complexity of her or his own sociocultural emergence or situatedness may exceed her or his capacity for narration—hence, if I am to claim agency, I also have the responsibility to account for my sociocultural whereabouts and its implications for myself and those with whom I work; I must be "dedicated to playing," dedicated to *doing* reflexivity even—and especially—while knowing it is never enough, never complete, never finished.

Performative Embodiment

I love performance most when I enter into it, when it calls me forward shamelessly, across those hard-edged maps into spaces where I must go, terrains that are foreign, scary, uninhabitable, but necessary. I must go to them to know myself more, to know you more.

—Soyini Madison (2006, p. 244)

Knowledge becomes constantly embodied and bodying forth through past, present and future practices, sensorily and emotionally shared with persons, objects, and institutions—actual or imagined, seen or unseen or never to be seen.

—Eleni Papagaroufali (2008, p. 121)

Having discussed the conceptualization of a performative-I disposition, I would focus further on the development of embodiment in performative-I and in performative autoethnography because the centrality of the body is what characterizes performance studies research and, subsequently, performative autoethnography (Jones, 2005). Papagaroufali's and Madison's thoughts reflect why, since communing with loss, I have been uncomfortable with the conceptualization of "I" in autoethnographic research. Though dialogue and performativity have become tenants in autoethnographic description, there was, within my own embodied experience, the absence of a connective tissue in theoretical descriptions. But in writing this chapter through a conflation of performance studies, autoethnography, and critical ethnography, I find a multiplicity and accountability in the performative-I, and a felt-sense of liminality, of circularity, of inbetweeness as reflected in Papagaroufali's and Madison's words. In "bodying forth" with others "seen or unseen or never to be seen" in an embodied communion (pleasant and difficult) with others, I find embodied presence in performative autoethnography. Different from an "ethnographic-I," this is a troubled, sensual, contingent embodiment of communitas. Pollock guides me:

> Entanglement, ravishment, love, writing: what I want to call performance writing does not project a self, even a radically destabilized one, as much as *a relation of being and knowing* that cuts back and forth across multiple "divisions" among selves, contexts, affiliations. . . . the self that emerges from these shifting perspectives is, then, *a possibility rather than a fact*, a figure of relation emerging from between lines of difference. (1998, pp. 86–87, emphasis mine)

A disposition,

a relation of being and knowing that cuts back,

a possibility

a figure of relation emerging

from difference

from entanglement

from ravishment,

from . . .

and it is here that I must stop; for to speak of love in relation to research, in an academic

context, an academic handbook, even one edited by the roaming free radical Norman Denzin

feels heresy,

not in the truth of my "un/learning body" (Madison, 2006),

because surely, truly, and ravishingly,

it is love and desire

for communitas, for Burke's consubstantiality, for articulation and interruption of the personally political pains that I inflict on others and that are inflicted upon me within the uneven, unjustified, and inequitable systems privileging some and disempowering others. Heresy or not, it is a disposition of love in autoethnographic research that has given me the courage to move into entanglements with others about race, gender, privilege, and more resulting in a different kind of knowing. Surely, it is only through a disposition of love, through love as an epistemological possibility, through a methodology of the heart that I have been able to fill the absence of loss with the linguistic presence of our child.

And so in speaking of an autoethnographic embodiment in research and in report, I am most comfortable in the liminal space of an embodied *disposition,* an embodied co-presence in conceptualizing the research and writing process of critical performative autoethnography, of the embodied performative-I. Engaging a liminal disposition then in performative autoethnography has *moved me out of* an autoethnographic, or "ethnographic–I." It has not moved me into the bodies of others; I do not "walk a mile in their shoes" or look at the world through their eyes. I do not *know* the Other through performing autoethnography, such would be hubris rather than pedagogy. Performance is not that innocent. Rather, through performance I come to know more fully what a painful and liberating process it is to try to represent the intimate nuances of a "hermeneutics of experience, relocation, co-presence, humility, and vulnerability" that Conquergood describes in conjunction with the texts of Frederick Douglass, where knowledge is located, engaged, and "forged from *solidarity with,* not separation from, the people" in research (Spry, 2006, p. 315). This is assuming, of course, that people would desire forging knowledge together, and, if not, that the autoethnographer has the humility to recognize resistance to collaboration and the courage to critically reflect upon why the resistance exists.

Seeking embodied knowledge through an embodied liminality with others has taken me out of my self, out of a singular self; it has taken me out of my body and pushed and pulled me into the liminal inbetweeness of meaning making with others, into the space where I am not me and not you, but where we are *us* in a place and time defined by the multiple and surely conflictual readings of our cultural situatedness and the meanings we make co-performatively/together. A performative-I embodied disposition has shifted me into the void between our materiality where we meet each other in dialogue, in argument, in communion, in anger, in the *mysterium tremendum* that performance studies scholar Leland Roloff (1973) describes where cultural presence is what our bodies in contexts are making together.

A liminally embodied disposition of performative-I autoethnography involves personal/political risk and vulnerability. A performative-I embodiment takes me out of my body and into the liminal space where I do not have the "safety" of "I," where I am vulnerable to contestation with others about what is what, about the critical readings and representations of bodies. Surely and without question, within a performative-I disposition, I am critically and unromantically aware that no matter how liminally out of my body I may move, it is a racially and financially privileged body that I live in. This material reality constitutes the reasons and need for the critical imagination (Denzin, 2005) needed to move out of body and into the rapture and entanglements of what Conquergood (1991) calls a "performance sensitive" way of knowing to articulate and transform dominant narratives that sediment systems of power.

We do not enter these spaces lightly or stumble into this process without strategy, without method, without a practice. The performative-I disposition wholly depends on performative embodiment where the body is the actor, agent, and text at once, where our views of the world are tested, refuted, and articulated through the negotiation of corporeal bodies in space and time. It is a "bodying forth" into critical social theorizing where the resulting text causes interruption, repudiation, and intervention on dominant narratives.

▣ CRITICAL FRAGMENTS OF CRAFT: AN ETHIC OF AESTHETICS IN PERFORMATIVE AUTOETHNOGRAPHY

Her vocabulary is so discriminating that each word wears the complex self-investigation that brought it into being.

—Carmine Starnino (2008, p. 149)

I only write or make art about myself when I am completely sure that the biographical paradigm intersects with larger social and cultural issues.

—Guillermo Gómez-Peña (2000, p. 7)

In the case of autoethnography, the two strands of barbed wire manifest as a demand to create knowledge (the epistemic) and a demand to create art (the aesthetic). While we need not see these demands as diametrically opposed, neither need we see them as synonymous. In any event, we leave the relationship between them unconsidered at our peril.

Craig Gingrich-Philbrook (2005, p. 303)

Performative autoethnography is a critical moral discourse (Conquergood, 1985, 1991; Denzin, 2003, 2008; Jones, 2005). It is grounded in "The performance paradigm [which] privileges particular, participatory, dynamic, intimate, precarious, embodied experience grounded in historical process, contingency, and ideology" (Conquergood 1991, p. 187). Consequently, in autoethnography's trajectory and development as a moral discourse it must be, at its foundation, epistemic. *All of the potentials and possibilities embodied in performative autoethnography depend upon the quality of its report, of its linguistic and aesthetic construction, of it its ability to make writing perform* (Alexander; 2006; Denzin, 2003, 2006b; Gingrich-Philbrook, 2001, 2005; Goodall, 2000, 2008;

Hamera, 2006; Pelias, 2004; Pollock, 1998; Spry 2008, 2009; Trinh T. Minh-ha, 1989). The moral imperative, then, of autoethnography is as much situated in its aesthetic craft as in its epistemological potential. The depth of knowledge generated (epistemology) by performative autoethnography is directly related to its aesthetic acumen, and just as autoethnography is a critical moral discourse, the aesthetic crafting of autoethnography is a sociocultural and political action. "Performance exposes aesthetics' social work," writes Judith Hamera, "as embodied, processual, rhetorical, and political and especially, as daily, as routine, a practice of everyday life" (2006a, p. 47). Here performative autoethnography, the performative-I disposition, operates as a movement of epistemologically embodied art crafted within and between the representations of power and powerlessness and is motivated by the desire for local embodied knowledge of how unequal power systems can be called up, called out, disassembled, reimagined, and reconfigured.

Our need at this point in the development of autoethnographic writing is, it seems to me, to write more about writing, in particular, about the politics, power, and privilege of it, the various styles, forms, embodiment, and aesthetic functions of it—in other words, a focus on the aesthetic representation of how the body co-performs with others in sociopolitical contexts. In his 2008 essay "Contemporary Fieldwork Aesthetics in Art and Anthropology," George Marcus calls for "an explicit rearticulating of its [ethnography's] aesthetic of method" because of the varying "norms and forms of knowledge" (p. 32) since his groundbreaking work with James Clifford in *Writing Culture.*

We must continue to develop writing from/with/of the co-performative body as co-present with others, the body as epistemologically central, heuristically inspirational, politically catalytic. "In performative writing," writes Madison, "we recognize that the *body* writes. Critical ethnography adheres to radical empiricism: the intersection of bodies in motion and space" (2005, p. 195). We must write from within the entanglements of co-presence, from the rapture of communion, from the un/comfortable risk and intimacy of dialogue, from the vulnerable and liminal inbetweenness of self/other/context.

But it is, of course, *through language* that we "body forth" in interpreting and articulating what the body "knows." In postmodern research, we sometimes like to think of the body as inherently "knowing" things without remembering that *the body knows what language constructs* (Spry, 2009). In conceptualizing performative writing, Gingrich-Philbrook lives "in body-language-body-language. My body makes language. It makes language like hair" (2001, p. 3). Embodied knowledge is generated from a body-language rapture, elation, conflation, each affecting (and sometimes abjecting) the other.

I want to spin the aesthetic crafting of language in autoethnography as an ethical imperative, as *a movement of embodied art crafted within a performative-I disposition of liminality*. In a foundational article, "Autoethnography's Family Values: Easy Access to Compulsory Experiences" (2005), Gingrich-Philbrook deftly articulates the double bind of the aesthetic and the epistemic in autoethnography as a response, in part, to what he argues is a regulatory valuing and valorizing of a compulsory hegemonic emotionality in autoethnography at the expense of developing alternative theory, aesthetic practices and epistemological products of autoethnography. He states, "However much one applauds autoethnography's artistic and social intentions, those intentions do not in themselves secure artistic results" (p. 308). Performance studies practitioners have always worked with the embodiment of emotion in the production of knowledge and are aware of the potential dangers when expecting the expression of emotion in research to stand-in for aesthetic acumen. Emotion is not inherently epistemic. I have tomes of writing expressing the emotional turmoil of loss during childbirth; as significant as that writing is to my own personal grief process, it is not performative autoethnography. Hamera helps clarify:

> *Experience is not scholarship.* . . . Performance links experience, theory, and the work of close critique in ways that make precise analytical claims about cultural production and consumption, and expose how both culture and our claims are themselves constructed things, products of hearts and souls, minds and hands. (2006a, p. 241, emphasis mine)

Many aspects of my own personal grief are not yet critically reflective of how this experience is personally part of sociopolitics, of cultural production. "This does not mean," writes Butler, "that I cannot speak of such matters, but only when I do, I must be careful to understand the limits of what I can do, the limits that condition any and all such doing. In this sense, I must become critical" (2005, p. 82). In my own aesthetic production of critique, in the engagement of hearts and souls, minds, and hands, I understand the limits of my own condition in offering epistemologies concerning dominant hegemonic structures of grief.

I think at this point in our development of autoethnography, we look at the epistemic as more sociopolitically and academically relevant, while viewing the aesthetic or literary as an added scholarly bonus, or worse, as ideologically benign. Gingrich-Philbrook employs the work of Murray Krieger (1992), who argues that the aesthetic "alerts us to the illusionary, the merely arbitrary claims to reality that authoritarian discourse would impose upon us; because, unlike authoritarian discourse, the aesthetic takes back the 'reality' it offers us in the very act of offering it to us" (quoted in Gingrich-Philbrook 2005, p. 310). Performative autoethnography is forged in the ontological tension between its epistemological potential and its aesthetic imperative. It is through language, after all, that we "give an account of ourselves." Language's propensity toward imperializing, toward "merely arbitrary claims to reality," makes this accounting a moral commitment, an ethical imperative.

Privileged peoples need not attend to imperializing aesthetics, as their words are framed by power. It is ethically imperative then, that the autoethnographer, who may certainly carry privilege into the research context, be acutely aware of the power dynamics involved in the aesthetics of performative autoethnography. Representation has risks (Denzin, 2003, 2006a, 2008; Denzin & Lincoln, 2007; Grande, 2008; Madison, 2005, 2009; Poulos, 2009; Smith, 1999; Spry, 2008). Aesthetics are not ideologically benign. Those risks can be negotiated by an ethic of care for aesthetic representation.

An ethic of aesthetic representation is illustrated through what Mindy Fenske calls an "ethic of answerability" where, in this case, the autoethnographer is responsible for and ethically liable for linguistic representations of the interpolations of self with others in contexts (2004, p. 8). Fenske argues that no hierarchy exists between craft and emotion, form and production, theory and practice, art and life. "Instead," she writes, "such relations are unified and dialogic. . . . Art and life are connected, one is not meant to transcend the other. Both content and experience, form and production . . . exist inside the unified act in constant interaction" (p. 9). In this dialogical ethic of care, emotion is not touted as the scholarly cure for realism, nor is aesthetic craft viewed as a mechanized technique handcuffing the raw essence of experience and emotion; rather, they are interdependent on one another, responsible to one another, liable to one another to represent the complex negotiations of meaning between selves and others in power-laden social structures. *Here art is not a reflection of life; they are, rather, answerable to one another.* "Form," writes Fenske "becomes a location inciting, rather than foreclosing, dialogue" (p. 11). The debilitating binary argument of craft over emotion, or practice over theory is deflated through Fenske's argument because these elements are mutually answerable to one another; knowledge is sought through their dialogic engagement suggesting a praxis, an ethical assembly, a resistance of hierarchy. Rather than a linear path from self to other, theory to practice, or emotion to craft, performative-I liminality is a dialogic space where experience and text effect and are effected by one another.

Clearly, embodiment is crucial in this ethics of aesthetics. Just as emotion and experience are not inherently epistemic, Fenske reminds us that "events are not ethical simply because they are embodied. . . . In order to achieve answerability, the embodied action must be responsible for its meaning, as well as liable to meaning" (2004, p. 12). The material body cannot be erased in composing autoethnography; rather the corporeal body is made fully present in performance and represented through critical reflections on the body's social constructions. In a recent student's performative autoethnography titled "Driving While Black," the truths of Anthony's life are not compromised by aesthetic craft, rather in the critical dialogic process of articulating, of crafting life, Anthony constructs and embodies knowledge that is subversive, pedagogical, and heuristic. Politically troubled aesthetics allow Anthony to read and re/write his social body as a transgressive text. To operate as if there is a hierarchy in art or life, in craft or emotion, in theory or practice is to engage, Fenske argues, "a type of aesthetic that lets the artist off the ethical hook" (2004, p. 13) by being tempted to offer simplistic notions of the "purity" (read "apolitical") of embodied experience in aesthetic composition. Any methodology, aesthetic or otherwise, that does not exercise as fundamental the critique of sociocultural systems and discourses of power sanitizes and imperializes critical reflexivity into a parlor game of identity construction where Self stands in front of a mirror trying on different cultural hats to see the "world" from the eyes of the Other. Epistemologically, these dialectics engaged in collaboration expand the depth and breadth of critical conversations and implementations. Autoethnography remains accountable by considering the political constructions of an "'I' that remains skeptical of authentic experience" (Jackson & Mazzei, 2008, p. 314) and, I would argue, of aesthetic purity.

Making writing perform. Making the story answerable to its own sociocultural emergence, to its own performance, to its own life as art and back again. Pelias writes, "Language is my most telling friend, my most fierce foe ... power lurks, will grab me at every turn. I must stare it down, write it down" (2007, p. 193). Power lurks as much in aesthetic construction as in epistemological construction because surely, they are inseparable. And though they are answerable to one another, the answers do not foreclose one another. Performative autoethnographic writing is about the continual questioning, the naming and renaming and unnaming of experience through craft, through heart, through the fluent body.

◼ CONCLUDING FRAGMENTS: ENTANGLEMENT, RAPTURE, AND WRITING

Did I actually reach out my arms

toward it, toward paradise falling, like

the fading of the dearest, wildest hope—

the dark heart of the story that is all

the reason for its telling?

—Mary Oliver (1986, p. 2)

There is a long time in me between knowing and telling.

—Grace Paley (1974, p. 127)

I write to show myself showing people who show me my own showing. I-You: not one, not two. In this unwonted spectacle made of reality and fiction, where redoubled images form and reform, neither I nor you come first.

—Trinh T. Minh-ha (1989, p. 22)

Oliver, Paley, and Trinh T. Minh-ha write of the passionate liminality, the inchoate corporeality, the continual redoubling where you and I are collaboratively present and singularly absent on the page. The aesthetic and epistemic collapsing into one another

with such desire that we cannot tell where one ends and the other begins.

It is a strange and alternate plain where we can, at once, feel and speak and hear,

where we can bleed and bleed and not die,

or where we can die and be resurrected—or not,

thankful that through performance, words make flesh.

◼ REFERENCES

Alexander, B. (2006). *Performing Black masculinity: Race, culture, and queer identity.* Lanham, MD: AltaMira Press.

Anderson, L. (2006). Analytic autoethnography. *Journal of Contemporary Ethnography, 35*(4), 373–395.

Anzuldúa, G. (2007). *Borderlands/La Frontera: The new mestiza* (3rd ed.). San Francisco: Aunt Lute Books.

Behar, R. (1997). *The vulnerable observer: Anthropology that breaks your heart.* Boston: Beacon Press.

Bhabha, H. (1993). *The location of culture.* New York: Routledge.

Butler, J. (2005). *Giving an account of oneself.* New York: Fordham University Press.

Clifford, J., & Marcus, G. E. (1986). *Writing culture: The poetics and politics of ethnography.* Berkeley: University of California Press.

Conquergood, D. (1985). Performing as a moral act: Ethical dimensions of the ethnography of performance. *Literature in Performance, 5,* 1–13.

Conquergood, D. (1991). Rethinking ethnography: Towards a critical cultural politics. *Communication Monographs, 58,* 179–194.

Davis, M., & Dibbs, M. (Producer/Director). (2001). *The Miles Davis story* [TV documentary]. New York: Columbia Music Video.

Denzin, N. K. (2003). *Performance ethnography: Critical pedagogy and the politics of culture.* Thousand Oaks, CA: Sage.

Denzin, N. K. (2005). Politics and ethics of performance pedagogy: Toward a pedagogy of hope. In D. S. Madison & J. Hamera (Eds.), *The SAGE handbook of performance studies* (pp. 325–338). Thousand Oaks, CA: Sage.

Denzin, N. K. (2006). Analytic autoethnography, or déjà vu all over again. *Journal of Contemporary Ethnography, 35*(4), 419–428.

Denzin, N. K. (2008). *Searching for Yellowstone: Race, gender, family, and memory in the postmodern West.* Walnut Creek, CA: Left Coast Press.

Denzin, N. K., & Giardina, M. D. (2007). Introduction: Ethical futures in qualitative research. In N. K. Denzin & M. D. Giardina (Eds.), *Ethical futures in qualitative research* (pp. 9–39). Walnut Creek, CA: Left Coast Press.

Denzin, N. K., & Lincoln, Y. S. (2007). *The landscape of qualitative research* (3rd ed.). Thousand Oaks, CA: Sage.

Ellis, C. (2004). *The ethnographic-I: A methodological novel about auto-ethnography.* Walnut Creek, CA: AltaMira Press.

Ellis, C. (2009). *Revision: Autoethnographic reflections on life and work.* Walnut Creek, CA: Left Coast Press.

Ellis, C., & Bochner, A. P. (2006). Analyzing analytic autoethnography: An autopsy. *Journal of Contemporary Ethnography, 35,* 429–449.

Fabian, J. (1983). *Time and the Other: How anthropology makes its objects.* New York: Columbia University Press.

Fabian, J. (2001). *Anthropology with an attitude.* Palo Alto, CA: Stanford University Press.

Fabian, J. (2007). *Memory against culture: Arguments and reminders.* Durham, NC: Duke University Press.

Fenske, M. (2004). The aesthetic of the unfinished: Ethics and performance. *Text and Performance Quarterly, 24*(1), 1–19.

Gale, K., & Wyatt, J. (2008). Becoming men, becoming-men? A collective biography. *International Review of Qualitative Research, 1*(2), 235–253.

Gale, K., & Wyatt, J. (2009). *Between the two: A nomadic inquiry into collaborative writing and subjectivity.* Newcastle upon Tyne, UK: Cambridge Scholars.

Gingrich-Philbrook, C. (2001). Bite your tongue: Four songs of body and language. In R. J. Pelias & L. C. Miller (Eds.), *The green window: Proceedings of the Giant City Conference on Performative Writing* (pp. 1–7). Carbondale: Southern Illinois University Press.

Gingrich-Philbrook, C. (2005). Autoethnography's family values: Easy access to compulsory experiences. *Text and Performance Quarterly, 25*(4), 297–314.

Gómez-Peña, G. (2000). *Dangerous border crossers: The artist talks back.* New York: Routledge.

Goodall, H. L. (2000). *Writing the new ethnography.* Walnut Creek, CA: AltaMira Press.

Goodall, H. L. (2008). *Writing qualitative inquiry: Self, stories, and academic life.* Walnut Creek, CA: Left Coast Press.

Grande, S. (2008). Red pedagogy: The un-methodology. In N. K. Denzin, Y. S. Lincoln, & L. T. Smith (Eds.), *Handbook of critical and indigenous methodologies* (pp. 233–254). Thousand Oaks, CA: Sage.

Hamera, J. (2006). Performance, performativity, and cultural poesies in practices of everyday life. In D. S. Madison & J. Hamera (Eds.), *The SAGE handbook of performance studies* (pp. 49–64). Thousand Oaks, CA: Sage.

hooks, b. (1994). *Teaching to transgress: Education as the practice of freedom.* New York: Routledge.

hooks, b. (1999). *Remembered rapture: The writer at work.* New York: Henry Holt.

Hunt, S. A., & Junco, N. R. (Eds.). (2006). Introduction to two thematic issues: Defective memory and analytical autoethnography. *Journal of Contemporary Ethnography, 35*(4), 1–3.

Jackson, A. Y., & Mazzei, L. A. (2008). Experience and "I" in autoethnography: A deconstruction. *International Review of Qualitative Research, 1*(3), 299–317.

Jones, A., & Jenkins, K. (2008). Rethinking collaboration: Working the indigene-colonizer hyphen. In N. K. Denzin, Y. S. Lincoln, & L. T. Smith (Eds.), *The handbook of critical and indigenous methodologies.* Thousand Oaks, CA: Sage.

Jones, S. H. (2005). Autoethnography: Making the personal political. In N. K. Denzin & Y. S. Lincoln (Eds.), *The SAGE handbook of qualitative research* (pp. 763–792). Thousand Oaks, CA: Sage.

Keller, L. (2009). The second wave: Return of the militias. Montgomery, AL: Southern Poverty Law Center. Retrieved August 8, 2009, from http://www.splcenter.org

Krieger, M. (1992). *Words about words about words: Theory, criticism, and the literary text.* Baltimore: Johns Hopkins University Press.

Langellier, K. (1999). Personal narrative, performance, performativity: Two or three things I know for sure. *Text and Performance Quarterly, 19,* 125–144.

Liakos, A. (2008). Canonical and anticanonical histories. In N. Panourgia & G. Marcus (Eds.), *Ethnographic moralia: Experiments in interpretive anthropology* (pp. 138–156). New York: Fordham University Press.

Madison, D. S. (2005). *Critical ethnography: Method, ethics, and performance.* Thousand Oaks, CA: Sage.

Madison, D. S. (2006). Performing theory/embodied writing. In J. Hamera (Ed.), *Opening acts: Performance in/as communication and cultural studies* (pp. 243–266). Thousand Oaks, CA: Sage.

Madison, D. S. (2009). Dangerous ethnography. In N. K. Denzin & M. Giardina (Eds.), *Qualitative inquiry and social justice* (pp. 187–197). Walnut Creek, CA: Left Coast Press.

Madison, D. S., & Hamera, J. (Eds.). (2006). *The SAGE handbook of performance studies.* Thousand Oaks, CA: Sage.

Marcus, G. E. (2008). Contemporary fieldwork aesthetics in art and anthropology: Experiments in collaboration and intervention. In N. Panourgia & G. Marcus (Eds.), *Ethnographic moralia: Experiments in interpretive anthropology* (pp. 29–44). New York: Fordham University Press.

Marsalis, W., with Hinds, S. S. (2005). *To a young jazz musician: Letters from the road.* New York: Random House.

Max, D. T. (2009, March 9). The unfinished. *The New Yorker.*

Mutua-Kombo, E. (2009). Their words, actions, and meaning: A researcher's reflection on Rwandan women's experience of genocide. *Qualitative Inquiry, 15,* 308–323.

Oliver, M. (1986). The chance to love everything. In *Dream work* (pp. 8–9). Boston: Atlantic Monthly Press.

Paley, G. (1974). Debts. In *Enormous changes at the last minute* (pp. 15–23). New York: Farrar, Straus & Giroux.

Panourgia, N., & Marcus, G. (Eds.) (2008). *Ethnographic moralia: Experiments in interpretive anthropology.* New York: Fordham University Press.

Papagaroufali, E. (2008). Carnal hermeneutics: From "concepts" and "circles" to "dispositions" and "suspense." In N. Panourgia & G. Marcus (Eds.), *Ethnographic moralia: Experiments in interpretive anthropology* (pp. 113–125). New York: Fordham University Press.

Pelias, R. (2004). *A methodology of the heart: Evoking academic and daily life.* Walnut Creek, CA: AltaMira Press.

Pelias, R. (2007). Performative writing: The ethics of representation in form and body. In N. K. Denzin & M. Giardina (Eds.), *Ethical futures in qualitative research* (pp. 181–196). Walnut Creek, CA: Left Coast Press.

Pineau, E. L. (2000). Nursing mother and articulating absence. *Text and Performance Quarterly, 20*(1), 1–19.

Pollock, D. (1998). Performing writing. In P. Phelan & J. Lane (Eds.), *The ends of performance* (pp. 73–103). New York: New York University Press, 73–103.

Pollock, D. (1999). *Telling bodies performing birth.* New York: Columbia University Press.

Poulos, C. (2009). *Accidental ethnography: An inquiry into family secrecy.* Walnut Creek, CA: Left Coast Press.

Richardson, L. (2007). *Last writes: A daybook for a dying friend.* Walnut Creek, CA: Left Coast Press.

Roloff, L. (1973). *The perception and evocation of literature.* New York: Scott Foresman.

Russell, L. (2004). A long way toward compassion. *Text and Performance Quarterly, 24*(3 & 4), 233–254.

Schechner, R. (1985). *Between theater and anthropology.* Philadelphia: University of Pennsylvania Press.

Smith, L. T. (1999). *Decolonizing methodologies: Research and indigenous peoples.* New York: St. Martin's Press.

Smith, S. (1998). Performativity, autobiographical practice, resistance. In S. Smith & J. Watson (Eds.), *Women, autobiography, theory: A reader* (pp. 108–115). Madison: University of Wisconsin Press.

Spry, T. (1998). Performative autobiography: Presence and privacy. In S. J. Dailey (Ed.), *The future of performance studies: Visions and revisions* (pp. 254–259). Annandale, VA: National Communication Association.

Spry, T. (2001a). Performing autoethnography: An embodied methodological praxis. *Qualitative Inquiry, 7,* 706–732.

Spry, T. (2001b). From Goldilocks to dreadlocks: Racializing bodies. In R. J. Pelias & L. C. Miller (Eds.), *The green window: Proceedings of the Giant City Conference on Performative Writing* (pp. 52–65). Carbondale, IL: Southern Illinois University Press.

Spry, T. (2003). Illustrated woman: Autoperformance in "Skins: A daughter's (re)construction of cancer" and "Tattoo stories: A postscript to 'Skins.'" In L. C. Miller, J. Taylor, & M. H. Carver (Eds.), *Voices made flesh: Performing women's autobiography* (pp. 167–191). Madison: University of Wisconsin Press.

Spry, T. (2004). Paper and skin: Bodies of loss and life. An autoethnography performed in various venues across the country.

Spry, T. (2006). A performance-I copresence: Embodying the ethnographic turn in performance and the performative turn in ethnography. *Text and Performance Quarterly, 26*(4), 339–346.

Spry, T. (2008). Systems of silence: Word/less fragments of race in autoethnography. *International Review of Qualitative Research, 1*(1), 75–80.

Spry, T. (2009). *Bodies of/and evidence.* Paper performed at the 2008 Congress of Qualitative Inquiry, University of Illinois, Champagne-Urbana, IL.

St.Pierre, E. A. (2008). Home as a site of theory. *International Review of Qualitative Research, 1*(2), 119–124.

Starnino, C. (2008). Five from Ireland. *Poetry* (November), 149–161.

Strine, M. S. (1998). Mapping the "cultural turn" in performance studies. In S. J. Dailey (Ed.), *The future of performance studies: Visions and revisions* (pp. 3–9). Annandale, VA: National Communication Association.

Swadener, B. B., & Mutua, K. (2008). Decolonizing performances: Deconstructing the global postcolonial. In N. K. Denzin, Y. S. Lincoln, & L. T. Smith (Eds.), *Handbook of critical and indigenous methodologies* (pp. 31–43). Thousand Oaks, CA: Sage.

Trinh, T. Minh-Ha. (1989). *Woman, native, other.* Bloomington: Indiana University Press.

Turner, V. (1986). *The anthropology of performance.* New York: PAJ Publications.

Visweswaran, K. (2006). Betrayal: An analysis in three acts. In I. Grewal & C. Kaplan (Eds.), *Scattered hegemonies: Postmodernity and transnational feminist practices.* Minneapolis: University of Minnesota Press.

Weston, K. (2008). "Real anthropology" and other nostalgias. In N. Panourgia & G. Marcus (Eds.), *Ethnographic moralia: Experiments in interpretive anthropology* (pp. 126–137). New York: Fordham University Press.

Wyatt, J. (2008). No longer loss: Autoethnographic stammering. *Qualitative Inquiry, 14,* 955–967.

Young, V. A. (2007). *Your average Nigga: Performing race, literacy, and masculinity.* Detroit, MI: Wayne State University Press.

31

THE METHODS, POLITICS, AND ETHICS OF REPRESENTATION IN ONLINE ETHNOGRAPHY

Sarah N. Gatson

▣ THE BOUNDARIES OF ETHNOGRAPHY AND THE INTERNET

Any study of Internet interactions is challenging because of the simultaneous dense interconnectedness of the Internet and the normal boundaries between networks and communities. Moreover, online community development is inherently a multisited enterprise, and each locale in an identified network and its interactions must be thoroughly investigated to best discover its salient boundaries. In this chapter, I address the methods, politics, and ethics of representing these complex arenas through ethnography. I will provide a brief overview of the history and types of online ethnographies. Grounded in this discussion of online ethnographic works, and speculation about the future of such endeavors, I discuss the value of two ways to envision online ethnography: (1) The extension of traditional collaborative ethnography, in which a network of participant observers in offline laboratories or networks, as well as online, work together (sometimes unknowingly) to produce ethnography. This speculation is grounded in my experiences with online communities, one rooted in television fandom,[1] one rooted in research on drug use discourse,[2] and one that uses the Internet for research, training, and educational purposes.[3] (2) Emerging from these experiences, I discuss autoethnographic network mapping, in which a researcher grounds an online network map on herself or himself. I speculate that, in particular, this method would be useful in pedagogical and public sociological projects where media literacy/citizenship is the specific aim. The chapter concludes with the argument that these sorts of ethnographic practices both ground themselves in the traditions of the method, and address issues of the empiricist critique by explicitly exploring the meaning of "empirical" versus the meaning of "objective" in the practice of the social sciences.

In the way I understand methods, ethics, and politics, each of these concepts overlaps with the others, as each is concerned with distributions of power. Although ethics and politics are concepts that connote concomitant ideas of power, methods may not be. Decisions regarding exactly what tool in one's methodological kit to employ often hinge on power—that of the researcher, the researched, and the shifting power relations between the two over time, perhaps especially in the intensely interactive method of ethnography (e.g., Ferguson, 1991, pp. 130–132; Kurzman, 1991, p. 261). The power in method is the power of representation of others (Markham 2005a). It's a basic power—you get to choose the questions and the boundaries of the field, and you write the narrative. The ethnographic texts produced are models of social relations (Smith, 1990). The social relations involved in online ethnographies ultimately reveal that my position as arbiter of textual reality is a rather precarious power (see Marcus, 1998, p. 97).

▣ ONLINE ETHNOGRAPHIC METHODS: EXTENDING THE CLASSICS

Generally, ethnographic methods may be divided into three areas, and each of these and their well-known mechanics and methods are easily adaptable to the research site that begins, merges with, or ends up in an online setting,

■ **Traditional field methods,** wherein a lone researcher enters a field site and becomes a covert or known participant observer. A subset of this classic type is collaborative ethnography, wherein pairs or teams of researchers, often a mentor plus field workers or students, engage the research site (e.g., Anderson, 1990; Burawoy, 1979; Drake & Cayton, 1945; Duneier, 1992; Geertz, 1973/2000; Hartigan, 1999; Kanter, 1977; Lynd & Lynd, 1927/1956; May & Patillo, 2000; Shostak, 1981; Tulloch & Jenkins, 1995).

- **Autoethnography,** wherein the researcher is the explicitly grounded native of a particular field site or social situation/ status (e.g., Bochner & Ellis, 2002; Ellis, 2004; Gatson, 2003; Hancock, 2007; Markham, 2005a; May, 2003).

- **Multisited/extended-case ethnography,** wherein the goal is to situate contexts within a dialogue between theory and the field, and the micro mundane world to the macro systems that structure those worlds (Burawoy, 1991, 2000), where "Empirically following the thread of cultural process itself impels the move toward multisited ethnography" (Marcus, 1998, p. 80; see also the Center for Middletown Studies, which, like the Chicago School, has been an ethnographic factory).

Each of these types may be said to be mainly about presentation of the data, as it is possible to tease out the autoethnographic self of the researcher(s), and locate and highlight the place in the macro system(s) of the ethnographic site.

Whether using any of the ethnographic methods outlined, the Internet is ideally situated to be a part of extending the reach of ethnography. Although the boundaries of Internet sites are inherently more permeable and less physically bounded (if no less graphically or cognitively bounded) than offline sites, it is possible to note the predominant way in which authors have presented particular online ethnographies, although many of these publications may be placed in all three categories, and all can be placed in at least two, simultaneously. The bleed between the categories is not necessarily unique to the online settings of these ethnographies, but noting these categories and where we might place particular analyses within them tells us something about the ethical and political place of the contemporary ethnographer.

Although first emerging within the last two decades, with its classics only about 15 years old, and despite the ongoing question of whether online ethnography is either advisable or possible (e.g., Ashton & Thorns, 2007; Derteano, 2006; Ethnobase, n.d.; Holström, 2005; Howard, 2001; Nieckarz, 2005; Watson, 1997/2003), the online ethnography already has a vast tradition from which to draw.

Traditional Field Methods Online

The earliest online ethnography is arguably Howard Rheingold's *The Virtual Community* (1993/2000), wherein the "homesteading" metaphor of the "frontier" of cyberspace took root (see De Saille, 2006, for a critique of this metaphor). Although Rheingold's text is grounded in his personal experiences as one of the creators of online communities, as well as in an analysis that explores the multiple and connected online and offline places and spaces in which his community ultimately exists, his presentation is, without formally using the tools of ethnography (see Rheingold, 2000 edition, pp. 54–55), a rich and useful account of the state of online community in its nascent years. Others in this vein include Ali (2009a, 2009b), Baumle (2009), Baym (1995a; 1995b, 1998, 2000), Davis (2008), DiSalvo and Bruckman (2009), Gatson and Zweerink (2000, 2004a, 2004b), Harmon and Boeringer (2004), Kendall (2002), Lu (2009), Markham (1998), Millard (1997), Mizrach (1996), Nieckarz (2005), O'Brien (1997, 1999), Parpart (2003), Sharf (1997), Shaw (1997/2002), Stivale (1997), and Turkle (1995).

Autoethnography Online

Unlike many works in the classical ethnographic tradition, online ethnographies are often written by consummate and acknowledged insiders in the communities of interest, often by individuals who start out as students, or indeed non-academics. Beginning again with Rheingold, these works explicitly ground the author(s) as a member (sometimes indeed as an architect) of the community of interest first and foremost, and they include Asim Ali (2009a, 2009b), Sarah Gatson and Amanda Zweerink (2004a, 2004b), Stacy Horn (1998), Jeffrey Ow (2000), Latoya Peterson (2009a, 2009b), Lisa Richards (2003), John Seabrook (1997), Sherry Turkle (1995), and Stephanie Tuszynski (2006).

Multisited Ethnography

One engages in this type of online ethnography by either exploring more than one online site, by including both online and offline sites, or building a multilayered narrative that develops the larger social context of a community under study (Marcus, 1998, pp. 84–88, 117–118, 241–242). Philip Howard's previous suggestion regarding the lack of appropriateness for straight ethnography in online settings did not hinge upon "real" versus "virtual" per se, but more specifically on a physical ideal of the field site, suggesting that most online sites are both non-physically bounded in any way, and "difficult to set . . . in a larger social context" (2001, p.565; see also Derteano, 2006; Nieckarz, 2005). This position however has given way to one that is multisited (Celeste, Howard, & Hart, 2009), and includes Ashton and Thorns (2007), Bakardjieva (2005), Bandy (2007), Blasingame (2006), Christian (2009), Connery (1997), Gatson (2007a, 2007b), Gatson and Zweerink (2004a, 2004b), Goodsell and Williamson (2008), Hampton and Wellman (2002, 2003), Heinecken (2004), Hine (2000), Islam (2008), Ito (1997), Kendall (2002), Knapp (1997), Komaki (2009), Leurs (2009), Mallapragada (2009), McPherson (2000), Mitchell (1999), Nakamura (2009), Ow (2000), Reid (2009), Richards (2003), Salaff (2002), Schmitz (1997/2002), Silver (2000), Stenger (2006), Stern and Dillman (2006), Tepper (1997),

Tuszynski (2006), Watson (1997/2003), Williams (2004), and Zickmund (1997/2002).

It is worth noting again the difficulty of categorizing online ethnography. If we take a set of published texts by a research team (Busher & James, 2007a, 2007b; James, 2007; and James & Busher, 2006, 2007) together, these works present something other than only the online interviewing techniques that are their main stated methods. Instead, we could re-categorize this set as an autoethnographic multisited ethnography in that the authors set the research within academia where they are members, having prior knowledge of the bulk of their participants. A second set of examples occurs in *Race in Cyberspace.* Several of the chapters are never called *ethnography,* but arguably are, and are multisited: Nakamura analyzes advertising texts that are "popular media narratives of commercial cyberspace" (Kolko, Nakamura, & Rodman, 2000, p. 9). Jennifer González (2000) presents her sites as mainly text about graphics, discussing sites where users may purchase avatars. These sites, among some others in this edited volume (as well as in Porter's 1997 *Internet Culture*), are treated mainly as texts, and institutional review board (IRB) and methodological considerations are not explicitly discussed. Similarly, Henry Bial's keynote presentation at the 2009 Texas A&M University Race and Ethnic Studies Institute Symposium moves the author from film studies and analyses of actor performances, to the performances of Jewish identity in several online arenas. Finally, Jeffrey Ow's statement is most instructive:

> As an Asian male cyborg in my own right, I choose to play my own intellectual game with the *Shadow Warrior* controversy, acknowledging the perverse pleasures of weaving an oppositional read of the controversy, creating much more horrid creatures of the game designers and gaming public than the digital entities on the computer screen. In each level of my game, the Yellowfaced Cyborg Terminator morphs into different entities, from the individual gamer, to company representatives, ending with the corporate entities. (2000, p. 54)

Thus, Ow, though grounded in his own participation in both gaming and a racist culture, reflects George Marcus, "Cultural logics . . . are always multiple produced, and any ethnographic account of these logics finds that they are at least partly constituted within sites of the so-called system (i.e., modern interlocking institutions of media, markets, states, industries . . .)" (Marcus, 1998, p. 81).

▣ POLITICIZING METHODS AND ETHICS IN THE ONLINE FIELD SITE: INHERENT MEMBERSHIP?

That none of the authors discussed in the previous paragraph called themselves ethnographers is generatively problematic. Max Travers (2009) argues that the newness and innovation claims of online ethnography are mainly political (see also

Hine, 2008). In a basic sense of the mechanics of what it is that an ethnographer does (goes to a site, observes the location, the interactions, the boundaries, talks to or observes the inhabitants, records or transcribes all such observations and interactions, reads one's transcriptions, observes or talks more, transcribes more, and finally prepares a narrative wherein theory emerges or is tested), he is correct.[4] However, Travers's dismissal of the new *methods* of online ethnography misses the possibilities of the new *field*—in the sense of field site(s)—of online ethnography. The site of the online ethnography necessarily pushes the definitional boundaries of generally accepted concepts such as self, community, privacy, and text.

Gary Fine argues, "Ethnography is nothing until inscribed: sensory experiences become text" (1993, p. 288). The online site is already text, already inscribed (even more graphic sites, such as YouTube, have text); researcher elicitation from subjects is often unnecessary. This seems to be one reason for Travers to dismiss the online ethnography as one that "usually results in a 'thinner' level of description" (2009; p. 173). Despite the few examples Travers cites to prove this assertion, in which his sense seems to be that online ethnographers read "a" posting by a subject, the dozens of examples wherein the researcher(s) rather reads hundreds, perhaps thousands, of posts, often by the same set of participants, over years or as an archive, are ignored. Online research can provide either the same level of depth as a one-shot, one-hour interview, or the same level of depth as that produced by the daily participating, embedded offline ethnographer. It may also provide the same level of in-depth analysis as any historical or comparative historical text-based analysis, wherein the text is gleaned from archival sources (see also Marcus, 1998, p. 84).

Perhaps because the site of entry is so often (assumed to be) a private space (home, office; privately held online account), the idea that the online field has special ethical boundaries is often taken for granted. However, when reading the ethics sections of just about any work presenting itself as ethnographic, we find the same sorts of boundary-establishing behaviors outlined; indeed they are not inherently different than those found in offline ethnographies. I started my first online ethnographic project before the Association of Internet Researchers' (AoIR) "Ethical Decision-Making and Internet Research" guidelines were written (2002). From the outset, I followed the practices therein, paying particular attention to how various Internet venues across the terrain of a single community established their own ethical expectations (AoIR, 2002, pp. 4–5; Gatson & Zweerink, 2004a, pp. 17–19). I didn't get my sense of ethical boundaries from the AoIR, but rather from having been trained in sociological research methods.

Ethical Guidelines for Online Ethnography: A Sketch

Dorothy Smith has asserted, "There is no such thing as non-participant observation" (1990, pp. 87). As acknowledged by

Judith Davidson and Silvana di Gregorio (Chapter 38, this volume), the higher scrutiny and surveillance of IRBs, coupled with the fact of "ordinary people ... actively engaged as indigenous qualitative researchers in the virtual world," further complicates the already somewhat fraught professionally defined understandings of informed consent, participation, observation, authoritative narratives, discourse, and scholarship. The online arena then, is a field in which defining insiders and outsiders is more explicitly complicated than in traditionally understood conceptions of ethnographic field sites.

In contrast to Annette Markham's assertion, the first step in online environments is reading (see 2005a, p. 794). As well, in contrast to Hugh Busher and Nalita James (2007), it may be argued that academics are always already in the audience or group being studied when the research site is grounded in online interactions (see Turkle, 1995, pp. 29–30). The Internet itself is one of the only definable fields with which the overwhelming majority of its researchers are already intimately familiar, at least in the mechanical sense (we read, we post, we e-mail, etc.). The content of any particular subfield site within the Internet may be unfamiliar, but the method of becoming an entrant will not be. In other words, lurking or reading online content *is* participant observation in a way that unobtrusive observation isn't in an offline ethnographic situation; if we're a reader of online spaces, we are already "in," in a real way because most online content is read (interpreted), and not necessarily interacted with by adding the reader's own post. But is it always participant observation for which one needs IRB permission to perform? When does reading become thinking become data gathering become data analysis? When is one a community member, a citizen, or a scholar? Does one need permission to read, or only to post or talk to others online? If, on the Internet, experience is already inscribed, already performed, and not in need of an ethnographer to validate it through scholarly revelation, we are again exposed as decision makers who arbitrate the definitions of the boundaries of appropriate interactions.

In a sense, all online ethnography is "disguised observation," but it is not also necessarily deceptive observation. The contemporary publicly accessible website carries with it an expectation of being under some level and type of observation, and it is questionable whether anyone participating in such sites has a reasonable or defensible expectation of being unobserved, or indeed of being able to control the observers' intentions or uses of such observation. The hegemonic bedrock of ethnographic ethics, however, involves both informed consent and an awareness of power differentials, both embedded in the historical excesses of human subjects research, as well as those of IRBs themselves. But, again, reading is its own form of interaction, and posting, submitting, and publishing one's text online invites readership and an audience, if not a community. Markham's concern with the researcher's "loss of

authority [or power] in the presentation of research, and diminishment of one's academic role as observer/interpreter/archivist of social life" mirrors the loss of control the everyday online writer has once he or she presses post/submit/publish (2005a, p. 800; see also Marcus, 1998, p. 97).

In noting the "10 lies of ethnography," Fine discusses the positives of power and information control, assuming that the ethnographer has the greater share of salient power and control (1993, p. 276). In online settings though, the researcher is hardly the lone ranger (an ethnographic character Fine impugns) controlling the information flows and representations of an isolated or previously unknown/ordinarily unknowable community; again this "lie" of offline ethnography is intensified online. One's colleagues may engage in open published critiques of one's work, but so may one's subjects, and *your* subjects are hardly yours alone.

The required online training manual for human subjects researchers states, "Researchers do not have the right to conduct research, especially research involving human subjects. Society grants researchers the privilege of conducting research. The granting of that privilege is based on the public's trust that research will be conducted responsibly. Erosion of that trust can result in the withdrawal of this privilege." I have discussed elsewhere (Gatson & Zweerink, 2004b) the complicated dynamics of who gives permission to who, and for what, when one's research site is a public venue with not even an unlocked door whose opening announces a certain basic level of entry, and at most slightly opaque windows that block certain kinds of participation. Those dynamics include the "privilege" of the researcher being just another subject in a way that classic offline ethnographers such as Clifford Geertz ([1973/2000] or even Joshua Gamson [1998]) did not have to confront. As well, Fine, citing Jack Douglas (1976), argues that the ethnographer has rights too. Given the very public nature of most online ethnography, those rights should be assessed under a model other than either a biomedical, or a 50-year-old social behavioral, one (see Stark, 2007). Rather, it should be one that takes the media literate citizen (including the online ethnographers themselves) into account (Fine, 1993, p. 271; see also Bassett & O'Riordan, 2002; Dingwall, 2007; Elm, Buchanan, & Stern, 2009; Feely, 2007a, pp. 766–770; 2007b; Johns, Hall, & Crowell, 2004; Katz, 2007; Kendall, 2004; Smith, 2004; Thomas, 2004). Online ethnographers have to engage in an exploration of our particular locations in connection to particular field sites, and it is fundamentally and qualitatively different exploring our place in the media-reading audience than it has been in exploring our place as outsiders to more or less bounded, easily identifiable, cultural or subcultural offline geographical locations. Thus, in a sense, we have to remake our guidelines for each online ethnography we decide to do, without at the same time abandoning our connections to professional and socio-legal ethics that we must simultaneously work under.

FINDING THE EDGE OF THE ETHNO: REPRESENTING ONLINE PLACES AND EXPERIENCES

How then does one go about using the political power one has as an online ethnographer? What is both useful and ethical in creating a representative narrative of an online site and its attendant identities and boundaries? I suggested earlier that one must understand one's place in the larger community and the ever-tightening circles that demarcate our memberships in groups and networks and thus understand one's multiple positions, identities, and power/resources, and how to ethically employ them.

The Interpenetration of Community Boundaries Online

'stina says:

(Mon Sep 14 10:27:21 1998)

QUESTION: This was in my local newspaper a couple of months ago, in the "letters" section:

As a high-school female student, it is troubling to me to think that my age group has been reduced to a simple stereotype of the child who has seen Titanic 18 times because Leonardo DiCaprio is "really really cute" and who religiously watches trashy, superficial programming like Buffy, the Vampire Slayer and Dawson's Creek.

I resent being told that I am so lucky to have role models like the short-skirted, thrift-store shopping, bubble-headed Buffy.... The empowerment of women is a long way off if the media think that portraying sexy, moronic, peroxided young girls will give us real teen-aged girls the power to take the world by storm. I have no problem with young—I am. I have no problem with sexy—that's cool. I have no problem with hair dye—that's between a girl and her hair-stylist. But what I do mind is moronic. To have "girl power," we must first learn to respect ourselves and others. And if the only images we see are those of heroines with no brains, how can we ever respect ourselves enough to believe we can be something more than Buffy, the vampire slayer?

Courtenay B. Symonds, Houston[5]

Why do you think that young women are consistently told it's "great" to have Buffy and Ally [McBeal] around as role models? Is it because t]here are so few role models out there for women, that the second [a] series that's not a sitcom comes out that's focused on women instead of men we're all supposed to follow that woman's lead?

Does this kid have a point?

'stina—who'd much rather be like Chris Carter's Dana Scully than Joss Whedon's Buffy Summers.

This textual excerpt from an online community—which generated an approximately hour-long analytical conversation about the gendered role-model appropriateness of Buffy Summers—in which I have spent more than a decade as a member, and more than 6 years as an autoethnographer presents an ideal framing device for a discussion of the methods, politics, and ethics of representation in online ethnography.[6] First, because **'stina** used a common strategy that in fact replicates an ethnographic technique—she made reference to a piece of conversation or text she'd overheard or read elsewhere to bring attention to a topic of concern to the community and herself as a member of it. Second, she quoted in full the words of someone else, whose original text appeared outside of the place of new inscription, for purposes of her own, again mimicking the technique of an ethnographer. Finally, by reproducing these texts in published form again (for the second time in Ms. Symonds' case, and for the first in **'stina**'s), I myself highlight methods, politics, and ethics—why these texts? Why these subjects? And, with whose permission?

Reuben May (2003) explored these issues in an article wherein he described himself as both a personal journal writer and a scholarly ethnographer. May's analysis is one of personal journals treated as archival textual evidence in and of themselves, not as field notes as his "statuses negate[d] the viability of formally studying college students' social behavior because his own behavior as a participant-observer would flirt with, rub against, or cross, social, administrative, or legal boundaries" (2003, p. 442). In exploring the difference between being an anonymous student participant-observer who "even wrote my first book about a neighborhood tavern because of my deep down urge to be around people and to share in their verbal games of sexual innuendo," and one who found it inappropriate to engage in the same techniques in exploring the nightlife in a college town wherein he was a non-anonymous and hypervisible Black professor, May tells us what he studied by telling us why he chose not to study it (2003, p. 443). In contrast to May's decisions, Markham (2005b), in a similarly contextualized article about sex/gender and class/race tensions and power dynamics, chose to present her narrative as a complex methodological piece that nevertheless presents findings from a formally defined ethnographic research project. Both of these authors were members of the contexts discussed but each made different choices about what could be defined as a legitimate research project, yet both published their accounts, and in the same journal.

Because of the very public nature of most World Wide Web sites (as opposed to online sites more generally, which includes more controlled access and private arenas from e-mail to newsgroups and bulletin boards, as well as intranet and Internet work- and education-based online arenas), the contentiousness over where the ethical boundaries are has been perhaps especially fraught because they raise issues perhaps thought settled already, as long as the formalities of the IRB process are followed. Is quoting from a blog the same thing as quoting from a newspaper, or a letter found in a historical archive? Is quoting from or reconstructing a face-to-face or overheard in-person conversation the same as quoting from or reconstructing a conversation or group discussion held through instant messaging or a bulletin/ posting board? All self-identified online ethnographers, like their offline counterparts, discuss these sorts of issues, and must figure out what exactly their particular online arena(s) replicate about offline, perhaps settled, situations so that they can defend their ethical choices (e.g., Barnes, 2004; Bassett and O'Riordan, 2002; Elm, Buchanan, & Stern, 2009; Sharf, 1999).

Jan Fernback argues that online space is "socially constructed and re-constructed . . . [and] is a repository for collective cultural memory—it is popular culture, it is narratives created by its inhabitants that remind us who we are, it is life as lived and reproduced in pixels and virtual texts. . . . Cyberspace is essentially a reconceived public sphere for social, political, economic, and cultural interaction. . . . [its] users are . . . authors, public rhetoricians, statesmen, pundits" (1997/2002, p. 37). Thus, does a multivocal or dialogic set of texts produce a consensus picture of a community, a fragmented sense of what is or was important to it as a whole, or a reproduction—or indeed a continuation—of the community itself? Ethnographic narrative inscription presents a holistic and often linear story of a people, their place, and their identity. This linear structure may be a necessity of adhering to standards of coherent presentation (see Markham, 2005b, for another perspective). However, communities, identities, and places are contested entities. They also change over time. One (or even two) ethnographers can't cover every facet of a community, and the production of a coherent narrative requires choice-making. Although we as ethnographers can and do produce consecutive (or concurrent), multifocal narratives of one community, we are still the inscribers, interpreters, and authorities. What happens when there are other scholarly inscribers in one's research field, as well as "lay" inscribers whose inscription is at least as analytical as one's own (again, see Davidson and di Gregorio, Chapter 38, this volume)? One outcome could be a collaborative multivocal ethnography that combines the practices of each of the three main areas of ethnographic methodology. This becomes *macro-ethnography.*

Kate Millet's *Prostitution Papers* presents us with an example of multivocal ethnography (1973; see also Davis & Ellis, 2008). Although Millet is presented as the book's author, she is really its editor, soliciting and organizing a dialogue between four women, herself included. Millet presented the approaches of each writer in the form of essays authored by women who were identified only by their initials. She herself was K., the scholar activist. J. was the former sex worker–current psychologist, M. was the former sex worker–current PhD student, and L. was the lawyer and policy advocate for the rights of prostitutes. Although there was an overarching consensus offered by the authors—sex work is work, sex workers are positioned by patriarchy and re-victimized by the criminal justice system that seeks to punish prostitutes but not (or rarely) the men they service—as each woman comes from a different status position, their narratives are quite divergent.

Millet called the work a "candid dialogue," and the project actually produced both the book and a film. Millet was arguably a liminal ethnographer. She took on the ethnographer's timeline of observation, inscription, and interpretation. By offering space for their own developed narratives, she invited her ostensible subjects to become ethnographers as well, auto-ethnographers in particular. Millet's book is thus one that combines ethnography as the experience of the ethnographer and the text(s) that ethnographers and their subjects co-create. In the online ethnography, there is never just "'one beginning [or] one ending'" (Bochner & Ellis, 2002, p. 11), and no one position of unequivocal power. James and Busher worry that even with the shift in the balance of power that comes with using e-mail for interviews, "researchers cannot escape the power they exert from structuring the rules of the process" (2006, p. 416). Busher and James also worry about the insecure environments and lack of privacy inherent in the field (2007a, p. 3), [but they do not see that this is true for the researcher as well. As Stevienna De Saille notes, without making the Internet a utopia, power and access are quite different online, "As previously disenfranchised people increasingly put up their own boards, pages, and blogs, thus defining their heterogeneous subjectivity to the world, can it be ignored that the technologies of the web do indeed allow the subaltern to speak?" (2006, p. 7).

The "engagement medi[a]" that are both television and the Internet (Askwith, 2007) potentially take us far beyond team-based collaborative ethnographies and allow us to further parse the boundaries, ethical and otherwise, of television audiences, and the communities embedded within those audiences (Bandy, 2007; Islam, 2008; Lotz & Ross, 2004; Shirky, 2002; Whiteman, 2009). Around the same time that I was making the decision to formally study The Bronze, the book, *Bite Me! An Unofficial Guide to the World of Buffy the Vampire Slayer* was published (Stafford, 1998/2002). It contained a section on the first major offline gathering of the community, the soon-to-be annual Posting Board Party, and included photographs of Bronzers, captioned with both their Bronze posting names and their offline, everyday names. Its 2002 edition contains even more easily identifiable information about several individual Bronzers, as well as thoroughly identifying information about The Bronze as a website, and some interviews with members (Stafford, 2002, pp. 113–156). As well, the site was well advertised, and well trolled by journalists, and through linked websites such as The Who's Who and What's What of The Bronze (where Bronzers themselves created a pre-Friendster/MySpace/Facebook place for ease of social networking, long before the mainstream discourse of "Web 2.0" appeared) self-promoted as the communal place to be.

The audience-author feedback loop (Kociemba, 2006) of The Bronze has always been complex and multileveled—it wasn't just Joss Whedon who got to be an author, and he wasn't the only one to hear from his audience and community. Jane Espenson, eventually part of the production team, started out as a linguistics graduate student and provided the introduction to Michael Adams's *Slayer Slang* (2004). Meredyth Smith (~**mere**~) went the other direction, from Bronzer first to Whedonverse writer.[7] Others were lurking and posting, and ruminating on the implications of being Bronzers. **elusio** did a couple of papers for undergraduate courses at a university in the United Kingdom, **Kenickie** was an undergraduate sociology major in the United Kingdom throughout most of the research, and **Psyche** and **seraphim** were graduate students in psychology and anthropology, respectively, who—like several other acafans—while doing no formal studies on The Bronze, nonetheless brought their intellectual interests to their offline and online fandom activities, as **Tamerlane** did, when he delurked after several years and became a visible member of the community,

Tamerlane says:
(Tue Apr 11 17:10:39 2000)

'stina, Jaan Quidam, Closet Buffyholic, SarahNicole, and others: This place really does interest me. More, I think, than the show. I think an awful lot of credit has to be given to **TV James** for instituting the format he did. One of my two long-time academic interests is Biology (the other of course being History). It always struck me as interesting the difference between your average college lecture class and a biology class with a field or even just a long, interactive lab component. Hanging around people for a three-hour open-ended lab, twice a week, where you spend the whole time wandering around looking at things and discussing them with others, led, I think, to a greatly increased connectivity (compared to even a three hour lecture). Not to mention the bonding that took place on long field excursions (one of my all time favorite camping trips remains a one-week Fire Ecology field-trip). Although I see communities develop at all the posting boards I visit (and I have been there at the beginning of a couple of others), I have never seen the cohesiveness of the Bronze duplicated elsewhere. This "message board" format, at least the way it has evolved, offers IMHO all of the best qualities of chatrooms … and threaded boards with few of the drawbacks. The ability to encompass a wide, free-flowing conversation that can be scrolled at leisure at any point within a one week period is pretty unique. The Usenet is similar of course, but the seemingly much higher level of anonymity and the much larger base of casual users seems to limit social interactions.

Here we see an example of how lurkers/readers have an acknowledged place as members of online communities and audiences—they may show up visibly at any time, and their observations/representations then become part of the communally produced text. Across the range of less formal photographic and inscribed observations, there were the Bronzers who wrote about their experiences on easily accessible websites (e.g., **Claris**'s site, www.NoDignity.com), wherein the ethnographer became the subject, where *sometimes* my informed consent was solicited before posting a quotation or a photo.

Other Bronzers (Ali, 2009a,[8] 2009b; Tuszynski, 2006) also wrote ethnographies, while Allyson Beatrice (2007) wrote a memoir that grounded her online communal experiences at The Bronze. Scholars who do not identify as Bronzers, but as aca-fans (*aca-fans* refers to academics who research the object of their fandom, or who identify as both fans and scholars) of *Buffy, the Vampire Slayer* (BtVS) to one degree or another, also wrote about Buffy fandom, and sometimes used The Bronze and Bronzers as data (Adams, 2004; Askwith, 2007; Bandy, 2007; Blasingame, 2006; Busse, 2002; Heinecken, 2004; Kem, 2005; Kociemba, 2006; Larbelestier, 2002; Parpart, 2003; Parrish, 2007; Richards, 2003; Stenger, 2006; Williams, 2004; see the Kirby-Diaz, 2009, collection for other broader BtVS fandom works). Dawn Heinecken and Michael Adams most explicitly use Bronzers as examples of audience members, with Adams acknowledging us/them as the hierarchical top of the fandom heap, whereas Heinecken does not draw a boundary around them as a community, but rather presents a division of fans that some Bronzers were a part of—Spike/Buffy 'shippers(short for "relationshippers"—fans of particular romantic pairings, either those actually appearing in a show/text, or those wished for)—without noting where these folks were on the Bronzer totem pole. Neither of these pieces nor most of the others include discussion of methods, or IRB issues. These authors generally do not position themselves as ethnographers (with the exceptions of Ali, Tuszynski, and Richards). In Adams's case, he is a linguist, accessing the publicly available development of slang stemming from a particular show and its fandom, without really drawing boundaries around the community(ies) that make up that fandom, and seeing his data as published text, part of the public flow of developing language.

The Networked Self and the Pedagogy of the Media Literate Citizen

As I have noted elsewhere (Gatson & Zweerink, 2004a), it is questionable how anonymous ethnographic sites have ever been. Thus, my comments in this section will mainly focus on field sites that are public arenas—not that there isn't a backstage, but they are backstages that one need not be a "professional" to gain access to. The (perhaps dirty) secret of online ethnography (all ethnography?)—like other mass media-based research, especially television studies—is that it exposes the lack of special knowledge needed to do it. It is intense, but its mechanics are fairly simple, and they're things we and our students are engaged in every day. All that is needed is the application of the sociological imagination to formalize the reading and posting I do at online communities where my membership ranges from the regular lurker, to the occasional poster, to the daily contributor, and where my online identity of **SarahNicole** remains a near-constant; I'm fairly certain that at the least one of my posts across these communities have already been incorporated into someone's thesis or dissertation at this point. Two examples demonstrate aspects of this experience of the Internet and create speculation of simultaneously teaching undergraduates responsible research methods and media literacy/citizenship.

In 2007, I did something common to academics; I read a review of my work. What was perhaps unusual but increasingly common was that I stumbled across the review online while "Googling myself" in a search for citations of my work (don't lie; you all do it too). What was truly unusual was that the review was written by some of the subjects of the research. This research was the outcome of a National Institute on Drug Abuse (NIDA)–funded study on raves and the use of "club drugs," which in our case focused on the online discourse surrounding these topics (Fire Erowid, 2007; Murguía, Tackett-Gibson, & Lessem, 2007). Early on, my co-principal investigator (PI) attended a conference at NIDA where he met Fire and Earth Erowid (Fire Erowid, 2002). Fire and Earth are the pseudonyms of the people who run Erowid, an online clearinghouse of information about drug use and alternative subcultures that was an important node in the network we were studying, so from nearly its inception, research subjects were involved in the research in ways that were previously unfamiliar to most of the research team. In their review, Fire characterized the work of Erowid thusly,

> Although our primary role is that of cultural documentarians rather than participants of the drug-using subculture, in the modern anthropological tradition we adhere to, the validity of one's understanding of a culture or community is based on whether one is a part of that culture or community. If there is too little connection between anthropologists, researchers, or documentarians and their subjects, the resulting research is likely to be inaccurate. Erowid was started out of our personal and academic interest in psychoactives, but we are only peripherally involved in many aspects of the field. We maintain connections and involvement with a variety of communities in order to better be able to serve their needs, represent their actions and viewpoints, and act as their trusted recorders and archivists. (2007).

These authors/subjects engage in a "contexting [of] the network" (Jones, 2004) of scholars and subjects that publicizes these connections in ways that go beyond publication in practically restricted-access journals read only by authoritatively legitimated experts.

The Internet exposes the fact that there's public, and then there's public; there's talking back (other scholars' letters to the editor, symposia/dialogue in journals; see Denzin, 2004), and then there's talking back (Borland, 2004; Chen, Hall, & Johns, 2004; Gatson & Zweerink, 2004b). The race and pop culture website Racialicious.com and some of the talk-back generated therein serve as examples of when subjects talk back. Racialicious is more moderated than many news outlets' comments section because it is partly meant to be a safe space to discuss issues related to race/ethnicity and racism (see http://www.racialicious.com/comment-moderation-policy/), but its capacity for generating talk-back with which many ethnographers may be unfamiliar is illustrative of the phenomenon wherein I think we can locate both scholars' place in the research network, as well as platform a teaching tool.

In 2009, as director of my university's Race and Ethnic Studies Institute, I hosted a symposium on race, ethnicity, and (new) media. Latoya Peterson, editrix of Racialicious, participated as both a keynote speaker (2009b) and research presenter (2009a). Peterson took copious notes on each presentation, and presented some of her synopses at Racialicious. The synopsis of Lisa Nakamura's keynote presentation and the article from which it was drawn (2009) drew sometimes angry commentary from fans of the game World of Warcraft, which, along with the gaming practice of creating machinima (animated music videos using images from the game), were the centerpiece of Nakamura's work. As well, danah boyd (2009) re-published at Racialicious some of her work looking at socioeconomic differences between MySpace and Facebook users, and some users of those social networking sites had much to critique about boyd's conclusions.[9]

These examples, along with my own experience with Erowid, demonstrate that both information and people, although theoretically having newly opened conduits, are also materially and ideologically embedded in truncated networks with less-than-permeable boundaries (Howard, 2004; Norris, 2004; Travers, 2000). The talk-back made possible by the Internet takes us beyond the professional deconstruction of our ethnographic pasts (Van Maanen, 2004; see also MacKinnon, 1997/2002), and pushes us as both scholars and teachers to explore our "distributed learning communities" (Haythornthwaite, 2002) and the "social context of user sophistication" (Hargittai, 2004). If it is important in the process of ethnographic research to locate the researcher as well as the subjects, and expose the connections between and among them, then the autoethnographic network mapping of particular research projects would be useful in pedagogical and public sociological projects where media literacy and online citizenship are the specific aims. In developing an undergraduate course on the sociology of the Internet, I use my own online network maps as an introduction to the major project of the class. The students will have to place themselves in their offline and online networks—where do they go and what do they read, watch, discuss, and publish? Do their experiences reflect the published works on their networked worlds? Why and why not?

Here, we have the inherent autoethnographic nature of Internet environments for the academic—being online is arguably (a large part of) our work environment, if not always our home or our third space. The Internet exposes us again in our quest for "unpolluted truth" (see Fine, 1993, p. 274)—if we ever were investigating isolated "primitives" outside of the macro system, we certainly aren't now. For every online interaction engaged in, every online observation logged, some other observer may be recording our actions, and observing us. Thus, we most especially "differ little from Erving Goffman's social actors" (Fine, 1993, p. 282), and increasingly our students are very likely to be more expert members of social worlds that they have the right to engage in and comment on, beyond their identities as nascent researchers. We should provide the tools for that commentary to be analytical and empirical, so that both they and we are aware of the political boundaries of our methods and ethics. Media-saturation is not the same thing as media-literacy, and navigating through representations across the field of mass media is an important skill, even if one chooses to opt out of much interaction with such media.

During the next decade and into the future as ethnography moves from face-to-face, to online textual and graphic communication, to the spaces of Second Life (Boellstorff, 2008) and such games as World of Warcraft, and back again to the offline context, the ethnographic experiences discussed herein both ground themselves in the traditions of the method and are generative of explicitly exploring the meaning of "empirical" versus the meaning of "objective" in the practice of the social sciences. We can tell empirically based trustworthy stories about human behavior online, perhaps especially because we as the ethnographers are eminently exposable as but one in a host of voices telling the stories, and we are un-removable subjects of those stories, perhaps waiting for someone else to tell our story for us. We could perhaps call this Ethnography 2.0, in an acknowledgement of the way in which this way of practicing ethnography "allows its [practitioners] to interact with other[s] or to change . . . content, in contrast to . . . [being] passive [subjects]" ("Web 2.0," 2/8/2010).[10] This visible and experiential reality does not remove our ethical responsibilities from us, but it does make the boundary surrounding the ostensibly objective outsider (the researcher, the lone scholar) especially permeable. It raises the question, can anyone be an ethnographer? If so, who watches the watchers?

▣ Notes

1. Gatson and Zweerink, 2000, 2004a, 2004b; Zweerink and Gatson, 2002. These works, though driven by my sociological research agenda, were coauthored with Amanda Zweerink, a "lay" ethnographer I met online within the community I ended up researching. She was in advertising at the time we began working together and at present is director of community at CurrentTV.

2. Gatson, 2007a, 2007b. These are two of the three chapters I wrote for an edited volume that emerged from a National Institutes on Drug Abuse collaborative grant that looked at drug use discourse online.

3. Coughlin, Greenstein, Widmer, Meisner, Nordt, Young, Gatson, et al., 2007; Coughlin, Greenstein, Widmer, Meisner, Nordt, Young, Quick, and Bowden, 2007; Desai et al., 2008; Gatson et al., 2005; Gatson et al., 2009; Nordt et al., 2007. These represent some of the work emerging from a multiyear project exploring the ways in which a group of researchers, students, teachers, and others have worked together toward a paradigm shift in the production and use of science. This approach sought to integrate research, teaching, and service—the traditional triumvirate of evaluation for academics in college and university systems—by combining the reintroduction of an old animal model with new and emerging technologies and the development of a new online/offline community model that incorporated the development of formal and informal networks.

4. For specifically methodological discussions of the processes of online ethnography, see Chen, Hall, and Johns, 2004; Dicks and Mason, 2008; Dicks, Soyinka, and Coffey, 2006; Gatson and Zweerink, 2004b; Hine, 2000; 2008; Hine, Kendall, and boyd, 2009; Kendall, 1999, 2004; LeBesco, 2004; Mann and Stewart, 2004; Markham, 2004; Markham and Baym, 2009; Walstrom, 2004.

5. The italics used here indicate that the poster was quoting directly from another source, in this case, the letters to the editor section of the *Houston Chronicle*.

6. My field site was The Bronze, a linear posting board (which originally also hosted a threaded posting board and chat room) located at the official website for the television series *Buffy the Vampire Slayer* (BtVS). It was called The Bronze after the club where the main characters often hung out, and the denizens of the online community dubbed themselves Bronzers.

7. Both of these women now have writing and producing credits outside the Whedonverse.

8. The **Jaan Quidam** addressed by **Tamerlane** earlier; I am **SarahNicole** addressed earlier.

9. See http://www.racialicious.com/2009/05/11/dont-hate-the-player-hate-the-game-the-racialization-of-labor-in-world-of-warcraft-conference-notes/ and http://www.racialicious.com/2009/07/09/the-not-so-hidden-politics-of-class-online/. For another example of such researcher/ researched engagement, see also the discussion of Boellstorff's ethnography of Second Life, in which the ethnographer participates: http://savageminds.org/2008/06/12/ethnography-of-the-virtual/#comment-392629

10. I don't actually think it's appropriate to apply the concept of Web 3.0 to the concept of online ethnography I am explicating herein, because as a concept, there are too many varied definitions of what this even means (see "Web 2.0," retrieved February 8, 2010, from Wikipedia, http://en.wikipedia.org/wiki/Web_2.0; and "Semantic Web," retrieved from February 8, 2010, from Wikipedia, http://en.wikipedia.org/wiki/Semantic_Web). However, especially because some consider Web 3.0 "as the return of experts and authorities to the Web" ("Web 2.0") and I think the Internet (both the technology and

the end users) have made the online ethnographic project one that is too inherently open to non-expert participation. Although we may continue to have easily separated reference groups relative to our particular ethnographic projects (e.g., our academic subjects versus our academic employers), and though to some degree we can still keep our finished products mainly to an academic audience if we choose, I think we have to deal with being no more necessarily experts in our online endeavors as our ostensible subjects, and I think this reality highlights and complicates the traditional ethnographic notion of participant observation in ways that "Web 3.0" doesn't.

◙ REFERENCES

Adams, M. (2004). *Slayer slang: A* Buffy the Vampire Slayer *lexicon.* Oxford, UK: Oxford University Press.

Ali, A. (2009a). "In the world, but not of it": An ethnographic analysis of an online *Buffy the Vampire Slayer* fan community. In M. K. Diaz (Ed.), *Buffy and Angel conquer the Internet: Essays on online fandom* (pp. 87–106). Jefferson, NC: McFarland.

Ali, A. (2009b). Community, language, and postmodernism at the mouth of hell. In M. Kirby-Diaz (Ed.), *Buffy and Angel conquer the Internet: Essays on online fandom* (pp. 107–126). Jefferson, NC: McFarland. (Original publication, 2000; available at http://terpconnect.umd.edu/~aali/buffnog.html)

Anderson, E. (1990). *Streetwise: Race, class, and change in an urban community.* Chicago: University of Chicago Press.

Ashton, H., & Thorns, D. C. (2007). The role of information communications technology in retrieving local community. *City & Community, 6*(3), 211–230.

Askwith, I. D. (2007). *Television 2.0: Reconceptualizing TV as an Engagement Medium.* Master's thesis, Massachusetts Institute of Technology, Boston, MA. Available at cms.mit.edu/research/theses/IvanAskwith2007.pdf

Association of Internet Researchers. (2002). Ethical decision-making and Internet research. Available at http://aoir.org/reports/ethics.pdf

Bakardjieva, M. (2005). *Internet society: The Internet in everyday life.* Thousand Oaks, CA: Sage.

Bandy, E. (2007, May 23). *From* Dawson's Creek *to "Dawson's Desktop": TV-web synergy in a multimedia world.* Paper presented at the annual meeting of the International Communication Association, San Francisco, CA. Available at http://www.allacademic.com/meta/p172730_index.html

Barnes, S. B. (2004). Issues of attribution and identification in online social research. In M. D. Johns, S. S. Chen, & G. J. Hall (Eds.), *Online social research: Methods, issues, ethics* (pp. 203–222). New York: Peter Lang.

Bassett, E. H., & O'Riordan, K (2002). Ethics of Internet research: Contesting the human subjects research model. *Ethics and Information Technology, 4*(3), 233.

Baumle, A. K. (2009). *Sex discrimination and law firm culture on the Internet.* New York: Palgrave Macmillan.

Baym, N. K. (1995a). The emergence of community in CMC. In S. G. Jones (Ed.), *CyberSociety: Computer-mediated communication and community* (pp. 138–163). Thousand Oaks, CA: Sage.

Baym, N. K. (1995b). From practice to culture on Usenet. In S. L. Star (Ed.), *The cultures of computing* (pp. 29–52). Oxford, UK: Blackwell/Sociological Review.

Baym, N. K. (1998). The emergence of online community. In S. G. Jones (Ed.), *CyberSociety 2.0: Revisiting computer-mediated communication and community* (pp. 35–63). Thousand Oaks, CA: Sage.

Baym, N. K. (2000). *Tune in, log on: Soaps, fandom, and online community.* Thousand Oaks, CA: Sage.

Beatrice, A. (2007). *Will the vampire people please leave the lobby?: True adventures in cult fandom.* Naperville, IL: Sourcebooks.

Bial, H. (2009, April 30–May). *Jew media: Performance and technology in the 58th century.* Paper presented at the Texas A&M University Race & Ethnic Studies Institute Symposium: Race, Ethnicity, and (New) Media, Texas A&M University, College Station, TX.

Blasingame, K. S. (2006). "I can't believe I'm saying it twice in the same century… but 'Duh … '": The evolution of the *Buffy the Vampire Slayer* sub-culture language through the medium of fanfiction in *Buffy the Vampire Slayer. Slayage: The Online Journal of Buffy Studies, 20.* Available at http://slayageonline.com/essays/slayage20/Blasingame.htm

Bochner, A. P., & Ellis, C. (2002). *Ethnographically speaking: Autoethnography, literature, and aesthetics.* Walnut Creek, CA: AltaMira Press.

Boellstorff, T. (2008). *Coming of age in Second Life: An anthropologist explores the virtually human.* Princeton, NJ: Princeton University Press.

Borland, K. (2004). "That's not what I said": Interpretive conflict in oral narrative research. In S. N. Hesse-Biber & P. Leavy (Eds.), *Approaches to qualitative research: A reader on theory and practice* (pp. 522–534). New York: Oxford University Press.

boyd, d. (2009). The not-so-hidden politics of class online. Racialicious.com. Available at http://www.racialicious.com/2009/07/09/the-not-so-hidden-politics-of-class-online/

Burawoy, M. (1979). *Manufacturing consent: Changes in the labor process under monopoly capitalism.* Chicago: University of Chicago Press.

Burawoy, M. (1991). *Ethnography unbound: Power and resistance in the modern metropolis.* Berkeley: University of California Press.

Burawoy, M. (2000). *Global ethnography: Forces, connections, and imaginations in a postmodern world.* Berkeley: University of California Press.

Busher, H., & James, N. (2007a, April 12–14). *Email communication as a technology of oppression: Attenuating identity in online research.* Paper presented at the Annual Conference of the British Sociological Association, University of London Available at http://hdl.handle.net/2381/439

Busher, H., & James, N. (2007b, September 5–8). *Building castles in the air: Colonising the social space in online qualitative research.* Paper presented at the British Educational Research Association Annual Conference, Institute of Education, University of London. Available at http://www.leeds.ac.uk/educol/documents/165971.htm

Busse, K. (2002). Crossing the final taboo: Family, sexuality, and incest in Buffyverse fan fiction. In R. Wilcox & D. Lavery (Eds.), *Fighting the forces: What's at stake in* Buffy the Vampire Slayer (pp. 207–217). Lanham, MD: Rowman & Littlefield.

Celeste, M., Howard, P. N., & T. Hart (2009, April 30–May 2). *(Con)Testing identities: Haitian and Indian women's use of social networking platforms.* Paper presented at the Texas A&M University Race & Ethnic Studies Institute Symposium: Race, Ethnicity, and (New) Media, Texas A&M University, College Station, Texas.

Center for Middletown Studies. Available at http://cms.bsu.edu/Academics/CentersandInstitutes/Middletown.aspx

Chen, S. S., Hall, G. J., & Johns, M. D. (2004). Research paparazzi in cyberspace: The voices of the researched. In M. D. Johns, S. S. Chen, & G. J. Hall (Eds.), *Online social research: Methods, issues, ethics* (pp. 157–173). New York: Peter Lang.

Christian, A. J. (2009, April 30–May 2). *YouTube: Black existentialism and network participation.* Paper presented at the Texas A&M University Race & Ethnic Studies Institute Symposium: Race, Ethnicity, and (New) Media, Texas A&M University, College Station, Texas.

Connery, B. (1997). IMHO: Authority and egalitarian rhetoric in the virtual coffeehouse. In D. Porter (Ed.), *Internet culture* (pp. 161–180). New York: Routledge.

Coughlin, D. J., Greenstein, E. E., Widmer, R. J. Meisner, J., Nordt, M. Young, M. F., Gatson, S. N., et al. (2007, April 29). *e-Research: A novel use of the Internet to perform live animal research from a laboratory distant from the site of animal care technicians and facilities.* Federation of American Societies for Experimental Biology meetings, Computers in Research and Teaching II poster session.

Coughlin, D. J., Greenstein, E. E., Widmer, R. J., Meisner, J., Nordt, M., Young, M. F., Quick, C. M., & Bowden, R. A. (2007). Characterization of an inflammatory response and hematology of the Pallid bat using "e-Research." *The FASEB Journal, 21,* 742.11.

Davis, J. L. (2008). *Presentation of self and the personal interactive homepage: An ethnography of MySpace.* Master's thesis, Texas A&M University, College Station, Texas.

Davis, C., & Ellis, C. (2008). Emergent methods in autoethnographic research: Autoethnographic narrative and the multiethnographic turn. In S. N. Hesse-Biber & P. Leavy (Eds.), *Handbook of emergent methods* (pp. 283–302). New York: Guilford Press.

Davis, J. L. (2008). *Presentation of self and the personal interactive homepage: An ethnography of MySpace.* Master's thesis, Texas A&M University, College Station, Texas.

Denzin, N. K. (2004). The art and politics of interpretation. In S. N. Hesse-Biber & P. Leavy (Eds.), *Approaches to qualitative research: A reader on theory and practice* (pp. 447–473). New York: Oxford University Press.

Derteano, P. F. M. (2006). Reflexiones para la reflexividad del investigador: Un acercamiento a través del estudio del fenómeno pornográfico (Reflections on the reflexivity of the investigator: An approach through the study of the pornographic phenomenon). Retrieved June 2006 from http://www.perio.unlp.edu.ar/question/nive12/articulos/ensayos/molinaderteano_1_ensayos_12primavera06.htm

Desai K. V., Gatson, S. N., Stiles, T., Laine, G. A., Stewart, R. H., & Quick, C. M. (2008). Integrating research and education at research-intensive universities with research-intensive communities. *Advances in Physiological Education, 32*(2), 136–141.

De Saille, S. (2006). A cyberian in the multiverse: Towards a feminist subject position for cyberspace. *Conference proceedings—Thinking gender—the NEXT generation.* UK Postgraduate Conference in Gender Studies, June 21–22, University of Leeds, UK, e-paper no. 19.

Dicks, B., & Mason, B. (2008). Hypermedia methods for qualitative research. In S. N. Hesse-Biber & P. Leavy (Eds.), *Handbook of emergent methods* (pp. 571–600). New York: Guilford Press.

Dicks, B., Soyinka, B., & Coffey, A. (2006). Multimodal ethnography. *Qualitative Research, 6*(1), 77–96.

Dingwall, R. (2007). "Turn off the oxygen . . .' *Law & Society Review, 41*(4), 787–796.

DiSalvo, B. J., & Bruckman, A. (2009, April 30–May 2). *Gaming manhood in African American culture.* Paper presented at the Texas A&M University Race & Ethnic Studies Institute Symposium: Race, Ethnicity, and (New) Media, Texas A&M University, College Station, Texas.

Douglas, J. (1976). *Investigative social research.* Beverly Hills, CA: Sage.

Drake, S., & Cayton, H. R. (1945). *Black metropolis: A study of Negro life in a northern city.* New York: Harcourt, Brace.

Duneier, M. (1992). *Slim's table: Race, respectability, and masculinity.* Chicago: University of Chicago Press.

Ellis, C. (2004). *The ethnographic I: A methodological novel about autoethnography.* Walnut Creek, CA: AltaMira Press.

Elm, M. S., Buchanan, E. A., & Stern, S. A. (2009). How do various notions of privacy influence decisions in qualitative Internet research? In A. N. Markham & N. K. Baym (Eds.), *Internet inquiry: Conversations about method* (pp. 69–98). Thousand Oaks, CA: Sage.

Ethnobase. (n.d.). Available at http://webdb.lse.ac.uk/ethnobase/bibliography.asp

Feely, M. M. (2007a). Legality, social research, and the challenge of institutional review boards. *Law & Society Review, 41*(4) 757–776.

Feely, M. M. (2007b). Response to comments. *Law & Society Review, 41*(4), 811–818.

Ferguson, A. A. (1991). Managing without managers: Crisis and resolution in a collective bakery. In M. Burawoy (Ed.), *Ethnography unbound: Power and resistance in the modern metropolis* (pp. 108–132). Berkeley: University of California Press.

Fernback, J. (2002). The individual within the collective: Virtual ideology and the realization of collective principles. In S. G. Jones (Ed.), *Virtual culture: Identity and communication in cybersociety* (pp. 36–54). Thousand Oaks, CA: Sage. (Original work published 1997)

Fine, G. A. (1993). Ten lies of ethnography: Moral dilemmas of field research. *Journal of Contemporary Ethnography, 22*(3), 267–294.

Fire Erowid. (2002). Face to face with NIDA: A conference on drugs, youth and the Internet. *Erowid Extracts, 3*(2).

Fire Erowid. (2007). Review of *Real Drugs in a Virtual World. Erowid Newsletter, 13.* Available at http://www.erowid.org/library/review/review.php?p=265

Gamson, J. (1998). *Freaks talk back: Tabloid talk shows and sexual nonconformity.* Chicago: University of Chicago Press.

Gatson, S. N. (2003). On being amorphous: Autoethnography, genealogy, and a multiracial identity. *Qualitative Inquiry, 9*(1), 20–48.

Gatson, S. N. (2007a). Assessing the likelihood of Internet information-seeking leading to offline drug use by youth. In E. Murguía, M. Tackett-Gibson, & A. Lessem (Eds.), *Real drugs in a virtual world: Drug discourse and community online* (pp. 99–120). Lanham, MD: Lexington Books.

Gatson, S. N. (2007b). Illegal behavior and legal speech: Internet communities' discourse about drug use. In E. Murguía, M. Tackett-Gibson, & A. Lessem (Eds.), *Real drugs in a virtual world: Drug discourse and community online* (pp. 135–159). Lanham, MD: Lexington Books.

Gatson, S. N., Meisner, J. K., Young, M. F., Dongaonkar, R., & Quick, C. M. (2005). The eBat project: A novel model for live-animal distance learning labs. *FASEB Journal, 19*(5), A1352.

Gatson, S. N., Stewart, R. H., Laine, G. A., & Quick, C. M. (2009). A case for centralizing undergraduate summer research programs: The DeBakey research-intensive community. *FASEB Journal, 633,* 8.

Gatson, S. N., & Zweerink. A. (2000). Choosing community: Rejecting anonymity in cyberspace. In D. A. Chekki (Ed.), *Community structure and dynamics at the dawn of the new millennium* (pp. 105–137). Stamford, CT: JAI.

Gatson, S. N., & Zweerink. A. (2004a). *Interpersonal culture on the Internet: Television, the Internet, and the making of a community.* Studies in Sociology Series, no. 40. Lewiston, NY: Edwin Mellen Press.

Gatson, S. N., & Zweerink. A. (2004b). "Natives" practicing and inscribing community: Ethnography online. *Qualitative Research, 4*(2), 179–200.

Geertz, C. (2000). *The interpretation of cultures.* New York: Basic Books. (Original work published in 1973)

González, J. (2000). The appended subject: Race and identity as digital assemblage. In B. E. Kolko, L. Nakamura, & G. B Rodman (Eds.), *Race in cyberspace* (pp. 27–50). New York: Routledge.

Goodsell, T. L., & Williamson, O. (2008). The case of the brick huggers: The practice of an online community. *City & Community, 7*(3), 251–272.

Hampton, K., & Wellman, B. (2002). The not so global village of Netville. In B. Wellman & C. Haythornthwaite (Eds.), *The Internet in everyday life* (pp. 345–371). Malden, MA: Blackwell.

Hampton, K., & Wellman, B. (2003). Neighboring in Netville: How the Internet supports community and social capital in a wired suburb. *City & Community, 2*(4), 277–312.

Hancock, B. H. (2007). Learning how to make life swing. *Qualitative Sociology, 30*(2), 113–133.

Hargittai, E. (2004). Informed web surfing: The social context of user sophistication. In P. N. Howard & S. Jones (Eds.), *Society online: The Internet in context* (pp. 256–274). Thousand Oaks, CA: Sage.

Harmon, D., & Boeringer, S. B. (2004). A content analysis of Internet-accessible written pornographic depictions. In S.N. Hesse-Biber & P. Leavy (Eds.), *Approaches to qualitative research: A reader on theory and practice* (pp. 402–408). New York: Oxford University Press.

Hartigan, J. (1999). *Racial situations: Class predicaments of whiteness in Detroit.* Princeton, NJ: Princeton University Press.

Haythornthwaite, C. (2002). Building social networks via computer networks: Creating and sustaining distributed learning communities. In K. A. Renninger & W. Shumar, *Building virtual communities: Learning and change in cyberspace* (pp. 159–190). Cambridge, UK: Cambridge University Press.

Heinecken, D. (2004). Fan readings of sex and violence in *Buffy the Vampire Slayer. Slayage, 11–12.* Available at http://slayageonline.com/Numbers/slayage11_12.htm

Hine, C. (2000). *Virtual ethnography.* Thousand Oaks, CA: Sage.

Hine, C. (2008). Internet research as emergent practice. In S. N. Hesse-Biber & P. Leavy (Eds.), *Handbook of emergent methods* (pp. 525–542). New York: Guilford Press.

Hine, C., Kendall, L., & boyd, d. (2009). How can qualitative researchers define the boundaries of their projects? In A. N. Markham & N. K. Baym (Eds.), *Internet inquiry: Conversations about method* (pp. 1–32). Thousand Oaks, CA: Sage.

Holström, J. (2005, retrieved). Virtuell etnografi—vad är det? (Virtual ethnography—What is it?). Retrieved 2005 from http://www.hanken.fi/portals/studymaterial/2005–2006/helsingfors/foretagsledningochorganisation/2235/material/handouts/virtuell_etnografi.pdf

Horn, S. (1998). *Cyberville: Clicks, culture, and the creation of an online town.* New York: Warner Books.

Howard, P. N. (2001). Network ethnography and the hypermedia organization: New organizations, new media, new methods. *New Media & Society, 4*(4), 551–575.

Howard, P. N. (2004). Embedded media: Who we know, what we know, and society online. In P. N. Howard & S. Jones (Eds.), *Society online: The Internet in context* (pp. 1–28). Thousand Oaks, CA: Sage.

Islam, A. (2008). *Television and the Internet: Enabling global communities and its international implications on society and technology.* Master's thesis, Communication and Leadership Studies, School of Professional Studies, Gonzaga University, Spokane, WA.

Ito, M. (1997). Virtually embodied: The reality of fantasy in a multi-user dungeon. In D. Porter (Ed.), *Internet culture* (pp. 87–110). New York: Routledge.

James, N. (2007). The use of email interviewing as a qualitative method of inquiry in educational research. *British Educational Research Journal, 33*(6), 963–976.

James, N., & Busher, H. (2006). Credibility, authenticity and voice: Dilemmas in online interviewing. *Qualitative Research, 6*(3), 403–420.

James, N., & Busher, H. (2007). Ethical issues in online educational research: protecting privacy, establishing authenticity in email interviewing. *International Journal of Research & Method in Education, 30*(1), 101–113.

Johns, M. D., Hall, G. J., & Crowell, T. L. (2004). Surviving the IRB review: Institutional guidelines and research strategies. In M. D. Johns, S. S. Chen, & G. J. Hall (Eds.), *Online social research: Methods, issues, ethics* (pp. 105–124). New York: Peter Lang.

Jones, S. (2004). Contexting the network. In P. N. Howard & S. Jones (Eds.), *Society online: The Internet in context* (pp. 325–334). Thousand Oaks, CA: Sage.

Kanter, R .M. (1977). *Men and women of the corporation.* New York: Basic Books.

Katz, J. (2007). Toward a natural history of ethical censorship. *Law & Society Review, 41*(4), 797–810.

Kem, J. F. (2005). *Cataloging the Whedonverse: Potential roles for librarians in online fan fiction.* Master's thesis, School of Information and Library Science of the University of North Carolina at Chapel Hill.

Kendall, L. (1999). Recontextualizing "cyberspace": Methodological considerations for online research. In S. G. Jones (Ed.), *Doing Internet research: Critical issues and methods for examining the Net* (pp. 57–74). Thousand Oaks, CA: Sage.

Kendall, L. (2002). *Hanging out in the virtual pub: Masculinities and relationships online*. Berkeley: University of California Press.

Kendall, L. (2004). Participants and observers in online ethnography: Five stories about identity. In M. D. Johns, S. S. Chen, & G. J. Hall (Eds.), *Online social research: Methods, issues, ethics* (pp. 125–140). New York: Peter Lang.

Kirby-Diaz, M. (Ed.) (2009). *Buffy and Angel conquer the Internet: Essays on online fandom*. Jefferson, NC: McFarland.

Knapp, J. A. (1997). Essayistic messages: Internet newsgroups as electronic public sphere. In D. Porter (Ed.), *Internet culture* (pp. 181–200). New York: Routledge.

Kociemba, D. (2006). "Over-identify much?": Passion, "passion," and the author-audience feedback loop in *Buffy the Vampire Slayer*. *Slayage: The Online Journal of Buffy Studies, 19*. Available at http://slayageonline.com/essays/slayage19/Kociemba.htm

Kolko, B. E., Nakamura, L., & Rodman, G. G. (2000). *Race in cyberspace*. New York: Routledge.

Komaki, R. (2009, April 30–May 2). *A Japanese social network site mixi and the imagined boundary of "Japan."* Paper presented at the Texas A&M University Race & Ethnic Studies Institute Symposium: Race, Ethnicity, and (New) Media, Texas A&M University, College Station, TX.

Kurzman, C. (1991). Convincing sociologists: Values and interests in the sociology of knowledge. In M. Burawoy (Ed.), *Ethnography unbound: Power and resistance in the modern metropolis* (pp. 250–270). Berkeley: University of California Press.

Larbelestier, J. (2002). *Buffy*'s Mary Sue is Jonathan: *Buffy* acknowledges the fans. In R. Wilcox & D. Lavery (Eds.), *Fighting the forces: What's at stake in* Buffy the Vampire Slayer (pp. 227–238). Lanham, MD: Rowman & Littlefield.

LeBosco, K. (2004). Managing visibility, intimacy, and focus in online critical ethnography. In M. D. Johns, S. S. Chen, & G. J. Hall (Eds.), *Online social research: Methods, issues, ethics* (pp. 63–80). New York: Peter Lang.

Leurs, K. (2009, April 30–May 2). *Be(com)ing cyber Mocro's: Digital media, migration and glocalized youth cultures*. Paper presented at the Texas A&M University Race & Ethnic Studies Institute Symposium: Race, Ethnicity, and (New) Media, Texas A&M University, College Station, TX.

Lotz, A. D., & Ross, S. M. (2004). Toward ethical cyberspace audience research: Strategies for using the Internet for television audience studies. *Journal of Broadcasting & Electronic Media Studies, 48*(3), 501–512.

Lu, J. (2009). *Software copyright and piracy in China*. Dissertation, Texas A&M University, College Station.

Lynd, R. S., & Lynd, H. M. (1956). *Middletown: A study in American culture*. New York: Harcourt, Brace. (Original work published in 1927)

MacKinnon, R. C. (2002). Punish the persona: Correctional strategies for the virtual offender. In S. G. Jones (Ed.), *Virtual culture: Identity and communication in cybersociety* (pp. 206–235). Thousand Oaks, CA: Sage. (Original work published 1997)

Mallapragada, M. (2009, April 30–May 2). *Desi webs: South Asian America, online cultures and the politics of race*. Paper presented at the Texas A&M University Race & Ethnic Studies Institute Symposium: Race, Ethnicity, and (New) Media, Texas A&M University, College Station, TX.

Mann, C., & Stewart, F. (2004). Introducing online methods. In S. N. Hesse-Biber & P. Leavy (Eds.), *Approaches to qualitative research: A reader on theory and practice* (pp. 367–401). New York: Oxford University Press.

Marcus, G. (1998). *Ethnography through thick and thin*. Princeton, NJ: Princeton University Press.

Markham, A. (1998). *Life online: Researching real experience in virtual space* (*Ethnographic Alternatives*, No. 6). Walnut Creek, CA: AltaMira Press.

Markham, A. (2004). Representation in online ethnographies: A matter of context sensitivity. In M. D. Johns, S. S. Chen, & G. J. Hall (Eds.), *Online social research: Methods, issues, ethics* (pp. 141–156). New York: Peter Lang.

Markham, A. (2005a). The methods, politics, and ethics of representation in online ethnography. In N. K. Denzin & Y. S. Lincoln (Eds.), *The SAGE handbook of qualitative methods* (3rd ed., pp. 793–820). Thousand Oaks, CA: Sage.

Markham, A. (2005b). "Go ugly early": Fragmented narrative and bricolage as interpretive method. *Qualitative Inquiry, 11*(6), 813–839.

Markham, A. N., & Baym, N. K. (2009). *Internet inquiry: Conversations about method*. Thousand Oaks, CA: Sage.

May, R. B. A. (2003). "Flirting with boundaries": A professor's narrative tale contemplating research of the wild side. *Qualitative Inquiry, 9*(3), 442–465.

May, R. B. A., & Patillo, M. (2000). Do you see what I see? Examining a collaborative ethnography. *Qualitative Inquiry, 6*(1), 65–87.

McPherson, T. (2000). I'll take my stand in Dixie-Net: White guys, the South, and cyberspace. In B. E. Kolko, L. Nakamura, & G. B. Rodman (Eds.), *Race in cyberspace* (pp. 117–132). New York: Routledge.

Millard, W. B. (1997). I flamed Freud: A case study in teletextual incendiarism. In D. Porter (Ed.), *Internet culture* (pp. 145–160). New York: Routledge.

Millet, K. (1973). *The prostitution papers*. New York: Avon.

Mitchell, W. J. (1999). *E-Topia: "Urban Life Jim, But Not as We Know It."* Cambridge: MIT Press.

Mizrach, S. (1996). Cyberanthropology. Retrieved August 18, 1999, from http://www.lastplace.com/page205.htm

Murguía, E., Tackett-Gibson, M., & Lessem, A. (Eds.). (2007). *Real drugs in a virtual world: Drug discourse and community online*. Lanham, MD: Lexington Books.

Nakamura, L. (2002). *Cyber types: Race, ethnicity, and identity on the Internet*. New York: Routledge.

Nakamura, L. (2009). Don't hate the player, hate the game: The racialization of labor in world of warcraft. *Critical Studies in Media Communication, 26*(2), 128–144.

Nieckarz, P. P., Jr. (2005). Community in cyber space?: The role of the Internet in facilitating and maintaining a community of live music collecting and trading. *City & Community, 4*(4), 403–424.

Nordt, M., Meisner, J., Dongaonkar, R., Quick, C. M., Gatson, S. N., Karadkar, U. P., & Furuta, R. (2007). eBat: A technology-enriched life sciences research community. *Proceedings of the American Society for Information Science & Technology, 43,* 1–25. Available at http://www3.interscience.wiley.com/journal/116327865/issue

Norris, P. (2004). The bridging and bonding role of online communities. In P. N. Howard & S. Jones (Eds.), *Society online: The Internet in context* (pp. 31–42). Thousand Oaks, CA: Sage.

O'Brien, J. (1997). Changing the subject. *Women and Performance: A Journal of Feminist Theory, 17.* Available at http://www.echonyc.com/~women/Issue17/

O'Brien, J. (1999). Writing in the body: Gender (re)production in online interaction. In M. A. Smith & P. Kollock (Eds.), *Communities in cyberspace* (pp. 76–104). New York: Routledge.

Ow, J. A. (2000). The revenge of the yellowfaced cyborg: The rape of digital geishas and the colonization of cyber-coolies in 3D realms' *Shadow warrior.* In B. E. Kolko, L. Nakamura, & G. B. Rodman, (Eds.), *Race in cyberspace* (pp. 51–68). New York: Routledge.

Parpart, L. (2003). "Action, chicks, everything": Online interviews with male fans of *Buffy the Vampire Slayer.* In F. Early & K. Kennedy (Eds.), *Athena's daughters: Television's new women warriors* (pp. 78–91). Syracuse, NY: Syracuse University Press.

Parrish, J. J. (2007). *Inventing a universe: Reading and writing Internet fan fiction.* PhD dissertation, University of Pittsburgh, Pittsburgh, PA.

Peterson, L. (2009a, April 30–May 2.). *Ewww—You got your social justice in my video game!* Paper presented at the Texas A&M University Race & Ethnic Studies Institute Symposium: Race, Ethnicity, and (New) Media, Texas A&M University, College Station, Texas.

Peterson, L. (2009b, April 30–May 2). *Talking about race in digital space.* Paper presented at the Texas A&M University Race & Ethnic Studies Institute Symposium: Race, Ethnicity, and (New) Media, Texas A&M University, College Station, Texas.

Porter, D. (1997). *Internet culture.* New York: Routledge.

Reid, R.A. (2009, April 30–May 2). *Harshin ur squeez: Visual rhetorics of anti-racist work in livejournal fandoms.* Paper presented at the Texas A&M University Race & Ethnic Studies Institute Symposium: Race, Ethnicity, and (New) Media, Texas A&M University, College Station, TX.

Rheingold, H. (2000). *The virtual community: Homesteading on the electronic frontier.* Reading, MA: Addison-Wesley. (Original work published 1993)

Richards, L. (2003). Fandom and ethnography. Available at http://www.searingidolatry.co.uk/lond/index2.html

Salaff, J. W. (2002). Where home is the office: The new form of flexible work. In B. Wellman & C. Haythornthwaite (Eds.), *The Internet in everyday life* (pp. 464–495). Malden, MA: Blackwell.

Seabrook, J. (1997). *Deeper: Adventures on the Net.* New York: Touchstone (Simon & Schuster).

Schmitz, J. (2002). Structural relations, electronic media, and social change: The public electronic network and the homeless. In S. Jones (Ed.), *Virtual culture: Identity & communication in cybersociety* (pp. 80–101). Thousand Oaks, CA: Sage. (Original work published 1997)

Sharf, B. (1997). Communicating breast cancer online: Support and empowerment on the Internet. *Women and Health, 26,* 65–84.

Sharf, B. (1999). Beyond netiquette: The ethics of doing naturalistic discourse research on the Internet. In S. G. Jones (Ed.), *Doing Internet research: Critical issues and methods for examining the Net* (pp. 57–74). Thousand Oaks, CA: Sage.

Shaw, D. F. (2002). Gay men and computer communication: A discourse of sex and identity in cyberspace. In S. G. Jones (Ed.), *Virtual culture: Identity and communication in cybersociety* (pp. 133–145). Thousand Oaks, CA: Sage. (Original work published 1997)

Shirky, C. (2002). Communities, audiences, and scale. Available at http://shirky.com/writings/community_scale.html

Shostak, M. (1981). *Nisa: The life and words of !Kung woman.* Cambridge, MA: Harvard University Press.

Silver, D. (2000). Margins in the wires: Looking for race, gender, and sexuality in the Blacksburg Electronic Village. In B. E. Kolko, L. Nakamura, & G. B. Rodman (Eds.) *Race in cyberspace* (pp. 133–150). New York: Routledge.

Smith, D. (1990). *Texts, facts, and femininity: Exploring the relations of ruling.* New York: Routledge.

Smith, K. M. C. (2004). "Electronic eavesdropping": The ethical issues involved in conducting a virtual ethnography. In M. D. Johns, S. S. Chen, & G. J. Hall (Eds.), *Online social research: Methods, issues, ethics* (pp. 223–238). New York: Peter Lang.

Stafford, N. (2002). *Bite me! An unofficial guide to the world of Buffy the vampire slayer.* Toronto, ON: ECW Press. (Original work published in 1998)

Stark, L. (2007). Victims in our own minds?: IRBs in myth and practice. *Law & Society Review, 41*(4), 777–786.

Stenger, J. (2006). The clothes make the fan: Fashion and online fandom when *Buffy the Vampire Slayer* goes to eBay. *Cinema Journal, 45*(4), 26–44.

Stern, M. J., & Dillman, D. A. (2006). Community participation, social ties, and use of the Internet. *City & Community, 5*(4), 409–424.

Stivale, C. J. (1997). Spam: Heteroglossia and harassment in cyberspace. In D. Porter (Ed.), *Internet culture* (pp. 133–144). New York: Routledge.

Tepper, M. (1997). Usenet communities and the cultural politics of information. In D. Porter (Ed.), *Internet culture* (pp. 39–54). New York: Routledge.

Thomas, J. (2004). Reexamining the ethics of Internet research: Facing the challenge of overzealous oversight. M. D. Johns, S. S. Chen, & G. J. Hall (Eds.), *Online social research: Methods, issues, ethics* (pp. 187–202). New York: Peter Lang.

Travers, A. (2000). *Writing the public in cyberspace: Redefining inclusion on the net.* Garland studies in American popular history and culture. New York: Garland Press.

Travers, M. (2009). New methods, old problems: A sceptical view of innovation in qualitative research. *Qualitative Research, 9*(2), 161–179.

Tulloch, J., & Jenkins, H. (1995). *Science fiction audiences: Watching Doctor Who and Star Trek.* New York: Routledge.

Turkle, S. (1995). *Life on the screen: Identity in the age of the Internet.* New York: Simon & Schuster.

Tuszynski, S. (2006). *IRL (in real life): Breaking down the binary of online versus offline social interaction.* PhD dissertation, Bowling Green State University, Bowling Green, OH.

Van Maanen, J. (2004). An end to innocence: The ethnography of ethnography. In S. N. Hesse-Biber & P. Leavy (Eds.), *Approaches to qualitative research: A reader on theory and practice* (pp. 427–446). New York: Oxford University Press.

Walstrom, M. K. (2004). "Seeing and sensing" online interaction: An interpretive interactionist approach to USENET support group research. In M. D. Johns, S. S. Chen, & G. J. Hall (Eds.), *Online social research: Methods, issues, ethics* (pp. 81–100). New York: Peter Lang.

Watson, N. (2003). Why we argue about virtual community: A case study of the Phish.Net fan community. In S. G. Jones (Ed.), *Virtual culture: Identity and communication in cybersociety* (pp. 102–132). Thousand Oaks, CA: Sage. (Original work published in 1997)

Whiteman, N. (2009). The de/stabilization of identity in online fan communities. *Convergence: The International Journal of Research into New Media Technologies, 15*(4), 391–410.

Williams, R. (2004). "It's about power": Executive fans, spoiler whores and capital in the *Buffy the Vampire Slayer* online fan community. *Slayage: The Online Journal of Buffy Studies,* 11–12.

Zickmund, S. (2002). Approaching the radical other: The discursive culture of cyberhate. In S. Jones *Virtual culture: Identity and communication in cybersociety* (pp. 185–205). Thousand Oaks, CA: Sage. (Original work published 1997)

Zweerink, A., & S. N. Gatson (2002). www.buffy.com: Cliques, boundaries, and hierarchies in an Internet community. In R. Wilcox & D. Lavery (Eds.), *Fighting the forces: What's at stake in Buffy the Vampire Slayer* (pp. 239–249). Lanham, MD: Rowman & Littlefield, pp. 239–249.

32

ANALYZING TALK AND TEXT

Anssi Peräkylä and Johanna Ruusuvuori

There are two much used but distinctively different types of empirical materials in qualitative research: interviews and "naturally occurring" materials. Interviews consist of accounts given to the researcher about the issues in which he or she is interested. The topic of the research is not the interview itself but rather the issues discussed in the interview. In this sense, research that uses "naturally occurring" empirical material is different; in this type of research, the empirical materials themselves (e.g., the tape recordings of mundane interactions, the written texts) constitute specimens of the topic of the research. Consequently, the researcher is in more direct touch with the very object that he or she is investigating.

Most qualitative research probably is based on interviews. There are good reasons for this. By using interviews, the researcher can reach areas of reality that would otherwise remain inaccessible such as people's subjective experiences and attitudes. The interview is also a very convenient way of overcoming distances both in space and in time; past events or far-away experiences can be studied by interviewing people who took part in them.

In other instances, it is possible to reach the object of research directly using naturally occurring empirical materials (Silverman, 2001). If the researcher is interested in, say, strategies used by journalists in interviewing politicians (e.g., Clayman & Heritage, 2002a), it might be advisable to tape-record broadcast interviews rather than to ask journalists to tell about their work. Or, if the researcher wants to study the historical evolvement of medical conceptions regarding death and dying, it might be advisable to study medical textbooks rather than to ask doctors to tell what they know about these concepts.

The contrast between interviews and naturally occurring materials should not, however, be exaggerated (see also Potter, 2004; Speer, 2002). There are types of research materials that are between these two pure types. For example, in *informal interviews that are part of ethnographic fieldwork*, and in *focus groups*,

people describe their practices and ideas to the researcher in circumstances that are much closer to "naturally occurring" than are the circumstances in ordinary research interviews. Moreover, even "ordinary" interviews can be, and have been, analyzed as specimens of interaction and reasoning practices rather than as representations of facts or ideas outside the interview situation. As Susan Speer (2002) recently put it, "The status of pieces of data as natural or not depends largely on what the researcher intends to 'do' with them" (p. 513). Margaret Wetherell and Jonathan Potter (1992), for example, analyzed the ways in which interviewees use different linguistic and cultural resources in constructing their relation to racial and racist discourses. On the other hand, as David Silverman (2001) put it, no data—not even tape recordings—are "untouched by the researcher's hands" (p. 159; see also Speer, 2002, p. 516); the researcher's activity is needed, for example, in obtaining informed consent from the participants. The difference between researcher-instigated data and naturally occurring data should, therefore, be understood as a continuum rather than as a dichotomy.

This chapter focuses on one end of this continuum. It presents some methods that can be used in analyzing and interpreting tape-recorded interactions and written texts, which probably are the types of data that come closest to the idea of "naturally occurring."

ANALYZING TEXTS

Uses of Texts and Variety of Methods of Text Analysis

As Dorothy Smith (1974, 1990) and Paul Atkinson and Amanda Coffey (1997) have pointed out, much of social life in modern society is mediated by written texts of different kinds. For example, modern health care would not be possible without patient records; the legal system would not be possible without laws and other juridical texts; professional training would not be possible without manuals and professional journals; and leisure would not be possible without newspapers, magazines, and advertisements. Texts of this kind have provided an abundance of material for qualitative researchers.

In many cases, qualitative researchers who use written texts as their materials do not try to follow any predefined protocol in executing their analysis. By reading and rereading their empirical materials, they try to pin down their key themes and, thereby, to draw a picture of the presuppositions and meanings that constitute the cultural world of which the textual material is a specimen. An example of this kind of informal approach is Clive Seale's (1998) small but elegant case study on a booklet based on a broadcast interview with the British playwright Dennis Potter (pp. 127–131). The interviewee was terminally ill at the time of the interview. Seale showed how the interview conveys a particular conception of death and dying, characterized by intensive awareness of the imminent death and special creativity arising from it.

An informal approach may, in many cases, be the best choice as a method in research focusing on written texts. Especially in research designs where the qualitative text analysis is not at the core of the research but instead is in a subsidiary or complementary role, no more sophisticated text analytical methods may be needed. That indeed was the case in Seale's (1998) study, in which the qualitative text analysis complemented a larger study drawing mostly on interview and questionnaire materials as well as on theoretical work. In projects that use solely texts as empirical materials, however, the use of different kinds of analytical procedures may be considered.

The researchers can choose from many methods of text analysis. The degree to which they involve predefined sets of procedures varies; some of them do to a great extent, whereas in others the emphasis is more on theoretical presuppositions concerning the cultural and social worlds to which the texts belong. Moreover, some of these methods can be used in the research of both written and spoken discourse, whereas others are exclusively fitted to written texts. In what follows, we briefly mention a few text analytical methods and then discuss two a bit more thoroughly.

Semiotics is a broad field of study concerned with signs and their use. Many tools of text analysis have arisen from this field. The most prominent of them may be *semiotic narrative analysis.* The Russian ethnologist Vladimir Propp (1968) and the French sociologist Algirdas Julien Greimas (1966) developed schemes for the analysis of narrative structures. Initially their schemes were developed in fairy tales, but later they were applied to many other kinds of texts. For example, by using Greimas's scheme, primordial structural relations (e.g., subject vs. object, sender vs. receiver, helper vs. opponent) can be distilled from the texts. Jukka Törrönen (2000, 2003) used and developed further Greimasian concepts in analyzing newspaper editorials

addressing alcohol policy, showing how these texts mobilize structural relations to encourage readers to take action to achieve particular political goals.

Another, more recent, trend in narrative analysis focuses on *narratives as practice within social interaction* rather than as text with an identifiable structure. In anticipation of the second half of this chapter (focusing on research on interaction), we will briefly introduce this new approach on narrative here. This new turn in narrative analysis lays emphasis on the multiple, fragmented, and situated nature of narrative (Hyvärinen et al., 2010). It investigates stories and storytelling as they operate within society. In this trend, context is not seen as a static setting but as multiple intersecting processes that are a resource for talk-in-interaction (De Fina & Georgakopoulou, 2008). Traditionally within the narrative field of study, the narratives that are investigated have been derived from interview data (Bamberg & Georgakopoulou, 2008). The focus has been on the *internal organization* of narratives—on the ways in which particular types of narrative organization are connected with factors such as gender, for example. Within the new trend, the focus has been turned more on the *external organization* of narratives, on the production of narratives in their immediate surroundings (Gubrium & Holstein, 2009, pp. vii–ix, 1–2). Narratives are analyzed as talk-in-interaction in varying contexts; on one hand, the focus is on the ways in which stories are told and shaped by other people and the surrounding context of the situation, and on the other hand, the focus is on the ways in which this context is shaped by the narrative tellings (De Fina & Georgakopoulou, 2008; Ochs & Capps, 2001, p. 2). To give an example, Michael Bamberg and Alexandra Georgakopoulou (2008) have analyzed the storytelling activities of 10-year-old boys in a group discussion as tools of identity work. Bamberg and Georgakopoulou's starting point is that narratives can be used as means to construct characters in space and further, positions relative to other participants of the situation (see also Sacks, 1974b). Thus, specific linguistic choices can be linked with larger social identities (Georgakopoulou, 2007, p. 13). Bamberg and Georgakopoulou (2008) show how 10-year-old boys juggle between two contrasting story lines (of being interested in girls and not being interested) in a focus group situation with an interviewer and three other boys. Their focus is on the discursive maneuvering of the boys between two master narratives that are dominant in the boys' peer group and on the development of their sense of self through this navigation process. The researchers point out that small stories that are told within changing situations can gradually amount to more constant ways of organizing life experience and result in life stories that form a sense of who we are.

The term *discourse analysis* (DA) may refer, depending on context, to many different approaches of investigation of written texts (and of spoken discourse as well). In the context of linguistics, DA usually refers to research that aims at uncovering the features of text that maintain coherence in units larger than the sentence (Brown & Yule, 1983). In social psychology, DA (or *discursive psychology,* as it has been called more recently) involves research in which the language use (both written and spoken) underpinning mental realities, such as cognition and emotion, is investigated. Here, the key theoretical presupposition is that mental realities do not reside "inside" individual humans but rather are constructed linguistically (Edwards, 1997; Potter, 2006; Potter & te Molder, 2005). *Critical discourse analysis* (CDA), developed by Norman Fairclough (1989, 1995) among others, constitutes yet another kind of discourse analytical approach in which some key concerns of linguistic and critical social research merge. Critical discourse analysts are interested in the ways in which texts of different kinds reproduce power and inequalities in society (see Wodak & Meyer, 2009). Liisa Tainio's (1999) study on the language of self-help communication guidebooks for married couples is one example of a CDA study. Tainio showed, for example, how in these texts the woman is expected to change for the communication problems to be solved, whereas the man is treated as immutable.

A *Foucauldian approach* to the analysis of texts, or *historical discourse analysis* (HDA) as it is sometimes called, focuses on tracing the interrelatedness of knowledge and power in studying historical processes through which certain human practices and ways of thinking have emerged. The term *analytics of government* (Dean, 1999; Meskus, 2009a; Rose, 1999) refers to a method of analysis where this type of research approach is in use. In the following, we will introduce an example of this approach.

Analyzing the Government of Human Heredity: A Research Example

Many scholars working with written texts have drawn insights and inspiration from the work of Michel Foucault. (For examples of his own studies, see Foucault, 1973, 1977, 1978. For examples of accessible accounts of his theories and methods, see Dean, 1999; Kendall & Wickham, 1999; McHoul & Grace, 1993; Rose, 1999.) Foucault did not propose a definite set of methods for the analysis of texts; hence, the ways of analyzing and interpreting texts of scholars inspired by him vary. For all of them, however, a primary concern is, as Potter (2004) aptly put it, how a set of "statements" comes to constitute objects and subjects. The constitution of subjects and objects is explored in historical context—or, in Foucault's terms, through *archeology* and *genealogy.*

A recent example of this kind of historical approach is offered by Mianna Meskus' (2009b) research on the ways in which the rationale and technologies concerning heredity have evolved in Finnish medicine and health care. Meskus focuses on the development starting in the early 20th century during which concepts such as *eugenics* and *racial hygiene* were gradually replaced by the idea of *risk,* and how the technologies for governing the sphere of heredity and reproduction changed respectively.

Meskus (2009b) investigates texts from the spheres of professional, political, and lay discourses (medical articles, policy documents, committee reports, guidebooks, and health magazines) tracing the interconnectedness of the advancements in genetics, the changes in national population policy and the practices of its implementation in health care. Her specific focus is on the technology of prenatal screening and the doctor-driven development during which it was gradually extended to encompass all pregnant women.

Meskus distinguishes three phases or periods in the government of human heredity. The first reaches from the beginning of the 20th century to the 1960s. During that period, it was thought that people with mental illness or cognitive impairment should be sterilized to enhance the "quality of population." In the second phase, in the 1970s and 1980s, the focus of policy turned from the quality of population to health. During this period, the state strongly invested in preventive health care, launching a nationwide system for health counseling. At the same time, the chromosomal diagnosis of congenital and hereditary diseases was implemented into clinical practices. At this stage, the concept of risk was attached to pregnancy, and technologies (such as amniocentesis) were developed and implemented to diagnose potential anomalies of the fetus in specific risk groups, such as mothers older than 40. If anomalies were found and future parents would so decide, abortion could be induced. In the third phase, starting in the 1990s, the development of genetics made available new tests that were relatively easy to implement clinically. Meskus shows how, in this latter phase, prenatal screening was adopted as a routine procedure for all pregnant mothers in Finland, but in Sweden, for example, fetal diagnostics were only targeted at specific risk groups. The rationale was presented as providing a possibility of *choice for parents* to control the health of their future baby. A central difference to the practice in Sweden was that whereas in Finland all pregnant mothers were routinely offered the test, and therefore had to say yes or no, such a routine offer was not made in Sweden, where the mothers were given information about hereditary diseases and could ask for the test on their own initiative, if they so decided.

Meskus points out how the development has advocated individual choice and at the same time covered the social and economical contexts within which prenatal screening has emerged as a routine practice. Through all the three periods, the procedures for managing the "quality" or health of the population with regard to pregnancy or childbirth were connected with the health policy interests in saving expenses of social and health care. However, the rationale of the doctors and geneticists who have advocated screening for all pregnant mothers has centered around future parents' increasing possibilities to know about the health of their future children and to choose whether they are willing to manage with a disabled child.

Meskus concludes by referring to new ethical problems that have arisen with this "freedom to choose." As the awareness of health risks among the public has increased and their possibilities to choose have been promoted, parents' expectations concerning the health and normality of their future children have also increased, and in some cases beyond the limits offered by medicine. In practice, however, the freedom to choose brings parents against a very difficult choice between abortion and taking the risk that their baby may be disabled. This freedom entails a heavy burden of responsibility for pregnant mothers and their partners in case anomalies are found. Thus, Meskus shows how adoption of a medical technology, such as prenatal screening, that is seemingly based on neutral medical knowledge is actually a result of various historical, social, and political underpinnings and may further result in unexpected ethical dilemmas.

Meskus' method is Foucauldian in the sense that she examines historical (textual) entities and ways of thinking through concepts that are typical for the period and for the texts under investigation (Meskus, 2009b, p. 232). Drawing on ethnographic ideas, she describes her research object in various, changing contexts: in different types of texts during different periods of time, to make a synthesis. Her versatile data include medical articles, administrative documents, memos, and guidebooks, with focus on issues that are presented as problematic, and on interests and debates around these concerns. Having arranged the data thematically, Meskus examines particular dimensions present in the expert texts: How are *the entities of interest* (scientific facts about heredity) defined and described, what are the *standpoints and styles of reasoning and argumentation* (how heredity is made problematic and what solutions are presented to the problems), and how are *the target groups* (particular sections of the population) defined. This analysis is then drawn together from the historical point of view by tracing the *continuities and turning points in the historical approach to the focus of interest* (heredity). The overarching idea is the intertwining of texts and practices. Meskus' study efficiently shows how the medical "facts" on heredity are produced in particular kinds of societal climate where particular policy ideas, values, and needs are present. These, then, are reflected on the practices of the government of heredity during each period.

Meskus' historical and Foucauldian way of analyzing and interpreting texts offers one compact alternative for qualitative text analysis. We now turn to a quite different way of reading texts in qualitative research, that is, *membership categorization analysis* (MCA).

Membership Categorization Analysis

Whereas Meskus' historical analysis was concerned with how issues are defined as problems in the texts and how the styles of reasoning are reformed or stabilized in time and across different types of data, MCA is concerned about *the descriptive apparatus* that makes it possible to say whatever is said.

Before we start to examine MCA, we want to remind the reader about the wide range of applications that this approach has. In addition to the analysis of written texts, it can be used in the analysis of interviews (e.g., Nikander, 2002; Roca-Cuerbes, 2008) and in the analysis of naturally occurring talk (e.g., Butler & Weatherall, 2006; Stokoe, 2003). In the following, however, we focus on the text analytical applications.

The idea of membership categorization came from the American sociologist Harvey Sacks (1974b, 1992). *Description* was a key analytical question for Sacks; he was concerned about the conditions of description, that is, what makes it possible for us to produce and understand descriptions of people and their activities. As Silverman (2001) aptly put it, Sacks was concerned about "the apparatus through which members' descriptions are properly produced" (p. 139). This interest led Sacks to examine categorization.

People are usually referred to by using categories. The point of departure for MCA is recognition that at any event, a person may be referred to by using many alternative categories. As the authors of this chapter, we may also be referred to as academics, Finns, parents, sociologists, Europeans, University of Tampere alumni, and so forth. MCA is about the selection of categories such as these and about the conditions and consequences of this selection.

Sacks's (1974b) famous example is the beginning of a story written by a child: *The baby cried. The mommy picked it up.* There are two key categories in this story: "baby" and "mommy." Why are these categories used, and what is achieved by them? If the mommy happened to be a biologist by profession, why would the story not go like this: *The baby cried. The scientist picked it up* (Jayyusi, 1991, p. 238)? Why do we hear the story being about a baby and *its* mother and not just about any baby and any mother? MCA provides answers to questions such as these and offers a toolkit for analyzing various kinds of texts.

Sacks (1992) noted that categories form sets, that is, collections of categories that go together. Family is one such collection, and "baby," "mother," and "father" are some categories of it. "Stage of life" is another collection; it consists of categories such as "baby," "toddler," "child," and "adult." Now, "baby" could in principle be heard as belonging to both collections, but in the preceding little story we hear it as belonging to the "family" collection. This is because in hearing (or reading) descriptions where two or more categories are used, we orient to a rule according to which we hear them as being from the same collection if they indeed can be heard in that way. Therefore, in this case we hear "baby" and "mommy" being from the device "family" (p. 247).

Categories also go together with *activities*. Sacks used the term *category-bound activities* in referring to activities that members of a culture take to be "typical" of a category (or some categories) of people. "Crying" is a category-bound activity of a baby, just as "picking a (crying) baby up" is a category-bound activity of a mother. In a similar fashion, "lecturing" is a category-bound activity of a professor. Activities such as these can be normative; it is appropriate for the baby to cry and for the mother to pick it up, but it is not appropriate for an adult to cry (like a baby) or for a mother to fail to pick her crying baby up. *Standardized relational pairs* consist of two categories where incumbents of the categories have standardized rights and obligations in relation to each other, with "mother and baby" clearly being one pair, just as "husband and wife" and "doctor and patient" are common pairs. Moreover, the receivers of descriptions can and do infer from actions to categories and vice versa. By knowing actions, we infer the categories of the agents; by knowing categories of agents, we infer what they do.

Even on the basis of these fragments of Sacks's ideas (for more thorough accounts, see Lepper, 2000; Schegloff, 2007b; Silverman, 1998), the reader may get an impression of the potential that this account offers for the analysis of texts. Sacks's ideas are resources for the analysis of texts as sites for the production and reproduction of social, moral, and political orders. Merely by bearing in mind that there is always more than one category available for the description of a given person, the analyst always asks "Why this categorization now?"

Let us examine a brief example of MCA. Marc Rapley, David McCarthy, and Alec McHoul (2003) report a social psychological analysis on the news coverage of a mass killing in Tasmania in 1996 (on MCA of an equivalent case, see Eglin & Hester, 2003). Rapley et al. focus on the public categorizations of the gunman both by lay people and professionals and pay attention to the tension that is created in between the candidate category memberships that are assigned to the gunman, in both lay and professional accounts of the incident.

The authors make use of Sacks' idea of methods of categorization, where particular ties are inferred between categories of person and their category-bound activities—including the moral accountability of these activities. The authors observe how in lay accounts of the event the gunman is presented as *a psycho* or as *schizophrenic*, but also as *a young man dogged by tragedy.* When assigned to the category of *mentally ill,* the man is supposed to lose his sense of reality and is thus regarded as capable of doing unexpected and abnormal things—he is not accountable for his actions. Conversely, as a member of society, *a young man,* the man's deviant actions can be judged as wrong and immoral. Thus, on the one hand, his actions are explained in terms of otherness, as the workings of a madman who is not responsible of his doings, but on the other hand, he is described as a member of a shared social order, and in this way, as morally accountable for his actions.

Interestingly, the authors find that professional explanations for the incident are no less incongruent. Some experts describe the killer as having little intellectual capability and not insane, whereas others refer to him as possibly schizophrenic. This way, a similar tension between the moral accountability and non-accountability of the gunman's actions is created as in lay explanations of the incident. The psychiatrists and psychologists who examined the gunman finally agreed that he suffered from a personality disorder and was in the borderline range between intellectual disability and a "dull normal individual," but did NOT suffer from a serious mental illness that would have prevented him from knowing the difference between right and wrong. Thus, the expert explanation offered made use of lay categories situating the gunman in between mad and not mad, abnormal and normal, "not-us" and "us," which then allowed the gunman to be held morally (and legally) accountable for his actions.

Following Sacks, Rapley et al. point out how the way in which we categorize people does the work of explanation, and how this categorization work is inherently moral, even when it is done by professional experts. According to Rapley et al.'s analysis, the categorizations used in the media were organized to produce an account of the gunman that retained his status as a moral and accountable actor. In the case presented, the psychiatric (expert) categories were also harnessed to accomplish this possibility. In terms of membership categorization, the categorizations of the actor were tied to moral types to accomplish practical moral judgments. The actual scientific grounds for choosing the particular categories were left aside. Rapley et al. conclude that categories (also psychiatric ones) should be regarded as resources that people use to accomplish things, in this case, a moral verdict, rather than treating them as neutral scientific facts: categorization as a method of describing events and thus for producing moral accounts precedes and grounds other "technical," "clinical," or "scientific" judgments.

Because all description draws on categorization, MCA has wide applicability in the analysis of texts. The analysis of categorization gives the researcher access to the cultural worlds and moral orders on which the texts hinge. Importantly, however, categorization analysis is not *only* about specific cultures or moralities. In developing his concepts, Sacks was not primarily concerned about the "contents" of the categorizations; rather, he was concerned about the ways in which we use them (Atkinson, 1978, p. 194). Therefore, at the end of the day, membership categorization analysis invites the qualitative researcher to explore the conditions of action of description in itself.

▣ ANALYZING TALK

Face-to-face social interaction (or other live interaction mediated by phones and other technological media) is the most immediate and the most frequently experienced social reality. The heart of our social and personal being lies in the immediate contact with other humans. Even though ethnographic observation of face-to-face social interaction has been done successfully by sociologists and social psychologists, video and audio recordings are what provide the richest possible data for the study of talk and interaction today. Such recordings have been analyzed using the same methods that were discussed previously in the context of interpretation of written texts. CDA, MCA, and even Foucauldian DA have all their applications in researching transcripts based on video or audio recordings. However, as Erving Goffman (1983) pointed out, to be fully appreciated, the face-to-face social interaction also requires its own specific methods. The interplay of utterances and actions in live social interaction involves a complex organization that cannot be found in written texts. *Conversation analysis* (CA) is presented as a method specialized for analyzing that organization.

Origins of Conversation Analysis

CA is a method for investigating the structure and process of social interaction between humans. As their empirical materials, CA studies use video or audio recordings made from naturally occurring interactions. As their results, these studies offer qualitative (and sometimes quantitative) descriptions of interactional structures (e.g., turn taking, relations between adjacent utterances) and practices (e.g., telling and receiving news, making assessments).

CA was started by Sacks and his coworkers, especially Emanuel Schegloff and Gail Jefferson (1977), at the University of California during the 1960s. At the time of its birth, CA was something quite different from the rest of social science. The predominant way of investigating human social interaction was quantitative, based on coding and counting distinct, theoretically defined actions (see especially Bales, 1950). Erving Goffman (e.g., 1955) and Harold Garfinkel (1967) had challenged this way of understanding interaction with their studies that focused on the moral and inferential underpinnings of social interaction. Drawing part of his inspiration from them, Sacks started to study qualitatively the real-time sequential ordering of actions—the rules, patterns, and structures in the relations between consecutive actions (Silverman, 1998). Schegloff (1992a) argued that Sacks made a radical shift in the perspective of social scientific inquiry into social interaction; instead of treating social interaction as a screen on which other processes (Balesian categories or moral and inferential processes) were projected, Sacks started to study the very structures of the interaction itself (Schegloff, 1992a, p. xviii).

Basic Theoretical Assumptions

In the first place, CA is not a theoretical enterprise but rather a very concretely empirical one. Conversation analysts make video or audio recordings of naturally occurring interactions, and they transcribe these recordings using a detailed notation system (see Appendix). They search, in the recordings

and transcripts, for recurrent distinct interactive practices that then become their research topics. These practices can involve, for example, specific sequences (e.g., news delivery sequence consisting of "news announcement," "announcement response," "elaboration," and "assessment" [Maynard, 2003]) or specific ways of designing utterances (e.g., "oh"-prefaced answers to questions [Heritage, 1998]). Then, through careful listening, comparison of instances, and exploration of the context of them, conversation analysts describe in detail the properties and tasks that the practices have (e.g., "oh"-preface as marking a change in the epistemic state of its speaker, see Heritage, 1998).

However, through empirical studies—in an "inductive" way—a body of theoretical knowledge about the organization of conversation has been accumulated. The actual "techniques" in doing CA can be understood and appreciated only against the backdrop of these basic theoretical assumptions of CA. In what follows, I try to sketch some of the basic assumptions concerning the organization of conversation that arise from these studies. There are perhaps three most fundamental assumptions of this kind (see also Heritage, 1984, Chapter 8; Hutchby & Wooffitt, 1998), namely that (a) talk is action, (b) action is structurally organized, and (c) talk creates and maintains intersubjective reality.

Talk Is Action

As in some other philosophical and social scientific approaches, in CA talk is understood primarily as a vehicle of human action (Schegloff, 1991). The capacity of language to convey ideas is seen as being derived from this more fundamental task. In accomplishing actions, talk is seamlessly intertwined with (other) corporeal means of action such as gaze and gesture (Goodwin, 1981). Some CA studies have as their topics the organization of actions that are recognizable as distinct actions even from a vernacular point of view. Thus, conversation analysts have studied, for example, openings (Schegloff, 1968) and closings (Schegloff & Sacks, 1973) of conversations, assessments and ways in which the recipients agree or disagree with them (Goodwin & Goodwin, 1992; Pomerantz, 1984), storytelling (Mandelbaum, 1992; Sacks, 1974a), complaints (Drew & Holt, 1988), telling and receiving news (Maynard, 2003), and laughter (Glenn, 2003; Haakana, 2001; Jefferson, 1984). Many CA studies have as their topic actions that are typical in some institutional environment. Examples include diagnosis (Heath, 1992; Maynard, 1991, 1992; Peräkylä, 1998, 2002; ten Have, 1995) and physical examination (Heritage & Stivers, 1999) in medical consultations, questioning and answering practices in cross-examinations (Drew, 1992), ways of managing disagreements in news interviews (Greatbatch, 1992), and advice giving in a number of different environments (Heritage & Sefi, 1992; Silverman, 1997; Vehviläinen, 2001). Finally, many important CA studies focus on fundamental aspects of conversational organization that make any action possible. These include turn taking (Sacks, Schegloff, & Jefferson, 1974), repair (Schegloff, 1992c; Schegloff, Jefferson, & Sacks, 1977), and the general ways in which sequences of action are built (Schegloff, 2007a).

Action Is Structurally Organized

In the CA view, the practical actions that comprise the heart of social life are thoroughly structured and organized. In pursuing their goals, the actors have to orient themselves to rules and structures that only make their actions possible. These rules and structures concern mostly the relations between actions. Single acts are parts of larger, structurally organized entities. These entities may be called *sequences* (Schegloff, 2007a).

The most basic and the most important sequence is called the *adjacency pair* (Schegloff & Sacks, 1973). It is a sequence of two actions in which the first action ("first pair part"), performed by one interactant, invites a particular type of second action ("second pair part") to be performed by another interactant. Typical examples of adjacency pairs include question–answer, greeting–greeting, request–grant/refusal, and invitation–acceptance/declination. The relation between the first and second pair parts is strict and normative; if the second pair part does not come forth, the first speaker can, for example, repeat the first action or seek explanations for the fact that the second action is missing (Atkinson & Drew, 1979, pp. 52–57; Merritt, 1976, p. 329).

Adjacency pairs often serve as a core around which even larger sequences are built (Schegloff, 2007a). So, a *pre-expansion* can precede an adjacency pair, for example, in cases where the speaker first asks about the other's plans for the evening and only thereafter (if it turns out that the other is not otherwise engaged) issues an invitation. An *insert expansion* involves actions that occur between the first and second pair parts and makes possible the production of the latter, for example, in cases where the speaker requests specification of an offer or a request before responding to it. Finally, in *post-expansion,* the speakers produce actions that somehow follow from the basic adjacency pair, with the simplest example being "okay" or "thank you" to close a sequence of a question and an answer or of a request and a grant (Schegloff, 2007a).

Talk Creates and Maintains the Intersubjective Reality

CA has sometimes been criticized for neglecting the "meaning" of talk at the expense of the "form" of talk (Alexander, 1988, p. 243; Taylor & Cameron, 1987, pp. 99–107). This is, however, a misunderstanding, perhaps arising from the impression created by the technical exactness of CA studies. Closer reading of CA studies reveals that in such studies, talk and interaction are examined as a site where intersubjective understanding about the participants' intentions is created and maintained (Heritage & Atkinson, 1984, p. 11). As such, CA gives access to the construction of meaning in real time where the methods or "vehicles" of this construction are inseparable from what is constructed (see also the example of MCA earlier). But it is important to notice that the conversation analytical "gaze" focuses exclusively on meanings and understandings that are made public through conversational action and that it remains "agnostic" regarding people's intrapsychological experience (Heritage, 1984).

The most fundamental level of intersubjective understanding—which constitutes the basis for any other type of intersubjective understanding—concerns *the understanding of the preceding turn displayed by the current speaker.* Just like any turn of talk that is produced in the context shaped by the previous turn, it also displays its speaker's understanding of that previous turn (Atkinson & Drew, 1979, p. 48). Thus, in simple cases, when producing a turn of talk that is hearable as an answer, the speaker also shows that he or she understood the preceding turn as a question. Sometimes these choices can be crucial for the unfolding of the interaction and the social relation of its participants, for example, in cases where a turn of talk is potentially hearable in two ways (e.g., as an announcement or a request, as an informing or a complaint) and the recipient makes the choice in the next turn. In case the first speaker considers the understanding concerning his talk to be incorrect or problematic, as displayed in the second speaker's utterance, the first speaker has an opportunity to correct this understanding in the "third position" (Schegloff, 1992c), for example, by saying "I didn't mean to criticize you; I just meant to tell you about the problem."

Another important level of intersubjective understanding concerns the *context* of the talk. This is particularly salient in institutional interaction, that is, in interaction that takes place to accomplish some institutionally ascribed tasks of the participants (e.g., psychotherapy, medical consultations, news interviews) (Arminen, 2005; Drew & Heritage, 1992; Heritage, 2004). The participants' understanding of the institutional context of their talk is documented in their actions. As Emmanuel Schegloff (1991, 1992b) and Paul Drew and John Heritage (1992) pointed out, if the "institutional context" is relevant for interaction, it can be observed in the details of the participants' actions—in their ways of giving and receiving information, asking and answering questions, presenting arguments, and so forth. CA research that focuses on institutional interactions explores the exact ways in which the performers of different institutional tasks shape their actions to achieve their goals.

Research Example

After these rather abstract considerations, let us consider a concrete example of CA research. A recent study by John Heritage and Geoffrey Raymond (2005) that focuses on the ways in which participants to an interaction manage their epistemic status, that is, their rights to know about the topic or target talked about. It has long been known in CA research that in describing events people also make explicit how they are able to know about the incident they are telling about, what sort of access they have to the incident (Sacks, 1992; Whalen & Zimmerman, 1990). Similarly, in telling stories or delivering news, people give primary rights to tell about an event to a person who has actually experienced the event (Maynard, 2003; Peräkylä, 1995; Pomerantz, 1984; Sacks, 1984). In their study "The Terms of Agreement," Heritage and Raymond (2005) describe how epistemic authority and subordination are constantly managed also in evaluating a common target in everyday talk and show some subtle and recurring methods with which this is done.

In CA terms, Heritage and Raymond's investigation concerns assessment sequences in conversation. Assessments are typically made in adjacency pairs, meaning that when one speaker assesses a target in conversation, the others orient to this first assessment as making relevant a second assessment.

Heritage and Raymond maintain that by making the first assessment, the speaker simultaneously claims to have a primary right to evaluate the target. Thus, in making the first assessment, speakers orient to the possibility that the other participants have a better access to, or a closer relationship with, the assessed target. Heritage and Raymond show various cases in which speakers regulate these epistemic rights, such as by downgrading them in making the first assessment or by upgrading them in making the second assessment. Reaching agreement thus requires careful management of the participants' epistemic status: It inherently involves negotiation on epistemic rights, authority, and subordination.

The following two extracts from Heritage and Raymond's article show unmarked assessment sequences, where both participants orient to their right to assess the target as unproblematic:

```
(1)  [VIYMC 1:4]

1 J:  Let's feel the water. Oh, it...

2 R:  -> It's wonderful. It's just right.

3     It's like bathtub water.
```

```
(2)  [NB:IV.7:-44]

1 A:  -> Adeline's such a swell [gal

2 P:                           [Oh God, whadda

3     gal. You know
```

In these cases, both speakers in both assessment sequences have similar access to the target that they are assessing and treat their rights to assess the target as equal. If this is not the case, speakers have various ways to make this clear. The following

two sequences show some ways in which the speakers of the *first assessment* may orient to their respective epistemic status. In extract 3, the speaker downgrades her epistemic rights with a *tag question*:

```
(3)  [Rah 14:2]

1 Jen: Mm [I: bet they proud o:f the fam'ly.=

2 Ver:    [Ye:s.

3 Jen:-> =They're [a luvly family now ar'n't [they.

4 Ver:            [°Mm:.°                     [They

5          are: yes ye[s.

6 Jen:              [eeYe[s::,

7 Ver:                   [Yes,

8 Jen: Mm: All they need now is a little girl

9      tih complete i:t.
```

It is evident in the first two lines of the sequence that Vera has more information on the family in question as she answers Jennie's question concerning the family. In line 3, Jennie assesses the family as lovely, and downgrades her assessment with a tag question *aren't they*. This way she indicates that her co-participant

has primary rights to assess the family, as she is the one who knows them better.

There are also methods to emphasize one's primary rights to assess a target. One such method is *negative interrogative* of which extract 4 shows a case:

```
(4) [SBL:2-1-8:5]
1 Bea: Wz las'night th'firs'time you met Missiz Kelly?
2        (1.0)
3 Nor: Me:t who:m?
4 Bea: Missiz Kelly?
5 Nor: ^Ye:s. hh[Yih kno] :w what<]
6 Bea:        [ Isn't ]she a cu]te little thi:ng?
```

In this extract, the interrogative syntax that Bea deploys in her first assessment at line 6 increases the relevancy of a response, the yes-no question structure predisposes the terms to be used in the response and that an agreeing response is expected. Through all these characteristics of the turn, Bea shows that her stance toward Mrs. Kelly is settled, she has an established acquaintance with her and has stronger rights to assess her than Norman.

Similarly, there are available for speakers of *the second assessment* to manage their epistemic status. One of these is the oh-preface. In the following extract (5) Ilene and Norman are talking about Norman's dog Trixie. The first assessment is in lines 9-10 and the second in line 11.

```
(5) [Heritage 1:11:4]
1 Ile: No well she's still a bit young though
2        isn't [she<ah me]an:=uh[:
3 Nor:       [She : :]      [She wz a year:
        la:st wee:k.
5 Ile: Ah yes. Oh well any time no:w [then.]
6 Nor:                       [Uh: : :] : [m
7 Ile:                               [Ye: s.=
8 Nor: =But she[:'s (  )          ]
9 Ile:        [Cuz Trixie started] so
10       early [didn't sh[e,
11 Nor:       [°O h : : [ye:s.°=
12 Ile: =°Ye:h°=
```

In line 11, we see how oh-prefacing of the second assessment indexes the speaker's independent access to the target. This is achieved with the oh-prefaces change-of-state characteristics, indicating that Ilene's first assessment has made it relevant for Norman to review his previous, preexisting experience of the target (see Heritage, 2002).

Thus, various methods can be deployed in asserting primary or secondary rights to assess a certain target in conversation. Heritage and Raymond show how through these methods, while agreeing and disagreeing with assessments, participants also negotiate who knows better about the target of the assessment. This work is sometimes subtle and the participants establish mutual alignment, but it can also involve competition and even conflict. Heritage and Raymond conclude by stating that their results point at "a dilemma at the heart of agreement sequences." People seek to know what others think about a certain target, but they at the same time have to pay heed to each other's epistemic rights. Especially when the question is of personal matters (assessing somebody else's grandchildren or pets for example), people may have to engage in complicated face-saving procedures to solve this basic dilemma. The analysis points out how involvement in or detachment from social relationships is an issue that is deeply practical and present in our everyday talk: This is an issue that we have to manage to some extent whenever we engage in the act of assessing a target.

Heritage and Raymond's findings are a good example of the sort of research that is capable of unraveling the fine-tuned logic of face-to-face interaction and to uncover the embedded norms of conduct that are oriented to by the participants in managing their social relations. Their article depicts some ways in which people encode and argue for their epistemic status in interaction. The authors' results, however, are relevant beyond the sphere of everyday interactions. Epistemic relations between people in different statuses are at the heart of many institutions—such as medicine and education. Heritage and Raymond's study provides a baseline in relation to which it is possible to analyze how epistemic rights are managed in many institutional encounters.

Rethinking the Place of Mental Realities

Some years ago, Martyn Hammersley (2003) instigated a debate concerning methodological foundations of conversation analysis. In particular, he criticized CA for refusing to acknowledge that various psychosocial features are not "observable" in the subjects' public actions or the immediate context of action, but nevertheless have bearing to these actions. Hammersley thus calls for more recognition for both psychological and social factors, which reside, as it were, "outside" the immediate interactional expression and context. The ways in which CA can address the social factors will be discussed at the conclusion of this chapter. Regarding the psychological realities, the recent research program outlined by N. J. Enfield and Stephen C. Levinson (2006) is of great interest. Levinson and his coworkers have brought together a key contemporary discussion in psychology on *theory of mind*, and the findings of conversation analysis. In result, they propose that the basic practices of social interaction involve a process of mutual "reading" of the mental states of the co-interactants.

Although CA traditionally has avoided references to mental states of the participants of interaction—for example by referring to *epistemic rights* rather than to *cognition*—for Enfield and Levinson (2006, p. 1), the interactants take part in a "shared mental world." This shared mental world involves the interactants' detailed expectations concerning each other's behavior and their understandings regarding each other's cognitions, intentions, and motives. It is a world that is shaped and maintained in and through the sequentially organized action.

Theory of mind is a cornerstone of conceptualization by Enfield and Levinson. This is not a "researcher's theory," but a basic competence of understanding the social world, shared by humans. It involves an ability to attribute to other persons a world of inner experience that is independent from the outer world and the observer's own experience—a world consisting of states such as beliefs, desires, and intentions (Premack, 1976).

According to Enfield and Levinson, theory of mind is in incessant use in social interaction. The use of theory of mind is normally automatic and unconscious. The interactants read each other's communicative intentions and respond to these (Enfield & Levinson, 2006, p. 5; Levinson, 2006a, p. 45). Interactants do not respond to other's behavior as such. Interaction requires interpretation of other's behavior: "mapping intentions or goals onto behaviour" (Levinson, 2006a, p. 45), whereby behavior is understood as intentional action. This process of interpretation, according to Levinson, involves "some kind of simulation of the other's mental world" (p. 45).

Levinson (2006a, 2006b) and Enfield (Enfield & Levinson, 2006), and the contributors to their recent collection (especially

Schegloff, 2006) show how the practices identified by CA—adjacency pairs, pre-sequences, recipient design, repair—involve reciprocal and reflexive simulation of the mental states of the participants. Through the integration of CA and the research tradition on theory of mind, suggested by Enfield and Levinson, we can thus arrive at a conceptualization of interaction that preserves the conversation analytical findings, and yet does not call into question the relevancy of mental processes.

The reinterpretation of conversation analytical findings in the light of the psychological research traditions on theory of mind involves a new turn in the conceptualization of social interaction. The coming years will show whether this new conceptualization yields new kinds of empirical research designs and research results in CA.

▣ CONCLUSION

In this chapter, we have introduced a number of qualitative research approaches that use language—text or talk—as data. Approaches like those that we have presented are sometimes criticized for their narrow focus, as investigating an arbitrary fracture of reality, a piece of text or a fragment of talk, that has no bearing on broader social issues (e.g., Hammersley, 2003). If we study language, do we neglect something else, which might be more important, at least in social and political terms? Does qualitative research on talk and text involve *merely* language, or can these approaches address broader social issues? To conclude this chapter, we compare some of the methods discussed for their relation to issues of *power* and *social change*. We focus on the three methods discussed most thoroughly: HDA, MCA, and CA. Our main conclusion will be that these methods are indeed potent in addressing broader social phenomena.

The HDA exemplified in this chapter by Meskus' work is most directly a method for investigating social change. Meskus showed us the evolvement of the management of human heredity in Finnish maternity health care. At the same time, her analysis of texts was about power—about the discourses given which certain decisions concerning the management of heredity were made and about the practices that were adopted as technologies of this management—as well as about the ways in which these developments concerned groups of individuals (in this case, mostly pregnant mothers). Meskus treated power here as a productive force—as something that calls realities into being rather than suppresses them.

The potential of MCA in dealing with questions pertaining to power and social change was well shown in Rapley et al.'s (2003) research where they demonstrated the deeply moral underpinnings of the use of categories that were neutral on the surface. The adoption and use of specific categories in social situations as well as in texts—the mere naming of a member as belonging to a certain category—simultaneously attributes specific obligations and refutations to the chosen category and thus also obliges the person in question. This was the case with the media struggle on the categorization of the gunman described by Rapley et al. MCA provides a method with which we can bring to the fore the subtle underpinnings of seemingly innocent language use: it shows how any categorization of a member or group in society involves their placement within certain moral space with regard to which their actions can be judged.

The relation of CA to broader social issues is more complex. CA that focuses on generic practices and structures of mundane everyday talk might seem irrelevant in power and social change. The research example we showed was about everyday casual conversations, the minute reality of which is perhaps far away from the large-scale questions of change in social, economic, and political structures. Michael Billig (1998) argued that this irrelevance may actually imply politically conservative choices. Even in researching institutional interaction, the fact that conversation analysts often focus on small details of video- or audio-recorded talk might seem to render their studies impotent for the analysis of social relations and processes *not* incorporated in talk (see also Hak, 1999).

From the CA point of view, two responses can be given to these criticisms. First, the significance of orderly organization of face-to-face (or other "live") interaction for *all* social life needs to be restated. No "larger-scale" social institutions could operate without the substratum of the interaction order. These institutions operate largely through questions, answers, assessments, accusations, accounts, interpretations, and the like. Hence, even when not focusing on hot social and political issues that we read about in the newspapers, CA is providing knowledge about the basic organizations of social life that make these issues, as well as their possible solutions and the debate about them, possible in the first place. The observation made by Heritage and Raymond, for instance, about the terms of agreement in social interaction makes it possible to suggest that such fine-tuned management of epistemic rights may lie behind various struggles for power and status at the workplace, in professional encounters, and so on. Further, CA research that is not explicitly framed around questions of power or status may, however, bring results that are relevant in discussing these topics. For instance, analyses of professional practices may bring forward covered ways of influencing clients to reach particular goals, which may then give reason to discuss the legitimacy or potential effects of these practices (see Clark, Drew, & Pinch, 2003, on sales encounters; Ruusuvuori, 2007, on homeopathic consultations).

Some CA research is more directly relevant for political and social concerns. For example, many CA studies have contributed to our understanding of the ways in which specific interactional practices contribute to the maintenance or change of the *gender system*. In these studies, gender and sexuality are treated as practical accomplishments rather than as "facts." Work by Candace West (1979) and Don Zimmerman (Zimmerman & West, 1975) on male–female interruptions is widely cited. More recently, Celia Kitzinger (2005) has shown how heterosexual speakers constantly allow their heterosexuality to be inferred in their talk and how this "both reflects and constructs heteronormativity" (p. 222; see also Kitzinger, 2000; Kitzinger & Kitzinger, 2007). In a somewhat more linguistic CA study, Tainio (2002) explored how syntactical and semantic properties of utterances are used in the construction of heterosexual identities in elderly couples' talk. Studies such as these (for an overview, see McIlvenny, 2002) also amply demonstrate the *critical* potential of CA. A different CA study on social change was offered in Steven Clayman and John Heritage's (2002b) work on question design in U.S. presidential press conferences. By combining qualitative and quantitative techniques, Clayman and Heritage showed how the relative proportions of different types of journalist questions, exhibiting different degrees of "adversarialness," have changed over time. As such, they explored the historical change in the U.S. presidential institution and media. A further example of a critical potential of a study that combines CA and statistical methods is Tanya Stivers and Asifa Majid's (2007) research on implicit race bias in asking questions in pediatric consultations. Stivers and Majid's study shows that parental race and education have a significant effect on whether doctors select children to answer questions. Thus, at least to scholars using CA or MCA in their analyses of the everyday world, these methods offer ample critical perspectives for inquiry of social life.

Dorothy Smith (among others) has criticized the Goffmanian approach (adopted in CA and MCA) to social interaction as a self-sufficient object of study of its own. She argues that treating the everyday world of social interaction as such an object isolates it from its context of broader forms of organization and makes it appear self-contained (Grahame, 1998). Smith (1987, pp. 152–154) maintains that local social organization is generated by social relations external to the local setting and that these social relations cannot be adequately grasped by investigating the local setting only. It seems to us, however, that the way in which CA is able to provide for detailed descriptions of the organization of the world of social interaction (such as the terms of agreement for instance) could rather be seen as one step further in uncovering the mechanisms through which social relations operate. Further, the new trends such as Kitzinger's feminist CA (2000, 2005) or Stivers & Majid's analysis of race bias in pediatric consultations (2007) show that in principle there is nothing in the actual method that would prevent combining its results with further investigations of the research object with other methods—to gain a more comprehensive view of it.

Thus, our conclusion is, qualitative research on text and talk is not only about language. The observations made by methods on text and talk provide one avenue to understanding social structures, as well as individual actions.

APPENDIX

THE TRANSCRIPTION SYMBOLS IN CA

[	Starting point of overlapping speech
]	End point of overlapping speech
(2.4)	Silence measured in seconds
(.)	Pause of less than 0.2 seconds
↑	Upward shift in pitch
↓	Downward shift in pitch
word	Emphasis
wo:rd	Prolongation of sound
°word°	Section of talk produced in lower volume than the surrounding talk
WORD	Section of talk produced in higher volume than the surrounding talk
w#ord#	Creaky voice
£word£	Smile voice
wo(h)rd	Laugh particle inserted within a word
wo-	Cut off in the middle of a word
word<	Abruptly completed word
>word<	Section of talk uttered in a quicker pace than the surrounding talk
<word>	Section of talk uttered in a slower pace than the surround talk
(word)	Section of talk that is difficult to hear but is likely as transcribed
( )	Inaudible word
.hhh	Inhalation
hhh	Exhalation
.	Falling intonation at the end of an utterance
?	Rising intonation at the end of an utterance
,	Flat intonation at the end of an utterance
word.=word	"Rush through" without the normal gap into a new utterance.
((word))	Transcriber's comments

Source: Adapted from Drew & Heritage (Eds.). (1992). *Talk at work: Interaction in institutional settings.* Cambridge, UK: Cambridge University Press.

◨   REFERENCES

Alexander, J. (1988). *Action and its environments: Toward a new synthesis.* New York: Columbia University Press.

Arminen, I. (2005). *Institutional interaction: Studies of talk at work.* Aldershot, UK: Ashgate.

Atkinson, J. M. (1978). *Discovering suicide: Studies in the social organization of sudden death.* London: Macmillan.

Atkinson, P., & Coffey, A. (1997). Analysing documentary realities. In D. Silverman (Ed.), *Qualitative research: Theory, method, and practice* (pp. 45–62). London: Sage.

Atkinson, J. M., & Drew, P. (1979). *Order in court: The organization of verbal interaction in judicial settings.* London: Macmlllan.

Bales, R. F. (1950). *Interaction process analysis: A method for the study of small groups.* Reading, MA: Addison-Wesley.

Bamberg, M., & Georgakopoulou, A. (2008). Small stories as a new perspective in narrative and identity analysis. *Text & Talk, 28*(3), 377–396.

Billig, M. (1998). Whose terms? Whose ordinariness? Rhetoric and ideology in conversation analysis. *Discourse & Society, 10,* 543–558.

Brown, G., & Yule, G. (1983). *Discourse analysis.* Cambridge, UK: Cambridge University Press.

Butler, C., & Weatherall, A. (2006). "No, we're not playing families": Membership categorization in children's play. *Research on Language and Social Interaction, 39*(4), 441–470.

Clark, C., Drew, P., & Pinch, T. (2003). Managing prospect affiliation and rapport in real-life sales encounters. *Discourse Studies, 5*(1), 5–31.

Clayman, S., & Heritage, J. (2002a). *The news interview: Journalists and public figures on the air.* Cambridge, UK: Cambridge University Press.

Clayman, S., & Heritage, J. (2002b). Questioning presidents: Journalistic deference and adversarialness in the press conferences of Eisenhower and Reagan. *Journal of Communication, 52,* 749–775.

De Fina, A., & Georgakopoulou, A. (2008). Introduction: Narrative analysis in the shift from texts to practices. *Text & Talk, 23*(3), 275–281.

Dean, M. (1999). *Governmentality: Power and rule in modern society.* London: Sage.

Drew, P. (1992). Contested evidence in courtroom cross-examination: The case of a trial for rape. In P. Drew & J. Heritage (Eds.), *Talk at work: Interaction in institutional settings* (pp. 470–520). Cambridge, UK: Cambridge University Press.

Drew, P., & Heritage, J. (1992). Analyzing talk at work: An introduction. In P. Drew & J. Heritage (Eds.), *Talk at work: Interaction in institutional settings* (pp. 3–65). Cambridge, UK: Cambridge University Press.

Drew, P., & Holt, E. (1988). Complainable matters: The use of idiomatic expression in making complaints. *Social Problems, 35,* 398–417.

Edwards, D. (1997). *Discourse and cognition.* London: Sage.

Eglin, P., & Hester, S. (2003). *The Montreal massacre: A story of membership categorization analysis.* Waterloo, ON: Wilfred Laurier University Press.

Enfield, N. J., & Levinson, S. (2006). Introduction: Human sociality as a new interdisciplinary field. In N. J. Enfield & S. C. Levinson (Eds.), *Roots of human sociality: Culture, cognition and interaction* (pp. 1–34). New York: Berg.

Fairclough, N. (1989). *Language and power.* London: Longman.

Fairclough, N. (1995). *Media discourse.* London: Edward Arnold.

Foucault, M. (1973). *The birth of the clinic: An archaeology of medical perception.* New York: Pantheon.

Foucault, M. (1977). *Discipline and punish: The birth of the prison.* London: Allen Lane.

Foucault, M. (1978). *The history of sexuality: Vol. 1. An introduction.* New York: Pantheon.

Garfinkel, H. (1967). *Studies in ethnomethodology.* Englewood Cliffs, NJ: Prentice Hall.

Georgakopoulou, A. (2007). *Small stories, interaction and identities.* Amsterdam: John Benjamins.

Glenn, P. (2003). *Laughter in interaction.* Cambridge, UK: Cambridge University Press.

Goffman, E. (1955). On face work. *Psychiatry, 18,* 213–231.

Goffman, E. (1983). The interaction order. *American Sociological Review, 48,* 1–17.

Goodwin, C. (1981). *Conversational organization: Interaction between speakers and hearers.* New York: Academic Press.

Goodwin, C., & Goodwin, M. H. (1992). Assessments and the construction of context. In A. Duranti & C. Goodwin (Eds.), *Rethinking context: Language as interactive phenomenon* (pp. 147–190). Cambridge, UK: Cambridge University Press.

Grahame, P. R. (1998). Ethnography, institutions, and the problematic of the everyday world. *Human Studies, 21,* 347–360.

Greatbatch, D. (1992). On the management of disagreement between news interviewees. In P. Drew & J. Heritage (Eds.), *Talk at work: Interaction in institutional settings* (pp. 268–302). Cambridge, UK: Cambridge University Press.

Greimas, A. J. (1966). *Semantique structurale.* Paris: Larousse.

Gubrium, J., & Holstein, J. (2009). *Analyzing narrative reality.* Thousand Oaks, CA: Sage.

Haakana, M. (2001). Laughter as a patient's resource: Dealing with delicate aspects of medical interaction. *Text, 21,* 187–219.

Hak, T. (1999). "Text" and "con-text": Talk bias in studies of health care work. In S. Sarangi & C. Roberts (Eds.), *Talk, work, and institutional order* (pp. 427–452). Berlin, Germany: Mouton de Gruyter.

Hammersley, M. (2003). Conversation analysis and discourse analysis: Methods or paradigms. *Discourse and Society, 14*(6), 751–781.

Heath, C. (1992). The delivery and reception of diagnosis in the general-practice consultation. In P. Drew & J. Heritage (Eds.), *Talk at work: Interaction in institutional settings* (pp. 235–267). Cambridge, UK: Cambridge University Press.

Heritage, J. (1984). *Garfinkel and ethnomethodology.* Cambridge, UK: Polity Press.

Heritage, J. (1998). Oh-prefaced responses to inquiry. *Language in Society, 27*(3), 291–334.

Heritage, J. (2002). Oh-prefaced responses to assessments: A method of modifying agreement/disagreement. In C. Ford, B. Fox, & S. Thompson (Eds.), *The language of turn and sequence* (pp.196–224). New York: Oxford University Press.

Heritage, J. (2004). Conversation analysis and institutional talk. In R. Sanders & K. Fitch (Eds.), *Handbook of language and social interaction* (pp. 103–146). Mahwah, NJ: Lawrence Erlbaum.

Heritage, J., & Atkinson, J. M. (1984). Introduction. In J. M. Atkinson & J. Heritage (Eds.), *Structures of social action* (pp. 1–15). Cambridge, UK: Cambridge University Press.

Heritage, J., & Raymond, G. (2005). The terms of agreement: Indexing epistemic authority and subordination in talk-in-interaction. *Social Psychology Quarterly, 68*(1), 15–38.

Heritage, J., & Sefi, S. (1992). Dilemmas of advice: Aspects of the delivery and reception of advice in interactions between health visitors and first-time mothers. In P. Drew & J. Heritage (Eds.), *Talk at work: Interaction in institutional settings* (pp. 359–417). Cambridge, UK: Cambridge University Press.

Heritage, J., & Stivers, T. (1999). Online commentary in acute medical visits: A method for shaping patient expectations. *Social Science and Medicine, 49,* 1501–1517.

Hutchby, I., & Wooffitt, R. (1998). *Conversation analysis: Principles, practices, and applications.* Cambridge, UK: Polity Press.

Hyvärinen, M., Hydén, L-C., Saarenheimo, M., & Tamboukou, M. (2010). Beyond narrative coherence: An introduction. In M. Hyvärinen, L-C. Hydén, M. Saarenheimo, & M. Tamboukou (Eds.) *Beyond narrative coherence: Studies in narrative 11.* Amsterdam: John Benjamins.

Jayyusi, L. (1991). Values and moral judgment: Communicative praxis as moral order. In G. Button (Ed.), *Ethnomethodology and the human sciences* (pp. 227–251). Cambridge, UK: Cambridge University Press.

Jefferson, G. (1984). On the organization of laughter in talk about troubles. In J. M. Atkinson & J. Heritage (Eds.), *Structures of social action* (pp. 346–369). Cambridge, UK: Cambridge University Press.

Kendall, G., & Wickham, G. (1999). *Using Foucault's methods.* London: Sage.

Kitzinger, C. (2000). Doing feminist conversation analysis. *Feminism & Psychology, 10,* 163–193.

Kitzinger, C. (2005). Speaking as a heterosexual: (How) does sexuality matter for talk-in-interaction. *Research on Language and Social Interaction, 38*(3), 221–265.

Kitzinger, C., & Kitzinger, S. (2007). Birth trauma: Talking with women and the value of conversation analysis. *British Journal of Midwifery, 15*(5), 256–264.

Lepper, G. (2000). Categories in text and talk. *A practical introduction to categorization analysis.* Introducing Qualitative Methods Series. London: Sage.

Levinson, S. (2006a). On the human "interaction engine." In N. J. Enfield & S. C. Levinson (Eds.), *Roots of human sociality: Culture, cognition and interaction* (pp. 39–69). New York: Berg.

Levinson, S. (2006b). Cognition at the heart of human interaction. *Discourse Studies, 8*(1), 85–93.

Mandelbaum, J. (1992). Assigning responsibility in conversational storytelling: The interactional construction of reality. *Text, 13,* 247–266.

Maynard, D. W. (1991). Interaction and asymmetry in clinical discourse. *American Journal of Sociology, 97,* 448–495.

Maynard, D. W. (1992). On clinicians co-implicating recipients' perspective in the delivery of diagnostic news. In P. Drew & J. Heritage (Eds.), *Talk at work: Interaction in institutional settings* (pp. 331–358). Cambridge, UK: Cambridge University Press.

Maynard, D. W. (2003). *Bad news, good news: Conversational order in everyday talk and clinical settings.* Chicago: University of Chicago Press.

McHoul, A. W., & Grace, A. (1993). *A Foucault primer: Discourse, power, and the subject.* Melbourne, Australia: Melbourne University Press.

McIlvenny, P. (2002). *Talking gender and sex.* Amsterdam: John Benjamins.

Merritt, M. (1976). On questions following questions (in service encounters). *Language in Society, 5,* 315–357.

Meskus, M. (2009a). Governing risk through informed choice: Prenatal testing in welfarist maternity care. In S. Bauer & A. Wahlberg (Eds.), *Contested categories: Life sciences in society* (pp. 49–68). Farnham, UK: Ashgate.

Meskus, M. (2009b). *Elämän tiede.* Tampere, Finland: Vastapaino.

Nikander, P. (2002). *Age in action: Membership work and stage of life categories in talk.* Helsinki: Finnish Academy of Science and Letters.

Ochs, E., & Capps, L. (2001). *Living narrative. Creating lives in everyday storytelling.* Cambridge, MA: Harvard University Press.

Peräkylä, A. (1995). *AIDS counselling; Institutional interaction and clinical practice.* Cambridge, UK: Cambridge University Press.

Peräkylä, A. (1998). Authority and accountability: The delivery of diagnosis in primary health care. *Social Psychology Quarterly, 61,* 301–320.

Peräkylä, A. (2002). Agency and authority: Extended responses to diagnostic statements in primary care encounters. *Research on Language and Social Interaction, 35,* 219–247.

Pomerantz, A. (1984). Agreeing and disagreeing with assessments: Some features of preferred/dispreferred turn shapes. In J. M. Atkinson & J. Heritage (Eds.), *Structures of social action: Studies in conversation analysis* (pp. 67–101). Cambridge, UK: Cambridge University Press.

Potter, J. (2004). Discourse analysis as a way of analysing naturally occurring talk. In D. Silverman (Ed.), *Qualitative research: Theory, method, and practice* (2nd ed., pp. 200–201). London: Sage.

Potter, J. (2006). Cognition and conversation. *Discourse Studies, 8*(1), 131–140.

Potter, J., & te Molder, H. (2005). Talking cognition: Mapping and making the terrain. In H. te Molder & J. Potter (Eds.), *Conversation and cognition* (pp. 1–54). Cambridge, UK: Cambridge University Press.

Premack, D. (1976). Language and intelligence in ape and man. *American Scientist, 64*(4) 674–683.

Propp, V. I. (1968). *Morphology of the folktale* (rev. ed., L. A. Wagner, Ed.). Austin: University of Texas Press.

Rapley, M., McCarthy, D., & McHoul, A. (2003). Mentality or morality? Membership categorization, multiple meanings and mass murder. *British Journal of Social Psychology, 42,* 427–444.

Roca-Cuerbes, C. (2008). Membership categorization and professional insanity ascription. *Discourse Studies, 10*(4), 543–570.

Rose, N. (1999). *Powers of freedom: Reframing political thought.* Cambridge, UK: Cambridge University Press.

Ruusuvuori, J. (2007). Managing affect: Integration of empathy and problem-solving in health care encounters. *Discourse Studies, 9*(5), 597–622.

Sacks, H. (1974a). An analysis of the course of a joke's telling in conversation. In R. Bauman & J. Sherzer (Eds.), *Explorations in the ethnography of speaking* (pp. 337–353). Cambridge, UK: Cambridge University Press.

Sacks, H. (1974b). On the analysability of stories by children. In R. Turner (Ed.), *Ethnomethodology* (pp. 216–232). Harmondsworth, UK: Penguin.

Sacks, H. (1992). *Lectures on conversation* (Vol. 1, G. Jefferson, Ed., with an introduction by E. Schegloff). Oxford, UK: Blackwell.

Sacks, H., Schegloff, E., & Jefferson, G. (1974). A simplest systematics for the organization of turn-taking for conversation. *Language, 50*, 696–735.

Schegloff, E. A. (1968). Sequencing in conversational openings. *American Anthropologist, 70*, 1075–1095.

Schegloff, E. A. (1991). Reflection on talk and social structure. In D. Boden & D. Zimmerman (Eds.), *Talk and social structure* (pp. 44–70). Cambridge, UK: Polity Press.

Schegloff, E. A. (1992a). Introduction. In G. Jefferson (Ed.), *Harvey Sacks: Lectures on conversation: Vol. 1. Fall 1964–Spring 1968.* Oxford, UK: Blackwell.

Schegloff, E. A. (1992b). On talk and its institutional occasion. In P. Drew & J. Heritage (Eds.), *Talk at work: Interaction in institutional settings* (pp. 101–134). Cambridge, UK: Cambridge University Press.

Schegloff, E. A. (1992c). Repair after next turn: The last structurally provided defense of intersubjectivity in conversation. *American Journal of Sociology, 98*, 1295–1345.

Schegloff, E. A. (2006). Interaction: The infrastructure for social institutions, the natural ecological niche for language, and the arena in which culture is enacted. In N. J. Enfield & S. C. Levinson (Eds.), *Roots of human sociality: Culture, cognition and interaction* (pp. 70–98). New York: Berg.

Schegloff, E. A. (2007a). *Sequence organization.* Cambridge, UK: Cambridge University Press.

Schegloff, E. A. (2007b). A tutorial on membership categorization. *Journal of Pragmatics, 39*, 462–482.

Schegloff, E. A., Jefferson, G., & Sacks, H. (1977). The preference for self-correction in the organization of repair in conversation. *Language, 53*, 361–382.

Schegloff, E. A., & Sacks, H. (1973). Opening up closings. *Semiotica, 8*, 289–327.

Seale, C. (1998). *Constructing death: The sociology of dying and bereavement.* Cambridge, UK: Cambridge University Press.

Silverman, D. (1997). *Discourses of counselling.* London: Sage.

Silverman, D. (1998). *Harvey Sacks: Social science and conversation analysis.* Cambridge, UK: Polity Press.

Silverman, D. (2001). *Interpreting qualitative data: Methods for analyzing talk, text, and interaction* (2nd ed.). London: Sage.

Smith, D. (1974). The social construction of documentary reality. *Sociological Inquiry, 44*, 257–268.

Smith, D. (1987). *The everyday world as problematic: A feminist sociology.* Toronto, ON: University of Toronto Press.

Smith, D. (1990). *The conceptual practices of power.* Toronto: University of Toronto Press.

Speer, S. (2002). "Natural" and "contrived" data: A sustainable distinction. *Discourse Studies, 4*, 511–525.

Stivers, T., & Majid, A. (2007). Questioning children: Interactional evidence of implicit bias in medical interviews. *Social Psychology Quarterly, 70*(4), 424–441.

Stokoe, E. (2003). Mothers, single women and sluts: Gender, morality and membership categorization in neighbour disputes. *Feminism & Psychology, 13*(3), 317–344.

Tainio, L. (1999). Opaskirjojen kieli ikkunana suomalaiseen parisuhteeseen. *Naistutkimus, 12*(1), 2–26.

Tainio, L. (2002). Negotiating gender identities and sexual agency in elderly couples' talk. In P. McIlvenny (Ed.), *Talking gender and sexuality* (pp. 181–206). Amsterdam: John Benjamins.

Taylor, T. J., & Cameron, D. (1987). *Analyzing conversation: Rules and units in the structure of talk.* Oxford, UK: Pergamon.

Törrönen, J. (2000). The passionate text: The pending narrative as a macrostructure of persuasion. *Social Semiotics, 10*(1), 81–98.

Törrönen, J. (2003). The Finnish press' political position on alcohol between 1993 and 2000. *Addiction, 98*(3), 281–290.

ten Have, P. (1995). Disposal negotiations in general practice consultations. In A. Firth (Ed.), *The discourse of negotiation: Studies of language in the workplace* (pp. 319–344). Oxford, UK: Pergamon.

Vehviläinen, S. (2001). Evaluative advice in educational counseling: The use of disagreement in the "stepwise entry" to advice. *Research on Language and Social Interaction, 34*, 371–398.

West, C. (1979). Against our will: Male interruption of females in cross-sex conversation. *Annals of the New York Academy of Science, 327*, 81–97.

Wetherell, M. (1998). Positioning and interpretative repertoires: Conversation analysis and post-structuralism in dialogue. *Discourse & Society, 9*, 387–412.

Wetherell, M., & Potter, J. (1992). *Mapping the language of racism: Discourse and the legitimation of exploitation.* London: Harvester.

Whalen, M., & Zimmerman, D. (1990). Describing trouble: Practical epistemology in citizen calls to the police. *Language in Society, 19*, 465–492.

Wodak, R., & Meyer, M. (2009). Critical discourse analysis: History, agenda, theory and methodology. In R. Wodak & C. Meyer (Eds.), *Methods of critical discourse analysis* (pp. 1–33). London: Sage.

Zimmerman, D. H., & West, C. (1975). Sex roles, interruptions, and silences in conversation. In B. Thorne & N. Henley (Eds.), *Language and sex: Difference and dominance* (pp. 105–129). Rowley, MA: Newbury House.

33

FOCUS GROUPS

Contingent Articulations of Pedagogy, Politics, and Inquiry

George Kamberelis and Greg Dimitriadis

As traditional research demarcations collapse and new questions and issues arise (as evidenced by the evolution of the *Handbook of Qualitative Inquiry* across four editions), focus groups offer a particularly fruitful method for "thinking through" qualitative research today. Basically, focus groups are collective conversations or group interviews. They can be small or large, directed or nondirected. Focus groups have been used for a wide range of purposes during the past century or so. The U.S. military (e.g., Robert Merton), multinational corporations (e.g., Proctor & Gamble), Marxist revolutionaries (e.g., Paulo Freire), literacy activists (e.g., Jonathan Kozol), and three waves of feminist scholar-activists (e.g., Esther Madriz), among others, have all used focus groups to help advance their concerns and causes.

In the last edition of this handbook, we argued that focus group research exists at the intersection of pedagogy, activism, and interpretive inquiry. We also argued that researchers are typically strategic in configuring these intersections. In this chapter, we build upon and extend that work: (a) by troubling the idea that the intersection of pedagogy, activism, and inquiry is always or primarily strategic and (b) by exploring both new possibilities of and new dangers faced by focus group research in the current social and political climate, especially in relation to debates around the politics of evidence. To accomplish these goals, we engage in two related pragmatic/rhetorical moves. First, we reimagine focus group work as almost always multifunctional. Second, we situate focus group work within a performative idiom. Before we revivify and

expand our earlier discussions of focus group work, we describe each of these moves.

MULTIFUNCTIONALITY AND FOCUS GROUPS

Multifunctionality has become an increasingly important construct used to explain complexity and contingency within many different disciplines. In linguistics, for example, it has long been known that many different linguistic functions (e.g., referential, conative, phatic, poetic) almost always operate simultaneously, with one function typically assuming a dominant role (e.g., Jakobson, 1960). In contemporary agriculture research, *multifunctionality* refers to the benefits beyond food production and trade that result from agricultural policies. Such benefits include such things as landscape preservation and increasing rural employment opportunities.

Inspired by Laurel Richardson's (2000) image of the crystal as productive for mapping the changing complexity of the lives of her research participants and her own life as a sociologist, we find the image of the prism to be useful for revisioning the primary functions of focus group work (pedagogy, politics, and inquiry), as well as the relations between and among them. A prism is a transparent optical element with flat, polished surfaces that can both refract and reflect light. The most common use of the term *prism* refers to a triangular prism—one that has a triangular base and three clear rectangular surfaces. When viewed from different angles, one can see more or less of each of its three surfaces. From some angles, one surface is completely and directly visible with the other two surfaces partially and obliquely visible. From other angles, two surfaces are completely visible with the third surface visible but obliquely and in the distance. Importantly, however, from every angle of vision, at least some of every surface is visible. Additionally, from some angles prisms break up light into its constituent spectral colors whereas from other angles they act like mirrors, reflecting all or almost all approaching light back toward its source. Similarly, all three focus group functions are always at work simultaneously, they are all visible to the researcher to some extent, and they all both refract and reflect the substance of focus group work in different ways.

Before saying more about the complex relations among the three primary functions of focus group work, and even though the meanings of each function may be (or appear to be) self-explanatory, we first briefly discuss how we have defined each function for our work here. The pedagogic function basically involves collective engagement designed to promote dialogue and to achieve higher levels of understanding of issues critical to the development of a group's interests and/or the transformation of conditions of its existence. In Paulo Freire's (1970/1993) *Pedagogy of the Oppressed,* for example, it is a matter of "reading the word" to better read the world. Among other things, this means asking and answering questions such as the following: What social facts are portrayed in a message as if they were perfectly "natural" or "normal"? Whose positions, interests, and values are represented in the message? Whose positions, interests, and values are absent or silent? Are any positions, interests, or values ridiculed, vilified, or demonized? How is the message trying to position its readers/viewers in relation to its messages? How does this message do its work through the use of specific textual features and specific arrangements of these features?

Although not necessarily, the political function of focus group work often builds on the pedagogic function. The primary goal of the political function is to transform the conditions of existence for particular stakeholders. Activism (or enacting the political function) can grow out of a wide variety of political orientations; it typically constitutes a response to conditions of marginalization or oppression; the goal of the political function is usually to transform these conditions, making them more democratic, and it may be enacted in a variety of ways including consciousness-raising activities; writing editorials or manifestos; participating in campaigns, public marches, or strikes; boycotting products or services; lobbying government agencies; or simply adjusting one's needs and desires or changing one's lifestyle. In a recent interview (Kreisler, 2002), linguist and political activist Noam Chomsky cited several key moments of political activism in history that he argues literally changed the world: the Lowell factory girls' protest in the 1850s that catalyzed the labor movement in the United States; the antiwar, civil rights, and feminist movements of

the 1960s; and the efforts of journalists, artists, and public intellectuals in Turkey fighting censorship today through acts of civil disobedience in their everyday professional work. Less obvious, more local forms of activism are also in evidence everywhere all the time—whether in neighborhoods, schools, or workplaces. Indexing both the nature and the importance of the political function, Chomsky noted that all effective political accomplishments "got there by struggle, common struggle by people who dedicated themselves with others, because you can't do it alone, and made [this] a much more civilized country. It was a long way to go, and that's not the first time it happened. And it will continue" (2002, Activism, para. 7).

Research (or inquiry) is perhaps the function most typically associated with focus group research. Yet, it is a slippery term with a long and contested history. At least since the Enlightenment, inquiry has been associated with the so-called hard sciences. From this perspective, reality and knowledge are *a priori* givens. The primary goals of inquiry are to explain, predict, and control both natural and social phenomena. And inquiry operates according to a correspondence theory of truth (one-to-one mapping of representations onto reality). Since the "interpretive turn," the nature and scope of inquiry have expanded considerably. Within this perspective, reality is considered to be (at least partially) socially constructed and thus changing and changeable. Knowledge is seen as partial and perspectival. The primary goal of inquiry within this view is to achieve richer, thicker, and more complex levels of understanding. And inquiry operates according to a logic of argumentation with the argument most well supported by evidence and warrants holding the day. Even more recently, inquiry has been shown to be messy, dirty, thoroughly imbricated within colonial and neocolonial impulses, and in need of retooling from the ground up to be more praxis-oriented and democratizing (Denzin & Lincoln, 2005). Based on the emergence of focus group research as a way to answer *how* and *why* questions that remained unanswered by positivistic quantitative methods, our working definition of the inquiry function is most closely aligned with that of the "interpretive turn" and especially the Chicago School of Sociology. The primary goal of inquiry from this perspective is to generate rich, complex, nuanced, and even contradictory accounts of how people ascribe meaning to and interpret their lived experience with an eye toward how these accounts might be used to affect social policy and social change. Clearly, we are already signaling that the boundaries between and among inquiry, pedagogy, and politics within focus group work are porous.

In this regard, these three functions are seldom related in simple, unproblematic ways. Political interventions, for example, do not necessarily emerge from inquiry. And when they do, this often occurs in unexpected or unintended ways. Similarly, pedagogy is not always central to activist work, but it can be and often is. Within any given project, these different dimensions or functions of focus group work emerge and interact in distinct and often disjunctive ways, eventually resulting in some unique interactive stabilization. We will touch on this idea of disjunctive emergence in our discussions of different lines of work within which one or another of these three functions has been framed as dominant. Toward the end of the chapter, we will address this issue more thoroughly and more directly when we discuss the performative nature of our own and our students' work. That said, we now discuss the efficacy of reframing focus group work within a performative idiom.

▣ FOCUS GROUPS AND/IN THE PERFORMATIVE TURN

Long connected with theater and elocution, performance has more recently emerged as basic, ontological, but inherently contested concept. "Performance is a contested concept because when we understand performance beyond theatrics and recognize it as fundamental and inherent to life and culture we are confronted with the ambiguities of different spaces and places that are foreign, contentious, and often under siege" (Madison & Hamera, 2006, p. xii). In moving across the academy, this turn to performance has posed new questions about our understandings of texts, practices, identities, and cultures. The turn to performance has also allowed us to see the world as always already in motion. Such a notion of performance gives us nowhere to hide in our responsibilities for the work we do. More specifically, we are forced to see the routine as ambiguous, as foreign, as contentious, and as often under siege.

This turn to performance has also created powerful spaces for thinking about emergent methodologies, those that "explore new ways of thinking about and framing knowledge construction," remaining ever-conscious of the links among epistemologies, methodologies, and the techniques used to carry out empirical work (Hesse-Biber & Leavy, 2006, pp. xi–xii). From this perspective, inquiry (and especially qualitative inquiry) is no longer a discrete set of methods we deploy functionally to solve problems defined *a priori*. Instead, we must question the reification of particular methods that has so marked the emergence of qualitative inquiry as a transdisciplinary field of inquiry (Kamberelis & Dimitriadis, 2005).

▣ NEW CHALLENGES FACING FOCUS GROUP WORK

Guided by commitments to multifunctionality and a performative idiom, our primary goal in this chapter is to reimagine the roles of focus groups in qualitative inquiry from the ground up. Our approach is thus conceptual and transdisciplinary. Only occasionally and in passing do we discuss procedural and practical issues related to selecting focus group members, facilitating focus group discussion, or analyzing focus group transcripts.

Many texts are available for readers who are looking for this kind of treatment (e.g., Barbour, 2008; Bloor et al., 2001; Krueger & Casey, 2008; Morgan, 1998; Schensul et al., 1999; Stewart, Shamdasani, & Rook, 2006). Instead, we both explore and attempt to move beyond recent historical and theoretical treatments of focus groups as "instruments" of qualitative research. In doing so, we try to show how, independent of their intended purposes, focus groups are almost always multivalent and contingent articulations of instructional, political, and empirical practices and effects. Focus groups thus offer unique insights into the possibilities of critical inquiry as deliberative, dialogic, democratic practice that is always already engaged in and with real-world problems and asymmetries in the distribution of economic, cultural, and social capital (e.g., Bourdieu & Wacquant, 1992).

Because the performative turn has decentered the research act—creating spaces that *de facto* collapse demarcations between and among research, pedagogy, and activism—reimagining qualitative inquiry largely involves seeing it more as a matter of asking and dwelling in new questions that are not necessarily answerable in finalizable ways. Although we look to start a discussion about the ways different strategies and practices can coalesce in productive and synergistic ways, we do not advocate an "anything goes" position. The specific competencies required by each dimension of focus group work—not to mention the articulation of multiple dimensions—demands specific sets of skills that are not easily or readily transferred from one dimension to another.

Our orientation to focus group work thus constitutes a productive challenge, encouraging a new angle of vision (among other things) on the politics of evidence. Mindful of the best impulses of the sociology of knowledge and the attendant co-implication of knowledge and power, the new politics of evidence, we believe, must attend to the specificity and autonomy of evidence in new ways. Recognizing that evidence never "speaks for itself," a key task today is finding ways to use evidence to challenge how we are situated and to help us develop new avenues of thought and practice. By locating focus groups within a performative idiom and at the intersection of research, pedagogy, and politics, we are provided no "alibis" for our work. Thus, for example, we cannot attend to the political dimension of our work without attending to the traditional empirical dimensions. And we cannot attend to either without attending to the ways our work circulates pedagogically. Focus group work is thus inevitably prismatic, with all three faces of the prism visible to some extent no matter which face we fix on or how direct our gaze.

Jean-Paul Sartre's notion of "bad faith" is instructive with regard to the reenvisaging the politics of evidence. By "bad faith," Sartre meant all the ways we refuse our basic, human freedoms through recourse to received and static ideas, beliefs, and roles. "Bad faith," Sartre wrote, "implies in essence the unity of a *single* consciousness" (2001, p. 208). Through bad faith, according to Stephen Priest, "we masquerade as fixed essences by the adoption of hypocritical social roles and inert value systems" (in Sartre, 2001, p. 204). To be in bad faith is to remain mired in either simple "facticity" or "transcendence"—to believe either that the empirical world is "as it is" and that we cannot move beyond it or to believe that we can move beyond our circumstances by simple force of will and imagination alone (Solomon, 2006). Either option lets us "off the hook" both for engaging our particular realities and circumstances and for imaginatively looking beyond them toward a broader "state" of human freedom. Either option robs us of our fundamental responsibilities to our fellow human beings.

The implications for how we live the politics of evidence are key here. As David Denter (2008) noted, a cornerstone of bad faith is the willingness to be "persuaded by weak evidence" (p. 84). That is, to arrive at conclusions *a priori* and then to search for evidence to support them. This means lowering one's "evidentiary standard[s]" in favor of one's preconceived dispositions (p. 84). In Sartre's own words, this approach to evidence is one that does not "*demand too much,*" that has a firm resolution "to count itself satisfied when it is barely persuaded, to force itself in decisions to adhere to uncertain truths" (Sartre quoted in Denter, 2008, p. 85). This impulse toward early and easy closure—to willfully "force" ourselves to accept evidence in service of positions we hold over time—is a cornerstone of "bad faith."

A new politics of evidence must avoid approaches to evidence that work either in the service of transcendent *a priori* ideals or brute empirical reductionism. We have seen both forms taken to extremes during the past decade or so, when evidence has been invalidated or discarded when not in service of neoconservative or progressive ideologies or when only the most reductive forms of "evidence-based" scholarship has been encouraged, funded, or published. In the end, we argue that focus groups have unique affordances that allow researchers to dwell in an evidentiary middle space, gathering empirical material while engaging in dialogues that help avoid premature consolidations of their understandings and explanations. This is a starting point for a new approach to evidence that respects the particularities and autonomies of evidence without assuming that evidence can ever speak for itself.

In short, the broader conception of focus group work we discuss in this chapter offers a useful intervention into debates around evidence. On the one hand, the collective nature of focus group discussions can help avoid the temptation to be too easily or quickly persuaded by weak evidence. Although groups can certainly reach consensus too easily—a point we will take up at the end of this chapter—focus group discussions allow us both to moderate and to calibrate this danger. On the other hand, a more expansive treatment of focus groups allows us to reflect continually on the particular limits of our methodological strategies. By destabilizing how we understand focus groups—locating them at contingent intersections of research, pedagogy,

and activism—we continually work against the tendency to reify our methodological strategies. Approaching focus groups in this way allows us to see our empirical material as always already refracted through multiple and often quite distinct prismatic faces, giving us more and more acute perspectives on the data we generate from them and how we interpret these data.

▣ FOCUS GROUPS AS PEDAGOGICAL, POLITICAL, AND RESEARCH PRACTICE

The Pedagogical Surface of Focus Group Work: Paulo Freire and Beyond

In this section, we highlight how focus groups have been important pedagogical sites and instruments throughout history. Acknowledging that there are a plethora of historical examples of pedagogically motivated focus groups—from dialogues in the 5th-century BC Athenian Square to early African American book clubs to union-sponsored "study circles" to university study groups—we analyze the focus groups cultivated by Freire in Brazil to illustrate their pedagogical dimensions. Through analyses of these exemplars, we show how collective critical literacy practices were used to address local politics and concerns for social justice. Among other things, we foreground the ways in which Freire worked *with* people and not *on* them, thus modeling an important praxis disposition for contemporary educators and qualitative researchers (e.g., Barbour & Kitzinger, 1999). Pedagogy, in Freire's work, was the dominant function of focus groups. However, inquiry always nourished pedagogy, and pedagogy was seen as useful only to the extent that it mobilized activist work.

Freire's work was intensely practical as well as deeply philosophical. His most famous book, *Pedagogy of the Oppressed* (1970/1993), can be read as equal parts social theory, philosophy, and pedagogical method. His claims about education are foundational, rooted both in his devout Christian beliefs and his commitment to Marxism. Throughout *Pedagogy of the Oppressed,* Freire argued that the goal of education is to begin to name the world, to recognize that we are all "subjects" of our own lives and narratives, not "objects" in the stories of others. We must acknowledge the ways in which we, as human beings, are fundamentally charged with producing and transforming reality together. He argued further that those who do not acknowledge this, those who want to control and oppress, are committing a kind of epistemic violence.

Freire often referred to oppressive situations as *limit situations*—situations that people cannot imagine themselves beyond. Limit situations naturalize people's sense of oppression, giving it a kind of obviousness and immutability. As particularly powerful ideological state apparatuses, schools, of course, play a big role in this naturalization process. Freire argued that most education is based on the *banking model* where educators see themselves as authoritative subjects, depositing knowledge into their students, their objects. This implies an Enlightenment worldview, where subject and object are *a priori* independent of each other, and where subjects are objectified and thus dehumanized. Among other things, the banking model of education implies that "the teacher teaches and the students are taught" and that "the teacher knows everything and the students know nothing" (1970/1993, p. 54). The model operates according to monologic rather than dialogic logics, serving the interests of the *status quo* and functioning to promote business as usual rather than social change. As problematic as it is, the banking model provides the epistemological foundation for most contemporary educational institutions and practices.

In the place of a banking model of education, Freire offered an alternative model that was based on the elicitation of words (and concomitant ideas) that are fundamentally important in the lives of the people for whom educational activities are designed. He called these words *generative words*. He spent long periods in communities trying to understand community members' interests, investments, and concerns to elicit comprehensive sets of generative words. These words were then used as starting points for literacy learning, and literacy learning was deployed in the service of social and political activism. More specifically, generative words were paired with pictures that represented them and then interrogated by people in the community who used the terms both for what they revealed and concealed with respect to the circulation of multiple forms of capital. Freire encouraged the people both to explore how the meanings and effects of these words functioned in their lives and to conduct research on how their meanings and effects did (or could) function in different ways in different social and political contexts. The primary goal of these activities was to help people feel in control of their words and to be able to use them to exercise power over the material and ideological conditions of their own lives. Thus, Freire's literacy programs were designed not so much to teach functional literacy but to raise people's critical consciousness (*conscientization*) and to encourage them to engage in "praxis" or critical reflection inextricably linked to political action in the real world. He was clear to underscore that praxis is never easy and always involves power struggles, often violent ones.

That Freire insisted that the unending process of emancipation must be a collective effort is far from trivial. Central to this process is a faith in the power of dialogue. Importantly, dialogue for Freire was defined as collective reflection and action. He believed that dialogue, fellowship, and solidarity are essential to human liberation and transformation. "We can legitimately say that in the process of oppression someone oppresses someone else; we cannot legitimately say that in the process of revolution, someone liberates someone else, nor yet that someone liberates himself, but rather that men in communion liberate each other" (1970/1993, p. 103). Only dialogue is capable of producing critical consciousness and praxis. Thus, all educational programs (and especially all language and literacy programs) must be dialogic. They must be spaces wherein "equally knowing subjects" (p. 31) engage in collective struggle in efforts to transform themselves and their worlds.

Within Freireian pedagogies, the development and use of generative words and phrases and the cultivation of *conscientization* are enacted in the context of locally situated "study circles" (or focus groups). The goal for the educator within these study circles is to engage, with people, in their lived realities, producing and transforming them. Again, for Freire, this kind of activity was part and parcel of literacy programs always already grounded in larger philosophical and social projects, those concerned with how people might more effectively "narrate" their own lives. In the context of these study circles, educators immerse themselves in the communities to which they have committed themselves. They try to enter into conversations, to elicit generative words and phrases together with their participants, and then to submit these words and phrases to intense reflection—presentation and re-presentation—to bring into relief lived contradictions that can then be acted upon.

To illustrate this kind of problem-posing education rooted in people's lived realities and contradictions, Freire created many research programs with participants, including one designed around the question of alcoholism. Because alcoholism was a serious problem in the city, a researcher showed an assembled group a photograph of a drunken man walking past three other men talking on the corner and asked them to talk about what was going on in the photograph. The group responded, in effect, by saying that the drunken man was a hard worker, the only hard worker in the group, and he was probably worried about his low wages and having to support his family. In their words, "he is a decent worker and a souse like us" (1970/1993, p. 99). The men in the study circle seemed to recognize themselves in this man, noting both that he was a "souse" but also situating his drinking in a politicized context. Alcoholism was "read" as a response to oppression and exploitation. The group went on to discuss these issues. This example of problem-posing pedagogy is quite different from (and we would argue much more effective than) a more didactic approach such as "character education," which would more likely involve "sermonizing" to people about their failings. Problem-posing education is proactive and designed to allow the people themselves to identify and generate solutions to the problems they face. The goal is to decode images and language in ways that eventually lead to questioning and transforming the material and social conditions of existence. Freire offered other examples as well, including showing people different (and contradictory) news stories covering the same event. In each case, the goal of problem-posing pedagogy is to help people understand the contradictions they live and to use these understandings to change their worlds.

In sum, focus groups have always been central to the kinds of radical pedagogies that have been advocated and fought for by such intellectual workers as Freire and his many followers (e.g., Henry Giroux, Joe Kincheloe, Jonathan Kozol, Peter McLaren). Organized around generative words and phrases and usually located within unofficial spaces, focus groups become sites of and for collective struggle and social transformation. As problem-posing formations, they operate locally to identify, interrogate, and change specific lived contradictions that have been rendered invisible by hegemonic power and knowledge regimes. Focus groups' operation also functions to reroute the circulation of power within hegemonic struggles and even to redefine what power is and how it works. Perhaps most importantly for our purposes here, the impulses that motivate focus groups in pedagogical domains or for pedagogical functions have important implications for reimagining and using focus groups as resources for constructing "effective histories" within qualitative research endeavors in the "seventh moment."

Although we have highlighted the pedagogical role of Freire's use of study circles, pedagogy is only a contingently situated function. That is, the three functions noted earlier can—and indeed, always have been—implicated in each other. Freire's study circles are excellent examples. Though designed as pedagogical activities, they were also and inextricably linked to activist-based and research-oriented activities. Recall that a starting point for Freire's work was to investigate and unearth the key generative phrases that indexed key social problems in the Brazilian communities in which Freire worked—clearly a form of research. And of course, his pedagogy was entirely entwined with an activist agenda—helping rural Brazilians learn to decode and then transform the world around them, especially its systemic oppressions.

In addition, the pedagogical uses of "focus groups" highlight several of the themes that run throughout this chapter, including those around the politics of evidence. The banking model of education, for example, draws broadly on what Sartre called "bad faith"—the recourse to static roles and stances defined *a priori*. As Freire repeatedly emphasized, such a model refuses our basic human freedoms, including the nature and effects of "authentic dialogue" which clearly is at the core of all of his work. When we adopt static roles, we refuse a transformative engagement with the world, opting instead for premature closure on key topics and issues. In contrast, the kind of pedagogically inflected focus groups advocated by Freire afford (even require) opportunities to engage with the world and with others in good faith.

The Political Surface of Focus Group Work: Consciousness-Raising Groups and Beyond

In this section, we offer descriptions and interpretations of focus groups in the service of radical political work designed within social justice agendas. In particular, we focus on how the consciousness-raising groups (CRGs) of second- and third-wave feminism have been deployed to mobilize empowerment agendas and to enact social change. This work complements and extends the explicitly pedagogical work we just discussed, especially in its active investment in community empowerment agendas and its commitment to praxis. It also provides important insights relevant for reimagining the possibilities of focus group activity within qualitative research endeavors. Where Freire's primary goal was to use literacy (albeit broadly defined) to mobilize oppressed groups to work against their oppression through praxis, the primary goal of the CRGs of second- and third-wave feminism was to build theory from the lived experiences of women that could contribute to their emancipation.

In our discussion of CRGs here, we draw heavily on Esther Madriz's retrospective analyses of second-wave feminist work as well as her own third-wave feminist empirical work. In both of these endeavors, Madriz focused on political (and politicized) uses of focus groups within qualitative inquiry, demonstrating that there has been a long history of deploying focus groups in consciousness-raising activities and for promoting social justice agendas within feminist and womanist traditions. Importantly, as forms of collective testimony, focus group participation has often been empowering for women, especially women of color (2000, p. 843). This is the case for several reasons. Focus groups decenter the authority of the researcher, allowing women safe spaces to talk about their own lives and struggles. These groups also allow women to connect with each other collectively, to share their own experiences, and to "reclaim their humanity" in a nurturing context (p. 843). Often, Madriz noted, women themselves take these groups over, reconceptualizing them in fundamental ways and with simple yet far-reaching political and practical consequences. In this regard, Madriz argued, "Focus groups can be an important element in the advancement of an agenda of social justice for women, because they can serve to expose and validate women's everyday experiences of subjugation and their individual and collective survival and resistance strategies" (p. 836). "Group interviews are particularly suited for uncovering women's daily experience through collective stories and resistance narratives that are filled with cultural symbols, words, signs, and ideological representations that reflect different dimensions of power and domination that frame women's quotidian experiences" (p. 839). As such, these groups constitute spaces for generating collective "testimonies," and these testimonies help both individual women and groups of women find or produce their own unique and powerful "voices."

As Madriz and others have noted, focus groups have multiple histories within feminist lines of thought and action. Soon after slavery ended in the United States, for example, churchwomen and teachers gathered to organize political work in the South (e.g., Gilkes, 1994). Similarly, early 20th–century "book clubs" were key sites for intellectual nourishment and political work (e.g., Gere, 1997). Mexican women have always gathered in kitchens and at family gatherings to commiserate and to work together to better their lives (e.g., Behar, 1993; Dill, 1994). And in 1927, Chinese women working in the San Francisco garment industry held focus group discussions to organize against their exploitation, which eventually led to a successful strike (e.g., Espiritu, 1997). Although we do not unpack these and other complex histories in this chapter, we do offer general accounts of the nature and function of focus groups within second- and third-wave feminism in the United States. These accounts pivot on the examination of several key original manifesto-like texts generated within the movement, which we offer as synecdoches of the contributions of a much richer, more complex, contradictory, and intellectually and politically "effective" set of histories.

Perhaps the most striking realization that emerges from examining some of the original texts of second-wave feminism are the explicitly self-conscious ways in which women used focus groups as "research" to build "theory" about women's everyday experiences and to deploy theory to enact political change. Interestingly but not surprisingly, this praxis-oriented work was dismissed by male radicals at the time as little more than "gossip" in the context of "coffee klatches." Ironically, this dismissal mirrors the ways in which qualitative inquiry is periodically dismissed for being "soft" or "subjective" or "nonscientific." Nevertheless, second-wave feminists persisted in building theory from the "standpoint" of women's lived experiences, and their efforts eventually became a powerful social force in the struggle for equal rights.

In many respects, the CRGs of second-wave feminism helped set the agenda for the next generation of feminist activism. As Hester Eisenstein (1984) noted, these groups helped bring personal issues in women's lives to the forefront of political discourse. Abortion, incest, sexual molestation, and domestic and physical abuse, for example, emerged from these groups as pressing social issues around which public policy and legislation had to be enacted. Importantly, these issues had previously been considered too personal and too intensely idiosyncratic to be taken seriously by men at the time, whether they were scholars, political activists, or politicians. By finding out which issues were most pressing in women's lives, CRGs were able to articulate what had previously been considered individual, psychological, and private matters to the agendas of local collectives and eventually to social and political agendas at regional and national levels.

Working within the movement(s) of third-wave feminism, Madriz used focus groups in powerful ways, some of which are evidenced in her 1997 book, *Nothing Bad Happens to Good Girls: Fear of Crime in Women's Lives*. In this book, Madriz discussed all the ways in which the fear of crime works to produce an insidious form of social control on women's lives. Fear of crime produces ideas and dispositions about what women "should"

and "should not" do in public to protect themselves, enabling debilitating ideas about what constitutes "good girls" versus "bad girls" and severely constraining the range of everyday practices available to women.

With respect to research methods, Madriz called attention to the fact that most research findings on women's fear of crime had previously been generated from large survey studies of both men and women. This approach, she argued, severely limits the range of thought and experience that participants are willing to share and thus leads both to inaccurate and partial accounts of the phenomenon. In other words, it is hard to get people—women in particular—to talk about such sensitive topics as their own fears of assault or rape in uninhibited and honest ways in the context of oral or written surveys completed alone or in relation to a single social scientist interviewer. This general problematic is further complicated by differences in power relations between researchers and research participants that obtain as a function of age, social class, occupation, language proficiency, race, and so on.

To work against the various alienating forces that seem inherent in survey research and to collect richer and more complex accounts of experience with greater verisimilitude, Madriz used focus groups, noting that these groups provided a context where women could support each other in discussing their experiences of and fears and concerns about crime. Indeed, these groups do mitigate the intimidation, fear, and suspicion with which many women approach the one-on-one interview. In the words of one of Madriz's participants, "When I am alone with an interviewer, I feel intimidated, scared. And if they call me over the telephone, I never answer their questions. How do I know what they really want or who they are?" (1997, p. 165). In contrast, focus groups afford women much safer and more supportive contexts within which to explore their lived experiences and the consequences of these experiences with other women who will understand what they are saying intellectually, emotionally, and viscerally.

This idea of safe and supportive spaces ushers in another important dimension of focus group work within third-wave feminist research, namely the importance of constituting groups in ways that mitigate alienation, create solidarity, and enhance community building. To achieve such ends, Madriz argued for the importance of creating homogenous groups in race, class, age, specific life experiences, and so on.

In relation to this point, CRGs of second-wave feminism suffered from essentializing tendencies (whether politically strategic or not) that ended up glossing the many, different, and even contradictory experiences of many women and groups of women under the singular sign of the homogenous "woman." More importantly and more problematically, this sign was constructed largely from the lived experiences of White middle-class women. Acknowledging the need to see and to celebrate more variability in this regard, third-wave feminist researchers refracted and multiplied the "standpoints" from which testimonies might flow and voices might be produced. Although many held onto the post-positivist ideal of "building theory" from lived experience, researchers such as Madriz pushed for theory that accounted more fully for the local, complex, and nuanced nature of lived experience, which is always already constructed within intersections of power relations produced by differences between and among multiple social categories (e.g., race, ethnicity, national origin, class, gender, age, sexual orientation, etc.). In the end, a primary goal of focus group activity within third-wave feminist research is not to offer prescriptive conclusions but to highlight the productive potentials (both oppressive and emancipatory) of particular social contexts (with their historically produced and durable power relations) within which such prescriptions typically unfold. In this regard, Madriz's work is a synecdoche for third-wave feminist work more broadly conceived—particularly work conducted by women of color such as Dorinne Kondo, Smadar Lavie, Ruth Behar, Aiwa Ong, and Lila Abu-Lughod.

The nature and functions of CRGs within second- and third-wave feminism offer many important insights into the potentials of focus group work. Building on Madriz's political reading of focus groups, and more specifically on the constructs of *testimony* and *voice,* we now highlight some of these potentials. One key purpose of focus groups within feminist work has been to elicit and validate collective testimonies and group resistance narratives. Such testimonies and narratives have been used by women (and could be used by any subjugated group) "to unveil specific and little-researched aspects of women's daily existences, their feelings, attitudes, hopes, and dreams" (Madriz, 2000, p. 836). Another key emphasis of focus groups within feminist and womanist traditions has been the discovery or production of *voice.* Because focus groups often result in the sharing of similar stories of everyday experience, struggle, rage, and the like, they often end up validating individual voices that had previously been constructed within and through mainstream discourses as idiosyncratic, selfish, and even evil. Because they foreground and exploit the power of testimony and voice, focus groups can become sites for overdetermining collective identity as strategic political practice—to create a critical mass of visible solidarity that seems a necessary first step toward social and political change.

Focus groups within feminist and womanist traditions have also mitigated the Western tendency to separate thinking and feeling, thus creating possibilities for reimagining knowledge as distributed, relational, embodied, and sensuous. Viewing knowledge in this light brings into relief the complexities and contradictions that are always involved in field work. It also brings into view the relations between power and knowledge and thus insists that qualitative research is always already political—implicated in social critique and social change.

Either from necessity or for strategic purposes, feminist work has always taken the constitutive power of *space* into

account. To further work against asymmetrical power relations and the processes of "othering," meetings are almost always held in safe spaces where women feel comfortable, important, and validated. This is a particularly important consideration when working with women who have much to lose from their participation such as undocumented immigrants, victims of abuse, or so-called deviant youth.

Finally, the break from second-wave to third-wave feminism both challenged the monolithic treatment of difference under the sign of "woman" that characterized much of second-wave thinking and highlighted the importance of creating focus groups that are relatively "homogeneous" with respect to life histories, perceived needs, desire, race, social class, region, age, and so forth because such groups are more likely to achieve the kind of solidarity and collective identity requisite for producing "effective histories" (Foucault, 1984). Although coalition building across more heterogeneous groups of women may be important in some instances, focused intellectual and political work is often most successful when enacted by people with similar needs, desires, struggles, and investments.

Such work does justice to the unique and often extremely complex vantage points of (often) subordinated groups. According to Patricia Hill Collins (1991), for example, African American women have often been expected to speak either from the vantage point of "woman" or "African American." Historically, these positions have spoken to the needs and concerns of "White women" and "Black men" respectively. For Collins, it is critical to do justice to and to acknowledge the ways Black women have their own needs, concerns, and experiences, which need to be addressed in their specificity. In this regard, she argued for what she calls "standpoint epistemologies" or knowledge frameworks generated from specific vantage points. The kind of focus group work envisioned and enacted by Madriz and other third-wave feminist researchers goes a long way toward imagining, enabling, and constructing such knowledge frameworks.

Yet, focus groups also allow us to challenge the limits of such knowledge claims. At their most reductive extreme, standpoint epistemologies essentialize the truth claims of particular subjects and create an imaginary kind of collective experience. Additionally, celebrating and generalizing situated individual experience can degenerate very quickly into a kind of uncritical relativism—an "anything goes" approach to knowledge that privileges individual biography and history in ways that reduce knowledge claims only to questions of power. Focus groups provide an important potential corrective here, allowing for both a collective articulation of particular subject positions, as discussed earlier, while opening them to contestation. In other words, focus groups allow people to speak in both collective and individual voices—creating space for traditionally marginalized groups to articulate their particular experience while allowing people to argue and disagree and ultimately produce what Michelle Fine calls "strong objectivity" (more on this later).

Politicized forms of focus group work are perhaps best evidenced today in various participatory action research (PAR) projects. In the United States, Fine has helped form various "research collectives" with youth at the City University of New York (CUNY) Graduate Center during the past several years, around several key contemporary issues (Cammarota & Fine, 2008). For example, Fine brought together multiethnic groups of suburban and urban high schools for *Echoes of Brown,* a study of the legacies of *Brown v. Board of Education.* Originally a study of the so-called achievement gap, the framework soon shifted, largely because of the focus group–like sessions that drew the participants together: "At our first session, youth from six suburban high schools and three urban high schools immediately challenged the frame of our research" (Torre & Fine, 2006, p. 273). After discussion, the framework changed from one of the "achievement gap"—a construct the youth felt put too much of the onus on themselves—to the "opportunity gap."

These research collectives create spaces for youth to challenge themselves and others in ongoing dialogue—a key affordance of focus group work, as noted earlier. "As we moved through our work, youth were able to better understand material, or to move away from experiences that were too uncomfortable, or to make connections across seemingly different positions" (Torre & Fine, 2006, p. 276). Ultimately, these youth were able to carry out both empirical projects around "push out rates" and disciplinary practices in schools as well as to produce powerful individual and collective testimonies about their own perspectives on and experiences of schooling, 50 years after *Brown.* Key here was the PAR "under construction" principle, the idea that opinions, ideas, beliefs, and practices are always expected to change and grow (Torre & Fine, 2006, p. 274). See Julio Cammarota and Michelle Fine (2008) for additional examples of this principle at work.

Fine and others have done their work largely in the United States, but Torre and Fine link PAR explicitly to histories of worldwide political struggle. "Based largely on the theory and practice of Latino activist scholars, PAR scholars draw from neo-Marxist, feminist, queer, and critical race theory ... to articulate methods and ethics that have local integrity and stretch topographically to site/cite global patters on domination and resistance" (2006, p. 271). They call particular attention to Colombian sociologist Fals Borda, whom many consider the "founder" of PAR. In so doing, Torre and Fine underscore the international scope of focus group work, particularly around their political uses.

Indeed, if we look elsewhere around the world, we quickly realize that the original impulses of feminist consciousness-raising work have been taken up and re-inflected in multiple ways for many purposes within various PAR initiatives. In Australia, for example, the "Deadly Maths Consortium" created by Thomas Cooper and Annette Barturo is a PAR project that

has been quite effective in improving mathematics learning among indigenous peoples (e.g., Cooper et al., 2008). In New Zealand, Russell Bishop has worked both tirelessly and very successfully to enact PAR initiatives designed to create positive change in classrooms, curricula, schools, and education policy based on caring relationships where power is shared among self-determining individuals within non-dominating relations of interdependence. Within his PAR focus groups, where cultural traditions are honored, learning is dialogic, and participants are connected to one another through the establishment of a common vision for what constitutes excellence in educational outcomes—all characteristics that are paramount to the educational performance of Maori students (e.g., Bishop et al., 2006). And indigenous scholars in the United States have used various forms of collaborative and participatory action research to work against the effects of colonization and to pursue their own liberation collectively and systematically (e.g., Grand, 2004; Wilson & Yellow Bird, 2005).

We have highlighted here the political function of focus groups—the ways these groups allow participants to coalesce around key issues, coproducing knowledges and strategies for transcending their circumstances. As noted throughout, however, other functions are embedded in these practices as well. For example, the theory building we emphasized relies on the kinds of critical, pedagogical practices associated with Freire and others. (Torre and Fine also explicitly link PAR to the work of Freire). The interactions between participants in consciousness-raising and other feminist groups, for example, are deeply pedagogical, as knowledge is co-created in situated and dialogic ways. Finally, these groups often engage in inquiry, especially inquiry focused on understanding asymmetrical power relations that render both women and particular groups of women as inferior in any number of ways. Although they have often been caricatured as uncritical support groups, consciousness-raising and other feminist groups have always been intensely concerned with producing new and useful knowledge about issues such as domestic violence or rape or workplace marginalization that were often broadly misconstrued within "official knowledges." And the PAR work we have mentioned continues to build on and extend the productive linkages among pedagogy, politics, and inquiry.

Fine's notion of "strong objectivity" is helpful for this discussion. According to Fine, we must work toward new forms of objectivity informed by the insights and advances of critical scholarship—particularly scholarship about the "situatedness" of all knowledge. For Fine, this reflection can be a source of better, more honest, and more "objective" accounts of our work. Drawing on the work of Sandra Harding, Fine argues that "strong objectivity" is "achieved when researchers work aggressively through their own positionality, values, and predispositions, gathering as much evidence as possible, from many distinct vantage points, all in an effort not to be guided, unwittingly, by predispositions and the pull of biography"

(Fine, 2006, p. 89). Such an approach helps researchers become more aware of potential "blind spots" that they may "import, wittingly and not, to their studies." Such work can be usefully done in "work groups," where empirical material can be discussed, pulled apart, and cleared of the "fog of unacknowledged subjectivities." These work groups seem to share the best impulses of focus groups, as participants forge new kinds of understandings and try to avoid premature closure.

In sum, the various practices and insights of second- and third-wave feminist and the various trajectories they have spawned (notably PAR) move us further down the road of imagining and enacting: (a) a commitment to morally sound, praxis-oriented research, (b) the strategic use of eclectic constellations of theories, methods, and research strategies, (c) the cultivation of dialogic relationships in the field, (d) the production of polyvocal non-representational texts, and (e) the conduct of mindful inquiry attuned to what is sacred in and about life and text.

▣ THE INQUIRY SURFACE OF FOCUS GROUP WORK: FROM POSITIVISM TO POSTSTRUCTURALISM AND BEYOND

Interest in focus groups in the social sciences has ebbed and flowed during the past 60 or so years. In many respects, the first really visible use of focus groups for conducting social science research may be traced back to the work of Paul Lazarsfeld and Robert Merton. Their focus group approach emerged in 1941, as the pair embarked on a government-sponsored project to assess media effects on attitudes toward America's involvement in World War II. Working within the Office of Radio Research at Columbia University, they recruited groups of people to listen and respond to radio programs designed to boost "morale" for the war effort (e.g., Merton, 1987, p. 552). Originally, the pair asked participants to push buttons to indicate their satisfaction or dissatisfaction with the content of the radio programs. Because the data yielded from this work could help them answer "what" questions but not "why" questions about participants' choices, they used focus groups as forums for getting participants to explain why they responded in the ways that they did. Importantly, Lazarsfeld and Merton's use of focus groups strategies for data collection always remained secondary to (and less legitimate than) the various quantitative strategies they also used. In other words, they used focus groups in exploratory ways to generate new questions that could be operationalized in quantitative work or simply to complement or annotate the findings yielded from their mostly large-scale survey studies.

In philosophy of science terms, the early use of focus groups as resources for conducting research was highly conservative in nature. This is not at all surprising when we consider that Lazarsfeld and Merton's work was funded by the military and included "interviewing groups of soldiers in Army camps about their responses to specific training films and so-called morale films" (Merton, 1987, p. 554). Their research also included many other media reception studies on topics such as why people made war bond pledges or how people responded to government-sponsored advertisements. The goal of most of this work was use knowledge about people's beliefs and decision-making processes to develop increasingly effective forms of propaganda—inquiry in the service of politics.

Although both their goals and the techniques merit harsh criticism (especially from progressive and radical camps), two key ideas from Lazarsfeld's and Merton's work have become central to the legacy of using focus groups within qualitative research: (a) capturing people's responses in real space and time in the context of face-to-face interactions and (b) strategically "focusing" interview prompts based on themes that are generated in these face-to-face interactions and that are considered particularly important to researchers.

The kind of focus group research conducted by scholars such as Lazarsfeld and Merton continued as a powerful force within corporate-sponsored market research, but it all but disappeared within the field of sociology in the middle-part of the 20th century, only to reemerge in the early 1980s in the form of "audience analysis" research. When it did reemerge, it was no longer wed to (or used in the service of) predominantly quantitative-oriented research, a fact that Merton bemoaned: "One gains the impression that focus-group research is being mercilessly misused as quick-and-easy claims for the validity of the research are not subjected to further, quantitative test . . ." (Merton, 1987, p. 557).

Criticisms such as these notwithstanding, audience analysis research was and is decidedly interpretive and increasingly dialogic and emancipatory. Its primary goal is to understand the complexities involved in how people understand and interpret media texts. Its methods are almost exclusively qualitative. In contrast to Lazarsfeld's and Merton's work, which focused on expressed content, audience analysis researchers typically focus on group dynamics, believing that the meanings constructed within groups of viewers are largely socially constructed. In a groundbreaking audience analysis study, for example, David Morley (1980) attempted to chart all the various ways in which groups of viewers from different social and economic classes responded to the popular television show, *Nationwide.* He conducted content analyses of many episodes of the show which included focus group interviews with people who had just watched these episodes, and he compared their responses with his analysis of the show's content. Working from within a social constructionist framework (e.g., Berger & Luckmann, 1966), Morley's use of focus groups was strategic: "The choice to work with groups rather than individuals . . . was made on the grounds that much individually based interview research is flawed by a focus on individuals as social atoms divorced from their social context" (1980, p. 97). For Morley and other scholars interested in audience reception practices, focus groups are invaluable because they afford insights into how meanings get constructed *in situ,* and thus allow researchers "to discover how interpretations were collectively constructed through talk and the interchange between respondents in the group situation—rather than to treat individuals as the autonomous repositories of a fixed set of individual 'opinions' isolated from their social context" (p. 97).

Janice Radway also used focus groups to great effect in her pioneering research on the reading practices of romance novel enthusiasts that resulted in her 1984 book, *Reading the Romance.* The research took place in and around a local bookstore, and Radway's participants included the storeowner and a group of 42 women who frequented the store and were regular romance readers. Like Morley, Radway developed a mixed-method research design that included text analysis and focus group interviews. Assisted by the bookstore owner (Dot), Radway was able to tap into the activity dynamics of existing networks of women who were avid romance novel readers. These women interacted frequently with Dot about newly published novels, and they interacted with each other as well. Radway simply "formalized" some of these ongoing social activities to generate a systematic and rich store of information about the social circumstances, specific reading practices, attitudes, reading preferences, and multiple and contradictory functions of romance reading among the women she studied. She took her cues about what books to read and what issues to focus discussions around from Dot and her other participants. She read all of the books that her participants read. She talked with many of them informally whenever they were at the bookstore together. And she conducted formal focus groups.

Among other things, Radway noted the importance of group dynamics in how different romance novels were interpreted and used. Even though the novels themselves were read privately, sharing their responses to the novels both in informal conversations and formal focus group discussions was very important to the women. Radway also came to understand how important belonging to a reading group was for mitigating the stigma often associated with the practice of reading romance novels: "Because I knew beforehand that many women are afraid to admit their preference for romantic novels for fear of being scorned as illiterate or immoral, I suspected that the strength of numbers might make my informants less reluctant about discussing their obsession" (1984, p. 252). Finally, the ways in which Radway positioned herself within the reading groups was crucial. She noted, for example, that when she was gently encouraging and when she backgrounded her own involvement, "the conversation flowed more naturally as the participants disagreed among themselves, contradicted one another, and delightedly discovered that they still agreed about many things" (p. 48).

All of the various strategies that Radway deployed helped to mobilize the collective energy of the group and to generate kinds and amounts of data that are often difficult, if not impossible, to generate through individual interviews and even observations. Additionally, these strategies—and participation in the focus groups themselves—helped to build a stronger and more effective collective with at least local political teeth. Radway concluded her book-length treatment of her romance novel project with a hopeful yet unfinalized call to praxis, noting, "It is absolutely essential that we who are committed to social change learn not to overlook this minimal but nonetheless legitimate form of protest.... and to learn how best to encourage it and bring it to fruition" (1984, p. 222)

If Radway began to outline the political, ethical, and praxis potentials of focus groups within qualitative inquiry, Patti Lather attempted to push the "limit conditions" of such work even further. In their book, *Troubling the Angels,* for example, Lather and Chris Smithies (1997) explored the lives, experiences, and narratives of 25 women living with HIV/AIDS. The book is filled with overlapping and contradictory voices that grew out of five years of focus group interviews conducted within different "support groups" in five major cities in Ohio. Lather and Smithies met and talked with their women participants at birthday parties and holiday get-togethers, hospital rooms and funerals, baby showers and picnics. Group dynamics among these women were unpredictable, emotionally charged (even ravaging), and changed constantly across the project. In what she calls a "postbook," Lather (2001, p. 210) acknowledged experiencing at least two "breakdowns" as she bore witness to the women's experiences and stories. Insofar as "breakdowns" are central to human understanding (Heidegger, 1927/1962), clearly these groups were surfaces of or for inquiry and pedagogy for researchers and research participants alike.

In both "strategic" and "found" ways, more organized occasions for "collecting data" constantly blurred into the "practices of everyday life" (deCerteau, 1984). Among other things, this social fact transformed the very nature of the focus groups these researchers conducted, rendering them more like rich and powerful conversations among people who cared deeply for each other. Yet Lather and Smithies were careful to work against the tendency to sentimentalize or romanticize their roles or their work by enacting what Lather (2001, p. 212) referred to as a "recalcitrant rhetoric" to counteract tendencies toward *verstehen* or simple empathy. Lather and Smithies tried to remain aware that the goals and rewards of their involvement in the groups were very different from the goals and rewards of their research participants. Their participants, for example, wanted to produce a "K-Mart" book, a collection of autobiographies or autoethnographies of "lived experience." Lather and Smithies were more interested in theorizing their participants' experiences and foregrounding the political (especially micropolitical) dimensions and effects of these experiences. According to Lather, these competing goals were constantly negotiated in focus groups. This pedagogic and political activity resulted in producing a book that embodies a productive, if uncomfortable/uncomforting tension between the two competing goals.

Although much of this book is devoted to troubling the waters of ethnographic representation, the experience of conducting fieldwork primarily through focus groups also troubled the waters of research practice. In this regard, Lather and Smithies integrated sociological, political, historical, therapeutic, and pedagogical practices and discourses in their work with the women they studied. In her postbook, for example, Lather claims to have looked constantly for "the breaks and jagged edges of methodological practices from which we might draw useful knowledge for shaping present practices of a feminist ethnography in excess of our codes but, still, always already forces already active in the present" (2001, pp. 200–201).

One of the most interesting sections of the book for our purposes in this chapter is one in which Lather and Smithies cultivate what they call a "methodology of getting lost":

> At some level, the book is about getting lost across the various layers and registers, about not finding one's way into making a sense that maps easily onto our usual ways of making sense. Here we all get lost: the women, the researchers, the readers, the angels, in order to open up present frames of knowing to the possibilities of thinking differently. (Lather & Smithies, 1997, p. 52)

Although these reflections refer to the book itself rather than the process of conducting the research that led to it, they apply equally well to working with research participants in the field in the sense that the reflections index the political, pedagogical, and ethical dimensions of all practices and all knowledge. For example, Lather and Smithies refused to position themselves as grand theorists and to interpret or explain the women's lives to them. Instead, they granted "weight to lived experience and practical consciousness by situating both researcher and researched as bearers of knowledge while simultaneously attending to the 'price' we pay for speaking out of discourses of truth, forms of rationality, effects of knowledge, and relations of power" (Lather 2001, p. 215). Through their tactical positioning, Lather and Smithies both challenged the researcher's right to know and interpret the experiences of others while they interrupted and got in the way of their participants' attempts to narrate their lives through a kind of innocent ethnographic realism where their voices simply spoke for themselves in some way (e.g., reading AIDS as the work of God's will). Additionally, Lather and Smithies acknowledged their impositions and admitted that a different kind of book—a K-Mart book—might have pleased their participants more. But such a book would have taken Lather and Smithies outside their own predilections and perhaps competencies as researchers, and is thus a task easier stated than accomplished. A K-Mart book has never been written, and Lather's (2007) follow-up to *Troubling the Angels* is even more theoretical than the original volume, highlighting the difficulties of accomplishing multiple interpretive tasks and embodying multiple voices simultaneously.

The various relational and rhetorical tactics enacted by Lather and Smithies bring to light the very complicated and sometimes troubling micro-politics that are part and parcel of research practice in the seventh, eighth, and ninth moments of qualitative inquiry (Denzin & Lincoln, 2005)—whether we are willing to see and enact these micro-politics in our own work. Lather and Smithies remind us constantly that there are no easy separations between "researcher" and "researched"; that research itself is always already relational, political, pedagogical, and ethical work; and that we have no alibis for thinking and acting otherwise lest we follow one or another path of "bad faith." There is no privileged place from which to experience and to report on experiences objectively. Only positions in dialogue. And even then, the issue of whose positions get foregrounded and whose get backgrounded remains a bit of a black hole that challenges the limits of dialogue, transparency, and self-reflexivity, which are inevitably and forever threatened by the possible (and usually invisible) reemergence of "bad faith." Notwithstanding our critique here, more than most other research of which we are aware, Lather and Smithies' work offers us ways to think about research that transcends and transforms the potentials of using focus groups for revisioning epistemology, interrogating the relative purchase of both lived experience and theory, reimagining ethics within research practice, and enacting field work in ways that are more attuned to its spiritual, even sacred, dimensions. And perhaps even more than this, the weaknesses and limitations of Lather and Smithies's work index the many experiential, epistemological, ethical, theoretical, and all-too-human challenges we still face in conducting seventh, eighth, and ninth moment qualitative inquiry (Denzin & Lincoln (2005). In this regard, focus groups surface the dialogic possibilities inherent in but often thwarted both in everyday social life and in research practice. What happens in focus groups can help researchers work against premature consolidation of their understandings and explanations, thus signaling the limits of reflexivity and the importance of intellectual/empirical modesty as forms of ethics and praxis. Such modesty can allow researchers to engage at least partially in "doubled practices" where we *both* listen to the attempts of others as they make sense of their lives *and* also resist the seductive qualities of "too easy" constructs such as "voice" or "faith" by recognizing and showing how experience itself (as well as our accounts of it) are always constituted within one or another "grand narrative" (Lather, 2001, p. 218). No less than life itself, doing social science has no guarantees.

▣ FOCUS GROUPS AS CONTINGENT, SYNERGISTIC ARTICULATIONS

We'd like now to highlight what we view as some of the more productive possibilities and some of the more serious dangers in conducting focus group research. Framing focus groups as contingent and often synergistic articulations of research, pedagogy, and politics seems to allow a clearer understanding of both.

First, a reconceptualized notion of focus groups can work to avoid premature closure at the institutional level. To return to the earlier discussion, if done in "bad faith," with an eye toward early closure, such groups scaffold "group think," registering and generating false notions of cohesion. For example, a certain narrow notion of focus group work has been used to construct notions of singular "publics" in popular political discussions. Recall, for example, the ways focus groups were used to gauge responses instantly during the 2008 presidential debates on CNN. This kind of brute empiricism—responses were gathered instantly and often immediately quantified on a modified sliding scale—was used to construct notions of what a supposed cross-section of Americans thought about the candidates. This impulse toward early closure and consolidation in focus group work can lead to the construction of construct *communities* that are facile and can be readily used toward any number of political ends. Understanding focus groups as contingent, synergistic articulations can help mitigate this problem.

Additionally, a more robust and complex version of focus groups can be used to resist local, institutional closure. Writing about the specificity of feminism (a movement we discussed above), Ellen Messer-Davidow (2002) writes, "To launch any feminist-studies project (whether it is to be located in the academy, the community, or both), individuals have to take collective action, and to take collective action they have to form a collective identity. Collective identity is not a conceptual product, like a job description, that individuals discuss, write up, and perform; it is the practical work of producing linkages among themselves while they negotiate their project's forms, objectives, and strategies" (p. 124). The practical work of producing and negotiating linkages was evidenced (in part) in the CRGs that were a part of second-wave feminism. As she shows, however, the academic incorporation of feminism was accomplished by a certain kind of institutional "formatting" that took up the smoother contours of the movement, producing an ever-proliferating meta-discourse with its own questions, rules, and feedback relays (p. 207). Keeping the multiple genealogies of focus groups in mind—including their co-implicated but distinct trajectories in research, pedagogy, and activism—might allow us to resist such institutional closure and reengage with a new and more productive version of the politics of evidence.

Second, although focus groups are becoming a larger part of public discourse, they are increasingly under new kinds of surveillance from within the academy (from institutional review boards [IRBs], funding agencies, publication venues, etc.). Ironically perhaps, the deployment of focus groups in the public sphere seems to have revivified and magnified the dangers of long-standing concerns about issues such as *anonymity*. Indeed, anonymity has been a long-standing cornerstone of academic research—in particular, the ability of research participants to choose not to be identified in research reports or to have sensitive,

personal information kept confidential. This kind of anonymity can usually be preserved in one-on-one interviews if the interviewer adheres to given protocols. Yet, in a group setting, trust and a commitment to confidentiality are more widely distributed. Indeed, the distribution of trust, as well as knowledge and experience, constitutes part of the power of focus groups. Yet, it also embodies their potential dangers. This paradox has surfaced repeatedly in our own work and the work of our students.

Greg Dimitriadis, for example, has served as major advisor and committee member for two doctoral students who used focus groups as a major tool for collecting empirical material. One student, Dr. Getnet Tizazu Fetene, studied attitudes toward HIV/AIDS among college-aged youth in Ethiopia. The second student, Dr. Touorouzou Some, investigated attitudes toward university cost-sharing among college-aged youth in Burkina Faso. Both projects were flagged by the IRB as potentially "high risk" endeavors, even though both students had been proactive in arguing that these very commonly discussed topics were not likely to cause undue stress among participants. The question of anonymity was central among the IRB's concerns. Originally, some suggested that one-on-one interviews should be conducted instead of focus groups. The logic was sound. Even if the researcher told members of the group that information shared during these sessions must remain confidential, they could not guarantee it. So why take the risk?

What was really at issue became clearer when Dimitriadis talked with the IRB. Unfamiliar with focus group work, some members thought focus groups were simply a way to conduct multiple interviews simultaneously—almost as a time-saving measure. Dimitriadis had to provide information about focus groups—including the fact that participants often feel more comfortable in groups, thereby diminishing the possibility of personal vulnerability and risk. He also had to make the point that some kinds of information are more likely to emerge from focus group discussions (as opposed to one-on-one interviews) including (and perhaps especially) information about sex and sexuality. Although the question of anonymity remained a thorny one to the IRB, it was reconstructed (and thus resolved) in the following way: The potential risk of participants breaking anonymity in focus groups was outweighed by the potential information that could only be gleaned from group interviews. Importantly, this demanded educating powerful administrators not entirely familiar with this methodological tool about its particular benefits—an interesting twist on the pedagogic function we have addressed in this chapter.

The point is worth underscoring. As is well documented by now (see this volume), IRBs can be problematic institutions for qualitative researchers because they are often targeted toward positivist and medical kinds of research. It is important to remember, however, that these boards are always balancing potential "risk/reward" ratios when approving this work. For example, such ratios are an explicit concern when, say, approaching clinical tests for a new drug. Qualitative researchers can often appropriate this language when "road bumps" such as this one around anonymity emerge. Such boards often have to be educated (quite literally) about different data collection techniques and have the potential benefits both to the study and the field explained. These can then be explicitly balanced against potential risks such as this one around anonymity. Such an approach, we think, is more useful than simply stating (again) that these boards "don't get it." It is often a question of moving across different language fields.

The rewards for this project on HIV/AIDS in Ethiopia were profound and highlight many of the issues thread throughout. To begin with, Fetene was *not* able to secure one-on-one interviews with his participants—a technique he hoped to use to complement his focus groups. As he reported, many of the students were reluctant (even unwilling) to discuss such sensitive material with a partial stranger in a close, interpersonal setting. This response defied the position of the IRB, which assumed one-on-one interviews would offer a more comfortable, less stressful interviewing context. Perhaps more importantly, Fetene's the empirical material yielded from the focus groups proved highly revealing, in ways that personally challenged him and his co-researcher Dr. Muluemebet Zenebe (who conducted the groups with young women). In particular, many of these young people—specifically, the young women—talked very openly about their romantic relationships, astonishing both researchers and forcing them to reevaluate their own beliefs and understandings about the culture in which they both grew up and lived. From a traditionally conservative culture, Fetene was initially surprised by how openly these young Ethiopian men talked about sex. His female co-researcher initially assumed these men were just trying to "show off" for each other—until she conducted focus groups with a group of young women. To her self-proclaimed astonishment, these discussions largely mirrored the discussions among the young men. This "evidence" forced both researchers to reconsider and revise long-held assumptions about their own culture and society.

Recall the discussions of "bad faith" and the politics of evidence we presented earlier in the chapter. Focus groups can be powerful sites or tools for interrupting and taking us beyond our own calcified positions—the temptation to be persuaded by weak evidence or *a priori* beliefs. In this case, focus group evidence proved transformative in moving both researchers past their own beliefs about Ethiopian youth and sexuality and the culture in which they were both so deeply immersed. Moreover, these discussions challenged the ways in which the global discourse around HIV/AIDS is presently constituting itself. Although much of this discourse has assumed that a lack of "knowledge" about safe sex is a key issue, these focus groups disrupted such assumptions, with young people stating again and again that they knew about the importance (for example) of condom use. These youth raised other key issues, including changing mores around sex. Again, and importantly, the focus group interviewing context was central to the emergence of

these knowledges and insights and for disrupting the kind of "commonsense" closure that often surrounds discussions of so-called sensitive topics. Also worth noting here is that a project that began focused primarily on *inquiry* quickly took on communitarian, political, and pedagogical valences for researchers and research participants alike.

Finally, we would like to say a bit more about the synergistic potential of focus group work. This potential was realized in highly visible ways in a project conducted by one of George Kamberelis's students, Graciana Astazarian. The project focused on the experiences and needs of recent Mexican American immigrants to a small community in the midwestern United States. Largely because it had an abundance of good manufacturing jobs, Mexican American immigrants were moving to this community in large numbers. However, because the influx of new immigrants was happening so fast, little or no infrastructure for supporting the health, education, transportation, employment, linguistic, and cultural needs of these people existed. Astazarian's project focused on understanding these needs. She conducted focus groups with small groups of women in a variety of contexts—homes, community centers, churches, and even the waiting line at the Latino Coalition office. Several issues emerged from the focus groups as particularly important problems for the women interviewed: learning English, transportation, language barriers in school and health care settings, and discrimination at work. We focus on the discussions of two of these problems—English language learning and transportation—for the ways in which they draw out the synergistic potential afforded by focus group work. Indeed, discussions about these (and other) problems almost always surfaced a good deal of complexity, nuance, and contradiction.

Many factors contributed to problems with English language learning among these new immigrants—from biological ones (learning disabilities) to life historical ones (bad school experiences) to social ones (linguistic and cultural imperialism) to educational ones (poor instructors and instructional materials) to economic ones (the cost of instruction). Discussing these various factors created tremendous solidarity among the participants. These discussions also surfaced some of the hidden, macro-level factors such as linguistic and cultural imperialism, which both surprised and gratified group members.

Astazarian's focus groups generated considerable political synergy around the issue of English language learning. Several participants created a manifesto of English language learning needs and desires that included: (a) having more opportunities for adult education; (b) having more financial aid for English classes; (c) having ESL classes at night and not just during the day; (d) having ESL classes that were dynamic, fun and encouraged oral communication; (e) creating different levels of classes to meet the different levels and needs of students; (f) having amnesty for undocumented immigrants; and (g) having more bilingual workers at most community agencies.

This group of participants also lobbied the Latino Coalition and other advocacy agencies to push for more translators in hospitals and clinics. Group members composed a letter to the editor of the local newspaper about the need for translation services in these sites. And they assembled and distributed an information sheet about the documentation required to open a bank account at most local banks. Within two years, and largely motivated by this focus group work, many positive changes occurred in the community. The nature and scope of ESL classes available in the community changed considerably and along the lines suggested. The Community Health Clinic hired three bilingual staff members. The Industrial Federal Credit Union hired five bilingual tellers.

Transportation to work, school, health care facilities, and so on was another key problem faced by these new immigrants. Public transportation was both scarce and unreliable. Underground car services run by entrepreneurial Mexican American men were expensive and unreliable. In discussing these problems, participants surfaced even more serious problems such as sexism and economic exploitation within the Mexican American community. They went on to talk about the fracturing effects these forces had on community solidarity. Within weeks of the initial discussion of this transportation problem, a group of women created a co-op ride sharing system. This system grew and became more efficient over time. Among other things, the co-op forced mercenary drivers to lower their fees.

Participants also instigated discussions of sexism that emerged in the initial focus group sessions, especially the oppression resulting because many husbands discouraged (even prevented) their wives from getting licenses. In about a year, women who had licenses began to teach women who didn't have licenses how to drive and arranged for them to take their driving tests.

As these examples show, a contingent and unpredictable potential for synergy exists within focus group work. This synergy often constitutes "breakdowns" (Heidegger, 1927/1962) that disclose complexities, nuances, and contradictions embodied in "lived experience." It often indexes social and economic forces such as linguistic imperialism, economic exploitation, and sexism that often get glossed or explained away by one or another cultural logic—thus introducing new versions of the politics of evidence that mitigate the effects of "bad faith" (Sartre, 2001) and allow neither researchers nor research participants any alibis for their actions. These positive potentials are, of course, accompanied by attendant dangers. Given how power operates within relations of dominance and oppression, naming and talking back to imperialism or sexism can have serious, even devastating, consequences. Indeed, the participants in this project faced such consequences in their marriages, in the Mexican American community, in the workplace, and in the community at large. Finally, and notwithstanding such dangers, synergy can motivate the kinds political and

pedagogic activity required for (a) connecting social science and social purpose; (b) encouraging and celebrating local, indigenous social sciences; and (c) decolonizing the academy within the contemporary imperatives of qualitative inquiry (Denzin & Lincoln, 2005, pp. 117–124).

▣ FINAL COMMENTS

Focus group research is a key site where pedagogy, politics, and inquiry intersect and interanimate each other. Because of their synergistic potentials, focus groups often produce data that are seldom produced through individual interviewing and observation and thus yield particularly powerful knowledges and insights. Specifically, the synergy and dynamism generated within homogeneous collectives often reveal unarticulated norms and normative assumptions. They also take the interpretive process beyond the bounds of individual memory and expression to mine historically sedimented collective memories and desires.

The unique potentials of focus group research are most fully realized when we acknowledge and exploit their multifunctionality. To return to the metaphor with which we opened this chapter, this kind of multifunctionality can best be imagined as a prism. That is, the three sides of this prism—pedagogy, politics, and research—are always implicated in and productive of each other. One surface of the prism may be most visible at any particular moment, but the others are always also visible, refracting what is brought to light in multiple and complex directions. How we choose to "hold" this imaginary prism at any moment in time has important consequences both for what we see and what we do with what we see.

In addition to enhancing the kinds and amounts of empirical material yielded from a qualitative study, focus group work also foregrounds the importance both of content and of expression because it capitalizes on the richness and complexity of group dynamics. Acting somewhat like magnifying glasses, focus groups induce social interactions akin to those that occur in everyday life but with greater intensity. More than observations and individual interviews, focus groups afford researchers access to social-interactional dynamics that produce particular memories, positions, ideologies, practices, and desires among specific groups of people. Focus groups also allow the researcher to see the complex ways people position themselves in relation to each other as they process questions, issues, and topics in focused ways. These dynamics, themselves, become relevant "units of analyses" for study.

If taken seriously, these dynamics help us to avoid premature closure on our understandings of the particular issues and topics we explore. They challenge us to avoid being persuaded too easily and too early by weak evidence. Although certainly not a simple solution to the complexities involved in the new politics of evidence we face, the way we have reconceptualized focus groups discourages us from ever being too comfortable in our understandings and insights. Understanding focus groups and their dynamics through multiple angles of vision forces us always to see the world in new and unexpected ways. We remain less tempted by the lures of simple facticity or transcendence—pulled always to see our empirical material in new and more rigorous ways.

In addition, focus groups function to decenter the role of the researcher. As such, focus groups can facilitate the democratization of the research process, allowing participants more ownership over it, and promoting more dialogic interactions and the joint construction of more polyvocal texts. These social facts were brought into relief by the feminist work conducted by Madriz, Radway, and Lather and Smithies that we discussed earlier. Although also functioning as sites for consolidating collective identities and enacting political work, focus groups allow for the proliferation of multiple meanings and perspectives, as well as interactions between and among them. Because focus groups get multiple perspectives on the table, they help researchers and research participants alike realize that both the interpretations of individuals and the norms and rules of groups are inherently situated, provisional, contingent, unstable, and thus changeable. In this regard, focus groups help us move toward constructing a "methodology of getting lost" and toward enacting "doubled practices" (Lather, 2001), which seem necessary first steps toward conducting eighth and ninth moment qualitative research.

Echoing Antonio Gramsci, we conclude that the "we" enabled by focus groups has "no guarantees." With no guarantees, focus groups must operate according to a hermeneutics of vulnerability (Clifford, 1988). Clifford developed the construct of a "hermeneutics of vulnerability" to discuss the constitutive effects of relationships between researchers and research participants on research practice and research findings. A hermeneutics of vulnerability foregrounds the ruptures of fieldwork, the multiple and contradictory positionings of all participants, the imperfect control of the researcher, and the partial and perspectival nature of all knowledge. Among the primary tactics for achieving a hermeneutics of vulnerability, according to Clifford, is the tactic of self-reflexivity, which may be understood in at least two senses. In the first sense, self-reflexivity involves making transparent the rhetorical and poetic work of the researcher in representing the object of her or his study. In the second (and, we think, more important) sense, self-reflexivity refers to the efforts of researchers and research participants to engage in acts of self-defamiliarization in relation to each other. In this regard, Elspeth Probyn (1993) discussed how the fieldwork experience can engender a virtual transformation of the identities of both researchers and research participants even as they are paradoxically engaged in the practice of consolidating them. This is important theoretically because it allows for the possibility of constructing a mutual ground between researchers and research participants even while recognizing that the ground is unstable and fragile. Self-reflexivity in this second sense is also important because it encourages reflection on interpretive research as the dual practice of knowledge gathering and self-transformation through self-reflection and mutual reflection with the other. Finally, as Lather (2001, 2007) has shown, even self-reflexivity has serious limits with respect to working against the triple crisis of representation, legitimation, and praxis. Indeterminacies always remain. Allowing ourselves to dwell in (and even celebrate) these indeterminancies—the truly prismatic nature of things worth studying—may be the best way to move down the roads of qualitative research practice and theory building at this particular historical juncture.

▣ REFERENCES

Barbour, R. (2008). *Doing focus groups.* Thousand Oaks, CA: Sage.

Barbour, R., & Kitsinger, J. (1999). *Developing focus group research.* Thousand Oaks, CA: Sage.

Behar, R. (1993). *Translated woman: Crossing the border with Esperanza's story.* Boston: Beacon Press.

Berger, P., & Luckmann, T. (1966). *The social construction of reality.* New York: Doubleday.

Bishop, R., Berryman, M., Cavanagh, T., Teddy, L., & Clapham, S. (2006). *Te Kotahitanga Phase 3 Whakawhanaungatanga: Establishing a culturally responsive pedagogy of relations in mainstream secondary school classrooms.* Wellington: New Zealand Ministry of Education.

Bloor, M., Frankland, J., Thomas, M., & Robson, K. (2001). *Focus groups in social research.* Thousand Oaks, CA: Sage.

Bourdieu, P., & Wacquant, L. J. D. (1992). *An invitation to reflexive sociology.* Chicago: University of Chicago Press.

Cammarota, J., & Fine, M. (Eds.). (2008). *Revolutionizing education.* New York: Routledge.

Chomsky, N. (2002). Activism, anarchy, and power: Noam Chomsky interviewed by Harry Kreisler. *Conversations with history,* March 22, 2002. Retrieved from http://www.chomsky.info/interviews/20020322.htm

Clifford, J. (1988). *The predicament of culture.* Cambridge, MA: Harvard University Press.

Collins, P. H. (1991). *Black feminist thought: Knowledge, consciousness and the politics of empowerment.* New York: Routledge.

Cooper, T. J., Baturo, A. R., Duus, E. A., & Moore, K. M. (2008). Indigenous vocational students, culturally effective communities of practice and mathematics understanding. In O. Figueras, J. L. Cortina, S. Alatorre, T. Rojano, & A. Sepulveda (Eds.), *Proceedings of the 32nd Annual Conference of the International Group for the Psychology of Mathematics Education* (pp. 378–384). Morelia, Mexico: PME.

deCerteau, M. (1984). *The practice of everyday life* (S. F. Rendall, Trans.). Berkeley: University of California Press.

Denter, D. (2008). *Sartre explained.* Chicago: Open Court.

Denzin, N. K., & Lincoln, Y. S. (2005). Epilogue: The eighth and ninth moments—qualitative research in/and the fractured future.

In N. K. Denzin & Y. S. Lincoln (Eds.), *The SAGE handbook of qualitative research* (3rd ed., pp. 1115–1126). Thousand Oaks, CA: Sage.

Dill, B. T. (1994). Fictive kin, paper sons, and compadrazgo: Women of color and the struggle for family survival. In M. B. Zinn & B. T. Dill (Eds.), *Women of color in U.S. society* (pp. 149–169). Philadelphia: Temple University Press.

Eisenstein, H. (1984). *Contemporary feminist thought.* New York: Macmillan.

Espiritu, Y. L. (1997). *Asian women and men: Labor, laws, and love.* Thousand Oaks, CA: Sage.

Fine, M. (2006). Bearing witness: Methods for researching oppression and resistance. *Social Justice Research, 19*(1), 83–108.

Foucault, M. (1984). Nietzsche, genealogy, and history. In P. Rabinow (Ed.), *The Foucault reader* (pp. 76–100). New York: Pantheon Books.

Freire, P. (1993). *Pedagogy of the oppressed.* New York: Continuum. (Original work published in 1970)

Gere, A. R. (1997). *Writing groups: History, theory, and implications.* Carbondale: Southern Illinois University Press.

Gilkes, C. T. (1994). "If it wasn't for the women . . .": African American women, community work, and social change. In M. B. Zinn & B. T. Dill (Eds.), *Women of color in U.S. society* (pp. 229–246). Philadelphia: Temple University Press.

Grand, S. (2004). *Red pedagogy: Native America social and political thought.* Lanham, MD: Rowman & Littlefield.

Heidegger, M. (1962). *Being and time* (J. Macquarrie & E. Robinson, Trans.). San Francisco: HarperSanFrancisco. (Original work published in 1927)

Hesse-Biber, S., & Leavy, P. (Eds.). (2006). *Handbook of emergent methods.* Thousand Oaks, CA: Sage.

Jakobson, R. (1960). Concluding statement: Linguistics and poetics. In T. A. Sebeok (Ed.), *Style in language.* Cambridge: MIT Press.

Kamberelis, G., & Dimitriadis, G. (2005). *On qualitative inquiry: Approaches to language and literacy research.* New York: Teachers College Press.

Kreisler, H. (2002). Activism, anarchism, and power: Conversation with Noam Chomsky, linguist and political activist. University of California at Berkeley, Institute of International Studies, Conversation with History Series. Retrieved August 1, 2009, from http://globetrotter.berkeley.edu/people2/Chomsky/chomsky-con0.html

Kreuger, R. A., & Casey, M. A. (2008). *Focus groups: A practical guide for applied research* (4th ed.). Thousand Oaks, CA: Sage.

Lather, P. (2001). Postbook: Working the ruins of feminist ethnography. *Signs: Journal of Women in Culture and Society, 27*(1), 199–227.

Lather, P. (2007). *Getting lost: Feminist efforts toward a double(d) science.* Albany: SUNY Press.

Lather, P., & Smithies, C. (1997). *Troubling the angels: Women living with HIV/AIDS.* Boulder, CO: Westview Press.

Madison, D., & Hamera, J. (2006). Introduction. In D. Madison & J. Hamera (Eds.), *The SAGE handbook of performance studies* (pp. xi–xxv). Thousand Oaks, CA: Sage.

Madriz, E. (1997). *Nothing bad happens to good girls: Fear of crime in women's lives.* Berkeley: University of California Press.

Madriz, E. (2000). Focus groups in feminist research. In N. K. Denzin & Y. S. Lincoln (Eds.), *Handbook of qualitative research* (2nd ed., pp. 835–850). Thousand Oaks, CA: Sage.

Merton, R. (1987). The focused group interview and focus groups: Continuities and discontinuities. *Public Opinion Quarterly, 51,* 550–566.

Messer-Davidow, E. (2002). *Disciplining feminism.* Durham, NC: Duke University Press.

Morgan, D. L. (1998). *The focus group guidebook.* Thousand Oaks, CA: Sage.

Morley, D. (1980). *The* Nationwide *audience.* London: British Film Institute.

Probyn, E. (1993). *Sexing the self: Gendered positions in cultural studies.* London: Routledge.

Radway, J. (1984). *Reading the romance: Women, patriarchy, and popular literature.* Durham, NC: University of North Carolina Press.

Richardson, L. (2000). Writing: A method of inquiry. In N. K. Denzin & Y. S. Lincoln (Eds.), *Handbook of qualitative research* (2nd ed., pp. 923–948). Thousand Oaks, CA: Sage.

Sartre, J. (2001). *Basic writings.* New York: Routledge.

Schensul, J. J., LeCompte, M. D., Nastasi, B. K., & Borgatti, S. P. (1999). *Enhanced ethnographic methods: Audiovisual techniques, focused group interviews, and elicitation techniques.* Walnut Creek, CA: AltaMira Press.

Solomon, R. (2006). *Dark feelings, grim thoughts: Experience and reflection in Camus and Sartre.* New York: Oxford University Press.

Stewart, D. W., Shamdasani, P. N., & Rook, D. (2006). *Focus groups: Theory and practice.* Thousand Oaks, CA: Sage.

Torre, M., & Fine, M. (2006). Researching and resisting. In S. Ginwright, P. Nogurea, & J. Cammarota (Eds.), *Beyond resistance* (pp. 269–283). New York: Routledge.

Wilson, W. A., & Yellow Bird, M. (2005). *For indigenous eyes only: A decolonization handbook.* Santa Fe, NM: School of American Research Press.

THE ART AND PRACTICES OF INTERPRETATION, EVALUATION, AND REPRESENTATION

In conventional terms, Part V of the *Handbook* signals the terminal phase of qualitative inquiry. The researcher and evaluator now assess, analyze, and interpret the empirical materials that have been collected. This process, conventionally conceived, implements a set of analytic procedures that produces interpretations that are then integrated into a theory, or put forward as a set of policy recommendations. The resulting interpretations are assessed in terms of a set of criteria, from the positivist or postpositivist traditions, including validity, reliability, and objectivity. Those interpretations that stand up to scrutiny are put forward as the findings of the research.

The contributors to Part V explore the art, practices, and politics of interpretation and evaluation, and representation. In so doing, they return to the themes of Part I—asking, that is, *how the discourses of qualitative research can be used to help create and imagine a free democratic society*. In returning to this question, it is understood that the processes of analysis, evaluation, and interpretation are neither terminal nor mechanical. They are like a dance—to invoke the metaphor used by Valerie Janesick (2010)—a dance informed at every step of the way by a commitment to this civic agenda. The processes that define the practices of interpretation and representation are always ongoing, emergent, unpredictable, and unfinished. They are always embedded in an ongoing historical and political context. As argued throughout this volume, in the United States, neoconservative discourse in the educational arena (No Child Left Behind, National Research Council) privileges experimental criteria in the funding, implementation, and evaluation of scientific inquiry. Many of the authors in this volume observe that this creates a chilling climate for qualitative inquiry.

We begin by assessing a number of criteria that have been traditionally (as well as recently) used to judge the adequacy of qualitative research. These criteria flow from the major paradigms now operating in this field, as well from standards set by governmental agencies.

EVIDENCE, CRITERIA, POLICY, AND POLITICS

Torrance (Chapter 34, this volume) reviews the debates surrounding qualitative research and social policy, especially in the United Kingdom, the United States, Australia, and New Zealand. Often this discourse has marginalized qualitative inquiry, claiming that it is of low quality, and holding up experimental design as the preferred scientific protocol. There is a worldwide movement to reassert empiricist, technicist approaches to the production of evidence for policy-making purposes. This move undercuts previous policies, which endorsed a hands-off approach to the public funding of university-based science. Today, in too many places, social science is expected to serve short-term government policy, economic development, and educational achievement.

Torrance reviews the major criticism of the experimental, RCT (randomized controlled trial) model. Too often there are not clear-cut effects that can be connected to the experimental treatment condition. In response, some investigators have moved to mixed method designs, while others resort to meta-reviews, arguing that evidence to inform policy should be accumulated across studies. Meta-reviews raise the issue of criteria of quality, and there are competing quality appraisal checklists that can be deployed.

In response to these governmental initiatives, various professional associations have developed their own criteria (see Denzin, Chapter 39, this volume, for a review). These discussions of quality revolve around issues of engagement, deliberation, ethics, and desires to reconnect critical inquiry to democratic processes.

We live in an age of relativism. In the social sciences today, there is no longer a God's-eye view that guarantees absolute methodological certainty; to assert such is to court embarrassment. Indeed, there is considerable debate over what constitutes good interpretation in qualitative research. Nonetheless, there seems to be an emerging consensus that all inquiry reflects the standpoint of the inquirer, all observation is theory-laden, and there is no possibility of theory-free knowledge. We can no longer think of ourselves as neutral spectators of the social world.

Consequently, few speak in foundational terms. Before the assault of methodological conservativism, relativists would calmly assert that no method is a neutral tool of inquiry, and hence the notion of procedural objectivity could not be sustained. Anti-foundationalists thought the days of naïve realism and naïve positivism were over. In their place stand critical and historical realism, and various versions of relativism. The criteria for evaluating research have become relative, moral, and political.

There are three basic positions on the issue of evaluative criteria: foundational, quasi-foundational, and non-foundational. There are still those who think in terms of a *foundational* epistemology. They would apply the same criteria to qualitative research as are employed in quantitative inquiry, contending that there is nothing special about qualitative research that demands a special set of evaluative criteria. As indicated in our introduction to Part II, the positivist and postpositivist paradigms apply four standard criteria to disciplined inquiry: internal validity, external validity, reliability, and objectivity. The use of these criteria, or their variants, is consistent with the foundational position.

In contrast, *quasi-foundationalists* approach the criteria issue from the standpoint of a non-naïve, neo- or subtle realism. They contend that the discussion of criteria must take place within the context of an ontological neorealism and a constructivist epistemology. They believe in a real world that is independent of our fallible knowledge of it. Their constructivism commits them to the position that there can be no theory-free knowledge. Proponents of the quasi-foundational position argue that a set of criteria unique to qualitative research needs to be developed. Hammersley (1992, p. 64; also 1995, p. 18; 2008; see also Wolcott, 1999, p. 194) is a leading proponent of this position. He wants to maintain the correspondence theory of truth, while suggesting that researchers assess a work in terms of its ability to (1) generate generic/formal theory; (2) be empirically grounded and scientifically credible; (3) produce findings that can be generalized, or transferred to other settings; and (4) be internally reflexive in terms of taking account of the effects of the researcher and the research strategy on the findings that have been produced.

Hammersley (2008) reduces his criteria to three essential terms: plausibility (is a claim plausible), credibility (is the claim based on credible evidence), and relevance (what is the claim's relevance for knowledge about the world). Of course, these terms require social judgments. They cannot be assessed in terms of any set of external or foundational criteria. Their meanings are arrived at through consensus and discussion in the scientific community. Within Hammersley's model, there is no satisfactory method for resolving this issue of how to evaluate an empirical claim.

For the non-foundationalists, relativism is not an issue. They accept the argument that there is no theory-free knowledge. Relativism, or uncertainty, is the inevitable consequence of the fact that as human beings we have finite knowledge of ourselves, and the world we live in. Non-foundationalists contend that the injunction to pursue knowledge cannot be given epistemologically; rather, the injunction is moral and political.

Accordingly, the criteria for evaluating qualitative work are also moral and fitted to the pragmatic, ethical, and political contingencies of concrete situations. Good or bad inquiry in any given context is assessed in terms of criteria that flow from a feminist, communitarian moral ethic of empowerment, community, and moral solidarity. Returning to Clifford Christians (Chapter 4), this moral ethic calls for research rooted in the concepts of care, shared governance, neighborliness, love, and kindness. Further, this work should provide the foundations for social criticism and social action.

In an ideal world, the anti- or non-foundational narrative would be uncontested. But such is not the case. We continue to live in dark days.

◼ INTERPRETIVE ADEQUACY IN QUALITATIVE RESEARCH

Altheide and Johnson (Chapter 35) call their approach to interpretive adequacy "analytical realism"; that is, there is a real world that we interact with. We create meaning in this world through interaction. They discuss how analytical realism can be used to enhance the credibility, relevance, and importance of qualitative methods and interpretive materials. All knowledge is contextual and partial. Evidence is a part of a communication process. This interactional process "symbolically joins an actor, an audience, a point of view, and . . . claims about the relations between two or more phenomena." This view of evidence-as-process is termed the "evidentiary narrative." It is shaped "by symbolic filters, including distinct epistemic communities, or collective meanings, standards, and criteria that govern sanctioned action."

They discuss how this view of evidence has been framed in clinical and policy studies, in action research, and in performance and autoethnography. Various forms of validity—successor, catalytic, interrogated, transgressive, imperial, ironic, situated—are discussed. They offer a hyphenated model—validity-as-culture, -as-ideology, -as-gender, -as-language, -as-relevance, -as-standards, -as-reflexive-accounting, and -as-marketable-legitimacy.

Their model of evidentiary narrative shows how evidence is not about facts, but about narrative. Their ethnographic ethic enacts this model of evidence, connecting it to relationships between the observer, the observed, the setting, the reader, and the written text. Their goal is not to offer a checklist for assessing quality or validity. They open their text with a quote from the artist Paul Klee—"A line is a dot that went for a walk." Their task, they contend, is "to continue pushing the line in new directions to illuminate our humanity and our communicative worlds."

▣ ANALYSIS AND REPRESENTATION

Laura Ellingson (Chapter 36) offers a continuum—right, left, middle—approach to the analysis and representation of qualitative materials. On the far right, there is an emphasis on valid, reliable knowledge generated by neutral researchers using rigorous methods to generate Truth. This is the space of postpositivism. At the left end of the continuum, researchers value humanistic, openly subjective knowledge—autoethnography, poetry, video, stories, narratives, photography, drama, painting. Truths are multiple, ambiguous; literary standards of truthfulness replace those of positivism. In the middle is work that offers description, exposition, analysis, insight, and theory, blending art and science, and often transcending these categories. First-person voice is used, scholars seek intimate familiarity with their textual materials, and grounded theory and multiple methods may be employed.

Ellingson offers a series of writing/stretching exercises, a process she calls "wondering." It asks the researcher to think seriously and freely about their empirical materials, their inquiry topics, the audiences for their work, the pleasures they derive from their project, their identities as writers, and the writing genres they feel most comfortable with.

Multigenre crystallization is Ellingson's postmodern-influenced approach to triangulation. Crystallization combines multiple forms of analysis and genres of representation into a coherent text. Crystallization seeks to produce thick, complex interpretation. It utilizes more than one writing genre. It deploys multiple forms of analysis, reflexively embeds the researcher's self in the inquiry process, and eschews positivist claims to objectivity. Crystallization features two primary types: those *integrated* into a single text, and those that are *dendritic,* involving multiple textual formations. Guerilla scholarship moves back and forth across both types of crystallization, and engages different methods, genres, paradigms, and ideologies, always in the name of social justice.

Ellingson predicts a sharp rise in the next decade in the number of researchers who are willing to take up her view of the qualitative continuum in pursuit of socially engaged programs. So do we.

▣ POST QUALITATIVE RESEARCH

St.Pierre (Chapter 37) calls for the resurgence of postmodernism, a philosophically informed inquiry that will resist calls for scientifically based forms of research (SBR). In so doing, she also offers a powerful postmodern critique of conventional humanistic qualitative methodology. (Her reading of the SBR discourse complements Torrance's critique of this movement.)

She convincingly argues that it is time for qualitative inquiry to reinvent itself, to put under erasure all that has been accomplished, so that something different can be done, a "rigorous reimagining of a capacious science that cannot be defined in advance and is never the same again." Thus does she take up the "posts"—postmodernism, poststructuralism—offering a valuable history of each discourse. She introduces the concepts of haecceity, assemblage, and entanglement to deconstruct the humanist concept of human being. She notes, though, that it is difficult to escape the concept of "I."

Drawing on her own research, she shows how deconstruction can work, from transgressive data, to understanding that writing is always analysis. Writers interpret as they write, so writing is a form of inquiry, a way of making sense of the world. Writing as a method of inquiry coheres with the development of ethical selves. St.Pierre troubles conventional understandings of ethics. Drawing on Derrida and Deleuze, she places ethics under deconstruction: "What happens when we cannot apply the rules?" We must not be unworthy of what happens to us. We struggle to be worthy, to be willing to be worthy. We seek a writing space that goes beyond the "posts," to reach into the future.

▣ QUALITATIVE RESEARCH AND TECHNOLOGY

Davidson and di Gregorio (Chapter 38) note that in the 1980s, as qualitative researchers began to grapple with the promise of computers, a "handful of innovative researchers created the first generation of what became known as CAQDAS (Computer Assisted Qualitative Data Analysis Software)," or QDAS (Qualitative Data Analysis Software), as they refer to it in their chapter. Initially, these software packages were used for simple text retrieval tasks. They quickly expanded into comprehensive all-in-one-packages that offered qualitative researchers a suite of digital tools that could be used to store, organize, analyze, represent, and transport qualitative materials.

Three decades later, with the explosion of the Internet and the emergence of web-based tools known as Web 2.0, or Web 3.0, QDAS is on the brink of a new wave of developments. These developments are multimodal, visually attractive, easier to learn, less expensive, and more socially connected. Yet despite this fact, QDAS is used by a small minority of qualitative researchers. There is still a lack of institutional understanding and support for the tools. The major initiatives in qualitative inquiry have not taken up these technologies.

But qualitative research and computer technology are in the midst of a revolution. This chapter sets the historical context for understanding these developments. Their six-stage model of QDAS development is fitted to our eight-stage model: traditional, modernist, blurred genres, crisis of representation, postmodernism, post-experimental, methodologically contested present, a fractured future. Key texts (grounded theory) and QDAS in each of their six stages are identified.

In 1989, the first international conference on qualitative computing was held at the University of Surrey, in the United Kingdom. In 1994, the Economic and Social Research Council (ESRC) in Great Britain funded the CAQDAS networking project. In 1995, Sage Publications agreed to market QSR: Qualitative Social Research International's NUD*IST package. By the end of the 1990s, three major QDAS packages came to dominate: ATLAS.ti, MAXQDA, and NVivo.

In the recent past, concern shifted to teaching with QDAS, developments of the E-project, and international conferences exploring new frontiers with Web 2.0. These frontiers include the electronic future, the rapid growth of media and social networking technologies on the Internet: wikis, YouTube, Flickr, Twitter, Facebook, and so forth. Ethical issues continue, as Gatson discussed in Chapter 31, many taken up by the Association of Internet Researchers. As we move to QDAS 2.0, how can we ensure that web-based storage systems have the security we need?

◨ THE POLITICS OF EVIDENCE

Denzin's chapter (Chapter 39) reviews the by now all-too-familiar arguments about policy, SBR, and the politics of evidence. He reviews state- and discipline-sponsored standards and criteria for qualitative work. He criticizes recent efforts by the American Education Research Association to offer a set of standards for reporting on humanities-oriented research. He notes the multiple points of tension within the qualitative inquiry community: Interpretivists dismiss postpositivists. Poststructuralists dismiss interpretivists, and postinterpretivists dismiss the interpretivists. Global efforts to impose a new orthodoxy on critical social science inquiry must be resisted.

◨ WRITING INTO POSITION

Pelias (Chapter 40) argues that the writerly self is a performance. In the moment of composition, the I comes into existence. Writing becomes a form of inquiry, a form of self-realization. Writing functions as way of moving the individual forward, into

poetic, narrative spaces. Evocative, reflexive, embodied practices allow the I to position him- or herself in partisan places. The three major sections of Pelias's chapter interrogate these strategies for composition and evaluation. Each writing position, each writing strategy, poeticizes the researcher's body, locates it in a story, a narrative, which is all that we have.

Reflexive writers write about their complicity in the problems they interrogate, inviting others to interrogate their own actions, seeking, perhaps, a new, more utopian democratic space. Qualitative researchers always write from a location of corporeal presence. As Spry argues in Chapter 30, they write from the site of the body, the body in pain, the abused body, the damaged body. They write to make the world a better place; they write in the hope of dialogue, of new possibilities. We sit at our desks trying, trying.

◨ POLICY AND QUALITATIVE EVALUATION

Program evaluation, as an applied science, is a major site of qualitative research. Evaluators are interpreters, and they do their work in socio-political contexts. Their texts tell stories. These stories are inherently moral and political, and relational. For Abma and Widdershoven (Chapter 41), evaluation is a relationally responsible practice. In their work, evaluators enact a shared understanding of what it means to be an evaluator and do evaluative work.

They offer a history of this field, which complements House's (2005) narrative. The field has moved from faddish experimental, social engineering, and quantitative evaluation studies (1960s), to small-scale qualitative studies, to meta-analyses and program theory. A move from a model of value-free inquiry to committed social justice projects, and back again, is also part of this history. In the 1980s, evaluation moved away from quantitative methods and value-free studies, toward qualitative studies focused on stakeholders, social justice concerns, and participatory techniques.

The qualitative evaluator makes judgments, based on values, ethics, methodologies, stakeholders' accounts, and contextual understandings. The evaluator develops evaluations that are in between advocacy and critique, midway between "antipathy and sympathy . . . an Aristotelian middle-ground position." Abma and Widdershoven's evaluator is a wise judge who understands that evaluation is a political practice; "it has unequal consequences for various stakeholders in the evaluation." There is a desire to empower people. They review feminist, transformative, democratic, participatory, critical, social justice, and fourth and fifth evaluation traditions—evaluation for understanding and evaluation for social critique and transformation—that advance these positions.

Qualitative evaluation is holistic, dialogical, and emergent. It evolves through closely connected stages or phases. It starts with hearing marginalized voices, then placing these voices in dialogue with one another. The evaluator empowers those who are heard, and creates a safe space for dialogue and inclusion. The authors present a case study from the field of psychiatry to illustrate their ideas, which they call "interactive evaluation." The challenge of ensuring that positive cultural change endures remains.

▣ CONCLUSION

The readings in Part V affirm our position that qualitative research has come of age. Topics that were contained within the broad grasp of the positivist and postpositivist epistemologies are now surrounded by multiple discourses. There are now many ways in which to write, read, assess, evaluate, and apply qualitative research texts. Even so, there are pressures to turn back the clock. This complex field invites reflexive appraisal, the topic of Part VI—the future of qualitative research.

▣ REFERENCES

Hammersley, M. (1992). *What's wrong with ethnography?* London: Routledge.

Hammersley, M. (1995). *The politics of social research.* London: Sage.

Hammersley, M. (2008). *Questioning qualitative inquiry: Critical essays.* London: Sage.

House, E. (2005). Qualitative evaluation and changing social policy. In N. K. Denzin & Y. S. Lincoln (Eds.), *The SAGE handbook of qualitative research* (3rd ed., pp. 1069–1082). Thousand Oaks, CA: Sage.

Janesick, V. (2010). *"Stretching" exercises for qualitative researchers.* Thousand Oaks, CA: Sage.

Wolcott, H. F. (1999). *Ethnography: A way of seeing.* Walnut Creek, CA: AltaMira Press.

34

QUALITATIVE RESEARCH, SCIENCE, AND GOVERNMENT

Evidence, Criteria, Policy, and Politics

Harry Torrance

The debate about how educational research and, more generally, social research, might better serve policy has been continuing for more than a decade now. It is not a new debate, and has been revisited many times since the inception of educational and social research as established university-based activities (e.g., Lagemann, 2000; Nisbet & Broadfoot; 1980; Weiss, 1972, 1980). However, it has been addressed with new vigor since the late 1990s as successive governments in the USA, the UK, and elsewhere have looked for better value for money from research, and more particularly looked for legitimating and supportive endorsements of their policies. The debate carries particular import for those working in what might be termed the broad field of "qualitative inquiry," since it has tended to privilege so-called 'scientific' approaches to educational and social research, by which is meant empirical investigations of educational activities and innovations, oriented to the identification of causality, explanation, and generalization (e.g., National Research Council, 2002). Implicitly, therefore, and sometimes quite explicitly, qualitative approaches to research are marginalized. The debate seems to reflect both long-term changes in what we might call the "terms of trade" between science and policy, along with more specific short-term jockeying for position amongst particular researchers and government officials/advisers at a particular point in time. This chapter will attempt briefly to review the background to the debate before examining some of its key elements in more detail and reflecting on its implications for the field of qualitative research and its relationship to policy.

The intensity and focus of the current debate in the UK can be dated from a speech in 1996 by David Hargreaves (then Professor of Education at Cambridge University) to the Teacher Training Agency (TTA—a government agency regulating teacher training). Hargreaves (1996) attacked the quality and utility of educational research, arguing that such research should produce an "agreed knowledge base for teachers" (p. 2) that "demonstrates conclusively that if teachers change their practice from X to Y there will a significant and enduring improvement in teaching and learning" (p. 5). Subsequent government-sponsored reviews and reports took their lead from this speech and produced what might be termed a mainstream policy consensus that the quality of educational research was low, particularly because so many studies were conducted on a small scale and employed qualitative methods, and therefore "something had to be done" (Hillage, Pearson, Anderson, & Tamkin, 1998; Tooley & Darby, 1998; Woodhead, 1998). That such claims were disputed need not detain us here (but see for example Hammersley, 1997, 2005; MacLure, 2003). It is worth noting, however, that subsequent analyses of papers published by the *British Educational Research Journal*, the leading UK journal of the British Educational Research Association, and of educational research projects funded by the UK Economic and Social Research Council (ESRC), demonstrated that critics had misrepresented the field and that in fact a wide range of methods were and are employed in British educational research, including-large scale quantitative analysis, experimental design, and mixed methods (Gorard & Taylor, 2004; Torrance, 2008).

The parallel intervention to Hargreaves in the USA is probably the National Research Council report (2002) "Scientific Research in Education," though this in turn was produced in response to already extant policy debate and legislation identifying what would be defined as "research" for purposes of federal funding—specifically the Reading Excellence Act, 1999, and the No Child Left Behind Act, 2001 (see Baez & Boyles, 2009, pp. 5 ff., for illustration and discussion of these acts). A huge literature has been prompted by this legislation, subsequent attempts to delineate the boundaries of "scientific research in education" and responses to those attempts, and again, it is not my purpose to review it further here (see, e.g., *Educational Researcher*, 2002, vol. 31, no. 8; *Qualitative Inquiry*, 2004, vol. 10, no. 1; *Teachers College Record*, 2005, vol. 107, no. 1). However, one quotation from this debate is worth highlighting, since in many respects it summarizes the "scientific" case, particularly the case for using not just a broadly quantitative empirical approach, but a specifically experimental design. Thus, Robert Slavin (2002), a leading proponent of the scientific method in the USA and recently appointed Director of the Institute for Effective Education at the University of York, UK, argues that "the experiment is the design of choice for studies that seek to make causal conclusions, and particularly for evaluations of educational innovations" (p. 18.). And, in a turn of phrase that is directly reminiscent of Hargreaves's (1996) speech, Slavin suggests that policy makers want to know "if we implement Program X instead of Program Y, or instead of our current program, what will be the likely outcomes for children?" (p. 18).

Here, then, is the essential focus of, and apparent justification for, the current debate. Educational research, and especially, qualitative approaches to educational research, has not provided a sufficiently cumulative and robust evidence base for the development of educational policy and practice, and in particular has not produced sufficient experimental data to allow policy makers to evaluate policy alternatives.

However, it is important to recognize that these criticisms are not restricted to the USA and/or UK policy contexts, nor indeed are they restricted to educational research. Reviews of and attacks on the quality of educational research, and particularly the quality of qualitative educational research, have similarly impacted debate in Australia and New Zealand (Cheek, 2007; Middleton, 2009; Yates, 2004), and are beginning to emerge in the European Union (Besley, 2009; Bridges, 2005, 2009; Brown, 2003). Overall, and to reiterate, the argument has been that educational research is too often conceived and conducted as a "cottage industry": producing too many small-scale, disconnected, non-cumulative studies that do not provide convincing explanations of educational phenomena or how best to develop teaching and learning. There is no cumulative or informative knowledge base in the field, and it is characterized as being of both poor quality and limited utility. Similar critiques have been leveled against social research more generally. In a speech to the ESRC in 2000, titled "Influence or Irrelevance" the then Secretary of State for Education, David Blunkett (2000), asserted that

> often in practice we have felt frustrated by a tendency for research . . . to address issues other than those directly relevant to the political and policy debate. . . . Many feel that too much social

science research is inward-looking, too piecemeal, rather than helping to build knowledge in a cumulative way, and fails to focus on the key issues of concern to policy-makers, practitioners and the public, especially parents. (p. 1)

In an attempt to address such political concerns, the Campbell Collaboration, a direct parallel to the Cochrane Collaboration in medical research, seeks to review and disseminate social science knowledge for policy makers through what it terms "systematic reviews" (Davies & Boruch, 2001; Wade, Turner, Rothstein, & Lavenberg, 2006). I will return to systematic reviewing below, but for the moment my point is that while the legislative concern to promote "scientific research in education" is a fairly specific American phenomenon, as is the particular focus on experimental design, this sits in a much broader international context of concern about the nature and purpose of social research and its relationship to policy. Such concerns reflect both the topics and methods of inquiry. Educational research, qualitative approaches to educational research, but also qualitative approaches to social research more generally have all come in for criticism and, taken together, suggest that qualitative inquiry is facing a global movement to reassert broadly empiricist and technicist approaches to the generation and accumulation of social scientific "evidence" for policy making. The focus, worldwide, is on seeking evidence to inform policy making, particularly evidence about "what works." Elements of such a movement will differ in their origins, orientations, and specific national aspirations. But equally they do seem to represent a concerted attempt to impose (or perhaps reimpose) scientific certainty and a form of center-periphery, research, development, dissemination (RDD) system management on an increasingly complex and uncertain social world.

▣ LONG-TERM TRENDS: WHITHER/ WITHER SCIENCE AND GOVERNMENT?

Part of the backcloth to the current debate is the uncertain status and legitimacy of both science and government at the present time. The role, purpose, and utility of science and scientific research is less agreed upon and less secure than it once was, and with respect to this, just as educational research can be seen to be situated in a wider debate about social research, so social research can be seen to be located in a wider debate about scientific research and the role of science in society. In the UK, for most of the 20th century, the relationship between science and government was determined by the so-called "Haldane principle" (after Viscount Haldane, an influential liberal politician who chaired the committee that articulated the principle in 1918). This settlement essentially resolved that university-based science would be funded from the public purse to pursue fundamental research, which would in turn produce unpredictable, but nevertheless substantial, long-term scientific and technical benefit—i.e., "basic" research would, over time, produce the platform for more "applied" technological developments and benefits. This was even characterized as the creation and operation of the "independent republic of science" by Michael Polyani (1962, as cited in Boden, Cox, Nedeva, & Barker, 2004). It has its direct parallel in the United States with the publication of Vannevar Bush's "Science: The Endless Frontier" (1945). This argued, on the back of scientific successes apparent in the Second World War, for the federal government to significantly expand support for scientific research on the basis of a similarly "arms length" linear model of "basic" research eventually leading to technological benefit (e.g., Greenberg, 2001, Chapter 3). More recently, however, government calls for much more short-term responsiveness and utility have pervaded policy debates and aspirations on both sides of the Atlantic and elsewhere—e.g., the Clinton focus on science and technology policy in the 1990s (Greenberg, 2001), and the current UK government concern to document and evaluate the "impact" of research through its new Research Excellence Framework (Department for Business, Innovations, and Skills [DBIS], 2009; Higher Education Funding Council for England [HEFCE], 2009), the successor to the Research Assessment Exercise (Torrance, 2006). Science in general, and social science in particular, is now expected to serve government policy and economic development very directly. This clearly begs questions about how to define quality and utility.

Equally, however, government itself is under pressure to "deliver," especially in areas of public policy. Since the first oil crisis of the 1970s put severe pressure on public spending, especially in the UK, and with the development and implementation of monetarist critiques of government spending in the 1980s, and the collapse of the Soviet Communist Bloc in 1989, there has developed a severe crisis of confidence and legitimation with respect to the role of government itself, especially with regard to the provision of public services: Are they really needed? If so, could they be better and more efficiently provided by other mechanisms and stakeholders? What reasons are there for state intervention in the lives of ordinary citizens? In this respect, government demand for "evidence" is as much a demand for material to justify its own existence, as it is a demand for the evaluation of particular policy alternatives. What is at stake is the legitimacy of policy intervention *per se.*

▣ EXPERIMENTALISM: PART OF THE SOLUTION OR PART OF THE PROBLEM?

Advocates of experimental design have inserted themselves into this uncertain nexus. Given such uncertainty, it is understandable that governments and policy makers will look to research for assistance. Research, or more generally, "science," is still largely regarded as independent of government and thus able, at least in principle, to provide disinterested evidence for both the

development and evaluation of policy, despite recent moves toward the development of a closer and more utilitarian relationship. To reiterate, "if we implement Program X instead of Program Y, or instead of our current program, what will be the likely outcomes for children?" (Slavin, 2002, p. 18). The attraction of the sort of evidence that Hargreaves (1996) and Slavin claim can and should be provided is easy to appreciate. It sounds seductively simple. When charged with dispensing large amounts of public money for implementing programs and supporting research, one can understand that policy makers might value this sort of help—at least as long as the answers to the questions posed are clear and not too radical or expensive.

But here's the rub—the answers to questions of public policy and program evaluation are often not very clear (nor indeed are the questions sometimes). More circumspect proponents of experimental methods, specifically randomized controlled trials (RCTs), acknowledge that in order for a causal relationship to be established, even within the narrow terms of an RCT, very specific questions have to be asked. Thus, for example, Judith Gueron (2002) argues that while "random assignment . . . offers unique power in answering the 'Does it make a difference?' question" (p. 15), it is also the case that "[t]he key in large-scale projects is to answer a few questions well" (p. 40). In the same edited volume of papers, produced from a conference convened to promote "Randomized Trials in Education Research," Thomas Cook and Monique Payne (2002) agree that

> most randomized experiments test the influence of only a small subset of potential causes of an outcome, and often only one . . . even at their most comprehensive, experiments can responsibly test only a modest number of the possible interactions between treatments. So, experiments are best when a causal question involves few variables [and] is sharply focused. (p. 152)

What these observations mean is that RCTs can be very good at answering very specific questions and attributing cause in a statistically descriptive (i.e., observable) way. What they cannot do is produce the questions in the first place: That depends on much prior, often qualitative, investigation, not to mention value judgments about what is significant in the qualitative data and what problem might be addressed by a particular program intervention. Nor can RCTs provide an explanation of *why* something has happened (i.e., the causal mechanisms at work). That, likewise, will depend on much prior investigation and, if possible, parallel qualitative investigation of the phenomenon under study, to inform the development of a theory about what the researchers think may be happening. Without a reasonable understanding of why particular outcomes have occurred, along with identifying the range of unintended consequences that will almost inevitably accompany an innovation, it is very difficult to generalize such outcomes and implement the innovation with any degree of success elsewhere. A good example of such problems is provided by California's attempt to implement smaller class sizes off the back of the apparent success of the Tennessee "STAR" evaluation. The Tennessee experiment worked with a sample, whereas California attempted statewide implementation, creating more problems than they solved by creating teacher shortages, especially in poorer neighborhoods in the state. There simply weren't enough well-qualified teachers available to reduce class size statewide, and those that were tended to move to schools in richer neighborhoods when more jobs in such schools became available (see Grissmer, Subotnik, & Orland, 2009).

Interestingly in this respect, Cook and Payne (2002) continue,

> The advantages of case study methods are considerable … we value them as adjuncts to experiments. … Case study methods complement experiments when … it is not clear how successful program implementation will be, why implementation shortfalls may occur, what unexpected effects are likely to emerge, how respondents interpret the questions asked of them, [and] what the causal mediating processes are … qualitative methods have a central role to play in experimental work. (p. 169)

One is tempted to ask, "So what's all the fuss about?" Why is some RCT advocacy so strident and exclusive? Of course, different researchers will vary in the importance they give to qualitative methods, and it is irritating to have qualitative methods reduced to an "adjunct" or a "complement" to experimental approaches, or as some activity to be undertaken before the "real" scientific work begins (see Shavelson, Phillips, Towne, & Feuer, 2003, p. 28). But it does seem as though those whose work actually involves the conduct of social science experiments have a well-informed view of the strengths of qualitative research, along with clear understandings of the limitations of experiments, as opposed to those who just engage in uninformed criticism of qualitative methods and advocacy for RCTs.[1]

There is not enough space here to go into all the potential problems of conducting randomized experiments in the "natural" (as opposed to laboratory) setting of the school or the classroom. Extensive philosophical and practical critiques (and rejoinders) about the nature of causality and the place of RCTs in understanding social interaction and evaluating human services have been published by Erickson and Gutierrez (2002), Howe (2004), and Maxwell (2004), among many others. Indeed, practitioners such as Gueron (2002) and Cook and Payne (2002), cited above, provide comprehensive accounts of the challenge of undertaking experiments "in the field." The real problem with experimental methods, however, is that even if conducted as effectively as possible, they often don't actually answer the "Does it make a difference?" question. Already, accounts of disappointing results are starting to appear in the press:

> Like a steady drip from a leaky faucet, the experimental studies being released this school year by the federal Institute of Education Sciences are mostly producing the same results: "No effects," "No effects," "No effects."

> The disappointing yield is prompting researchers, product developers, and other experts to question the design of the studies, whether the methodology they use is suited to the messy real world of education, and whether the projects are worth the cost, which has run as high as $14.4 million in the case of one such study. (Viadero, 2009, p. 1)

We should not be surprised. It was precisely the confounding problems of diverse implementation and interaction effects that produced so many "no significant difference" results in the 1960s in the context of curriculum evaluation studies. Reflections on such results prompted the development and use of qualitative methods in evaluation studies in the first place, in the1970s and 1980s (Cronbach, 1975; Cronbach & Associates, 1980; Guba & Lincoln, 1981, 1989; Hamilton, Jenkins, King, MacDonald, & Parlett, 1976; Stake, 1967, 1978; Stenhouse, 1975; Stenhouse, Verma, Wild, & Nixon, 1982). Indeed, in one mixed method study of the "problems and effects of teaching about race relations" (as issues of race were called in the UK in those days), it was reported that 60% of the sample student population became less racially prejudiced as measured by attitude tests after following a particular program, but 40% became *more* prejudiced—as the author himself mused, what on earth is one supposed to do with such a result (Stenhouse et al., 1982)?

◼ BEYOND SINGLE STUDIES: SYSTEMATIC REVIEWING

The response of those interested in unpacking the sort of dilemma highlighted above would probably be to conduct further detailed investigation of the program as implemented, and indeed, Stenhouse and his team (Stenhouse et al., 1982) did conduct other investigations, including an "action research" approach to the development of the program. Similar mixed methods evaluation studies are often funded in the UK as those stakeholders with an interest in the development—local authorities (i.e., school districts), head teachers (principals), school governors, and so forth—seek maximum information about the effects of an intervention, not just a one-off research result about whether or not it "works" (e.g., Somekh et al., 2007).

However, a different approach has been advanced by those committed to experimental design but who acknowledge the potential weakness of relying on single studies—that of so-called "systematic reviewing." Advocates of systematic reviewing argue that evidence to inform policy should be accumulated across studies, but not just any studies, rather, only those that pass strict tests of quality. And those tests of quality have until relatively recently involved focusing on large-scale samples and, ideally, experimental designs (Gough & Elbourne, 2002; Oakley 2000, 2003). The case for developing systematic reviewing is based on transparency of process and clear criteria for including and excluding studies from the review. The case derives from critiques of so-called "narrative reviewing," which, it is claimed, focuses on summarizing findings, in relation to a particular argument, rather than reviewing the whole field dispassionately and "systematically" so that the reader can be confident that all relevant prior knowledge in a field has been included and summarized. Arguments in favor of conducting such reviews reflect the critiques of social and educational research outlined earlier: that the findings of empirical studies are often too small-scale, non-cumulative, or contradictory to be useful. Advocates are closely associated with the Cochrane Collaboration in medical and health care research and the Campbell Collaboration in social science, both of which favor the accumulation and dissemination of research findings based on scientific methods, particularly randomized controlled trials. As such, systematic reviewing is very much located within the international "evidence-based policy and practice" movement (Davies, 2004; Davies & Boruch, 2001; see also Mosteller & Boruch, 2002, p. 2, for evidence of the close networking of this international movement).

The original criteria of quality employed by systematic reviewing clearly derived from the medical model, but it is interesting to note that even as some researchers started to argue the relevance of an RCT-based medical model to educational and social research in the UK, it was already being subject to criticism in the field of medicine itself. Medical researchers understood that many issues of patient treatment and care require the design of qualitative as well as quantitative studies, and substantial subsequent developments have tried to find ways of integrating the findings of qualitative studies into systematic reviews (e.g., Barbour & Barbour, 2003; Dixon-Woods, Booth, & Sutton, 2007; Dixon-Woods, Fitzpatrick, & Roberts, 2001). Sometimes this has led to the rather absurd deployment of Bayesian statistics to incorporate qualitative data into quantitative estimates of effects—essentially transforming expert judgment of qualitative studies into numerical indicators by rank-ordering the importance of key variables discernable from qualitative studies. The rank-ordered expert judgments of quality are then rendered into probabilities (of which variables are likely to be most important) and included in quantitative meta-analyses. This seems to do little more than add a spurious mathematical accuracy (to three decimal points in Roberts, Dixon-Woods, Fitzpatrick, Abrams, & Jones, 2002) to what would be far better left as "expert" judgment. At least we can be appropriately skeptical of expert judgment, precisely because it is usually expressed in narrative form, even if we might also regard it as the best available evidence in the circumstances. Nevertheless, such developments indicate that qualitative data are appreciated as important in understanding the conduct and impact of medical processes and treatments.

The original "hard line" position of systematic reviewing in social research has now been significantly modified, as it has encountered considerable skepticism over the last 10 years in the UK (Hammersley, 2001; MacLure; 2005; cf. also Oakley's 2006 response, and Hammersley's 2008 rejoinder). Work is now underway to integrate different kinds of research findings, including those of qualitative research, into such reviews. This involves attempts to appraise the quality and thus the "warrant" of individual qualitative research studies and their findings: Are they good enough to be included in a systematic review or not? Once again, however, this can lead toward absurdity rather than serious synthesis as the complexity of qualitative work is rendered into an amenable form for instant appraisal. Thus, for example, Attree and Milton (2006) report on a "Quality Appraisal Checklist . . . [and its associated] quality scoring system . . . [for] "the quality appraisal of qualitative research" (p. 125). Studies are scored on a 4-point scale:

A No or few flaws

B Some flaws

C Considerable flaws, study still of some value

D Significant flaws that threaten the validity of the whole study (p. 125)

Only studies rated A or B were included in the systematic reviews that the authors conducted, and in the paper they attempt to exemplify how these categories are operationalized in their work. But their descriptions beg many more questions than they answer. The above scale simply provides a reductionist checklist of mediocrity. Even the most stunning and insightful piece of qualitative work can only be categorized as having "No or few flaws."

To try to be fair to the authors, they indicate that

> the checklist was used initially to provide an overview of the robustness of qualitative studies . . . to balance the rigor of the research with its importance for developing knowledge and informing policy and practice. (Attree & Milton, 2006, p. 119)

But this is precisely the point at issue with respect to using research to inform policy: Standards and checklists *cannot* substitute for informed judgment when it comes to balancing the rigor of the research against its potential contribution to policy. This *is* a matter of judgment, both for researchers and for policy makers.

◼ IMPACT ON QUALITATIVE RESEARCH: SETTING STANDARDS TO CONTROL QUALITY

Many other criticisms could be directed at systematic reviewing in addition to its apparent disdain for qualitative evidence. For example, it is also very expensive and inefficient in terms of time and material resources, given the little it often delivers in terms of actual "findings." The results of systematic reviews can take many months to appear, and policy makers in England are as likely to ask for very rapid reviews of research to be conducted over a few days or weeks, and possibly assembled via an expert seminar, as to commission longer-term systematic reviews (Boaz, Solesbury, & Sullivan, 2004, 2007). However, the more general issue for this chapter is the impact of the "scientific evidence movement" on qualitative research, and the above checklist produced by Attree and Milton (2006) well illustrates the contortions that some qualitative researchers are starting to go through, in order to maintain the visibility of their work in the context of this movement.

A major response to the evidence movement has been for organizations and associations to start trying to "set standards" in qualitative research, and indeed in educational research more generally, to reassure policy makers about the quality of qualitative research and to reassert the contribution that qualitative research can (and should) make to government-funded programs. However, the field of qualitative research, or qualitative inquiry, is very broad, involving large numbers of researchers working in different countries, working in and across many different disciplines (anthropology, psychology, sociology, etc.), different applied research and policy settings (education, social work, health studies, etc.), and different national environments with their different policy processes and socioeconomic context of action. It is not at all self-evident that reaching agreement across such boundaries is desirable, even if it were possible. Different disciplines and contexts of action produce different readings and interpretations of apparently common literatures and similar issues. It is the juxtaposition of these readings, the comparing and contrasting within and across boundaries, that allows us to learn about them and reflect on our own situated understandings of our own contexts. Multiplicity of approach and interpretation, and multivocalism of reading and response, are the basis of quality in the qualitative research community and, it might be argued, in the advancement of science more generally. The key issue is to discuss and explore quality across boundaries, thereby continually to develop it, not fix it, as at best a good recipe and at worst a narrow training manual.

Nevertheless, various attempts at "setting standards" are now being made, often, it seems, with the justification of "doing it to ourselves, before others do it to us" (Cheek, 2007; see also the discussion by Moss et al., 2009). In England, independent academics based at the National Centre for Social Research (a not-for-profit consultancy organization) were commissioned by the Strategy Unit of the UK government Cabinet Office to produce a report on "Quality in Qualitative Evaluation: A Framework for Assessing Research Evidence" (Cabinet Office, 2003a). The rationale seems to have been that UK government departments are increasingly commissioning policy evaluations in the context of the move toward evidence-informed policy and practice and that guidelines for judging the quality of qualitative approaches and methods were considered to be necessary.

The report is in two parts: a 17-page summary, including the "Quality Framework" itself (Cabinet Office, 2003a), and a 167-page full report (Cabinet Office, 2003b), including discussion of many of the issues raised by the framework. The framework is a guide for the commissioners of research when drawing up tender documents and reading reports, but it is also meant to influence the conduct and management of research and the training of social researchers (Cabinet Office, 2003a, p. 6). However, the short "Quality Framework" begs many questions, while the full 167-page report reads like an introductory text on qualitative research methods. Paradigms are described and issues rehearsed, but all are resolved in a bloodless, technical, and strangely old-fashioned counsel of perfection. The reality of doing qualitative research and indeed of conducting evaluation, with all the contingencies, political pressures, and decisions that have to be made, is completely absent. Thus, in addition to the obvious need for "Findings/conclusions [to be] supported by data/evidence" (Cabinet Office, 2003b, p. 22), qualitative reports should also include

> Detailed description of the contexts in which the study was conducted; (p. 23)
>
> Discussions of how fieldwork methods or settings may have influenced data collected; (p. 25)
>
> Descriptions of background or historical developments and social/organizational characteristics of study sites; (p. 25)
>
> Description and illumination of diversity/multiple perspectives/alternative positions; (p. 26)
>
> Discussion/evidence of the ideological perspectives/values/philosophies of the research team. (p. 27)

And so on and so forth, the document continues across a total of six pages and 17 quality "appraisal questions."

No one would deny that these are important issues for social researchers to take into account in the design, conduct, and reporting of research studies. However, simply listed as such, they comprise a banal and inoperable set of standards that beg all the important questions of conducting and writing up qualitative fieldwork. Everything cannot be done; *choices* have to be made: How are they to be made, and how are they to be justified?

To be more positive for a moment, and note the arguments that might be put forward in favor of setting standards, it could be argued that if qualitative social and educational research is going to be commissioned, then a set of standards that can act as a bulwark against commissioning inadequate or underfunded studies in the first place ought to be welcomed. It might also be argued that this document at least demonstrates that qualitative research is being taken seriously enough within government to warrant a guidebook being produced for civil servants. This might then be said to confer legitimacy on civil servants who want to commission qualitative work; on qualitative social researchers bidding for such work; and indeed on social researchers more generally, who may have to deal with local research ethics committees (RECs; IRBs in the USA), which are predisposed toward a more quantitative natural science model of investigation. But should we really welcome such "legitimacy"? The dangers on the other side of the argument, as to whether social scientists need or should accede to criteria of quality endorsed by the state, are legion. In this respect, it is not at all clear that, *in principle,* state endorsement of qualitative research is any more desirable than state endorsement of RCTs.

Similar guidelines and checklists are starting to appear in the USA. Thus, for example, Ragin, Nagel, and White (2004) report on a "Workshop on Scientific Foundations of Qualitative Research," conducted under the auspices of the National Science Foundation and with the intention of placing "qualitative and quantitative research on a more equal footing . . . in funding agencies and graduate training programs" (p. 9). The report argues for the importance of qualitative research and thus advocates funding qualitative research *per se,* but equally, by articulating the "scientific foundations" it is arguing for the commissioning of not just qualitative research, but "proper" qualitative research. Thus, for example, they argue that

> Considerations of the scientific foundations of qualitative research often are predicated on acceptance of the idea of "cases." . . . No matter how cases are defined and constructed, in qualitative research they are studied in an in-depth manner. Because they are studied in detail their number cannot be great. (pp. 9–10)

This is interesting and provocative with respect to the idea of standards perhaps acting as a professional bulwark against commissioning inadequate or underfunded studies: A quick and cheap survey by telephone interview would not qualify as high-quality "scientific" qualitative research. But when it comes to the basic logic of qualitative work, Ragin et al. (2004) do not get much further than arguing for a supplementary role for qualitative methods:

> Causal mechanisms are rarely visible in conventional quantitative research . . . they must be inferred. Qualitative methods can be helpful in assessing the credibility of these inferred mechanisms. (p. 15)

In the end, Ragin et al.'s (2004) "Recommendations for Designing and Evaluating Qualitative Research" also conclude with another counsel of perfection:

> These guidelines amount to a specification of the *ideal* qualitative research proposal. A strong proposal should include as many of these elements as feasible. (p. 17, emphasis original)

But again, that's the point: What is *feasible* (and relevant to the particular investigation) is what is important, not what is ideal. How are such crucial choices to be made? Once again, "guidelines" and "recommendations" end up as no guide at all; rather, they are a hostage to fortune whereby virtually any qualitative proposal or report can be found wanting.

A potentially much more significant example of this tendency is the American Educational Research Association (AERA) *Standards for Reporting on Empirical Social Science Research in AERA Publications* (2006). The *Standards* comprise eight closely typed double-column pages and include "eight general areas" (p. 33) of advice, each of which is subdivided into a total of 40 subsections, some of which are subdivided still further. Yet only one makes any mention of the fact that research findings should be interesting or novel or significant, and that is the briefest of references under "Problem Formulation," which we are told should answer the question of "why the results of the investigation would be of interest to the research community" (p. 34). Intriguingly, whether the results might be of interest to the *policy* community is not mentioned as a criterion of quality.

As is typical of the genre, the *Standards* include an opening disclaimer that

> The acceptability of a research report does not rest on evidence of literal satisfaction of every standard. . . . In a given case there may be a sound professional reason why a particular standard is inapplicable. (p. 33)

But once again, this merely restates the problem rather than resolves it. The *Standards* may be of help in the context of producing a book-length thesis or dissertation, but no 5,000-word journal article could meet them all. Equally, however, even supposing that they could all be met, the article might still not be worth reading. It would be "warranted" and "transparent," which are the two essential standards highlighted in the preamble (p. 33), but it could still be boring and unimportant.

It is also interesting to note that words such as *warrant* and *transparency* raise issues of trust. They imply a concern for the very existence of a substantial data set as well as how it might be used to underpin conclusions drawn. Yet the issue of trust is only mentioned explicitly once, in the section of the *Standards* dealing with "qualitative methods": "It is the researcher's responsibility to show the reader that the report can be trusted" (AERA, 2006, p. 38). No such injunction appears in the parallel section on "quantitative methods" (p. 37); in fact, the only four uses of the actual word *warrant* in the whole document all occur in the section on "qualitative methods" (p. 38). The implication seems to be that quantitative methods really are trusted—the issue doesn't have to be raised—whereas qualitative methods are not. Standards of probity are only of concern when qualitative approaches are involved.

▣ CAPACITY BUILDING, PROFESSIONALIZATION, AND THE RETREAT INTO "SCIENCE"

One response to the above examples of standards and guidelines is simply to accept them at face value. As I have already noted, in many respects they are unremarkable, and one of the key weaknesses that I have identified (i.e., their attempt at comprehensiveness) could even be used as a teaching device—for example, by asking students to identify which issues might *actually* be more or less important in the design of a particular study. And yet such documents carry more import than this— they also legitimate a particular delineation and control of the discourse surrounding qualitative research. In so doing, and in combination with other interventions such as the increasing reach of ethics committees and government regulation of research activity (Department of Health, 2005; Lincoln & Tierney, 2004; Torrance, 2006), they are beginning to change the very social relations of research and the ways in which issues of research quality have hitherto been addressed. Pursuing and developing quality in qualitative research has involved reading key sources iteratively and critically, in the context of designing and conducting a study, and discussing the implications and consequences with doctoral supervisors, or colleagues or project advisory groups. *Setting* standards in qualitative research, however, is a different enterprise. It implies the identification of universally appropriate and applicable procedures, which in turn involves documentary and institutional realization and compliance.

Much of the activity associated with such moves also now goes under the heading of "capacity building," certainly in the UK. As the government seeks to concentrate research resources in a smaller number of universities and extract maximum economic and social value from them, "centers of excellence" are being promoted, along with a concomitant obligation for the centers to link with and train in standard procedures those left stranded outside them (Department for Business, Innovations, and Skills, 2009; ESRC, 2005, 2009; National Centre for Research Methods, n.d.; Torrance, 2006). Similar aspirations also seem to be emerging in the USA (Eisenhart & DeHaan, 2005; NRC, 2005). It seems, then, that what is going on here is a struggle over the political economy and bureaucratic institutionalization of social research. What we are witnessing is a crucial moment in the continuing professionalization of social research. Governments are looking to control and quality-assure the process of social research and in so doing are treating researchers as an almost directly employed category of government worker in the "nationalized industry" of knowledge production. This in turn provides threats and opportunities for researchers as they seek to position themselves as both independent and autonomous sources of disinterested (i.e., scientific) advice, but nevertheless trustworthy professionals who can be relied upon to focus on topics of interest to policy and deliver a high-quality product.

Thus, some researchers are attempting to respond to the pressure of policy and the evidence movement by producing defensive documents that emphasize the need for professional standards and self–regulation (i.e., the AERA *Standards* above). In so doing, they appeal to and attempt to reassert the independence of "science" and the scientific community as a self-regulating group which, while broadly inclusive, nevertheless has clear boundaries and not only can define and protect standards, but *will*. Other researchers are seeing opportunities to redefine

the field and their place within it (i.e., their status and access to research funding). This is similarly being pursued by an appeal to science, but it involves a much more exclusive, elitist, and static interpretation of science—defined by method, rather than broad approach, and by association with other more specifically social science disciplines such as psychology, political science, and economics. However, this latter group seems increasingly out of step with government demands for utility (namely, the problems created by results that simply show "no effects"), and thus it would appear that they are deploying the rhetoric of science as part of an internal struggle with other researchers, rather than in any direct response to the supposed needs of government (see also Baez & Boyles, 2009, for a longer-term analysis of such trends).

■ SCIENCE IS NOT ENOUGH: TOWARD A DIFFERENT APPROACH

Interestingly, just as we've been here before with respect to 1960s/1970s disillusionment with research results that constantly showed "no significant difference," so we've been here before with respect to the response of the research community. Barry MacDonald (1974/1987) identified similar tensions over what role the research community should play in evaluating educational innovations. He identified three ideal types of approaches to evaluation—*autocratic, bureaucratic,* and *democratic,* aligning autocratic with scientific research, bureaucratic with confidential technical collaboration, and democratic with providing information for the widest possible public audience:

> Autocratic evaluation is a conditional service to . . . government. . . . It offers external validation of policy in exchange for compliance with its recommendations . . . the evaluator . . . acts as expert adviser. . . .
>
>   Bureaucratic evaluation is an unconditional service to . . . government. . . . The evaluator . . . acts as a management consultant [and] the report is owned by the bureaucracy and lodged in its files. . . . Democratic evaluation is an information service to the whole community about the characteristics of an educational program. . . . The democratic evaluator recognises value pluralism and seeks to represent a range of interests . . . techniques of data gathering and presentation must be accessible to non-specialist audiences. (pp. 44–45)

Of course, times change and the parallels with current debates are not exact. In particular, the obviously favored stance of "democratic evaluation" still presupposes that data can be gathered and interests represented in a fairly straightforward, realist fashion. Such aspirations would be more complex to accomplish now. Yet such a formulation also resonates with contemporary issues around stakeholder involvement, voice, and the engagement of a wider community in deciding which research questions are important to ask and how best to try to answer them.

It is now widely recognized from many different perspectives, including that of the empowerment of research subjects on the one hand, and policy relevance and social utility on the other, that an assumption of scientific disinterest and independence is no longer sustainable. Other voices must be heard in the debate over scientific quality and merit, particularly in an applied, policy-oriented field such as education. Thus, for example, Gibbons el al. (1994) distinguish between what they term Mode 1 and Mode 2 knowledge, with Mode 1 knowledge deriving from what might be termed the traditional academic disciplines, and Mode 2 knowledge deriving from and operating within "a context of application":

> [I]n Mode 1 problems are set and solved in a context governed by the, largely academic, interests of a specific community. By contrast, Mode 2 knowledge is carried out in a context of application. (p. 3)

Such knowledge is "transdisciplinary ... [and] involves the close interaction of many actors throughout the process of knowledge production" (p. vii). In turn, quality must be "determined by a wider set of criteria which reflects the broadening social composition of the review system" (p. 8).

The language employed by Gibbons et al. (1994) and the assumed context of operation very much reflect an engineering/technology-transfer type set of activities, but they also mirror a far wider set of concerns with respect to redefining the validity and social utility of research. There is a clear orientation toward the co-creation of knowledge through collaborative problem-solving action—rather than the discovery of knowledge through centralized, "expert" experimental investigation, which then gets disseminated to "practitioners" at the periphery. Ideas about the co-creation of knowledge link with deliberative and empowerment models of evaluation (Fetterman, 2001; House & Howe, 1999), which in turn owe something to MacDonald's (1974/1987) original notion of "democratic evaluation" (explicitly so, in House & Howe's case). The concept of "Mode 2 knowledge" also reflects something of the arguments around indigenous knowledge (Smith, 2005) and the many articulations and interrogations of how to identify and represent different "voices" in research (e.g., Alcoff, 1991; Fielding, 2004; Goodley, 1999; Jackson & Mazzei, 2009). Such arguments, coalescing into a diverse, contested, but nevertheless highly provocative and promising constellation of issues around the validity, utility, and ethics of social research, also bring us to the very limit of what it is currently possible to think about the relationship of qualitative inquiry to science, policy, and democracy. The challenge we face is how to sustain the tension between interrogating and reconceptualizing problems—"thinking the new"—while also addressing the "here and now" of the enduring social and political issues that face our society (see Lather, 2004, and Lather's contribution to Moss et al., 2009). The issue is how to reconcile the (research) need to investigate and comprehend complexity with the (policy) urge to simplify and act. To invert Marx, policy

makers seek to change the world, but first they need to try to understand it, while involving others in both processes.

The scholarly retreat into trying to define the "scientific" merit of qualitative research simply in terms of theoretical and methodological standards, rather than in wider terms of social robustness and responsiveness to practice, seems to betray a defensiveness and loss of nerve on the part of the scholarly community. We need to acknowledge and discuss the imperfections of what we do, rather than attempt to legislate them out of existence. We need to embody and enact the deliberative process of academic quality assurance, in collaboration with research participants, not subcontract it to a committee. Assuring the quality of research, and particularly the quality of qualitative research, must be conceptualized as a vital and dynamic process that is always subject to further scrutiny and debate. The process cannot be ensconced in a single research method or a once-and-for-all set of standards. Furthermore, it should be oriented toward risk taking and the production of new knowledge, including the generation of new questions (some of which may derive from active engagement with research respondents and policy makers) rather than supplication, risk aversion, and the production of limited data on effectiveness for a center-periphery model of system maintenance ("what works").

What this means for the actual conduct of social research, particularly qualitative research, over the medium to long term is still difficult to say, but various modest examples are emerging in the UK. These involve designing studies with collaborating sponsors and participants, including policy makers and those "on the receiving end" of policy, and talking through issues of validity, warrant, appropriate focus, and trustworthiness of the results, rather than trying to establish all of the parameters in advance (see, e.g., James, 2006; Pollard, 2005; Somekh & Saunders, 2007; Somekh et al., 2007; Torrance et al., 2005; Torrance & Coultas, 2004). It can also involve new forms of dissemination and intellectual engagement with participants, rather than the simple reporting of "research findings" (MacLure, Holmes, MacRae, & Jones, 2010).

The process is not without its problems or critics, especially with respect to issues of co-option into a too closely defined "bureaucratic" agenda—policy makers and sponsors usually being rather more powerful than research participants. But in essence, the argument is that if research is to engage critically with policy and practice, then research and policy making must progress, both theoretically and chronologically, in tandem. Neither can claim precedence in the relationship. Research should not simply "serve" policy; equally, policy cannot simply "wait" for the results of research. And just as participant and practitioner perspectives (often called research "end-users" by policy makers) may be used by policy to attempt to discipline the research agenda pursued by researchers, equally, such perspectives can be used to critically interrogate policy. Research will encompass far more than simply producing policy-relevant findings; policy making will include far more than simply disseminating and acting upon research results. Where research and policy do cohere, the relationship should be pursued as an iterative one, with gains on both sides.

Ultimately, the issue revolves around whether or not quality is protected and advanced by compliance with a particular set of standards, or by the process of open democratic engagement and debate. Governments, and some within the scholarly community itself, seem to be seeking to turn educational research into a technology that can be applied to solving short-term educational problems, thereby also entrenching the power of the expert in tandem with the state. An alternative vision proposes research as a system of reflective and engaged enquiry that might help practitioners and policy makers think more productively about the nature of the problems they face and how they might be better addressed. And in fact, the latter process will be as beneficial to policy as to research. Producing research results takes time, and, as we have seen above, such results are unlikely to be unequivocal. Drawing policy makers and practitioners into a discussion of these issues will improve the nature of research questions and research design, while also signaling to them that the best evidence available is unlikely ever to be definitive—it should inform and educate judgment, but it cannot supplant judgment, nor should it.

Both the concept and the practice of science and government are under severe pressure at present, and ironically, despite all the recent criticisms of qualitative research, it is qualitative research that is best placed to recover and advance new forms of science and government, precisely because it rests on direct engagement with research participants. Many recent discussions of quality in qualitative research revolve around issues of engagement, deliberation, ethical process, and responsiveness to participant agendas, along with the need to maintain a critical perspective on both the topic at hand and the power of particular forms of knowledge (Lincoln, 1995; Schwandt, 1996; Lather, 2004; Smith, 2005). It is these strengths of a qualitative approach that are needed to reinvigorate the research enterprise and reconnect it with democratic processes.

▣ NOTE

1. See Grissmer, Subotnik, and Orland (2009) for another illustration of the significance of qualitative data in focusing research questions and modifying the analyses of an experimental study of housing provision.

▣ REFERENCES

Alcoff, L. (1991, Winter). The problem of speaking for others. *Cultural Critique*, 5–32.

American Educational Research Association. (2006). Standards for reporting on empirical social science research in AERA publications. *Educational Researcher, 35*(6), 33–40.

Attree, P., & Milton, B. (2006). Critically appraising qualitative research for systematic reviews: Defusing the methodological cluster bombs. *Evidence and Policy, 2*(1), 109–126.

Baez, B., & Boyles, D. (2009). *The politics of inquiry: Education research and the "culture of science."* Albany: State University of New York Press.

Barbour, R., & Barbour, M. (2003). Evaluating and synthesizing qualitative research: The need to develop a distinctive approach. *Journal of Evaluation in Clinical Practice, 9*(2), 179–185.

Besley, T. (Ed.). (2009). *Assessing the quality of educational research in higher education: International perspectives.* Rotterdam, The Netherlands: Sense Publishers.

Blunkett, D. (2000). Influence or irrelevance: Can social science improve government? Speech to the Economic and Social Research Council (ESRC). (Reprinted in *Research Intelligence, 71,* British Educational Research Association, and *Times Higher Education* 2000, February 4, 2000.) Available at http://www.timeshighereducation.co.uk/story.asp?storyCode=150012&sectioncode=26

Boaz, A., Solesbury, W., & Sullivan, F. (2004). *The practice of research reviewing 1: An assessment of 28 review reports.* London: UK Centre for Evidence-Based Policy and Practice, Queen Mary College.

Boaz, A., Solesbury, W., & Sullivan, F. (2007). *The practice of research reviewing 2: Ten case studies of reviews.* London: UK Centre for Evidence-Based Policy and Practice, Queen Mary College.

Boden, R., Cox, D., Nedeva, M., & Barker, K. (2004) *Scrutinising science: The Changing UK government of science.* London: Palgrave.

Bridges, D. (2005, December 16). *The international and the excellent in educational research.* Paper prepared for the Challenges of the Knowledge Society for Higher Education Conference, Kaunas, Lithuania.

Bridges, D. (2009). Research quality assessment in education: Impossible science, possible art? *British Educational Research Journal, 35*(4), 497–517.

Brown, S. (2003, September 17). *Assessment of research quality: What hope of success?* Keynote address to European Educational Research Association annual conference, Hamburg, Germany.

Bush, V. (1945, July). *Science: The endless frontier.* A report to the president by Vannevar Bush, Director of the Office of Scientific Research and Development. Washington, DC: U.S. Government Printing Office. Available at http://www.nsf.gov/od/lpa/nsf50/vbush1945.htm

Cabinet Office. (2003a). *Quality in qualitative evaluation: A framework for assessing research evidence* [Summary]. London: Author.

Cabinet Office. (2003b). *Quality in qualitative evaluation: A framework for assessing research evidence* [Full report]. London: Author.

Cheek J. (2007). *Qualitative inquiry, ethics, and the politics of evidence. Qualitative Inquiry, 13*(8), 1051–1059.

Cook, T., & Payne, M. (2002). Objecting to the objections to using random assignment in educational research. In F. Mosteller & R. Boruch (Eds.), *Evidence matters: Randomized trials in education research* (pp. 150–178). Washington, DC: Brookings Institution Press.

Cronbach, L. (1975). Beyond the two disciplines of scientific psychology. *American Psychologist, 30,* 116–127.

Cronbach, L., & Associates. (1980). *Toward reform of program evaluation*. San Francisco: Jossey-Bass.

Davies, P. (2004). Systematic reviews and the Campbell Collaboration. In G. Thomas & R. Pring (Eds.), *Evidence-based practice in education* (pp. 21–33). Maidenhead, UK: Open University Press.

Davies, P., & Boruch, R. (2001). The Campbell Collaboration. *British Medical Journal, 323,* 294–295.

Department for Business, Innovations, and Skills. (2009). *Higher ambitions: The future of universities in a knowledge economy*. Available at http://www.bis.gov.uk/policies/higher-ambitions

Department of Health. (2005). *Research governance framework for health and social care* (2nd ed.). London: Author.

Dixon-Woods, M., Booth, A., & Sutton, A. (2007). Synthesizing qualitative research: A review of published reports. *Qualitative Research, 7*(3), 375–422.

Dixon-Woods, M., Fitzpatrick, R., & Roberts, K. (2001). Including qualitative research in systematic reviews: Opportunities and problems. *Journal of Evaluation in Clinical Practice, 7*(2), 125–133.

Economic and Social Research Council. (2005). *Postgraduate training guidelines*. Available at http://www.esrcsocietytoday.ac.uk/ESRCInfoCentre/Images/Postgraduate_Training_Guidelines_2005_tcm6-9062.pdf

Economic and Social Research Council. (2009). *Capacity building clusters*. Available at http://www.esrcsocietytoday.ac.uk/ESRCInfoCentre/research/CapacityBuildingClusters/index.aspx

Eisenhart, M., & DeHaan, R. (2005). Doctoral preparation of scientifically based education researchers. *Educational Researcher, 34*(4), 3–13.

Erickson, F., & Gutierrez, K. (2002). Culture, rigor, and science in educational research. *Educational Researcher, 31*(8), 21–24.

Fetterman, D. (2001). *Foundations of empowerment evaluation*. Thousand Oaks, CA: Sage.

Fielding, M. (2004). Transformative approaches to student voice: Theoretical underpinnings, recalcitrant realities. *British Educational Research Journal, 30*(2), 295–311.

Gibbons, M., Limoges, C., Nowotny, H., Schwartzman, S., Scott, P., & Trow, M. (1994). *The new production of knowledge*. Thousand Oaks, CA: Sage.

Goodley, D. (1999). Disability research and the "researcher template": Reflections on grounded subjectivity in ethnographic research. *Qualitative Inquiry, 5*(1), 24–46.

Gorard, S., & Taylor, C. (2004). *Combining methods in educational and social research*. Maidenhead, UK: Open University Press.

Gough, D., & Elbourne, D. (2002) Systematic research synthesis to inform policy, practice, and democratic debate. *Social Policy and Society, 1*(3), 225–236.

Greenberg, D. (2001). *Science, money, and politics* Chicago: University of Chicago Press.

Grissmer, D., Subotnik, R., & Orland, M. (2009). *A guide to incorporating multiple methods in randomized controlled trials to assess intervention effects*. Available at http://www.apa.org/ed/schools/cpse/activities/mixed-methods.aspx

Guba, E., & Lincoln, Y. (1981). *Effective evaluation: Improving the usefulness of evaluation results through responsive and naturalistic approaches*. San Francisco: Jossey-Bass.

Guba, E., & Lincoln, Y. (1989). *Fourth generation evaluation*. Newbury Park, CA: Sage.

Gueron, J. (2002). The politics of random assignment: Implementing studies and affecting policy. In F. Mosteller & R. Boruch (Eds.), *Evidence matters: randomized trials in education research* (pp. 15–49). Washington, DC: Brookings Institution Press.

Hamilton, D., Jenkins, D., King, C., MacDonald, B., & Parlett, M. (1976). *Beyond the numbers game*. London: Macmillan.

Hammersley, M. (1997). Educational research and teaching: A response to David Hargreaves' TTA lecture. *British Educational Research Journal, 23*(2), 141–161.

Hammersley, M. (2001). On systematic reviews of research literature: A narrative response. *British Educational Research Journal 27*(4), 543–554.

Hammersley, M. (2005). The myth of research-based practice: The critical case of educational inquiry. *International Journal of Social Research Methodology, 8*(4), 317–330.

Hammersley, M. (2008). Paradigm war revived? On the diagnosis of resistance to randomized controlled trials and systematic review in education. *International Journal of Research and Method in Education, 31*(1), 3–10.

Hargreaves, D. (1996). *Teaching as a research-based profession*. Teacher Training Agency 1996 Annual Lecture. London: Teacher Training Agency.

Higher Education Funding Council for England. (2009). *Research Excellence Framework*. Bristol, UK: Author.

Hillage, J., Pearson, R., Anderson, A., & Tamkin, P. (1998). *Excellence in research on schools* (DfEE Research Report 74). London, Department for Education and Employment.

House, E., & Howe, K. (1999). *Values in evaluation and social research*. Thousand Oaks, CA: Sage.

Howe, K. (2004). A critique of experimentalism. *Qualitative Inquiry, 10*(1), 42–61.

Jackson, A., & Mazzei, L. (Eds.). (2009). *Voice in qualitative inquiry*. London: Routledge.

James, M. (2006). Balancing rigor and responsiveness in a shifting context: Meeting the challenges of educational research. *Research Papers in Education, 21*(4), 365–380.

Lagemann, E. (2000). *An elusive science: The troubling history of education research*. Chicago: University of Chicago Press.

Lather, P. (2004). This IS your father's paradigm: Government intrusion and the case of qualitative research in education. *Qualitative Inquiry, 10*(1), 15–34.

Lincoln, Y. (1995). Emerging criteria for quality in qualitative and interpretive research. *Qualitative Inquiry, 1*(3), 275–289.

Lincoln, Y., & Tierney, W. (2004). Qualitative research and institutional review boards. *Qualitative Inquiry, 10*(2), 219–234.

MacDonald, B. (1987). Evaluation and the control of education. In R. Murphy & H. Torrance (Eds.), *Evaluating education: Issues and methods*. London: Harper & Row. (Reprinted from *Innovation, evaluation, research and the problem of control*, pp. 9–22 [SAFARI Interim Papers], by B. MacDonald & R. Walker, Eds., 1974, Norwich: UK: University of East Anglia, Centre for Applied Research in Education.

MacLure, M. (2003). *Discourse in education and social research*. Maidenhead, UK: Open University Press.

MacLure, M. (2005). Clarity bordering on stupidity: Where's the quality in systematic review? *Journal of Education Policy, 20*(4), 393–416.

MacLure, M., Holmes, R., MacRae, C., & Jones, L. (2010). Animating classroom ethnography: Overcoming video-fear. In L. Mazzei & K. McCoy (Eds.), Thinking with Deleuze in qualitative research [Special issue]. *International Journal of Qualitative Studies in Education, 23*(5), 543–556.

Maxwell, J. (2004). Causal explanation, qualitative research, and scientific enquiry in education. *Educational Researcher, 33*(2), 3–11.

Middleton, S. (2009). Becoming PBRF-able: Research assessment and education in New Zealand. In T. Besley (Ed.), *Assessing the quality of educational research in higher education: International perspectives* (pp. 193–208). Rotterdam, The Netherlands: Sense Publishers.

Moss, P., Phillips, D., Erickson, F., Floden, R., Lather, P., & Schneider, B. (2009). Learning from our differences: A dialogue across perspectives on quality in education research. *Educational Researcher, 38*(7), 501–517.

Mosteller, F., & Boruch, R. (Eds.). (2002). *Evidence matters: Randomized trials in education research.* Washington, DC: Brookings Institution Press.

National Centre for Research Methods. (n.d.). *A strategic framework for capacity building within the ESRC National Centre for Research Methods (NCRM).* Available at http://www.ncrm.ac.uk/TandE/capacity/documents/NCRMStrategicFrameworkForCapacityBuildingMain.pdf

National Research Council. (2002). *Scientific research in education.* Washington, DC: Author.

National Research Council. (2005). *Advancing scientific research in education.* Washington, DC: Author.

Nisbet, J., & Broadfoot, P. (1980). *The impact of research on policy and practice in education.* Aberdeen, Scotland: Aberdeen University Press.

Oakley, A. (2000). *Experiments in knowing.* Cambridge, UK: Polity Press.

Oakley A. (2003). Research evidence, knowledge management and educational practice: Early lessons from a systematic approach. *London Review of Education, 1*(1), 21–33.

Oakley, A. (2006). Resistances to new technologies of evaluation: Education research in the UK as a case study. *Evidence and Policy, 2*(1), 63–88.

Pollard, A. (2005). Challenges facing educational research. *Educational Review, 58*(3), 251–267.

Ragin, C., Nagel, J., & White, P. (2004). *Workshop on scientific foundations of qualitative research.* Available at http://www.nsf.gov/pubs/2004/nsf04219/start.htm

Roberts, K., Dixon-Woods, M., Fitzpatrick, R., Abrams, K., & Jones, D. (2002). Factors affecting uptake of childhood immunization: A Bayesian synthesis of qualitative and quantitative evidence. *The Lancet, 360,* 1596–1599.

Schwandt, T. (1996). Farewell to criteriology. *Qualitative Inquiry 2*(1), 58–72.

Shavelson, R., Phillips, D., Towne, L., & Feuer, M. (2003). On the science of education design studies. *Educational Researcher, 32*(1), 25–28.

Slavin, R. (2002). Evidence-based education policies: Transforming educational practice and research. *Educational Researcher, 31*(7), 15–21.

Smith, L. (2005). On tricky ground: Researching the native in the age of uncertainty. In N. K. Denzin & Y. S. Lincoln (Eds.), *The SAGE handbook of qualitative research* (3rd ed., pp. 85–107). Thousand Oaks, CA: Sage.

Somekh, B., & Saunders, L. (2007). Developing knowledge through intervention: Meaning and definition of "Quality" in research into change. *Research Papers in Education, 22*(2), 183–197.

Somekh, B., Underwood, J., Convery, A., Dillon, G., Jarvis, J., Lewin, C., et al. (2007). *Final report of the evaluation of the ICT Test Bed Project.* Coventry, UK: Becta.

Stake, R. (1967). The countenance of educational evaluation. *Teachers' College Record, 68,* 523–540.

Stake, R. (1978). The case study method in social inquiry. *Educational Researcher, 7*(2), 5–8.

Stenhouse, L. (1975). *An introduction to curriculum research and development.* London: Heineman.

Stenhouse, L., Verma, G., Wild, R., & Nixon, J. (1982). *Teaching about race relations: Problems and effects.* London: Routledge.

Tooley, J., & Darby, D. (1998). *Educational research: A critique.* London: Office for Standards in Education.

Torrance, H. (2006). Research quality and research governance in the United Kingdom: From methodology to management. In N. K. Denzin & M. Giardina (Eds.), *Qualitative inquiry and the conservative challenge* (pp. 127–148). Walnut Creek, CA: Left Coast Press.

Torrance, H. (2008). *Overview of ESRC research in education: A consultancy commissioned by ESRC: Final report.* Available at http://www.sfre.ac.uk/uk/

Torrance, H., Colley, H., Ecclestone, K., Garratt, D., James, D., & Piper, H. (2005). *The impact of different modes of assessment on achievement and progress in the learning and skills sector.* London: Learning and Skills Research Centre.

Torrance, H., & Coultas, J. (2004). *Do summative assessment and testing have a positive or negative effect on post-16 learners' motivation for learning in the learning and skills sector?* London: Learning and Skills Research Centre.

Viadero, D. (2009, April 1). "No effects" studies raising eyebrows. *Education Week.* Available at http://www.projectcriss.com/newslinks/Research/MPR_EdWk--NoEffectsArticle.pdf

Wade, C., Turner, H., Rothstein, H., & Lavenberg, J. (2006). Information retrieval and the role of the information specialist in producing high-quality systematic reviews in the social, behavioral, and education sciences. *Evidence and Policy, 2*(1), 89–108.

Weiss, C. (1972). *Evaluating action programs.* Boston: Allyn & Bacon.

Weiss, C. (1980). *Social science research and decision-making.* New York: Columbia University Press.

Woodhead, C. (1998, March 20). Academia gone to seed. *New Statesman,* pp. 51–52.

Yates, L. (2004). *What is quality in educational research?* Buckingham, UK: Open University Press.

35

REFLECTIONS ON INTERPRETIVE ADEQUACY IN QUALITATIVE RESEARCH

David L. Altheide and John M. Johnson

A line is a dot that went for a walk.

—Paul Klee

Over 15 years ago, we published "Criteria for Assessing Interpretive Validity in Qualitative Research" (Altheide & Johnson, 1994). In this work, we continued the development of our ideas about qualitative research, many of which were developed during a long series of professional meetings in the 1980s. Our animating questions were how interpretive methodologies should be judged by readers (audiences) who share the perspective that how knowledge is acquired, organized, interpreted, and presented is relevant for the substance of those claims. We presented our ideas about how to make the claims and narratives of qualitative research more trustworthy to readers and audiences. We called our approach "analytic realism," to identify how reflexive and interpretive methods could be presented to enhance their credibility, relevance, and importance.

Much has happened in the world of qualitative methods during the last two decades, and important questions and issues now span many new disciplines, venues, arenas, perspectives, theories, and problem areas. Methodological issues once considered relevant for only a minority of anthropologists and sociologists are now discussed (and disputed) in many other disciplines, especially education, policy studies, the health sciences, gender studies, communication, cultural studies, justice studies, and others. Many new models of representation and interpretation have arisen during this period, including those from advocates for linking qualitative research to justice values, issues, and agendas. Performative writing and performance ethnography emerge as new positions, and advocates for standpoint epistemologies cross disciplinary boundaries. A robust debate about using cyberspace for research generates new issues and perspectives. In the context of this creative flux, conservative countermovements emerge to standardize or normalize the practice and evaluation. In the United States, a scientifically based research (SBR) countermovement has arisen, also called scientific inquiry in education (SIE), and this is joined by similar movements in the United Kingdom, called the research assessment exercise (RAE), or in Australia the research quality framework (RQF). Two reports by the U.S. National Science Foundation have proven especially troublesome and problematic for American scholars (Lamont & White, 2009; Ragin, Nagel, & White, 2004). These movements have generated heated debates about "the politics of evidence," and inspired claims about qualitative inquiry being "under fire" by these state-sponsored efforts to constrain and control the criteria of scientific inquiry (Denzin, 2009).

Our advocacy of analytic realism in the 1994 paper was intended to align our efforts with philosophical realism, arguably the dominant philosophical position in the social sciences for many decades. The basic idea of realism is that there is a real world with which we act and interact (an "obdurate" world in the words of George Herbert Mead), that individuals and groups create meaning in this world, and that while our theories, concepts, and perspectives may approach some kind of valid understanding, they cannot and do not exhaust the phenomena of our interest. All theories, concepts, and findings are grounded in values and perspectives; all knowledge is contextual and partial; and other conceptual schemas and perspectives are always possible. We are heartened that many other scholars have advanced some related version of this perspective, such as "critical realism" (Bhaskar, 1979; Harré & Madden 1975; Manicas & ebrary, Inc. 2006; Maxwell 2008), "experimental realism" (Lakoff, 1987), "subtle realism" (Emerson, Fretz, & Shaw, 1995; Hammersley, 1992), "ethnographic realism" (Lofland, 1995), "innocent realism" (Haak, 2003), "natural realism" (Putnam, 1999), and "emergent realism" (Henry, Julnes, & Mark, 1998). These different versions of realism share certain basic ideas: that human social life is meaningful, and that it is essential to take these meanings into account in our explanations, concepts, and theories; furthermore, to grasp the importance of the values, emotions, beliefs, and other meanings of cultural members, it is imperative to embrace an interpretivist approach in our scientific and theoretical work. According to Maxwell (2008), these versions of realism reflect an ontological realism while simultaneously accepting a form of epistemological constructionism and relativity. They oppose the radical constructivist view, which denies the existence of any reality apart from our constructions of it (or them). This kind of realist perspective has proven very valuable in grasping the relationship(s) between the meanings and perspectives of cultural members and the social contexts in which they are embedded, and especially for understanding conflicts or differences in meanings for actors located in the same situation or context.

When knowledge and evidence are viewed from a symbolic interactionist perspective, evidence is seen *as part of a communication process that symbolically joins an actor, an audience, a point of view, assumptions, and claims about the relations between two or more phenomena.* This view of evidence-as-process is termed the "evidentiary narrative," and draws attention to the ways in which credible information and knowledge are buffeted by symbolic filters, including distinctive "epistemic communities," or collective meanings, standards, and criteria that govern sanctioned action (see Altheide, 2008). It is heartening to see that numerous qualitative researchers in a number of these epistemic communities have made impressive strides in addressing their validity issues within their own perspectives, and have developed many useful ideas to create more trustworthy knowledge to be shared with other audiences. We briefly review some of these developments as a prelude to saying how this has changed our views in recent years.

◫   FRAMING VALIDITY ISSUES IN INTERPRETIVE RESEARCH

There are many ways to use, practice, promote, and claim qualitative research, and in each there is a proposed or claimed relationship between some field of human experience, a form of representation, and an audience. Researchers and scholars in each of these areas have been grappling with issues of truth, validity, verisimilitude, credibility, trustworthiness, dependability, confirmability, and so on. What is valid for clinical studies or policy studies may not be adequate or relevant for ethnography or autoethnography or performance ethnography. We return to this point later.

Clinical Studies

In recent years, researchers in several clinical fields have utilized qualitative methods to grasp and articulate invisible or taken-for-granted realities they experience in clinical settings. Miller and Crabtree (2005) write,

> Qualitative clinical researchers bring several power perspectives to the clinical encounter that help surface the unseen and the unheard and also add depth to what is already present. These include understanding disease as a cultural construction . . . possessing knowledge of additional medical models such as biopsychosocial and humanistic models, homeopathy, and non-Western models that include traditional Chinese, Ayurvedic, and shamanism, and recognizing the face and importance of spirituality in human life. (p. 612)

Clinical practitioners who utilize observational, narrative, or discourse methods are seeking to articulate standards of validity (or truthfulness) that can be shared with others in their field, and hence are subject to independent tests and verification. Of these efforts, Rolfe (2004) writes,

> [For some,] validity and reliability [in clinical studies] are achieved when the researcher rigorously follows a number of verification strategies in the course of the research process. "Together, all these verification strategies incrementally and interactively contribute to and build reliability and validity, thus ensuring rigor. Thus, the rigor of qualitative inquiry should be beyond question, beyond challenge, and provide pragmatic *scientific evidence* that *must* be integrated into our developing knowledge base" [Morse et al., 2002, emphasis added by Rolfe]. This statement of intent exemplifies very strongly the aspirations of some qualitative researchers to the values, approaches, terminologies, and hence, to the certainties of the "hard" sciences. Rigor is clearly the key to success. . . . [But others] argue that issues of validity in qualitative studies should be linked not to "truth" or "value," as they are for the positivists, but rather to "trustworthiness," which becomes a matter of persuasion whereby the scientist is viewed as having made those practices visible, and therefore, auditable. (p. 305)

These comments illustrate the ongoing debates of the *evidence-based practice movement* in clinical studies, where for many years now qualitative practitioners and researchers have been discussing the applicability of such ideas as truth, validity, reliability, trustworthiness, and so on. While many anthropologists or sociologists who practice qualitative research might be primarily motivated to make fundamental contributions to the basic knowledge of their disciplines, or contributions to a substantive problem area, those who practice qualitative methods in clinical studies typically intend a different audience, those interested in advancing effective clinical practice. These comments illustrate that there is a diversity of purposes that animate qualitative research, and that criteria of usefulness in research are tied to these practical purposes and disciplinary/occupational values.

Policy Studies

Qualitative research is increasingly used in policy studies, where the intention is to study how various actors bring and make meaning in actual concrete settings, and the consequences of these actions. The goals of this type of research are crisply stated by Hammersley (2005): "Qualitative policy research is aimed at having an impact on current programs and practices" (p. 3). The focus here can be on the impact or consequences of policy, but additionally the processes of how official law or policies are translated and interpreted, from the heights of inception down to the points of implementation, to the "street-level" realities. There is little doubt that qualitative research can be more flexible than traditional quantitative research, and has the potential to adjust research agendas to meet changing demands in the field. While the focus of quantitative research is usually on "outcome measures," or metrics, qualitative research has the potential to study the complex social and bureaucratic processes whereby laws and policies are actually implemented in daily life. In the United Kingdom, central policy makers have proposed their own set of standards for what they expect of qualitative research in these areas (Cabinet Office, 2003). These should include a detailed description of the contexts in which the study was conducted, a discussion of how fieldwork settings or methods may have influenced data collection, descriptions of background or historical developments and social/organizational characteristics of study sites, description and illumination of diversity/multiple perspectives or positions, and discussion/evidence of the ideological perspectives/values/philosophies that guide the researcher or research team (see Torrance, 2007, pp. 55–79).

The tensions and debates in policy studies mirror those in other areas; some feel that validity or truth is better served by affirming a set of research standards (such as those offered by the UK Cabinet Office above), whereas others think that truth is better served by making comparative assessments between studies, over time, and between diverse settings. Harry Torrance (2007) writes,

> Assuring the quality of research, and particularly the quality of qualitative research in the context of policy making, must be conceptualized as a vital and dynamic process that is always

subject to further scrutiny and debate. The process cannot be ensconced in a single research method or a once-and-for-all set of standards. (p. 73)

Others in the policy field also affirm the necessity for comparing research reports with prior reports, and for comparing settings with other settings (Hammersley, 1992). This is a common theme in what is now called "the new public management movement," which seeks greater transparency and clarity of the policy-making and policy-implementation processes. This suggests a dual approach to the issue of validity: certain expectations for the researcher or research team to show readers the grounds for trusting their report, on the one hand, in conjunction with a measured and realistic skepticism by readers, on the other hand, to place the claims of any given research report in a context of many other reports, even one's life experiences.

Action Research

Action research or participatory action research is another emergent form of qualitative research, which usually involves one researcher or a research team in the field, participating with societal members to produce social change or implement a social policy or organized response to a problem. Kemmis and McTaggart (2005) propose their vision of this kind of research:

> Through participatory action research, people can come to understand that—and how—their social and educational practices are located in, and are the product of, particular material, social, and historical circumstances that *produced* them and by which they are *produced* in everyday social interaction in a particular setting. By understanding their practices as the product of particular circumstances, participatory action researchers becomes alert to clues about how it may be possible to *transform* the practices they are producing and reproducing through their current ways of working. (p. 565, emphasis original)

Social action researchers do not deny the relevance or importance of foundational or basic knowledge, but often insist that they seek experiential knowledge as well, often expressing the hope that theory and pragmatics together can achieve a whole that is greater than its parts. To provide grounds for trusting participatory action research, Ladkin (2004) proposes that research done in this vein should include accounts to demonstrate emergence and enduring consequences of actions or policies, accounts of how the research dealt with pragmatic issues of practice and practicing, accounts of how the research deals with questions of significance, and accounts showing how the research considers a number of different ways of knowing. While it would be easy to criticize Ladkin's proposals as being very abstract and insubstantial, or for emphasizing certain things (like the consequences) that are not within the control of the researcher or research team, our main point here is to show that those working in this vein are struggling with issues of truthfulness and validity, and seek to engage this debate within the confines of their own expertise.

Autoethnography and Expressive Frames

A diversity of current research seeks to break down the prior barriers between subject and object, between the knower and the known, between the self and the social, between the spiritual and the empirical, and between the writer and the audience. Autoethnography is only one of several names given to this emerging enterprise, and autoethnographers commonly seek to integrate the storyteller and the story. Laurel Richardson (1997) says that "writing stories about our 'texts' is a way of making sense of and changing our lives" (p. 5). Carolyn Ellis (2009) seeks to "open up conversations about emotions in romantic and family relationships" (p. 17). She additionally says,

> Thus, reexamining the events we have lived through and the stories we have told about them previously allows us to expand and deepen our understandings of the life we have led, the culture in which we have lived, and the work we have done. This review provides new possibilities for understanding ourselves and keeps us from remaining stuck in the interpretations we have settled on in the past. (p. 13)

On many occasions, autoethnographers grapple with the issue of "memory," and how to contextualize or re-contextualize people and events from long ago. This issue is larger than the vagaries of remembering empirical facts, because it includes the many issues of interpretation and perspective. On the standards involved in this kind of research, Bochner (2007) states,

> Of course, my gravest obligation is not to lie. But the space between lying and telling the truth can be vast. If telling the truth is merely saying what I remember, then I have set the bar of obligation extremely low. Once the past was there, now it is gone. I want to be faithful to the past, but what I remember of my history is anchored by what summons me *now* to remember, and my memory is, in part, a response to what inspired my recollections. (p. 198, emphasis original)

For many of its practitioners, autoethnography becomes a disciplined way to interrogate one's memory, to contextualize or re-contextualize empirical facts or memories within interpretations or perspectives that "make sense" of them in new or newly appreciated ways. Many of these studies deal with intimate and family relations, and commonly seek to make explicit what is usually taken for granted within these relationships, the many seen-but-unspoken or known-but-not-acknowledged complexities of our lives. In some autoethnographic studies, the time frame of interest is longer than that of a traditional, observational ethnography, which might take as long as 10 years between the time of inception and the time of the final report;

many autoethnographers seek to elucidate the changes in meaning or perspective over many years, even decades, where the long passage of time itself produces new or altered understandings of past "facts."

Performance ethnography and expressive artists often share the autoethnographer's desire to explore and communicate the deeply personal or taken-for-granted aspects of personal or daily life, but in addition they often seek to engage their audience in a more direct manner, often seeking to evoke an emotional response. This may be done with acted performances, poems, photography, multimedia collages, or readings. Szto, Furman, and Langer (2005) write,

> The photographer is an ethnographer in this sense [of trying to capture a subject's reality]. You try to capture the context. You have to take poetic license and select context.... In the role of researcher, the poet must engage in conscious and constant self-exploration. When he (or she) writes about a subject in front of himself (herself), or when he (she) is reducing data from narratives, he (she) has to be very clear to stay faithful to the data. His (her) notes serve as both data to be worked with, as well as ethnographic notes that explore their reactions. Many times, these biases should be presented so the readers can decide for themselves how to interpret the poem. The first allegiance of the researcher, as poet, has to be to the subject's experience. In a sense, there are two types of poems for the researcher. There are poems in which they merely present the subjects' experience as accurately as possible, hopefully utilizing their words, and then there are interpretive poems, in which they deconstruct the meaning of the experience and consciously allow for interpretation. (p. 139)

There is great diversity of qualitative research, and there is diversity in the ways to justify or legitimize each of the above approaches. While these approaches differ, they also share an ethical obligation to make public their claims, to show the reader, audience, or consumer why they should be trusted as faithful accounts of some phenomenon. Moreover, each of these approaches reflects the context and purposes for the practitioners and audiences, including clients. The pragmatic utility of validity as "good for our present intents and purposes" cuts through all of the methodological approaches and authoritative claims. In other words, whether it is truthful, accurate, on the mark, and so forth is framed by an ecology of knowing tied to practices and intentions, and ultimately, "our justifications" for using this method. What is common to each of these approaches, and by implication all forms of inquiry, is a *process* of acquiring information, organizing it as data, and then analyzing and interpreting those data with the help of refractive (conceptual, theoretical, perhaps political) lenses.

We have noted that validity has been referred to many ways, including successor validity, catalytic validity, interrogated validity, transgressive validity, imperial validity, simulacra/ironic validity, situated validity, and voluptuous validity (see, e.g., Atkinson, 1990; Atkinson, 1992; Guba, 1990; Hammersley, 1990, 1992; Lather, 1993; Wolcott, 1990). Our effort to clarify the logic-in-use by many qualitative researchers suggested that a heuristic view of "hyphenated validity" could help clarify the methodological discourse at the time (early 1990s).

Types of Validity

Validity-as-culture (VAC) is well known to social science students. A basic claim is that the ethnographer reflects, imposes, reproduces, writes, and then reads their cultural point of view for the "others." Point of view is the culprit in validity. The solution includes efforts to include more points of view, including reassessing how researchers view the research mission and the research topic. Atkinson (1992, p. 34 ff.) suggests that ethnographies can be mythologized: "But the sense of class continuities is hardly surprisingly stronger in the British genre than in the American which is more preoccupied with a sense of place."

Validity-as-ideology (VAI) is very similar to VAC, except the focus is on the certain specific cultural features involving social power, legitimacy, assumptions about social structure, e.g., subordinate/superordinate.

Validity-as-gender (VAG), like the previous two, focuses on taken-for-granted assumptions made by "competent" researchers in carrying out their conceptual and data collection tasks, including some issues about power and domination in social interaction. One concern is that these asymmetrical aspects of social power may be normalized and further legitimated.

Validity-as-language/text (VAL) resonates with all that have come before, particularly how cultural categories and views of the world, as implicated in language, and more broadly, "discourse," restricts decisions and choices by how things are framed.

Validity-as-relevance/advocacy (VAR) stresses the utility and "empowerment" of research to benefit and uplift those groups often studied, relatively powerless people, e.g., the poor, peasants, etc.

Validity-as-standards (VAS) asserts that the expectation about a distinctive authority for science, or the researchers legitimized by this "mantle of respectability," is itself suspect, and that truth-claims are so multiple as to evade single authority or procedure. In the extreme case, science ceases to operate as a desirable model of knowledge, because it is, after all, understanding rather than codified, theoretically integrated information—as knowledge—that is to be preferred. (Altheide & Johnson, 1994, p. 488)

These approaches to validity, while certainly not definitive, reflected the purpose and audiences for research approaches and applications. The subtext was openness and engagement. Notwithstanding intense debates that occurred among the practitioners, the tone was inclusive and the spirit was to not overlook important segments of, say, audiences that might not be so well served by unintended limitations of the generation of knowledge. Informed by our basic assumption that *the social world is an interpreted world, not a literal world, always under symbolic construction, with emphasis on awareness of the process of the ethnographic work*, we offered another inclusive view:

validity-as-reflexive-accounting (VARA), which places the researcher, the topic, and the sense-making process in interaction.

We identified these hyphenated validities (above) as illustrations of the range of attention the "problem of validity" has received. But another standard for validity has appeared. The SBR, SIE, and RAE movements noted above represent examples of an expanded context of control involving disparate audiences with oversight interests in corralling and regulating qualitative research in accordance with conventional formats of communication and regulation, and ultimately legitimacy associated with more positivistic methodology. This has led to a new version of validity, validity-as-marketable-legitimacy (VAML), which refers to the negotiated order of socially sanctioned (and respectable) research methodology.

This latest version of validity is being promoted for bureaucratic, rational, and organizational purposes and not to enhance inquiry, creativity, or discovery, but rather, accountability, as in; Funding this is warranted according to our "guidelines"; therefore no individual will be accountable for any errors, and so forth (Denzin, 2010; Kvale & Brinkmann, 2008). An unintended consequence of VAML is to dampen the creative search for varied forms of truth and relevant search. Openness and the tremendous success of the explorations of various approaches to qualitative work across many disciplines have, paradoxically, contributed to its utility, use, and imposed limitations for practical purposes. In recent years, the debate has intensified and changed, as qualitative research has become more accepted for funding and "practical" applications, e.g., policy research. One interpretation is that qualitative research is now in the marketplace of ideas, where the coinage is not just intellectual prowess, but actual coin. The approaches to validity that we delineated were consistent with the pursuit of truth and a logic of discovery. However, the success of varied approaches to qualitative research has opened up market possibilities not only for publication and teaching, but also for funding and sponsored projects that are accountable to administrators and overseers of agencies and organizations that must answer to other scientific and political constituencies. The standardization and reduction of an array of approaches leads to more than smoothing out sharp edges; domains become sacrificed for the sake of an established lexicon, rhetoric, and narrative of authorized knowing, ultimately as "objective." The push to the linear criteria and decision making—recall Klee's dot that went for a walk!—means that approved criteria and checklists of acceptability and standardization matter. We will say a bit more about this later in the essay after discussing the complexities of evidence and the importance of tacit knowledge in understanding the social construction of reality. For now, we wish to emphasize that the standardization of qualitative work is also risk avoidance and demonstrates a "risk society" approach to solving the problem of evaluating research, and so forth (Erikson & Doyle, 2003). However, evidence is not that simple.

The Problem of Evidence

Much of the foregoing discussion rests on an understanding of "evidence," or agreed-upon—or potentially agreeable—information that would serve as a basis or foundation. There is a rich literature on the control of information, research subjects, and topics (Van den Hoonaard, 2002), as well as the politics of evidence (Altheide, 2008; Denzin & Giardina, 2008). Evidence and facts are similar but not identical. We can often agree on facts, e.g., there is a rock, it is harder than cotton candy. Evidence involves an assertion that some facts are relevant to an argument or claim about a relationship. Since a position in an argument is likely tied to an ideological or even an epistemological position, evidence is not completely bound by facts, but is more problematic and subject to disagreement. Indeed, until the 1990s, most qualitative research was not taken seriously by many sociologists, and was regarded as second-rate social science; editors and reviewers for the discipline's major journals would rarely publish qualitative reports. The basic problem, of course, was that qualitative research, with exceptions, was not regarded as being based on data; quotes and observations were not regarded as appropriate evidence, especially if there were not a large enough number or "N" of these, thus rendering qualitative claims as akin to quantitative estimates. The situation was so dire that several groups of qualitative researchers started their own journals, including *Urban Life* (later, *Journal of Contemporary Ethnography*), *Symbolic Interaction,* and *Qualitative Sociology.* In 2011—as this handbook attests—the situation has much improved, as qualitative methods, data, and "evidence" have come to be more accepted. Yet, as the foregoing discussion indicates, there remains a kind of impatience with the wide-ranging epistemologies that certain interest groups want to compartmentalize and regulate.

The problem that sociologists have with conflicts over evidence, however, is minimal compared to the nonacademic settings. When the president of the United States uses rhetoric and photographs to demonstrate/show that Iraq had weapons of mass destruction (WMD) and therefore should be invaded in order to keep the world safe, and when this "proof" turns out to be false, then it is even more important that evidence be examined and critically analyzed. Accordingly, qualitative researchers have focused on evidence and its social contexts (Denzin & Giardina, 2008).

We wish to stress that communication strategies, formats, and paradigmatic boundaries can cloud our vision. The symbolic meaning filters that are called forth all stem from various memberships. Ultimately, evidence is bound up with our identity in a situation. The multiple memberships we hold in various epistemic communities are situationally shuffled and joined for a particular purpose (e.g., when an assumption or value is challenged or called into question). An "evidentiary narrative" emerges from a reconsideration of how knowledge and belief systems in everyday life are tied to epistemic communities that provide perspectives, scenarios, and scripts that reflect symbolic social and moral orders.

> An "evidentiary narrative" symbolically joins an actor, an audience, a point of view (definition of a situation), assumptions, and a claim about a relationship between two or more phenomena. If any of these factors are not part of the context of meaning for a claim, it will not be honored, and thus, not seen as evidence. Moreover, only the claim is discursive, or potentially problematic, but it need not be so. (Altheide, 2009, p. 65)

The idea is that evidence is not about facts per se, but is about an argument, a narrative that is appropriate for the purpose-at-hand. That means it is contextualized and part of a bounded project, with accompanying assumptions, criteria, rules of membership, participation, and so on.

From a sociology-of-knowledge perspective, the active reception of a point "of information" is contingent on the "media logic" of legitimacy (acceptability) of the information source, the technology, medium, format and logic through which it is delivered (Altheide & Snow, 1979). What is meant by "evidence" can be viewed as "information that is filtered by various symbolic filters and nuanced meanings compatible with membership" (Altheide, 2009, p. 65). Only then can the information be interpreted as evidence in juxtaposition with an issue, problem, or point of contention. Conversely, information that is not suitably configured and presented is likely to be resisted, if not rebuffed, within a prevailing discourse.

Earlier we argued (Altheide & Johnson, 1994) that a more encompassing view of the ethnographic enterprise would take into account the process by which the ethnography occurred, which must be clearly delineated, including accounts of the interaction between the context, researcher, methods, setting, and actors. The broad term that we offered, "analytic realism," is based on the view that the social world is an interpreted world, not a literal world, always under symbolic construction (even deconstruction!). We can also apply this perspective to understand how situations in everyday life are informed by social contexts and uses of evidence. This application illuminates the process by which evidence is constituted. We can now see how any effort to standardize and limit qualitative research criteria is doomed to failure and irrelevance. As long as the core of qualitative research is extended to more specific audiences and uses, criteria for validity will be linked with the evidence appropriate for specific tastes and uses. It is useful, then, to consider the following elements of what we termed "an ethnographic ethic" (Altheide & Johnson, 1994, p. 489) when trying to understand evidence that is stated or affirmed in a situation:

(1) The relationship between what is observed (behaviors, rituals, meanings) and the larger cultural, historical, and organizational contexts within which the observations are made (the substance);

(2) The relationship between the observer, the observed, and the setting (the observer);

(3) The issue of perspective (or point of view), whether that of the observer or the member(s), used to render an interpretation of the ethnographic data (the interpretation);

(4) The role of the reader in the final product (the audience); and

(5) The issue of representational, rhetorical, or authorial style used by the author(s) to render the description or interpretation (the style).

Each of these areas includes questions or issues that must be addressed and pragmatically resolved by any particular observer in the course of his or her research. As originally formulated, these five dimensions of qualitative research include problematic issues pertaining to validity. Indeed, we argued that the "ethnographic ethic" calls for ethnographers to substantiate their interpretations and findings with a reflexive account of themselves and the process(es) of their research (Altheide & Johnson, 1993).

The evidentiary narrative is built on several arrays of meaning: what we know, who we are, and what we consider as evidence of either our most basic assumptions (e.g., that there is order in the world), or a specific claim about part of that order (e.g., my beliefs are legitimate and truthful). Prior to delving into a solution, let's attempt an overview of the problem. We live as social beings and are accountable to some people but not others. Why do we accept certain claims but reject others, and what would lead us to change our minds? In the modern era, this involved scientific authority about certain empirical truths that were based on "data" as evidence, or as a kind of fact-guided proof. The modernist project relied on rationality, including some formal rules of logic grounded in an objective view of things. This view has been seriously questioned by extensive research and writings encompassing such approaches as ethnomethodology, phenomenology, existential sociology, symbolic interactionism, feminism, literary criticism, performance studies, and autoethnography, to name a few (Denzin & Lincoln, 1994). These approaches contributed to the "reflexive turn" in the social sciences, and the examination of how the research process, including the "act of writing" partially produces the research result (Marcus & Clifford, 1986; Van Maanen, 1988).

The reflexive turn is central to the problem of evidence, not just why people believe "crazy things" and won't easily consider evidence that would lead them to reject such beliefs, but more basically, why—and how—do researchers and scientists accept information as evidence for a particular matter? The subject matter under investigation still matters, but mainly as a "product" that is socially constructed.

Efforts like SBR and SIE referred to above attempt to formulate what qualitative research should look like, in terms of criteria

and checklists, but such efforts are risky, since this is likely to be mainly regulative but not constitutive of the process that gave rise to the innovations in the first place. The upshot is nothing less than treating ethnography and qualitative research as a commodity—a product—to be bought and sold in a market, but this market, like all markets, is reflexive of the process and interest that gives rise to all markets. Nico Stehr's (2008) insights about knowledge markets are instructive:

> [W]hen viewed from a contemporary sociological perspective—and this is the much more common and developed critique—markets are not so much responsible for widespread affluence; rather, they represent a rather harsh, impersonal institution that has put into practice what major classical sociological theorists have always anticipated and, of course, feared. From this perspective market relations are nothing but power relations. Market relations are a form of pure power relationship, pitting the owners of the means of production against the owners of labor power. (p. 85)

The power relations inherent in funding agencies and sanctioning boards that sanctify appropriate methodology can, perhaps unintentionally, purge qualitative research of the subtle but important distinctions noted above.

The creative logic-in-use of scientific discovery that has shaped much of qualitative inquiry and development is pushed aside as qualitative research becomes recast as "expertise" and becomes a resource for knowledge producers to employ for practical purposes and for diverse audiences with limited understanding (and interest) in the complexities of qualitative research, including varieties of validity. Steven Fuller (2008) argues that historically, science and expertise have been antithetical forms of knowledge, with the former associated with creativity and contemplation, and in a sense, examining the nature and realms of order and possibilities, while the latter was more associated with application, and "doing" practical things. Focusing on research in science technological studies (STS), he asserts,

> Science and expertise are historically opposed ideas: The former evokes a universalistic ideal meant to be pursued in leisure, while the latter consists of particular practices pursued to earn a living. However, expertise can serve the universalistic ideal of science by undermining the authority of other expertises that would cast doubt on the viability of this ideal. Put bluntly, expertise is "progressive" only when it serves as the second moment of a Hegelian dialectic.... I see the modern university—specifically through its teaching function—as the place where this moment most often happens. (p. 115)

Fuller's concern is to save inquiry, exploration, and discovery as process from a steamrolling rhetoric and organizational push for completion and results that can flatten subtleties as useful products and procedures. This issue is apparent in current discussions about the nature and uses of evidence.

◨ THE EVIDENTIARY NARRATIVE AS PROCESS

We began with the quote from Klee, about a dot that went for a walk, and before one realizes, there is a line. The qualitative researcher shows us a "line" by describing and telling about the meanings in order to make them visible. Because of the reflexivity of all research and the indexicality of all communication, this is often problematic. The nature and meaning of a person's experience is not isomorphic with the researcher's account of that experience. As Schutz (1967) noted, the lifeworld is a world interpreted by social actors. These are first-order cognitions and constructs; the social scientist or qualitative research must interpret the actors' meanings and provide second-order constructs and accounts. As noted above, these second-order constructs are made within a social/cultural/historical context, with an intended audience. Our emphasis here is on the ways audience members might critically assess research reports or other representations in order to apply their criteria of validity, adequacy, or truthfulness. Some of the dimensions we will here examine are basic, yet they provide tools to assist us in assessing qualitative research. This raises the issue of the transparency of qualitative research, the ways used by researchers to "connect the dots" of their efforts, from inception to final report or representation. It is not possible to have complete or total transparency, again because of the inherent reflexivity and indexicality of the research process itself, yet most qualitative reports contain guidance about the relatedness of the observations, findings, claims, explanations, or conclusions.

Practitioners of qualitative methods routinely encounter certain problems and issues in the conduct of their observations, experiences, or research. While many of these problems and issues are legendary in traditional social science ethnography, rather than drawing on these well-known examples from classical observational research, we will use illustrations from an exceptional recent autoethnography, H. Lloyd (Bud) Goodall's book *A Need to Know: The Clandestine History of a CIA Family* (2006), which reports his experiences as a child growing up in a family governed by secrecy and paranoia, and routinely under the surveillance of the U.S. government. Only upon the death of his father did Goodall learn that he had been a CIA operative. Bud's subsequent impassioned quest for answers about his father's work, life, and commitments provided some clarity about family history, his mother's illness, and numerous moves, as well as documenting the institutional and organizational contexts for real family events, personal lives, and a young boy's fears and insecurities. Bud Goodall's problem was solved through careful weaving of subtle and tacit knowledge of real-life events that were constructively revealed by replaying narratively shaped memories, and checking these against new interpretations gleaned from various documents about his family history, including a Bible, a diary, and his father's agency codebook, *The Great Gatsby*.

We draw on this extraordinary project to illuminate the sometimes circular path of the dimensions of qualitative inquiry.

We argue that in one form or another, qualitative research, especially ethnography, involves some data collection, analysis, and interpretation, although these may not always be as apparent and transparent as a reader may desire. Indeed, a distinctive aspect of qualitative research designs that emerge from lived experience is the blurring of data collection and analysis, since the latter often informs what new data or examples/comparisons to seek out, clarify, and compare. For example, Professor Goodall (2006) began with his personal experiences, but later came to rely on various documents, including photographs, family and military records, and so forth, for more information. All research involves collecting, organizing, analyzing, and interpreting data. Sometimes this involves recognizing that we have something worthwhile, and then finding out what it is. The most basic concern is perhaps an epistemological one: How do we know? But this involves several features of the "knowing process," including what is it that we know, how did we learn or come to know something about it, and how do we make sense of this? We are not suggesting any particular approach, mode, or representation for these activities, but only that they help distinguish a social science "telling" or "accounting" from other genres.

Qualitative research should provide a window for a critical reading, or at the very least, permit an informed reader's queries about what is being read, or seen, or heard. Our position is that any claim for veracity, validity, adequacy, or truthfulness turns on the transparency of these dimensions, and their personal relevance, pertinence, and significance for the audience member (e.g., reader, listener, viewer). Transparency promotes empathic and sympathetic understanding and participation between the author and the audience. We wish to stress, then, that it is not enough to "like" a research account/narrative, but that for social science purposes, such "liking" can be linked to these dimensions.

The power of extensive involvement with the lifeworld one is experiencing is apparent from journalistic accounts as well. Journalists do not have a method as such. They inquire and search out leads for a story. Their main connection to a kind of realism is to have "facts checked" by someone else, which often involves someone finding out if a person actually told the journalist something, but not whether what that person said was actually true. While social scientists operate with a theoretical orientation that guides data collection and interpretation, being close to the data, the actual experience, is critical for understanding. Journalists are often closer to the action, and their descriptions provide great insight, even though their rendering is less theoretically guided and informed. This can be seen with journalist Roberto Saviano's (2007) statement of how he knows about the corruption of the Camorra, Italy's brutal criminal organization that dominates commerce in southern Italy. In describing his familiarity with the criminal control of the cement industry, Saviano states,

> I know and I can prove it. I know how economies originate and where they get their odor. The odor of success and victory. I know what sweats of profit. I know. And the truth of the war takes no prisoners because it devours everything and turns everything into evidence. It doesn't need to drag in cross-checks or launch investigations. It observes, considers, looks, listens. . . . The proofs are not hidden in some flash drive concealed in a hole in the ground. I don't have compromising videos hidden in a garage in some inaccessible mountain village. Nor do I possess copies of secret service documents. The proofs are irrefutable because they are partial, recorded with my eyes, recounted with words, and tempered with emotions that have echoed off iron and wood. I see, hear, look, talk, and in this way I testify, an ugly word that can still be useful when it whispers, "It's not true," in the ear of those who listen to the rhyming lullabies of power. The truth is partial; after all, if it could be reduced to an objective formula, it would be chemistry. I know and I can prove it. And so I tell. About these truths. (p. 213)

We wish to emphasize that those portions of experience that are the basis for a particular project, argument, or account need to be available or at least be a point of reference for the reader or audience. What, for example, was the basis for a claim or explanation or rendering or narrative with which we are engaged? Is it personal experience (e. g., biographical), a critical event/experience shared with others, an observation—or series of observations such as an extended ethnography, informal or formal interviews, reflections on documentary "evidence" or accounts, and so on? Even if the critical insight is personal, emergent from a creative consciousness, is the referent in any sense actual, and if not, in what ways is it connected to the lifeworld presumably shared by readers, viewers, or listeners? An example is insights about metaphors of life or how identities are shaped by popular culture. Saviano (2007) reports how dominant Camorra leaders mimic Hollywood bad guys as part of a lifestyle of domination and control:

> Camorra villas are pearls of cement tucked away on rural streets, protected by walls and video cameras. There are dozens and dozens of them. Marble and parquet, colonnades and staircases, granite fireplaces with the boss's initials. One, the most sumptuous, is particularly famous, or perhaps it has merely generated the most legends. Everyone calls it Hollywood. Just saying the word makes you understand why. . . . Walter Schiavone's villa really does have a link to Hollywood. People in Casal di Principe say the boss told his architect he wanted a villa just like Tony Montana's, the Miami Cuban gangster in *Scarface.* He'd seen the film countless times and it had made a deep impression in him, to the point that

he came to identify with the character played by Al Pacino. With a bit of imagination, Schiavone's hollowed face could actually be superimposed on the actor's. The story has all the makings of a legend. People say Schiavone even gave his architect a copy of the film; he wanted the Scarface villa, exactly as it was in the movie. (pp. 344–345)

Many insights found in qualitative research originate in a researcher's personal experience. A bit later in the chapter, we discuss the importance of tacit knowledge. As we move from insights or hunches to more robust claims-making, we run into the issue of sampling, that is, how cases were selected. Qualitative researchers do not discount research from small samples, even a sample of one—a case—but it is helpful if this information is available. Indeed, we could say that most ideas—no matter what method is used to develop them—are not disconnected from personal experience. But there is more to it. Unless it is clear that the account being offered is completely idiosyncratic or unique, then we also would like to know how a "case" was selected, as well as whether other cases were examined, other comparisons made, and if not, are others implied or suggested? In other words, qualitative researchers are trying to make statements about the world of experience such as how things are organized and the consequences of this action. We do not know of a single piece of research in which the researcher avowed that the work being examined was not relevant for anything, any situation, context, application, and the like. The relevance might be limited, but the work matters because it is assumed to shed some light on a specific or related problem that goes beyond the particular case being illuminated, referred to, or scrutinized. Again, Roberto Saviano's (2007) account of the banality of fear and intimidation illustrates the taken-for-granted moral order dictated by the Camorra. Here, he discusses the violation of the code of morality by a witness to a street killing, who was willing to testify. And there were consequences for this brave woman:

> It wasn't testifying in itself that generated such fear, or her identifying a killer that caused such a scandal. The logic of Omerta isn't so simple. What made the young teacher's gesture scandalous is that she considered being able to testify something natural, instinctive, and vital. In a land where [lying] is considered to be what gets you something and [truth] what makes you lose, living as if you actually believe truth can exist is incomprehensible. So the people around you feel uncomfortable, undressed by the gaze of one who has renounced the rules of life itself, which they have fully accepted. (pp. 279–280)

Saviano's gathering of various examples for his story provides data for other investigators, criminal and sociological.

The expanded focus to other types of data is well illustrated with Goodall's (2006) experience, which began with his own family, but he subsequently branched out to explore how CIA operatives were dealt with by the government, how their families were treated, and so on.

> I didn't go into this research project believing that I would be able to discover the "truth" about my father's clandestine career, or even about the meaning of my family's cold war life. I hoped for something less grand. I hoped to find an adequate, even if partial, explanation for what happened to us, and why. (p. 24)

This work involved interviews, historical records and other documents, and even connections with other social science literature, but he was able to relate a unique case—his family—with other cases, and was able to track how organizational practices and cultures—including the unique culture of the CIA—and how it contributed to family dysfunction, on the one hand, but also foreign policy missteps, on the other.

Moreover, it is clear from Goodall's (2006) excellent monograph what the different sources of data were, how they were interwoven, and the reader is given a good understanding of how his subsequent interpretations and conclusions are more or less closely tied to the various data, while other considerations may be a bit more speculative and less data bound.

Saviano (2007) understands what any ethnographer might, but the way that he knows is not transparent, partly because, as a journalist, he answers to a different epistemological canon. Yet, it is compelling enough to motivate researchers to know as much as he does, but to be able to offer an account of how we know what we claim. Listen to his testament of authenticity in the final pages of his riveting book:

> I was born in the land of the Camorra, in the territory with the most homicides in Europe, where savagery is interwoven with commerce, where nothing has value except what generates power. Where everything has the taste of a final battle. It seemed impossible to have a moment of peace, not to live constantly in a war where every gesture is a surrender, where every necessity is transformed into weakness, where everything needs to be fought for tooth and nail. In the land of the Camorra, opposing the clans is not a class struggle, an affirmation of a right, or a expropriation of one's civic duty. It's not the realization of one's honor or the preservation of one's pride. . . . To set oneself against the clans becomes a war of survival, as if existence itself—the food you eat, the lips you kiss, the music you listen to, the pages you read—were merely a way to survive, not the meaning of life. Knowing is thus no longer a sign of moral engagement. Knowing—understanding—becomes a necessity. (pp. 300–301)

While we are convinced, ethnography must provide more to maintain authority.

We have addressed mainly personal experiences and very close—if not biographical—observations and recollections to this point. But the same logic about the veracity and relevance of a report resonates with various kinds of interviews, additional observations, and even documents. Again, it is not our intent here to say which is better or which ones you should trust more than the others, but we mainly want to emphasize that the reader/listener/viewer should be able to discern what was used in general, as well as in specific instances. Any audience member will discern his or her own criteria for being "convinced" or "skeptical" of the connection between what is being reported and its avowed source. But the main concern is that the connection be apparent, and to the extent possible, transparent. This can even be aided by adding a methodological appendix in which parts of the research process are delineated, which we address shortly below, or even an occasional footnote.

These general principles of transparency can be applied to data analysis as well. Data analysis seems self-evident in some first-person qualitative reports, but it is seldom as straightforward as it seems. The popular applications of "grounded theory" provide one rationale for coding and comparison, but there are other modes of comparison as well. The skilled writer can weave in the creative processes of data explication, comparison, and triangulation that are usually involved in qualitative reports, while in other instances it is helpful to have a section (e.g., footnote) delineating "what I did" and "how I came up with this." Again, Goodall's (2006) use of official government records about his father's bogus job descriptions clearly illustrated the pervasive bureaucratic duplicity and lies, even when he was trying to understand his deceased father's life, troubles, and perspective. His use of these materials helped make an organizational bureaucratic process of duplicity more visible. The challenge of making our research approach visible underlies validity issues.

▣ TACIT KNOWLEDGE AND
 AN ECOLOGY OF UNDERSTANDING

Good ethnographies display tacit knowledge. We focus on the dimensions of "an ecology of understanding." Contextual, taken-for-granted, "tacit knowledge" plays a constitutive role in providing meaning. Goodall (2006) reports some bewilderment as a child about what his father did for a living, and where he would disappear to for weeks at a time. His parents' answer, "It's complicated," would be given to other queries over the years. Social life is spatially and temporally ordered through experiences that cannot be reduced to spatial boundaries, as numerous forms of communication attempt to do, especially those based on textual and linear metaphors. More specifically, experience is different from words and symbols about those experiences. Words are always poor representations of the temporal and evocative lifeworld. Words and texts are not the primary stuff of the existential moments of most actors in what Schutz (1967) termed the "natural attitude." They are very significant for intellectuals and wordsmiths who claim to represent such experiences. Yet, as those word-workers have come to rely on and substitute words and other texts for the actual experiences, their procedures of analysis have been reified to stand for the actual experience. Therein lies much of the problem that some have termed the "crisis in representation."

Capturing members' words alone is not enough for ethnography. If it were, ethnographies would be replaced by interviews. Goodall (2006) could not simply have told his story to an interviewer, because part of the story was not clear to him until he reflected more on his past, sought other information sources, compared different versions of events, and pieced the meanings together into a coherent narrative. His book, in other words, is not the foundation of his story, but is actually part of the method he used in discovering, selecting, and interpreting experiences. Good ethnographies, like Goodall's, reflect tacit knowledge, the largely unarticulated, contextual understanding that is often reflected in nods, silences, humor, and naughty nuances. This is the most challenging dimension of ethnography, and gets to the core of the members' perspective, or for that matter, the subtleties of membership itself. As Laura Nader (1993) suggests, this is the stuff of ethnography: "Anthropology is a feat of empathy and analysis" (p. 7). But, without doubting the wisdom of Professor Nader, it is necessary to give an accounting of how we know things, what we regard and treat as empirical materials—the experiences—from which we produce our second (or third) accounts of "what was happening."

▣ REFLEXIVE ACCOUNTING FOR SUBSTANCE

As we learn more about other significant and essentially invariable dimensions of settings, such as hierarchical organization, these are added. In order to satisfy the basic elements of the ethnographic ethic, the following "generic" topics can be found in ethnographic reports. Goodall's (2006) descriptions of his mother's practiced attempts to appear more sophisticated and the tensions and problems that this created are apt illustrations. After noting that his mother's modest West Virginia childhood hardly prepared her for the duties and cultural performances of a vice consul's wife, often before the critical eye of Ambassador to Italy Clare Boothe Luce, we read the following:

> Embassy social gatherings were an American cultural performance. They relied heavily on appearances and the careful cultivation of approved patterns of perceptions. My mother was given detailed instruction on appropriate behavior at social functions during a State Department orientation for spouses.... She also had been coached on conversation. She knew how to appear interested in a topic when she was uninterested in it; how to deflect talk when confronted with an unwanted question or merely something tedious or dull; how to appear happy and energetic when she felt tired and unhappy; and how to appear helpless or clueless to get out of trouble if the case required it. (p. 141)

There is a distinction to be drawn between interesting, provocative, and insightful accounts of ethnographic research, on the one hand, and high-quality ethnographic work. Given our emphasis on the reflexive nature of social life, it will not surprise the reader that we prefer those studies that enable the ethnographic audience to symbolically engage the researcher and enter through the research window of clarity (and opportunity). While no one is suggesting a "literal" accounting, our work and that of many others suggests that the more a reader (audience member) can engage in a symbolic dialogue with the author about a host of routinely encountered problems that compromise ethnographic work, the more our confidence increases. Good ethnographies increase our confidence in the findings, interpretations, and accounts offered.

▣ ACCOUNTING FOR OURSELVES

A key part of the ethnographic ethic is how we account for ourselves. Good qualitative research—and particularly ethnographies—shows the hand of the ethnographer. The effort may not always be successful, but there should be clear "tracks" that the attempt has been made. We are in the midst of a rediscovery that social reality is constructed by human agents—even social scientists—using cultural categories and language in specific situations or contexts of meaning. This interest is indeed welcomed because it gives us license to do yet another elucidation of the "concept of knowing."

Our collective experience in reading a literature spanning more than 50 years, along with our own work on numerous topics and projects, suggests that there is a minimal set of problem areas that are likely to be encountered in most studies. We do not offer a solution to the problems that will follow, but only suggest that these can offer a focus for providing a broader and more complete account of the reflexive process through which something is understood (Altheide, 1976; Denzin, 1991; Douglas, 1976; Johnson, 1975). Such information enables the reader to engage the study in an interactive process that includes seeking more information and contextualizing findings, reliving the report as the playing out of the interactions between the researcher, the subjects, and the topic in question.

The idea for the critical reader of an ethnography is to ask whether or not any of the basic issues of data collection and analysis were likely to have been relevant problems, were they explicitly treated as problematic by the researcher, and if so, how were they addressed, resolved, compromised, avoided, and so forth. Because these dimensions of ethnographic research are so pervasive and important for obtaining truthful accounts, they should be implicitly or explicitly addressed in the report. Drawing on such criteria enables the ethnographic reader to approach the ethnography interactively and critically, and to ask what was done, how it was done, what are the likely and foreseen consequences of the particular research issue, and how was it handled by the researcher. These dimensions represent one range of potential problems likely to be encountered by an ethnographer.

No study avoids all of these problems, although few researchers give a reflexive account of their research problems and experience. One major problem is that the phenomenon of our interest reflects multiple perspectives; there are usually a multiplicity of modes of meanings, perspectives, and activities, even in one setting. Indeed, this multiplicity is often unknown to many of the official members of the setting. Thus, one does not easily "become the phenomenon" in contemporary life. As we strive to make ourselves, our activities and claims, more accountable, a critical feature is to acknowledge our awareness of a process that may actually impede and prevent our adequate understanding of all relevant dimensions of an activity.

Our experience suggests that the subjects of ethnographic studies are invariably temporally and spatially bounded. That the range of activities under investigation occurs in time and space (which becomes a "place" when given a meaning) provides one anchorage, among many others, for penetrating the hermeneutic circle. One feature of this knowledge, of course, is its incompleteness, its implicit and tacit dimensions. The qualitative researcher seeks to draw on the tacit dimension in order to make meanings and order more explicit, and in a sense, more visible. Our subjects always know more than they can tell us, usually even more than they allow us to see; likewise, we often know far more than we can articulate. Even the most ardent social science wordsmiths are at a loss to transform the nuances, subtleties and the sense of the sublime into symbols! For this reason, we acknowledge the realm of tacit knowledge, the ineffable truths, unutterable partly because they are between meanings and actions, the glue that separates and joins human intentionality to more concretely focused symbols of practice. As we have stressed, the key issue is not to capture the informant's voice, but to elucidate the experience that is implicated by the subjects in the context of their activities as they perform them, and as they are understood by the ethnographer. Harper's (1987) explanation of how he used photography in a study of a local craftsman illustrates this intersection of meaning:

> The key, I think, is a simple idea that is the base of all ethnography. I want to explain the way Willie has explained to me. I hope to show a small social world that most people would not look at very closely. In the process I want to tell about some of the times between Willie and me, thinking that at the root of all sociology there are people making connections, many like ours. (p. 14)

One approach to making ourselves more accountable and thereby sharing our experience and insights more fully with readers, is to locate inquiry within the process and context of actual human experience. Our experience suggests that researchers should accept the inevitability that all statements are reflexive, and that the research act is a social act. Indeed, that is the essential rationale for research approaches grounded in the contexts of experience of the people who are actually involved in their settings and arenas.

Studies of popular culture and symbolic reality construction are, for some people, ethnographically on the edge; the life-worlds under investigation are experiential, but representational of entertainment-oriented media personas and styles. Considerable care must be taken to show something of the process, the contexts of understanding and the social relationships connecting the symbolic to the actual or the real, as in everyday life discipline, language, and actions involving enforcement and resistance.

Recent work employing *qualitative document analysis* (also referred to as "ethnographic content analysis") illustrates the application of an ethnographic ethic and tacit knowledge where the emphasis is on discovery and description, including searching for contexts, underlying meanings, patterns, and processes, rather than mere quantity or numerical relationships between two or more variables. Michael Coyle's (2007) study of "The Language of Justice" examined numerous documents, including dictionaries, to trace the etiology of the terms *victim*—especially *innocent victim*—and *evil*. His focus was the penetration of these terms into public discourse about crime and punishment, including the popular phrase "tough on crime" that was widely used by moral entrepreneurs and politicians.

Chris Schneider (2008) drew on a lifelong interest and study of popular music to track how digitized and embedded technologies can simultaneously resist and assist authority and social control. His examination of a range of documents, especially popular music (e.g., rap music), as well as school contexts for controlling students' use of interactive portable technology (e.g., cell phones, iPods), illuminated how the communication of control is embedded in products marketed for entertainment.

Similarly, as Tim Rowlands (2010) has masterfully shown, the world of virtual experience and reality can be investigated, described, and rendered theoretically through intensive fieldwork in cyberspace. As with Coyle's and Schneider's studies, Rowlands' personal experience was significant in providing resources and questions to uncover data; years of experience as a serious player of various "computer games" opened up experiences with other players and provided opportunities for informal interviews to clarify the semiotic scenarios of, for example, *EverQuest,* that could be conceptually treated as exploration of one virtual utopia. What were the rules, Rowlands asked, what were the underlying assumptions of order, and above all, what does justice look like in virtual space? He found violence and the iconography of violence, but also a technologically rendered cosmology of alienated confrontations as task completion for role fulfillment that players recognized and pursued as pastimes of boredom, fun, work, and as an inevitable feature of the computer environment that reflected a capitalistic order. Examining the evidence within the confines of interpretive validity suggests, "MMOs [massive multiplayer online games] such as *EverQuest* serve in important ways to shape our understanding of what virtual worlds currently are" (p. 369).

Let us bring the parts of the argument closer together. Evidence is a feature of the interaction between an audience, a claim, and practical epistemologies of everyday life. Whether evidence is convincing follows along the same lines; is the evidence "good enough," or is there enough evidence relevant to the topic? The key question is, what is relevant for the topic or question under investigation? A reader's (listener's) perspective about what is "evidence" is informed by biography, culture, and so forth.

◙   CONCLUSION

The great artist Paul Klee, who stated that a line is a dot that went for a walk, could enjoy the task of trying to capture the intersection of many lines—borne of numerous dots—that are heading in different directions. We seek, among other things, to understand the nature, process, and consequences of social interaction, on the one hand, and how this promotes individual and joint renderings of a definition of a situation, on the other hand. A positivistic view of validity works fine in a different social universe where there are not multiple perspectives, vastly different methods and materials with which to work, and myriad uses and audiences. But that is not our social research world. The social world and its human actors make and interpret meanings through an interaction process that contributes to the construction, reification, and resistance of social reality. Any method that obliterates the essential role of emergence, negotiation, and tacit knowledge will not be valid. And any effort to remake this world to comply with an idealized model is folly, will lack credibility, and is doomed to failure. Our overview of validity issues in qualitative research, with a focus on ethnographic reports, suggests that a proper set of standards or criteria for assessing validity entails considering the place of evidence in an interaction process between the researcher, the subject matter (phenomenon to be investigated), the intended effect or utility, and the audience for which the project will be evaluated and assessed.

We continue to hold that focusing on the process of investigation and communicating that process, the problems and solutions encountered in accessing, collecting, analyzing, and interpreting data—to the best of our ability—is quite consistent with analytic realism, or the general notion that the social world is an interpreted one. We are not opposed to using evidence, but prefer that the spirit of the evidentiary narrative be considered.

We do not intend to leave the question of validity completely open-ended. There are parameters, and the criteria and process can become tighter within the community of scholars who employ certain methods and criteria. While we do not recommend "recipe" methods, we have offered several "lists" of items to consider when focusing on the presentation of research problems and solutions in a given project (Altheide & Johnson, 1994). For example, many approaches to interviewing (e.g., focused, life history, etc.), ethnography (e.g., grounded theory), and document analysis (qualitative content analysis) have developed criteria and procedures for optimal work. But our task is not to refine and squeeze the novelty and richness out of experience in favor of some bygone notion of rigor and efficiency, nor is it to make sure that creative problem solving and discovery are compromised in order to dot the i's and cross the t's. We want to see more dots flourish and evolve into creative insights. Our task is to continue pushing the line in new directions to illuminate our humanity and communicative worlds.

◙   REFERENCES

Altheide, D. L. (1976). *Creating reality: How TV news distorts events.* Beverly Hills, CA: Sage.

Altheide, D. L. (2008). The evidentiary narrative: Notes toward a symbolic interactionist perspective about evidence. In N. K. Denzin & M. D. Giardina (Eds.), *Qualitative inquiry and the politics of evidence* (pp. 137–162). Walnut Creek, CA: Left Coast Press.

Altheide, D. L. (2009). *Terror post 9/11 and the media.* New York: Peter Lang.

Altheide, D. L., & Johnson, J. M. (1993). The ethnographic ethic. In N. K. Denzin (Ed.), *Studies in symbolic interaction* (pp. 95–107). Greenwich, CT: JAI Press.

Altheide, D. L., & Johnson, J. M. (1994). Criteria for assessing interpretive validity in qualitative research. In N. K. Denzin & Y. S. Lincoln (Eds.), *Handbook of qualitative research* (pp. 485–499). Newbury Park, CA: Sage.

Altheide, D. L., & Snow, R. S. (1979). *Media logic.* Beverly Hills, CA: Sage.

Atkinson, P. (1990). *The ethnographic imagination: Textual constructs of reality.* New York: Routledge.

Bhaskar, R. (1979). *The possibility of naturalism: A philosophical critique of the human sciences.* Brighton, UK: Harvester.

Bochner, A. P. (2007). Notes toward an ethics of memory in autoethnographic inquiry. In N. K. Denzin & M. D. Giardina (Eds.), *Ethical futures in qualitative research: Decolonizing the politics of knowledge* (pp. 197–208). Walnut Creek, CA: Left Coast Press.

Cabinet Office. (2003). *Quality in qualitative evaluation: A framework for assessing research evidence* [Full report]. London: Author.

Coyle, M. J. (2007). *The language of justice: Exposing social and criminal justice discourse.* Unpublished doctoral dissertation, School of Justice and Social Inquiry, Arizona State University, Tempe.

Denzin, N. K. (2009). *Qualitative inquiry under fire.* Walnut Creek, CA: Left Coast Press.

Denzin, N. K. (2010). A qualitative stance: Remembering Steinar Kvale (1938–2008). *International Journal of Qualitative Studies in Education, 23*(2), 125–127.

Denzin, N. K., & Giardina, M. D. (Eds.). (2008). *Qualitative inquiry and the politics of evidence.* Walnut Creek, CA: Left Coast Press.

Denzin, N. K., & Lincoln, Y. S. (Eds.). (1994). *Handbook of qualitative research.* Newbury Park, CA: Sage.

Douglas, J. D. (1976). *Investigative social research: Individual and team field research.* Beverly Hills, CA: Sage.

Ellis, C. (2004). *The ethnographic I.* Walnut Creek, CA: AltaMira Press.

Ellis, C. (2009). *Revision: Autoethnographic reflections on life and work.* Walnut Creek, CA: Left Coast Press.

Emerson, R. M., Fretz, R. I., & Shaw, L. L. (1995). *Writing ethnographic fieldnotes.* Chicago: University of Chicago Press.

Erikson, R. V., & Doyle, A. (2003). *Risk and morality.* Toronto, ON, Canada: University of Toronto Press.

Fuller, S. (2008). Science democratized—Expertise decommissioned. In N. Stehr (Ed.), *Knowledge and democracy* (pp. 105–117). New Brunswick, NJ: Transaction.

Goodall, H. L., Jr. (2006). *A need to know: The clandestine history of a CIA family.* Walnut Creek, CA: Left Coast Press.

Guba, E. G. (1990). Subjectivity and objectivity. In E. W. Eisner & A. Peshkin (Eds.), *Qualitative inquiry in education* (pp. 74–91). New York: Teachers College Press.

Haak, S. (2003). *Defending science—within reason.* Amherst, NY: Prometheus.

Hammersley, M. (1990). *Reading ethnographic research.* London: Longman.

Hammersley, M. (1992). *What's wrong with ethnography? Methodological explorations.* London & New York: Routledge.

Hammersley, M. (2005). The myth of research-based practice: The critical case of educational inquiry. *International Journal of Social Research Methodology, 8*(4), 317–330.

Harper, D. (1987). *Working knowledge: Skill and community in a small shop.* Berkeley: University of California Press.

Harré, R., & Madden, E. (1975). *Causal powers.* Oxford, UK: Basil Blackwell.

Henry, G., Julnes, G. J., & Mark, M. (Eds.). (1998). *Realist evaluation.* San Francisco: Jossey-Bass.

Johnson, J. M. (1975). *Doing field research.* New York: Free Press.

Kemmis, S., & McTaggart, R. (2005). Participatory action research: Communicative action in the public sphere. In N. K. Denzin & Y. S. Lincoln (Eds.), *The SAGE handbook of qualitative research* (3rd ed., pp. 599–603). Thousand Oaks, CA: Sage.

Kvale, S., & Brinkmann, S. (2008). *InterViews: Learning the craft of qualitative research interviewing.* Thousand Oaks, CA: Sage.

Ladkin, D. (2004). Action research. In C. Sale, G. Gobo, J. F. Gubrium, & D. Silverman (Eds.), *Qualitative research practice* (pp. 536–48). New York: Routledge.

Lakoff, G. (1987). *Women, fire, and other dangerous things: What categories reveal about the mind.* Chicago: University of Chicago Press.

Lamont, M., & White, P. (2009). *Workshop on interdisciplinary standards for systematic qualitative research.* Washington, DC: National Science Foundation.

Lather, P. (1993). Fertile obsession: Validity after poststructuralism. *Sociological Quarterly, 34,* 673–93.

Lofland, J. (1995). Analytic ethnography. *Journal of Contemporary Ethnography, 24*(1), 30–67.

Mancias, P. T., & ebrary, Inc., 2006. *A realist philosophy of social science explanation.* Cambridge, UK: Cambridge University Press.

Marcus, G. E., & Clifford, J. (1986). *Writing culture: The poetics and politics of ethnography.* Berkeley: University of California Press.

Maxwell, J. A. (2008). The value of a realist understanding of causality for qualitative research. In N. K. Denzin & G. D. Giardina (Eds.), *Qualitative inquiry and the politics of evidence* (pp. 163–181). Walnut Creek, CA: Left Coast Press.

Miller, W. L., & Crabtree, B. F. (2005). Clinical research: Participatory action research. In N. K. Denzin & Y. S. Lincoln (Eds.), *The SAGE handbook of qualitative research* (3rd ed., pp. 605–639). Thousand Oaks, CA: Sage.

Nader, L. (1993). Paradigm busting and vertical linkage. *Contemporary Sociology, 33,* 6–7.

Putnam, H. (1999). *The threefold cord: Mind, body, and the world.* New York: Columbia University Press.

Ragin, C., Nagel, J., & White, P. (2004). *Workshop on scientific foundations of qualitative research.* Washington, DC: National Science Foundation.

Richardson, L. (1997). *Fields of play.* New Brunswick, NJ: Rutgers University Press.

Rolfe, G. (2004). Validity, trustworthiness, and rigor: Quality and the idea of qualitative research. *Journal of Advanced Nursing, 53*(3), 304–310.

Rowlands, T. E. (2010). *Empire of the hyperreal.* Unpublished doctoral dissertation, School of Social Transformation, Arizona State University, Tempe.

Saviano, R. (2007). *Gomorrah.* New York: Farrar, Straus and Giroux.

Schneider, C. J. (2008). *Mass media, popular culture and technology: Communication and information formats as emergent features of social control.* Unpublished doctoral dissertation, School of Justice and Social Inquiry, Arizona State University, Tempe.

Schutz, A. (1967). *The phenomenology of the social world.* Evanston, IL: Northwestern University Press.

Stehr, N. (2008). *Moral markets.* Boulder, CO: Paradigm.

Szto, P., Furman, R., & Langer, C. (2005). Qualitative research in sociology in Germany and the U.S. *Focus: Qualitative Social Work, 4*(2), 135–156.

Torrance, H. (2007). Building confidence in qualitative research: Engaging the demands of policy. In N. K. Denzin & M. D. Giardina (Eds.), *Qualitative inquiry and the politics of evidence* (pp. 55–79). Walnut Creek, CA: Left Coast Press.

Van den Hoonaard, W. C. (2002). *Walking the tightrope: Ethical issues for qualitative researchers.* Toronto, ON, Canada: University of Toronto Press.

Van Maanen, J. (1988). *Tales of the field: On writing ethnography.* Chicago: University of Chicago Press.

Wolcott, H. F. (1990). On seeking—and rejecting—validity in qualitative research. In E. W. Eisner & A. Peshkin (Eds.), *Qualitative inquiry in education: The continuing debate* (pp. 121–152). New York: Teachers College Press.

36

ANALYSIS AND REPRESENTATION ACROSS THE CONTINUUM

Laura L. Ellingson

[Researchers] do jump across traditions, we do straddle metatheoretical camps, and (unfortunately) we do let paradigmatic "definitions" constrain our work. . . . [I want to] allow for comfortable jumps and straddles and to loosen some of these constraints.

—K. I. Miller (2000, p. 48)

In the interest of loosening some of the unproductive methodological constraints highlighted by K. I. Miller (2000), I encourage qualitative researchers to consider jumping and straddling multiple points across the field of qualitative methods—consciously, actively, and creatively. Large qualitative projects provide ample opportunity for producing a series of analyses and interpretations that subdivides findings, not only according to logical topic segments, but also with regard to embodying a range of artistic, expository, and social scientific writing genres or other representational media. Yet too often researchers learn and embrace a handful of strategies and settle into comfortable methodological ruts. Such complacency is not limited to practitioners of more traditional, social scientific forms of qualitative research; singular devotion to autoethnographic, narrative, and other artistic approaches may be just as slavish. I go beyond supporting multiple methods research strategies to advocate the use of multiple methods of analysis and representation that span artistic and scientific epistemologies, or ways of knowing.

For the purposes of this chapter, analysis of data or other assembled empirical materials will be understood as the process of separating aggregated texts (oral, written, or visual) into smaller segments of meaning for close consideration, reflection, and interpretation. Forming representations will mean rendering intelligible accounts of analyses, such as through construction of themes or patterns; transformation of journal entries or transcripts into narratives; or explication of an individual account using a particular theoretical lens. Of course, the processes of analysis and representation overlap throughout the duration of a qualitative project; for example, the production of ethnographic fieldnotes involves both selection of details of an encounter or setting to document (i.e., analysis) and generation of a representation of that analysis (the written notes).

In this chapter, I make the case for qualitative methods to be conceptualized as a continuum anchored by art and science, with vast middle spaces that embody infinite possibilities for blending artistic, expository, and social scientific ways of analysis and representation (Ellis & Ellingson, 2000; Potter, 1996). Such an approach moves past the tendency to understand art and science as dichotomies (i.e., mutually exclusive, paired opposites) by illuminating research and representational options that fall between these two poles. To begin, I review traditional and current ways of straddling points of the continuum within research projects. I then offer a guide to being conscious about the interpretive process of selecting methods and genres, and describe three promising strategies for deliberate endeavors to traverse the qualitative continuum.

▣   A CONTINUUM APPROACH

Embracing a Continuum

Dichotomous thinking remains pervasive within methodological debates: "[D]ifferences have been cast in terms of binaries.... All are distinguished by virtue of what they are not" (Gergen, 1994, p. 9). Nowhere are dualisms evidenced more strongly than in the quantitative/qualitative divide. Even within the qualitative field itself, polarities mark the differences between interpretivists and realists (Anderson, 2006; Atkinson, 2006; Ellis & Bochner, 2006). On the one hand, qualitative social scientists often rely on appeals to an authoritative tradition to disparage what they characterize as the subjective, messy, nongeneralizable work of the "navel-gazers" to their left, reifying the natural primacy of "hard" over "soft" ways of knowing. On the other hand, artistic/interpretive researchers may construct traditional social scientists as deluded positivists obsessed with the mythical gods of objectivity, validity, and reliability, in order to legitimate creative analytic approaches to analysis and representation (Richardson, 2000). Moving beyond defining artistic approaches as "not science" and social science as "not art" takes some generative thinking. Many qualitative researchers, if not most, will affirm that art and science are ends of a continuum, not a dichotomy; yet in practice, the use of the methodological "other" to legitimate our particular methodological and paradigmatic preferences leads to a reification of dichotomous ways of conceptualizing and writing about qualitative research practices.

A continuum approach to mapping the field of qualitative methodology constructs a nuanced range—or broad spectrum— of possibilities to describe what traditionally have been socially constructed as dichotomies such as art/science, hard/soft, and qualitative/quantitative (Potter, 1996). Significantly, the continuum is made up primarily of a vast and varied middle ground, with art and science representing only the extreme ends of the methodological and representational range, rather than each constituting half of the methodological ground. Such middle-ground approaches need not represent compromise or a lowering of artistic or scientific standards. Rather, they can signal innovative approaches to sense making and representation.

Building upon Ellis's (2004) representation of the two ends of the qualitative continuum (i.e., art and science) and the analytic mapping of the continuum developed in Ellis and Ellingson (2000), I envision the continuum as having three main areas, with infinite possibilities for blending and moving among them (Ellingson, 2009). As exemplified in Figure 36.1, the goals, questions posed, methods, writing styles, vocabularies, role(s) of researchers, and criteria for evaluation vary across the continuum as we move from a realist/positivist social science stance on the far right, through a social constructionist middle ground, to an artistic/interpretive paradigm on the left. Each of these general approaches offers advantages and disadvantages, and they are not mutually exclusive. Moreover, no firm boundaries exist to delineate the precise scope of left/middle/right; these reflect ideal types only. Furthermore, terms of demarcation and description used throughout the continuum (e.g., interpretive, postpositivist) are suspect and contestable; use of key terminology in qualitative methods remains dramatically inconsistent across disciplines, paradigms, and methodological communities, with new terms arising continually (Gubrium & Holstein, 1997). At any point on the qualitative continuum, a set of assumptions about epistemology (i.e., about what knowledge is and what it means to create it) influences choices surrounding the collection of empirical materials and analysis methods, which in turn tend to foster (but do not require) particular forms of representation.

I now sketch the right, left, and middle areas of the continuum, along with a brief review of both traditional and more contemporary approaches to spanning the boundaries among those areas.

Right/Science

At the far right of the qualitative continuum (see Figure 36.1), emphasis on valid and reliable knowledge, as generated by neutral researchers utilizing the scientific method to discover universal Truth, reflects an epistemology commonly referred to as positivism (Warren & Karner, 2010). Historically, social scientists understood positivism as reflected in a "realist ontology, objective epistemology, and value-free axiology" (K. I. Miller, 2000, p. 57). Few, if any, qualitative researchers currently subscribe to an absolute faith in positivism, however. Many postpositivists, or researchers who believe that achievement of objectivity and value-free inquiry are not possible, nonetheless embrace the goal of production of generalizable knowledge through realist methods and minimization of researcher bias, with objectivity as a "regulatory ideal" rather than an attainable goal (Guba & Lincoln, 1994). In short, postpositivism does not embrace naive belief in pure scientific truth; rather, qualitative research conducted in a strict postpositivist tradition utilizes precise, prescribed processes and produces social scientific reports that enable researchers to make generalizable claims about the social phenomenon within particular populations under examination.

Postpositivists commonly utilize qualitative methods that bridge quantitative methods, in which researchers conduct an inductive analysis of textual data, form a typology grounded in the data (as contrasted with a preexisting, validated typology applied to new data), use the derived typology to sort data into categories, and then count the frequencies of each theme or category across data. Such research typically emphasizes validity of the coding schema, inter-coder reliability, and careful delineation of procedures, including random or otherwise systematic sampling of texts. Content analyses of media typify this

Figure 36.1   Qualitative Continuum

Writing

Use of first-person voice
Literary techniques
Stories
Poetry/poetic transcription
Multivocal, multigenre texts
Layered accounts
Experiential forms
Personal reflections
Open to multiple interpretations

Use of first-person voice
Incorporation of brief narratives in research reports
Use "snippets" of participants' words
Usually a single interpretation, with implied partiality and positionality
Some consideration of researcher's standpoint(s)

Use of passive voice
"View from nowhere" (Haraway, 1998)
Claim single authoritative interpretation
Meaning summarized in tables and charts
Objectivity and minimization of bias highlighted

Researcher

Researcher as the main focus, or as much the focus of research as other participants

Participants are main focus, but researcher's positionality is key to forming findings

Researcher is presented as irrelevant to results

Vocabularies

Artistic/Interpretive: inductive, personal, ambiguity, change, adventure, improvisation, process, concrete details, evocative experience, creativity, aesthetics

Social Constructionist/Postpositivist: inductive, emergent, intersubjectivity, process, themes, categories, thick description, co-creation of meaning, social construction of meaning, standpoint, ideology (e.g., feminism, postmodernism, Marxism)

Positivist: deductive, tested, axioms, measurement, variables, manipulation of conditions, control, predication, generalizability, validity, reliability, theory driven

Criteria

Do stories ring true, resonate, engage, move?
Are they coherent, plausible, interesting, aesthetically pleasing?

Flexible criteria
Clarity and openness of processes
Clear reasoning and use of support
Evidence of researcher's reflexivity

Authoritative rules
Specific criteria for data, similar to quantitative
Proscribed methological processes

(Continued)

597

Figure 36.1 (Continued)

	Artistic/Interpretive	Social Constructionist/Postpositivist	Positivist
Writing	Use of first-person voice; Literary techniques; Stories; Poetry/poetic transcription; Multivocal, multigenre texts; Layered accounts; Experiential forms; Personal reflections; Open to multiple interpretations	Use of first-person voice; Incorporation of brief narratives in research reports; Use "snippets" of participants' words; Usually a single interpretation, with implied partiality and positionality; Some consideration of researcher's standpoint(s)	Use of passive voice; "View from nowhere" (Haraway, 1998); Claim single authoritative interpretation; Meaning summarized in tables and charts; Objectivity and minimization of bias highlighted
Researcher	Researcher as the main focus, or as much the focus of research as other participants	Participants are main focus, but researcher's positionality is key to forming findings	Researcher is presented as irrelevant to results
Vocabularies	Artistic/Interpretive: inductive, personal, ambiguity, change, adventure, improvisation, process, concrete details, evocative experience, creativity, aesthetics	Social Constructionist/Postpositivist: inductive, emergent, intersubjectivity, process, themes, categories, thick description, co-creation of meaning, social construction of meaning, standpoint, ideology (e.g., feminism, postmodernism, Marxism)	Positivist: deductive, tested, axioms, measurement, variables, manipulation of conditions, control, predication, generalizability, validity, reliability, theory driven
Criteria	Do stories ring true, resonate, engage, move? Are they coherent, plausible, interesting, aesthetically pleasing?	Flexible criteria; Clarity and openness of processes; Clear reasoning and use of support; Evidence of researcher's reflexivity	Authoritative rules; Specific criteria for data, similar to quantitative; Proscribed methological processes

approach. For example, Feng and Wu (2009) derived a typology of values represented in advertisements in a Chinese Communist Party newspaper, which they used to code their samples from the years 1980 and 2002. They then compared the frequency of messages reflecting values considered "hedonistic" with those messages reflecting utilitarian values, thus evaluating the overall appeals in the newspaper (see also Kuperberg & Stone, 2008). This type of analysis can also utilize textual data generated by researchers and participants, such as transcribed interviews, focus groups, or participant journals. Bruess and Pearson's (1997) study of interpersonal rituals between heterosexually married couples and both single-sex and cross-sex adult friends is illustrative. In interviews and in written accounts solicited from participants, researchers collected descriptions of everyday rituals that affirm our bonds to significant others. They conducted an inductive analysis, then counted frequencies and used chi-square tests to determine significance of differences between friends and married couples (see also Cousineau, Rancourt, & Green, 2006; Güven, 2008).

Other postpositivist qualitative research modeled similar research processes, but without quantification of themes or categories. Glaser and Strauss (1967) supported their original formulation of grounded theory methods with positivist claims of inductively deriving themes that "emerge" from the data, ensuring validity as long as researchers correctly employed the method. Adopting the language of science, they did not require counting of themes, but they emphasized detailed analytical procedures and a detached, scholarly voice, generating research reports quite similar to statistical reports (e.g., Kwok & Sullivan, 2007).

Another way postpositivist researchers support ideals of objectivity is to highlight issues of data collection while glossing over details of analysis and omitting any acknowledgment of researchers' role in sense making. Such an approach assumes a shared understanding of data analysis procedures and a relatively impartial researcher. Contemporary realist ethnographers sometimes follow this model of understanding and reporting their analytic processes with few or no details in favor of devoting more space to documentation of findings; see for example Brooks and Bowker's (2002) study of play in the workplace (see also Meyer, 2004).

Left

At the left end of the continuum, researchers value humanistic, openly subjective knowledge, such as that embodied in stories, poetry, photography, and painting (Ellis, 2004). Researchers may study their own lives and/or those of intimate others, community members, or strangers. Among the artistic/interpretivists, truths are multiple, fluctuating, and ambiguous. Autoethnographers, performance studies scholars, and others engaged in such practices embrace aesthetics and evocation of emotion and identification as equally or even more important than illumination of a particular topic (Richardson, 2000). Literary standards of truthfulness in storytelling (i.e., verisimilitude) replace those of social scientific truth (Ellis, 2004). Like all art, creative social science representations enable us to learn about ourselves, each other, and the world through encountering the unique lens of a person's (or a group's) passionate rendering of reality into a moving, aesthetic expression of meaning. At its best, art sparks compassion and inspires people to nurture themselves, their communities, and the world.

Analysis and representation merge to a great degree as art is produced, based upon experiences or empirical materials. A tremendous variety of artistic practices exists, with new forms continually arising; I review some of the most common here. *Autoethnography* is research, writing, story, and method that connect the autobiographical to the cultural, social, and political through the study of a culture or phenomenon of which one is a part, integrated with relational and personal experiences (Ellingson & Ellis, 2008). Autoethnographers typically produce emotionally evocative accounts, such as Lee's (2006) autoethnography of her grief following a loved one's suicide (see also Defenbaugh, 2008; Kiesinger, 2002; Lindemann, 2009; Rambo, 2005, 2007; Secklin, 2001). *Narratives* constructed from fieldnotes, interview transcripts, personal experiences, or other empirical materials enable readers to think with and feel with a story, rather than explicitly analyzing its meaning (Frank, 1995). Parry (2006) constructed short stories based upon her interviews with women negotiating pregnancy, birth, and midwifery (see also Abu-Lughod, 1993; Drew, 2001; Tillmann-Healy, 2001; Trujillo, 2004). The creative analytic practice of *poetic representation* of findings provides rich modes for artistic expression of research (Faulkner, 2007, 2010; Richardson, 1992a, 1992b, 1993, 2000). Variations include research poetry that represents a fieldwork experience in India (Chawla, 2006), investigative poetry critiquing the U.S. prison-industrial complex (Hartnett, 2003), and hybrid poems exploring issues of culture and identity (Prendergast, 2007; see also Austin, 1996; González, 1998). Many *video* representations of empirical materials involve participatory methods, ideally empowering participants to act on their own behalf, as reflected in a project designed to highlight the voices of parents of children in Head Start programs (McAllister, Wilson, Green, & Baldwin, 2005; see also Carlson, Engebretson, & Chamberlain, 2006; Nowell, Berkowitz, Deacon, & Foster-Fishman, 2006; Singhal, Harter, Chitnis, & Sharma, 2007; White, 2003). Many researchers create live *performances* based on autoethnography, ethnographic fieldnotes, or interviews that engage audiences and invite them to connect to and empathize with others (Spry, 2001). One research team developed a dramatic production exploring personhood in patients with Alzheimer's disease that is used as a teaching tool with medical students (Kontos & Naglie, 2007; see also Gray & Sinding, 2002; Mienczakowski, 1996, 2001).

Middle

Enlarging, illuminating, and bolstering the middle ground of qualitative inquiry has been and remains my particular quest as a qualitative methodologist. I am most at home in what I have come to understand as the middle ground—not a fence-sitting, ambivalent, or commitment-phobic place, but a rich, varied, and complex location. In the middle sits not merely work that is "not art" or "not science," but also work that offers description, exposition, analysis, insight, theory, and critique, blending elements of art and science or transcending the categories.

In the middle ground, qualitative researchers adopt social constructionist or postmodernist-influenced perspectives of meaning as intersubjective and co-created, but also as retaining emphasis on the significance of commonalties and connections across the objects of analysis (Ellingson, 2009). Middle-ground work often concerns the construction of patterns, e.g., themes, categories, and portrayals, as well as practicalities, e.g., applied research, recommendations for action, empowerment of marginalized groups (Ellis & Ellingson, 2000). First-person voice departs from the passive voice that characterizes positivist work, moving away from objectivity and toward intersubjectivity, but not all the way to avowed subjectivity of art. Researchers acknowledge claims of truth as contingent upon (among other things) the politics of research funding, indeterminacy of language in which authors express claims, and fallibility of human sense making. Rather than apologizing for being subjective or ignoring the ways in which researchers construct findings, middle-ground qualitative researchers often reflect upon their standpoints to shed light on how their race, class, gender, dis/ability, sexuality, and other identities and experiences shape research processes and results (Ellingson, 1998). Rigor, depth of analysis, and reflexivity constitute important criteria for evaluating claims of middle-ground qualitative research (Fitch, 1994).

Qualitative researchers in the middle ground achieve "intimate familiarity" with their textual materials by rereading them many times, making notes on emergent trends, and then constructing themes or patterns concerning aspects of the culture (Strauss & Corbin, 1998; Warren & Karner, 2010). Charmaz (2000, 2005, 2006) situates grounded theory methods within social constructionist theory, developing the constructed nature of all knowledge claims as arising out of relationships; thus, meaning resides not in people or in texts, but *between* them. One such study parses the meanings of emotional strain on palliative cancer nurses (Sandgren, Thulesius, Fridlund, & Petersson, 2006; see also Bergen, Kirby, & McBride, 2007; Ellingson, 2007; Larsen, 2006; Low, 2004; Miller-Day & Dodd, 2004; Montemurro, 2005; Sacks & Nelson, 2007; Wilson, Hutchinson, & Holzemer, 2002; Zoller, 2003). Similar middle-ground forms of analysis include "deriving themes" (Apker, 2001; Meyer & O'Hara, 2004) and narrative analysis (Goodier & Arrington, 2007; Riessman, 2008; Vanderford, Smith, & Olive, 1995). These middle-ground forms of analysis rely almost exclusively on traditional research report format for representation, but some researchers have begun to find unique ways of complementing reports with creative analytic representations, as I discuss below.

Straddling the Continuum: Traditional Approaches

Traditionally, researchers triangulate by employing multiple quantitative and qualitative methods or measures to capture more effectively the truth of a social phenomenon (Lindlof & Taylor, 2002). In positivist and (some) postpositivist research, triangulation or multiple methods design involves an attempt to get closer to the truth by bringing together multiple forms of data and analysis to clarify and enrich a report on a phenomenon (e.g., Creswell & Clark, 2006). Hence, while methods may complement and even contrast in terms of procedures (e.g., interviews and numerical surveys), the epistemological underpinnings are consistent, with both upholding positivist or postpositivist goals of generalization and prediction. While such work often includes both qualitative and quantitative data or a range of different qualitative data or statistical measurements combined into a single report, manuscripts tend to reflect traditional writing conventions and do not include artistic or creative genres. Scott and Sutton (2009), for instance, utilized open-ended and quantitative surveys and interviews to explore relationships between teachers' emotions and practice change during and following participation in a series of professional development workshops (see also Castle, Fox, & Souder, 2006; Hodgkin, 2008; O'Donnell, Lutfey, Marceau, & McKinlay, 2007).

Straddling the Continuum: Contemporary Approaches

Endless numbers of innovative ways to blend or transcend art and science in qualitative collection of empirical materials, analysis, and representation exist, with more being developed all the time. Hybrids may integrate inductive/grounded theory analysis with other methods of representation, such as photography in a study of quality of life among African American breast cancer survivors (López, Eng, Randall-David, & Robinson, 2005) or participatory action research techniques employed by researchers exploring female childhood sexual abuse survivors' experiences with physical therapy (Teram, Schachter, & Stalker, 2005). Another way to overtly blend the voices of art and science is to weave them into a single representation. *Layered accounts* move back and forth between academic prose and narrative, poetry, or other art, revealing their constructed nature through the juxtaposition of social science and artistic ways of knowing (Ronai, 1995). Layered accounts often connect personal experiences to theory, research, and cultural critique or to discussions of methodological matters. As such, Jago's

(2006) essay incorporates narratives to illustrate the experience of growing up with an absent father (see also Markham, 2005; Saarnivaara, 2003; Tracy, 2004).

Another current approach to straddling methodological camps is to reach out to multiple audiences outside of the academy with research findings (Fine, Weis, Weseen, & Wong, 2000). For example, Hecht and Miller-Day (2007) detailed their project that collected narratives from adolescents about their substance abuse experiences and conducted systematic analyses, which were published in an academic research journal. At the same time, the authors transformed their findings into dramatic performances and into a substance abuse prevention curriculum that clearly reflected artistic and pedagogical forms of representation far different from their scholarly report. I continue my discussion of reaching multiple audiences later in this chapter. At this point, I engage the decision-making processes involved in considering opportunities for analysis and representation across the qualitative continuum.

◙ ENGAGING INTERPRETIVE PROCESSES: SOME STRATEGIES FOR SUCCESSFULLY NAVIGATING THE CONTINUUM

Now that I have stated my case for constructing qualitative methods as a continuum and urged researchers to navigate the possibilities for analysis and interpretation in a fluid and diverse way, I offer some suggestions for how to think about and move through a qualitative project. This is an answer (albeit not *the* answer) to the perennial question by students and practitioners of qualitative methods during and after they have collected a rich, intriguing set of empirical materials: What do I *do* with all this *stuff*? No substitute exists for wading through the interpretive process oneself, and each project holds unique opportunities and constraints. Yet qualitative researchers and students share many common experiences, obstacles, and decision points as we navigate our projects. For those who decide to cross the qualitative continuum instead of limiting themselves to one location, I offer some suggestions for how to do it, either when moving within a single project or across different projects. Many of these strategies were inspired by and overlap with reflexive exercises offered by other methodologists (see Janesick's [2006] "stretching exercises" and Richardson's [2000] writing exercises). I tuned these strategies to the challenges and opportunities of deliberately choosing to adopt more than one epistemological, analytical, and/or representational position on the continuum (Ellingson, 2009). Please consider the order to be flexible, and freely pick and choose which seem applicable to a particular project.

I encourage researchers to wonder, target audiences, strategically select material, consider format, keep the forest and the trees, acknowledge mutual influence, make each piece count, be pragmatic, and own the process.

Wonder

In addition to reading relevant methodological, theoretical, and topic-specific materials, I also suggest that researchers explore their goals in a process I call *wondering*. To prepare, explore the answers you generate to the following questions in deciding which points of the continuum will manifest within a project. I heartily encourage journaling or freewriting on these topics, using writing as a method of inquiry to open up possibilities (Richardson, 2000).

Empirical Materials/Analysis

- What cases, events, stories, or details come to mind immediately when I think about my data or other empirical materials?
- What have I learned about my materials by immersing myself in them?
- What contradictions, inconsistencies, or exceptions to the rules do I notice in my empirical materials?
- How does my identity relate to my work? How do my age, gender, race/ethnicity, nationality, abilities and disabilities, special talents, formative experiences, etc., shape how I understand my participants?
- How do I think my participants perceive me?
- What have my participants taught me about their worlds? About mine?
- How is power revealed and concealed in my empirical materials?
- How am I complicit with systems of power in my empirical materials and analyses?
- What truths seem to be missing from the preliminary analyses and accounts I have worked on?

Topics

- What are the key content claims I want to make about my topic?
- What patterns do I wish to explore?
- What is/are my thesis statement(s) for this project?
- What political implications of my project do I want to explore?
- What pragmatic suggestions for improving the world have I developed? Or in what areas do I detect a need for improvement that I might be able to shed light on with my study?
- What questions do I still have about my setting, participants, and processes?

Audiences

- What academic audiences do I want to reach with my work?
- What community, lay, or popular audiences could benefit from my findings?
- What would my favorite auntie [insert friend/relative of your choice] want to know about this topic?

- What nonprofit or government agencies could benefit from my project?
- What policies could be improved using ideas from my project?
- Sharing with which audiences would bring me the most satisfaction? Why?

Researcher Desire

- What is my favorite thing about my empirical materials? What makes me smile when I think of it? What makes me cry? What makes me angry?
- What would be fun to write?
- What process issues or ideas come up in my journaling that intrigue me?
- What strong emotions do I have about my participants, their stories, and our relationships?
- Whose research do I admire? Why?
- What about my study embarrasses me or makes me feel self-conscious? Why?
- What am I most proud of in my empirical materials?
- If one of my mentors asked me about my project, what would I want to tell her or him?

Genres

- What new forms would I like to experiment with?
- What types of writing am I good at?
- What non-written forms could I use to collect or represent my empirical materials?
- What genres do I enjoy reading? Why?
- With what genres are my participants most familiar and comfortable?
- What texts could I produce that would benefit my participants?
- How do accounts I have written or produced (e.g., fieldnotes, transcripts, photographs, e-mails, memos) shape each other? (adapted from Ellingson, 2009, pp. 75–77)

Any manuscript or representation requires innumerable decisions about content, language, and style. The freedom of moving beyond a single method or form of representation makes this process exponentially more complex, but also invigorating. Wondering enables researchers to explore options throughout the duration of qualitative projects as new opportunities, insights, and relationships develop. I urge researchers to set aside time for wondering and (re)answer questions throughout their collection of materials, analysis, and writing and producing art in other media (Fine et al., 2000).

Target Audiences

One way to divide up findings at multiple points along the continuum is to judge what would most appeal to different audiences to whom a project has relevance, and then to produce pieces accordingly. Qualitative researchers segment audiences based on methodologies, (sub)topics, disciplines, publication

outlets (e.g., academic journals, newsletters, websites, newspapers), policy makers, practitioners, community organizations, and other stakeholders. For instance, a researcher could publish an artistic, evocative, layered account in *Qualitative Inquiry* to reach an audience of ethnographers and interpretive qualitative researchers from sociology, communication, anthropology, and education; a grounded theory analysis in *Journal of Applied Communication Research* to reach organizational, interpersonal, and group communication scholars; a content analysis or a case study in *Qualitative Health Research* to reach qualitative research scholars in nursing, medicine, social work, medical sociology, medical anthropology, and health communication; and an article summarizing her or his research and offering recommendations to benefit practitioners who read a relevant professional newsletter or trade magazine.

Strategically Select Material

Any qualitative researcher blessed with a rich body of empirical materials faces the challenge of deciding which theories and research to cite, of selecting examples with which to illustrate analytic themes, and of choosing specific incidents about which to write narratives. First, consider the research questions or issues to be addressed and the genre(s) to be used in the project. Reflect on the main point or thesis statement, and then make sure that every example or incident chosen clearly embodies one or more of these messages. Second, select particularly telling or evocative moments, quotes, or examples for lengthier representation in narrative or poetic writing, and keep the exemplars and instances that are representative but more easily broken into snippets to illustrate themes for analytic accounts. Planning ahead about which stories or incidents to develop narratively avoids boring and unhelpful repetition of the same incidents in both artistic representations and more social scientific pieces. Next, think in terms of which stories (or poems, photographs, etc.) can be shared and commented upon with a manageable amount of background information. Oftentimes qualitative researchers find themselves in the position of not being able to include a narrative or scene in an article because it would not make sense to readers unfamiliar with other aspects of their empirical materials, and insufficient space exists to fully contextualize the interaction. Similarly, some examples, taken out of context and serving as one of only a few representations of participants in a manuscript, may characterize participants in ways not intended or supported. Thus, choice of examples should be based in part on their comprehensiveness and their transportability. Finally, when writing systematic qualitative analyses, endeavor to illustrate themes with examples from different types of materials (e.g., transcripts, fieldnotes, organizational documents) and from a range of participants. Incorporating variety in this way when possible adds richness and interest to findings.

Consider Format

Perhaps the primary formatting concern for those who want to straddle the continuum within a single representation is to ensure that chosen formats embody the purpose of the article or other representation. That is, the form should reinforce the content, providing another kind of evidence or support for the main argument or another answer to a research question. A second formatting concern centers on choosing a structure that enables both showing *and* telling about a phenomenon in an effective manner. These goals are not mutually exclusive, of course; stories can also tell, and analysis can show in the midst of explication. However, accounts that combine more than one approach along the continuum should embody balanced portrayals: show *and* tell, talk *and* listen, move forward *and* step back, portray the personal *and* the political. Try to formulate a text or series of texts that provides not just multiple perspectives but a range of perspectives—group, societal, individual, dyadic, critical, appreciative, and so on. Resist limited notions of abstract or universal "fairness" or "equity" and instead think about what balance best serves the goals of the particular project.

Many authors now rely on some variant of the layered account that alternates narrative and academic prose in order to straddle constructionist (middle ground) and artistic epistemologies (Ronai, 1995). The best layered accounts show a world through particularly evocative accounts while also telling analytical arguments about theory and research. Magnet (2006), for example, alternates narratives of conversations from her life with theoretically informed critiques of her own White privilege as it informs and is informed by a dominant culture of racial oppression. The author presents a balance of showing and telling that does not necessarily accord equal length to each style but instead ensures a symbiotic relationship between the styles that serves to illuminate the central point of the article. One final formatting caution: Researchers should keep in mind the risks of reader fatigue; some very fragmented or overly complex text formats, while effective for highly motivated readers, can lose their luster over the course of a lengthy text. Avoid making mere novelty of format its own goal and instead make choices that enable readers to comprehend and engage material as fully as possible.

Keep the Forest *and* the Trees

As researchers make decisions about where on the continuum to situate different parts of a project, in what form, and for what purposes, it is wise to keep in mind the project's larger picture. I find this wider purpose difficult to keep in my head when immersed in a study; moving back and forth among the concrete details of narratives, the analytic perspective involved in constructing patterns, and paradigmatic/ideological goals presents a significant challenge. Yet that big picture view

remains critical to the construction of a meaningful project involving multiple epistemologies. To illustrate, consider several aspects of my big picture for the ethnographic exploration of a dialysis unit (Ellingson, 2007, 2008): my overall paradigm of social constructionism, my feminist commitments to praxis and critique of power, and my pragmatic goal of the transformation of the medical establishment into a more humane and just environment for employees, patients, and their loved ones. Collectively, these goals constitute a big picture of my project, and I work to ensure that each piece I produce adds to the picture—aesthetically, conceptually, and ethically.

Acknowledge Mutual Influence

I advise researchers traversing the qualitative continuum to continually reflect on how writing or creating in one genre impacts representation in other genres; multidimensional thinking may arise, as each genre offers a different way of knowing about a topic. As we move back and forth among (for instance) constructing narratives, writing personal reflections, and inductively deriving analytical categories and processes, we play with the constraints of various genres and epistemologies by allowing each to inspire and shape the others. That is, we place the modes of thinking and writing into conversation with one another. After composing narrative, we could go back to a typology of themes and rethink how a narrative does and does not fit neatly within the categories. Narratives also focus attention on some events and divert attention from others. Ordering events constructs meaning(s) for them, and these subsequent meanings may affect the analytical process as well. The reflexive relationship among different forms of analysis and artistic representation is one of the benefits of exploring multiple points on the continuum, and I encourage others to explore how their representational practices (writing, filming, painting, performing, and so on) and analysis processes mutually influence one another.

Make Each Piece Count

Another helpful strategy for crossing the continuum is to make certain that each piece of a qualitative project provides some rich material and unique argument that differs in important ways from any other representations. That does not mean we should not refer to any of the same ideas, literature, methods, or theory in more than one representation, but that we be able to point to and articulate (ideally in a single sentence) the unique contribution of each piece. Moreover, I urge readers to take advantage of the particular opportunities and constraints of each genre, outlet, or medium by choosing wisely which points to make in which form. For example, trying to accomplish the goal of showing individuals' suffering, while (probably) possible within the boundaries of a traditional scholarly journal article, may be better suited to a performance or a series of poems. Tracy's (2000) study of communication and emotion provides an excellent illustration: In a mainstream journal, she published a scholarly analysis of emotional labor of social directors on a cruise ship. This same project yielded a case study included in a volume of organizational cases intended for use by students in organizational communication courses (Tracy, 2006). Then, she authored a script, performed, and acted as a consultant to the director of an ethnodrama, presented as a live performance that offered an embodied representation of the personal and professional rewards and costs of such work for her participants (Tracy, 2003). Each of these forms represented different points on the qualitative continuum, aptly reflected their content, and offered a significant contribution to the body of knowledge on emotion work.

Be Pragmatic

In service of the goal of completing pieces of projects that reflect multiple points of the continuum, be strategic; divide materials in ways that fit with opportunities to complete work in a timely manner. The great ideas for articles that we never get around to finishing do no service to anyone; looming deadlines prompt us to complete representations. I urge researchers to shape pieces of their work to calls for submissions to edited collections, conference panels, special issues of journals, or other fora, if/as suitable calls appear. While I concede the possibility of going to extremes by "retreading" work or making the same basic argument over and over without adding any new insights, I find such intellectual laziness relatively rare. Most of us segment our qualitative studies and adapt portions of our analyses and representations to multiple audiences because we spent years and countless hours collecting and conducting them, and they are *rich*. We cannot hope to come close to exhausting our empirical materials, and so we produce legitimate and often highly valuable scholarship by drawing fresh water from already drilled wells. Pragmatism underlies the decision making of prolific qualitative researchers, and upholding this value may encourage researchers to complete pieces reflecting diverse points on the qualitative continuum.

Own the Process

A final suggestion for crossing the continuum is that as researchers we should be absolutely clear about what we did (and did not) do in our projects. This includes collection of empirical materials, analysis techniques, and especially choices about constructing representations. Terminology and practices vary widely among qualitative methodologists, even within the same discipline, and detailed methodological accounts are crucial (Potter, 1996). For example, I explained how I constructed the "day in the life" ethnographic narrative of an interdisciplinary geriatric oncology clinic:

> [T]o construct a day, I have taken liberties with chronology, condensing into a single day events that actually happened at different times during my fieldwork. I used the narrative convention of time frame (a day) to provide a sense of plot movement and improve clarity for readers. While faithfully representing the interactions I observed, I altered minor details of an interaction in service of constructing a view of the clinic that reflects the team and the people it serves in an intelligible or comprehensible manner. In addition to the chronology changes, I made two other types of changes. (Ellingson, 2005, p. 16)

By explaining my process, I help alleviate suspicions that I took an "anything goes," sloppy attitude toward constructing my representation. While some colleagues may not like or approve of some methods or genres regardless of how they are explained, concise, explicit details of analytical and representational processes make it more difficult for them to dismiss choices as careless or random. Accounting for each element of our research processes (possibly in an appendix or endnote) constitutes an important nod toward methodological rigor. As many have posited, engaging in creative or other boundary-blurring work should be no less rigorous, exacting, and subject to standards of peer evaluation (e.g., Denzin & Lincoln, 2005).

Moreover, such methodological road maps assist others who may seek to follow. One of my persistent criticisms of qualitative researchers is their unwillingness or inability (due to space constraints) to represent the mistakes and misdirections that characterize real research; we neaten it up to sound credible and get published. I believe that few qualitative researchers deliberately lie, but we often commit the sin of omission. I find this representational practice particularly problematic when some of the audience for our work—students and those new to our methodologies or a particular genre, in particular—find themselves shocked when in the field or conducting an interview and all goes much less smoothly than it was "supposed to go." While these processes inevitably reflect individual personalities and circumstances, many commonalities of experience exist and deserve to be shared among the qualitative community. Researchers need models of how to negotiate uncomfortable circumstances, whether in the field or in our offices writing (or producing) representations.

◙ WHERE TO FROM HERE?

Jumping, straddling, and loosening barriers among qualitative methodologies constitute high priorities for me. Straddling the continuum in creative ways provides a means of enlarging the possibilities and impact of qualitative research. In the final section of this chapter, I discuss three promising paths to navigating the qualitative continuum—engaging a multigenre crystallization framework (Ellingson, 2009), pursuing social justice work (Denzin & Giardina, 2009), and adopting "guerilla scholarship" strategies to enlarge editorial and generic boundaries (Rawlins, 2007)—and conclude with a look toward the next decade of traversing the qualitative continuum.

Possibilities of Multigenre Crystallization

In order to encourage boundary-spanning work along the qualitative continuum, I champion a postmodern-influenced approach to triangulation I term *crystallization,* building upon Richardson's (2000) work. Richardson invoked the crystal as an alternative metaphor to the two-dimensional, positivist image of a triangle as the basis for methodological rigor and validity. I further articulated this alternative to triangulation, which I defined as follows:

> Crystallization combines multiple forms of analysis and multiple genres of representation into a coherent text or series of related texts, building a rich and openly partial account of a phenomenon that problematizes its own construction, highlights researchers' vulnerabilities and positionality, makes claims about socially constructed meanings, and reveals the indeterminacy of knowledge claims even as it makes them. (Ellingson, 2009, p. 4)

Crystallization thus serves to promote multiple perspectives on topics, while destabilizing those same claims, yielding a postmodern form of validity (see also Janesick, 2000, p. 392; Saukko, 2004, p. 25). Several principles further clarify the approach.

First, as with any qualitative approach, crystallization seeks to *produce knowledge about a particular phenomenon through generating a deepened, complex interpretation.* All good qualitative research should provide an in-depth understanding of a topic through "thick description" (Geertz, 1973). But crystallization provides another way of achieving depth, through not only the compilation of many details but also the juxtaposition of contrasting modes of organizing, analyzing, and representing those revealing details. Second, crystallization *utilizes forms of analysis or ways of producing knowledge across multiple points of the qualitative continuum,* generally including at least one middle-ground (constructivist) or middle-to-right (postpositivist) analytic method and one interpretive, artistic, performative, or otherwise creative approach. Third, crystallized texts *include more than one genre of writing or representation.* Crystallization depends upon segmenting, weaving, blending, or otherwise drawing upon two or more genres, media, or ways of expressing findings. The slipperiness of generic categories notwithstanding, crystallized texts (and crystallized series of separate texts) draw much of their strength from their willful crossing of epistemological boundaries. A fourth principle is that crystallized texts *feature a significant degree of reflexive consideration of the researcher's self* in the process of research design, collection of empirical materials, and representation. Depending upon the researcher's goals, explicit evidence of authorial reflexivity may be subtle,

explicit, or creatively manifested. Fifth, crystallization *eschews positivist claims to objectivity and a singular, discoverable truth and embraces, reveals, and even celebrates knowledge as inevitably situated, partial, constructed, multiple, and embodied.* It brings together multiple methods *and* multiple genres simultaneously to enrich findings *and* to demonstrate the inherent limitations of all knowledge; each partial account complements the others, providing pieces of the meaning puzzle but never completing it, marking the absence of the completed image. At the same time that we surrender objectivity and singular truth, we can still make claims to know, recommendations for action, pragmatic suggestions for improvement, and theoretical insights. Although the terminology of crystallization is not yet widely used, some researchers already engage in practices that reflect these principles.

Crystallization features two primary types: integrated and dendritic. *Integrated crystallization* refers to multigenre texts that reflect the above principles in a single, coherent representation (e.g., a book, a performance) and take one of two basic forms: woven, in which small pieces of two or more genres are layered together in a complex blend; or patched, in which larger pieces of two or more genres are juxtaposed to one another in a clearly demarcated, sequential series. An outstanding exemplar of woven crystallization is Thorp's (2006) book about a participatory action project involving creating and caring for a garden at a diverse, under-resourced primary school that serves urban children, the majority of whom live in or skirt the edges of poverty. The book combines photographs, fieldnotes, poems, and analytic prose generated by the researcher with digital images of children's journals, drawings, and diagrams, artfully jumbled together in a pastiche. In an exploration of backstage teamwork among members of an interdisciplinary geriatric oncology team, I highlighted the constructed nature of accounts via patched crystallization, placing genres next to each other in a series of chapters—ethnographic narrative, grounded theory analysis, autoethnography, and feminist critique—in order to show how all accounts inevitably invoked authorial power (Ellingson, 2005; see also Bach, 2007; Lather & Smithies, 1997; D. L. Miller, Creswell, & Olander, 1998).

Dendritic crystallization refers to the ongoing and dispersed process of making meaning through multiple forms of analysis and multiple genres of representation without (or in addition to) combining genres into a single text. A particular benefit of conceptualizing the production of a series of separate representations as collectively constituting a form of postmodern methodological triangulation is scholarly legitimacy and support for academics to reach multiple audiences within and outside the academy while earning scholarly credit for work often considered to be "only" professional service. A compelling description of a research project incorporating dispersed representations that reflect multiple points on the continuum is found in Miller-Day's (2008) essay describing her study of low-wage working mothers and the personal and family challenges facing households classified as living in poverty, which included a community performance that Miller-Day argues persuasively is as significant as academic representations of findings (see also Lieblich, 2006).

Promoting Social Justice Across the Qualitative Continuum

Much of the practice of ethnography and other qualitative methods remains rooted in the passion for exposing and addressing injustice that characterized the Chicago School, many members of which explored marginalized groups in urban settings (see Lindlof & Taylor, 2002; Warren & Karner, 2010). Conquergood (1995) argued that research is always political, potentially revolutionary, and never neutral: Researchers "must choose between research that is 'engaged' or 'complicit.' By engaged I mean clear-eyed, self-critical awareness that research does not proceed in epistemological purity or moral innocence" (p. 85). Researchers cannot remain uninvolved—to refuse to advocate or to assist is to reinforce existing power relations, not to remain impartial. Calls to socially engaged work proliferate across the social sciences (e.g., Denzin & Giardina, 2009; Denzin & Lincoln, 2005; Frey & Carragee, 2007; Harter, Dutta, & Cole, 2009), often under the rubric of applied (Frey & Cissna, 2009), translational (Zerhouni, 2005), participatory action (Wang, 1999), or feminist (Hesse-Biber, 2007) research processes. I encourage readers to think of their work as always already political in its practices and implications and to use multiple modes of analysis and representation to highlight the material and ideological implications of their research practices and findings.

By consciously crossing the qualitative continuum to make the most of our rich empirical materials, we can produce written, oral, visual, and multimedia accounts that meet specific needs and interests of diverse audiences. To reach practitioners, policy makers, social commentators, and other stakeholders, we must engage in meaningful dialogue—a process that requires us to listen as much as (or more than) we speak. When we bring our ideas and willingness to collaborate to divergent academic disciplines (Parrott, 2008) and to the general public, we act as scholars and as public intellectuals who "embody and enact moral leadership" (Papa & Singhal, 2007, pp. 126–127; see also Brouwer & Squires, 2003; Giroux, 2004). When we speak out, we move beyond the important work of knowledge creation and theory building to apply our scholarly resources to benefit people more directly. The more varied our methodological toolbox, the more opportunities we have to creatively address social inequities and work for positive change—for example, through mixed methods research design (Mertens, 2007; Sosulski & Lawrence, 2008), visual and participatory methods such as photovoice (Singhal et al., 2007), and multigenre representation of research (Ellingson, 2009).

Enlarging Editorial Boundaries: The Practice of Guerilla Scholarship

Those of us who feel passionately that our work holds the potential to help people, to promote social justice, to shed light on complex problems, and to significantly influence our disciplines need to make sure important work that serves those goals gets done and published (or otherwise shared). If that goal requires being subtle (or even sneaky), so be it. In order to reach our intended audiences, some work may have to accommodate conventions that may not fit comfortably with postmodern, feminist, or narrative sensibilities, but which currently constitute the cost of admission to disciplinary ground. Rawlins (2007) explains that he resorted to fitting his work within the boundaries of traditional social scientific journal conventions by

> aping its trappings, writing style, and subdivisions . . . in order to *pass* as a serious researcher. I call such activity *guerilla scholarship*. It is necessary when certain ways of knowing are stringently enforced to the exclusion and neglect of others. The stated and unstated regimes of certain journals require these kinds of accoutrements. (p. 59, emphasis original)

Conversely, some work that closely follows traditions of (post)positivism may find that journals devoted to artistic/interpretive or "creative analytic" work (Richardson, 2000) reject the work for its apparent adherence to values that they reject. Likewise, practitioners may have jargon and communication norms that must be met to gain entry to their trade publications, or community organizations may request to have all theory, concepts, and findings translated into laypeople language. Given that no innocent or neutral portrayal is possible in any genre or medium, I consider adopting varied presentational norms that span the qualitative continuum in order to access particular outlets to be well within the boundaries of ethically sound practice. Indeed, to refuse to adapt to such conventions may just as easily reflect arrogance as integrity.

Another form of guerilla scholarship I suggest is to cite within an article multiple other works that reflect different methods, genres of representation, ideologies, or even paradigms (Ellingson, 2009). For example, Harter, Norander, and Quinlan (2007) dared to cite their research group's narratives written for a nonprofit agency's website (Harter, Norander, & Young, 2005) and a newspaper article (Novak & Harter, 2005) in a scholarly article on public intellectualism and activism published in a mainstream journal. Who knows what unsuspecting reader might follow the trail of cites straight into a forest of new practices that could broaden her or his horizons. Reference sections fulfill this function, of course, but footnotes, epigraphs, highlighted quotes in an essay or report, and interludes or other disruptive discourses can all be places to invoke the methodological or representational Other and lure the audience into new fields of play. Methods sections also offer good places to highlight connections; when I discuss analysis, I sometimes engage in guerilla scholarship by discussing the mutual influence of other pieces of my larger project with the one I documented in that method section. I urge qualitative scholars to share (or at least hint at) work from disparate areas of the qualitative continuum—that is, mention performances of findings when reporting the process of producing an inductive typology; in a performance playbill or flyer, add a note about reports based on the same data. We can also push the envelope through footnotes and the occasional provocative or creative phrase here and there that challenges the standards of acceptability in a publication outlet and thus subversively broaden the horizons of the publication and, by extension, our disciplines.

🔲 CONCLUSION

My passion for meandering across the qualitative continuum grows every year as I encounter the outstanding work of colleagues across the social sciences, education, and health and human services fields, and as I continue to stretch my own capabilities and interests into new areas. In the coming decade, I foresee a sharp rise in the number of researchers who willingly employ methods that span significant areas of the qualitative continuum in their efforts to address critical issues through socially engaged research programs. By this I mean not simply that some researchers will add some quantitative measures to highly structured qualitative content analyses or complement narratives with performances, but that the methodological lions and lambs shall lie down together peacefully in the form of stories tied to statistics, typologies interwoven with photographs, and community reports accompanied by poststructuralist critiques. I sense that such practices shall be, if not widespread, at least not uncommon. Structures and institutions change more slowly, and I hold out somewhat less hope that doctoral training programs will readily adapt to methodological and representational plurality that truly crosses epistemological boundaries within a given candidate's program of study in the same period of time. Still, the link between nimble navigation of the qualitative continuum and the pursuit of positive social change is clear, and it will undoubtedly prove irresistible to more and more researchers.

I am not naive; of course no researcher can achieve excellence in all forms of qualitative analysis and representation. We all have analytical strengths and weaknesses, individual standpoints and disciplinary predispositions, ideological commitments and artistic impulses. I encourage respect for the complexities and rigors of every art form and analytical technique, but I reject fear of the unknown and methodological prejudice inherited through our academic family lineages. K. I. Miller (2000) warned that too much emphasis on categorizing types of researchers or research

orientations can serve to constrain researchers into thinking and acting in accordance with their perceptions of their researcher type rather than pursuing whichever important research questions interest them through whichever methods (and representational forms) best enable the pursuit of answers. I have thus provided what I hope are some useful questions and strategies for those who resist or want to resist unproductive limits on the ongoing navigation of the qualitative continuum.

🔲 REFERENCES

Abu-Lughod, J. L. (1993). *Writing women's worlds: Bedouin stories.* Berkeley: University of California Press.

Anderson, L. (2006). Analytic autoethnography. *Journal of Contemporary Ethnography, 35*(4), 373–395.

Apker, J. (2001). Role development in the managed care era: A case of hospital-based nursing. *Journal of Applied Communication Research, 29,* 117–136.

Atkinson, P. (2006). Rescuing autoethnography. *Journal of Contemporary Ethnography, 35,* 400–404.

Austin, D. A. (1996). Kaleidoscope: The same and different. In C. Ellis & A. P. Bochner (Eds.), *Composing ethnography* (pp. 206–230). Walnut Creek, CA: AltaMira Press.

Bach, H. (2007). *A visual narrative concerning curriculum, girls, photography, etc.* Walnut Creek, CA: Left Coast Press.

Bergen, K. M., Kirby, E., & McBride, M. C. (2007). "How do you get two houses cleaned?" Accomplishing family caregiving in commuter marriages. *Journal of Family Communication, 7,* 287–307.

Brooks, L. J., & Bowker, G. (2002). Playing at work: Understanding the future of work practices at the Institute for the Future. *Information, Communication & Society, 5,* 109–136.

Brouwer, D. C., & Squires, C. R. (2003). Public intellectuals, public life, and the university. *Argumentation and Advocacy, 39,* 201–213.

Bruess, C. J. S., & Pearson, J. C. (1997). Interpersonal rituals in marriage and adult friendship. *Communication Monographs, 64,* 25–45.

Carlson, E. D., Engebretson, J., & Chamberlain, R. M. (2006). Photovoice as a social process of critical consciousness. *Qualitative Health Research, 16,* 836–852.

Castle, S., Fox, R. K., & Souder, K. O. (2006). Do professional development schools (PDSS) make a difference? A comparative study of PDS and non-PDS teacher candidates. *Journal of Teacher Education, 57,* 65–80.

Charmaz, K. (2000). Grounded theory: Objectivist and constructivist methods. In N. K. Denzin, & Y. S. Lincoln (Eds.), *The SAGE handbook of qualitative research* (2nd ed., pp. 509–535). Thousand Oaks, CA: Sage.

Charmaz, K. (2005). Grounded theory in the 21st century: A qualitative method for advancing social justice research. In N. K. Denzin & Y. S. Lincoln (Eds.), *The SAGE handbook of qualitative research* (3rd ed., pp. 507–535). Thousand Oaks, CA: Sage.

Charmaz, K. (2006). *Constructing grounded theory: A practical guide through qualitative analysis.* Thousand Oaks, CA: Sage.

Chawla. D. (2006). The bangle seller of Meena Bazaar. *Qualitative Inquiry, 12*(6), 1135–1138.

Conquergood, D. (1995). Between rigor and relevance: Rethinking applied communication. In K. N. Cissna (Ed.), *Applied communication in the 21st century* (pp. 79–96). Mahwah, NJ: Lawrence Erlbaum.

Cousineau, T. M., Rancourt, D., & Green, T. C. (2006). Web chatter before and after the Women's Health Initiative results: A content analysis of on-line menopause message boards. *Journal of Health Communication, 11,* 133–147.

Creswell, J. W., & Clark, V. L. P. (2006). *Designing and conducting mixed methods research.* Thousand Oaks, CA: Sage.

Defenbaugh, N. (2008). "Under erasure": The absent "ill" body in doctor–patient dialogue. *Qualitative Inquiry, 14,* 1402–1424.

Denzin, N. K., & Giardina, M. D. (2009). *Qualitative inquiry and social justice.* Walnut Creek, CA: Left Coast Press.

Denzin, N. K., & Lincoln, Y. S. (2005). Introduction: The discipline and practice of qualitative research. In N. K. Denzin & Y. S. Lincoln (Eds.), *The SAGE handbook of qualitative research* (3rd ed., pp. 1–32). Thousand Oaks, CA: Sage.

Drew, R. (2001). *Karaoke nights: An ethnographic rhapsody.* Walnut Creek, CA: AltaMira Press.

Ellingson, L. L. (1998). "Then you know how I feel": Empathy, identification, and reflexivity in fieldwork. *Qualitative Inquiry, 4,* 492–514.

Ellingson, L. L. (2005). *Communicating in the clinic: Negotiating front-stage and backstage teamwork.* Cresskill, NJ: Hampton Press.

Ellingson, L. L. (2007). The performance of dialysis care: Routinization and adaptation on the floor. *Health Communication, 22,* 103–114.

Ellingson, L. L. (2008). Patients' inclusion of spirituality within the comprehensive geriatric assessment process. In M. Wills (Ed.), *Spirituality and health communication* (pp. 67–85). Cresskill, NJ: Hampton Press.

Ellingson, L. L. (2009). *Engaging crystallization in qualitative research: An introduction.* Thousand Oaks, CA: Sage.

Ellingson, L. L., & Ellis, C. (2008). Autoethnography as constructionist project. In J. A. Holstein, & J. F. Gubrium (Eds.), *Handbook of constructionist research* (pp. 445–465). New York: Guilford Press.

Ellis, C. (2004). *The ethnographic I: A methodological novel about autoethnography.* Walnut Creek, CA: AltaMira.

Ellis, C., & Bochner, A. P. (2006). Analyzing analytic autoethnography: An autopsy. *Journal of Contemporary Ethnography, 35*(4), 429–449.

Ellis, C., & Ellingson, L. L. (2000). Qualitative methods. In E. F. Borgatta & R. J. V. Montgomery (Eds.), *Encyclopedia of Sociology* (2nd ed., Vol. 4, pp. 2287–2296). New York: Macmillan Library Reference.

Faulkner, S. L. (2007). Concern with craft: Using Ars Poetica as criteria for reading research poetry. *Qualitative Inquiry, 13*(2), 218–234.

Faulkner, S. L. (2010). *Poetry as method: Reporting research through verse.* Walnut Creek, CA: Left Coast Press.

Feng, J., & Wu, D. D. (2009). Changing ideologies and advertising discourses in China: A case study of *Nanfang Daily. Journal of Asian Pacific Communication, 19,* 218–238.

Fine, M., Weis, L., Weseen, S., & Wong, L. (2000). For whom? Qualitative research, representation, and social responsibilities. In N. K. Denzin, & Y. S. Lincoln (Eds.), *Handbook of qualitative research* (2nd ed., pp. 107–132). Thousand Oaks, CA: Sage.

Fitch, K. L. (1994). Criteria for evidence in qualitative research. *Western Journal of Communication, 58,* 32–38.

Frank, A. W. (1995). *The wounded storyteller: Body, illness, and ethics.* Chicago: University of Chicago Press.

Frey, L. R., & Carragee, K. M. . (2007). *Communication activism, vol. 1: Communication for social change.* Cresskill, NJ: Hampton Press.

Frey, L. R., & Cissna, K. (Eds.). (2009). *Handbook of applied communication research.* New York: Routledge.

Geertz. C. (1973). *The interpretation of cultures.* New York: Basic Books.

Gergen, K. J. (1994). *Realities and relationships: Soundings in social construction.* Cambridge, MA: Harvard University Press.

Giroux, H. A. (2004). Cultural studies, public pedagogy, and the responsibility of intellectuals. *Communication and Critical/ Cultural Studies, 1*(1), 59–79.

Glaser, B., & Strauss, B. (1967). *The discovery of grounded theory: Strategies for qualitative research.* Chicago: Aldine.

González, M. C. (1998). Painting the white face red: Intercultural contact presented through poetic ethnography. In J. Martin, T. Nakayama, & L. Flores (Eds.), *Readings in cultural contexts* (pp. 485–495). Mountain View, CA: Mayfield.

Goodier, B. C., & Arrington, M. I. (2007). Physicians, patients, and medical dialogue in the NYPD Blue prostate cancer story. *Journal of Medical Humanities, 28*(1), 45–58.

Gray, R., & Sinding, C. (2002). *Standing ovation: Performing social science research about cancer.* Walnut Creek, CA: AltaMira Press.

Guba, E. G., & Lincoln, Y. S. (1994). Competing paradigms in qualitative research. In N. K. Denzin & Y. S. Lincoln (Eds.), *Handbook of qualitative research* (pp. 105–117). Thousand Oaks, CA: Sage.

Gubrium, J. F., & Holstein, J. A. (1997). *The new language of qualitative method.* New York: Oxford University Press.

Güven, B. (2008). Experience, instruction, and social environment: Fourth and fifth grade students' use of metaphor. *Social Behavior and Personality, 36,* 743–752.

Harter, L. M., Dutta, M., & Cole, C. (Eds.). (2009). *Communicating for social impact.* Cresskill, NJ: Hampton Press.

Harter, L. M., Japp, P. M., & Beck, C. (Eds.). (2005). *Narratives, health, and healing: Communication theory, research, and practice.* Mahwah, NJ: Lawrence Erlbaum.

Harter, L. M., Norander, S., & Quinlan, M. M. (2007). Imaginative renderings in the service of renewal and reconstruction. *Management Communication Quarterly, 21,* 105.

Harter, L. M., Norander, S., & Young, S. (2005). *Collaborative art: Cultivating connections between self and other.* Available at http:// www.passionworks.org/articles

Hartnett, S. J. (2003). *Incarceration nation: Investigative prison poems of hope and terror.* Walnut Creek, CA: AltaMira Press.

Hecht, M. L., & Miller-Day, M. (2007). The Drug Resistance Strategies Project as translational research. *Journal of Applied Communication Research, 35,* 343–349.

Hesse-Biber, S. N. (Ed.). (2007). *Handbook of feminist research: Theory and praxis.* Thousand Oaks, CA: Sage.

Hodgkin, S. (2008). Telling it all: A story of women's social capital using a mixed methods approach. *Journal of Mixed Methods Research, 2*(3), 296–316.

Jago, B. (2006). A primary act of imagination: An autoethnography of father-absence. *Qualitative Inquiry, 12,* 398–426.

Janesick, V. J. (2000). The choreography of qualitative research design: Minuets, improvisations, and crystallization. In N. K. Denzin &

Y. S. Lincoln (Eds.), *Handbook of qualitative research* (2nd ed., pp. 379–399). Thousand Oaks, CA: Sage.

Kiesinger, C. (2002). My father's shoes: The therapeutic value of narrative reframing. In A. Bochner & C. Ellis (Eds.), *Ethnographically speaking: Autoethnography, literature, and aesthetics* (pp. 95–114). Walnut Creek, CA: AltaMira Press.

Kontos, P. C., & Naglie, G. (2007). Expressions of personhood in Alzheimer's disease: An evaluation of research-based theatre as a pedagogical tool. *Qualitative Health Research, 17*(6), 799–811.

Kuperberg, A., & Stone, P. (2008). The media depiction of women who opt out. *Gender & Society, 22,* 497–517.

Kwok, C., & Sullivan, G. (2007). The concepts of health and preventive health practices of Chinese-Australian women in relation to cancer screening. *Journal of Transcultural Nursing, 18*(2), 118–126.

Larsen, E. A. (2006). A vicious oval. *Journal of Contemporary Ethnography, 35,* 119–147.

Lather, P., & Smithies, C. (1997). *Troubling the angels: Women living with HIV/AIDS.* Boulder, CO: Westview Press.

Lee, K. V. (2006). A fugue about grief. *Qualitative Inquiry, 12,* 1154–1159.

Lieblich, A. (2006). Vicissitudes: A study, a book, a play: Lessons from the work of a narrative scholar. *Qualitative Inquiry, 12,* 60–80.

Lindemann, K. (2009). Cleaning up my (father's) mess: Narrative containments of "leaky" masculinities. *Qualitative Inquiry, 16,* 29–38.

Lindlof, T. R., & Taylor, B. C. (2002). *Qualitative communication research methods* (2nd ed.). Thousand Oaks, CA: Sage.

López, E., Eng, E., Randall-David, E., & Robinson, N. (2005). Quality-of-life concerns of African American breast cancer survivors within rural North Carolina: Blending the techniques of photovoice and grounded theory. *Qualitative Health Research, 15,* 99–115.

Low, J. (2004). Managing safety and risk: The experiences of people with Parkinson's disease who use alternative and complementary therapies. *Health, 8,* 445–463.

Magnet, S. (2006). Protesting privilege: An autoethnographic look at whiteness. *Qualitative Inquiry, 12,* 736–749.

Markham, A. N. (2005). "Go ugly early": Fragmented narrative and bricolage as interpretive method. *Qualitative Inquiry, 11*(6), 813–839.

McAllister, C. L., Wilson, P. C., Green, B. L., & Baldwin, J. L. (2005). "Come and take a walk": Listening to Early Head Start parents on school-readiness as a matter of child, family, and community health. *American Journal of Public Health, 95,* 617–625.

Mertens, D. M. (2007). Transformative paradigm: Mixed methods and social justice. *Journal of Mixed Methods Research, 1,* 212–225.

Meyer, M. (2004). From transgression to transformation: Negotiating the opportunities and tensions of engaged pedagogy in the feminist organizational communication classroom. In P. M. Buzzanell, H. Sterk, & L. H. Turner (Eds.), *Gender in applied communication contexts* (pp. 195–213). Thousand Oaks, CA: Sage.

Meyer, M., & O'Hara, L. S. (2004). When they know who we are: The National Women's Music Festival comes to Ball State University. In P. M. Buzzanell, H. Sterk, & L. H. Turner (Eds.), *Gender in applied communication contexts* (pp. 3–23). Thousand Oaks, CA: Sage.

Mienczakowski, J. (1996). An ethnographic act: The construction of consensual theatre. In C. Ellis & A. P. Bochner (Eds.), *Composing ethnography: Alternative forms of qualitative writing* (pp. 244–264). Walnut Creek, CA: AltaMira.

Mienczakowski, J. (2001). Ethnodrama: Performed research—Limitations and potential. In P. Atkinson, A. Coffey, S. Delamont, J. Lofland, & L. Lofland (Eds.), *Handbook of ethnography* (pp. 468–476). Thousand Oaks, CA: Sage.

Miller, D. L., Creswell, J. W., & Olander, L. S. (1998). Writing and retelling multiple ethnographic tales of a soup kitchen for the homeless. *Qualitative Inquiry, 4*(4), 469–491.

Miller, K. I. (2000). Common ground from the post-positivist perspective: From "straw-person" argument to collaborative coexistence. In S. R. Corman & M. S. Poole (Eds.), *Perspectives on organizational communication: Finding common ground* (pp. 47–67). New York: Guilford Press.

Miller-Day, M. A. (2008). Performance matters. *Qualitative Inquiry, 14*(8), 1458–1470.

Miller-Day, M. A., & Dodd, A. H. (2004). Toward a descriptive model of parent–offspring communication about alcohol and other drugs. *Journal of Social and Personal Relationships, 21*(1), 69–91.

Montemurro, B. (2005). Add men, don't stir. *Journal of Contemporary Ethnography, 34,* 6–35.

Novak, D. R., & Harter, L. M. (2005, June 14–18). Blues fest showcase world's best. *StreetWise,* pp. 1–2.

Nowell, B. L., Berkowitz, S. L., Deacon, Z., & Foster-Fishman, P. (2006). Revealing the cues within community places: Stories of identity, history, and possibility. *American Journal of Community Psychology, 37,* 29–46.

O'Donnell, A. B., Lutfey, K. E., Marceau, L. D., & McKinlay, J. B. (2007). Using focus groups to improve the validity of cross-national survey research: A study of physician decision making. *Qualitative Health Research, 17,* 971–981.

Papa, M. J., & Singhal, A. (2007). Intellectuals searching for publics: Who is out there? *Management Communication Quarterly, 21,* 126–136.

Parrott, R. (2008). A multiple discourse approach to health communication: Translational research and ethical practice. *Journal of Applied Communication Research, 36,* 1–7.

Parry, D. C. (2006). Women's lived experiences with pregnancy and midwifery in a medicalized and fetocentric context: Six short stories. *Qualitative Inquiry, 12,* 459–471.

Potter, W. J. (1996). *An analysis of thinking and research about qualitative methods.* Mahwah, NJ: Lawrence Erlbaum.

Prendergast, M. (2007). Thinking narrative (on the Vancouver Island ferry): A hybrid poem. *Qualitative Inquiry, 13,* 743.

Rambo, C. (2005). Impressions of Grandmother: An autoethnographic portrait. *Journal of Contemporary Ethnography, 34,* 560–585.

Rambo, C. (2007). Handing IRB an unloaded gun. *Qualitative Inquiry, 13,* 353–416.

Rawlins, W. K. (2007). Living scholarship: A field report. *Communication Methods and Measures, 1,* 55–63.

Richardson, L. (1992a). The consequences of poetic representation: Writing the other, rewriting the self. In C. Ellis & M. G. Flaherty (Eds.), *Investigating subjectivity: Research on lived experience* (pp. 125–140). Thousand Oaks, CA: Sage.

Richardson, L. (1992b). The poetic representation of lives: Writing a postmodern sociology. *Studies in Symbolic Interaction, 13,* 19–29.

Richardson, L. (1993). Poetics, dramatics, and transgressive validity: The case of the skipped line. *Sociological Quarterly, 35,* 695–710.

Richardson, L. (2000). Writing: A method of inquiry. In N. K. Denzin & Y. S. Lincoln (Eds.), *Handbook of qualitative research* (2nd ed., pp. 923–943). Thousand Oaks, CA: Sage.

Riessman, C. K. (2008). *Narrative methods for the human sciences.* Thousand Oaks, CA: Sage.

Ronai, C. R. (1995). Multiple reflections on childhood sex abuse: An argument for a layered account. *Journal of Contemporary Ethnography, 23,* 395–426.

Saarnivaara, M. (2003). Art as inquiry: The autopsy of an [art] experience. *Qualitative Inquiry, 9*(4), 580–602.

Sacks, J. L., & Nelson, J. P. (2007). A theory of nonphysical suffering and trust in hospice patients. *Qualitative Health Research, 17,* 675–689.

Sandgren, A., Thulesius, H., Fridlund, B., & Petersson, K. (2006). Striving for emotional survival in palliative cancer nursing. *Qualitative Health Research, 16*(1), 79–96.

Saukko, P. (2004). *Doing research in cultural studies: An introduction to classical and new methodological approaches.* Thousand Oaks, CA: Sage.

Scott, C., & Sutton, R. E. (2009). Emotions and change during professional development for teachers. *Journal of Mixed Methods Research, 3*(2), 151–171.

Secklin, P. L. (2001). Multiple fractures in time: Reflections on a car crash. *Journal of Loss and Trauma, 6*(4), 323–333.

Singhal, A., Harter, L. M., Chitnis, K., & Sharma, D. (2007). Participatory photography as theory, method, and praxis: Analyzing an entertainment-education project in India. *Critical Arts, 21*(1), 212–227.

Sosulski, M. R., & Lawrence, C. (2008). Mixing methods for full-strength results. *Journal of Mixed Methods Research, 2*(2), 121–148.

Spry, T. (2001). Performing autoethnography: An embodied methodological praxis. *Qualitative Inquiry, 7*(6), 706–732.

Strauss, A., & Corbin, J. (1998). *Basics of qualitative research: Techniques and procedures for developing grounded theory* (2nd ed.). Thousand Oaks, CA: Sage.

Teram, E., Schachter, C. L., & Stalker, C. A. (2005). The case for integrating grounded theory in participatory action research: Empowering clients to inform professional practice. *Qualitative Health Research, 15,* 1129–1140.

Thorp, L. (2006). *Pull of the earth: Participatory ethnography in the school garden.* Walnut Creek, CA: AltaMira Press.

Tillmann-Healy, L. (2001). *Between gay and straight: Understanding friendship across sexual orientation.* Walnut Creek, CA: AltaMira Press.

Tracy, S. J. (2000). Becoming a character for commerce: Emotion labor, self subordination, and discursive construction of identity in a total institution. *Management Communication Quarterly, 14,* 90–128.

Tracy, S. J. (2003). *Navigating the cruise—A trigger script ethnodrama.* Tempe: The Hugh Downs School of Human Communication's Empty Space Theater, Arizona State University.

Tracy, S. J. (2004). The construction of correctional officers: Layers of emotionality behind bars. *Qualitative Inquiry, 10*(4), 509–533.

Tracy, S. J. (2006). Navigating the limits of a smile: Emotion labor and concertive control on a cruise ship. In J. Keyton & P. Shockley-Zalabak (Eds.), *Case studies for organizational communication: Understanding communication processes* (2nd ed., pp. 394–407). Los Angeles: Roxbury.

Trujillo, N. (2004). *In search of Naunny's grave: Age, class, gender, and ethnicity in an American family.* Walnut Creek, CA: AltaMira Press.

Vanderford, M. L., Smith, D. H., & Olive, T. (1995). The image of plastic surgeons in news media coverage of the silicone breast implant controversy. *Plastic and Reconstructive Surgery, 96*(3), 521–538.

Wang, C. C. (1999). Photovoice: A participatory action research strategy applied to women's health. *Journal of Women's Health, 8*(2), 185–192.

Warren, C. A. B., & Karner, T. X. (2010). *Discovering qualitative methods: Field research, interviews, and analysis* (2nd ed.). Los Angeles: Roxbury.

White, S. A. (2003). Introduction: Video power. In S. A. White (Ed.), *Participatory video: Images that transform and empower* (pp. 17–30). Thousand Oaks, CA: Sage.

Wilson, H. S., Hutchinson, S. A., & Holzemer, W. L. (2002). Reconciling incompatibilities: A grounded theory of HIV medication adherence and symptom management. *Qualitative Health Research, 12*(10), 1309–1322.

Zerhouni, E. A. (2005). Translational and clinical science: Time for a new vision. *New England Journal of Medicine, 353,* 1621–1623.

Zoller, H. M. (2003). Health on the line: Identity and disciplinary control in employee occupational health and safety discourse. *Journal of Applied Communication Research, 31*(2), 118–139.

37

POST QUALITATIVE RESEARCH

The Critique and the Coming After

Elizabeth Adams St.Pierre

This chapter about *post* qualitative research may come both too late and too soon. I know it comes too late because I began to write it over 20 years ago, in 1995, in my first qualitative research report, my dissertation (St.Pierre, 1995). It may come too soon because qualitative research is still under a deliberate, naive, and crude attack as it has been since the beginning of the 21st century. With that in mind, I acknowledge the slippery politics of my critique. However, I am weary both of defending an overdetermined qualitative inquiry I find increasingly limited and also of the always already failed romance of trying to "talk across differences" (see, for example, Moss et al., 2009) with people who haven't kept up; are "paradigms behind" (Patton, 2008, p. 269); and have for over half a century, it seems, not read and/or engaged the linguistic turn, the cultural turn, the interpretive turn, the narrative turn, the historical turn, the critical turn, the reflexive turn, the rhetorical turn, the postmodern turn, and others. With Spivak (1993), I cannot see why "people who do not have the time to learn should organize the construction of the rest of the world" (p. 187).

Before providing a postmodern critique of what I call in this chapter *conventional humanist qualitative methodology*, I will sketch the political context in which this chapter is written, one in which both postmodernism and qualitative methodology have been rejected by "science."

The Context of Scientifically Based Research

Qualitative research in education in the United States has taken a beating since 2002 when the U.S. No Child Left Behind Act took effect and the National Research Council (NRC) (2002) published its report, *Scientific Research in Education* (SRE). By establishing experimental research and, preferably, randomized controlled trials as the gold standard for high-quality research,

those two documents—one a federal law that mandated research methodology—exemplified the positivist and conservative restoration in the larger audit and accountability culture that privileges an instrumental, engineering model of social science that feeds on metrics to establish "what works." Notwithstanding claims of inclusiveness and in the fervor of a new scientism, qualitative research was rejected as not rigorous enough to count as high-quality science.

As time passed, it became clear that postmodernism had also become a whipping boy, a *codeword* for critiques of positivist tendencies offered by all those "turns" listed above and for critiques, more specifically, by feminist, race, Marxist, queer, postcolonial, and other theories. One of the first examples of how postmodernism was used to stand in for what was non-experimental and non-positivist was the following statement in the 2002 NRC report:

> We assume that it is possible to describe the physical and social world scientifically so that, for example, multiple observers can agree on what they see. Consequently, we reject the postmodernist school of thought when it posits that social science research can never generate objective or trustworthy knowledge. (p. 25)

A footnote to that statement is even more perplexing:

> This description applies to an extreme epistemological perspective that questions the rationality of the scientific enterprise altogether, and instead believes that all knowledge is based on sociological factors like power, influence, and economic factors. (p. 25)

Both statements assume views of reality and reason about which all the "turns," not just postmodernism, are skeptical. But even law enforcement officials experience social constructionism when eyewitnesses differently describe/construct what they "saw"; and many social *and* natural scientists acknowledge that power, politics, and economic factors influence knowledge production in the sciences.

The NRC report cited as its authority for those two statements rejecting postmodernism a book by Phillips (a member of the NRC committee) and Burbules (2000). But my careful reading of that book found almost no discussion of postmodernism or citations to postmodern scholars except Lyotard, so it cannot serve as a warrant for the committee's claims about postmodernism. Nonetheless, Phillips (2006) continued to critique postmodernism without citing postmodern scholars, calling postmodernism "extreme" and "at the left-hand end or pole" of some "continuum," claiming for himself a moderate, temperate position. Later, Phillips (2009) seemed surprised that researchers would object to the rampant orthodox positivism in the scientifically based research (SBR) debates and admonished resisters to SBR—those in a "diseased condition" (p. 164)—for being unruly, recalcitrant, querulous, vituperative, and filled with "postmodernist contumely" (p. 193).

Other examples of unsubstantiated, under-sourced critiques of postmodernism followed, especially from Grover Whitehurst, the former Director of the U.S. Institute of Education Sciences (IES). In his 2003 Presidential Invited Session at the American Educational Research Association, he proclaimed that we need less theory, naming postmodernism in particular, and more of "what works." Whitehurst (IES, 2008) continued to attack postmodernism in his final report to Congress in 2008, reporting that during his tenure at IES he had had to distinguish the institute's work "from what had become the dominant forms of education research in the latter half of the 20th century: qualitative research grounded in postmodern philosophy and methodologically weak quantitative research" (p. 5). In the report, he referred to "the ascendance of postmodern approaches to education research" (p. 6). Who knew? I, for one, was unaware that postmodern qualitative research had ever been dominant in any field. True to form, Whitehurst did not cite postmodern scholars or texts in either case, but the link between postmodernism and qualitative research was, nonetheless, established in their rejection.

Whitehurst's counterpart in England, Anne Oakley, Director of the Social Science Research Unit at the Institute of Education at the University of London, had been rejecting postmodernism for some time. In an odd twist, she called critiques of scientifically based research "resistance texts" and "conservative responses to real or imagined threats, including that of 'new' technology [the randomized controlled trial and the systematic review (SR); see MacLure, 2005, for a postmodern critique of SRs], and its ability to reveal *previously concealed features of academic work*" (Oakley, 2006, p. 64, emphasis added). Oakley laid the blame for poor social science research on postmodernism, chiding it for "fashionable nonsense and word games" (p. 78) though, like Whitehurst, she cited none of the key postmodern scholars.

Perhaps Henig's (2008) comment summed up the positivist concern about research that is not positivist. He wrote that educational research had become a fragmented field that was "overly abstract (e.g., neo-Marxist, post-structuralism, gender identity, or critical race studies) and seen as insufficiently rigorous—leaning toward qualitative over quantitative research and less concerned with causal mechanisms than telling convincing stories" (p. 51). Of course, it is only positivism that claims to be non-abstract, objective, and both theory- and value-free. Qualitative inquiry has always carried on its strong, supple back theory-, value-, power-, and politically-laden investigations of the lived experiences of people in the midst of living—hardly a neutral arena. It has never claimed science can be nowhere, disentangled from the humans who produce it.

The examples above illustrate how, in the SBR debates, postmodernism—without citational authority or warrant—became a code word for any philosophical approach that is not positivist. Linked to qualitative methodology, which could never be rigorous because it is not experimental, postmodernism was doubly damned. In writing about the resistance to Derrida's work, which has profoundly informed postmodern theories, Lamont (1987) suggested that

> this large opposition was related to Derrida's attack on the basic tenets of the humanist tradition and interpretive activity. The very violence of these attacks contributed to the institutionalization of deconstruction; it indicated that Derrida had become a force to be contended with. (p. 612)

Hodkinson (2004) pointed out that postmodernism's critique of foundationalism in educational research has been significant though both ignored and rejected in the new audit culture with its "increasing dominance of procedures of target setting, outcomes measurement, and the focus on effectiveness and efficiency, rather than purpose or values (p. 17). But a conservative, positivist restoration in social science research methodology has not been secured; postmodern and other approaches in the social sciences are entrenched and proliferate.

The Resurgence of Postmodernism

This chapter is my contribution to the *resurgence of postmodernism,* a body of critique especially useful in times like these as positivism once again falters across the social sciences as evidenced in education, for example, by the failure of the "what works" mentality of the IES in too many "no effect" research results (Viadero, 2009) and, in economics, the hardest of the soft social sciences, by the recent failure of supposed rational free markets and consumers (see Cullenberg, Amariglio, & Ruccio, 2001). Postmodern critiques, along with those of the other "turns," emerged half a century ago in response to the excesses of the positivisms; produced a sea change in the social sciences and humanities; and are needed once again to open up structures being disciplined, regulated, and normalized.

I wish to make it clear here at the beginning that the "posts" (e.g., post-colonialism, post-critical, post-humanist, post-Fordist, post-positivist, post-feminist, post-foundational, post-emancipatory, post-memory, post-subjective, post-everything—henceforth, I will abbreviate using "posts") have never offered alternative structures and that *I do not and cannot offer an alternative methodology*—a recipe, an outline, a structure, for post qualitative research—another handy "research design" in which one can safely secure oneself and one's work. The "posts" do not offer a corrective or a fix.

Instead, I use the occasion of writing this chapter to summarize and extend my continuing concerns with what I refer to as conventional humanist qualitative inquiry and to call for a renewed commitment to a *reimagination of social science inquiry* enabled by postmodernism in spite of the current positivist orthodoxy. My critique is not that qualitative research is unscientific; it is My critique is that, to a great extent, it has been so disciplined, so normalized, so centered—especially because of recent assaults by SBR—that *it has become conventional,* reductionist, hegemonic, and sometimes oppressive and has lost its radical possibilities "to produce different knowledge and produce knowledge differently" (St. Pierre, 1997, p. 175). I am well aware that work accomplished under the umbrella of "qualitative research" is also radical, unconventional, and exciting. I do not believe, however, that that is the qualitative inquiry described in most textbooks or taught in most university courses. I think it's time, therefore, to move away from a centering, defensive mode and get on with the invention of science.

I argue that the concepts/categories that structure conventional, humanist qualitative inquiry have increasingly tightened since 1985, for example, when Lincoln and Guba wrote one of the first qualitative research texts, *Naturalistic Inquiry.* We now have thousands of textbooks, handbooks, and journal articles that have secured *qualitative methodology* by repeating that structure in book after book with the same chapter headings so that we now believe it is true and real. *We've forgotten we made it up.*

The "post" in the title of this chapter can be thought of both chronologically—what comes after conventional humanist qualitative research—and, more importantly, deconstructively. I will discuss deconstruction in detail later in the chapter, but here I explain that I envision at least two deconstructive approaches. The first follows Derrida in putting a structure *sous rature,* or under erasure. In this approach, we retain the structure of qualitative research methodology—its structuring concepts and categories—because it appears necessary and, at the same time, cross it out because it is inaccurate. Thus, we could write ~~qualitative research methodology~~ to signal the opening up—not the rejection—of the structure. "Persistently to critique a structure that one cannot not (wish to) inhabit is the deconstructive stance" (Spivak, 1993, p. 284). Working within

the enclosure of conventional humanist qualitative inquiry while troubling it is what I have done for almost 20 years, but I seem no longer content with that approach.

During the last decades, we've deconstructed many of qualitative methodology's concepts/categories: e.g., *interview* (Scheurich, 1995), *validity* (Lather, 1993), *data* (St. Pierre, 1997), *voice* (Jackson & Mazzei, 2009), *reflexivity* (Pillow, 2003). The deconstruction of even one concept/category disrupts other related structuring concepts/categories, as I illustrate later in this chapter, and initiates the cascading collapse of methodology's center, its failure in the wake of the "posts." The difficulty for the poststructural researcher lies in trying to function in the ruins of the structure after the theoretical move that authorizes its foundations has been interrogated and its limits breached so profoundly that its center no longer holds.

Of course, *the structure had always already been ruptured, ruined.* In self-defense, some wrote textbooks to convince us it was coherent and asked researchers to organize their work into inadequate existing concepts (e.g., *research design, data, data collection, data analysis, interview, observation, representation*) even though they could not contain it. Much, therefore, has been unintelligible, and science has, accordingly, been impoverished.

A second deconstructive approach helps here. Derrida also explained that deconstruction is more than working within and against a structure. It is also the overturning and displacement of a structure so that something(s) different can be thought/done. In this second approach, one is no longer "residing within the closed field [of the structure] thereby confirming it" (Derrida, 1972/1981, p. 41) but is "overturning and displacing a conceptual order, as well as the nonconceptual order with which the conceptual order is articulated" (Derrida, 1971/1982, p. 329). After that displacement, we are in play as we radically de-naturalize what we've taken for granted. Here, we refuse alternatives and pursue the *supplement,* what always already escapes the structure.

This is the science that beckons—this is the lure. And this is deconstruction at its finest—the rigorous reimagining of a capacious science that cannot be defined in advance and is never the same. This science is *différance,* not repetition. Borrowing from Deleuze and Guattari (1980/1987), it is not *is;* it is *becoming.* We move away from Plato's poisoned gift of ontological determination, a logic of identity and prediction—*Science is this; science is not that*—toward a logic of the "and"—*This and this and this and this. . . .* Thinking and doing that science is the invitation we risk accepting.

Reading Theory

Before one can put postmodernism or any theoretical approach to work, one must read and study it. Given the impossibility of "talking across differences" that I, and others, experienced in the SBR debates, I am convinced that the study of philosophy should precede the study of research methodology so that, for example, the typical social science researcher would understand the epistemological and ontological assumptions that structure positivist, interpretive, critical, postmodern, and other methodologies in the social sciences. Attempts to disentangle science and philosophy are always dangerous.

If such study were the norm, most readers of the 2002 National Research Council report, *Scientific Research in Education,* would recognize the following sample statements from that report as positivist (there are many others): "cumulative knowledge" (p. 1), "at its core, scientific inquiry is the same in all fields" (p. 2), "replicate and generalize across studies" (p. 4), and "multiple observers can agree on what they see" (p. 25). That many readers of the 2002 NRC report could not do so is the result of a failure to read, not necessarily a failure to teach. Students do not have to take doctoral courses in theory in order to read and study positivism or social constructionism or critical race theory or postmodernism, though such courses would surely help researchers understand that *science* is not one thing but a highly contested concept whose meaning and practices shift across philosophical approaches and historical and political moments.

Unfortunately, we hesitate to read outside our comfort areas and too casually reject texts that seem too hard to read. It's doubtful we would expect to quickly understand an advanced physics text, yet we expect a philosophy text to be welcoming and accessible. Why should we expect to understand Derrida or Foucault or Deleuze and Guattari on first reading? Perhaps it's arrogant to think we should quickly understand concepts that have age-old and contentious histories such as *knowledge, truth, reason, reality, power,* and *language* and new concepts such as *différance,* the *rhizome,* and Foucaultian *archaeology.*

But the idea that language should be clear is not only deeply embedded in our anti-intellectual culture but also in positivism. For example, Ayer (1936), a logical positivist, wrote in his mid-20s,

> For we shall maintain that no statement which refers to a "reality" transcending the limits of all possible sense-experience can possibly have any literal significance; from which it must follow that the labours of those who have striven to describe such a reality have all been devoted to the production of nonsense. (p. 17)

This view of language is commensurate with the positivist determination to eschew metaphysics, which is speculative, deals with concepts that can't be verified with empirical evidence, and are thus senseless according to Ayer. It also represents positivism's "search for certainty" (Reichenbach, 1951) echoed in the call for clear language. Following Ayer, Maxwell (2010), for example, could dismiss Deleuze and Guattari, who introduced new language that might enable new realities, and claim they are "simply 'running their mouths'"

(p. 6), which, of course, one could say about Ayer, Carnap, Reichenbach, Husserl, Marx, Einstein, Neils Bohr, Grigori Perelman, and many other scholars, including Maxwell, if one were so inclined.

I advise students to take seriously Lacan's (as cited in Ulmer, 1985) advice, "to read does not obligate one to understand. First it is necessary to read . . . avoid understanding too quickly" (p. 196). I have little sympathy with excuses not to read difficult texts, and I advise students to read harder when the text seems too hard to read, to just keep reading, letting the new language wash over them until it becomes familiar. I encourage them to develop "reading management strategies," for example, dictionaries of concepts they don't understand. The dictionary I began as a doctoral student is now over 700 pages long. The entry on *subjectivity,* the focus of my work, is over 30 pages. Still, I don't know what subjectivity *means.*

If we don't read the theoretical and philosophical literature, we have nothing much to think with during analysis except normalized discourses that seldom explain the way things are. However, when we study a variety of complex and conflicting theories, which I believe is the purpose of doctoral education, we begin to realize, as Fay (1987) suggested, that *we have been theorized,* that we and the world are products of theory as much as practice, and that putting different theories to work can change the world. History, of course, tells us that.

I don't care whether students make the postmodern or any other turn. I do expect them to read hard. By the time they write their dissertation proposals, I expect them to have studied several bodies of high-level theory (e.g., feminist theories, race theories, phenomenology, postmodernism, social constructionism), the theories of their disciplines that have taken up those high-level theories (e.g., in English education, reader response theory that is thinkable because of social constructionism), and the methodological literature.

In all this, I am increasingly interested in *readiness,* and find Butler's (1995) question apropos: "how is it that we become available to a transformation of who we are, a contestation which compels us to rethink ourselves, a reconfiguration of our 'place' and our 'ground'?" (p. 132). I expect we're inclined toward certain interests but, no doubt, that's as much a matter of experience and education as disposition.

Clearly, what I describe here is not the "training model" of the natural sciences. My desire is for students to go into a study immersed in a field of complex and contradictory theory rich enough to address the complex and contradictory nature of whatever they encounter in fieldwork and analysis. They desperately need theories, interpretive frameworks, for analyzing data rather than more and/or better methods for collecting it, else they produce poorly conceived and theorized work. Hurworth (2008), in her study of the teaching of qualitative inquiry, concurred and wrote that theory and practice should be taught together. And Neumann, Pallas, and Peterson (2008) wrote that we should focus research preparation on the challenge of the "management and use of epistemological diversity in research" (p. 1478). The following three sections of this chapter describe poststructural theories I have found particularly helpful, which, once read, studied, and taken up, will change the way one reads the world.

Postmodernism and Poststructuralism (the "posts")

Postmodern and poststructural analyses include diverse and contradictory critiques that resist, subvert, and refuse any structural formation. Rajchman (1987) wrote that postmodernism "does not comprise a School of Thought" but refers to a "motley and elastic range of things" (p. 49). While taking account of the "turns" mentioned in this chapter, the "posts" announce a radical break with the humanist, modernist, imperialist, representationalist, objectivist, rationalist, epistemological, ontological, and methodological assumptions of Western Enlightenment thought and practice.

In the "posts," the "epistemological point of departure in philosophy is inadequate" (Butler, 1992, p. 8) and, some might argue, incommensurable. *Epistemology*—the branch of philosophy concerned with what counts as knowledge and how knowledge claims are justified as true—assumes a certain kind of subject, the sovereign, knowing subject secured in advance of knowing and always separate from it—the originating subject of knowledge and history. The "posts" deconstruct that subject as I will discuss later in the chapter. *Ontology* is the branch of metaphysics concerned with what exists (what "is"), with being and reality and how entities are organized. In the "posts," the ontological point of departure is also inadequate. In conventional philosophy, it is important to keep epistemological and ontological issues separate, and, if they are confused, we say a "category mistake" has been made. But postmodern theories disrupt the distinction between epistemology and ontology, as does physics, for example, "where the rise of quantum theory with its interpretational problems was one of the first major challenges to the ontic/epistemic distinction" (Atmanspacher, 2002, p 50).

The terms *postmodernism* and *poststructuralism* are often used interchangeably; however, there are acknowledged differences in their meaning. Thus far, I have used the term *postmodern* in this chapter, though I would describe my own work as *poststructural*. Lather (1993) differentiated these two terms as follows: postmodernism "raises issues of chronology, economics (e.g., post-Fordism) and aesthetics whereas poststructural[ism] is used more often in relation to academic theorizing 'after structuralism'" (p. 688).

Postmodernism has been used to refer to "the new stage of multinational, multiconglomerate consumer capitalism, and to all the technologies it has spawned" (Kaplan, 1988, p. 4) as well as to the avant garde in the arts, "the erosion of the older

distinction between high culture and so-called mass or popular culture" (Jameson, 1988, p. 14). Flax (1990) wrote that

> postmodern discourses are all deconstructive in that they seek to distance us from and make us skeptical about beliefs concerning truth, knowledge, power, the self, and language that are often taken for granted within and serve as legitimation for contemporary Western culture. (p. 41)

Poststructuralism is a French term that represents the European, particularly French, avant garde in critical theory. Peters (1999) explained that poststructuralism, inspired by Nietzsche, is a "specifically philosophical response to the alleged scientific status of structuralism" (p. 1) in all its guises. "In philosophy," Harvey (1989) wrote,

> the intermingling of a revived American pragmatism with the post-Marxist and poststructuralist wave that struck Paris after 1968 produced what Bernstein calls "a rage against humanism and the Enlightenment legacy." This spilled over into a vigorous denunciation of abstract reason and a deep aversion to any project that sought universal human emancipation through mobilization of the powers of technology, science, and reason. (p. 41)

Poststructuralism offers critiques of the scientistic pretensions of structural tendencies in all disciplines—linguistics, anthropology, psychology, economics, and so forth.

Interestingly, the word *postmodernism* first appeared in architecture to reflect a new way of thinking about space: Harvey (1989) credited Jameson for calling it "contrived depthlessness" (p. 58). This is consistent with the general postmodern critique of foundationalism, the idea that there are absolute truths propping up everyday human activity. Classical foundationalism encourages empiricism, the idea that all knowledge is derived from sensory experience. In his general introduction to postmodernism, Harvey (1989) cited the postmodern position of the editors of the architectural journal *PRECIS*, who in 1987 summarized postmodernism quite elegantly as follows:

> postmodernism as a legitimate reaction to the "monotony" of universal modernism's vision of the world. "Generally perceived as positivistic, technocentric, and rationalistic, universal modernism has been identified with the belief in linear progress, absolute truths, the rational planning of ideal social orders, and the standardization of knowledge and production." Post-modernism, by way of contrast, privileges "heterogeneity and difference as liberative forces in the redefinition of cultural discourse." Fragmentation, indeterminacy, and intense distrust of all universal or "totalizing" discourses (to use the favoured phrase) are the hallmark of postmodernism thought. The rediscovery of pragmatism in philosophy (e.g., Rorty, 1979), the shift of ideas about the philosophy of science wrought by Kuhn (1962) and Feyerabend (1975), Foucault's emphasis upon discontinuity and difference in history and his privileging of "polymorphous correlations in place of simple or

complex causality," new developments in mathematics emphasizing indeterminacy (catastrophe and chaos theory, fractal geometry), the re-emergence of concern in ethics, politics, and anthropology for the validity and dignity of "the other," all indicate a widespread and profound shift. (p. 9)

As Eagleton (1987) announced, "we are now in the process of wakening from the nightmare of modernity, with its manipulative reason and fetish of the totality" (p. 9), a modernity that elevated science to a religion, appealed to a "surface rationality of proof, logic, axiom, explicitness" (McCloskey, 2001, p. 103).

Postmodernism argues against the values, practices, and goals of Enlightenment humanism as they play out in Modernism: "History, Progress, Freedom, Reason, Transcendence and Man" (Finn, 1993, p. 134). Such totalizing discourses are meta-theories "through which all things can be connected or represented" (Harvey, 1989, p. 45), including the idea of a unified theory of science (the positivist's claim), or a grand unified theory (GUT) of physics, or one all-embracing, deterministic principle such as social class found in Marxism. In his report on scientific knowledge, Lyotard (1979/1984) described postmodernism as a "condition," wrote that he found it in America, and defined it simply as an "incredulity toward meta-narratives"(p. xxiv), grand, totalizing theories like Harvey's listed above.

The catastrophic events of World War II interrupted centuries-old Western traditions. Spivak (1993) explained the French response as follows:

> The critique of humanism in France was related to the perceived failure of the European ethical subject after the War. The second wave in the midsixties, coming in the wake of the Algerian revolution, sharpened this in terms of disciplinary practice in the humanities and social sciences because, as historians, philosophers, sociologists, and psychologists, the participants felt that their practice was not merely a disinterested pursuit of knowledge, but productive in the making of human beings. It was because of this that they did not accept unexamined human experience as the source of meaning and the making of meaning as an unproblematic thing. And each one of them offered a method that would challenge the outlines of a discipline: archaeology [Foucault], genealogy [Foucault], power/knowledge reading [Foucault], schizo-analysis [Deleuze & Guattari], rhizo-analysis [Deleuze & Guattari], nonsubjective psychoanalysis [Lacan], affirmative deconstruction [Derrida], paralogic legitimation [Lyotard]. (p. 274)

There is much to read about each of these and other "methods" used in the "posts," but here I provide a brief description only of Derrida's affirmative deconstruction, which I mentioned earlier. It is deconstruction that made me aware early on that any concept/category is a structure attempting to contain and close off meaning and, at the same time, that that concept/category is available to rupture and rethinking.

Derrida's Deconstruction

Derrida (1988) made it clear that "deconstruction is not a method" and cannot "be reduced to some methodological instrumentality or to a set of rules and transposable procedures" (p. 4). Further, "in spite of appearances, deconstruction is neither an *analysis* nor a *critique*" (p. 4). He cautioned that "all sentences of the type 'deconstruction is X' or 'deconstruction is not X' a priori miss the point, which is to say that they are at least false" (p. 4). He continued,

> Deconstruction is neither a theory nor a philosophy. It is neither a school nor a method. It is not even a discourse, nor an act, nor a practice. *It is what happens* [emphasis added], what is happening today in what they call society, politics, diplomacy, economics, historical reality, and so on and so forth. (Derrida, 1990, p. 85)

To acknowledge the ethical nature of deconstruction, Derrida (as cited in Caputo, 1993) wrote simply, "deconstruction is justice" (p. 86). Spivak (1989) added to Derrida's list of what deconstruction is not, commenting that "Deconstruction is not an essence. It's not a school of thought; it is a way of reading" (p. 135). But Derrida (1989/2002) wrote that deconstruction is not something done to a text: "This deconstruction does not apply itself to such a text. It never applies itself to anything from the outside. It is in some way the operation or rather the very experience that this text, it seems to me, first does itself, by itself, on itself" (p. 264). Again, deconstruction is not something done to a text or any other structure; *the text, the concept, the structure undoes itself.* More generally, Spivak (1993) claimed that "deconstruction has always been about the limits of epistemology" (p. 123) and, in particular, an argument against the Western metaphysics of *presence.*

If it now seems impossible to understand what deconstruction *is,* what it *means,* that is precisely Derrida's point, that the meaning of any signifier—*deconstruction, reason, truth, science*—cannot be secured, cannot be *present,* but is constantly deferred, absent. Given this, we can no longer ask, "What is science? What exactly does it mean, once and for all?" because the sign (e.g., *science*) has no center, no constant, essential meaning that holds across time and all instances of its use. Science is always already different from itself.

Deconstructive discourses critique *essentialism,* then, an ontological concept that characterizes the work of thinkers such as Plato, Aristotle, Descartes, Spinoza, Leibniz, Kant, and Husserl. According to Fuss (1989), "essentialism is classically defined as a belief in true essences—that which is most irreducible, unchanging, and therefore constitutive of a given person or thing" (p. 2). In humanist thought, without that essence—the *center*—that unique and unchanging core that guarantees the thing its meaning, the thing would cease to exist. Essentialism pervades Western thought. For example, Derrida (1972/1981) questioned the idea in phenomenology of an *a priori* "layer of pure meaning, or a pure signified . . . a layer of prelinguistic or presemiotic (preexpressive, Husserl calls it) meaning " (p. 31), some preexisting meaning that language brings to light, to presence, and expresses.

A companion example is the phrase, "the thing in itself," as opposed to its appearance, which much concerned Plato and, later, Kant. Nietzsche (1887/1967) argued that "Kant no longer has a right to his distinction between 'appearance' and 'thing-in-itself' (p. 300) because there are no things-in-themselves. "If we were to remove all the relationships and actions of a thing, the thing does not remain" (p. 302). Things—and some will say people as well—exist not by themselves but only in relations. In deconstruction, "the thing itself always escapes" (Derrida, as cited in Spivak, 1974, p. lxix). Meaning appears only fleetingly and then begins to decay as it misfires and re-forms within the play of language. The sign, then, is not a structure of identity, of presence, but a radical structure of difference, absence. Derrida used the concept *différance* (meaning both "to differ" and "to defer") to account for the endless differences in and the deferral of the meaning of a sign. As Spivak explained, "word and thing or thought never in fact become one" (p. xvi). Because of this, the linguistic system cannot be totalized, and language, then, is always "being born" and exists between the "already" and the "not yet" (Derrida, 1967/1974, p. 244); meaning is always "to come" [*avenir*].

In an early lecture about deconstruction, Derrida (1966/1970), in his careful way, explained the problem of essence, the presumed centered structure:

> Thus it has always been thought that the center, which is by definition unique, constituted that very thing within a structure which while governing the structure, escapes structurality. This is why classical thought concerning structure could say that the center is paradoxically, *within* the structure and *outside it.* The center is at the center of the totality, and yet, since the center does not belong to the totality (is not part of the totality), the totality *has its center elsewhere.* The center is not the center. The concept of centered structure—although it represents coherence itself, the condition of the *episteme* as philosophy of science—is contradictorily coherent. And as always, coherence in contradiction expresses the force of a desire. The concept of centered structure is in fact the concept of a freeplay based on a fundamental immobility and a reassuring certitude, which is itself beyond the reach of the freeplay. With this certitude anxiety can be mastered. (p. 248)

The notion of a centered structure (presence, essence, core), then, is the illusion, the ruse, the cheat, the grounding mistake of Western metaphysics and positivist philosophy of science. Because there is no pure center, *presence,* "the structure of the sign is determined by the trace or track of that other which is forever absent" (Spivak, 1974, p. xvii). The authority for the coherence of the structure lies not in its center, in the interior of the structure (presence), but elsewhere—in exteriority (absence). The structure cannot (it never could) authorize itself and comes undone (has always already been undone).

Yet deconstruction demonstrates that that idea—presence, essence—is but a *description* of the world that, one might argue, releases us from responsibility, from the difficult intellectual, political, and ethical struggles of dealing with the ambiguity and contingency of human existence. As Keenan (1997) wrote, responsibility comes not with certainty but with "the removal of grounds, the withdrawal of the rules or the knowledge on which we might rely to make our decisions for us. No grounds means no alibis, no elsewhere to which we might refer the instance of our decision" (p. 1). Once we give up appeals to transcendental/foundational truth, essence, and originary meaning (the universal, the eternal), responsibility and justice assume their full weight. As Mouffe (1996) explained, "the absence of foundation 'leaves everything as it is,' as Wittgenstein would say, and obliges us to ask the same questions in a new way" (p. 38). Along those lines, Scott (1988) suggested we stop asking essentializing questions like "What does it mean?" and ask instead, "How do meanings change? How have some meanings emerged as normative and others been eclipsed or disappeared? What do these processes reveal about how power is constituted and operates?" (p. 35).

Deconstruction not only decenters structures that presume foundational/transcendent meaning, it also deconstructs the structure of binary oppositions (e.g., Self/Other, identity/*différance*) also organized by *presence.* The first term of the binary represents "presence and the logos; the inferior serves to define its status and mark a fall" (Spivak, 1974, p. lxix). The privileged term in the binary can only be thought in opposition to the other term; that is, the unmarked term depends on the marked term for its meaning. For example, Said (1978) illustrated that the Occident had to create the Orient in order to define itself, and to define itself as superior.

Derrida wrote that the binary oppositions of metaphysics are violent hierarchies because those on the wrong side of the binary can be brutalized for their difference. For that reason, the binary can't simply be neutralized. It must first be reversed— "fight violence with violence" (Spivak, 1974, p. lxxvii). The first step in deconstruction, then, is to reverse the binary. So in the binary, heterosexual/homosexual, for example, homosexual should hold the privileged position so that heterosexuals can feel the violence of being called abnormal, deviant, sinful, evil, and so on. In the next step, the winning term is displaced to make room for a new concept that can't be understood in terms of the old structure, a concept that not only undoes the binary but also encourages entirely different thinking about sexuality. This is the affirmative move of deconstruction, the overturning and opening up of a violent structure so that something different might happen.

Some call deconstruction, and other poststructural "methods," nihilistic, relativist, anarchist, antipolitical, deliberately obfuscatory, and on and on, but for many they are ethical practices of freedom, as Culler (1982) described below:

> If "sawing off the branch on which one is sitting" seems foolhardy to men of common sense, it is not so for Nietzsche, Freud, Heidegger, and Derrida; for they suspect that if they fall there is no "ground" to hit and that the most clear-sighted act may be a certain reckless sawing, a calculated dismemberment or deconstruction of the great cathedral-like trees in which Man has taken shelter for millennia. (p. 149)

If one has been on the wrong side of binaries and trapped in essentialist structures that control and shut down meaning and lives, the persistent critique of deconstruction against their founding violence can, indeed, be liberating. In conclusion, deconstruction is not just attention to language but to the very material structures we create through language and social practice, including that very material structure we call the *human being.*

Entanglement/Haecceity/Assemblage

In my own work, I have used Deleuze and Guattari's concepts *haecceity* and *assemblage* and the notion of *entanglement* from quantum physics to deconstruct one of the most powerful legacies of Enlightenment humanism—the *human being,* the *individual,* the *self,* the *person.* Once that master concept ruptures, every single related structure fails because we humans are at the center of them all—or, at least, we've been led to believe we are. In this section, I briefly describe the following: the humanist description of the human being, *entanglement* from quantum physics, and a Deluzoguattarian reconception of the human being. The failure of the humanist subject produces the failure of humanist methodology as I illustrate later.

Enlightenment humanism produced a particular description of the human being—an epistemological subject. Descartes' (1637/1993) foundationalism provided one description of that human being, a knowing subject who knows chiefly through rational deduction—"I think, therefore I am" (p. 18). Locke (1690/1924) would later dispute Descartes—Locke argued that the mind is a blank slate with no innate ideas—and describe a "self," an "individual," with a personal identity based on consciousness. Locke, the first of the British empiricists, believed that knowledge is determined only by experience derived from the senses, the chief tenet of empiricism. The human beings these two Enlightenment thinkers, in particular, described (invented) have had remarkable staying power in modernism and in conventional social science.

Elsewhere (St.Pierre, 2000) I have discussed and will not repeat here a postmodern critique of the liberal individual of Enlightenment humanism—a sovereign, lucid, transparent,

free, agentive, self-sufficient, rational, knowing, meaning-giving, conscious, stable, coherent, unified, self-identical, reflective, autonomous, intentional, and ahistoric individual who is "endowed with a will, a freedom, an intentionality, which is then subsequently "expressed" in language, in action, in the public domain" (Butler, 1995, p. 136). But both Descartes' and Locke's beliefs, their descriptions of the human, prevail in the human sciences, and we believe the human is, indeed, separate from everything else and, usually, master of the universe. Many binary oppositions follow from that assumption: Self/Other, subject/object, knower/known, man/nature, and so forth.

It is the *principle of individuation,* the criterion of identity, that enables us to organize the undifferentiated into identity, to determine when/where one thing ends and another begins, to divide and separate. Typically, we individuate by establishing an essence and then claim that everything that has that essence is identical, the same. At some point, it became possible to think that one human being could be individuated/divided from other human beings—that each human has a center, an identity (an "inner self," "inner voice"). We also individuated the human from everything else, from all that is not human. Clearly, individuation (the creation of categories such as *man* and *nature*) is an act of power.

In postmodernism, however, the aim is to de-individualize, to disrupt individuations we believe are real, and, in that work, Deleuze and Guattari are most helpful. They borrowed from Duns Scotus an old concept, *haecceity,* that he borrowed from Aristotle to describe a non-subjective assemblage of humans, time, space, physical objects, and everything else: "It should not be thought that a haecceity consists simply of a décor or a backdrop that situates subjects. . . . It is the entire assemblage in its individual aggregate that is haecceity . . . that is what you are, and . . . you are nothing but that" (Deleuze & Guattari, 1980/1987, p. 262). Rajchman's (2000) examples of this impersonal assemblage follow: "An hour of a day, a river, a climate, a strange moment during a concert can be like this—not one of a kind, but the individuation of something that belongs to no kind" (Deleuze & Guattari, 1980/1987, p. 85). Deleuze (1990/1995) rethought the idea of the human being as *assemblage:* "Felix [Guattari] and I, and many others like us, don't feel we're persons exactly. Our individuality is rather that of *events* [emphasis added] . . . a philosophical concept, the only one capable of ousting the verb 'to be' and its attributes" (p. 141).Haecceity is not stable but is always becoming. It is not *is* but *and.* "A haecceity has neither beginning nor end, origin nor destination; it is always in the middle. It is not made of points, only of lines. It is a *rhizome*" (Deleuze & Guattari, 1980/1987, p. 263).

I think, then, of *haecceity* as mingling, assemblage, as relation, as becoming, perhaps as Benjamin's (1999) *constellation* or *entanglement.* In quantum physics, when "two entities interact, they entangle. . . . No matter how far they move apart, if one is tweaked, measured, observed, the other seems to instantly respond. . . . And no one knows how" (Gilder, 2008, p. 3). Barad (2007) explained entanglement in quantum theory as follows:

> To be entangled is not simply to be intertwined with another, as in the joining of separate entities, but to lack an independent, self-contained existence. Existence is not an individual affair. Individuals do not preexist their interactions; rather, individuals emerge through and as part of their entangled intra-relating. Which is not to say that emergence happens once and for all, as an event or as a process that takes place according to some external measure of space and of time, but rather that time and space, like matter and meaning, come into existence, are iteratively reconfigured through each intra-action, thereby making it impossible to differentiate in any absolute sense between creation and renewal, beginning and returning, continuity and discontinuity, here and there, past and future. (p. ix)

Quantum physics seems quite Nietzschean and Deleuzean, and here we see that Science and Philosophy do not have to be at odds.

Experimental physics also disrupted our conventional understanding of space and time in which we believe objects and people exist in a stable space that moves through time so that, for example, one can supposedly observe the same phenomena as they pass through time and change, progress, or "develop"—linearity. With entanglement in mind, Massey (1994) argued, "it is not that the interrelations between objects occur in space and time; it is these relationships which create/define space and time" such that "the spatial is social relations 'stretched out'" (p. 263). It is only manmade history that brings time into being and establishes clear connections. The shift from the vertical (history, depth) to the horizontal (simultaneity, surface) characterizes the shift from modernism to postmodernism.

Space-time from physics is dynamic, fractured, porous, paradoxical, and non-individual with sets of space-time relations existing simultaneously, rhizomatically and overlapping, interfering with each other. "There is no choice between flow (time) and a flat surface of instantaneous relations (space)" (Massey, 1994, p. 265) because time is not linear and space is not flat. The human desire to measure and control everything extended to time itself (we invented clocks), but time is out of joint and always has been. "In the theory of relativity," Hawking (1988) wrote, "there is no absolute time. Each observer has his own measure of time" (p. 87), so we all live different times.

Even so, some descriptions have great purchase in the social sciences, especially positivist approaches that mimic ideas no longer supported in the natural sciences. Ideas of absolute time, linearity, and sequenced progression enable positivist ideas such as cause and effect, the accumulation of knowledge, and so on. But such linearity is unintelligible in space-time because "people are everywhere conceptualizing and acting on different spatialities" (Massey, 1994, p. 4). Thus, there cannot be a fixed point from which a fixed individuation can observe another fixed individuation at another fixed point. And the positivist statement in the 2002 NRC report that the world can be described so that "multiple observers can agree on what they see" (p. 25) is not thinkable in space-time because everything is entangled and always already overlapping, dynamic, contested, multiple, antagonistic, becoming, in process.

So the positivist social science ideas of replication and generalizability are unthinkable in space–time, along with the subjectivity/objectivity distinction and related concepts such as *bias* and *brute fact/data*. In a similar fashion, the interpretive and critical social science idea of, for example, *culture*—a coherent set of people traveling together in space and time—is also unthinkable. In all those examples, we attempt to "stabilize the meaning of particular envelopes of space-time" (Massey, 1994, p. 5) and study them.

Quantum physics' entanglement and space–time require a different description of the human being. On the one hand, it might be difficult to give up the "I" one is familiar with; on the other, why should we rank Descartes' "I" and Locke's "self" (both inventions) above Deleuze and Guattari's "assemblage" or "haecceity" (another invention) or, indeed, some other description of the "human" we may have forgotten and/or not yet thought?

Clearly, language strains and stutters here. I have not yet learned to write easily without saying "I," "me," "myself," "one," "oneself," though I do not need them so much anymore for living. But as Deleuze and Guattari (1980/1987) explained, the goal is not to reach the "point where one no longer says I, but the point where it is no longer of any importance whether one says I" (p. 3). They did advise that to resist subjectification, "You have to keep enough of the organism for it to reform each dawn and you have to keep small supplies of significance and subjectification, if only to turn them against their own systems when the circumstances demand it" (p. 160). That is the work of politics, "to refuse what we are" (Foucault, 1982, p. 216).

The implications of entanglement are staggering. If one no longer thinks of oneself as "I" but as entangled with everyone, everything else—as haecceity, as assemblage—what happens to concepts in social science research based on that "I"—the *researcher,* the *participant, identity, presence, voice, lens, experience, positionality, subjectivity, objectivity, bias, rationality, consciousness, experience, alienation, reflexivity, freedom, transformation, dialogue?* In space–time, how does one think of *research design, research process, timeline, narrative, cause and effect, accumulation of knowledge, generalizability, replicability, predictability, scaling up?* In entanglement, how does one think about "face-to-face" methods like *interviewing* and *observation,* methods that privilege *presence,* Derrida's bane?

It is indeed difficult to escape the "I." Even those who've studied the "posts," and, in particular, poststructural theories of subjectivity, seem unable *not* to write the humanist human being. For example, they may include a "subjectivity statement" to describe themselves even though *subjectivity* in the "posts" is incommensurable with that practice. They may write lengthy, rich, thick descriptions of individual participants to whom they've assigned pseudonyms, producing them as autonomous, coherent, intentional, knowing, speaking subjects. But in the posts, participants are not an "epistemological dead-end" (Sommer, 1994, p. 532)—an object of knowledge—but rather a line of flight that takes us elsewhere—participants as provocateurs. It should be clear that if we no longer believe in a disentangled humanist self, individual, person, we have to rethink qualitative methods (interviewing and observation) grounded in that human being as well as humanist representation.

Deconstruction Happens

Putting deconstruction and entanglement to work in conventional humanist qualitative methodology renders it not only incomprehensible but also without an alternative that can be described in advance of the study in a proposal or captured at the end in representation. Each researcher who puts the "posts" to work will create a different *articulation* (e.g., Hall, 1986/1996; Laclau & Mouffe, 1985), *remix, mash-up, assemblage,* a *becoming* of inquiry that is not *a priori,* inevitable, necessary, stable, or repeatable but is, rather, created spontaneously in the middle of the task at hand, which is always already *and, and, and. . . .* I argue that this has always been the case but that researchers have been trained to believe in and thus are constrained by the pre-given concepts/categories of the invented but normalized structure of "qualitative methodology," its "designs" and "methods," that are as positivist as they are interpretive, often more so.

In general, qualitative inquiry, especially after SBR, comes with so many instructions and limits that rigor seems impossible. By rigor, I mean the demanding work of freeing oneself from the constraints of existing structures, what Foucault (1966/1970) called the "order of things," so that one can think the unthought. Rigor is the work of différance, not repetition, work that bell hooks (1989) called "too deep" and the poet Rilke, "too large." I can barely think into that space though I do find myself there, most often when I write, because, for me, writing, that old technology (absence) Plato feared, "can free thought from what it silently thinks and enable it to think differently" (Foucault, 1984/1985, p. 9).

I don't believe the rigorous inquiry I desire can be taught in a sequence of research courses or described in textbooks, but I've thought a great deal about how one might become available to deconstructive work and the transformations it can enable. I cannot explain how deconstruction happens, but it does if one has read enough and puts it to work.

I have written elsewhere about how I came to theory (St. Pierre, 2001) in my own doctoral program. Briefly, I quickly exhausted the mid-level theories of my field, English education, which

seemed mostly focused on preservice teacher education, did not situate itself within larger theoretical frameworks, and therefore seemed to come from nowhere. I had studied philosophy as an undergraduate when the "turns" were being thought and written but missed them entirely after I became a high school English teacher and, later, a librarian. Still, I knew there were larger conversations I was missing as a doctoral student.

I took as many qualitative research courses as I could find and studied methodology with feminist professors in education and sociology making the postmodern turn. Their casual references in class to Foucault, Butler, Trinh, Said, Derrida, Omi and Winant, Habermas, and other theorists were my only guidance to the "turns."

I never had a theory course in my doctoral program; instead, as scholars have done for centuries, I followed the citational trail from one text to another, finding and reading on my own, for example, feminism, Frankfurt School critical theory, interpretive anthropology, postcolonialism, critical race theory, social constructionism, poststructuralism, postmodernism, and theories of space–time and memory. I felt far behind some of my fellow students and read harder to catch up. Much of the work I read had been written during my own lifetime, and it spoke eloquently to me. Loving language, I reveled in Butler's and Derrida's rich, complex, perfectly formed sentences. Deleuze and Guattari's new concepts (e.g., *bodies without organs, becoming animal, smooth space*) spilled over me, immediately useful. I formed and then shed attachments to scholars and their theories as I read, and those shifts taught me there would always be another sentence in another book that might well shatter my life again. Reading guaranteed transformation, and I couldn't get enough.

I began to understand that *theory produces people.* Theory was no longer an abstract, sometimes impenetrable discourse but a powerful, personal tool I needed to study for my own good. I realized I and my culture had been theorized by Enlightenment humanism whose projects seldom served women well and that the "posts," in particular, combined with other theories, offered many analyses to resist humanism's oppressive structures, especially its foundational structure, the *human being.*

To study the construction of subjectivity (including my own), I dutifully designed and accomplished a qualitative study that was a combination of an interview study with older women who lived in my hometown and an ethnography of their community. Until I began to write my dissertation, the two bodies of literature I had diligently studied—theory and qualitative research methodology—mostly remained separate. But as I wrote the dissertation, qualitative methodology, so clearly grounded in the humanism (and positivism) I no longer believed, ruptured.

No doubt the nature of my study intensified that breakdown. I returned to my hometown in Essex County where my family still lived 20 years after I had left, to study women who had taught me how to be a woman. As I interviewed them, they were simultaneously old women happy to help a hometown girl get her doctorate *and* the lovely, formidable young women I remembered from childhood. I never knew who/when/where I was during fieldwork. Subjectivity, space–time, and "reality" exploded and overwhelmed me. It was not simply that I "had multiple subjectivities" or "moved among subject positions" but that I was always already a simultaneity of relations with humans and the nonhuman (I could no longer think/live that dichotomy)—the "women" and "me" in all times and places; my father, long dead, loving me; the streets and storefronts of the town; all us cousins catching lightning bugs on a summer evening; Essex County's red clay tobacco fields; my beloved aunt whose smile saved everyone who met her; all of us, everything, de-individualized, de-identified, *dis-individuated*. A rhizome, assemblage, haecceity, my life. *A life.* Theory produced me differently, and I am not the same. I never was.

But I could not have thought those thoughts by thinking alone. That work about subjectivity (an inadequate concept) required a simultaneity of living, reading, and writing. I needed living (*experiences* is inadequate) for which humanist individuations no longer worked (me with the women in space–time simultaneously); theories that provided language to think living differently (the "posts" and theories of space–time and memory), and the setting-to-work of writing that forced the rupture and demanded I move on. When writing the next word and the next sentence and then the next is more than one can manage; when one must bring to bear on writing, in writing, what one has read and lived, that is thinking that cannot be taught. *That is analysis.*

So writing became a field of play in which the study took place, a space (never just textual) as important as Essex County. *In the thinking that writing produced,* the humanist subject was the first humanist concept/category to fail, but many others did as well.

Here, I briefly discuss the failure of one other key concept, *data* (and data collection and data analysis and others). I have written in detail elsewhere (St.Pierre, 1997) about my troubles with data. Until I began analysis—the thinking that writing enables—I believed the textbook definition of qualitative data: that which is textualized, fixed, and made visible in words in interview transcripts and fieldnotes. That is a foundationalist description that enables us to treat words as brute data—transparent, neutral, independent of theory, *waiting to be analyzed*. However, words are always thinkable, sayable, and writable (in interviews and fieldnotes) only within particular grids of intelligibility, usually dominant, normalized discursive formations. What this means is that words we collect in interviews and observations—data—are always already products of theory.

The researcher's first task is to recognize the theory(ies) that enabled others' words in interviews, and she can do that only if she's studied theory. Her next task is to theorize those already theorized words that reflect "experiences," living. Of course, the theory(ies) she uses determines whether those words even count as data because words (or anything else) become data only when theory acknowledges them as data. In this way, theory is deeply imbricated in the concept *data* and in every related concept, e.g., "methods" of collecting data, data analysis. I believe the understanding of data in conventional humanist qualitative methodology, strongly influenced now by SBR and scientism in general, is increasingly positivist because, first, it must be fixed and visible in words, and, second, because we increasingly treat words as brute, uninterpreted data rather than as already interpreted data we must explain. It should not be surprising, then, that one who thinks with poststructural theories of language would have trouble with humanist qualitative methodology's description of data.

Here's a brief description of how deconstruction happened for me. As I wrote about subjectivity, I realized I was thinking/writing not only with textualized data in interview transcripts and fieldnotes, but also with data that were not textualized, fixed, and visible. I therefore decided to claim and name some of those other data so they might be accounted for. For one thing, I believed that if participants' words about subjectivity in interview transcripts counted as data, so did theorists' words about subjectivity in books I had read—why discount their expertise? I thought with everyone's words as I wrote—Foucault's and my former Latin teacher's—and it no longer made sense to separate those data into chapters called "literature review" and "interpretation." It has since occurred to me that that artificial separation actually encourages researchers *not* to theorize data from interviews and observations.

I also identified what I called "transgressive data" clearly at work in my study—*emotional data, dream data, sensual data, memory data,* and *response data*—data that were not visible and that disrupted linearity, consciousness, and the mind/body dichotomy. (Recently, several of my students have, unaware of each other's work, identified *spectral data.*) Much data—*what we think with when we think about a topic*—were identified *during* analysis and not before. Until one begins to think, one cannot know what one will think with. In that sense, data are collected during thinking and, for me, especially during writing.

I wrote earlier in this chapter and elsewhere (Richardson & St.Pierre, 2005) that, for me, writing is analysis. Twenty years ago, I did not believe that coding data was analysis, and I haven't changed my mind. It's certainly something one can do with data—label and sort (and count)—but I'm not sure one would do it if one had never heard of it. There are many ways to "stay close to the data" (conventional data, that is), for example, reading transcripts and fieldnotes and listening to tapes repeatedly. But if those data are only a portion of the data we use in thinking, in analysis, coding makes even less sense, e.g., how/why would one code sensual data?

I strongly advise my own doctoral students not to code data because I have seen too many students, even those who've done their hard theoretical reading, become exhausted after months of tedious coding and never do the theoretical analysis they could. Their findings are pedestrian, and they produce low-level, insignificant themes; untheorized stories; or extended descriptions that do not get to the intellectual problem of explaining why things are as they are. I do, however, ask them to explain what they did when they thought they were "doing analysis," and they describe a multitude of activities—washing the car and weeding the garden (*the physicality of theorizing*), making charts and webs, talking with friends, writing, listening to music, reading transcripts, reading more theory, dozing on the couch, and so forth. The positivism imbedded in qualitative research quickly fails—audit trails can't capture that work, it can't be triangulated, and it is never saturated.

If we agree with Derrida and many others that language cannot contain and close off meaning and cannot transport meaning from one person to another, it's difficult to understand why we believe that isolating and labeling a word or group of words (a chunk) with another word (a code) is scientific or rigorous or "analysis"—even if we do that work with computer software, which makes the process seem detached, objective, systematic, and even more scientific. If a word is data, isn't a code (a word) data as well? Do we code codes? I expect some do because I have read entire books on coding and discovered that there are, for example, supercodes. Some even count the occurrence of codes, giving weight to codes that appear most frequently even though we know that the most significant data in a study might occur only once. I argue that coding is a positivist practice, a relic from the positivist social science of the 1920s and 1930s when qualitative data were handled in a quasi-statistical fashion, when words were considered *brute data*, when it seemed best to treat words as numbers and even to turn them into numbers for the sake of clarity and simplicity, for the sake of scientific analysis—the pathology of quantification.

I expect we teach coding because we don't know how to teach thinking. But I will always believe that if one has read and read and read, it's nigh onto impossible *not* to think with what others have thought and written. (If one has not read much, perhaps one needs to code.) I imagine a cacophony of ideas swirling as we think about our topics with all we can muster—with words from theorists, participants, conference audiences, friends and lovers, ghosts who haunt our studies, characters in fiction and film and dreams—and with our bodies and all the other bodies and the earth and all the things and objects in our lives—the entire assemblage that is *a life* thinking *and, and, and. . . .* All those data are set to work in our thinking, and we think, and we work our way somewhere in thinking. My advice is to read, and analysis, whatever it is, will follow. (*Do tell me what you think you are thinking with when you think—what are your data? And do tell me what you think you do when you think—when you do analysis? Do that.*) In the end, it is impossible to disentangle *data, data collection,* and *data analysis.* Those individuations no longer make sense. We could just give them up.

Any concept/category can be deconstructed as I've deconstructed data here. This work is not only playful in the Derridaean sense but also necessary if we're to move out of the structures that prevent us from thinking differently. Here we do not repeat the same—this work is *différance,* not repetition.

A Return to Philosophy

What can be said here at the end of this chapter? I have reviewed some poststructural theories, stressed the importance of studying those theories if one wants to use them, and then used them to deconstruct several concepts/categories that ground conventional humanist qualitative inquiry: language, the human being, data, and related structuring concepts. I suspect that some familiar, naturalized concept becomes troublesome in every study, but deconstruction requires that instead of suppressing or ignoring that trouble, we "take it with the utmost seriousness, with literal seriousness, so that it transforms itself" (Spivak, 1989, p. 129) and overturns the structure it helps to organize. What happens next is not predictable, and, for that reason, the "posts" do not and cannot offer an alternative methodology.

It is critical at this deconstructive moment that is always already a political moment that we do not, in fear or confusion, submit to the positivist resurgence or any other foundationalist legitimating authority that promises to secure science and save us. The urge to create new structures of comfort is almost insurmountable, but we must work against the current conservative restoration in favor of a time to come, a future that is "no more historical than it is eternal: it must be untimely, always untimely" (Deleuze, 1995/2001, p. 72).

The "posts" and interpretive and critical theories were described half a century ago in a time much like this one in response to crippling foundationalist, especially positivist, dogma. The conventional humanist qualitative inquiry described in textbooks—the structure we call a "methodology" with "research designs" and "methods" that still mimics a simulacrum of the natural sciences—emerged as one project of the interpretive turn, the critique of positivist social science, which itself began in earnest in the 1920s with the logical positivism and logical empiricism of the Vienna Circle and continued well into post–World War II social science. But it is difficult to escape what one critiques, and qualitative inquiry has always been organized as much by positivism as by interpretivism. Its latent positivism was clearly evident when it almost immediately hardened in response to the scientific- and evidence-based movements and accommodated, for example, epistemologically and ontologically incommensurable "mixed methods." In any case, it is entirely humanist.

Conventional humanist qualitative inquiry organized and established itself mainly through a proliferation of publications that convinced us it was real. The scattered post critiques, never well-organized into rival publications, nonetheless gnawed away at it throughout the social sciences, even economics, and ruptured and ruined it beyond recognition. Those working those ruins deconstructively (working within/against the structure) who paused—for too long, I suspect—to defend qualitative inquiry from the resurgence of positivism are now countering with a *resurgence of postmodernism,* calling it out to continue work it began in the 1960s.

I believe qualitative inquiry is more vulnerable than ever now that its positivist tendencies have been outed. More importantly, I suggest that we've worked within/against the ruined structure long enough. We could now, if we wish, give up conventional humanist qualitative inquiry and its structuring concepts and categories—just let it go.

We can now do something different from the beginning. Am I saying "anything goes"? Well, anything always goes until someone who has some power draws a line. But many of us are weary of all the lines drawn around social science inquiry these days. I believe inquiry should be provocative, risky, stunning, astounding. It should take our breath away with its daring. It should challenge our foundational assumptions and transform the world. We must, even so, be vigilant in analyzing the consequences of human invention and the structures it endlessly creates. Humanism's projects created spectacular failures that the "turns" identified half a century ago. Why not try something different?

But what happens next, is happening now, and has always happened, cannot be predicted or controlled. People everywhere always re-think, deconstruct, invent, and we theorists and researchers are always catching up, trying to make sense of their work. We individuate, we order, we name, we try to control, we draw lines. Nonetheless, they resist structural boundaries as they create entanglements that may initially seem incongruous. I propose we worry less—so much angst—about what might happen if we give up exhausted structures and attend to what *is* happening. *Deconstruction has already happened; it is happening at this moment, everywhere.*

My desire is for post inquiry to remain *unstable* as we create different articulations, assemblages, becomings, mash-ups of inquiry given the entanglement that emerges in our different projects. Whether it will be called "science" depends, as always, on who has the power to make such decisions. Some will work to keep science capacious, others won't, and still others won't care much at all.

I am reminded of Foucault's (1984) questions about subjectivity here at the end because whether we, each of us, is available to a transformation of inquiry, much less of ourselves (the "I") and our world, is always already personal *and* philosophical: "How are we constituted as subjects of our own knowledge? How are we constituted as subjects who exercise or submit to power relations? How are we constituted as moral subjects of our own actions" (p. 49). Foucault's projects focused on how

power relations constitute human beings within discursive formations that are only *descriptions,* albeit sometimes powerful descriptions with very real material effects. Rorty (1986) explained that those descriptions are not necessarily rational, intentional, ethical, or progressive:

> The urge to tell stories of progress, maturation and synthesis might be overcome if we once took seriously the notion that we only know the world and ourselves *under a description.* For doing so would mean taking seriously the possibility that we just *happened* on that description—that it was not the description which nature evolved us to apply, or that which best unified the manifold of previous descriptions, but just the one which we have now *chanced* to latch onto. (p. 48)

At this very moment, we are latched onto descriptions that are producing us and the world, descriptions that, over time, have become so transparent, natural, and real that we've forgotten they're fictions. We accept them as truth.

What philosophy can do, what the "posts" can do, is reach "for the speculative possibilities that exceed our present grasp, but may nevertheless be our future" (Rorty, 1986, p. 48). The resurgence of positivism evidenced in SBR once again attempted to detach science from philosophy as positivism always does; to reduce knowledge to that produced by science; and to reduce science to systematic procedures and protocols, mechanistic technique, statistical manipulation, and causal structures. The call for the resurgence of postmodernism in this chapter has, of course, all along, been a call for philosophically informed inquiry accomplished by inquirers who have read and studied philosophy. It seems we have to keep on learning that philosophy and science are not individuated but always already entangled. The most important task of post qualitative inquiry is to attend to that false and grievous distinction.

◼ REFERENCES

Atmanspacher, H. (2002). Determinism is ontic, determinability is epistemic. In H. Atmanspacher & R. Bishop (Eds.), *Between chance and choice: Interdisciplinary perspectives on determinism* (pp. 49–74). Thoveton, UK: Imprint Academic.

Ayer, A. J. (1936). *Language, truth, and logic.* London: Victor Gollancz.

Barad, K. (2007). *Meeting the universe halfway: Quantum physics and the entanglement of matter and meaning.* Durham, NC: Duke University Press.

Benjamin, W. (1999). *The arcades project* (H. Eiland & K. McLaughlin, Trans.). Cambridge, MA: Harvard University Press.

Butler, J. (1992). Contingent foundations: Feminism and the question of "postmodernism." In J. Butler & J. W. Scott (Eds.), *Feminists theorize the political* (pp. 3–21). New York: Routledge.

Butler, J. (1995). For a careful reading. In S. Benhabib, J. Butler, D. Cornell, & N. Fraser (Eds.), *Feminist contentions: A philosophical exchange* (pp. 127–143). New York: Routledge.

Caputo, J. D. (1993). *Against ethics: Contributions to a poetics of obligation with constant reference to deconstruction.* Bloomington: Indiana University Press.

Cullenberg, S., Amariglio, J., & Ruccio, D. F. (Eds.). (2001). *Postmodernism, economics and knowledge.* London: Routledge.

Culler, J. (1982). *On deconstruction: Theory and criticism after structuralism.* Ithaca, NY: Cornell University Press.

Deleuze, G. (1995). *Negotiations: 1972–1990* (M. Joughin, Trans.). New York: Columbia University Press. (Original work published 1990)

Deleuze, G. (2001). *Pure immanence: Essays on a life* (A. Boyman, Trans.). New York: Zone Books. (Original work published 1995)

Deleuze, G., & Guattari, F. (1987). *A thousand plateaus: Capitalism and schizophrenia* (B. Massumi, Trans.). Minneapolis: University of Minnesota Press. (Original work published 1980)

Derrida, J. (1970). Structure, sign, and play in the discourse of the human sciences. In R. Macksey & E. Donato (Eds. & Trans.), *The structuralist controversy: The languages of criticism and the sciences of man* (pp. 247–272). Baltimore: Johns Hopkins University Press. (Lecture delivered 1966)

Derrida, J. (1974). *Of grammatology* (G. C. Spivak, Trans.). Baltimore: Johns Hopkins University Press. (Original work published 1967)

Derrida, J. (1981). *Positions* (A. Bass, Trans.). Chicago: University of Chicago Press. (Original work published 1972)

Derrida, J. (1982). Signature, event, context. In J. Derrida, *Margins of philosophy* (A. Bass, Trans.) (pp. 307–330). Chicago: University of Chicago Press. (Original work published 1971)

Derrida, J. (1988). Letter to a Japanese friend. In D. Wood & R. Bernasconi (Eds.), *Derrida and différance* (pp. 1–5). Evanston, IL: Northwestern University Press.

Derrida, J. (1990). Some statements and truisms about neologisms, newisms, positisms, parasitisms, and other small seismisms. (A. Tomiche, Trans.). In D. Caroll (Ed.), *The states of "theory": History, art, and critical discourse.* New York: Columbia University Press.

Derrida, J. (2002). Force of law: The "mystical foundation of authority." In J. Derrida, *Acts of religion* (G. Anidjar, Ed.) (pp. 230–298). New York: Routledge. (Original work circulated 1989)

Descartes, R. (1993). *Discourse on method and Meditations on first philosophy* (4th ed., D. A. Cress, Trans.). Indianapolis, IN: Hackett Publishing Company. (*Discourse on Method* first published 1637 and *Meditations on First Philosophy* first published 1641)

Eagleton, T. (1987, February). Awakening from modernity. *Times Literary Supplement, 20,* 6–9.

Fay, B. (1987). *Critical social science: Liberation and its limits.* Ithaca, NY: Cornell University Press.

Feyerabend, P. K. (1975). *Against method* (3rd ed.). London: Verso.

Finn, G. (1993). Why are there no great women postmodernists? In I. Taylor (Ed.), *Relocating cultural studies: Developments in theory and research* (pp. 123–152). New York: Routledge.

Flax, J. (1990). Postmodernism and gender relations in feminist theory. In L. J. Nicholson (Ed.), *Feminism/Postmodernism* (pp. 39–62). New York: Routledge.

Foucault, M. (1970). *The order of things: An archaeology of the human sciences* (A. M. S. Smith, Trans.). New York: Vintage Books. (Original work published 1966)

Foucault, M. (1982). The subject and power. *Critical Inquiry, 8*(4), 777–795.

Foucault, M. (1984). What is enlightenment? (C. Porter, Trans.). In P. Rabinow (Ed.), *The Foucault reader* (pp. 32–50). New York: Pantheon Books.

Foucault, M. (1985). *The history of sexuality. Volume 2. The use of pleasure* (R. Hurley, Trans.). New York: Vintage Books. (Original work published 1984)

Fuss, D. (1989). *Essentially speaking: Feminism, nature & difference.* New York: Routledge.

Gilder, L. (2008). *The age of entanglement.* New York: Knopf.

Hall, S. (1996). On postmodernism and articulation: An interview with Stuart Hall. In D. Morley & K.-H. Chen (Eds.), *Stuart Hall: Critical dialogues in cultural studies* (pp. 131–150). London: Routledge. (Reprinted from *Journal of Communication Inquiry, 10*(2), pp. 45–60, 1986)

Harvey, D. (1989). *The condition of postmodernity: An enquiry into the origins of cultural change.* Cambridge, MA: Blackwell.

Hawking, S. (1988). *A brief history of time: From the big bang to black holes.* New York: Bantam Books.

Henig, J. R. (2008). The evolving relationship between researchers and public policy. In F. M. Hess (Ed.), *When research matters: How scholarship influences education policy* (pp. 41–62). Cambridge, MA: Harvard Education Press.

Hodkinson, P. (2004). Research as a form of work: Expertise, community and methodological objectivity. *British Educational Research Journal, 30*(1), 9–26.

hooks, b. (1989). *Talking back: Thinking feminist, thinking black.* Boston: South End Press.

Hurworth, R. E. (2008). *Teaching qualitative research: Cases and issues.* Rotterdam, The Netherlands: Sense Publishers.

Institute of Education Sciences. U.S. Department of Education. (2008). *Rigor and relevance redux: Director's biennial report to Congress* (IES 2009–6010). Washington, DC.

Jackson, A. Y., & Mazzei, L. A. (2009). *Voice in qualitative inquiry: Challenging conventional, interpretive, and critical conceptions in qualitative research.* New York: Routledge.

Jameson, F. (1988). Postmodernism and consumer society. In E. A. Kaplan (Ed.), *Postmodernism and its discontents: Theories, practices* (pp. 13–29). New York: Verso.

Kaplan, E. A. (1988). Introduction. In E. A. Kaplan (Ed.), *Postmodernism and its discontents: Theories, practices* (pp. 1–9). New York: Verso.

Keenan, T. (1997). *Fables of responsibility: Aberrations and predicaments in ethics and politics.* Stanford, CA: Stanford University Press.

Kuhn, T. S. (1970). *The structure of scientific revolutions* (2nd ed.). Chicago: University of Chicago Press. (Original work published 1962)

Laclau, E., & Mouffe, C. (1985). *Hegemony and socialist strategy: Towards a radical democratic politics.* London: Verso.

Lamont, M. (1987). How to become a dominant French philosopher: The case of Jacques Derrida. *American Journal of Sociology, 93*(3), 584–622.

Lather, P. (1993). Fertile obsession: Validity after poststructuralism. *Sociological Quarterly, 34*(4), 673–693.

Lincoln, Y. S., & Guba, E. G. (1985). *Naturalistic inquiry.* Newbury Park, CA: Sage.

Locke, J. (1924). *An essay concerning human understanding.* Oxford, UK: Clarendon Press. (Original work published 1690)

Lyotard, J-F. (1984). *The postmodern condition: A report on knowledge* (G. Bennington & B. Massumi, Trans.). Minneapolis: University of Minnesota Press. (Original work published 1979)

MacLure, M. (2005). "Clarity bordering on stupidity": Where's the quality in systematic review? *Journal of Education Policy, 20*(4) 393–416.

Massey, D. (1994). *Space, place, and gender.* Minneapolis: University of Minnesota Press.

Maxwell, J. A. (2010, January 19). Review of the book, *Theory and Educational Research: Toward Critical Social Explanation,* by J. Anyon. Available at http://edrev.asu.edu/reviews/rev882.pdf

McCloskey, D. (2001). The genealogy of postmodernism: An economist's guide. In S. Cullenberg, J. Amariglio, & D. F. Ruccio (Eds.), *Postmodernism, economics, and knowledge* (pp. 102–128). London: Routledge.

Moss, P. A., Phillips, D. C., Erickson, F. D., Floden, R. E., Lather, P. A., & Schneider, B. L. (2009). Learning from our differences: A dialogue across perspectives on quality in educational research. *Educational Researcher, 38*(7), 501–517.

Mouffe, C. (1996). Radical democracy or liberal democracy? In D. Trend (Ed.), *Radical democracy: Identity, citizenship, and the state* (pp. 19–26). New York: Routledge.

National Research Council. (2002). *Scientific research in education* (R. J. Shavelson & L. Towne, Eds.). Committee on Scientific Principles for Education Research. Washington, DC: National Academies Press.

Neumann, A., Pallas, A. M., & Peterson, P. L. (2008). Exploring the investment: Four universities' experiences with the Spencer Foundation's research training grant program: A retrospective. *Teachers' College Record, 110*(7), 1477–1503.

Nietzsche, F. (1967). *The will to power* (W. Kaufman, Ed.; W. Kaufman & R. J. Hollingdale, Trans.). New York: Vintage Books. (Original work published 1887)

Oakley, A. (2006). Resistances to "new" technologies of evaluation: Education research in the UK as a case study. *Evidence and policy: A journal of research, debate and practice, 2*(1), 63–87.

Patton, C. (2008). Finding "fields" in the field: Normalcy, risk, and ethnographic inquiry. *International Review of Qualitative Research, 1*(2), 255–74.

Peters, M. (1999). (Posts-) modernism and structuralism: Affinities and theoretical innovations. *Sociological Research Online, 4*(3). Available at http://www.socresonline.org.uk

Phillips, D. C. (2006). A guide for the perplexed: Scientific educational research, methodolatry, and the gold versus platinum standards. *Educational Research Review, 1,* 15–26.

Phillips, D. C. (2009). A quixotic quest? Philosophical issues in assessing the quality of education research. In P. B. Walters, A. Lareau, & S. H. Ranis (Eds.), *Education research on trial: Policy reform and the call for scientific rigor* (pp. 163–195). New York: Routledge.

Phillips, D. C., & Burbules, N. C. (2000). *Postpositivism and educational research.* Lanham: Rowman & Littlefield.

Pillow, W. S. (2003). Confession, catharsis, or cure? Rethinking the uses of reflexivity as methodological power in qualitative research. *International Journal of Qualitative Studies in Education, 16*(2), 175–196.

Rajchman, J. (1987, November/December). Postmodernism in a nominalist frame: The emergence and diffusion of a cultural category. *Flash Art, 137,* 49–51.

Rajchman, J. (2000). *The Deleuze connections.* Cambridge: MIT Press.

Reichenbach, H. (1951). *The rise of scientific philosophy.* Berkeley: University of California Press.

Richardson, L., & St.Pierre, E. A. (2005). Writing: A method of inquiry. In N. K. Denzin & Y. S. Lincoln (Eds.), *The SAGE handbook of qualitative research* (3rd ed., pp. 959–978). Thousand Oaks, CA: Sage.

Rorty, R. (1979). *Philosophy and the mirror of nature.* Princeton, NJ: Princeton University Press.

Rorty, R. (1986). Foucault and epistemology. In D. C. Hoy (Ed.), *Foucault: A critical reader* (pp. 41–49). Cambridge, MA: Basil Blackwell.

Said, E. W. (1978). *Orientalism.* New York: Vintage Books.

Scheurich, J. J. (1995). A postmodernist critique of research interviewing. *International Journal of Qualitative Studies in Education, 8*(3), 239–252.

Scott, J. (1988). Deconstructing equality-versus-difference: Or, the uses of poststructuralist theory for feminism. *Feminist Studies, 14*(1), 33–50.

Sommer, D. (1994). Resistant texts and incompetent readers. *Poetics Today, 15*(4), 523–551.

Spivak, G. C. (1974). Translator's preface. In J. Derrida *Of Grammatology* (G. C. Spivak, Trans.) (pp. ix–xc). Baltimore: Johns Hopkins University Press.

Spivak, G. C. (1989). In a word: Interview (E. Rooney, Interviewer). *Differences, 1*(2), 124–156.

Spivak, G. C. (1993). *Outside in the teaching machine.* New York: Routledge.

St.Pierre, E. A. (1995). *Arts of existence: The construction of subjectivity in older, white southern women.* Unpublished doctoral dissertation, The Ohio State University, Columbus.

St.Pierre, E. A. (1997). Methodology in the fold and the irruption of transgressive data. *International Journal of Qualitative Studies in Education, 10*(2), 175–189.

St.Pierre, E. A. (2000). Poststructural feminism in education: An overview. *International Journal of Qualitative Studies in Education, 13*(5), 477–515.

St.Pierre, E. A. (2001). Coming to theory: Finding Foucault and Deleuze. In K. Weiler (Ed.), *Feminist engagements: Reading, resisting, and revisioning male theorists in education and cultural studies* (pp. 141–163). New York: Routledge.

Ulmer, G. L. (1985). *Applied grammatology: Post(e)-Pedagogy from Jacques Derrida to Joseph Beuys.* Baltimore: Johns Hopkins University Press.

Viadero, D. (2009). "No effect" studies raising eyebrows. *Education Week, 28*(27), 1 & 14.

Whitehurst, G. J. (2003). *The Institute of Education Sciences: New wine and new bottles.* Paper presented at the annual meeting of the American Educational Research Association, Chicago.

38

QUALITATIVE RESEARCH AND TECHNOLOGY

In the Midst of a Revolution

Judith Davidson and Silvana di Gregorio

In the early 1980s, as qualitative researchers began to grapple with the promise and challenges of computers, a handful of innovative researchers brought forth the first generation of what would come to be known as CAQDAS (Computer Assisted Qualitative Data Analysis Software), or QDAS (Qualitative Data Analysis Software) as we will refer to it here. These stand-alone software packages were developed, initially, to bring the power of computing to the often labor-intensive work of qualitative research. While limited in scope at the beginning to text-retrieval tasks, for instance, these tools quickly expanded to become comprehensive all-in-one packages that offered qualitative researchers the following:

(1) A convenient digital location in which to organize all materials related to one study;

(2) A suite of linked digital tools that could be applied to those materials, including the ability to store and organize data as well as fragment, juxtapose, interpret, and recompose that same material;

(3) Easy portability; and

(4) A remarkable new form of transparency that allowed the researcher, and others, the opportunity to view and reflect upon the materials (di Gregorio & Davidson, 2008).

QDAS packages were flexible enough that they could be used by qualitative researchers from diverse disciplinary and methodological perspectives (although many early developers and users were from sociology). Over time, as the technological context advanced, these tools came to incorporate possibilities for working with multimodal data, as well as to provide new ways for multiple researchers to work together on the same project. A small but robust literature on the use, applications, and implications of QDAS also burgeoned.

Close to 30 years later, QDAS packages are comprehensive, feature-laden tools of immense value to many in the qualitative research world. However, with the advent of the Internet and the emergence of web-based tools known as Web 2.0, QDAS is now challenged on many fronts as researchers seek out easier-to-learn, more widely available and less expensive, increasingly multimodal, visually attractive, and more socially connected technologies to support their qualitative research endeavors (Anderson, 2007).

Truly, qualitative research and technology is in the midst of a revolution. The purpose of this chapter is to set a historical context for understanding the development of QDAS and to bring that discussion into alignment with the broad stream of methodological discussions in qualitative research, examine the challenges that face qualitative researchers at this pivotal moment, and attempt to predict how the current technological dilemma of our field will be resolved.

DEFINITIONS AND DELIMITATIONS

Our primary aim is to examine those tools that support qualitative researchers in the organization and analysis of qualitative research projects. Silver's (2009) adaptation of Lewins and Silver's (2007) model presents a particularly robust view of these functions.

Figure 38.1 The Basic Idea of QDAS Packages

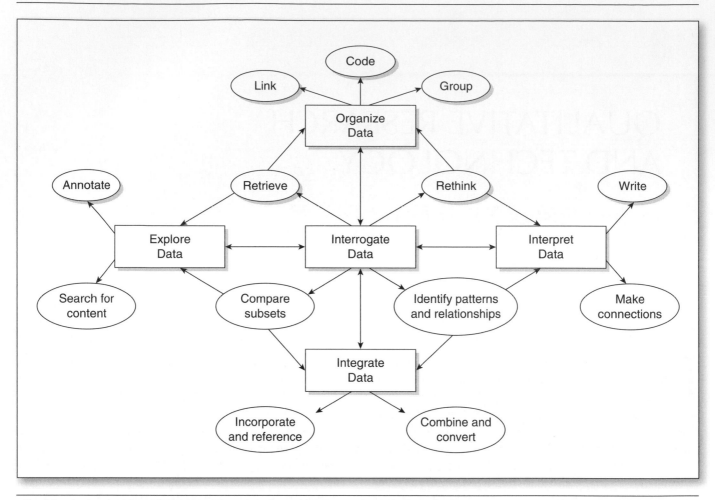

Adapted from Lewins, A., & Silver, C. (2007), *Using Software in Qualitative Research: A Step-by-Step Guide.* London: Sage.

As illustrated above, qualitative analysis is a process that requires the exploration, organization, interpretation, and integration of research materials (data). These four components require that researchers retrieve, rethink, compare subsets, and identify patterns and relationships. Various QDAS program features support analysis tasks including linking and grouping, annotating and searching, writing and making connections, and incorporating references and combining or converting findings.

These are the same tasks researchers using traditional methods perform except that without the power of the computer, it is difficult to retrieve data, so you are limited in comparing subsets and identifying patterns, which then have a limiting effect on your ability to rethink data.

In the earliest versions of such software, data preparation was an important concern. Materials had to be prepared in particular ways so that they could be entered into the program. As computers and QDAS have become more sophisticated, less special preparation is required. MS word formats, jpegs, and video formats can be imported with ease. However, to take advantage of certain affordances of the software (such as automatic coding), you still need to prepare the materials in specified ways (although the preparations are simple and minimal in nature). It could be argued that these affordances may have an impact on how researchers—particularly novice researchers—would design their projects, attending to the affordances rather than the research questions.

In this piece, our goal is to focus on tools providing specific support for the functions described above in the Lewins and Silver (2007) model. QDAS programs are comprehensive stand-alone packages providing a kind of one-stop shop for the technologically savvy qualitative researcher. In this article, we will use the term QDAS to refer to stand-alone software tools that perform the functions described above. It is true, however, that with the explosion of the Internet and Web 2.0, tools with these capacities come in a variety of forms (stand-alone and net-based), opening up uncharted territory for the qualitative researcher. We will introduce the term *QDAS 2.0* to describe this hybrid state of affairs.

In setting parameters for this piece, we have decided not to include the area of "Internet Research," where the focus is on the collection of Internet data or the Internet as a virtual site of study (Hine, 2008). For more information on this topic, we would direct you to Sarah Gatson's article in this volume "The Methods, Politics, and Ethics of Representation in Online Ethnography" (Chapter 31).

We have also shied away from those discussions of technologies for qualitative research that are primarily focused on the presentation of research findings (Dicks & Mason, 2008).

◩ QDAS: An Overview

Discussions of QDAS rely upon a range of frameworks for understanding the development of QDAS or the nature of the tools (see, for instance, Fielding, 2008; Hesse-Biber & Crofts, 2008; Kelle, 1995; T. Richards & Richards, 1994; Tesch, 1990; Weitzman & Miles, 1995). For the most part, these discussions have focused on the developments of the technology itself and less on the relationship of the changing technology to the changing methodological currents of qualitative research. For that reason, we have chosen to anchor our discussion of the development of QDAS around Denzin and Lincoln's (2003, 2008) eight critical moments in the chronology of qualitative research. In using the Denzin and Lincoln chronology, we are aware of its possible ethnocentric limitations (Cisneros, 2008a, 2008b, 2009), but we feel that this is a broadly recognized stage structure among qualitative researchers.

Taking a historical approach to qualitative research and QDAS should help the reader to see how technology has always been a part of qualitative research (di Gregorio & Davidson, 2008; see Table 38.1).

Stage I: Pre-QDAS

| Traditional Period—early 1900s to WWII | Stage I: Pre-QDAS: Notebooks; typewriters and carbon paper |
| Modernist—post WWII to 1970 | Stage I: Pre-QDAS continues: McBee Keysort cards, InDecks Information Retrieval cards, manual to electric portable typewriters, photocopying |

In Stage I, a period that spans more than half the 20th century, the classical set of qualitative research technologies was established. The lionized anthropologist of the "traditional moment" in qualitative research was pictured with a pad and pen in hand, or perhaps in a tent slaving away on his or her manual typewriter, eyes on a notebook full of notes, or a pile of note cards. A significant technological development of this era was the advent of carbon paper. This enabled the researcher to type multiple copies of an interview transcript or observation notes. The copies could be cut up and sections of text could be filed according to themes. The development of QDAS-like tools—notched index cards such as McBee Keysort Cards and InDecks Information Retrieval Cards (Kelly, 2008) and tabs (Tenner, 2005)—represented important technological steps forward in the field. Each index card contained notes from an interview. A mastercard would have descriptors associated with

Table 38.1 Lincoln and Denzin's Stages of Qualitative Research Tweaked and Integrated With Davidson and di Gregorio's Stages of QDAS Development

Moments in Qualitative Research	Stages in the Development of QDAS
1. Traditional Period—Early 1900s to WWII	Stage I: Pre–Qualitative Data Analysis Software notebooks; typewriter and carbon paper
2. Modernist—Post WWII to 1970	Stage I continues . . . McBee Keysort Cards; InDecks Information Retrieval cards; manual to electric portable typewriters; photocopying
3. Blurred Genres—1970 to 1986	Stage II: QDAS begins
4. Crisis of Representation—Mid-1980s to early 1990s	Stage III: The Typology Era: Matching program to project
5. Postmodernism—Early to mid-1990s	Stage IV: Focus on developing similar and competing features: Experimentation with generic tools for qualitative research (QR) functions
6. Post-Experimental Inquiry—1995–2000	Stage IV continues . . .
7. Methodologically Contested Present—2000–2008	Stage V: Development of meta-perspectives on the use of QDAS
8. A Fractured Future	Stage VI: Development of Web 2.0/3.0. Networked technologies move to the fore. QDAS 2.0 comes into being.

the holes around its periphery. Each card was "coded" by punching out the relevant hole(s) in a card. Retrieval was done by inserting a needle in the stack of cards in the hole for a particular descriptor—those cards that were not lifted out by the needle were the ones relevant for the query. Using two needles simultaneously would achieve a Boolean AND.

Toward the end of the modernist period of qualitative research, important changes were taking place in the computer arena that would make QDAS possible. The early mainframe computers were developed (1965) with a focus on the conduct of quantitative content analysis. In 1968, the University of Chicago released the quantitative analysis program—SPSS—and McGraw-Hill published the first user's manual for this program in 1970 (http://www.spss.com/corpinfo/history.htm).

While quantitative researchers were quick to integrate the use of software in their practice, qualitative researchers were not. It may be that statistical software packages mapped more readily onto already existing practices. Computers were associated with counting and mathematical calculations that were consistent with statistical analysis. Qualitative researchers, on the other hand, could only see computers contributing to a quantitative reduction of their analysis.

Toward the end of this period, Glaser and Strauss published *The Discovery of Grounded Theory* (1967). It would have a deep and reverberating effect on the future of QDAS and its place within the field of qualitative research (Fielding, 2008). In this work, Glaser and Strauss challenged the dominant sociological perspectives of the time as exemplified in the works of Talcott Parsons, Robert Merton, and Peter Blau. Glaser and Strauss provided the rationale for the use of qualitative data in capturing "the unfolding nature of the meanings, interpretations and processes that the sociologist is studying" (Layder, 1993). Their book inspired an upsurge of interest in the use of qualitative data in sociology.

Our discussion of noncomputer qualitative research technologies may seem unrelated to the birth of QDAS, but in truth they have great significance for the emergence of QDAS. The concept of *skeuomorphs* is critical in explaining the connection between earlier technologies and QDAS: "A skeuomorph is a design feature that is no longer functional in itself, but that refers back to a feature that was functional at an earlier time.... Skeuomorphs visibly testify to the social or psychological necessity for innovation to be tempered by replication" (Hayles, 1999, p. 17).

In Stage I, qualitative researchers struggled to find effective means of organizing and analyzing research materials. In that process, they developed techniques from card sorts and indexing to annotations and searching that informed the structure and form of QDAS in Stage II. These earlier forms lived on as skeuomorphs embedded in the design and structure of QDAS.

Stage II: QDAS Begins

Blurred genres—1970 to 1986	Stage II: QDAS begins

QDAS was born as qualitative researchers struggled with the issue of blurred genres and as the boundaries between academic areas become fuzzy, allowing for new kinds of intersection and overlap between the social sciences and humanities (Denzin & Lincoln, 2003). As qualitative research struggled with this internal argument, it also struggled with an external context within universities and funding agencies that emphasized science-like approaches to research, which was to have implications for qualitative research and QDAS.

The 1980s are described by Fielding and Lee (1998) as a period of experimentation in computer use in general—databases, quantitative content analysis, word processors, and QDAS. NUD*IST (Non-numerical Unstructured Data Indexing, Searching, and Theorizing) was released in 1981; the Ethnograph was released in 1984. Both are forms of QDAS.

Seidel (1998), developer of the Ethnograph, describes how he identified the analysis processes of pre-QDAS qualitative researchers, as he developed the components of his QDAS tool. Noticing and collecting, which he identified as key analysis processes, were then translated into the computer method. A criticism of the Ethnograph and similar tools was that noticing and collecting could be mistakenly seen as the sum total of the researcher's work (Seidel, 1998).

In research computing in general, the trend was toward programs that were operated from individual computers, as opposed to the large, immoveable mainframe machines.

What also characterizes the beginnings of QDAS is the relative isolation of the various developers. Programs were developed with little knowledge of what other software developers were doing to develop related solutions. The early QDAS programs were developed in noncommercial environments either by social scientists themselves or by social scientists working with programmers (Fielding, 2008).

Stage III: The Typology Era—Matching Program to Project

Crisis of Representation—Mid-1980s to early 1990s	Stage III: The Typology Era: Matching program to project

The "crisis of representation" raised its head in the mid-1980s as qualitative researchers "made research and writing more reflexive and called into question the issues of gender, class, and race" (Denzin & Lincoln, 2003, p. 26).

The corresponding development in QDAS we have dubbed "The Typology Era," which represents the period in which users were concerned with methodological alignment of tool and project. As a consequence, much thought was given to the paradigmatic or methodological perspectives of the developer, as it was believed that these were embedded in the software design and would shape the work of users in specific directions congruent with the developer's methodological bias, regardless of the user's intent.

QDAS came to the fore at the same moment that Strauss and Corbin's (1990) step-by-step guide for doing grounded theory was published. While Glaser and Strauss (1967) provided the justification for qualitative analysis and developing theory from analysis grounded in empirical data, their book did not elaborate on how to conduct such an analysis. Strauss and Corbin produced a text that represented grounded theory as a highly structured set of procedures. Indeed, their text overpowered many voices within the field, and it is only recently that the full breadth of grounded theory possibilities are being discussed (Bryant & Charmaz, 2010). The result was that newcomers from traditionally quantitative disciplines were attracted to the method, as it provided a "cookbook" for conducting qualitative analysis. At the same time, their book also created great antipathy within the field from researchers who felt that alternative perspectives on grounded theory were not represented.

QDAS emerged at this moment and was quickly drawn into the fray. QDAS had been developed as generic tools to support researchers working with unstructured data through providing computerized assistance to the common tasks required when working with this kind of data. QDAS developers were quick to state that their tools were certainly compatible with grounded theory approaches. This apparent association between grounded theory and QDAS led to concerns about the epistemologies of the software developers and the influence their methodological perspectives might have had on the shape of the tool (Coffey, Holbrook, & Atkinson, 1996; for a response to this critique, see Kelle, 1997, and Lee & Fielding, 1996). As a result, QDAS came to be seen as more prescriptive in nature than traditional (non-computerized) methods of qualitative research. This was the state of affairs in QDAS as qualitative research entered the crisis of representation.

In 1989, the Department of Sociology at the University of Surrey, United Kingdom, organized the first international conference on qualitative computing where, for the first time, a dialogue was established among developers and early adopters (Fielding & Lee, 2007). While this group of QDAS enthusiasts was attentive to the developments in the wider world of qualitative analysis, the same cannot be said for the wider world of qualitative research, which paid little attention to the developments in QDAS. QDAS users were a kind of subcultural movement, and their concerns were not incorporated in the contemporary debates around qualitative analysis.

Another important landmark occurred in 1990 when Renata Tesch published the first book to provide an overview of types of qualitative analysis and software tools. While Tesch identifies 46 "brands" of qualitative research, she argues that there are not 46 approaches to analysis. Instead, she describes 10 principles that are common across most types of qualitative research. She makes a distinction between *structural* analysis (event structure analysis, discourse analysis, ethnoscience, ethnography of communication, and structural ethnography) and *interpretational* analysis (which includes the bulk of types of qualitative analysis; they can be subdivided into theory-building and

interpretive/descriptive analysis). The structural analysis approaches differ in that their goal is to create a model—the organizing system common to all approaches is not a means to an end but the end itself. For this approach to analysis, she recommends text retrievers and database managers, which are general tools that can be adapted. However, at the time there were two tools that specifically supported structural analysis—ETHNO and TAP. Interpretational analysis was best supported by the qualitative analysis programs that were evolving at the time—QUALPRO, the Ethnograph, TEXTBASE ALPHA, and HyperQual. AQUAD, NUD*IST, and HyperResearch are briefly mentioned as supporting theory building.

It is important to note two things from Tesch's (1990) seminal work: (1) Grounded theory is featured as only 1 of the 46 types of qualitative analysis, and (2) apart from a couple of Mac-based programs, the rest were for the DOS operating system. The software developed at the time was restricted by the possibilities offered by a DOS system. And the DOS system was soon to be replaced by the Windows operating system. Those software programs that did not make the transition to Windows quickly were left behind. In the next period, software that became dominant had made that transition quickly. NUD*IST (later NVivo), the Ethnograph and HyperResearch, and a few software packages that did not feature in Tesch's review—namely, ATLAS.ti and WinMax (later MAXQDA), became prominent in the field.

Tesch's (1990) book was a key text for the new group of qualitative software users. However, unlike Strauss and Corbin (1990), it did not have the same impact on the wider qualitative research community. The common principles of analysis across most approaches to qualitative analysis that software supported were lost in the discussion and justification of the epistemological bases of these different approaches.

Tesch's (1990) points are amplified by Fielding and Lee in a 2007 presentation overviewing the history of QDAS:

> There was also an assumption that we were seeking to establish some kind of orthodoxy around the analytic process, particularly in relation to grounded theory. This is an assumption we have always strongly resisted. For us, identification of the coding features found in many qualitative data analysis programs with grounded theory has tended to elide program features, analytic procedures and methodological approaches (p. 10)

Stage IV: Similar and Different—An Era of Competition

Postmodernism—Early to mid-1990s	Stage IV: Focus on developing similar and competing features: Experimentation with generic tools for qualitative research (QR) functions
Post-Experimental Inquiry—1995–2000	Stage IV continues

The postmodern turn in qualitative research refers to the "triple crisis of representation, legitimation, and praxis ... in the human disciplines" (Denzin & Lincoln, 2003, p. 28). New forms of representation emerged in the field at the same time that social justice concerns were ardently pushed forward.

Following on the heels of the University of Surrey's conference on qualitative computing, Stage IV represents a radical shift away from methodological alignment of software and project to a broader perspective on the common qualities expected from a basic QDAS package. This stage is illustrative of the increasing sophistication of experienced QDAS users, with the skills and knowledge to look critically across packages, compare features, and explain similarities and differences. A mark of this era is the emergence of key texts on the topic of QDAS; of particular note are the volumes by Fielding and Lee (1991, 1998), Miles and Weitzman (1994; Weitzman & Miles, 1995), and Kelle (1995).

In discussing the publication of their 1991 volume, Fielding and Lee (2007) commented, "It took three proposals before our publisher agreed but the book based on the 1989 conference became their best seller" (p. 5). Fielding and Lee's 1998 book is one of the few studies based on empirical work that looks at how researchers use software to support qualitative data analysis.

Miles and Weitzman (1994; Weitzman 2002; Weitzman & Miles, 1995) built on the work of Tesch (1990) but hardened the distinction between different types of software—text retrievers, textbase managers, code-and-retrieve programs, code-based theory builders, and conceptual network builders. Their focus on comparing technical features of each package had the unfortunate consequence of reinforcing the procedural, scientific image of these programs. While this image attracted users from disciplines without a qualitative analysis tradition, it pushed away many experienced qualitative analysts. In addition, many of the DOS programs they covered transitioned into Windows packages or ceased to exist, as did many of the Apple Macintosh programs—following a downturn in Apple's fortunes during the 1990s.

Kelle's (1995) edited volume, which is dedicated to Tesch, contains a rich collection of articles by the small but hardy group that was creating the software and trying to figure out how to apply it to practical research questions. The contributors included developers and other academics in a relatively egalitarian mix, far different from the more commercial and distanced relationships that were to prevail in a later period. This volume makes a significant contribution to pushing the theoretical perspectives necessary to bring QDAS into the qualitative research field.

In this era, there were many notable and significant developments that supported and exploited this growing knowledge of QDAS. In 1994, the Economic and Social Research Council (ESRC) in Great Britain funded the CAQDAS Networking Project (which emerged out of the awareness raised by the 1989 conference). Based in the Sociology Department at the University of Surrey, it provided (and still provides) advice and training in multiple QDAS packages (http://caqdas.soc.surrey.ac.uk/). Freelance trainers with specialization in QDAS begin to emerge at this time, including one of the authors of this article (di Gregorio). Many of these trainers have specialized in one specific tool, and a smaller number have offered support in multiple tools.

Sage Publications played a key role in disseminating information about and marketing these packages. In 1995, they agreed to market the QSR's newly formed software package—NUD*IST. This was followed subsequently by marketing a number of other qualitative packages, such as ATLAS.ti, Hyper Research, and WinMAX (later MAXQDA). Sage's involvement as a market center for QDAS was short-lived, but their brief involvement did provide important visibility for many of the packages. QSR's NVivo was released in 1999 and marks the close of this era of QDAS development and comparison. NVivo is unusual in the annals of QDAS in that it has been the focus of several book-length texts by the developer and independent writers (Bazeley, 2007; Bazeley & Richards, 2000; Gibbs, 2002; L. Richards, 1999).

As QDAS users compared packages, other qualitative researchers used on-the-shelf software such as Word or Excel to support their analysis. Hahn (2008) has illustrated how Word can be used to support qualitative analysis, while Ritchie and Lewis (2003) have documented how Excel can be used to support their approach to qualitative analysis. It is interesting to note that the National Centre for Social Research (where Ritchie and Lewis were based) have since developed their own QDAS—FrameWork (http://www.framework-natcen.co.uk/).

By the end of this period, three major QDAS packages had come to dominate the scene: ATLAS.ti (www.atlasti.com), MAXQDA (www.MAXQDA.com), and NVivo (www.qsrinternational.com). While the dominant packages of this era were generic tools, other packages emerged with enhanced features that supported particular types of analysis, e.g., Transana (http://www.transana.org/) for video analysis and QDAMiner (http://www.provalisresearch.com/QDAMiner/QDAMinerDesc.html) for mixed methods.

While QDAS made remarkable growth during this period, seeds of contention had been sown. Qualitative researchers debated postmodernism and theories of representation, and QDAS developers and friends, who were also qualitative researchers, listened and examined the tools in light of these arguments. However, the reverse could not be said to be true. Qualitative researchers who had not been introduced in a positive way to QDAS did not gravitate toward using these tools. Some believed the rumors they had heard about the epistemological problems, others did not have access to the tools and were not motivated to find access, and still others did not know that such tools existed. Fear of technology itself was, undoubtedly, an important factor in the failure of senior researchers, in particular, to integrate these new tools into their practices.

On this point, Fielding and Lee (2007) have pointed out, "Since qualitative researchers traditionally learn their craft in apprenticeship mode, the approach we were suggesting [QDAS] seemed to take some mystique out of the analytic process, potentially challenging in consequence the charismatic authority of the teacher" (p. 10).

Stage V: Development of Meta-Perspectives

Methodologically Contested Present—2000–2008	Stage V: Development of meta-perspectives on the use of QDAS, 2000–2008

The methodologically contested seventh period of qualitative research continues to explore the issues of form and justice, legitimation and representation.

Moment seven corresponds to our Stage V, the development of meta-perspectives on the use of QDAS. In Stage V, QDAS adherents sought to develop strategies for research analysis grounded in a cohesive theoretical perspective on QDAS. The Strategies Conferences in the UK (1999–2006), international conferences focused on strategies for using the two QDAS packages developed by QSR International—NUD*IST and NVivo—were important in this process (Fielding & Lee, 2007). These conferences enabled a small and dedicated group of users to discuss ways of working with these packages, including using them for literature review (di Gregorio, 2000), teamworking (di Gregorio, 2001; Gilbert & di Gregorio, 2004; L. Richards, 2006), ways of managing analysis—in general (di Gregorio, 2003a, L. Richards, 2004; T. Richards, 2004) and for specific types of research—including narrative analysis (Gibbs, 2004), and evaluation research (Kaczynski & Miller, 2004; Richter & Clary, 2004), and grounded theory (diGregorio, 2003b). While the focus was on using NUD*IST and NVivo, it became clear to some participants that many of the principles of use under discussion were relevant regardless of which QDAS package was used. Some of the papers from these conferences are still accessible from www.qual-strategies.org.

The software package MAXQDA has also been the focus of a series of annual conferences beginning in 2005 (Computergestutzen Analyse Qualitatirer Daten, or CAQD; see http://www.caqd.de/).

The development of a meta-perspective was crystalized by Lewins and Silver (2007). Their book not only incorporates a step-by-step guide for working with the three major packages of the time (ATLAS.ti, NVivo, and MAXQDA), but it is also organized around the processes and tasks involved in qualitative analysis (rather than the features and functions of each software package).

Our work (di Gregorio & Davidson, 2008) looked across software packages to offer a common framework for representing a qualitative research design when setting up a project in QDAS regardless of the software package adopted. In this work, the term *E-Project* is introduced, meaning the electronic container that is used for storing and organizing all materials related to one project. The E-Project is described for the first time in the QDAS literature as a research genre, and theories of the literary or social science genre are applied to discussion of standards and procedures that cut across individual software packages (see also Davidson 2005a, 2005b, 2005c, 2005d; Davidson & di Gregorio, 2007; di Gregorio, 2005, 2006a, 2006b, 2007; di Gregorio & Davidson, 2007).

This new foundation (the e-project as a generic term for qualitative research projects in any QDAS package, and the notion of the e-project as a genre) serves as the basis for a new approach to research design. The approach described in *Qualitative Research Design for Software Users* (di Gregorio & Davidson, 2008) presents the notion of (1) a software shell representing the research design of a study, (2) the development of an interpretive system through disaggregation and re-contextualization of the data, and (3) thorough reliance on an interactive practice that sustains dialogue between technology and methodology. Whereas earlier QDAS writers had given greater weight to the methodological fit of the tool to the methodological perspective of the researcher, the E-Project approach assumes all tools have limitations and that tools for research are in constant flux and development. Therefore, researchers must engage in an active dialogue between methodology and technology in order to craft the appropriate fit for their work.

Other developments that characterize this period include the expanding use of QDAS in diverse disciplines and various kinds of institutions—academia, business, and other sectors. Sectors that are starting to look at the use of QDAS include market research (di Gregorio, 2008b; Rettie, Robinson, Radke, & Ye, 2007; Vince & Sweetman, 2006), law firms (Coia, 2006), research institutes, and the public sector (di Gregorio & Davidson, 2008). This expansion of QDAS use brought increasing demand for information on best methods for research design with these tools (di Gregorio, 2005; di Gregorio, 2006b; di Gregorio & Davidson, 2008). However, expansion in the market research and commercial sectors has been slow and piecemeal with suspicion of QDAS being associated with quantitative methods (Ereaut, 2002) or not understanding the benefits of its use (di Gregorio, 2008b) despite efforts to demonstrate its value (Ereaut & di Gregorio, 2002, 2003).

The increased awareness and use of QDAS during this time led inevitably to discussions about the teaching of qualitative research using QDAS (Bringer, Johnston, & Brackenridge, 2004; Davidson, 2004, 2005c; Davidson & Jacobs, 2008; Davidson, Siccama, Donohoe, Hardy-Gallagher, & Robertson, 2008; di Gregorio & Davidson, 2008; Gilbert, 1999; Jackson, 2003; Kuhn & Davidson, 2007). In 2003 and 2005, the University of Wisconsin campus was the venue for two conferences on teaching qualitative research with QSR products (Davidson, 2005b; di Gregorio 2003b). The 2003 conference resulted in a special issue of the *Qualitative Research Journal* devoted to the topic of teaching qualitative research with QSR software (*Teaching Qualitative Research With QSR Software,* 2003). Concern about the instruction of QDAS has led software developers to increasing training activities and developing licensing incentives for university settings as well as to developing materials specifically for trainers and teachers.

The majority of discussion about teaching with QDAS has focused on doctoral-level work in higher education. Di Gregorio and Davidson (2008) spend considerable time discussing how QDAS can change pedagogical practices related to qualitative research instruction and advisement for the dissertation. For instance, the new transparency and portability afforded by QDAS allow advisors and doctoral students to engage in interpretive work in ways that were not possible before. Texts, coding systems, and the very coding itself are easily shared through projects sent as file attachments or the joint viewing of the e-project in class or supervision sessions. This has led to the need for new standards for the writing and reading of the e-project as a genre, and the need for faculty to gain fluency in this new literary form.

Issues related to scaling up QDAS use within and across institutions also began to be discussed more widely in Stage V, including issues related to the role of professional organizations in setting standards for qualitative research conducted in QDAS and the ethical issues that QDAS raises for researchers in this era of institutional review boards (Davidson & Jacobs, 2007; Davidson et al., 2008; di Gregorio & Davidson, 2008, 2009; ESRC, 2005; Office of Public Sector Information, 1998; Strike et al., 2002). See the subsequent section on ethics for a fuller discussion of this issue.

In Europe, the high point of Stage V was the CAQDAS 2007 Conference organized by the CAQDAS Networking Project to celebrate 13 years of their work in this area. This international conference drew developers and users together to discuss their use of and experiments with different forms of media and software tools. Innovative papers pointing to future developments included Dario Da Re's (2007) study on the use of video, multimedia, and the web as a new form of qualitative representation (see www.raccontiditerra.it); Parmegianni (2007) on visual sociology; and Tutt and Shaukat's (2007) presentation on MiMeG—a tool for remote collaborative video analysis (see conference materials at http://caqdas.soc.surrey.ac.uk/Resources/Caqdas07conference/caqdas07conferenceintro.html).

In the United States, the conclusion of Stage V might well be the May 2008 "A Day in Technology in Qualitative Research," a preconference day of the Fourth International Congress of Qualitative Inquiry, where an international group of scholars came together to share perspectives on the use of technology in qualitative research. Presentations offered opportunities to explore many of the new frontiers of the field, from the integration of QDAS and geospatial concerns (Cisneros, 2008a; Kwan, 2008) and discussions of e-portfolios as a form of qualitative research (Arndt, 2008), to issues related to teaching QDAS (Davidson et al., 2008), deeper integration of QDAS across disciplines (Gilbert et al., 2008), and the application of QDAS in new liberatory ways (Lapadat, 2008), as well as discussions of directions beyond QDAS, including the use of wikis (Bhattacharya & McCullough, 2008).

As this stage of work began to come to a close, the divisions between the mainstream of qualitative research and the world of QDAS had become increasingly apparent to those in the QDAS world. Although humanistic and arts-based forms of qualitative research were proliferating in this period, it was next to impossible to find examples of qualitative research of this nature employing QDAS (Davidson, 2009; Davidson et al., 2009). New possibilities for QDAS were arising in the world of the Internet, but it was unclear how this would affect the stand-alone packages (di Gregorio, 2009, 2010).

Stage VI: Networked Tools to the Fore

A Fractured Future	Stage VI: Development of Web 2.0–networked tools to the fore: QDAS 2.0

As we enter Stage VI, a period characterized by powerful new movements within qualitative research, from autoethnography and performance ethnography to indigenous methodologies and the rise of discussion about mixed methods, QDAS use stands at an uncomfortable juncture. The concerns come from within and without—from QDAS developers and qualitative researchers; to institutions of higher education, nonprofits, and commerce; to the world of rapid Internet development.

Among many qualitative researchers, there continues to be a sense of unease with QDAS, and a lingering problem is the close association of grounded theory and QDAS (Fielding, 2008; Hesse-Biber & Crofts, 2008). For some, QDAS seems to embody the divide between scientific and humanistic or artistic approaches that qualitative researchers have struggled so hard to sort out, and QDAS is allied with the scientific side in their minds. For those who seek to conduct qualitative research in more humanistic ways, there appears to be a resistance to the transparency afforded by QDAS, and an unspoken belief that their interpretive strategies should not be put on the public chopping block in this manner (Jackson, 2009). Here, too, the fear of failing with the use of technology may be an important factor.

Some of the difficulty with adopting QDAS seems to be related to the complexity of the comprehensive software packages. Early QDAS packages were limited in functionality and had a fairly simple format with only a few menus and drop-down selections. As the programs became more powerful, they also became richer with features, and menus and drop-downs have become more deeply buried and complicated to the average user. Beginning users can easily feel overwhelmed by the choices available (Mangabeira, Lee, & Fielding, 2004).

Not surprisingly, we find that many of those using the packages do so in a very limited way and are often not aware of the features that could assist them. Some researchers have tried these packages, and lacking adequate support, failed in their attempts. Lack of technical support for these tools has been an ongoing problem since their inception. As a result, they warn off other potential users as they share experiences of their QDAS disasters. Newcomers can also be put off by the lack of an intuitive interface and, with some packages, a visually uninteresting interface.

QDAS has not become the practice of most senior researchers in the field of qualitative research, and many rising researchers still lack exposure to QDAS use in their graduate training. Technologically savvy graduate students have forged ahead using QDAS for their research, but in some cases must proceed without adequate technical support, and with advisors who cannot make sufficient sense of QDAS use. For those graduate students working with advisors with knowledge of QDAS, other faculty with whom they work may not be equally skilled. To date, there have been few exemplars of integrated QDAS use at the institutional level from which to learn (Davidson & Jacobs, 2007; Davidson et al., 2008; di Gregorio & Davidson, 2008b). Some developers are taking innovative steps to change these circumstances, such as QSR's recent Teaching Grant Award program (QSR International, 2009).

The lack of institutional support for QDAS in higher education has been a significant issue. While there are signs that QDAS support is increasing within selected institutions, overall support remains uneven within and across institutions. Support also varies depending upon the national context. The problems with building institutional capacity in QDAS are multifaceted. Until recently, universities had been slow to provide researchers with licenses for these products, technical support for their use, and acknowledgment for researchers and teachers who explore these tools. On many campuses, support for QDAS lags far behind support for basic quantitative technologies. Moreover, requests from qualitative researchers for these basic technological tools may be greeted with greater skepticism from administrators, requiring greater justifications than what quantitative researchers face making similar requests (Davidson & Jacobs, 2007; Davidson et al., 2008). As di Gregorio and Davidson (2008) report, "lacking a critical mass of users and adequate paths of knowledge distribution, users are often isolated within organizations or fields" (p. 2).

The lack of institutional support could be attributed to several issues. First, senior faculty are not advocating for it because they are less likely to be users. Newer faculty and graduate students, who are users, may lack the institutional clout to acquire and implement QDAS use. Second, information technology

departments have a significant say in technology purchases despite the fact that they may have limited knowledge about higher education curricular needs. Given their technical backgrounds, it may be that quantitative tools have more sway with them than qualitative research tools. Third, the entry point for QDAS into an institution is usually an individual faculty member, so there may be many hives of QDAS activity within an institution that are not aware of each other's activities. In other words, QDAS implementation does not fall under any unified policy direction of the institution.

Institutional support for QDAS differs significantly based upon the national context. The UK has been at the forefront in providing national leadership for the development of digital capacity for qualitative researchers. The ESRC National Centre for Research Methods and the Joint Information Systems Committee (JISC) have provided leadership and resources for the development of qualitative research methodology in many ways, including the provision of training and the development of e-resources.

At the same time that many universities have proven resistant to QDAS, there is significant expansion of the use of QDAS in governmental, commercial, and other nonprofit settings (di Gregorio, 2006b; di Gregorio & Davidson, 2008). These include such diverse fields as health, criminal justice, law, and social policy. Virtually every field in which unstructured data are present, and there is the need for fine-grained studies of small pockets of such data, is discovering that QDAS could help ease the burden. Many of these studies fall into the arena of evaluation or evaluation-like research, with little use being made of the capacity of QDAS to support less scientific and more humanistic forms of qualitative research. In some cases, new converts to QDAS are coming without training or experience in qualitative research and, lacking other reference points, develop their QDAS use with only quantitative research training as background. Such studies only exacerbate some qualitative researchers' beliefs that QDAS is really quantitative research in disguise. And, indeed, discussions regarding the possibilities of using computer capacities to integrate qualitative and quantitative data have also been on the rise (Fielding, 2008; Nasukawa, 2006).

QDAS developers are closely attuned to the market possibilities for their wares. In addition to expanding disciplinary markets, they are also looking to expand language markets as they make their tools available in different language formats. Developers have also expanded their training and educational forces, and developed new and attractive licensing strategies. Each new version of the software packages demonstrates attentiveness to user demands, from better tools for visualization, geospatial work, and multiple media modalities to the capacity for greater collaboration and teamwork and integration of quantitative data.

To date, there has been little discussion among academic and nonacademic users of these tools. Thus, we know little about the ways these different audiences are using these tools, the unique challenges they have encountered, or the standards for use that are developing in these different arenas. However, the Merlien Institute, an independent organization founded in 2006, offers a unique platform for connecting various communities interested in qualitative research and the use of QDAS through its mini-conferences on these topics (http://merlien.org/).

The Challenge of the Internet

At the same moment that QDAS has come to this pass, qualitative researchers have encountered another challenge. With the rapid growth of the Internet, we are faced with an avalanche of unstructured data—words, images, video, and even sensory data. The rise of Web 2.0 applications, such as YouTube, Flickr, Twitter, and Facebook, allows anyone with access to the Internet to upload their own data—be it photos, video, or chat—to share with others, to comment, and to organize.

These opportunities come paired with their own set of challenges for qualitative researchers: How can we make best use of this information? What are the tools of the future for qualitative researchers? What are the ethical guidelines for qualitative research in QDAS 2.0? How can we ensure the security of data in online environments?

Software developers with no relationship to the world of qualitative research and its rich traditions in the arena of interpretation of unstructured data have created tools that will help users to organize and analyze unstructured information in this virtual world. In so doing, they must ask and answer the same kinds of questions that have long been discussed among qualitative researchers: How do we deal with unstructured data of many formats? How does one best organize, search, sort, pattern, and manipulate this kind of data? An example of the most recent innovations at the time of this writing is the development of Web 3.0—moving from a web of documents to a web of data (Berners-Lee, 2009).

A feature of this amazing amount of unstructured data is that it is available everywhere to just about anyone. Not only is the data readily available in the forms of texts, images, and video, but also the tools and services with which one can work with the data. There has been a huge growth in technologies to decipher and search among these forms of unstructured data. Digital text searches are essential in today's world, and one can also search sound and visual images (di Gregorio, 2010).

It is becoming increasingly apparent that many kinds of people want to research and work with various kinds of unstructured texts (Greif, 2009). The things they do to the texts they examine (search, tag, index, annotate, memo, interpret, represent) bear close resemblance to the things qualitative researchers have long done to unstructured data. These ordinary people are actively engaged as indigenous qualitative researchers in the virtual world. Examples of these indigenous

researchers include teenagers who tag and organize YouTube videos and crafters of all sorts engaging in their community of practice (di Gregorio, 2009, 2010). They use a range of tools, pulled off the web, to achieve their personal research goals. The analysis of this phenomenon of tagging items on the Internet and the development of *folksonomies,* a user-created, bottom-up categorical structure development with an emergent thesaurus (Vander Wal, 2007), has been dominated by those in library science, information architecture, and new media (di Gregorio, 2008a).

Unlike the restricted world of QDAS, the developers of these new QDAS-like Web 2.0 tools represent a broad cross-section of the information community. There are the major computer companies that survived the shift into the Internet age, such as Microsoft, IBM, Xerox, and Apple, all of which are conducting important research leading to new tool possibilities. There are also web-based companies that emerged from the Web 2.0 revolution, including such well-known names as Google, Amazon, and Apple iTunes, who have also made important contributions to the possibilities of QDAS-like tools. The new forms of the telephone/Internet provider companies are also serious contenders. AT&T, Verizon, Apple—creator of the iPhone—and Telenor are a few of the names in this marketplace. Governmental agencies have also played a role in the development of new QDAS-like tools. In the UK, the Joint Information Systems Committee and the Economic and Social Research Council have been particularly active in this area with the funding of projects that will develop new digital tools and training available to qualitative researchers. Finally, there are thousands of independent Web 2.0 developers scattered across the globe that are working on their own to create new products that can do the kinds of things that QDAS has done. They are so prolific that even the most tuned-in leaders in the field complain that it is hard to keep abreast of these developments (Greif, 2009). An example of this work can be found in the emergence of an application for iPhone use designed specifically for qualitative market researchers—EverydayLives (www.everydaylives.com)—which allows the researcher to document activities in the field, tagging or coding the text, audio, or visual records and sharing them with researchers in other locations from a mobile (cell) phone.

For qualitative researchers, an important portal to the world of these new tools has been DiRT (the Digital Research Tools wiki), which provides a comprehensive list of QDAS and Web 2.0 research tools (http://digitalresearchtools.pbworks.com). There are a few promising net-based tools emerging that appear to have capabilities very close to QDAS. A good example of this development is A.nnotate, a tool created by developers in the UK, with financial support from the Scottish government, that allows researchers to annotate or tag text and visual data, and index and organize these tags. The annotations can be viewed and worked upon by collaborative groups (A.nnotate.com).

Qualitative researchers have also begun to experiment with new net-based tools that have special affinity with the ways qualitative researchers work. Wikis are very important in this category. Dicks and Mason (2008) experimented with qualitative hyperlinking in their Ethnographic Hypermedia Environment project. They used StorySpace to do this but see wikis as a way to turn this kind of an analysis into an interactive endeavor, "which allows informants and participants, other academics and indeed the general public to make contributions, allowing the work to grow and develop organically over time" (p. 584). Melanie Hundley (2009a) of Vanderbilt University has used wikis in qualitative research courses as a collective e-project site to house multimodal data collected by students and has explored the ways the hypertext capacities of a wiki support the notion of "data as event." Hundley (2009b) has also used a website for her autoethnography—*The Bard on the Digital Porch*—which invites the reader to decide how to traverse the hyperlinks she constructed so that the reader can participate in the co-construction of the interpretation of meaning. Kakali Bhattacharya (2009) has also explored the possibilities of wikis for qualitative research teaching; while more oriented toward the use of wikis for theoretical study, her work offers possibilities for thinking about wiki's as a collaborative tool for qualitative research in general. Di Gregorio has developed an online course on qualitative analysis based in a Wetpaint wiki, which incorporates videos, collaborative work on analysis, a discussion board, and a chat area (http://qdas01 .wetpaint.com). Again, while focused on teaching, it illustrates how researchers can reflect and share insights on each other's work. Bennett (2008) has produced a YouTube video on how to use wikis for qualitative analysis (www.youtube.com/watch?v=Jwfce BwNmuk&feature=related). We (Davidson and di Gregorio) wrote this chapter in a PBWorks wiki—where we constructed collaboratively the initial outline of the chapter, created a resource section where we uploaded documents and links to references, and used the comments feature to communicate and keep a record of our developing thinking. We saw strong parallels between the ways we use QDAS e-projects and the process of collaborative writing in the wiki, and for this reason we believe wikis are very worthy of closer study by qualitative researchers concerned with the development of digital tools for qualitative research analysis.

Di Gregorio (2010) has related QDAS tools to existing Web 2.0 tools, as exemplified in Table 38.2. As can be seen in the table, there are equivalent Web 2.0 tools for the core QDAS tools. Unsurprisingly, Web 2.0 tools are stronger with hyperlinking, visualizing, and collaborating, while QDAS tools are stronger in fine coding and sophisticated searching. However, as there are considerable investments in Web 2.0 tools from large companies such as Google, IBM, and Microsoft, the capabilities of Web 2.0 tools in fine coding and searching could soon surpass the QDAS tools.

Table 38.2 Comparison of Web 2.0 and QDAS tools

	Web 2.0	QDAS
Organizing	Tagging	Coding
	Grouping	Sets, families
	Hyperlinking	Hyperlinking
Reflective tools	Blogging	Memoing
	Annotating	Annotating
	Mapping	Mapping
Exploring tools	Visualizing	Model, map, network
	Searching	Text search, coding search
Integrating tools	Blogging with hyperlinks	Memoing with hyperlinks
	Collaborating through wikis	Merging projects

In the United Kingdom, as mentioned earlier, a strong push for development of research tools has come from JISC, the ESRC and their associated National Centre for E-Social Science (NCeSS), as well as the National Centre for Research Methods (NCRM). Grants from the NCRM have funded Qualitative Innovations in CAQDAS (QUIC), the University of Surrey's work on expanding qualitative research computing into the areas of mixed methods, visualization and geospatialization, and large-scale collaboration (http://caqdas.soc.surrey.ac.uk/QUIC/ quicheader.html). In addition, through QUIC, Surrey will expand their training program to encompass online training including on interactive protocols and software exemplars. NCeSS funded the development of MiMeg, a collaborative video analysis tool (www.ncess.ac.uk/tools/mimeg/) and DRS, a tool to integrate multimodal data (audio and visual records) with GPS and sensory probe data, which is billed as the "next-generation CAQDAS" (http://caqdas.soc.surrey.ac.uk/PDF/DRSdistinguishingfeatures.pdf). NCeSS was closed in 2009, and its key work is being carried forward by the Manchester eResearch Centre (MeRC) (www.merc.ac.uk/)

These new developments emerging at the intersection between QDAS and the Internet raise important issues for qualitative researchers as they seek out digital tools that will be beneficial to their work.

Moving From QDAS to QDAS 2.0

The next movement in the evolution of technology in qualitative research will include both QDAS and QDAS-like tools available on the Internet. For this reason, we would like to offer the term *QDAS 2.0* to refer to this hybrid state of affairs that will probably persist for some time. If this new movement is to successfully integrate QDAS 2.0 with the work of qualitative researchers, we need to develop tools that meet basic user requirements, create a shared vision of the qualitative analysis process that will allow us to make the best use of QDAS 2.0, and to engage in conversation with the new field of developers.

Basic Requirements

Both older tools (QDAS) and newer tools (QDAS 2.0) will have to meet evolving user standards. In addition to meeting the standards for one or more components of QDAS in the realm of data organization and analysis, it will be essential that these new tools

- Possess an intuitive and visually attractive interface;
- Are easily accessible;
- Have powerful, intuitive, and contextualized search tools;
- Are easily combined to create new user-specialized tools;
- Provide opportunities for visualization and spatialization;
- Offer ease of integration with quantification tools; and
- Offer strong functionality for collaborative work.

Other issues that QDAS 2.0 must grapple with as it moves forward are the new challenges to privacy and ethics that web-based tools raise—for instance, housing data on a third-party server controlled by a commercial interest (see discussion below).

Creating a Shared Vision of the Qualitative Analysis Process

In order to mend the rift between qualitative research and QDAS 2.0, we feel it is necessary to re-present the critical discussion that those engaged in QDAS 2.0 have been making about the nature of QDAS 2.0 in relationship to the nature of the qualitative analysis process. In her landmark book on QDAS, Tesch (1990) concluded that while there are many brands of software programs, they provide a similar range of features for the conduct of qualitative research analysis. In other words, qualitative analysis processes, regardless of the methodological approach of the researcher, must basically do a range of similar things. While they may vary in particulars, the overall range of things one does in analysis is similar across methodological approaches. Thus, QDAS is a generic tool tailored to do these basic tasks that must occur in qualitative research analysis.

T. Richards and Richards (1994) support this conclusion when they point to the importance of the researcher's skills, perspectives, and ability to work with the tools. Morse and Richards (2002) add to this discussion, emphasizing the importance of the researcher's epistemological stance in shaping a research project. They demonstrate how different questions dictate different approaches, but also how analysis processes are composed of a similar range of possibilities. Lewins and Silver (2007) are essentially making the same point; that is, the research process is composed of a range of common tasks, and QDAS are generic tools designed to support the researcher to do these tasks. Hesse-Biber and Crofts (2008), in a similar vein, discuss the commonalities of the qualitative research process as the backdrop to understanding the generic nature of QDAS packages. Fielding (2008) also makes the point that these software packages provide generic support that works with the range of qualitative research methodologies, emphasizing that users have to make the decisions about what the computer will be asked to do. Di Gregorio and Davidson (2008) stipulate the importance of ongoing dialogue between technology and methodology. There is always, we believe, a dialectic between researcher and technology, regardless of whether the technology is a software program or a note card. It is the researcher, not the technology, that decides the question that will guide the research and how it will be approached, as well as deciding critical methodological stances toward gender, power, and voice. This point is similar to the stand taken by L. Richards (2005).

Many of the arguments we have heard made against the use of technology (it imposes unnecessary hierarchy, it is variable dependent, it leads to artificial understanding, it presumes an objective observer, etc.) are arguments about issues that are in the realm or control of the researcher and are not a function of the technology itself. These are arguments that could be applied to many technologies employed by qualitative researchers, not just those of the digital sort.

Following Tesch's (1990) lead, we would describe qualitative analysis as distinct because of the nature of the process (iterative, flexible, and reflexive), the characteristic tasks (disaggregation and reaggregation driven by the search for patterns and comparison of categories), and the role of the researcher (responsible for integrating methodology and substance).

Expanding the Conversation: Addressing Ethical Concerns

QDAS has, until recently, flown under the radar of ethical discussions in qualitative research. QDAS was considered simply a place in the computer where the data was stored, and, for that reason was no more or less problematic than any kind of electronic data stored on a stand-alone computer. Similar restrictions were applied to QDAS as to documents in a word processing software, that is, ethical bodies sought to make sure that adequate safeguards of privacy and security were in place so that information would not be inadvertently released or inappropriately available to prying eyes.

Because of software restrictions, e-projects have not been stored on the Internet. However, this restriction is on the verge of change as increasing numbers of users need to be able to work with data in an Internet environment, as opposed to a stand-alone computer environment. IRBs are also increasingly sophisticated about the technological issues that can infringe on the privacy of research participants, and they will need to be assured that the new cycle of digital tools can be used in a way that will protect the safety of human subjects. While these discussions will be challenging, we have no doubt that these issues are solvable.

A greater ethical dilemma, however, looms on the horizon for qualitative researchers working with QDAS 2.0. In regard to Internet research ethics, Bassett and O'Riordan (2002) argue that

> the use of spatial metaphors in descriptions of the Internet has shaped the adoption of the human subjects research model. Whilst this model is appropriate in some areas of Internet research such as email communication, we feel that researchers, when navigating the complex terrain of Internet research ethics, need also to consider the Internet as cultural production of texts. (p. 233)

In this volume, Gatson (Chapter 31) discusses the ethical complexities facing Internet researchers and points to the guidelines of the Association of Internet Researchers as a starting point for designing policies (http://aoir.org).

Buchanan, Delap, and Mason (2010) identify seven issues that fuel concern regarding ethics with research and new digital technologies. These are "malleability of technology; black box nature of systems; increasing complexity of systems; potential scope and magnitude of impact; difficulty of anticipating consequences; potential irreversibility; [and] rapid pace of technical development . . . lack of precedents" (n.p.).

These challenges have led, in some cases, to creative responses. As an example, researchers working on the VIBE (Virtual Information Behavior Environments) project at the University of Washington Information School have developed a "consent bot, named Harvey, to allow them to gain consent from participants in Second Life where they are conducting research" (Lin, Eisenberg, & Marino, 2010, n.p.).

The ethical dilemmas facing researchers working with QDAS 2.0 cannot be underestimated. This issue will require strong attention from qualitative researchers and other information technology specialists.

▣ CONCLUSION

QDAS 2.0 offers spectacular possibilities to qualitative researchers. From their inception as simple text-retrieval programs to their current state as comprehensive stand-alone packages and what is emerging—Web 2.0 tools with various capacities—they have been a controversial subject among qualitative researchers. As we move more deeply into the digital age, their use, which was once a private choice, will become a necessity.

We recognize that there is strong resistance to the notion that elements of analysis are common among diverse methodological approaches to qualitative research. This resistance is a residue, we believe, of the tough battles of legitimacy qualitative researchers fought to gain a position in academic and other circles. It is time, however, to put this one to rest. As the pressure for participation in the digital world increases, it is critical that qualitative researchers get beyond these artificial and self-imposed barriers they have erected and get on with more important tasks.

The possibilities and development of QDAS 2.0 will depend in large part upon the capacity of qualitative researchers to enter discussions about QDAS and Web 2.0. It will require our field to initiate or join conversations taking place, not only among academic colleagues and QDAS developers, but also with the wider world of Internet entrepreneurs, from the IBMs and Amazons of the world to the range of small, savvy developers creating exciting new tools for use with fine-grained analysis of unstructured data in all its forms.

▣ REFERENCES

Anderson, P. (2007, February). What is Web 2.0? Ideas, technologies and implications for education. *JISC Technology and Standards Watch.* Available at: http://www.jisc.ac.uk/publications/reports/2007/twweb2.aspx

Arndt, A. (2008, May). *Artifacts and assemblages: Electronic portfolios in educational research.* Paper presented at "A Day in Technology in Qualitative Research," a preconference day of the Fourth International Congress on Qualitative Inquiry, University of Illinois, Urbana-Champaign.

Bassett, E., & O'Riordan, K. (2002). Ethics of Internet research: Contesting the human subjects research model. *Ethics and Information Technology, 4*(3), 233–247.

Bazeley, P. (2007). *Qualitative data analysis with NVivo.* Thousand Oaks, CA: Sage.

Bazeley, P., & Richards, L. (2000). *The NVivo qualitative project book.* Thousand Oaks, CA: Sage.

Bennett, N. (2008, April 5). *Using wikis to conduct qualitative research* [YouTube video], Available at http://www.youtube.com/watch?v=JwfceBwNmuk

Berners-Lee, T. (2009, March 13). The next web of open, linked data [YouTube video]. Available at http://www.youtube.com/watch?v=OM6XIICm_qo

Bhattacharya, K. (2009). *Portal to three wiki spaces developed by K. Battacharya's qualitative research classes.* Available at http://kakali.org/memphiswebsite/kakaliorg1/community.html

Bhattacharya, K., & McCullough, A. (2008, May). *De/colonizing democratic digital learning environments: Carving a space for wiki-ology in qualitative inquiry.* Paper presented at "A Day in Technology in Qualitative Research," a preconference day of the Fourth International Congress on Qualitative Inquiry, University of Illinois, Urbana-Champaign.

Bringer, J., Johnston, L., & Brackenridge, C. (2004). Maximizing transparency in a doctoral thesis: The complexities of writing about the use of QSR*NVIVO within a grounded theory study. *Qualitative Research, 4*(2), 247–265.

Bryant, A., & Charmaz, K. (2010). *The SAGE handbook of grounded theory.* Thousand Oaks, CA: Sage.

Buchanan, E., Delap, A., & Mason, R. (2010, January). *Ethical research and design in cyberspace.* Paper presented at the 43rd Hawaii International Conference on Systems Science, Koloa.

Cisneros, C. (2008a, May). *Emergent approaches on linking qualitative software to qualitative geography.* Paper presented at "A Day in Technology in Qualitative Research," a preconference day of the Fourth International Congress on Qualitative Inquiry, University of Illinois, Urbana-Champaign.

Cisneros, C. (2008b). On the roots of qualitative research. In J. Zelger, M. Raich, & P. Schober (Eds.), *Gabek III: Organisationen und ihre Wissensnetze* (pp. 53–75). Innsbruck, Austria: StudienVerlag.

Cisneros, C. (2009, May). *Qualitative data analysis software: Challenges from the periphery.* Paper presented as part of a panel titled Humanistic Issues Regarding Qualitative Data Analysis Software (QDAS): Teaching, Learning, and the Representation of Data in a Digital Age, at the Fifth International Congress on Qualitative Inquiry, University of Illinois, Urbana-Champaign.

Coffey, A., Holbrook, B., & Atkinson, P. (1996). Qualitative data analysis: Technologies and representations. *Sociological Research Online, 1*(1). Available at http://www.socresonline.org.uk/1/1/4.html

Coia, P. (2006, June). How a global law firm works with NVivo 7. *Nsight, 29.*

Da Re, D. (2007, April 18–20). *Research results showed by a video and by a website.* Paper presented at the CAQDAS 2007 Conference: Advances in Qualitative Computing, Royal Holloway, University of London, Egham, UK. (website for the project discussed in this paper is http://www.raccontiditerra.it/)

Davidson, J. (2004, September 1–3). *Grading NVivo: Making the shift from training to teaching with software for qualitative data analysis.* Paper presented at the Fifth International Conference on Strategies in Qualitative Research: Using QSR Nivo and NUD*IST, University of Durham, UK.

Davidson, J. (2005a, April). *Genre and qualitative research software: The role of "the project" in the post-electronic world of qualitative research.* Paper presented at the American Educational Research Association Annual Meeting, Montreal, Quebec, Canada.

Davidson, J. (2005b, April). *Learning to "read" NVivo projects: Implications for teaching qualitative research.* Paper presented at the Second Teaching Qualitative Research Using QSR Products Conference, University of Wisconsin, Madison.

Davidson, J. (2005c, Spring). Learning to think as a teacher within the NVivo container. *QSR Newsletter.*

Davidson, J. (2005d, April). *Reading "the project": Qualitative research software and the issue of genre in qualitative research.* Paper presented at the First International Congress of Qualitative Inquiry, University of Illinois, Urbana-Champaign.

Davidson, J. (2009, May). *Autoethnography/self-study/arts-based research/qualitative data analysis software: Mixing, shaking, and recombining qualitative research tools in the act of recreating oneself as qualitative researcher, instructor, and learner.* Paper presented at the Fifth International Congress on Qualitative Inquiry, University of Illinois, Urbana-Champaign.

Davidson, J., & di Gregorio S. (2007, May). *Research design in qualitative research software.* Paper presented at the Third International Congress on Qualitative Inquiry, University of Illinois, Urbana-Champaign.

Davidson, J., Donohoe, K., Tello, S. Christensen, L., Steingisser, G., & Varoudakis, C. (2009, May). *Initiating qualitative inquiry: Report on an experiment with a cluster of powerful tools—autoethnography, arts-based research, and qualitative data analysis software.* Poster session presented at the Fifth International Congress on Qualitative Inquiry, University of Illinois, Urbana-Champaign.

Davidson, J., & Jacobs, C. (2007, May). *The qualitative research network: Working cross-campus to support qualitative researchers at the University of Massachusetts-Lowell.* Paper presented as part of a panel titled Institutionalizing Qualitative Research: Emerging Models, at the Third International Congress on Qualitative Inquiry, University of Illinois, Urbana-Champaign.

Davidson, J., & Jacobs, C. (2008). The implications of qualitative research software for doctoral work: Considering the individual and institutional context. *Qualitative Research Journal, 8*(2), 72–80.

Davidson, J., Siccama, C., Donohoe, K., Hardy-Gallagher, S., & Robertson, S. (2008, May). *Teaching qualitative data analysis software (QDAS) in a virtual environment: Team curriculum development of an NVivo training workshop.* Paper presented at the Fourth International Congress on Qualitative Inquiry, University of Illinois, Urbana-Champaign.

A Day in Technology in Qualitative Research. (2008, May 17–21). Preconference day at the Fourth International Congress on Qualitative Inquiry, University of Illinois, Urbana-Champaign.

Denzin, N. K., & Lincoln, Y. S. (2003). Introduction: The discipline and practice of qualitative research. In N. K. Denzin & Y. S. Lincoln (Eds.), *The landscape of qualitative research* (2nd ed., pp. 1–46). Thousand Oaks, CA: Sage.

Denzin, N. K., & Lincoln, Y. S. (2008). Introduction: The discipline and practice of qualitative research. In N. K. Denzin & Y. S. Lincoln (Eds.), *The landscape of qualitative research* (3rd ed., pp. 1–44). Thousand Oaks, CA: Sage.

Dicks, B., & Mason, B. (2008). Hypermedia methods for qualitative research. In S. Hesse-Biber & P. Leavy (Eds.), *Handbook of emergent methods* (pp. 601–612). New York: Guilford Press.

di Gregorio, S. (2000, September 29–30). *Using NVivo for your literature review.* Paper presented at the Strategies in Qualitative Research: Issues and Results From Analysis Using QSR NVivo and NUD*IST conference, Institute of Education, London.

di Gregorio, S. (2001, November). Teamwork using QSR N5 software: An example from a large-scale national evaluation project. *NSight Newsletter.*

di Gregorio, S. (2003a, May 8–9). *Analysis as cycling: Shifting between coding and memoing in using qualitative software.* Paper presented at the Strategies in Qualitative Research: Methodological Issues and Practices Using QSR NVivo and NUD*IST conference, Institute of Education, London.

di Gregorio, S. (2003b). Teaching grounded theory with QSR NVivo [Special issue]. *Qualitative Research Journal,* 79–95. Available at http://www.latrobe.edu.au/aqr

di Gregorio, S. (2005, May 11–13). *Software tools to support qualitative analysis and reporting.* Paper presented at the Business Intelligence Group Conference, The New B2B: A Widening Horizon, Chepstow, UK.

di Gregorio, S. (2006a, June). The CMS Cameron McKenna Project—How it looks in NVivo 7. *Nsight, 29.*

di Gregorio, S. (2006b, September 13–15). *Research design issues for software users.* Paper presented at the Seventh International Strategies in Qualitative Research Conference, University of Durham, UK.

di Gregorio, S. (2007). Qualitative Analysesoftware. In R. Buber & H. Holzmuller (Eds.), *Qualitative Marktforschung: Konzpete, Methoden, Analysen.* Wiesbaden, Germany: Gabler.

di Gregorio, S. (2008a). *Folksonomies: A tool to learn from others?* [Online wiki]. Available at http://folksonomiesanddelicious. pbworks.com/

di Gregorio, S. (2008b, Fall). Is technophobia holding back advances in the analysis of qualitative data? *QRCA Views, 7, 1.*

di Gregorio, S. (2009, June 4–5). *Qualitative analysis and Web 2.0.* Paper presented at the Second International Workshop on Computer-Aided Qualitative Research, Utrecht, The Netherlands.

di Gregorio, S. (2010, January 5–8). *Using Web 2.0 tools for qualitative analysis: An exploration. Proceedings of the 43rd Annual Hawaii International Conference on System Sciences (CD-ROM).* Washington, DC: IEEE Computer Society Press.

di Gregorio, S., & Davidson, J. (2007, February 13). *Research design, units of analysis and software supporting qualitative analysis.* Paper presented at the CAQDAS 2007 Conference: Advances in Qualitative Computing, Royal Holloway, University of London, Egham, UK.

di Gregorio, S., & Davidson, J. (2008). *Qualitative research design for software users.* London: Open University Press/McGraw-Hill.

di Gregorio, S., & Davidson, J. (2009, May 20–23). *Research design and ethical issues when working within an e-project.* Paper presented at the Fifth International Congress of Qualitative Inquiry at the University of Illinois, Urbana-Champaign.

Economic and Social Research Council. (2005). *Research ethics framework.* Available at http://www.esrcsocietytoday.ac.uk/ESRCInfo Centre/opportunities/research_ethics_framework

Economic and Social Research Council. (n.d.). *Our research.* Available at http://www.esrcsocietytoday.ac.uk/ESRCInfoCentre/research

Ereaut, G. (2002). *Analysis and interpretation in qualitative market research.* London: Sage.

Ereaut, G., & di Gregorio, S. (2002, June). *Qualitative data mining.* Presentation at Association for Qualitative Research Conference, London.

Ereaut, G., & di Gregorio, S. (2003, June 6). *Can computers help analyse qualitative data?* Presentation at Association for Qualitative Research Conference, London.

Fielding, N. (2008). The role of computer-assisted qualitative data analysis: Impact on emergent methods in qualitative research. In S. Hesse-Biber & P. Leavy (Eds.), *Handbook of emergent methods* (pp. 675–695). New York: Guilford Press.

Fielding, N., & Lee, R. (1991). *Using computers in qualitative research.* London: Sage.

Fielding, N., & Lee, R. (1998). *Computer analysis and qualitative research.* Thousand Oaks, CA: Sage.

Fielding, N., & Lee, R. (2007, April 18–20). *Honouring the past, scoping the future.* Plenary paper presented at CAQDAS 07: Advances in Qualitative Computing Conference, Royal Holloway, University of London, Egham, UK.

Gibbs, G. (2002). *Qualitative data analysis: Explorations with NVivo.* Buckingham, UK: Open University Press.

Gibbs, G. (2004, September 1–3). *Narrative analysis and NVivo.* Paper presented at the Fifth International Strategies in Qualitative Research Conference, University of Durham, UK.

Gilbert, L. (1999). *Reflections of qualitative researchers on the uses of qualitative data analysis software: An activity theory perspective.* Doctoral dissertation, University of Georgia: Athens, GA.

Gilbert, L., Boudreau, M., Coverdill, J., Freeman, M., Harklau, S. L., Joseph, C., et al. (2008, May). *Faculty learning community: Experiences with qualitative data analysis software.* Paper presented at "A Day in Technology in Qualitative Research," a preconference day of the Fourth International Congress on Qualitative Inquiry, University of Illinois, Urbana-Champaign.

Gilbert, L., & di Gregorio, S. (2004, September 1–3). *Team research with QDA software: Promises and pitfalls.* Paper presented at the Fifth International Strategies in Qualitative Research Conference, University of Durham, UK.

Glaser, B., & Strauss, A. (1967). *The discovery of grounded theory.* Chicago: Aldine.

Greif, I. (2009). *Web 2.0 Expo NY: Irene Greif (IBM), what ManyEyes knows* [YouTube video]. Available at http://www.youtube.com/watch?v=nXSOM7WUNaU

Hahn, C. (2008). *Doing qualitative research using your computer: A practical guide.* Thousand Oaks, CA: Sage.

Hayles, N. K. (1999). *How we became posthuman: Virtual bodies in cybernetics, literature, and informatics.* Chicago: University of Chicago Press.

Hesse-Biber, S., & Crofts, C. (2008). User-centered perspectives on qualitative data analysis software: Emergent technologies and future trends. In S. Hesse-Biber & P. Leavy (Eds.), *Handbook of emergent methods* (pp. 655–674). New York: Guilford Press.

Hesse-Biber, S. N., & Leavy, P. (2007). *The practice of qualitative research.* Thousand Oaks, CA: Sage.

Hine, C. (2008). Internet research as emergent practice. In S. Hesse-Biber & P. Leavy (Eds.), *Handbook of emergent methods* (pp. 525–541). New York: Guilford Press.

Hundley, M. (2009a, May). *Data as event.* Paper presented at the Fifth International Congress on Qualitative Inquiry, University of Illinois, Urbana-Champaign.

Hundley, M. (2009b, May). *Gilding the lily: Creating the Bard on the digital porch.* Paper presented at the Fifth International Congress on Qualitative Inquiry, University of Illinois, Urbana-Champaign.

Jackson, K. (2003). Blending technology and methodology: A shift toward creative instruction of qualitative methods with NVivo [Special issue]. *Qualitative Research Journal, 15.*

Jackson, K. (2009, May 20–23). *Troubling transparency: Qualitative data analysis software and the problems of representation.* Paper presented at the Fifth International Congress of Qualitative Inquiry, University of Illinois, Urbana-Champaign.

Kaczynski, D., & Miller, E. (2004, September 1–3). *Evaluation team design considerations using NVivo.* Paper presented at the Fifth

International Strategies in Qualitative Research Conference, University of Durham, UK.

Kelle, U. (1995). Introduction: An overview of computer-aided methods in qualitative research. In U. Kelle (Ed.), *Computer-aided qualitative data analysis: Theory, methods and practice* (pp. 1–17). London: Sage.

Kelle, U. (1997). Theory building in qualitative research and computer programs for the management of textual data. *Sociological Research Online, 2*(2).

Kelly, K. (2008). One dead media. *The Technium Blog*. Available at http://www.kk.org/thetechnium/archives/2008/06/one_dead_media.php

Kuhn, S., & Davidson, J. (2007). Thinking with things, teaching with things: Enhancing student learning in qualitative research through reflective use of things. *Qualitative Research Journal. 7*(2), 63–75.

Kwan, M. (2008, May 17–21). *Geo-narrative: Extending Geographic Information Systems for narrative analysis in qualitative research.* Keynote presentation at "A Day in Technology in Qualitative Research," a preconference day of the Fourth International Congress on Qualitative Inquiry, University of Illinois, Urbana-Champaign.

Lapadat, J. (2008, May). *Liberatory technologies: Using multimodal literacies to connect, reframe, and build communities from the bottom up.* Paper presented at "A Day in Technology in Qualitative Research," a preconference day of the Fourth International Congress on Qualitative Inquiry, University of Illinois, Urbana-Champaign.

Layder, D. (1993). *New strategies in social research.* Cambridge, UK: Polity Press.

Lee, R., & Fielding, N. (1996). Qualitative data analysis: Representations of a technology: A comment on Coffey, Holbrook, and Atkinson. *Sociological Research Online, 1*(4).

Lewins, A., & Silver, C. (2007). *Using software in qualitative research: A step-by-step guide.* Thousand Oaks, CA: Sage.

Lin, P., Eisenberg, M., & Marino, J. (2010). *"Hi! I'm Harvey, a consent bot": How automating the consent process in SL addresses challenges of research online.* Poster session at the February 2010 iConference, University of Illinois, Urbana-Champaign.

Mangabeira, W., Lee, R. M., & Fielding, N. G. (2004). Computers and qualitative research: Adoption, use, and representation. *Social Science Computer Review, 22,* 167.

Miles, M., & Weitzman, E. (1994). Appendix: Choosing computer programs for qualitative data analysis. In M. B. Miles & M. A. Huberman (Eds.), *Qualitative data analysis* (2nd ed., pp. 311–317). Thousand Oaks, CA: Sage.

Morse, J. M., & Richards, L. (2002). *Readme first for a user's guide to qualitative methods.* Thousand Oaks, CA: Sage.

Nasukawa, T. (2006). *TAKMI (text analysis and knowledge mining) and sentiment analysis, IBM research, Tokyo Research Laboratory.* Paper presented at agenda-setting workshop, Bridging Quantitative and Qualitative Methods for Social Science Using Text Mining Techniques, at National Centre for e-Social Science, Manchester, UK.

Office of Public Sector Information. (1998). *Data Protections Act of 1998.* London: Her Majesty's Stationery Office.

Parmeggiani, P. (2007, April 18–20). *Using computer-assisted qualitative data analysis software for visual sociology.* Paper presented at the CAQDAS 2007 Conference: Advances in Qualitative Computing, Royal Holloway, University of London, Egham, UK.

QSR International. (2009, August 12). *What's new? Recipients of NVivo teaching grants announced.* Available at http://www.qsrinternational.com/news_whats-new_detail.aspx?view=168

Rettie, R., Robinson, H., Radke, A., & Ye, X. (2007, April 18–20). *The use of CAQDAS in the UK market research industry.* Paper presented at the CAQDAS 2007 Conference—Advances in Qualitative Computing, Royal Holloway, University of London, Egham, UK.

Richards, L. (1999). *Using NVivo in qualitative research.* Victoria, Australia: Qualitative Solutions and Research.

Richards, L. (2004, September 1–3). *Validity and reliability? Yes! Doing it in software.* Paper presented at the Fifth International Conference on Strategies in Qualitative Research: Using QSR NVivo and NUD*IST, University of Durham, UK.

Richards, L. (2005). *Handling qualitative data.* Thousand Oaks, CA: Sage.

Richards, L. (2006, September 13–15). *Farewell to the Lone Ranger? What happened to qualitative=small?* Paper presented at the Sixth International Strategies in Qualitative Research Conference, University of Durham, UK.

Richards, T. (2004, September 1–3). *Not just a pretty node system: What node hierarchies are really all about.* Paper presented at the Fifth International Conference on Strategies in Qualitative Research: Using QSR NVivo and NUD*IST, University of Durham, UK.

Richards, T., & Richards, L. (1994). Using computers in qualitative research. In N. K. Denzin & Y. S. Lincoln (Eds.), *Handbook of qualitative research* (pp. 445–462). Thousand Oaks, CA: Sage.

Richter, D., & Clary, L. (2004, September 1–3). *Using NVivo in the analysis of data from a site visit program.* Paper presented at the Fifth International Strategies in Qualitative Research Conference, University of Durham, UK.

Ritchie, J., & Lewis, J. (2003). *Qualitative research practice: A guide for social science students and researchers.* London: Sage.

Seidel, J. (1998). *Qualitative data analysis.* Available at http://www.qualisresearch.com (originally published as Qualitative data analysis, in *The Ethnograph v5.0: A Users Guide,* Appendix E, 1998, Colorado Springs, CO: Qualis Research)

Silver, C. (2009). *Choosing the right software for your research study: An overview of leading CAQDAS packages.* Paper presented at the 2009 Computer Assisted Qualitative Research Conference, Utrecht, The Netherlands.

Strauss, A., & Corbin, J. (1990). *Basics of qualitative research: Grounded theory procedures and techniques.* Newbury Park, CA: Sage.

Strike, K., Anderson, M., Curren, R., van Geel, T., Pritchard, I., & Robertson, E. (2002). *Ethical standards of the American Educational Research Association: Cases and commentary.* Washington DC: American Educational Research Association.

Teaching qualitative research with QSR software. (2003). *The Journal of the Association for Qualitative Research* [Special issue].

Tenner, E. (2005, February). Keeping tabs: The history of an information age metaphor. *Technological Review.*

Tesch, R. (1990). *Qualitative research: Analysis types and software tools.* Basingstoke, UK: Falmer.

Tutt, D., & Shaukat, M. (2007, April 18–20). *Evaluation of MiMeG in use: Technical and social issues in remote collaborative video analysis.* Paper presented at the CAQDAS 2007 Conference: Advances in Qualitative Computing, Royal Holloway, University of London, Egham, UK.

Vander Wal, T. (2007, February 2). *Folksonomy coinage and definition.* Available at: http://www.vanderwal.net/folksonomy.html

Vince, J., & Sweetman, R. (2006, September 29). *Managing large scale qualitative research: Two case studies.* Paper presented at Words Instead of Numbers: The Status of Software in the Qualitative Research World, the Association for Survey Computing, Imperial College London.

Weitzman, E. (2000). Software and qualitative research. In N. K. Denzin & Y. S. Lincoln (Eds.), *Handbook of qualitative research* (2nd ed., pp. 803–820). Thousand Oaks, CA: Sage.

Weitzman, E., & Miles, M. (1995). *Computer programs for qualitative data analysis.* Thousand Oaks, CA: Sage.

Qualitative Data Analysis Software

ATLAS.ti. *ATLAS.ti home page,* http://www.atlasti.com/

MAXQDA. http://www.MAXQDA.com/

NVIVO. A product of QSR International, http://www.qsrinternational.com/

QDAMiner. http://www.provalisresearch.com/QDAMiner/QDAMiner Desc.html

QSR. *QSR International home page,* http://www.qsrinternational.com/

Transana. http://www.transana.org/index.htm

Technologies, Technology Companies, and Research Resources

Amazon. *About Amazon,* http://www.amazon.com/Careers-Homepage/

A.nnotate. *About A.nnotate,* http://a.nnotate.com/about.html

Apple. *Apple science,* http://www.apple.com/science/

AT&T Labs, Inc. *Research,* http://www.research.att.com

CAQDAS. *CAQDAS Networking Project,* http://caqdas.soc.surrey .ac.uk/

DiRT. *Digital Research Tools wiki,* http://digitalresearchtools.pbworks .com

DReSS. http://www.esrcsocietytoday.ac.uk/esrcinfocentre/viewaward page.aspx?awardnumber=RES-149–25–0035

EverydayLives. http:// www.everydaylives.com

Google. *Google Labs,* http://www.googlelabs.com

IBM. *IBM Center for Social Research,* http://www.research.ibm.com/ social/index.html

Microsoft. *Microsoft Live Labs,* http://livelabs.com

MiMeg. http://www.esrcsocietytoday.ac.uk/esrcinfocentre/viewaward page.aspx?awardnumber=RES-149–25–0033

PARC (Palo Alto Research Center). http://www.parc.com

Telenor. *Telenor Research and Innovation,* http://www.telenor.com

39

THE POLITICS OF EVIDENCE[1]

Norman K. Denzin

There is a current dispute between qualitative and quantitative research. It is international, acrimonious, and there are elements of state-sponsored support "in the West" for a return to a kind of neopositivist quantitative inquiry.

—I. Stronach (2006, p. 758)

To serve evidence-based policymaking we probably need to invent a . . . myth for qualitative work, that is we too have clear-cut guidelines and criteria, maybe not randomized control trials, but we have our criteria.

—M. Hammersley (2005a, p. 4)

Qualitative researchers are caught in the middle of a global conversation concerning emerging standards and guidelines for conducting and evaluating qualitative inquiry (St.Pierre, 2006). This conversation turns on issues surrounding the politics and ethics of evidence, and the value of qualitative work in addressing matters of equity and social justice (Lather, 2006). In some senses, this is like old wine in new bottles, 1980s battles in a new century.

Like an elephant in the living room, the evidence-based model is an intruder whose presence can no longer be ignored. Within the global audit culture,[2] proposals concerning the use of Cochrane and Campbell criteria,[3] experimental methodologies, randomized controlled trials, quantitative metrics, citation analyses, shared data bases, journal impact factors, rigid notions of accountability, data transparency, warrantablity, rigorous peer-review evaluation scales, and fixed formats for scientific articles now compete, fighting to gain ascendancy in the evidence-quality-standards discourse (Feuer, Towne, & Shavelson, 2002; Lather, 2004; NRC, 2002; Thomas, 2004).

The interpretive community must mount an articulate critique of these external threats to our "collective research endeavor" (Atkinson & Delamont, 2006, p. 751; Freeman, deMarrais, Preissle, Roulston, & St.Pierre, 2007). We must create our own standards of quality, our own criteria.

I want to read the controversies surrounding this discourse within a critical pedagogical framework, showing their contradictions, their overlaps, the gaps that stand between them (Denzin, 2003). Standards for assessing quality research are pedagogies of practice, moral, ethical and political institutional apparatuses that regulate and produce a particular form of science, a form that may be no longer workable in a trans-disciplinary, global, and postcolonial world. Indeed, within the evidence-based community, there is the understanding that qualitative research does not count as research unless it is embedded in a randomized controlled trial (RCT)! Further, within this community, there are no agreed-upon procedures, methods, or criteria for extracting information from qualitative studies. These interpretations must be resisted.

In reviewing these multiple discourses, I hope to chart a path of resistance. Because the qualitative research community is not a single entity, guidelines and criteria of quality need to be fitted to specific paradigmatic and genre-driven concerns, e.g., grounded theory studies versus performance ethnographies. I favor flexible guidelines that are not driven by quantitative criteria. I seek a performative model of qualitative inquiry, a model that enacts a performance ethic based on feminist, communitarian assumptions.

I align these assumptions with the call by first and fourth world scholars for an indigenous research ethic (Bishop, 1998;

Rains, Archibald, & Deyhle, 2000; L. T. Smith, 1999). This call opens the space for a discussion of ethics, science, causality, and trust and a reiteration of moral and ethical criteria for judging qualitative research (Denzin, 2003, 2007; Denzin, Lincoln, & Giardina, 2006). I will conclude with a set of recommendations concerning review panels, scholarly associations, journals, and criteria for evaluating qualitative research.

▣ THE ELEPHANT IN THE LIVING ROOM

I agree with Atkinson and Delamont (2006) who state, "We are appalled by the absurd proposal that interpretive research should be made to conform to inappropriate definitions of scientific research. . . . Equally disturbing is the argument that qualitative research should not be funded if it fails to conform to these criteria" (p. 751; see also Erickson & Gutierrez, 2002, p. 221). Hammersley (2005a), in turn, observes that "[q]ualitative research tends to suffer by comparison with quantitative work because there is a myth that quantitative researchers have clear-cut guidelines which are available for use by policymakers (Was it a randomized controlled trial? Was there a control group?)" (p. 3).

Morse (2006a) extends the argument: " Indeed, qualitative inquiry falls off the positivist grid. Why it barely earns a Grade of C- on the Cochrane scale! It gets worse! It receives the 'does not meet evidence standard' on the 'What Works Clearinghouse' (WWC) Scale" (p. 396; Cheek, 2005, 2006).

Feuer et al. (2002) offer the counterargument:

Although we strongly oppose blunt federal mandates that reduce scientific inquiry to one method . . . we also believe that the field should use this tool in studies in education more often than is current practice. . . . Now is the time for the field to move beyond particularized views and focus on building a shared core of norms and practices that emphasize scientific principles. (p. 8)

A report by the National Center for Dissemination of Disability Research (2007) states, "We need criteria for comparing research methods and research evidence, we need terms like credibility (internal validity), transferability (external validity), dependability (reliability), confirmability (objectivity)" (n.p.).

A skeptic must ask, "Whose science? Whose scientific principles?"

▣ TWO OTHER ELEPHANTS

The elephant wears two other garments, the cloak of meta-analysis and the disguise of mix methods research. The meta-analysis disguise invites the production of systematic reviews that incorporate qualitative research into meta-analyses (Dixon-Woods et al., 2006). The mixed methods disguise revisits the concept of triangulation, asking how qualitative and quantitative methods can be made to work together (Moran-Ellis et al., 2006).

There are problems with both disguises. Meta-analyses of published articles hardly count as qualitative research in any sense of the term. The return to mixed methods inquiry fails to address the incommensurability issue—the fact that the two paradigms are in contradiction (Smith & Hodkinson, 2005). Any effort to circumvent this collision, through complementary strengths, single-paradigm, dialectical, or multiple-paradigm, mixed methods approaches seems doomed to failure (see Teddlie & Tashakkori, 2003, pp. 19–24).[4]

▣ WHOSE CRITERIA? WHOSE STANDARDS?

Extending J. K. Smith and Deemer (2000), within the qualitative inquiry community there are three basic positions on the issue of evaluative criteria: foundational, quasi-foundational, and non-foundational (see also Creswell, 2007; Guba & Lincoln, 1989, 2005; Lincoln & Guba, 1985; Spencer, Ritchie, Lewis, & Dillon, 2003). *Foundationalists,* including those that apply the Cochrane and Campbell Collaborations, are in this space, contending that *research is research,* quantitative or qualitative. All research should conform to a set of shared criteria (e.g., internal, external validity, credibility, transferability, confirmability, transparency, and warrantability (see Dixon-Woods, Shaw, Agarwal, & Smith, 2004; Dixon-Woods et al., 2006; Teddlie & Tashakkori, 2003).

Quasi-foundationalists contend that a set of criteria, or a guiding framework unique to qualitative research, needs to be developed. These criteria may include terms like reflexivity, theoretical grounding, iconic, paralogic, rhizomatic, and voluptuous validity (Eisner, 1991; Lather, 1993; Lincoln & Guba, 1985). In contrast, *non-foundationalists* stress the importance of understanding, versus prediction (Denzin, 1997; Wolcott, 1999). They conceptualize inquiry within a moral frame, implementing an ethic rooted in the concepts of care, love, and kindness (see also Christians, 2005).

▣ POLICY AND PRAXIS

Evaluative criteria, as pedagogical practices, are shaped by what is regarded as the proper relationship between qualitative inquiry and social policy. Within the critical qualitative inquiry community, at least four pedagogical stances, or identities, can be distinguished, each with its own history: (1) discipline-based qualitative research focused on accumulating fundamental knowledge about social processes and institutions; (2) qualitative policy research aimed at having an impact on current programs and practices; (3) critical qualitative approaches that disrupt and destabilize current public policy or social discourse; and (4) public intellectuals, public social scientists, and cultural critics who use qualitative inquiry and interpretive work to address current issues and crises in the public arena (Hammersley, 2005a, p. 3).

Hammersley (2005a,) cautions that, "We should not allow the close encounters promised by the notion of evidence-based policymaking, or even 'public social science,' to seduce us into illusions about ourselves and our work" (p. 5).

Torrance (2006) is quite assertive:

This new orthodoxy seems perversely and willfully ignorant of many decades of debate over whether, and if so in what ways we can conduct enquiry and build knowledge in the social sciences, pausing only to castigate educational research for not being more like … medical research. (p. 127)

▣ THE POLITICS OF EVIDENCE

The term *politics* (and ethics) *of evidence* is, as Morse (2006a) observes, an oxymoron in more than one way. Evidence "is something that is concrete and indisputable, whereas politics refers to 'activities concerned with the … exercise of authority [and power]'" (p. 395). Evidence in a countable or measurable sense is not something that all qualitative researchers attend to. Few critical ethnographers (Madison, 2005) think in a language of evidence; they think instead about experience, emotions, events, processes, performances, narratives, poetics, and the politics of possibility.

Moreover, evidence is never morally or ethically neutral. But, paraphrasing Morse (2006a), who quotes Larner (2004, p. 20), the politics and political economy of evidence is not a question of evidence or no evidence. It is rather a question of who has the power to control the definition of evidence, who defines the kinds of materials that count as evidence, who determines what methods best produce the best forms of evidence, whose criteria and standards are used to evaluate quality evidence. On this, Morse is quite clear: "Our evidence is considered soft … it is considered not valid, not replicable, not acceptable! We have failed to communicate the nature of qualitative evidence to the larger scientific community … we have failed to truly understand it ourselves" (pp. 415–416). The politics of evidence cannot be separated from the ethics of evidence.

▣ STATE- AND DISCIPLINE-SPONSORED EPISTEMOLOGIES

This ethical, epistemological, and political discourse is historically and politically situated. It plays out differently in each national context (see Atkinson & Delamont, 2006; Cheek, 2006; Gilgun, 2006; Morse, 2006a, 2006b; Preissle, 2006). In the United States, the United Kingdom, Continental Europe, New Zealand, and Australia, the conversation criss-crosses audit cultures, indigenous cultures, disciplines, paradigms, and epistemologies, as well as decolonizing initiatives. Depending on the nation-state, the discourse goes by various acronyms. In the United States, it is called SBR (scientifically based research), or SIE (scientific inquiry in education). In the United Kingdom, the model goes by the letters RAE (the British research assessment exercise), and in Australia, it goes by RQF for the research quality framework. All of these models are based, more or less, on the assumption that since medical research is successful, and randomized experimental designs are used and appreciated in medical science, this should be the blueprint for all good research (but see Timmermans & Berg, 2003).

There is not a single discourse. In the postpositivist, foundational, and quasi-foundational U.S. communities, there are multiple institutions (and conversations) competing for attention, including (1) the Institute of Education Science (IES) within the U.S. Department of Education; (2) the What Works Clearinghouse (WWC), funded by the IES; (3) the Cochrane-Campbell Collaboration (CCC), which contracts with the WWC; (4) the National Research Council-SBR framework (2002), which implements versions of CCC and WWC; (5) the recently IES-funded ($850,000) Society for Research on Educational Effectiveness (SREE); and (6) the 2006 standards for reporting adopted by the American Education Research Association (AERA), which explicitly addresses standards for qualitative research, some of which are contained in documents prepared by members of the Cochrane Qualitative Methods Group (Briggs, 2006).[5]

▣ NATIONAL RESEARCH COUNCIL

The federally funded National Research Council (NRC) scientifically based research (SBR), or evidence-based movement, argues that educational, health care, and other social problems can be better addressed if we borrow from medical science, and upgrade our methods and create new gold standards for evaluating evidence (National Research Council, 2002; NRC, 2005).

For this group, quality research is scientific, empirical, and linked to theory; it uses methods for direct investigation and produces coherent chains of causal reasoning based on experimental or quasi-experimental findings, offering generalizations that can be replicated and used to test and refine theory. If research has these features, it has high quality, and it is scientific (NRC, 2005).

In the United States, such research must also conform to the Office of Human Subject Research definition of scientific inquiry—namely, that scientific research is any activity designed to test an hypothesis, permit conclusions to be drawn, and thereby to develop or contribute to generalizable knowledge expressed in theories, principles, and statements of relationships. Research is described in a formal protocol that sets forth an objective and a set of procedures designed to reach that objective. (U.S. Code of Federal Regulations, Title 45, Part 46, as quoted in American Association of University Professors [AAUP], 2001, p. 55; see also AAUP, 1981, 2002, 2006)

Hand in glove, ethics and models of science now flow into one another. IRB panels can simultaneously rule on research that is ethically sound and of high quality. If these assumptions are allowed, we have lost the argument even before it starts. Cannella and Lincoln (2004) are clear on this point:

The NRC report is a U.S. government–requested project designed to clearly define the nature of research that is to be labeled as representing quality. . . . Accurately referred to as methodological fundamentalism . . . contemporary conservative research discourses . . . have ignored critical theory, race/ethic studies, and feminist theories and silenced the voices and life conditions of the traditionally marginalized. (p. 165; see also Feuer, 2006; Freeman et al., 2007; Hammersley, 2005a; St.Pierre, 2006; St.Pierre & Roulston, 2006)

▣ IMPLEMENTING THE NRC MODEL

Thirteen recommendations for implementing the NRC model are directed to federal funding agencies, professional associations and journals, and schools of education. These recommendations state that

Research agencies should

- Define and enforce better-quality criteria for peer reviewers;
- Ensure peer reviewer expertise and diversity;
- Create infrastructures for data sharing.

Publishers and professional associations should

- Develop explicit standards for data sharing;
- Require authors to make data available to other researchers;
- Create infrastructures for data sharing;
- Develop standards for structured abstracts;
- Develop a manuscript review system that supports professional development.

Schools of education and universities should

- Enable research competencies;
- Ensure that students develop deep methodological knowledge;
- Provide students with meaningful research experiences.

There are several problems with these NRC formulations and recommendations. I start with Maxwell (2004a, 2004b). He

unravels and criticizes the centrally linked assumptions in the model. His six points constitute powerful criticisms of SBR. He argues that the model assumes a narrow, regularity view of causation; privileges a variable-oriented, as opposed to a process-oriented, view of research; denies the possibility of observing causality in a single case; neglects the importance of context, meaning, and process as essential components of causal and interpretive analysis; erroneously asserts that qualitative and quantitative research share the same logic of inference; and presents a hierarchical ordering of methods for investigating causality, giving priority to experimental and other quantitative methods (2004b, p. 3).

Feuer et al. (2002) attempt to finesse this criticism, creating a special place for qualitative research, suggesting it can be used to capture the complexities involved in teaching, learning, and schooling, that is,

when a problem is poorly understood, and plausible hypotheses are scant—qualitative methods such as ethnographies . . . are necessary to describe complex phenomena, generate theoretical models and reframe questions. . . . We want to be explicit . . . [that] we do not view our strong support for randomized field trials and our equally strong argument for close attention to context . . . as incompatible. Quite the contrary: When properly applied, quantitative and qualitative research tools can both be employed rigorously and together. (p. 8)

Finessing aside, the NRC is clear on this point, that "a randomized experiment is the best method for estimating [causal] effects" (Feuer et al., 2002, p. 8).

Flashback to 1926. Déjà vu all over again. Lundberg (1926), sociology's archpositivist, is arguing against the use of the case method:

The case method is not in itself a scientific method at all, but merely the first step in the scientific method. . . . [T]he statistical method is the best, if not the only scientific method. . . . [T]he only possible question . . . is whether classification of, and generalizations from the data should be carried out by random, qualitative, and subjective method . . . or through the systematic, quantitative, and objective procedures of the statistical method. (p. 61)

Fast forward to 1966, to Howard S. Becker:

The life history method has not been much used by contemporary sociologists, a neglect which reflects a shift in the methodological stance of the researcher. Rigorous, quantitative, and (frequently) experimental designs have become the accepted modes of investigation. This situation is unfortunate because the life history, when properly conceived and employed, can become one of the sociologist's most powerful observational and analytic tools. (p. xviii)

The presumption that only quantitative data can be used to identify causal relationships is problematic. Maxwell (2004b) shows how the SBR model neglects meaning, context, and process. He demonstrates that causality can be identified (after Hume) in the single case; that is, multi-case, variable-based causal arguments are just one form of causal interpretation. Other causal, or quasi-causal, models of course are based on multi-variant, process, contextual, and interactionist-based assumptions. Further, causality as a type of narrative is only one form of interpretation. Autoethnographic, performative, arts-based, ethnodramatic, poetic, action-based, and other forms of narrative representation are equally powerful methods and strategies of analysis and interpretation.

In addition to Maxwell's six basic criticisms, I add the following. First, amazingly, there is little attention given to the process by which evidence is turned into data. This is not a simple process, and is not accomplished by waving a wand over a body of observations. Second, there is also no detailed discussion of how data are to be used to produce generalizations, test and refine theory, and permit causal reasoning. It is clear, though, that data become a commodity that does several things. That is, third, evidence as data carries the weight of the scientific process. This process works through a self-fulfilling, self-validating process. You know you have quality data that are scientific when you have tested and refined your theory. How you have addressed problems in the real world remains a mystery.

Fourth, the focus on data sharing is critical, and of central concern. It is assumed that quality data can be easily shared. But complex interpretive processes shape how evidence is turned into data, and how data, in turn, are coded, categorized, labeled, and assembled into data banks (Charmaz, 2005). Data are not silent. Data are commodities, produced by researchers, perhaps owned by the government or by funding agencies. What would it mean to share my data with you? Why would I want to do this? If I own my data, I want to have ownership over how it is used, including what is published from it. The injunction to engage in data sharing requires amplification. Data sharing involves complex moral considerations that go beyond sending a body of coded data to another colleague.

Fifth, money and concerns for auditing from the audit culture seem to drive the process. This is evidenced in the emphasis placed on funding and quality peer reviews. If quality data can be produced and then shared, then granting agencies get more science for less money. However, in order for greater data sharing to occur, more quality projects need to be funded. For this to happen, granting agencies need a better peer review system with better-trained reviewers, who are using more clearly defined rating scale levels. Reviewers will be helped if researchers write proposals that use rigorous methodologies and the very best research designs. Such projects will surely have high standards of evidence. Thus does the self-fulfilling process reproduce itself. We know we are getting quality science of the highest order because we are using methods of the

highest order. Reviewers can easily identify such work. The blind peer review, based on presumptions of objectivity, is the key to this system.[6]

The peer review system is not immune to political influence. Kaplan (2004) has demonstrated that the George W. Bush administration systematically stacked federal advisory and peer review committees with researchers whose views matched the president's on issues ranging from stem-cell research to ergonomics, faith-based science, AIDS, sex education, family values, global warming, and environmental issues in public parks (see also Monastersky, 2002).

▣ SREE

The Society for Research on Educational Effectiveness (SREE) extends the federally sponsored NRC agenda. It appears to oppose recent efforts within AERA to soften NRC guidelines (see below). The code words for SREE, which plans its own journal (*Journal of Research on Educational Effectiveness—* JREE), handbook (*Handbook of Research on Educational Effectiveness*), and electronic journal (*Research Notes on Educational Effectiveness*), are rigorous research design and randomized control experiment. The mission of SREE is

> to advance and disseminate research on the causal effects of education interventions, practices, programs, and policies. As support for researchers who are focused on questions related to educational effectiveness, the Society aims to: 1) increase the capacity to design and conduct investigations that have a strong base for causal inference, 2) bring together people investigating cause-and-effect relations in education, and 3) promote the understanding and use of scientific evidence to improve education decisions and outcomes. (www.sree-net.org; see also Viadero, 2006)[7]

There is no place in SREE here for qualitative research. This is hardcore SBR: evidence-based inquiry. Scientific research becomes a commodity to be sold in a new journal, a commodity that serves and embodies the interests of educational science as narrowly defined.

▣ THE COCHRANE, CAMPBELL, WHAT WORKS CLEARING HOUSE COLLABORATIONS

The Cochrane, Campbell, and What Works Clearinghouse Collaborations are inserting themselves into the qualitative research conversation. All three represent state-sponsored projects. All three are dedicated to producing so-called scientific peer reviews of quality (evidence-based) research that can be used by policy makers. The Cochrane Qualitative Methods Group focuses on methodological matters arising from the inclusion of findings from qualitative studies into systematic reviews of evidence-based inquires. The Campbell Methods Group focuses on methodological issues associated with process evaluations, which use mixed methods, while including evidence gathered via qualitative methods. It is understood that qualitative research can help in understanding how an intervention is experienced, while providing insight into factors that might hinder successful implementation.

Randomized controlled trials are central to all three collaborations. Hence, qualitative evidence is of primary interest only when it is included as a data gathering technique in an experimental, or quasi-experimental, study (Briggs, 2006). There is some debate on this point, that is, whether "only qualitative research embedded within relevant RCTs should be included" (Briggs, 2006). The Campbell Collaboration only includes qualitative materials if they are part of controlled observations (Davies, 2004). However, there is no consensus on how to include qualitative evidence in such work—namely, how to identify, record, appraise, and extract data from qualitative studies.

◨ APPRAISAL TOOLS

Enter CASP—the Critical Appraisal Skills Program (Briggs, 2006), which was developed in conjunction with the Cochrane Qualitative Research Methods Group (CQRMG). The Cochrane Group (Briggs, 2006) has a broad, but conventional definition of qualitative research, encompassing specific methods (interviews, participant and nonparticipant observation, focus groups, ethnographic fieldwork), data types (narrative), and forms of analysis (ethnography, grounded theory, thematic categories).

CASP, like any number of other checklists (Dixon-Woods, et al., 2004; Jackson & Waters, 2005; Popay, Rogers, & Williams, 1998; Spencer et al., 2003), is an assessment tool developed for those unfamiliar with qualitative research. The tool presents a series of questions focused around three broad issues: rigor, credibility, and relevance. Ten questions concerning aims, methodology, design, subject recruitment, data collection, researcher–participant relationship, ethics (IRBs), data analysis, statement of findings, and value of research are asked. The reviewer of a study writes comments on each of these issues.

CASP implements a narrow model of qualitative inquiry. Methods are not connected to interpretive paradigms (e.g., feminism, critical theory). Multiple strategies of inquiry and analysis (case or performance studies, narrative inquiry, critical ethnography) go unidentified. Nor is the complex literature from within the interpretive tradition on evaluating qualitative research addressed (see Christians, 2005). Thus, CASP offers the reviewer a small, ahistorical tool kit for reading and evaluating qualitative studies.

Checklists

Here, Hammersley (2005a) is again relevant. This is the myth of the checklist, the myth of the guideline. Consider the guidelines prepared for the British Cabinet Office (Spencer et al., 2003). This is another checklist with 16 categories (scope, timetable, design, sample, data collection, analysis, ethics, confirmability, generalizability, credibility, etc.), 80 specific criteria (clearly stated hypotheses, outcomes, justification of analysis methods, triangulation, etc.), and 35 broad criteria (explicit aims, appropriate use of methods, assessment of reliability and validity, etc.).

This is old-fashioned postpositivism, applying a soft quantitative grid (confirmability, hypotheses, credibility) to qualitative research. But there is more going on. Like CASP, the Spencer et al. (2003) tool kit introduces the notion of credibility, that is, whether the findings can be trusted. If they can be trusted, they must be confirmable, valid, and reliable, which means they can be generalized. If they are not credible, the whole house of cards falls down.

Torrance (2006) exposes the underlying theory at work here, noting that "it is a traditional, positivist model, that is the truth is out there to be discovered" (p. 128). Yet, as he observes, "these scholars still cannot solve the problem of epistemological incommensuration . . . but . . . this is little more than experts 'rating' qualitative evidence on an agreed scale so it can be included in meta-analyses of effect sizes" (p. 140).

◨ AERA

The American Education Research Association (AERA, 2006, 2008) has recently added its collective voice to the conversation, supplementing and departing from the NRC recommendations. Two sets of guidelines, one for empirical research, the other for humanities-based work, have been offered. Both sets are intended to help authors and journal editors and reviewers who may not be familiar with expectations guiding such work. They are also intended to foster excellence in the production of high-quality research.

Standards for Empirical Social Science Research

Two global standards are offered for reporting empirical research: warrantability and transparency (AERA, 2006).[8,9] Reports of research should be warranted, that is, adequate evidence, which would be credible (internal validity), should be provided to justify conclusions. Reports should be transparent, making explicit the logic of inquiry used in the project. This method should produce data that have external validity; reliability; and confirmability, or objectivity. Like the NRC guidelines, these standards are to be used by peer reviewers, research scholars, and journal publishers, and in graduate education programs where researchers are trained.

There is extensive discussion of quantitative procedures (AERA, 2006, pp. 6–10), but trust is not an issue.

Trust

Trust *is* an issue for qualitative researchers. The AERA (2006) report is explicit, asserting that

> It is the researcher's responsibility to show the reader that the report can be trusted. This begins with the description of the evidence, the data, and the analysis supporting each interpretive claim. The warrant for the claims can be established through a variety of procedures including triangulation, asking participants to evaluate pattern descriptions, having different analysts examine the same data (independently and collaboratively), searches for disconfirming evidence and counter-interpretations. (p. 11)

This is all clear enough, but these validating procedures and standards are not held up for quantitative researchers. When qualitative evidence does not converge, the report recommends that

> critical examination of the preexisting perspective, point of view, or standpoint of the researcher(s), of how these might have influenced the collection and analysis of evidence, and of how they were challenged during the course of data collection and analysis, is an important element in enhancing the warrant of each claim. (AERA, 2006, p. 11)

Here is the heart of the matter. The perspective of the qualitative researcher can influence the collection of evidence in such a way as to introduce a lack of trust into the research process. That presence potentially undermines the credibility and warrantability of the report. But why would the qualitative researcher's effects on the research process be greater or less than the effects of the quantitative researcher? Doesn't the quantitative researcher have an effect on the collection, analysis, and interpretation of evidence, including deciding what is evidence?

The 2006 AERA recommendations call for the responsible use of quasi-foundational tools; that is, threats to trust can be overcome. Transparency—that is, trust—is increased by clearly discussing the process of interpretation, highlighting the evidence and alternative interpretations that serve as a warrant for each claim, providing contextual commentary on each claim. When generalizations extend beyond a specific case, researchers must clearly indicate the sampling frame, population, individuals, contexts, activities, and domains to which the generalizations are intended to apply (external validity). The logic supporting such generalizations must be made clear.

A sleight of hand is at work in the AERA recommendations. The intent of the report is now clear. Two things are going on at once—a familiar pattern. Qualitative research is downgraded to the status of a marginal science, second-class citizenship. Since it lacks trustworthiness, it can be used for discovery purposes, but not for the real work of science, which is verification. Only under the most rigorous of circumstances can qualitative research exhibit the qualities that would make it scientific, and even then, trust will be an issue. Trust becomes a proxy for quality; transparency and warranted evidence function as proxies for objectivity.

Clearly, AERA wants a space for qualitative research that is not governed by the narrow NRC experimental and quasi-experimental guidelines. We all want this. To its credit, AERA wants a broad-based, multimethod concept of quality. But they falter in asserting that empirical research reports should be warranted and transparent. These are criteria for doing business as usual. No wonder SREE was created. AERA's educational science does not require randomized controlled experiments. SREE's does.

Rereading Trust and Ethics

Trust in this discourse resurfaces as a proxy for more than quality. It spills over to the researcher who does research that lacks trust. Untrustworthy persons lie, misrepresent, cheat, engage in fraud, or alter documents. They are not governed by measurement and statistical procedures that are objective and free of bias. They may not be shady characters; they may be well-intended, gifted actors, poets, fiction writers, or performers, but they are not scientists! Qualitative researchers are not to be trusted because their standpoints can influence what they study and report. Somehow quantitative researchers are freed from these influences. This of course is a sham!

By implication, qualitative scientists are being charged with fraud, with misrepresenting their data. This may be because many qualitative researchers do not have data and findings, tables and charts, statistics and numbers. We have stories, narratives, excerpts from interviews. We perform our interpretations and invite audiences to experience these performances, to live their way into the scenes, moments, and lives we are writing and talking about. Our empirical materials can't be fudged, misrepresented, altered, or distorted because they are life experiences. They are ethnodramas.

Apples Turned Into Oranges: Turning Interpretations Into Data

Like the NRC, AERA's ethical guidelines focus on issues relevant to reporting results. Authors have an obligation to address the ethical decisions that shaped their research, including how the inquiry was designed, executed, and organized. Incentives for participating, consent waivers and confidentiality agreements, and conflicts of interest should be presented and discussed. Reporting should be accurate, free of plagiarism, fully accessible to others, and without falsification or fabrication of data or results. Data should be presented in such a way that any qualified researcher with a copy of the relevant data could reproduce the results.

Thus are interpretive materials turned into data. The interpretive process becomes an exercise in seeking patterns of evidence, presenting evidence in a way that will engender trust on the part of the reader, while avoiding charges of misrepresentation or fabrication (more on ethics below). But this is not how qualitative researchers work.

Standards for Reporting on Humanities-Oriented Research in AERA Publications[10]

The 2008 Draft of Standards for Humanities-Oriented Research extends the place of qualitative inquiry in educational research.[11] The document recognizes that traditional social science standards for empirical research cannot be automatically applied to humanities-oriented research. The document focuses on five genres of humanities-linked inquiry: philosophy, history, arts-based educational research (ABER), literary studies, and studies of the politics of knowledge.[12] Space prohibits a discussion of all five genres. I will focus on ABER because of its overlap with experimental forms of qualitative inquiry (see Barone, 2001; Cahnmann-Taylor & Siegesmund, 2008; Eisner, 1991; Finley, 2008; Leavy, 2009; Richardson, 2000a, 2000b).

Two strands of arts-based inquiry, the humanistic or traditional, and the activist, critical pedagogical, can be identified. The traditional strand, the one emphasized in the AERA (2008) report, contrasts empirical and artistic approaches to qualitative research. Dance, film, poetry, drama, and the plastic arts are used to explore various facets of the human condition: the relationship between reason and emotion, the ethical life, self, identity, and meaning (p. 3; see also Finley, 2008). Activist, radical, performative, ethical, and revolutionary forms of arts-based work, projects which disrupt, interrupt and challenge structures of oppression, are not taken up (see Finley, 2008).

The report defers to those forms and methods of humanities-oriented research that are empirical and use interpretive methods in the analysis of texts, text analogues, and textual artifacts (AERA, 2008, p. 4). It is asserted that such work is inextricably empirical, which means it can be counted, assessed, and evaluated in terms of a politics of evidence. This means there is overlap between empirical and humanities-oriented research (p. 4).

Accordingly, the standards for evaluating humanistic work overlap with those applied to empirical work. Seven standards, each with a series of substandards[13] that elaborate the major standard, are offered: (1) significance, (2) conceptualization, (3) methods, (4) substantiation, (5) coherence, (6) quality of communication, and (7) ethics.[14] (These could have been included in the 2006 empirical standards document.)

Substantiation and coherence are key standards, and they are intertwined. Together they establish the warrant for the arguments in a text, the adequacy or credibility of its interpretations, the quality and use of evidence, its transparency, and critical self-awareness. A warrantable humanities-based text, like its empirical counterpart, uses evidence that justifies its conclusions. Such a text demonstrates internal and external coherence, offering compelling confirming and disconfirming evidence, and an awareness of competing, external perspectives.

What if a work is deliberately not empirical? What if it disrupts the concept of the empirical. What if it disallows the concept of the text, and turns the text into a performance, into a site where meaning is multiple, plural, and unclear? In such a case, an empirical-textual model no longer applies, and the standards of coherence and substantiation no longer apply.

Reading the New Standards

As with the discussion of qualitative research, it is clear that AERA wants a space for humanities-based inquiry that is not governed by narrow SBR guidelines. But the window it creates for this form of inquiry is quite narrow. AERA wants to hold humanistic inquiry to a modified set of evidence-based standards. Underneath its claim for inclusiveness, it brings the same criteria—transparency, coherence, evidence, trust—to the humanities that it applied to qualitative inquiry.

Its discussion of arts-based educational research (ABER) ignores, as it did in the discussion of qualitative inquiry, a large methodological and interpretive literature concerning empowerment discourses, critical performance ethnography, art-for-social-action purposes, dialogic spaces, public art, censorship, and neoliberal forms of governmental regulation (see Finley, 2003, 2005). It seems that this document was produced outside the discourse it was intended to regulate.

The effect, however, is disarming. There is the impression that we are one big happy family, with different people doing different things. That is not the case. In fact, we are better described as a "house divided."[15] Accordingly, we should resist the "new orthodoxy." By asserting that everything we do is inextricably empirical, the AERA seeks to diminish, if not erase, hard-fought distinction, and all in the name of science!

◼ ◼ ◼

It is as if the NRC, SREE, and AERA guidelines were written in a time warp. Over the last three decades, the field of qualitative research has become a interdisciplinary field in its own right. The interpretive and critical paradigms, in their multiple forms, are central to this movement. Complex literatures are now attached to research methodologies, strategies of inquiry, interpretive paradigms, and criteria for reading and evaluating inquiry itself. Sadly, little of this literature is evident in any of the recent national documents. It seems that the qualitative community is hemmed in from all sides. But before this judgment is accepted, the "for whom" question must be asked—that is, high-quality science, or evidence, for whom? (Cheek, 2006). NRC, AERA, and SREE's umbrellas are too small. We need a larger tent.

▣   THE QUALITATIVE INQUIRY COMMUNITY

There are tensions over the politics of evidence within the interpretive community: (1) Interpretivists dismiss postpositivists, (2) poststructuralists dismiss interpretivists, and now (3) the postinterpretivists dismiss the interpretivists (Preissle, 2006, p. 692; see also Hammersley, 2005b; Hodkinson, 2004; MacLure, 2006). Some postpositivists are drawn to the SBR standards movement, seeking to develop mixed or multiple methodological strategies that will conform to the new demands for improving research quality. Others reject the gold standard movement, and argue for a set of understandings unique to the interpretive, or postinterpretive, tradition (St.Pierre & Roulston, 2006). Atkinson and Delamont (2006) call for a return to the classics in the Chicago School tradition. The American Education Research Association (2006) aims to strike a middle ground, neither too postpositivist nor too interpretivist.

The immediate effects of this conversation start at home, in departments and in graduate education programs where PhD's are produced and tenure for qualitative research scholars is granted. Many fear that the call for SBR will drown out instruction, scholarship, and the granting of tenure in the qualitative tradition, or confine them to a narrow brand of interpretive work (Eisenhart, 2006). Worse yet, it could lead to a narrow concept of orthodoxy.[16]

▣   RESISTANCE

We must resist the pressures for a single gold standard, even as we endorse conversations about evidence, inquiry, and empirically warranted conclusions (Lincoln & Cannella, 2004). We cannot let one group define the key terms in the conversation. To do otherwise is to allow the SBR group to define the moral and epistemological terrain that we stand on. Neither they nor the government own the word *science*. Habermas (1972) anticipated this nearly 40 years ago:

The link between empiricism, positivism and the global audit culture is not accidental and it is more than just technical. Such technical approaches deflect attention away from the deeper issues of value and purpose. They make radical critiques much more difficult to mount . . . and they render largely invisible partisan approaches to research under the politically useful pretense that

judgments are about objective quality only. In the process human needs and human rights are trampled upon and democracy as we need it is destroyed. (p. 122)

Bourdieu (1998) elaborates:

The dominants, technocrats, and empiricists of the right and the left are hand in glove with reason and the universal. . . . More and more rational, scientific technical justifications, always in the name of objectivity, are relied upon. In this way the audit culture perpetuates itself. (p. 90)

There is more than one version of disciplined, rigorous inquiry—counter-science, little science, unruly science, practical science—and such inquiry need not go by the name of science. We must have a model of disciplined, rigorous, thoughtful, reflective inquiry, a "postinterpretivism that seeks meaning but less innocently, that seeks liberation but less naively, and that . . . reaches toward understanding, transformation and justice" (Preissle, 2006, p. 692). It does not need to be called a science, contested or otherwise, as some have proposed (Eisenhart, 2006; Preissle, 2006; St.Pierre & Roulston, 2006).

Lather (2006) extends the argument:

The commitment to disciplined inquiry opens the space for the pursuit of 'inexact knowledges' (p. 787), a disciplined inquiry that matters, applied qualitative research . . . that can engage strategically with the limits and the possibilities of the uses of research for social policy (p. 789). The goal is a critical "counter-science." . . . that troubles what we take for granted as the good in fostering understanding, reflection and action (p. 787). We need a broader framework where such key terms as science, data, evidence, field, method, analysis, knowledge, truth, are no longer defined from within a narrow policy-oriented, positivistic framework. (pp. 787 & 789)

▣   A NEW TERRAIN: TROUBLE WITH THE ELEPHANT

Let's return to the elephant in the living room. Consider the parable of the blind men and the elephant. Lillian Quigley's children's book, *The Blind Men and the Elephant,* is a retelling of an ancient fable about six blind men who visit the palace of the Rajah. There, the men have their first encounter with an elephant. As each man touches the animal in turn, he reports to the others what he feels:

The first blind person touches the side of the elephant and reports that it feels like a wall. The second touches the trunk and says an elephant is like a snake. The third man touches the tusk and says an elephant is like a spear. The fourth person touches a leg and says it feels like a tree. The fifth man touches an ear and says it must be a fan, while the sixth man touches the tail and says how thin, an elephant is like a rope.

There are multiple versions of the elephant in this parable, multiple lessons. We can never know the true nature of things. We are each blinded by our own perspective. Truth is always partial.

To summarize,

Truth One: The elephant is not one thing. If we call SBR the elephant, then according to the parable, we can each know only our version of SBR. For SBR advocates, the elephant is two things: an all-knowing being who speaks to us, and a way of knowing that produces truths about life. How can a thing be two things at the same time?

Truth Two: For skeptics, we are like the blind persons in the parable. We only see partial truths. There is no God's view of the totality, no uniform way of knowing.

Truth Three: Our methodological and moral biases have so seriously blinded us that we can never understand another blind person's position. Even if the elephant called SBR speaks, our biases may prohibit us from hearing what she says. In turn, her biases prevent her from hearing what we say.

Truth Four: If we are all blind, if there is no God, and if there are multiple versions of the elephant, then we are all fumbling around in the world just doing the best we can.

▣ TWO OTHER VERSIONS OF THE ELEPHANT

The version above is the blind person's version of the elephant. There are at least two other versions, 2.1 and 2.2. Both versions follow from the version above, but now the elephant refers to a painfully problematic situation, thing, or person in one's life space. Rather than confront the thing, and make changes, people find that it is easier to engage in denial, to act like the elephant isn't in the room. This can be unhealthy because the thing may be destructive. It can produce codependency. We need the negative presence of the elephant in order to feel good about ourselves.

This cuts two ways at once, hence versions 2.1 and 2. 2. In **Fable 2.1,** SBR advocates treat qualitative research as if it were an elephant in their living room. They have ignored our traditions, our values, our methodologies; they have not read our journals, or our handbooks, or our monographs. They have not even engaged our discourses about SBR. Like the six blind men, they have acted as if they could create us in their own eye. They say we produce findings that cannot be trusted, we are radical relativists, we think anything goes. They dismiss us when we tell them they only know one version of who we are. When we tell them their biases prevent them from understanding what we do, they assert that we are wrong and they are right.

In **Fable 2.2,** the elephant is located in our living room. With notable exceptions, we have tried to ignore this presence. Denial has fed codependency. We need the negative presence of SBR to

define who we are. For example, we have not taken up the challenge of better educating policy makers, showing them how qualitative research and our views of practical science, interpretation, and performance ethics can positively contribute to projects embodying restorative justice, equity, and better schooling (Preissle, 2006; Stanfield, 2006). We have not engaged policy makers in a dialogue about alternative ways of judging and evaluating quality research, nor have we engaged SBR advocates in a dialogue about these same issues (but see St.Pierre, 2006). And, they have often declined the invitation to join us in a conversation. As a consequence, we have allowed the SBR elephant to set the terms of the conversation.

If we are to move forward positively, we have to get beyond Fable 2.2, beyond elephants, blind persons, and structures of denial. We must create a new narrative, a narrative of passion and commitment, a narrative that teaches others that ways of knowing are always already partial, moral, and political. This narrative will allow us to put the elephant in proper perspective. Here are some of the certain things we can build our new fable around:

1. We have an ample supply of methodological rules and interpretive guidelines.

2. They are open to change and to differing interpretation, and this is how it should be.

3. There is no longer a single gold standard for qualitative work.

4. We value open peer reviews in our journals.

5. Our empirical materials are performative. They are not commodities to be bought, sold, and consumed.

6. Our feminist, communitarian ethics are not governed by IRBs.

7. Our science is open-ended, unruly, and disruptive (MacLure, 2006; Stronach, Garratt, Pearce, & Piper, 2007).

8. Inquiry is always political and moral.

9. Objectivity and evidence are political and ethical terms.

We live in a depressing historical moment of violent spaces, unending wars against persons of color, repression, the falsification of evidence, the collapse of critical, democratic discourse, and repressive neoliberalism disguised as dispassionate objectivity prevails. Global efforts to impose a new orthodoxy on critical social science inquiry must be resisted; a hegemonic politics of evidence cannot be allowed. Too much is at stake.

▣ NOTES

1. This chapter revises and extends arguments in Denzin (2009).

2. Audit culture refers to a technology and a system of accounting that measures outcomes and assesses quality in terms of so-called

objective criteria such as test scores. Some argue that the global audit culture implements conservative, neoliberal conceptions of governmentality (Bourdieu, 1998; Habermas, 1972, 2006).

3. Lather (2004) offers a history and critical reading of this alphabet soup of acronyms: CC (Cochrane Collaboration), C2 (Campbell Collaboration), AIR (American Institutes for Research), WWC (What Works Clearinghouse), IES (Institute of Education Science). There has been a recent move within CC and C2 to create protocols for evaluating qualitative research studies (see Briggs, 2006; National CASP Collaboration, 2006; see also Bell, 2006, and below).

4. Over the past four decades, the discourse on triangulation, multiple operationalism, and mixed methods models has become quite complex and nuanced (see Saukko, 2003, and Teddlie & Tashakkori, 2003, for reviews). Each decade has taken up triangulation and redefined it to meet perceived needs.

5. The common thread that exists between WWC and C2 is the No Child Left Behind (NCLB) and Reading First Acts. These acts required a focus on identifying and using scientifically based research in designing and implementing educational programs (What Works Clearinghouse).

6. Ironically, the blind peer review recommendation flies in the face of a recent CC study, which argues that there is little hard evidence to show that blind peer reviews improve the quality of research (Jefferson, Rudin, Brodney Folse, & Davidoff, 2003; White, 2003; see also Judson, 2004, pp. 244–286). Indeed, the Cochrane Collaboration researchers found few studies examining this presumed effect.

7. Their first annual conference (March 2–4, 2008) was outcomes based, calling for rigorous studies of reading, writing, and language skills; mathematics and science achievement; social and behavioral competencies; and dropout prevention and school completion.

8. Warrantability and transparency are key terms in the new managerialism, which is evidence based, and audit driven; that is, policy decisions should be based on evidence that warrants policy recommendations, and research procedures should be transparently accountable (Hammersley, 2004). Transparency is also a criterion advanced by the Cochrane Qualitative Methods Group (Briggs, 2006).

9. The reporting standards are then divided into eight general areas: problem formation, design, evidence (sources), measurement, analysis and interpretation, generalization, ethics, title and abstract.

10. I thank Kenneth Howe for his comments on this section. He was a member of this AERA committee.

11. The revised and finalized version of these standards is published in *Education Researcher, 38*(6), August/September 2009, 481–486.

12. The report reduces interpretive work to three generic categories, or kinds of objects: texts, text analogues (reports, narratives, performances, rituals), and artifacts (works of art).

13. For example, the significance standard has four levels, involving topic and scholarly contribution. The methods standards have three levels, conceptualization has five levels, and so forth.

14. Ethically, humanities research, as with empirical research, should be carried out in accordance with IRB approval. Scholars should announce their values and discuss any conflicts of interest that could influence their analysis.

15. I thank Ken Howe for this phrase.

16. In the last two decades, qualitative researchers have gone from having fewer than 3 journals dedicated to their work to now having 20 or more (Chenail, 2007).

▣ REFERENCES

American Association of University Professors. (1981). Regulations governing research on human subjects: Academic freedom and the institutional review board. *Academe, 67,* 358–370.

American Association of University Professors. (2001). Protecting human beings: Institutional review boards and social science research. *Academe, 87*(3), 55–67.

American Association of University Professors. (2002). Should all disciplines be subject to the common rule? Human subjects of social science research. *Academe, 88*(1), 1–15.

American Association of University Professors, Committee A. (2006). *Report on human subjects: Academic freedom and the institutional review boards.* Available at http://www.aaup.org/AAUP/About/committees/committee+repts/CommA/

American Education Research Association. (2006). *Standards for reporting on empirical social science research in AERA publications.* Available at http://www.aera.net/opportunities/?id =1480

American Education Research Association. (2008, August/September). Standards for reporting on humanities-oriented research in AERA publications. *Educational Researcher, 38*(6), 481–486.

Atkinson, P., & Delamont, S. (2006, November/December). In the roiling smoke: Qualitative inquiry and contested fields. *International Journal of Qualitative Studies in Education, (19)*6, 747–755.

Barone, T. (2001). *Touching eternity: The enduring outcomes of teaching.* New York: Teachers College Press.

Becker, H. S. (1966). Introduction. In C. Shaw, *The jack-roller* (pp. v–xviii). Chicago: University of Chicago Press.

Bell, V. (2006). *The Cochrane Qualitative Methods Group.* Available at http://www.lancs.ac.uk/fass/ihr/research/public/cochrane.htm

Bishop, R. (1998). Freeing ourselves from neo-colonial domination in research: A Maori approach to creating knowledge. *International Journal of Qualitative Studies in Education, 11,* 199–219.

Bourdieu, P. (1998). *Practical reason.* Cambridge, UK: Polity.

Briggs, J. (2006). *Cochrane Qualitative Research Methods Group.* Available at http://www.joannabriggs.eduau/cqrmg/role.html

Cahnmann-Taylor, M., & Siegesmund, R. (Eds.). (2008). *Arts-based research in education: Foundations for practice.* New York: Routledge.

Cannella, G. S., & Lincoln, Y. S. (2004, April). Dangerous discourses II: Comprehending and countering the redeployment of discourses (and resources) in the generation of liberatory inquiry. *Qualitative Inquiry, 10*(2), 165–174.

Charmaz, K. (2005). Grounded theory in the 21st century: A qualitative method for advancing social justice research. In N. K. Denzin & Y. S. Lincoln (Eds.), *The SAGE handbook of qualitative research* (3rd ed., pp. 507–535). Thousand Oaks, CA: Sage.

Cheek, J. (2005). The practice and politics of funded qualitative research. In N. K. Denzin & Y. S. Lincoln (Eds.), *The SAGE handbook of qualitative research* (3rd ed., pp. 387–410). Thousand Oaks, CA: Sage.

Cheek, J. (2006, March). What's in a number? Issues in providing evidence of impact and quality of research(ers). *Qualitative Health Research, 16*(3), 423–435.

Chenail, R. J. (2007). Qualitative research sites. *The Qualitative Report: An Online Journal.* Available at http://www.nova.edu/sss/QR/web.html

Christians, C. (2005). Ethics and politics in qualitative research. In N. K. Denzin & Y. S. Lincoln (Eds.), *The SAGE handbook of qualitative research* (3rd ed., pp. 139–164). Thousand Oaks, CA: Sage.

Creswell, J. W. (2007). *Qualitative inquiry and research design: Choosing among five approaches* (2nd ed.). Thousand Oaks, CA: Sage.

Davies, P. (2004). Systematic reviews and the Campbell Collaboration. In G. Thomas & R. Pring (Eds.), *Evidence-based practice in education* (pp. 21–33). New York: Open University Press.

Denzin, N. K. (1997). *Interpretive ethnography.* Thousand Oaks, CA: Sage.

Denzin, N. K. (2003). *Performance ethnography: Critical pedagogy and the politics of culture.* Thousand Oaks, CA: Sage.

Denzin, N. K. (2007). The secret Downing Street memo, the one percent doctrine, and the politics of truth: A performance text. *Symbolic Interaction, 30*(4) 447–461.

Denzin, N. K. (2009). The elephant in the living room: Notes on the politics of inquiry. *Qualitative Research, 9*(1), 139–160.

Denzin, N. K., & Giardina, M. D. (2006). Qualitative inquiry and the conservative challenge. In N. K. Denzin & M. D. Giardina (Eds.), *Qualitative inquiry and the conservative challenge* (pp. ix–xxxi). Walnut Creek, CA: Left Coast Press.

Denzin, N. K., & Lincoln, Y. S. (2005). The discipline and practice of qualitative research. In N. K. Denzin & Y. S. Lincoln (Eds.), *The SAGE handbook of qualitative research* (3rd ed., pp. 1–32). Thousand Oaks, CA: Sage.

Denzin, N. K., Lincoln, Y. S., & Giardina, M. D. (2006, November/December). Disciplining qualitative research. *International Journal of Qualitative Studies in Education, 19*(6), 769–782.

Dixon-Woods, M., Bonas, S., Booth, A., Jones, D. R., Miller, T., Sutton, A. J., et al. (2006, February). How can systematic reviews incorporate qualitative research? A critical perspective. *Qualitative Research, 6*(1), 27–44.

Dixon-Woods, M., Shaw, R. L., Agarwal, S., & Smith, J. A. (2004). The problem of appraising qualitative research. *Quality & Safety in Health Care, 13,* 223–225.

Eisenhart, M. (2006, November/December). Qualitative science in experimental time. *International Journal of Qualitative Studies in Education, 19*(6), 697–708.

Eisner, E. W. (1991). *The enlightened eye.* New York: Macmillan.

Erickson, F., & Gutierrez, K. (2002, November). Culture, rigor, and science in educational research. *Educational Researcher, 31*(8), 21–24.

Feuer, M. J. (2006). Response to Bettie St. Pierre's "Scientifically Based Research in Education: Epistemology and Ethics." *Adult Education Quarterly, 56*(3), 267–272.

Feuer, M. J., Towne, L., & Shavelson, R. J. (2002, November). Science, culture, and educational research. *Educational Researcher, 31*(8), 4–14.

Finley, S. (2008). Arts-based research. In J. G. Knowles & A. L. Cole (Eds.), *Handbook of the arts in qualitative research* (pp. 71–81). Thousand Oaks, CA: Sage.

Freeman, M., deMarrais, K., Preissle, J., Roulston, K., & St. Pierre, E. A. (2007). Standards of evidence in qualitative research: An incitement to discourse. *Educational Researcher, 36*(1), 1–8.

Gilgun, J. F. (2006, March). The four cornerstones of qualitative research. *Qualitative Health Research, 16*(3), 436–443.

Guba, E., & Lincoln, Y. S. (1989). *Fourth-generation evaluation.* Newbury Park, CA: Sage.

Guba, E., & Lincoln, Y. S. (2005). Paradigmatic controversies, contradictions, and emerging confluences. In N. K. Denzin & Y. S. Lincoln (Eds.), *The SAGE handbook of qualitative research* (3rd ed., pp. 191–216). Thousand Oaks, CA: Sage.

Habermas, J. (1972). *Knowledge and human interests* (2nd ed.). London: Heinemann.

Habermas, J. (2006). *The divided West.* Cambridge, UK: Polity.

Hammersley, M. (2004). Some questions about evidence-based practice in education. In G. Thomas & R. Pring (Eds.), *Evidence-based practice in education* (pp. 133–149). New York: Open University Press.

Hammersley, M. (2005a, December). Close encounters of a political kind: The threat from the evidence-based policy-making and practice movement. *Qualitative Researcher, 1,* 2–4.

Hammersley, M. (2005b, April). Countering the "New Orthodoxy" in educational research: A response to Phil Hodkinson. *British Educational Research Journal, 31*(2), 139–156.

Hodkinson, P. (2004, February). Research as a form of work: Expertise, community and methodological objectivity. *British Educational Research Journal, 30*(1), 9–26.

Jackson, N., & Waters, E. (2005). Criteria for the systematic review of health promotion and public health interventions. *Health Promotion International, 20*(4), 367–374.

Jefferson, T., Rudin, M., Brodney Folse, S., & Davidoff, F. (2006). Editorial peer review for improving the quality of reports of biomedical studies. *Cochrane Database of Methodology Reviews, 1.*

Judson, H. F. (2004). *The great betrayal: Fraud in science.* New York: Harcourt Brace.

Kaplan, E. (2004). *With God on their side: How the Christian fundamentalists trampled science, policy, and democracy in George W. Bush's White House.* New York: New Press.

Larner, G. (2004). Family therapy and the politics of evidence. *Journal of Family Therapy, 26,* 17–39.

Lather, P. (1993). *Getting smart: Feminist research and pedagogy with/in the postmodern.* New York: Routledge.

Lather, P. (2004). This is your father's paradigm: Government intrusion and the case of qualitative research in education. *Qualitative Inquiry, 10*(1), 15–34.

Lather, P. (2006, November/December). Foucauldian scientificity: Rethinking the nexus of qualitative research and educational policy analysis. *International Journal of Qualitative Studies in Education, 19*(6), 783–792.

Lather, P. (2007). *Getting lost: Feminist efforts toward a double(d) science.* Albany: SUNY Press.

Lincoln, Y. S., & Cannella, G. S. (2004, February). Dangerous discourses: Methodological conservatism and governmental regimes of truth. *Qualitative Inquiry, 10*(1), 5–10.

Lincoln, Y. S., & Guba, E. (1985). *Naturalistic inquiry.* Beverly Hills, CA: Sage.

Lundberg, G. (1926, October). Quantitative methods in sociology. *Social Forces, 39,* 19–24.

MacLure, M. (2006, November/December). The bone in the throat: Some uncertain thoughts on baroque method. *International Journal of Qualitative Studies in Education, 19*(6), 7239–7746.

Madison, D. S. (2005). *Critical ethnography: Methods, ethics, and performance.* Thousand Oaks, CA: Sage.

Maxwell, J. A. (2004a). Causal explanation, qualitative research, and scientific inquiry in education. *Educational Researcher, 23*(2), 3–11.

Maxwell, J. A. (2004b, August). Using qualitative methods for causal explanation. *Field Methods, 16*(3), 243–264.

Monastersky, R. (2002, November 25). Research groups accuse education department of using ideology in decisions about data. *Chronicle of Higher Education, 2.*

Moran-Ellis, J., Alexander, V. D., Cronin, A., Dickenson, M., Fielding, J., Sleney, J., et al. (2006, February). Triangulation and integration: Processes, claims, and implications. *Qualitative Research, 6*(1), 45–60.

Morse, J. M. (2006a, March). The politics of evidence. *Qualitative Health Research, 16*(3), 395–404.

Morse, J. M. (2006b, March). Reconceptualizing qualitative inquiry. *Qualitative Health Research, 16*(3), 415–422.

National CASP Collaboration. (2006). *10 questions to help you make sense of qualitative research, Critical Appraisal Skills Program (CASP).* Milton Keynes Primary Care Trust. Available at http://www.pdptoolkit.co.uk/Files/Critical%20Appraisal/casp.htm

National Center for Dissemination of Disability Research. (2007). Available at http://www.ncddr.org/kt/products.focus.focus9/

National Research Council. (2002). *Scientific research in education.* Committee on Scientific Principles for Education Research (R. J. Shavelson & L. Towne, Eds.). Washington, DC, National Academies Press.

National Research Council. (2005). *Advancing scientific research in education.* Committee on Scientific Principles for Education Research (L. Towne, L. Wise, & T. M. Winters, Eds.). Washington, DC, National Academies Press.

Popay, J., Rogers, A., & Williams, G. (1998). Rationale and standards for the systematic review of qualitative literature in health services research. *Qualitative Health Research, 8,* 341–351.

Preissle, J. (2006, November/December). Envisioning qualitative inquiry: A view across four decades. *International Journal of Qualitative Studies in Education, 19*(6), 685–696.

Quigley, L. (1996). *The blind men and the elephant.* New York: Scribner.

Rains, F. V., Archibald, J., & Deyhle, D. (2000). Introduction: Through our eyes and in our own words—The voices of indigenous scholars. *International Journal of Qualitative Studies in Education, 13*(4), 337–342.

Richardson, L. (2000a). Evaluating ethnography. *Qualitative Inquiry, 6*(2), 253–255.

Richardson, L. (2000b). Writing: A method of inquiry. In N. K. Denzin & Y. S. Lincoln (Eds.), *Handbook of qualitative research* (2nd ed., pp. 923–948). Thousand Oaks, CA: Sage.

Saukko, P. (2003). *Doing research in cultural studies: An introduction to classical and new methodological approaches.* London: Sage.

Smith, J. K., & Deemer, D. K. (2000). The problem of criteria in the age of relativism. In N. K. Denzin & Y. S. Lincoln (Eds.), *Handbook of qualitative research* (2nd ed., pp. 877–896). Thousand Oaks, CA: Sage.

Smith, J. K., & Hodkinson, P. (2005). Relativism, criteria and politics. In N. K. Denzin & Y. S. Lincoln (Eds.), *The SAGE handbook of qualitative research* (3rd ed., pp. 915–932). Thousand Oaks, CA: Sage.

Smith, L. T. (1999). *Decolonizing methodologies: Research and indigenous peoples.* Dunedin, NZ: University of Otago Press.

Spencer, L., Ritchie, J., Lewis, L., & Dillon, L. (2003). *Quality in qualitative evaluation: A framework for assessing research evidence.* London: Government Chief Social Researcher's Office, Crown Copyright.

Stanfield, J. H. (2006, November/December). The possible restorative justice functions of qualitative research. *International Journal of Qualitative Studies in Education, 19*(6), 723–728.

St.Pierre, E. A. (2006). Scientifically based research in education: Epistemology and ethics. *Adult Education Quarterly, 56*(3), 239–266.

St.Pierre, E. A., & Roulston, K. (2006, November/December). The state of qualitative inquiry: A contested science. *International Journal of Qualitative Studies in Education, 19*(6), 673–684.

Stronach, I. (2006, November/December). Enlightenment and the "Heart of Darkness": (Neo) imperialism in the Congo, and elsewhere. *International Journal of Qualitative Studies in Education, 19*(6), 757–768.

Stronach, I., Garratt, D., Pearce, C., & Piper, H. (2007, March). Reflexivity, the picturing of selves, the forging of method. *Qualitative Inquiry, 13*(2), 179–203.

Teddlie, C., & Tashakkori, A. (2003). Major issues and controversies in the use of mixed methods in the social and behavioral sciences. In A Tashakkori & C. Teddlie (Eds.), *Handbook of mixed methods in social and behavioral research* (pp. 3–50). Thousand Oaks, CA: Sage.

Thomas, G. (2004). Introduction: Evidence: Practice. In G. Thomas & R. Pring (Eds.), *Evidence-based practice in education* (pp. 1–20). New York: Open University Press.

Timmermans, S., & Berg, M. (2003). *The gold standard: The challenge of evidence-based medicine and standardization in health care.* Philadelphia: Temple University Press.

Torrance, H. (2006). Research quality and research governance in the United Kingdom. In N. K. Denzin & M. Giardina (Eds.), *Qualitative inquiry and the conservative challenge* (pp. 127–148). Walnut Creek, CA: Left Coast Press.

Viadero, D. (2006). New group of researchers focuses on scientific study. *Education Week, 25*(21), 1 & 16.

White, C. (2003, February 1). Little evidence for effectiveness of scientific peer review. *British Medical Journal, 326*(7383), 241.

Wolcott, H. F. (1999). *Ethnography: A way of seeing.* Walnut Creek, CA: AltaMira Press.

40

WRITING INTO POSITION

Strategies for Composition and Evaluation

Ronald J. Pelias

When I write it feels like I'm carving bone. It feels like I'm creating my own face, my own heart.

—G. Anzaldúa (1999, p. 95)

I'm sitting at my desk considering how the self is positioned in scholarly writing, how the self commands attention even when the self is not seemingly central to the discussion. As this argument gathers in my head, I remember Joan Didion's initial claim in her essay, "Why I Write" (2000):

> In many ways writing is the act of saying I, of imposing oneself upon other people, of saying *listen to me, see it my way, change your mind.* It's an aggressive, even hostile act. You can disguise its aggressiveness all you want with veils of subordinate clauses and qualifiers and tentative subjunctives, with ellipses and evasions— with the whole manner of intimating rather than claiming, of alluding rather than stating—but there's no getting around the fact that setting words on paper is the tactic of a secret bully, an invasion, an imposition of the writer's sensibility on the reader's most private space. (pp. 17–18, emphasis original)

I am drawn to Didion's insight, seduced by its logic. It seems, in part, to write my experience of writing, positioning me in a productive place, a place where I might see myself writing. Being pulled in by Didion, I transcribe her comment in my journal and jot the following note:

> When I write, I am asserting a self, insisting that I matter. In general, I would argue, research is a way of claiming space. It takes an extended turn with the implication that one's writing merits attention. Such a commitment is a call to arrogance and to significance.

Research cannot exist without a belief in its seriousness; it cannot prosper in the belief in its singular truth. Research lives in possibility and in promise.

In that moment of composition, I come to see what I believe, what I did not know before I started writing. I arrive at a place of resonant articulation. I move toward clarity, toward, as Robert Frost (1963) would suggest his poems are, "a momentary stay against confusion" (p. 2). I perform myself into being and I emerge within and through the consequences of my always political and material assertions. I am engaging in a process that Laurel Richardson (2000; see also Richardson & St.Pierre, 2005) would call "writing as a method of inquiry" (p. 923).

Following Richardson, in this chapter I build on the idea that writing is a "method of inquiry." I write myself into a position that identifies what I have come to understand by that claim as well as what I presently believe about qualitative writing. Part I argues that writing functions as both a realization and a record. As a realization, it locates itself on a continuum of possibility to certainty, of the subjunctive to declarative, as well as situates itself on a continuum of personal discovery to public argument. As a record, realizations find their form as poetic, narrative, and dramatic utterances and emerge as descriptive, deconstructive, and critical claims. Part II argues that the evocative, reflexive, embodied, partial and partisan, and material characteristics often associated with qualitative research encourage certain compositional strategies. Part III, using the device of juxtaposition, outlines how the more and less effective essay establishes itself.

🔲 **PART I: POSITIONING SELF, POSITIONING WRITING**

I am sitting at my desk trying to remember some of the creative writers I have read that suggest in some way that writing is a method of inquiry. I open my journal and quickly find Stephen Dunn's (1993) claim that an essayist might best be seen as "a person who believes there's value in being overheard clarifying things for himself" (p. ix). On another page, I discover Lee Smith (2007) echoing Dunn's suggestion:

> Whether we are writing fiction or nonfiction, journaling or writing for publication, writing itself is an inherently therapeutic activity. Simply to line up words one after another upon a page is to create some order where it did not exist, to give recognizable shape to the sadness and chaos of our lives. (p. 41)

This clarifying function of writing I see on another page of my journal when Theodore Roethke (2001) notes that a poem is "one more triumph over chaos" (p. 77). My search continues with Natalie Goldberg (1986): "Writing is the act of burning through the fog in your mind" (p. 86). What these and other creative writers have come to understand is that writing is a strategy of circling, of making present what might have slipped away, of calling into focus through an attentiveness to and negotiation with language. It is a process, as Marvin Bell (2002) would have it, where the writer "listens" to the writing "as it goes" (p. 13). In that way, writing is, as M. L. Rosenthal (1987) explains, "the unfolding of a realization, the satisfying of a need to bring to the surface the inner realities of the psyche" (p. 5). As Don Geiger (1967) says of the lyric poem, writing "records the process of the speaker's realization" (p. 152). With such surrounding discourse, it is not surprising that Richardson would up the stakes by arguing for writing's methodological status and that her argument would be embraced by many scholars who engage in qualitative research.

Joining those who align themselves with Richardson, I want to outline how writing might function as a realization and as a record. These terms—*realization* and *record*—point toward the writer's process and completed text. Writers come to realize what they believe in the process of writing, in the act of finding the language that crystallizes their thoughts and sentiments. It is a process of "writing into" rather than "writing up" a subject. When writing *up* a subject, writers know what they wish to say before the composition process begins. When writing *into* a subject, writers discover what they know through writing. It is a process of using language to look at, lean into, and lend oneself to an experience under consideration. This "languaging" unearths the writer's articulate presence. It positions, marks a place, a material stance in the world. In short, languaging matters.

As writers proceed, their realizations come forward on a continuum between the declarative and the subjunctive. These realizations might be called "is-ness" or "perhaps-ness" utterances. "Is-ness" claims assert, "This is . . ."; "perhaps-ness" statements reside in "may be." For writers, then, a realization may carry considerable authority, materialize without doubt, feel certain, or may exist tentatively, appearing as a possibility among many, contingent upon circumstances. Realizations settle into writers in different ways, often guiding the stance they take toward their subjects. Moving with assurance or caution, believing they possess overwhelming evidence or only perhaps a small piece of a larger puzzle, writers stand behind their ideas, sometimes pushing them forward as points that seem obvious and worthy of attention and sometimes holding them close by, keeping them from making too much noise. On the "is-ness" side of the continuum, arguments come forward as definitive and, at times, sufficient for public advocacy. On the "perhaps-ness" side, claims invite further dialogue, call for further research, live in their questions.

Realizations, whether emerging as "is-ness" or "perhaps-ness" claims, also unfold on a continuum from the personal to the public. Personal realizations inform writers about themselves as individuals. They place them in touch with their own attitudes, beliefs, and feelings; with their own relational attachments and political investments; with their own sense of the world. Such writing might be located in personal identity (e.g., Alexander, 2006; Myers, 2008; Trujillo, 2004; Warren, 2001; Young, 2007); in trauma, illness, and loss (e.g., Defenbaugh, 2011; Ellis, 1995; Rambo Ronai, 1996; Richardson, 2007; Watt, 2005, 2008); or in relational dynamics (e.g., Adams, 2006; Poulos, 2009; Tillmann-Healy, 2001). In short, personal realizations tell writers how they might see themselves, how they might make sense of their experiences. When sharing their insights, they invite readers to acknowledge their perspectives and perhaps to identify with them. Their writing becomes a location for readers' consideration.

Public realizations place writers in contact with the social or cultural sphere. They come to understand how the social world unfolds, highlighting how structural schemes guide human behavior, how institutional practices control human desires and dictate entitlements, how cultural understandings privilege some but not others. Such writings might focus on colonial and postcolonial logics (e.g., Anzaldúa, 1999; Bhabha, 1994), corporate and governmental behaviors (e.g., Goodall, 1989, 2006; Tracy, 2003), or social and political injustices (e.g., Denzin, 2008; Lockford, 2008). They often carry an implicit or explicit call to action. They may come forward as a utopian dream, as a location of hope, or as an ethical imperative. They may call for a reordering in the name of social justice. They may appear radical, even anarchist, when change from within seems insufficient or impossible. They may excite; they may incite.

The familiar feminist insight that the personal is political is a quick reminder of the dangers of separating the personal and the public. Personal utterances are revelatory in the public sphere, particularly when previously silenced or minimized, and public pronouncement and legislation find their most profound articulation as they impact individual bodies. By contemplating the personal, public realizations emerge; by considering the public, personal insights become apparent. The personal/public distinction is useful, however, for noting what seems rhetorically foregrounded in a given work.

Realizations find their form by becoming a record of what a writer has come to discover. They take shape as a poetic, narrative, or dramatic record. While these forms share much in common, scholars tend to gravitate to one form or another in their work and, in so doing, their scholarly efforts unfold differently. The poetic record comes forward as poem, an inquiry that depends upon the power of poetic devices (e.g., figurative language, prosody, lining) to structure its insights. Whether turning to their own or others' experiences as data, as a source for their evidentiary claims, researchers (e.g., Brady, 2003; Hartnett, 2003; Prendergast, Leggo, & Sameshima, 2009) working with poetic form use the condensed emotional intensity of poetry as well as careful research to render their subjects. As Hartnett explains, turning to the poetic merges "the evidence-gathering force of scholarship with the emotion-producing force of poetry" (p. 1). In this sense, such efforts stand as what some have called "investigative poetry" (e.g., Hartnett, 2003; Hartnett & Engels, 2005; Sanders, 1976).

The narrative record is the form most frequently found in scholarly circles. It is a formed tale, told by a narrator, relying upon point of view, plot, and character. In most cases, the narrators in scholarly research are reliable; that is, they share the same values as their authors. As readers engage these tales, they expect authors to operate under a contract of truth-telling. They assume that authors will try to render their stories as honestly as they can, recognizing, of course, that no account is the final word. "Storying" is a self-making process. It helps authors construct their perceptions of the world, see new terrain, and live with alternative views. It is also a culture-making process. Stories often carry a sense of social responsibility, a need to tell to further social justice. Such stories come forward as acts of witnessing, as testimony on behalf of others. Ellis (2009a) offers a telling description of the power of story for scholars:

> Stories are what we have, the barometers by which we fashion our identities, organize and live our lives, connect and compare our lives to others, and make decisions about how to live. These tales open our hearts and eyes to ourselves and the world around us, helping us change our lives and our world for the better. (p. 16)

The dramatic record taps into theatrical practice, highlighting conflict and dialogue. It may emerge as a script on the page (e.g., Ellis & Bochner, 1992) or on the stage (e.g., Gray & Sinding, 2002; Pineau, 2000; Saldana, 2005). Relying upon the artistic techniques of performance, scholars working with dramatic forms write their research findings as scripts designed to display multiple speakers engaged in interaction with one another or multiple voices reflecting varying perspectives. The labels most commonly associated with the dramatic record are ethnodrama (e.g., Mienczakowski, 2001; Saldana, 2005) and performance ethnography (e.g., Alexander, 2005; Denzin, 2003; Madison, 2005) The aim of such renderings is to offer embodied representations, portrayals that bring to life research findings, often with the desire to participate in "a cultural politics of hope" (Denzin, 2003, p. 24).

Whether realizations take shape as poetic, narrative, or dramatic utterances, they also take form as interpretive, deconstructive, or critical utterances. Realizations, in an interpretive form, rely most heavily on description; in deconstructive form, they open possibilities; in critical form, their investigative pulse pushes toward social action. To think of writing as a realization recorded in a given form is to suggest that writing is a performative act. It is a speech act that participates in the world, a material utterance that matters. It may function to simply reinforce current construction or it may provide alternative ways of seeing the world. It is always political. The speaker is always, for better or worse, positioned. To say that writing is a method of inquiry best understood as realization recorded in a given form, however, does not offer the specific strategies that qualitative scholars often employ. The next section turns to some of these compositional strategies.

▣ ## PART II: COMPOSITIONAL STRATEGIES

I am sitting at my desk trying to identify how qualitative researchers make their cases, how they shape words on the page, how they bring readers into their essays. I remember some of the key descriptions of qualitative work, descriptions that point in similar directions across numerous arguments on behalf of qualitative work (e.g., Colyar, 2009; Denzin & Lincoln, 2005; Ellis, 2004; Goodall, 2000, 2008; Pollock, 1998). I begin by noting how qualitative scholars often indicate that their work is evocative, reflexive, embodied, partial and partisan, and material, and that each of these dimensions pushes these scholars toward certain compositional strategies. Although separated here for explanatory purposes, I should note that readers are likely to discover writers using multiple strategies within their essays. All essays are necessarily partial and partisan as well as material. Writing, for better or worse, matters. Essays vary in their degree of evocativeness, reflexivity, and embodiment. Readers, however, often have a feel for what strategy or strategies are being privileged in a given essay.

Evocative

Qualitative researchers who employ the evocative do so to enrich or disrupt normative understanding. Their work relies on the *literary* and *possibilizing*. In calling upon the literary, they use literary devices (e.g., figurative language, dialogue, rhythm) to create an experience for the reader. They see their work as aesthetic, borrowing from various literary traditions and believing that the affective has a place in scholarly writing. Tillmann-Healy (2003), for example, offers a compelling rendering of her family relationships during a time of loss through the use of metonymy. She describes her own and other family members' hands to establish the situation's relational and emotional force. In a particularly poignant moment, she tells the reader her thoughts as she looks at her grandfather's hands:

> I stare at the eighty years ingrained in my grandfather's hands. Culinary hands that kept enlisted men fed in the Second World War. Supple hands that stroked my grandmother's raven hair. Strong hands that repaired the dam restraining the eager Mississippi. Calloused hands that constructed my father's childhood home on Elm Street. Proud hands that cradled three boys, and later, eight grandchildren. Paternal hands that carved holiday turkeys. Nurturing hands that cultivated the garden soil that burst open in spring in symphonies of crimson and marigold. Tired hands soothed by sweating glasses of lime Kool-Aid. Forgetful hands that rattled the cup of Yahtzee dice for one too many turns. Aging hands stained with the burgundy of exploded blood vessels. Incorrigible Parkinson's hands that played invisible pianos as he sat in his napping chair, watching "wrastling." (p. 176)

This passage finds its power in the rhythmic repetition, in the carefully selected details, and in the chronological structure that delays the telling hands of her grandfather's illness. Through the literary, the reader, guided by Tillmann-Healy, learns about loss.

Possibilizing works by putting on display multiple readings and alternative actions that the reader can consider. It strives for an escape from the crisis of representation, fully aware that it cannot fully satisfy its own desire. It values writing, to use Derrida's term, in excess in order to create a space for dialogue and different ways of being. As Pollock (1998) suggests, "Performative, evocative writing confounds normative distinctions between critical and creative (hard and soft, true and false, masculine and feminine), allying itself with logics of possibility rather than of validity or causality" (p. 81). Madison (2005) further explains,

> In a performance of possibilities, the possible suggests a movement culminating in creation and change. It is active, creative work that weaves the life of the mind with being mindful of life, of merging the text with the world, of critically traversing the margin and the center, and of opening more and different paths for enlivening relations and spaces. (p. 172)

Writers who rely on possibilizing might present conflicting narratives of the same event, generate a proliferation of readings of a given episode, identify various actions that could be taken, call attention to the inadequacy of their rendering, and so on. They write cautious of language's hegemony, cautious that their advocacy does not become another gesture of power. They write for the possible.

Often, both literary and possibilizing strategies take form by deploying multiple speakers. Some calling upon these strategies aspire to the aesthetic standards of dramatic scripts and find their complete articulation in full theatrical staging. Such is typically the case with ethnodrama, performance texts, and performance ethnography. Spry's "From Goldilocks to Dreadlocks: Hair-Raising Tales of Racializing Bodies" (2001a), Smith's *Fires in the Mirror* (1993) and *Twilight: Los Angeles, 1992* (2000), and Kaufman's *The Laramie Project* (Kaufman & Members of the Tectonic Theater Project, 2001) are clear examples of this type of work. Others using these strategies are content to have their drama on the page, using multiple speakers to establish various perspectives, juxtapositions, and collaborations. Ellis's "Telling Tales on Neighbors: Ethics in Two Voices," (2009b), Denzin's *Searching for Yellowstone* (2008), and Gale and Wyatt's "Two Men Talking: A Nomadic Inquiry Into Collaborative Writing" (2008) serve as instructive examples.

Reflexive

Reflexive writing strategies allow researchers to turn back on themselves, to examine how their presence or stance functions in relationship to their subject. Reflexive writers, ethically and politically self-aware, make themselves part of their own inquiry. Reflexive writing strategies include indicating how the researcher emerged as a *contaminant,* how the researcher's *insider* status was revelatory or blinding, and how the researcher is *implicated* in the problem being addressed. Researchers who see themselves as a contaminant might argue that their own positionality or procedures negatively influenced the study. In such cases, researchers offer a cautionary note suggesting that their claims be read in full awareness of the researcher's influence. Rhetorically, this strategy often proves effective because the reader trusts that the researcher is sensitive to the issue and is likely to temper his or her arguments accordingly.

Researchers who claim insider status indicate that they share cultural membership with the group under investigation. Often, insider status comes unquestioned (e.g., "as a Japanese immigrant living in the United States," "as a single mother of three children," "as a person with cancer"). Other times, insider status may become a point of discussion. In such cases, researchers describe their relationship to the group they are studying to argue that they have spent sufficient time or been accepted by others warranting their claim of insider membership. Once insider status is established, researchers may assert that their insider position allowed them to have insights that outsiders could not or, conversely, that their insider status may have kept them from seeing operative cultural logics.

Chawla (2003) offers an instructive example. She presents a deeply reflexive essay about her research on East Indian arranged marriage. As a woman of Indian background, Chawla first notes that she wrote her own story before collecting tales from other Indian women. Then she questions her procedure: "I began to ask myself why I wrote my story before I listened to my participants. I wondered if I had imposed my story on them by writing first" (p. 276). Later, she troubles her own insider status:

> While I do have incomplete stories from my memories of the different arranged marriages that surround me, these are stories about other people in these marriages, and not stories about *myself*. In these memories, I remain an observer, albeit an observer to myself. My lack of a direct involvement in the reality of an arranged marriage makes me question my legitimacy to do this research. I am still worried that I am too much on the *outside* in this existential and intellectual displacement. (p. 277, emphasis original)

Such arguments are particularly useful writing strategies because they acknowledge that researchers cannot be separated from their research, that the researchers' relationship to those they study as well as their procedures influence their findings, and that researchers who reflect about their stance offer more trustworthy and honest accounts.

Researchers who see themselves as implicated write about their complicity in the problem they are trying to address. In short, they position themselves as contributing to the predicament. For example, one might argue that he or she, as a meat eater, has some responsibility for deforestation; that he or she, although arguing for racial equality, has participated in racist speech; or that he or she served in a war he or she did not support. Structurally, this strategy often first acknowledges one's mistakes and ends with a pledge to alternative conduct. It has the advantage of pointing one's finger at oneself instead of at others. In doing so, it invites others who may have committed similar "sins" to join with the researcher in enlightened behaviors. Denzin (2008) offers a powerful example. He implicates himself as he tracks his own family history with Native Americans. Following an introductory chapter, he begins his discussion, "In the 1950s my brother, Mark, and I spent our summers, until we were young teenagers, with our grandparents on their farm south of Iowa City, Iowa. Saturday nights were special. Grandpa loved those 'cowboy and Indian' movies, and so did I" (p. 25). In the next paragraph, he writes, "In fourth grade I was Squanto in the Thanksgiving play about the pilgrims" (p. 25). Throughout, Denzin places his previous behavior under suspicion as he writes "our way into a militant democratic utopian space, a space where the color line disappears, and justice for all is more than a dream" (p. 23). In a similar fashion, Myers (2008) uses the trope

of dentistry to uncover his straight and White privilege. His line, "When I was about 12 years old, it was obvious to my parents and my peers (and myself whenever I looked in the mirror) that I was simply not straight enough" (p. 161), offers a flavor of how he implicates himself throughout his essay.

Embodied

Qualitative researchers who value embodiment write from a location of corporeal presence. They see the body as a site of knowledge demanding scholarly attention (Conquergood, 1991; Madison, 1999, 2005; Spry, 2001b, 2009). As Spry (2001a) explains, "Coaxing the body from the shadows of academe and consciously integrating it into the process and production of knowledge requires that we view knowledge in the context of the body from which it is generated" (p. 725). To do so, researchers write into the *mind/body split* as a corrective to cognitive renderings, call upon the *sensuous* body, and tap into *bodily experiences.*

Writing into the mind/body split, researchers make the body present. Instead of privileging mind over the body, they insist that the body provides flesh to sterile, distant, cognitive accounts. They proceed by writing from affective space, often with the desire to provide a more complete picture of human experience. For them, purely cognitive descriptions of human behavior fail to give rich and nuanced portrayals, erase the individual in the name of generalizability, and lack resonant validity. Closely associated is sensuous writing. As a writing strategy, it asks researchers to speak from the senses, to recognize how the body takes in the world around it, to allow the body to be alive in research. Stoller (1997) explains that sensuous scholarship "is an attempt to reawaken profoundly the scholar's body by demonstrating how the fusion of the intelligible and the sensible can be applied to scholarly practices and representations" (p. xv). Another related strategy is writing from bodily experience. In this case, researchers speak about some incident where the body was the site of the incident's happening. Scholars have used this strategy to write about such topics as illness, trauma, violence, grief, race, sexuality, gender, ethnicity, and so on. Common to such studies is that the body becomes a location of knowledge, a place where the researcher speaks from felt experience, from an awareness of what the body endured, from a sense of self. In doing so, the vulnerable body gains agency by asserting its history and living presence. It turns victim into survivor and the voiceless into a cultural worker. It carries the potential for identification and for social change.

Spry (2001b) presents a telling example of how her body becomes deployed on behalf of scholarly inquiry. Writing about her decision as a White woman to grow dreadlocks, she says,

> Their time had come.
>
> And as they emerged, they evoked many comments from many people.

A most interesting theme of comments emerged from White women:

"Tami, aren't you afraid of offending Black people

by wearing dreads?" "I mean, what will they think?"

"Aren't you 'taking something away' from Black people

by growing dreads?"

As if I could

As if I were in   racial drag

As if I were   drag racing   to the finish line

of an essentialized,   homogenized   *Blackness.*

But what began to emerge for me

were essentialized, homogenized images of

Whiteness.

And I began to see the ways

that I had been living much of my life

In White   racial   drag.   (p. 724)

Spry, in this brief excerpt from a longer work, writes about and from her body, allowing the reader to consider how White privilege and, more generally, race functions in U.S. culture. Readers sense her resistance to those who through their questions would essentialize race, feel her struggle to make sense of her own racialized body, and question their own racial positioning and understanding. Spry gives readers a performative encounter with race, fully embodied, felt, and emotionally honest.

Partial and Partisan

Researchers proceed with the knowledge that their work is always partial and partisan. That is, they understand that they can never say everything about anything (partial) and that everything they say carries ideological weight (partisan). Given such awareness, researchers may elect to write into the heart of the matter by calling attention to *linguistic limitations,* by highlighting how a given argument is *ideologically laden,* or by *uncovering* the hidden. Using the strategy of noting linguistic limitations permits researchers to acknowledge how language is slippery, never in perfect correspondence to the subject it attempts to capture. Noting linguistic limitations draws attention to the elusive and inexpressible that haunts all qualitative inquiry. It reminds readers what is at stake in any rendering. Researchers might put this strategy into play by discussing the implications of a given word choice, by calling into question their own representations (e.g., specifying what is missing, suggesting

why their writing is inadequate), by offering multiple narratives of the same event, or by continuously adding to a previous account. Such maneuvers remind the reader to stay alert, to accept any claim with caution, to be suspicious of language's ability to represent human experience.

Holman Jones (2002) offers a poignantly expressed example:

Some days I wonder if, after all the hours I've spent in seminar rooms and alone in front of my computer writing, I have been reduced to making lists of words, to scripting fragments. Unable to express in finely wrought sentences the injustices of oppression or the beauty of a solution, I make lists that signify worlds. Words that set off explosions of thought and feeling. (p. 187)

In this moment, Holman Jones offers the reader a glimpse into her struggle to write in a way that satisfies her. She hints that her academic position and her isolation as a writer may be contributing factors. She does not have the words to capture what she desires. She has only fragments, lists, that point to her subject, but she is unable to write what troubles her most. As her essay progresses, the reader sees that she does, in part through the use of fragments and lists, write her desire. This turn is not uncommon in qualitative inquiry. Readers often encounter researchers who lament language's limitations while demonstrating its power.

Researchers, recognizing that all utterances are ideologically laden, point to language itself as well as their own language practices. When discussing language itself, they might demonstrate how language privileges some at the expense of others, how a word's etymology points to its unproductive connotations, or how language conceals as much as it reveals. Such writing shows how language as a discursive system is never innocent, without bias. When turning on their own language practices, researchers might make explicit their ideological commitments and their operative hidden assumptions. They often do so by revealing their personal beliefs, noting their private agendas, or specifying their ethical stance. Closely associated to the strategy of making ideology explicit is the tactic of *uncovery.* Uncovery is a deconstruction, an act of revealing the absent in a given assertion. Its primary approach is proliferation, excess, with the aim, at times, of reconstruction.

Russel y Rodriguez (2002) argues that her own discipline, anthropology, requires "static and uncomplicated single identities of its subjects and theoreticians" (p. 347) that are at odds with her own positionality:

The Chicana among feminists, the feminist among Chicanos. The Chicano nationalist among Euro-Americans, the bridge builder among Chicano nationalists. Half Anglo, half Chicana. The newcomer in academe, the overeducated at home. The minority among the majority, the mainstream among raza. (pp. 347–348)

Russel y Rodriguez recognizes how each of these descriptors places her in differing ideological currents. Uncovering these currents and marking their force, she writes to free herself and others from a space where "both normative and oppositional stances contribute to a silencing praxis in anthropology" (p. 348).

Material

As I hope I made clear earlier in this essay, writing as a realization and record is a performative act, a material manifestation of a writer's labor and ideology, an enunciation that carries weight in the social world. Material texts often demonstrate what matters through personal, scholarly, and social articulations. Among the strategies available to researchers foregrounding the material are the *curative,* the *citational corrective,* and the *socially consequential.* Writing curatively, researchers work therapeutically in the desire to heal their own and others' wounds. They may be struggling, for instance, to come to terms with illness, relational tensions, oppression, or physical violence. Poulos (2009) gives a seductive explanation of why researchers call upon the curative:

> As I inquire into the depth and contours and possibilities of the secret world of families, I necessarily encounter some of the "darker" moments of the human spirit. But, along the way, I discover—in the eruption of a story, in the soft reminiscent light of accidental talk, in a burst of memory overstepping memory—a world of hope. (p. 15)

The hope functions not only for the authors of curative texts but also for those who might see themselves in the account and for those whose understanding of a given problem is enlarged.

Kiesinger (2003), writing of her sexual abuse and of her resulting binge eating and purging in her powerful poem, "He Touched, He Took," stands as a useful example. The reader is first given a description of the assault. Here is a small sample:

> She will put language to the images and sensations as they come to her.
>
> His breath, hot on her neck.
>
> His scent—an odd mixture of dirt and Lysol.
>
> The feel of his hands, rough and calloused, as he places candy in hers—
>
> soft and small. (p. 177)

The poem next moves to her descriptions of binging and purging—binging to fill the void he left, purging to rid herself of him. The poem ends with the following: "She writes from and through her memories with the hope of recovering what he took when he touched—with the hope of rediscovering what was fundamentally her own" (p. 184). Kiesinger's text becomes another purging. She writes of her own horrific experience, an experience that, unfortunately, others have lived. She writes with hope for her own healing and marks a space where others might find comfort in seeing their own story told. She writes to teach us all.

Researchers who use citational corrective as a writing strategy start from dissatisfaction with a given scholarly claim. They proceed by writing against what stands as an adequate account. They challenge ongoing logics by such tactics as juxtaposing their own personal experience against disciplinary claims, speaking from the margins, and adding the affective to cognitive accounts. My desire, for example, to write "Confessions of an Apprehensive Performer" (Pelias, 1999) initiated with the belief that despite the considerable amount of research on communication apprehension, none captured the felt experience of the apprehensive speaker. I placed myself in dialogue with research findings to demonstrate how as an apprehensive I seldom saw myself in these cognitive claims, or when I did, the claims rang empty. In one section, I quote several scholars on the topic and offer flippant responses to their points. Here is one instance:

> Watson, Monroe, and Atterstrom:
>
> Communication apprehension (CA), the fear of oral communication with another person or persons, is found to affect many individuals negatively by inhibiting amounts of communication and interfering with effectiveness in life experiences. (*Communication Quarterly, 37,* 1989, p. 67)
>
> No shit. (p. 83)

My aim was not to denigrate other scholars but to put on display my emotional response as an apprehensive when reading such points and to suggest that research on apprehension might benefit by attending to apprehensives' personal accounts.

The writing strategy that foregrounds the socially consequential puts on display alternative social constructions and practices. Scholars who are committed to the idea that research should matter to everyday lives take as their charge to articulate new ways of being. Their work comes forward as an emancipatory pedagogy, ethically charged, calling for action. They strive to move beyond description, to become critically engaged, to create utopian spaces. They are cultural workers, laboring on behalf of social justice. Denzin and Giardina (2009) offer an eloquent statement of how socially consequential writing would ideally unfold:

> Inquiry grounded in critical indigenous pedagogy should meet multiple criteria. It must be ethical, performative, healing, transformative, decolonizing, and participatory. It must be committed to dialog, community, self-determination, and cultural autonomy. It must meet peoples' perceived needs. . . . It seeks to be unruly, disruptive, critical, and dedicated to the goals of justice and equity. (p. 29)

I am sitting at my desk knowing that any listing of writing strategies is incomplete. Language, despite its limitations, is too rich and generative to be nailed down into simple categories. There are no limits to the productive procedures a researcher might employ. There is no simple formula for writing success. Likewise, writing strategies are not stable; they overlap, slip away, change. They may prove effective for one study but not for another. There are no magic techniques that, abracadabra, turn an essay into a "must read." Yet, if employed with rhetorical skill; with sensitivity to the task at hand and with an eye toward writing evocatively, reflexively, and in an embodied fashion; deploying language's partial and partisan nature for the material, for social justice, then certain writing strategies may prove to be a helpful way in.

▣ PART III: AN EVALUATIVE POSITION

I am sitting at my desk trying to contemplate what qualitative work I want to applaud and what efforts seem lacking. I'm curious why I am seduced by some work but not others, why the best work seems to engage and the weaker work seems to fall flat and leaves me cold. I wish to articulate what I like and what I don't without imposing my evaluative stance but acknowledging that I have one that guides my practice as a reviewer, teacher, and writer. I leave open the possibility of other evaluative and more productive schemes, but I believe that the following assessments live in the company of other scholars' values (e.g., Bochner, 2000; Ellis, 2004; Goodall, 2000, 2008; Richardson, 2000). I am sitting at my desk ready to consider other readings, but I continue, putting an evaluative self forward.

The flat piece, a cold dinner, is forced down, taken in with little pleasure. It lacks the heat of the chef's passions, the chef's sensuous self who knows, without spice, all is bland. The engaging piece makes each mouthful worthy of comment, encourages lingering, savoring, remembering. In its presence, I want to invite my colleagues and students to enjoy its flavors.

The flat piece does not know its place. It carries on as if no one has ever spoken before. It once again invents the wheel. The engaging piece knows it place, bows in the direction of the previous as it takes its conversational turn. It does so, not with the perfunctory review, but with the respectful incorporation of others' ideas. At times, it works as a process of sense making, of letting others carry it along, but it is always as interesting as those it quotes. It proceeds without creating a killing field.

The flat piece clothes itself in fancy garb, often hiding its essential simplicity. The engaging piece knows just how to dress for the occasion, perhaps adorning itself in the unfamiliar but always with an eye toward making any flourish an expressive part of the whole. Instead of an ostentatious display, its design encourages further conversation.

The flat piece lives in the abstract, speaks across and above the individual. The engaging essay resides in the precise. It puts leaves on bare limbs, but never bowling balls or toasters. It realizes what leaves are needed to create its tree. And as the leaves accumulate, become thick, each one counts.

The flat piece structures its ideas with familiar models. It learned too well the lessons from first-year English composition and scientific reasoning. It forgets the vibrant relationship of form and content; instead, it unfolds in a predetermined manner. The engaging work allows structure to emerge, guided by the necessity of its subject. When its form is fulfilled, it satisfies. It knows the imperatives of its content.

The flat piece has easy and ready answers. It says what it knows, finding its way to where it began. It works to prove itself right. The engaging essay exists in struggle, searching for what it may come to realize. It becomes a small, nervous solution, offered with humility. Its ideas slide into cautious claims, noting its limitations.

The flat piece thinks its author can be invisible, above it all, bodiless. It is a megaphone speaking from on high, producing pronouncements. The engaging essay has its author own his or her bodily presence. Knowing that all bodies are historically, culturally, and individually saturated, it brings forward its own situatedness to suggest a context for reading and to demonstrate how its account is tainted. It understands how it is fettered. It speaks from the body to those who wish to listen.

The flat piece thinks its head can be separated from its heart. Its passion is buried; its politics is denied. It is a doctor with a scalpel. The engaging essay joins the head and the heart. It resists Descartes' logic, but if trapped by his binary, it turns "think" into "feel." It works best when it speaks from the heart through the power and passion of its heady constructions to render human experience, albeit never completely, in its lived complexity. It is the nurse with your chart by her side telling a comforting and clear tale.

The flat piece accepts the given, trusts in the already forged. It ornaments the already built, decorates the status quo. The engaging piece plays, opens closed doors, discovers hidden passageways, creates new spaces. It is mischievous, utopian, saying the unsayable, the forbidden, the dangerous. It knows the master's house can be rebuilt. It believes there should be no master.

The flat piece errs by thinking it has the truth. It believes it has covered all the bases, exhausted all the angles, handled all the counterarguments. Its smugness sticks to the page. The engaging essay situates itself in the conditional, aware of language's slippery slide, aware that today's certainty is tomorrow's joke, aware that it is always located in its author.

The flat piece turns politics into sloganeering, easy platitudes. It rails against known enemies, pointing its finger in a familiar direction. It is full of fury, often signifying a liberal education, but nothing more. Too often, it is merely correct. The engaging piece locates politics in the body, shows how politics matters to individual lives, speaks of and to the heart. It positions readers in another's life, brings forward their empathic selves, even while at times implicating them, so that they want to work on behalf of social justice.

The flat piece proceeds unaware of its moral consequences. It rushes in, claims space, forgetting others may be present. Its emissions pollute the social environment. The engaging essay looks both left and right. It finds its ethical stance with its hand in another's hand.

▣ A Summary Positioning

I am sitting at my desk, positioned, having taken an extended turn, asking perhaps more of my reader than I am entitled to request or that my prose might bear. My narrative is, of course, only one story that might be told about writing qualitative research. I invite the reader to decide if it is flat or engaging. In either case, it is a record of what I've come to realize, a statement that lobbies for space, a material gesture on behalf of qualitative inquiry designed to enhance our efforts at writing the social world, of making the social world a more habitable and just place. I offer it in the hope of dialogue and of new possibilities. I am sitting at my desk, trying.

▣ References

Adams, T. (2006). Seeking father: Relationally reframing a troubled love story. *Qualitative Inquiry, 14*(4), 704–723.

Alexander, B. K. (2005). Performance ethnography: The reenacting and inciting of culture. In N. K. Denzin & Y. S. Lincoln (Eds.), *The SAGE handbook of qualitative research* (3rd ed., pp. 411–442). Thousand Oaks, CA: Sage.

Alexander, B. K. (2006). *Performing Black masculinity: Race, culture, and queer identity.* Walnut Creek, CA: AltaMira Press.

Anzaldúa, G. (1999). *Borderlands/la frontera: The new mestiza* (2nd ed.). San Francisco: Aunt Lute Books.

Bell, M. (2002). Thirty-two statements about writing poetry (work-in-progress). *The Writer's Chronicle, 13.* Available at http://www.coppercanyonpress.org/400_opportunities/430_gettingpub/bell.cfm

Bhabha, H. K. (1994). *The location of culture.* New York: Routledge.

Bochner, A. P. (2000). Criteria against ourselves. *Qualitative Inquiry, 6,* 266–272.

Brady, I. (2003). *The time at Darwin's reef. Poetic explorations in anthropology and history.* Walnut Creek, CA: AltaMira Press.

Chawla, D. (2003). Rhythms of dis-location: Family history, ethnographic spaces, and reflexivity. In R. P. Clair (Ed.), *Expressions of ethnography: Novel approaches to qualitative methods* (pp. 271–279). Albany: State University of New York Press.

Colyar, J. (2009). Becoming writing, becoming writers. *Qualitative Inquiry, 15*(2), 421–436.

Conquergood, D. (1991). Rethinking ethnography: Towards a critical cultural poetics. *Communication Monographs, 58,* 179–194.

Defenbaugh, N. (2011). *Dirty tale: The chronically ill journey.* Creskill, NJ: Hampton Press.

Denzin, N. K. (2003). *Performance ethnography: Critical pedagogy and the politics of culture.* Thousand Oaks, CA: Sage.

Denzin, N. K. (2008). *Searching for Yellowstone: Race, gender, family and memory in the postmodern West.* Walnut Creek, CA: Left Coast Press.

Denzin, N. K., & Giardina, M. D. (2009). Qualitative inquiry and social justice: Toward a politics of hope. In N. K. Denzin & M. D. Giardina (Eds.), *Qualitative inquiry and social justice* (pp. 11–50). Walnut Creek, CA: Left Coast Press.

Denzin, N. K., & Lincoln, Y. S. (Eds.). (2005). *The SAGE Handbook of qualitative research* (3rd ed.). Thousand Oaks, CA: Sage.

Didion, J. (2000). Why I write. In J. Sternburg (Ed.), *The writer on her work* (pp. 17–25). New York: W. W. Norton.

Dunn, S. (1993). *Walking light: Essays and memoirs.* New York: W. W. Norton.

Ellis, C. (1995). *Final negotiations: A story of love, loss, and chronic illness.* Philadelphia: Temple University Press.

Ellis, C. (2004). *The ethnographic I: A methodological novel about autoethnography.* Walnut Creek, CA: AltaMira Press.

Ellis, C. (2009a). *Revision: Autoethnographic reflections on life and work.* Walnut Creek, CA: Left Coast Press.

Ellis, C. (2009b). Telling tales on neighbors: Ethics in two voices. *International Review of Qualitative Research, 2*(1), 3–28.

Ellis, C., & Bochner, A. P. (1992). Telling and performing personal stories. The constraints of choice in abortion. In C. Ellis & M. Flaherty (Eds.), *Investigating subjectivity: Research on lived experience* (pp. 79–101). Newbury Park, CA: Sage.

Frost, R. (1963). The figure a poem makes. In *Selected poems of Robert Frost* (pp. 1–4). New York: Holt, Rinehart & Winston.

Gale, K., & Wyatt, J. (2008). Two men talking: A nomadic inquiry into collaborative writing. *International Review of Qualitative Research, 1*(3), 361–380.

Geiger, D. (1967). *The dramatic impulse in modern poetics.* Baton Rouge: Louisiana State University Press.

Goldberg, N. (1986). *Writing down the bones: Freeing the writer within.* Boston: Shambhala.

Goodall, H. L. (1989). *Casing the promised land: The autobiography of an organizational detective.* Carbondale: Southern Illinois University Press.

Goodall, H. L. (2000). *Writing the new ethnography.* Walnut Creek, CA: AltaMira Press.

Goodall, H. L. (2006). *A need to know: The clandestine history of a CIA family.* Walnut Creek, CA: Left Coast Press.

Goodall, H. L. (2008). *Writing qualitative inquiry: Self, stories, and academic life.* Walnut Creek, CA: Left Coast Press.

Gray, R., & Sinding, C. (2002). *Standing ovation: Performing social science research about cancer.* Walnut Creek, CA: AltaMira Press.

Hartnett, S. J. (2003). *Incarceration nation: Investigative prison poems of hope and terror.* Walnut Creek, CA: AltaMira Press.

Hartnett, S. J., & Engels, J. D. (2005). "Aria in time of war": Investigative poetry and the politics of witnessing. In N. K. Denzin & Y. S. Lincoln (Eds.), *The SAGE handbook of qualitative research* (3rd ed., pp. 1043–1068). Thousand Oaks, CA: Sage.

Holman Jones, S. (2002). Torch. In N. K. Denzin & Y. S. Lincoln (Eds.), *The qualitative inquiry reader* (pp. 185–215). Thousand Oaks, CA: Sage.

Kaufman, M., & Members of the Tectonic Theater Project. (2001). *The Laramie Project.* New York: Vintage Books.

Kiesinger, C. E. (2003). He touched, he took. In R. P. Clair (Ed.), *Expressions of ethnography: Novel approaches to qualitative methods* (pp. 177–184). Albany: State University of New York Press.

Lockford, L. (2008). Investing in the political beyond. *Qualitative Inquiry, 14,* 3–12.

Madison, D. S. (1999). Performing theory/embodied writing. *Text and Performance Quarterly, 19,* 107–124.

Madison, D. S. (2005). *Critical ethnography: Method, ethics, and performance.* Thousand Oaks, CA: Sage.

Mienczakowski, J. (2001). Ethnodrama: Performed research—limitations and potential. In P. Atkinson, A. Coffey, S. Delamont, J. Lofland, & L. Lofland (Eds.), *Handbook of ethnography* (pp. 468–476). Thousand Oaks, CA: Sage.

Myers, B. (2008). Straight and White: Talking with my mouth full. *Qualitative Inquiry, 14,* 160–171.

Pelias, R. J. (1999). *Writing performance: Poeticizing the researcher's body.* Carbondale: Southern Illinois University Press.

Pelias, R. J. (1999). Confessions of an apprehensive performer. In R. J. Pelias, *Writing performance: Poeticizing the researcher's body* (pp. 79–87). Carbondale: Southern Illinois University Press.

Pineau, E. (2000). Nursing mother and articulating absence. *Text and Performance Quarterly, 20,* 1–19.

Pollock, D. (1998). Performative writing. In P. Phelan & J. Lane (Eds.), *The ends of performance* (pp. 73–103). New York: New York University Press.

Poulos, C. (2009). *Accidental ethnography: An inquiry into family secrets.* Walnut Creek, CA: Left Coast Press.

Prendergast, M., Leggo, C., & Sameshima, P. (Eds.). (2009). *Poetic inquiry: Vibrant voices in the social sciences.* Rotterdam, The Netherlands: Sense Publishers.

Rambo Ronai, C. R. (1996). My mother is mentally retarded. In C. Ellis & A. Bochner (Eds.), *Composing ethnography: Alternative forms of qualitative writing* (pp. 109–310). Walnut Creek, CA: AltaMira Press.

Richardson, L. (2000). Writing: A method of inquiry. In N. K. Denzin & Y. S. Lincoln (Eds.), *Handbook of qualitative research* (2nd ed., pp. 923–948). Thousand Oaks, CA: Sage.

Richardson, L. (2007). *Last writes: A daybook for a dying friend.* Walnut Creek, CA: Left Coast Press.

Richardson, L., & St. Pierre, E. A. (2005). Writing: A method of inquiry. In N. K. Denzin & Y. S. Lincoln (Eds.), *The SAGE handbook of qualitative research* (3rd ed., pp. 959–978). Thousand Oaks, CA: Sage.

Roethke, T. (2001). *On poetry and craft.* Port Townsend, WA: Copper Canyon Press.

Rosenthal, M. L. (1987). *The poet's art.* New York: W. W. Norton.

Russel y Rodriguez, M. (2002). Confronting anthropology's silencing praxis: Speaking of/from Chicana consciousness. In N. K. Denzin & Y. S. Lincoln (Eds.), *The qualitative inquiry reader* (pp. 347–376). Thousand Oaks, CA: Sage.

Saldana, J. (2005). *Ethnodrama: An anthology of reality theatre.* Walnut Creek, CA: AltaMira Press.

Sanders, E. (1976). *Investigative poetry.* San Francisco: City Lights Books.

Smith, A. D. (1993). *Fires in the mirror.* New York: Doubleday.

Smith, A. D. (2000). *Twilight: Los Angeles, 1992.* New York: Random House.

Smith, L. (2007). A life in books. *The Writer's Chronicle, 40*(2), 37–41.

Spry, T. (2001a). From Goldilocks to dreadlocks: Hair-raising tales of racializing bodies. In L. C. Miller & R. J. Pelias (Eds.), *The green window: Proceedings of the Giant City Conference on Performative Writing* (pp. 52–65). Carbondale: Southern Illinois University.

Spry, T. (2001b). Performing autoethnography: An embodied methodological practice. *Qualitative Inquiry, 7,* 706–732.

Spry, T. (2009). Bodies of/as evidence in autoethnography. *International Review of Qualitative Research, 1,* 603–610.

Stoller, P. (1997). *Sensuous scholarship.* Philadelphia: University of Pennsylvania Press.

Tillmann-Healy, L. M. (2001). *Between gay and straight: Understanding friendship across sexual orientation.* Walnut Creek, CA: AltaMira Press.

Tillmann-Healy, L. M. (2003). Hands. In R. P. Clair (Ed.), *Expressions of ethnography: Novel approaches to qualitative methods* (pp. 175–176). Albany: State University of New York Press.

Tracy, S. J. (2003). Watching the watchers: Making sense of emotional constructions behind bars. In R. P. Clair (Ed.), *Expressions of ethnography: Novel approaches to qualitative methods* (pp. 159–172). Albany: State University of New York Press.

Trujillo, N. (2004). *In search of Naunny's grave: Age, class, gender, and ethnicity in an American family.* Walnut Creek, CA: AltaMira Press.

Warren, J. T. (2001). Absence for whom? An autoethnography of White subjectivity. *Cultural Studies <=> Critical Methodologies, 1,* 36–49.

Watt, J. (2005). A gentle going? An autoethnographic short story. *Qualitative Inquiry, 11,* 724–732.

Watt, J. (2008). No longer loss: Autoethnographic stammering. *Qualitative Inquiry, 14,* 955–967.

Young, V. A. (2007). *Your average Nigga: Performing race, literacy, and masculinity.* Detroit, MI: Wayne State University Press.

41

EVALUATION AS A RELATIONALLY RESPONSIBLE PRACTICE

Tineke A. Abma and Guy A. M. Widdershoven

Evaluation is an applied science, and evaluators do their work amidst the sociopolitical dynamics of the practice evaluated. Although evaluation requires methodological steps, in this chapter we emphasize that evaluation is first of all a relationally responsible endeavor. In their work, evaluators enact a shared understanding of what it means to be an evaluator and do evaluative work. We argue that evaluators should engage with the world around them, taking fully into account their relationship with those in the practice being evaluated. This engagement includes a responsibility for the relationships developed as well as a shared responsibility for the development of socially just practices. A case example is presented from the field of psychiatry to illustrate the ideas in the chapter.

▣ EVALUATION AS A PRACTICE INVOLVING IMPLEMENTATION AND EDUCATION

Evaluation is first of all a practice, meaning that evaluators are workers, acting and conducting all kinds of activities like negotiating contracts, talking to people, typing out transcriptions, and writing reports. Much of their actions are based on a shared understanding of what it means to be an evaluator. "Shared" refers here to a way of doing things that is common in the evaluation field. Defining evaluation as a practice stresses that evaluation is much more complex than just the application of textbook knowledge; evaluators also have to deal with the sociopolitical complexity of their practice (impatient clients, bureaucratic managers, invisible users, etc).

The practice of evaluation is not inwardly focused, but very much grounded in the problems arising in the real world. Evaluation should be useful for practice. Therefore, implementation has always been an issue of concern in the field. From their beginnings, it was clear that evaluation studies have to inform and be useful for decision making and practice, whether these findings influence ways of thinking gradually, and as part of other information sources in an incremental policy-making process, or are directly used to adjust less complicated decision-making processes at the micro level (Patton, 1988; Weiss, 1988). Evaluation work is thus part of a sociopolitical process (Abma & Schwandt, 2005; Greene, 1994, 2000; House, 1981; Palumbo, 1987). This sociopolitical context creates a complicated dynamic; evaluative work often needs to be done within a short time frame, there may be a sponsor who may claim a special say, or the evaluator may encounter resistance or interest groups trying to influence the outcomes. The responsibilities of the evaluator are therefore demanding; there are many actors with conflicting interests and perspectives. In addition, programs or policies can be stopped as a result of the evaluation and people can be sanctioned. Implementation thus requires a political awareness and sensitivity on the part of the evaluator.

An important dimension of implementation is the pedagogical one. Evaluation fosters learning processes and can do so in various ways. A didactic approach to learning will strive for the application of findings after the evaluation has been conducted. It follows a linear conception of knowledge production: dissemination—translation—application. This transmissional view of information processing understands learning as a cognitive act, something that occurs in the mind of an individual and separate from the rest of one's activities. Acquiring knowledge and applying it are considered as distinct steps. A pedagogical approach to learning will place more emphasis on the learning that occurs in the action and interaction between practitioners. Learning is then understood as a social and collective process, and directly related to the working context of participants. This means that the evaluation process itself will foster learning experiences when it challenges participants to reflect on their practice and to ask themselves critical questions regarding the goodness of their practice. Qualitative evaluations foster such processes through face-to-face interactions. Deliberately engaging participants in a learning process concerning their practice will create co-ownership for practice improvements, and facilitates the implementation right from the start of the evaluation. Reasoning from a pedagogical stance, one should not consider the evaluation as an add-on after the intervention or program has been developed, but rather as a reflective learning process that is part of the development of practice. Action, reflection, and evaluation will then become intensively intermingled. Such a process is not the sole responsibility of the evaluator, we will argue, but a shared responsibility of all involved. The evaluator will engage all and foster interactions and relationships of trust in the interest of a more inclusive society.

The purpose of this chapter is to argue for an approach that takes the relational responsibility of evaluation seriously, both in term of relations as well as in terms of the normative horizon of social justice. Let us first go back into history, without pretending to cover all, to gain an insight into the intellectual history of the field, and qualitative evaluation in particular.

▣ HISTORY OF THE FIELD

Evaluation is a young, typically modernist discipline and profession. It was just after the Second World War that evaluation as a formal way of thinking and discipline came to life in the United States, closely related to developments in the field of education. Its birth is marked by the U.S. frontier ethos and federal attempts to improve the situation of poor people, concurrent with the launch of Sputnik by the Russians and a movement toward developing and enhancing the knowledge of children (Guba, personal communication, summer 1994). Massive educational programs were federally sponsored, with the obligation to evaluate their effectiveness. Evaluation was connected to social engineering; evaluators had to find out which programs worked best, what causal relations explained the program's effectiveness in order to generalize, predict, and control the performance of these programs elsewhere, as a basis for social betterment. Since evaluating programs had not been done before, social scientists and other professionals had to develop ways to measure, monitor, describe, control, and judge practices. Initially, scholars and practitioners in the field focused on the development of appropriate methodologies, which were initially predominantly quasi-experimental in design. In such a design, hypotheses—the articulated policy goals and assumptions—are a priori formulated (program theory) and then tested with the use of quantitative methods and techniques.

Qualitative evaluation can be defined as inquiry that establishes the value and goodness of a practice based on insiders' and contextual knowledge. The rise of qualitative evaluation dates back to the 1970s when Cronbach (Cronbach et al., 1980) proposed that the evaluation field needed good social anthropologists, a call that inspired many evaluators to become naturalistic or field researchers with a special focus on social relations (Stake, 1991; personal communication, summer 1994). In the UK, Parlett and Hamilton (1972) further elaborated the metaphor of the evaluator as social anthropologist. The authors proposed an approach to evaluation that is adaptive to situations. Central features were that the evaluator familiarizes him- or herself "thoroughly with the day-to-day reality" in order to "unravel the complexity." Being present and recording observations were the predominant activities. The evaluator was to record discussions with and between participants, and keep track of the language, jargon, and metaphors in order to gain insight into implicit pre-assumptions and interpersonal relations. In the USA, Stake also proposed a broader conceptualization of evaluation that included social interactions (Abma & Stake, 2001; Stake, 1967). Evaluators should not only gather data about program input and outcomes, but also describe its context and judge the quality of intermediary processes. Guba and Lincoln (1981) refer explicitly to Stake's work in the development of their "naturalistic" approach to evaluation.

These qualitative approaches have in common that they aim to holistically understand the evaluated program from the insider perspectives of the program participants and other stakeholders. This holistic understanding implies that the evaluator pays attention to many mutually influencing "factors" that shape the program and its context, such as the history of the program, the organization and culture in which it is embedded, the persons and personalities that take the lead, the political dynamics, and the social interactions and relations among stakeholders. All these factors are important in evaluation, because they are integral to and partly constitutive of the quality and effectiveness of the program being evaluated. Stake (1991) and others mention "personalities to be determining factors" (p. 12), but other aspects of social relations also count: leadership and charisma, caring and respect, reciprocity and collaboration, safety and structure, opportunities to learn and grow, but also hierarchy, conflict, envy, suspicion and mistrust, and many more. These are all aspects of stakeholders' relationships with one another that become interwoven in the fabric of the program being evaluated and thereby are integral to program quality and effectiveness. A program or policy should therefore be understood as a social practice; it is never just an intervention, but always a socially, historically, and culturally determined pattern of relations, interactions, and values. An example can be found in the work of Mabry (1991) who evaluated the arts education in an inner-city school. Here is a vignette based on her work:

> Alexandre Dumas is an elementary school located in the south side of Chicago. The evaluator aims to see how they teach the arts at Dumas. The evaluation is part of a larger study of arts education in elementary schools. A single step in the foyer and the evaluator realizes Dumas doesn't fit the generalized descriptions of a school in a poor neighborhood. It doesn't fit the expectations she has not tried to preconceive. Ushering in a sea of attentive black faces the evaluator suddenly realizes she is white. Having talked to many teachers and attended several classroom lessons, she concludes that at Dumas the arts are seen as a way out, a way to acknowledge and celebrate being African American as well as an entrée into mainstream culture, an opportunity to experience the good things in life. With no exception everyone attributes Dumas' success to the charisma and dedication of the principal, Ms. Silvia Peters. Her leadership makes Dumas a dream amidst the nightmare of many of Chicago's all–African American public schools. The evaluator presents a rich portrayal in order to show us who Sylvia Peters is and how she did it: She made the arts fun; she made them accessible; and she made them important.

Within qualitative approaches, like the one presented above, there is an awareness that evaluators should attend to the plurality of values and interests of those participating in the program and those of other stakeholders. Stake (1975) was among the first to draw attention to the one-sidedness and shortcomings of an approach that just took the program goals as the criterion for evaluation. This would lead to a management bias, and Stake therefore enlarged the scope of evaluation to include the issues of all possible stakeholders. This was based on the idea that a phenomenon has various, sometimes conflicting meanings for different stakeholders (Abma & Stake, 2001; Stake & Abma, 2005). Being responsive to the issues of stakeholders assumes an appreciation of their experiential knowledge. Methodologically, the acknowledgment of plurality implies that the "design" gradually emerges in conversation with the stakeholders. Metaphorically one may compare the designing process in such an evaluation with improvisational dance (Janesick, 2000). Whereas the minuet prescribes the definite steps, definite turns, and foot and arm movements, improvisation is spontaneous and reflexive of the social condition. The evaluator charts the progress and examines the route of the study as it proceeds by keeping track of his or her role in the research process. Since the design is not preordained, important methodological decisions have to be made along the way, like determining a point of saturation and selecting issues that require further exploration.

Also, the users of these approaches are well aware of the interpretive character of the evaluative work. Issues of stakeholders are not there to be discovered or revealed, they are not ready-made, but have to be elicited by the evaluator as a midwife. The birth of meaning is never just a matter of demonstration. Human beings, evaluators included, are interpreters. In order to make sense of our world and endow our experiences or those of others with meaning, we bring in our own background, training, prior experiences, desires, and standpoints. Every description is laden with interpretation. Yet, qualitative evaluation studies try to stay as close to the stakeholders' accounts and narratives as possible and are skeptical about the use of conceptual frameworks to prevent foreclosure or reduction of the data. In order to illuminate the quality of the practice, the evaluator has to use his or her judgment. This should not be understood as judgment in terms of calculation against preordained criteria. The qualitative evaluator does not predefine a set of evaluation criteria, but uses the stakeholder issues and their experiences as well as his or her observations as a source to come to an assessment of the program's quality. Stake explains that this is partly an intuitive process; one develops an understanding of the quality and later rationalizes what makes the practice good (personal communication, summer 1994; Stake & Schwandt, 2006). Schwandt (2005) refers to the Aristotelian virtue of *phronesis*, or wise judgment, to indicate what it means to evaluate the quality of a practice.

Wise judgment is a kind of ordinary, empirical, quasi-aesthetic, contextual knowing. Berlin (as quoted in Schwandt, 2005) aptly describes it as follows:

> Capacity for integrating a vast amalgam of constantly changing, multicolored, evanescent, perceptually overlapping data, too many, too swift, too intermingled to be caught and pinned down and labeled like so many individual butterflies. . . . To seize a situation in this sense one needs to see, to be given a kind of direct, almost sensuous contact with the relevant data, and not merely to recognize their general characteristics, to classify them or reason about them, or analyze them, or reach conclusions and formulate theories about them. (p. 325)

Wise judgment requires a power to attend to the particulars of a situation, to discriminate, and to see relevant details. Part of the evaluator's wisdom is also that he or she finds a balanced, middle-ground position in between antipathy and sympathy, between emotional and rational, and does justice to all stakeholders. The Aristotelian middle-ground position is also crucial in describing the practice. Stake (1982) notes, "The wisdom of the evaluator's findings will be little appreciated if couched in words that hurt too little or too much" (p. 80). Developing such wisdom is a process that never ends in the scholarly life of an evaluator; it is fostered among novices by a process of *Bildung*, which entails more than learning about qualitative methods and techniques. It requires a safe and friendly context where exploration and reflection on the self-as-evaluator (and of his or her authority, responsibility, obligation, and so forth) are stimulated.

◼ NEW INTERPRETATIVE PARADIGMS

The qualitative evaluation approaches described above emphasize the contextual understanding of programs by the evaluator as a wise judge. Wise judgment is detached. The evaluator as a wise judge wants to understand the world of the practitioners without taking into account the relevance for him- or herself and the input of the evaluator. This is quite different from interactive and dialogical evaluation approaches; here the relations between the evaluator and practitioners are central.

The difference between old and new interpretative approaches can be clarified by the work of the hermeneutic philosopher Hans Georg Gadamer (1960). According to Gadamer, there are three ways of understanding the world. The first stance is objectivist: The knower is detached from the outside world, stands above or outside that world (at least claims such a stance), and tries to explain the world with the use of universal laws. The perspectives of others in that world can be wrong, and their incorrect vision can be explained by underlying causalities or motives. We recognize this stance in the objectivist judgment approaches following a quantitative design. The second stance is subjectivist. Now the knower tries to understand the world by placing him- or herself in the footsteps of practitioners living in that world. The perspectives of other people are not wrong, but different. This empathic understanding of the uniqueness of the other from within can be recognized in the qualitative approaches described above. It can lead to relativism. Both the objectivist and subjectivist stance are problematic according to Gadamer. In both these stances, the knower does not relate to world; he does not take into account what the world means for himself nor what his input means for the world. The third position goes beyond objectivism and relativism (Bernstein, 1983) and is dialogical: Now the knower engages with the world around, taking fully into account what the world means for him, and vice versa. The knower listens to the other from a readiness to accept the other's input as being relevant for himself. Knower and known engage in a joint learning process in which both change in identity and new horizons emerge. The knower no longer just looks at the world, but interacts and takes responsibility for the process of development in the world. This intersubjective position can be found in interactive, dialogical approaches to evaluation. In these approaches, the evaluator and the participants in practice learn from each other and are jointly responsible for the outcome of this learning process.

Relevant evaluative traditions that embody the idea that the evaluator has a relationship with and thus a responsibility for the evaluated practice include those that promote transformation and change in the interest of human flourishing, such as feminist and transformative approaches to evaluation (Mertens, 2002, 2009; Whitmore et al., 2006), empowerment evaluation (Fetterman, 1994), democratic evaluation (MacDonald, 1977; Murray, 2002), participatory evaluation (Greene, 1997, 2006; Suarez-Herrera, Springett, & Kagan, 2009; Themessl-Huber & Grutsch, 2003), critical evaluation (Segerholm, 2001), social justice promoting traditions in evaluation (House, 1981, 1993), and fourth- and fifth-generation evaluation (Guba & Lincoln, 1989; Lincoln, 1993). The evaluator establishes particular relationships in his or her evaluation as a way of challenging such relationships—especially of power—in the context outside the evaluation. The evaluative purpose is the establishment of equal and just relations in society and the empowerment of marginalized groups in society, and therefore engagement and ownership is valued. In order to help effect the kinds of transformations desired, the evaluator purposefully uses the relational dimensions of evaluation. He or she purposefully establishes certain kinds of relationships—those that are accepting, respectful, and reciprocal—in order to help promote the overall social changes desired.

There is thus more emphasis on the evaluator's responsibility to foster the interactions among practitioners as a way to jointly develop socially responsible practices. Active partnership and participation are central values. In traditional qualitative approaches, the evaluator does all the interpretive and judgmental work alone; in an interactive approach, this is a joint responsibility of the evaluator and all the practitioners (including clients, patients, citizens, etc.). In interactive evaluation, the social relations between the evaluator and various practitioners and among practitioners are therefore central. Interactions and relationships always matter because these shape the evaluative knowledge that is generated in an evaluation and because these relationships convey what particular norms and values are being advanced in the evaluation. It matters, for example, that evaluators kneel in the mud alongside psychiatric patients because relationships thus formed are respectful, equitable, and accepting. With this action, they communicate the values of respect, attentiveness, and engagement (as opposed to the more distanced and hierarchic relation in objectivist evaluation approaches). In interactive evaluation, the social relations in the setting are not just an object of study, but there is an active engagement with the people in the setting. This broader responsibility stems from the critical consciousness and awareness that social practices are often marked by inequalities and social injustice, and the desire to create more responsible practices through evaluative work.

The underlying ontological notion is that human beings are fundamentally relational. Our social world is a product of social interactions and relations. Understanding of our socially constructed world can only be generated by developing a relation and dialogue with and between the inhabitants of this world. Epistemologically these traditions are grounded in the notion that object and subject mutually influence each other. There is a dialogical relationship: Instead of two independent entities standing in front of each other, knower and known are now engaged in a conversation. In this conversation, participants may change. Mertens (2002) describes the relationship between object and subject, knower and known as follows: "interactive, sensitive to those with the least power, and empowering to those involved in the process" (p. 106). Others point out that relations between evaluator and program participants become more democratic: The locus of control shifts from the evaluator as an expert to program people and other stakeholders who also have a say in the process (Greene, 1997; MacDonald, 1977; Murray, 2002; Themessl-Huber & Grutsch, 2003). Instead of being objective and impartial, evaluators working in these traditions show a "multiple partiality"; they identify with all people involved, which enables them to act as teachers who can explain the various experiences to the groups. At the same time, they are open to learning from each of the participants.

Evaluators working in these traditions understand evaluation as a political practice; it has unequal consequences for various stakeholders in the evaluation. They ask themselves *whose* interests they want to serve (Schwandt, 1997; Segerholm, 2001). Social relations and societal structures are not taken for granted, but critically examined and transformed. The evaluator criticizes power imbalances and the status quo, often on the grounds of critical theories. He or she acts as a social critic, arguing against domination, oppression, exploitation, cruelty, and violence (Segerholm, 2001; Mertens, 2002), and as an advocate of a particular silenced and marginalized group (Lincoln, 1993), not just to promote the interests of this group, but to allow the group to participate equally in the overall learning process. The intention to pay attention to social relations is driven by emancipatory and democratic ideals and a human rights agenda (Mertens, 2009). There is an engagement to empower people; to enlarge people's abilities to govern their own lives. On an individual level, empowerment refers to the development of voice, to more creative ways of interacting with one's social environment and the ability to take action in a more self-conscious way. On the collective level, empowerment refers to obtaining more influence and power in organizations and, eventually, in the processes of decision making in organizations and institutions.

Fostering Action, Interaction, and Dialogue

Qualitative evaluation aims to develop a rich and holistic understanding of the complexity of social practices from the insider's perspective of various stakeholder groups. This is taken a step further in interactive evaluation where the aim is to enhance the mutual understanding between practitioners. The interactive evaluator helps the practitioners to develop new and richer ways of dealing with actual problems in their practice.

They are engaged in dialogues about real problems. This distinguishes a dialogue from a theoretical debate. A dialogical approach implies a crucial role for action, experience, and learning. A dialogue presupposes that the participants already have some interest in and insight into the matter at hand. It also presupposes that the participants can elaborate their interest and knowledge through an exchange of perspectives.

In qualitative evaluation, practitioners are approached as meaning-making persons. They are invited to share their experiences with the evaluator. Interactive evaluation turns the process of sharing information into a two-sided relationship in which both the evaluator and the practitioner are active. The role of the practitioners goes beyond that of providing information. In interactive approaches, practitioners are actively involved in the evaluation process from beginning to end. Ways to engage practitioners may range from being an advisor to the process or a full member of the evaluation team to being participant in a learning community. Each time anew the evaluator will search for the best ways to engage people in the process. This means that data are gathered, not about participants in practice, but together with practitioners (professionals, clients, and other stakeholders). Likewise, the interpretation of the data in light of their consequences requires a dialogue with participants in practice. Methodological decisions made along the way will also be part of the negotiation with practitioners. The idea of emergence—not planning everything ahead—is an important prerequisite for the development of voice and inclusion. Generally, the emerging character of the process fosters the feeling of co-ownership.

Phases in the Process

Step 1: Ideally, the interactive evaluation process starts with the group least heard to ensure a balanced and fair process (Abma, Nierse, & Widdershoven, 2009; Baur, Abma, & Widdershoven, 2010). Their voice is unknown and needs to be articulated and elaborated. Also along the way, deliberate attention should be paid to the identification of "victims" or "silenced voices," those whose interests are at stake but remain unheard (Lincoln, 1993). Such voices may be hard to find, for example, because people want to remain anonymous or because they fear sanctions. To fully articulate the voice of those least heard often involves an intensive process. The evaluator needs to establish contact and develop a trusting relationship. Patience is an important virtue, and sometimes one needs to work with "cultural brokers" to be acceptable as an evaluator and to develop sensitivity for the specific cultural values and norms of the marginalized group.

Step 2: To prepare these groups for a dialogue with other groups often entails a process of empowerment where individuals first need to develop their own intimate voice, and then are brought into contact with each other to develop a shared agenda. Enclave deliberation among those with the same interests helps to turn an intimate voice into a political voice. Methods such as surveys and experiments are often not very appropriate to reach those who are marginalized (Mertens, 2002). Evaluators should therefore consider other methods to gain access to experiences, for example, in-depth interviews, focus groups, or storytelling workshops. Via interviews, people gain personal acknowledgment for their experiences (Koch, 2000). Sometimes other methods are more appropriate, such as focus groups. This depends on the population, so flexibility is needed with an attentive eye for the material conditions (timing, location, restitution for traveling costs, etc.) that create a comfortable space to speak up. The issues of established groups are often easier to determine. Their issues and agenda are often already documented, and persons from these groups are more articulate and able to provide focused information. Hence, this takes less time and energy.

Step 3: Having identified the issues for each stakeholder group, the next step is to create conditions and to organize dialogues and interactions between groups of stakeholders whose interests may diverge. Interaction between stakeholder groups is a deliberative process. *Deliberation* refers to the interaction and dialogue between participants. They will not just accept each other's beliefs and persuasions, but will explore these. Listening, probing, and questioning characterize this process, rather than confronting, attacking, and defending. Central features of dialogue are openness, respect, inclusion, and engagement (Abma et al., 2001; Greene, 2001). Dialogue may lead to consensus. Absence of consensus is, however, not problematic; on the contrary, differences stimulate a learning process (Widdershoven, 2001). Conditions for dialogue involve the willingness of stakeholders to participate, to share power, and to change in the process (Abma et al., 2001; Widdershoven & Abma, 2007).

Dialogue requires respect and openness, and that the evaluator should create a social infrastructure to facilitate participation and stakeholder communication. Deliberate attention should be paid to the power relations (Koch, 2000; Mertens, 2009). If a face-to-face encounter is impossible given asymmetries between stakeholder groups, one may organize a virtual meeting to stimulate a learning process between participants (Widdershoven, 2001). Experiences that have been exchanged in the safe environment of homogeneous groups are then introduced as issues in other stakeholder groups. By presenting such issues through stories, a climate of open discussion and dialogue may be fostered (Abma & Widdershoven, 2005). Active engagement of as many stakeholders as possible and deliberation minimize the chance of bias and domination of one party. Of course, bringing people together does not imply that everyone is heard. The moderator of the dialogues should therefore be alert for subtle mechanisms of exclusion. Afterward, evaluators should assess whether the dialogical process was really open. A careful reading of the transcript and an evaluation of the deliberative process with participants can give insight into this. During the process, the evaluator should also be prepared for the fact that those who are established and who feel superior may not be willing to participate in the process (They may think, what is there to learn?), or may join the process but find it hard to listen to the voices of other groups. Those who are not established might feel uncertain and need extra support to bring in their voice.

Dialogue and deliberative forums can be explicitly used to give voice to program participants who don't usually have a voice (House & Howe, 1999). This experience hopefully can affect participants' political power outside the evaluation context. Segerholm (2001) notes, however, that inclusion of stakeholders does not guarantee that values like equity and social justice will be prominent. There is always the risk of pseudo-participation. Evaluators should therefore ask who will profit from an inclusive deliberation.

Roles of the Evaluator

The roles of the interactively working evaluator include those of facilitator, teacher, and Socratic guide, and indicate an equal, friendly, and cordial relationship. The evaluator should engage with the people in the setting (Kushner, 2000). The evaluator has a "relational responsibility" and should pay deliberate attention to the "social relations of inquiry" (Greene, 2002). This means that the evaluator will try to create a "safe space" for participants and stakeholders "to learn how to relate and communicate equally and justly, a space that is unfettered by outside status and role differentials and animated by norms of reciprocity, parity, and respect" (Greene, 1997, p. 176; Greene, 2001). The evaluator must be sensitive to subtle mechanisms of exclusion. This is in line with what Aristotle (1997) said about wise judgment. The Aristotelian view of wise judgment goes beyond the detached view described above; it recognizes a hermeneutic and dialogical relationship between judge and world (Gadamer, 1960). The wise evaluator not only has a good perception of the practice and an adequate moral relation with the practice (unbiased, respectful), but he also strives toward just relations in the practice and is sensitive to power relations and helps practitioners to solve moral problems, not by placing him- or herself outside or above the practice, but by engaging in that practice.

▣ CASE EXAMPLE: REDUCTION OF COERCION IN PSYCHIATRY

In this part, we present a story from the field to illustrate our idea of evaluation as socially responsible practice. It concerns an evaluation as part of a movement to reduce coercion and restraint in psychiatry in The Netherlands. The purpose of the evaluation was to engender a dialogue about the application of coercive interventions and to develop normative guidelines for professionals in practice. In the first phase (1999–2001), six psychiatric hospitals participated in the process. The emphasis was on developing guidelines. The second phase also took 3 years (2002–2005). At this point, 12 psychiatric hospitals were involved. In this phase, the focus was on implementation of the guidelines. A learning community was set up among the project coordinators to foster active learning and connect experiments.

Inclusion, Voice, Learning Community, and Ownership

The evaluation processes were set up in a responsive way. In line with the notion that one needs to start with the group having least influence, a lot of time and energy was invested to contact patients, and to gain an insight into their experiences with coercion. It proved, however, to be difficult to include patients for in-depth interviews. They did not feel comfortable enough to tell their story to a professional who was an outsider, and the issue touched on a sensitive period in their lives. Therefore, we started working along with a research partner from an advocacy group who helped us to set up a series of communal meetings with client representatives. In the safe environment of this group, participants began to tell what had happened to them. The testimony of Jenny (pseudonym), a woman with a bipolar disorder, was recognized as expressing well the patients' perspective on coercion. The responses and recognition turned her personal testimony into a collective story witnessing the experiences inside the institutions. The story aptly revealed the interactional character of crises, how professionals were part of the problem situation without seeing it. Sharing and recognizing each other's stories helped the patients to transform personal issues into collective ones, and to build support among themselves as a group to place their issues on the agenda.

Jenny's Testimony, or Developing a Political Voice

Jenny was taken into care in a psychiatric hospital where she felt ill at ease and was hot-tempered. She says, *I could be very angry, because I felt powerless. I would be warned, and this would make me feel rebellious. If a nurse would say, "Do not extinguish your cigarette in that plant," I would not see this as reasonable, and I would become more and more obstruse.*

As a result, Jenny was often put in an isolation cell. The cell makes her feel miserable: *It is not humane to be in a cell without a toilet. These cardboard chamber pots are very awkward.*

After a while the medication starts to work. Yet, she is anxious about the risk of becoming numbed by it. Luckily, this does not occur: *I was really worried,*

that I would not have any feelings any more. Well, it did not turn out that way.

She is hesitant about leaving the hospital and returning home. In the end, she is forced to go. At home she often feels down and lonely: *I stayed in the hospital for three years, and then I had to go home. I went home one day, and never returned to the hospital. I experienced a lot of anxiety and loneliness. I felt left alone by the hospital. I often called the emergency phone.*

Subsequently, in order to foster the interaction between stakeholder groups, heterogeneous dialogue groups (managers, nurses, psychiatrists, patients, and family) were organized to discuss the issues derived earlier in the process. In these mixed groups, the focus was on ethical problems regarding coercion. We were not so much interested in the question of when coercion is legitimate (a legal question), but how coercion can be prevented and, if needed, be carried out in a careful and responsible manner (an ethical question). Analyzing the material, we (the research team) were able to establish ethical problems about coercion and restraint, and to develop guidelines for action, which we called quality criteria. These criteria were again discussed within dialogue groups to validate them and to facilitate stakeholder interaction and mutual learning. The criteria changed in this iterative process (see Table 41.1 for an overview of the most important criteria). To make the quality criteria assessable for those working in practice, a short brochure was composed and widely disseminated. In addition, a longer document was crafted that grounded the quality criteria in concrete experiences and theoretical insights (Berghmans, Elfahmi, Goldsteen, & Widdershoven, 2001).

Table 41.1 Quality Criteria for Coercion and Compulsion

- Be aware of contradictory obligations in handling situations of coercion.
- Create room for emotions, reflect upon them, and discuss them.
- Pay attention to the process character of coercion: Anticipate and evaluate incidents.
- Pay attention to communication: Be attentive and open toward the patient; reflect upon goals and means.

After the drafting of the document, the second phase of the project started. In 12 institutions, implementation activities were set up. These were different for each institution, based upon local context and expertise. Staff from the institutions coordinated the activities, while the evaluators focused on giving feedback and organizing meetings among the local organizers to exchange experiences and learn from one another. Also, a learning community was set up among the project coordinators in each institution. The learning community aimed to foster "situated learning" (versus learning from material abstracted from context) in relationships between people. It was based on the assumption of an intimate connection between knowledge, relationships, and action. The learning process of the project coordinators was intimately connected with and embedded in their practice. The knowledge produced was context-specific, practice- and experience-based, and interactively derived. Below, an example is given of the knowledge generated among the project leaders in response to a problematic situation presented and discussed in one of the collegial meetings. It concerns the question of whether or not the institution should build extra seclusion rooms concentrated in a specialized unit. The problem was brought in by a project coordinator, called Larry (pseudonym), who was also the head of the closed units in one of the participating hospitals.

Larry's Problem, or Generating Support

Larry, project coordinator of the implementation of the quality criteria, has been asked to give his Board of Directors advice about the number of rooms required in the new seclusion unit. There appears to be a growing group of youngsters with aggressive behavior. Seclusion is considered to be the only option for them. A specialized seclusion unit is proposed as the solution for the problems at hand.

The participants in the meeting ask questions about Larry's power. He responds that he has at least some influence on the formation of the seclusion unit and the number of seclusion rooms. In response to a question concerning the vision of care underlying the plans, Larry explains, *The management reasons, How should these bricks be piled and what is the most logical and cheap way to do that? There is not much talk about our vision of quality of care.*

Participants recognize the case and bring several negative experiences with specialized seclusion units to the fore, such as that patients may experience it as a punishment to go to another unit; the seclusion unit easily becomes a sort of internal police service within the organization and the availability of seclusion rooms will create a demand. One of the participants calls it a "logistics story" and misses the underlying ideas about what quality of care means. Participants also question the necessity of the reorganization: *For whom is this? It is certainly not in the interest of the staff and patients!* The group gives the advice to pay more attention to the means to prevent seclusion, such as creating "healing environments" and enhancing communication and interaction with patients.

Larry's story shows how work experiences were shared within the learning community. Participants shared stories about failures, hard-earned knowledge that if not narrated might not be known. These stories were used by Larry, who presented the perceived problems and alternative solutions in discussions with his Board of Directors. In the end, the Board decided to build four instead of the planned six seclusion rooms.

Evaluators and local staff wrote down their experiences in a book, describing both the tensions in the project and the positive outcomes (Abma, Widdershoven, & Lendemeijer, 2005). During the implementation, the formulation of the guidelines did not change. Yet the meaning of the guidelines for practice was further specified through experiments and the development of so-called "good practices." (It was agreed between organizing staff and ethicists that we should not use the notion of "best practices." At that time, the hospitals had just started to experiment with new approaches, and we did not expect to gain a shared understanding of a "best" practice.)

Perspective, Experiential Knowledge, Dialogue, Shared Understanding

One of the key assumptions of interactive evaluation regards the notions of perspective and exchange of perspective, or, in terms of Gadamer (1960), fusion of horizons. Qualitative evaluation aims to articulate and explore the various, sometimes conflicting perspectives on a situation under consideration and to foster communication and dialogue among these perspectives.

In the project, we deliberately explored the perspectives of all stakeholder groups on coercion; managers, psychiatrists, nurses, patients, and family were consulted so we could understand their experiences and point of view. It soon became clear that their perspectives varied. Patients like Jenny focused on the negative side of coercion, and expressed their feelings of dehumanization, powerlessness, anger, loss of control, and fear. They brought to the fore that seclusion, the preferred method of coercion in The Netherlands, was often experienced as a punishment. They also complained about the fact that such incidents were never discussed afterward.

The nurses took another perspective, focusing on the constraining organizational conditions (lack of staff, shortage of time, too many administrative tasks, large caseload) that complicated their work. They also reported about the difficulties involved in assessing the risk of danger, and mentioned problems in the decision-making process. Dialogue about coercion assisted the nurses to articulate their identity; instead of being a "guardian," they began to redefine their identity in terms of being a "coworker" of the patient, developing a new, more cordial relationship (Landeweer, Abma, & Widdershoven, 2010). Psychiatrists focused on the means of coercion and lacking the legal conditions to substitute seclusion by (enforced) medical treatment. Family members felt that the communication with professionals was not always optimal: Why were they not informed about the situation of patients and consulted about proper courses of action? Finally, the project coordinators and managers like Larry reported about the culture of control (versus negotiation) in their institutions and resistance to change. Each of these perspectives was articulated, explored, and placed in the context of the background, training, experiences, and position of each of the groups. In other words, meanings and (moral) judgments were associated with the concrete context and the positions of stakeholders.

Interactive evaluation conceives dialogue as a learning process. Stakeholders were engaged in genuine dialogues. This was unique, since in public debates on coercion these parties tended not to talk, but to discuss and debate with each other and act strategically toward each other. In public, parties would bring their standpoints to the fore, taking different positions without engaging in dialogue. In the project, however, parties gained a name and face and became persons who were starting to talk with each other, exchanging perspectives.

An example concerns the conversation about the role of the patient's history in decisions about coercion. Professionals stressed that one should place a patient's behavior in the larger context of former ways of behaving. In order to judge whether a certain action is dangerous, one should take into account previous experiences with the patient. This obligation was stressed by psychiatric nurses. On the other hand, patients brought to the fore that the patient should get a chance to show different behavior. If the staff intervenes because they predict problems, based on prior events, they do not trust the patient to be able to handle the situation better this time. As a consequence, the patient is "locked up" in his or her history. Nurses and patients agreed that there is no simple solution to this dilemma. The best thing to do is to recognize that it is unfair not to take the history into account (since that would take away possibilities of prevention) and to see the patient's history as a causal determinant for the current situation. This example shows that nurses and patients learned from each other and developed new and enriched insights. They not only learned about the concrete perspectives of the others, but they also became aware of the more general fact that concrete situations are multi-interpretable and that certain methods or communication strategies (like dialogue) bring out these various interpretations.

The example also shows the relevance of practical rationality and wise judgment that guides interactive qualitative evaluation, as a characteristic not just of the evaluator, but of all the participants in the process. Our conversations always started from practice, with practical case examples, appreciating and using the experiences of participants as a valid source of moral wisdom. The practical knowledge of the participants was articulated and further developed by dialogical exchanges. By these means, participants gained more general knowledge of the situation and

developed rules to deal with it in the form of quality criteria. They were able to develop wise judgment, not as an individual quality, but as a group capacity.

The shared understanding between parties—all supported the quality criteria—was largely due to the fact that people listened to each other's experiences around coercion and restraint, and came to understand uncertainties and worries. The inclusion of many voices and (disciplinary) perspectives enhanced the insight into the moral problems encountered in the care for psychiatric patients and led to a fuller understanding of what is required to improve the quality of coercion and care in general. For example, the input of the patients helped to further unravel the complexities in communication between patients and professionals. We also noted that people with various disciplinary backgrounds learned from each other. For example, nurses and psychiatrists came to appreciate each other's view of the situation. The dialogues enhanced the mutual understanding among participants and gave a voice to those who often have no say in policy processes. Moreover, the continuous dialogue on and reworking of the documents enabled the participants to influence the process and secured their commitment.

An important characteristic of the process was the commitment to and responsibility for creating change and improving practice, shared by the evaluators and the participants. Everybody was convinced that the existing way of dealing with coercion was unsatisfactory, and that reduction and quality improvement were urgent. Thus, the project had a political nature. By seeing this as a shared responsibility, strategic action by individuals was prevented. All parties recognized that they were unable to develop solutions individually, and needed each other to create a better practice. Both the evaluators and the participants showed their vulnerability: Since they would not be satisfied with simple but inadequate solutions, they would sometimes show emotions like frustration and even anger. By being open and sincere about this, the participants created an atmosphere of trust and mutual support, which was experienced as unique and helped to overcome problems along the way.

▣   DISCUSSION AND CONCLUSION

In this chapter, we presented the basic notions of an interactive approach to evaluation. Interactive evaluation is relational and socially responsible. It goes beyond objective approaches to evaluation in that it takes into account the meanings of participants in practice. In this sense, interactive evaluation is part of the qualitative tradition in evaluation. It differs from classical qualitative approaches in that it does not regard the interpretation of meanings in practice as an activity of the evaluator, listening to the participants as a wise judge, but as a dialogical process; both the evaluator and the participants in practice are active in the process of meaning making and interpretation;

wise judgement is an activity shared by all. Interactive evaluation is relationally responsible in that the evaluator engages in the practice and stimulates the participants to be active and responsible themselves for the quality of their interactions, communication, and relations. In addition, the evaluator tries to develop a shared responsibility with the participants to jointly work toward the normative ideal of a more inclusive society. We described how this worked in the case of reduction of coercion in psychiatry. The evaluation was based on the political idea that the practice of coercion should be improved. This presupposition was shared among all the participants. The role of the evaluator was to foster a dialogue on experiences around coercion and to stimulate the parties involved to take and share responsibility for improvement.

One may question whether the evaluator, in giving voice to the group least heard, does not take sides, and thus strategically influences the evaluation process. In order to create conditions for dialogue, the evaluator does have to support weaker parties and thus empower them. This process of empowerment is, however, not dictated by the evaluator. The participants in the weaker group themselves become confident by listening to each other and sharing emotions and knowledge. In the case of Jenny, the story was developed in interaction with and through the responses of the other patients. They shared her indignation, and as a group became *self*-conscious. The group helped Jenny to articulate her intimate voice, and in the chorus with other voices the group developed a political voice and agenda. This prepared them for the dialogue with other, more established parties. Likewise, Larry was empowered, not by the evaluators, but by the group of colleague project leaders. In the political science literature, developing a voice within a group of participants with the same interests is known as "enclave deliberation" (Karpowitz, Raphael, & Hammond, 2009). Enclave deliberation helps to deal with power differences between groups and prevents domination of established groups. It serves as an alternative to proportional deliberation, a process with equal amounts of participants with diverging interests. Creating preconditions for dialogue is not a strategic activity. It requires trust in the participants and respect for their ability to express their experiences. The evaluator can only succeed if the members of the group take responsibility for helping each other in making explicit issues and concerns.

By emphasizing that interactive evaluation is relationally responsible, we do not deny that other approaches to evaluation can and should be responsible. Objective approaches to evaluation have to be fair and measure outcomes in a valid way. This is an example of being responsible. Interpretive approaches presuppose wise judgment of the evaluator. Interactive evaluation underlines the importance of validity and wisdom, but defines and realizes these notions in a new and radical way. In interactive evaluation, validity and wisdom are regarded as social and relational products. What counts as valid measurements should be agreed upon by all parties involved. Practical wisdom should be shared between participants and developed in a joint learning process. The responsibilities as interactive evaluators define them also to have consequences for the quality criteria that guide their work. Many years ago, Guba and Lincoln (1989) provided a set of trustworthiness, authenticity, and fairness criteria. We find these still to be very helpful. Interactive evaluation is not only good when it results in better understandings, it should also lead to an enhanced insight among all participants, so that they are better equipped to interact creatively with their surroundings. The process toward such shared understanding and responsibility should be fair and just. Striving for one codified set of quality criteria does not acknowledge that various approaches have diverse sets of responsibilities and goals.

What are the challenges for interactive evaluation as a socially responsible practice? One challenge is to develop ways of making cultural change enduring. Through dialogue, participants in a practice can be stimulated to become more open toward each other, to understand various perspectives and give up traditional oppositional views. This openness is a precondition for practice change. Yet, developing shared understandings is not enough to make change enduring. In the case of reduction of coercion in psychiatry, we saw a large change in views. Whereas coercion used to be regarded as a normal element of psychiatric practice, it is now seen as problematic and unwanted. In many institutions, initiatives to reduce coercion have been set up, resulting in a decrease in the number and the length of seclusions. These successes are, however, fragile. It may become difficult to develop new ideas because so much has already been achieved. Thus, new input is constantly needed. When project leaders move to new jobs, the project is vulnerable. This shows that the project is still too dependent on individual people, and structural implementation is weak.

A further challenge and new territory is the trend toward intercultural diversity. Over the years, it has become more common for nurses, patients, and families to have a voice in health care policy; these groups have become more empowered and accepted as equal participants. Yet, professionals, patients, and families from a non-Western cultural background are less involved in processes of policy making. The participants in the project of reduction of coercion in psychiatry were mainly White. Over the years, the population of institutions of mental health care in The Netherlands has become more diverse, with a larger proportion of patients coming from other cultures. A challenge for the future will be to ensure their participation, giving them voice and stimulating them to be responsible partners in interactive evaluation. Evaluators may seize the opportunity to develop and evaluate diversity-sensitive practices with the communities concerned (Burlew, 2003). Working in a globalized context may then actually result in an enrichment of our practice, as it will confront us with new values and norms and place our own ideas in perspective.

▣ REFERENCES

Abma, T. A., Greene, J., Karlsson, O., Ryan, K., Schwandt, T. S., & Widdershoven, G. (2001). Dialogue on dialogue. *Evaluation, 7*(2), 164–180.

Abma, T. A., Nierse, C., & Widdershoven, G. A. M. (2009). Patients as research partners in responsive research. Methodological notions for collaborations in research agenda setting. *Qualitative Health Research, 19*(3), 401–415.

Abma, T. A., & Schwandt, T. S. (2005). The practice and politics of sponsored evaluations. In B. Somekh & C. Lewis (Eds.), *Research methods in the social sciences* (pp. 105–112). London: Sage.

Abma, T. A., & Stake, R. (2001). Stake's responsive evaluation: Core ideas and evolution. *New Directions for Evaluation, 92,* 7–22.

Abma, T. A., & Widdershoven, G. A. M. (2005). Sharing stories: Narrative and dialogue in responsive nursing evaluation. *Evaluation and the Health Professions, 28*(1), 90–109.

Abma, T. A., Widdershoven, G. A. M., & Lendemeijer, B. (Eds.). (2005). *Dwang en drang in de psychiatrie, De kwaliteit van vrijheidsbeperkende maatregelen.* Utrecht, The Netherlands, Lemma.

Aristotle. (1997). *Ethica Nicomachea* (Book IV). Amsterdam: Kallias.

Baur, V., Abma, T. A., & Widdershoven, G. A. M. (2010). Participation of older people in evaluation: Mission impossible? *Evaluation and Program Planning, 33*(3), 238–245.

Berghmans, R., Elfahmi, D., Goldsteen, M., & Widdershoven, G. A. M. (2001). *Kwaliteit van dwang en drang in de psychiatrie.* Utrecht/ Maastricht: GGZ Netherlands/Universiteit Maastricht.

Bernstein, R. J. (1983). *Beyond objectivism and relativism.* Oxford, UK: Oxford University Press.

Burlew, A. K. (2003). Research with ethnic minorities: Conceptual, methodological, analysis issues. In G. Bernal, J. E., Trimble, A. K. Burlew, & F. T. L. Leong (Eds.), *Handbook of racial and ethnic minority psychology.* Thousand Oaks, CA: Sage.

Cronbach, L. J., et al. (1980). *Toward reform of program evaluation.* San Francisco: Jossey-Bass.

Fetterman, D. (1994). Empowerment evaluation. *Evaluation Practice, 15,* 1–6.

Gadamer, H. G. (1960). *Wahrheit und Methode.* Tübingen, Germany: J.C.B. Mohr.

Greene, J. C. (1994). Qualitative program evaluation: Practice and promise. In N. K. Denzin & Y. S. Lincoln (Eds.), *Handbook of qualitative research* (pp. 530–544). Thousand Oaks, CA: Sage.

Greene, J. C. (1997). Participatory evaluation. In L. Mabry (Ed.), *Evaluation and the post-modern dilemma. Advances in program evaluation* (Vol. 3, pp. 171–189). Greenwich, CT: JAI Press.

Greene, J. C. (2000). Understanding social programs through evaluation. In N. K. Denzin & Y. S. Lincoln (Eds.), *Handbook of qualitative research* (2nd ed., pp. 981–1000). Thousand Oaks: Sage.

Greene, J. C. (2001). Dialogue in evaluation: A relational perspective. *Evaluation, 7*(2), 181–203.

Greene, J. C. (2002, October). *Evaluation as education.* Paper presented at the European Evaluation Society, Seville, Spain.

Greene, J. C. (2006). Evaluation, democracy, and social change. In I. F. Shaw, J. C. Greene, & M. M. Mark (Eds.), *The SAGE handbook of evaluation* (pp. 141–160). London: Sage.

Greene, J. C., & Abma, T. A. (Eds.). (2001). Responsive evaluation. *New Directions for Evaluation, 92.*

Guba, E. G., & Lincoln, Y. S. (1981). *Effective evaluation.* San Francisco: Jossey-Bass.

Guba, E. G., & Lincoln, Y. S. (1989). *Fourth-generation evaluation.* Newbury Park, CA: Sage.

House, E. R. (1981). *Evaluating with validity.* Beverly Hills, CA: Sage.

House, E. R. (1993). *Professional evaluation.* Newbury Park, CA: Sage.

House, E. R., & Howe, K. R. (1999). *Values in evaluation and social research.* Thousand Oaks, CA: Sage.

Janesick, V. J. (2000). The choreography of qualitative research design: Minuets, improvisations, and crystallization. In N. K. Denzin & Y. S. Lincoln (Eds.), *Handbook of qualitative research* (2nd ed., pp. 379–400). Thousand Oaks, CA: Sage.

Karpowitz, C. F., Raphael, C., & Hammond, A. S. (2009). Deliberative democracy and inequality: Two cheers for enclave deliberation among the disempowered. *Politics & Society, 37,* 576–615.

Koch, T. (2000). Having a say: Negotiation in fourth-generation evaluation. *Journal of Advanced Nursing, 31,* 117–125.

Kushner, S. (2000). *Personalizing evaluation.* London: Sage.

Landeweer, E. G. M., Abma, T. A., & Widdershoven, G. A. M. (2010). The essence of psychiatric nursing: Redefining nurses' identity through moral dialogue about reducing the use of coercion and restraint. *Advances of Nursing Science, 33*(4), E1–E12.

Lincoln, Y. S. (1993). I and thou: Method, voice, and roles in research with the silenced. In D. McLaughlin & W. Tierney (Eds.), *Naming silenced lives* (pp. 29–47). London: Routledge.

Mabry, L. (1991). Alexandre Dumas Elementary School, Chicago, Illinois. In R. Stake, L. Bresler, & L. Mabry (Eds.), *Custom & cherishing: The arts in elementary schools* (pp. 137–176). Chicago: University of Illinois, National Arts Education Research Center.

MacDonald, B. (1977). A political classification of evaluation studies. In D. Hamilton, D. Jenkins, C. King, B. MacDonald, & M. Parlett (Eds.), *Beyond the numbers game* (pp. 224–227). London: Macmillan.

Mertens, D. (2002). The evaluator's role in the transformative context. In K. E. Ryan & T. S. Schwandt (Eds.), *Exploring evaluator role and identity* (pp. 103–118). Greenwich, CT: IAP.

Mertens, D. M. (2009). *Transformative research and evaluation.* New York: Guilford Press.

Murray, R. (2002). Citizens' control of evaluations: Formulating and assessing alternatives. *Evaluation, 8*(1), 81–100.

Palumbo, D. J. (Ed.). (1987). *The politics of program evaluation.* Newbury Park, CA: Sage.

Parlett, M., & Hamilton, D. (1972). Evaluation as illumination: A new approach to the study of innovatory programs. In G. Glass (Ed.), *Evaluation review studies annual* (Vol. 1, pp. 140–157). Beverly Hills, CA: Sage.

Patton, M. Q. (1988). The evaluator's responsibility for utilization. *Evaluation Practice, 9,* 5–24.

Schwandt, T. S. (1997). Whose interests are being served? Program evaluation as conceptual practice of power. In L. Mabry (Ed.), *Evaluation and the post-modern dilemma: Advances in program evaluation* (Vol. 3, pp. 89–104). Greenwich, CT: JAI Press.

Schwandt, T. S. (2005). On modeling our understanding of the practice fields. *Pedagogy, Culture and Society, 13*(3), 313–332.

Segerholm, C. (2001). Evaluation as responsibility, conscience, and conviction. In K. E. Ryan & T. S. Schwandt (Eds.), *Exploring evaluator role and identity* (pp. 87–102). Greenwich, CT: IAP.

Stake, R. E. (1967). The countenance of evaluation. *Teachers College Record, 68,* 523–540.

Stake, R. E. (1975). To evaluate an arts program. In R. E. Stake (Ed.), *Evaluating the arts in education: A responsive approach* (pp. 13–31). Columbus, OH: Merrill.

Stake, R. E. (1982, August). How sharp should the evaluator's teeth be? *Evaluation News,* pp. 79–80.

Stake, R. E. (1991). Retrospective on "The Countenance of Educational Evaluation." In M. W. McLaughlin & D. C. Phillips (Eds.), *Evaluation and education: At quarter century, ninetieth yearbook of the National Society for the Study of Education (NSSE)* (pp. 67–88). Chicago: University of Chicago Press.

Stake, R. E., & Abma, T. A. (2005). Responsive evaluation. In S. Mathison (Ed.), *Encyclopedia of evaluation* (pp. 376–379). Thousand Oaks, CA: Sage.

Stake, R. E., & Schwandt, T. S. (2006). On discerning quality in evaluation. In I. F. Shaw, J. C. Greene, & M. M. Mark (Eds.), *The SAGE handbook of evaluation* (pp. 404–418). London: Sage.

Suarez-Herrera, J. C., Springett, J., & Kagan, C. (2009). Critical connections between participatory evaluation, organizational learning and intentional change in pluralistic organizations. *Evaluation, 15*(3), 321–342.

Themessl-Huber, M. T., & Grutsch, M. A. (2003). The shifting locus of control in participatory evaluations. *Evaluation, 9*(1), 92–111.

Weiss, C. (1988). If program decisions hinged only on information: A response to Patton. *Evaluation Practice, 9,* 15–28.

Whitmore, E., Gruijt, I., Mertens, D. M., Imm, P. S., Chinnan, M., & Wandersman, A. (2006). In I. F. Shaw, J. C. Greene, & M. M. Mark (Eds.), *The SAGE handbook of evaluation* (pp. 340–359). London: Sage.

Widdershoven, G. A. M. (2001). Dialogue in evaluation: A hermeneutic perspective. *Evaluation, 7*(2), 253–263.

Widdershoven, G. A. M., & Abma, T. A. (2007). Hermeneutic ethics between practice and theory. In R. E. Ashcroft, A. Dawson, H. Draper, & J. R. McMillan (Eds.), *Principles of health care ethics* (pp. 215–222). West Sussex, UK: Wiley.

Part VI

THE FUTURE OF QUALITATIVE RESEARCH

And so we come to the end, which is only the starting point for a new beginning, the contested future. Several observations have structured our arguments to this point. The field of qualitative research continues to transform itself. The changes that took shape in the first decade of this new century are gaining momentum, even as they confront multiple forms of resistance. A new generation is making its presence felt. Scholars trained in the postmodern and experimental moments take for granted what early generations fought to establish.

The indigenous, gendered, narrative turn has been taken. Foundational epistemologies, what Schwandt (2007) calls epistemologies with the big E, have been replaced by post-postconstructivist, hermeneutic, feminist, poststructural, pragmatist, critical race, and queer theory approaches to social inquiry. Epistemology with a small e has become normative, displaced by discourses on ethics and values, conversations on and about the good, and about the just and moral society.

Qualitative inquiry is under assault from three sides. First, on the *political right,* are the methodological conservatives who are connected to neoconservative governmental regimes. These critics support evidence-based, experimental methodologies, or mixed methods. This stance consigns qualitative research to the methodological margins. Second, on the *epistemological right,* are neotraditionalist methodologists who look with nostalgia at the golden age of qualitative inquiry. These critics find in the past all that is needed for inquiry in the present. Third, on the *ethical right,* are mainstream, biomedical scientists and traditional social science researchers who invoke a single ethical model for human subject research. The ethical right refuses to engage the arguments of those researchers who engage in collaborative, consciousness raising, empowering inquiry.

Qualitative researchers in the seventh and eighth moments must navigate between these three oppositional forces, each of which threatens to deny the advances in qualitative research over the past three decades. These critics do not recognize the influences of indigenous, feminist, race, queer, or ethnic border studies. We need to protect ourselves from these criticisms. We also need to create spaces for dialogue and public scholarly engagement of these issues.

The chapters in this volume collectively speak to the great need for a compassionate, critical, interpretive, civic social science. This is an interpretive social science that blurs both boundaries and genres. Its participants are committed to politically informed action research, inquiry directed to praxis, and social change. Hence, as the reformist movement called qualitative research gains momentum, its places in the discourses of a free democratic society become ever clearer. With the action researchers, we seek a disciplined set of interpretive practices that will produce radical democratizing transformations in the public and private spheres of the global postcapitalist world. Qualitative research is the means to these ends. It is the bridge that joins multiple interpretive communities. It stretches across many different landscapes and horizons, moving back and forth between the public and the private, the sacred and the secular.

Paradigm shifts and dialogues have become a constant presence within and across the theoretical frameworks that organize both qualitative inquiry and the social and human sciences. The move to standpoint epistemologies has accelerated. No one any longer believes in the concept of a unified sexual subject, or indeed, any unified subject. Epistemology has come out of the closet. The desire for critical, multivoiced, postcolonial ethnographies increases as capitalism extends its global reach.

We now understand that the civic-minded qualitative researcher uses a set of material practices that bring the world into play. These practices are not neutral tools. This researcher thinks historically and interactionally, always mindful of the structural processes that make race, gender, and class potentially repressive presences in daily life. The material practices of qualitative inquiry turn the researcher into a methodological (and epistemological) *bricoleur.* This person is an artist, a quilt maker, a skilled craftsperson, a maker of montages and collages. The interpretive *bricoleur* can interview; observe; study material culture; think within and beyond visual methods; write poetry, fiction, and autoethnography; construct narratives that tell explanatory stories; use qualitative computer software; do text-based inquiries; construct *testimonios* using focus group interviews; and even engage in applied ethnography and policy formulation.

It is apparent that the constantly changing field of qualitative research is defined by a series of tensions and contradictions, as well as emergent understandings. These tensions and understandings have been felt in every chapter in this volume. Here, as in previous editions, we list many of them for purposes of summary only. They take the form of questions and assertions:

1. Will the performance turn in qualitative inquiry lead to performances that decolonize theory, and help deconstruct that global, postcolonial situation?

2. Will critical, indigenous interpretive paradigms, epistemologies, and pedagogies flourish in the eighth moment?

3. Will critical, indigenous interpretive paradigms, epistemologies, and pedagogies lead to the development and use of new inquiry practices, including counternarratives, autoethnographies, cultural poetics, and arts-based methodologies?

4. Can indigenous qualitative researchers take the lead in decolonizing the academy?

5. Will the emphasis on multiple standpoint epistemologies and moral philosophies crystallize around a set of shared understandings concerning the contributions of qualitative inquiry to civil society, civic discourse, and critical race theory?

6. Will the criticisms from the methodological, political, and ethical conservatives stifle this field?

7. Will the performance turn in ethnography produce a shift away from attempts to represent the stream of consciousness and the world of internal meanings of the conscious subject?

8. How will feminist, communitarian, and indigenous ethical codes change IRBs? Will the two- and three-track IRB model become normative?

9. Will a new interpretive paradigm, with new methods and strategies of inquiry, emerge out of the interactions that exist between the many paradigms and perspectives we have presented in this volume?

10. How will indigenous, ethnic, queer, postcolonial, and feminist paradigms be fitted to this new synthesis, if it comes?

11. Will the postmodern, anti-foundational sensibility begin to form its own foundational criteria for evaluating the written and performed text?

12. When all universals are gone, including the postmodern worldview, in favor of local interpretations, how can we continue to talk and learn from one another?

There are no definitive answers to any of these questions. Examined from another angle, the 12 questions listed above

focus on the social text, history, politics, ethics, the Other, and interpretive paradigms more broadly.

▣ INTO THE FUTURE

The chapter by Judith Preissle (Chapter 42) reflexively moves qualitative inquiry into the future. In so doing, she draws on her own experiences of more than 40 years as a qualitative researcher and teacher. (She established QUALRS-L, a listserv discussion group on qualitative research, in 1991.) She uses multiple metaphors—bramble bush, tapestry, umbrella, confederacy—to describe this sprawling discourse. It has multiple roots—like a bramble bush. It is like a tapestry, with many different threads, colors, and patterns. It is like an umbrella in that it covers a variety of traditions. It is also like a confederacy in the Native American Iroquois manner—it extends across time and space; its membership is fluid.

She reads qualitative research's future through the lens of these metaphors, seeing a future that will be more just, more free, more open to alternative perspectives, and more secure. She imagines a world where young scholars will bring new understandings of critical inquiry into the classroom and into their written work.

▣ RIGHT AND LEFT POLE ACTIVIST METHODOLOGIES[1]

Eisenhart and Jurow (Chapter 43) enter this pedagogical space. They review the current literature on qualitative pedagogy, and offer extended discussion of the two-semester Introduction to Qualitative Research course required of all entering doctoral students in the College of Education at the University of Colorado, Boulder.

They reinforce the argument that the literature on teaching qualitative research continues to reflect the 1980 paradigm disputes. With Phillips (2006), they see two pedagogical camps, or two poles on a continuum, a right pole and a left pole. On the right pole are the traditionalists who view methods as objective tools. Traditionalists focus their teaching on questions of design, technique, and analysis. This is "qi" in small letters.

As expected, the experimentalists are on the left pole; this is QI in big letters! Those on the left pole take a more avant garde, activist view of method and pedagogy. They adopt a subjective, interpretive approach to inquiry. They concentrate on method as praxis, or method as a tool for social action. Performance ethnographers, action researchers, and community organizers are all in the left pole group. They want to change the world by creating texts that move persons to action. They want texts that move from personal troubles to public institutions. They want to teach students how to do this.

There is a third pole; this is the space of social justice. Right- and left-pole methodologists can be united around social change issues. Traditional methodologists, like left-pole activists, can teach students how to do ground-level social justice inquiry. This is inquiry that is indigenous, collaborative, and community-based.

This initiative combines pedagogy and methodology to

- Help clarify competing definitions of a social problem;
- Collect and use narratives, life stories, statistics, numbers, and facts to expose the limits of official ideologies and to dramatize the extent of injustice operating in a crisis situation;
- Isolate points of intervention;
- Suggest alternative moral points of view;
- Articulate questions concerning "how," not "why" the problem was created;
- Connect personal troubles with public issues;
- Secure multiple instances of injustice through interview, observation, archives, and personal experience;
- Collaborate with community members to produce and perform ethnodramas that dramatize the situations that have been uncovered;
- Interpret and publicize audience feedback.

In these and other interpretive ways, traditional and experimental interpretive methods can be combined in projects that are committed to advancing social justice agendas. More is involved, as elaborated below.

◙ TEACHING TO THE LEFT POLE FOR BRICOLEURS

Eisenhart and Jurow (Chapter 43) observe that teaching to the left pole involves more then technique. It centers on postmodern epistemological, philosophical principles, including the politics of knowing, as well as issues surrounding objectivity, performance, reflexivity, writing and the first-person voice, complicity with the Other, ethics, values, and truth. In order to travel to the place of the experimental text, students obviously need instruction in a large literature that has traditionally not been regarded as central to methodology. This is qualitative research that is messy, performative, poetic, political, and reflexive. It is autoethnographic, inquiry shaped by the call to social action, by a commitment to undo pedagogies of oppression.

So conceived, there are three attitudes to be enacted, or goals to be pursued. First, teaching, understood as critical pedagogy, is the practice of making the political and the ideological visible through the act of performance itself. Teaching is a performative act. Second, this act is an invitation for students to use their own experiences as vehicles for pushing back against structures of racial, sexual, and class oppression, an invitation to become agents in their own biographies.

Third, in order to realize the first two goals, students become autoethnographers, authors of dramas about their own lives. This performance format presumes that all playwrights are ethnodramatists (Saldaña, 2005, p. 33). This is emancipatory theatre, critical performance ethnography for the oppressed, a theatre that exposes the lies we tell one anther about race and ourselves.

At the same time, students on the left pole need instruction on right-pole methodologies. Critical scholars need to, at some point in their careers, be deeply immersed in the methodological classics of their discipline. They need to know how to interview, do fieldwork, work in archives, do participant observation, write autoethnography, do case studies, engage the various forms of participatory action research (PAR), do focus groups, and write grounded theory. As *bricoleurs,* they need all of these methodologies in their social action tool kit. Traditional qualitative research can be used as a tool to leverage social change. This can happen when participation, action, experience, and inquiry are joined under a critical format.

Language is a key. If new forms of social text are created, then new voices are heard—the voices of change, the voices of resistance. Research is thereby connected to political action. Systems of language and meaning and paradigms of knowing are changed. When this happens, the world changes.

The collapse of foundational epistemologies has led to emerging innovations in methodology. These innovations reframe what is meant by validity. They have shaped the call for increased textual reflexivity, greater textual self-exposure, multiple voicing, stylized forms of literary representation, and performance texts. These innovations shade into the next issues, surrounding representation.

Representational issues involve how the other will be presented in the text. Representational strategies converge with a concern over the place of politics in the text. We can no longer separate ideology and politics from methodology. Methods always acquire their meaning within broader systems of meaning, from epistemology to ontology. These systems are themselves embedded in ethical and ideological frameworks as well as in particular interpretive communities. Our methods are always grafted into our politics.

Scientific practice does not stand outside ideology. As argued in the first and second editions, a poststructural social science project seeks its external grounding not in science, but rather in a commitment to post-Marxism, and an emancipatory feminism. A good text is one that invokes these commitments. A good text exposes how race, class, and gender work their ways into the concrete lives of interacting individuals.

We foresee a future where research becomes more relational, where working the hyphen becomes easier, and more difficult, for researchers are always on both sides of the hyphen. We also see a massive spawning of populist technology. This technology will serve to undermine qualitative inquiry as we know it, including disrupting what we mean by a stable subject (Where is the cyberself located?). The new information technologies also increase the possibilities of dialogue and communication across time and space. We may be participating in the reconstruction of the social sciences. If so, qualitative inquiry is taking the lead in this reconstruction.

Finally, we predict that there will be no dominant form of qualitative textuality in the seventh and eighth moments; rather, several different hybrid textual forms will circulate alongside one another. The first form will be the classic, realist ethnographic text, redefined in poststructural terms. We will hear more from the first-person voice in these texts. The second hybrid textual form will blend and combine poetic, fictional, and performance texts into critical interventionist presentations. The third textual form will include *testimonios* and first-person (autoethnographic) texts. The fourth form will be narrative evaluation texts, which work back and forth between first-person voices and the *testimonio*. These forms will be evaluated in terms of an increasingly more sophisticated set of local, indigenous, anti-foundational, moral, and ethical criteria.

Variations on these textual forms will rest on a critical rethinking of the notion of the reflexive, self-aware subject. Lived experience cannot be studied directly. We study representations of experience: stories, narratives, performances, dramas. We have no direct access to the inner psychology and inner world of meanings of the reflexive subject. The subject in performance ethnographies becomes a performer. We study performers and performances, persons making meaning together, the how of culture as it connects persons in moments of co-creation and co-performance.

▣ HISTORY, PARADIGMS, POLITICS, ETHICS, AND THE OTHER

Many things are changing as we write our way out of writing culture, and move into the eighth moment of qualitative research. Multiple histories and theoretical frameworks, when before there were just a few, now circulate in this field. Today, foundationalism and postpositivism are challenged and supplemented by a host of competing paradigms and perspectives. Many different applied action and participatory research agendas inform program evaluation and analysis.

We now understand that we study the Other to learn about ourselves, and many of the lessons we have learned have not been pleasant. We seek a new body of ethical directives fitted to postmodernism. The old ethical codes failed to examine research as a morally engaged project. They never seriously located the researcher within the ruling apparatuses of society. A feminist, communitarian ethical system will continue to evolve, informed at every step by critical race, postcolonial, and queer theory sensibilities. Blatant voyeurism in the name of science or the state will continue to be challenged.

Performance-based cultural studies and critical theory perspectives, with their emphases on moral criticism, will alter the traditional empiricist foundations of qualitative research. The dividing line between science and morality will continue to be erased. A postmodern, feminist, poststructural, communitarian science will move closer to a sacred science of the moral universe.

As we edge our way into the twenty-first century, looking back, and borrowing Max Weber's metaphor, we see more clearly how we were trapped by the twentieth century and its iron cage of reason and rationality. Like a bird in a cage, for too long we were unable to see the pattern in which we were caught. Co-participants in a secular science of the social world, we became part of the problem. Entangled in the ruling apparatuses we wished to undo, we perpetuated systems of knowledge and power that we found, underneath, to be all too oppressive. It's not too late to get out of the cage. Today, we leave that cage behind.

And so do we enter, or leave, the eighth moment, moving into an uncertain future.

▣ NOTE

1. This section reworks material in Denzin (2010, pp. 55–57).

▣ REFERENCES

Denzin, N. K. (2010). *The qualitative manifesto.* Walnut Creek, CA: Left Coast Press.

Phillips, D. C. (2006). A guide for the perplexed: Scientific educational research, methodolatry, and the gold versus platinum standards. *Educational Research Review, 1*(1), 15–26.

Saldaña, J. (2005). *Ethnodrama.* Walnut Creek, CA: AltaMira Press.

Schwandt, T. C. (2007). *Qualitative inquiry* (3rd ed.). Thousand Oaks, CA: Sage.

42

QUALITATIVE FUTURES

Where We Might Go From Where We've Been

Judith Preissle[1]

In the early 1980s, coming off some success in publishing journal articles on qualitative research methods, my colleague Marki LeCompte and I put together a prospectus for a qualitative research methods book in education and sent it to two publishers. An editor at one of them expressed an interest that led to two editions of the text (Goetz & LeCompte, 1984; LeCompte & Preissle, 1993) and a Spanish translation (Goetz & LeCompte, 1988). From the other, we received a polite refusal accompanied by a vitriolic review of our material that ended with the assertion that no one would be doing qualitative research anymore. The reviewer said that qualitative research had been a frivolous endeavor, its time had come and gone, and the projected manuscript would have no market.

I begin my consideration of the future of qualitative research with this cautionary tale because what is ahead of us is always uncertain, some of our expectations will not be met, and some of what happens is rarely anticipated. Although my undergraduate preparation in history convinced me that the present and future reflect the past, that reflection is itself based in hindsight and is partly made up of what was not expected. Furthermore, scholars and academics awake each day to make their worlds— to teach, study, and write with the confidence that what they are creating from the past and the present will contribute to scholarship in the coming years. The nature of research and publication are such that we are always engaged now in preparation for later, and later may be a year, 5 years, or 50 years away.

In this chapter, I base projections for and speculations about the future of qualitative research on three sources. First is my own individual experience of 40 years of doing and teaching qualitative research. Second is the view that experience provides of the history and development of qualitative research methods and design. Third is the work of emerging scholars along with other expert projections of qualitative futures (e.g., Loseke & Cahill, 1999a, 1999b; Denzin & Lincoln's commentaries in previous handbooks, 1994, 2000, 2005; and such). The emphasis is on what changes and what remains the same in practicing qualitative scholarship and how we even recognize the difference. Finally, my focus is on the future of research practice and methodology rather than on broader social and economic trends, although my projections assume that globalization, technical upheavals, and cultural strife will not succumb to apocalypse.

In this material, I consider the dangers of prognostication, as illustrated in the anecdote that begins the chapter, and I acknowledge that my own relationship with the future has been ambivalent. Engagement in the present generally has distracted me from envisioning what is to come. Consequently, I review some of these relevant "presents" that have culminated in my professional history as a qualitative methodologist in the United States. Understanding our present and future based on the concrete experiences of individuals has multiple, overlapping traditions in qualitative scholarship: life history research (e.g., R. Atkinson, 1998; Goodson & Sikes, 2001; Langness & Frank, 1981), biographical and autobiographical work (e.g., Bertaux, 1981; Goodson & Walker, 1991; Kridel, 1998; Okely & Callaway, 1992), narrative representations of experience (Hatch & Wisniewski, 1995; Hinchman & Hinchman, 1997; Josselson & Lieblich, 1993; Polkinghorne, 1988), and autoethnography (Ellis, 2009; Reed-Danahay, 1997). Feminists have contributed illuminating accounts of their lives as scholars (Reinharz, 1979/1984; Richardson, 1997), a practice arguably pioneered earlier in the 20th century by fieldworkers such as Malinowski (1967/1989) and Mead (1972). What follows is less detailed than autoethnography and less comprehensive than life history. They are anecdotes from an academic memoir, they are intended to illuminate a past and project a future, and they invite readers to reflect on their own similar and different experiences with scholarship.

◨ JUDE'S HISTORY

Many children grow up confident of what they want to be and do for a living. My only vision was to be independent. No one was more surprised than I was when practicing education became even more important than achieving and maintaining independence.

Becoming an Educator

In 1962, when I was an undergraduate at Grinnell College in Iowa, my father began pressing me for my plans for independence—a livelihood to support arriving at adulthood. He assured me that my ideas of taking the civil service exam and working for the U.S. federal government in some capacity would lead me nowhere. He said most policy analysts were projecting large reductions in government workers for the coming years. This conversation occurred before John Kennedy was assassinated; before Lyndon Johnson assumed the presidency; and before Johnson's Great Society programs initiated an expansion, rather than a contraction, in the federal government workforce. My father's mistaken ideas may account for my skepticism about my own and anyone else's abilities to project the future, including the qualitative research future.

What routes other than civil service and law make sense for undergraduates majoring in history? My father pressed for teaching certification so I would have "something to fall back on." Having settled for a history major because an anthropology degree was not available at my college at that time, I feared education would take me even further afield from my interest in studying cultures. Reginald Archambault (1963), a Dewey scholar who taught education at Grinnell during those years, set me straight. Archambault introduced his students to Dewey's formulation of the teacher as a mediator between "the child and the curriculum": the teacher who studies child and subject matter to engage the child in learning about the world. This was a life-changing revelation for me. I myself had had few experiences of such teachers. My assumption had been that these gifted individuals were just "born" with those talents. I had not realized that people could learn to become progressive educators if they set themselves that goal. That became my goal.

In the following years, I brought the scientific approach to history taught at Grinnell together with some preparation in English and later study of anthropology at the University of Minnesota to a career teaching social studies and language arts to 12-year-olds. Most of those years, the social studies I taught was an experimental anthropology curriculum developed as one of the federally funded Project Social Studies programs, part of the broader effort by the U.S. government to reform precollegiate curriculum in mathematics, science, language, and social studies. Studying anthropology at the university and studying the youngsters in my classes provided an orientation to doing research deeply lodged in *practice* framed by *theory*.

My own instructors in history, anthropology, and some sociology were insistent on addressing the tensions in these fields between a science and a humanities orientation to conducting research. These scholars represented the different perspectives as sources of creative thought stimulating new insights and methods of study. My education courses, however, presented a much narrower view of scholarship: The ideal taught while I was pursuing graduate study from 1966 to 1975 was experimental design, with survey research a sorry second choice. The proper subject of such scholarship was the improvement of public education in the United States. The research I completed for a master's degree was a quasi-experimental comparison of two approaches to teaching about minority groups in the United States. Although the instruments I used indicated significant differences in results, I never published the study because my observations of the two approaches made me skeptical of those results. At the time, I had no idea of how to combine those observations into a formal report of the research. Three more years of graduate study began to address that limitation.

Becoming an Ethnographer

Ethnographic activity was something I understood early on as a graduate student as occurring at some other remote time and place. I read widely in the cultural anthropology and qualitative sociology literature of research done in the past by people who had become experts in the present when I was reading them. Like my earlier history curriculum, this fieldwork literature was dominated by Western scholars, Europeans and North Americans, but their purview was global rather than centered on the North Atlantic. Even so, the global world was interpreted from Western frameworks.

I read Margaret Mead, Howard Becker and his colleagues, and anyone offering qualitative analyses in my field of education: Elizabeth Eddy, Estelle Fuchs, Jules Henry, Lou Smith, George and Louise Spindler, and Harry Wolcott. All of these scholars were established researchers at the time, and some had much good work ahead of them—although Henry died in 1969 while I was teaching junior high school full-time and studying for a master's degree part-time. Nevertheless, these people were reporting work done in the 1960s and before, some of it outside the United States, and most were individuals I would later meet. For someone from a family with little experience of higher education, much less scholarship, these direct, face-to-face encounters with practicing researchers made fieldwork accessible and possible.

My first formal experiences doing qualitative research were interviews with rural children about their views of city life and close observations of youngsters and their novice teachers whom I was supervising. Trying to bring fieldwork and other qualitative endeavors into my own present was an effort. What I read about had been done in the past. My instructors in education who had ongoing research projects were running surveys or experiments. The anthropologists and qualitative sociologists with whom I studied were between field projects, and I did not begin to meet students returning from their own fieldwork until late in my doctoral career. Just bringing qualitative scholarship from other people's pasts into my present felt audacious. The future, my own or that of qualitative research, was the last thing on my mind. Of course, an education is supposed to be preparation for the future, but it did not help that being a full-time graduate student was lots of fun and something I found myself reluctant to give up.

However, it was 1972, and I had to make a decision about a dissertation. The future was intruding on the present. Just what was I preparing to do? Who was I intending to be? I decided that I wanted to be the Margaret Mead of education in the United States. Although many of the people I was reading had conducted qualitative studies in U.S. schools, the daily lives of teachers and youngsters in classrooms in Western nations had been documented only in incomplete fragments. What I wanted to do was reveal the rich complexities of what goes on among teachers and students as they interact to teach and learn. Maturing in the midst of the U.S. civil rights movement and the succeeding second wave of feminism radicalized many of us in an occupation otherwise characterized by conventionality (Lortie, 1975). As noted previously, I had taught middle grades children from 1965 to 1971, and I team–taught the last 4 of those 6 years. My colleagues and I had made changes every year in an attempt to improve our students' experiences, but as a teacher educator I saw a gap—a rupture—between the changes educational policy makers and innovators planned and what went on in classrooms. Part of my motivation to contribute to what I envisioned as an ethnology of schools and classrooms was to inform the public, the politicians, and educational researchers of what goes on among teachers and students—"the good, the bad, and the ugly." I thought we should learn how things *are* before we intervene to make things as they *ought to be*. I had seen my own well-intentioned efforts to challenge the status quo lead to mixed results.

After several false starts, I conducted for a dissertation study what I called, after Burnett (1973), a microethnography of a third-grade classroom in the rural Midwest (Goetz, 1975). The ethnology has and has not happened. My teacher, Judith Friedman Hansen, did draw on my dissertation and other educational ethnographies for her 1979 synthesis of educational anthropology. LeCompte and I attempted a chapter-length educational ethnology about a decade later (1992). Most importantly,

however, the growth in qualitative research studies in education across its subfields has been such that a single synthesis, one ethnology, is no longer feasible. What we have attained by the first decade of the 21st century is the possibility of many different ethnologies of education. We not only know more about how teachers and students experience schooling, but we also have more detailed accounts of the effect on these experiences of race, ethnicity, gender, class, religion, ability, citizenship status, and sexual orientation.

As for me, my projected career in teacher education and even in educational anthropology was diverted elsewhere. What characterized the abandoned career was its preoccupation with schooling and education in the United States, as much of my initial reading indicated. What characterized the developing career, however, was a continuing ambivalent relationship with the future. I learned to plan ahead for the next hurdle and then to formulate goals for the next year, but long-term planning remains a challenge because I continue to enjoy the present so much.

Becoming a Qualitative Methodologist

In 1975, I took a position in elementary social studies education at the University of Georgia. The faculty believed my dissertation experience and preparation in anthropology were sufficient for developing and offering the first qualitative course to be taught at the university, and that was part of my assigned load. Now, four decades later, I teach qualitative research full-time for a graduate certificate program housed in a college of education, but serving programs across the university. Those enrolled in the program come from all parts of the globe, with some qualitative research classes composed of predominantly international students. Over this period, my instruction and scholarship have been enriched by a year's study in philosophy, by the necessity to keep reading across the social and professional sciences to address my students' concerns, and by the adventure of exploring studies of education elsewhere. Examples of these have been Willis's (1977) critical ethnography of working-class males in the United Kingdom, the inspired work on gender regimes in Australia by Kessler and her colleagues (Kessler, Ashendon, O'Connell, & Dowsett, 1985), and the exploration of schooling and child labor in India by the Dutch anthropologist Nieuwenhuys (1994).

By the middle of the 1980s, other faculty members assumed my responsibilities in social studies teacher education as the student demand for qualitative courses grew from one course to many courses. Although I continue to research and write on topics in educational anthropology, my recent instructional assignment has been almost entirely devoted to the Qualitative Research Program at the University of Georgia. The irony is that I myself never took a course in qualitative research. To the extent that such material was available at Indiana University and most other research institutions in the mid-20th century, it was embedded as a brief topic in courses like the sociological research methods I did take or in anthropological field methods courses focusing on such crucial tasks as acquiring a new language on site.

Summary

Our futures as qualitative practitioners depend in part on decisions each of us makes in the present(s). Each qualitative scholar has a unique history developed over time from particular experiences, although many methodologists of my generation also came to their expertise through variant routes. My particular stance on qualitative research reflects a practice–theory balance acquired as a classroom teacher and an interdisciplinary orientation developed from framing social sciences in liberal arts environments.

Our futures as qualitative practitioners also require visions of what we hope to accomplish and why. I have summarized some of my own hopes in the preceding pages. These hopes are more likely to be achieved if they are articulated. What changes and what stabilities do we intend, expect, and care about? Who are we working with to make these things happen? What are we bringing from our pasts to create our presents and our futures? I turn next to considering the past of qualitative research activity.

▣ QUALITATIVE RESEARCH'S HISTORY

The history of any scholarly subject is the accumulated histories of the people who create and contribute to the subject. The multiple histories of qualitative research are grounded in the practices of those who work from and study qualitative traditions. These histories are specific to times, places, and disciplinary fields. They contribute to the confusion and aggravation experienced by many who expect the research methods and design they study to be conceptually clear, well-organized, and rationally sensible.

In this section, I draw from and summarize material published elsewhere (Preissle, 2006) about what it means to be a qualitative researcher and do qualitative work. In attempting to synthesize an interdisciplinary specialty with contributions from scholars around the world, I organize the discussion with two different metaphors. These draw attention to different facets of qualitative methodologies and are my attempts to make sense of such a complex area of study. The first metaphor, the *qualitative confederacy,* represents the community of qualitative practitioners and theorists. The second metaphor, the *qualitative tapestry,* represents what we produce or make as we practice and theorize. I begin by discussing who we methodologists are.

The Qualitative Confederacy

To the extent that people have been watching, listening, asking, and collecting material (Wolcott, 1992) to understand and interpret the world within considered frameworks, they have been doing qualitative research for millennia. Herodotus and Thucydides in Greece and Sima Qian in China left accounts of studying and interpreting the world and even assessing the quality of the information they gathered. What is more recent is documenting these activities in detail, reflecting on the purposes they serve, and connecting the acts with the thinking that directs and is directed by engagement with the world. If the acts and activities are methods, then the reflections are the methodology (see Lather, 1992). Qualitative methodology has been invented, reinvented, and borrowed around the world and in many substantive fields of study. It is based on conceptually framed records of our direct sensory experiences of the world, reported so as to preserve as much of that experience as possible. Qualitative methods are practiced in most social and behavioral sciences, across the professional fields such as education, social work, clinical study, law, library science, health and medicine, and in journalism and the humanities. Figure 42.1 represents a third metaphor, the *bramble bush* of disciplinary roots, stems, and branches contributing to qualitative methods and methodologies.

This bramble bush of qualitative research is sometimes referred to as an *umbrella* term that encompasses the variety of traditions involved, but I prefer to consider qualitative methods and methodologies as a *confederation* of practitioners who share a commitment to theoretically and conceptually formulating an engagement with the world that produces vivid descriptive accounts of human experience. This is confederacy in the Native American Iroquois manner: periodic gatherings of equals to discuss matters of common concern characterized by

Figure 42.1 The Qualitative Disciplinary Bramble Bush

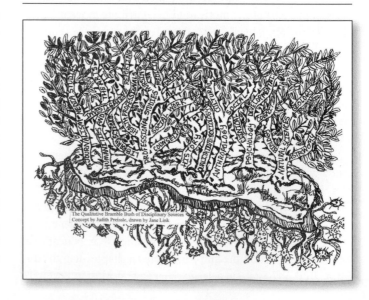

extended discussion and by candid, but amiable disagreements. The confederacy extends across time and space from the ancients and into the future we envision. Its membership is fluid as new scholars are inducted into the work and even as thinkers long dead may be newly recognized for their contributions to qualitative methodologies. Viewed also as a bramble bush, the confederacy clumps into fields of study that draw, borrow, or rob from one another in sometimes prickly fashion.

The Qualitative Tapestry

Most social science and professional disciplines have some tradition of qualitative methods, although the methodologies may be more recent. In sociology, anthropology, and history, where the methods can be traced to the origins of the fields, qualitative methods and methodology are deeply integrated with content. Conversations about *how* are intertwined with conversations about *what*. The subject matter and the research method implicate each other such that they may be difficult to separate. Anthropologists, for example, customarily assume that an ethnography is a study of a group's culture. In contrast, other social scientists may use the term as a synonym for fieldwork or qualitative research generally. This appears to be what is meant by ethnography in the title of the periodical, *Journal of Contemporary Ethnography*. Even in considering what may be generic approaches like interviewing, many scholars begin with what they want to learn and how others have studied it. The research question, the subject matter, and the theoretical frameworks clearly direct the research design in these situations.

The earliest fieldwork manual I have encountered is the French *philosophe* Degérando's (1800/1969) guide for French sailors on how to properly study the indigenous peoples they would encounter in their explorations of the South Coast of Australia (Preissle, 2004). However, what qualitative methodologists now consider classic texts are likewise concerned with *how* to study *what*. Later in the 19th century, the English sociologists Harriet Martineau (1838/1989) and Beatrice Webb (1926) documented the social lives of ordinary people in the United States and Great Britain. The Polish-American Florian Znaniecki's 1934 discussion of analytic induction focuses on discovering sociological categories and theories, as does Glaser and Strauss's 1967 qualitative text. Powdermaker (1966), Williams (1967), and Wax (1971) all offered guidelines for fieldwork directed to the anthropological study of cultures (see also Mead & Métraux, 1953/2000). Bloch's discussion (1953) of observation, composed in France under the 1940s German occupation, is qualified as historical observation and the particular problems posed for historians of observing human "tracks." During this same period, as Riessman (2008) emphasizes, what has become the qualitative research methodology of narrative analysis was being explored in language and literature studies. My point here is that the history of qualitative research is in part tied to scholars from around the world who practice specific disciplines and fields of study.

From relatively early on in the history of qualitative research methodology, however, more interdisciplinary approaches developed. One example is John Dollard's (1935) anthology of exemplary life histories with a discussion of how the life history should be developed. He draws not only from across the social sciences, but also from the humanities. Dollard, trained in sociology and cited generously in anthropology, later turned to psychology in what became a thoroughly interdisciplinary career. This contrasts with the approach taken by Gottschalk, Kluckhohn, and Angell (1945) in discussing personal documents in research; their treatment is organized by the three disciplines the three scholars represent: history, anthropology, and sociology. Some scholars have worked both within and across disciplines. The English sociologist John Madge, whose academic preparation was in architecture and economics, wrote both about interdisciplinary social science practice (1953) and about sociology as a scientific practice (1962). By the middle years of the 20th century, disciplinary and interdisciplinary treatments of observation (e.g., Adams & Preiss, 1960), interviewing (e.g., Merton, Fiske, & Kendall, 1956), case study approaches (e.g., Foreman, 1948), and fieldwork (e.g., Junker, 1960) were providing the beginning literature of what would become qualitative research methodologies. Adler and Adler (1999) discuss this diversification of participation in what they call the "ethnographers' ball," and Wolcott (2009) illustrates some of the diversification with his tree of qualitative practices (see Figure 42.2).

Threaded, with something like an occasional metallic glint, through the mid-century literature—disciplinary and interdisciplinary—were epistemological challenges to the objectivist-realist or positivist premises of most methodological scholarship. Smith (1989, 1993) and others (e.g., Heshusius & Ballard, 1996) have offered accounts of some of these challenges. Smith and Heshusius (1986) detail, for example, the disagreements among German scholars at the turn of the 20th century between the hermeneutically oriented Wilhelm Dilthey (1883/1989) and the more empirically oriented Max Weber (1903–1917/1949) about whether the physical and the social world can and ought to be studied with similar assumptions. Smith and Heshusius themselves concluded that they cannot and ought not. Phenomenology, reformulated by another German scholar, Edmund Husserl (1893–1917/1999), from earlier philosophical ideas to provide what he hoped would be a foundation for the empirical sciences, was transformed by later continental thinkers into alternatives to positivist thinking (e.g., Berger & Luckmann, 1966; Schutz, 1962) that form the roots of current-day constructionism. By the last decade of the 20th century, reflection on epistemological and ontological assumptions had become central issues in the various methodological traditions. My colleague Linda Grant and I, for example, have

Figure 42.2 Portraying Qualitative Research Strategies Graphically

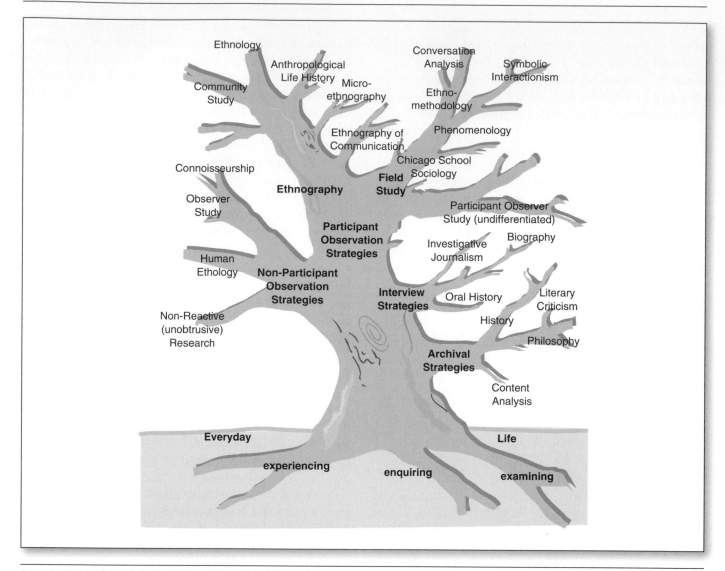

Source: Reprinted with permission from Wolcott, 2009, as adapted from Wolcott, 1992.

discussed some of the different epistemological approaches underlying observation (Preissle & Grant, 2004), and we attempted to represent this graphically (see Figure 42.3).

Having introduced the metaphor of a tapestry for qualitative methodologies, where some of the threads are metallic for philosophies and others are organic materials for methods, I turn to colors to represent purpose. Purposes are, of course, tied to both reflecting-theorizing-philosophizing and practicing-acting-doing, but purposes are also riddled with priorities and values. To weave purpose into this tapestry, I borrow Lather's 1992 framework, itself an elaboration of research purposes formulated by the German philosopher and sociologist Jürgen Habermas (1973). This borrowing of ideas to build different ideas is one of the functions of the intellectual communities of practice I am claiming here. So one purpose for (or color of) inquiry is to find out how the world works and what is going on in it. I think of it as the Pandora impetus, and it appears to be strong in the human species. People want to know, and people

value knowing for its own sake. People pass along what they know and do not know as gossip, rumor, assertions, speculations, claims. As the Pandora myth indicates, knowledge, misinformation, and disinformation can be dangerous, but people still seek to *understand*. For some people, understanding how things are is a stepping-stone to the future, to *predicting* what is to come and to controlling how things will be. For other people, the priority is to improve human experience, to *emancipate*. Freeing people from pain, suffering, abuse, and oppression is another age-old desire, associated especially with the practice of the world religions. Secular emancipation, however, developed with modernism: Marxism, feminism, race-consciousness, and postcolonialism. Secular emancipation is grounded in assumptions about human equity, about distributions of power and resources, and about the potential for communities directed by deliberative choice rather than by traditions and ideologies. Whereas predicting and understanding have often been associated with disinterest and neutrality, emancipating is fueled by

Figure 42.3 Philosophies and Fieldwork Conduct

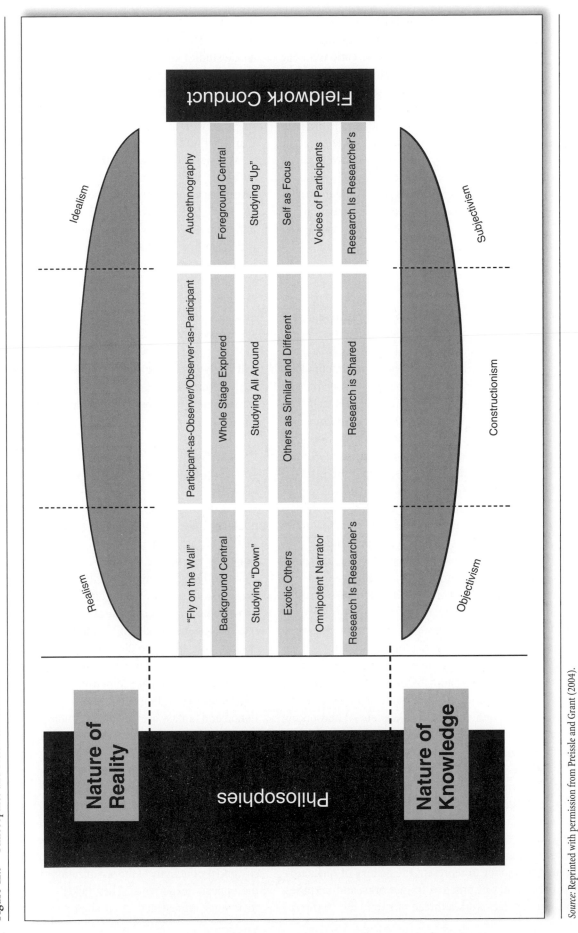

Source: Reprinted with permission from Preissle and Grant (2004).

Note: Philosophical framework adapted from Crotty, M. (1999). *The Foundations of Social Research: Meaning and Perspective in the Research Process.* London: Sage.

care (see, for example, He & Phillion, 2008). Care for others has been inextricably tied to qualitative research traditions, motivating, for example, Beatrice Webb's (1926) observations among the urban poor in Great Britain and Harriet Martineau's (1838/1989) investigations of 19th-century United States.

Lather (1992) extends Habermas's three categories of predicting, understanding, and emancipating to include a fourth: *deconstructing*. Deconstruction seeks "to produce an awareness of the complexity, historical contingency, and fragility of the practices we invent to discover the truth about ourselves" (p. 88). Although drawing from philosophical explorations and arguments, analysts add reflections on language and on the particularities of time and space when they deconstruct human experiences, arguments, and knowledge claims.

Summary

I am combining two organizing metaphors here to represent qualitative research. One is that of a confederation. I have proposed that the various methodological approaches we call qualitative form a confederation of methods and methodologies. What does this mean? What are the constituent parts? I believe we are made up of networked communities of scholarly practice—intellectual communities. Most of us are members of several such communities, so the groups are fluid and highly permeable and they have overlapping memberships. Unlike geographically identified disciplinary groups such as the American Educational Research Association (AERA) or the Hong Kong Sociological Association, the qualitative methodology confederation is made up, for example, of focus group practitioners, field study researchers, phenomenological researchers, constructivist-constructionist scholars, and poststructural feminists. Just as important are the international composition and global context of this scholarship. We are communities because we share a scholarly practice that we communicate about with one another through books, through journal articles, among professional networks formed initially in face-to-face situations like classrooms, in subgroups at professional conventions such as the International Congresses of Qualitative Inquiry, and across the increasingly diverse venues offered by the Internet: listservs, online journals, blogs, social networking sites, and such. We meet and debate, we read and write back to one another, we engage both our deceased colleagues and those still living, and we change our practices as a result of these encounters. We learn some new ideas from an Internet discussion group, reformulate them as we talk about them to colleagues we meet face to face, try them out in publications directed to completely different audiences, and respond to challenges posted on someone's blog.

The second metaphor to represent the history of qualitative research is the tapestry of practice, frameworks, and purposes. Although some scholars work a thread or two in any particular

project, most recent scholarship draws self-consciously from mixes of color, metals, and organic materials in constructing single studies and designing careers of many studies. As I talk to people whose initiation to qualitative research was a social science or the arts or practitioners' self-study, I am struck by how we all associate that initiation with normal qualitative research, with a qualitative practice somehow more authentic than what we later encounter. That is the hold of the past on us.

My goal in juxtaposing the history of qualitative research with my own scholarly history is to prompt all qualitative researchers and methodologists to reflect on our educations and developments as human beings who pursue knowledge and understanding and to connect ourselves to the communities in which we practice. Each of us entered into a world made by predecessors, but we make and remake that world as we live in/ on/through it. Each individual experience is different, depending on who we are, where we have been educated, and how we have developed our practice. The particularities of time and space may make our challenges specific to now, but we continue to share facets of those challenges with those predecessors and will surely pass along similar issues to our successors. Our networked communities of practice, our global confederacy, live across time. I turn now from past to future.

◪ LOOKING INTO/TOWARD/ FOR QUALITATIVE RESEARCH FUTURES

Having discussed some varied paths scholars have taken to the qualitative, through purpose, method, and philosophy, in this section of the chapter I reflect on my own and others' projections of where these communities of qualitative practice may be going. Although many science fiction stories represent people living in a future not much differently from how they live in a present, the images I propose are less static. I believe people live *into* futures, much like falling into water. They live *toward* their futures, as if driving to destinations. Scholars especially live *for* their futures: planning, studying, and reporting in cycles intended to accumulate something they may consider knowledge or understanding. What I have borrowed here from science fiction and fantasy are the ideas of multiple, alternative futures. First, people clearly project different ideas of what is to come, even in an intersubjective world. Second, people recall different pasts and experience different presents, so even similar projections may vary significantly.

Here is an example. Some years ago, I listened to several former leaders of a large professional organization describe the current state of their field of study and project what they would like to see happen next. Some talked about the intimacy, intensity, and productivity of the organization's past. Others described the opposite experience. They talked about struggling to get their work represented in an alien and rejecting environment until more recently when the organization had become more diverse and hence more welcoming. The first group projected a future allowing them to return to their past. The second group embraced the present and projected a future allowing them to continue it. Neither group considered an inclusive future: a future welcoming a variety of activities and open even to what is not yet envisioned.

Such a future requires that scholars recognize how porous are their intellectual traditions and how much they are members of multiple, overlapping associations and traditions. I have conceptualized qualitative scholars as a confederacy, not just because I believe the metaphor fits how we are currently operating, but also because it is an image I endorse. In a recent commentary (Freeman, deMarrais, Preissle, Roulston, & St.Pierre, 2007), for example, on standards for qualitative research, my colleagues and I at the University of Georgia strongly advocated for control of standards to be as local as possible. We urged communities of practice, however they define themselves, to begin developing standards tailored to, for example, narrative analysis, social phenomenology, and poststructural feminist interviewing. Some of these will be shared across traditions because of common membership and similar approaches, but standards formulated by practicing scholars are likely to be both more applicable to what people are doing and more amenable to change as new conditions and understandings warrant.

A clear disadvantage of local determination is duplication and confusion. Common language for our practices and shared meanings for ideas may promote better understanding. Certainly a discourse available and accessible to all has its advantages, and I am not mandating a cacophony here. I have contributed to and benefited from discussions of such hegemonic terms as *validity*. I have no objection to the idea of *truth*, for example, so long as truth can be singular or plural and relative to the community practicing it. The problem is that the meaning of constructs such as validity and truth is rarely unitary. In some respects, that is why they are so powerful. Recognizing the different meanings and examining how their facets play out across situations permit flexibility, creativity, and adapting to the particulars of research situations (for an example of this, see Scriven's 1972 treatment of objectivity and subjectivity). What my colleagues and I stressed in our commentary on standards (Freeman et al., 2007) was that most methodologies from a variety of persuasions used the term validity, but their discussions of what *constitutes* validity were richly informed by scholars proposing substitute language.

A task that we as qualitative scholars have, going forward, is to propose the words and ideas we are going to draw on or draw up to communicate about our practice. What are we going to preserve and what are we going to change? This is important not only for how we communicate among ourselves, but also for how we bring in new scholars and how we reach out to the public.

I have taught qualitative research methods since 1976. Much has changed since that first class. The wealth of methodological materials and examples of qualitative research accumulated since then is remarkable. It is overwhelming. In my program, instructors take seriously a charge to introduce students to the range and variation of qualitative research practices. We may teach our own specialties more confidently, but we are committed to representing respectfully others' traditions. This is a challenge, and some students resist. These students often begin knowing what they want to learn, and it is *not* everything.

We also teach research design informed by theory and epistemology. Intellectual meltdowns by the end of the first month of the first course are common among our students. They press us for singular language, and we resist. Our intention is partly to represent what we know to be the diverse and confusing history of qualitative methodologies; just as important is that these students are the future. To produce and to respond to what is to come, they need as much of the qualitative resources provided by others as we can manage. Many of them develop innovations that surprise their teachers, as I discuss subsequently.

Some of what is to come we can anticipate and plan for, but much of it is neither controllable nor foreseeable. For example, growing up in the 1950s, my parents worried about the effect of the growing popularity of television on the newspapers they read avidly. What actually developed was a more symbiotic relationship where newspapers and television news complemented each other (Heflin, in press). Who knew then that the actual threat to newspapers would be a technology developed by the military called the Internet (Jones, 2009)?

Another unexpected development is the result of the obesity epidemic in youth. People who grew up in the United States and other western nations in the 20th century have benefitted from improved public health and medical treatment, being assured by health experts that they would live healthier and longer lives than their parents and grandparents. However, such adults alive today may be the last group to enjoy this privilege because of the increased health risks associated with obesity in youth (Olshansky et al., 2005). Although the individual consequences of excess weight have been recognized for many years, many people are shocked to learn that their children may be denied their own longevity. What does this have to do with the future of qualitative research? It is a caution, I believe. At one time, the worry was that the world's food supply would not support the world's population. Agricultural technologies have been used to address that issue, but now the world is faced with two challenges. One is the distribution problem of some people getting too much food and others getting too little. The other is the quality problem of inadequate nutrition and overloads of sugars, fats, and such.

Likewise, throughout most of human history, information and knowledge have been limited. The recent intensification of globalization and other developments have produced what can only be considered a glut of information and knowledge. In the past, people lacked the information necessary to understand one another. In the 21st century, many people are inundated by information, but its quality and distribution, like that of food, are uneven. Misinformation and disinformation circulate as rapidly as reliable information and well-vetted knowledge. Some people have the challenge of too much information, while others are deprived of what they need to know to survive: locations of bombs from wars long over, safe routes home during heavy rains, and so forth. Qualitative scholars rarely begin research with no knowledge or information about what they want to study. The mass media, the Internet, and university libraries and archives provide so much relevant material that the task becomes how to handle it: What is pertinent and what is not? What sources are dependable and which are not? When should we stop browsing the literature and start getting our own data? *When* should we seek *what* information? Does beginning with no review of the literature really prevent biases? Glaser and Strauss (1967), contrary to what is attributed to them, did not prescribe reviewing literature after data gathering; they said only that scholars should think about and weigh the consequences of when literature is to be reviewed (p. 253).

Most, but not all, qualitative research scholars have ready and plentiful access to producing, distributing, and applying knowledge, but these privileges are not universal. As I have suggested, many people lack such access, and this constitutes an ethical challenge to be addressed into the future. However, the democratization of higher education, the increasing participation of professionals in studying and evaluating their own situations, and the social networking and information-sharing software that permits people to make their private lives public—a kind of exhibitionism referred to as *celebritizing*—all function to widen access. In a previous consideration of the future, for example, I speculated about whether increasing numbers of people will become researchers of their own lives, activities, and experiences (Preissle, 1999). As insiders to our own organizations, activities, and communities, how will we address the ethics of exposing others as we expose ourselves? Likewise, how may this break down the boundaries between the ivory tower of academia and the so-called real world if everyone can learn and practice the creation of knowledge? On the one hand, people gain access to what has been the domain of the experts, and they can make the knowledge they want and believe to be useful. On the other hand, what is produced or learned may be of uneven quality and may serve questionable purposes. Already the hegemony of government-supported research (Lather, 2010) has been challenged by funds for research provided by philanthropic foundations, by nongovernmental organizations, and by a variety of special interest groups (e.g., deMarrais, 2006).

Qualitative scholars of all varieties and housed both inside and outside academia are challenged and pressed by technological developments, such as those I have just described, that permit a constant data stream documenting people's lives. On the one hand, the ease and accessibility of audio and video recording devices and the simplicity of reproducing and storing the records may seem like a qualitative dream come true. On the other hand, ordinary people may have means to create their own ethnographies if they can learn to document and interpret their own lives and distribute those accounts on Internet sites like YouTube. Alternatively, ordinary people may be even more reluctant to want their privacy threatened by such records. For example, the local school district where my university is located had for several years a ban on video recording for research purposes because of the threat to student privacy. However, presuming we have willing participants and no such restrictions, our recording technology presents other issues such as how to use the material ethically, how to analyze mountains of audio and video files, and how to interpret the recorded life.

In another example of a technological challenge, for many years now, people new to qualitative inquiry have posed the same question on QUALRS-L, the listserv discussion group on qualitative research that I founded in 1991. They want to know what technology will automatically transcribe their audio files so that they do not have to suffer the tedium of manual transcription or the cost of hiring it out. The answer to that question has not changed in 10 years. Transcription software must be trained to particular voices, cannot easily handle records with two and more voices, and must be carefully vetted for inevitable inaccuracies. Furthermore, the better the software, the more expensive it is. I believe these answers are changing, but I am myself surprised that this is taking so long.

What has surprised me even more has been a proposal by a former student (Markle, 2008) to give up transcription altogether for what he proposes is an analysis of the audio record that is much closer to the interview experience. Using software developed for music editing, Markle shows how to generate an analysis in text that accompanies the audio record. I think of the audio file sound as the music and the analysis as lyrics, but the lyrics are an analysis of the spoken language, and they are layered so that any segment of conversation may have multiple lyrics that comment on the talk. Markle argues that the digital sound file is a more accurate and complete record of what was said than any written transcription can provide and that transcription is unnecessarily time consuming. Furthermore, the music editing software permits scholars to embed portions of an audio or video record right into research manuscripts. Voices of participants would be heard rather than just read. Markle notes that transparency, for those who value it, is improved because data are closer to what the researcher experienced.

Another surprise I have experienced recently is how novice scholars are challenging, even transgressing, epistemological and theoretical boundaries. Most methodologists have little difficulty with the idea of mixing methods of data selection, collection, and analysis, and the recent reinvigoration of mixed methodology is a testament to this acceptance (Greene, 2007). In their classic formulation of research epistemologies, for example, Guba and Lincoln (1994) say, "both qualitative and quantitative methods may be used appropriately with any research paradigm" (p. 105), but caution that "a resolution of paradigm differences can occur only when a new paradigm emerges that is more informed and sophisticated than any existing one" (p. 116). As I have noted previously, however, others argue for a more-or-less complete incommensurability among paradigms. My own position is far less doctrinaire. I have even been accused by some of philosophical schizophrenia. Research approaches are mental models that we formulate for our study of the world. Although I value coherence, consistency, and other such elements under some circumstances, I also believe that innovation and creative problem solving often mean finding ways of putting things together that convention has separated.

One such example comes from another student in the program where I teach. Van Cleave (2008) positions herself as a post-humanist scholar who deconstructs how qualitative interviewing is commonly presented in a variety of introductory texts. She cites many of the questionable assumptions about shared interviewer–respondent meanings and intentions so brilliantly criticized by Scheurich (1997). Then she makes the audacious move of applying Schacter's (2001) cognitive and neuroimaging research on memory, especially what he calls false memory, to raise additional crucial questions about the content of what people report in interviews. She uses clearly postpositivist research work to support her post-humanist claims about how we ought to make sense of interview data. Van Cleave affirms the value of the interview while she insists on reinterpreting what the content might mean: Interviews ought not to be considered true representations of human experience, but rather representations of human sense making.

Another such paradigm-challenging study, by my colleague Chávez (2004), combined analyses of molecular genetics and brain functioning, psychometrics, and phenomenological interviewing to establish links among genetic profiles, neurological activity, and creativity. Now practicing psychotherapy in the United States, Chávez is a Mexican psychiatrist who studied 40 well-known scholars and artists, 30 psychiatric outpatients, and 30 healthy but not famous individuals to demonstrate consistent associations among these three elements. Chávez, an accomplished pianist and poet, brings an eclectic stance to the humanities, the sciences, and the human study she pursues. She and many of her generation of scholars appear little affected by "the repeated, and indeed long-standing, tensions between scientific and interpretive inquiry, between realist and experimental texts, between [the] impersonal and experiential" (P. Atkinson, Coffey, & Delamont, 1999, p. 470). Behar's (1999) observation that "ethnography is reinvented with every journey" (p. 477) projects a future in which research itself may be reconceptualized with every endeavor. Reconceptualization

occurs for method, for design, for methodology, for theory, and for philosophy. St.Pierre (2010), for example, is challenging the construct of qualitative itself in her projections for a post-qualitative scholarship.

▣ CONCLUSION

Our futures are *serial*. They are composed of events particular to each of us in the coming years, decades, and centuries. Denzin and Lincoln's formulation of change over time in qualitative research is an example of chronological history moving into a serial future: sequential moments. These are "the traditional (1900–1950); the modernist or golden age (1950–1970); blurred genres (1970–1986); the crisis of representation (1986–1990); the postmodern, a period of experimental and new ethnographies (1990–1995); [and] postexperimental inquiry (1995–2000)" (2000, p. 2). These six were succeeded by "the methodologically contested [seventh moment] (2000–2004) . . . of conflict, great tension, and, in some quarters, retrenchment [and] . . . the eighth moment (2005–) . . . confronting the methodological backlash associated with 'Bush science' and the evidence-based social movement" (2005, p. 20). Denzin and Lincoln emphasize differences in their formulation. When Margaret Mead (1962) cautioned that adults are always immigrants to the present, she too was emphasizing the differences between what we have become accustomed to and what we must now and in the future adjust to. Mead's is also a serial understanding of time.

Hammersley (1999) objects to this serial analysis. He disagrees with the time divisions, seeing similarities rather than differences between some periods, and he posits our futures as *recurrent* rather than serial, worrying that the dogma of positivism has been replaced with new dogmas. Others may view the evidence-based eighth moment as another recurrence, an attempt to return to the modernist or golden age, when the federal government funded large scientific projects to solve poverty, racism, and educational inequities. However, different views of time and the future illuminate different understandings of human experiences. Recurrent conceptualizations of time have their own issues:

> All the turning that has been done over the past thirty years—the interpretive turn, the linguistic turn, the constructionist turn, the rhetorical turn, the narrative turn. Indeed, all this turning makes a person dizzy. . . . But isn't life itself uncertain, contingent, and paradoxical? Aren't human beings steeped in ambiguity and contradiction? (Bochner & Ellis, 1999, p. 488)

What are the alternatives to viewing the future through serial, recurrent, or chaotic prisms? My preference is to incorporate them all into *recursive* visions of what is to come that concede the unexpected but recognize both what is new and what is the same old experience. We plan for a future that is somewhat recognizable. We plan for a future we want to make more just, more free, more secure than our presents and pasts. We prepare our students for the varieties of practices called qualitative methodologies while urging them to think and do beyond these practices.

◨ NOTE

1. Judith Preissle has also published as Judith Preissle Kasper and Judith Preissle Goetz.

◨ REFERENCES

Adams, R. N., & Preiss, J. J. (Eds.). (1960). *Human organization research: Field relations and techniques.* Homewood, IL: Dorsey Press and the Society for Applied Anthropology.

Adler, P. A., & Adler, P. (1999). The ethnographer's ball—revisited. *Journal of Contemporary Ethnography, 28*(5), 442–450.

Archambault, R. D. (1963). Introduction. In R. D. Archambault (Ed.), *John Dewey on education: Selected writings* (pp. xiii–xxx). New York: Modern Library.

Atkinson, P., Coffey, A., & Delamont, S. (1999). Ethnography: Post, past, and present. *Journal of Contemporary Ethnography, 28*(5), 460–471.

Atkinson, R. (1998). *The life story interview.* Thousand Oaks, CA: Sage.

Behar, R. (1999). Ethnography: Cherishing our second-fiddle game. *Journal of Contemporary Ethnography, 28*(5), 472–484.

Berger, P. L., & Luckmann, T. (1966). *The social construction of reality: A treatise in the sociology of knowledge.* Garden City, NY: Doubleday.

Bertaux, D. (Ed.). (1981). *Biography and society: The life history approach in the social sciences.* Beverly Hills, CA: Sage.

Bloch, M. (1953). *The historian's craft.* New York: Random House.

Bochner, A. P., & Ellis, C. S. (1999). Which way to turn? *Journal of Contemporary Ethnography, 28*(5), 485–499.

Burnett, J. (1973). Event description and analysis in the microethnography of urban classrooms. In F. A. J. Ianni & E. Storey (Eds.), *Cultural relevance and educational issues* (pp. 287–303). Boston: Little, Brown.

Chávez, R. A. (2004). *Evaluación integral de la personalidad creativa: Fenomenología, clínica y genética (Integral evaluation of the creative personality: Phenomenology, clinic and genetics).* Unpublished doctoral dissertation, Facultad de Medicina, National Autonomous University of Mexico UNAM, Mexico City.

Degérando, J.-M. (1969). *The observation of savage peoples* (F. C. T. Moore, Ed. & Trans.). Berkeley: University of California Press. (Original work published 1800)

deMarrais, K. (2006). The haves and the have mores: Fueling a conservative ideological war on public education. *Educational Studies, 39*(3), 203–242.

Denzin, N. K., & Lincoln, Y. S. (1994). Introduction: Entering the field of qualitative research. In N. K. Denzin & Y. S. Lincoln (Eds.), *Handbook of qualitative research* (pp. 1–17). Thousand Oaks, CA: Sage.

Denzin, N. K., & Lincoln, Y. S. (2000). Introduction: The discipline and practice of qualitative research. In N. K. Denzin & Y. S. Lincoln (Eds.), *Handbook of qualitative research* (2nd ed., pp. 1–28). Thousand Oaks, CA: Sage.

Denzin, N. K., & Lincoln, Y. S. (2005). Introduction: The discipline and practice of qualitative research. In N. K. Denzin & Y. S. Lincoln (Eds.), *The SAGE handbook of qualitative research* (3rd ed., pp. 1–32). Thousand Oaks, CA: Sage.

Dilthey, W. (1989). *Introduction to the human sciences* (R. A. Makkreel & F. Rodi, Eds.). Princeton, NJ: Princeton University Press. (Original work published 1883)

Dollard, J. (1935). *Criteria for the life history, with analyses of six notable documents.* Freeport, NY: Books for Libraries Press.

Ellis, C. (2009). *Revision: Autoethnographic reflections on life and work.* Walnut Creek, CA: Left Coast Press.

Foreman, P. B. (1948). The theory of case studies. *Social Forces, 26*(4), 408–419.

Freeman, M., deMarrais, K. D., Preissle, J., Roulston, K, & St. Pierre, E. A. (2007). Standards of evidence in qualitative research: An incitement to discourse. *Educational Researcher, 36*(1), 25–32.

Glaser, B. G., & Strauss, A. L. (1967). *The discovery of grounded theory: Strategies for qualitative research.* Chicago: Aldine.

Goetz, J. P. (1975). *Configurations in control and autonomy: A microethnography of a rural third-grade classroom.* Unpublished doctoral dissertation, Indiana University, Bloomington.

Goetz, J. P., & LeCompte, M. D. (1984). *Ethnography and qualitative design in educational research.* New York: Academic Press.

Goetz, J. P., & LeCompte, M. D. (1988). *Etnografia y diseno cualitativo en investigacion educative* (A. Ballesteros, Trans.). Madrid, Spain: Morata.

Goodson, I., & Sikes, P. (2001). *Life history research in educational settings: Learning from lives.* Buckingham, UK: Open University Press.

Goodson, I., & Walker, R. (1991). *Biography, identity and schooling: Episodes in educational research.* London: Falmer Press.

Gottschalk, L., Kluckhohn, C., & Angell, R. (1945). *The use of personal documents in history, anthropology, and sociology.* New York: Social Science Research Council.

Greene, J. C. (2007). *Mixed methods in social inquiry.* San Francisco: Jossey-Bass.

Guba, E. G., & Lincoln, Y. S. (1994). Competing paradigms in qualitative research. In N. K. Denzin & Y. S. Lincoln (Eds.), *Handbook of qualitative research* (pp. 105–117). Thousand Oaks, CA: Sage.

Habermas, J. (1973). *Theory and practice* (J. Viertel, Trans.). Boston: Beacon Press.

Hammersley, M. (1999). Not bricolage but boatbuilding: Exploring two metaphors for thinking about ethnography. *Journal of Contemporary Ethnography, 28*(5), 574–585.

Hansen, J. F. (1979). *Sociocultural perspectives on human learning: An introduction to educational anthropology.* Englewood Cliffs, NJ: Prentice Hall.

Hatch, J. A., & Wisniewski, R. (Eds.). (1995). *Life history and narrative.* London: Falmer Press.

He, M. F., & Phillion, J. A. (2008). *Personal-passionate-participatory inquiry into social justice in education.* Charlotte, NC: IAP.

Heflin, K. (in press). The future will be televised: Newspaper industry voices and the rise of television news. *American Journalism.*

Heshusius, L., & Ballard, K. (Eds.). (1996). *From positivism to interpretivism and beyond: Tales of transformation in educational and social research (the mind–body connection).* New York: Teachers College Press.

Hinchman, L. P., & Hinchman, S. K. (Eds.). (1997). *Memory, identity, community: The idea of narrative in the human sciences.* Albany: State University of New York Press.

Husserl, E. (1999). *The essential Husserl: Basic writings in transcendental phenomenology* (D. Welton, Ed.). Bloomington: Indiana University Press. (Original work published 1893–1917)

Jones, A. S. (2009). *Losing the news: The future of the news that feeds democracy.* New York: Oxford University Press.

Josselson, R., & Lieblich, A. (Eds.). (1993). *The narrative study of lives.* Newbury Park, CA: Sage.

Junker, B. H. (1960). *Field work: An introduction to the social sciences.* Chicago: University of Chicago Press.

Kessler, S., Ashendon, D., O'Connell, R. W., & Dowsett, G. W. (1985). Gender relations in secondary schooling. *Sociology of Education, 58*(1), 34–48.

Kridel, C. (1998). *Writing educational biography: Explorations in qualitative research.* New York: Garland.

Langness, L. L., & Frank, G. (1981). *Lives: An anthropological approach to biography.* Novato, CA: Chandler & Sharp.

Lather, P. (1992). Critical frames in educational research: Feminist and post-structural perspectives. *Theory Into Practice, 31*(2), 87–99.

Lather, P. (2010). *Engaging science policy: From the side of the messy.* New York: Peter Lang.

LeCompte, M. D., & Preissle, J. (1992). Toward an ethnology of student life in schools and classrooms: Synthesizing the qualitative research tradition. In M. D. LeCompte, W. L. Millroy, & J. Preissle (Eds.), *The handbook of qualitative research in education* (pp. 815–859). New York: Academic Press.

LeCompte, M. D., & Preissle, J. (1993). *Ethnography and qualitative design in educational research* (2nd ed.). New York: Academic Press.

Lortie, D. C. (1975). *Schoolteacher: A sociological study.* Chicago: University of Chicago Press.

Loseke, D. R., & Cahill, S. E. (1999a). Ethnography: Reflections at the millennium's turn—Part 1 [Special issue]. *Journal of Contemporary Ethnography, 28*(5), 437–585.

Loseke, D. R., & Cahill, S. E. (1999b). Ethnography: Reflections at the millennium's turn—Part 2 [Special issue]. *Journal of Contemporary Ethnography, 28*(6), 597–723.

Madge, J. (1953). *The tools of social science.* London: Longmans, Green.

Madge, J. (1962). *The origins of scientific sociology.* New York: Free Press of Glencoe.

Malinowski, B. (1989). *A diary in the strict sense of the term.* Palo Alto, CA: Stanford University Press. (Original work published 1967)

Markle, D. T. (2008). *Beyond transcription: Promoting alternative qualitative data analysis.* Paper presented at the 2008 SQUIG Conference in Qualitative Research, Athens, GA.

Martineau, H. (1989). *How to observe morals and manners.* New Brunswick, NJ: Transaction. (Original work published 1838)

Mead, M. (1962). *Coming of age in America* [Audiotape]. Guilford, CT: Audio-Forum Sound Seminars.

Mead, M. (1972). *Blackberry winter: My earlier years.* New York: Simon & Schuster.

Mead, M., & Métraux, R. (Eds.). (2000). *The study of culture at a distance.* New York: Berghahn Books. (Original work published 1953)

Merton, R. K., Fiske, M., & Kendall, P. L. (1956). *The focused interview: A manual of problems and procedures.* Glencoe, IL: Free Press.

Nieuwenhuys, O. (1994). *Children's lifeworlds: Gender, welfare and labour in the developing world.* London: Routledge.

Okely, J., & Callaway, H. (Eds.). (1992). *Anthropology and autobiography. Association of Social Anthropologists Monograph 29.* London: Routledge.

Olshansky, S. J., et al. (2005). A potential decline in life expectancy in the United States in the 21st century. *New England Journal of Medicine, 352*(11), 1138–1145.

Polkinghorne, D. E. (1988). *Narrative knowing and the human sciences.* Albany: State University of New York Press.

Powdermaker, H. (1966). *Stranger and friend: The way of an anthropologist.* New York: W. W. Norton.

Preissle, J. (1999). An educational ethnographer comes of age. *Journal of Contemporary Ethnography, 28*(6), 650–659.

Preissle, J. (2004, April). *A rhizomposium on neglected figures in qualitative research: Joseph-Marie de Gérando.* Roundtable paper presented at the meeting of the American Educational Research Association, San Diego.

Preissle, J. (2006). Envisioning qualitative inquiry: A view across four decades. *International Journal of Qualitative Research in Education, 19*(6), 685–695.

Preissle, J., & Grant, L. (2004). Fieldwork traditions: Ethnography and participant observation. In K. B. deMarrais & S. D. Lapan (Eds.), *Foundations for research: Methods of inquiry in education and the social sciences* (pp. 161–180). Mahwah, NJ: Lawrence Erlbaum.

Reed-Danahay, D. E. (Ed.). (1997). *Auto/ethnography: Rewriting the self and the social.* Oxford, UK: Berg.

Reinharz, S. (1984). *On becoming a social scientist: From survey research and participant observation to experiential analysis.* New Brunswick, NJ: Transaction. (Original work published 1979)

Richardson, L. (1997). *Fields of play: Constructing an academic life.* New Brunswick, NJ: Rutgers University Press.

Riessman, C. K. (2008). *Narrative methods for the human sciences.* Thousand Oaks, CA: Sage.

Schacter, D. (2001). *The seven sins of memory: How the mind forgets and remembers.* Boston: Houghton Mifflin.

Scheurich, J. J. (1997). *Research method in the postmodern.* London: Falmer Press.

Schutz, A. (1962). *Collected papers 1: The problem of social reality.* (M. Natanson & H. L. van Breda, Eds.). The Hague, The Netherlands: Martinus Nijhoff.

Scriven, M. (1972). Objectivity and subjectivity in educational research. In Lawrence G. Thomas (Ed.), *Philosophical redirection of educational research* (pp. 94–142). Chicago: National Society for the Study of Education.

Smith, J. K. (1989). *The nature of social and educational inquiry: Empiricism versus interpretation.* Norwood, NJ: Ablex.

Smith, J. K. (1993). *After the demise of empiricism: The problem of judging social and educational inquiry.* Norwood, NJ: Ablex.

Smith, J. K., & Heshusius, L. (1986). Closing down the conversation: The end of the quantitative–qualitative debate among educational inquirers. *Educational Researcher, 15*(1), 4–25.

St.Pierre, E. A. (2010, May 29). *Resisting the subject of qualitative inquiry.* Paper presented at the Sixth International Congress of Qualitative Inquiry, University of Illinois, Urbana-Champaign.

Van Cleave, J. (2008). *Deconstructing the conventional qualitative interview.* Paper presented at the 2008 SQUIG Conference in Qualitative Research, Athens, GA.

Wax, R. H. (1971). *Doing fieldwork: Warnings and advice.* Chicago: University of Chicago Press.

Webb, B. (1926). *My apprenticeship* (Vols. 1 & 2). London: Longmans, Green.

Weber, M. (1949). *The methodology of the social sciences.* New York: Free Press. (Original work published 1903–1917)

Williams, T. R. (1967). *Field methods in the study of culture.* New York: Holt, Rinehart & Winston.

Willis, P. E. (1977). *Learning to labour: How working class kids get working class jobs.* Farnborough, UK: Saxon House.

Wolcott, H. F. (1992). Posturing in qualitative research. In M. D. LeCompte, W. L. Millroy, & J. Preissle (Eds.), *The handbook of qualitative research in education* (pp. 3–52). New York: Academic Press.

Wolcott, H. F. (2009). *Writing up qualitative research* (3rd ed.). Thousand Oaks, CA: Sage.

Znaniecki, F. (1934). *The method of sociology.* New York: Farrar & Rinehart.

43

TEACHING QUALITATIVE RESEARCH

Margaret Eisenhart and A. Susan Jurow

All programs to develop qualitative researchers must grapple with at least seven key questions: What should students be taught about qualitative research? Should explicit instruction in research methods be required? How much of the program should be devoted to methods versus foundational or topic-specific coursework? Should prospective researchers be prepared in one method or several? How should research ethics be covered? How should research competence be assessed? And, how can the curricular goals, whatever they are, be designed into instruction? (See Page, 2001, and Preissle & Roulston, 2009, for some additional questions.)

Answering these questions can be daunting. Qualitative research communities, whether they be sociologists, anthropologists, educational researchers, psychologists, nurses, or others, do not agree on research priorities. Scattered around the world, they do not face the same research problems or questions about their work. Whether they be discipline-, field-, or practice-based, qualitative researchers do not share one approach. Instructors have different areas of expertise, and understandably, they want to teach in ways consistent with their expertise. Yet at least in compulsory research courses, they have a responsibility to prepare students to conduct research on a range of topics, use various research designs, and work with diverse groups. Further, there is always a limited amount of time for courses in a degree program, even less for so-called "methods" courses, and all course activities have to fit, at least generally, the requirements of university schedules and rules. The proverbial pedagogical question can be a harsh master: What is a qualitative research instructor to do on Monday morning? Surprisingly few qualitative researchers have written in any detail about the teaching decisions they make.

In this article, we first present an overview of the literature that bears on teaching qualitative research. Then, we describe in some detail the pedagogical approach we (Eisenhart and Jurow) took to teaching a two-course introduction to qualitative

research required of all entering doctoral students in the School of Education at the University of Colorado, Boulder. In the final section, we reflect on what we have achieved and not achieved in our teaching and suggest some directions for future reflections on and studies of teaching qualitative research.

LITERATURE ON TEACHING QUALITATIVE RESEARCH

The literature on qualitative research focuses mainly on how to do research. Numerous books and articles cover processes and procedures for conducting qualitative research. They range from compendia of different traditions, approaches, and techniques (e.g., Creswell, 2002; Denzin & Lincoln, 1994, 2000, 2005; Green, Camilli, & Elmore, 2006; LeCompte, Millroy, & Preissle, 1992; Schensul & LeCompte, 1999); to extended discussions of a single approach, such as grounded theory (Charmaz, 2006; Glaser & Strauss, 1967), ethnography (Agar, 1996; Hammersley & Atkinson, 1995; Wolcott, 2009), qualitative evaluation (Patton, 2002), or participatory action research (McIntyre, 2008); to guides to specialized techniques, such as ethnographic interviewing and participation observation (Spradley, 1979, 1980), discourse analysis (Phillips & Hardy, 2002), qualitative media analysis (Altheide, 1996), interpretive policy analysis (Yanow, 2000), and systematic self-observation (Rodriguez & Ryave, 2002). Others have focused on particular phases of research practice, e.g., Wolcott on analysis and interpretation (1994) and writing up qualitative materials (2008) and Miles and Huberman on data management and analysis (1994). A few have written from the perspective of novices (or students) as they struggled to learn how to do qualitative research (Heath & Street, 2008; Lareau & Shultz, 1996), and a few others, about the relationships that form between qualitative novice and more experienced mentor (Lee & Roth, 2003; Minichiello & Kottler, 2009). Certainly this body of literature, even more extensive than listed here, serves as a basis for what is taught (and the variety of what is taught) as qualitative research, but it rarely focuses on pedagogical approaches or strategies themselves.

Two recent publications, Hurworth's *Teaching Qualitative Research* (2008) and Garner, Wagner, and Kawulich's *Teaching Research Methods in the Social Sciences* (2009), directly address this lack of attention to pedagogy. Finding little guidance in the literature to answer her questions about how to teach qualitative research, Hurworth conducted case studies in qualitative research classes in seven Australian and English universities. Garner et al., concerned about the lack of information about how to teach both qualitative and quantitative research methods, compiled articles from researchers around the world in hopes of stimulating the development of a "pedagogical culture in research methods education." In both cases, the authors were shocked at how little has been written on this topic.

Not surprisingly, what literature there is on teaching qualitative research reflects the divide that has fractured the qualitative research community since the 1980s—the divide between those who take a more conventional social science view of methods and concentrate (at least in what they write about teaching) on research designs (case study, ethnography, narrative research, etc.) and techniques (participant observation, open-ended interviewing, etc.) and those who take a more critical or "avant garde" view and concentrate on epistemological and ontological principles.

Phillips (2006) caricatures this divide as two poles of a continuum—a right pole and a left pole. In Phillips's view, researchers at the right pole want to do "rigorous" qualitative research—research that is systematic and accessible to scrutiny by others, and they believe that such research can be done. This pole has also been referred to as "conventional social science," "Old Guard," "modernist," and "scientific." At this end of the continuum, clearly specified, transparent "methods" are what are critically important. This is "qi" in small letters (as labeled by the left pole)—where the focus is on methods per se, almost always multiple methods, for gathering and analyzing or interpreting qualitative data that can serve as evidence for warranted knowledge claims.

The left pole views all research as inherently subjective; therefore, it cannot be and should not try to be systematic and transparent in the style of conventional social science. The point of qualitative research is to lead readers or listeners to understand their taken-for-granted worlds in a new light that is generally sympathetic to the plight of nondominant groups and critical of dominant forms of privileging. In consequence, the left pole is highly skeptical of conventional ways of doing, talking, and thinking, including the conventional ways of doing research. This pole is also referred to as "nonscientific," "postmodern," "discursive" (where different discourses are put into conversation with each other, e.g., a dominant and nondominant one), the "linguistic turn in social science" (where texts,

rather than actions, are the object of study), or a "moral discourse" (where morality-in-context is stressed, and universal morality is usually denied). At this end of the continuum, what's important is the force with which a researcher confronts or undermines the taken-for-granted and the status quo. This is "QI" in capital letters (again, as labeled by the left pole), where the focus is on making a statement, telling a story, initiating an action, or catalyzing a movement that promotes "social justice" by putting diverse voices in "conversation" with each other. This position implies at least a different *use* of methods and the materials ("data") derived from them, if not new methods themselves.

Teaching Toward the Right Pole

Qualitative research instructors who lean toward the right pole tend to write about ways to teach conventional qualitative methods and conventional social science habits of mind.[1] Prior to 1990, most qualitative research instructors leaned in this direction. In 1992, Webb and Glesne published results of interviews and questionnaires from 75 qualitative research instructors and reviews of 55 course syllabi from colleges and schools of education in the United States. They reported that most instructors aimed to develop an appreciation for social science theories and an understanding of how qualitative methods have been used in conventional social science. Common methodological issues addressed were finding a site, identifying appropriate research questions, taking fieldnotes and conducting an interview, keeping a research journal, doing data analysis, coding, finding patterns or themes, and writing a research report.

Keen (1996), for example, recommended an approach outlined by sociologists Lofland and Lofland (1995):

> [We] outline a constellation of "thinking topics" that enable students to focus on patterns of social interaction and social organization in the setting. These topics are a set of conceptual categories that identify a series of units of analysis, beginning with the micro and moving progressively toward the macro. They include practices, episodes, encounters, roles, relationships, groups, organizations, social worlds, and lifestyles or subcultures. With the thinking topics in hand, students can begin to transform themselves from general spectators into theoretical observers and can identify a variety of practices, roles, relationships, groups ... in the setting.... To further facilitate the development of analysis, we also introduce coding.... We discuss how to use the thinking topics we've identified as coding labels to create a coding scheme, which is to be used in organizing and sorting the data (Lofland and Lofland, 1995, 186–193).... To lay the final foundations for our theory construction, and to further focus the gathering of data, the last two weeks of participant observation are accompanied by investigating the relationships between the various units of analysis we've identified through asking questions of the data. To facilitate this process, [we] present eight basic questions for social analysis of the thinking topics: type, frequency,

magnitude, structure, process, cause, consequence, and agency of the participants involved (1995, 123–148). (as quoted in Keen, 1996, pp. 170–171)

In 2008, Hurworth, troubled by still-limited information about teaching qualitative research, reviewed what she could find (see her Chapter 2 for a useful review and her Chapter 12 for a list of resources) and then conducted her own intensive case studies of seven qualitative research courses at Australian and English universities. One of her major findings from the literature is that while authors often list what they include in their courses (finding a site, identifying research questions, etc.), they almost never discuss their curriculum design or pedagogical decisions (Keen was apparently an exception). She writes,

> Often authors just list what they include [in their courses]. . . . [T]here is little indication of how [they] have made or evaluated their choices. It has been a matter of "This is what I do." Therefore, any curriculum issue tends to be trivialised or glossed over. For example, when questioned about how he devised his course, [one author] just said somewhat naively: "The topics I teach are similar to those in any qualitative or field methods course." (p. 159)

From her observations of the seven courses, Hurworth (2008) found six topics commonly included in the curriculum (although to varying degrees): examinations of particular qualitative methodologies (ethnography, case study, etc.), discussions of paradigm differences (positivist, interpretivist, etc.), components of research design and a research proposal, participant observation, interviewing, and fieldwork practices (gaining rapport and entry, etc.). Other topics, including document analysis, data analysis, credibility and rigor, ethics, history of qualitative research, writing about results, and the use of qualitative methods in multiple methods designs, were covered in some but not all of the courses and usually received rushed and minimal attention at the end of the course.

One clear finding from both reviews and surveys (Webb & Glesne [1992] and Hurworth [2008], spanning 1990–2007) is the common use of a research project as a frame for teaching qualitative research. Webb and Glesne found that 39 of the 55 syllabi they collected required a major piece of qualitative research and 6 more required mini-projects. They suggest that doing a research project leads students to important insights about qualitative research:

> The act of doing qualitative research forces most students to question their own assumptions. Observing and interviewing puts students in close contact with the experiences of others. They soon learn that the research methods serve as guides for intelligence, not as a technical substitute for thought. The act of data analysis rids most students of the naïve assumption that the data will somehow speak for themselves and that researchers can avoid interpreting what they have seen and heard. (pp. 776–777)

Hurworth also found that in most courses students are expected to conduct some kind of research project, and the project counted for 50–75% of the course grade.

In fact, the student research project seems to have become a "signature pedagogy" of qualitative research courses. This term was coined by Shulman (2005) to describe characteristic

> types of teaching that organize the fundamental ways in which future practitioners are educated for their new professions. In these signature pedagogies, the novices are instructed in critical aspects of the three fundamental dimensions of professional work—to *think,* to *perform,* and to *act with integrity.*" (p. 52, italics original)

Examples of signature pedagogies include the case dialogue method of teaching in law schools, bedside teaching on daily clinical rounds in medical education, and—we would add—the small-scale research project in qualitative research education.

Others have noted the pedagogical advantages of having students conduct qualitative research projects. Preissle and Roulston (2009) write that "practical engagement in field-related exercises and authentic research activities is integral" to qualitative research courses (p. 16). Strayhorn (2009) suggests that engaging students in authentic research activities is the most appropriate pedagogy for developing higher-order cognitive skills such as application of concepts and strategies to new situations, analysis, synthesis, and evaluation. In fact,

> [A] consensus has emerged that qualitative methods are best taught through "hands-on practice" (Blank 2004), "active learning techniques" (Crull and Collins 2004), "real-world" research (Potter, Caffrey, and Plante 2003), or "learning-by-doing" (Rifkin and Hartley 2001). Evidence specific to teaching qualitative methods shows that experiential pedagogies enhance students' enjoyment of learning (Rohall, Moran, Brown and Caffrey 2004); heighten students' awareness of the complexity of research choices and the philosophies that inform them (Hopkinson and Hogg 2004); impress on students that qualitative research is a process shaped by the particular social context in which it unfolds (Winn 1995); impart to students specific skills such as in-depth interviewing (Roulston, deMarrais, and Lewis 2003); give students confidence in applying qualitative research techniques (Walsh 2003); and help students appreciate the value of qualitative research (Rifkin and Hartley 2001). (Raddon, Nault, & Scott, 2007, n.p.)

Teaching Toward the Left Pole

Those who lean toward the left pole tend to stress teaching about critical or postmodern epistemological principles and habits of mind. These discussions center on beliefs, values, and ethics rather than research design or techniques. Lather (2006),[2] for example, argues that teachers of research should concentrate on the "politics of knowing and being known" and the "logics of inquiry and philosophies and histories of knowledge" (p. 47). She offers a pedagogical approach that challenges conventional assumptions about the world and how researchers can know it. Her approach focuses on five *aphorias* (impasses or complications) that expose students to the messiness, inconclusiveness, and partiality of current research practice: aphorias of objectivity (How can any researcher truly be objective?), complicity (How can any researcher avoid making research political?), difference (How can any researcher avoid static categorizations of difference?), interpretation (How can researchers deal with differences in constructions of reality among participants, social groups, researchers, and stakeholders?), and legitimization (Who has the authority to make and evaluate knowledge claims in the face of multiple, sometimes competing claims, and what are the effects of this exercise of power?). Lather's pedagogical intent is not to produce answers to these questions but to explore them for what they teach about the limits of research practice and the multiple ways of producing and being produced by research.

Somewhat similarly, Preissle and deMarrais (2009) discuss five key principles that guide their qualitative methods instruction. These principles include responsiveness (qualitative researchers must interact with and respond appropriately to research participants), reflexivity (qualitative researchers must study themselves studying topics, participants, and settings), recursiveness (qualitative researchers must work back and forth across the phases of research, such that data collection leads to refined research questions and more informed data collection, etc.), reflectivity (qualitative researchers should bring multiple concepts and theories from various disciplines to bear on collected materials in order to understand the data), and contextual (qualitative researchers assume that settings, participants, time, and place are integral to understanding human experience and behavior).

Phillips (2006) and others (e.g., Atkinson, Coffey, & Delamont, 2003) have criticized both left and right poles for their excesses—for example, researchers on the left who accuse those on the right of being closet positivists, having physics envy, or selfishly wanting to preserve the status of social scientists as experts, and those on the right who accuse those on the left of being wild-eyed radicals, extreme relativists, incoherent theorists, and shrill critics of everything. Arguments focused on the extremes have tended to drown out the voices of those in the middle. As Phillips writes,

> Situated in the middle of the continuum are a variety of moderate or temperate positions. . . . [H]ere research is seen as a fallible enterprise that attempts to construct viable warrants or chains of argument that draw upon diverse bodies of evidence and that support any assertions that are being made. (p. 17)

Phillips continues, and we agree,

> The most that reasonably can be expected is that research results ought to be *constrained* by the evidence [a major concern of those

on the right]; and that evidence indicating differential harms and benefits ought to be given due weight [a major concern of those on the left]. . . . [Further, it has] to be acknowledged that *without* careful observation, testing, measurement, construction of ingenious apparatus, designing questionnaires, making models, doing calculations, drawing implications, playing hunches, and so forth, scientific inquiry (however characterized) would not be able to get off the ground. (pp. 22, 24, emphasis original)[3]

We find ourselves near the middle of Phillips's (2006) continuum with respect to our teaching of qualitative research methods, and more toward the right than the left. On the one hand, we require that students engage in specific activities to scaffold their introduction to the practices of qualitative research, and we teach specific skills and techniques that qualitative researchers have conventionally used. On the other hand, we encourage certain interpretive dispositions (habits of mind) that we believe are central to the practice of qualitative research. In the following section, we describe our approach.

◨ QUALITATIVE RESEARCH IN EDUCATION AT THE UNIVERSITY OF COLORADO, BOULDER

The qualitative methods courses we teach are framed by a larger faculty-led initiative, begun in 2002, to strengthen the doctoral program as part of the Carnegie Foundation's Initiative on the Doctorate (www.carnegiefoundation.org/previous-work/professional-graduate-education). The reform was motivated in part by faculty concern that the doctoral curriculum had been virtually unchanged for 10 years and partly by national concerns about the quality of PhD graduates in education research (Burkhardt & Schoenfeld, 2003; Lagemann, 2000; Neumann, Pallas, & Peterson, 1999; Schoenfeld, 1999). Our School of Education is known for teacher education and graduate programs that emphasize research-based classroom practice, research methodology, and educational policy. The special character of the school is its shared commitment to equal educational opportunity, diversity, research-based reform, and collaborative research (http://www.colorado.edu/education). When first implemented in the late 1980s, our existing doctoral program had been ahead of many others in its requirements for coursework in foundational issues in educational research and in both quantitative and qualitative research methods. But over time, concerns had mounted about the datedness of some course content, the balance and extent of training in qualitative and quantitative methods, and the link between research and practice.

As we reviewed other schools' programs and debated alternatives, we returned again and again to the idea of a common core. Although our survey of highly ranked schools of education showed that few required common courses for all doctoral students, the faculty decided that establishing a common core was desirable for three interconnected reasons: (1) Many of our doctoral students are former teachers and lack a discipline-based research background from their undergraduate or master's preparation, i.e., they enter our program with little or no previous research training; (2) prior experience as teachers and lack of research training sometimes produce students who are skeptical of the importance of research for educational improvement—in some cases, this skepticism leads to a rift between students primarily interested in research and those with teaching or activist orientations; and (3) students' inconsistent patterns of course taking were hampering our ability to offer advanced courses (because almost every course included some novices).

A faculty committee set out to design a core set of courses that would establish a common language and shared discourse about education research; present common norms and standards for the conduct of education research; and build an intellectual and methodological foundation for advanced, specialized coursework. After months of discussion, the committee settled on a concrete proposal for the core: a set of courses to be required for all entering doctoral students in the school and taken as a cohort. It would include two 2-semester courses in the "big (or foundational) ideas" of education and educational research, two 2-semester courses in quantitative methods, and two 2-semester courses in qualitative methods. One course of each type (one big ideas course, one quantitative course, one qualitative course) would run concurrently each semester of the students' first year.

(See Page, 2001, for a description of a similar program design at the University of California, Riverside.) One additional core course in multicultural education was scheduled for the first semester of the second year. In making this proposal, we were doubling the research methods courses required (from one course to two in each methodology), but by incorporating required material from old courses with new material in the core courses, we were able to eliminate some old courses and hold the total increase in required courses to only one. See Figure 43.1 for an overview of the new course design.

The new core went into effect in the fall of 2004 with a first-year cohort of 16 students. Although surprised by the existence of a core when they arrived on campus, the students were mostly enthusiastic about the new courses. At the beginning of the semester, several students commented that they were thrilled to have a clear program of study laid out for them and were eager to begin. Since then, the new courses have received mostly good reviews from both students and faculty.

Teaching Qualitative Research by Doing Real Research

The qualitative courses were further constrained by Eisenhart's decision, made in 2006, to organize the first of the two-course qualitative sequence around a "real" research and outreach project, the Learning Landscapes Initiative (LLI), which was already underway. LLI is a collaborative project of community groups,

Figure 43.1   Overview of the New Doctoral Program at the University of Colorado, Boulder

YEAR ONE: THE CORE	
First Semester	*Second Semester*
Big Ideas: Perspectives on Classroom Teaching and Learning (3 hrs) Qualitative Methods I (3 hrs) Quantitative Methods I (3 hrs) Specialty Seminar (1 hr)	Big Ideas: Education Research and Social Policy (3 hrs) Qualitative Methods II (3 hrs) Quantitative Methods II (3 hrs) Specialty Seminar (1 hr)

YEAR TWO: INTERMEDIATE	
First Semester	*Second Semester*
Multicultural Educ (3 hrs) Specialty Area Courses/Advanced Methods Courses (3 or 6 hrs)	Specialty Area Courses/Advanced Methods Courses (3 or 6 hrs)

YEAR THREE: INTERMEDIATE/CAPSTONE	
First Semester	*Second Semester*
Specialty Area Courses/Advanced Methods Courses/ Capstones (3 or 6 hrs)	Specialty Area Courses/Advanced Methods Courses/ Capstones (3 or 6 hrs)

school leaders, and university faculty to build new playgrounds at schools in Denver (Brink & van Vliet, 2004). Architects, landscapers, urban planners, and community members have designed, raised money for, and built new playgrounds at 48 elementary schools since 1998; efforts to redo all the city schools' playgrounds continue. The new playgrounds include age-appropriate play structures, artwork, grassy playing fields, and gardens. The new playgrounds replaced aging ones that had old or unsafe equipment and surfaces. Research accompanied the opening of the new playgrounds; the original LLI research team studied how the redesigned playgrounds affected children's physical activity levels and the community's pride and involvement with the school.

Although Eisenhart was committed to the context of LLI as the starting point of her pedagogy for the first-semester course, she also wanted the students to gain experience formulating their own small-scale research projects. Thus, for the final assignment in her class, she asked students to develop a new study based on their experience of LLI. They were required (by institutional review board [IRB] constraints) to conduct their research at LLI schools (if studying children or schools), but they could choose the topic. The new topics became the focus of students' "real" research in Jurow's second-semester course.

The decision to locate students' research projects in the context of LLI and its participating schools was inspired in large part by our commitment to a social practice theory of learning (Bourdieu, 1977; Lave & Wenger, 1991; Wenger, 1998). From this perspective, in order to learn something new, novices must engage in authentic "practices" of a new field while receiving guidance and support from experts and context. The practices of a field, as we refer to them here, are the activities in which regular participants in the field engage and that direct them to perceive the world related to that field in a particular way. Through gradually increased participation in activities with the continuing guidance of more experienced participants and the social organization of the context, novices are expected to learn techniques, strategies, and perspectives for engaging appropriately in the field. In our courses, students had opportunities to participate in activities that were part of an ongoing research endeavor in the actual context of that endeavor, receive feedback on their skills from experts, share their findings with a research team and community members, and eventually take on responsibilities related to conducting qualitative research. Students' research experiences were intended to be "authentic" in the sense that they would have "use value" (by contributing to a community of research practice) as well as "exchange value" (by contributing to course requirements and earning a grade).

From this perspective, novices (here: first-year doctoral students in education) encountered a pedagogy of qualitative research that we (two experienced qualitative researchers) structured, consciously and unconsciously (via our dispositions), as a means of teaching and learning qualitative research. Our decision to organize the courses around a real research project was extraordinarily consequential for the learning environment. Demands of entrée, rapport, ethics, communication, scheduling, observing, interviewing, and reviewing literature for research about children on playgrounds and in elementary schools provided much of the structure for topic coverage and sequence in class, as well as for class assignments and deadlines.

By structuring student participation in these aspects of a real research project, we hoped to cultivate dispositions that we value in qualitative researchers. We hoped students would grasp the distinctiveness of qualitative research, particularly its commitment to learn from participants' perspectives, values, and context through open-ended and flexible inquiry. We hoped they would see that research methods should be used to answer specific research questions (and not simply because one likes or dislikes a method). We hoped they would come to appreciate the importance of careful analysis and thoughtful interpretation of qualitative materials. We hoped they would appreciate how qualitative methods can reveal unexpected insights about people and places thought to be known or understood already.

We also hoped to cultivate ethical dispositions consistent with good research practice. Because students would be observing young children on school playgrounds; interviewing teachers, students, and parents; and using the material they collected and analyzed to make presentations and contribute to LLI, their behavior and activities had to meet the requirements for research with human subjects, including those of the university's institutional review board. Although it can be useful for students to prepare their own IRB materials as part of a class project, we did not want to delay the start of the research; thus, Eisenhart filed the IRB materials, contacted the schools, and received approval before the beginning of the first semester course.[4] During class time, she shared the IRB materials with the students and reviewed ways of minimizing risk: obtaining consent and assent, maintaining confidentiality, refraining from direct involvement except in an emergency, reporting problems and results, and being clear and honest about research purposes with everyone involved.

The decision to focus the students' activities in these ways also was made on pedagogical and logistical grounds, based on past experience. Both instructors had previously taught qualitative research methods in which students chose their own sites and research questions from the start. Although this approach has the advantage of allowing students to select a research topic based on interest, in our experience it also meant that students proceeded at different paces and had different needs at any given time during the course. This situation made it impossible to match class topics and discussion with students' immediate research needs (see also Keen, 1996).

Overview of the Qualitative Research Methods Sequence

In the two-course qualitative sequence, students were introduced to qualitative methods in the first course and revisited these ideas as they completed their own research in the second course. Both courses were framed by a view of interpretive, qualitative research as it developed in cultural anthropology and fieldwork sociology. Students read books and articles focused on qualitative research design and interpretation (e.g., Michael Agar's *The Professional Stranger* [1996], Joseph Maxwell's *Qualitative Research Design* [2004], and Phil Carspecken's *Critical Ethnography in Educational Research* [1996]) and book-length ethnographic studies (e.g., Barrie Thorne's *Gender Play* [1993] and Annette Lareau's *Unequal Childhoods* [2003]). As students read and discussed these texts in class, they simultaneously engaged in supervised fieldwork.

In the first semester, teams of students conducted observations and interviews at nine Denver playgrounds to understand children's social experiences during recess. (Details on the project and the students' work will be discussed later.) Students coded and reconstructed their data and wrote short technical reports on their results. Drawing on their experiences in the field and in class, students (individually or in pairs) developed proposals for small-scale studies they would complete in the second-semester course.

In the second course (taught by Jurow), students focused more on the interpretive and critical epistemology of qualitative research; read additional exemplars of qualitative research; and studied texts on how to collect, organize, and analyze data. As in the first semester, the students completed course readings as they conducted supervised fieldwork. Course readings and assignments were coordinated with and to some extent developed in response to students' research activities. Students participated in workshop-style discussions focused on data collection and analysis strategies as well as emergent research dilemmas. Toward the end of the semester, students created posters to share their research in a School of Education poster session where they received feedback from faculty and other students. Building on their posters, they wrote final research papers in which they presented their studies.

Activities, Skills, and Dispositions of Qualitative Research

In this section, we discuss the activities we arranged, the skills we hoped students would develop, and the dispositions we wanted to cultivate. Although we discuss activities, skills, and dispositions separately for analytic ease and clarity, in both theory and practice these features overlap and interrelate.

Research Activities

The central activities that we organized for the students included (1) activities for answering (and later developing) research questions with qualitative data, (2) activities for communicating about research, and (3) activities for making social and cultural comparisons.

Answering research questions with qualitative data. To help students understand how qualitative research methods are used to answer research questions, Eisenhart organized students' early project work around a set of research questions. Rather than asking students to enter the field with no particular focus, which can quickly lead to student frustration, or to develop their own, which is difficult for most novices, Eisenhart structured the students' fieldwork so they would experience (rather than design) qualitative research questions and the kinds of data that can answer such questions. The questions she proposed complemented the LLI's ongoing work. While LLI focused on comparing old and new playgrounds in terms of quantified measures of children's physical activity level, Eisenhart's research questions focused on comparing old and new playgrounds in terms of social behavior. She provided the students with three research questions to begin their research:

1. What kinds of social interactions—among children and between children and adults—take place on the playground?

2. How do these interactions vary depending on gender and ethnicity?

3. Why do these patterns of social interaction occur; i.e, what is a credible explanation for the patterns observed?

She emphasized that the students' contribution to LLI would come in providing tentative answers to these questions for the LLI research team. The three research questions then framed the first-semester readings; class discussions; and assignments on observation, interviewing, and analysis—the main research skills emphasized in the course sequence (described in more detail under "skills" below). Students' observations, interviews, analysis and write-up of their first-semester fieldnotes and transcripts were directed toward providing answers to the three questions.

Eisenhart's questions focused the students' attention on local interactions and encouraged them to interpret the interactions in terms of social and cultural significance. She hoped that the form and focus of the questions, as well as the content of the answers, would serve as a guide to the kinds of qualitative research questions that students would want to ask and be able to answer in their second-semester projects. For the most part, though focused on a range of topics, students in the second semester proposed research questions about participants' social experiences and how these were constrained and afforded by interactional, societal, cultural, and personal dimensions of their lives. Some of the research questions students asked included the following: "What are African American and Mexican American fifth graders' perceptions about race?" "What kinds of interracial interactions do boys and girls have on the playground?" "What meanings/messages might be conveyed to the jumpers through lyrics in jump rope songs?" "What influences formation of a child's identity on the playground?" and "Are home and school separate spheres with respect to the acquisition and valuing of cultural capital?" Sub-questions to these primary questions allowed students to identify and investigate patterns and variations in the children's playground interactions.

Communicating qualitative research. In order to help students learn how to share their research with others, both Eisenhart and Jurow organized opportunities for students to talk and write about qualitative research. Lave and Wenger (1991) point out that for newcomers to a community of practice, "the purpose is not to learn *from* talk . . . it is to learn *to* talk" (pp. 108–109, emphasis original).

In our courses, talk *about* qualitative methods was frequent and took the form of mini-lectures and class discussions about conducting a study. In both courses, students read, discussed, and critiqued book-length ethnographies. In the first semester, Eisenhart divided the class in half for key readings twice during the first semester: once when each half read portions of a methods text (Agar, 1996, or Carspecken, 1996) and once when each half read an ethnography (Lareau, 2003, or Carter, 2005). Students were given study questions focused on key ideas in each book and came to class prepared to describe their book to those who had not read it. Discussions in which students focus on a text to identify themes and offer critiques are what doctoral students do in most of their courses. This is a valuable form of discourse, allowing students to develop a shared language for communicating, but it does not allow students to engage directly in trying out their own strategies for discussing and justifying qualitative work.

Both instructors also created opportunities for students to practice talking and writing *as* qualitative researchers. At the end of the first course, students wrote technical reports that addressed Eisenhart's research questions; these reports were sent to the LLI researchers. Toward the end of the second semester, students were required to be presenters in a schoolwide poster session. Prior to the session, students developed short written and verbal descriptions of their studies for the poster session. They considered what their most intriguing finding was, what was important about their results, and what still puzzled them. Explaining their study at the poster session verbally positioned students as qualitative researchers in the eyes of the wider school community.

In Jurow's second-semester class, the students' final course papers followed a standard format for publishing qualitative research with an introduction, conceptual framework, literature review, methods section, results, discussion, and conclusion. Jurow encouraged the students to revise their course papers for presentation at conferences and possible publication. Six of the

students took this advice and presented their findings in a symposium at the 2008 American Educational Research Association (AERA) meetings.

These multiple and genuine opportunities for talking about and sharing qualitative research were purposefully designed into our courses. Students developed common language for discussing and evaluating qualitative research (others' and their own) and for positioning themselves as qualitative researchers. They also developed an expectation to share the stories, patterns, and potential implications of their qualitative research findings with audiences beyond their immediate peers and instructors.

Social and cultural comparisons. LLI allowed us to organize students' research experiences to encourage comparison of data in and across sites. In the first semester of the 2006–2007 and 2007–2008 doctoral cohort, students (n=12/each cohort) were divided into three groups of four. Each group was assigned three schools to study: one with an old playground, one with a recently developed new playground, and one with an older new playground. Each student was required to conduct six observations with fieldnotes, thus assuring that each group had 24 sets of fieldnotes spanning three playgrounds. To write the technical report in which they identified and tried to explain patterns they had observed, the students needed to work together to organize, analyze, and develop shared interpretations of their data from across the three sites. Further, in class discussions of the research process and emerging results, students had to consider the findings and interpretations of those who had worked at different sites.

In the second semester, when students chose their own research topics, opportunities for comparison across sites shifted. Students were no longer required to work together at their field sites, although the majority chose to do so.[5] Jurow designed classroom activities to capitalize on the fact that students were studying and facing similar issues at the playground sites. She required students to comment on each others' emerging analyses using standards for evaluating qualitative research that had been previously discussed in class. She purposefully paired individual students/groups that had something in common to read and comment on each others' preliminary claims. Students were then encouraged to consider their emerging interpretations in light of data that others had collected or conceptual frameworks that others were using. These class activities were meant to deepen students' understandings of what was happening in and across playground sites. They also aimed to help students review their peers' analyses of social and cultural patterns in critical and constructive ways.

The LLI group research project with its multiple playground sites enabled us to emphasize the value of social and cultural comparison in qualitative research. For example, students came to consider how different schools' views and policies on diversity affected children's interracial interactions, how playgrounds with spare play structures versus those with more elaborate structures affected the kinds of games children could play and with whom they played, and how children's arguments were resolved differently when peer mediators were present and when they were not. When students in prior years had conducted individual research studies at sites of their own choosing, the possibility for such comparison was limited. In the context of the LLI group research project, students had to look beyond their own data and their own sites in order to answer research questions and complete assignments. We also found that collaboration happened frequently and organically as students learned about each other's work.

Qualitative Research Skills

In the context of the three central activities, we assigned readings, held class discussions, and made assignments to encourage students to develop particular skills that we think necessary (though not sufficient) for doing qualitative research. We focused on having students learn to collect data through relatively unobtrusive methods; write fieldnotes; conduct interviews; and interpret these data by coding, narrating, and reconstructing the material collected. Students practiced these techniques as they collected data to answer research questions (first Eisenhart's and then their own) in the context of the LLI project.

In the first semester, students focused on practicing observational and fieldnote skills. They were required to conduct an open-ended, then a focused, and then a structured observation (each student conducted a minimum of six observations). Fieldnotes from each observation type were turned in to the instructor for feedback on amount of detail, level of concreteness, and likelihood of contributing answers to the research questions that she posed to them.

Using their own fieldnotes and those of two other students who had observed at different playground sites (for a total of 24 fieldnote observations from three sites), student teams began data analysis. Eisenhart introduced students to two different approaches to data analysis: Spradley's (1979, 1980) coding by domain and Carspecken's (1996) reconstructive analysis. Following Spradley, she asked students to develop a coding scheme based on the research questions (e.g., kinds of interactions, ways of displaying gender), and then to add codes that emerged from the data. Each team developed its own scheme, applied it to its observation data, and then received feedback from the instructor on its coding.

Building on these codes, Eisenhart then used readings on Carspecken's (1996) "reconstructive analysis" to encourage interpretations of the patterns. Carspecken focuses his analytic approach on identifying "meaning fields" in interactions, i.e., all possible meanings that any actor could give to an interaction, and then considering the social and historical "horizons of

meaning" that give substance to the meaning fields. For example, in a playground interaction in which a girl hits a boy, the analyst identifies meaning fields by speculating about what this interaction could mean to all the actors. From the perspective of the girl, it could mean she likes the boy, so she hits him to show affection; or, she hates the boy, so she hits him to express distaste, and so on. From the perspective of the boy, it could mean that the girl likes him and wants to get his attention; or, the girl is mad at him, and so forth. Once salient meaning fields have been identified, then the analyst considers the logic or narrative that binds the fields together (Do these interactions/meaning fields represent the social norm for children's gender relations in this community or something new and different? Do these interactions/meaning fields represent a historical legacy of gender relations or something new and different?). This approach, which involves systematically considering multiple observations of interactions and their potential meanings for different participants in an activity, allows the analyst to identify which interpretations or meaning fields capture most completely the logic of the actions. The intention behind using Carspecken's approach was to give the students a concrete means for developing a social and cultural analysis of their observations.

Drawing on their own interests and what they were learning through participation at their sites during the first semester, students were asked to develop their own research foci for the second-semester course. To help them locate their interests in the literature, Eisenhart had each student identify 10 relevant sources, read and briefly annotate each source, and then post the results on the class discussion board for all to share. When it came time to write the conceptual framework for the second-semester proposal, students were advised to consult the annotated list, choose three to four primary sources to focus their research interests, and use these to propose a tentative conceptual framework for their own study.[6] The students then wrote research proposals in which they articulated their emerging conceptual framework, research questions, and study design.

Students thus entered the second semester with some foundational knowledge of qualitative research, some strategies for collecting and analyzing data, and a sense of prior research on their research topic. As in the first semester, students began their studies by conducting observations at their sites. This allowed them to develop a sense of what was happening in the local context, what had changed since the first semester (if they were studying the same site), or how this site differed from the one they had studied in the first semester. With this orienting information, students were expected to refocus their research question (if necessary) and then complete initial interviews with site participants who might shed light on or extend their observations.

It was both expected and acceptable for students to modify their research foci in the second semester. In one case, a student who was interested in studying children's interracial interactions switched playground sites from the first to the second semester so that she could conduct her study at a school that had greater numbers of children from different racial groups. At her new site, she quickly noted that the children used what she described as "sophisticated language to talk about race." Her curiosity about the children's language use led her to expand her original research focus from looking only at the children's playground interactions to include an analysis of the children's participation in a character education program run by the school. In other cases, students who ended the first semester with a particular research focus changed their research questions altogether after noticing practices that they found more personally, theoretically, or practically compelling at their sites in the second semester. For example, a pair of students initially interested in how children use rules on the playground shifted their study to messages about gender norms and romantic relations conveyed in jump rope rhymes. This shift allowed the students to narrow what they felt was too broad of a focus to a topic on which they could gather more specific and richer data. In line with authentic qualitative research practices, students were encouraged to view their research investigations not as set in stone, but as responsive to their interests, what they learned from their fieldwork, and their efforts to collect usable data.

In the second semester, the students engaged in approximately 8 weeks of fieldwork. During this time, they also participated in class discussions and activities that increased their skill at writing fieldnotes, designing interview protocols, conducting interviews, and interacting with study participants.

Building on their first-semester experiences with data analysis, in the second semester, students were introduced to a further variety of approaches to thinking about qualitative data analysis. All the course readings emphasized an inductive approach, but were selected to offer the students a range of complementary techniques to make sense of their data (e.g., grounded theory, thematic analysis). As students read about approaches to analyzing data, they wrote about their emerging understandings of their data. Following an interpretive approach to qualitative research methods, a focus of the students' writing was on understanding children's interpretations of actions within their meaning systems. There were a number of analytic writing assignments in the second-semester course: a series of short memos written while students collected their data, a rough and then a final poster draft, and a rough and then a final course paper. In addition to receiving feedback from Jurow on their emerging analyses and writing, peers critiqued each other's writing, and faculty and students at the schoolwide poster session provided feedback on students' work. Through this repeated process of feedback and critique, students often reconsidered original ways of framing their studies, identified the need for additional conceptual tools, and refined their research foci.

The students' approaches to analysis, writing, and presentation of findings were developed in relation to professional standards for sharing qualitative research in education. Students read and referred back to texts including the American Educational Research Association's (2006) standards for reporting empirical research as they reviewed research articles and ethnographies in class. Jurow and the students used the standards to have critical and grounded discussions about issues such as the links among the conceptual framework, the research design, and the interpretations; the adequacy of evidence presented in a text; and the validity of a study's findings. Jurow encouraged the students to use the AERA recommendations as a shared framework for evaluating their own and each other's research posters and papers. In this way, the class developed a common vocabulary for discussing and debating the strengths and weaknesses of qualitative studies.

In summary, we intended for students to learn and practice skills related to participant observation, interviewing, interpreting, and writing up analyses in our courses. Although we focused on these foundational skills for doing qualitative research, our commitment to social practice theory led us to believe that students would learn more than this through their engagement in the LLI group research project. Indeed, this was the case. Students faced ethical issues at their sites related to ensuring children's safety and rights, needed to revise their research questions in light of the data they collected, and shifted their research roles as they came to value children as active participants rather than simply objects of research. As these kinds of problems emerged out of the students' research, we discussed them in class, read texts related to them, and helped students develop skills for managing them in real-time research contexts.

Dispositions for Qualitative Research

In addition to the specific activities we scaffolded and the skills we taught, we wanted our students to develop a set of dispositions, or a more general sense of "how to do" qualitative research. One disposition we certainly hoped to encourage was a commitment to careful and systematic qualitative research. Our attention to the detail of conventional qualitative skills, designs, and sequencing was intended to support this goal. But this was not the only disposition we wanted to nurture in our students. We also hoped they would develop commitments to open-ended and flexible inquiry that leads to new insights about what research participants care about and mean by their actions; a reflective stance on one's subjectivity and position as a researcher (and other social identities), including their effects on research questions, design, and outcomes; and ethical behavior in the conduct of research.

We used several strategies to encourage open-endedness and flexibility. Eisenhart presented observations as proceeding from general guidelines such as "what's happening here" to more focused and then structured observations, where each successive type is developed from the preceding one. She presented reconstruction (following Carspecken, 1996) as a process of identifying and considering many possible interpretations and only then deciding on and justifying "final" interpretations. When students began their own research in the second semester, Jurow emphasized that they should think about their study proposals as resources for thinking about what they would do in the field as opposed to using a script. She recommended that students set aside their proposals for the first few visits to the site so that they could familiarize themselves with the current happenings at the site and how these might affect their research plans. These early observations typically led to a refocusing or sometimes a complete change of students' research questions or data collection strategies. As a case in point, a pair of students, who had regularly observed girls and boys playing foursquare during their first-semester fieldwork, did not see as many mixed-gendered interactions during the second semester. As a result, the students revised their research question from one focused on gender differences to one that considered how children negotiate rules. This was a more appropriate question to ask given what the children were doing, and it required the students to rethink their conceptual framework, their research questions, and the purpose of their observations and interviews.

Also in the second semester, students read Erickson's (1986) chapter in which he introduces the concept of a natural history of inquiry. This method assumes that qualitative research will take turns that one cannot predict at the start, and from these turns, one will learn new things about the topic under study, but mostly one will learn about him- or herself as a researcher. To illustrate this idea, students read qualitative studies (e.g., Whyte's 1955 book, *Street Corner Society*) in which the researcher wrote about the development of his or her research and the unexpected directions the inquiry took based on early data analysis, relationships with participants, challenges in getting access to a particular site, funding and time constraints, and changing commitments related to the purpose of the research.

Throughout both semesters, the students were asked to keep track of how what they were learning affected the course of their project. One student, with some experience doing experimental and survey research, thought that he should not interact with the children on the playground because it would taint his data. Over the course of a few weeks, however, he started to realize that he would get a much deeper understanding of what the children were dealing with by talking and playing with them. This student began engaging with the children, who were eager to have his attentions, and this allowed him to see that they were not passive objects of gender socialization, but active agents in making sense of themselves as gendered beings.

By talking with students explicitly about the value of being responsive to study participants' experiences, we tried to convey that this was a desirable quality for a qualitative researcher to have. By reading qualitative studies in which an unplanned-for

refocusing of the research study was prominent in the author's account, we wanted students to see that this was not a disreputable practice. Further, when students needed to refocus their projects, we tried to use these as "teachable moments" to discuss how to decide on appropriate changes to conceptualizing the research project, methods of data collection, and the approach to analysis.

In order to cultivate a reflective stance in our students, we created occasions for them to share stories about their field experiences in class and in writing so that they could consider the ethical, social, and personal dimensions of their research. Some former teachers shared that it was hard for them to watch passively as children acted cruelly to one another on the playground. Should they intervene or not? If so, what kinds of interventions might be good ideas or bad ideas? Others who had been teachers found that they took a decidedly more active role in their research with the children because of their former experiences. Was this good or bad? Some students struggled with how their involvement with students might bias their data; others worried that their failure to interact with students might bias their data. In one situation, student researchers learned about the possibility of criminal behavior at school. What should they do? Should they report the allegation or protect their sources? Who was likely to be harmed and in what ways? Was it ethical to continue to conduct research at this school? Was it ethical to continue to use these data or not? How could they finish the course if they could not continue at the school or use the data?

These professional, ethical, and personal issues were explored throughout the courses in readings, in-class writing, and whole-class discussions. As part of a workshop-style format, Jurow asked two students to talk about problems or questions they were facing in their research as part of every class session. The volunteers talked about their struggles and how they affected their relationships in the field, approaches to data collection and analysis, and their opinions about the import of their research. These discussions were organized as opportunities for students to consider their values, experiences, and commitments alongside their positions as women, men, and members of different cultural communities. Students were invited to reflect on their subjectivities and the reasons they were asking the questions they asked, the perspectives they wanted to investigate or make known, and the purposes of their research beyond fulfilling a course requirement. Subjectivities were examined not as an end in themselves, but as part and parcel of doing qualitative research.

Summary

Consistent with Lave and Wenger, our pedagogy of research emphasized novices' exposure to and participation in authentic research practice—doing the actual work, for the purposes of contributing to an ongoing research project (as well as completing class work), in hopes of nurturing valued dispositions of qualitative inquiry. The pedagogical decisions we made were intended to give students access to and practice with conventional qualitative skills of data collection, analysis, write-up, and presentation, and to encourage them to develop a "sense of the game" of qualitative research.

Student Responses to Our Teaching

Although we did not systematically collect data about our students' responses to our teaching, we do have some evidence of their learning in our courses (see Jurow & Eisenhart, in preparation, for a more detailed discussion). In general, most of the students kept up with the readings, participated in class discussions, and followed the guidelines for setting up and conducting their research projects. Along the way, they had lots of questions and concerns, especially about the LLI project. Some did not want to spend so much time doing research, even for practice, on playgrounds. As they began the observations, they worried about not being able to record everything, not being able to figure out what was going on because they couldn't hear or understand what the children were saying. As the observation data came in and they started preliminary analysis, they worried that their fieldnotes did not contain anything interesting, that the instructions had not been clear enough for them to know how to "do it right." We took these concerns, which are typical of novice qualitative researchers, to suggest that the students were trying to engage with the tasks we set forth.

All the students engaged in some practices associated with qualitative research and tried to get better at them. These practices included taking responsibility for developing a research focus, collecting and analyzing some data, writing and presenting results, collaborating with others to do research, and using discourses of qualitative research. As the students worked on getting better at qualitative research—and they did get better—they also encountered difficulties that stalled their progress. Some challenges were due to the nature of qualitative research; more were due to time constraints imposed by the course structure and organization. Data analysis and interpretation proved especially challenging for all of them (see also Keen, 1996). Our efforts to scaffold students' access to strategies for analysis and interpretation were not sufficient, in the time period we had, to overcome students' lack of experience with conceptual frameworks as used in qualitative studies and with the skills and traditions of qualitative analysis and interpretation.

The students' responses suggest that participating in the qualitative course sequence led them to recognize some of the skills and dispositions associated with being a qualitative researcher. The responses also show the students' ability to use the discourse of qualitative research emphasized in our courses. And, they reveal the possibility of multiple learning trajectories.

Some students began to think of themselves as qualitative researchers before the end of the courses. Others needed more experience and greater responsibility. Interestingly, among all the things the students did in the context of our pedagogy of qualitative research, the defining feature of being a qualitative researcher, to them, was being able to analyze qualitative data, arguably the thing they struggled with the most and for which we think they most needed more time.

◨ CONCLUSION

Clearly our approach to teaching qualitative methods is only one of many approaches that might be taken. Our courses were constrained by the fact that they are compulsory courses for graduate students, many of whom have had limited research experience. They also were constrained by our need to contribute to a core curriculum that groups together all new students in the School of Education, regardless of their specialty area or interest, for introductory methods training. They also were constrained by our theoretical interest in social practice theory as a useful guide to teaching research and our previous dissatisfaction with class-initiated research projects in which each student chose his or her own topic and site.

The pedagogical approach we took in this teaching context was to organize much of our two-course sequence in terms of an ongoing research project to which our students contributed. This allowed us to engage in a form of recursive instruction in which we introduced and then spiraled back to strategies and techniques for observing, participating, interviewing, collaborating, analyzing, interpreting, writing up, and presenting qualitative research. It enabled the class as a whole to progress through all the phases of qualitative research in two semesters. It highlighted the value of collaborations and cross-site comparisons in qualitative research and built a community of researchers—including students and faculty from two universities, as well as teachers, administrators, students, and parents from nine urban schools—with some shared knowledge, language, and experience.

But it did not do everything. We did not systematically introduce all types of qualitative research (e.g., case study, ethnography, grounded theory). We did not encourage a necessarily activist approach to qualitative research (e.g., critical, antiracist, participatory action). We left the students on their own to learn qualitative data analysis software if they wanted. Hopefully, the students developed a foundation on which these things can and will be added.

We also did not do as much as we could have to study our students' learning in our courses. We kept journals about our teaching, collected some student journals, evaluated student assignments, surveyed students at the end of each course, and organized focus group discussions (in which we did not participate) after the completion of our courses in an effort to collect information about the students' learning. Although useful, this information was not systematically collected or analyzed.

Perhaps most salient given other articles in this handbook, we did not spend much time on epistemological debates about positivism vs. interpretivism, interpretivism vs. constructivism, or critical or postmodern stances on qualitative research.[7] Clearly our pedagogical decisions stressed research practice—student opportunities to design, conduct, and complete a research study and to make them aware of professional standards in relation to which they could evaluate their efforts. Students thus practiced strategies for taking fieldnotes, interviewing participants, analyzing data, and reviewing their own and others' research reports according to conventional standards for reporting empirical research. In this sense, we gravitate toward the right pole, the "qi" end, of Phillips's (2006) continuum and have perhaps compromised our students' ability to participate in the epistemological debates that are so prominent in U.S. educational research circles at this time.

On the other hand, we gave attention to left pole issues, too. We located students' research projects within a group project (the LLI) that was selected, among other reasons, for its affordances for investigating issues of equity. The LLI provided opportunities for investigating the lives of children from racial and linguistic minority backgrounds who have been historically underserved by public schools. By locating the students' research in Denver public schools that served these populations, we wanted our students to recognize the specificity of children's lives and move beyond deficit models of thinking about urban, minority children that are pervasive in educational research. Eisenhart's initial set of research questions provided a concrete way for students to investigate issues related to children's experiences of race and gender on the playgrounds and how these might vary across sites.

Our pedagogical practices, including the readings we selected, the feedback and guidance we gave students on their emerging analyses, and class exercises, were intended to teach students how to collect data systematically and make valid claims *as well as* engage them in discussions of the meaning of the patterns they observed and critiques of these patterns. For instance, when students read Thorne's (1993) *Gender Play* and Lareau's (2003) *Unequal Childhoods,* they considered the researchers' analyses of gender, race, and class hierarchies and discussed issues such as whether the arguments were backed by the evidence provided and if they went far enough in their critiques of present social structures.

Our attention to issues toward the left pole was enhanced because our students were simultaneously taking a first-year course on perspectives on classroom teaching and learning that covered sociocultural perspectives, feminist theory, and critical race theory. In that course, students read studies conducted from these perspectives, and these perspectives in turn influenced

how some students framed their research questions, thought about the purposes of their research, and considered their roles as researchers in our courses. We welcomed and capitalized on these connections across courses. Especially in the second-semester course, students' interests in avant garde theories and their implications for how researchers engage with study participants, include participants' voices, and act on what they learn through their research, influenced how we discussed their research practice. For example, when a pair of students who were interested in studying young children's ways of talking about race decided to shift from passive roles as observers to being more involved and directive in their interactions with children on the playground, this was encouraged. Jurow talked with the students about the tensions they felt between advocating for a particular view of race, which came out of their commitment to a critical perspective on race and their own experiences, and understanding and documenting the children's perspectives. These students then had to rethink how they would write up their study and evaluate its validity. Their shift led to a discussion with the entire class around these issues.

In this way, we encouraged students to discuss objectivity, complicity, difference, interpretation, and legitimization as they arose in the context of their research projects and readings. Unlike Lather (2006), we did not make these issues the centerpiece of our teaching, but we could not teach qualitative research without attending to them.

There were also ways in which we did not emphasize right-pole priorities. For example, we gave limited attention to anthropological and sociological theories that have been so important to the development of qualitative research in the disciplines. In a previous version of our school's qualitative research offerings, doctoral students were required to take at least one theory-oriented course in anthropology or sociology. They could choose from courses in the Anthropology or Sociology Departments, or a course in Anthropology and Education or Sociology of Education offered by the School of Education, and they had to complete this course before they could enroll in Ethnographic (or Qualitative) Methods. Under this arrangement, students typically spent one semester learning about discipline-based social science theories and writing a theoretically inspired research proposal for implementation in the second-semester methods course. In the methods course, they conducted a small-scale qualitative study of their own choosing. This course arrangement highlighted the place of social science theory as a precursor to designing a qualitative research study, but gave the students very little time to actually practice doing research. At the time, we offered no other courses in qualitative methods.

Another omission was the lack of attention to research designs employing multiple methods. This has been of special concern to the students in our program, who complain that both qualitative and quantitative researchers tout the value of multiple methods but do not teach how to do it. Our position has been that in order to use multiple methods effectively, students must first learn the strengths and weaknesses of specific traditions or designs. Our position would be more tenable if our program offered a subsequent course in multiple methods (which is now in development), although we will face resistance to adding another even quasi-required methodology course to the curriculum.

In our doctoral program, once students have taken our two courses, further coursework and practice in qualitative research is optional. We hope that students will take more advanced qualitative courses before beginning their (qualitative) dissertations, or work with professors on qualitative research projects. We hope that our faculty colleagues advise them to do so. Some students do take more courses and have opportunities for additional practice. But others will conduct and complete qualitative dissertations with no further methodological training.

The practice-oriented project work that was so consequential for our pedagogy was demanding for both the students and for us. Hours outside of class went into setting up the project work; designing the components; conducting the studies; providing individual feedback; doing revisions; and developing posters, final papers, and AERA presentations. But following the development of the students' work over time and listening to them talk about their experiences have been gratifying. They have had opportunities for practice, advice, critique, and revision. They have struggled with issues that are important to us in qualitative research: how to open yourself to perspectives and possibilities different from your own, how to find out about meaning and context, how to think through and rethink a research design, how to grasp and represent the complexity of social and cultural phenomena, how to position oneself in a research study, and how to collaborate and compare so as to go both deep and wide in a research study. At this point, we do not know whether attentiveness to these issues can be taught and learned in other ways, nor do we know what else could be achieved with other pedagogical strategies. We hope that our intentions, successes, and failures will inspire other qualitative researchers to do more research and writing about their teaching intentions, their pedagogical strategies, and their students' learning.

▣ NOTES

1. In saying that these instructors focus on method, we do not mean that they have a formulaic or recipe-like approach to teaching methods, as has sometimes been assumed.

2. Lather is talking specifically about educational research, in which she includes both "qualitative" and "quantitative" research.

3. Atkinson, Coffey, and Delamont (2003) take a somewhat different position with respect to what they call "classic" or "Old Guard" ethnographers—those whose careful work during 1950–1970 in sociology and anthropology served to more clearly articulate the methodological approaches used by qualitative researchers, versus "avant garde" or "critical" or "postmodern" ethnographers who came to have more and more influence from 1980 on. Their main point is that classic ethnographers laid the groundwork for the avant garde and thus the sharp distinction between the two groups is inaccurate and misleading. They do not see a paradigm shift but an evolution of the field in which the issues have become more complex and the answer strategies more diverse.

4. In our experience, the semester time needed to secure IRB approval for students' individual projects substantially limits time for students to practice other aspects of research. Also, because individual approvals can take different amounts of time—some are approved quickly while others take weeks—the instructor's ability to schedule class topics and activities to match students' progress is compromised.

5. Thirteen of 24 students collaborated with at least one other person for their second-semester research project. Others (11 of 24) worked on their own, but 4 of them chose to collect data as a "quasi-team" at one school site (i.e., they shared fieldnotes, transcripts, and a survey that they all used to answer their own research questions).

6. In our courses, we make a distinction between a literature review and a conceptual framework. We define the literature review as a summary of important previous research; we define the conceptual framework as a skeletal structure for organizing or guiding a new study (Eisenhart, 1991). We expect students to use the results of the literature review, in conjunction with guidance from theoretical perspectives, to develop a conceptual framework. Because developing a conceptual framework takes time and is difficult, Eisenhart shortened the process in her course.

7. We should note that our students do have other opportunities to cover this material in their coursework. One of the other core courses, Perspectives on Teaching and Learning, gives considerable attention to critical, feminist, and postmodern perspectives. Another course that is not required but many students take, Philosophical Issues in Educational Research, also covers this material in depth.

▣ REFERENCES

Agar, M. (1996). *The professional stranger: An informal introduction to ethnography.* San Diego, CA: Academic Press.

Altheide, D. L. (1996). *Qualitative media analysis.* Thousand Oaks, CA: Sage.

American Educational Research Association. (2006). Standards for reporting on empirical social science research in AERA publications. *Educational Researcher, 35*(6), 33–40.

Atkinson, P. A., Coffey, A., & Delamont, S. (2003). *Key themes in qualitative research.* Walnut Creek, CA: AltaMira Press.

Blank, G. (2004). Teaching qualitative data analysis to graduate students. *Social Science Computer Review, 22*(2), 187–196.

Bourdieu, P. (1977). *Outline of a theory of practice.* Cambridge, UK: Cambridge University Press.

Brink, L., & van Vliet, W. (2004). *If they build it, will they come? An evaluation of the effects of the redevelopment of inner-city school grounds on the physical activity of children.* Denver: University of Colorado.

Burkhardt, H., & Schoenfeld, A. (2003). Improving educational research: Toward a more useful, more influential, and better-funded enterprise. *Educational Researcher, 32*(9), 3–14.

Carspecken, P. (1996). *Critical ethnography in educational research.* New York: Routledge.

Carter, P. (2005). *Keepin' it real: School success beyond black and white.* Oxford, UK: Oxford University Press.

Charmaz, K. (2006). *Constructing grounded theory: A practical guide through qualitative analysis.* Thousand Oaks, CA: Sage.

Creswell, J. W. (2002). *Research design: Qualitative, quantitative, and mixed methods approaches* (2nd ed.), Thousand Oaks, CA: Sage.

Crull, S. R., & Collins, S. M. (2004). Adapting traditions: Teaching research methods in a large class setting. *Teaching Sociology, 32*(2), 206–212.

Delgado-Gaitan, C. (1993). Researching change and changing the researcher. *Harvard Educational Review, 63*(4), 389–411.

Denzin, N. K., & Lincoln, Y. S. (Eds.). (1994). *Handbook of qualitative research.* Thousand Oaks, CA: Sage.

Denzin, N. K., & Lincoln, Y. S. (Eds.). (2000). *Handbook of qualitative research* (2nd ed.). Thousand Oaks, CA: Sage.

Denzin, N. K., & Lincoln, Y. S. (Eds.). (2005). *The SAGE handbook of qualitative research* (3rd ed.). Thousand Oaks, CA: Sage.

Eisenhart, M. (1991). Conceptual frameworks for research circa 1991: Ideas from a cultural anthropologist; implications for mathematics education researchers. *Proceedings of the thirteenth annual meeting of psychology of mathematics education, North America* (pp. 202–219). Blacksburg, VA: Psychology of Mathematics Education.

Erickson, F. (1986). Qualitative methods in research on teaching. In M. Wittrock (Ed.), *Handbook of research on teaching* (3rd ed., pp. 119–161). New York: Macmillan.

Garner, M., Wagner, C., & Kawulich, B. (Eds.). (2009). *Teaching research methods in the social sciences.* Burlington, VT: Ashgate.

Glaser, B. G., & Strauss, A. L. (1967). *The discovery of grounded theory: Strategies for qualitative research.* New York: Aldine.

Green, J. L., Camilli, G., & Elmore, P. B. (2006). *Handbook of complementary methods in education research.* Mahwah, NJ: Lawrence Erlbaum.

Hammersley, M., & Atkinson, P. (1995). *Ethnography: Principles in practice* (2nd ed.). London: Tavistock.

Heath, S. B., & Street, B. V. (with Mills, M.). (2008). *Ethnography: Approaches to language and literacy research.* New York: Teachers College.

Hopkinson, G. C., & Hogg, M. K. (2004). Teaching and learning about qualitative research in the social sciences: An experiential learning approach amongst marketing students. *Journal of Further and Higher Education, 28*(3), 307–320.

Hurworth, R. E. (2008). *Teaching qualitative research: Cases and issues.* Rotterdam, The Netherlands: Sense.

Jurow, A. S., & Eisenhart, M. (with Eyerman, S., Gaertner, M., Roberts, S., Seymour, M., Spindler, E., & Subert, A.). (2011, manuscript in preparation). *Learning to be a qualitative researcher in education.*

Keen, M. F. (1996). Teaching qualitative methods: A face-to-face encounter. *Teaching Sociology, 24,* 166–176.

Lagemann, E. C. (2000). *An elusive science: The troubling history of education research.* Chicago: University of Chicago Press.

Lareau, A. (2003). *Unequal childhoods: Class, race, and family.* Berkeley: University of California Press.

Lareau, A., & Shultz, J. (1996). *Journeys through ethnography: Realistic accounts of fieldwork.* Boulder, CO: Westview.

Lather, P. (2006). Paradigm proliferation as a good thing to think with: Teaching qualitative research as a wild profusion. *Qualitative Studies in Education, 19*(1), 35–57.

Lave, J., & Wenger, E. (1991). *Situated learning: Legitimate peripheral participation.* Cambridge, UK: Cambridge University Press.

LeCompte, M. D., Millroy, W. L., & Preissle, J. (Eds.). (1992). *Handbook of qualitative research in education.* New York: Academic Press.

Lee, S., & Roth, W.-M. (2003). Becoming and belonging: Learning qualitative research through legitimate peripheral participation. *Forum: Qualitative Sozialforschung [Forum: Qualitative Social Research]* [Online serial], *4*(2). Available at http://www.qualitative research.net/index.php/fqs/article/view/708

Lofland, J., & Lofland, L. (1995). *Analyzing social settings: A guide to qualitative observation and analysis.* Belmont, CA: Wadsworth.

Maxwell, J. (2004). *Qualitative research design: An interactive approach* (2nd ed.). Thousand Oaks, CA: Sage.

McIntyre, A. (2008). *Participatory action research: Qualitative research methods series, 52.* Thousand Oaks, CA: Sage.

Miles, M. B., & Huberman, M. A. (1994). *Qualitative data analysis: An expanded sourcebook* (2nd ed.). Thousand Oaks, CA: Sage.

Minichiello, V., & Kottler, J. A. (2009). *Qualitative journeys: Student and mentor experiences with research.* Thousand Oaks, CA: Sage.

Neumann, A., Pallas, A., & Peterson, P. L. (Eds.). (2008, July). Investment in the future: Improving education research at four leading schools of education: Campus experiences of the Spencer Foundation's Research Training Grant Program [Special issue]. *Teachers College Record, 110*(7).

Page, R. N. (2001). Reshaping graduate preparation in educational research methods: One school's experience. *Educational Researcher, 30,* 19–25.

Patton, M. Q. (2002). *Qualitative evaluation and research methods* (3rd ed.). Thousand Oaks, CA: Sage.

Phillips, D. C. (2006). A guide for the perplexed: Scientific educational research, methodolatry, and the gold versus platinum standards. *Educational Research Review, 1*(1), 15–26.

Phillips, N., & Hardy, C. (2002). *Discourse analysis: Investing processes of social construction.* Thousand Oaks, CA: Sage.

Potter, S. J., Caffrey, E. M., & Plante, E. G. (2003). Integrating service-learning into the research methods course. *Teaching Sociology, 31,* 38–48.

Preissle, J., & deMarrais, K. (2009, May 23). *Qualitative pedagogy: Teaching ethnography and other qualitative traditions.* Paper presented at the fifth international congress of qualitative inquiry, University of Chicago, Urbana-Champaign.

Preissle, J., & Roulston, K. (2009). Trends in teaching qualitative research: A 30-year perspective. In M. Garner, C. Wagner, & B. Kawulich (Eds.), *Teaching research methods in the social sciences* (pp. 13–21). Surrey, UK: Ashgate.

Raddon, M., Nault, C., & Scott, A. (2007, August 11). *"Learning by doing" revisited: The complete research project approach to teaching qualitative methods.* Paper presented at the annual meeting of the American Sociological Association, New York. Available at http://www.allacademic.com/meta/p182602_index.html

Rifkin, S. B., & Hartley, S. D. (2001). Learning by doing: Teaching qualitative methods to health care personnel. *Education for Health, 14*(1), 75–85.

Rodriguez, N., & Ryave, A. L. (2002). *Systematic self-observation.* Thousand Oaks, CA: Sage.

Rohall, D. E., Moran, C. L., Brown, C., & Caffrey, E. (2004). Introducing methods of sociological inquiry using living-data exercises. *Teaching Sociology, 32*(4), 401–407.

Roulston, K., deMarrais, K., & Lewis, J. (2003). Learning to interview: The student interviewer as research participant. *Qualitative Inquiry, 9*(4), 643–668.

Schensul, J. J., & LeCompte, M. D. (1999). *The ethnographer's toolkit.* Walnut Creek, CA: AltaMira Press.

Schoenfeld, A. (1999). The core, the canon, and the development of research skills: Issues in the preparation of education researchers. In E. C. Lagemann & L. S. Shulman (Eds.), *Issues in education research: Problems and possibilities* (pp. 166–202). San Francisco: Jossey-Bass.

Shulman, L. S. (2005). Signature pedagogies in the disciplines. *Daedalus, 134*(3), 52–59.

Spradley, J. P. (1979). *The ethnographic interview.* New York: Holt, Rinehart & Winston.

Spradley, J. P. (1980). *Participant observation.* New York: Holt, Rinehart & Winston.

Strayhorn, T. L. (2009). The (in-)effectiveness of various approaches to teaching research methods. In M. Garner, C. Wagner, & B. Kawulich (Eds.), *Teaching research methods in social sciences* (pp. 119–130). Burlington, VT: Ashgate.

Thorne, B. (1993). *Gender play: Girls and boys in school.* New Brunswick, NJ: Rutgers University Press.

Walsh, M. (2003). Teaching qualitative analysis using QSR NVivo. *The Qualitative Report, 8*(2), 251–256. Available at http://www.nova.edu/ssss/QR/QR8-2/walsh.pdf

Webb, R. B., & Glesne, C. (1992). Teaching qualitative research. In M. D. LeCompte, W. L. Millroy, & J. Preissle (Eds.), *The handbook of qualitative research in education* (pp. 771–814). San Diego, CA: Academic Press.

Wenger, E. (1998). *Communities of practice: Learning, meaning, and identity.* Cambridge, UK: Cambridge University Press.

Whyte, W. F. (1955). *Street corner society: The social structure of an Italian slum* (4th ed.). Chicago: University of Chicago Press.

Wolcott, H. F. (1994). *Transforming qualitative data: Description, analysis, and interpretation.* Thousand Oaks, CA: Sage.

Wolcott, H. F. (2008). *Writing up qualitative research* (3rd ed.). Thousand Oaks, CA: Sage.

Yanow, D. (2000). *Conducting interpretive policy analysis.* Thousand Oaks, CA: Sage.

EPILOGUE

Toward a "Refunctioned Ethnography"

Yvonna S. Lincoln and Norman K. Denzin

If the Industrial Age was built on people's backs, and the Information Age on people's left hemispheres, the Conceptual Age is being built on people's right hemispheres. We've progressed from a society of farmers to a society of factory workers to a society of knowledge workers. And now we're progressing yet again—to a society of creators and empathizers, pattern recognizers, and meaning makers.

—D. H. Pink (2005, n.p.)

nd so it is we come to another punctuation point in the history of qualitative research and qualitative methods. The changes in even the past 6 years since the third edition of *The SAGE Handbook of Qualitative Research* have been enormous, and they are markers along the trajectory of qualitative and interpretive histories that these mature and new authors have written here. The "pattern recognizers" and "meaning makers" have not, despite considerable denigration, disparagement, and deprecation from some parts of the positivist camp, lost any particular ground, unless it is in funding. And even there, mixed methods proposals have frequently won the day (and the external funding).

We are at an interesting crossroads. On the one hand, qualitative methods as a field has been able to propose methods far beyond those of original fieldwork experts and texts (e.g., McCall and Simmons's [1989] original textbook, wherein they proposed that all fieldwork rested in interviewing and observation). So, for instance, methodologists have been prompt to adapt their methods to emerging technologies and to explore what kinds of data about social life emerge when they examine cultural artifacts—particularly the artifacts of popular culture—or what kinds of data are created by online communities. On the other hand, methodologists have also interrogated deeply classical methods, such as interviewing, and how such methods work at the intersections of race, class, gender, and nationality/hybridity. Our understandings about, for example,

interviewing as well as observation is far more nuanced, sophisticated, and sensitive than it was even 20 years ago. There are several such profound changes in the field. This epilogue will cover a number of the more important, including the turn to social justice; the turn, in interpretive work, to critical stances; the rise of mixed methods; the possibilities for cumulating knowledge and understanding for qualitative research; and the upcoming struggle to design even more contemporary methods that address the effects of late capitalism and its reshaping of economies, cultural structures and mores, and social life across the global community.

SOCIAL JUSTICE

A mere two decades ago, only a handful of scholars were talking about the impact of their work on issues of social justice, by which they meant the ability of social science to be put to policy objectives with the purpose of redressing a variety of historically reified oppressions in modern life: racism, economic injustice, the "hidden injuries of class," discrimination in the legal system, gender inequities, and the new oppressions resulting from the restructuring of the social welfare system to "workfare." Today, many scholars, positivists and interpretivists alike, purposefully direct their own research toward uncovering such injustices, exposing how historic social structures reify and reinvent discriminatory practices, and proposing new forms of social structures that are less oppressive. The turn toward social justice, of course, is directly linked with the turn toward more critical stances in interpretation and representation (Cannella & Lincoln, 2009; Denzin, 2009, 2010; Denzin & Giardina, 2006, 2007, 2008, 2009).

◨ THE TURN TO CRITICAL STANCES

The call for more critical stances in interpretive inquiry is not new (Lather, 1986, 1991, 1992, 2004, 2007); it has a long history, marching alongside qualitative researchers like a gentle moral conscience on the road trip to somewhere, destination unknown. However, until qualitative researchers were able to understand the connection between social science writ broadly and the quest for better policy, for a more just and democratic society, for a more egalitarian distribution of goods and services, critical perspectives were, in some cases, simply a companion, not a conjoined voice. That has changed.

While it is the case that not all qualitative researchers aim for social justice explicitly—some being focused on simply describing some phenomenon, or helping to create deep *verstehen* of some hitherto unexplored situation—it is the case that many now ask themselves what the outcomes of their research will produce in terms of more extended equality and less domination and discrimination (Mertens, 1998). Ruth Bleier (1984, 1986) makes a similar point: With resources for social science extremely limited, can we afford to engage in such scientific work without having it embody some larger social purpose, principally the amelioration of some social ill? She observes that

> Science is a powerful tool for good as well as for evil, for *emancipation* as well as for exploitation. How scientists use their time and talent, their training at public expense, their *public research funds, and the public trust are not matters to be brushed aside lightly.* Many fascinating research problems wait for attention. Residing, as we do, inside a universe filled with enigmas, many of which lend themselves to research approaches congenial to our personal styles, many with applications beneficial to segments of society that are due for some benefits, how do we justify working on research whose applications threaten to be deeply destructive of natural resources, of human life, of the dignity and self-respect of a racial or ethnic or gender group? (Bleier, 1986, p. 35, emphasis added)

This sense that something is owed back for the education and its privileges that scholars enjoy, but also that there are serious issues to which attention should be turned—issues which bear a strong relationship to redistributing the benefits of society more equitably—is a feeling that has come to full flower at the millennium. However much critics and disparagers of qualitative research criticize interpretive work as "advocacy" or as arguably political—since all knowledge is ultimately political—it is unlikely that issues of social justice, or

the more equal distribution of goods and services, or the elimination of discrimination and injustice, will go away. How can we ever return to our former naivete? We know too much, we understand too deeply, to go back now.

◨ MIXED METHODS

Nowhere can the theoretical and methodological tumult of the last decade be seen more clearly than in the crucible of mixed methods. As we wrote in the introduction to this volume, we—the two of us—share some misgivings about the paradigmatic (epistemological, ontological, axiological) state of mixed methods research *at this point in time.* The second edition of Tashakkori and Teddlie's *The SAGE Handbook of Mixed Methods in Social and Behavioral Research* (2010) deals with multiple issues that we and others have raised (Teddlie, personal correspondence, 2009), including epistemological concerns. We are uncomfortable with the stance that some have taken that epistemology doesn't matter. We feel strongly that such a declaration essentially negates the work of dozens (if not hundreds) of feminist theorists; critical theorists; race and ethnic studies theorists; queer and other "embodied" theorists; as well as theorists of postcolonialism, hybridities, borders, Latino/a and Latino critical studies, and others. Epistemology matters. Standpoint matters. Each of these things—one philosophical, one embodied and sociocultural—gives meaning and inflection to both the beginning (the research question) and the ending (the findings) of any inquiry. To deny their influence is to miss most of the major debates of the last quarter-century of qualitative research and indeed, the social sciences more broadly. The apotheosis of mixed methods at the expense of epistemology debates and ontological concerns appears to us to be misplaced, given the attention devoted to the careful plumbing of these issues and the delicate, razor-edged exploration of their implications by scholars for many decades.

Pragmatism may not, either, be the answer to the ontological, epistemological, and axiological concerns raised by the gentle voices of dissent. As we point out, too, the pragmatism that has been captured thus far is neither the classical pragmatism of Peirce or William James, nor is it the revised neopragmatism of Rorty or Habermas. A deep revision of the paradigmatic claims for the foundations of mixed methods seems to us to be called for at this time.

The original criticisms of mixed methods research seem to endure. While no classical methodologists appear to have anything particular against the possibility of mixing methods (see, for instance, Guba & Lincoln, 1981, 1989; Lincoln & Guba, 1985), it is with the caveat that such mixing should be done under the aegis of a single paradigm, such that the ontological, epistemological, and axiological concerns are coherent and resonant. Indeed, there are times when both kinds of data can illuminate different aspects of the same phenomenon so that a sharper and bolder picture emerges. Until such time as the paradigmatic issues are resolved, however, we welcome the ongoing debate and trust that the new mixed methods journal *(Journal of Mixed Methods Research)* will contribute clarity and strength to the conversation.

▣ THE CUMULATION OF QUALITATIVE KNOWLEDGE

One of the many myths surrounding qualitative research is that policy formulation utilizing such research is either difficult or impossible (Lincoln, in progress). Frequently dismissed as "anecdotal" by its detractors, qualitative research has often turned inward, addressing its own community of believers, who choose their own, less global, more locally focused means to effect social change. The major issue appears to be that of aggregating data, much as quantitative researchers rely on meta-analytic techniques for discovering strands of data and meaningful findings that can be "translated" into policy arguments. Aggregation, or cumulation of findings, is not a technique (or set of techniques) often taught or even discussed in doctoral research preparation, so it is not surprising that even qualitative researchers can and do accept the myths about non-aggregatability. There are, however, possibilities for cumulating and aggregating qualitative data, and there are techniques for secondary analyses of data that are beginning to be more widely known (Heaton, 2004), and there are historical efforts—developed for and utilized by the federal government—at case study aggregation analysis (Lucas, 1972, 1974). There are additional techniques being suggested that parallel quantitative meta-analytic techniques. Both secondary analyses and meta-analytic techniques are being made possible by computer-aided archival of data; such large-scale storage permits comparisons of data, reusing and "reworking" of qualitative data in a fraction of the time it might have taken a decade ago, and permits the kinds of comparisons of findings that allow policy construction. An exploration of these options, with a direct focus on their applicability for policy purposes, is the centerpiece of new and future efforts at addressing the cumulation issue.

▣ NAVIGATING THE CONTEMPORARY

Perhaps the most serious current issue confronting qualitative inquirers is generating a contemporary history and ethnography of the rapid changes characterizing the global community today. Davis and Marquis (2005) comment that "we are constantly reminded that we live in a world in which large organizations have absorbed society and vacuumed up most of social reality" (p. 332), and in which "MNCs [multinational corporations] dominate the world economy (and thus society) through their concentrated control of capital. It hardly seems a fair fight, as large organizations continue their drive to vacuum up whatever is left of social life." Davis and Marquis are not the only observers to note this phenomenon. Perrow (1991) commented on the same issue more than a decade earlier, and closer to this year, Faubion and Marcus (2009) and Westbrook (2008) addressed the same issues of contemporariness. Michael Fischer's (2009) powerful foreword to the Faubion and Marcus volume lays out a project of enormous scope: a beginning set of dialogues that "do not satirize older anthropologies and instead build upon, and extend into a new era, a recursive set of intellectual conversations and experiments" (p. ix), including their ability to guide readers into "the mediations of guarded, packaged, and traded elusive information" that serve to help us "understand the structure of the circuits as [well as] to challenge or guesstimate the veracity in the information packets" (p. viii). Much of the six dissertations reported on in this "Rice project" (dissertations completed at Rice University under the direction of Marcus and Fischer, with particular foci in mind) have to do with faked data (from informants), "made-up statistics and corruption stories" (p. viii) emanating from fieldwork projects, principally in postcolonial settings, but also in governmental and NGO (nongovernmental organization) environments. We have virtually no ethnographic tools to deal with wide-scale corruption, with seriously faked data, with governmental efforts to "vacuum up" what multinational corporations have not managed to "capture," or with information that is "elusive," deliberately hidden, ethically "unusable" (Hamilton, 2009), treated as corporate intellectual property, or carefully packaged for extra-organizational consumption.

Westbrook (2008), too, argues for an "ethnography of present situations," a "refunctioned ethnography" wherein "critical reflexivity" is operationalized "so that self-consciousness is not merely deployed as a critique of texts and stances after the fact, but is instead a part of the design and performance of anthropological work from the beginning" (p. 111). Westbrook reminds us, as well, of the first great principle of the postmodern: "[R]ather than a description or representation in the ordinary sense, which is in principle replicable, the expressions of ethnography for present situations are *in principle* unique" (p. 65, italics original). The function of these reimagined, repurposed, refunctioned ethnographies of "present situations" is to account for how we chose this future over some other; how we moved from the ravages of modernism to the globalized, late-capitalism neoliberalism sowing its seeds around the globe; and how we can examine the "navigator"—the ethnographer herself—how we examine multiple respondents "in some relation to one another"; and what Westbrook calls "liaisons," or connections, junctures, and nexuses that provide alliance between ethnographers and those who can and will supply the raw data of present-situation ethnographies. Faubion and Marcus (2009) would add to these three ingredients "circuits," or the routes by which information, data, stories, narratives, and other dialogic and textual material flows both around and to the ethnographer—or is prevented from reaching the ethnographer.

The call here is for an ethnography that moves beyond the "sin" and "guilt" of modernity, and that attempts to unravel the complexity of the milieu which is, according to Davis and Marquis (2005), "vacuuming up" social life and social reality. In the call for the practical, the pragmatic, or "what works," we may have lost sight of the fact that we are rapidly losing the means to socially construct *any* worlds, let alone one that is more just, more socially, economically, and culturally equitable.

▣ CONCLUSION: A REAL EPILOGUE

The foregoing are some of the more troubling reasons we term this fourth edition merely a punctuation point in the history of qualitative methods, not a period, not the end of the page. We have work to do, important work, and we must do it fast and well. While keeping our eyes on issues of social justice, we must also contrive how to represent multiple findings from multiple studies in order to achieve presence and voice at the policy table. We must learn to talk with those who speak quantitatively and those who speak qualitatively, but do so with consonance, coherence, and suasion. We must likewise research and make transparent the changes that are overtaking the world, so that we understand the futures we have chosen and are empowered to choose others if we so wish. Far from being some imaginary endpoint, we are in fact at the edge of a new colonialism, a new era, one that we did not fully choose, and one that we must begin to understand more fully than we have to this point. The only meaningful method for that understanding is a refashioned, refunctioned, repurposed imaginary for ethnography and ethnographers. And that is yet to be invented.

▣ REFERENCES

Bleier, R. (1984). *Science and gender: A critique of biology and its theories on women.* Oxford, UK: Pergamon Press.

Bleier, R. (Ed.). (1986). *Feminist approaches to science.* Oxford, UK: Pergamon.

Cannella, G. S., & Lincoln, Y. S. (2009). Deploying qualitative methods for critical social purposes. In N. K. Denzin & M. D. Giardina, (Eds.), *Qualitative inquiry and social justice: Toward a politics of hope* (pp. 53–72). Walnut Creek, CA: Left Coast Press.

Davis, G. E., & Marquis, C. (2005, July/August). Prospects for organizational theory in the early twenty-first century: Institutional fields and mechanisms. *Organization Science, 16*(4), 332–343.

Denzin, N. K. (2009). *Qualitative inquiry under fire: Toward a new paradigm dialogue.* Walnut Creek, CA: Left Coast Press.

Denzin, N. K. (2010). *The qualitative manifesto: A call to arms.* Walnut Creek, CA: Left Coast Press.

Denzin, N. K., & Giardina, M. D. (Eds.). (2006). *Qualitative inquiry and the conservative challenge: Confronting methodological fundamentalism.* Walnut Creek, CA: Left Coast Press.

Denzin, N. K., & Giardina, M. D. (Eds.). (2007). *Ethical futures in qualitative research: Decolonizing the politics of knowledge.* Walnut Creek, CA: Left Coast Press.

Denzin, N. K., & Giardina, M. D. (Eds.). (2008). *Qualitative inquiry and the politics of evidence.* Walnut Creek, CA: Left Coast Press.

Denzin, N. K., & Giardina, M. D. (Eds.). (2009). *Qualitative inquiry and social justice: Toward a politics of hope.* Walnut Creek, CA: Left Coast Press.

Faubion, J. D., & Marcus, G. E. (Eds.). (2009). *Fieldwork is not what it used to be: Learning anthropology's method in a time of transition.* Ithaca, NY: Cornell University Press.

Fischer, M. M. J. (2009). Foreword: Renewable Ethnography. In J. D. Faubion & G. E. Marcus (Eds.), *Fieldwork is not what it used to be: Learning anthropology's method in a time of transition* (pp. vii–xiv). Ithaca, NY: Cornell University Press.

Guba, E. G., & Lincoln, Y. S. (1981). *Effective evaluation.* San Francisco: Jossey-Bass.

Guba, E. G., & Lincoln, Y. S. (1989). *Fourth-generation evaluation.* Thousand Oaks, CA: Sage.

Hamilton, J. A. (2009). On the ethics of unusable data. In J. D. Faubion & G. E. Marcus (Eds.), *Fieldwork is not what it used to be: Learning anthropology's method in a time of transition* (pp. 73–88). Ithaca, NY: Cornell University Press.

Heaton, J. (2004). *Reworking qualitative data.* Thousand Oaks, CA: Sage.

Lather, P. (1986). Research as praxis. *Harvard Educational Review, 56*(3), 257–277.

Lather, P. (1991). *Getting smart: Feminist research and pedagogy within the postmodern.* New York: Routledge.

Lather, P. (1992). Critical frames in educational research: Feminist and poststructural perspectives. *Theory Into Practice, 31*(2), 1–13.

Lather, P. (2004). Scientific research in education: A critical perspective. *Journal of Curriculum and Supervision, 20*(1), 14–30.

Lather, P. (2007). *Getting lost: Feminist efforts toward a double(d) science.* Albany: State University of New York Press.

Lincoln, Y. S. (in progress). *Policy from prose: The perfect adequacy of policy formulation from qualitative research.* (Paper accepted for presentation, American Educational Research Association, New Orleans, LA, April 8–12, 2011)

Lincoln, Y. S., & Guba, E. G. (1985). *Naturalistic inquiry.* Thousand Oaks, CA: Sage.

Lucas, W. K. (1972). *The case survey method: aggregating case experience.* Santa Monica, CA: The Rand Corporation.

Lucas, W. K. (1974). *The case survey and alternative methods for research aggregation.* Washington, DC: The Rand Corporation.

McCall, G. J., & Simmons, J. L. (Eds.). (1989). *Issues in participant observation.* New York: Random House.

Mertens, D. M. (1998). *Research methods in education and psychology: Integrating diversity with quantitative and qualitative approaches.* Thousand Oaks, CA: Sage.

Perrow, C. (1991). A society of organizations. *Theory of society, 20,* 725–762.

Pink, D. H. (2005, February). Revenge of the right brain, *Wired.* Available at http://www.wired.com/wired/archive/13.02/brain.html

Tashakkori, A., & Teddlie, C. (2010). *The SAGE handbook of mixed methods in social & behavioral research* (2nd ed.). Thousand Oaks, CA: Sage.

Westbrook, D. A. (2008). *Navigators of the contemporary: Why ethnography matters.* Chicago: University of Chicago Press.

AUTHOR INDEX

SUBJECT INDEX

ABOUT THE EDITORS

Norman K. Denzin is Distinguished Professor of Communications, College of Communications Scholar, and Research Professor of Communications, Sociology and Humanities, at the University of Illinois, Urbana-Champaign. He is the author, editor, or coeditor of numerous books, including *The Qualitative Manifesto; Qualitative Inquiry Under Fire; Flags in the Window: Dispatches From the American War Zone; Searching for Yellowstone: Identity, Politics and Democracy in the New West; Performance Ethnography: Critical Pedagogy and the Politics of Culture; Screening Race: Hollywood and a Cinema of Racial Violence; Performing Ethnography;* and *9/11 in American Culture.* He is past editor of *The Sociological Quarterly*; coeditor of *The SAGE Handbook of Qualitative Research,* Fourth Edition; coeditor of *Qualitative Inquiry*; editor of *Cultural Studies <=> Critical Methodologies*; editor of *International Review of Qualitative Research,* editor of *Studies in Symbolic Interaction,* and founding President of the International Association of Qualitative Inquiry.

Yvonna S. Lincoln is Ruth Harrington Chair of Educational Leadership and Distinguished Professor of Higher Education at Texas A&M University, where she also serves as Program Chair for the higher education program area. She is the coeditor, with Norman K. Denzin, of the journal *Qualitative Inquiry,* and of the first and second, third and now fourth editions of *the SAGE Handbook of Qualitative Research* and the *Handbook of Critical and Indigenous Methodologies.* As well, she is the coauthor, editor, or coeditor of more than a half dozen other books and volumes. She has served as the President of the Association for the Study of Higher Education and the American Evaluation Association, and as the Vice President for Division J (Postsecondary Education) for the American Educational Research Association. She is the author or coauthor of more than 100 chapters and journal articles on aspects of higher education or qualitative research methods and methodologies. Her research interests include development of qualitative methods and methodologies, the status and future of research libraries, and other issues in higher education. And, she's fun.

ABOUT THE CONTRIBUTORS

Tineke A. Abma is Professor in Client Participation in Elderly Care at the department of Medical Humanities, VU University Medical Center and senior researcher at the EMGO+ Institute for Health and Care Research at the same university in Amsterdam. Her scholarly work concentrates on the methodology of interactive approaches to qualitative research and program evaluation, including the participation and empowerment of patients and vulnerable groups. More recently she works in the field of bioethics with a special interest in ethics of care and hermeneutic ethics. Practice fields include chronic care and disabilities, elderly care, and psychiatry.

David L. Altheide is Regents' Professor in the School of Justice and Social Inquiry at Arizona State University, where he has taught for 36 years. A sociologist who uses qualitative methods, his work has focused on the role of mass media and information technology in social control. Dr. Altheide received the 2005 George Herbert Mead Award for lifetime contributions from the Society for the Study of Symbolic Interaction (SSSI). He is also a three-time recipient of the SSSI's Cooley Award, for the best book of the year in 2007 for *Terrorism and the Politics of Fear* (2006); in 2004 for *Creating Fear: News and the Construction of Crisis* (2002); and in 1986 for *Media Power* (2005). His most recent book is *Terror Post 9/11 and the Media* (2009). His teaching efforts were recognized with the SSSI's Mentor Excellence Award in 2007.

Michael Angrosino is Professor Emeritus of Anthropology at the University of South Florida. He is an applied medical anthropologist who specializes in mental health policy and service delivery issues and in studies of religion in secular society. His most recent books are *Exploring Oral History: A Window on the Past* (2008) and *How Do They Know That? The Process of Social Research* (2010).

Mary Brydon-Miller, PhD, directs the University of Cincinnati's Action Research Center and is Professor of Educational Studies and Urban Educational Leadership in the College of Education, Criminal Justice, and Human Services. She is a participatory action researcher who engages in both community-based and educational-action research. She coedited the volumes *Traveling Companions: Feminism, Teaching, and Action Research* (with Patricia Maguire and Alice McIntyre), *From Subjects to Subjectivities: A Handbook of Interpretive and Participatory Methods* (with Deborah Tolman), and *Voices of Change: Participatory Research in the United States and Canada* (with Peter Park, Budd Hall, and Ted Jackson). Her other publications include work on participatory action research methods, academic writing in the social sciences, refugee resettlement, elder advocacy, and disability rights. Her current scholarship focuses on ethics and action research.

Gaile S. Cannella holds the Velma E. Schmidt Endowed Chair in Early Childhood Studies in the Department of Teacher Education and Administration, University of North Texas, and focuses on critical social science and qualitative research methodologies in her teaching and research projects. She has written several books, including *Deconstructing Early Childhood Education: Social Justice and Revolution; Kidworld: Childhood Studies, Globalization and Education* (with J. Kincheloe)*; Childhood and (Post)colonization* (with R. Viruru)*;* and *Childhoods: A Handbook* (with L. Diaz Soto). Cannella is past Vice President of the International Association of Qualitative Inquiry.

Kathy Charmaz is Professor of Sociology and Director of the Faculty Writing Program at Sonoma State University, a program she designed to help faculty complete their research and scholarly writing. She has written, coauthored, or coedited nine books including *Constructing Grounded Theory: A Practical Guide Through Qualitative Analysis,* which has been translated into Chinese, Japanese, Polish, and Portuguese and received a Critics' Choice Award from the American Educational Studies Association. Among her recent writings are two multiauthored books, *Developing Grounded Theory: The Second Generation,* and *Five Ways of Doing Qualitative Analysis: Phenomenological Psychology, Grounded Theory, Discourse Analysis, Narrative Research, and Intuitive Inquiry.* She served as the 2009–2010 president of the Society for the Study of Symbolic Interaction and has been an officer in several other professional societies. Kathy Charmaz gives many workshops on qualitative methods, grounded theory, and writing for publication around the globe.

Susan E. Chase is Professor of Sociology at the University of Tulsa. Her most recent book is *Learning to Speak, Learning to Listen: How Diversity Works on Campus.* Using the methods of narrative ethnography and narrative analysis, she presents a case study of how undergraduates engage diversity issues at City University (a pseudonym). The book portrays how some students at this predominantly white university learn to speak and listen to each other across social differences, especially race. She is also the author of *Ambiguous*

Empowerment: The Work Narratives of Women School Superinten-dents, and *Mothers and Children: Feminist Analyses and Personal Narratives* (with Mary Rogers).

Julianne Cheek is Professor at Atlantis Medical College, Oslo, Norway. She is an associate editor of *Qualitative Health Research,* a past coeditor of *Health: An Interdisciplinary Journal for the Social Study of Health, Illness and Medicine,* and an editorial board member of a number of journals related to qualitative inquiry and/or social health. Julianne has attracted funding for many qualitative health-related projects. She has also reviewed funding applications for a number of funding schemes including the Australian Research Council and National Health and Medical Research Council. Her books, book chapters, and journal arti-cles reflect her ongoing interest in qualitative inquiry and the politics of that inquiry as it relates to the world of health and social care. In addi-tion, she has a long interest in the mentoring and development of qualitative inquirers, including the development of postdoctoral pro-grams. In 2010, she had the honor of serving as the vice president of the International Association of Qualitative Inquiry, as well as on the advisory board of the International Congress of Qualitative Inquiry, held annually at the University of Illinois.

Clifford G. Christians is a Research Professor of Communications Emeritus at the University of Illinois, Urbana-Champaign, where he was Director of the Institute of Communications for 16 years. He has been a visiting scholar in philosophical ethics at Princeton University, in social ethics at the University of Chicago, and a PEW fellow in ethics at Oxford University. He completed the third edition of Rivers and Sch-ramm's *Responsibility in Mass Communication,* has coauthored *Jacques Ellul: Interpretive Essays* with Jay Van Hook, and has written *Teaching Ethics in Journalism Education* with Catherine Covert. Christians is also the coauthor, with John Ferre and Mark Fackler, of *Good News: Social Ethics and the Press.* Christians' *Media Ethics: Cases and Moral Reason-ing,* with Mark Fackler, Kathy McKee, Peggy Kreshel, and Robert Woods, is now in its ninth edition. He has authored or coauthored *Com-munication Ethics and Universal Values, Handbook of Mass Media Ethics, Moral Engagement in Public Life: Theorists for Contemporary Ethics, Ethical Communication: Moral Stances in Human Dialogue,* and *Normative Theories of the Media.* He was editor of *Critical Studies in Media Communication* and currently edits *The Ellul Forum.*

John W. Creswell is a Professor of Educational Psychology and teaches courses and writes about mixed methods research, qualitative meth-odology, and general research design. He has been at the University of Nebraska at Lincoln for 30 years and has authored 12 books, many of which focus on alternative types of research designs, comparisons of different qualitative methodologies, and the nature of and use of mixed methods research. His books are read around the world by audiences in the social sciences, education, and the health sciences. In addition, he has codirected for the last five years the Office of Qualita-tive and Mixed Methods Research at Nebraska that provides support for scholars incorporating qualitative and mixed methods research into projects for extramural funding. He serves as the founding coedi-tor for the *Journal of Mixed Methods Research,* and he has been an Adjunct Professor of Family Medicine at the University of Michigan and assisted investigators in the health sciences and education on the research methodology for National Institutes of Health and National Science Foundation projects. He is a Senior Fulbright Scholar to South

Africa and lectured during 2008 at five universities to faculty in educa-tion and the health sciences.

Judith Davidson is an Associate Professor in the Graduate School of Education at the University of Massachusetts-Lowell (UML). Her major interest is in the area of qualitative research methodology, with an emphasis on technologies for qualitative research. She has served as a qualitative researcher on projects investigating a diverse range of areas, from early childhood education for at-risk students and elemen-tary arts education to K–12 technology planning, math/science instruction for in-service teachers, and preservice teacher recruit-ment. She is a founder of the cross-campus Qualitative Research Net-work and has worked with faculty, staff, and graduate students to implement qualitative research technologies in diverse disciplinary areas on the UML campus. She is the author of *Living Reading: Explor-ing the Lives of Reading Teachers* (2005), and coauthor of *Qualitative Research Design for Software Users* (2008) and *Adolescent Literacy: What Works and Why* (1998). She was a leader in the organization of the "A Day in Technology in Qualitative Research" preconference day at the International Congress on Qualitative Inquiry 2008 conference. She has presented and published widely on topics related to qualitative research and its technologies. Currently she is working on a project that blends a cluster of qualitative research techniques—document analysis, self-study, arts-based research, and use of qualitative data analysis software—to create a new approach to artful computing in qualitative research. She received a PhD from the University of Illinois College of Education, an MS from Bank Street College of Education, and her BA from Antioch University.

Silvana di Gregorio received her PhD in social policy and adminis-tration from the London School of Economics and Political Science. She has worked in several applied research settings including the Nuffield Centre for Health Services Studies, University of Leeds, and the Department of Social Policy, Cranfield University, where she was involved in numerous practitioner-research studies. During the 1990s, she was Director of Graduate Research Training at Cranfield School of Management, where she developed her interest in methodological issues, particularly looking at the affordances of software to support the analysis of qualitative data. In 1996, she resigned her position at Cranfield to focus on consulting and teaching on a range of packages that support qualitative analysis. SdG Associates is her consulting business. She is coauthor with Judith Davidson of *Qualitative Research Design for Software Users* (2008), which addresses both methodologi-cal and practical issues related to working with qualitative data analy-sis software packages—regardless of which brand of package is used. She is currently updating her skills by doing an MSc in E-Learning at the University of Edinburgh. She is also exploring the use of Web 2.0 tools to support the analysis of qualitative data. She is on the Advisory Board of the European Chapter of the Merlien Institute, which pro-motes innovations in qualitative research.

Cynthia B. Dillard (Nana Mansa II of Mpeasem, Ghana, West Africa) is Professor of Multicultural Teacher Education in the School of Teach-ing and Learning at The Ohio State University. Her major research interests include critical multicultural education, spirituality in teach-ing and learning, and African/African American feminist studies. She has published numerous book chapters and articles in journals including *International Review of Qualitative Research; Race, Ethnicity* *and Education; The Journal of Teacher Education;* and *The International Journal of Qualitative Studies in Education and Urban Education.* Her first book, *On Spiritual Strivings: Transforming an African American Woman's Academic Life* was published in 2006 by SUNY Press and was selected for the 2008 Critics' Choice Book Award by the American Educational Studies Association (AESA). Most recently, her research and service is focused in Ghana, West Africa, where she has established a preschool, is building a new elementary school, and is enstooled/crowned Nana Mansa II, Queen Mother of Development, in the village of Mpeasem, Ghana.

Greg Dimitriadis is Professor of Sociology of Education in Graduate School of Education at the State University of New York at Buffalo. He is interested in urban education and the policies that serve urban youth. More specifically, he is interested in the potential value and importance of nontraditional educational curricula (e.g., popular culture), programs (e.g., arts-based initiatives), and institutions (e.g., community centers) in the lives of disenfranchised young people. His most recent work has dealt with the contemporary complexities of qualitative inquiry, including its history and philosophical and theoretical underpinnings, as well as the ways "theory" generated outside the field of education can be brought to bear on the questions and concerns facing educational researchers and practitioners today. He is author or editor (alone and with others) of more than 10 books and 50 articles and book chapters. His books include *Performing Identity/Performing Culture: Hip Hop as Text, Pedagogy, and Lived Practice; Friendship, Cliques, and Gangs: Young Black Men Coming of Age in Urban America;* and *Studying Urban Youth Culture.* His work has appeared in journals, including *Teachers College Record, Anthropology and Education Quarterly,* and *British Journal of Sociology of Education.* He edits the book series Critical Youth Studies and coedits Key Ideas and Education, both published by Routledge.

Margaret A. Eisenhart is University Distinguished Professor and Charles Professor of Educational Anthropology and Research Methodology at the University of Colorado, Boulder. Her research focuses on culture and gender in education, and women in science and engineering. She has conducted research in elementary and secondary schools, colleges, universities, and workplaces. Her most important works include *Educated in Romance: Women, Achievement, and College Culture* (with Dorothy Holland); *Women's Science: Learning and Succeeding From the Margins* (with Elizabeth Finkel); and *Designing Classroom Research* (with Hilda Borko). In her current research project, "Female Recruits Explore Engineering" (the FREE Project), she developed and delivered a program to encourage high school minority girls' interest in engineering and IT and studied how the program and its goals fit into the context of the girls' lives. She is a fellow of the American Anthropological Association and the American Educational Research Association, and a member of the National Academy of Education.

Laura L. Ellingson, PhD, University of South Florida, is Associate Professor of Communication and Women's & Gender Studies at Santa Clara University. Her research focuses on gender, qualitative methodology, extended and chosen family, and interdisciplinary communication in health care organizations. She is author of *Communicating in the Clinic: Negotiating Frontstage and Backstage Teamwork* (2005) and *Engaging Crystallization in Qualitative Research* (2009), and coauthor (with Patricia J. Sotirin) of *Aunting: Cultural Practices That Sustain*

Family and Community Life (2010). She is the Senior Editor for Qualitative, Interpretive, and Rhetorical Methods at the journal *Health Communication.*

Frederick Erickson is George F. Kneller Professor of Anthropology of Education and Professor of Applied Linguistics at the University of California, Los Angeles (UCLA). His contributions to the field of anthropology of education have earned him numerous honors and awards, including fellowships from the Spencer Foundation and Annenberg Institute for Public Policy, a Fulbright Award, the Spindler Award for Scholarly Contributions to Educational Anthropology from the American Anthropological Association, and a Lifetime Achievement Award for Research on the Social Context of Education from Division G of the American Educational Research Association (AERA). Erickson's writings on the video-based microethnographic study of classroom and family interaction, and on qualitative research methods more generally, are widely cited. His recent book, *Talk and Social Theory: Ecologies of Speaking and Listening in Everyday Life* (2004) received an Outstanding Book Award for 2005 from AERA. He serves on the editorial boards of *Research on Language and Social Interaction, Discourse and Communication, the International Review of Qualitative Research,* and *Teachers College Record.* In 1998–1999 and again in 2006–2007, he was a fellow at the Center for Advanced Study in the Behavioral Sciences at Stanford University. In 2000, he was elected a member of the National Academy of Education, and in 2009, he was elected a fellow of the American Educational Research Association.

Susan Finley—researcher, teacher, artist, and author—is an activist who has implemented community-based educational efforts with people living in tent communities, with street youths, and among economically poor children and their families, housed and unhoused. Her artworks have been exhibited in galleries and in professional research venues, and have been published as covers and illustrations of research texts. Her original poetry and drama, as well as coauthored theater scripts, have been performed in arts venues, on campuses, and at professional research meetings. Curricular innovations for K–8 students and preservice teachers have been developed through the At Home At School Community Education effort that Susan designed and directs (2002–).

Bent Flyvbjerg is BT Professor and Chair of Major Programme Management at Oxford University's Saïd Business School and Founding Director of Oxford's BT Centre for Major Programme Management. He works for better management of megaprojects and cities. He also writes about phronetic social science and case study research. Bent Flyvbjerg was twice a Visiting Fulbright Scholar to the United States, where he did research at UCLA, the University of California, Berkeley, and Harvard University. His books include *Making Social Science Matter, Rationality and Power, Megaprojects and Risk,* and *Decision-Making on Mega-Projects.* His books and articles have been translated into 18 languages and his research covered by *Science, The Economist,* the *Financial Times,* the *New York Times,* the BBC, NPR, and many other media. Bent Flyvbjerg has served as adviser to the United Nations, the EU Commission, and government and business in many countries.

Sarah N. Gatson is Associate Professor of Sociology at Texas A&M University. Her research interests include race/ethnicity, gender, culture, legal studies, identity, community, citizenship, media and technology, and qualitative methods. Her work has appeared in *Law & Social Inquiry, Research in Community Sociology, Qualitative Research, Qualitative Inquiry, Contemporary Sociology, Feminist Media Studies,* and *Advances in Physiology Education,* and as book chapters in *Fighting the Forces: What's at Stake in Buffy the Vampire Slayer* (edited by David Lavery and Rhonda Wilcox, 2002), *Faculty of Color Teaching in Predominantly White Institutions* (edited by Christine Stanley, 2006), and *Real Drugs in a Virtual World: Drug Discourse and Community Online* (edited by Edward Murguia, Melissa Tackett-Gibson, and Anne Lessem, 2007). Gatson is the coauthor of the book *Interpersonal Culture on the Internet—Television, the Internet, and the Making of a Community* (2004).

Michael D. Giardina (PhD, University of Illinois) is Assistant Professor in the College of Education at Florida State University, where he teaches courses on physical cultural studies, qualitative research methods, and popular media. He is the author of *Sporting Pedagogies: Performing Culture & Identity in the Global Arena* (2005), which received the 2006 Most Outstanding Book award from the North American Society for the Sociology of Sport, and *Sport, Spectacle, and NASCAR Nation: Consumption and the Cultural Politics of Neoliberalism* (2011, with Joshua I. Newman). Giardina is also the editor of *Youth Culture & Sport: Identity, Power, & Politics* (2007, with Michele K. Donnelly), *Globalizing Cultural Studies: Ethnographic Interventions in Theory, Method, and Policy* (2007, with Cameron McCarthy et al.), and a series of seven books with Norman K. Denzin on cultural studies and interpretive research in the post–9/11 era, including *Contesting Empire/Globalizing Dissent: Cultural Studies after 9/11* (2006) and *Qualitative Inquiry and Social Justice: Toward a Politics of Hope* (2009). His work has similarly appeared in such journals as *Qualitative Inquiry, American Behavioral Scientist, Harvard Educational Review, International Journal of Qualitative Studies in Education,* and *Cultural Studies <=> Critical Methodologies.* He is Associate Editor of the *Sociology of Sport Journal* and a member of the editorial board of *Cultural Studies <=> Critical Methodologies.*

Davydd J. Greenwood is the Goldwin Smith Professor of Anthropology at Cornell University where he has served as a faculty member since 1970. He was elected Corresponding Member of the Spanish Royal Academy of Moral and Political Sciences in 1996. He served as the John S. Knight Professor and Director of the Mario Einaudi Center from 1983 to 1995 and as Director of the Cornell Institute for European Studies from 2000 to 2008. His work centers on action research, political economy, ethnic conflict, community and regional development, and neoliberal reforms of higher education. He has worked in the Spanish Basque Country, Spain's La Mancha region, and the Finger Lakes region of Upstate New York. The author of 8 books and more than 40 articles on agricultural industrialization, the use of biological ideologies to support political economic regimes, industrial cooperatives, and action research, he is currently working on the role of action research in developing creative responses to neoliberal reforms in higher education.

Egon Guba (1924–2008) was Professor Emeritus of Education, Indiana University. He received his PhD from the University of Chicago in quantitative inquiry (education) in 1952, and thereafter served on the faculties of the University of Chicago, the University of Kansas, the Ohio State University, and Indiana University. Over a two-decade period, he studied paradigms alternative to the received view and espoused a personal commitment to one of these: constructivism. He

was the coauthor of *Effective Evaluation* (1981), *Naturalistic Inquiry* (1985), and *Fourth Generation Evaluation* (1989), all with Yvonna S. Lincoln, and he is the editor of *The Paradigm Dialog* (1990), which explores the implications of alternative paradigms for social and educational inquiry. He is the author of more than 150 journal articles and more than 100 conference presentations, many of them concerned with elements of new-paradigm inquiry and methods.

Jaber E. Gubrium is Professor and Chair of Sociology at the University of Missouri. He has had a long-standing program of research on the social organization of care in human service institutions and pioneered in the reconceptualization of qualitative methods and the development of narrative analysis. His publications include numerous books and articles on aging, the life course, medicalization, and representational practice in therapeutic context. Collaborating for over 25 years, Gubrium and James Holstein have authored and edited dozens of books, including *Analyzing Narrative Reality, The New Language of Qualitative Method, The Active Interview, Handbook of Constructionist Research, Handbook of Interview Research, The Self We Live By, Constructing the Life Course,* and *What is Family?*

Judith Hamera is Professor and Head of the Department of Performance Studies at Texas A&M University. Her scholarship is interdisciplinary, contributing to American, communication, and cultural studies, as well as performance studies. Her most recent book is the *Cambridge Companion to American Travel Writing* (2009), coedited with Alfred Bendixen; and she has completed a monograph examining the cultural work of the American home aquarium, to be published by the University of Michigan Press. She is the author of *Dancing Communities: Performance, Difference and Connection in the Global City* (2007), which received the Book of the Year award from the National Communication Association's Ethnography Division. Other books are *Opening Acts: Performance In/As Communication and Cultural Studies* (2006); and, coedited with D. Soyini Madison, *The SAGE Handbook of Performance Studies* (2006). Her essays have appeared in *Communication and Critical/Cultural Studies, Cultural Studies, TDR: The Drama Review, Modern Drama, Text and Performance Quarterly, Theatre Topics,* and *Women and Language.*

James A. Holstein is Professor of Sociology in the Department of Social and Cultural Sciences at Marquette University. His research and writing projects have addressed social problems, deviance and social control, family, and the self, all approached from an ethnomethodologically informed, constructionist perspective. His publications include numerous books and articles on qualitative research methods. Collaborating for over 25 years, Holstein and Jaber Gubrium have authored and edited dozens of books, including *Analyzing Narrative Reality, The New Language of Qualitative Method, The Active Interview, Handbook of Constructionist Research, Handbook of Interview Research, The Self We Live By, Constructing the Life Course,* and *What is Family?*

John M. Johnson is Professor of Justice Studies at Arizona State University, where he has taught for almost 40 years with his close friend David Altheide. They have changed in many ways during their lives, including many of their views about qualitative methods.

A. Susan Jurow is an Assistant Professor of Educational Psychology and Research on Teaching and Teacher Education at the University of Colorado, Boulder. Her research focuses on the relations among learning, teaching, and communication. She has conducted research in elementary and middle schools, professional development programs, and universities. In these projects she has focused on (1) the development of practice-linked identities to describe how people identify with particular ways of knowing, acting, and valuing, and are positioned to participate in social practices, and (2) individuals' engagement in long-term projects of learning located in classrooms and out-of-school settings. Her multidisciplinary approach to studying learning and teaching draws on the learning sciences, anthropology, and discourse analysis.

George Kamberelis is a Wyoming Excellence Chair of Literacy Education at the University of Wyoming. He received a PhD in Education and Psychology and an MS in Psychology from the University of Michigan, an MA in Literature and Religion from the University of Chicago, and a BA in Philosophy and Religion from Bates College. Most of his research and writing has focused on the philosophical foundations of qualitative inquiry, methods of qualitative inquiry, and sociocultural dimensions of literacy practice. He has coauthored two books with Greg Dimitriadis (*On Qualitative Inquiry* and *Theory for Education*) and contributed chapters to many edited volumes. His work has also been published in various journals including *Qualitative Inquiry, Annals of the American Academy of Political and Social Science, Journal of Russian & East European Psychology, Journal of Contemporary Legal Issues, Reading Research Quarterly, Research in the Teaching of English,* and *Linguistics and Education.* Over the years, Kamberelis has taught courses on social theory, the logics of inquiry, qualitative research methods, literacy and society, classroom discourse, and media literacy.

Joe L. Kincheloe (1950–2008) was a researcher's researcher, a teacher's teacher, and the quintessential embodiment of critical pedagogy. A tireless champion for socially just pedagogy, he authored and edited over 60 books and hundreds of articles underpinned by his commitment to engagement, authenticity, and cultural work. His notions of teacher as researcher, critical constructivism, research bricolage, postformal thinking, and critical cultural studies are internationally recognized. Joe was the supervisor and chair for scores of doctoral students from Pennsylvania State University, CUNY Graduate Center, and McGill University. His work and legacy continue to make a difference in faculties, schools, and communities from Barcelona, Spain, to Utrecht, Netherlands, to Daejong, Korea, to Melbourne, Australia, to Winnipeg, Manitoba, and back to where his work was grounded, Sao Paulo, Brazil. Joe was a father, a husband, a musician, a teacher, a friend, and a researcher; his life was a bricolage of hyperreality, Tennessee Volunteers football, epistemological quandaries, and radical love. His curiosity and wonder informed his work, and he filled his life with questioning the unquestioned and naming the unnamed.

Michael Kral, PhD, is Assistant Professor of Psychology and Anthropology at University of Illinois at Urbana-Champaign, and in the Department of Psychiatry, University of Toronto. He works primarily with Inuit in Arctic Canada on suicide prevention, community wellness, and community-based participatory action research. He has coedited *Suicide in Canada* (with Antoon Leenaars, Susanne Wenkstern, Isaac Sakinofsky, Ron Dyck, and Roger Bland), *About Psychology: Essays at the Crossroads of History, Theory, and Philosophy* (with Darryl Hill), and a special issue of the *American Journal of Community Psychology* (with Mark Aber and Jorge Ramirez Garcia) on culture and community

psychology. Michael's current research is on indigenous youth resilience in the circumpolar north, indigenous community action and success stories, and urban indigenous well-being.

Antjie Krog is teaching at the University of Western Cape, South Africa. She has published 12 poetry volumes and 3 nonfiction books in English: *Country of my Skull,* on the South African Truth and Reconciliation Commission; *A Change of Tongue,* about the transformation in South Africa after 10 years of democracy; and recently *Begging to be Black,* about living with a Black majority. Krog has been awarded all the main prizes available in South Africa for poetry and nonfiction, as well as the Hiroshima Foundation Award for Peace and Culture (2000); the Open Society Prize (2006) from the Central European University (previous winners were Jürgen Habermas and Vaclav Havel); and a research fellowship at Wissenschaftskolleg zu Berlin in 2007–2008.

Morten Levin is a Professor at the Department of Industrial Economics and Technology Management, the Faculty of Social Sciences and Technology Management, at the Norwegian University of Science and Technology in Trondheim, Norway. He holds graduate degrees in engineering and in sociology. Throughout his professional life, he has worked as an action researcher with a particular focus on processes and structures of social change in the relationships between technology and organization. His action research has taken place in industrial contexts, in local communities, and in university teaching, where he has developed and directed three PhD programs in action research. He is author of a number of books and articles, including *Introduction to Action Research: Social Research for Social Change,* and serves on the editorial boards of *Systemic Practice and Action Research, Action Research International, Action Research,* and *The Handbook of Action Research.*

James H. Liu is Professor of Psychology at Victoria University of Wellington and Deputy Director of its Centre for Applied Cross Cultural Research. He was born in Taiwan and grew up in a small town in the midwestern United States. He obtained a bachelor's degree in computer science from the University of Illinois and worked as an aerospace engineer for Hughes Aircraft Space and Communications. He completed a PhD in social psychology in 1992 at University of California, Los Angeles, and a postdoctoral fellowship at Florida Atlantic University. He has been teaching at Victoria University of Wellington since 1994. His research is at the intersection of cross-cultural psychology and intergroup relations. He specializes in the study of social identity and representations of history. He has more than 100 academic publications, and his edited volumes include *New Zealand Identities: Departures and Destinations* (2005), *Restorative Justice and Practices in New Zealand* (2010), *Ages Ahead: Promoting Intergenerational Relationships* (1998), and *Progress in Asian Social Psychology,* Volumes 2 (1997) and 6 (2006). He was Secretary General of the Asian Association of Social Psychology from 2003 to 2007, Treasurer from 1999 to 2003, and is now editor of the *Asian Journal of Social Psychology.* A naturalized citizen of two countries, he describes himself as a "Chinese-American-New Zealander." He is married to Belinda Bonzon Liu, and they have a daughter who is even more hyphenated: a Chinese-American-Filipino-New Zealander. His father is an eminent neo-Confucian philosopher living in Taiwan.

Susan A. Lynham is Associate Professor in the Research Methodology, and Organizational Performance and Change PhD programs at Colorado State University. Her scholarship focuses on strategic human resource development, leadership in complex and diverse environments, and applied theory building research methods. A past board member of the Academy of Human Resource Development, Susan serves as Editor-in-Chief of the *Advances in Developing Human Resources* journal. She obtained her PhD from the University of Minnesota, in May 2000.

Patricia Maguire, EdD, Professor of Education and Counseling, is the Chairperson of Western New Mexico University, Gallup Graduate Studies Center. She has worked as a school and mental health counselor, international development trainer (Africa, Jamaica), and community activist. Since 1987, Maguire has been a member of a collaborative team developing transformative-oriented graduate education programs in a rural community on the edge of the Navajo Nation and Pueblo of Zuni. Her networking, research, and publication interests include the interface between feminisms and participatory action research and teacher action research, building on her 1987 book, *Doing Participatory Research: A Feminist Approach* and *Traveling Companions: Feminism, Teaching, and Action Research* (2004), coedited with Mary-Brydon Miller and Alice McIntyre. Maguire's activist interests include working against sexual and interpersonal violence, supporting community efforts to address hunger and food insecurity, and promoting transformative teacher action research.

Peter McLaren is Professor of Urban Schooling, the Graduate School of Education and Information Studies, University of California, Los Angeles. He is the author and editor of more than 45 books on various subjects within the pedagogy of liberation. His writings have been translated into 25 languages. McLaren has been honored in numerous countries for his international scholarship and political activism. Among these honors, a Catedra McLaren has been established at Bolivarian University, Venezuela, and La Fundacion McLaren and Instituto Peter McLaren have been created in Northern Mexico by teachers, researchers and activists. As an award-winning author and educational activist, McLaren lectures worldwide.

Donna M. Mertens is Professor of Research Methodology and Program Evaluation at Gallaudet University in Washington, D.C. Her work focuses on the development of transformative research and evaluation methodologies with a goal of increasing social justice and furthering human rights. Her publications include *Research and Evaluation in Education and Psychology: Integrating Diversity with Quantitative, Qualitative, and Mixed Methods* (3rd ed., 2010), *Transformative Research and Evaluation* (2009), and *The Handbook of Social Research Ethics* (edited with Pauline Ginsberg, 2009).

Janice M. Morse, PhD (Nursing), PhD (Anthropology), FAAN, is a Professor and Presidential Endowed Chair at the University of Utah College of Nursing, and Professor Emeritus, University of Alberta, Canada. She was the founding director of the International Institute for Qualitative Methodology (IIQM, 1997–2007), University of Alberta; founding editor for the *International Journal of Qualitative Methods;* and since 1991 has served as the founding editor for *Qualitative Health Research.* Morse is the recipient of the Episteme Award (Sigma Theta Tau) and honorary doctorates from the University of Newcastle (Australia) and Athabasca University (Canada). She is the author of 370 articles and 15 books on qualitative research methods, suffering, comforting, and patient falls. Her most recent book (with Linda Niehaus) is *Mixed Method Design: Principles and Procedures* (2009).

Joshua I. Newman (PhD, University of Maryland) lectures in the areas of sport and physical culture, qualitative research, cultural studies, and critical pedagogy at the University of Otago's (New Zealand) School of Physical Education. Broadly speaking, his research, teaching, and supervision are committed to interrogating the intersections of late capitalism, identity, and cultural politics of the body. In recent years, his work has focused on the politics of embodiment and enfleshed performance within the U.S. South. He is the author of *Embodying Dixie: Studies in the Body Pedagogics of Southern Whiteness* (2010) and *Sport, Spectacle, and NASCAR Nation: Consumption and the Cultural Politics of Neoliberalism* (2011, with Michael D. Giardina). Newman has also published numerous research articles in journals such as *Cultural Studies <=> Critical Methodologies; American Behavioral Scientist; The Review of Education, Pedagogy, and Cultural Studies; Sociology of Sport Journal;* and *International Review for the Sociology of Sport.*

Susan Noffke is Associate Professor of Curriculum & Instruction at the University of Illinois at Urbana-Champaign. She taught in public elementary and middle schools for a decade, and has led preservice and inservice teachers and doctoral students in courses in action research for the past 25 years. She has also worked with teachers in school-based action research projects, and has used action research to study her own practice as a teacher educator. Her major scholarship has addressed the history of action research, its existing and theoretical foundations, international variations, as well as studies using action research in community school settings.

Chinwe Okpalaoka is Director of Special Programs in the College of Arts and Sciences at The Ohio State University. In this position, she oversees the recruitment and retention of targeted student populations as well assists in the development and management of freshman seminar courses for the university. Her major research interests include immigrant education, immigrant ethnic identity development, and African/African American feminist studies. She is currently working with other scholars on a collaborative piece that forwards the notion of transnational Black feminisms and corresponding methodologies. The most recent focus of her work is an examination of how West African immigrant girls' experiences in the United States might be interpreted through a transnational Black feminist lens.

Virginia Olesen, Professor of Sociology (Emerita), Department of Social and Behavioral Sciences, University of California, San Francisco, continues her long time exploration of feminist qualitative research with writing on the temporal dynamics of critical interpretive research in an age of despair (*Critical Studies <=> Critical Methodologies*, Vol. 9, pp. 52–55, 2009), the limits of reflexivity in qualitative research ("Reflexive Anthropology and Research Ethics: Elvi Whittaker's Contributions" in *Ethnography, Epistemology, Ethics: Essays in Honour of Elvi Whittaker,* G. V. Loewen, Ed., forthcoming) and feminist qualitative research and grounded theory ("Feminist Qualitative Research and Grounded Theory: Complexities, Criticisms and Opportunities," pp. 417–435 in A. Bryant and K. Charmaz, Eds., *The SAGE Handbook of Grounded Theory,* Sage, 2007). A pioneer in feminist qualitative studies of women's health, she recently worked with an interdisciplinary group (nurses, physicians, anthropologists, sociologists, social workers) that examined issues around the health of incarcerated women ("Social Capital: A Lens for Examining Health of Incarcerated and Formerly Incarcerated Women," pp. 13–22 in D. C. Hatton and A. A. Fisher, Eds., *Women Prisoners and Health Justice, Perspectives, Issues and Advocacy for an International Hidden Population,* 2009).

Ronald J. Pelias, Professor, teaches performance studies in the Department of Speech Communication at Southern Illinois University, Carbondale. His most recent books are *Writing Performance: Poeticizing the Researcher's Body* (1999), *A Methodology of the Heart: Evoking Academic & Daily Life* (2004) and *Leaning: A Poetics of Personal Relations* (2011).

Anssi Peräkylä is Professor of Sociology at the University of Helsinki. His research interests include medical communication, psychotherapy, emotional communication, and conversation analysis. His books include *AIDS Counselling* (1995) and *Conversation Analysis and Psychotherapy* (coedited, 2008), both published by Cambridge University Press. His work has appeared in journals such as *Sociology, Sociology of Health and Illness, Social Psychology Quarterly, Research on Language and Social Interaction,* and *Psychotherapy Research.* In his current research project, he is exploring the linkages between interactional management of emotion, and psychophysiological processes in the body.

Ken Plummer is Emeritus Professor of Sociology at the University of Essex. His first book was *Sexual Stigma* (1975) and his most recent is *Sociology: The Basics* (2010). He has written on critical humanism and life stories, gay and lesbian life, sexual storytelling, narrative work, labeling theory, symbolic interactionism, queer theory, inequalities, rights, and citizenship. He was the founder editor of the journal *Sexualities* in 1996.

Judith Preissle is the 2001 Distinguished Aderhold Professor for the College of Education at the University of Georgia (UGA) where she has taught since 1975. She is an affiliate faculty member of the Institute for Women's Studies, who honored her with the 2006 Women's Studies Faculty Award. As an interdisciplinary scholar, she studies qualitative research methods and design; the anthropology of education; and gender, ethnic, and immigration studies. Recently, she has been studying ethics and the philosophy of social science. Her chapter, "Feminist Research Ethics" appeared in 2007 in *The Handbook of Feminist Research: Theory and Praxis.* She is the coauthor of *Ethnography and Qualitative Design in Educational Research* (1984, 1993), translated into Spanish in 1988, and coeditor of *The Handbook of Qualitative Research in Education* (1992). Her most recent book, coauthored with Xue Lan Rong, is the second edition of *Educating Immigrant Students* (2009). In addition to numerous book chapters, she has authored articles in such journals as the *American Educational Research Journal,* the *Anthropology of Education Quarterly,* the *Journal of Contemporary Ethnography,* the *Review of Educational Research,* the *Educational Researcher,* the *International Journal of Qualitative Studies in Education,* and *The Elementary School Journal.* She e-manages QUALRS-L (Qualitative Research for the Human Sciences), established in 1991 and the oldest such forum online, and QualPage, a web page on qualitative research. She founded the qualitative research program at UGA that now offers a graduate certificate program. In 2009 she was made a Fellow of the American Educational Research Association.

Jon Prosser is Director of the International Education Management program and a member of the Leeds Social Science Institute at Leeds University, United Kingdom. He was project leader for the Economic and Social Research Council's Building Capacity in Visual Methods program, which was part of the UK Researcher Development Initiative. He was involved as a visual methodologist in the Real Life Methods project based at Leeds and Manchester universities. Currently he is contributing to the groundbreaking Realities program based at the Morgan Centre, University of Manchester, and a study of Visual Ethics, both funded by the National Centre for Research Methods, University of Southampton. He is also involved in the Campaigning for Social Change: Understanding the Motive and Experiences of People With Dementia project with Bradford University. He is perhaps best known for editing *Image-based Research: A Sourcebook for Qualitative Researchers* (1998), which was the first book in the field to present visual research not as a stand-alone strategy taking one particular form or perspective, but as a theoretically and methodologically varied approach that drew on other approaches to conducting research. He sees his current work, which involves taking photographs for the local Riding for the Disabled Association calendar, as challenging and important.

Judith Rosenberg works as a nurse practitioner and ethnographic researcher in the Department of Pediatrics at the University of South Florida's College of Medicine. Her research has focused on adolescents, particularly those with disabilities and those at risk for domestic violence. She earned her doctorate in applied anthropology at the University of South Florida.

Johanna Ruusuvuori is Senior Researcher at the Finnish Institute of Occupational Health. Her research interests include professional–client interaction in health care encounters (general practice, homeopathy, maternity health care, psychotherapy), interaction in multiprofessional meetings, emotion in social interaction, intertwine of facial expression and spoken interaction, qualitative methodology and conversation analysis. She has published in journals such as *Social Psychology Quarterly, Social Science & Medicine,* and *Journal of Pragmatics.* In her current research, she develops evaluation methods for occupational health promotion.

Anu Sabhlok, PhD, is Assistant Professor of Humanities and Social Sciences at the Indian Institute of Science Education and Research in Mohali, India. She is also deeply involved with a local NGO, CEVA: Center for Education and Voluntary Action where she integrates her work on gender, education, and urban environments with community concerns related to social and economic development. As part of her dissertation research on disaster-relief, she volunteered with SEWA (Self-Employed Women's Association) in Gujarat, India, and conducted community-based research in order to address issues of identity and justice. Her previous training is as an architect and her design works to embody the participatory ethic.

Linda Shopes is a freelance developmental editor and consultant in oral and public history. She has written widely on oral and public history with a focus on interpretive issues, community history, and ethics, including institutional review board (IRB) review of oral history. She coedited *Oral History and Public Memories* (2008) and *The Baltimore Book: New Views of Local History* (1991), served as contributing editor for oral history to the *Journal of American History,* and is co-general editor of Palgrave's *Studies in Oral History* series. She is a past president of the U.S. Oral History Association.

Tami Spry is a Professor of Performance Studies in the Communication Studies Department at St. Cloud State University in Minnesota. Using autoethnographic writing and performance as a critical method of inquiry into culture and communication, Spry's national and international performance work, publications, directing, and pedagogy focuses on the development of cultural critique that engenders dialogue about difficult sociocultural issues in our everyday communal and global lives. Her publications appear in *Text and Performance Quarterly, Critical Studies <=> Critical Methodologies, Qualitative Inquiry, International Review of Qualitative Research, Women and Language, The SAGE Handbook of Qualitative Research,* and various anthologies. Her book, *Body, Paper, Stage: Writing and Performing Autoethnography* is available in Spring 2011.

Hilary Stace is a Research Fellow at the School of Government, Victoria University of Wellington, New Zealand. Her PhD research is on autism and public policy, specifically how the expertise of policy consumers—people with autism and their families—can be brought into the policy process. Her lived experience as a parent of an adult autistic son has fuelled her autism advocacy. She is also interested in collaborative approaches to disability research.

Shirley R. Steinberg is the incoming Director and Chair of the Werklund Foundation Youth Leadership Education, and Professor of Youth Studies in Education at the University of Calgary, and was Research Professor at The University of Barcelona. The author and editor of many books in critical pedagogy, urban and youth culture, and cultural studies, some of her recent books include *Critical Qualitative Research* (with Gaile Cannella, 2012); the third edition of *Kinderculture: The Corporate Construction of Childhood* (2011); *Teaching Against Islamophobia* (edited with Joe Kincheloe and Christopher Stonebanks, 2010); *19 Urban Questions: Teaching in the City* (2010); *Christotainment: Selling Jesus Through Popular Culture* (edited with Joe Kincheloe, 2009); and *Diversity and Multiculturalism: A Reader* (2009). With Joe Kincheloe, she founded The Paulo and Nita Freire International Project for Critical Pedagogy (freireproject.org). A regular columnist for CTV and CBC Radio, she speaks internationally on critical pedagogy, youth studies, leadership, qualitative research, and social justice.

Elizabeth Adams St.Pierre is Professor of Language and Literacy Education and Affiliated Professor of both the Qualitative Research Program and the Women's Studies Institute at the University of Georgia. Her interests focus on the work of language in the construction of subjectivity, a poststructural critique of conventional qualitative inquiry, and a critique of "scientifically based research."

Martin Sullivan, PhD, QSO, is a senior lecturer at the School of Health and Social Services, Massey University, Palmerston North, New Zealand, where he coordinates postgraduate programs in disability studies and social policy. He coedited with Patricia O'Brien *Allies in Emancipation. Shifting From Providing Service to Being of Support* (2005), was part of the expert panel that reviewed and set guidelines for the disability research funded by the Health Research Council of New Zealand, and was a ministerial appointment to the National Ethics Advisory Committee for 6 years. He is currently working on a longitudinal study into the first two years of transition from spinal unit to community for people with spinal cord injury.

Abbas Tashakkori (PhD, Social Psychology, University of North Carolina at Chapel Hill) is a Professor and Chair of Educational Psychology Department at the University of North Texas, Denton. He has been a Mellon Post-Doctoral Fellow of the Carolina Population Center and a faculty member at various universities in the United States and abroad

for almost three decades. He has published more than 70 articles and book chapters, and authored or edited five books, including the *Handbook of Mixed Methods in the Social & Behavioral Research* (2nd ed., 2010, with Charles Teddlie). He has a rich and diverse history of research, program evaluation, and writing on attitude change, self-perceptions and efficacy, planning and evaluation of school improvement programs, and utilization of mixed methods in educational, social, and behavioral research.

Charles Teddlie (PhD, Social Psychology, University of North Carolina at Chapel Hill) is a Distinguished Professor Emeritus in the College of Education at Louisiana State University. He has been an investigator on several mixed methods research studies internationally and in the United States. Professor Teddlie has taught research methods courses for over 25 years, including statistics, qualitative methods, and mixed methods and was awarded the Excellence in Teaching Award from the LSU College of Education. He has been the co-owner and evaluation director of K. T. Associates, a consulting company specializing in mixed methods research and evaluation, since 1986. He has produced numerous articles and chapters in education, psychology, and evaluation and coauthored or coedited a dozen books including two editions of the *Handbook of Mixed Methods Research in Social & Behavioral Research* (2003, 2010) and *Foundations of Mixed Methods Research* (2009) with Abbas Tashakkori.

Barbara Tedlock is Distinguished Professor of Anthropology at the State University of New York at Buffalo. She served as editor-in-chief of the *American Anthropologist* (1993–1998). She received the 2003 SUNY Chancellor's Research Recognition Award for "Overall Excellence of Research in the Social Sciences." Tedlock is a former president of the Society for Humanistic Anthropology and a member of PEN (Poets-Essayists-Novelists). Her publications include six books and more than 120 articles and essays.

Harry Torrance is Professor of Education and Director of the Education and Social Research Institute, Manchester Metropolitan University, UK. His research interests include the interrelation of assessment, teaching, and learning; testing and educational standards; the role of assessment in educational reform and policy development; qualitative research methodology; the development of applied research; and the relationship between research and policy, research governance, and research management. He has undertaken many research projects investigating these topics, funded by a wide range of sponsors. He is editor of the four-volume *SAGE Handbook of Qualitative Research Methods in Education* (2010), a former editor of the *British Educational Research Journal,* and an elected member of the UK Academy of Social Sciences.

Guy A. M. Widdershoven is Professor in Medical Philosophy and Ethics and head of the department of Medical Humanities, VU University Medical Center, and senior researcher at the EMGO+ Institute for Health and Care Research at the same university in Amsterdam. He has published on hermeneutic ethics and its application in empirical ethics, moral deliberation, and the ethics of chronic care (psychiatry and elderly care). He is Scientific Director of the Netherlands School of Primary Care Research (CaRe) and former president of the European Association of Centers of Medical Ethics (EACME).

SAGE Research Methods Online

The essential tool for researchers

**Sign up now at
www.sagepub.com/srmo
for more information.**

An expert research tool

- An **expertly designed taxonomy** with more than 1,400 unique terms for social and behavioral science research methods

- **Visual and hierarchical search tools** to help you discover material and link to related methods

- Easy-to-use navigation tools
- Content organized by complexity
- Tools for citing, printing, and downloading content with ease
- Regularly updated content and features

A wealth of essential content

- The most comprehensive picture of quantitative, qualitative, and mixed methods available today

- More than **100,000 pages of SAGE book and reference material** on research methods as well as editorially selected material from SAGE journals

- More than **600 books** available in their entirety online

Launching 2011!

SAGE research methods online